Catalysis from A to Z

A Concise Encyclopedia

Edited by
Boy Cornils, Wolfgang A. Herrmann
Robert Schlögl, Chi-Huey Wong

WILEY-VCH

Weinheim · New York · Chichester
Brisbane · Singapore · Toronto

Prof. Dr. B. Cornils
Kirschgartenstraße 6
D-65719 Hofheim
Germany

Prof. Dr. R. Schlögl
Fritz-Haber-Institut
der Max-Planck-Gesellschaft
Faradayweg 4–6
D-14195 Berlin (Dahlem)
Germany

Prof. Dr. Dr. mult. h.c. W. A. Herrmann
Anorganisch-Chemisches Institut
Technische Universität München
Lichtenbergstraße 4
D-85747 Garching
Germany

Prof. Dr. C.-H. Wong
Department of Chemistry
The Scripps Research Institute
10550 N. Torrey Pines Road
La Jolla, CA 92037
USA

Library of Congress Card No. applied for

A catalogue record for this book is available from the British Library

Die Deutsche Bibliothek – CIP Cataloguing-in-Publication-Data
A catalogue record for this publication is available from Die Deutsche Bibliothek

ISBN 3-527-29855-X

Composition: Kühn & Weyh Software GmbH, D-79111 Freiburg
Printing: betz-druck GmbH, D-64291 Darmstadt
Bookbinding: Wilh. Osswald + Co., D-67433 Neustadt
Printed in the Federal Republic of Germany

Preface

> "Wir bekommen begründeten Anlaß zu vermuten,
> daß in lebenden Pflanzen und Tieren Tausende
> von katalytischen Prozessen zwischen den
> Geweben und Flüssigkeiten vor sich gehen."
>
> *J. J. Berzelius (1836)**

Catalysis occurs always and everywhere. Catalysis is not a separate scientific discipline of limited scope, but rather a cross-linking science that embraces all chemistry and biology. There is hardly any scientist in these and related areas who does not encounter catalytic phenomena in his work. Beyond that, many catalytic processes are known to the educated member of the public, to the politician as well as to any other informed person. Anyone who knows about Mozart, Ravel, and Gershwin should know about catalysis as well. There is no life without the miracles of catalytic reactions in plants, animals, and human beings.

Because of this broad importance of catalysis both in science and in daily life, four scientists covering the main fields of catalysis have edited the present encyclopedia, which is the first comprehensive "Catalysis from A to Z". Approximately 3000 keywords have been explained by 165 authors and coauthors, who are all active in their respective areas. It was our aim to edit a concise handbook which is quickly at hand if one wants to know the meaning of certain catalytic or catalysis-related terminology. We hope that the present book proves a useful compendium for everyone whose occasional or permanent interest is catalysis – scholars, high-school teachers, professors, chemists, bioscientists, and physicists, both within and beyond the academic world.

We have been fortunate to have conferred on a multiplicity of topics with experts. Nevertheless we might have overlooked keywords and entries which are important for catalysis. Please keep the editors informed of any such cases.

We thank the team of Wiley-VCH, Dr. F. Sagi from Lyon (France) for supporting the French translations, and especially our freelance copyeditor, Mrs. Diana P.E. Boatman, M.A., M.R.S.C, C. Chem, from Redhill, Surrey (UK), for their endless patience and help during the difficult process of completion.

Frankfurt, München, Berlin, La Jolla
January 2000

B. Cornils
W. A. Herrmann
R. Schlögl
H.-C. Wong

---

* "We have good reasons to assume that thousands of catalytic processes take place in living organisms between the tissues and the body fluids" (*Ann. Chim. et Phys.* **1836**, 146–151).

List of Contributors

Dr. Roger Alberto, Paul-Scherrer-Institut Würenlingen, CH-5232 Villigen-PSI Aargau/Switzerland

Dipl.-Chem. I. Albers, Universität Hamburg, D-20146 Hamburg/Germany

Dr. Markus Albrecht, Universität Karlsruhe, D-76131 Karlsruhe/Germany

Prof. Dr. Markus Antonietti, Max Planck Institute of Colloids and Interfaces, D-14424 Potsdam/Germany

Dr. Reiner Anwander, Technische Universität München, D-85747 Garching/Germany

Dr. Michael Arndt-Rosenau, Universität Hamburg, D-20146 Hamburg/Germany

Dipl.-Chem. David Arrowsmith, Universität Hamburg, D-20146 Hamburg/Germany

Prof. Dr. Didier Astruc, Université de Bordeaux I, F-33405 Talence Cédex/France

Prof. Dr. Torsten Bach, Universität Marburg, D-35032 Marburg/Germany

Prof. Dr. Jan-E. Bäckvall, University of Uppsala, S-75121 Uppsala/Sweden

Prof. Dr. Manfred Baerns, Institut für Angewandte Chemie, D-12484 Berlin/Germany

Dr. Helmut Bahrmann, Celanese AG/Werk Ruhrchemie, D-46128 Oberhausen/Germany

Prof. Dr. Joachim Bargon, Universität Bonn, D-53115 Bonn/Germany

Prof. Dr. Jean-Marie Basset, Laboratoire COMS, CPE-Lyon, F-69616 Villeurbanne/France

Prof. Dr. Peter Behrens, Universität Hannover, D-30167 Hannover/Germany

Prof. Dr. Matthias Beller, Institut für Organische Katalyseforschung, D-18055 Rostock/Germany

Dipl.-Chem. Bettina Bems, Fritz-Haber-Institut der MPG, D-14195 Berlin/Germany

Dipl.-Chem. Inken Beulich, Universität Hamburg, D-20146 Hamburg/Germany

Dr. Claudio Bianchini, Istituto per lo Studo della Stereochimica ed Energetica dei Composti di Coordinazione, I-50132 Firenze/Italy

Dipl.-Chem. Volker Böhm, Technische Universität München, D-85747 Garching/Germany

Prof. Dr. Helmut Bönnemann, MPI für Kohlenforschung, D-45470 Mülheim/Germany

Dr. Armin Börner, Institut für Organische Katalyseforschung, D-18055 Rostock/Germany

Dipl.-Ing. Ingolf Boettger, Fritz-Haber-Institut der MPG, D-14195 Berlin/Germany

Prof. Dr. G.C. Bond, Brunel University, Uxbridge, Middlesex, UB8 3PH/United Kingdom

Prof. Dr. Henri Brunner, Universität Regensburg, D-93053 Regensburg/Germany

Dr. O.V. Buyevskaya, Institut für Angewandte Chemie, D-12484 Berlin/Germany

Dr. Jürgen Caro, Institut für Angewandte Chemie, D-12484 Berlin/Germany

Prof. Dr. Yves Castanet, Université das Sciences et Technologies de Lille, F-59652 Villeneuve d'Ascq Cédex/France

Dr. Peter Claus, Institut für Angewandte Chemie, D-12484 Berlin/Germany

Prof. Dr. Gabriele Centi, Universitá di Bologna, I-40136 Bologna/Italy

Prof. Dr. Boy Cornils, Hoechst AG, D-65926 Frankfurt/Germany

Dr. Gerd Collin, DECHEMA, D-60486 Frankfurt/Germany

C. Coperit, Laboratoire COMS, CPE-Lyon, F-69616 Villeurbanne/France

Prof. Dr. Mark E. Davis, California Institute of Technology, Pasadena, CA 91125/USA

Dr. J. Paul Day, Corning Incorporated, Corning, NY 14831/USA

Prof. Dr. Eckehard V. Dehmlow, Universität Bielefeld, D-33615 Bielefeld/Germany

Dipl.-Chem. Martin Dieterle, Fritz-Haber-Institut der MPG, D-14195 Berlin/Germany

Prof. Dr. Eckhardt Dinjus, Forschungszentrum Karlsruhe, D-76021 Karlruhe/Germany

Dr. Manfred Döring, Universität Jena, D-07743 Jena/Germany
Dr. Eite Drent, Shell Research and Technology Centre, NL-1030 BN Amsterdam/
The Netherlands
Dr. Birgit Driessen-Hölscher, RWTH Aachen, D-52056 Aachen/Germany
Dr. Anette Eckerle, Wiley-VCH, D-69469 Weinheim/Germany
Dipl.-Chem. Robert Eckl, Technische Universität München, D-85747 Garching/Germany
Prof. Dr. Rudi van Eldik, Universität Erlangen-Nürnberg, D-91058 Erlangen/ Germany
Dr. Markus Enders, Universität Heidelberg, D-69120 Heidelberg/Germany
Dr. Jörg Eppinger, Technische Universität München, D-85747 Garching/Germany
Prof. Dr. Alexander C. Filippou, Humboldt-Universität Berlin, D-10115 Berlin/Germany
Dipl.-Chem. Armin Fischer, Fritz-Haber-Institut der MPG, D-14195 Berlin/Germany
Dr. Richard W. Fischer, Celanese AG/Werk Ruhrchemie, D-46128 Oberhausen/Germany
Dipl.-Chem. Frank Freidanck, Universität Hamburg, D-20146 Hamburg/Germany
Prof. Dr. Gernot Frenking, Universität Marburg, D-35043 Marburg/Germany
Dipl.-Chem. Jörg Fridgen, Technische Universität München, D-85747 Garching/Germany
Dr. Carl-Dieter Frohning, Celanese AG/Werk Ruhrchemie, D-46128 Oberhausen/Germany
Dr. Alois Fürstner, MPI für Kohlenforschung, D-45470 Mülheim/Germany
Dr. Holger Geissler, Clariant AG/Werk Griesheim, D-65916 Frankfurt/Germany
Dr. Gisela Gerstberger, Technische Universität München, D-85747 Garching/Germany
Dipl.-Chem. Dieter Gleich, Technische Universität München, D-85747 Garching/Germany
Dipl.-Chem. Hans W. Görlitzer, Technische Universität München, D-85747 Garching/Germany
Dipl.-Chem. Mattis Gosmann, Universität Hamburg, D-20146 Hamburg/Germany
Prof. Dr. Helmut Greim, GSF-Forschungszentrum für Umwelt und Gesundheit GmbH,
D-85764 Neuherberg/Germany
Dipl.-Chem. Marco M. Günter, Fritz-Haber-Institut der MPG, D-14195 Berlin/Germany
Dr. Peter Härter, Technische Universität München, D-85747 Garching/Germany
Dipl.-Chem. Michael Haevecker, Fritz-Haber-Institut der MPG, D-14195 Berlin/Germany
Prof. Brian E. Hanson, Virginia Polytechnic Institute and State University,
Blacksburg VA 24061-0212/USA
Dr. Detlef Heller, Institut für Organische Katalyseforschung, D-18055 Rostock/Germany
Dr. Jochem Henkelmann, BASF AG, D-67056 Ludwigshafen/Germany
Prof. Dr. Dietmar Hesse, Universtät Hannover, D-30167 Hannover/Germany
Prof. Dr. Dr. mult. h.c. Wolfgang A. Herrmann, Technische Universität München,
D-80333 München/Germany
Prof. Dr. Andreas Heumann, Université d'Aix-Marseille, F-13397 Marseille Cédex 13/France
Dipl.-Chem. Wolfgang Hieringer, Technische Universität München,
D-85747 Garching/Germany
Dr. Charles C. Hobbs, Jr., Celanese Ltd., Corpus Christi, TX 78469-9077/USA
Prof. Dr. D. Hönicke, Technische Universität Chemnitz, D-09107 Chemnitz/ Germany
Prof. Dr. István T. Horváth, Eötvös Loránd University, H-1117 Budapest/Hungary
Prof. Dr. Ronald Imbihl, Universität Hannover, D-30167 Hannover
Dr. Friederike Jentoft, Fritz-Haber-Institut der MPG, D-14195 Berlin/Germany
Dr. Rolf Jentoft, Fritz-Haber-Institut der MPG, D-14195 Berlin/Germany
Prof. Dr. Zilin Jin, Dalian University of Technology, 116012 Dalian/PR China
Dipl.-Chem. Yvonne Joseph, Fritz-Haber-Institut der MPG, D-14195 Berlin/Germany
Prof. Dr. Philippe Kalck, Ecole Nationale Supérieure de Chimie de Toulouse,
F-31077 Toulouse Cedex 4/France
Prof. Dr. Walter Kaminsky, Universität Hamburg, D-20146 Hamburg/Germany

Prof. Dr. Helmut Knözinger, Universität München, D-80333 München/Germany

Dr. Christian W. Kohlpaintner, Celanese AG/Werk Ruhrchemie, D-46128 Oberhausen/ Germany

Prof. Dr. Klaus Köhler, Technische Universität München, D-85747 Garching/Germany

Dr. Randolf D. Kohn, Technische Universität Berlin, B-10623 Berlin/Germany

Prof. Dr. Udo Kragl, Universität Rostock, D-18055 Rostock/Germany

Dr. Roland Kratzer, Aventis GmbH & Co KG, D-65926 Frankfurt/Germany

Dr. Jürgen G.E Krauter, Degussa-Hüls AG, D-63457 Hanau/Germany

Dr. Steffen Krill, Degussa-Hüls AG, D-63403 Hanau/Germany

Dr. Fritz Kühn, Technische Universität München, D-85747 Garching/Germany

Dipl.-Chem. Christian Kuhrs, Fritz-Haber-Institut der MPG, D-14195 Berlin/Germany

Dr. Jürgen Kulpe, Celanese AG/Werk Ruhrchemie, D-46128 Oberhausen/Germany

Dr. Walter Leitner, MPI für Kohlenforschung, D-45470 Mülheim/Germany

F. Lefebre, Laboratoire COMS, CPE-Lyon, F-69616 Villeurbanne/France

Prof. Dr. Heiner Lieske, Institut für Angewandte Chemie, D-12484 Berlin/Germany

Prof. Dr. Torsten Linker, Universität Stuttgart, D-70569 Stuttgart/Germany

Prof. Dr. Hanfan Liu, Chinese Academy of Sciences, Beijing 1000 80/PR China

Dr. Egbert Lox, Degussa-Hüls AG, D-63403 Hanau/Germany

Prof. Dr. Bernhard Lücke, Institut für Angewandte Chemie, D-12484 Berlin/ Germany

Prof. Dr. Wilhelm F. Maier, MPI für Kohlenforschung, D-45470 Mülheim/Germany

Prof. Dr. Bogdan Marciniec, Adam Mickiewicz University Poznań, PL60–780 Poznan/Poland

Dr. Gerhard Mestl, Fritz-Haber-Institut der MPG, D-14195 Berlin/Germany

Dr. Dimitrios Mihalios, Technische Universität München, D-85747 Garching/Germany

Prof. Dr. Ilya I. Moiseev, Russian Academy of Sciences, 117907 Moscow GSP-1/Russia

Prof. Dr. Eric Monflier, Université d'Artois, F-62307 Lens Cédex/France

Prof. Dr. André Mortreux, Université des Sciences et Technologies de Lille, F-59652 Villeneuve d'Ascq Cédex/France

Dr. J.P. Müller, Institut für Angewandte Chemie, D-12484 Berlin/Germany

Prof. Dr. Klaus Müller-Dethlefs, University of York, GB-York Y010 5DD/United Kingdom

Prof. Dr. Martin Muhler, Ruhr-Universität Bochum, D-44780 Bochum/Germany

Prof. Dr. Ronny Neumann, The Hebrew University of Jerusalem, Jerusalem/Israel 91904

Prof. Dr. A.F. Noels, Université de Liège, B-4000 Sart Tilman/Belgium

Prof. Dr. Günther Oehme, Universität Rostock, D-18055 Rostock/Germany

Prof. Dr. Tamon Okano, Tottori University, J-680 Tottori/Japan

Prof. Dr. Jun Okuda, Johannes-Gutenberg-Universität Mainz, D-55099 Mainz/Germany

Prof Dr. Hélène Olivier, Institut Francais du Pétrole, F-92852 Rueil-Malmaison Cédex/France

Dr. Peter Panster, Degussa-Hüls AG, D-63403 Hanau/Germany

Dr. U. Prüsse, Bundesforschungsanstalt für Landwirtschaft, D-38116 Braunschweig/Germany

Dipl.-Chem. Oliver Pyrlik, Universität Hamburg, D-20146 Hamburg/Germany

Dr. Florian Rampf, Technische Universität München, D-85747 Garching/Germany

Prof. Dr. Manfred T. Reetz, MPI für Kohlenforschung, D-45470 Mülheim/Germany

Prof. Dr. Oliver Reiser, Universität Regensburg, D-93053 Regensburg/Germany

Prof. Dr. Albert Renken, École Polytechnique Fédérale de Lausanne, CH-1015 Lausanne/ Switzerland

Dr. Peter Roesky, Universität Karlsruhe, D-76128 Karlsruhe/Germany

Prof. Dr. Ingo Romey, Universität GH Essen, D-45141 Essen/Germany

Dr. Karola Rück-Braun, Johannes Gutenberg-Universität Mainz, D-55099 Mainz/Germany

Dipl.-Chem. Emilio Sanchez-Cortezon, Fritz-Haber-Institut der MPG, D-14195 Berlin/Germany

Dr. Gerd Sandstede, Sandstede Technologie Consulting, D-60598 Frankfurt/Germany
Dr. Thomas Schedel-Niedrig, Fritz-Haber-Institut der MPG, D-14195 Berlin/Germany
Dipl.-Chem. Andreas Scheybal, Fritz-Haber-Institut der MPG, D-14195 Berlin/Germany
Dr. Annette Schier, Technische Universität München, D-85747 Garching/Germany
Prof. Dr. Robert Schlögl, Fritz-Haber-Institut der MPG, D-14195 Berlin/Germany
Prof. Dr. Hubert Schmidbaur, Technische Universität München, D-85747 Garching/Germany
Prof. Dr. A. Dieter Schlüter, Freie Universität Berlin, D-14195 Berlin/Germany
Dipl.-Chem. Volker Scholz, Universität Hamburg, D-20146 Hamburg/Germany
Dr. Detlef Schröder, Technische Universität Berlin, D-10623 Berlin/Germany
Prof. Dr. Ferdi Schüth, MPI für Kohlenforschung, D-45470 Mülheim/Germany
Dr. Michael Schulz, BASF AG, D-67056 Ludwigshafen/Germany
Dr. Rudolf Schwabe, GSF – Forschungszentrum für Umwelt und Gesundheit GmbH,
D-85764 Neuherberg/Germany
Dipl.-Chem. Constantin Schwecke, Universität Hamburg, D-20146 Hamburg/Germany
Prof. Dr. Pam Sears, The Scripps Research Institute, La Jolla CA 92037/USA
Dr. Christian Seel, Universität Bonn, D-53121 Bonn/Germany
Prof. Dr. Denis Sinou Université Claude Bernard Lyon I, F-69622 Villeurbanne Cedex/France
Prof. Dr. Roger A. Sheldon, Delft University of Technology, NL-2628 BL Delft/
The Netherlands
Dr. Brad L. Smith, Celanese Ltd., Wilmington NC 28402/USA
Prof. Dr. Othmar Stelzer, Bergische Universität, D-42097 Wuppertal/Germany
Dr. Thomas Straßner, Technische Universität München, D-85747 Garching/Germany
Dipl.-Chem. Christian Strübel, Universität Hamburg, D-20146 Hamburg/Germany
Prof. Dr. Masato Tanaka, National Institute of Materials and Chemical Research Tsukuba,
Ibaraki 305–8565/Japan
Prof. Dr. Rudolf Taube, Technische Universität München, D-85747 Garching/Germany
Sir John Meurig Thomas, SCD, FRS, The Royal Institution of Great Britain, London W1X 4BS/
United Kingdom
Dr. Werner R. Thiel, Technische Universität München, D-85747 Garching/Germany
Dr. Paull Torrence, Celanese Ltd., Corpus Christi, TX 78469-9077/USA
Dr. Stephan Trautschold, Clariant GmbH/Werk Gersthofen, D-86005 Augsburg/Germany
Dipl.-Chem. Marc Vathauer, Universität Hamburg, D-20146 Hamburg/Germany
Prof. Dr. Fritz Vögtle, Universität Bonn, D-53121 Bonn/Germany
Prof. Dr. Klaus-Dieter Vorlop, Bundesforschungsanstalt für Landwirtschaft,
D-38116 Braunschweig/Germany
Dr. Matthias Wagner, Technische Universität München, D-85747 Garching/Germany
Dipl.-Chem. Ulrich Weingarten, Universität Hamburg, D-20146 Hamburg/Germany
Dipl.-Chem. Ralf Werner, Universität Hamburg, D-20146 Hamburg/ Germany
Dr. Thomas Weskamp, Technische Universität München, D-85747 Garching/Germany
Prof. Dr. John M. Winfield, University of Glasgow, Glasgow G12 8QQ/United Kingdom
Dipl.-Chem. Hans-Jörg Wölk, Fritz-Haber-Institut der MPG, D-14195 Berlin/Germany
Prof. Dr. Chi-Huey Wong, The Scripps Research Institute, La Jolla, CA 92037/USA
Dr. Noriaki Yoshimura, Kuraray Co. Ltd., J-710 Okayama/Japan
Dipl.-Chem. Jochen Zoller, Technische Universität München, D-85747 Garching/Germany

How to Use this Encyclopedia

Abbreviations Used

abbr.	abbreviation
acc.	according to, accordingly
calcn(s).	calculation(s)
calc(d)(ng).	calculat(ed)(ing)
cat(s).	catalyst(s), catalytical(ly), catalysis, catalyzing, catalyze(s)
comb(s).	combination(s)
comp(s).	component(s)
conc(s).	concentration(s)
cond(s).	condition(s)
conv(s).	conversion(s)
cpd(s).	compound(s)
3D	three dimensional
DE	German patent
ee	enantiomeric excess
EP	European patent
equ(s).	equation(s)
equil(s).	equilibrium, equilibria
equiv.	equivalent
FR	French patent
GB	British Patent
H	hydrogen
hc(s).	hydrocarbon(s)
het.	heterogen, heterogeneous(ly)
hom.	homogen, homogeneous(ly)
manuf(d)(ng).	manufacture(d)(ing)
m/m	molar ratio
N	nitrogen
O	oxygen
prep(d)(n).	prepared, preparation
press.	pressure
proc(s).	process(es)
prod(s).	production, product(s)
®	trade mark
rac.	racemic
rt	room temperature
S	sulfur
sc.	supercritical
spec(s).	specification(s)
synth.	synthesis, synthethic, synthesize, synthetically
temp(s).	temperature(s)
UHV	ultra high vacuum

US	US patent
UV	ultra violet light
w/o	with or without
wt.%	weight%
w/w	by weight (e.g., %)
→	cross reference, cf.
⇀	reaction

Plurals of noun abbreviations are formed by adding "s" to the singular abbr. (such as abbrs.).

References

The references at the end of most of the entries illustrate developments or important aspects of the keyword, being mainly from recent publications. References which are mentioned often are given with only the name of the author and are compiled under "General references cited" (cf. below). Other information is cited mainly from recent publications in journals; in this case we dispensed with the names of the authors. Of course, no attempt has been made to provide a complete literature review of the topics covered by each keyword.

By courtesy of Marcel Dekker, Inc. (New York), Kluwer Academic Publishers (Dordrecht), and Gulf Publishing, Comp. (Houston, TX) various Figures have been reproduced from their respective publications.

Cross References

Cross references to other keywords are given by → or by *.

Translations

Each keyword is translated into French (**F**) and German (**G**); identical translations are indicated by **E=F=G**.

Trademarks, Patents, etc.

Trademarks (®, ™) are normally not marked, simply to refrain from using the encyclopedia for product advertising purposes. Patents are cited according to international convention (US for the United States, FR for France, GB for Great Britain, DE for Germany, EP European Patent, etc.).

General References Cited

E. W. Abel, F. G. A. Stone, G. Wilkinson (Eds.), *Comprehensive Organometallic Chemistry*, Pergamon Press, Oxford 1995.

R. D. Adams, A. F. Cotton, *Catalysis by Di- and Polynuclear Metal Cluster Compounds*, Wiley, Chichester 1998.

J. R. Anderson, M. Boudart (Eds.), *Catalysis – Science and Technology*, Springer, Berlin 1981.

Anonymous, *Modern Petroleum Technology*, 2 Vols., Wiley, New York 1999.

D. Astruc, *Electron Transfer and Radical Processes in Transition-Metal Chemistry*, VCH, New York 1995.

J. D. Atwood, *Inorganic and Organometallic Reaction Mechanisms*, Wiley-VCH, Weinheim 1997.

R. L. Augustine, *Heterogeneous Catalysis for the Synthetic Chemist*, Marcel Dekker, New York 1996.

D. R. Barton, A. E. Martell, D. T. Sawyer (Eds.), *The Activation of Dioxygen and Homogeneous Catalytic Oxidation*, Plenum Press, New York 1993.

J. M. Basset, B. C. Gates, J. P. Candy, A. Choplin, M. Leconte, F. Quignard, C. C. Santini, *Surface Organometallic Chemistry*, in: *Molecular Approaches to Surface Catalysis*, Kluwer, Dordrecht 1988.

E. R. Becker, C. J. Perreira (Eds.), *Computer-Aided Design of Catalysts*, Marcel Dekker, New York 1993.

M. Beller, C. Bolm (Eds.), *Transition Metals for Organic Synthesis*, 2 Vols., Wiley-VCH, Weinheim 1998.

A. Benninghoven, F. G. Rüdenauer, H. W. Werner, *Secondary Ion Mass Spectrometry*, Wiley, New York 1987.

H. Beyer (Ed.), *Catalysis by Microporous Materials*, Elsevier, Amsterdam 1995.

A. Bielanski, J. Haber, *Oxygen in Catalysis*, Marcel Dekker, New York 1991.

J. R. Blackborrow, D. Young, *Metal Vapor Synthesis in Organometallic Chemistry*, Springer, Berlin 1979.

G. C. Bond (Bond-1), *Catalysis by Metals*, Academic Press, London 1962.

G. C. Bond (Bond-2), *Homogeneous Catalysis*, Oxford Science Publ., Clarendon Press, Oxford 1987.

K. H. Büchel, H.-H. Moretto, P. Woditsch, *Industrielle Anorganische Chemie*, 3rd Ed., Wiley-VCH, Weinheim 1999.

W. Büchner, R. Schliebs, G. Winter, K. H. Büchel, *Industrial Inorganic Chemistry*, VCH, Weinheim 1989.

J. Burgess, *Ions in Solution*, Horwood, Chichester 1988.

J. B. Butt, E. E. Petersen, *Activation, Deactivation, and Poisoning of Catalysts*, Academic Press, San Diego 1988.

G. Centi, F. Trifiro (Eds.), *New Developments in Selective Oxidation*, Elsevier, Amsterdam 1990.

L. Cerveny (Ed.), *Catalytic Hydrogenation*, Elsevier, Amsterdam 1986.

J. Chastain (Ed.), *Handbook of X-ray Photoelectron Spectroscopy*, Perkin-Elmer Co., Eden Prairie 1992.

F. Ciardelli, E. Tsuchida, D. Wöhrle (Eds.), *Macromolecule-Metal Complexes*, Springer, Heidelberg 1996.

J. H. Clark, *Catalysis of Organic Reactions by Supported Inorganic Reagents*, VCH, New York 1994.

A. N. Collins, G. N. Sheldrake, J. Crosby (Eds.), *Chirality in Industry*, Wiley, Chichester 1992.

H. M. Colquhoun, D. J. Thompson, M. V. Twigg, *Carbonylation*, Plenum Press, New York 1991.

K. A. Connors, *Chemical Kinetics*, VCH, Weinheim 1990.

B. Cornils, W. A. Herrmann (Eds.)(Cornils/Herrmann-1), *Applied Homogeneous Catalysis with Organometallic Compounds*, VCH, Weinheim 1996.

B. Cornils, W. A. Herrmann (Eds.)(Cornils/Herrmann-2), *Aqueous-Phase Organometallic Catalysis*, Wiley-VCH, Weinheim 1998.

R. H. Crabtree, *The Organometallic Chemistry of the Transition Metals*, 2nd Ed., Wiley, New York 1994.

M. E. Davis, S. L. Suib (Eds.), *Selectivity in Catalysis*, ACS Symposium Series 517, ACS, Washington 1993.

B. Delmon, G. F. Froment (Eds.), *Catalyst Deactivation*, Elsevier, Amsterdam 1987.

F. Diederich, P. J. Stang, *Metal-catalyzed Cross-coupling Reactions*, Wiley-VCH, Weinheim 1998.

B. Dietrich, P. Viout, J.-M. Lehn, *Macrocyclic Chemistry*, VCH, Weinheim 1993.

T. J. Dines, C. H. Rochester, J. Thomson, *Catalysis and Surface Characterization*, Royal Society of Chemistry, Cambridge 1992.

H. G. Elias, *Makromoleküle*, 5 th ed., Hüthig und Wepf, Basel 1990/1992.

G. Ertl, H. Knözinger, J. Weitkamp (Eds.), *Handbook of Heterogeneous Catalysis*, 5 Vols., Wiley-VCH, Weinheim 1997.

J. Falbe (Ed.) (Falbe-1), *New Syntheses with Carbon Monoxide*, Springer, New York 1980.

J. Falbe (Ed.) (Falbe-2), *Chemierohstoffe aus Kohle*, Thieme, Stuttgart 1977; English version: *Chemical Feedstocks from Coal*, Wiley, New York 1982.

J. Falbe (Falbe-3), *Synthesen mit Kohlenmonoxid*, Springer, Berlin 1967; English version 1970 (Falbe-3a).

J. Falbe, U. Hasserodt (Eds.) (Falbe-4), *Katalysatoren, Tenside und Mineralöladditive*, Thieme, Stuttgart 1978.

R. J. Farrauto, C. H. Bartholomew, *Fundamentals of Industrial Catalytic Processes*, Blackie Academic & Professional, Chapman & Hall, London 1997.

G. Fink, R. Mülhaupt, H.H. Brintzinger, *Ziegler Catalysts – Recent Scientific Innovations and Technical Improvements*, Springer, Berlin 1995.

A. Fürstner, *Active Metals*, Wiley-VCH, Weinheim 1995.

B.C. Gates (Gates-1), *Catalytic Chemistry*, Wiley, New York 1992.

B.C. Gates, L. Guczi, H. Knözinger (Gates-2), *Metal Clusters in Catalysis*, Elsevier, Amsterdam 1986.

B. Giese, *Radicals in Organic Synthesis: Formation of Carbon-Carbon Bonds*, Pergamon, Oxford 1986.

R. van Grieken, A. Markowicz, *Handbook of X-Ray Spectrometry*, Dekker, New York 1992.

M. Guisnet, J. Barrault, C. Bouchoule, D. Duprez, C. Montassier, G. Perot, *Heterogeneous Catalysis and Fine Chemicals*, Studies in Surface Science and Catalysis *41* (1988), *59* (1990), *78* (1993) and *108* (1997), Elsevier, Amsterdam.

D.C. Harris, M.D. Bertolucci, *Symmetry and Spectroscopy*, Dover Publications, New York 1989.

T.R. Hartley, *Supported Metal Complexes*, Reidel, Dordrecht 1985.

F.R. Hartley, *The Chemistry of Organophosphorus Compounds*, Wiley, New York 1990.

R.F. Heck, *Palladium Reagents in Organic Synthesis*, Academic Press, London 1985.

R.M. Heck, R.J. Farrauto, *Catalytic Air Pollution Control – Commercial Technology*, van Nostrand-Reinhold, New York 1995.

L. Hegedus (Ed.), *Catalyst Design – Progress and Perspectives*, Wiley, New York 1987.

C.L. Hill, *Activation and Functionalization of Alkanes*, Wiley, New York 1989.

I.T. Horváth, F. Joó, *Aqueous Organometallic Chemistry and Catalysis*, NATO ASI Series, Kluwer, Dordrecht 1995.

H. Ibach, D.L. Mills, *Electron Energy Loss Spectroscopy and Surface Vibrations*, Academic Press, New York 1982.

B. Imelik, J.C. Vedrine, *Catalyst Characterization*, Plenum Press, London 1994.

K.J. Ivin, J.C. Mol, *Olefin Metathesis and Metathesis Polymerization*, Academic Press, San Diego 1997.

Y. Iwasawa (Ed.), *Tailored Metal Catalysts*, D. Reidel, Dordrecht 1986.

P.W. Jolly, G. Wilke, *The Organic Chemistry of Nickel*, Academic Press, London 1974.

W. Kaminsky, H. Sinn (Eds.), *Transition Metals and Organometallics as Catalysts for Olefin Polymerization*, Springer, New York 1988.

Kirk/Othmer-1, *Encyclopedia of Chemical Technology*, 4th Ed., Wiley, New York 1991–1998.

Kirk/Othmer-2, *Concise Encyclopedia of Chemical Technology*, Wiley-Interscience, New York 1985.

H. Krauch, W. Kunz, *Reaktionen der organischen Chemie*, 6th Ed., Hüthig, Heidelberg 1997.

K.J. Laidler, *Chemical Kinetics*, 3rd Ed., Harper and Row, New York 1987.

T. Laue, A. Plagens, *Namens- und Schlagwort-Reaktionen der Organischen Chemie*, Teubner, Stuttgart 1994.

N.J. Long, *Metallocenes – An Introduction to Sandwich Complexes*, Blackwell Science, Oxford 1997.

J. March, *Advanced Organic Chemistry*, 4th Ed., Wiley, New York 1992.

L.N. Mander, *Stereoselektive Synthese*, Wiley-VCH, Weinheim 1998.

H.E. Mark, N.M. Bikales, C.G. Overberger, G. Menges (Eds.), *Encyclopedia of Polymer Sciences and Engineering*, Wiley-Interscience, New York 1985–1989.

A.E. Martell, R.D. Hancock, *Metal Complexes in Aqueous Solution*, Plenum, New York 1996.

J.J. McKetta, W.A. Cunningham (Eds.) (McKetta-1), *Encyclopedia of Chemical Processing and Design*, Marcel Dekker, New York 1976–1998.

J.J. McKetta (Ed.) (McKetta-2), *Chemical Processing Handbook*, Marcel Dekker, New York 1993.

Methoden der Organischen Chemie/Methods of Organic Chemistry/Methodicum, 4th Ed., Thieme, Stuttgart 1978–1999.

A. Mortreux, F. Petit, *Industrial Aspects of Homogeneous Catalysis*, Riedel, Dordrecht 1988.

F. Montanari, L. Casella (Eds.), *Metalloporphyrin Catalyzed Oxidations*, Kluwer, Dordrecht 1994.

K. Morokuma, J.H. van Lenthe, P.W.N.M. van Leeuwen, *Theoretical Aspects of Homogeneous Catalysis*, Kluwer, Dordecht 1995.

J.A. Moulijn, P.W.N.M. van Leeuwen, R.A. van Santen (Eds.), *Catalysis*, Elsevier, Amsterdam 1993.

J. Mulzer, H. Waldmann (Eds.), *Organic Synthesis Highlights III*, Wiley-VCH, Weinheim 1998.

S.-I. Murahashi, S.G. Davies, *Transition Metal Catalyzed Reactions*, Blackwell Science, 1999.

S. Neufeldt, *Chronologie Chemie 1800–1980*, 2nd Ed., VCH, Weinheim 1987.

J.W. Niemantsverdriet, *Spectroscopy in Catalysis*, VCH, Weinheim 1993.

R. Noyori, *Asymmetric Catalysis in Organic Synthesis*, Wiley, New York 1994.

I. Ojima, *Catalytic Asymmetric Synthesis*, VCH, Weinheim 1993.

G.A. Oláh (Ed.)(Oláh-1), *Friedel-Crafts and Related Reactions*, Wiley, New York 1964.

G.A. Oláh (Oláh-2), *Friedel-Crafts Chemistry*, Wiley-Interscience, New York 1973.

Organic Reactions, Wiley, New York 1942–1999.

Organic Syntheses, Vols. 1–69, Wiley, New York 1932–1993.

G.W. Parshall, D.I. Ittel, *Homogeneous Catalysis*, 2nd Ed., Wiley, New York 1993.

S. Patai, Z. Rappoport (Eds.) (Patai-1), *The Chemistry of Organic Silicon Compounds*, Wiley, Chichester 1989.

S. Patai (Ed.) (Patai-2), *The Chemistry of Organophosphorus Compounds*, Wiley, New York 1990.

K.R. Payne (Ed.), Chemicals from Coal: New Processes, in: *Critical Reports on Applied Chemistry*, Vol. 14, Wiley, Chichester 1987.

M.J. Perkins, *Radical Chemistry*, Ellis Horwood, Chichester 1994.

H. Pignolet (Ed.), *Homogeneous Catalysis with Metal Phosphine Complexes*, Plenum Press, New York 1983.

R.P. Quirk (Ed.), *Transition Metal Catalyzed Polymerizations*, Cambridge University Press, Cambridge 1988.

C. Reichardt, *Solvents and Solvent Effects in Organic Chemistry*, VCH, Weinheim 1988.

P.N. Rylander, *Catalytic Hydrogenation over Platinum Metals*, Academic Press, New York 1967.

D. Schomberg, M. Salzmann, *Enzyme Handbook*, Springer, Heidelberg 1993.

G.M. Schwab, *Catalysis from the Standpoint of Chemical Kinetics*, Macmillan, London 1936, and van Nostrand, New York 1937.

R.A. Sheldon, J.K. Kochi (Sheldon-1), *Metal-catalyzed Oxidations of Organic Compounds*, Academic Press, New York 1981.

R.A. Sheldon (Ed.) (Sheldon-2), *Metalloporphyrins in Catalytic Oxidation*, Marcel Dekker, New York 1994.

L. Simandi, *Catalytic Activation of Dioxygen by Metal Complexes*, Kluwer Academic Publishers, Boston 1992.

M. Sinnott (Ed.), *Comprehensive Biological Catalysis*, Academic Press, London 1998.

G.A. Somorjai, *Principles of Surface Chemistry and Catalysis*, Wiley, New York 1994.

G.G. Stanley, *Catalysis of Organic Reactions*, Dekker, New York 1995.

A.B. Stiles (Ed.), *Catalyst Supports and Supported Catalysts*, Butterworths, Boston 1987.

H.H. Storch, N. Golumbic, R.B. Anderson, *The Fischer-Tropsch and Related Syntheses*, Wiley, New York 1951.

G. Strakul (Ed.), *Catalytic Oxidations with H_2O_2 as Oxidant*, Kluwer, Dordrecht 1992.

W. Stumm (Ed.), *Aquatic Chemical Kinetics*, Wiley, New York 1990.

B.P. Sullivan, K. Krist, H.E. Guard, *Electrochemical and Electrocatalytic Reactions of Carbon Dioxide*, Elsevier, Amsterdam 1993.

J.M. Thomas, W.J. Thomas, *Heterogeneous Catalysis*, Wiley-VCH, Weinheim 1997.

A. Togni, R.L. Halterman (Eds.) (Togni-1), *Metallocenes*, Wiley-VCH, Weinheim, 1998.

A. Togni, T. Hayashi (Eds.) (Togni-2), *Ferrocenes*, 2 Vols., Wiley-VCH, Weinheim 1994.

H. Topsøe, B.S. Clausen, F.E. Massoth, *Hydrotreating Catalysis*, Springer, Berlin 1996.

D.L. Trimm, S. Akashah, M. Absi-Halabi, A. Bishara (Eds.), *Catalysts in Petroleum Refining 1989*, Elsevier, Amsterdam 1990.

B.M. Trost, I. Fleming (Eds.), *Comprehensive Organic Synthesis*, Pergamon, Oxford 1991.

J. Tsuji, *Transition Metal Reagents and Catalysts*, Wiley, Chichester 2000.

M.V. Twigg, *Catalyst Handbook*, 2nd Ed., Wolf, London 1989.

R. Ugo (Ed.), *Aspects of Homogeneous Catalysis*, Kluwer Academic Publishers, Dordrecht 1990.

Ullmann's Encyclopedia of Industrial Chemistry, 5th Ed., Verlag Chemie (VCH), Weinheim 1985–1996.

Ullmann's Encyclopedia of Industrial Chemistry, 6th Ed., 1999 electronic release (fully network-able database); Wiley-VCH, Weinheim.

R. van Eldik, C. D. Hubbard (Eds.), *Chemistry under Extreme or Non-Classical Conditions*, Wiley-Spectrum, New York/Heidelberg 1997.

R. A. van Santen, *Theoretical Heterogeneous Catalysis*, World Scientific, Singapore 1991.

R. A. van Santen, P. W. N. M. van Leeuwen, J. A. Moulijn, B. A. Averill, *Catalysis – An Integrated Approach*, 2nd Ed., Elsevier, Amsterdam 1999.

K. Weissermel, H.-J. Arpe, *Industrial Organic Chemistry*, 4th Ed., VCH, Weinheim 1993 (5th Ed. in 1998).

H. Werner, W. Schreier, *Selective Reactions of Metal-Activated Molecules*, Vieweg, Wiesbaden 1998.

R. I. Wijngaarden, A. Kronberg, K. R. Westerterp, *Industrial Catalysis*, Wiley-VCH, Weinheim 1998.

R. G. Wilkins, *Kinetics and Mechanism of Reactions of Transition Metal Complexes*, VCH, Weinheim 1991.

Winnacker-Küchler (H. Harnisch, R. Steiner, K. Winnacker, Eds.), *Chemische Technologie*, Hanser, Munich 1981.

C.-H. Wong, G. M. Whitesides, *Enzymes in Synthetic Organic Chemistry*, Pergamon, New York 1994.

Yu. I. Yermakov, Y. A. Zakharov, B. N. Kuznetsov, *Catalysis by Supported Metal Complexes*, Elsevier, Amsterdam 1981.

S. Yoshida, S. Sasaki, H. Kobayashi, *Electronic Processes in Catalysis*, VCH, Weinheim 1994.

A

AA 1: → acetic acid; 2: → acrylic acid; 3: → adipic acid; 4: → acrylamide; 5: atomic absorption, → atomic absorption spectroscopy (AAS)

AAA → acetic acid anhydride

AAS → atomic absorption spectroscopy

ABB process → Lummus procs.

ab-initio calculations Nonrelativistic quantum mechanical ab-initio calcns. (aics.) aim at approximate solutions of the *Schrödinger equ. without making use of experimental values. The first step in an aic. is the solution of the one-electron *Hartree–Fock (HF) equs., which approximate the many-electron wave function as a prod. of one-electron functions. The prod. of the one-electron functions has the form of a determinant, called the Slater determinant, because a simple prod. would violate the Pauli principle. The integro-differential HF equs. are usually approximated by the *LCAO, which expands the one-electron functions as a set of basis functions called the basis set. The LCAO approximation leads to linear equs. which can be solved iteratively in the *SCF procedure. Most aics. also make use of the Born–Oppenheimer approximation, i.e., the electrons move in the field of fixed nuclei. The quality of an HF-LCAO aic. is determined by the size of the basis set. Since the HF-LCAO equs. are variational, the calcd. energy is always higher than the true energy. The difference between the correct (nonrelativistic) energy and the HF energy with an infinite basis set is called the *correlation energy. There are several methods for calcng. correlation energy in aics. The most popular ones are Møller–Plesset perturbation theory, configuration interaction (CI), multi-config-

uration SCF (MCSCF), and coupled-cluster theory. Aics. can thus be carried out at many levels of theory, the quality of which depends on the size of the basis set and the accuracy of the correlation treatment. High-level aics. may challenge the accuracy of experimental values, but they can become very expensive. Relativistic aics. try to solve the Dirac equs., which give relativistic total energies. Solving the Dirac equs. is substantially more complicated than solving the Schrödinger equ., because the former equs. involve a four-comp. Hamiltonian. Most aics. circumvent this problem by using *pseudopotentials. Aics. can also predict physical properties of atoms and molecules such as ionization potentials, electron affinities, equil. geometries, bond dissociation energies, activation and reaction enthalpies, rotational and vibrational spectra, NMR chemical shifts, coupling constants, etc. In this respect they are interesting for cat. as well. The theoretical properties are obtained from calcn. of first or second derivatives of the energy with respect to coordinates or electromagnetic fields. **F** calculs d'ab-initio; **G** Ab-initio-Rechnungen.   G. FRENKING

Ref.: Thomas/Thomas, p. 400.

ABS 1: alkylbenzene sulfonate, manuf. by *alkylation of benzene and subsequent *sulfonation; 2: a co- or graft-*copolymer of acrylonitrile, butadiene, and styrene, where the disperse phase formed by a butadiene elastomer is distributed in a continuous phase consisting of a styrene–acrylonitrile copolymer.

B. CORNILS

abzymes → catalytic antibodies

ac, Ac acetyl (MeCO–) or acyl (MeCO–O–) groups in *complex cpds. or as *ligands. Acc. to the IUPAC rules for nomenclature abbrs. for *ligands should be written with small letters.

B. CORNILS

Ac₂O → acetic (acid) anhydride, AAA

accelerator → activator, *promoter, *co-catalyst

acceptor ligands are *ligands which usually form stable *complexes with *transition metals in low positive, zero, or negative oxidation states. These ligands bear low-energy, empty orbitals (acceptor orbitals) and are able to accept electrons (ligand acidity) from a metal center. This occurs upon interaction with filled metal-centered orbitals of the same symmetry (metal–ligand *back-bonding) and reduces the electron density at the metal center brought about by the ligand-to-metal σ- or π-donation of electrons. Depending on the symmetry of the *overlapping orbitals with respect to the metal–ligand axis, als. can be classified into π-acceptors (π-acids, e.g. CO, N_2, *phosphines, alkenes, *carbenes) and δ-acceptors (e.g. *arenes, *cp). Another classification differentiates between single-faced and double-faced als. π-Als. exert a strong ligand field leading usually to complexes with a low-spin *configuration. **E=F=G.**

A.C. FILIPPOU

Ref.: DeKock, Gray, *Chemical Structure and Bonding*, University Science Books, Sausalito 1989.

accessibility of het. cats. means the facility of diffusion of heat and mass into and out of their *surfaces. The physicochemical background is a considerable topic in cat. reaction engineering. Important factors to be taken into account are, inter alia, gradients of temps. and concs., and *diffusion coefficients, reaction charateristics, *porosity, oscillations (steady/unsteady states), spatial distribution of the *active sites (*distribution of metals), etc. In many cases, *selectivity, *stability, and a. must be considered together; retroactive consequences to the choice of reactors and

their design are not excluded. **F** accessibilité; **G** Zugänglichkeit.

B. CORNILS

Ref.: Anderson/Boudart, Vol. 8, Chapter 3.

acac, ACAC acetylacetone anions in *complex cpds. or as *ligands

acetamidocinnamate amidohydrolase
This *enzyme cat. the *hydrolysis of acetamidocinnamic acid to phenylpyruvic acid and acetamide. This transformation has been used synth. in the prod. of L-phenylalanine from acetamidocinnamic acid by *coupling this reaction with enzymatic *transamination. **E=F=G.**

P.S. SEARS

Ref.: *J.Biotechnol.* **1986**, *4*, 293.

acetate kinase (EC 2.7.2.1) cat. the synth. of adenosine 5-triphosphate (*ATP) from acetyl phosphate (AcP) and adenosine 5′-diphosphate (*ADP).

The reaction is in favor of *ATP formation, and the *enzyme also accepts other *NDPs. The reaction has been used in the *regeneration of ATP for enzymatic synth. **E=F=G.**

C.-H. WONG

acetic acid (AA, AcOH) is an important intermediate, manuf. preferably by cat. procs. such as *oxidation of hydrocarbons (procs. of *Celanese LPO, *Bayer, *BP, *Hüls), acetaldehyde *oxidation (*Hoechst AA, *Daicel, *Rhône-Poulenc), or *carbonylation of methanol (procs. of *BASF, *Monsanto). **F** acide acétique; **G** Essigsäure.

B. CORNILS

Ref.: Kirk/Othmer-1, Vol. 1, p. 121; Ullmann, Vol. A1, p. 45; Cornils/Herrmann-1, p. 104; McKetta-1, Vol. 1, p. 216; *Chem.Tech.* **1971**, 600.

acetic acid anhydride (AAA, Ac₂O, acetic anhydride) important intermediate, manufd. by oxidative *dehydration of acetaldehyde (*Hoechst AAA proc.) or by *carbonylation of *methyl acetate or dimethyl ether (procs. of *Eastman, *Halcon, *Hoechst). **F** anhydride acétique; **G** Essigsäureanhydrid (Acetanhydrid).   B. CORNILS

Ref.: Kirk/Othmer-2, p. 10; Ullmann, Vol. A1, p. 65; Cornils/Herrmann-1, p. 104; Weissermel/Arpe, p. 180; McKetta-1, Vol. 1, p. 258.

acetoacetyl CoA reductase → poly-β-hydroxybutyrate

acetoacetyl CoA thiolase → poly-β-hydroxybutyrate

acetoxylation is the introduction of the acetoxy group –O–CO–CH$_3$ into organic cpds. such as arenes (benzene → phenyl acetate), alkenes (ethylene → *vinyl acetate; cf. reactions of *Treibs and *Moiseev), dienes (butadiene → 1,4-diacetoxybutene), or ketones (cyclohexane → acetoxycyclohexane). In the presence of *acetic acid (source for the acetoxy group) and strong oxidants like Pb(OAc)$_4$ (LTA), NaNO$_2$, NaNO$_3$, KMnO$_4$, K$_2$Cr$_2$O$_7$, CuBr$_2$/MnO$_2$, a. of aromatic cpds. leads to arene acetates. Alkylaromatic cpds. like toluene yield benzyl acetate (allylic-like oxidation). As. are important in the *Wacker proc. In acetic acid in presence of alkali acetates ethylene reacts with Pd salts to give vinyl acetate utilizing oxygen as oxidant and applying CuCl$_2$ as *co-cat. for the re-oxidation of the *intermediate Pd(0) species. 1,4-Diacetoxybutene is formed by a. of butadiene with Pd(II)/Pd(0) cats. and oxygen as oxidant (commercialized by *Mitsubishi BDO proc.). Other established oxidants are organic *peroxides, *hydroperoxides, and esters of peroxy acids (Kharasch–Sosnovsky reaction) likewise in the presence of *transition metals (Pb, Hg, Pd) in acetic acid as solvent and reactant. Other commercial procs. via a. are those of *Bayer AA, *ChemSystems PO, *Bayer–Hoechst VAM, *Halcon VAM. **E=F; G** Acetoxylierung.   R.W. FISCHER

Ref.: W. J. Mijs et al. (Eds.), *Organic Synthesis by Oxidation with Metal Compounds*, Plenum Press, New York and London 1986; p. 785; Cornils/Herrmann-1, Vol. 1, p. 394; Ertl/Knözinger/Weitkamp, Vol. 5. p. 2295.

acetylation is the introduction of acetyl groups CH$_3$CO– into other organic cpds. and thus a special case of acylation. In general, an acetyl donor (acetyl halides, *AAA, *AA) react directly with a hydrogen active cpd. (e.g., alcohols, amines, thiols, C–H active cpds.) to yield the acetylated derivatives. The reaction is commercially important. In some cases the reaction of *AA with alcohols or amines is cat. by acids (H$_2$SO$_4$), solid acids (*zeolites), and *Lewis acids (e.g., ZnCl$_2$). The reaction of acid chlorides with NH or OH groups in the presence of stoichiometric amounts of base (pyridine) is called the Schotten–Baumann reaction. A. of organometallic cpds. such as *Grignard, Cu, Sn, Zn, or Li reagents to yield methyl ketones proceeds with acetyl chloride or *AA; ketones can be acylated by *AAA in the presence of BF$_3$ as cat. to give β-diketones. The most important method for the synth. of acetylbenzenes is the *Friedel–Crafts acylation. **F** acétylation; **G** Acetylierung.   M. BELLER

Ref.: March.

acetylcholine esterase (EC 3.1.1.7) is a serine-type *esterase which contains the Glu-Ser-His *catalytic triad and cat. the *hydrolysis of the neurotransmitter acetylcholine. The *enzyme from electric eel has been used in the *enantioselective hydrolysis of various esters (e.g., the esters of *meso*-diols) used in organic synth. Many sulfonyl and phosphonyl fluoride cpds. are mechanism-based inactivators (*inhibitors) of the enzyme. **E=F=G**.

                                      C.-H. WONG

acetyl coenzyme A (acetyl CoA) is a *cofactor used in acetyl transfer reactions as in the synth. of fatty acids (cf. *fatty acid biosynthesis), steroids, terpenoids, and macrolides.

Acetyl CoA is synth. in Nature via several procs. 1: By reaction of acetyl phosphate with CoA, cat. by *phosphotransacetylase, which also releases inorganic phosphate. 2: By reaction of acetate with CoA in the presence of *ATP cat. by *acetyl CoA synthetase. The by-products are AMP and inorganic phosphate. 3: Reaction of acetylcarnitine with CoA cat. by *carnitine acetyltransferase gives acetyl CoA and carnitine. 4: By reaction of citrate, *ATP, and CoA cat. by ATP citrate *lyase gives acetyl CoA, ADP, oxaloacetate, and inorganic phosphate. 5: Reaction of pyruvate, *NAD, and CoA cat. by *pyruvate dehydrogenase gives acetyl CoA, NADH, and CO_2. 6: Stepwise *catabolism of fatty acids, in which a β-ketoacyl CoA intermediate and CoA are converted by the enzyme β-ketoacyl thiolase to acetyl CoA and a β-ketoacyl CoA that is shorter by two carbons (*fatty acid degradation). The *phosphotransacetylase reaction is the most favorable and has been used in the enzymatic synth. of acetyl CoA and its regeneration. **E=F=G.**   C.-H. WONG

Ref.: *J.Am.Chem.Soc.* **1992**, *114*, 7287.

acetyl CoA regeneration Enzymatic reactions that utilize *coenzyme A (CoA) thioesters as substrates are involved in the *biosynthesis of many natural prods. Due to the high cost of CoA, however, these *enzymes can only be used practically in organic synth. if the CoA thioester is recycled. Many examples of CoA recycling have been demonstrated using the synth. of citric acid from oxaloacetate and acetyl CoA. Alternatively, *phosphotransacetylase (EC 2.3.1.8) cat. the transfer of the acetyl group to CoA from acetyl phosphate. Propanoyl phosphate, butanoyl phosphate, and others are all accepted by phosphotransacetylase.

Acetyl CoA can also be recycled with *carnitine acetyltransferase (EC 2.3.1.7) and acetylcarnitine. Propanoylcarnitine and others were also accepted as substrates. Another enzymatic approach to regenerate acetyl CoA uses acetyl CoA synthetase (EC 6.2.1.1) which requires *ATP (Figure 1).

A non-enzymatic regeneration process for acetyl CoA utilizes a *phase transfer catalyst in an *aqueous two-phase system. This meth-

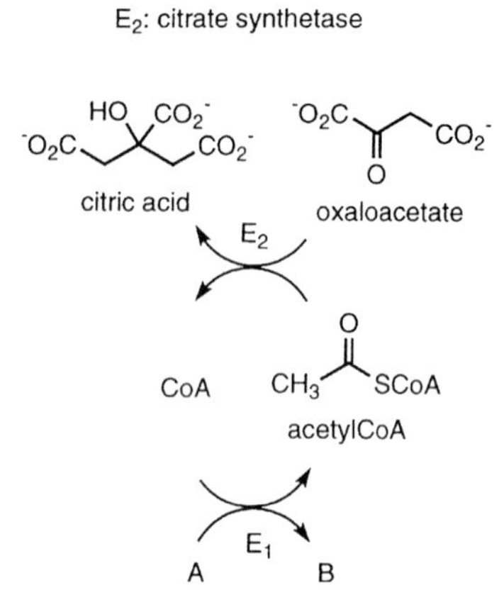

E_2: citrate synthetase

citric acid   oxaloacetate

acetylCoA

od can also be used to prepare many different acyl CoA derivatives for use as substrates for CoA-dependent enzymes (Figure 2). **F** régénération d'acétyl CoA; **G** Acetyl-CoA-Regenerierung.   C.-H. WONG

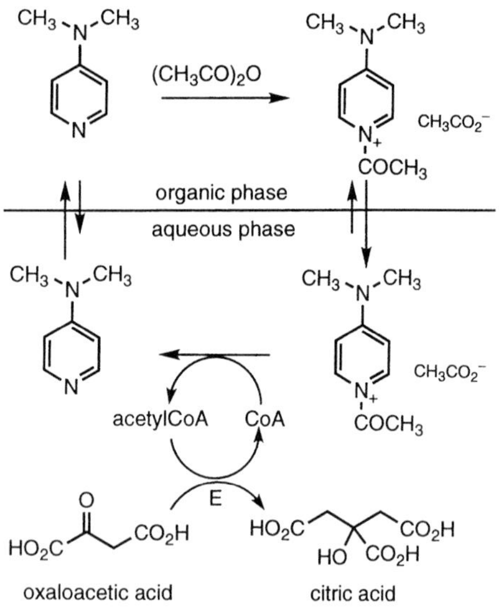

organic phase
aqueous phase

acetylCoA   CoA

oxaloacetic acid   citric acid

E = citrate synthetase

Ref.: *J.Am.Chem.Soc.* **1992**, *114*, 7287; *Bioorg.Chem.* **1989**, *17*, 1; *J.Org.Chem.* **1986**, *21*, 2842.

acetyl coenzyme A synthetase → acetyl coenzyme A

acetylene hydration → Kutscheroff reaction, *Nieuwland reaction

N-acetylglucosaminyltransferases

(GnTs) cat. the transfer of a GlcNAc residue from a donor such as *UDP-GlcNAc or N-acetylglucosaminylphosphoryldolichol to mannose or another acceptor. Different GlcNAc transferases cat. the addition of GlcNAc residues to the core pentasaccharide of asparagine glycoproteins. These and other GlcNAc transferases have been used for oligosaccharide synth. and have also been utilized to transfer other sugar residues onto oligosaccharides (Figure). In addition to transferring GlcNAc, N-acetylglucosaminyl transferase from human milk cat. the transfer of 3-, 4-, or 6-deoxy-GlcNAc from the respective *UDP derivative to Manα1,3(Manα1,6)ManβO-$(CH_2)_8CO_2CH_3$. The 4- and 6-deoxy-GlcNAc analogs can also be transferred by GlcNAc transferase II, but UDP-3-deoxy-GlcNAc is not a substrate for this enzyme. **E=F=G.**

C.-H. WONG

N-acetylneuraminate (NeuAc) aldolase

(EC 4.1.3.3) also known as *sialic acid aldolase, cat. the reversible *condensation of pyruvate with D-N-acetylmannosamine (ManNAc) to form N-acetyl-5-amino-3,5-dideoxy-D-*glycero*-D-*galacto*-2-nonulosonic acid (NeuAc or sialic acid). Although the β-anomer predominates in solution, the α-anomer of NeuAc is the substrate for the *enzyme, and the initial prods. of aldol cleavage are α-D-ManNAc and pyruvate. The enzyme has a catabolic function in vivo. For synth. purposes, however, an equil. favoring the aldol prod. can be achieved by using excess pyruvate. NeuAc aldolase from both bacteria and animals is a *Schiff base-forming type I aldolase.

Extensive substrate *specificity studies have been carried out on this enzyme. Of the many NeuAc derivatives synth. so far, most give

prods. with *S*-configuration at C 4. Under thermodynamically controlled reactions with certain sugars, the *stereochemistry at C 4 can become reversed. For example, in the NeuAc aldolase catalyzed synth. of KDO, a mixture of (*S*)-C 4 and (*R*)-C 4 products were isolated when D-arabinose was the substrate. Also, NMR studies with several other sugars showed that, over time, prods. with a C 4 equatorial group predominated in some cases. These examples violate the normal stereochemical preference of the enzyme. Apparently, pyruvate attacks the acceptor sugar to give the thermodynamically more stable prod., and the facial selectivity is merely a consequence of the preference to form a C 4 equatorial prod. Several biologically interesting L-sugars were synthesized using this method, including L-NeuAc, L-KDO and L-KDN. Leftover pyruvate can be destroyed with pyruvate decarboxylase to simplify the prod. isolation.

A facile synth. of 9-O-acetylNeuAc was improved by *regioselective irreversible acetylation of ManNAc cat. by *subtilisin followed by NeuAc aldolase-cat. condensation of the resulting 6-O-acetylManNAc with pyruvate (yield 80 %). Some other 9-O-acylated NeuAc derivatives may also be prepared in this fashion. **E=F=G.**

C.-H. WONG

Ref.: *New J.Chem.* **1988**, *12*, 733; *J.Am.Chem.Soc.* **1992**, *114*, 10138; *J.Am.Chem.Soc* **1988**, *110*, 6481.

N-acetylneuraminic acid synthetase → N-acetylneuraminate aldolase

acetyl phosphate/acetate kinase (aph./

ack.) Aph. is useful as the ultimate source of phosphate in *ATP regeneration. With this system, *acetate kinase (EC 2.7.2.1) is used as the cat. for conv. of *ADP to ATP.

Aph. is easily prepared in multimolar quantities by reaction of *AAA with phosphoric acid. This ease of synth. makes aph./ ack. the most economical system for large-scale prod. in many cases. A major drawback to the use of the aph./ ack. system is the instability of acetyl phosphate. Aph. *hydrolyzes rapidly in solution. The phosphoryl donor strength of acetyl phosphate is good but not as strong as that of phosphoenolpyruvate (cf. *adenosine triphosphate regeneration, *acetate kinase). **E=F=G.**   C.-H. WONG

Ref.: *J.Am.Chem.Soc.* **1979**, *101*, 5828.

achiral   are non-*chiral cpds. They can be converted via *enantioselective synth. into enantiomers.

acid–base catalysis   is a well developed part of solution cat. Generally, the essential steps of aqueous abc. involve proton transfer reactions (*Brønsted acid–base concept) as distinguished between *general* and *specific* abc. In *specific* acid cat., solvated protons SH^+ (H_3O^+ in water) are the proton donors for substrate B (cf. Figure) and the rate of the acid cat. is only proportional to the conc. of SH^+ that is actually present in the solution (in water the rate correlates to the pH). Analogously, specific base cat. occurs with the proton acceptor S^- (OH^- in water). Generally, specific abc. does not involve proton transfer in the rate-determining step:

step 1: SH^+ + B $\rightleftharpoons$ $(SHB^+)^{\ddagger}$ $\rightleftharpoons$ BH^+ + S
step 2: BH^+ + S $\longrightarrow$ products + SH^+

specific acid catalysis: $k_{obs.} = k_{SH^+}[SH^+]$

general acid catalysis: $k_{obs.} = k_0 + k_{SH^+}[SH^+] + k_1[HA^1] + k_2[HA^2] + ...$

Brønsted catalysis law: $\log k_n = \alpha \log K_{a(n)} + \log C$ or $k_n = CK_{a(n)}^{\alpha}$

In *general* acid cat., the rate is increased not only by an increase in $[SH^+]$ but also an increase in the conc. of other (non-dissociated) acids. Each individual acid (H_3O, HA^1, HA^2, etc.) present in the system can act as a proton donor in the rate-determining step 1. The analogous situation for bases is general base cat. In general acid cat. the strongest acids catalyze best. The *Brønsted cat. equation (law) relates to the acidity and the cat. ability of the cat., i.e., the individual cat. constants k_n should be related to the equil. acidities. The

Brønsted slope α is a measure of the sensivity of the reaction to the acid strengths of the various cats. The *Brønsted law is an empirical linear free-energy function which is closely related to the *Hammond postulate. In *Lewis acid cat. metal ions or their *complexes also function as cats. by bonding to organic substrates and polarizing them.

Important cat. procs. in aqueous solutions are *hydrolyses, *decarboxylations, and phosphate cleavage reactions. Lewis acids, derived from hard main group (B, Al, Ga) and d- (Ti, Zn, Cu, Ru) and f- (rare earth series) transition metal cations acc. to the *HSAB concept are particularly effective cats. in carbonyl *additions and other basic C–C bond forming reactions, e.g., with *asymmetric cats. from organometallic complexes in *aldol, *Diels–Alder, or *Michael reactions. Lewis acid cat. by metal ions is also an important feature in *enzyme cat. (cf. *general acid–base cat.). The prod. distributions of *dehydration reactions are often taken for the qualitative characterization of acid–base properties.

Heterobimetallic complexes can act as acid–base *bifunctional cats., e.g. by displaying both Lewis acidic and Brønsted basic behavior. For abc. in solids cf. *solid acid–base cats. **F** catalyse acido-basique; **G** Säure–Base-Katalyse.   R. ANWANDER

Ref.: Gates-1; Trost/Fleming; Ertl/Knözinger/Weitkamp, Vol. 2, p. 689.

acid catalysis $\rightarrow$ acid–base catalysis

acid protease $\rightarrow$ aspartyl protease

ACM (AM) $\rightarrow$ acrylamide

ACN (AN) $\rightarrow$ acrylonitrile

ACR   advanced cracking reactor, $\rightarrow$ catalytic cracking

acrolein   is the simplest unsaturated aldehyde, manuf. mainly by *oxidation of propylene on *multimetallic cats. (e.g., the procs. of *BP/Ugine, *Degussa, *Shell, *SOHIO).

B. CORNILS

acronyms relevant for catalysis are given under the appropriate keyword.

Ref.: D. and H. Noether, *Encyclopedia of Chemical Technology*, VCH, Weinheim 1993; *Appl.Catal.* **1996**, *135*, N2; Anderson/Boudart, Vol. 10, p. 177; Ertl/Knözinger/Weitkamp, Vol. 2, p. 772; Farrauto/Bartholomew, p. 119.

acrylamide acrylic acid amide (AM, ACM), manuf. by partial *hydrolysis of acrylonitrile, important monomer for polyacrylamide, e.g., procs. of *Mitsui Toatsu, *Nitto AM. **F=G=E**.   B. CORNILS

acrylic acid is the simplest unsaturated carboxylic acid, manufd. by *Reppe synth. from acetylene or by cat. gas-phase *partial oxidation of propene (procs. of *BASF, *Mitsubishi, *Nippon Shokubai). **F** acide acrylique; **G** Acrylsäure.   B. CORNILS

Ref.: Kirk/Othmer-1, Vol. 1, p. 287; Kirk/Othmer-2, p. 24; Ullmann, Vol.A1, p. 161; Weissermel/Arpe, p. 289; McKetta-1, Vol. 1, p. 401.

acrylonitrile (ACN, AN) is a cat. manuf. intermediate, cf. *ammonoxidation procs. of *BP(Distillers)/Ugine, *DuPont, *Montedison, *Snamprogetti, *SOHIO; cf. *Nieuwland cat. **E=F; G** Acrylnitril.   B. CORNILS

ACT advanced cleavage technology, → Lummus phenol process

activated carbon (charcoal) Activated coal or carbon is the most important carbon material for cat. applications. It is mainly used as a *support for metals such as Pt and Pd in *hydrogenation reactions. The origin of ac. can be either natural or synth. The natural origin causes the appearance of large quantities of impurities, most of them silicates. The carbonization of diverse seeds like olive stones, coconuts, and peaches offers the possibility of creating excellent acs. The first step is the carbonization of the carbonaceous precursor in an inert atmosphere. In this step volatile material and water are evolved from the carbon precursor. The second step is the *activation* step. C is selectively burned by chemical and/ or physical means. The goal of this step is to create the porosity and a large *surface area. The materials obtained are excellent support materials due to high surface areas and a well developed *porosity which are essential for achieving large metal *dispersions. The inertness of the ac. can be used as an advantage in reactions where strong *metal support interactions (*MSIs) are not desired. At the same time, the so-called inertness of C is rather misleading. The surface of ac. has a proportion of *active sites, constituted by unsaturated valences at the edges and defects of the graphene layers. Ac. is one of the C materials with the highest surface area. Furthermore, the presence of heteroatoms such as N, H, and O introduces *active sites on the carbon surface. The oxygen content of the ac. present as oxygenated carbon groups may be enhanced with *selective oxidation steps. Depending of the oxidation proc., basic or acid groups can be created on the carbon surface. These groups (together with the pH of the metal solution) condition the dispersion of the metal precursor. Ac. can act as a *bulk cat., e.g., for the manuf. of *sulfur chlorides or phosgene.

A relatively new group of carbonaceous materials included in the acs. are the carbon *molecular sieves. These materials possess much larger surface areas where the porosity can be regulated by relatively simple methods as with *zeolites. The distinction between *shaping and size can be more easily conducted with these supports than with other high surface area supports. **F** charbon active; **G** Aktivkohle.

R. SCHLÖGL, E. SANCHEZ-CORTELON

Ref.: Ullmann, Vol.A6, p. 157.

activated complex → transition state

Activated Metals & Chemicals a group of companies headquartered in Sevierville, Tennessee (USA), developing, manufacturing, and recycling base and precious metal cats. AMC is an important producer of *sponge metal cats. by the *Raney proc.

J. KULPE

activation energy   1: represents the energy barrier that a system must pass on its way from the reactant to the prod. state (Figure; cf. also Figure under keyword *potential energy diagrams, *thermodynamics in cat.).

First determined empirically by *Arrhenius (cf. *Arrhenius plot), the ae. is central for the understanding of het. cat. because the rate enhancement through a cat. can only be due to a higher collision frequency of the reactants or to a lowering of the ae. The ae. measured acc. to Arrhenius represents an *apparent ae. because cat. reactions typically involve several mechanistic steps and only in special cases when a single rate-limiting step exists can the ae. be assigned to this step.

R. VAN ELDIK, R. IMBIHL

2-Ae. in *enzyme cat.: at its simplest, enzyme cat. occurs in at least two discrete steps: substrate binding with a binding equil. constant K_S and substrate conv. with (first-order) *rate constant k_{cat}. For an enzymatic reaction that follows the classical *Michaelis-Menten mechanism, K_s is equal to K_m, the *Michaelis constant. These are discrete energetic events, and the equil. constant and first-order rate constant can be related to the free energies of binding and activation using the thermodynamic and *transition state relation (*Eyring equ.). If the enzyme substrate *complex is greatly stabilized (i.e. the free energy of E:S is lowered) without a corresponding drop in the energy of the transition state, the first-order (maximal) reaction rate will suffer because the difference in the energies of the transition state and the E:S complex will be increased. Enzymes frequently overcome unfavorable entropic terms by preorganizing the substrate (and cat. groups) within the *active site: the unfavorable entropic term is then incorporated into the binding constant instead of the cat. constant. Favorable binding energy is provided by *hydrogen bonding, *salt bridges, and *hydrophobic interactions. **F** énergie d'activation; **G** Aktivierungsenergie.

P.S. SEARS

Ref.: Wilkins; Connors; Atwood; van Eldik/Hubbard; Burgess.

activation volume $\Delta V^{\neq}$ represents the partial molar volume change of the reactants in solution on going to the *transition state, acc. to the *transition state theory. It is used to construct a volume profile for reactions, e.g. A + B → AB, which describes changes in the partial molar volume along the reaction coordinate (Figure) and is used to analyze the nature of the transition state, and thereby elucidate the underlying reaction mechanism.

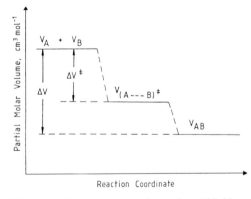

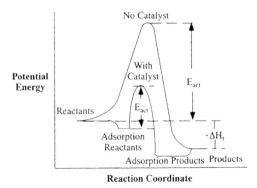

Reaction Coordinate

[by courtesy reproduced from Farrauto/Bartholomew, Kluwer Academic Publ., Figure 1.3]

The reaction volume, given by $\Delta V = V_{AB} - (V_A + V_B)$ represents the overall change in partial molar volume that occurs during the reaction. $\Delta V^{\neq}$ is in general analyzed in terms of intrinsic (as a result of changes in bond lengths and angles) and solvational (as a result of changes in charge and dipole moment that affect solvent electrostriction) volume contributions. **F** volume d'activation; **G** Aktivierungsvolumen.   R. VAN ELDIK

Ref.: Wilkins; van Eldik/Hubbard; Burgess.

activator   general name for cpds. or measures (e.g. *radicals, UV) that activate a given system, thus acting as accelerators or *promoters – often synergistically. In cat., the a. (the *co-catalyst) conventionally is either an *additive that converts the *precatalyst (*precursor) to its active form (active species generation) or that strengthens a cat. additively or synergetically. For example, the Pd(II) *complex **1** must be reduced to the active Pd(0) species **2** by activator cpds. such as N_2H_4 or Na formate to start *Heck-type C–C coupling reactions. The a. is thus required in at least stoichiometric amounts with regard to the cat. (Figure 1).

1
inactive

2
active

An activation of cats. may even occur through dissociation (e.g., in the case of *phosphine-modified Ni cats. [*ligand-modified cats.]) (Figure 2). In some cases the *modifiers of transition metal cpds. such as *phosphines are called co-cats. or as.

$$Ni[P(C_6H_5)_3]_4 \rightleftharpoons Ni[P(C_6H_5)_3]_3 + P(C_6H_5)_3$$

inactive   active

By way of contrast, a cat. that activates a molecule to become more reactive is *not* called an a. since the *terminus technicus* catalyst has the higher priority, including activating, accelerating, and otherwise reaction-improving qualities. **F** activateur; **G** Aktivator.

W.A. HERRMANN
Ref.: Butt/Petersen; *J.Organomet.Chem.* **1995**, *500*, 149.

active site density   describes the relation between the total number of potential reaction sites and the number of *active sites (ass.) in het. cat. The determination of this important parameter is difficult experimentally as only in a few cases are probe molecules known which count the abundance of ass. A typical example is the selective *chemisorption of NO over metallic Cu. On ass. the molecule is split into N and O. The oxygen remains on the Cu whereas the nitrogen is released in the gaseous phase. The selective *chemisorption of *ammonia on acid sites in *zeolites and other oxides is often used to determine the number of ass. The relation of this number to the total *surface area (TSA) gives the as. density. This number usually indicates that only a small fraction of the cat. *surface is active (typically 1 %). **F** densité des sîtes actives; **G** Dichte der aktiven Stellen.

R. SCHLÖGL

active site-directed irreversible inhibitors   are cpds. that covalently modify the *active site of an *enzyme and destroy the enzyme's activity. This group of *inhibitors includes mechanism-based *inactivators (suicide inhibitors, suicide substrates), cpds. that only inhibit the enzyme upon activation by that enzyme (and thus require enzyme activity for covalent modification), and affinity labels (or "Trojan horse" inhibitors), cpds. that are naturally labile but can be targeted to the active site by their resemblance to substrate, product, or *transition state.

An example of a mechanism-based inactivator is phenylmethylsulfonyl fluoride, an inhibitor used for inactivating the *serine proteases (Figure). The nucleophilic active site serine of this class of proteases attacks the sulfonyl fluoride, releasing fluoride and forming a relatively stable sulfonate ester. **E=F=G**.

P.S. SEARS

active site model → active sites

active sites (active centers) 1–general: The terms (introduced by H.S. Taylor) distinguish between the geometrical *surface area of a het. cat. available for *physisorption, and specific sites representing only a fraction of the total *surface area where cat. conv. actually takes place. The concept is based on the idea that atoms which belong to certain structural defects of the surface such as, for example, the edge atoms (cf. *single crystal) of an atomic step will have properties different from atoms belonging to the regular structural elements and therefore exhibit potentially an enhanced cat. *activity. This picture of ass. is highly problematic because firstly no strict definition exists and secondly because no method is available with which the number of ass. can be determined reliably. Titration of a surface with CO or other reactive molecules (cf. *active site titration) yields the number of *chemisorption sites, which will be smaller than the geometrical surface determined with the *BET method, but these sites may not represent the bottleneck with respect to a cat. reaction. A second reason why the idea that structural defects govern the cat. activity has lost ground came from the results of surface science studies showing that the regular structural elements of a surface typically account for most of the cat. *activity and that surface defects play a role only under specific conds.

R. IMBIHL

2–in enzyme cat.: The as. is the region, often a cleft, where substrate is bound and cat. takes place. In order for any *enzyme to be a useful cat. for synth., it must be predictable whether a particular cpd. will be a substrate and if so, the *stereo- and *regiospecificity of the reaction. As. models are developed to predict the substrate preferences of enzymes. These models are sometimes based on available crystal structures of enzymes complexed with substrate, prod. or *inhibitor, but in cases where such data are not available, they may be based on exhaustive studies of the enzyme's substrate preferences (e.g. models of *horse liver alcohol dehydrogenase, HLADH). In this model the enzyme as. is represented as layers of cubes of arbitrary size, depending on the accuracy needed. Certain cubes are considered forbidden and others allowed or limited for occupation by the substrate. A similar model has been constructed for *pig liver esterase (cf. Figure under that keyword), an enzyme frequently used for the selective *hydrolysis of *chiral esters. **F** sîtes actifs; **G** aktive Stellen. P.S. SEARS

Ref.: Ertl/Knözinger/Weitkamp, Vol. 1, p. 5; Fürstner; Thomas/Thomas; *J.Am.Chem.Soc.* **1990**, *112*, 4946, **1982**, *104*, 4666, and **1982**, *104*, 4659; Augustine, p. 27.

active site titration When determining the kinetic parameters k_{cat} or k_{cat}/K_m for an *enzyme, it is necessary to estimate the exact number of active molecules of enzyme. Simple calcns. based on the molecular weight of the enzyme and the weight of enzyme added to a solution will give an upper limit to the number of active sites present, but this number will be greater than the true value if the enzyme prepn. is impure or a portion of the enzyme is inactive. Accurate determination of the number of active sites can be accomplished by ast., which often requires the use of a labeled (fluorescent, radioactive, or spin-labeled) mechanism-based *inactivator. For an enzyme with a two-step reaction mechanism, a substrate with a rapid first step and a very slow second step (resulting in so-called "*burst kinetics"), and for which the first step produces an easily detectable prod. or *intermediate, can be used. It is preferable that the reaction of the inhibitor with the enzyme can be monitored spectrophotometrically or fluorimetrically. **F** titration des sîtes actifs; **G**=E.

P.S. SEARS

active surface area (ASA) supplies kinetic data (cf. *adsorption, chemisorption)

activity → catalytic activity

ACV δ-(L-aminoadipyl)-L-cysteinyl-D-valine

ACV synthetase   cat. the synth. of δ-(L-aminoadipyl)-L-cysteinyl-D-valine   (ACV) from individual *amino acids. The *enzyme from *Streptomyces clavuligerus* has been studied extensively.

It also accepts unnatural amino acids as substrates to give ACV analogs. ACV is a precursor to the antibiotics penicillin and cephalosporin. **E=F=G**.   C.-H. WONG

acyclic diene metathesis (ADMET) → metathesis

acylase   Acylase I (aminoacylase, EC 3.5.1.14) cat. the *enantioselective *hydrolysis of N-acyl-L-amino acids and analogs, especially the N-acetyl and the chloroacetyl, trifluoroacetyl, and methoxyacetyl derivatives. N-Acyl-α-methyl-α-amino acids are substrates for the kidney acylases, but not for the *Aspergillus* *enzyme. The enzyme contains a zinc ion. The acylases selective for N-acyl-D-proline and N-acyl-D-amino acids also exist. **E=F=G**.   C.-H. WONG
Ref.: *J.Am.Chem.Soc.* **1989**, *111*, 6354; *Angew.Chem. Int.Ed.Engl.* **1990**, *29*, 417; *Appl.Environ.Microbiol.* **1991**, *57*, 1259.

acylation   is the introduction of acyl groups such as –CO–R or –SO₂–R in organic cpds. containing C–H acid groups such as –OH, –SH, or –NH (cf. reactions of *Darzens, *Friedel–Crafts, *Nenitzescu). The treatment of acyl chlorides or carboxylic acid anhydrides (*AAA, *PAA) with alcohols is a commercially important reaction which yields phthalic esters (used as *plasticizers), terephalic acid esters (*DMT), etc. The reaction of carboxylic acids with alcohols or amines is cat. by inorganic acids (H₂SO₄), *zeolites, *aluminosilicates, or *Lewis acids such as

ZnCl₂ or *p*-toluenesulfonic acid. Ketones can be acylated by acid anhydrides in the presence of BF₃ as cat. to give β-diketones. Other a. reactions or carbon nucleophiles include the prep. of acyl cyanides, diazo ketones, or β-keto esters. **F** acylation; **G** Acylierung.
                                    M. BELLER
Ref.: March; Ertl/Knözinger/Weitkamp, Vol. 5, p. 2358.

acyl enzyme   Many *enzymes, such as the *thiol and *serine proteases, form covalent adducts to the substrate. When the reaction involves acyl transfer, the adduct is called an acyl enzyme. In the serine and thiol proteases, for example, the *active site serine or cysteine attacks the carbonyl of the peptide substrate to form an ester (or thioester). This acyl enzyme is then hydrolyzed. **E=F=G**.   P.S. SEARS

acyllactone rearrangement   is a proton-cat. *alcoholysis of α-acylated lactones (*acid–base cat.) which yields tetrahydrofurans or pyrans. **F** réarrangement d'acyllactones; **G** Acyllacton-Umlagerung.   B. CORNILS
Ref.: Krauch/Kunz.

acyloin condensation   1–general: the cat. *dimerization of aldehydes to hydroxyketones. Cats. are, e.g., cyanide ions.   B. CORNILS
2–enzymatic: the same *condensation as under 1, cat. by *pyruvate decarboxylase. The proc. may also be carried out with yeast. **F** condensation   acyloine   (enzymatique); **G** Acyloinkondensation.   C.-H. WONG
Ref.: Krauch/Kunz; *J.Chem.Soc.Chem.Commun.* **1993**, 341.

N-acylproline acylase, L-specific   Many *amidases and *proteases will not *hydrolyze the amides of secondary amines. *HIV protease, which hydrolyzes the Tyr-Pro bond, is one exception, and this *specificity has been used in the design of very specific *inhibitors of that *enzyme. Another amidase that accepts N-dialkylated amides is acyl proline acylase (proline acylase), which can hydrolyze acyl groups such as acetyl and chloroacetyl from the amino group of proline and other N-disubstituted amines. This enzyme has been used in the resolution of proline and other *N*-dialkyl amino acids. **E=F=G**.   P.S. SEARS

Ref.: *Angew.Chem.Int.Ed.Engl.* **1992**, *31*, 195; *Biochim. Biophys.Acta* **1983**, *744*, 180.

acyl transfer, enzymatic. *Enzymes such as *esterases, *amidases, *proteases, *lipases, and *acyltransferases cat. an acyl transfer reaction from one substrate to another. Reactions include the transfer of an acyl group from an ester to an alcohol, amine, thiol, or water, or the transfer of an acyl group from an amide or thioester to another acceptor. The transfer of an acyl group from an acyl *CoA or acetyl phosphate is also called an at. reaction. **F** transfert d'acyl, enzymatique; **G** enzymatischer Acyltransfer.   C.-H. WONG

AD asymmetric dihydroxylation, cf. *dihydroxylation

Adam and Eve principle is a recycle proc. as a combination of 1: an endothermal methane *cracking with water into CO/H_2 (via nuclear heated helium, EVA), 2: transportation of the *syngas to the very distant user, and 3: the exothermically cat. *methanation (ADAM) of the syngas on the user's site. The resulting CH_4 returns 4: via pipeline to the conv. site which might be located (as well as the nuclear reactor) in remote areas. This principle makes possible the local separation of energy supplier and energy user. **F** principe d'Adam et d'Eve; **G** Adam-und-Eva-Prinzip.   B. CORNILS

Ref.: Anderson/Boudart, Vol. 1, Chapter 1; *Appl.Catal.* **1981**, *1*, 125; Winnacker/Küchler, Vol. 5, p. 445.

Adam's catalyst liquid phase *Pt-Pt oxide cat., prepared by the *fusion of chloroplatinic acid with sodium nitrate or nitrite; mainly used for fine chemicals production. **F** catalyseur d'Adams; **G** Adams-Katalysator.   B. CORNILS

Ref.: Augustine, p. 231.

adatoms are one type of point defect on a real *surface. In contrast to *vacancies they represent additional atoms on the surface (cf. *islands, *overlayer). These point defects are present in most surfaces and are important participiants of atom transport along the surface, although their equil. conc. is much less than 1 % of a monolayer even at the melting point. The number of neighboring atoms of an a. is highly reduced compared to those of an atom located in an atomic plane. Therefore surface reactions often take place at these highly uncoordinated sites. The relative conc. of an a. can be varied, depending on the method of surface preparation. **F=E**; **G** Adatome.   R. SCHLÖGL, Y. JOSEPH

addition of ammonia to double bonds → ammonia lyases

addition reactions 1–general: reactions between reactants with C–C or with C–heteroelement single, double, or triple bonds and reactive cpds. such as alkenes or alkynes *without* formation of byproducts such as H_2O, CO_2, etc. as in *condensation reactions. Most ars. are cat., e.g., *hydroformylation, *hydrocyanations, *hydrophosphinylation, *Markovnikov reaction, *Diels–Alder reaction, *Thorpe nitrile add., *oligo- or *polymerization, *cycloaddition, *etherification etc. Examples are the procs. of *Atochem, *BASF BDO; *Knapsack–Griesheim ACN, *IFP isoprene, *MTBE syntheses, *Kriewitz–Prins reaction, *Ultee. 2: *radical additions, cf. *polymerizations, *Kharasch add.   B. CORNILS

3–enzymatic: Many *enzymes form adducts (*complexes) with their substrates. Enzymes for ars. include *aldolase cat. *aldol addition reactions, *transaldolase-cat. transaldol reactions, *transketolase-cat. transketol reactions, *oxynitrilase-cat. addition of a cyanide to an aldehyde to form the cyanohydrin, *fumarase cat. addition of water to fumarate and analogs to form L-malate and analogs, ammonia *lyase cat. addition of ammonia to alkenes, etc. **F** réactions d'addition; **G** Additionsreaktionen.   C.-H. WONG

Ref.: Wong/Whitesides, Vol. 12, p. 312; Ertl/Knözinger/Weitkamp, Vol. 5, p. 2370.

additives are substances (*dopants) added to other cpds. or cpd. mixtures which influence even in small amounts their course of manufg. or their properties, → activators, *co-catalysts, *doping, *inhibitors, *promotion. A possible genesis is given in the Figure. **F** additives; **G** Additive.   B. CORNILS

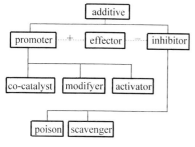

Ref.: McKetta-1, Vol. 2, p. 77; Farrauto/Bartholomew, p. 59.

adenosine deaminase (EC 3.5.4.4) cat. the hydrolytic release of ammonia from the adenine base of adenosine to give inosine. **E=F=G**.   P.S. SEARS

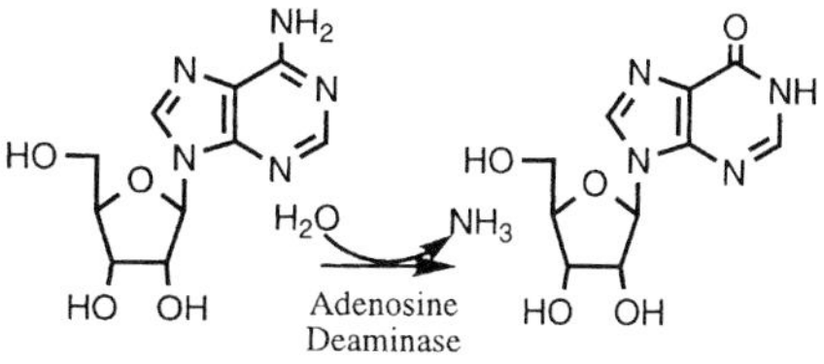

adenosine diphosphate (ADP) → adenosine phosphates

adenosine kinase (EC 2.7.1.20) uses *ATP as a high-energy phosphate source to phosphorylate the 5′-hydroxyl group of adenosine to give adenosine monophosphate (*AMP) (Figure). **E=F=G**.   P.S. SEARS

adenosine monophosphate (AMP) → adenosine phosphates

adenosine phosphates (ATP, ADP, AMP) This class of cpds. includes adenosine monophosphate (AMP), adenosine diphosphate

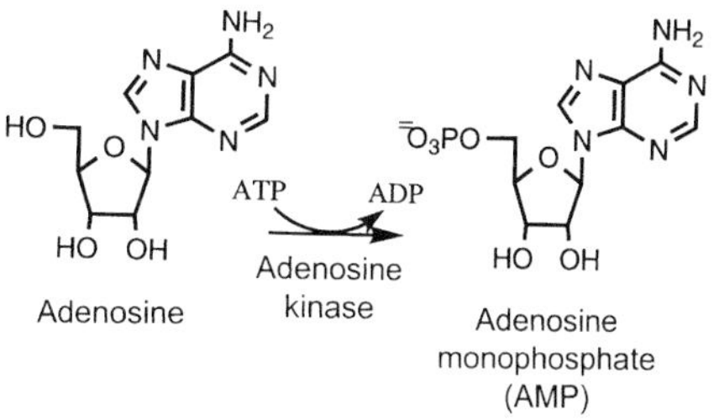

adenosine kinase

(ADP), and *adenosine triphosphate (ATP). They act as *cofactors in numerous *enzymatic reactions, especially those cat. by *kinases. A kinase cat. the transfer of the γ-phosphoryl group from ATP to another acceptor and generates ADP as a byproduct. A pyrophosphoryl transfer reaction will generate AMP as a byproduct. Many kinase-cat. reactions have been used in the synth. of organic phosphates

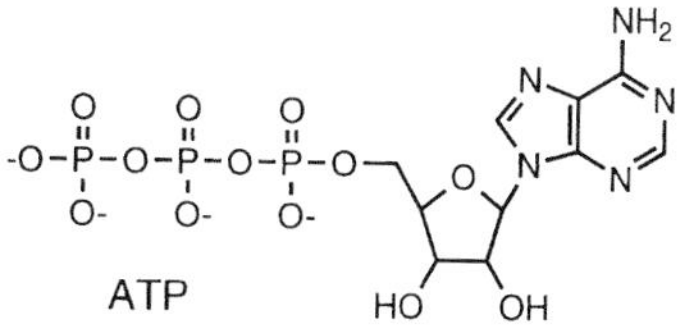

Many enzymes couple thermodynamically unfavorable reactions to the *hydrolysis of ATP to drive the reaction forward. The cofactor ATP may be regenerated from ADP using acetyl phosphate + acetate kinase, phosphoenolpyruvate + pyruvate kinase, or several other schemes. **E=F=G**.   C.-H. WONG

adenosine-5′-phosphosulfate (APS) → 3′-phosphoadenosine   5′-phosphosulfate (PAPS), → adenosine phosphosulfuryl kinase

adenosine phosphosulfuryl kinase (APS kinase) cat. the 3′-phosphorylation of adenosine phosphosulfate to PAPS (*3′-phosphoadenosine 5′-phosphosulfate). **E=F=G**.   P.S. SEARS

adenosine triphosphate (ATP) is *chiral at the α-,β-, or γ-P atom. The α- and β-phosphorus groups of *nucleoside triphosphates are prochiral and the γ-phosphorus group is prochiral. To understand the stereochemical course of nucleoside triphosphate-dependent phosphoryl transfer reactions, the phosphorus group involved in the reaction must be chiral. Introduction of the chirality required for mechanistic studies can be accomplished by replacing one (or two) of the oxygen atom(s) with S or isotopic oxygen (^{17}O or ^{18}O). The synth. of nucleoside triphosphates that are chiral at phosphate have been well developed. Of particular interest are the *enzymatic synth. of phosphorothioate ATP analogs including ATPαS, ATPβS, and ATPγS. ATPαS with S-configuration at the α-position (S_P-ATPαS) can be prepared from AMPαS via adenylate *kinase and pyruvate kinase reactions, and R_P-ATPαS can be prepared from ADPαS in a reaction cat. by *creatine kinase. *pyruvate kinase also cat. the synth. of S_P-ATPβS from ADPβS. The minor byproduct R_P-ATPβS can be decomposed selectively with *hexokinase in the presence of glucose. R_P-ATPβS can be prepared from ADPβS in a reaction cat. by *acetate kinase; the minor product S_P-ATPβS that is formed can be selectively decomposed with myosin. Both R_P- and S_P-AMP containing S and ^{18}O can be prep. from diadenosine pyrophosphate using a proc. cat. by nucleotide pyrophosphatase.

Several forms of ATP chiral at γ-phosphorus have been prepared (e.g.[γ-^{18}O]ATPγS in 1978). Pyruvate kinase coupled with Met-^{t}RNA synthetase was used for the preparation of (R_P)-γ^{17}O, γ^{18}O-ATPγS, and glyceraldehyde 3-phosphate dehydrogenase coupled with phosphoglycerate kinase was used for the synth. of ATPγS containing ^{17}O and/or ^{18}O.

[α-^{32}P]-ATP was also prepared from [γ-^{32}P]-ATP via adenosine kinase, adenylate kinase, and *creatine kinase-cat. reactions. **E=F; G** Adenosintriphosphat. C.-H. WONG

Ref.: *Annu.Rev.Biochem.* **1985**, *54*, 367; *Methods Enzymol.* **1982**, *87*, 179, 213, 279.

adenosine triphosphate regeneration

from ADP is used in large- scale enzymatic phosphorylations. The *enzyme systems for ATPr. are well developed and efficient. ATPr. is based on the selective transfer of phosphate from a high-energy donor, such as *phosphoenolpyruvate, to ADP by use of a *kinase enzyme. The enzymes that are involved should have sufficient *activity and *stability to achieve practical *turnover numbers throughout the reaction, and should be inexpensive.

There is also a need for ATPr. from AMP. AMP is not a substrate for acetate or *pyruvate kinase, and a different method is required. *Adenylate kinase (EC 2.7.4.3) cat. the equilibration of AMP, ADP, and ATP. This enzyme can, therefore, be used in conjunction with acetate or pyruvate kinase to generate ATP from AMP. Adenylate kinase has also been found to accept other nucleoside monophosphates, including *CMP. Note that the system for regeneration of CTP from CMP requires the use of a cat. amount of ATP and AMP in order to initiate the cycle. **F** régéneration d'ATP; **G** ATP-Regenerierung. C.-H. WONG

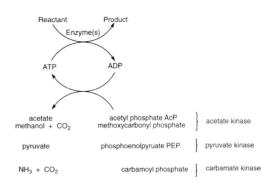

Ref.: *Tetrahedron Lett.* **1988**, *29*, 1123; *Appl.Biochem. Biotechnol.* **1987**, *16*, 95.

S-adenosylhomocysteine (SAH) is the byproduct of methyl transfer from S-adenosylmethionine (SAM) to an acceptor, and can be used in several pathways. Following *hydrolysis of the adenosine, the resulting homocysteine may be recycled to methionine via a *vitamin B_{12}-dependent *enzyme that transfers a methyl group from N^5-methyl-THF. Alternatively, it may be combined with serine to form cystathionine, which can be conv. to α-ketobutyrate and L-cysteine by the action of cystathionine lyase. **E=F=G.**   P.S. SEARS

S-adenosylmethionine (SAM) (–)-S-Adenosyl-L-methionine (AdoMet,SAM) is a *cofactor required in many *enzyme reactions, particularly as an electrophilic methyl source in transmethylation reactions. Enzymatically active SAM has the S configuration at the *chiral amino acid and at the sulfonium center. Inversion of the sulfonium center produces an inactive cofactor, making the stereocontrol in synth. SAM reactions crucial to their utility. SAM is labile to both alkaline and acid *hydrolysis, so a rapid turnover is a requirement for economical use of SAM in organic synth. Chemical methods involving methylation of *S-adenosylhomocysteine with methyl iodide produce a slight excess of the inactive (+)-isomer. SAM is synth. in nature from L-methionine and ATP, in a reaction cat. by SAM synthetase. SAM is a strong inhibitor of SAM synthetase, and the reaction must therefore be carried out in dilute solutions. The best methods for the prod. of SAM are fermentative techniques using microbial cells. The product of transmethylation, *SAH, is not remethylated enzymatically; it is degraded by

ROH   ROCH₃

SAM

SAH

E_1

ATP + methionine

E_1:   SAM synthetase (EC 2.5.1.6)
E_2:   Methyltransferase
SAH: S-adenosylhomocysteine

SAH *hydrolase to L-homocysteine and adenosine. L-Homocysteine may be methylated to give methionine and SAM (Figure). **E=F; G** S-Adenosylmethionin.   C.-H. WONG

Ref.: *Biotech.Appl.Biochem.* **1987**, 9, 39; *Trends Biotech.* **1984**, 2, 37.

adenylate kinase → nucleoside diphosphate synth.

adipic acid is a hexane diacid, manuf. preferably from cyclohexane (CH) by multi-step *oxidative cleavage. The first step includes the cat. *oxidation of CH at 125–165 °C/0.8–1.5 MPa over Mn or Co salts to *KA oil; boric acid as a cat. *additive is possible (*Bashkirov reaction, *IFP variant). The next stage, the oxidation of the KA oil, proceeds either over Cu/Mn acetate or with HNO_3 in presence of NH_4 metavanadate and Cu nitrate at 50–80 °C (* SD proc.). Other procs.: cf. *Monsanto, *BASF. **F** acide adipique; **G** Adipinsäure.   B. CORNILS

Ref.: Kirk/Othmer-1, Vol. 1, p. 466; Kirk/Othmer-2, p. 37; McKetta-1, Vol. 2, p. 146; Ullmann, Vol. A1, p. 269.

adiponitrile is adipic acid dinitrile (ADN, ADP), a cat. manuf. intermediate, cf. procs. of *Butachem, *Monsanto. **F** nitrile d'acide adipique; **G** Adipinsäurenitril, Adipodinitril.
    B. CORNILS

Ref.: McKetta-1, Vol. 2, p. 162.

Adkins catalyst is an heterogeneous-phase cat. on the basis of Cu supported chromium oxide ("copper chromite", doped with Ba, Mn), mainly for *hydrogenations, e.g., of *fats, fatty acids, or fatty acid amides or esters (cf., e.g., procs. of *Eastman BHMC, *Henkel Direct Hydrogenation, *Lurgi). **F** catalyseur d'Adkins; **G** Adkins-Katalysator.   B. CORNILS

Ref.: *J.Am.Chem.Soc.* **1931**, 53, 1091; *Organic Reactions*, Vol. 8, p. 1.

adlayers Condensed, *physisorbed, or *chemisorbed molecules or atoms on a *surface constitute the adlayers. They are investigated for changes in the electronic structure with respect to the free molecule or atom, their chemical nature, or geometric structure.

If there exists a strong and specific interaction between the substrate atoms and the adsorbed molecules or atoms (chemisorption), a *commensurate structure may be formed. If only physisorption or condensation of the adsorbate occurs on the surface because of weak adsorbate–substrate interactions, an *incommensurate structure could be found. **E=F=G.**   R. SCHLÖGL, Y. JOSEPH

ADMET acyclic diene metathesis, → metathesis

ADN, ADP → adiponitrile

ADP 1: → adenosine diphosphate; 2: → adiponitrile

adparticles general term for *adatoms, molecules, *adlayers, etc.

adsorbate-adsorbate interactions → coadsorption

adsorbate inhibition → substrate inhibition

adsorbate-surface interactions → Temkin isotherm

adsorption the fundamental proc. occurring at het. interfaces. For cats. the most relevant are a. procs. of molecules from the gaseous phase or from the liquid phase to solid *surfaces. A. requires the existence of attractive forces between the molecules (*adsorbate*) and the solid (*adsorbent*). These forces can be weak and of a dispersive nature (*van der Waals attaction) or strong, leading to directed chemical bonds. Acc. to the type of interaction, a. procs. are discriminated into *physisorption, when weak undirected interactions dominate the attraction, and *chemisorption, when directed forces govern the interaction between adsorbate and adsorbent. The absolute magnitude of the interaction potential is a less well-suited variable by which to distinguish chemisorption (more than 40 kJ/mol) from physisorption (less than 40 kJ/mol),

which was, however used frequently in the past. The determination of the interaction potential or *adsorption energy* is an important task in describing a interface proc. The other important variable to determine is the amount of adsorbed molecules per unit mass of the adsorbate. Here two fundamentally different cases can be distinguished. A. can occur until all the available solid surface is covered with one *monolayer of adsorbate. At this point the specific *coverage, which is the ratio between empty and filled sites on the *surface, is unity. For chemisorption procs. this is the limiting a. capacity. For nonspecific physisorption procs., however, more than one monolayer can be adsorbed on a solid. This multilayer a. ends with the condensation of the adsorbate onto the adsorbent forming a film of liquid or of solid "ice". In the context of het. cat. chemisorption is relevant, as in multilayer a. systems no chemical contact exists between the solid (cat.) and the molecules. A. and its reversal, *desorption, are fundamental elementary steps in each cat. reaction. Before a cat. can act upon a substrate this has be brought into contact with the cat. within the dimensions of the length of a chemical bond (adsorption). After the action of the cat. the resulting prod. molecules have to leave the cat. (desorption) in order to free the *active sites for a new *turnover cycle. It is thus evident that an exact knowledge of the fine details of a. *kinetics and *thermodynamics of educts and prods. in a cat. reaction is of fundamental importance for its description.

A. reactions can be investigated under conds. of equil. between the molecular phase and the solid. Using barometric techniques the amount of adsorbed gas is determined as a function of its press. at a fixed temp. The result is an *adsorption isotherm* in which the amount of adsorbed molecules is plotted against the press. of the gas. This press. is usually given in the relative unit p/p_0 where p_0 designates the saturation press. of the substance under study at the temp. of the isotherm. The same treatment holds for ions in solution and molecules of a liquid. Here concs.

of the adsorbate are determined and the abscissa of the isotherm is given as the absolute conc. of the adsorbate in an inert diluent (*solvent). From a series of a. isotherms at different temps. two important results can be obtained. The amount of adsorbate as a function of its temp. under equil. conds. can be obtained as an *adsorption isobar*. In such a graph the amount of adsorbate is plotted against its temp. If the press. required to reach the same amount of adsorbed molecules at various temps. is plotted against $1/T$, from the slope of this graph the *isosteric heat of a.* for a given amount of adsorbate is obtained. To determine the coverage of a solid with adsorbed molecules one has to distinguish between two cases. In the case of small diatomic molecules and metal surfaces a rough estimation can be done by assuming that each surface metal atom adsorbs one molecule. This assumption for the case of CO molecules is often incorrect as, besides linear coordinations, bridged a. geometries also occur which reduce the coverage by a factor of two. Close-packed metals exhibit typically 100 μmol a. sites/m^2 geometric *surface area. This geometric constant can either be used to estimate the coverage when the surface area is known or to determine the surface area when a coverage of unity is established. The latter situation is one of the most frequently used methods to determine the surface area of an irregularly shaped cat. surface. The level of information about complicated surfaces of, e.g., a *supported metal cat. can be enormously enhanced when the surface area is determined with two molecules, one of which is adsorbed non-specifically on all sites of the surface (often N at 77 K), and the other one (H or CO at 270 K) is specifically adsorbed on the metal sites of a supported metal cat. The surface area measured with the unspecific probe molecule is called *total surface area* (TSA), whereas the surface area measured with the specific adsorbate is called *active surface area* (ASA).

The ability of a material to adsorb a molecular or atomic species is often described in terms of the *sticking coefficient S (being the ratio between the number of adsorbed molecules per unit time and the number of attempts of the molecules to land on the surface per unit time). The latter quantity can be calc. from kinetic gas theory and the former can be measured from the uptake kinetics of the adsorbate after a jump in its pressure. These coefficients can vary from unity (CO on *transition metals at 300 K) down to 10-12 (N$_2$ on Ru). The sticking coefficient does not scale with the adsorption energy. S describes the complex proc. of approximation of a molecule to the surface, its reorientation in the proper a. geometry and the release and dissipation of the a. energy. Often the chemisorption of molecules on surfaces occurs via a weakly bound *precursor species which can more easily be desorbed than convert into a strongly adsorbed species. In such cases the application of external press. enhances the overall sticking coefficient.

The *adsorption probability* of a given species depends not only on the material of the adsorbent (substrate) but also on the crystallographic orientation of the surface. The sticking coefficient can vary over orders of magnitude for the same system in different geometries (S for O on silver (110), 10^{-2}; for silver (111), 10^{-7}). This phenomenon of *structure sensitivity was identified by surface science and can explain the extreme sensitivity of the performance of a cat. to the morphological shape of its active solid comp.

The experimental a. isotherms have been described by various theoretical models in order to relate the kinetic and geometric variables of the a. proc. The simplest sorption model is the *Langmuir isotherms which describes chemisorption in the limit of a given number of a. sites which are all equal in a. energy under equil. conds. In addition, no interaction between a neighboring adsorbed species should exist and no more than one monolayer should be adsorbed. Then the coverage Θ is given by $\Theta = Kp/(1+Kp)$, with K designating the ratio of the velocity constants for a. and desorption. This simple model often holds in practical cases and can be analyzed by plotting the adsorbed amount/adsorbate press. vs.

the adsorbate press. A linear relationship indicates the validity of this model. If all conds. of the Langmuir model hold but the a. of more than one monolayer is allowed, then the elaborate *BET isotherm* (*Brunauer-Emmett-Teller) is a suitable model. This model describes a. as a sequence of successive Langmuir a. procs. with an additional parameter C describing the difference between the velocity constants for a. and desorption for the first layer and all successive layers. An additional parameter is the amount of adsorbate which fills the first monolayer, V_{mono}. The useful representation of the BET isotherm for practical examination of an a. isotherm is then

$$z/[(1-z)V] = 1/C \cdot V_{mono} + (C-1)z/C \cdot V_{mono}$$

with z designating p/p_0 . A plot of the left-hand side of the equ. against z yields a straight line when the BET isotherm is a valid model and the parameters C and V_{mono} can be determined from the ordinate intercept and the slope of the graph. **E=F=G**.

R. SCHLÖGL

Ref.: Kirk/Othmer-1, Vol. 1, p. 493; J. Gregg, K.S.W. Sing, *Adsorption, Surface Area, and Porosity*, Academic Press, London 1982;V. Ponec et al., *Adsorption on Solids*, Butterworth, London 1974.

adsorption energy → adsorption

adsorption isobars → adsorption

adsorption isosteric heat → adsorption

adsorption isotherm describes the *adsorption/*desorption equil. of a gas or molecules dissolved in a liquid in contact with a *surface at variable partial press. and fixed temp. Depending on the type of substrate and on the type of *adsorption various ais. have been proposed with empirical formulas describing the *coverage or adsorbed mass as a function of the partial press.: the *Langmuir isotherm valid for sub-monolayer adsorption and energetically equivalent sites, the *Henry isotherm as a special case of the Langmuir i. in the range of small press., the *Freundlich isotherm when the adsorption sites are not

energetically equivalent but exhibit an exponential distribution with respect to the heat of adsorption, the *Temkin isotherm when the heat of adsorption drops linearly with coverage, and the *Brunauer–Emmett–Teller (BET) isotherm valid for multilayer adsorption (cf. also *heat of adsorption). **F** isotherme d'adsorption; **G** Adsorptionsisotherme. R. IMBIHL

Ref.: Somorjai; Gates-1, p. 329.

adsorption probability → adsorption

Advanced Cleavage Process (ACT) → Lummus phenol process

Advanced Oxidation Process (AOP) → catalytic waste water treatment

AEM → analytical electron microscopy

aerobe an organism that uses oxygen in its metabolic pathways, primarily as a terminal electron acceptor in cellular respiration but also as an oxidant in a variety of *oxidation reactions involving organic molecules (cf. *monooxygenase). Many organisms are obligate aerobes, with an absolute requirement for oxygen in order to survive. Others are facultative aerobes, and can survive in the absence of oxygen but grow more slowly. **F** aérobies; **G** Aerobier. P.S. SEARS

aerogel → sol–gel catalyst, *silica

Aerosil → silica

AES 1: → Auger electron spectroscopy; 2: → atomic emission spectroscopy

affinity constant a general term for association or dissociation constants

affinity labeling of enzymes → active site-directed inhibitors

AFI *zeolite structural code for *ALPO-five *zeolites; cf. *aluminophophate

AFM → atomic force microscope

AFS atomic fluorescence spectroscopy, → electron spectroscopy for chemical analysis

afterburning (catalytic afterburning) proc. for the removal of pollutants such as *VOCs by treatment under thermal (afterburning) or cat. conditions (*catalytic combustion). Combinations of both techniques (cf. Figure) are also known. **G** Nachbrennen.   B. CORNILS

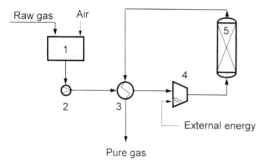

Raw gas   Air

Pure gas

1 mixer and filter; 2 blower; 3 heat exchanger; 4 burner; 5 catalytic reactor.

aging a reduction of *catalyst lifetime by slow transformations initiated by the reaction conds. other than the presence of the reactants or *poisons (*fouling, *deactivation). Temp. (*sintering, recrystallization), solvents, high press., or the mechanical forces during stirring or motion of the cat. bed (agglomeration) can cause structural changes, fragmentation of the solid, formation of stable complexes, or loss of active phase by friction which all lead to deactivation. **F** vieillir; **G** Altern.   B. CORNILS

agostic interactions Electron-deficient metal *complexes sometimes show the behavior that a σ-bond of the *ligand donates electronic charge to the metal through weak interactions which are called ais (Figure).
There may be α-ais. (a), β-ais. (b), or γ-ais. (c), which differ by the position of the X–Y σ-donor bond. Ais. may become manifest by acute bond angles α, lengthening of the X–Y σ-bond, characteristic signals in the NMR and IR spectra, and restricted conformational flexibility of the ligand. Agostic interactions

can be found in main group and transition metal *complexes. A more common expression for agostic interactions in main group cpds. is hyperconjugation.   G. FRENKING

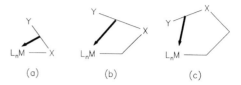

(a)   (b)   (c)

Ref.: Moulijn/van Leeuwen/van Santen.

AIBN → azoisobutyronitrile

Air Products Catfin process for the *dehydrogenation of propane to propylene over Cr_2O_3-alumina cat. The required heat of reaction is supplied by storing heat in the cat. bed during the decoking period. **F** procédé Catfin d'Air Products; **G** Air Products Catfin-Verfahren.   B. CORNILS
Ref.: McKetta-2, p. 806.

Akzo Nobel catalysts Akzo Nobel N.V. at Arnheim (Netherlands) has two business units dealing with polymer chemicals and cats. Akzo is the global leader in *polymerization cats. such as organic *peroxides, aluminum alkyls, and *Ziegler–Natta cats. and one of the top players in *FCC and *hydroprocessing cats.; other business areas are *isomerization and *oxychlorination cats. Akzo Nobel Catalysts has joint ventures in *FCC cats. (e.g., F.C.C. S.A. in Santa Cruz/Brazil), hydroprocessing cats. (e.g., Nippon Ketjen Co. Ltd. in Tokyo/Japan), and cat. services (e.g., *Eurecat S.A.).   J. KULPE

Alan Wood ore is a iron oxide cat. for the *Fischer–Tropsch synth. (*Hydrocol proc.).

ALBENE process developed in India for the manuf. of ethylbenzene from fermentation ethanol (as the *alkylation agent) and benzene. The *zeolite (ZSM-5) cat. tolerates up to 40 % of water. **F** procédé d'ALBENE; **G** ALBENE-Verfahren.   B. CORNILS

Alcalase   the Novo Nordisk trade name for its formulation of the *serine protease "Subtilisin Carlsberg" (EC 3.4.21.14), commonly used in laundry detergents for removing proteinaceous stains (see also *subtilisin).

P.S. SEARS

alcoholates   → alkoxides

alcohol dehydrogenases (EC 1.1.1.1) a class of *enzymes that cat. the reversible *oxidation of alcohols to aldehydes or ketones using the *cofactor *nicotinamide adenine dinucleotide (NAD) or its 2'-phosphate (*NADP). The cofactor is reduced to its 1,4-dihydronicotinamide derivative during the *hydride transfer reaction. The stereochemical course of the reaction depends on the *stereospecificity of the alcohol *dehydrogenase. The enzymes from yeast, *horse liver (both NAD-specific) and *Thermoanaerobacterium brockii* (NADP-specific) cat. the transfer of pro-(R) hydride from the cofactor to the *re* face of the carbonyl substrate (usually an aldehyde for the yeast and liver enzymes and a ketone for the *Thermoanaero-bacterium* enzyme). Those from *Pseudomonas* (NAD-specific) and *Lactobacillus kefir* (NADP-specific) cat. the transfer of pro-(R) hydride from the reduced cofactor to the *si* face of the carbonyl substrate. The enzyme from *Mucor javanicus* (NADP-specific) cat. the transfer of pro-(S) hydride to the *si* face of the carbonyl substrate (cf. also entries on the individual hydrogenases). **E=F=G**.

C.-H. WONG

alcohol dehydrogenases from microorganisms   In addition to (baker's and brewer's) *yeast alcohol dehydrogenase (ad.), the ad. from *Candida parapsilosis* has quite broad substrate specificity, transferring the pro-(R) hydride of NADH to yield (S)-alcohols. *Mucor javanicus* ad. (MJADH) has unusual *stereospecificity, delivering the hydride to the *si* face of the carbonyl to give an (R)-alcohol. The natural substrate of MJADH is dihydroxyacetone, but the *enzyme will ac-

cept a variety of substrates. The ads. from *Pseudomonas sp.* (ATCC 49688) cat. the transfer of the pro-(R) hydrogen from NADH to the *si* face of a number of acyclic ketones to give (R)-alcohols. The NADPH-dependent ad. from *Lactobacillus kefir* (ATCC 35411) cat. the reduction of acetophenone and other ketones to the corresponding (R)-alcohols. **E=F=G**.

C.-H. WONG

Ref.: *J.Org.Chem.* **1992**, *57*, 1532 and **1992**, *57*, 1526; *Biocatal.* **1993**, *8*, 31.

alcohol oxidase   an *enzyme which cat. the *oxidation of an alcohol to an aldehyde using the *cofactor *flavin mononucleotide (FMN) or *flavin dinucleotide (FAD). **E=F=G**.

C.-H. WONG

alcohols as motor fuels   Cat. manuf. MeOH or EtOH exclusively or mixtures thereof with petroleum-based gasoline are used as motor fuels (*gasohol, *synfuels).

B. CORNILS

Ref.: Kirk/Othmer-1, Vol. 1, p. 826; McKetta-1, Vol. 2, p. 357, Vol. 20, pp. 1, 11.

alcoholysis   a reaction analogous to *hydrolysis in which the OH group of an alcohol, instead of water, cleaves the C-element bond (e.g., *acyllactone rearrangement). **F** alcoolyse; **G** Alkoholyse.

B. CORNILS

aldol addition (aldolization)   is the base-cat. *addition of aldehydes or ketones containing α-hydrogen atoms with aldehydes or ketones yielding β-hydroxy carbonyl cpds. (the aldols). Often *dehydration of the initial aldol gives α,β-unsaturated aldehydes or ketones (via aldol *condendsation):

$$R-CH_2-COR^1+R^2-CO-R^3 \rightarrow$$
$$R-CH(COR^1)-CH(OH)-R^2R^3 \rightarrow$$
$$RC(COR^1)=CR^2R^3 + H_2O$$

The aa. is part of a class of reactions whereby a stabilized carbanion adds to carbonyl cpds.; other examples include reactions of *Darzens, *Knoevenagel, *Perkin, *Thorpe, *Wittig (*olefination). Aas. are performed in the presence of base cats. (aqueous alkali, Li amides, NaH, *tert*. amines), but the *condensation can also take place with acid cats., in which case *dehydration usually follows. The aa. may create two new stereogenic centers. Using classical methods all four stereoisomers were often obtained. In order to control *diastereo- and *enantioselectivity of aldol condensations, reactions are performed with pre-formed silyl enol-ethers (*Mukaiyama reaction), boron, Mg, Sn, or Zr enolates in the presence of *Lewis acids such as $TiCl_4$, $SnCl_4$, $La(OTf)_3$, etc. By using chiral Lewis acids cat. asymmetrical condensations have also been achieved. The self-condensation is of commercial interest with the aa. of acetaldehyde/acetone (to form crotonaldehyde; *Union Carbide sorbic acid proc., proc. of *Deutsche Texaco to MIBK) and with *n*-butyraldehyde (to 2-ethylhexenal and subsequently to *2-EH). The aa. of *iso*-butyraldehyde and formaldehyde yields hydroxypivalaldehyde (→ neopentylgylcol; cf. also *Degussa acrolein proc.). The combination of oxo and aldol reaction is the *aldox proc. For optically active aa. with Au(I) cf. the ligand *BPPFA. **F** addition du type aldol; **G** Aldoladdition (Aldolisierung).   M. BELLER

Ref.: March; *Organic Reactions*, Vol. 16, p. 1.

aldolases   a class of *enzymes which cat. *aldol additions in a reversible or irreversible manner, usually using a ketone donor and an aldehyde acceptor. More than twenty as. have been identified so far, and they can be classified into two types based on the reaction mechanism. In type I as., an *active site lysine residue forms a *Schiff base with the donor, which in turn adds *stereospecifically to the acceptor. Type I as. are found primarily in animals and higher plants; type II as. are found primarily in microorganisms where they use a zinc *cofactor as a *Lewis acid to activate the substrate. **E=F=G**.   C.-H. WONG

Ref.:   *Tetrahedron Organic Chemistry Series* **1994**, Vol. 12, p. 195.

aldolization   → aldol addition

Aldox process   a special variant of the *oxo synthesis in which C_{2n+2} alcohols are obtained from C_n olefins via *hydroformylation, followed by controlled *aldol reaction of the resulting aldehydes, in-situ *crotonization of the hydroxyaldehydes, and finally by *hydrogenation of the unsaturated aldehydes to the alcohols. The reaction needs special *promoters for the oxo cat. such as Zn or Mg salts. The purity of the products is unsatisfactory. **F** procédé d'Aldox; **G** Aldox-Verfahren.

B. CORNILS

Ref.: Falbe-1, p. 71, 80, McKetta-1, Vol. 5; *J. Mol. Catal. A:* **1998**, *132*, 189.

Alfen, Alfol → Ethyl procs.

aliphatic aldehyde oxidation   Aliphatic aldehydes undergo easily *oxidation reactions, even at ambient temps. Therefore, in aged aldehyde samples significant amounts of the corresponding carboxylic acid can often be found. With air as oxidant (*autoxidation) the first oxidation prod. of aldehydes is a carboxylic peroxy acid R–C(O)–O–O–H. This acid, which is normally stable at low temps., reacts with a further aldehyde molecule to two molecules of the corresponding carboxylic acid (*Baeyer–Villiger oxidation). The oxidation of acetaldehyde is used for the manufacture of peracetic acid, *AA, and *AAA. The latter is produced from acetaldehyde with air at 55–65 °C and increased press. in presence of cat. amounts of Mn(II) or Co(II) acetate. The metal cats. initiate the oxidation via a one-electron process by transfer of an electron from the aldehyde to an M(III) species (formed from radicals which are generated via autoxidation procs.). The M(III) ion is regenerated subsequently by peracid or acetylperoxy radicals. The oxidation of aldehydes with O, cat. by Mn or Co salts, is commercialized for the prod. of various carboxylic acids such as butyric, valeric, or 2-ethylhexanoic acid. **F** oxidation des aldehydes aliphatiques; **G** Oxidation aliphatischer Aldehyde.   R.W. FISCHER

Ref.: Sheldon-1, p. 359.

alkane oxidation transformation of alkanes in the presence of oxygen, air, or other oxidants (e.g., H_2O_2, N_2O) to O-containing prods. or olefins. The reactions can be performed catalytically or non-cat. Depending on the target prods. there are two main types of ao. reactions: *selective (or partial) *oxidation and *total oxidation (*combustion). The target prods. of *selective oxidation are alkenes, alcohols, aldehydes, ketones, acids, ethers, epoxides, or anhydrides while total oxidation results in formation of carbon oxides as only C-containing prods.

Partial oxidation of low alkanes to valuable chemicals is of great interest because of their low cost as feedstocks. The reactions are generally carried out in the presence of *mixed-oxide or *supported metal cats. Conv. of n-butane to maleic anhydride using V-phosphorous (VPO) cat. is the only het. cat. alkane-selective oxidation in commercial use (cf. *maleic acid, proc. of *Alusuisse/Lummus ALMA). Other procs. such as the synth. of ethylene and acetic acid from ethane having a potential as alternatives to present technologies are still in the development stage. For the selective transformation of higher alkanes, liquid-phase oxidations are used. A research target for high alkane oxidation is the replacement of hom. by het. cats. (cf. the procs. of *BP, *Bayer, *Celanese).

The *cat. combustion of alkanes has considerable potential relating to the development of cat. combustors for power prod. and to industrial exhaust gas stream purification. **F** oxidation des alkanes; **G** Alkanoxidation.

O.V. BUYEVSKAYA, M. BAERNS

Ref.: *Hydrocarb.Proc.* **1985**, *No. 9, 64*, 123; *Catal. Today* **1998**, *45*, 13; R.J. Brotherton, H. Steinberg (Eds.), *Progress in Boron Chemistry*, Vol. 3, Pergamon Press, 1970.

Alkar process → UOP procs.

alkoxides are alcoholates, i.e., cpds. consisting of a metal cation and an anion of an alcohol; e.g., $CH_3OH + Na \rightarrow CH_3ONa$ (Na methanolate) $+ 0.5\ H_2$. Used as cat. (*Neber

rearrangement, *Thorpe nitrile add., *Bamford–Stevens reaction, *Reppe synth., *atactic polymers, etc.) and as *condensation agents, e.g., *aldol addition, *Claisen condensation, *hydride shift reactions, *carbonylations (*Halcon/SD or *Leonard formic acid), etc. **F** alcoolates; **G** Alkoholate.

B. CORNILS

alkoxycarbonylation is the Pd-cat., hom. reaction (*oxidative carbonylation) between alkenes, CO, and alcohols, yielding unsaturated esters or diesters. **E=F**; **G** Alkoxycarbonylierung.

B. CORNILS

Ref.: Falbe-1, p. 106, *J.Mol.Catal. A:* **1999**, *143*, 263, 325.

alkoxylation proceeds while reacting unsaturated cpds. like alkenes with alcohols, generally in alcoholic solutions of Pd salts, yielding acetals or ketals. This corresponds to the *oxidation of the alkene with $Pd(II)X_2$ to form the acetal or ketal and the successive generation of Pd(0) and hydrogen halide. With diols as substrates cyclic acetals are obtained, e.g., dioxolanes from 1,3-dioxanes from 1,3-glycols. An example of a alkoxypalladation is the Murahashi synth. of γ,δ-unsaturated alcohols using O_2 as oxidant and $Pd(OAc)_2/Cu(OAc)_2$. Intermolecular alkoxypalladations are used for the formation of O-heterocycles, e.g., the synth. of tetrahydropyran derivatives. If alkoxypalladations are performed under a CO atmosphere (cat. $PdCl_2$, excess $CuCl_2$), *alkoxycarbonylations are observed. **E=F**; **G** Alkoxylierung. R.W. FISCHER

Ref.: Cornils/Herrmann-1, Vol. 1, p. 404; Beller/Bolm, Vol. 2, p. 311.

alkyd resins → Baekeland–Lederer–Manasse condensation

alkylates alkylated hydrocarbons, manufd. by cat. *alkylation of suitable raw materials via, e.g., alkenes, 1: for gasoline production, or 2: intermediates for petrochemicals (e.g., *ethylbenzene). **F** alkylate; **G** Alkylat.

Ref.: McKetta-1, Vol. 2, p. 414. B. CORNILS

alkylation insertion of alkyl groups R such as methyl, ethyl, or propyl by *addition or *substitution reactions (e.g., reactions of *Friedel–Crafts, *Meerwein, *Bahlson, etc.) into molecules. The alkyls may be bonded via C, hetero- (O, N, S, Si), or metal atoms (*organometallic cpds.). During *petroleum processing and in *petrochemistry, *alkylates are manufd. cat. (over $AlCl_3$, *SPA, sulfuric acid, *ion exchange resins, hydrofluoric acid, or *zeolites) for the prod. of either gasolines or of *intermediates (cf. manuf. of *EB and *cumene, procs. of *BASF, *CdF, *CDTech, *Exxon, *Kellogg cascade, *Mobil/Badger, *Lummus/UOP, *Monsanto/Lummus, *Phillips, *Stratco, *UOP). The alkylating agents are alkenes or lower alcohols (*ALBENE proc.). **E=F; G** Alkylierung.   B. CORNILS

Ref.: Kirk/Othmer-1, Vol. 2, p. 85; Kirk/Othmer-2, p. 71; McKetta-1, Vol. 2, p. 414; McKetta-2, p. 80; Ullmann, Vol. A1, p. 185; *CHEMTECH* **1998**, *(6)*, 40 and **1998**, *(7)*, 46; Ertl/Knözinger/Weitkamp, Vol. 4, p. 2039 and Vol. 5, p. 2123; Farrauto/Bartholomew, p. 565.

alkyne metathesis a thermoneutral reaction where alkylidene moieties are statistically redistributed over het. cats. (WO_3–SiO_2 at 350 °C) or hom. cats. ($Mo(CO)_6$–ArOH at 110 °C or by UV/rt). Applied to functionalized alkynes in hom. cat. (R^1 or R^2 = CH_2[COR, CO_2R, NR_2, halogen]; Figure)

$$2 \; R^1 C \equiv C R^2 \; \rightleftharpoons \; R^1 C \equiv C R^1 + R^2 C \equiv C R^2$$

The mechanism is interpreted via metallacyclobutadiene intermediates using high oxidation state carbyne species as well defined *initiators. The metathesis of terminal acetylenes is followed by *polymerization. **F** metathèse des alcynes; **G** Alkin-Metathese.

A. MORTREUX

Ref.: *J.Chem.Soc.Chem.Commun.* **1968**, 1548 and **1974**, 786; *J.Mol.Catal.* **1982**, 75, 93 and **1995**, 96, 95.

Allied cyclohexanone oxime process is conducted in the vapor phase, reacting cyclohexanone, ammonia and oxygen over Porasil A, a high surface area *silica.   B. CORNILS

Ref.: *J.Catal.* **1981**, *70*, 66, 72 and 84.

Allied malic acid process manuf. of malic acid by cat. *addition (*hydration) of water to *MAA. **F** procédé Allied pour l'acide malique; **G** Allied Äpfelsäure-Verfahren.   B. CORNILS

allosteric interactions In enzymatic cat., the binding of a molecule (an allosteric effector) at a site that may be far from the *active site can affect the activity of the *enzyme. *Cooperativity between different active sites of a multimeric enzyme is an example of an allosteric interaction. This is observed, for example, in the $c_6 r_6$ heterododecameric enzyme *aspartate transcarbamoylase (ATCase) from *E. coli*. The binding of a molecule of the substrate analog *N*-(phosphonacetyl)-L-aspartic acid (PAA) to one subunit induces a very large shift of the subunits with respect to each other, placing them in a more cat. active arrangement. The *conformational shift is symmetric; that is, one subunit cannot shift to the more active state without a corresponding shift in the other subunits. Thus, the addition of a small amount of PAA actually activates the enzyme. Shifting of subunits relative to each other is a very common mechanism for allostery. It has been observed in a large number of enzymes, including *chorismate mutase. **F** interactions allostériques; **G** allosterische Beziehungen.   P.S. SEARS

Ref.: *J.Mol.Biol.* **1987**, *193*, 527; **1987**, *196*, 853, and **1988**, *263*, 18583.

alloy catalysts Alloying (bimetal formation) of metals in het. cats. may influence both the *activity and *selectivity, stronger than proportional. Examples are Sn-Pd alloys for more selective *low temp. *hydrogenation of alkynes, Pt-Rh alloys for higher NO yields in *oxidation of ammonia, or Ag-Au alloys for improved *selectivity in ethylene *oxidation. There are also examples of alloying active and inactive metals (cf. *catalyst poisons). The distinction from *bimetallic cat. or bimetallic *clusters may be difficult (cf. *ensemble, *Weisz criterion, *Vegard's rule). Amorphous alloys are called *metallic glasses and are active for, e.g., oxidations. Alloys may also be used as *metal film cats. **F** catalyseurs d'alliage; **G** Legierungskatalysatoren.

B. CORNILS

Ref.: Ertl/Knözinger/Weitkamp, Vol. 2, p. 803 and 1009; F. Habashi (Ed.), *Alloys*, Wiley–VCH, Weinheim 1998; Moulijn/ van Leeuwen/van Santen, p. 166.

allylation   Reaction of allylic substrates with C, N, S, or O nucleophiles Nu (Tsuji–Trost reaction, allylic amination) is cat. mainly by a Pd(0) *complex stabilized by a P or N donor *ligand L or by Ni(0), Ir(I), Rh(I), W(0), or Mo(0) *complexes to give the C-, N-, S-, or O-allylated cpds., respectively (Figure).

Pd(0) sources are, e.g., Pd(PPh$_3$)$_4$ cats., although the active Pd(0) cat. can be formed *in situ* from Pd(OAc)$_2$ and a phosphine ligand. The reaction can also be performed in water with Pd(OAc)$_2$/*TPPTS.

Allylic acetates are the most often used allylic substrates. The mechanism occurs via a π-allyl intermediate; the *regioselectivity is generally determined by steric factors with the formation of the new bond at the less hindered site. The *selectivity can also be controlled by *ligand effects and *electronic factors. Enantioselective allylation occurs with *chiral ligands (diphosphines, diamines, or bisheteroatomic ligands such as P/N) in association with Pd(0) to give *ee higher than 90 %. The reaction of allylic silanes and stannanes with aldehydes, ketones, orthoesters, acetals, or ketals is cat. by *Lewis acids such as Me$_3$SiI, Me$_3$SiOTf, etc., to give the corresponding protected allylic alcohols (*Sakurai reaction). **F**=**E**; **G** Allylierung

D. SINOU

Ref.: Abel/Stone/Wilkinson, Vol. 8, p. 799 (1982) and p. 797 (1995); Trost/Fleming, Vol. 4, p. 585; *Tetrahedron Asymmetry* **1992**, *3*, 1089; *Chem. Rev.* **1996**, *96*, 395; *Tetrahedron Lett.* **1976**, *16*, 1295; Cornils/Herrmann-2, pp. 221 and 401; Mulzer/Waldmann, p. 8; *J. Mol. Catal. A:* **1999**, *144*, 473.

allylic amination   → allylation (Tsuji–Trost reaction)

ALMA process   → Lummus procs.

Alphabutol process   → IFP procs.

alpha-olefins, α-olefins (AO)   linear alkenes with a double bond in the α-position, manuf. via the procs. of *Ethyl, *Exxon, *Gulf, *Shell, etc.; cf. *LAO.

B. CORNILS

ALPO   aluminum phosphate, → aluminophosphate

alternating copolymerization   The discovery of highly active and selective Pd cats. in the early 1980s at Shell turned polyketones by ac. of ethylene and CO (which was described in early work of *Reppe) from lab. to commercial reality. The active cats. comprise *cis*-coordinated cationic Pd(II) species associated with weakly or non-coordinating anions. The mechanism of *polymerization involves the migratory *insertion of CO into a Pd–alkyl bond followed by a migratory insertion of ethylene into the Pd–acyl bond. The *chelating *bidentate *ligand (e.g., *phosphines, amines, etc.) ensures a *cis* orientation of the growing polymer chain and the incoming *monomer needed to facilitate migratory insertion. The perfectly alternating *insertion of ethylene and CO monomers is due to a combination of kinetics and thermodynamics. Chain transfer during *polymerization, i.e., via chain termination and initiation, can proceed via *alcoholysis of Pd acyl species. The *polyketones described are produced by Shell under the tradename Carilon. The new family of cats. is also active for ac. or *terpolymerization with higher alkenes, both aliphatic or heteroatom-functionalized. Under certain conds. the polymers can be transformed into a polyspiroketal structure, isomeric with polyketones. **F** copolymérisation alternante; **G** alternierende Copolymerization.

E. DRENT

Ref.: *Chem. Rev.* **1996**, *96*, 663; Mark/Bikales/Overberger/Menges, Vol. 10, p. 369.

Altman Sidney (born 1939), professor of molecular biology at Yale, discovered that not only *enzymes but also RNA could function as cats. (*ribozymes). A. was awarded the 1989 Nobel prize (together with Cech).

B. CORNILS

Ref.: *Angew.Chem.Int.Ed.Eng.* **1990**, *29*, 749; *Nature (London)* **1990**, *342*, 391.

alumina the most widely used single cpd. in het. cat. The origin of its usefulness is the *redox stability of pure a. (Al_2O_3) in conjunction with a very wide variation in its *acid-base properties as a solid. This the reflection of the amphoteric character of the Al^{3+} ion in aqueous solution. The specific *surface area, the *particle size, and the reactivity with water in the gas phase can be varied over orders of magnitude for each parameter and a wide selection of combinations of these parameters are possible. A. is used as a *support material for metal particles and oxides. The tunable properties of a. render it the first choice for almost any proc. where the residual *Lewis acidity of a. is not negative. Its *redox stability and its high thermal stability give the supported particles a very high degree of *stability and ensure the *lifetime of many cat. systems. A. is also used as an acid cat. system in many applications. Here either the intrinsic acidity of the OH groups at its surface or *anchored mineral acids are used as *active centers. The OH ions occur from *defects (see below) or from reaction of the surface Al atoms with water from the atmosphere. Hydroxyl chemistry on a. is very complex and always related to defects in the *bulk or on its *surface.

The very good stability of a. in very acidic media and even under steam load ensure the stability and durability of such cats. The acidity function of a. is often moderated by forming crystalline or amorphous ternary oxides, preferably with *silica. Amorphous *aluminosilicates, *zeolites, or *mesoporous solids and *ceramic cats. are very important classes of functional materials where a. creates the desired chemical activity. The versatility of a. is based upon a rich structural chemistry which can be well controlled synthetically. The starting point is always Al hydroxide, which can be obtained as gibbsite or bayerite. Careful *dehydration by heating in air and control of the heating rates can produce Al oxyhydroxide (*boehmite) or one of the six modifications of anhydrous oxide, Al_2O_3. The stuctures of these oxides can be described as close-packed layers of oxoanions and Al^{3+} cations in tetrahedral and octahedral vacancy positions. The three families of alumina referred to as α, β, and γ forms result from stacking variations of the oxoanions. The most common is γ-alumina, which is often referred to as a "defective *spinel structure".
E=F; **G** Aluminiumoxid.

R. SCHLÖGL

Ref.: Farrauto/Bartholomew, p. 60.

aluminophosphate (ALPO, AlPO, alum-(in)ophosphate) The large family of as. ("AlPOs") are similar to *zeolites in that they contain a crystalline host framework with *micropores. The framework is built from corner-sharing tetrahedra, wherein the tetrahedrally coordinated atoms (*T atoms) are connected by bridging O atoms. In contrast to the strict (but outdated) definition of a *zeolite as having an *aluminosilicate framework, the T atoms in a. are Al and P. They usually occur in a 1:1 ratio, so that the framework of composition [$AlPO_4$] is neutral. Like zeolites, as. are prepared by hydrothermal reactions involving structure-directing agents or *templates.

As. were first synth. by researchers from Union Carbide and the laboratory codes have been transferred to the scientific literature. Important as. of this group are: $AlPO_4$-5 (read "alpo-five"), $AlPO_4$-11 and $AlPO_4$-34. Another a., VPI-5, was at the time of its discovery (1988) the zeolite-like material with the largest pore diameter. VPI-5 contains a unidimensional system of parallel 18-membered ring (see *zeolites) channels with a diameter of ca. 12 A.

Similarly to zeolites, AlPOs are synth. under mild hydrothermal conds. Due to the possibility of using templating molecules (typically amines and alkylammonium ions) in these synth., many different framework topologies have been obtained. These span a wider range of pore sizes than zeolites do. AlPOs exhibit very good thermal stability (ca. 1000 °C), and are stable in steam (600 °C); these values are similar to those of zeolites. The *surface characteristic of AlPOs is moderately hydrophilic. The use of pure as. in cat. is restricted by the fact that these materials are not acidic. Acidity can be introduced by replacing part of the Al^{3+} by divalent metal atoms (e.g., Mg^{2+}, Co^{2+}, Zn^{2+}) leading to *metal aluminophosphates ("MeAPOs" such as VAPO with vanadium, MnAPO, CrAPO or CoAPO), or by partial substitution of phosphorus (formally P^{5+}) by Si^{4+}. In the *silicoaluminophosphates ("SAPOs"), however, Si usually replaces part of the Al as well. In addition, cat. active species can be deposited in the voids of the open frameworks of AlPOs. **E=F=G.**   P. BEHRENS

Ref.: Thomas/Thomas.

aluminosilicates (alumosilicates) are ternary oxides of Al and Si which contain Al and Si in fourfold tetrahedral coordination. Among the crystalline ass., *microporous *zeolites are the most important materials for het. cat. (e.g., for *acylations, *thermofor procs., *transalkylation [*Toray/UOP Tatoray proc.]). They can act as cat. themselves due to their strong acidity when in the protonated form, or they can be used as cat. *supports. Amorphous as. which can be prepared by *precipitation from solution (for example via a *sol–gel proc.) are less strongly acidic. **E=F; G** Alumosilikate.

P. BEHRENS

Ref.: Ertl/Knözinger/Weitkamp, Vol. 5, p. 2360.

aluminum chloride $AlCl_3$, catalyst for *Friedel–Crafts reactions

Aluminum Co. Selexsorb process   purification of polymer grade ethylene, inter alia removing oxygen by a bed (*guard bed) of reduced Cu cat. **F** procédé Selexsorb d'Aluminum Co.; **G** Aluminum Co. Selexsorb-Verfahren.   B. CORNILS

Ref.: Hydrocarb.Proc. **1996**, (4), 137.

aluminum in catalysis   Al plays a considerable role in catalysis: *alumina as a cat. (e.g., for *dehydrations or for the *Leonard amine proc., *Mobil LTI proc., *Oppenauer oxidations; cf. also *chlorinated alumina) or as a *support; $AlCl_3$ as a *Lewis acid (bulk or supported, e.g., on *bauxite, cf. *Shell or *Phillips vapor-phase butane isomerization) for widespread reactions (*acylations acc. to *Nenitzescu, *chlorinations, *dehydrations, *Friedel-Crafts, *isomerizations), or AlF_3 for *fluorinations). Al *alkoxides serve for *Tishchenko or *Meerwein-Ponndorf-Verley reactions. See also *aluminophosphates and *aluminosilicates.

Al alkyls have great importance in *Ziegler-Natta cat. for alkene *polymerization, "aufbau" reaction, di- or *oligomerization (e.g., procs. of *IFP Alphabutol, *IFP Dimersol, etc.), *Simmons-Smith reaction, etc. Cf. also *alumina, *methylalumoxane (MAO). **F** aluminium en catalyse; **G** Aluminium in der Katalyse.   B. CORNILS

Ref.: *Organic Reactions*, Vol. 6, p. 469, Vol. 32, p. 375, Vol. 34, p. 1, Vol. 36, p. 249; *Houben-Weyl/Methodicum Chimicum*, Vol. XIII/4 and E 18.

aluminum phosphate   → aluminophosphate

Alusuisse LAR process   This "Low Air Ratio" proc. converts o-xylene or naphthalene via *oxidation to phthalic anhydride at 400 °C in tubular reactors under air/reactant (mass) ratios between 10/1 and 22/1. The cat. is V_2O_5 (K and Mo as *promoters) on TiO_2 (cf. *titania).   B. CORNILS

Ref.: Hydrocarb.Proc. **1984**, (11), 83.

Alusuisse/Lummus ALMA process   → Lummus procs.

Alusuisse maleic anhydride process manuf. of *MAA by *dehydroxidation of butane over het. cats. on the basis of *transition metals. At conv. of 15 % per pass the selectivities reach 60 %. **F** procédé d'Alusuisse pour

anhydride maléique; **G** Alusuisse Maleinsäureanhydrid-Verfahren.   B. CORNILS

α-lytic protease   → cf. under entry "L"

AM (**ACM**)   → acrylamide

Amadori rearrangement   comprises the
reaction of D-glucose with *p*-toluidine to 1-des-
oxy-1-(*p*-toluidino)-D-fructose in addition to
the expected D-glucosylamine. The prods. are
formed by rearrangement of the glucosylamine
in the presence of cat. amounts of acids. In general Amadori cpds. are obtained by the reaction
of the corresponding aldose with a slight excess
of amine in the presence of water and cat.
amounts of acids. Apart from arylamines, primary and secondary aliphatic amines such as
piperidine, morpholine, etc., undergo the ar. to
yield the corresponding 1-amino-2-ketones.
F réarrangement d'Amadori; **G** Amadori-
Umlagerung.   M. BELLER
Ref.: H. Paulsen, K.W. Pflughaupt, *The Carbohydrates*
(Eds. Horton, Pigman), Kluwer, Dordrecht 1980, Vo-
l.Ib, p. 899.

Amberlyst   synth. *ion exchange resins

ambidentate ligands   have different modes of bonding to the metal (e.g., M–NO_2 or M–
ONO).   R.D. KOHN

**American Cyanamide anthraquinone
process**   cat. *carbonylation of benzene and
*CO with metal or *transition metal chlorides
at 220 °C/2.5–7 MPa for the manuf. of anthraquinone. **F** procédé American Cyanamide
pour l'anthraquinone; **G** American Cyanamide Anthrachinonerfahren.   B. CORNILS
Ref.: Weissermel/Arpe, p. 327.

amidases   are *enzymes which cat. the *hydrolysis of amides, and include *acylases,
*aminopeptidases, *hydantoinases, *penicillin
acylases, *carboxypeptidases, peptidases, and
*proteases. Acylase I (aminoacylase, EC
3.5.1.14), a commercially available *enzyme
from porcine kidney (PKA) or *Aspergillus*
(AA), cat. the hydrolysis of acylated L-α-ami-

no acids, and is commonly used in the
*resolution of *enantiomers. β-Amino acids,
N-alkylated, and *N*-acylated amino acids (including proline) are not substrates. D-Ami-
noacylases are known, as well. A D-amino-
acylase isolated from *Alcaligenes faecalis* cat.
the hydrolysis of large, neutral *amino acids
such as D-methionine and D-phenylalanine
but accepts small or polar amino acids poorly.
*Penicillin acylase (EC 3.5.1.11) cat. the
hydrolysis of the *N*-benzoyl group of penicillin to make 6-aminopenicillanic acid, but also
cat. the hydrolysis of the phenylacetyl protecting group from a variety of cpds. including
amino acids, alcohols, amines, and sugars. See
also *proteases. **E=F=G**.   C.-H. WONG

amidocarbonylation   three-comp.   reaction for the synth. of *N*-acyl α-amino acids
starting with an aldehyde, an acid amide, and
CO under Co or Pd cat. The formal sequence
is given in the Figure.

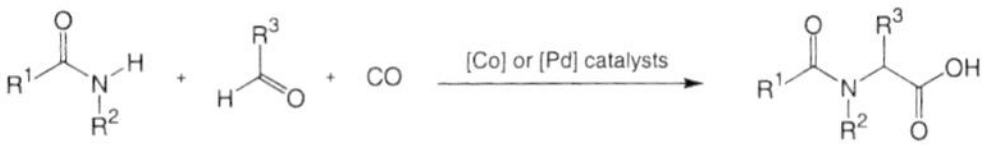

Originally, the a. was performed with
$Co_2(CO)_8$ as cat. at 120 °C/15 MPa. Today's
accepted mechanism involves the nucleophilic
attack by the amide N on the aldehyde carbonyl C atom resulting in an α-hydroxy
amide. The hydroxy group is substituted by
the $[Co(CO)_4]^-$ anion to give an alkyl Co
*complex. After CO *insertion the acyl com-
plex reacts with water yielding the *N*-acyl
amino acid. The prods. are used as detergents
(sarconisates). Besides Co, Pd systems (e.g.,
$PdBr_2$-LiBr-H^+) can also be used as cats.
F=E; **G** Amidocarbonylierung.
                                                 M. ECKERT, M. BELLER
Ref.: *J.Chem.Soc.,Chem.Commun.* **1971**, 1540; Beller/
Bolm, Vol. 2, p. 79.

aminating hydrogenation   → reductive
amination

amination   is the general term for the cat.
introduction of an amino group into an organ-

ic cpd. As. can be subdivided into *reductive, oxidative, and normal aminations. Special reviews describe the het. cat. reactions of *ammonia and amines with a wide variety of organic reactants. The following may be used as substrates for this reaction: alcohols and phenols (*ammonolysis, a well established proc. for the commercial manuf. of amines), carbonyl cpds. (*aminating hydrogenation), carboxylic acids and esters, alkenes, halogen cpds., ethers, hydrocarbons, CO, and CO_2. The main reaction routes are shown in the Figure.

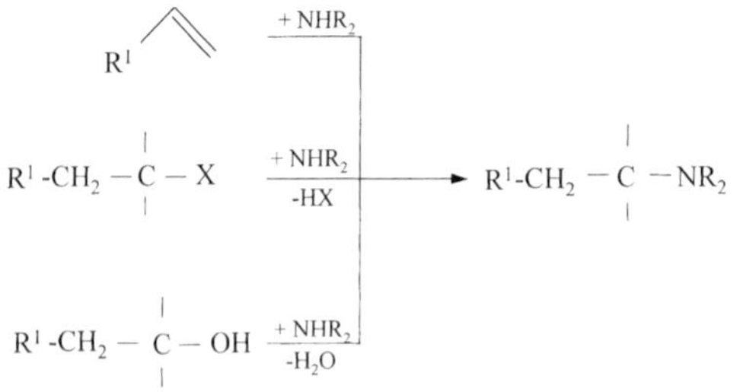

Examples are given by the procs. of *ATO, *Berol Kemi, *Jefferson, *Leonard, etc. A special review concerning the metal-initiated amination of alkenes and alkynes is also available. **F**=E; **G** Aminierung.   B. DRIESSEN-HÖLSCHER

Ref.: Ertl/Knözinger/Weitkamp, Vol. 5, p. 2334; *Chem. Rev.* **1998**, *98*, 675; Kirk/Othmer-1, Vol. 2, p. 482.

aminative peroxidation cat. procedure of simultaneous formation of –NH– and –O–O– bonds, e.g., with *Ube's 12-aminododecanoic acid proc. **F** peroxidation aminatif; **G** aminierende Peroxidation.   B. CORNILS

Ref.: Weissermel/Arpe, p. 263.

amine oxidases → copper amine oxidases, *amino acid oxidases

amino acid esterases *enzymes that cat. the *hydrolysis of amino acid esters. Many esterases, *lipases, and *proteases have this property and often exhibit strong *enantioselectivity. *Proteases, especially the *serine proteases, are usually specific for L-amino acid esters. A D-amino acid esterase was recently discovered and used in the synth. of D-peptides. **F** esterases de l'acide aminé; **G** Aminosäureesterasen.   C.-H. WONG

Ref.: *Angew.Chem.Int.Ed.Engl.* **1989**, *28*, 450.

amino acid oxidases The amino acid oxidases (aaos., L-amino acid oxidase, EC 1.4.3.2; D-amino acid oxidase, EC 1.4.3.3) cat. the *oxidation of amino acids to their respective imino acids, which then deaminate in water to the α-keto acids. They are flavoenzymes (*FAD or *FMN), and use O_2 as an oxidant to regenerate the reduced *cofactor, producing *hydrogen peroxide.
In *eukaryotes, the aaos. are sequestered in *peroxisomes, where the H_2O_2 produced is rapidly destroyed by *catalase. **F** oxidase de l'acide aminé; **G** Aminosäureoxidasen.

P.S. SEARS

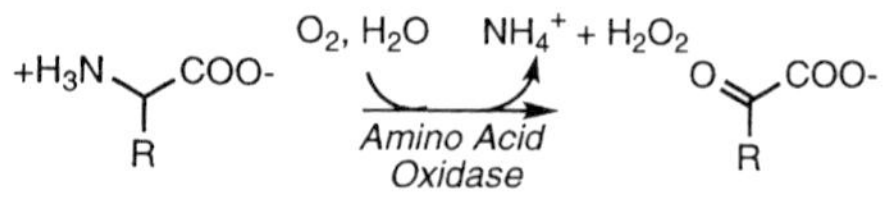

Ref.: D. Schomberg, M. Salzmann, *Enzyme Handbook*, Springer, Heidelberg 1993, Vol. 6.; *Proc.Natl. Acad.Sci. USA* **1971**, *68*, 987.

amino acid resolution → acylase, *amino acid esterase, *amidase, *amino acid oxidase, *penicillin acylase

amino acids (abbreviations) Among those common in biocat. are Gly (glycine), Ala (alanine), Phe (phenylalanine), Val (valine), Tyr (tyrosine), Thr (threonine), Asp (aspartic acid), Arg (arginine), Pro (proline), etc. **F** acides aminés; **G** Aminosäuren. B. CORNILS

aminoacyl-tRNA synthetases cat. the formation of aminoacyl tRNA through the formation of aminoacyl AMP from an amino acid and *ATP. See also *non-ribosomal peptide bond formation. **E**=**F**=**G**.   C.-H. WONG

amino alcohols as ligands represent bi- or polydentate *ligands for hom. cats. Common routes to aas. include *reduction of amino acids, amino carbonyl cpds., or hydroxy oximes, treatment of *N*-boc-amino esters with *Grignard reagents and many other synth. A large number of natural prods. contains amino alcohol functionalities. In particular, optically active aas. derived from chiral natural cpds. are applied in *asymmetric cat. and synth. Chiral β-aa. react with $CH_3B(OH)_2$ to form 1,3,2-oxazaborolidines which are efficient and highly selective cat. for the *reduction of ketones with H_3B^*L [L = *THF or $S(CH_3)_2$]. Various *chiral aas. are used as cats. for, e.g., the *enantiomeric nucleophilic *addition of dialkylzinc reagents to aldehydes. Mo and V complexes with chiral ligands such as ephedrine are cats. in asymmetric *epoxidations of allylic alcohols with *TBHP. **F** aminoalcools commes ligands; **G** Aminoalkohole als Liganden.   J. FRIDGEN

Ref.: *Chem.Rev.* **1996**, *96*, 835 and **1992**, *92*, 935; *Angew. Chem.Int.Ed.Engl.* **1994**, *33*, 497.

α-**aminoalkylation** a variant of the *Morita-Baylis-Hillman reaction; cf. *hydroxyalkylation

aminohydroxylation (*hydroxyamination) 1,2-*Amino alcohols can be prep. by the *sharpless *asymmetric aminohydroxylation of alkenes, which is closely related to *asymmetric *dihydroxylation. Os cat. the suprafacial *addition of a nitrogen atom, coming from an *N*-acyl- or *N*-sulfonyl chloramine salt and an oxygen atom to C–C double bonds. The active cat. species is best described as $OsO_3(NR)(L)$ with different chiral *ligands L. (*Os as catalyst metal). **F=E**; **G** Aminohydroxylierung.   T. STRASSNER

Ref.: Mulzer/Waldmann, p. 57; *Angew.Chem.Int. Ed.Engl* **1999**, *111*, 339, 1080.

S-**aminolevulinic acid dehydratase** cat. the formation of porphobilinogen, in a pathway leading to *vitamin B_{12} through a multienzyme-cat. reaction that includes the sequential methylation cat. by *S-adenosylmethionine-dependent methyltransferase. **F** déhydratase de l'acide *S*-amino-5-oxo-4-valérique; **G** *S*-Aminolävulinsäure-Dehydratase.   C.-H. WONG

S-aminolevulinic acid dehydratase

aminolysis Enzymatic synth. of peptides can be accomplished with *proteases by one of two approaches. In the thermodynamic approach, the hydrolytic reaction of proteases is reversed by conducting the reaction in nearly anhydrous organic *solvents, by the selective precipitation of the product, or by a similar technique that draws the reaction in favor of synth. In the kinetic approach, peptides are synth. via aminolysis of esters in aqueous buffers or aqueous/organic mixtures by an amine nucleophile, so that there is competition between the amine and water as nucleophiles for the acyl *enzyme intermediate. Inclusion of water-miscible organic co-solvents, increasing the conc. of amine nucleophile, and certain *active site modifications, such as methylation of the active site histidine of *serine proteases or conv. of serine proteases to *thiolproteases, can improve the aminolysis: hydrolysis ratio. **F** aminolyse enzymatique; **G** enzymatische Aminolyse.   P.S. SEARS

Ref.: *Angew.Chem.Int.Ed.Engl.* **1991**, *30*, 1437; *Biotechnol.Prog.* **1996**, *12*, 423.

aminopeptidase → proteases

aminophosphine phosphinites (AMPPs) are synth. by phosphinylation of parent *chiral amino alcohols using $ClPPh_2$ and/or $ClPR_2$ acc. to:

$$HNR^1CHR^2CHR^3OH + 2\ ClP(Ar/R)_2 \rightarrow$$
$$(Ar/R)_2PNR^1CHR^2CHR^3OP(Ar/R)_2$$

Amidophosphine phosphinites are prepd. similarly from amido alcohols, mixed AMPPs from 1 eq. $ClPAr_2$ and 1 eq. $ClPR_2$. The AMPPs are useful *chiral bi- or tridentate *ligands for *asymmetric cat., e.g., the *hydroformylation of styrene (30–65 % *ee*) and *hydrovinylation of dienes (>90 % *ee*). Very high *activity and *ees are obtained for the *hydrogenation of ketones, ketolactones (>99 % ee), ketoesters, etc. **E=F=G**.

A. MORTREUX

Ref.: Coord.Chem.Rev. **1998**, *180*, 1615.

aminophosphinites are synth. by monophosphinylation of amino alcohols acc. to HNMeCHMeCHPhOH + ClPPh$_2$ → HNMeCHMeCHPhOPPh$_2$. They are useful *chiral *ligands for linear *dimerization of dienes, e.g., from butadiene to octa-1,3,6-triene, with hom. Ni cats. (*TON > 5000). 1,3-Pentadiene gives 4,5-dimethyloctatrienes (*ee >90 %) in head-to-head reaction. Homodimerization of dienic esters, including methyl sorbate, is possible, as well as butadiene–dienic ester *codimerization to new trienic esters. **E=F=G**.

A. MORTREUX

Ref.: Coord.Chem.Rev. **1998**, *180*, 1615.

aminotransferases *pyridoxal 5'-phosphate dependent *enzymes that cat. the reversible transfer of the amino group of an amino acid donor to a keto acid acceptor. The reaction comprises two half-reactions: the first involves transfer of the donor amino group to pyridoxal 5'-phosphate to give a 2-keto acid prod. (which is released from the enzyme) and an enzyme-bound pyridoxamine 5'-phosphate; the second is the binding of another 2-keto acid to the enzyme and transfer of the amino group from pyridoxamine 5'-phosphate to the 2-keto acid to produce an L-amino acid and regenerate the *cofactors. Of the many known transaminases, aspartate a. (EC 2.6.1.1) from *E. coli* is the most useful (Figure 1).

A D-amino acid a. from *Bacillus sp.* has been isolated that cat. *transamination between var-

ious *D*-amino acids and α-keto acids. The aspartate as. from pig heart and bacteria have been studied with regard to their enantioselectivity for the *amination of 4-hydroxy-4-methyl, and 4-ethyl-2-ketoglutaric acids. Transaminases used in synth. include L-lysine: 2-oxoglutarate 6-aminotransferase (EC 2.6.1.36) and 4-aminobutyrate:2-ketoglutarate transaminase (EC 2.6.1.19); the latter was used in the synth. of L-phosphinothricin (L-homoalanin-4-yl(methyl)-phosphinic acid) (Figure 2).

In an enzymatic approach to the synth. of L-phenylalanine, *acetamidocinnamate amidohydrolase (EC 3.5.1.-) was used to cat. the *hydrolysis of acetamidocinnamic acid to phenylpyruvic acid, which was in turn converted to L-phenylalanine via enzymatic *transamination. **E=F=G**.

C.-H. WONG

ammo(n)dehydrogenation is the cat. *condensation reaction under *dehydrogenation of *hydrocarbons or aromatics in presence of ammonia acc. to CH$_4$ + NH$_3$ → HCN + 3 H$_2$. Cats. are rare metals (*Degussa BMA proc.), γ-Mo$_2$N, or Ni-NiO (*DuPont aniline proc.). A transition to *ammoxidation is the *Lummus proc. to *PTA. **F** ammo(n)déshydrogénation; **G** Ammondehydrierung.

B. CORNILS

ammonia lyases The *enzymes that cat. the reversible *addition of ammonia to alkenes (cf. *amination) are named ammonia lyases. They include aspartate al. (aspartase, EC 4.3.1.1), 3-methylaspartate ammonia lyase (EC 4.3.1.2), and other amino acid als. L-Aspartase from *E. coli* cat. the *addition of ammonia to the C-2 *si* face of fumarate to form L-aspartate. The enzyme is specific for its amino acid substrate. Addition of ammonia to mesaconic acid is cat. by 3-methylaspartate al. (EC 4.3.1.2). Replacement of the methyl group with H, Cl, or Br is also acceptable, and gives the corresponding 3-substituted aspartic acid. **F** lyases d'ammoniaque; **G** Ammoniak-lyasen.

C.-H. WONG

X = Me, H, Cl, Br

ammonia oxidation $\rightarrow$ oxidation of ammonia

ammonia slip secondary ammonia emissions resulting from the SCR (*selective catalytic reduction) or *DeNOx procs.

ammonia synthesis As. by the *Haber–Bosch proc. involves the synth. of the NH_3 molecule from the constituent elements in a high press./high temp. reaction. The commercial proc. is carried out over a promoted iron metal cat. at 673 K/approx. 15 MPa. Reactors with capacities up to 1000 tons per day are used. The reaction is exothermic. It is limited by an equil. which is under conds. of practical reaction rates always on the side of the educts, requiring loop operation with recovery of the easily condensable product gas. The feed gases are prepared from air (nitrogen) and from a H source (*hydrogen manuf., *syngas).

The high press. is macroscopically required to shift the equil. to the prod. since a 50 % decrease in volume (acc. to $N_2 + 3\,H_2 \rightarrow 2\,NH_3$) is associated with the reaction (Le Chatelier principle). In an atomistic picture the press. is required to ensure a sufficient *coverage of the cat. with the weakly binding dinitrogen molecule which is highly inert ($N\equiv N$ triple bond, -911.13 kJ/mol dissociation energy). The activation of the di-nitrogen into *chemisorbed atomic N is the kinetic bottleneck of the proc. over Fe cats. The following stepwise reaction of atomic N with pre-adsorbed H atoms is facile and gradually lowers the bonding interaction between the $(NH)_x$ species

and the cat. The activation of di-H over iron is facile under the conds. of the as. The cat. has to be chosen so that it activates N but does not form either stable nitrides or hydrides which prevent the reaction to ammonia. Only Ru has been found to be practically useful besides Fe, although many thousands of systems have been tested over the years. Fe is only active in a special metastable *texture which can be obtained by a critical *reduction proc. (*activation) of pre-melted (*fused) Fe oxide (*magnetite). *Promoters such as *alumina and potassium are useful to stabilize the active Fe (alumina) and to support the as. (potassium). Ru is only active in the form of small particles supported on graphite or MgO. It also requires promotion by Cs in order to reach its full activity. The much higher price of Ru is compensated by a higher tolerance of ammonia in the gas phase, allowing higher press. and lower temps. and thus reducing the cost of recirculating the non-converted gas feed.

The mechanism of the as. is the best known in het. cat. The reaction served as a prototype for the scientific strategy to understand the complexity of het. cat. reactions. It is now possible to predict the technical reactor performance based on parameters of the individual elementary step reactions. The benefit of studying elementary step reactions on well-defined *single crystals as models for the structurally complex technical cats. has thus been demonstrated. The other possible mechanism, of successive *hydrogenation of dinitrogen to diimine and hydrazine with the N–N bond being broken only in the last step of the synth., is definitively ruled out from working in the commercial reactions. It is, however, the mechanism of biological ammonia synth., which is achieved by bacteria fixing N_2 from air into NH_3 and using the energy prod. by this proc. (11 kJ/mol) supporting their life. The energetics of this proc. is comparable to that of the commercial reaction and the concs. which can be reached (dissolved diluted NH_3 in water) are much inferior to those resulting from the *Haber–Bosch proc. (pure liquefied

NH_3). As most of the ammonia is required for further processing there is little incentive to replace the high temp. proc. by the low press. and low temp. biological proc.

The most important single application of ammonia is the prod. of artificial fertilizers. Major prod. lines requiring NH_3 are explosives, dyestuffs, plastics and many life-science prods. For modern variants cf. the procs. of *Haldor Topsøe, *ICI AMV, *Kellogg KAAP, or *Linde LAC. **F** synthèse d'ammoniaque; **G** Ammoniaksynthese.   R. SCHLÖGL

Ref.: A. Mittasch, *Geschichte der Ammoniaksynthese*, VCH, Weinheim 1951; A. Nielsen (Ed.), *Ammonia, Catalysis and Manufacture*, Springer, Berlin 1995; Ertl/Knözinger/Weitkamp, Vol. 4, p. 1697; M. Appl, *Ammonia*, Wiley–VCH, Weinheim 1999.

ammonolysis the slightly exothermal conversion of alcohols into amines by reaction with ammonia, primary, or secondary amines in the presence of oxidic or metallic cats. (cf. *amination). The prod. pattern is governed by the thermodynamic equil. which in turn is determined by the ratio between the alcohol and the aminating agent. Thus, an excess of NH_3 is applied if primary amines are the target prods. Unwanted amines can be recycled in order to increase the yield of the desired product.

The a. of lower alcohols (methanol, ethanol) is carried out in the presence of oxidic cats. with dehydrating properties (*silica, *aluminosilicates, *zeolites); in this case it is only by gas-phase operation at 260–340 °C/1–4 MPa. *Shape selectivity of *zeolites improves the yield of dimethylamine, the most important of the methylamines. Alcohols in the C_3–C_8 range are usually converted in gas-phase procs.; for higher (fatty) alcohols liquid-phase operation is preferred. Ni- and Cu-containing cats. are usually applied. In order to preserve the metallic state of the cat. hydrogen must be present. Temps. between 120 and 180 °C and press. up to 2 MPa are common. Whereas the a. of monohydric alcohols proceeds smoothly and with high selectivity, glycols (and polyols) need severe reaction conds. (160–220 °C/8–30 MPa) and only partial conversion to diamines is

achieved (e.g., *Celanese HMDA proc.), **F** ammoniolyse; **G** Ammonolyse.

C.D. FROHNING

amm(on)oxidation (oxidative ammonolysis) is the selective, cat. *oxidation of hydrocarbons (particularly alkenes, toluenes, or methyl heterocycles) to nitriles in presence of ammonia acc. to $R–CH_3 + 3\ O_2 + NH_3 \rightarrow R–CN + 3\ H_2O$.

More than 4 MM tpy of *acrylonitrile are produced by the *SOHIO proc. (fluid-bed technology at 400–500 °C/slightly increased press.; procs. of *BP/Ugine, *Montedison, *Snamprogetti, etc.) on *multicoponent, *multifunctional cats. based on Bi-Mo oxides (*bismuth molybdate). For the specific activation an optimum *redox activity (substrate oxidation by the cat., reoxidation by oxygen; cf. *redox reactions, het.) is necessary. The mechanism is well studied. In the cat. cycle the interaction with ammonia forms an a. center for the coordination of the alkene. The key step for the alkene activation is the H abstraction in the allylic position. In the reoxidation of the cat. the communication of the corresponding *active sites and necessary anion *vacancies in the lattice play a role (cf. Figure under the keyword SOHIO). Further interesting a. reactions are: isobutene to methacrylonitrile (*multicomponent Bi-Mo-O cat.) and toluene to benzonitrile (V-Ti-O cat.). *Antimonates are a. cats., too. The cats. may be received from *precursors such as *VPO. When ammonia in ammoxidations of alkenes is absent, the oxidation leads to the unsaturated aldehydes, e.g. acrolein from propylene (*BP/Ugine proc.). In the same way substituted toluenes or methyl heterocycles are ammoxidized to their corresponding nitriles; xylenes can be converted by reaction of one or of both methyl groups to mono- or dinitriles (e.g., p-tolunitrile, terephthalonitrile [*Lummus proc.], o-tolunitrile, phthalodinitrile, etc.). Other prods. are chlorobenzonitriles, nicotinonitrile, 4-cyanothiazole, etc.

Alkanes are less reactive and give lower yields. The reaction of methane (or methanol) with NH$_3$ and oxygen yields HCN and can be regarded as a. or *ammondehydrogenation (at 1100–1200 °C on Pt-Rh *gauze cats. or with Ni-NiO cats.; procs. of *Andrussov, *Degussa BMA, *Sumitomo). The a. of pyridine side-chains is described in *Lonza's nicotin-amide proc. **E=F=G**.   B. LÜCKE

Ref.: Rylander in Anderson/Boudart, p. 28; Ertl/Knözinger/Weitkamp, Vol. 5, pp. 2253, 2302, and 2326.

ammoximation   the cat. formation of ox-imes from ketones, ammonia and oxygen. Of particular interest is the formation of cyclo-hexanone oxime which is manuf. by reaction in the gaseous phase on *silica based cats. or in the liquid phase (with H$_2$O$_2$ or O$_2$ as oxi-dants on Ti-ZSM-5 *zeolites). Procs. are li-censed by *Allied or *Enichem. **E=F=G**.
                                                      B. LÜCKE

Ref.: Ertl/Knözinger/Weitkamp, Vol. 5, p. 2326.

Amoco/Mid Century TPA process

manuf. of *terephthalic acid (*TPA) by cat. *oxidation of *p*-xylene with air (cf. *DMT). The liquid-phase *oxidation in *AA proceeds with Co and Mn acetates together with *co-catalysts such as NH$_4$Br/CBr$_4$ at 175–230 °C/ 1–2 MPa. **F** procédé Amoco pour l'acide té-rephthalique; **G** Amoco Terephthalsäure-Ver-fahren.   B. CORNILS

Ref.: Cornils/Herrmann-1, p. 543.

Amoco trimellitic acid process   two-step
manuf. of *trimellitic acid anhydride (TMA) by 1: cat. *carbonylation of *m*-xylene with CO (cat. HF·BF$_3$) and 2: *oxidation of the intermediate 2,4-dimethylbenzaldehyde over MnBr$_2$/HBr). **F** procédé Amoco pour l'acide trimellitique; **G** Amoco Trimellithsäure-Verfahren.

Ref.: Weissermel/Arpe, p. 316.   B. CORNILS

Amoco UltraCat process   special *regen-
eration method for *FCC cats.

amorphous structures   are solids where a loss in the periodicity of the crystal is present. This loss of crystallinity can be explained by the introduction of *defects in the periodicity.

Those defects can be two- and three-dimen-sional, breaking the two- and/or three-periodic sequence of the crystal atoms. Nevertheless, the definition of amorphous structures is rather am-biguous since a material can show itself as amor-phous with X-ray diffraction, while *HREM or *HRTEM can still show areas where nanocrys-talline materials are present. Therefore, a dis-tinction between X-ray amorphous and nano-crystalline must be considered. Ass. present in catalysis can be cat. *supports such as *silica, *aluminum oxides and *carbon supports (*acti-vated carbon and *carbon blacks). Further-more, oxide-based cats. also present the active phase as being amorphous. Defects are thought to provide one of the *active sites during re-action where *chemisorption and activation procs. are accelerated. **F** structures amorphes; **G** amorphe Strukturen.
                     R. SCHLÖGL, E. SANCHEZ-CORTELON

AMP   1: *aminophosphinite; 2: adenosine monophosphate

amplification   the term a. is used in cat. in enzymatic *cascade reactions and in *asym-metric cat. for the amplification of signals or informations.

AMPP   → aminophosphine phosphinite

AMV process   → ICI procs.

amylase, amyloglucosidase   Amylases are *glycosidases that *hydrolyze glucose polymers. They may be retaining or inverting, and there are amylases that accept either anomeric linkage. α-Amylase, an *enzyme found in saliva, hydrolyzes the Glc-α1,4-Glc unit of starch (amylose). It is a retaining en-doglycosidase, which will hydrolyze starch at random positions within the molecule. β-Amylase, on the other hand, hydrolyzes the β1,4-Glc linkages found in cellulose. It is an inverting exoglycosidase, which hydrolyzes a disaccharide (maltose, Glcβ1,4Glc) from the non-reducing end of the polymer. **E=F=G**.
                                                        P.S. SEARS

AN (ACN)   → acrylonitrile

anabolism   refers to *biosynthetic pathways designed to build molecules that are useful for the cell from common intermediates. The prod. of lipids from acetyl *CoA, proteins from amino acids, sugars from glyceraldehyde, or polysaccharides from sugars are examples of anabolic procs. Anabolism is contrasted with *catabolism, pathways that destroy molecules for excretion, recycling, or energy generation. **F** anabolisme; **G** Anabolismus.   P.S. SEARS

anaerobe   is an organism that does not use oxygen. If the organism *requires* an oxygen-free environment, it is called an obligate (or "strict") anaerobe. If it can survive in oxygen-containing environments, it is called aerotolerant. **F** anaérobies; **G** Anaerobier.
                                              P.S. SEARS

analytical electron microscopy (AEM) permits the determination of elemental compositions of solid cats. by energy dispersive detection of the electron induced X-ray emission (probe: electrons; response: photons). The ex-situ technique is non-destructive; the data are local and bulk- sensitive.   R. SCHLÖGL

anatase   → titania

anchor(ing) 1: *immobilization of hom. cats. on suitable *supports (*carriers) acc. to any bonding; 2: some authors distinguish between *immobilization and anchoring in that respect that "anchored" ought to mean *complexes which are chemically bonded to the support surface; 3: groups connecting *guests in *template catalysis.   B. CORNILS

ancillary ligands synonymously termed auxiliary or spectator *ligands, comprise the portion of the ligand sphere of a *heteroleptic *complex $MX_nY_mZ_o$ which is generally not the site of reactivity, i.e., als. are not involved in basic cat. steps such as *ligand exchange or *insertion reactions. However, the reactivity of an electronically and coordinatively unsaturated metal–ligand fragment is often manipulated by *tailoring/fine-tuning of the al., thus imparting mononuclearity, rigidity, solubility, or kinetc stability to the complex (cf. *ligand accelerated cat.). Als. can be classified acc. to their valency (charge) and coordination mode: neutral als. comprise *phosphines and *carbene ligands, while tied-back *cp, *salen, and *BINOL ligands represent divalent als. **F** ligand ancillaire; **G** Steuerligand.   R. ANWANDER

Ref.: *New J.Chem.* **1995**, *19*, 525; *Top.Organomet. Chem.* **1999**, *2*, 1.

Anderson–Emmett–Kölbel mechanism
This proposed mechanism of *Fischer-Tropsch synth. explains the initiation and chain growth of the *hydrogenation of CO to *hydrocarbons (hcs.). The assumed first step is the simultaneous chemisorption of hydrogen and CO to a cat. center, leading to an enolic primary complex (Figure 1):

$$ \underset{M}{\overset{O}{\underset{\|}{\overset{\|}{C}}}} \xrightarrow{2H} \underset{M}{\overset{H}{\underset{|}{C}}\diagdown OH} $$

Chain growth is initiated by splitting off water from two adjacent primary complexes (Figure 2):

$$ \underset{M}{\overset{H}{C}\diagdown OH} + \underset{M}{\overset{H}{C}\diagdown OH} \longrightarrow \underset{M\ \ M}{\overset{H}{C}-\overset{OH}{C}} \xrightarrow{2H} \underset{M}{\overset{H_3C}{C}\diagdown OH} $$

Chain growth continues via *condensation and hydrogenation steps, leading to linear hc. chains. Termination occurs via *dehydrogenation/hydrogenation of the growing molecule to alkenes, w/o subsequent hydrogenation to alkanes. **F** mécanisme reactionnel selon A–E–K; **G** A–E–K-Mechanismus.   C.D. FROHNING

Andrussov process for the manuf. of HCN by *amm(on)oxidation of methane and NH_3 and O_2 with Pt-Rh cats. (in *gauze reactors) at 1000–1200 °C. **F** procédé d'Andrussow; **G** Andrussow-Verfahren.   B. CORNILS

Anfinsen, Christian B. (1916–1995), professor of biochemistry at Harvard and Johns Hopkins University in Baltimore/MD. Worked on the biochemistry of proteins and the relations between chemical structure and cat. *activity of *ribonuclease. Nobel laureate in 1972 (together with Moore and Stein).

B. CORNILS

Ref.: *Angew.Chem.* **1973**, *85*, 1065.

angular overlap model (AOM) is a simple *MO method to calculate the splitting of the valence d orbitals of a *transition metal under the influence of *ligands in a *complex. It gives the interaction energy ε between a transition metal d orbital ϕ_d and a ligand orbital ϕ_L as a function of the squared overlap integral S^2_{dL} as being $\varepsilon = KS^2_{dL}$. The constant K is proportional to the inverse of the energy difference between the non-interacting orbitals. Since the radial part of the overlap integral at a given distance is constant, the strength of the interaction ε depends only on the angular factor between ϕ_d and ϕ_L. Angular overlap factors are tabulated and can be used to estimate electron configurations and excitation energies. **E=F=G**.   G. FRENKING

Ref.: *Adv.Inorg.Chem.Radiochem.* **1978**, *21*, 1978.

ANIC/ENI MTBE cracking process for the manuf. of pure isobutene by cracking of *MTBE in fixed-bed reactors over boron pentasil *zeolites.   B. CORNILS

Ref.: US 4.656.016 (1987); *Appl.Catal.* **1994**, *115*, 180.

ansa-**metallocenes**   → cf. under "M"

anthrahydroquinone (anthraquinol) is manuf. by cat. *hydrogenation of alkylanthraquinone over Ni or Pd cats. at 30–35 °C. With oxygen, 2-alkylanthrahydroquinone undergoes extremely rapid *autoxidation via the *endo*-peroxide with formation of H_2O_2 and regeneration of anthraquinone, the *reaction carrier. This exchange between anthra- and anthrahydro-quinone is the basis for various procs. for the manuf. of *hydrogen peroxide

(*anthraquinone proc.). **F=E**; **G** Anthrahydrochinon.   B. CORNILS

Ref.: Kirk/Othmer-2, p. 627; Büchner/Schliebs/Winter/Büchel, p. 24.

anthraquinone processes for the manuf. of H_2O_2 via *anthrahydroquinone (former procs. of IG Farben, DuPont, etc.).

antibodies, antibody catalysis → catalytic antibodies

anticatalysis Corresponding to the definition of *catalysis, anticatalysis (*negative cat.*) involves a *decrease* in the rate of approach to the equil. of a chemical reaction without the anticatalyst being substantially consumed itself (cf. *inhibitor). **F** anticatalyse; **G** Antikatalyse.   B. CORNILS

Ref.: Ertl/Knözinger/Weitkamp, Vol. 1, p. 22.

antigen-binding fragments (Fragment: antigen binding, F_{ab}) F_{ab}s. are the portions of immunoglobulins responsible for antigen recognition. Immunoglobulins (IgGs) are composed of four polypeptide chains, two heavy and two light, where the heavy chains are roughly twice the size of the light chains. These chains form a Y-shaped structure as shown below, which consists of variable regions (V_L and V_H), which determine the binding specificity, and several "constant" regions (C_L and C_H). The "arms" are F_{ab}s and can be produced by limited digestion with the protease *papain. Alternatively, a dimeric F_{ab}, called an F_{ab}', can be separated from the base, called the F_c region (Fragment: complement), by digestion with pepsin.

F_{ab}s may also be produced at the genetic level via deletion of the C-terminal half of the heavy chain, or the fragment may be trimmed even further to the variable regions only (V_L and V_H) to form the F_V fragment, which is often produced as a single chain (scF_V) through linkage of the two domains via a flexible polypeptide tether. **F=F=G**.   P.S. SEARS

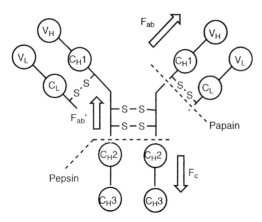

antigen-binding fragments

anti-Markovnikov addition → Markovnikov addition

antimonates The second class of *multicomp. cats. for *ammoxidation (besides Bi molybdates) are antimonates, mainly uranium antimonate (as commercialized by *SOHIO) on SiO_2, or Fe antimonate (*UCB in Belgium, *Nitto in Japan). Sn antimonates were introduced (and are now being phased out) by *BP. **F=E**; **G** Antimonate.   B. CORNILS
Ref.: Ertl/Knözinger/Weitkamp, Vol. 5, p. 2318.

antimony in catalysis For halogenations (*chlorination, *fluorination, *Swarts fluorination) Sb chlorides and fluorides are used. $SbCl_3$ is used as a *Lewis acid in *Friedel–Crafts reactions. Sb_2O_3 is the most common *Lewis acid cat. for the polycondensation of *BHET to polyethylene terephthalate (PET). Sb is said to be a *promoter for the *contact masses of the *Rochow synth. *Antimonates with Fe, U, Sn, Mn, or Ce are an important group of *ammoxidation cats. **F** antimoine en catalyse; **G** Antimon in der Katalyse.
                                                      B. CORNILS
Ref.: S. Patai (Ed.), *The Chemistry of Organic Arsenic, Antimony & Bismuth Compounds*, Wiley, Chichester 1994; *Houben-Weyl/Methodicum Chimicum*, Vol. XIII/8.

antioxidants cf. *auto(o)xidation, *catalase, *superoxide dismutase

anti/syn insertion concerns the *stereospecificity of the butadiene insertion reaction into the allyl–metal bond. Depending on the *cis* or *trans* configuration of the η^4- or η^2-coordinated butadiene in the cat. *complex, the new butenyl group which is generated by C–C bond formation exhibits the *anti* or *syn* configuration, respectively, in accordance with the *principle of least structure variation. **F** insertion anti ou syn; **G** Anti/syn-Einschiebung.   R. TAUBE
Ref.: Cornils/Herrmann-1, p. 285.

anti/syn isomerism indicates the different configurations in the η^3-coordinated allyl group concerning the position of the substituent R at the C(3) atom in reference to the hydrogen atom at the C(2) atom (Figure).

$$\eta^3\text{-}anti \qquad \eta^1\text{-}/\sigma\text{-}C(3) \qquad \eta^3\text{-}syn$$

The transformation of one isomer into the other proceeds via an η^1- or σ-C(3) coordination of the allyl group. The given definition of *anti/syn* isomerism is inconsistent with the IUPAC *E/Z* nomenclature, but is generally used in the literature. **F** isomérisation anti/syn; **G** Anti/syn-Isomerisierung.   R. TAUBE
Ref.: *Organometallics* **1995**, *14*, 4132, and **1999**, *18*, 3045.

AO → alpha-olefin

AOM → angular overlap model

AOP advanced oxidation process, → cat. waste water treatment

APAP acetyl *p*-aminophenol (acetaminophen), cf. *profens

Apel tradename of a *COC polymer from Mitsui

aperture angle → metallocenes

°API (American Petroleum Institute), measure in degrees API for the density of petroleum-derived prods. at 60 °F (15,56 °C).

B. CORNILS

apoenzyme As compared to the *holoenzyme, the a. is solely the polypeptide portion of an *enzyme, without the *cofactors, *coenzymes, metal ions, or metal *complexes found in the active cat.

P.S. SEARS

APP → atactic polypropylene, *polypropylene

apparent activation energy If the logarithm of the *rate* of a cat. reaction be plotted versus $1/T$, the slope of the linear plot equates to E_a/R, where E_a is termed the *apparent activation energy* (*Arrhenius plot). The reason for the prefix "apparent" is that, in principle, as the temp. is raised the *surface concs. of the adsorbed reactants will decrease, because their adsorptions are exothermic. The rate therefore does not rise as quickly as it would if the reactants did not desorb, and the value of the measured *activation energy therefore depends on how quickly the surface coverages decrease as the temp. is raised. In theory, applying the Arrhenius equation to a rate constant (which is what ought to be done) will give the true activation energy E_t, but obtaining *rate constants depends on finding a satisfactory and valid rate expression.
If the Arrhenius plot for the simple unimolecular reaction A → B made with results obtained using such a pressure of A, and such a range of temp., that the coverage θ_A remains very close to unity, then E_t is obtained. If however θ_A is low throughout, then it is E_a that is measured, and since the temp. dependence of the adsorption coefficient b_A is given by $(d\ln b_A/dT) = \Delta U^\theta/RT^2$, therefore $E_a = E_t + \Delta U^\theta$. Under the former cond., the reaction will be zero-order in A; under the latter, it will be first-order: therefore the relation first proposed by M.I. Temkin in 1935 (*Temkin isotherm), viz. $E_a = E_t + n_A\Delta U^\theta$, where n_A is the order of reaction, describes both conds.

F l'énergie apparente d'activation; **G** scheinbare Aktivierungsenergie.

G.C. BOND

Ref.: Laidler.

apparent density of catalysts → bulk density

APS → 3'-phosphoadenosine 5'-phosphosulfate (PAPS)

APXPS appearance potential X-ray photoemission spectroscopy, → X-ray spectroscopy

aqueous-phase catalysis a technique (aqueous biphasic cat., abc.) and a special variant of *two-phase cat. Abc. obeys *Manassen's principle. In abc., the active cat. for the reaction is (and remains) dissolved in water, so that the reactants and reaction prods., which are ideally organic and relatively nonpolar, can be separated off after the reaction is complete by simply decanting the second phase from the cat. solution, thus making it easy to recirculate the latter (Figure). The fact that *selectivity- and yield-reducing procedures (such as thermal stresses caused by chemical cat. removal or distillations) for separating prod. and cat. are avoided makes it possible to use sensitive reactants and/or obtain sensitive reaction products by hom. cat.

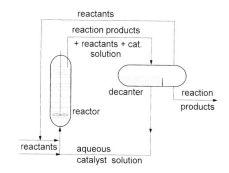

Originally, there was little academic research work and only some industrial effort. It was not until the work at Ruhrchemie AG that development led to the first large-scale utilization of the abc. technique at the beginning of the 1980s, viz. in *hydroformylation (*Ruhrchemie/Rhône-Poulenc proc.). Abc. is used commercially for the hydroformylation of lower alkenes C_3–C_5. A total of about 600 000 tpy of preferably linear C_4 and C_5 aldehydes are being produced at two locations. This corresponds to approx. 10 % of the world capacity.

The proc. is based on water-soluble *TPPTS, viz. triply *m*-sulfonated triphenylphosphine (*TPP). The actual oxo cat. is $[HRh(CO)\{(m\text{-}SO_3NaC_6H_4)_3P\}_3]$ and thus bears nine sulfonate substituents (i.e., three per P atom) and is accordingly readily soluble in H_2O (approx. 1.1 kg/L). The proc. is extremely economic and environmentally benign. The partition coefficients between aqueous and organic phases follow the *Hansch equ. For the relation between aqueous and non-aqueous solvents in *enzyme cat., cf. *enzyme cat. in organic solvents. **F** catalyse en phase aqueuse; **G** wäßrige Zweiphasenkatalyse. B. CORNILS

Ref.: *Proc. 4th Int.Symp.Homog.Cat.*, Leningrad, 1984, p. 487; Horváth/Joó; Cornils/Herrmann-1, p. 577; Cornils/Herrmann-2; *Adv. Organomet.Chem.* **1992**, *34*, 219; *Chem.Ing.Tech.* **1994**, *66*, 916; *J.Organomet.Chem.* **1995**, *502*, 177; Grieco (Ed.), *Organic Synthesis in Water*, Blackie Academic, London 1998; Reichardt; Martell/Hancock; Stumm.

aqueous solvents for enzyme catalysis
→ enzyme catalysis in organic solvents

ARAES
resolved Auger electron spectroscopy, → Auger electron spectroscopy

ARCO
*Atlantic Richfield Corp.; see also their procs.

ARCO butanediol process
manuf. of *butanediol (*BDO) by *hydroformylation of allyl alcohol with *ligand-modified Rh cat. **F** procédé ARCO pour le butandiol; **G** ARCO Butandiolverfahren. B. CORNILS

ARCO ethyleneglycol process
via *acetoxylation of ethylene with acetic acid and oxygen in the presence of a cat. system consisting of TeO_2/HBr and subsequent *hydrolysis of the diacetoxyethylene. The plant was shut down because of severe corrosion problems. **F** procédé ARCO pour l'éthylèneglycol; **G** ARCO Ethylenglykolverfahren. B. CORNILS

ARCO ethylurethane process
converts nitrobenzene in the presence of EtOH and CO directly via *reductive carbonylation into ethylcarbamate (ethylurethane). Elemental Se is used as the cat. to generate hydrogen in a *WGSR. The proc. was never applied commercially, probably due to the highly toxic intermediates H_2Se and Se=C=O. **F** procédé ARCO pour l'uréthane d'éthyl; **G** ARCO Ethylurethan-Verfahren. W.R. THIEL

ARCO Magnaforming process
a variant of the *platforming proc. to achieve high-*RON gasolines. The cat. proc. (Pt) at 1–2.5 MPa may directed toward high contents of aromatics, *LPG, or isobutane. **F** procédé Magnaforming d'ARCO; **G** ARCO Magnaforming-Verfahren. B. CORNILS

Ref.: Winnacker/Küchler, Vol. 5, p. 99.

ARCO Octafining process
*hydroisomerizes *ethylbenzene (a C_8 hydrocarbon from pyrolysis gasoline prod.) to xylenes as well as the *m*-xylene to the *p*-isomer over a *dual-function cat. with Pt on amorphous *silica-*alumina at 500 °C/1–2.5 MPa. **F** procédé Octafining d'ARCO; **G** ARCO Octafining-Verfahren. B. CORNILS

Ref.: *Petr.Refiner* **1958**, *37*(8), 208; Ertl/Knözinger/Weitkamp, Vol. 5, p. 2127.

ARCO propylene oxide process
Manuf. of *propylene oxide (PO) by combined *radical cat. *epoxidation of propylene and co-oxidation of an auxiliary cpd. (*reaction carrier, commercially isobutane or ethylbenzene). The proc. was jointly developed by the former *Halcon and by Atlantic Richfield, *oxirane). Depending on the reaction carrier, a second prod. is co-produced, e.g., PO/*tert.*-butanol (a development of *Texaco) or PO/1-phenylethanol (*styrene). The overall reaction when started from propylene and ethylbenzene (EB) includes the formation of EB

hydroperoxide by *radical-initiated oxidation (at 130 °C/3.5 MPa) and the coupled transfer of oxygen to propylene by hom. cat. (Mo, W) at 90–130 °C/2–4 MPa at residence times of approx. 2 h, thus eventually yielding PO in high selectivity and 1-phenylethanol as byproduct which is dehydrated to styrene. **F** procédé ARCO pour l'oxyde de propylène; **G** ARCO Propylenoxyd-Verfahren.

B. CORNILS

Ref.: Kirk/Othmer-1, Vol. 20, p. 271; McKetta-1, Vol. 45, p. 88; Ullmann. Vol. A22, p. 239.

areal reaction rate
the reaction rate divided by the surface area of the cat.

arene–aromatics oxidation
comprises the *oxidation of alkyl-substituted arenes to aromatic carboxylic acids (alkyl side chain = methyl) or ketones (alkyl side chain = ethyl or higher homologs). The oxidation is performed in acetic acid at 100–220 °C/2–3MPa with air. The applied cats. are Co-Mn salt mixtures (mixing ratio from 10/1 to 1/10 or higher) with HBr as promoting *co-cat. In special cases other cat. metals like Ni or Fe or *promoters like Zr are added to provide high cat. *activity. Fine-tuned cats. allow extremely high *STY (>1 kgL^{-1}h^{-1}) if applied at high reaction temps. This type of cat. oxidation reaction using air (oxygen) as oxidant is of high importance for the synth. of *terephthalic acid and other aromatic dicarboxy systems (cf. procs. of *Amoco, *Bofors, *Eastman, *IFP, *Hüls, *Lummus, *Mitsubishi, *Toray; cf. also *DMT synth.). The oxidation reaction runs via a *radical chain proc., started by *autoxidation of the alkyl–arene substrates. Thus, any substituents are tolerated except those which act as a radical trap (*radical scavenger). The great advantage of the proc. is its extraordinary high *selectivity (up to 99 %) in spite of the drastic reaction conds. Most of the arene carboxylic acids are very insoluble in the solvent *AA and thus easily separated and purified by simple washing with the solvent and consecutive drying.

R.W. FISCHER

Ref.: Cornils/Herrmann-1, Vol. 1, pp. 439, 541.

arene coupling
Bi-, ter-, oligo-, and polyaryls are important as *chelating *ligands (e.g., *BINAP, *BISBI). Homocoupling of aryl halides can be achieved by Ni(II)/Zn(0); *Ullmann-type coupling with Pd(0) and stoichiometric amounts of reducing agents such as Zn(0) or isopropanol. Arenes are also coupled by *Lewis or mineral acids (*Scholl reaction) but yields are low. Cu(I) cats. in the presence of oxygen are useful for the *oxidative coupling of phenols by a *radical type mechanism to form polyethers (p-phenylene oxide, PPO). Unsymmetrical and symmetrical bi-, ter-, oligo-, and polyaryls are selectively prepared by Pd(0)- or Ni(0)-cat. *cross-coupling of aryl halides or triflates with aryl organometallic cpds. The elements most commonly used are B (*Suzuki c.), Sn (*Stille c.), Zn (*Negishi c.), and Mg (*Grignard c.). Other metals such as Li, Al, Zr, Si, or Cu have also been used. The mechanism of all of these reactions (Figure) consists of the basic organometallic reaction types such as *oxidative addition, *transmetallation, and *reductive elimination. **F** couplage des arènes; **G** Aren(Aromaten)kupplung.   V. BÖHM

Ref.: *Tetrahedron* **1998**, *54*, 263.

arene dioxygenase
Since the 1970 enantioselective *oxidation of toluene to *cis*-dihydrotoluenediol and benzene to *cis*-cyclohexadienediol by a mutant of *Pseudomonas putida*, many arenes have been shown to yield diols through microbial techniques (Figure). All reactions were cat. by *dioxygenases using molecular O$_2$. *cis*-Dihydrotoluenediol has been used for synth. of terpene and prostanoid synthons and cyclohexadienediol has been used in the synth. of polybenzene. Cy-

clohexadienediol was also used in the synth. of carbocyclitols such as pinitol, conduritols, and hydroxyenones. A dioxygenase from *Pseudomonas putida* also cat. the *oxidation of styrene, chlorobenzene and derivatives to the corresponding cyclohexadiene *cis*-diols, which were used for synth. of natural products. **F** dioxygenase aromatique; **G=E**.

C.-H. WONG

Ref.: *Biochem.* **1970**, *9*, 1626; *J.Org.Chem.* **1990**, *55*, 4683.

ARGE process one of three technically and economically successful variants of the *Fischer–Tropsch synth., developed 1954 by the ARGE (consortium) Ruhrchemie/Lurgi. Core of the proc. is a water-cooled tube bundle reactor (i.d. 46 mm, length 12.000 mm) which allows efficient removal of the reaction enthalpy, thus avoiding hot spots and related problems. A gas recycle is necessary. The synth. design targets maximum yields of diesel range motor fuels. The reaction conds. are 220–260 °C/2–3 MPa, ratio H_2/CO 1.3–2.0, *GHSV of syngas 500–700, conv. of syngas 65–70 %. Precipitated Fe cat., doped with K_2O and SiO_2, yields a prod. composition (in % w/w) of approx. C_1/C_2 7, C_3/C_4 10, b.p. 30–165 °C 17, 165–230 °C 8.5, 230-320 °C 12, 320–460 °C 23, > 460 °C 18, oxygenates 4. The diesel and the *wax fraction are of excellent quality. The yield of motor fuels can be increased by common procs. One plant at *SASOL-One still operates the ARGE process. **F** procédé d' ARGE; **G** ARGE-Verfahren. C.D. FROHNING

Ref.: Falbe-1, 2; Anderson/Boudart, Vol. 1; Storch/Golumbic/Anderson.

arginine kinase, arginine phosphate
Aph. is a high-energy phosphate source among some invertebrates, such as the molluscs. Arginine kinase (EC 2.7.3.3), usually

isolated from lobster tail, cat. the (reversible) transfer of phosphate from aph. to *ADP. In this way, aph. acts as a high-energy phosphate reservoir for the cell, much like *creatine phosphate in mammalian cells. Ak. has been used in the large-scale production of aph. **E=F=G**.

P.S. SEARS

Ref.: *Bioorg.Chem.* **1984**, *12*, 170.

ARIS process → VEB Leuna procs.

arm-off dissociation the normally reversible dissociation of *ligands to generate *vacant coordination sites for other incoming ligands (Figure). O. STELZER

Ref.: *Organometallics* **1991**, *10*, 3798.

Arndt–Eistert reaction a *homologation of preferably alkyl carboxylic acid chlorides via diazoketones and their metal cat. rearrangement to *carbenes and eventually to the homologous acid. **F** réaction d'Arndt et Eistert; **G** Arndt–Eistert-Reaktion. B. CORNILS

Arofining process → Howe-Baker procs.

aromatic oxidation is the *oxidative conv. of aromatic systems into hydroxy-substituted derivatives or quinones. Phenol is produced by the *Hock reaction. Other older methods (synth. starting from benzene via benzenesulfonic acid or chlorobenzene and consecutive *hydrolysis, cf. *Raschig proc.) are of less economic significance. Oxidation of aromatic systems with *Fenton's reagent gives hydroxy-substituted arenes in fair to good yields. Substituted hydroquinone derivatives can be synth. by *hydrogenation of the corresponding quinones derived from the arene precursors by *oxidation. The prepn. of quinones from arenes is performed by selective oxidation applying various oxidants such as CrO_3/H_2SO_4, Ce ammonium nitrate or sulfate, MnO_2, per-iodic acid, supported Pd(II) (*electrochemically or with H_2O_2 os oxidant), or peroxyacetic acid. To apply H_2O_2 as oxidant, activation with Re-based cats. is required (cf. *Re as catalyst metal). Highly efficient for the H_2O_2-mediated formation of quinones are organorhenium cats. like CH_3ReO_3, its higher homologs or simple inorganic Re oxides like ReO_3 and Re_2O_7. The system $AcOH/H_2O_2$, cat. with, (CH_3ReO_3) transforms 2-methylnaphthalene to 2-methyl-naphthoquinone (menadione, vitamin K_3) with high conv. (>80 %) and high *regioselectivity (>85 %). The Re-based cats. can be efficiently applied to the oxidation of a broad variety of aromatic substrates. **F** oxidation aromatique; **G** Aromatenoxidation.

R.W. FISCHER

Ref.: Cornils/Herrmann-1, Vol. 1, p. 430.

aromatic substitution The electrophilic aromatic *substitution (Ar-S_E) is generally referred to as the *Friedel–Crafts reaction. Under *Lewis acid cat., a H atom is replaced either by a carboxylic acid halide to form an aromatic ketone; by an alkene, alkyl halide, or by alcohol to form an alkylated aromatic cpd. (e.g., *ethylbenzene, *cumene), or by Br_2 or Cl_2 to form an aryl halide. (*halogenation). Most frequently *Lewis acids like $AlCl_3$ or mineral acids (or *superacids) are used. In the latter case *sulfonation or *nitration may occur. The attack of the electrophile El proceeds via σ-complexes (Wheland complexes, **1**).

1   2

Benzene and alkylbenzenes can undergo *formylation by CO and HCl with Lewis acid cats. (*Gattermann synth.). Furthermore, electrophilic substitution can be achieved by

Pd(0) cats. at the *ipso* position of aryl halides or triflates (*arene coupling, *arylation). The nucleophilic *ipso* substitution takes place at activated positions, e.g., at methoxy groups, via Meisenheimer complexes **2**. An important example is the replacement of a hydroxy by an amino group with NH_3 in the presence of $NaHSO_3$ as cat. (*Bucherer reaction). Aryldiazonium salts are activated in the *ipso* position for nucleophilic substitution (Ar-S_N1) by heating (Schiemann reaction) or via *radicals (Ar-$S_{RN}1$) by cat. amounts of Cu(I) (*Sandmeyer reaction). **F** substitution aromatique; **G** aromatische Substitution.   V. BÖHM

Ref.: March.

aromatization the conv. of alkanes into aromatic hydrocarbons, e.g., C_7H_{16} (heptane) → C_6H_5-CH_3 (toluene) + 4 H_2 (*dehydrocyclization; e.g., procs. of *Catarol, *IFP BTX, *UOP Cyclar, etc.). Unlike this C_{6+} a., which is not connected with a change of the C number of the reactant molecule, the a. of lower alkanes necessarily requires a preceding lengthening of the C chain, at least a *dimerization (*dehydrodimerization). Because of these complicated skeletal reactions the *selectivity of the a. of lower alkanes is mostly inferior. The a. of lower alkanes proceeds on *bifunctional cats., e.g., Ga, Zn, or Pt, supported on *zeolites. The reaction sequence is alkane → *dehydrogenation (on the Ga, Zn, Pt species) to alkenes → *oligomerization of the alkenes on the H^+ function of the zeolite, and finally the *dehydrocyclization. Cf. also *C_2/C_3 hydrocarbon a. **F** aromatisation; **G** Aromatisierung.   H. LIESKE

Ref.: Ertl/Knözinger/Weitkamp, Vol. 4, p. 2069.

Arosat process   → Lummus procs.

ARPES angle-resolved photoelectron spectroscopy,   → electron spectroscopy

Arrhenius Svante (1859–1927), professor of physics in Stockholm, worked on electrolytes, reaction kinetics (*A. plot). Nobel laureate 1903.   B. CORNILS

Arrhenius plot expresses the temp. dependence of the standard internal energy change ΔU^{θ} as $(\mathrm{d}\ln K_c/\mathrm{d}T)_p = \Delta U^{\theta}/RT^2$, where K_c is the equil. constant in conc. units. K_c is the quotient of *rate constants for forward and reverse reactions (k_f and k_b), and therefore $(\mathrm{d}\ln k_f/\mathrm{d}T) = E_f/RT^2$ and $(\mathrm{d}\ln k_b/\mathrm{d}T) = E_b/RT^2$. Thus $(E_f - E_b) = \Delta U^{\theta}$, and providing the E terms are not functions of temp., $k = A \exp(-E/RT)$, where A is a constant known logically as the *pre-exponential factor: this equ. describes forward and reverse reactions equally. In 1889 *Arrhenius gave an interpretation of the above equ., saying that molecules only reacted after acquiring an *activation energy E, such molecules being in equilibrium with non-activated molecules. This equation is associated with his name, and a plot of ln k versus $1/T$ should afford a straight line of slope $-E/R$.

From both theoretical and practical standpoints it is important to know how temp. affects the rates of cat. reactions. However, because of the difficulty (hinted at elsewhere, cf. *rate constants) of obtaining valid rate constants, it is usual to plot ln (rate) versus $1/T$, and such *pseudo-Arrhenius* plots are often linear at least over some limited range of temp., the slope giving an *apparent activation energy. A possible source of confusion is that cats. may suffer *deactivation at higher temps. and thus lead to curvature of the plot. **F** équation d'Arrhenius; **G** Arrhenius-Gleichung.

Ref.: Laidler. G.C. BOND

ARS process → Stone & Webster procs.

Arton trade-name of *Japan Synthetic Rubber

ART process asphalt residual treating, cf. *Engelhard FCC proc.

arylation means the introduction of aromatic groups into organic cpds. Most versatile is the *Heck–Mizoroki vinylation via Pd(0) cats. such as Pd(PPh₃)₄ or a mixture of Pd(OAc)₂ and tri-o-tolylphosphine. After *oxidative addition of aryl halides, triflates, or diazonium salts to Pd(0), the *complexes react with unhindered alkenes to result in arylalkenes by replacing one of the vinyl hydrogens. For the *stability of the intermediate organo-Pd cpds. the lack of β-sp³-hydrogens is crucial, thus making the method inapplicable for the alkylation of alkenes.

Established transformations of aryl halides by Pd(0) cats. include the formation of aryl alkynes (*Sonogashira reaction), arylamines (Buchwald–Hartwig reaction, *amination), aryl *phosphines (*Stelzer reaction), and aryl ethers by substitution of an H atom on the reaction partner. For the arylation of arenes cf. *arene coupling. For the arylation of nucleophiles or electrophiles see *aromatic substitution. **F=E**; **G** Arylierung. V. BÖHM

aryl sulfatases catalyze the *hydrolysis of aryl sulfates to the aryl alcohol + sulfate.

C.-H. WONG

aryl sulfotransferase → sulfotransferase

as. asymmetrical, → asymmetric cat.

AS → atom scattering

ASA → active surface area

Asahi methacrylic acid process(es)
Two-step proc. for the manuf. of *methacrylic acid by het. cat. *oxidation of *tert.*-butanol to methacrolein (over Mo-Fe-Ni oxides at $420\,°C/0.1–0.3$ MPa) and subsequent cat. oxidation, using the same *multicomponent metal oxides (*MCM cat.), of the acrolein to *methacrylic acid. Between both steps the methacrolein must be purified. Another Asahi proc. starts from isobutyric acid methyl ester, which is cat. *sulfodehydogenated by H₂S/S. **F** procédé Asahi pour l'acide méthacrylique; **G** Asahi Methacrylsäureverfahren.

Ref.: Weissermel/Arpe, p. 284. B. CORNILS

Asahi methanediphenyl diisocyanate (MDI) process is a multi-step proc. for the manuf. of MDI by *oxidative carbonylation of aniline at 150–180 °C/5–8 MPa with metallic Pd and alkali iodide as *promoter to N-phenylurethane, followed by *condensation with HCHO (acid cat.) and *thermolysis at 250 °C/3 MPa to MDI. **F** procédé Asahi pour MDI; **G** Asahi MDI-Verfahren.   B. CORNILS

Ashland RCC (reduced crude conversion) process   method for the *cat. cracking of high-boiling feeds such as atmospheric reduced crude oil. **F** procédé RCC d'Ashland; **G** Ashland RCC-Verfahren.   B. CORNILS
Ref.: McKetta-1, Vol. 47, p. 34.

aspartame   an artificial sweetener (Asp–Phe–OMe). The N-protected *precursor can be made enzymatically from the protected amino acids with the *protease thermolysin. The reaction in water is driven virtually to completion due to precipitation of the dipeptide product. **F** synthèse enzymatique d'Aspartame; **G** enzymatische Aspartamsynthese.
                                                         P.S. SEARS

CBz-L-Asp + L-PheOMe $\underset{}{\overset{\text{Thermolysin}}{\rightleftharpoons}}$ H₂O + CBz-L-Asp-L-PheOMe ↓
                                                                (96% yield)

Ref.: *Tetrahedron Lett.* **1979**, *28*, 2611.

aspartate ammonia lyase   → ammonia lyases

aspartokinase   from *E. coli* (EC 2.7.2.4) cat. the *phosphorylation of the β-CO₂ group of aspartate to form β-aspartyl phosphate.

$$\text{⁻O}_2\text{C} \overset{\text{NH}_3^+}{\underset{}{\diagup}} \text{CO}_2^- \xrightarrow{\text{ATP}} \text{⁻O}_2\text{C} \overset{\text{NH}_3^+}{\underset{}{\diagup}} \overset{\text{O}}{\underset{}{}}\text{OPO}_3^=$$

↓

MeO₂C-, H₂NOC-

Studies of substrate *specificity indicate that the α-amino group is essential for substrate recognition. The 1-carboxyl group is not required for substrate recognition: both the 1-amide and 1-esters are competent alternative substrates. In addition, β-derivatized structural analogs such as β-hydroxamate, the β-amide, or β-esters are phosphorylated at the carboxyl group through a reversal of *regioselectivity. **E=F=G**.   C.-H. WONG
Ref.: *Biochem.* **1992**, *31*, 799.

aspartyl   proteases   *Proteases cat. the *hydrolysis of amide bonds in proteins or peptides. Many can also accept simple ester and amide substrates. The aps., also called acid proteases or carboxyl proteases, use a pair of carboxylic acids at the *active site that are currently thought to act as a general acid and base.

Though some early studies indicated that there might be a covalent acyl or amino *enzyme, repetition or reinterpretation of the work indicates that none probably exists. These enzymes are typically active under acidic conds., though several are active at neutral pH. Examples are the digestive enzyme pepsin and HIV protease. **E=F=G**.
                                                         P.S. SEARS

***Aspergillus*, expression system**   *Aspergillus niger* is a filamentous fungus commonly used for high-yield prod. of proteins and other biochemicals (e.g., antibiotics). This organism can secrete proteins to very high yields, and has the additional advantage that, since it is *eukaryotic, it can carry out any necessary post-translational modifications unique to eukaryotes. Moreover, *Aspergillus* has *GRAS status   and therefore can be used for prod. of proteins targeted for human consumption such as *lactase. **E=F=G**.   P.S. SEARS

association constant   → Scatchard plot

asymmetric catalysis   is an *enantioselective synth. in which a *chiral cat. is used. Compared to the substrate to be refined, the *chiral cat. is present in substoichiometric quantities. Therefore, ac. results in an economical multiplication (*amplification) of the chiral information contained in a small amount of cats. to produce a large amount of prod. Multiplication factors up to millions are possible. *Enzymatic cat. is mostly ac. **F** catalyse asymétrique; **G** asymmetrische Katalyse. H. BRUNNER

Ref.: E.N. Jacobsen (Ed.), *Comprehensive Asymmetric Catalysis*, Springer, Heidelberg, 1999; Collins/Sheldrake/Crosby; Mander; Noyori; Ojima.

asymmetric dihydroxylation (AD) → dihydroxylation

asymmetric site (enantioselective site)   the essential element in the *enantioselective synth. of molecules via het. cats. Ass. are *surface sites for which the *transition states to two enantiomers exhibit different activation barriers. Surfaces with ass. are generated by pre-adsorption of large *optically active organic molecules which steer the *adsorption and reaction toward the desired enantiomer. **F** site asymétrique; **G** asymmetrische Stellen. R. IMBIHL
Ref.: Thomas/Thomas.

asymmetric synthesis → enantioselective synthesis, *enantioselectivity in enzymatic catalysis

atactic polymers   poly-1-alkenes and vinylic polymers (p.) with a statistically random configuration at their tertiary C-atoms (*stereoselectivity, *tacticity). They have an amorphous structure. All *radical-initiated polymerizations give atactic p. (polyvinyl chloride, polystyrene, polyacrylonitrile, etc.). Free radical cats. such as peroxodisulfate, organic *peroxides, or *AIBN are often used for polymerization in bulk, solution, or in aqueous emulsion and suspension. Anionic polymerization of *styrene by butyllithium, of *acrylonitrile by *alkoxides, or of *propylene oxide by *ROMP cats. give aps. BF_3, $AlCl_3$, etc. are used for the cationic polymerization of isobutene, vinyl ethers, etc. to aps. Atactic *polypropylene is produced by simple

*Ziegler–Natta cats. such as $TiCl_4/AlEt_3$; pure atactic polyolefins are obtained by *metallocene cats. They are highly viscous liquids. **F** polymères atactiques; **G** ataktische Polymere.
W. KAMINSKY

Ref.: *Angew.Chem.* **1957**, 69, 213; P. Rempp, E.W. Merril, *Polymer Synthesis*, Hüthig und Wempf, Basel 1986.

ATC → atom transfer chain (catalysis)

ate complexes → Lewis acid–base concept, *homoleptic complexes

Atelier de Vitrification Marcoule process   converts (vitrifies) wastes from radioactive fuel elements or from cat. into a permanently storable form by incorporating (embedding) it in glassy matrixes. The proc. takes place in electrically heated rotary furnaces with ground glass frit (cf. *final disposal of cats.). **F** procédé AVM; **G** AVM-Verfahren.
B. CORNILS

Ref.: Büchner/Schliebs/Winter/Büchel, p. 591.

Atgas coal gasification   consists of a molten iron bath at approx. 1500 °C (cf. *molten salt media). Coal, steam, and oxygen are introduced together with limestone by injection through lances. The cat. coal gasification yields mainly CO. **F** Atgas gaséification de charbon; **G** Atgas Kohlevergasungsverfaren.
B. CORNILS

Ref.: J.G. Speight, *The Chemistry and Technology of Coal*, Dekker, New York 1994.

Atlantic Richfield → cf. also *ARCO

Atlantic Richfield Duotreat process manuf. of food-grade white oils meeting rigid FDA specs. by two-stage, fixed-bed *hydrogenation of naphthenic and dewaxed raffinates, thus reducing their aromatics content. **F** procédé Duotreat d'Atlantic Richfield; **G** Atlantic Richfield Duotreat-Verfahren. B. CORNILS
Ref.: *Hydrocarb.Proc.* **1978**, (9), 127.

Atlantic Richfield/HR *HDA process thermal *dealkylation (*hydrodealkylation, *HDA) of aromatic cpds. to yield benzene, toluene, xylenes, or naphthalene. One major application is benzene from toluene at 650–760 °C/3.5–7 MPa. Other procs. use cats. such as *supported Cr, Co, or Mo. **F** procédé d'Atlantic Richfield/HR de HDA; **G** Atlantic Richfield/HR HDA-Verfahren.   B. CORNILS

Atlantic Richfield methanediphenyl diisocyanate (*MDI) process

three-step proc. for the manuf. of *MDI by cat. (SeO$_2$, KOAc) *carbonylation of nitrobenzene to *N*-phenylurethane, *condensation with HCHO, and *thermolysis to MDI. **F** procédé d'Atlantic Richfield pour MDI; **G** Atlantic Richfield MDI-Verfahren.   B. CORNILS

Atlantic Richfield Pentafining process

for the upgrading of light naphtha (C$_5$/C$_6$ hydrocarbons) by treatment with *bifunctional cats. *alumina/*silica, thus avoiding the problems with HCl and continuous chloride addition (as, e.g., with *Standard Oil's Isomate proc.). **F** procédé Pentafining d'Atlantic Richfield; **G** Atlantic Richfield Pentafining-Verfahren.   B. CORNILS
Ref.: *Petr.Refiner* **1956**, *35*(4), 138 and **1957**, *36*(5), 172.

Atlantic Richfield Xylene-Plus process

cat. *transalkylation or *disproportionation of, e.g., toluene to benzene and xylenes, or mesitylene and toluene to xylenes. Typically *isomerization cat. such as *Lewis acids (HF, BF$_3$, AlCl$_3$) or CoO-MoO$_2$ on *aluminosilicates are used. **F** procédé Xylene-Plus d'Atlantic Richfield; **G** Atlantic Richfield Xylene-Plus-Verfahren.   B. CORNILS
Ref.: Winnacker/Küchler, Vol. 5, p. 211.

ATOCHEM ω-aminoundecanoic acid process

three-step proc. for the manuf. of ω-aminoundecanoic acid (a presursor for nylon-11) by 1: *hydrolysis of undecenoic acid methyl ester (*ATO heptaldehyde proc.), 2: *peroxide-cat. *addition of HBr to the unsaturated acid in a bubble column, and 3: *amination of the undecanoic acid bromide with ammonia. **F** procédé d'ATO pour l'acide ω-aminoundecanoique; **G** ATO ω-Aminoundecansäure-Verfahren.   B. CORNILS

ATOCHEM Atol-1 process

manuf. of *HDPE with *Ziegler–Natta cats. on the basis of Mg/AlR$_3$ at 60–100 °C/1–3 MPa in gasphase fluidized-bed technology. **F** procédé Atol-1 d'ATO; **G** ATO Atol-1-Verfahren.
Ref.: *Appl.Catal.* **1994**, *115*, 180.   B. CORNILS

ATOCHEM heptaldehyde process

manuf. of *n*-heptaldehyde as byprod. of the thermal cracking of castor oil dimethyl ester at 300 °C to undecenoic acid ethyl ester (*ATO ω-aminoundecanoic acid proc.). **F** procédé d'ATO pour heptaldéhyde; **G** ATO Heptaldehyd-Verfahren.   B. CORNILS

ATOCHEM LLDPE process

manuf. of *LLDPE by oxygen-cat. ethene polymerization in tubular reactors at 180 MPa. The conv. rate is 22 % per pass. **F** procédé d'ATO pour le LLDPE; **G** Atochem-LLDPE-Verfahren.
Ref.: *Hydrocarb.Proc.* **1977**, *56*(11), 206.   B. CORNILS

Atol process → Atochem procs.

atom economy is a conception coined by Trost, introducing (besides synthetic efficiency as the main target) the question of "how much of the reactants end up in the product?". This is atom economy. Thus, the primary goal is the evaluation of synth. methods offering 100 % conv. at 100 % *selectivity – a problem of the reaction type and of cats. as "activators". It is obvious that in terms of ae. *cycloadditions, cyclo- and other *isomerizations, rearrangements, *disproportionations, and *addition reactions are preferred over *substitutions, eliminations, or *condensation reactions. *Cross-couplings are another possibility, having the disadvantage of compulsory co-production of other components. In this respect Sheldon's concept of the *E factor is more far-ranging and progressive. **F** économie atomique; **G** Atomökonomie.
          B. CORNILS

Ref.: *Science* **1991**, *254*, 1471; *Angew.Chem.Int. Ed.Engl.* **1995**, *34*, 259.

atom efficiency   is atom economy related to whole plant sites. Values of 1 kg of waste per tonne of prod. are excellent.

Ref.: *Green Chem.* **1999**, *1*, G3; J.H. Clark (Ed.), *Chemistry of Waste Minimisation*, Chapman and Hall, London 1995.

atomic absorption spectroscopy (AAS) This photon-based destructive method (ex situ) is well suited for elementary compositions. The data are integral and bulk-sensitive.

R. SCHLÖGL

Ref.: Ullmann, Vol. B5, p. 559; K.W. Jackson, *Electrothermal Atomization for Analytical Atomic Spectroscopy*, Wiley, Chichester 1999.

atomic emission spectroscopy (AES)   is less used in cat. research.

atomic force microscopy (AFM)   The AFM yields a view on virtually any surface in atomic detail and thus delivers local, *surface-sensitive (capable of collection in situ) data about morphology and atomic structures. The samples must be flat; artefacts can also be investigated (non-contact AFM, NC-AFM, through strong forces between sample and tip).

Ref.: Niemantsverdriet, p. 176.   R. SCHLÖGL

atom scattering (AS)   for the determination of *surface atomic structures (probe/ resonance: atoms). The UHV, ex-situ method is non-destructive but needs *single crystals. The data received are local and surface-sensitive.   R. SCHLÖGL

atom transfer chain (ATC) **catalysis** chain reaction proc. like *ETC but the particle transfer in the chain is now an atom instead of an electron or electron hole. ATC cat. can be considered as an inner-sphere version of ETC cat. since there is a net *redox change upon atom transfer. In the same way that ETC procs. are usually favorable when the cross-redox step is exergonic, ATC procs. are best designed if the bond broken is weak-

er than the bond formed in the cross-atom transfer step.

The initiation step is similar to ETC; it can involve 1: an electron transfer provided by electrochemistry or by redox reagents; 2: generation using *radicals; 3: any means of introducing radicals (e.g., for the addition of cat. amounts of a radical such as *AIBN). The two main types of ATC cat. reactions are halogen ATC and hydrogen ATC, inter alia the *autoxidation of hcs. or *Kharasch type halogen additions. Another well known example is the *oxidative addition of alkyl iodides to 16-electron square Ir(I) *complexes. H-ATC reactions are found for the exchange of a carbonyl with a *phosphine in carbonyl–monohydride complexes, in heterobimetallic *reductive elimination of methane, *insertion of alkenes, *coenzyme B_{12} dependent rearrangements, etc. **F** catalyse par transfert d'atome en chaîne; **G** ATC-Katalyse.   D. ASTRUC

Ref.: Astruc; *Chem.Rev.* **1983**, *87*, 425; *Angew.Chem. Int.Ed.Engl.* **1982**, *21*, 1; *Inorg.Chem.* **1980**, *19*, 3230 and 3236; *J.Am.Chem.Soc.* **1972**, *94*, 4043, **1973**, *95*, 7908, and **1977**, *79*, 2527; *J.Organomet.Chem.* **1980**, *190*, 1533.

atom transfer method   *radical chain reaction in which the propagation step involves a *homolytic substitution of a prod. radical with the *precursor (e.g., an alkyl halide). Aside from an initiator, these reactions need no other added reagents and are real cat. *radical reactions. This method has the advantage that no other reaction for the radical R· can compete and that only cat. amounts of toxic tin reagents initiate the reactions. However, the key atom transfer step must be fast enough to propagate the chain. There are examples of H atom transfer, but halogen atom transfer reactions are of larger synth. usefulness. The applications range from intermolecular radical *additions to *cyclizations. **F** méthode de tranfer atomique; **G** Atomtransfermethode.   T. LINKER

Ref.: *Acc.Chem.Res.* **1988**, *21*, 206.

atom transfer radical polymerization (ATRP)   → radical catalysis (polymerization)

ATP   adenosine 5'-triphosphate, → adenosine phosphates

ATP sulfurylase → 3'-phosphoadenosine 5'-
phosphosulfate

ATRP   atom transfer radical polymerization,
→ radical catalysis (polymerization)

attrition loss   indicates the mechanical stability of cat. *pellets (spheres, rings, cylinders,
extrudates; cf. *forming of cats.) and of cat.
grains. The methods of determining the al. are
not standardized; therefore a number of
different procedures are in use: tumbling to
simulate the loading into a reactor, or linearly
increasing forces applied to the diameter or
the axis of tablets in specially constructed devices which are commercially available. Procs.
with fluidized or entrained cats. generate fines
mainly due to mechanical abrasion. The simulation of the mechanical stress in fluidized
test units is used where either the formation
of fines is measured directly (as *carryover)
or the shift in particle size distribution is monitored. **F** abrasion; **G** Abrieb   C.D. FROHNING
Ref.: Ertl/Knözinger/Weitkamp, Vol. 2, p. 589; US
5.821.192.

"aufbau" reaction   The "aufbau" (growing) reaction is the   *insertion of ethylene
into aluminum hydride or AlEt$_3$ at temps. of
100–120 °C. The prods. are mixed aluminum
trialkyls AlR$_3$ in which the alkyl groups consist of up to 200 ethylene units. *Oxidation
and *hydrolysis of these prods. lead to alcohols with an even number of carbon atoms
(e.g., *Ethyl Alfol proc.). At 200 °C the "aufbau" reaction becomes a cat. reaction due to
elimination of n-alkenes and the simultaneous
formation of active alkylaluminum hydrides.
The ar. cannot be performed with *α-olefins,
which react with AlR$_3$ with the formation of
dimers (*Shell SHOP proc.). The *addition of
TiCl$_4$ to the ar. led to the discovery of *Ziegler–Natta cat. **F** réaction d'Aufbau; **G** Auf-
baureaktion.   W. KAMINSKY, D. ARROWSMITH
Ref.: Angew.Chem. **1952**, *64*, 323.

Auger electron spectroscopy (AES)   is
based on the relaxation of photoionized
atoms to ions of lower energy. It yields integral, surface-sensitive information about *surface compositions and chemical bonding. The
UHV method is non-destructive and ex-situ
(cf. *SAM).   R. SCHLÖGL
Ref.: Niemantsverdriet, p. 37,68; Ullmann, Vol.B6,
p. 45; Ertl/Knözinger/Weitkamp, Vol. 2, p. 632.

Aurivillius phases   a family of complex oxides, derived from the structure of *perovskite, CaTiO$_3$. Bismuth titanates form a
homologous series represented by $(Bi_2O_2)^{2+}$
$(A_{n-1}B_nO_{3n+1})^{2-}$. In general A is a large cation, typically Ca, Sr, Ba, Pb, etc., that can
readily fit into the interstices in the perovskite
layers (composed of corner-shared BO$_6$ octahedra), B is the octahedral cation (typically
Ta, Nb, Ti, W, Mo) and n the number of layers
of corner-shared octahedra in each perovskite
slab. Usually n ranges from 1 to 5, and a well-
known example is Bi$_4$Ti$_3$O$_{12}$. In a sense the
phases may be regarded as regular intergrowths of pure octahedral layers. Bi$_2$MoO$_6$ is
an efficient cat. for the selective oxidation of
propene to acrolein (cf. *bismuth molybdates
and the procs. of *BP/Ugine, *Shell, or
*SOHIO). A whole family of structures may
be formed by recurrent intergrowths at the
sub-unit cell level (Sillen phases). Cpds. like
PbBi$_3$WO$_8$Cl are good cats. for the *oxidative
coupling of methane to C$_2$ prods. **F** phases
d'Aurivillius; **G** Aurivillius-Phasen.
                                              J.M. THOMAS
Ref.: L. Mandelkorn (Ed.), *Nonstoichiometric Compounds*, Academic Press, New York 1964; *Faraday Dis-
cuss.Chem.Soc.* **1989**, *87*, 33; C.N.R. Rao, B. Raveau,
Transition Metal Oxides, VCH, Weinheim 1995.

autocatalysis 1-general: A. is applied when
a prod. of a cat. reaction acts to accelerate the
reaction; the rate in a closed system thus increases continuously, unless or until diminu-
tion of the reactant concs. counteracts the
effect. A. is usually discussed in the context of
the hom. cat. *Belousov–Zhabotinsky reaction, where interaction of several autocatalytic steps leads to *oscillatory behavior.

There are few examples of a. in het. cat., but oscillatory phenomena occur widely, especially in cat. *oxidations (CO, H_2, etc.): autocatalytic reactions are therefore implicated. It is now recognized that the state of the *surface, presence of *adsorbed species, etc., must be regarded as a "product of the reaction", and when metal *single crystals are used as cat. it is possible to follow alternating surface structures in parallel with rate changes. Analysis of what is occurring is complicated by simultaneous variations in reaction enthalpy and cat. temp. For a. in supramolecular chemistry cf. *self replication. Asymmetric a. with amplification of *chirality is also known

G.C.BOND

2-in enzyme catalysis: Many of the *proteases are not active until partial proteolytic digestion (they are secreted in their inactive *proenzyme, or "zymogen" form), but once activated they may go on to activate other proenzymes. This is true for many of the *serine proteases such as the bacterial protease *subtilisin or the mammalian enzymes *trypsin and *chymotrypsin. **F** autocatalyse; **G** Autokatalyse. P.S. SEARS

Ref.: Gates-1, p. 62; Thomas/Thomas; Mulzer/Waldmann, p. 79; Ertl/Knözinger/Weitkamp, Vol. 1, p. 22.

autoclave chemical reactor used in batch, semibatch, or continuous operation, usually made of steel (in labs.: of glass) and able to withstand high press. and temps. (*catalytic reactors, *laboratory reactors). Batch or semibatch reactors (BR) provide species conc. data (reactant-product) as a function of real time and show the character of a stirred-tank reactor (*CSTR, cf. *catalytic reactors). In batch operation the charge is brought to the desired conds. (temp. and press.). Semibatch operation can be done in the same way; however, one or more reactants are continuously fed and prods. are withdrawn. The first so-called rocking bomb autoclaves were used in early studies of het. cat. *hydrogenations in a liquid phase. Modern as. are propeller-stirred and internally buffered. Therefore, *suspended solid cats. are maintained in suspension by shaking (rocking) or stirring, providing greater contacting efficiencies between gas, liquid, and cats. Sites of as. vary from lab. units to commercial size, which may be over 15 m long and >1 m in diameter. The chemical and petrochemical industry uses various types of as. for the manuf. of fine and bulk chemicals, pharmaceuticals, and dyes. **E**=**F**; **G** Autoklav. P. CLAUS, D. HÖNICKE

Autofining process → BP procs.

automotive exhaust catalysts are used to decrease the content of pollutants in automobile exhaust gases to meet legislative emission standards. Their development was triggered by legislation in California (1966) and the US Federal Clean Air Act of 1970 (cf. *environmental cat.). While early technology was centered around the Otto engine, cat. exhaust-gas purification is now widely applied also with diesel engines (*diesel engine emissions). Due to different residual oxygen concs., there are fundamental differences in the cat. exhaust-gas technologies for Otto (air/fuel ≈ stoichiometric) and diesel engines (air excess, "lean" regime). Aecs. operate under conds. that may undergo wide and rapid changes dependent on the driving regime. In Otto engine exhaust converters, cat. temps. vary between 250 and 900 °C and reactant *space-times between 10 and 100 000 kg s mol^{-1}. The exhaust contains typically an average of 750 ppm hcs., 1000 ppm NOx, 0.7 vol% CO, 0.2 vol% H_2, 0.5 vol% O_2, 12–14 vol% H_2O and CO_2. Diesel exhaust temps. are mostly between 200 and 300 °C, rarely >600 °C. In addition to NOx and hcs., the exhaust contains particulates of varying composition, oxygen and SO_2, while CO is almost absent. Exhaust of gasoline-fueled lean-burn engines resembles that of diesel engines except for the absence of particulates, a low SO_2 content and somewhat higher temps.

Cats. for Otto engine exhaust are referred to as *three-way catalysts (TWC): they provide *activity for three simultaneous reactions: CO and hc. *oxidation, and NOx *reduction. Early technology refrained from NOx reduction (oxidation catalysts) or performed reduction in an extra-converter, with air injection prior to the oxidation cat. (dual-bed converter, engine operated in fuel-rich regime). The TWC is bound to stoichiometric engine operation controlled by an oxygen sensor (lambda sensor). It fails to reduce NOx under lean conds., and to oxidize CO and hcs. under fuel-rich ("rich") conds. The TWC is able to accumulate O or hcs., which permits its operation under the cyclic variation of the exhaust-gas O content imposed by the feedback control of the engine: after accumulating oxygen, it provides high oxidation convs. during excursion into the rich region, after accumulating hcs. it is able to reduce NOx during subsequent lean excursion. Overall, application of the TWC reduces the emissions of CO and hydrocarbons by $\approx 95\,\%$, that of NOx by $\approx 75\,\%$. Since the cat. is *poisoned by lead, TWC technology has had a strong impact on the phaseout of leaded gasoline. To minimize press. drop, the TWC is usually applied in *monolith design. The cat. contains noble-metal particles on a high-area *support modified by *additives to improve *stability and storage properties. Pt and Rh are often employed ($1–2$ g L^{-1} cat., Pt:Rh = 5–20:1). Recently, the use of Pd ($2–10$ g L^{-1}) has strongly increased, mostly to replace the Pt comp., but all-Pd cats. are also in use. The basic *support material is γ-Al_2O_3, with increasing use of ZrO_2 to support the Rh comp. Typical support additives that increase oxygen storage capacity and texture stability are CeO_2, BaO, and La_2O_3 (up to 30 wt-%); cf. *ceramic cats. Recent developments of TWC technology aim at the improvement of cold-start behavior and cat. lifetime. The former is achieved by placing a robust *pre-catalyst (often of Pd type) near the engine outlet or by trapping cold-start hc. emissions in a *zeolite, which releases them onto the cat. at higher temps. The latter is attempted by improving *metal–support interactions and the flow regime (radial distribution) in the monoliths, but further changes in fuel and lubricants may also be required for that purpose.

For diesel engines, present emission standards are met by oxidation cats. (DOC) that accelerate combustion of unburned hcs. and soft particulate comps. The cats. contain Pt and/or Pd on high-area supports, with additives suppressing SO_2 oxidation activity. Future legislation will require removal of NOx and of particulate soot matter. Particles can be withheld by traps, e.g., monoliths with porous walls, through which the gas is forced by appropriate channel plugging. To avoid application of such traps in regenerative schemes, which often leads to breakdown due to thermal stress during *regeneration, research aims at the self-regenerating trap capable of oxidizing the soot at the relatively low exhaust temps. This oxidation may be achieved by O in the diesel exhaust (cat. trap with oxidation cat.), but better results are reported for reduction with NO/NO_2 (cat. trap with NO reduction cat., or inert trap with oxidation pre-catalyst). Considerable research effort has been put into implementing the selective reduction of NOx in lean exhaust streams by hcs. (*DeNOx) with cats. that typically contain zeolite-supported Cu, Co, Ce, or Pt, or alumina-supported metals (Ag, Pt). Commercialization of this approach has been prevented so far mainly by insufficient cat. durability. Pt systems, which combine high activity with acceptable durability, produce N_2O as a byproduct in intolerable amounts. In an alternative approach, the TWC is modified by addition of an NOx-storage comp. (e.g., Ba cpds.). The accumulated nitrates are periodically reduced by fuel injection. In real exhaust, the storage media suffer, however, from competitive sulfate adsorption. Recently, a technology performing NO reduction with urea has been commercialized for heavy diesel engines. This approach utilizes V_2O_5-WO_3-TiO_2 cats. known from *DeNOx with *stationary sources; the reductant NH_3 is formed by thermal decomposition of the injected urea. **F** catalyseurs pour le gaz d'échappement d'automobiles; **G** Kraftfahrzeug-Abgaskatalysatoren.   M. MUHLER

Ref.: Ertl/Knözinger/Weitkamp, Vol. 4, p. 1559; R.M. Heck, R.J. Farrauto, *Catalytic Air Pollution Control*, van Nostrand Reinhold, New York 1994; Ullmann, Vol.A3, p. 189; *Chem. Tech.* (Leipzig) **1995**, *47*, 205; Farrauto/Bartholomew, p. 581; Thomas/Thomas, p. 576.

aut(o)oxidation is the *oxidation of organic cpds. with molecular oxygen, either by *autocatalysis or catalyzed by *radical *initiators. Molecular O (triplet character) reacts very fast with most radicals. Thus the rate of a. is determined by the rate of radical formation (H-abstraction). A., specially the radical formation step, is favored by heat and light and catalyzed in the radical formation proc. by metal ions like Co, Mn, Cu, Ni, Fe. The primary prods. of a. are *hydroperoxides and *peroxides and their radical forms respectively. They are followed by their secondary prods. ketones, epoxides, or alcohols. These convs. are often guided by chemoluminescence (cf. *Belousov-Zhabotinsky reaction). Electron-rich cpds. or cpds. which easily form stable radicals (like allylic radicals or *t*-butyl radicals) undergo fast a. A prominent technical example of direct use of an a. is the highly selective conv. of isopropylbenzene (cumene) to hydroperoxoisopropyl benzene, which reacts further to phenol and acetone (*Hock proc.).

Not always is a. a desirable pathway: it causes aging of rubber and plastic, resinifying of oils and lubricants, the rancidity of fat and fatty oils. The formation of explosive α-hydroperoxo ethers from ethers demonstrates, besides the autoignition of organic material, the potential risk of a. To avoid undesired (aut)oxidation reactions, *radical scavengers are added to prevent the start of *radical chain reactions. Examples of such antioxidants or stabilizers are organic cpds. such as phenols, hydroquinones, aromatic amines, tocopherols (vitamin E), ascorbic, lactic, or citric acid. In the living cell free O radicals (superoxide, hydroperoxide, hydroxide radical) are formed by non-enzymatic reduction of oxygen. A denaturation of the cell membranes can be caused by such radicals, which is severe damage to the living system. This is normally avoided by enzymatic protection by *enzymes such as *superoxide dismutase, *catalase, or *glutathione peroxidase, which reduce lifetime and conc. of the oxidative active radicals to a minimum. However, all cell damage caused by a. contribute to the aging process of living materials (oxidative stress). Cf. also *peroxides. **E=F=G.** R.W. FISCHER

Ref.: Cornils/Herrmann-1, Vol. 1, p. 439; A. Streitwieser, C. H. Heathcock, *Organische Chemie*, VCH, Weinheim 1980; F.A. Carey, R.J. Sundberg, *Organische Chemie*, VCH, Weinheim 1995, p. 668.

auxiliary compounds *optically active cpds. such as *solvents, *additives, reactants, or cats. (even circularly polarized light), which transform prochiral (*achiral) precursors to one of the two *enantiomers preferentially. Example: *Oppolzer's sultam. **G** optisch aktive Hilfsverbindungen.

Ref.: Mander; *Chem.Ber.* **1989**, 268; *Chem.Rev.* **1996**, *96*, 835; K. Rück-Braun, H. Kunz, *Chiral Auxiliaries in Cycloadditions*, Wiley-VCH, Weinheim 1999; A.J. Pearson, W.J. Roush, *Handbook of Reagents for Organic Synthesis*, Wiley-VCH, Weinheim 1999.

auxiliary ligand → ancillary ligand

avidity refers to the increase in binding affinity observed between polyvalent species as compared to the affinity of the monomeric species. For example, vancomycin (V) binds to the dipeptide D-Ala-D-Ala with a moderate affinity, $K_D = 10^{-6}$ M, but the trivalent compounds (RV_3 and $R'[D-Ala-D-Ala]_3$) have a binding affinity for each other equal to the cube of the monovalent dissociation constant, $K_D = 10^{-18}$M. **F** aviditée; **G** Avidität.

Ref.: *Science* **1998**, *280*, 708. P.S. SEARS

AVM process → Atelier de Vitrification Marcoule process

azaisobutyronitrile (AIBN) 2,2′-Azobis (2-methylpropionitrile) decomposes upon heating into two organic *radicals and nitrogen. It is used as an *initiator for radical *polymerizations and organic transformations which proceed via radical mechanisms. M. BELLER

azasugars → iminocyclitol

B

back-biting reactions The reactive center (e.g., a *radical, cation) of a chain molecule can be transferred internally through an intramolecular back-biting attack. In the case of *alkane *oxidation which proceeds via *radical mechanisms, a sufficiently long hydrocarbon chain enables a back-biting reaction, i.e., after addition of oxygen the $R^1CH_2–CH_2–CHR^2–OO\cdot$ radical abstracts one hydrogen atom in the γ-position. Similar procs. occur in the course of radical *polymerizations and lead to chain branching, whereas intramolecular back-biting *transesterifications are side-reactions of the polyester formation, producing a mixture of both linear and cyclic molecules. Back-biting transfer of cationic centers was observed during the cat. decomposition of *polyethylene. **E=F=G**.

D. GLEICH

Ref.: *J.Am.Chem.Soc.* **1979**, *101*, 7574; *Macromolecules* **1990**, *23*, 1640; *Bull.Chem.Soc.Jpn.* **1991**, *64*, 3585.

back-bonding (back-donation) widely used concept of the theory of chemical bonding, particularly in *coordination chemistry, to explain the strength of bonds between *transition metals in low oxidation states and *acceptor ligands, e.g., the M–C bonds in metal carbonyls. B. in these cpds. results from the *overlap of filled metal-centered orbitals with empty, energetically low *ligand-based orbitals of π-symmetry. The excess negative charge of the metal brought about by the *ligand-to-metal electron donation is thereby reduced. Experimental evidence for the existence of b. is given by different techniques, e.g., IR, NMR, photoelectron spectroscopy, or single-crystal X-ray diffraction studies. B. is also suggested to exist in hypervalent cpds. of main group elements (e.g., Xe fluorides and oxides), influencing charge distribution and binding energy of the cpds. (cf. also *Chatt–Duncanson model). **E=F; G** Rückbindung.

A.C. FILIPPOU

Ref.: F. Cotton, G. Wilkinson, *Advanced Inorganic Chemistry*, Wiley, New York 1988; Gates-1, p. 74; W. Kutzelnigg, *Einführung in die Theoretische Chemie*, Bd. 2, VCH, Weinheim, 1994.

bacterial display For selection of proteins from a bacterial *library of protein mutants or from a cDNA library, it is most convenient to have the protein on the outside of the bacterium, either as a secreted protein or displayed on the bacterial surface as a fusion with a transmembrane protein. Display on the surface is particularly convenient if a *ligand exists that binds only the desired protein: if the ligand can be conjugated to a fluorophore, it provides a means of fluorescently labeling only those cells producing the desired protein. Fluorescence-activated cell sorting (FACS) can then select those clones (cf. *molecular recognition). Alternatively, the ligand can be attached to a solid *support. Passing a bacterial suspension over the top of the *immobilized ligand will select out those members that display the protein of interest. This technique can be used in the selection of *enzymes if the affinity ligand is a mechanism-based inactivator or other *inhibitor (e.g., a *transition state analog). **E=F=G**.

P.S. SEARS

Ref.: *Protein Eng.* **1998**, *11*, 825.

baculovirus diverse category of rod-shaped viruses that infect primarily insect cells. Those most commonly used as expression vectors are the *Autographa californica* nuclear polyhedrosis virus (AcMNPV) and the *Bombyx mori* nuclear polyhedrosis virus (BmNPV), with *Spodoptera frugiperda* (fall armyworm) Sf9 and Sf21 or *Bombyx mori* (silkworm) BMN4 cells as the hosts. The advantage of the baculovirus system over a prokaryotic (bacterial) expression system is that the host insect cell line is *eukaryotic, and thus many proteins that require posttranslational modifications common to eukaryotes (such as glycosylation) for activity will be processed correctly in these cell lines.

The term "polyhedrosis virus" refers to the fact that, in addition to budding from the cell, the virus also forms occlusion bodies in the nucleus of the cell. The occlusion bodies consist of the viral DNA enclosed in a gel composed mainly of a single protein, polyhedrin. The *promoter for polyhedrin is very strong: at very late stages after infection, 20 % of the mRNA produced may be polyhedrin mRNA. Due to the strength of the promoter, it is commonly used for expression of the desired heterologous gene. **E=F=G.**   P.S. SEARS

Ref.: D.R. O'Reilly, L.K. Miller, V.A. Lucklow, *Baculovirus Expression Vectors*, W.H. Freeman and Co., New York 1992.

Badger EB process   → BASF procs.

Baekeland–Lederer–Manasse condensation   cat.-influenced *condensation of phenols and formaldehyde (*hydroxymethylation) between the carbonyl group and *o*- or *p*-substituted hydroxyaromatics.   Acc. to the type of cat. (*acid, *base, or metal salts), the temp., and the ratio phenol/HCHO, different condensation products (PF resins) are obtained (Bakelite, novolacs, resols, resitols, resites) which are used as raw materials for resins and lacquers, coatings, or varnishes. **F** condensation de B–L–M., **G=E.**

B. CORNILS

Ref.: Kirk/Othmer-1, Vol. 2, p. 53; Kirk/Othmer-2, p. 867; McKetta-1, Vol. 48, p. 116; Ullmann, Vol. A1, p. 409 and Vol. 19, p. 371; Mark/Bikales/Overberger/Menges, Vol. 1, p. 644.

Baeyer–Villiger oxidation (BV rearrangement) 1–chemical: *oxidation of ketones (acyclic or cyclic) with peracids to yield carboxylic acid esters or lactones. The first step of the rearrangement is the proton-cat. *addition of a nucleophilic peroxo oxygen to the ketone carbonyl giving an intermediate tetrahedral perhemiketal $R^1R^2C(OH)-O-O-$ $C(=O)R^3$. The consecutive, rate-determining step involves the rearrangement of one of the two substituents to the next O atom with simultaneous O–O bond cleavage to give the ester and the acid moieties: $R^2C(=O)OR^1$ + R^3COOH. The migratory propensity decreases in the order tertiary> secondary > primary > methyl. The rearrangement occurs with retention of configuration at the migrating substituent. Besides organic peracid reagents, nucleophilically acting *transition metal peroxo *complexes, activating organic peroxides (alkyl peroxides, bis[trimethylsilyl]-peroxide), or H_2O_2 can also transform ketones via BVo. to the corresponding esters and lactones. Examples of active cats. are simple *Lewis acids like $SnCl_4$ or $BF_3 \cdot Et_2O$, *transition metal peroxo complexes like (1,5-dicarboxypyridyl)Mo(O)-(μ-O$_2$), $CH_3Re(O)$ $(μ-O_2)_2 \cdot H_2O$, or cationic peroxo Pt complexes like $[(dppe)Pt(CF_3)]^+$. Besides H_2O_2, oxygen can also be used as oxidant if transition metal cats. ($Ni^{2+/3+}$, Fe_2O_3, Cu^{2+}) and aldehydes ($RCHO$, C_6H_5CHO) are applied as sources for the formation of peroxo acids. *Asymmetric variations of the BVo. are known by applying asymmetric copper (oxygen/co-oxidant: peracid) or titanium complexes (*TBHP) or *enzymes (see below).

R.W. FISCHER

2–enzymatic: BVos. are cat. by flavoenzymes in the presence of *NAD(P)H and oxygen. Mechanistic studies of the *enzyme cyclohexanone monooxygenase from *Acinetobacter* NCIB 9871 indicate that the *FAD-4α-hydroperoxide intermediate acts as a nucleophile reacting with the carbonyl group. In addition to cyclohexanone, many other cyclic ketones, acyclic ketones, aldehydes, and boronic acids are also substrates, and proceed with stereochemistry similar to that of the chemical reaction. The enzyme also cat. *enantioselective transformation of substituted cyclic ketones which may or may not contain enantiotopic substituents, a valuable process from the synth. point of view. In large-scale reactions, the required NADPH *cofactor must be regenerated from NADP. The enzyme, however, is unable to oxidize olefins. Epoxide formation has not been observed. The flavin peroxide also functions as an electrophile, oxidizing heteroatoms such as S, N, Se, and P. *Pseudomonas putida* NCIMB 10007 contains a *monooxygenase which cat. BVo. of some bicyclic ketones. **F** oxidation de Baeyer–Villiger; **G** Baeyer–Villiger-Oxidation.

C.-H. Wong

Ref.: Beller/Bolm, Vol. 2, p. 213; G. Strukul, *Catalytic Oxidations with Hydrogen Peroxide as Oxidant*, Kluwer, Dordrecht 1992, p. 210; Sheldon/Kochi, p. 94; *Biochemistry* **1982**, *21*, 2644; *J.Am.Chem.Soc.* **1988**, *110*, 6892; *Angew.Chem.* **1998**, *110*, 1256; *Organic Reactions*, Vol. 43, p. 251.

Balsohn benzene olefin addition

basic reaction for *alkylation of aromatic hydrocarbons with alkenes by *acid cats., e.g., the *addition of benzene and ethylene to *ethylbenzene (cf. *Friedel–Crafts reaction, e.g., procs. of *BASF, *Mobil for *EB). **F** addition Balsohn de benzène et d'oléfine; **G** Balsohn Benzol-Olefin-Addition.

B. Cornils

Bamford–Steven reaction

cleavage of tosylhydrazones to alkenes under the cat. influence of bases such as *alkoxides. **F=G=E**.

band bending

distortion of the electron bands (*band theory) in a solid due to electrostatic interaction with charges at the *surface or interface of a solid. While in a metal such electrostatic interactions are screened over very short (atomic) distances, due the much lower density of free electrons in a semoconductor bb. can extend up to 1000 Å inside the bulk. Bb. is important when electron–hole pairs are created at the interface/surface through lightening and the charges, instead of recombining, become separated and take part in a cat. reaction, i.e., in *photocatalysis and photoelectrocatalysis. Experimentally bb. can be determined with *photoelectron spectroscopy by a shift of the core levels of the atoms in the bb. region. **E=F=G**.

R. Imbihl

Ref.: H. Lüth, *Surfaces and Interfaces of Solid Materials*, Springer, Berlin 1995; Thomas/Thomas, p. 392.

band gap

the energy gap between the *conduction band (cb.) and the *valence band (vb.) with the term "gap" meaning that no electronic states exist for this energy range (cf. *band theory). If the bg. between a fully occupied vb. and an empty cb. is much larger than the thermal energy kT then no thermal promotion of electrons into the cb. is possible and the material is an insulator; if the bg. is comparable to kT then some electrons are promoted into the cb. and a *semiconductor results. If impurities are present in the solid then local energy states can exist in the band gap which are termed according to their chemical nature and their position relative to the *Fermi level as acceptor or donor levels. The width of the bg. depends on the size of a *cluster and can thus be tailored to fit the the the excitation energy of visible light. **E=F=G**.

R. Imbihl

Ref.: Thomas/Thomas, p. 598.

band theory

According to the band theory, bands of energies permitted in a solid are related to the discrete allowed energies – the energy levels – of single, isolated atoms. When the atoms are brought together to form a solid, these discrete energy levels become perturbed interactions termed "quantum mechanical effects", and the many electrons in the collection of individual atoms occupy a band. The highest energy band occupied by electrons is the *valence band* (vb.). In a conductor, the vb. is partially filled, and since there are numerous empty levels, the electrons are free to move under the influence of an electric field. In an insulator, electrons completely fill the vb.; and the gap between it and the next band, which is the *conduction band* (cb.), is large. The electrons cannot move under the influence of an electric field unless they are given enough energy to cross the large energy gap to the cb. In a semiconductor, the gap to the cb. is smaller than in an insulator. In *photocatalysis this property is used for generation of electon-hole pairs by absorption of light. In metals there is an overlap of vb. and cb.

The next highest band to the vb. is the cb., which is separated from the vb. by an energy gap. Just as electrons at one energy level in an individual atom may transfer to another empty energy level, so electrons in the solid may transfer from one energy level in a given band to another in the same band or in another band, often crossing an intervening gap of forbidden energies. The excitation of electrons by absorption of light causes the generation of electron–hole pairs in semiconductors. **F** théorie de bande; **G** Bändertheorie.

R. SCHLÖGL, B. BERNS

Ref.: Thomas/Thomas, p. 598.

bare metal ions   In the gas phase, elementary steps and key procs. of *transition metal cats. can be examined with a first-principle approach. Therefore, putative reactants, intermediates, and prod. complexes are examined by sophisticated MS experiments. For example, the gas-phase *oxidation of methane by the metal oxenoid FeO^+ can be divided into three steps. 1: *coordination of the reactants to yield an encounter *complex, $CH_4 \cdot FeO^+$; 2: C–H bond *activation to an *insertion intermediate, $CH_3–Fe–OH$; and 3: *reductive elimination to afford $(CH_3OH)Fe^+$ which subsequently liberates methanol. These different species can be generated and characterized using MS. The net reaction is $FeO^+ + CH_4 \rightarrow Fe^+ + CH_3OH$. A cat. sequence can be realized in the gas phase, if the bare Fe^+ ion formed is reoxidized to FeO^+ in a second reaction, e.g., $Fe^+ + N_2O \rightarrow FeO^+ + N_2$. Kinetic analysis provides *rate constants and *TONs of the *cat. cycles and also uncovers possible sinks due to irreversible procs. A key aspect of these gas-phase studies is that theory and experiment can be compared with each other without any assumptions due to the presence of *ligands, *solvents, counterions, etc. Thus, *isotope labeling in conjunction with kinetic studies and *ab-initio calcns. allow one to determine rate-determining steps in the oxidation of *hcs. by *transition metal oxides. Similar studies were made with other couplings such as CH_4/O_2, CH_4/NH_3,

and CH_4/CO_2, alkyne *oligomerizations, *regio- and *stereoselective C–H *bond activation, and *transfer hydrogenation. For the use of bare Ni complexes cf. *cyclooligomerization. **F** métal ions nus; **G** nackte Metallionen.

D. SCHRÖDER

Ref.: *Angew. Chem. Int. Ed. Engl.* **1995**, *34*, 1973 and **1998**, *37*, 829; *J. Mass Spectrom.* **1996**, *37*, 829; *Pure Appl. Chem.* **1997**, *69*, 273; *Acc. Chem. Res.* **1995**, *28*, 430.

barium in catalysis   There are few applications of Ba: in *washcoats for *automotive exhaust cats. (BaO/*ceria on *alumina) or in *catalytic combustion (Ba hexaaluminate), and as a *promoter of *Adkins cats. for *fat hydrogenation. $BaCO_3/BaSO_4$ have been proposed as *supports.   B. CORNILS

barrels   as production units, cf. *bbl., *bpcd, *bpsd

Barton reaction   *photolytic/*radical reaction for the conv. of γ-$CH_3–$ into $–CHO$ groups (*catalytic radical reactions). **F** réaction de Barton; **G** Barton-Reaktion.

B. CORNILS

base catalysis   → acid–base catalysis

base reactions   → Hieber base reaction

BASF acetic acid (AA) process   first proc. for the *carbonylation of methanol with CO to *actic acid (AA) under high press., developed by *von Kutepow. The cat. *precursor was CoI_3, the actual cat. methyl Co carbonyls. The proc. has been substituted by the Rh-based *Monsanto AA process. **F** procédé BASF pour l'acide acetylique **G** BASF Essigsäure-Verfahren.   B. CORNILS

Ref.: Winnacker/Küchler, Vol. 6, p. 82.

BASF acrylic acid process   former manufng. proc. of *acrylic acid by *Reppe synth. from acetylene, CO, and water with a cat. system consisting of $NiBr_2$/CuI (*hydrocarboxylation). **F** procédé BASF pour l'acide acrylique; **G** BASF Acrylsäureverfahren. B. CORNILS

BASF adipic acid process a proposed three-step synth. based on the *hydroesterification of butadiene, CO, and methanol (Figure).

The initially formed 3-pentenoate is *isomerized in situ to the 4-pentenoic acid ester in order to afford dimethyl adipate in the second hydroesterification step. Adipic acid is obtained after *hydrolysis of the ester with an overall *selectivity of approx. 70 %. **F** procédé BASF pour l'acide adipique; **G** BASF-Adipinsäurevefahren.

W. MÄGERLEIN, M. BELLER
Ref.: *Appl.Catal.* **1994**, *115*, 188.

BASF anthraquinone process multi-step proc. of Li_3PO_4-cat. *dimerization of styrene to 1-methyl-3-phenylindane and its *oxidation to anthraquinone. **F** procédé BASF pour l'anthraquinone; **G** BASF Anthrachinon-Verfahren.   B. CORNILS

BASF–Badger ethylbenzene process for the manuf. of *ethylbenzene (EB) by *alkylation of benzene with ethylene (*Balsohnbenzene olefin addition), reacting both comps. with finely ground $AlCl_3$ in the liquid phase and removal of the cat. by alkaline washing. **F** procédé BASF-Badger pour l'EB; **G** BASF–Badger EB-Verfahren.   B. CORNILS
Ref.: Winnacker/Küchler, Vol. 5, p. 207,221.

BASF butanediol process is based on the *Reppe *ethynylation. The reaction sequence comprises the *addition reaction of acetylene and formaldehyde to butynediol (cat.: Bi-*promoted Cu acetylenide) in a trickle bed (*catalytic reactors) at 110 °C/0.5–2 MPa and the subsequent *hydrogenation in the liquid phase at 70–100 °C/25–30 MPa to *butane-1,4-diol (*BDO). **F** procédé BASF pour le

BDO; **G** BASF Butandiol-Verfahren.

B. CORNILS
Ref.: Weissermel/Arpe, p. 99; Falbe-1, p. 151; Falbe-3, p. 112.

BASF butanol process acc. to *Reppe's *hydrocarbonylation by reaction of propylene, CO, and water *without* intermediate aldehyde formation (cf. *hydroformylation). Modified Fe pentacarbonyls are used as cats. for the preceding *WGSR and the *hydrocarbonylation. **F** procédé BASF pour le butanole; **G** BASF Butanolverfahren.   B. CORNILS
Ref.: Weissermel/Arpe, p. 200.

BASF *tert*-butylamine process manuf. from isobutene and ammonia in "fixed-bed reactor gas-phase reaction over highly active het. cats." (BASF).   B. CORNILS
Ref.: *Appl.Catal.* **1994**, *115*, 184.

BASF γ-butyrolactone (GBL) process for the manuf. of *γ-butyrolactone by cat. gasphase *dehydrogenation (*dehydrocyclization) of *BDO over Cu cats. at 200–250 °C. **F** procédé BASF pour γ-butyrolactone; **G** BASF-Gamma-butyrolacton-Verfahren.   B. CORNILS

BASF catalysts BASF AG produces more than 140 different cats., half of them only for captive use and licensees. Main areas are *oxidation (*sulfuric acid, *acrylic acid, *phthalic and *maleic anhydrides), *hydrogenation (oil industry: i.e., *hydrotreating, aromatics hydrogenation, *selective hydrogenation) as well as other proc. cats. (*oxychlorination, *styrene), environmental applications (NO_x; Dioxin abatement), and *fuel cells.   J. KULPE

BASF Catasulf process Special variant of the *Claus process for H_2S/CO_2 removal from *natural gas or *syngas. **F** procédé Catasulf; **G** Catasulf-Verfahren.   B. CORNILS

BASF hydrocracking process for *hydrocracking of heavy sour feedstocks (virgin gas oil, deasphalted vacuum residues, etc.) into lower boiling comps. by single (or two)-

stage operation including *desulfurization, *denitrification, and *cracking over proprietary cats. **F** procédé hydrocracking de BASF; **G** BASF Hydrocracking-Verfahren.

Ref.: *Hydrocarb.Proc.* **1978**, 57(5), 122.   B. CORNILS

BASF hydroquinone process   older proc. based on the *Reppe *cyclocarbonylation of acetylene and CO under high press. Cats. used were Co, Fe, Rh, or Ru carbonyls. **F** procédé BASF pour l'hydroquinone; **G** BASF Hydrochinonverfahren.   B. CORNILS

Ref.: Falbe-3, p. 91

BASF hydroxylamine process   comprises the *reduction of NO with a Pt cat. (supported on C), suspended in sulfuric acid acc. to $2 NO + 3 H_2 + H_2SO_4 \rightarrow (NH_3OH)_2SO_4$. The side-reaction of NH_3 formation is suppressed by high H_2 press., low temp., and low H^+ ion conc. **F** procédé de BASF pour l'hydroxylamine; **G** BASF Hydroxylamin-Verfahren.   B. CORNILS

Ref.: Büchner/Schliebs/Winter/Büchel, p. 54.

BASF/IFP hydrocracking process   conv. of heavy sour feedstocks into lower boiling prods. through single or two stage operation (one through or liquid recycle), combined with different cats. In the first stage the heavy stock is desulfurized, denitrified (*HDS, *HDN) and partly cracked, in the second the cracking is completed. **F** procédé d'hydrocracking de BASF/IFP; **G** BASF/IFP-Hydrocrackingverfahren.   B. CORNILS

Ref.: *Hydrocarb.Proc.* **1972**, (9), 143.

BASF/IFP residue hydrodesulfurization improvement of heavy feedstocks by removal of S, N, and metallic contaminants (*HDS, *HDN, *HDM) by preheating the feedstock together with H-rich gas plus recycle and then pass through a fixed-bed reactor. *HDS reaches 85 %. **F** procédé HDS de BASF/IFP; **G** BASF/IFP-HDS-Verfahren.

Ref.: *Hydrocarb.Proc.* **1972**, (9), 176.   B. CORNILS

BASF maleic acid anhydride (MAA) process   for the isolation of *maleic acid an-

hydride from the wastewater of phthalic acid anhydride plants by cat. *oxidation of *p*-xylene or naphthalene. **F** procédé BASF pour l'anhydride maléique; **G** BASF Maleinsäureanhydrid-Verfahren.   B. CORNILS

BASF methanol process   As early as 1913, methanol (besides other oxygenates) was synth. via *hydrogenation of CO; about 10 years later the first commercial production unit was started. The high-press. proc. used ZnO-Cr_2O_3-based cats. which required rather drastic reaction conds. (25–35 MPa/320–380 °C). By this proc. MeOH rapidly became one of the most important chemicals and was produced in bulk quantities.

ZnO alone shows convenient activity for the synth. of MeOH, but deactivates within a short period of time (recrystallization). Cr(III) oxide is a weak cat. for this purpose, but stabilizes the ZnO structure and thus enables the technical application of the cat. $ZnCr_2O_4$ (zinc chromite; *spinel structure) has good activity and longevity and is frequently used as a *precursor of the final cat. The Zn/Cr ratio is of some importance. *Precipitation is widely applied for the generation of the cats. The suitability of the ZnO-Cr_2O_3 cats. is mainly due to their thermal stability and their resistance against cat. *poisons which are introduced via the *syngas. S has very little influence on the lifetime of the cat. and only Fe (as the carbonyl) has a detrimental effect (formation of methane and oxygenates) as a deposit on the *surface of the cat. Only when syngas of sufficient purity became economically available did the importance of the high press. synth. cease and the so-called low-press. synth. (5–10 MPa) were successfully commercialized (procs. of *ICI, *Lurgi, *Haldor). Today only a few older methanol high-press. units are on stream, and new plants use exclusively the low-press. technology. **F** procédé BASF pour le méthanol; **G** BASF Methanolverfahren.

C.D. FROHNING

Ref.: DE 293.787 (1913) and DE 415.686 (1923); Winnacker/Küchler, Vol. 5, p. 503.

BASF methyl methacrylate (MMA) process
a modern three-step proc. for the manuf. of methyl methacrylate (MMA) by 1: *condensation of formaldehyde and propionaldehyde to methacrolein under the influence of *tert*-amines as cats. at 160–210 °C/4–8 MPa; 2: *oxidation to methacrylic acid (MA), and 3: *esterification of MA with methanol to methyl methacrylate (MMA). **F** procédé BASF pour le méthyl méthacrylate; **G** BASF Methylmethacrylatverfahren. B. CORNILS
Ref.: DE 3.213.681 (1983); DE 3.211.901 (1983); *Appl. Catal.* **1994**, *115*, 182.

BASF oxo process
1-Co cat.: BASF technology of the *hydroformylation proc. by Co carbonyl cat. *addition of CO and H_2 (*syngas) to the double-bonds of alkenes at 150–170 °C/25 MPa total press. Based on mutual origins of this variant as originally developed by *Ruhrchemie, the characteristics of this proc. are the application of aqueous Co salt solutions as cat. *precursors and the *decobalting of the oxo crudes via oxidation of their Co carbonyl content from Co^{-1} to water-soluble Co^{2+} in acidic solution. The BASF technology processed C_3 to C_{12} olefins. 2-Rh cat.: The new BASF oxo process uses (similar to Union Carbon's *LPO process) *TPP-modified Rh carbonyls for the hydroformylation of propylene. **F** procédé BASF d'oxo; **G** BASF-Oxoverfahren. B. CORNILS
Ref.: Falbe-1, p. 164; *Hydrocarb.Proc.* **1977**, 56(11), 135.

BASF phenylacetone process
conv. of 3-phenylpropanal to phenylacetone, het. catalyzed in fixed bed operation over *zeolites at 400 °C (conv. >90 %, selectivity >95 %). B. CORNILS
Ref.: *ECN Specialty Chemicals* **1991**, (2), 13.

BASF phthalic anhydride process
manuf. of PA by *oxidation of *o*-xylene with air at 375–410 °C over doped het. V_2O_5 cats. in tubular reactors (*catalytic reactors). **F** procédé de BASF pour l'anhydride phthalique;

G BASF Phthalsäureanhydrid-Verfahren. B. CORNILS
Ref.: *Hydrocarb.Proc.* **1977**, 56(11), 194.

BASF process to S-1-phenylethylamine
from phenylethylamine by *resolution of rac . phenylethylamine over a *lipase (from *Pseudomonas* sp.) with the *acylation reagent ethyl methoxyacetate in >95 % *ee* (100 tpy). B. CORNILS
Ref.: *J.Prakt.Chem.* **1997**, *339*, 381.

BASF process to vitamin A
14-step synth. starting with the *hydroformylation of 1-vinylethylene diacetate using unmodified Rh cat. at high temps. and press. resulting in a pronounced *selectivity towards the *branched* aldehyde (low *n/iso* ratio). After elimination of AcOH *trans*-2-formyl-4-acetoxybutene is obtained which subsequently is combined with a PPh_3 derivative of vinyl-β-ionone in a *Wittig-type coupling. The C_{15} building block vinyl-β-ionol derives from isobutene and formaldehyde in a multi-step synth. with citral and β-ionone. **F** procédé BASF pour la vitamine A; **G** BASF Vitamin A-Verfahren. R. KRATZER
Ref.: Cornils/Herrmann-1, p. 8.

BASF propionic acid process
*hydrocarboxylation of ethylene acc. to *Reppe synthesis which since 1951 has only been used by BASF (current capacity 80 000 tpy). Ethylene is reacted with CO and water at 270–310 °C/20–25 MPa in the presence of 0.5–1 % Ni cat. The active species is believed to be $NiH(O_2CEt)(CO)_2$ formed by reaction of pa. with $Ni(CO)_4$ obtained under reaction conds. from the cat. *precursor $Ni(O_2CEt)_2$. The *cat. cycle consists of the classical steps: *insertion of ethylene into the Ni–H bond followed by migratory insertion of the ethyl ligand into an Ni–CO bond to give an acyl complex which is then nucleophilically attacked by water to give pa. and to regenerate the Ni–H species. The proc. is highly selective for pa. (98 % based on ethene) and gives typical yields of 96 %. The main side-reaction is the formation of H_2 by the *WGSR yielding ethane and diethyl ketone as byprods. The Ni cat. is almost quantitatively recirculated. Major disad-

vantage of the proc. is the corrosivity which makes it necessary to plate the reactor with silver (Figure). **F** procédé BASF pour l'acide propionique; **G** BASF Propionsäureverfahren.

M. SCHULZ

Ref.: Cornils/Herrmann-1, p. 138; Ullmann, Vol. A22, p. 223.

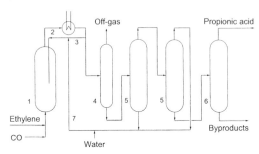

1 reactor; 4 gas separator; 5–6 distillation; 7 catalyst recycle.

BASF route to citral is important for the manuf. of the C_{15} building block in *BASF's synth. of vitamin A. The synth. starts with the reaction of isobutene and formaldehyde yielding 3-methyl-2-buten-1-ol (prenol) and 3-methyl-2-butenal, a mixture which can be converted to 3-methyl-2-butenal diprenyl acetal.

In the presence of an acid cat. at 150 °C an enol is first formed. The intermediate further reacts in a sequence of *Claisen and Cope rearrangements to citral (3,7-dimethyl-2,6-octadien-1-al). **F** procédé de BASF pour citrale; **G** BASF Citral-Verfahren. R. KRATZER

Ref.: EP 0.021.074; Ullmann, Vol. A11, p. 159; *Appl. Catal.* **1994**, *115*, 202.

BASF tetrahydrofuran (THF) process by *dehydration of *butanediol (BDO) in the presence of acid cats. such as H_3PO_4, sulfuric acid, or acid *ion exchange resins at 110–125 °C at rt. or at 300 °C/10 MPa. **F** procédé BASF pour le THF; **G** BASF-THF-Verfahren. B. CORNILS

BASF–Wyandotte allyl alcohol process by *isomerization of *propylene oxide over Cr_2O_3 cat. in the liquid phase. **F** procédé BASF–Wyandotte pour l'alcool allylique; **G** BASF–Wyandotte-Allylalkohol-Verfahren. B. CORNILS

Bashkirov oxidation metal ion (Co, Mn) cat. *oxidation of paraffins (e.g., from kerosenes or *FT synthesis) with air to secondary alcohols (e.g., *Nippon Shokubai proc.); the oxidation is also important for the manuf. of cyclohexanol (*KA oil) as a *precursor of *adipic acid and for C_{12}-acid (*Hüls proc.). The *selectivity may be enhanced by *promoters such as boric acid or amines. **F=G=E**.

Ref.: Weissermel/Arpe, p. 207 B. CORNILS

basis set In the *Hartree–Fock method each electron is described by a one-electron function which is called the *molecular orbital (MO). The MOs. are usually approximated by a linear combination of atomic orbitals (LCAO), which form the basis set for the *ab-initio calcns.

The mathematical functions for atomic basis sets are usually Gaussian-type functions $e^{-\zeta r}$, because the integrals over Gaussian functions are much easier to calc. than integrals over Slater functions $e^{-\zeta r^2}$, although the latter give a better description of the spacial distribution of the electrons. Bss. have been developed for all atoms of the periodic system. They are obtained by optimization of the exponents ζ with respect to the energy of the atom. Several Gaussian functions may be used to describe a single atomic orbital, and the coefficients of the functions may be kept constant during the calcn. This yields a contracted Gaussian function, while a single function is called a primitive Gaussian. The quality of a bs. is given by the number and by the contraction of the Gaussian functions. A minimal bs. has one contracted Gaussian function for each orbital. A double-zeta bs. has two contracted Gaussian functions, a triple-zeta basis set has three, etc. Bss. may contain polarization functions, which have a higher angular momentum than the valence orbitals. Thus, p, d, f, and higher functions are polarization functions for H and He; d, f, and higher functions are polarization functions for the main-group elements, etc. Bss. are also used to solve the Kohn–Sham equations in *density functional theory. Once a method is chosen, then the choice of the bs. is the most important factor for the quality of an ab-initio or DFT calcn. **E=F=G**.

G. FRENKING

batch reactor (BR) → catalytic reactors

bauxite Al hydroxide mineral, occasionally used as cat. *support (e.g., for AlCl$_3$ in procs. of *Shell's former vapor-phase butane isomerization and *Phillips cat. isomerization) or as cat. for the refinery of lube oils. **F=E**; **G** Bauxit.   B. CORNILS

Bayer acetic acid (AA) process former proc. by indirect liquid-phase *oxidation of n-butenes. By proton-cat. *addition of *AA to butene, 2-acetoxybutane is formed at 110 °C/1.5–2 MPa which is oxidized with air at 195 °C/6 MPa to AA (Bayer–Krönig proc.). **F** procédé Bayer pour l'acide acétique; **G** Bayer Essigsäure-Verfahren.   B. CORNILS

Bayer anthraquinone process
cat. *Diels–Alder reaction of 1,4-naphthoquinone and butadiene. **F** procédé Bayer pour l'anthraquinone, **G** Bayer Anthrachinon-Verfahren.

B. CORNILS

Bayer benzene chlorination   former proc. for the *chlorination of benzene in the liquid phase with Fe(III) chloride as cat. (cf. *Dow proc.). **F** Bayer chloration de benzène; **G** Bayer Benzolchlorierung.   B. CORNILS

Bayer cold hydrogenation process for the removal of propyne and propadiene (approx. 4 %) from cracker C$_3$ fractions. Over Pd cats. (in isothermal fixed-bed operation in tubular reactors, cooled with boiling propylene) the content of the C$_3$ byprods. is reduced to below 10 ppm; cf. *low-temp. hydrogenation, *Lummus CDHydro technology). **F** procédé

Bayer pour l'hydrogénation froide; **G** Bayer-Kalthydrierung.   B. CORNILS
Ref.: *Erdöl, Kohle-Erdgas-Petrochem.* **1983**, *6*, 249.

Bayer–Degussa propylene oxide (PO) process manuf. of *PO from propylene and perpropionic acid (from H$_2$O$_2$ and recycled propionic acid) at 60–80 °C. Perpropionic acid is extracted from the H$_2$O$_2$ solution by benzene (cf. *Interox proc.). **F** procédé Bayer/Degussa pour le PO; **G** Bayer/Degussa PO-Verfahren.

B. CORNILS

Bayer dimethyl carbonate process reaction of ethylene oxide with CO, followed by *transesterification with methanol. As cats. ammonium or alkali metal salts at 120–150 °C/3–5MPa are proposed (proc. presumably not realized).   B. CORNILS
Ref.: *Appl.Catal.* **1994**, *115*, 182.

Bayer double-contact (double-absorption) process manuf. of sulfuric acid by V$_2$O$_5$-cat. *oxidation of SO$_2$ to SO$_3$. Characteristic features are the removal of SO$_3$ after the fourth cat. rector and at SO$_2$ convs. of 98 % and the afterreaction of SO$_2$ to conv. rates of 99.7 %. **F** procédé contact double de Bayer; **G** Bayer Doppelkontakt-Verfahren.

B. CORNILS
Ref.: Kirk/Othmer-1, Vol. 23, p. 363, 385; Ertl/Knözinger/Weitkamp, Vol. 4, p. 1780.

Bayer–Hoechst vinyl acetate (VAM) process economical realization of the Pd-cat. *Moiseev reaction, the gas-phase *acetoxylation of ethylene, *acetic acid, and oxygen at 175–200 °C/0.5–1 MPa. The most important proc. variants of *Celanese, *USI, and *Bayer–Hoechst differ in the *promoters of the base-cat. metal Pd and the *support. **F** procédé Bayer–Hoechst pour l'acétate de vinyle; **G** Bayer-Hoechst Vinylacetat-Verfahren.

B. CORNILS
Ref.: Weissermel/Arpe, p. 229; *Hydrocarb.Proc.* **1977**, 56(11), 234; *Chem.Ing.Tech.* **1968**, *40*, 781.

Bayer hydrazine process reaction between NaOCl and NH$_3$ in the presence of

ketones. The initially obtained methyl ethyl ketoneazine is *hydrolyzed to hydrazine. **F** procédé de Bayer pour l'hydrazine; **G** Bayer Hydrazin-Verfahren.   B. CORNILS
Ref.: R. Schliebs, 193rd ACS Meeting Abstracts, Denver 1987.

Bayer isobutene oligomerization process separation of *iso-* from *n*-butenes by *oligomerization of the branched isomer, proton-cat. by acid *ion exchange resins to polymer olefins. **F** procédé Bayer pour l'oligomérisation d'isobutène; **G** Bayer Isobutenoligomerisierungs-Verfahren.   B. CORNILS

Bayer–Krönig process → Bayer AA proc.

Bayer maleic anhydride process by *selective *oxidation of C_4 cuts over special cats. at 400–440 °C; the tubular fixed-bed reactor is heated by a salt melt (Figure). **F** procédé Bayer pour MAA; **G** Bayer MAA-Verfahren.
                                                    B. CORNILS
Ref.: *Hydrocarb.Proc.* **1977**, 56(11), 180.

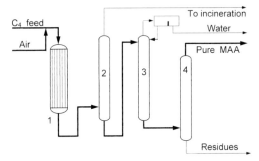

1 reactor; 2 scrubber; 3 dehydration; 4 distillation.

Bayer process to hydroxylamine → hydroxylamine

bbl. barrel (42 US liquid gallons = 5.61 ft³ = 0.1590 m³), a production measure in the *petroleum processing industry and *petrochemistry; as daily unit cf. *bpcd and *bpsd.
                                                    B. CORNILS

BBP butyl benzyl phthalate, a *plasticizer based on butanol, benzyl alcohol, and phthalic anhydride. **F** phthalate de butyle-benzyl; **G** Butylbenzylphthalat.

BDO → butane-1,4-diol (1.4-butanediol)

BEA *zeolite structural code for zeolite-beta

Beavon sweetening Proc. for the cat. conv. of sulfur cpds. in gas streams (SO_2, S_x, COS, CS_2) to H_2S and their removal. **F** procédé Beavon; **G** Beavon-Verfahren.
Ref: *Hydrocarb. Proc.* **1996**, (4), 106   B. CORNILS

Beckmann rearrangement Oximes rearrange in the presence of acids or anhydrides to substituted amides. A multitude of cats. have been used, inter alia, PCl_5, H_2SO_4, SO_2, HCOOH, HCl/AcOH/ Ac_2O, polyphosphoric acid, etc. The Ba. is accompanied by salt formation, which can be prevented by using liquid HF. The Ba. of cyclohexanone oxime is used commercially for the manuf. of ε-caprolactam, the precursor for polyamides. **F** réarrangement de Beckmann; **G** Beckmann'sche Umlagerung.   M. BELLER
Ref.: March; *Organic Reactions*, Vol.11, p.1 and Vol.35, p.1; *J.Mol.Catal.A:* **1999**, *149*, 25.

beidellite belongs to the groups of *clay minerals. It is a 2:1 trilayer phyllosilicate, i.e., its layers are built up from one sublayer with octahedrally coordinated cationic species and two sublayers with tetrahedrally coordinated cationic species. B. is closely related to *montmorillonite, and both structures can be derived from pyrophyllite, $Al_2(OH)_2[Si_4O_{10}]$. A typical formula of a beidellite is $A_{0.4}\{Al_2 (OH)_2[Al_{0.4}Si_{3.6}O_{10}]\} \cdot n\ H_2O$ with A typically H^+, Na^+, K^+. E=F=G.   P. BEHRENS

Belousov–Zhabotinsky reaction Ce ion cat. *oxidation of malonic acid by bromate ions in aqueous H_2SO_4. The interaction of several *autocatalytic steps leads to chemical *oscillations.   B. CORNILS
Ref.: Tyson, *The Belousov-Zhabotinskii Reaction*, Springer, Berlin 1976.

Bender sweetening → Petrolite process

bentonite swellable *clay, containing Na containing *montmorillonite. Ca bentonites are called Wyoming b. Both can be used as *supports for het. cats.   B. CORNILS

benzoin condensation → acyloin condensation

Bergius, Friedrich (1884–1949), professor of physical chemistry at Hannover (Germany), discovered, inter alia, in 1912/1913 the *coal liquefaction (*Bergius coal hydrogenation, *Bergius–Pier or IG process). Nobel laureate in 1931 (together with *Bosch).   B. CORNILS
Ref.: E. von Schmidt-Pauli, *Friedrich Bergius*, Mittler & Sohn, Berlin 1943; *Chemie in unserer Zeit* **1985**, 59.

Bergius coal hydrogenation Proc. for *coal liquefaction by *coal hydrogenation at high press. and high temps., originally *without* cats. The fundamental development work was done 1912/1913 by *Bergius et al. in the labs of the Technical University Hannover (Germany). The proc. was later commercialized by BASF as the *Bergius–Pier proc. (cf. *German technology). **F** hydrogénation Bergius du charbon; **G** Bergius Kohlehydrierung.
Ref.: DE 301.231, DE 304.348 (1913).   B. CORNILS

Bergius–Pier process → German technology

Berol Kemi ethylene diamine process manuf. of ethylene diamine by reaction of monoethanolamine and ammonia on *promoted Ni cats. at 300 °C/25 MPa in the gaseous phase.   **F** procédé Berol Kemi pour l'éthylènediamine; **G** Berol Kemi Ethylendiamin-Verfahren.   B. CORNILS

Berry pseudorotation → phosphoranes

Berty reactor internal recycle reactor with a stationary packed bed, for a closer approach of lab. reactors to commercial *cat. reactors. This is important for fast or exothermic cat. reactions. In a Br. commercial cats. can be tested at the same prod. rates as will be required from them in commercial units and no extrapolation

of performance is needed. Many variations of this differential reactor type are known.
In general, a Br. consists of a basket, in which a variable amount of cat. can be placed, and an impeller for the internal circulation of the gas in the reactor. The blower is driven magnetically and the gas from the turbine flows through a draft tube to the top of the cat. bed. The measurement of the press. drop over the bed detects indirectly the flow rate through the bed. For the cat. reaction, the press. drop versus flow relationship has to be measured in separate equipment. The whole reactor is as well isolated as an *autoclave system to make possible testing under high press.
                                      R. SCHLÖGL, H.J. WÖLK
Ref.: *Catal.Rev.-Sci.Eng.* **1979**, *20*(1), 75; Ertl/Knözinger/Weitkamp, Vol. 3, pp. 1190, 1360; Farrauto/Bartholomew, p. 240.

Berzelius Jöns Jacob, Freiherr von (1779–1848), professor of chemistry in Stockholm (Sweden). Besides other important discoveries he coined in 1836 the term "catalysis".
Ref.: Neufeldt.   B. CORNILS

BET *Brunauer–Emmett–Teller and their *adsorption model. The BET method is applied for *surface areas, porosities, and *texture. The very flexible method pursues press. changes (cf. *chemisorption, *total surface area) and is non-destructive. The data are locally and *surface sensitive.   R. SCHLÖGL

Bextol process → Shell procs.

BF → bright field imaging

BHET → bis(hydroxyethyl)terephthalate

BHMC bis(hydroxymethyl)cyclohexane (1,4-dimethylol cyclohexane), → Eastman procs.

bidentate phosphines are *ligands containing two P atoms linked by a chain of carbon or hetero atoms, being capable to *chelate or bridge one or two cat. active metals. Similar to monodentate phosphines, the ligand properties are dependent on the electron-donating or -withdrawing character of the substituents and on the geometry and chain length of the bridging units. Apart from their tremendous *coordination chemistry, the large number of tailormade bps. (cf. *tailoring of cats.) allows the control of *chemo-, *regio-, and *enatioselectivities in hom. cat. reactions. *Chiral bps. play a dominant role as *co-cats. in almost all cat. asymmetric reactions. Some enantioselective *hydrogenation, *reduction, or *isomerization reactions are performed industrially. Bps. are expensive and sensitive towards *P–C bond cleavage. **F** phosphines bidentaire; **G** zweizähnige Phosphane. R. ECKL

Ref.: Pignolet; Hartley; Gates-1, p. 73,88.

bifunctional catalysts, also called dual-functional cats., carry two different cat. *active sites, acting in a synergistic manner to cat. a complex reaction (cf.*ensemble). An example of this type is the classical Pt/*alumina cat. in the prod. of gasoline. In the naphtha *reforming proc., the metal cpd. of that cat. favors *dehydrogenation as well as *hydrogenation, and the acid function of the alumina *surface is responsible for skeletal *isomerization via carbenium ion intermediates (→ dehydroisomerization, *dehydrocyclization). In order to pass through this sequence of reaction steps without diffusional limitations, the mean distance between the two different active sites has to been adjusted acc. to the *Weisz criterion. The synergism and the advantages of bcs. are best demonstrated on reaction systems containing a reversible reaction followed by an irreversible one. Since the second reaction prevents the first from reaching thermodynamic equil. the educt conv. is much higher than that obtained with the monofunctional cats. acting in separate reactors. Mo, V, W, etc. are often constituents of bc. (*ammoxidation, *Mo as cat. metal). Bc. plays a role in cat. layers of *fuel cells, too. **F** catalyseurs bifonctionels; **G** bifunktionelle Katalysatoren. D. HESSE

Ref.: Adv.Chem.Eng. **1964**, 5, 37; Thomas/Thomas, p. 427; Gates-1, p. 210,396.

bifunctional reactions of a given species are transformation reactions of that cpd. cat. by *bifunctional cats. Examples are the *dehydrocyclization of n-hexane to benzene over Pt-Al$_2$O$_3$ cat. and the *ammoxidation reaction. **F** réactions bifonctionelles; **G** bifunktionale Reaktionen. D. HESSE

bimetallic catalysts cats. containing atoms of two different metals. The addition of a second metal to the active comp. follows the aim to influence the properties (*selectivity, *activty, *stability) of the cat. In practice, this type of cat. can be het. or hom. (cf. also *ensemble).

The bimetallic aggregates of a het. cat. can consist of a binary *alloy of an active element (e.g., Ni) and a second metal (e.g., Cu, cf. *alloy cats.). They can, however, also be made of two immiscible comps. like Os and Cu. The observation that immiscible metal systems show the same cat. behavior as completely miscible ones demonstrates that the change in *selectivity is expected only if the undesired reaction needs an ensemble of sites (multiplet) in order to proceed. It is, thus, a *structure-sensitive reaction. Adding a second metal can increase not only the selectivity of the cat., but also its *stability. This effect plays an important role in modern prod. of high-octane gasoline, e.g., acc. to Chevron's *Rheniforming proc. over Pt-Re cats. Contrary to the het. bimetallic cats., hom. ones (e.g., *bimetallic clusters) have so far not led to such a tremendous breakthrough. Some papers deal with hom. bimetallic *hydroformylation cats. (e.g. Sn-Pt systems), but their performance is mediocre. Other examples are mentioned under the keywords *BINOL, *BP Ferrofining, *Cyclopol, *Daicel picoline, *Houdriforming, procs. of *hydrocracking or *hydrotreating, *IFP cat. reforming, or *Hyval procs., *Mukaiyama aldol reaction, *Union Oil Unidak, etc. *Chemical vapor deposition has also been proposed for the prepn. of bcs. **F** catalyseurs bimétallique; **G** Bimetallkatalysatoren. D. HESSE

Ref.: J.H. Sinfelt, Bimetallic Catalysts, Wiley, New York 1983; R.D. Adams, F.A. Cotton (Eds.), Catalysis by Di- and Polynuclear Metal Cluster Complexes, Wiley, New York 1998; M.G. Scaros, M.L. Prunier (Eds.), Catalysis of Organic Reactions, Marcel Dekker, New York 1995, p. 363; Ertl/Knözinger/Weitkamp, Vol. 1, p. 257, Vol. 2, pp. 814, 866; J.Mol.Catal. 1999, 142, 147.

bimetallic clusters term introduced by Sinfelt for small aggregates in the range 10 to 60 Å consisting of two metals. Bcs. can also exist with two metals which do not *alloy in the bulk. Bcs. are of interest in het. cat. because of the high dispersion of the metal and due to the fact that *bimetallic cats. often exhibit a performance superior to that of the constituents. **F** clusters bimétalliques; **G** bimetallische Cluster.   R. IMBIHL

Ref.: J. H. Sinfelt, Bimetallic Catalysts, Wiley, New York 1983.

BINAP, binap one of the outstanding C_2 symmetric *ligands for hom. cat. (R)-BINAP may be converted into the *water-soluble tetrasulfonated BINAP$_{TS}$ by *sulfonation with oleum.

(S)-BINAP   (R)-BINAP$_{TS}$

Synth. by reaction of the *Grignard agent derived from 2,2'-bibromo-1,1'-binaphthyl with $(C_6H_5)_2P$ (O)Cl and reduction of the resolved (R,S)-2,2'-bis (diphenylphosphinyl)-1,1'-binaphthyl with HSiCl$_3$/triethylamine. Rh(I) *complexes of (S)-BINAP are cats. for the asymmetric *hydrogenation of prochiral α-acylamino acrylic acids and esters (up to 100 % *ee). Other *complexes cat. *asymmetric *isomerizations and *Heck couplings; Ru derivatives the asymm. *hydrogenation, e.g., of α,ß-unsaturated acids as for the synth. of naproxen (*profenes).   O. STELZER

Ref.: Tetrahedron Asymmetry 1993, 4, 2461; J.Org. Chem. 1986, 51, 629; Tetrahedron 1984, 40, 1245; Cornils/Herrmann-1, Vol. 1, p. 546; Synthesis 1992, 503.

BINAP$_{TS}$ → tetrasulfonated *BINAP

BINAPHOS, binaphos *phosphine/*phosphite *ligand with two axially asym. moieties for the hom. cat., existent as (R,S) and (R,R) forms.

The Rh/(R,S)-BINAPHOS *complex cat. the *enantioselective *hydroformylation of p-iso-butylstyrene to give a *chiral (S)-(+)-aldehyde in 92 % *ee, which serves as a *precursor for ibuprofen (*profenes). The *alternating copolymerization of propylene with CO is effected by applying [CH$_3$Pd {(R,S)-BINA-PHOS}(CH$_3$CN)]$^+$.   O. STELZER

Ref.: J.Am.Chem.Soc. 1993, 115, 7033 and 1995, 117, 9911; Cornils/Herrmann-1.

binaphthyl the *chiral 1,1'-binaphthyl group in *complex cpds. or as a *ligand. **F=G=E.**

BINAS, binas well investigated *ligand for hom. cat. It is prepared by direct *sulfonation of *NAPHOS with oleum. (S)-(–)-BINAS is formed in addition to small amounts of BINAS-6 and –7 on reaction of (S)-(–)-NAPHOS in the presence of boronic acid.
BINAS is the most active ligand system in Rh-cat. *aqueous-phase *hydroformylation of propene due to its large *natural bite angle (Figure). The n/iso ratio and the activity are

(X = SO$_3$Na)

higher when compared with *TPPTS. The *optical yields of biphasic *hydroformylation of styrene are lower than in the mono-phase. Pd/BINAS *complexes are cats. in *reductive carbonylation of nitrobenzenes and in *carbonylation of benzyl chlorides.　　O. STELZER

Ref.: *Angew.Chem.Int.Ed.Engl.* **1995**, *34*, 811; *J.Mol. Catal.A:***1995**, *97*, 65; *J.Organomet.Chem.* **1997**, *532*, 243; Cornils/ Herrmann-2, p. 441.

binder general term for substances which interconnect equal or different materials, thereby increasing the consistency of the mixture of particles to improve the mechanical properties and maintain the physical integrety of the *support particles. Bs. can be useful in several forming steps of cat. manuf. (*pelletizing, extrusion, tableting). Bs. can be organic or inorganic substances. Organic bs. are mostly added as *monomers, homogenized with the particles, and subsequently polymerized (e.g., acrylates or vinyl alcohol) exhibiting glue-like properties. If desired, the organic bs. can be removed by *calcination after forming is complete. Inorganic bs. are either hydraulic (calcium and/or *aluminum silicates, cements) or non-hydraulic ($Ca(OH)_2$, MgO, *clays) mortars. Acids and caustics in small amounts may also serve as bs., dissolving part of the material and thereby increasing the interparticle cohesion (extrusion of oxide mixtures). **F** liant; **G=E**.　　C.D. FROHNING

binding energy the energy of electron orbitals of atoms or molecules. The be. is defined as the difference between the energy of the *Fermi level (highest electron occupied states) and that of the orbital. The unit of be. is the electronvolt (eV). The highest be. have the core level electrons (electron levels near the atom core), the lowest be. have the valence electrons (electrons in the orbitals nearest to the Fermi level). The be. can be measured with photoelectron spectroscopy (*ESCA or *XPS).

The be. of the core level electrons is specific for each element. Furthermore, they vary within a span of a few eVs (chemical shift) if an atom has chemical bonds to other atoms. The chemical shift depends on the chemical environment of

the atom and may make it possible to identify chemical substances with XPS.

The bes. of the valence electrons also depend on the identity of atoms but they are similar for all atoms. The photoelectron spectrum of the valence electrons (*UPS) is characteristic for a chemical substance and is much more surface-sensitive than the XPS spectrum of the core levels. **F** énergie de liaison; **G** Bindungsenergie.　　R. SCHLÖGL, I. BOETTGER

BINOL, binol important oxygen donor and C_2-symmetric *ligand for hom. cat. with the binaphthyl skeleton. BINOL is prepared by Fe(III)-cat. *dimerization of 2-naphthol and *resolution of the *racemate with *N*-benzylcinchonidinium chloride (*cinchona alkaloids). Due to its flexible structure, BINOL forms stable *complexes with main group (B, Al, Sn) and *transition metals (Ti, Zr) including lanthanoids (La, Yb) (Figures).

(R = H; X = CF_3SO_3; base = *cis*-1,2,6-trimethylpiperidine)

$(*\text{C}_{\text{O-H}}^{\text{O-H}}$ = 6,6'-disubstit. (R)-(+)-BINOL)

Some of these complexes are cats. for *asymmetric C-C coupling reactions. 3-(2-Butenoyl)-1,3-oxazolidin-2-one reacts with cyclopentadiene in presence of a Yb complex of (*R*)-(+)-BINOL and a base to afford the *Diels–Alder adduct with high *ee. La–Na-BINOL complexes employed in *Michael additions are *bifunctional cats. Substitition in the 6,6'-position of BINOL leads to superior *bimetallic cats., e.g., the La-Li species in the *Henry aldol addition.　　O. STELZER

Ref.: *Org.Synth.* **1988**, *67*, 1; *Tetrahedron* **1994**, *50*, 11623; *J.Am.Chem.Soc.* **1995**, *117*, 6194; *J.Org.Chem.* **1995**, *60*, 7388; Beller/Bolm, Vol. 1, p. 313.

biocatalysis requires that the substrate in *biotransformation initially binds noncovalently to the *active site of the cat. as a *Michaelis complex with the proper alignment of reactants and cat. groups in the *active site. The chemical step then takes place, through a covalent or noncovalent intermediate or through reaction with a *cofactor, followed by release of the prod. and *regeneration of the cat. (*catalytic cycle). Acc. to the thermodynamic cycle and *transition state theory, the cat. binds to the reaction *transition state more strongly than to the ground state by a factor approx. equal to the rate acceleration, and the factors contributing to the enormous transisition state stabilization often include *acid–base cat., nucleophilic–electrophilic cat., and cat. by *proximity and strain. B. is increasingly occupied for the manuf. of fine chemicals. **F** biocatalyse; **G** Biokatalyse.

C.-H. WONG

Ref.: Sinnott; S.M. Roberts, *Biocatalysts for Fine Chemicals Synthesis*, Wiley–VCH, Weinheim 1999; W.-D. Fessner (Ed.), *Biocatalysis*, Springer, Berlin 1999; K. Faber, *Biotransformations in Organic Chemistry*, 3rd Ed., Springer, Berlin 1997.

biomimetic catalysts are small molecules that are designed to imitate the acceleration and *selectivities of *enzyme-cat. reactions.

C.-H. WONG

bioreactor reactor used for a chemical transformation accomplished via the use of *enzymes or cells (cf. *catalytic reactors). There are many designs available, depending on the number of liquid *phases and whether the enzyme is soluble or *immobilized. The simplest design is a *STR with the substrate(s) and enzyme in solution. This is useful for small-scale reactions, and for cases in which the enzyme is cheap and the substrate is somewhat soluble in water. It does not need to be totally soluble, as long as a moderate amount of substrate is in solution at any given time. To have a reasonable reaction rate and high *TON, the *cofactor conc. is usually set at around its K_m value (approx. 0.1 mM in most cases) or somewhat higher and the substrate conc. is in

the range of 0.1–0.5 M, unless *substrate or *product inhibition is a problem.

In many cases, a continuous-flow reactor (CFR) is desired. This will be the case for very large-scale reactions, or for industrial procs. where stopping and starting batch reactions is not ideal. CFRs are available in a number of different geometries. A simple *PFR may require a second stage to separate the enzyme from the product stream for recycling. In addition, such a reactor setup will not alleviate prod. inhibition. Other bs. restrain the enzyme, either by filtering the prod. stream through a *membrane with pores too small for the enzyme to pass through, or by *immobilizing the enzyme on a solid *support. Fixed-bed reactors with immobilized enzyme are frequently used. If either the substrate or the prod. is insoluble in water, it will be necessary for an organic phase to flow through the reactor as well. Adequate *mass transfer between the aqueous and organic phases can be accomplished via agitation or constraining the liquids within a contact system of very high *surface area, as in *hollow fibre reactors. One commonly used setup is to have the enzyme immobilized on the surface of the hollow fibers, and pass the aqueous phase through the inside and the organic phase through the outside. **F** bioréacteur; **G** Bioreaktor.

P.S. SEARS

biosynthesis Biosynthetic pathways are the *enzymatic routes used in vivo for the synth. of cpds. B., the synth. of cpds. by cells (or *enzymes), can provide an inexpensive route to the prod. of many natural prods.; cf. *biotransformation. **F** biosynthèse; **G** Biosynthese.

P.S. SEARS

biotin vitamin found in peanuts, chocolate, and eggs. B. is usually synth. in adequate amounts by intestinal bacteria. It is a *cofactor used in carboxyl transfer reactions, during which it becomes transiently carboxylated at N-1 (Figure).

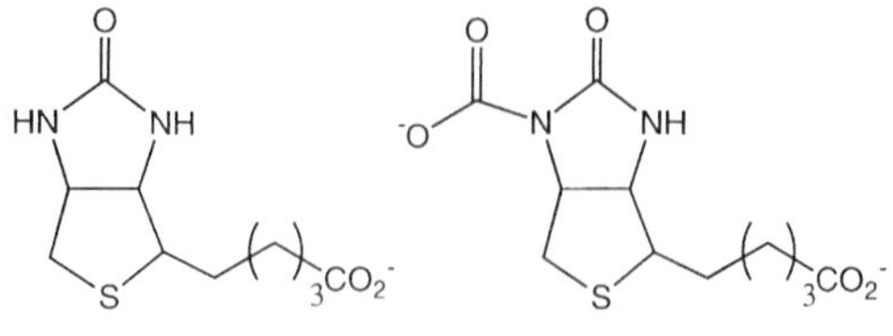

Biotin   N-carboxybiotin

It is bound very strongly by avidin ($K_D \sim 10^{-15}$ M) and streptavidin ($K_D \sim 10^{-17}$ M). The strength of this binding, coupled with the ease of derivatization at the carboxylate, has made this assocation very useful for the tight *complexation and/or *immobilization of a variety of compounds. **E=F=G**.   P.S. SEARS

biotransformation typically refers to the transformation of one cpd. into another via a living system or with biological cats. such as dried cells or *enzymes (cf. *biocatalysis).
               P.S. SEARS

biphasic catalysis → two-phase catalysis

BIPHEMP, biphemp first *optical active bis(triarylphosphine) with the axially disymmetric biphenyl moiety. It is available as (R)-(−)- and (S)-(+)-species and is prepd. by *Ullmann reaction of 2-iodo-3-nitrotoluene to 2,2′-dimethyl-6,6′-dinitrobiphenyl, which is reduced to the amino derivative. *Resolution, *Sandmeyer reaction, and subsequent halogen–metal exchange with t-BuLi after treatment with $(C_6H_5)_2PCl$ yields BIPHEMP (Figure).

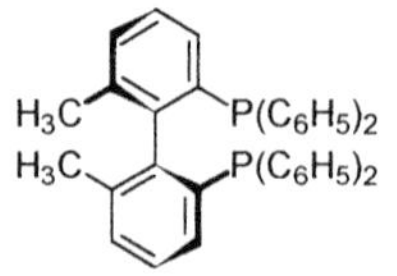

It induces *asymmetry in Rh-cat. *isomerization of N,N-diethylnerylamine to the corresponding enamine, which after *hydrolysis yields (S)-citronellal. BIPHEMP behaves very similarly to *BIPHEP. O. STELZER
Ref.: Helv.Chim.Acta **1988**, 71, 897 and **1991**, 74, 370; Tetrahedron Asymmetry **1991**, 2, 51.

BIPHEP, biphep closely related to *BIPHEMP, both methyl groups of which are substituted by methoxys. In a *water-soluble derivative of BIPHEP the PPh_3 substituents are sulfonated. It is prep. by ortho-lithiation/iodination of 3-methoxyphenyldiphenylphosphine oxide and *Ullmann reaction of the 2-iodophenydiphenylphosphine oxide which gives the racemic BIPHEP oxide. After resolution and *reduction the enantiomerically pure (R)- and (S)-BIPHEP are obtained. Ru *complexes of the (S)-isomer are used in chemo- and *enantioselective *hydrogenations. Sulfonated BIPHEP can serve as *ligand for the Ru-cat. *hydrogenation of unsaturated acids (*ee in water: 99%). Cationic Rh(I) complexes of the type [Rh–BIPHEP-(diene)]$^+$BF$_4^-$ (diene = *COD, *NBD) afford high *enantioselectivities in *isomerizations.
           O. STELZER
Ref.: Helv.Chim.Acta **1991**, 74, 370; Pure Appl.Chem. **1996**, 68, 131; Tetrahedron Asymmetry **1991**, 2, 51.

BIPY, bipy, dipy 2,2′-bipyridine group in *complex cpds. or as a *ligand.

Birch reaction the partial *hydrogenation or *transfer hydrogenation from benzene to cyclohexadiene with Na-Hg in the presence of alcohols. **F=E=G**.   B. CORNILS

BISA → bisphenol A

BISBI, bisbi flexible *bidentate diphosphine 2,2′-bis(diphenylphosphinomethyl)-1,1′-bi phenyl (Figure).

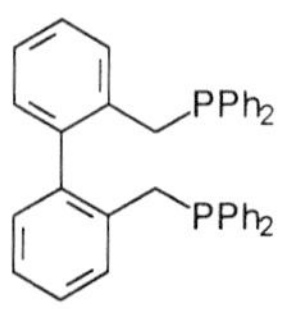

Chelating complexation with transition metals yields nine-membered rings with a wide range of possible *natural bite angles (113°). BISBI is described as a *regioselective *co-cat. for Rh-cat. *hydroformylations. X-ray crystal structures and NMR studies prove that BISBI and

related *ligands favor bis-equatorial coordination geometries in trigonal bipyramidal metal *complexes. BISBI provides the backbone for the sulfonated derivative *BISBIS.   R. ECKL

Ref.: *J.Am.Chem.Soc.* **1992**, *114*, 11817; *Inorg.Chem.* **1991**, *30*, 4271.

BISBIS, bisbis sulfonated derivative of the *ligand BISBI, obtained by direct *sulfonation by means of oleum (H_2SO_4/SO_3). BISBIS is a mixture of constitutional isomers with 4 to 6 sulfo groups. The very high solubility in water (>1 kg/L) makes BISBIS valuable as *ligand for *aqueous-phase cat. When used for the *hydroformylation of propylene, BISBIS shows a comparable n/i-ratio but excellent *activity at low P/Rh ratios when compared with *TPPTS.   R. ECKL

Ref.: *J.Mol.Catal.* **1992**, *73*, 191 and **1995**, *97*, 65.

bishydroformylation Monoalkenes C_nH_{2n} are converted to linear and branched aldehydes $C_nH_{2n+1}CHO$ by reaction with H_2 and CO (*hydroformylation), cat. by metal carbonyls (Rh, Co) which may be *ligand-modified. The bh. of dienes with isolated double bonds (e.g., hexa-1,5-diene) yields a mixture of linear and branched dialdehydes, e.g., isomeric octanedials. Conjugated dienes yield exclusively monoaldehydes when reacted over unmodified Co or Rh cats. When Rh and an excess of *PPh₃ are used, even these dienes (butadiene, isoprene) can be bishydroformylated to a mixture of the corresponding mono- and dialdehydes (in the case of butadiene at about 80 % conv. approx. 50:50 mono- and dialdehydes). **E=F**; **G** Dihydroformylierung.

C.D. FROHNING

bis(hydroxyethyl)terephthalate (BHET) The mutual *intermediate BHET for the prod. of *TPA or *DMT [and the "monomer" for the prod. of *poly(ethylene terephthalate), *PET] is manuf. by either *esterification of *TPA with glycol or by glycolysis of *DMT.   B. CORNILS

Ref.: Cornils/Herrmann-1, p. 545.

bishydroxylation   → hydroxylation

bismuth in catalysis Bi plays a decisive role in *ammoxidation chemistry and *partial *oxidations (*bismuth molybdate, procs. of *Asahi to methyl methacrylate, *Degussa or *Shell to acrolein, *SOHIO to acrylonitrile or acrolein, *Sumitomo to HCN). The *addition reaction of acetylene and formaldehyde to butynediol (*BASF butanediol proc.) is cat. by a Bi-*promoted Cu acetylenide in a trickle-bed operation. For Bi titanates and molybdates cf. *Aurivillius phases. B. CORNILS

Ref.: *Houben-Weyl/Methodicum Chimicum*, Vol. XIII/8.

bismuth molybdates   multicomp. cats. for *oxidation and *amm(o)oxidation reactions, e.g., $Bi_xMo_yO_z$, $M_2^{3+}(MoO_4)_3$, or $M^{2+}MoO_4$, where M^{3+} and M^{2+} are tri- or divalent *transition metals (cf. *Aurivillius phases). Fe and Co are used as molybdates. Another class of ammoxidation cats. are *antimonates.

B. CORNILS

Ref.: Ertl/Knözinger/Weitkamp, Vol. 5, pp. 2253,2311.

bis(oxazoline)   → oxazoline

bisphenol A (BPA, BISA) important *intermediate made from 1 mol acetone and 2 (*bis*) mol phenol, manuf. by proton-cat. *condensation (e.g., procs. of *Chiyoda, *Hooker, *Lummus). **F** bisphénol A, **G=E**. B. CORNILS

Ref.: McKetta-1, Vol. 4, p. 406.

bis(sulfonamide)   → sulfonamide

bisupports mixture of *supports such as $SiO_2/MgCl_2$, *alumina/ *silica, or different modfications of *titania as carrier for *metallocene or *Ziegler-type *polymerization.

Ref.: *J.Mol.Catal. A:* **1999**, *144*, 61.   B. CORNILS

bite angle   → natural bite angle

Blaser–Heck reaction Pd-catalyzed variant of the *decarbonylation of benzoic acid chlorides in the presence of activated alkenes yielding arylated alkenes.   M. BELLER

Ref.: *J.Organomet.Chem.* **1982**, *233*, 267.

bleeding (off) → leaching

block polymers Homopolymers with a block structure are generated from (pro)-*chiral *monomers and feature long sequences of monomer base units showing identical *stereochemical configuration (*stereo*-block polymers) or homopolymers featuring long sequences showing the same *tacticity. Stereoblock polymers are usually produced if stereoregulation is due to *chain end control, while changes in tacticity during *chain growth are the result of enantiomorphic *site control and cats. which may change their structure. Block *copolymerization yields *copolymers containing long sequences of the same configurational base unit. The prod. of block copolymers is usually based on a *living or quasi-living character of the active species. Thus the polymer chain is growing for a long time compared to the time needed for exchanging the monomer feed from monomer A to B. Generally two different methods for changing the monomer feed can be applied: one usually needs a real living *polymerization, which is rarely achieved in cat. polymerization, while the other method is widely employed for the prod. of impact-modified polypropenes from *Ziegler–Natta cats. In the latter case in a first reactor propylene is polymerized in a gas-phase process to yield a homopolymer *PP. Then the polymer granules containing the active cat. are transferred to a second reactor where ethylene is added for the formation of an ethylene–propylene copolymer in the *PP granules which results in the impact modification of the polypropene. **F** bloc polymères; **G** Blockpolymere.

W. KAMINSKY, M. ARNDT-ROSENAU
Ref.: Mark/Bikales/Overberger/Menges, Vol.Suppl., p. 380; *CHEMTECH* **1999**, *29*(June), 38.

Blyholder model qualitative molecular orbital model devised by Blyholder to explain experimental IR-band positions and intensities for the stretching vibration of *chemisorbed *CO acc. to the electron-donating properties of the adsorbent metal *surface. **F** modèle de Blyholder; **G** Blyholder-Modell. W. HIERINGER
Ref.: *J.Chem.Phys.* **1964**, *68*, 2772.

BMA process → Degussa procs.

BNPPA, bnppa a strong acid and useful for the *resolution of *racemic amines or as a *ligand for *hydroxycarboxylation. Both isomers are prep. by cyclocondensation of 1,1'-binaphthyl-2,2'-diol with $POCl_3$ in presence of triethylamine or pyridine, subsequent treatment with H_2O/*THF, and resolution of the racemic mixture with *(+)-cinchonine (Figure).

Optically active ibuprofen (*profens) is obtained by $PdCl/CuCl_2$- cat. *hydrocarboxylation of *p*-isobutylstyrene in the presence of (*R*)- or (*S*)-BNPPA under mild conds. The Rh-mediated tetrakisbinaphthol phosphate (Figure) may be employed as a cat. for *asymmetric dipolar *cycloaddition of diazo cpds. to furan or 2,3-dihydrofuran (50 % *ee). O. STELZER
Ref.: *Org.Synth.* **1989**, *67*, 1; *Synthesis* **1992**, 503; *J.Am. Chem.Soc.* **1990**, *112*, 2803; *Tetrahedron Lett.* **1992**, *33*, 5987.

boc, BOC butoxycarbonyl as a protective group in peptides.
Ref.: T.W. Greene, *Protective Groups in Organic Synthesis*; Wiley-Interscience, New York 1981.

BOC process → UOP procs.

BOC Sure process variant of the *Claus process for the recovery of S from H_2S-containing gas streams. **F** procédé Sure de BOC; **G** BOC Sure-Verfahren. B. CORNILS
Ref.: *Hydrocarb. Proc.* **1996**, (4), 144.

Bodenstein Max Ernst August (1871–1942), professor of physical chemistry. Made important contributions to chemical equilibria, reaction kinetics, chain reactions, etc. (*Bodenstein number).   B. CORNILS

Bodenstein number (*Bo*)   the characteristic parameter of the so-called dispersion model and is used to characterize the deviation of the residence time distribution of real reactors from ideal reactor performance by axial mixing. Thus, *Bo* represents the ratio of convection to dispersion which is given by equ. (1),

$$Bo = \frac{u\,L}{D_{ax}} = Pe_{ax}\,\frac{L}{d} \qquad (1)$$

where *u* is the flow rate (m/s), *L* is the reactor length, D_{ax} is the axial diffusion coefficient (m^2/s), *d* is a characteristic length (e.g., diameter of spheric catalyst particles), and Pe_{ax} is the Peclet number for axial dispersion. If dispersion is negligible, the reactor behavior turns to that of an ideal *PFR (plug-flow reactor). In the case of increasing mixing, the axial conc. profil disappears and the reactor behavior corresponds to that of an ideal *CSTR. The degree of conversion which it is possible to obtain in a tubular flow reactor depends on two dimensionless numbers, the *Damköhler number I (*DaI*) and the *Bo*, acc. to equ. (2),

$$\frac{c_A^{RFR}}{c_A^{PFR}} = \exp\left[\frac{(DaI)^2}{Bo}\right] \qquad (2)$$

where c_A^{RFR} and c_A^{PFR} are the educt concs. in the real-flow and plug-flow reactor, respectively. The axial dispersion for a single gas-phase flow (at high Reynolds numbers) through a packed bed of solid non-porous spheres is characterized by a Bodenstein number of 2. For practical purposes, axial dispersion effects can be neglected under certain conds. **F** nombre de Bodenstein; **G** Bodenstein-Zahl.

P. CLAUS, D. HÖNICKE

boehmite   common bauxite mineral in which there are disseminated grains or pea-like masses. B. consists of O, OH double layers in which the anions are in cubic packing. The Al ions are octahedrally coordinated. These layers are composed of chains of [AlO(OH)]$_2$ extending in the direction of the a-axis. The double layers are linked by hydrogen bonds between the hydroxyl ions in adjacent planes. B. crystals exhibit perfect cleavage parallel to the (010) plane. The crystal system is orthorhombic (D_{2h}^{17}). At about 375 K Al(OH)$_3$ converts to AlO(OH). The conv. temp. appears to be the same for all three γ-hydroxides: bayerite (monoclinic), b., and hydrargillite (monoclinic pseudo hexagonal). **E=F=G**.   R. SCHLÖGL, H.J. WÖLK

Bofors trimellitic anhydride process manuf. of trimellitic anhydride (TMA) by proton-cat. *oxidation of 1,2,4-trimethylbenzene with dilute nitric acid at 170–190 °C. **F** procédé Bofors pour l'anhydride trimellitique; **G** Bofors Trimellithsäureanhydrid-Verfahren.   B. CORNILS

Bohn–Schmidt anthraquinone hydroxylation   a multiple *hydroxylation of 1-hydroxyanthraquinone by means of sulfuric acid/S or sulfuric acid/boric acid and cat. amounts of Hg or Se. **F** réaction de Bohn et Schmidt; **G** Bohn–Schmidt Anthrachinonhydroxylierung.   B. CORNILS

borohydrides (boranates)   B. have a broad potential for *reduction reactions. MBH$_4$ (M = Li, Na, K, Ca, Zn) and BH$_3$ *complexes with amines, pyridine, *THF, Me$_2$S, and others are widely used for *selective hydrogenations (e.g. –CHO to –CH$_2$OH groups in the presence of C–C double bonds), because of costs mainly on the lab scale. In the presence of cats. such as B(OMe)$_3$, halides of Co, Ni, Os, Cu, Pt, and Ti borohydrides as well as the LiBH$_4$/Me$_3$SiCl system enable the reduction of other functionalities. The *enantioselective reduction of prochiral α,β-unsaturated esters can be achieved using MBH$_4$ in the presence of cat. amounts of a *chiral Co *complex (97 % *ee). For the enantioselective reaction of prochiral ketones cf. the *Itsuno reagent.   H.-J. KREUZFELD, M. BELLER

Ref.: A. Pelter, K. Smith, H.C. Brown, *Borane Reagents*, Academic Press, New York 1988; Noyori; J. Fuhrhop, G. Penzlin, *Organic Synthesis*, VCH, Weinheim 1994; *J.Chem.Soc.,Chem.Commun.* **1983**, 469; *Angew.Chem.Int.Ed.Engl.* **1989**, *28*, 60: *J.Org.Chem.* **1986**, *51*, 4512; *Chem.Rev.* **1998**, 98, 2685.

boron in catalysis BF_3 is (besides $AlCl_3$ or $ZnCl_2$) a well-known *acid cat. for a series of reactions such as *acylation, *alkylation/ *hydroalkylation (e.g., Alkar or cumene procs. of *UOP), *Diels-Alder, *Friedel-Crafts, *isomerization (*Mobil LTI proc.), or the *Koch synth. $HF \cdot BF_3$ serves as a carbonylation cat. (cf. *Mitsubishi trimellitic acid). Boric (boronic) acids are cats. for the metal ion-cat. *oxidation of paraffins (*Bashkirov reaction, e.g., *Nippon Shokubai proc.) and for the *Suzuki reaction. B. CORNILS

boron trifluoride (BF_3) is used as addition cpd. with ethers, alcohols, etc., mainly as *Friedel–Crafts cat.

Bosch Carl (1874–1940), president and CEO of BASF, co-founder of I.G. Farbenindustrie AG. Made important contributions to *ammonia synth. (*Haber–Bosch process) and to *coal hydrogenation. Nobel laureate 1931 (together with *Bergius). B. CORNILS
Ref.: K. Holdermann, *Im Banne der Chemie*, Econ, Düsseldorf 1953.

Boudouard equilibrium At high temps., particularly in the presence of metal cats., CO undergoes reversible *disproportionation acc. to $2\ CO \rightleftharpoons C + CO_2$; $\Delta H° = -172.5$ kJ/mol.
The equil. conc. of CO is 10 % at 550 °C and 99 % at 1000 °C. As the forward reaction involves a reduction of gaseous molecules in the system it is accompanied by a large decrease in entropy. This implies that the reverse reaction becomes more favored at higher temp. Commercial relevance of this equil. is in the procs. of *coal gasification and other carbonaceous materials with CO_2. Furthermore, the Be. has to be considered during cat. procs. in which carbon can be produced, e.g., in the form of coke (*coking)

during *steam reforming of hydrocarbons. *Deposited carbon can affect the performance of the cat. by *poisoning and reducing heat transfer. This causes so-called hot bands and finally damages the reformer tube. One option to reduce C formation in the steam reforming reaction of methane is to operate at higher steam/methane ratios. **F** équilibre de Boudouard; **G** Boudouard-Gleichgewicht.
R. SCHLÖGL, M.M. GÜNTER

BPA → bisphenol A

BP acetic acid process → BP Cativa acetic acid technology

BP Autofining process to *desulfurize light distillates incl. *LNG concentrates, naphthas, kerosenes, etc., and to *dehydrogenate simultaneously naphthenes without an external source of hydrogen over a fixed cat. bed at 370–430 °C/2 MPa. **F** procédé Autofining de BP; **G** BP Autofining-Verfahren.
Ref.: *Hydrocarb.Proc.* **1978**, (9), 123. B. CORNILS

BP Butane Isom process manuf. of isobutene by cat. *isomerization (fixed-bed, vaporphase, Pt on acid Al_2O_3) of *n*- to *i*-butane at 160 °C and its *dehydrogenation. The outlet for isobutene is *alkylation and *MTBE. **F** procédé BP butane isom; **G** BP Butanisom-Verfahren. B. CORNILS
Ref.: *Hydrocarb.Proc.* **1982**, *61*(9), 170.

BP Catalytic Dewaxing process selective *hydrocracking by cat. deparaffination (*dewaxing) on Pt/H-*mordenite cat. in the gaseous phase. **F** procédé Catalytic Dewaxing de BP; **G** BP Catalytic Dewaxing-Verfahren.
Ref.: McKetta-1, Vol. 15, p. 346. B. CORNILS

BP Cativa acetic acid technology a low-press., liquid-phase *carbonylation of MeOH, cat. by Ir cpds., *promoted by iodine and *co-promoted by variuous *transition metal and group IIIA salts (i.e., Ru, Os, Re, W, Cd, Zn, Hg, Ga, In) at 150–200 °C/ 1.5–5 MPa. The technology is implemented at two commercial plants to replace Rh-cat. *Monsanto AA procs. For comparable reaction conds., the Ir-cat. system provides improved cat. *stability and reduced acetaldehyde-derived unsaturate and alkyl iodine impurities. *Reaction rate is directly proportional to Ir and CO and inversely dependent on I conc. The rate determining step is the migratory insertion of CO to form the anionic acyl intermediate $[Ir(CH_3CO)(CO)_2I_2]^-$ (cf. mechanism as given with *Monsanto's AA proc.). Metal ion co-promoters facilitate the elimination of ionic iodide from the Ir *coordination sphere. **F** procédé Cativa de BP; **G** BP Cativa-Verfahren.   P. TORRENCE

Ref.: D.J. Watson, *Catalysis of Organic Reactions* **1998**, 75, 369; EP 0.749.948 A1 (1996); US 5.877.347.

bpcd   barrels per calendar day as a measure for the production per *calendar* day and not per *stream* day (*bpsd).

bpd   barrels per day as a measure for production figures of the *petroleum processing industry (*bpcd,*bpsd).

BP (Distillers)/Ugine acrylonitrile process   variant of the *amm(on)oxidation technology converting propene to *acylonitrile. In a two-step procedure propene is firstly oxidized to *acrolein on Se-CuO cats., followed by *amm(on)oxidation on MoO_3 fixed-bed cats. (between 1960 and 1980 over Sn *antimonates) with air/ammonia. **F** procédé BP (Distillers)/ Ugine pour le nitrile acrylique; **G** BP (Distillers)/Ugine Acrylnitril-Verfahren.   B. CORNILS

BPE, bpe   bis(phospholano)ethane, a *bidentate *ligand for, e.g., *enantioselective cat. *hydrogenation.

BP/Erdoelchemie Etherol process   The proc., jointly developed, comprises the manuf. of C_4–C_6 alkenes from *FCC effluents and methanol. The cat. (Pd on very acidic sulfonic *ion exchange resin) is trifunctional for *esterification, *isomerization, and *hydrogenation. A *catalytic distillation is incorporated.

F procédé   Etherol   de   BP/Erdölchemie; **G** BP/Erdölchemie Etherol-Verfahren.
                                                                                    B. CORNILS

Ref.: EP 0.048.893 (1982); *Appl.Catalysis* **1994**, *115*, 182.

BP Ferrofining process   a mild *hydrotreating (*hydrofinishing) of lube oils over *bimetallic cats. with fresh H_2. **F** procédé Ferrofining de BP; **G** BP Ferrofining-Verfahren.
                                                                                    B. CORNILS

Ref.: *Hydrocarb.Proc.* **1972**, (9),153; Winnacker/Küchler, Vol. 5, p. 121.

BP/Hercules phenol process   the BP/ Hercules variant of *Hock's process via *cumene hydroperoxide (120 °C, metal salts as cat.) and its acid cat. cleavage in homogeneous solution with dilute sulfuric acid at 60–65 °C into phenol and the co-prod. acetone. **F** procédé BP/Hercules pour le phénol; **G** BP/Hercules Phenol-Verfahren. B. CORNILS

BP hydrocarbon oxidation   former British Distillers proc. for the liquid-phase naphtha *oxidation with air in *radical reaction at 200 °C/5 MPa to yield mainly *acetic and *propionic acids. **F** procédé BP pour l'oxydation d'hydrocarbures; **G** BP Kohlenwasserstoffoxidations-Verfahren.   B. CORNILS

BP hydrofining process   to remove sulfur from a wide range of distillate feedstocks by a treatment over fixed-bed *HDS cats. Depending on the feedstock, the proc. can be combined with BP's *Autofining proc. to minimize H consumption. **F** procédé hydrofining de BP; **G** BP Hydrofining-Verfahren.

Ref.: *Hydrocarb.Proc.* **1978**, (9), 138.   B. CORNILS

BP hydrocracking process   to lower the S content and to improve the cold properties of *VGO prods. such as gas oils, diesel oils, lube oil basestocks, etc. by fixed-bed, one-through, single-stage cat. operation with cold recycle gas injection between the different cat. beds. **F** procédé hydrocracking de BP; **G** BP Hydrocracking-Verfahren.   B. CORNILS

Ref.: *Hydrocarb.Proc.* **1978**, (9), 118.

BP Innovene HDPE/LLDPE process

BP Innovene HDPE/LLDPE process to produce *LLDPE or *HDPE using the gas-phase Innovene technology (fluid-bed reactor) with either *Ziegler–Natta or *Cr cats. injected directly into the reactor (Figure). **F** procédé BP pour l'HDPE/LLDPE; **G** BP HDPE/LLDPE-Verfahren. B. CORNILS

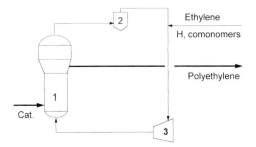

1 fluidized-bed reactor; 2 cyclone; 3 compressor.

BP C₄ isomerization process

BP C₄ isomerization process for the *isomerization of n-butane to isobutane over fixed-bed Pt-*alumina cats. (same cat. as in *BP's C_5/C_6 isom. proc.) at 150–200 °C/1–2 MPa. The *LHSV is approx. 2–3. Organic chloride cpds. are used for make-up (cf. *chlorinated alumina). **F** procédé C₄ isomérisation de BP; **G** BP C₄-Isomerisationsverfahren. B. CORNILS
Ref.: *Hydrocarb.Proc.* **1972**, (9), 123.

BP C₅/C₆ isomerization process

BP C₅/C₆ isomerization process to increase the octane ratings of pentane and hexane fractions by *isomerization over a specially chloride-treated fixed-bed Pt cat. on *alumina (cf. *chlorinated alumina). Losses of HCl are made up by a small injection of organic chloride to the feedstock (*make-up). **F** procédé C_5/C_6 isomérisation de BP; **G** BP C_5/C_6-Isomerisationsverfahren. B. CORNILS
Ref.: *Hydrocarb.Proc.* **1978**, (9), 170; *Petrochemical Int.* **1971**, *11*(12), 52.

BPO benzoyl peroxide, a starter for *radical reactions.

BPPFA, bppfa

BPPFA, bppfa a C_1-symmetric *bidentate *phosphine *ligand combining elements of planar and central chirality. The *R,S*-isomer is prep. by stepwise lithiation of (*R*)-**1** (a *ferrocene derivative) with n-BuLi and treatment of the lithiation prods. with chlorodiphenylphosphine (Figure).

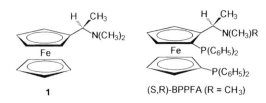

(*S,R*)-BPPFA may be applied in Rh-cat. *enantioselective *hydrogenation of olefinic substrates (e.g., (*Z*)-α-acetaminocinnamic acid, 93 % *ee*) and other cpds. *Aldol reactions of methyl-α-isocyanocarboxylates with aldehydes in presence of gold(I) *complexes of (*R,S*)-BPPFA (*gold as catalyst metal) derivatives yield optically active 4-methoxycarbonyl-4,5-dialkyl-2-oxazolines. O. STELZER
Ref.: *Bull.Chem.Soc.Japan* **1980**, *53*, 1138; *Pure Appl. Chem.* **1996**, *68*, 29; *Acc.Chem.Res.* **1982**, *15*, 395; *Tetrahedron Asymmetry* **1995**, *6*, 2503; *Tetrahedron* **1988**, *44*, 5253.

BPPM, bppm

BPPM, bppm 1-*tert.*-Butoxycarbonyl-(2*S*,4*S*)-2-[(diphenylphosphino)methyl]-4-(diphenylphosphino)pyrrolidine is an optical active *ligand for hom. cat. It is prep. by reaction of 1-(*tert.*-butoxycarbonyl)-4-hydroxy-L-prolinol di-*p*-toluene sulfonate with $(C_6H_5)_2PNa$. It is used for *hydrogenations and *hydroformylations (e.g., of styrene with *bimetallic Pt/ $SnCl_2$) with moderate selectivities (*iso/n* = 0,55; 77 % *ee*). Higher *ees* (>96 %) are achieved by trapping the aldehydes with triethyl *o*-formate. *Transfer hydrogenation of itaconic acid with $HCOOH/NR_3$ is cat. by Rh/BPPM *complexes. O. STELZER
Ref.: *J.Org.Chem.* **1981**, *46*, 2954; *Organometallics* **1991**, *10*, 1183; *J.Mol.Catal.* **1983**, *21*, 203; Cornils/Herrmann-1, Vol. 1, p. 210.

BP protein process

BP protein process *biocat. manuf. of high-protein yeast (trademark toprina) from n-alkanes by *fermentation in *bioreactors. The proc. is run continuously consuming air and ammonia, mineral nutrients, and growth factors at a special pH and at 30 °C. The proc. (and the prod.) are now obsolete. **F** procédé BP pour des protéines; **G** BP-Proteinverfahren.   B. CORNILS

Ref.: *Hydrocarb.Proc.* **1977**, 56(11), 222.

BP residue desulfurization
of high sulfur content atmospheric residues over selective, robust, and inexpensive cats. designed to accomodate deposited metals (*HDM) and reduce S (*HDS) to the desired level. The cycle period between cat. change is normally 6 months. **F** procédé BP de désulfurisation de résidue; **G** BP Rückstandsentschwefelung.
                                      B. CORNILS

Ref.: *Hydrocarb.Proc.* **1972**, (9), 175; *Oil&Gas J.* **1971**, 69(Oct.), 75.

bpsd   a measure of production per *stream* day (and not per *calendar* day, → bpcd)

BP selective hydrogenation process
for the refinement of gasoline byprods. from *steam cracking by cat. *hydrogenating *selectively the diolefins to olefins to produce a stable motor gasoline (single stage proc. at 180 °C/2–8 MPa). The comb. with desulfurization (*HDS, avoiding aromatics hydrogenation) yields suitable feedstocks for aromatics extraction (second stage at 270–340 °C/6–9 MPa). **F** procédé BP pour l'hydrogénation sélective; **G** BP-Selektivhydrierungsverfahren.

Ref.: *Hydrocarb.Proc.* **1972**, (9), 177.   B. CORNILS

BP wax hydrofinishing process
similar to *BP's Ferrofining proc. by treatment over Ni and Ni-Mo cats. **F** procédé wax hydrofinishing de BP; **G** BP Wachs-Hydrofinishing-Verfahren.   B. CORNILS

Ref.: Winnacker/Küchler, Vol. 5, p. 122.

bpy, BPY   *barrels per year

BR   1: batch reactor, → catalytic reactors; 2: butadiene rubber, → polybutadiene

Brassard's diene
1,3-dimethoxy-1-(trimethylsilyloxy)-1,3-butadiene, an activated, conjugated diene, comparable to *Danishef-

sky's diene, also applied in *Lewis acid cat. hetero *Diels–Alder reactions with ketones or aldehydes to δ-lactones and dihydropyrans (Figure).

With $Eu(hfc)_3$, high *stereoselectivity can be obtained. **F** diène de Brassard; **G=E**.
                                      G. GERSTBERGER

Ref.: *J.Am.Chem.Soc.* **1984**, *106*, 4294.

Bravais lattices   → crystallography of surfaces

Brichima hydroquinone process
manuf. of pyrocatechol and hydroquinone by *oxidation of phenol with *Fenton's reagent (Fe/Co and H_2O_2) at 40 °C. The conv. rate is 20 % per pass; the *selectivity 90 %. **F** procédé Brichima pour l'hydroquinone; **G** Brichima Hydrochinon-Verfahren.   B. CORNILS

Ref.: Ertl/Knözinger/Weitkamp, Vol. 5, p. 2331.

Briggs–Haldane kinetics/mechanism
(cf. also *Michaelis–Menten (MM) kinetics, mechanism). In the simplest *enzymatic reaction, given as $E + S \rightleftharpoons E:S \rightarrow E + P$ (with →: k_1, ←: k_{-1})   the MM mechanism assumes that the first step is essentially at equil., thus $[E:S] = (k_1/k_{-1})[E][S]$ or $[E:S] = [E][S]/K_s$, where $K_s = k_{-1}/k_1$ (the dissociation constant of the *Michaelis complex). This is only true if the cat. step is slow with respect to the binding equilibration between E, S, and E:S. The Briggs-Haldane (BH) mechanism does not require this assumption. Instead, it is assumed that the conc. of the Michaelis complex stays nearly constant, and so for a limited time, $d[E:S]/dt = 0 = k_1[E][S] - k_{-1}[E:S] - k_2[E:S]$. This is called the *steady-state (or pseudo steady-state) approximation. Combining this equ. with the mass balance equation for enzyme ($[E:S] = [E_o] - [E]$) (where E_o = total enzyme added) and the rate equ. $\upsilon = k_2[E:S]$

(where υ is the reaction velocity), yields the relation $\upsilon = k_2[E_o][S]/(K_m + [S])$, where K_m equals $(k_2 + k_{-1})/k_1$, not K_s. It clearly has the same form as the equ. for the MM mechanism, but the constant K_m no longer has the same meaning. Note that in the limit $k_2 \ll k_{-1}$, the relation collapses back to that for the MM mechanism, where $K_m = K_s$. At the opposite limit, where $k_2 \gg k_{-1}$, the rate becomes dominated by the assocation of enzyme and substrate, and the velocity is given by $\upsilon = k_1[E_o][S]$. **F** cinétique/mécanisme de Briggs–Haldane; **G** Briggs-Haldane-Kinetik/Mechanismus. P.S. SEARS

bright field imaging (BF) is an electron contrast formation mode using all-diffracted beams. The method is non-destructive.

R. SCHLÖGL

Brintzinger-type metallocenes group 3, 4, and 5 *ansa*-metallocenes. They are important as so-called *stereorigid *metallocene cats. in *stereoselective reactions, most notably in *polymerization of propylene. J. OKUDA

bromelain (EC 3.4.22.4) a relatively nonspecific *thiol endoprotease typically isolated from pineapple stem. It prefers Lys, Arg, Ala, Phe, or Tyr in the P_1 position, and has a preference for hydrophobic amino acids in the P_2 position. The pH optimum is between 6 and 8. Activity is enhanced in the presence of reducing agents and chelators such as EDTA. **E=F=G.** P.S. SEARS

Ref.: Schomberg/Salzmann, Vol. 5.; *Methods Enzymol.* **1970**, *XIX*, 273.

bromoperoxidase → peroxidase

Brønsted Johannes Nicolaus (1879–1947), professor of physical chemistry in Copenhagen (Denmark). Developed the *Brønsted *acid–base concept. B. CORNILS

Brønsted acid–base concept was independently developed by *Lowry and *Brønsted for aqueous solutions and was later extended to non-aqueous systems by Brønsted. An acid HA is defined as a proton donor (HA → H^+ + A^-), and a base B is a proton acceptor (B + H^+ → BH^+). Brønsted acid–base reactions always involve corresponding acid–base pairs, the equil. of which is called a protolytic system acc. to HA + B $\rightleftharpoons$ A^- + BH^+. A^- is the conjugate base of the acid HA and BH^+ is the conjugate acid of the base B. Brønsted acids and bases can be of neutral, cationic, and anionic nature. Examples of B. acids include HCl, NH_4^+, or HSO_4^-; B. bases comprise NH_3 or ClO_4^-. Molecules which behave both are amphoteric (e.g., H_2O). In a *solvent S, protons are coordinated to solvent molecules to yield SH^+, referred to as the lyonium ion of the solvent. Water is the standard solvent for setting up an acidity scale. The strength of B. acids and bases can be measured by the magnitude of the equil. constant for dissociation K_a (K_b), using activities. Strong acids are completely dissociated in water (pK_a = pH; pK_b = pOH; pH + pOH = 14). Since a universal scale of acidity is not possible for different reasons, various scales were introduced; e.g., the acid–base behavior of weak B. bases can be measured by the Hammett acidity function (cf. *Hammett constants).

The B. acid–base concept also applies for *solid acid–base cats. Predominant B. acid sites of solids are OH groups (e.g., in *zeolites). Their determination is difficult because a clear distinction of B. and *Lewis acid sites is difficult to achieve (methods of characterization include titration, thermochemistry, spectroscopy, calorimetry, etc.). A variety of organic reactions are cat. by B. acids, e.g., *dehydration, ether cleavage, *addition reactions, *rearrangements, etc. Cf. also *Brønsted equation. **F** concept acido-basique de B.; **G** B. Säure–Basen-Prinzip nach Brønsted.

R. ANWANDER

Ref.: *J.Am.Chem.Soc.* **1950**, *72*, 1164; H.L. Finston, A.C. Rychtman, *A New View of Current Acid–Base Theories*, Wiley, New York 1982; *J.Catal.* **1990**, *124*, 97.

Brønsted equation a linear free energy relation between the log of the rate constant (proportional to the activation energy, $\Delta G^{\neq}$) of a reaction involving an acid–base as a cat., nucleophile, or leaving group, and the strength (pK_a, proportional to the energy of deprotonation) of the acid–base involved. For a base the equation is $\log k_2 = A + \beta pK_a$ where A is a constant that depends on the particular reaction and β is the Brønsted β value for that base. For acid–base cat., the B. acid and base values are usually between 0 and 1, with 0 denoting no and 1 denoting complete proton transfer. For nucleophilic displacement reactions, the sign and magnitude of the β value can give information regarding the charge developed in the *transition state (cf. also *Brønsted acid–base concept, *Hammett equation). **F** équation de Brønsted; **G** Brønsted-Gleichung.

P.S. SEARS

Brønsted law   → *acid–base catalysis

Brønsted value   → Brønsted equation

Brunauer–Emmett–Teller (BET) isotherm
→ adsorption

BTX   collective term for benzene, toluene, xylenes in petrochemistry

bu, Bu   butyl group

i-bu, Bu   iso-butyl (branched butyl) group

n-bu, Bu   normal-butyl (unbranched butyl) group

t-bu, Bu, tert.-bu, Bu   tertiary butyl group

bubble column reactor   → catalytic reactors

Bucherer reaction $NaHSO_3$-cat. substitution of aromatic OH groups by amino groups (cf. *aromatic substitution).   B. CORNILS

Buchner   Eduard (1860–1917), professor of chemistry, inter alia at Würzburg (Germany). Worked on yeasts and discovered the *zymase-cat. (cell-free) *fermentation, thus being one of the fathers of *biotechnology. Nobel laureate in 1907.   B. CORNILS
Ref.: *Ber.Deut.Chem.Ges.* **1917**, *50*, II.

Bürgi–Dunitz trajectory   → Felkin–Anh model

BuLi, n-BuLi   n-butyllithium

bulk (carrier-free) catalysts   cats. which consist only of cat. active ingredients (unlike *supported cats. which consist of cat. active constitents *and* support). *Fused cats., *skeletal cats., *precipitate cats. (under certain limitations), *sol–gel cats. cats. by *flame hydrolysis technique, *heteropolyacids, cat. active *carbons, *gauze cats., or metal blacks (*Pt black, *Ru black) belong to the bulk cats. **F** catalyseurs de masse; **G** Vollkatalysatoren

B. CORNILS

bulk density of catalysts   Bd. (or apparent density, unit kg/L or lb/cft) is the mass of cat. filling a unit space determined by a standardized procedure (e.g., ASTM C 29/29M-97, DIN 53466). This measure allows to calc. the amount of cat. necessary for a given reactor volume. Bd. depends on the shape of the cat. particles (tablets, *pellets, powder, rings, lumps, spheres, etc.) which leads to the free interspace (void volume) between the particles, and on the density of the cat. particles. Common ranges for the bd. are 0.5–1.5 for shaped and 0.2–0.8 for powdered cats.

C.D. FROHNING

bulk-to-surface migration   → Tammann temperature

bulky ligands   are *ligands exhibiting high steric requirements (sterically encumbered ligands), expressed in large values of their *Tolman cone angles (*Θ values). Bls. are applied to achieve a kinetic stabilization of electronically highly unsaturated coordination cpds. by providing a sterically protected environment (*steric effects). Typical reactions that can be suppressed in the presence of bls.

are the coordination of donor molecules to *Lewis acidic metal centers resulting in low *coordination numbers, associative decomposition pathways involving reactions with the solvent or with another molecule of the cpd., and intramolecular insertion reactions such as *C–H activation. Bls. as a rule impart mononuclearity and concomitantly a better solubility of their *complexes, which is often a prerequisite for the generation of cat. active cpds. offering *vacant coordination sites. In general, an increased steric bulk of a ligand can be achieved by replacing its H atoms with a bulkier substituent such as *tert.*-Bu, SiMe$_3$, phenyl, or mesityl. Typical bls. are alkyl ligands like [HC(SiMe$_3$)$_2$]$^-$ (e.g., for the synth. of donor-free Ln[HC(SiMe$_3$)$_2$]$_3$) or 1-norbornanid (Co(1-norb)$_4$ is a low-spin tetrahedral complex), cp ligands like [C$_5$Me$_5$]$^-$ (the standard ligand in lanthanidocene chemistry to prevent complex agglomenzation), amide ligands like [N(SiMe$_3$)$_2$]$^-$, *phosphine ligands like PPh$_3$ or P(*t*-Bu)$_3$, *alkoxide ligands {like [OC(*t*-Bu)$_3$]$^-$ (tritox)}, siloxide ligands like [OSi(*t*-Bu)$_3$]$^-$ (silox), and boraoxide ligands like {OB-[C$_6$H$_2$(*i*-Pr)$_3$-2,4,6]$_3$}$^-$. In general the bulkiness increases with the number of substituents at the α-atom, i.e., CR$_3^-$ > PR$_3$ > NR$_2^-$ > OR$^-$. **F** ligands écartés; **G** sperrige Liganden.

J. EPPINGER

Ref.:*Adv.Organomet.Chem.* **1991**, *33*, 291; *Inorg.Chem.* **1989**, *328*, 2380.

burst kinetics If an *enzyme has more than one step following the initial substrate binding, and if the first step is much faster than subsequent steps, then when an overwhelming amount of substrate ([S] >> K_m) is added to the enzyme, all of the enzyme will be rapidly converted to the intermediate ES', and then slowly react further to generate product and free enzyme.

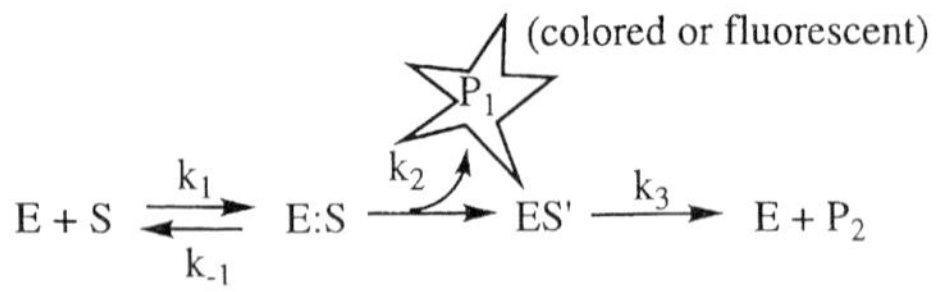

$$E + S \underset{k_{-1}}{\overset{k_1}{\rightleftharpoons}} E{:}S \xrightarrow{k_2} ES' \xrightarrow{k_3} E + P_2$$

(colored or fluorescent) P$_1$

If the first step yields a colored prod. (e.g., P$_1$ in the Figure is colored, or ES' is colored), then there will be a colorimetric "burst". This is commonly observed for *serine protease *hydrolysis of *p*-nitrophenylesters, in which the acyl enzyme is rapidly formed and deacylation proceeds more slowly. This burst can be used to titrate the number of enzyme *active sites. If the substrate conc. is less than K_m, more time will be required for the enzyme to be converted to ES', and if k_3 is sufficiently fast, some of the ES' intermediate will proceed through the second step before all of the enzyme has made it through the first step. This will cause the magnitude of the burst to be decreased. See also *active site titration. **F** cinétique de "burst"; **G** "Burst"-Kinetik.

P.S. SEARS

Buss hydrogenation system developed by Buss AG for discontinuous reactions (e.g., *hardening of fats) in a *slurry-phase loop reactor. The reactants and the dispersed cats. are circulated by a special centrifugal pump through a venturi tube-based injection mixing nozzle. Through this device both the fresh and the recycled H are introduced resulting in a very efficient *mass transfer. **F** système Buss pour l'hydrogénation; **G** Buss-Hydriersystem.

B. CORNILS

Ref.: *cav-High Tech.* **1987**, October; *J.Amer.Oil Chem. Soc.* **1987**, *60*, 1257.

Butachimie adiponitrile process manuf. of *adiponitrile by hom. cat. *hydrocyanation of butadiene with Rh(0)/*phosphine/ phosphite. **F** procédé Butachimie pour l'ADN, **G** Butachimie ADN-Verfahren.

B. CORNILS

Butakem tartaric acid process two-step proc. for the manuf. of tartaric acid by 1: *epoxydation of *maleic anhydride (MA) with H$_2$O$_2$ over Mo or W cats., and 2: the *hydrolysis of the epoxytartaric acid. **F** procédé Butakem pour l'acide tartrique; **G** Butakem Weinsäure-Verfahren.

B. CORNILS

Butamer process → UOP procs.

1,4-butanediol (butane-1,4-diol), BDO) is
an important *intermediate, inter alia for
*GBL and *THF, cat. manuf. by *hydrofor-
mylation of allyl alcohol (*ARCO BDO),
*ethynylation (*BASF, *GAF), or other
procs. (*Lurgi/ BP, *Mitsubishi, *Toyo)

B. CORNILS

Ref.: Ullmann, Vol. A4, p. 447.

Butane isom process → BP procs.

bz benzoyl group in *complex cpds. or *ligands

C

C₁ as a building block The modern chemical industry bases on degrading and on synth. reactions. Among the synth. procs. building blocks such as *carbon monoxide (CO), *syngas (CO + H₂), *formaldehyde, *methanol, and *methane (natural gas) gain importance since they may be links between different feedstocks (such as *coal, *wood, shale, peat, *petroleum) and a subsequent chemistry. Independently of the raw material, CO and syngas can be produced from, e.g., coal (*coal gasification) or petrochemical raw materials by existent procs., thus being the basis for downstream intermediates, fine, or bulk chemicals such as *AA, *AAA, *SNG, etc. (*petrochemistry, *petroleum processing). Nearly all of the converting procs. are cat., most of them homogeneously (e.g., synth. of *oxo prods., *AA, *AAA, *acrylic acid, *Fischer-Tropsch, *vinyl acetate, etc.). **F** C₁ comme matière première chimique; **G** C₁ als Chemiebaustein. B. CORNILS

Ref.: Kirk/Othmer-1, Vol. 5, p. 97; Kirk/Othmer-2, p. 1287; Payne, McKetta-1, Vol. 7, p. 246, Vol. 29, p. 484; McKetta-2, p. 2; Winnacker/Küchler, Vol. 5, p. 502f;

C₂/C₃-hydrocarbon aromatization The *aromatization of C₂ and C₃ hydrocarbons can be considered as an additional source for aromatics as chemical feedstocks. *Zeolites (mostly ZSM-5 type) are used as base cats. for the a. of short-chain hcs. For cat. *doping, Pt, Ga, and Zn cpds. are used. The a. reaction comprises numerous reaction pathways. For alkenes, *oligomerization, *cyclization, and subsequent *dehydrogenation occur besides *isomerization and *cracking. The reactions proceed via classical *carbenium ion chemistry on the *Brønsted acid sites of zeolites. For the a. of ethane and propane, the primary de-hydrogenation step facilitated by doping of the zeolites is important.

The conv. of an alkane into aromatic hcs. becomes more favorable with increasing C number. Because of the high thermodynamic barrier ethane has often been considered as a non-aromatizable prod. while the a. of propane and higher alkanes is much easier to achieve. Only a few data on the direct transformation of ethane into aromatics are known. Propane a. developed by *BP and *UOP is presently put into practice as the *Cyclar proc. (formerly *Catarol). **F** aromatisation des hydrocarbures C₂ et C₃; **G** Aromatisierung von C₂- und C₃-Kohlenwasserstoffen. O.V. BUYEVSKAYA, M. BAERNS

Ref.: J.Catal. **1970**, *17*, 205 and 216; Catal.Rev.Sci.Eng. **1992**, *34*(3), 179; Appl. Catal. A: **1995**, *131*, 347.

C₂-symmetric ligands are important for *enantioselective synth. for three major reasons. 1: In most cases metal *complex cats. offer more than one *coordination site for their substrate. As it is usually impossible to block undesired coordination site(s), diastereomeric *transition states cannot be ruled out. In the majority of scenarios for absolute stereochemical control, the presence of a C₂ symmetry axis within the chiral cat. reduces the number of possibly competing diastereomeric transition states. 2: Distinguishable coordination sites, e.g., equatorial and apical positions in a trigonal pyramid (cf. e.g., *phosphoranes) have to be considered in the case of *bidentate cats. Whereas C₁-symmetric ligands may coordinate to a metal center in two different manners, the exchange of equatorial and apical positions of a C₂-symmetric *ligand renders the cat. unchanged. 3: C₂-symmetric ligands are synth. very easily by either linking two chiral building blocks (e.g., amino acids) or by modifying already C₂-symmetric cpds. (e.g., tartaric acid), both readily available from the *chiral pool. Another source of chiral C_2-symmetric educts are binaphthylic and similar cpds. (e.g., *BINAP, *BINAPHOS, *BINAS, *BINOL) which show axial chirality due to a hindered rotation about a C-C bond. There are numerous examples of this concept. Commercially used are ligands such as *DIOP, *CHIRAPHOS, or *DIPAMP and some *phosphites. Other donor atoms are important for scientific use like *TADDOL (based on tartaric acid, used for the *Sharpless epoxidation), binaphthols (as cited earlier), or *salens (for the *Jacobsen epoxidation), nitrogen donor ligands like *sulfonamides or *bis(oxazolines), as well as *carbenes. For reasons of clarity it should be noted that C_2-symmetric *ansa-metallocenes in most cases do not contain C_2-symmetric ligands. Their chirality originates from the metal coordination and thus results in racemic mixtures known as *rac*-forms. **F** ligands avec C_2 symétrie; **G** C_2-symmetrische Liganden.

H.W. GÖRLITZER

Ref.: M. Nógrádi; *Stereoselective Synthesis*, 2nd Ed., VCH, Weinheim 1995; *Chem.Rev.* **1989**, *89*, 1581; Cornils/Herrmann-1.

C₂ symmetry has gained considerable importance in cat. with reference to *C_2-symmetric *ligands. Molecules showing this type of symmetry contain the symmetry element C_2 which stands for a rotatory axis of the order of two. In general, a symmetry axis of the order of n is an axis such that rotation of a molecule by an angle of $360°/n$ about such an axis produces a superposable entity. This means that rotation by an angle of 180° about a C_2 axis leaves a C_2-symmetric molecule apparently unchanged. **F** symétrie C_2; **G** C_2 Symmetrie.

H.W. GÖRLITZER

Ref.: B. Testa, *Principles of Organic Stereochemistry*, Dekker, New York 1979.

C₃-symmetric ligands N-containing acyclic and cyclic C_3-symmetric cpds. such as hydrotris(pyrazlolyl)borates or 1,4,7-triazacyclononanes have been applied as *ligands in model Cu and Fe *complexes, to clarify the function of *active sites of non-heme oxidizing *enzymes and oxygen-carrier metalloproteins, e.g., *methane monooxygenase or hemocyanin. Accordingly, biomimetic Fe, Cu, and Mn *complexes with achiral and chiral C_3-symmetric *ligands were developed for cat. *oxidations with 30 % H_2O_2 or *tert*-butyl perbenzoate (allylic oxidation of cycloalkanes, *epoxidation of olefins, *oxidation of alcohols to aldehydes). With *chiral Cu-based cats., derived from a chiral tris(*oxazoline) ligand, up to 88 % *ee was achieved in allylic oxidation. For C_3-symmetric ligands of higher order rotational symmetry, advantages in *enantioselective metal-mediated organic reactions were proposed. Cu-based systems with polypyrazolyl ligands were reported to be efficient cats. for *carbene and nitrene transfer to form *cyclopropanes/-propenes and aziridines. Chiral, tetradentate trialkanolamines were used successfully in enantioselective sulfide oxidations with Ti cats. Recently, dimeric Zr complex ligands proved to be highly selective cats. for the desymmetrization of *meso* epoxides in the presence of azidotrimethylsilane. **F** ligands avec C_3 symétrie; **G** C_3-symmetrische Liganden. K. RÜCK-BRAUN

Ref.: *Angew.Chem.Int.Ed.Engl.* **1998**, *37*, 248; *Nach. Chem.Tech.Lab.* **1998**, *46*, 646; *Synlett* **1995**, 1245; *Organometallics* **1998**, *17*, 1984; *Pure Appl.Chem.* **1998**, *70*, 1041.

Cabosil (Cab-O-Sil) → silica

CACD → computer-aided (assisted) catalyst design

CAD computer-aided (assisted) design, → computer-assisted catalyst design, → computational catalysis.

CAE computer-assisted (aided) engineering

calcining (calcination) a heat treatment of cat. *precursors in an oxidizing atmosphere for a couple of hours. The temp. is placed at a level to fully convert the precursor into oxides, i.e., to remove all anions except oxygen, and all physical or chemically bound *solvent molecules (mostly water). C. thus transfers the primary structure, generated, e.g., by *precipitation, into a composition of higher thermodynamic stability. **F** calcination; **G** Calcinierung.   C.D. FROHNING

Ref.: McKetta-1, Vol. 5, p. 440; Farrauto/Bartholomew, p. 100; Ullmann, Vol. A5, p. 351.

calix[n]arenes are torus-shaped *macrocyclic cpds. based on *meta*-cyclophanes, *n* being the number of monomers. Similar to *cyclodextrins, cs. can bind reactive or cat. groups, *ligands as inclusion molecules in *host cavities, or *enzyme precursors to one or both of the toroidal faces. Water-soluble or *PTC- active c. are also known. **E=F=G**.

B. CORNILS

Ref.: *Acc.Chem.Res.* **1983**, *16*, 161; *Organic Synthesis, Coll.Vol. 8*, Wiley, New York 1993, p. 75; Vicens, Böhmer, *Calixarenes – A Versatile Class of Macrocyclic Compounds*, Kluwer, Dordrecht, 1991; Ertl/Knözinger/Weitkamp, Vol. 2, p. 901; C.D. Gutsche, *Calixarenes*, RSC No. 242, Cambridge 1998.

calnexin As an endoplasmic reticulum (ER) membrane-resident protein-folding *chaperone, calnexin binds to the terminal glucose residues of glycoproteins, retaining them in the ER until they are fully folded. Glucosidases in the ER remove the terminal glucose residues, but unfolded glycoproteins can be reglucosylated by an ER-resident glucosyltransferase that recognizes unfolded glycoproteins. **E=F=G**.   P.S. SEARS

Ref.: *Trends Cell Biol.* **1997**, *7*, 193.

calorimetry method for the determination of heat effects involved in a known change in state of a material using a calorimeter. This can involve a change in *phase, temp., press., volume, chemical composition, phase transition, boiling, sublimation and vaporization, crystallite structure inversion, chemical reaction, or any other property of the material which is associated with the change in heat content. A calorimeter measures heat with reference to the change of temp. or the state of a calorimeter substance. Calorimeters are

classified as isothermal, adiabatic and heat flow calorimeters.

In catalysis, differential thermal analysis (DTA) and *differential scanning calorimetry (DSC) are used mainly. Both are normally coupled with *thermogravimetry (TG) resp. differential thermogravimetry (DTG). For measurement of very small changes in heat, exchange methods of microcalorimetry are used. In DTA the temp. of the sample is measured as a function of time, and a heating or cooling curve is recorded. The sample temp. is continuously compared with a reference material temp., the difference in temp. being recorded as a function of furnace temp. or time (assuming that the furnace temp. rise is linear with respect to time). Experimentally, this is accomplished by employing a furnace that contains a sample holder or block containing two symmetrically located and identical chambers. Each chamber contains an identical thermocouple and the temp. difference between them is recorded. The DTA signal then show a exothermic or endothermic signal relative to furnace temp. or time, usually giving only qualitative information about heat exchange of the sample. Quantitative analysis is possible with quantitative DTA, which works with defined thermic resistance in the experimental setup.

DSC uses in principle the same experimental setup as DTA. Unlike to DTA the sample temp. is maintained isothermal with a reference substance by supplying heat to the sample or reference material. The amount of heat required to maintain these isothermal conds. is then recorded as a function of time or temp. (enthalpy curve). The parallel measurement of DTG (measurement of weight changes of the cat. sample) enables to decide from where a DTA signal originates, e.g., from reaction heat or from a change in phase of the cat. In addition to DTA, DSC, and DTG it is possible to measure gas composition with a *mass spectrometer or *gas chromatograph (cf. also *temp.-programmed reaction spectroscopy [TPRS]). **E=F**; **G** Kalorimetrie.   R. SCHLÖGL, I. BOETTGER

Ref.: W.W. Wendlandt, *Thermal Methods of Analysis*, Wiley, New York 1974.

Calvin cycle The *photosynthetic pathway in plants occurs in two stages, 1: the so-called "light reactions," in which light is harvested and used to generate *ATP and *NADPH; and 2: the "dark" reactions, in which CO_2 is reduced to carbohydrate. The central pathway of the dark reactions is the Cc. (reductive pentose phosphate cycle). The carbon-fixing *enzyme of this cycle is ribulose bisphosphate carboxylase (RuBP carboxylase, Rubisco), which converts the C_5 sugar ribulose 1,5-bis-phosphate plus CO_2 into two molecules of the C_3 molecule 3-phosphoglycerate. This glycerate is converted to *G3P, which requires 1 ATP and 1 NADPH. A complex system of *disproportionation reactions follows, and in sum, six molecules of CO_2 provide one hexose phosphate. This also requires 18 ATP and 12 NADPH per hexose, which are provided by the "light reactions" of *photosynthesis (see *electron transport). **E=F=G**. P.S. SEARS

CAM computer-assisted (aided) manufacture

Candida sp. lipases The lipases from different _Candida_ yeast species (including _C. cylindracea_, CCL) and the recombinant form of _C. antarctica_ lipase (SP435) are commonly used in organic synth. The *enzyme CCL was also used in organic *solvents for the *resolution of racemic open-chain secondary alcohols via *transesterification. This *enzyme was also applied to the resolution of menthol in aqueous and organic solvents. The enzyme accommodates bulky esters in its active site. For transesterifications the length of the alkyl chain of the acyl group influences the efficiency and *selectivity, and a clear minimum was observed for transesterification or *esterification of six-carbon acyl moieties. The enzyme showed better *enantioselectivity in hydrophobic organic solvents than that in hydrophilic solvents. High selectivity of the enzymes for bulky substrates can be seen in a number of CCL-cat. enantioselective transformations. In the acylation of other polyols, CCL is selective for the less hindered primary hydroxy, or the benzylic hydroxyl group.

H_2O_2 is also a substrate for CCL, and fatty acids have been converted to peracids for use in *epoxidation. **E=F=G**. C.-H. WONG

$R = CH_3(CH_2)_n-$
$n = 5 - 13$
$R' = C_{13}H_{27}$

Canmet process processing of heavy Canadian oils and residues over cats. (coal or metal doped coal), cf. *VEBA proc., *coal liquefaction, *coal hydrogenation. B. CORNILS
Ref.: Ullmann, Vol. A 18, p. 76.

Cannizzaro reaction (C. dismutation, C. disproportionation) is the base-cat. reaction of two molecules of aldehydes without α-hydrogen (otherwise *aldol addition occurs). One molecule of aldehyde is *oxidized to the corresponding acid, while the other is *reduced to the primary alcohol, acc. to R-CHO + R-CHO + OH$^-$ → RCH$_2$OH + RCOO-. Normally, best yields of alcohol and acid are 50 % each, however by reacting formaldehyde with other aldehydes (crossed Cr.) high yields of alcohols can be obtained. Such crossed Crs. are of some importance in the *enzymatic prod. of ethanol. The mechanism of the Cr. involves a *hydride shift of the former carbonyl C-H bond, which is facilitated after the attack of hydroxide ions. Of commercial interest are Crs. for the prep. of furfuryl alcohol or of neopentylglycol. The *Tishchenko reaction proceeds with Na or Al *alkoxides as cats. and yields esters. **F** réaction de Cannizzaro; **G** Cannizzaro-Reaktion.
Ref.: March. M. BELLER

capillary condensation The effect manifests itself in a sorption *hysteresis on *mesoporous *adsorbents. The first section of the sorption *isotherm curve is identical with the corresponding isotherm given by a chemically similar nonporous solid having the same *surface area. Thus, *monolayer *coverage and the initial stages of multilayer adsorption occur on the mesopore walls in the same manner as on the open surface. At a certain partial press. p the adsorption curve deviates from the normal behavior and is shifted upwards to higher partial press. This upward swing is associated with the filling of mesopores by *capillary condensation*. Its occurrence is due to the formation of a curved liquid-like meniscus with surface tension σ between layers of condensed gas in opposite walls of pores of particular size and shape. According to the *Kelvin equ. the gaseous *phase which is in equil. with the concave surface of the liquid has a lowered vapor press. compared to the free liquid surface. As a consequence condensation occurs at lower vapor press. Therefore upon reducing the relative press. p/p_0 for the desorption branch, desorption occurs at a lower vapor press. Together with the disappearance of the liquid meniscus, both isotherms then coincide. **F** condensation capillaire; **G** Kapillarkondensation.   R. Schlögl, M.M. Günter

capillary impregnation a special procedure to generate impregnated cats., by which the active comp. is deposited preferentially on the inner *surface of the carrier (*support). The *porosity of a carrier is based on a system of pores of different length and diameter, the walls of which represent the inner *surface. Generally the inner surface (determined, e.g., by the *BET method) by far exceeds the outer (geometrical) surface of a carrier, even in powder form, and can be as high as 10^3 m^2/g, provided that the fraction of pores with small diameter is sufficiently high. The *micropore volume diffusion follows three regimes: restricted or configurational diffusion (below 1 nm), Knudsen (1–100 nm), bulk (>100 nm). Cf. also *impregnation, *incipient wetness method, *pore volume impregnation, *SLPC. **F** imprégation capillaire; **G** Kapillarimprägnierung.
Ref.: *CHEMTECH* **1973**, *3*, 498.   C.D. Frohning

carbamoyl methyl viologen (CAV) → enoate reductase

carbamoyl phosphate synth. from bicarbonate, glutamine, and *ATP. It is an activated one-carbon cpd. and a key intermediate in pyrimidine and urea *biosynthesis. It is also a phosphoryl donor for ATP regeneration from *ADP.   C.-H. Wong

carbenes as ligands Divalent C species :CR^1R^2 exhibit σ-donor and π-acceptor properties upon binding to *transition metals. In principle, a double bond analogue structure results, depending on the nature of R^1 and R^2. With regard to the reactivity of the metal-bonded C fragment, electrophilic (Fischer) and nucleophilic (Schrock) carbenes are discussed. In many cases the carbene (alkylidene) ligand can be transferred to other substrates; *cyclopropanation and alkene *metathesis are prominent examples. However, metal-carbene *complexes often react with cleavage of the metal-C bond. For this reason common carbenes with R^1,R^2 = H, alkyl, aryl, alkoxy, or amino are not considered as ligands that survive the standard reaction conds. of organometallic cat. The classification into *Fischer and *Schrock carbenes does not hold for N-heterocyclic carbenes of the imidazole and imidazolidine series, which were proven to be good σ-donors and virtually not π-acceptors and resemble electron-rich *phosphines regarding their spectroscopic data. N-heterocyclic carbenes exhibit promising properties such as directing ligands in hom. cat., e.g., for alkene metathesis, *Heck olefination, *hydroformylation, *hydrogenation, *isomerization, and (asymmetric) *hydrosilylation. **F** carbènes comme ligands; **G** Carbene als Liganden.   T. Weskamp
Ref.: *Angew.Chem.Int.Ed.Engl.* **1997**, *36*, 2162; F.Z. Dörwald, *Metal Carbenes in Organic Synthesis*, Wiley-VCH, Weinheim 1998.

carbenium cation, carbonium cation → carbocation

Carberry type reactor special variant of lab. reactors, restricted to larger cat. particles.

The flow rate of the fluid through the cat. bed is not known (and the temp. cannot be measured); various basket devices (containing the het. cat. in baskets which are part of the stirrer) have been proposed to optimize the fluid-cat. contact. The cr. is restricted to low endo- or exothermal reactions. **F** réacteur selon Carberry; **G** Carberry-Reaktor. B. CORNILS

Ref.: Anderson/Boudart, p. 131; *AIChE J.* **1994**, *40*, 862; G.B. Tatterson, *Scaleup and Design of Industrial Mixing Processes*, McGraw-Hill, New York 1994.

carbocations   cations with an even number of electrons with the positive charge formally located on one or more C atoms. Two types are distinguished. *Carbenium ions* (cs., IUPAC 1983, formerly: carbonium ions), are "classical" cs., with a tri-coordinated, sp^2-hybridized C-atom. C. stability is strongly dependent on the substituents (hyperconjugation, field effect); the stability being $R_3C^+ > R_2HC^+ > R_1H_2C^+$. Heteroatoms also stabilize the cs. (resonance). Cs. are stable in *superacid solutions in solvents of low nucleophilicity; some cs. give stable salts. Spectroscopic analysis is performed with, e.g., *NMR, *IR, *XPS. The ^{13}C-NMR chemical shift provides excellent evidence for positively charged carbon atoms (Me_3C^+: $\delta = -135.4$). Reactions of the electrophilic cs. can be classified as *additions (to simple nucleophiles such as halide, hydride, hydroxide, but also to double bonds), cleavage (with the simplest case being elimination of a proton to give an unsaturated cpd., but also cleavage into alkenes and cs.), and rearrangements. These reaction types are integral to many commercial procs., such as *alkylation, *acylation, *cracking, and *isomerization. Cs. are formed either when a leaving group takes a pair of electrons away or by protonation of an unsaturated system, so that *Lewis or Brønsted acids are suitable reagents to obtain cs. *Acid cat. (hom. or het.) is thus used in the above-mentioned procs. Many reactions of cs. yield cs. as prods., and they can be considered as chain carriers, making it sometimes difficult to distinguish between catalytic and initiation reactions.

Alkanium ions (as., IUPAC 1995, formerly carbonium ions) are "non-classical" carbocations with a penta-coordinated carbon and a 2-electron-3-center bond; as. are detectable in the gas phase by *MS. As. rapidly decay to alkenes (or hydrogen) and cs., and their formation requires a very strong *Brønsted acid in order to protonate an alkane. Their existence as *transition states in cat. reactions has been postulated from prod. distributions. Theory-based calcns. of structure and stability of small (C_1-C_4) cs. and as. are reported. **E=F**; **G** Carbocationen.

F. JENTOFT

Ref.: March; *Angew.Chem.Int.Ed.Engl.* **1995**, *34*, 1393.

carbometallation   C-C bond forming reaction in which organometallics are added to alkenes or alkynes with simultaneous formation of a new M-C bond. Besides that, the term c. is most often used for reactions in which the *addition of the organometallics to the alkenes or alkynes is controlled. Important are *addition reactions involving alkynes, since it is easier to obtain discrete alkenyl metal intermediates rather than alkyl metal species produced by addition to alkenes. The use of organo Al, Mg, and Zn cpds. as addition reagents is widespread. In the presence of zirconocene dichloride as cat., they are useful systems to functionalize alkenes. Specially, alkenyl metal species, which are able to form carbocycles in a subsequent reaction step, are interesting in organic synth. The reaction of diynes or enynes with zirconocene dichloride/*n*-BuLi (*Negishi reagent) produces zircona-cyclopentadienes and -cyclopentenes which can be transformed into cyclic butadienes and cyclopentenones, respectively. Carbopalladation (cf. *palladacycles) is of importance (e.g., *Heck reaction). Pd-cat. cascade (*tandem) reactions allow for the formation of cyclic polyenes. The *Ziegler-Natta type *polymerization of alkenes is a c. reaction where the addition of the organometallics is uncontrolled and therefore the termination is statistical. **E=F**; **G** Carbometallierung.   P. HÄRTER

Ref.: Trost-Fleming, Vol. 4, p. 965; *Acc.Chem.Res.* **1987**, *20*, 65; *Synthesis* **1988**, 1; *Chem.Rev.* **1996**, *96*, 365.

carbon The following examples illustrate the multiple role of carbon in cat. There are two major fields where C is present: first, most of the industrial reactions (except the NH_3 and H_2SO_4 synth.) are based on the *hydrogenation, *reforming, and/or the *oxidation of *hydrocarbons; secondly, C materials are used as carbon *supports and for some cases as cat. itself. The materials which are mostly used as C supports are *activated carbon (charcoal) and *carbon blacks, mainly due to their high *surface areas, a well developed *porosity, and an inert behavior. Pt and/or Pd are normally deposited on these supports for hydrogenation of *hcs. They are not recommended in oxidation reactions because the combustion limits their use. The inertness of C supports is an advantage for some reactions like *HDS where it is known that their use improves the reaction compared to the *alumina conventional supports.

Carbon is mainly regarded as a deactivation agent, specially in reforming and cracking reactions conducted with *zeolites. The excessive cracking of the hcs. causes the appearance of a dense carbonaceous layer which deactives the cat. In some *oxidation reactions, a similar layer is found to be formed on oxide based cats., e.g., in the *oxidative *dehydrogenation of ethylbenzene. The carbonaceous layer is oxidized with air and/or steam and thus regenerated for further use of the cat.

Another minor application of these materials is their use as *bulk cats. themselves. A few examples are the prod. of phosgene, the oxidation of SO_2 and NO, or the manuf. of *sulfur chlorides. Other reactions are oxidative dehydrogenations of hydrocarbons such as the above-mentioned reaction of ethylbenzene to styrene, where the carbonaceous layer is thought to be actually the active site. **F** carbone; **G** Kohlenstoff.   R. SCHLÖGL, E. SANCHEZ-CORTELON
Ref.: *Ind.Cat.News* **1998**, (9), 7; Augustine; Ullmann, Vol. A5, p. 95; Kirk/Othmer-1, Vol. 4, p. 949; Kirk/Othmer-2, p. 203.

carbon black Cbs. are solids industrially synthesized by the combustion of hydrocarbons in oxygen poor atmospheres. Depending on the reaction conds. a variety of materials with different *surface areas and *porosities can be achieved. Although the main applications of these materials are still in printer ink and tire prod., metals *supported on carbon blacks can also be used for cat. *hydrogenation. A disadvantage of these materials is the small particle size (from 8 to 500 nm), but the possibility of producing pellets confers them their potential use as cat. supports. **F** noir de carbone; **G** Ruß.
                        R. SCHLÖGL, E. SANCHEZ-CORTELON
Ref.: Augustine, p. 166; Kirk/Othmer-2, p. 207; McKetta-1, Vol. 6, p. 187.

carbon dioxide as a building block

CO_2, a cheap and abundant C-containing raw material, has attracted attention in recent years. Its multipurpose and coincident application possibilities as substrate, as solvent (*supercritical phase cat.), as reactant (*carbon dioxide reforming), and as extracting agent are a particularly intriguing prospect. CO_2 is a potential building block for C-C chains or as a competitive source of C in the chemical industry (e.g. in *CO_2 reforming) and appears very attractive, although the molecule is rather inert and its reactions are energetically highly unfavorable. Cats. play a crucial role in CO_2 convs., and suitable cats. can realize interactions between CO_2 and a substrate S on a *transition metal centre M. CO_2 has demonstrated surprising versatility by exhibiting a great variety of *coordination and reaction modes in its homo- and polynuclear metal *complexes. Most cat. reactions involving CO_2 activation proceed via formal *insertion of CO_2 into highly reactive M-X bonds with formation of new C-X bonds. Coordinated CO_2 as in stable complexes is not necessarily required. Reactions are generally initiated by nucleophilic attack of X at the *Lewis acidic C atom of CO_2. Procs. based on CO_2 are known for the synth. of hydrocarbons (via *Fischer-Tropsch synth.), energy-rich C_1 molecules (CH_4, CH_3OH, HCOOH, etc.), and fine chemicals containing functionalities such as -COO- (acids, esters, lactones), -O-COO- (organic carbonates), -N-COO- (carbamates), or -N-CO-

(ureas, amides). *Electro- and *photocatalytical reductions of CO_2 can also be cat. by metal complexes. **F** dioxide de carbone comme matière première; **G** Kohlendioxid als Rohstoff.

E. DINJUS

Ref.: T. Inui et al. (Eds.), *Studies in Surface Sciences and Catalysis*, Elsevier, Amsterdam 1998; van Eldik/Hubbard; *Energy & Fuels* **1996**, *10*, 305; *Coord.Chem .Rev.* **1996**, *153*, 257; Cornils/Herrmann-1, p. 1048 and -2, p. 486; *Adv.Inorg.Chem.* **1995**, *43*, 409; *J.Mol. Catal.A:* **1999**, *141*, 193.

carbon dioxide reforming

reaction of CO_2 with aliphatic hydrocarbons acc. to $C_nH_{2n+2} + CO_2 \rightleftharpoons 2n\ CO + (n+1)\ H_2$. The reaction is more endothermic than *steam reforming. Whereas the steam reforming of methane yields *syngases with an H_2/CO ratio of 3, CO_2 reforming of CH_4 results in a ratio of 1, which may be of interest for some syngas-consuming reactions. The equil. composition is influenced by the *WGSR. Cdr. needs higher temps. (with the danger of *coke formation); advantageously cats. may be used (Ni on *supports such as *alumina or MgO). Cdr. has been used in the *reduction of iron ores (Midrex, Purofer procs.). Cf. also *cat. reforming, *steam reforming. **E=F**; **G** CO_2 Reformierung.   B. CORNILS

Ref.: *Chem.Eng.Sci.* **1988**, *43*, 3049; Ullmann, A14, p. 558f; *CHEMTECH*, **1999**, *29*(Jan.), 37.

carbonic anhydrase

a *metalloenzyme (Zn^{2+}) that cat. the *hydration of CO_2 to carbonic acid.

$$\text{CO}_2 + \text{H}_2\text{O} \underset{\text{Carbonic anhydrase}}{\rightleftharpoons} \text{HCO}_3^- + \text{H}^+$$

It is an extremely efficient *enzyme, with a k_{cat} of approx. $10^6\ s^{-1}$ and k_{cat}/K_m of almost $10^8\ s^{-1}M^{-1}$ in the hydration direction. This puts the enzyme near the *diffusion controlled limit. **E=F=G**.   P.S. SEARS

carbon molecular sieves → carbon supports

carbon monoxide an important *ligand in hom. cat., commercially manuf. either by *partial oxidation of coal, gaseous, or liquid fossil feedstocks (mostly hydrocarbons) or by separation from *syngas (the mixture of CO and H_2). Careful purification of CO is required if used for cat. purposes (e.g., by *Claus procs.). Some technical procs. substitute CO or syngas by formic acid or formaldehyde (*liquid syngas).

The high π-acceptor capacity of CO causes the stability of carbonyl *complexes of low valent late *transition metals, many of which are cats. in different *carbonylation reactions, e.g., $Ni(CO)_4$, $Fe(CO)_5$, $Co_2(CO)_8$, $Rh_6(CO)_{16}$. *IR spectroscopy of C-O moieties is important for the analysis of bond order and symmetry of carbonyl complexes.

A large number of large-scale commercial procs. use CO in cat. reactions involving the C-C bond formation between substrate and CO as, e.g., in *hydroformylation, *hydrocarboxylation, *carbonylation, alcohol *homologation, *Fischer-Tropsch synthesis (*FTS), *Reppe syntheses, etc. (*CO reactions). The CO content of syngas influences *rate and *selectivity of the reactions, e.g., a high partial press. of CO increases the selectivity for *n*-products in the *hydroformylation reaction (*n/iso* ratio). In the CO *hydrogenation (FTS) the $CO:H_2$ ratio, together with the choice of cat., press., and temp., determines the prod. spectrum ranging from methane over alcohols to hydrocarbons and waxes. Most of the cat. procs. involving CO use temps. from 100 to 350 °C and press. between 0.5 and 50 MPa. **F** monoxyde de carbone; **G** Kohlenmonoxid.   F. RAMPF

Ref.: Falbe-1 through –4.

carbon monoxide reactions are mostly *hom.* cat. reactions which use CO as a *C_1 building block (cf. *carbon monoxide). Due to the properties of CO as *ligand in *complexes (e.g. with Ni, Co, Fe, Ru, Rh, Pd) they are particularly used for *carbonylations, which can formally be described by equs.

$$RX + CO \rightarrow R\text{-}CO\text{-}X \tag{1}$$
$$H_2C{=}CH_2 + CO + HX \rightarrow H_3C\text{-}CH_2\text{-}CO\text{-}X \tag{2}$$
$$R\text{-}X + CO + Nu\text{-}H \rightarrow R\text{-}CO\text{-}Nu + HX \tag{3}$$
$$R\text{-}X + CO + Y^- \rightarrow R\text{-}CO\text{-}Y + X^- \tag{4}$$

Some reactions are applied commercially, i.e., the carbonylation of methanol (equ.1), the *hydroformylation (2), and the *carbonylation of organic halides (equs. 3,4). Aromatic nitro cpds. can be cat. carbonylated to isocyanates. Aniline derivatives are obtained by cat. *reduction of aromatic nitro cpds. with CO. Recently the use of CO as building block for *alternating copolymers has been introduced by Shell. In organic and fine chemical synth. the *Pauson-Khand reaction represents a method to obtain substituted cyclpentenones from alkenes, alkynes, and CO. In *het.* cat. the *Fischer-Tropsch synth., the *methanation, and the *water gas shift reaction play important roles. **F** réactions de l'oxyde de carbone; **G** CO-Reaktionen.

B. ZIMMERMANN, M. BELLER

Ref.: Weissermel/Arpe, p. 137; Cornils/Herrmann-1,2; Falbe-1 through -4.

carbon supports

carbon supports are included in the family of cat. *supports. The main application of a cat. support is to maintain the cat. active phase in a highly dispersed state without intervening in the reaction. Nevertheless, it is known that cat. supports are not so inert and that *interactions between the active phase and the support (cf. *MSI) take place. This can have both positive and negative influences in the cat. reaction. The most used carbon supports are *activated carbon and *carbon blacks. These materials, consisting mainly of C and heteroatoms like H, O, and N, are excellent supports. The metals which are normally supported are Pt and Pd in *hydrogenation reactions. The carbon supports are oxidized prior to anchoring the metal *precursors. A variety of oxygenated basic and acid groups are formed. Normally, this is conducted by burning selectively a part of the carbon with oxygen or with acids or basic aqueous media. The aqueous metal solutions then have a better wettability and (depending of the basic-acid character of the predominant carbon groups) the pH of the metal solution and the ionic character of the metal solution better control the degree of metal dispersion. In some cases, the degree of dispersion achieved is reported to be near 99 %. A special group of activated carbons are the carbon *molecular sieves. These materials show the largest surface areas of any commercial cat. support. The main porosity in these materials consists of *micropores. The amorphous structure of the usual carbon supports limits their use to non-oxidizing atmospheres. The combustion of the support can be an advantage when recovering the precious metal active sites. Cf. *supports. **F** supports de charbon; **G** Kohlenstoffträger.

R. SCHLÖGL, E. SANCHEZ-CORTELON

carbonylation

carbonylation introduces the C=O group into organic molecules by hom. cat. reactions. In general, CO is used as a building block (*carbon monoxide, *carbon monoxide reactions). In addition are known cs. of organic halides (Ar-X, Bn-X, etc.: X = halogen) to the corresponding aldehydes, acids, or esters, *methoxycarbonylations, *amidocarbonylations to *N*-acylamino acids from aldehydes or alkenes and acetamide, cs. of aromatic and aliphatic amines to *N*-formyl derivatives and ureas as well as *oxidative cs. of alcohols, alkenes, and amines (dialkylcarbonates, carbamates, and dialkyloxalates; cf. the corresponding procs. of *Ube). Aromatic nitro cpds. may be carbonylated cat. to isocyanates in a one-step proc. or via carbamate intermediates. Examples are the procs. of *American Cyanamide (benzene), *Asahi (aniline), *ARCO (ethylurethane), *BASF (hydroquinone), *Halcon (acetic or formic acid, vinyl acetate), *Hoechst or *Monsanto (AAA), *Leonard (formic acid), *Monsanto (adipic acid), *Mitsubishi, etc. Cf. other keywords such as *cyclocarbonylation, *oxidative c., and others. **E=F, G** Carbonylierung.

B. ZIMMERMANN, M. BELLER

Ref.: Falbe-1 and -3; Weissermel/Arpe; Colquhoun/ Thompson/Twigg; Cornils/Herrmann-1, p. 29; Cornils/ Herrmann-2, p. 271, 373; Ullmann, Vol. A5, p. 217; Kirk/Othmer-1, Vol. 5, p. 104; *Chem. Tech.* **1971**, 600; *J. Mol. Catal. A:* **1997**, *127*, 33,95.

carbonylation of alkynes was applied after *Reppe's work and *BASF's acrylic acid proc., starting with acetylene. Methacrylic acid and its esters could in principle be manuf. by a similar c. of propyne. A class of highly active hom. Pd cats., developed at Shell, made the c. of higher alkynes efficient. Thus methyl methacrylate (MMA) can be produced from propyne (from the C_3 stream of naphtha crackers), CO, and methanol (cf. *methoxy-carbonylation). Characteristics of the new proc. are the cat.'s high *activity and *selec-tivity to MMA. The Pd cat. consists of a cat-ionic Pd(II) center ligated with a phosphine comprising a 2-pyridylphosphine moiety and being associated with weakly coordinating an-ions. Typically, reaction rates of 50 000 mol of propyne converted per mol of Pd and per h can be observed with 99 % selectivity. (50 °C/ 1 MPa). With Fe or Rh cats. alkynes yield aro-matics (*cyclization). **F** carbonylation des al-cynes; **G** Alkincarbonylierung. E. DRENT

Ref.: *J. Organomet. Chem.* **1994**, *457*, 57; *Chem. Ind. (London)* **1968**, 1732; Falbe-3a, p. 94.

carbonylative cyclization → cyclocar-bonylation, *cyclization

carbonyl reductase → oxidoreductases

carboxylation introduction of a carboxylic group -COOH into an organic cpd., i.e., the synth. of carboxylic acids derived from unsub-stituted cpds. such as the reaction of metalor-ganics like *Grignard reagents or Li organic cpds. with CO_2 or the Kolbe-Schmitt reaction. Oxidative c., the oxidative transformation of aliphatic aldehydes or ketones (first-step oxida-tion prod. from alkane oxidation with air, cf. *oxidation of hcs.) to carboxylic acids is com-mercialized. In the case of longer-chain deriva-tives like fatty acids (C_9-C_{15}) or the *oxidation of keto derivatives of *waxes (C_{30+}) *selectivity drops and chain degradation can occur. Analo-gous alkyl aromatic cpds. are converted to their aromatic carboxylic acid derivatives via oxida-tion of methyl or alkyl groups with oxygen and Co/Mn acetate cat. This reaction can be applied to synth. arylpropionic or fluorinated acids, si-lylated esters, and also β-amino acids as pro-ducts derived from functionalized alkenes. An-other important reaction is the hydrocarboxyl-ation of butadiene. For *hydrocarboxylation cf. the appropriate keyword. **E=F**; **G** Carboxylie-rung. R.W. FISCHER

Ref.: Cornils/Herrmann-1, Vol. 1, pp. 169, 187.

carboxyl proteases → aspartyl proteases

carboxypeptidases exopeptidases that cat. the release of C-terminal L-amino acid from peptides. Three c. have been used in protein chemistry: c. A (a Zn^{2+} *enzyme that is specific for neutral or acidic amino acids); c. B (specific for basic amino acids), and c. Y (a serine-type enzyme from Bakers yeast that accepts a broad spectrum of substrates, in-cluding Pro). A c. C from orange leaves has been shown to have a broad substrate specifi-city useful for the *kinetically controlled synth. of peptides. These cs. may also be used in the *resolution of unnatural amino acids. For example, c. Y cat. the *hydrolysis of trifluoroacetyl 7-fluorotryptophan to give the corresponding L-amino acid. **E=F**; **G** Carboxypeptidase. C.-H. WONG

Carilon tradename of an *alternating eth-ylene-carbon monoxide *copolymer from Shell.

carnitine acetyltransferase → acetyl CoA regeneration

Car-Parrinello (CP) method In order to gain detailed insight into the *reaction mech-anism of cat. procs., it is necessary to study the time evolution of the system using *molecular dynamics (MD) simulation. Because MD calcns. must be carried out at time steps of the order of femtoseconds (10^{-15} s), a large number of energy and gradient calcns. are necessary to study, e.g., the reaction profile of a cat. reaction.

This is not possible for *ab-initio and *DFT calcns., which are necessary, however, to treat electronic procs. that involve *transition metals. The CP method takes the converged wavefunction at the starting point and treats the orbitals as variables with fictive masses, which makes the calcn. of the forces (first derivatives) much less time-consuming. Although the wavefunction at each time step (except in the beginning) is only approximate, the error which is introduced in this way can be controlled by suitable MD parameters and thus becomes negligible. CP simulations in conjunction with DFT methods are currently used to gain insight into the mechanism of transition metal cat. reactions, which would otherwise not be possible. **F** méthode de Car-Parinello; **G** Car-Parinello-Methode.

Ref.: *Phys.Rev.Lett.* **1985**, *55*, 2471.   G. FRENKING

carrier   1: in general catalysis, a carrier is a *support; 2: in molecular biology a carrier is a *vector.

carrier-free catalyst   → bulk catalyst

carry-over   → leaching

Casale process   a special variant of NH_3 synth. The reaction is exothermic and accompanied by a decrease in volume (at constant press.): $N_2 + 3/2 H_2 \rightarrow NH_3$ (ΔH^0_{700K}= –52.5 kJ/mol). The value of the equil. constant therefore increases as the temp. is lowered, and the equil. NH_3 conc. increases with increasing press. (cf. *ammonia synth.). Passing through a cat. bed, the temp. of the NH_3 synth. gas rises, diminishing the conv. The construction of any converter therefore aims at a limitation of the peak temp. by an efficient removal of the heat of reaction. A number of solutions of the problem have been of-

fered, and several of them have been technically applied.

When the control of temp. is not achieved via the stepwise addition of cold gas (axial flow quench converter), an internal heat exchanger is used (indirect cooling). The heat exchanger in the Casale reactor is placed centrally along the axis of the reactor in the middle of three cat. beds. By special construction of the cat. bed *supports and the gas distribution, a mixture of radial and axial gas flow distribution is achieved (Figure).

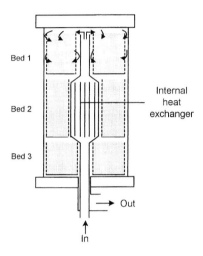

The result of the design is a tall vessel of relatively small diameter. The Casale reactors are made of a special steel, allowing press. up to 100 MPa (cf. *Linde LAC proc.). **F** procédé de Casale; **G** Casaleverfahren.   C.D. FROHNING

cascade molecules   part of the nomenclature for *dendrimers.

Ref.: *J.Polym.Sci.Part A:Polymer Chem.* **1993**, *31*, 641; Mulzer/Waldmann, p. 391.

cascade reaction   1: → domino reaction; 2: enzymatic cascaded reactions are a series of reactions in which the prod. of the first cat. the second reaction, and the prod. of the second cat. the third reaction, and so forth. The primary advantage of such a scheme is that it allows for the tremendous amplification of a small signal: suppose each *enzyme E_1 cat.

the production of 10 molecules of product, E_2. Suppose also that E_2 is also an *enzyme, and that each enzyme E_2 goes on to cat. the production of 10 molecules of product P_3. Then for each enzyme E_1, 100 molecules of prod. P_3 have been produced, and the more levels to the cascade, the greater the signal amplification. Blood clotting is an example of a multilevel cascade reaction, as are the *protein kinase cascades, commonly used in cellular signal transduction pathways. **F** réactions en cascade; **G** Kaskadenreaktionen.   P.S. SEARS

caspases   *thiolproteases that cleave specifically at aspartic acid residues. Due to their homology to the interleukin β-converting *enzyme (ICE, caspase 1), they are also called ICE-like proteases. Unlike most thiolproteases, which contain an *active site triad of Cys, His, and Asn, the cs. have an active site dyad of only Cys and His: the active site histidine hydrogen bonds to a backbone carbonyl oxygen in the c., rather than to a sidechain carbonyl of Asn as in other thiolproteases. The cs. are highly specific for aspartic acid in the P_1 position (the carboxyl side of the scissile amide bond), but their *specificity at other positions varies. The cs. are known as key mediators of inflammation and apoptosis (organized cell death). They are produced in an inactive *proenzyme form and, in all cases, cleavage at specific aspartate residues generates the active enzyme. Caspase *cascades are believed to be involved in apoptosis. **E=F=G**.
P.S. SEARS

Ref.: *Biochem.J.* **1997**, *326*, 1; *Chem.Biol.* **1998**, *5*, R97.

cat.   abbr. for "catalyst", "catalytic", "catalytically", "catalyzed", or "catalysis" in keywords, reaction schemes, or formulae. **F=G=E**.

catabolism   refers to *biosynthetic pathways designed to degrade biomolecules for excretion, recycling, or energy generation. Nucleoside degradation to the sugar and base, lipid degradation to *acetyl CoA, and protein *hydrolysis to amino acids are catabolic processes. C. is contrasted with *anabolism. **F** catabolisme; **G** Katabolismus.   P.S. SEARS

Catacarb process   removes H_2S/CO_2 from gases or gas mixtures by absorption with hot solutions of potassium carbonate or borate, containing an amine cat. (similar: UOP Benfield proc.). **F** procédé Catacarb; **G** Catacarb-Verfahren.   B. CORNILS

Catacol technology   → IFP's *catalytic distillation technique, e.g. *IFP ether proc.

Catadiene   → Houdry procs.

catalases   ferri-Fe(III)-hemoproteins that cat. the *dismutation of two molecules of H_2O_2 to O_2 and water acc. to $2 H_2O_2 \rightarrow O_2 + 2 H_2O$. Although H_2O_2 is preferred, cs. will also accept alkyl *peroxides to a lesser extent. One of the more remarkable facts about the cs. studied is that the reaction rate is close to the *diffusional limitation: k_{cat}/K_m (the apparent second-order rate constant of the enzymatic reaction) is $>10^7$ $M^{-1}sec^{-1}$, and is near the calcd. diffusion-controlled encounter rate between substrate and *enzyme. Cf. also *auto(o)xidation. **E=F**; **G** Katalasen.
P.S. SEARS

catalysis   acc. to *Ostwald, c. is the phenomenon in which a small quantity of substance, the *catalyst, increases the rate of a chemical reaction (or: the rate of approaching the equil. of a chemical reaction) without itself being substantially consumed. Only *thermodynamically feasible reactions can be accelerated by cats. Since reversible reactions are accelerated in both directions, forward and reverse, cats. cannot influence their equilibria. In a bulk of various possible reactions a given cat. will not catalyze equally all or several of the possible pathways, thus offering an elegant method for directing convs. in the desired course and velocity (*partial or *selective convs.). Energy profiles along the *reaction coordinate represent the pathways of uncat. (*intermediates with higher activation barrier) and cat. convs. with activated intermediates lower in energy. The tremendous acceleration of reactions and the possibility of directive and selective action of c. and cats. are the reasons for their importance in the chemical industry. Thus, approx. 80 % of all chemicals (valued at some US$ 10^{12}) are produced in contact with cats. in one or more proc. steps (cf. *cat. market). C.is therefore central to the chemical industry and for technologies for *environmental protection. For the history of c., cf. the timetable (*history of cat.).

Although introduced in 1836 by *Berzelius the term c. was scientifically defined firstly in 1894 by *Ostwald. It was not until the start of the 20th century that scientific observation, border-crossing, i.e., interdisciplinary occupation (e.g., organometal chemistry, material sciences, physical chemistry, chemical engineering), and surface sciences enlarged the knowledge of c. and cats. considerably. The lowering of the rate of reaction is called *anticatalysis. In *autocatalytic reactions the cat. is formed by the reaction products themselves.

Despite the general definition, cs. are classified roughly acc. to their *phase behavior. In *heterogeneous c. the cat. phase is different from the reactant phase. Usually the cat. is a solid, whereas the reactants (and the reaction prods.) are gases or liquids. The reaction occurs in presence of the cat. (*"Kontakt", contact) in multiphase mixtures. In the case of *supported cats. the solids are typically robust porous materials with high *surface areas; in the form of *oxide cats. they are non-porous. Thus, surface sciences are very important for het. c. The separation of prods. and het. cats. is easy. Approx. 80 % of the most economically important reactions are cat. heterogeneously, e.g., *ammonia synth., *sulfuric acid prod., *oil processing, manuf. of *polymers, *petrochemicals, *auto exhaust cats., etc.

In *homogeneous c. (with *homogeneous or *molecular cats.) the well-defined cat. is molecularly dispersed with the reactants and the reaction prods. Both combinations, liquid/liquid/gas and gas/gas are well known. Examples of hom. c. are the *lead chamber proc. (gas/gas), *hydroformylation, *carbonylations, *oxidation of *hcs., or the manuf. of *DMT (all liquid/liquid). Borderline cs. are those with *clusters, *colloids, *radicals, *surface organometallic chemistry, *phase transfer or *three-phase cats., and c. with *immobilized (anchored) cats.

Likewise important for chemical research and industry are het. and hom. c. as decisive procs. or methods to manuf. fuels, heavy chemicals, polymers, fibers, pharmaceuticals, dyes, pesticides, or resins. Bioc. is the basis for procs. in living cells. A comparison between both types of c. is given in the Table (below).

In *biocatalysis, the cats. are proteins (*enzymes) which cat. most biological reactions in vivo. They also cat. reactions using both nat-

	Homogeneous	Heterogeneous
	Catalysis	
Activity (rel. to content of active material)	High	Variable
Selectivity	High	Variable
Reaction conditions	Mild	Harsh
Service life of catalysts	Variable	Long
Sensivity toward catalyst poisons	Low	High
Diffusion problems	None	High
Catalyst recycling	Expensive	Unnecessary
Variability of steric properties of catalysts	Possible	Not possible
Variation of electronic props.	Possible	More or less empirical
Mechanistic understanding	Plausible under random conditions	More or less impossible

ural and unnatural substrates *in vitro* (cf. *history of catalysis). Ribonucleic acids can also act as cats. (*ribozymes), involved in RNA splicing in vivo and many novel reactions in vitro. *Antibodies with cat. properties (*abzymes) can also be elicited through immunization with a reaction *transition state analog or a reactive intermediate such as *hapten. Bioc. is most effective in aqueous solution and is classified as hom. cat., with the exception that some enzymatic reactions occur at the lipid-water interphase (e.g., lipase cat. reactions). When biocats. are used in living-cells or resting-cells, as *immobilized form, or *suspended in organic solvents, the proc. becomes het. cat.

Despite the general opinion ("the cat. ...remains unaffected") and acc. to the above-mentioned definition ("the cat. ...is not substantially consumed") it is now well recognized that cats. function by forming chemical bonds with the reactants (or by bringing the substrates to close proximity), thus opening up pathways to their conv. into prods. with regeneration of the cat. C. is thus a cyclic proc. including bonding of reactants with one species of the cat. and decoupling of prods. from another while regenerating the initial cat. (*turnover number, *turnover frequency, *catalytic cycle, *regeneration). In this respect, hom. cat. cycles are specially prominent, although there are many examples of het. cycles, too. Unlike hom. cats., the intermediary states of the het. cats. are highly reactive and difficult to prove, to indicate, and to measure.

The Table mentions various aspects of the characterization and assesment of different cs. For the rate of conv. and for the relative rate of formation of each prod. from two or more competing pathways the *cat. activity and *selectivity are responsible. The *stability is influenced by *deactivation and is denoted as *catalyst lifetime. Because cat. stability may be affected by cat. *poisons the sensivity of cats. against poisons is important. The recovery of cats. after their lifetime is finished or the possibility of recycling them may be economic criteria (cf. *reclamation of cats., *regeneration). The same is true for the *tailoring of cats. acc. to the mechanistic understanding and the variability of their steric or electronic properties. Since *biocats. are *chiral, bioc. is often *stereo- and *regioselective, and the *rate of reaction and the stability of cats. are often affected by temp., pH, *solvents, and ionic strength. The efficiency of biocat. is measured by its cat. *turnover number or specific activity, indicated by k_{cat} (*catalytic constant), and the *specificity of bioc. is defined by the *specific constant, i.e., the ratio of of k_{cat} to K_m (the *affinity constant for a substrate binding to the cat.) for a given cat. and substrate. The specificity (e.g., substrate sp. or stereosp.), stability, and specific activity of a biocat. can be altered by random or *site-directed mutagenesis or *directed evolution (cf. also *in vitro evolution of biocats.).

Specially for *het. c.*, *diffusion, *mass transport, and *heat transfer play a decisive role, because external and internal conc. and temp. gradients can build up in the fluid (liquid or gas) boundary layer around the cat. particles and inside the pores. Such effects determine the regimes (kinetic or diffusive) and the overall reaction rate (or selectivity) by influencing the *interphase mass or heat transfer, the solubilities, or the intrinsic *kinetics in solid/gas/liquid (and to a considerably lesser extent in gas/gas, liquid/liquid, or gas/liquid/liquid reactions)(cf. *effectiviness factor, *Thiele modulus).

The *reaction mechanisms of het. reactions involve *chemisorption and *cat. activation as the first steps, depending on the valence forces. In *hom. c.* the mechanisms are controlled by *heterolytic or by *homolytic bond splitting, and by *insertion. In organometallic c. (cf. below) *metal substrate *coordination, *ligand tuning, *ligand exchange, and *oxidative addition (or *reductive elimination) play an important role.

The cats. for *het. c.* are frequently *supported cats. Thus, the principal comps. of typical het. cats. (*supports, *metals, *metallic oxides or sulfides, *zeolites, etc.) determine the course of the c. Special favorable effects may be observed in the presence of or as a consequence of *activators, *alloying ingredients, *binders, *intermetallic cpds., *precursors, *promoters, or superior *distribution or mixture of metals (e.g., *bimetallic cats., *bifunctional c.). The *preparation method for het. cat. may also be decisive, thus underlining the empirical background of het. c. The function of all cats. arises from their ability to change their geometric and electronic structures dynamically in the presence of educt and prod. molecules. This dynamic response is restricted in hom. cats. to a number of atoms and bonds (the molecular nature of the *active site), whereas in het. cats. the *cooperative effects of *surfaces and of the *bulk create a much more complex mode of action. The advent of *model systems and of *in situ characterization techniques allow to understand that the deviations from the ideal solid state structure of a het. cat. have a decisive influence on its function. This is the reason for the "chemical memory" of a cat. material for its thermal and preparation pre-history. The lack of understanding of solid-state dynamics as compared to *molecular dynamics is the most serious obstacle in gaining insight into the mode of operation of het. cats. Greater knowledge of the common mode of operation of all categories of cats. is a successful result of overcoming the artificial barriers between the three areas of hom., het., and bio cat.

The cats. for *hom. c.* may be gases (NO in *lead chamber or *Deacon proc.), acids or bases (*acid-base cat., e.g. *hydrolyses), or organometallic *complexes (as in *carbonylations, *hydroformylations).

Cs. involve nearly all chemical convs., such as *hydrogenation, *oxidation, *halogenation, *addition or *condensation reactions, *alkylations, *cracking/eliminations, etc. Cat. procs. need special technologies and reactors (*catalytic reactors). **F** catalyse; **G** Katalyse.

B. CORNILS, W.A. HERRMANN,
R. SCHLÖGL, C.-H. WONG

Ref.: Anderson/Boudart; Basset et al.; Bond-1; Cornils/Herrmann-1; Dines/Rochester/Thomson; Ertl/Knözinger/Weitkamp; Farrauto/Bartholomew; Fürstner; Gates-1; Guisnet et al.; Iwasawa; Kirk/Othmer-1, Vol. 5, p. 320; Kirk/Othmer-2, p. 224; McKetta; Moulijn/van Leeuwen/van Santen; Mulzer/Waldmann; Parshall/Ittel; Schwab; Stanley; Ullmann, Vol. A5, p. 313; Wijngaarden/ Westerterp/Kronberg; van Santen et al. (Eds.), *Catalysis*, Elsevier, Amsterdam 1999.

catalysis in the gas phase special chapter of (hom.) cat. with catalyst and reactants (at least the cat.) remaining in the gas phase, e.g., the *dehydration of alcohols in the presence of HBr or the *lead chamber proc. for sulfuric acid with NO. Other examples cf. *hom. cat. **F** catalyse en phase gazeuse; **G** Gasphasenkatalyse.   B. CORNILS

catalysis literature Catalysis and cat. literature are documented in many and varied sources such as books (cf. References, p. XIIIff.), patents, journals, and trade literature. There is much primary literature with examples of application and cat. performance data over a broad range of different cats. and reaction conds. In most cases this primary literature – e.g., patents – is difficult to read because companies and inventors are reluctant to make their progress known and their propietary technology needs to be protected: in most cases the company that discovered a new cat. or technology first uses it commercially. In some cases their information is deliberately misleading.

Primary literature from academia in journals (original contributions) or books suffer from the fact that the researchers often use pure comps. and ideal cats. so their data are extremely theoretical and do not meet requirements with impure and multicomponent, commercial feedstocks under industrial conds. *Catalyst lifetimes and other performance data are mostly not recorded.

Trade literature and information from cat. suppliers and from licensors may be extremely helpful. Specially with potential customers they share confidential information about their cats. under the customer's conds. and requirements. The same is true for companies that license and market cats. and cat. technologies. Information under secrecy agreements is always reliable because the loss of trust and confidence may be punished with severe penalties and the loss of "good name". **F** littérature du catalyse: **G** Katalyse-Literatur.

B. CORNILS

Ref.: Kirk/Othmer-1, Vol.5, p.369; Kirk/Othmer-2, pp.642,839; Ullmann, Vol.B1.

catalysis of the Earth's atmosphere → environmental catalysis

catalyst A cat. is a substance which alters the speed of a chemical reaction by its presence although not being consumed significantly. In a cat. reaction the cat. generally enters into chemical comb. with the reactants but is ultimately regenerated so that the amount of cat. remains unchanged. Cats. may be classified generally acc. to their physical state (gas, solid, liquid), by their chemical nature (metals, oxides, proteins), by the nature of the reactions that they cat. (*oxidation, *reduction, *cracking, etc.) or by the number and kind of *phases present (hom., het., cf. *catalysis.

A cat. accelerates a chemical reaction in both directions equally. Therefore, a cat. does not affect the *position* of equil. of a chemical reaction; it affects only the *rate* at which equil. is attained. Apparent exceptions to this generalization are those reactions in which one of the prods. is also a cat. for the reaction. Such reactions are termed *autocatalytic. Where a given substance or a comb. of substances undergoes two or more simultaneous reactions that yield different prods., their distribution may be influenced by the use of a cat. that selectively accelerates one reaction relative to the other(s). By choice of the appropriate cat., a particular reaction may occur at the expense of the other. Many important applications of cats. are based on *selectivity differences of this kind. *Enzymes play an essential role in living organisms, where they accelerate reactions that otherwise would require temps. that would destroy most of the organic matter.

Many cat. procs. are known in which the cat. and the reactants are not present in the same phase. These are known as *het.* cat. reactions.

They include reactions between gases or liquids or both at the *surface of a solid cat. Since the surface is the place at which the reaction occurs, it is generally prepared in ways that produce large surface areas per unit of cat.; finely divided metals, metal *gauzes, metals incorporated into *supporting matrices, and metallic films have all been used in modern het. cat. The metals themselves are used, or they are converted to oxides, sulfides, or halides. Together with *hom.* cats. they are used for a variety of industrial procs., mostly in the *petroleum processing, bulk and fine chemical, and *polymer industries, in which organic molecules are *isomerized, built up from simple molecules, *oxidized, *hydrogenated, or caused to polymerize. **F** catalyseur; **G** Katalysator.

R. SCHLÖGL, M. DIETERLE

Ref.: Gates-1,p.430; Twigg; Ullmann, Vol.A5, p.313; H.F.Rase, *Handbook of Commercial Catalysts*, CRC Press, Boca Raton FL 1999.

catalyst additives → additives

catalyst alliance Driven by increased global competition the need for rapid commercialization of new cats. and cat. technologies by comb. of different competencies from various sources (or companies) facilitates cat. alliances and joint ventures, such as, e.g., *Haldor/*Kellog or *Catalytica/ *Conoco/*Neste Oy. Well known is *Criterion Catalyst Company (CCC) with its joint ventures with La Roche, ABB Lummus, and Zeolyst Int. to commercialize (among others) the *SynSat catalyst (Figure). **F** alliance du catalyse; **G** Katalyse-Allianz.

B. CORNILS

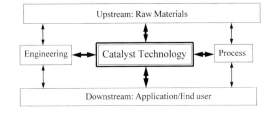

Ref.: *Cattech* **1997**, *1*, 5.

catalyst carrier → support

catalyst characterization includes IU-PAC definitions and recommendations concerning the terminology of cats. Referring to this, a manual includes cat. formulation and *preparation (including methods of prepn., *precursors, morphology), physical properties (*pore structure, *density, *mechanical strength), fine structure (*surface structure and topography, *active sites), and cat. properties (*activity, *selectivity, *inhibition, *regeneration). **F** caractérisation des catalyseurs; **G** Katalysatorcharakterisierung.   B. CORNILS

Ref.: Ertl/Knözinger/Weitkamp, Vol. 3, pp. 1503, 1516; *Pure Appl.Chem.* **1985**, *57*, 603 and **1991**, *63*, 1227; Imelik/Vedrine; Dines/Rochester/Thomson.

catalyst cycle → catalytic cycle

catalyst deactivation → deactivation

catalyst forming is the formation of the cats.' size and *shape acc. to the latter type of reaction, the operation, or the reactor employed. Thus, e.g., small particles or powder cats. are used in liquid-phase operation or in fluid-bed reactors. Various spheres act in moving bed, ebullated-bed, or in fixed-bed reactors. Size and shape of the cats. have to be adjusted to the reaction and the reactor (for details, cf., *catalytic reactors and *forming of cats.). **F** mise en forme des catalyseurs; **G** Katalysatorformung.   B. CORNILS

Ref.: A.B. Stiles, T.A. Koch, *Catalyst Manufacture*, Marcel Dekker, New York 1995; Ertel/Knözinger/ Weitkamp, Vol. 1, pp. 264, 412.

catalyst forms comprise *pellets, pills, rings, extrudates, spheres, granules, tablets, powders, particulates, or *monoliths, depending on manufacture and latter application (cf. *forming, *shape, *pellets).   B. CORNILS

catalyst fouling → deactivation, *fouling

catalyst lifetime of each cat. is limited. Lifetime (long-term stability) can be defined as the period of time during which a predetermined bandwith of reaction conds. can be kept without inacceptable loss in *activity or *selectivity. The term bandwith of conds. mainly refers to temp., to a smaller extent to *space velocity in continuous procs. or residence time in discontinuous procs., respectively. The tolerable temp. band may be as narrow as a few degrees Celsius in selectively demanding reactions (e.g. *hydroformylation, *fat hardening), or stretch over some tens to even a hundred degrees Celsius (e.g., *catalytic combustion) in cases where conv. is dominant over selectivity (see Table below).

For obvious reasons, the preservation of the cat. activity over an elongated period of time is highly desirable. However, even in case of very rapid *deactivation (e.g., *zeolites in *FCC procs. by *coke *deposition and blocking of the cat. *surface and *pore system), the comb. reaction with *deactivation/*regeneration renders the overall proc. favorable and economically viable.

Besides activity and selectivity the third important target in the development of cats. is

catalyst lifetime

Reaction	Catalyst	Temp. range [°C]	Lifetime span [years]
Ammonia synthesis	$Fe-K_2O-Al_2O_3$	450–600	8–10
Methanol synthesis	$Cu-ZnO-Al_2O_3$	280–350	6–8
Hydrogenation of oxo-aldehydes	$Cu-SiO_2$	140–180	3–6
EO synthesis	Ag	180–200	3–5
Hydroformylation	Rh/phosphine	95–110	2–4
Hydrodesulfurization	Ni-Mo	350–450	1–2
Hydrogenation of crude fatty acids	$Ni-SiO_2$	190–220	0.01
Catalytic cracking	zeolites	500–550	0.00001

an acceptable lifetime. When the influence of external *poisons is negligible, the loss of cat. active centers by chemical, thermal, or mechanical influences is the determining factor for the lifetime performance of a cat. Examples of chemical influences are the decomposition of *ligands (Rh-*phosphine *complexes in *hydroformylation) or the volatilization of active cat. ingredients (formation of $Ni(CO)_4$ in *methanation). Chemically induced deactivation may be accelerated by thermal influences, e.g., by recrystallization or *sintering of the active phases of cats. Finally the mechanical structure of cats. may be damaged by mechanical influences, e.g., agitation of cat. *suspensions or mechanical shocks in cat. beds, or simply by disintegration of shaped cats. under reaction conds. with time.

The cl. mirrors the true lifetime of the cats. only, when the loading of the cat. is reflected. Therefore, the cl. parameter, expressed as total weight of feed per weight of cat. per unit total time on stream $(kg\ h^{-1}\ kg^{-1} = h^{-1})$, is the better and more meaningful dimension. **F** durée de vie du catalyseur; **G** Katalysatorlebensdauer (-standzeit).   C.D. FROHNING

catalyst market   in 1995 was estimated to be about US$ 8.6 billions and is predicted to be US$ 10.7 billions in 2001:

Sector	US$ billion 1995	US$ billion 2001
Refining	1.9	2.4
Chemicals	2.2	2.5
Polymerization	1.4	1.8
Environmental	3.1	4.0
Total	8.6	10.7

Ref.: *Catal. Today* **1999**, *51*, 535; *ECN* **1996**, 3–9 June, p. 22.

catalyst particle size   A particle is a small solid object of any size from the atomic scale $(10^{-10}$ m) to the macroscopic scale $(10^{-3}$ m). Particles larger than 10^{-6} m are called grains (*zeolites, *carbons, *Raney metals) and particles smaller than 2 nm are called aggregates (metals) or *clusters (metals, oxides). A crys-

tallite is a small single crystal and a particle could be formed by one or more crystallites. For the complete characterization the following physical properties are needed: linear dimension, *surface area, volume, mass, setting rate, and response of electrical, optical, or acoustical field.

Cat. particles present an indeterminate size distribution which can be very narrow (metal particles in *zeolite cages) or broad with possibly two or more maxima (*particle size distribution). The particle form is not as well defined as the size. Particles are usually not spherical and their shape is not homogeneous. A combination of two or more physical methods can describe the cps. and *shape. In general, particles are assumed to be spherical, especially for metals, because of the difficulty of establishing size and shape distribution together. The measurement of the cps. is possible by gas *chemisorption (adsorption-desorption methods, *hydrogen-oxygen titration) or *X-ray diffraction line broadening analysis, *SAXS, *TEM, *SEM, and magnetic methods. **F** diamètre des particules du catalyseur; **G** Katalysatorkorngröße.

R. SCHLÖGL, H.J. WÖLK

Ref.: Ertl/Knözinger/Weitkamp, Vol. 2, p. 439; Ullmann, Vol. B2, Chapters 2–1 and 8–1; Farrauto/Bartholomew, p. 134.

catalyst pellets   → pellets, *catalyst forms

catalyst poisoning   In contrast to *deactivation, which may be caused by chemical, thermal, or physical effects, poisoning of cats. is a chemical process. A cat. *poison therefore is a substance which lessens or destroys the effectiveness of a cat. under normal conds. of use. The poison is irreversibly *chemisorbed by the active comp. of the cat. under reaction conds. and alters the chemical state of this comp. making it cat. inactive. This definition does not exclude the *regeneration of the cat. under conds. different from reaction conds. Several classes of cats. and respective poisons can be defined. Metals (predominantly *transition metals) containing cat. are sensitive to small amounts of impurities in the feed which possess free electron pairs which are bound via d-orbitals and form stable metal cpds. Typical examples of poisons are cpds. of groups V and VI of the periodic table, namely NH_3, PH_3, AsH_3, H_2S, thiophene, thioethers, Se, Te, which poison Ni, Co, Cu, Fe, Pd and Pt-containing cats. by chemical reaction or strong chemisorption. For metallic silver in the synth. of *EO, chlorine or chlorine-containing cpds. are poisons due to the formation of AgCl. Hg, Zn, and Pb can form *alloys with a number of metals, making them cat. inactive, e.g., Pt or Pd cats..

Cats. with acidic or basic properties are sensitive to basic or acidic substances, respectively. The *cat. cracking of hcs. requires *aluminosilicates or *zeolites with acidic functions which are blocked by chemisorption of ammonia, amines, organic bases, or metal ions. Coordinatively unsaturated molecules can also exhibit strong interaction with active centers of cats. forming stable intermediates. CO and CN^- are well-known poisons for *hydrogenation cats. Prods. of the cleavage of cat. *ligands (e.g., *P-C bond cleavage of *phosphines), namely phosphide-bridged dimers or *clusters, may also act as poisons. Asphaltenes or N cpds. may be poisonous to *hydrotreating cats. Some poisons can be removed by special treatment of the inactivated cats. (cf. *regeneration/*reclamation). Cf. also *directed poisoning, *sulfided cats. **F** empoisonnement des catalyseurs; **G** Katalysatorvergiftung.   C.D. FROHNING

Ref.: J.B.Butt, E.E.Petersen, *Activation, Deactivation, and Poisoning of Catalysts*, Academic Press, New York 1988; Farrauto/ Bartholomew, p. 266.

catalyst rejuvenation   → regeneration

catalyst support   → support

catalyst-support interactions   → metal-support interactions

catalyst surface   The cs. in a wider sense is the total *surface of the cat. It comprises the *external* surface as well as the *internal* surface

(*pores). In a narrower sense, the term cs. means the surface of the active phase of the cat. The latter definition is especially useful since it is on the active phase's surface that the cat. reaction takes place. Assuming that there are no transport limitations (*mass transport) the *rate of reaction of a het. cat. reaction should be accurately proportional to the surface area of the active phase of the cat. The characterization of the cs. includes, besides the above-mentioned geometric features (i.e., the total surface area, the area of the active phase and the *pore structure of the cat.), its chemical composition also. The total surface area can be determined experimentally via the *adsorption of chemically inert gases whereas the area of the active phase can in some cases be measured by selective gas *chemisorption. Adsorption *isotherms may also contain information on the pore structure. This is the case when *hysteresis loops occur in the isotherms, which are due to retarded condensation and re-evaporation of the gas in the fine pores. The *pore size distribution can be measured by *mercury porosimetry. The chemical composition of the cs. can be determined using surface analysis techniques such as *XPS, *UPS, *AES, or *ISS.

Also, the cs. can be investigated with microscopy techniques giving a direct image of the cs. Whereas *scanning electron microscopy (SEM) allows resolutions up to 20 nm, with scanning tunneling and atomic force microscopy (*STM and AFM) it is possible to reach atomic resolution. In combination with energy-dispersive X-ray analysis (*EDX), SEM allows an elemental mapping of the cs., which is used for the identification of the different phases and especially of the active phase of the cat. **F** surface de catalyseur; **G** Katalysatoroberfläche.

R. SCHLÖGL, A. SCHEYBAL

Ref.: Dines/Rochester/Thomson; Somojai; Gates-1, p. 220; Farrauto/Bartholomew, p. 154.

Catalytica Inc.   originally founded as a non-profit company to develop and sell cats. and cat. procs. Today C. is a public company.

Subsidiaries are Catalytica Pharmaceuticals (for the manuf. of pharmaceuticals or fine chemicals) and Catalytica Combustion Systems (generally for the reduction of hazardous emissions and specially for the development and marketing of combustors for flameless combustion). C.D. FROHNING

catalytic acronyms → acronyms

catalytic activity This term is used qualitatively to describe how effective a given cat. is in carrying out a desired reaction. Statements such as "a cat. shows high activity if it achieves its objective under mild conds. of temp. and press.", or that "its activity is low if very vigorous conds. have to be used" must of course be made in the context of the reaction's inertia. To be meaningful the expression "activity" should refer to specified conds. of temp. and reactant concs. but it is usually taken to imply a quality of the cat. that is not too much affected by the precise conds. of the experiment. For most purposes it is preferable to state the *rate at which reactants are transformed into desired prods. under defined conds. Highly active cats. are sometimes unstable and liable to rapid *deactivation, because activity is due to a small number of extraordinary sites that quickly become *poisoned: a low but stable activity is often better.

The general pattern of cat. activity shown by various types of solid for various classes of reaction is by now well established, although surprises do still happen. Some general principles are given here. In seeking to locate or devise an active cat. for a target reaction, one must first ensure that the reactants can be *chemisorbed on its *surface, and the prod. is not so strongly adsorbed as to constitute a *cat. poison. The cat. has to be chemically and physically stable in the presence of the reactants and prods. at the reaction temp., and must not be unduly sensitive to traces of poisons in the feedstock. A very helpful general theorem, applicable when trying to choose the best member of a group of similar materials, is the *volcano principle.

The selection of a likely cat. from the whole class of cat. materials requires first the use of the concept of compatibility between reactants and cat. Thus *oxidations are usually best cat. by oxides; the number of metals that can withstand the necessary reaction conds. without themselves being oxidized is very limited, and these are to be found in the second and third rows of groups 8–11 of the periodic table. Reactions requiring dissociation of the H_2 molecule on the other hand proceed well on many metals, but only on a very few oxides. Those reactions involving S-containing molecules (e.g., *hydrodesulfurization) need a sulfide as cat., because even if the cat. starts life as an oxide it soon becomes sulfided. Reactions of polar molecules, and those which involve ionic intermediates such as protons or *carbocations, go best on the *surfaces of acidic oxides: thus for the *dehydration of an alcohol, a cat. is needed that can chemisorb the product water molecule before releasing it into the fluid phase. **F** activité catalytique; **G** katalytische Aktivität. G.C. BOND

Ref.: Ertl/Knözinger/Weitkamp, Vol. 1, p. 1; Farrauto/Bartholomew, p. 40; Ullmann, Vol. A5, p. 322.

catalytic antibodies (abzymes) Proteins which can bind and stabilize the *transition state of a reaction will accelerate the reaction. *Antibodies are proteins specifically designed to bind foreign molecules that enter the body, and the use of *transition state analogs as *haptens for the induction of as. that could bind the transition state and thus cat. the reaction has been proposed. This has proven to be a moderately effective approach, although the as. induced have not reached the rate accelerations observed with *enzymes; the ratio of the cat.:uncat. rate is on the order of 10^3–10^5 for the as., as compared to 10^6–10^8 (or even greater) for *enzymes. With appropriate design of the antigens, specific functional groups can be induced in the binding site of an a. to perform *general acid-base or nucleophilic-electrophilic cat. With this new technique, new protein cats. can be designed and prepared for reactions that may be disfavorable or not attainable otherwise, or have different reaction *mechanisms or *specificities compared to the corresponding enzyme-cat. reactions. New strategies using reactive intermediates such as haptens have been used to elicit more efficient cas.

Reactions that are rare or have no counterpart in nature, including a variety of *pericyclic reactions, have been accelerated with cas. Other antibody-cat. reactions include ester *hydrolysis, peptide synth. (from esters), glycoside hydrolysis, *transesterifications, and *stereospecific reductions (cf. also *phage display, *chorismate mutase, *catalytic triad, *pericyclic reactions. **F** anticorps catalytique; **G** katalytische Antikörper.

P.S. SEARS, C.-H. WONG

Ref.: *Science* **1994**, *265*, 1059 and **1997**, *389*, 271; *Acc. Chem. Res.* **1997**, *30*, 115.

catalytic coal gasification   → coal gasification

catalytic combustion (afterburning) flameless combustion proc. in which combustible cpds. (e.g., contaminated fuels from gas turbines) react with oxygen on the *surface of het. cats. thus providing CO_2, H_2O, and thermal energy. The temps. applied (1000–1300 °C) are much lower than for conventional combustion (up to 2000 °C) so avoiding higher NO_x levels (below 5 ppm, cf. temp. profiles in Figure).

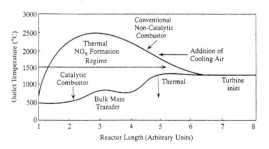

[by courtesey reproduced from Farrauto/Bartholomew, Kluwer Academic Publ., Figure 11.21]

Especially the combustion of alkanes has considerable potential relating to the develop-

ment of cat. combustors for power prod. and to the industrial exhaust gas stream purification. The pollutants may be hydrocarbons, *volatile organic cpds. (VOCs), or even *CFCs (cf. *CFC destruction). The cats. involved are *ceramic or metal *honeycombs with *washcoats based on *cordierite, *mullite, or *perovskites such as $LaCoO_3$ or Sr-doped $LaCoO_3$. Conventional cats. contain Ba-stabilized *alumina plus Pt and Pd. **F** combustion catalytique; **G** katalytische Verbrennung.

B. CORNILS

Ref.: Farrauto/Bartholomew, p. 649; Ertl/Knözinger/Weitkamp, Vol. 4, p. 1668; P.L. Silverstone, *Composition Modulation of Catalytic Reactors*, Gordon and Breach, London 1998; R.E. Hayes, S.T. Kolaczkowski, *Introduction to Catalytic Combustion*, Gordon and Breach, London 1997; Wm.R. Seeker, C.P. Koshland. *Incineration of Hazardous Waste*, Gordon and Breach 1994.

catalytic constant (k_{cat})   *Enzymes display saturation *kinetics: as the conc. of substrate increases, a point will be reached at which the enzyme *active site is always occupied (unless the reaction is diffusion limited). The reaction velocity at saturation is called V_{max} (units of M s^{-1}), and V_{max} divided by the enzyme conc. is the kinetic parameter k_{cat} (s^{-1}), the first-order rate constant for an enzymatic reaction. k_{cat} is also called the catalytic constant, or the *turnover number (TON).

In the case of a simple, one-step reaction where binding is fast compared to the reaction step (*Michaelis-Menten mechanism), cc. represents the first-order *rate constant of the conv. of the enzyme-substrate complex to the enzyme-prod. complex. In multi-step reactions, it represents a combination of rate constants (though it may be dominated by one; see *rate-determining step in enzymic reactions). **F** constant catalytique; **G** katalytische Konstante.

P.S. SEARS

catalytic converter   → automotive exhaust catalysts

catalytic cracking   *Petroleum processing includes cracking, i.e., transforming heavy *hydrocarbons to lower molecular weight cpds. Actual worldwide prod. by cc. is over 500×10^6 tonnes/y. In a typical modern *refinery, cat. crackers contribute from 20 to 50 % of the blending comps. in the gasoline pool. The original cracking cats. were acid-treated *clays, replaced by synth. silico-*alumina in the 1940s and then by *zeolite in the 1960s. The original Y-type *faujasite *zeolites were then modified by further dealumination or rare earth exchange (cf., e.g., *alumino phosphates). These modified zeolites are formulated into matrices of different *activities in order to provide a wide array of cat. properties, which can be *tailored to the specific characteristics of the feedstock.

A typical commercial *FCC cat. is composed of about 5 to 40 % of a Y-type *faujasitic zeolite, a *silica/*alumina *binder or matrix, and a clay filler. The matrix provides porosity and may contain a cat. active silica/alumina comp. The zeolite is either a rare earth exchanged Y or an ultra- stable dealuminated Y. The zeolite typically forms agglomerates of 1 to 2 μm in size and may be loosely or tightly embedded in the matrix. The cat. must meet severe requirements, such as resistance to *poisoning by metals (V and Ni), hydrothermal stability at high *regeneration temps., low rate of *coke and gas formation (especially in a high- metals environment) and good activity for very large molecules. In addition, due to a high rate of cat. *make-up (*cat. lifetime is of the order of 30 days), its cost must be low.

Cc. was the first large-scale application of circulating-bed reactors (CBR, cf. *FCC and *cat. reactors). During the cracking proc., extensive coke *deposition takes place, resulting in loss of *activity. Typically, the cat. loses 90 % of its activity within one second. Continuous *regeneration is thus needed, and the combustion of the coke provides the heat required for the endothermic cracking reactions. The introduction of the CBR technology, with the cat. moving from the reaction zone (riser or *riser-tube reactor) to the regeneration zone, allows this continuous regeneration of the cat. and the transfer by the solid cat. itself of the heat released in the regeneration zone to the reaction zone.

In cc. many reactions take place simultaneously. Cracking occurs by C-C bond cleavage of alkanes, *dealkylation, etc. *Isomerization and even *condensation reactions also take place. These reactions occur via positively charged hc. ions (*carbocations), formed by various possible mechanisms. For alkanes, it is usually assumed that carbenium ions formed by acid-cat. activation (hydride abstraction) are the first intermediate. The carbenium ion may undergo one of several basic reactions. Thermal, *radical-type cracking reactions may also occur, making the kinetic modeling of the proc. quite complex. The elucidation of the factors controlling the relative rates in the complex reaction network is still not complete.

Examples of commercial realizations are given under the procs. entries of *Ashland, *Chevron, *Houdry, *ICI, *Standard Oil, *Stone & Webster (*deep catracking), *UOP, *VEBA, etc. **F** "cracking" catalytique; **G** katalytisches Cracken, Katcracking.   G. CENTI

Ref.: P.B. Venuto, E.T. Habib, *Fluid Catalytic Cracking with Zeolite Catalysts*, Dekker, New York 1979; *Catal. Rev.-Sci.Eng.* **1984**, *26*, 525; **1989**, *31*, 215; **1989**, *24*, 1; **1994**, *36*, 405; McKetta-1, Vol. 13, p. 1; Farrauto/Bartholomew, p. 535.

catalytic cycle 1: visual interpretation of a complex reaction mechanism by subdividing the overall reaction into a series of ad- and desorption steps to and from a cat. active center and arranging the intermediates in a logical sequence to form a closed cycle (cf. examples *homogeneous cat., *hydroformylation). Frequently applied to explain the course of a reaction cat. by *ligand-modified metal atoms taking into account the *conformational requirements of the *central atom and the *ligands (e.g., square planar arrangements in fourfold, trigonal bipyramids into fivefold *coordination) and the formation of intermediates characterized by electronically stable configurations (e.g., *Tolman's 18–16-electron rule for the configurations of *transition metal atoms). Usually the cc. comprises an *oxidative addition and a *reductive elimination. A typical example is the *Heck-Breslow interpretation of the mechanism of *hydroformylation.

2: Recirculation of used or made-up/regenerated cats. within a single commercial proc. (*make-up, *recycling, *regeneration). **F** cycle catalytique; **G** Katalysatorcyclus.

C.D. FROHNING

Ref.: E. Riedel, *Moderne Anorganische Chemie*, W. de Gruyter, Berlin 1999.

catalytic dewaxing   → dewaxing

Catalytic Dewaxing process   → BP procs.

catalytic distillation   → catalytic (reactive) disitillation

catalytic membrane   → membranes

catalytic microreactor   1: In classical cat. science microreactors (ms.) are used as lab. instruments to screen substances (*catalytic pilot plant, *miniplant). Reactor set-ups resemble those used in conventional synth. chemistry in dimension and involve reaction sizes in the millimole range and cat. amounts in the milligram range. Such reactor dimensions are easy to operate but represent only a microscopic section of any commercial proc. Under such conds. the *heat and *mass flow conds. are different from those in large systems, leading to significant discrepancies in *kinetic parameters for large and small reactors (scale-up problem). Other problems may evolve with prod. and recycle streams and with the simulation of recovery steps under real conds. Rarely procs. are investigated in cycling operations and cats. are often only studied in single-pass operation. This avoids handling and contamination problems but precludes meaningful analysis of *lifetime and *deactivation problems in ms. 2: Modern ms. are *catalytic reactors with dimensions on the micrometer scale holding microscopic

amounts of cat. In such devices (manufactured from microsystem technology, no longer usual with discrete elements) the problems of *heat and *mass transport are almost eliminated and the immediate response of the reaction set-up to changes in reaction conds. allows operation of exergonic procs. in regimes of gas flow and compositions different from those accessible for larger reactors. Such devices can be used to produce on-line small amounts of hazardous chemicals, avoiding all the risks of large plants and eliminating all handling problems. Typical applications are *polymerizations with the prod. of starter cpds. Other areas include biological and analytical applications.

The recent field of high-throughput testing and *combinatorial cats. development would be impossible without the existence of microreactor technology. Only through the small amount of cat. and the immediate contact of reactants with the cat. a quick response (seconds instead of hours) of a system to changing reaction conds. can be expected. Formation reactions are quick and extensive diffusion limitations are excluded. This reduces by orders of magnitude the time required for analyzing a given cat. A common problem with all ms. is their sensitivity to *poisoning as there is not enough cat. in the reactor to act as a *guard bed for the main fraction of the cat. as occurs naturally in larger systems. For the micromechanical prepn. of cms. the *LIGA technique is used. **F** microréacteur catalytique; **G** katalytischer Mikroreactor.

R. SCHLÖGL

Ref.: W. Ehrfeld, V. Hessel, *Microreactors*, Wiley-VCH, Weinheim 1999.

catalytic monoliths (reactors)   cat. converters for cleaning exhaust gas streams (e.g., *automotive exhaust cats.). They have to fulfill several requirements: low press. drop at varying and extremely high linear gas velocities, mechanical resistance to thermal shocks (steep temp. gradients), high specific thermal load, and large *surface for the cat. active material. The comb. of requirements is best ful-

filled by monoliths, constructed of material with excellent thermal and mechanical properties and consisting of an integral bundle of *ceramic tubes, the walls of which are coated with cat. active material (cf. *monolytic honeycomb). The cmr. is thus a single block of typically ceramic material containing an array of parallel, uniform, straight channels having a density of 20 to 80 cm^{-2} related to equivalent diameters of 0.8 to 2.5 mm. The channel walls are covered with the cat. active comp. (cf. Figure (g) on page 105). One of the best base materials available is *cordierite, a synth. produced mineral, corresponding to the composition $Mg_2Al_3(AlSi_5O_{18})$, having an extremely low thermal expansion coefficient. The wet *precursor is extruded forming honeycomb channels surrounded by an outer shell rendering mechanical stability. As the dried material only owns geometrical surface, a *washcoat consisting of *alumina or *aluminosilicate is applied which after drying and *calcination is *impregnated with the active elements. The monoliths are mostly used for automotive exhaust cats. Several units can be combined in parallel and in series in a single apparatus to attain high throughputs, e.g., for cat. control of NO_x in power plant effluent gases (*stationary sources). For automotive emission control Pt-Rh are preferred. **F** monolith catalytique; **G** katalytischer Monolith.
Ref.: Thomas/Thomas.   C.D. FROHNING

catalytic nitrate reduction   proc. for the cat. decrease of nitrate to nitrogen (*hydrodenitrification). The reaction proceeds via several intermediates (nitrite, nitric and nitrous oxide). Ammonium is formed as a byprod. due to over-hydrogenation. For nitrate reduction a *bimetallic cat. is necessary, whereas the *reduction of nitrite can be performed by monometallic cats. Suitable *supported metal cats. are Pd-Sn, Pd-Cu, or Pd-In for nitrate and Pd for nitrite reduction. The *selectivity to N_2 ranges from 80 to 99 % depending on the reaction conds. (e.g., pH). H_2 or formic acid can serve as reductants (cf. *transfer hydrogenation). Generally, formic acid provides higher selectivities. **F** réduction catalytique de nitrate; **D** katalytische Nitratreduktion.

U. PRÜSSE, K.-D. VORLOP
Ref.: F.I.I.G. Janssen, R.A. van Santen, *Environmental Catalysis*, Imperial College Press, London 1999, p. 195.

catalytic oxidation   includes procs. of the petrochemical industry (*selective oxidation, oxidative *dehydrogenation or *amm(on)oxidation, *oxychlorination, *epoxidation, and *ammoximation), inorganic chemicals industry (oxidation of SO_2 and NH_3 for the manuf. of H_2SO_4 and HNO_3, and HCN by oxidation of CH_4), fine chemicals industry (*hydroxylation, *acetoxylation, oxidation of carbohydrates), *refinery industry, energy production (*methanation, *cat. combustion in gas turbines, and *fuel cells), and technologies for protection of the environment (*automotive exhaust cats.; *SCR, cat. elimination of *volatile organic cpds. [VOC], *cat. combustion, and the co. of organic matter in water). The growth of oxidation cats. is about 11 % per year, twice as high as the average of all cats.

Different classes of cats. are used in these processes – 1: *transition metal oxides or *mixed oxides (*bulk type or *supported; for example Fe molybdate for methanol oxidation, V oxide supported on *titania for phthalic anhydride synth. from *o*-xylene); 2: supported noble metals (Ag on *alumina for *EO synth., Pt on *alumina for co.); 3: *zeolites containing metal cations (in isomorphic substitution or in cationic positions – e.g., Ti *silicalite for phenol *hydroxylation or cyclohexanone *ammoximation) or metal complexes; 4: metal *complexes in solution (Co for terephthalic acid synth. from *p*-xylene); and 5: other types (e.g., Pt-Rh *gauzes for NH_3 oxidation, cf. *metal gauze reactors). Most of these co. procs. occur in the gas phase over solid cats., but there are significant examples of hom. procs. A third class of growing interest are procs. in the liquid phase, but using solid cats. such as *Ti silicalite. In most oxidation reactions molecular oxygen (air, oxygen-enriched air, oxygen) is used as the oxidizing agent (in the vapor or liquid phase), while a few other procs. use organic peroxides, H_2O_2, or HNO_3 as the oxidants (in the liquid phase). In the vapor phase for oxide-type cats. a *redox mechanism, in which the substrate is oxidized by the cat. and then re-oxidized by molecular oxygen (e.g., cf. the *Wacker proc.), predominates. In the liquid phase, either *radical chain or oxygen transfer mechanisms operate, apart from specific cases such as in the *Wacker proc., where a mediated *redox mechanism is effective. New types of *selective oxidation procs. are also under development; cf. also *alkane oxidation. **F** oxidation catalytique; **G** katalytische Oxidation.

G. CENTI

Ref.: J.M. Thomas, K.I. Zamaraiev (Eds.), *Perspective in Catalysis*, Blackwell, Oxford 1992, p. 371; *Catal. Today*, **1994**, *19*, 215; **1997**, *34*, 269; **1998**, *41* (New Concepts in Selective Oxidation issue).

catalytic pilot plant A catalytic pp. is the special version of a conventional pp., the semitechnical (semicommercial) experimental station between lab/miniplant development and large-scale plant operation, aiming, inter alia, at proc. development, confirming the feasibility of technical concepts, experiences with process engineering and handling of equipment, collecting data for scaling-up and for prod. specifications, and chemical, technical, and economical optimization. Special duties of cat. pps. are the layout of cat. procs., manuf. procs. for cats., the variation of different combs. of cats. and reaction apparatuses, securing of *cat. lifetimes and simulation of aging conds. of cats., long-term behavior of *supports, cat. metals, and *ligands, the influence of load variations, *poisons, *recycle of side-streams, etc. The scale-up factor for cat. pp. is generally between 10^2 and 10^4. Due to high costs of operating labor, in future pps. will be substituted by microunits (*miniplants, *catalytic microreactors), automated test reactors, or even by stand-alone computers. **F** installation pilote, **G** Katalyse-Technikum.

B. CORNILS

Ref.: Kirk/Othmer-1, Vol. 19, pp. 79, 144; Kirk/Othmer-2, pp. 892, 954; Ullmann, Vol. B4, pp. 454, 474; *Chem. Eng. Progr.* **1991**, *87*(1), 21; Anderson/Boudart, Vol. 1, p. 43 and Vol. 8; Augustine, p. 97; *Catalysis Today* **1999**, *51*, 535; W. Hoyle, *Pilot Plants*, RSC No. 236, Cambridge 1999.

catalytic radical reactions Many *radical reactions proceed via *radical chain reactions, started by *radical initiators. For synth. applications, the formation of C-C bonds is the most interesting topic, which can be conveniently realized by the addition of alkyl halides to electron-poor alkenes in the presence of $HSnBu_3$. Only cat. amounts of radicals are involved and are regenerated in the reaction cycle (Figure). The disadvantage of this strategy is the high toxicity of tin cpds., which are furthermore difficult to separate.

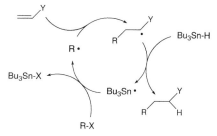

(R = alkyl, aryl; X = I, Br, SePh; Y = aryl, electron acceptor)

To overcome this problem, less toxic silicon hydrides or the thiohydroxamate method (*Barton esters) were employed. Furthermore, modern radical reactions were developed which require only cat. amounts of Sn cpds., such as the *atom transfer or the *Giese/Stork method. Very recently, *Lewis acid cat. radical reactions became attractive for *stereoselective radical reactions. **F** réactions radicalaires catalytiques; **G** katalytische Radikalreaktionen.

T. LINKER

Ref.: Giese; D.P. Curran, N.A. Porter, B. Giese, *Stereochemistry of Radical Reactions*, VCH, Weinheim 1996.

catalytic (reactive) distillation cat. chemical reaction occurring during distillation and thus a single simultaneous unit operation. Although invented much earlier, cd. was developed during the development of *MTBE from methanol and C_4-hydrocarbons containing the desired reactant isobutene (Figure 1; cf. *Hüls/UOP Ethermax proc.).

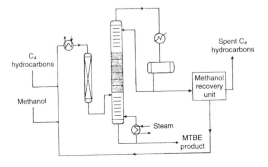

The cats. (beds, bundles, bales, structured packings such as *Katapak structures, *gauze cats., etc.) are stacked into the distillation tower to fill the cross-section between the rectification and the stripping section. Cat. reaction and separation of comps. occur within the single unit operation; heat evolved from the exothermic cat. reaction is utilized to manage the energy requirements of the column. The position of the cat. within the distillation tower (and thus the relationship between the rectifying and stripping section) depends on the relative volatilities of reactants and prods. Advantageously, the prod. can be withdrawn from the reaction zone as soon as it is formed. Examples are the procs. of *CDTech (cumene, ethylbenzene, ethers), *BP Etherol, *IFP (ether, *Catacol tech-

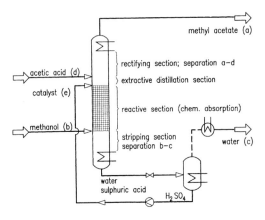

nique), *Hüls/UOP Ethermax, *Neste NEx). Another example is the synthesis of *methyl acetate. The simultaneous implementation of reaction and distillation in a counter-current operated column is shown in Figure 2. **F** distillation catalytique; **G** katalytische Destillation.

B. CORNILS

Ref.: Ertl/Knözinger/Weitkamp, Vol. 3, p. 1479, Vol. 4, p. 1994; Thomas/Thomas, p. 567; *Trans. Inst. Chem. Eng.* **1992**, *A 70*, 448; *Hydrocarb. Proc.* **1994**, (9), 27; *Chem. Eng. Technol.* **1999**, *22*, 95.

catalytic reactors 1: devices in which cat. chemical reactions take place. The reactors (rs.) may be operated in batch, semibatch, or continuous (cont.) modes. Operating in a *batch* mode includes charging the r. with reactants and cats., which are brought to the desired temp. and press., then reaction, cooldown and discharging. The batch operation is not applicable for gas-phase reactions. In *semibatch* r. operation, reactants are in the batch mode but meanwhile a co-reactant (mostly a gas, e.g., in order to maintain constant press.) is fed and withdrawn cont. In a r. for *cont.* operation the reactants are cont. fed and the prods. and unconverted reactants are removed in a corresponding steady stream. Different *phases of reactants can be used: gaseous reactants over solid cats., liquid reactants or liquid-gaseous reactants, both in the presence of dissolved or dispersed solid cats. The cont. mode of operation can be used over a wide range and for high throughputs at low operation costs. Various types of *laboratory and commercially used rs. may be classified, but all of them can be assigned to three basic ideal r. types with individual characteristics, viz. discont. or batch stirred-tank r. (*BR), cont. stirred-tank r. (*CSTR), and tubular r. (*plug-flow reactor, *PFR). The *discont. reactor* can be a tank with an internal stirrer (batch stirred-tank r., Figure (a)) or with an external circulating pump (b).

In addition, baffles are occasionally fitted into the tank to assist mixing. Cooling or – less often – heating of the reaction mixture may be effected by a r. jacket or internal coils. As-

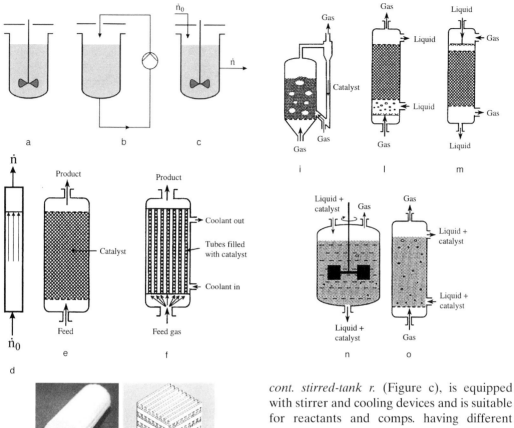

suming that a complete mixing on macrosopic and microscopic scales takes place, the r. is referred to as ideal. The mixing behavior is an important criterion for the reactor efficiency, especially when the reaction system contains comps. having different phases (gas-liquid, liquid-solid cat., liquid-gas-solid cat.) or nonmiscible liquids. The true reaction time exposed to all reactant molecules is equivalent to the time span of the reaction, if the reactants have been ideally mixed (cf. *mass transport in cat.).

From the mass balance for a batch r., the required reaction time can be calc. The batch r. can be used in semi-batch mode too. The

cont. stirred-tank r. (Figure c), is equipped with stirrer and cooling devices and is suitable for reactants and comps. having different phases as specified for the discont. reactor. The CSTR is characterized by a perfect mixing which is infinite in the case of an ideal reactor acc. to the description of the BR. As a consequence, concs. and temp. are uniform throughout the tank r.; however, the *residence times are distributed around the mean value. In CSTRs, the residence time distribution is very broad. Accordingly, the final conv. degree is lower than in other basic r. types due to the fact that unconverted reactants can get into the product stream, i.e., larger tank rs. are needed to attain same conv. degrees. The mass balance for a CSTR leads to the mean residence time of the reactants at a given conv. degree X.

Tubular reactors (*plug flow reactors, PFR, d) are used for cont. operations. In the ideal tubular reactor, ideal mixing takes place in radial direction, but no mixing in a axial direction, i.e., plug flow is assumed. Therefore, all reactant molecules experience the same, well-defined residence time as in a batch reactor. The mass balance for a PFR is equivalent to that of a batch r. From that, the same r. volume and reaction time at constant conv. degree resulted. Various phases can be converted in PFRs such as gaseous or liquid reactants, and gas-liquid reactant mixtures over solid cats. as packed-bed or cat. active tube walls. The most important advantage of PFRs is the favorable conds. for temp. control due to heat removal or supply. Various reactors of the same or different types can be connected in a system of reactors in order to increase the throughput or to optimize conv. and prod. yield. Types of interconnections are parallel, series and recycle (e.g., CSTRs in series). A survey of commercial cat. rs. and the reacting systems and cats. is given in the Table (below).

Cat. liquid-phase reactions (reaction system I in the Table) are carried out as hom. cat. reactions in stirred-tank rs. in batch, semi-batch, or cont. mode of operation with dissolved organometal *complexes as cats. The reactions of commercial interest are *hydroformylations, *carbonylations, *polymerizations, *hydrogenations, and *oxidations. In addition, tubular rs. are also used for liquid-phase reactions like polymerization, hydrogenation, *hydrolysis, or *dehydrochlorination. The basic type of PFR can carry the cat. particles. A novel type of tubular r. possesses micro-

channels with cat. active channel walls (cf. *cat. microreactors).

Cat. gas-phase reactions (reaction system II) play a major role in *refinery, *petrochemistry, and industrial organic technology. Various bulk prods. and intermediates are manuf. in this way using solid cats. The reactions take place in tubular rs. which are designed as fixed-bed or fluidized-bed rs., depending upon the heat of reaction and the *thermodynamic stability of the prod. formed. The least expensive kind to build is an *adiabatic r.* (Figure e) with a fixed bed of cat. and without internals for transferring heat. They are generally most practical for large-scale and relatively slow reactions without large heat effects (cf. *heat transfer in cat.). Several beds may be used in series (multistage reactor) so that the reacting gas can be cooled between beds or a cool reactant gas is injected as quench gas between them (e.g., *methanol synth.).

Adiabatic fixed-bed rs. are also used for highly exothermic reactions which occur extremely rapidly at high temp. (e.g., *formaldehyde synth.). The very short reaction times needed, on the order of milliseconds, are realized by a very thin layer of cats. (≤1 cm) or by using *gauzes (also called shallow-bed rs., *metal gauze rs.). For more exothermic reactions which are temp.-sensitive relating to the prod. *selectivity, *multitube fixed-bed rs.* (Figure f) with external cooling are used. The coarse cat. particles (2 to 8 mm in diameter)

Reaction system	Reactor type	Catalyst
I. Liquid-phase reactant	• Batch reactor	Dissolved homogeneous catalyst
	• Continuous stirred-tank reactor (CSTR)	
	• Tubular reactor (PFR)	Coarse catalyst particles or microchannel structure
II. Gas-phase reactant	• Tubular reactor (PFR)	Coarse catalyst particles or monolithic structure or microchannel structure
	• Fluidized-bed reactor	Fine catalyst particles
III. Gas- and liquid-phase reactants	• Bubble column	Fine catalyst particles or coarse catalyst particles or dissolved homogeneous catalyst
	• Trickle-bed reactor	
	• Fluidized-bed reactor	
	• Batch and continuous stirred-tank reactor	

are used to fill in several thousands (up to 27 000) of tubes having diameters from 2 to 5 cm and lengths from 1 to 10 m. A variety of reactions such as *partial oxidations (e.g., *maleic anhydride from butane, *EO), *partial hydrogenations (*low-temp. hydrog.), *de-hydrogenation (of *EB to styrene), as well as *oxichlorination, *isomerization, and *cracking reactions, take place in multitube rs. Special types of fixed-bed rs. are build with *cat. mono-liths or microstructured wafers (Figure g; *monolitic honeycomb). Another type of fixed-bed r. is the *microchannel r., designed with a stack of microstructured, typically metal-lic, wafers (Figure h).

For very exothermic reactions which cannot be controlled with a multitube fixed-bed r. or when the cat. must be removed or replaced frequently, a *fluidized-bed r.* (i) is used (e.g., *ammoxidation of propene to AN). Cat. *par-ticle sizes from 10 to 200 µm are desirable for efficient fluidization caused by the force of gas flow (≤ 0.6 ms^{-1}) leading to an excellent uniformity of temp. This is achievable throughout the bed because of the cat. move-ment which results in the excellent *heat ex-change between solid and gas. However, the r. needs attrition-resistant, non-agglomerating cats. Increasing gas velocities (≥ 6 ms^{-1}) cause cat. ejection, which is applied in a riser reac-tor, e.g., the *fluid catalytic cracker (see Fig-ure under *FCC process). The cat. is con-stantly recycled between the reaction zone A and *regeneration zone B. The *riser reactor is a kind of moving-bed reactor, in which part of the cat. is moved by gravity and the rest by lifting force from the fluid (cf. circulating-bed r. in *cat. cracking, *FCC, *riser-tube r.).

Cat. gas- and liquid-phase reactions (reaction system III) take place either on solid cats. in *three-phase* rs. over packed or suspended cats. or in two-phase rs. (cf. *two-phase cat.). Three-phase rs. are *packed bubble col-umns* or *slurry reactors (l), and *trickle-bed rs.* (m). Both have a cat. fixed bed with gas and liquid flowing through it. Sometimes a pulsing flow of the liquid through the trickle bed is used in order to increase the gas-liquid

interfacial area (cf. *unsteady-state procs. in cat.). Reactions over suspended cats. are per-formed in BRs or CSTRs (n) or in *bubble col-umns* (o). However, sophisticated techniques are needed to separate the fine solid cat. par-ticles from the liquid phase. By use of coarser cats., fluidized-bed rs. may be applied. They are characterized by the formation of a well-defined agitated cat. bed within the liquid phase. In two-phase rs. the feed is converted in the presence of a dissolved hom. cat. The reaction takes place typically in the same rs. as the three-phase reactions and additionally also in special types like spray rs., thin-film rs., or cat. *membrane reactors (CMRs). 2: for biomimetic reactions cf. *bioreactor. **F** réacteurs catalytiques; **G** katalytische Reak-toren.   P. CLAUS, D. HÖNICKE

Ref.: Farrauto/Bartholomew, p. 199; Augustine, p. 97; Wijngaarden/Kronberg/Westerterp; P.L. Silverstone, *Composition Modulation of Catalytic Reactors*, Gor-don and Breach, London 1998; K.D.P. Nigam, A. Schumpe, *Three-Phase Sparged Reactors*, Gordon and Breach 1996; S.S.E.. Elnashaie, S.S. Elshishini, *Model-ling, Simulation and Optimization of Industrial Fixed Bed Catalytic Reactors*, Gordon and Breach, London 1994; Ullmann, Vol. A5, p. 359.

catalytic reforming This term describes procs. in gasoline refining but it can be used in a wider sense for all commercial procs. that yield valuable prods. in a *refinery proc. The cr. process has been used in petroleum refin-ing for approximately half a century. It con-verts naphthas to gasoline comps. of high anti-knock quality. These comps., mainly aromatic hydrocarbons (*hcs.), are less vulnerable to the preignition procs. responsible for the "knocking" that decreases the power output of an internal combustion engine. The naphtha fraction of crude oil is composed pre-dominantly of saturated *hcs. (alkanes and cy-cloalkanes) which have boiling points within the approximate range of 320–479 K. The ob-jective in cr. is to convert the alkanes and cy-cloalkanes to aromatic hcs. as selectively as possible (cf. also *aromatization). Reaction temps. of 700–800 K and press. of 500–3000 kPa are employed. H formed in the reactions is recycled to the proc. to suppress formation of carbonaceous residues (*coking) on the cat. The cat. naphtha reforming is the second largest refinery process.

The important reactions in cr. are 1: *dehydrognation of cyclohexanes to aromatics; 2: *dehydroisomerization of alkylcyclopentanes to aromatics; 3: *dehydrocyclization of alkanes to aromatics; 4: *isomerization of n-alkanes to branched alkanes; and 5: fragmentation reactions (*hydrocracking or *hydrogenolysis) yielding prods. of lower C number than the reactant. In order to complement the dehydrogenation reaction with an acid-catalyst reaction, which permits isomerization in addition *bifunctional cat. were employed. Since the 1950s Pt and *alumina or *silica-alumina as acid cat. were employed. The original proc. operated at 3–4 MPa press. in a fixed-bed mode. The introduction of *bimetallic cats. in 1967 accomplished a lowering of the operating press. The bimetallic cat. contained Re or Ir, in addition to Pt and acidic alumina. These cats. maintained activity much better than the traditional Pt cat.

Reforming units typically consist of several fixed-bed reactors in series. In such a unit, the naphtha feedstock is vaporized and heated to reaction temp. before it enters the first reactor. As the hcs. in the naphtha undergo reaction during passage through the cat. bed, the temp. of the vapor stream decreases continuously due to the endothermicity of the reaction. The predominant reaction occurring in the first reactor is the dehydrogenation of cyclohexanes to aromatics. Noteworthy are the proc. variants of *Catarol, *Chevron (Isocracking, *Rheniforming), *Exxon Powerforming, *Haldor Topsøe, *Houdry (Houdryflow, Houdryforming), *Howe-Baker, *IFP, *Kellogg KRES, *Phillips STAR, *Standard Oil Ultraforming, *UOP, etc. For reasons of completeness the term cr. has to be expanded to some other refinery proc., e.g., the *steam reforming of hcs. to yield CO and H (e.g., the *ICI syngas proc.) acc. to $C_nH_m + m\ H_2O \rightarrow n\ CO + (n+m/2)\ H_2$. The steam reforming proc. has found broad application in the chemical industry for the prod. of H for *ammonia synth. or the prod. of *syngas for the manufacture of *methanol and *oxo alcohols. Another reforming process that has to be mentioned is the steam reforming of alcohols which generates hydrogen and CO_2. Nowadays this reaction has attracted interest because of the possibility of producing hydrogen "on board" from, e.g., methanol for *fuel cell applications. Cf. also *carbon dioxide reforming. **E=F**; **G** katalytische Reformieren.   R. SCHLÖGL, M.M. GÜNTER

Ref.: Ertl/Knözinger/Weitkamp, Vol. 4, p. 1939; Gates 1, p. 396; Farrauto/Bartholomew, p. 553.

catalytic selectivity   → selectivity

catalytic shaping   → shaping

catalytic site control   another expression for *site control mechanism (*enantiomorphic site control).

catalytic support   → support

catalytic triad   A large number of *enzymes that cat. the *hydrolysis of esters and amides work via nucleophilic cat. where deprotonation of the nucleophile, either serine or cysteine, occurs via the participation of another *active site residue, a histidine. The protonated form of the histidine is usually stabilized by H bonding to a third residue, which may be an aspartic acid (*serine proteases), glutamic acid (*esterases such as *acetylcholine esterase), or asparagine (*thiol proteases). Mutations of any of the three residues to alanine causes a dramatic drop in the cat. efficiency of the enzyme. Interestingly, a particularly active ester-cleaving cat. *antibody that was elicited with a phosphonate *hapten was crystallized and found to have two of the three residues of a cat. triad: the serine and histidine. **G** katalytische Triade.   P.S. SEARS

Ref.: Science **1994**, 265, 1059; Nature (London) **1988**, 332, 564; Angew.Chem.Int.Ed.Engl. **1999**, 38, 2348.

catalytic turnover number → turnover number (TON)

catalytic waste water treatment cat. procs. to remove pollutants from waste water. The nature of the pollutants depends on the waste water source. Frequent pollutants are C-, halogen-, P-, S-, and N-containing cpds. Most of them are efficiently removed by the established biocat. treatment steps (biological waste water treatment). Chemical, preferably cat., procs. have to be applied for the treatment of non-biodegradable and/or toxic cpds. or to achieve high degrees of removal. In cwwt. oxidative procs., aspiration methods to achieve a complete mineralization of the pollutants (formation of CO_2, H_2O, etc.) are predominant and already used for commercial wwt. These procs. are divided into wet air oxidation (WAO), supercritical water oxidation procs. (SCWO), photocat. oxidation, and advanced oxidation procs. (AOPs; cf. *titania photocatalysis). Depending on the pollutant and proc., air, oxygen, ozone, or H_2O_2 can serve as source for the oxidizing species, which is in the vast majority the hydroxyl *radical, and both hom. (Fe or Mn salts, e.g., *Fenton's reagent) and het. cats. (broad variety of metal oxides, supported metals on metal oxides) can be applied. Reductive procs., i.e., *hydrogenation procs., for wwt. like *dehydrochlorination; chlorinated cpds. → non-chorinated cpds. + HCl) or *hydrodenitrification (*cat. nitrate reduction → nitrogen) are emerging. For the use of solvents under supercritical conds. in waste water treatment cf. *supercritical phase cats. **G** katalytische Abwasserbehandlung. U. PRÜSSE, K.-D. VORLOP
Ref.: Ullmann, Vol.B8, p. 1; *Ind.Eng.Chem.Res.* **1998**, 37, 309.

Catarol process older *reforming proc. for the prod. of aromatics from aliphatic fractions of refineries by thermal treatment over Cu cats. Further developments are based on *Pt and on *Re cats. **F** procédé Catarol; **G** Catarol-Verfahren. B. CORNILS

Catasulf process → BASF procs.

catcracking → catalytic cracking

catechol 1,2-dioxygenase (EC 1.13.11.1) a non-heme Fe containing *oxidoreductase that cat. the ring-opening *oxidation of pyrocatechol to *cis,cis*-muconate. A variety of substituted pyrocatechols are also accepted. **F=E**; **G** Brenzcatechin 1,2-dioxygenase: P.S. SEARS

Ref.: *J.Biol.Chem.* **1983**, 258, 14422; Schomberg/Salzmann, Vol. 8.

Catfin process → Air Products procs.

cathepsins a diverse group of lysosomal *proteases: cathepsins (cs.) B, S, K, H, and L are *thiolproteases, for example, while cs. C, D, and E are aspartic acid *proteases and c. G is a *serine protease. Although their primary function is to degrade proteins within the lysosome, cs. are found in a number of other subcellular locations, including the cytoplasm and the plasma membrane, and procathepsins (inactive *precursors) have also been found in the extracellular space. A large amount of interest in the c. has resulted from the discovery that certain cs. are elevated in a variety of disease states. For example, the levels of several cs. (esp. B and D) are elevated in many cancer cell lines, and correlate with their metastatic potential. C. K is elevated in osteoclasts, the cells that reabsorb bone, and thus its inhibition may be a potential route for treatment of osteoporosis. **E=F=G**. P.S. SEARS
Ref.: *Biochem.Cell.Biol.* **1996**, 74, 799; *Expert Opin. Invest.Drugs* **1997**, 6, 1199; *Eur.J.Cancer* **1992**, 28A, 1780.

Cativa technology → BP procs.

Catofin process → Houdry procs., *Lummus procs.

CAT-resins *hydrocarbon resins (mostly epoxy resins) in which *transition metal (TM) cpds. either increase the rate of curing of amino- or anhydro-cured epoxy systems or initiate the homopolymerization of epoxy resins. These metal *complexes often respresent so-called *thermally or photochemically latent acelerators* in one-pot compositions. TM 1,3-diketonates exhibit latent cat. properties in the curing of bisphenol-A epoxy resins with tetrahydrophthalic anhydride; matrices formed by *cross-linking of those epoxy resins with metal carboxylates, *chelated with polyamine *ligands, show higher strength and lower heat deformation. To cat. the anionic homopolymerization of epoxy resins latently, the original curing amines or nitrogen hetrocycles (e.g., imidazoles, *tert.* amines) are cordinated in TM *complexes which have excellent stability at rt. and which cure rapid at elevated temps.

TM-imidazole and metal-amine complexes (CATs) are very suitable for one-pot applications in industrial practice for the manuf. of composite materials (e.g., prepregs, resin transfer molding, resin injection, etc.). The complex nature of these curing cat. (e.g., by variation of the metals) influences not only the gel-time but also determines the physical properties of the final epoxy material (e.g., glass transition temp.) mainly because of suitable morphology and cross-link density. **E**=**F**; **G** CAT-Harze. M. DÖRING

Ref.: W.R. Ashcroft, *Chemistry and Technology of Epoxy Resins*, Blackie Academic, London 1993, p. 58; Parshall/Ittel.

CAV carbamoyl methyl viologen, → enoate reductase

cavities geometrical caves in, e.g., *pores, *cyclodextrins, *template cats., proteins (*chaperones), or *zeolites; cf. also *pore models. B. CORNILS

Ref.: E.G. Derouane, M.S. Whittingham, A.J. Jacobson (Eds.), *Intercalation Chemistry*, Academic Press, New York 1982; Gates-1, p. 254.

CBR circulating bed reactor; cf. *cat. cracking, *catalytic reactors

cbz, CBZ carbobenzoxy as protective group in peptides

C-C activation The aliphatic C-C single bond is thermodynamically and kinetically inert with regard to fission because the sp^3 orbitals along the C-C axis are inaccessible to metals and thus show little reactivity. Therefore, C-C bond fusion is often coupled to auxiliary reactions (e.g., *hydrogenation, *C-H activation prior to C-C bond breaking). Active metal sites are necessary to induce such reactions, the simplest of which corresponds to an oxidative addition (Figure 1).

Depending on the electron density at the metal M, C-C bond cleavage can be thermodynamically more favorable than C-H fusion. The main challenge is, however, the consecutive functionalization of the metal-attached hydrocarbon group. C-C activation can be facilitated by steric strain (Figure 2).

L = P(C₆H₅)₃
R = C₆H₅
X = C₂H₅O, C₆H₅

The *dismutation of saturated hydrocarbons (*metathesis of alkanes) can be achieved by structurally well-defined surface-attached organometallic (*SOMC) cats. Also the degradation of polymers such as *PE or *PP under hydrogen to give mainly ethane and methane opens certain chances to recycle polymers. **F** activation C-C; **G** C-C-Aktivierung. W.A. HERRMANN

Ref.: Cornils/Herrmann-1, p. 1179; *Nature (London)* **1993**, *364*, 699; *Science* **1994**, *265*, 359 and **1996**, *271*, 966; FR 9.508.552 (1995).

C-C bond formation 1: general: is a class of reactions in which C-C bonds are newly formed. Moreover, C-C bond forming reactions are the key steps for the building of nearly every *complex molecule. Commercially extremely important C-C bfs. include, among others, *transition metal cat. *polymerizations of alkenes (*polyolefins such as *HDPE, *LDPE, or *PP), *carbonylations (especially *Monsanto's AA proc. and *hydroformylation), *hydrocyanation, *aldol and similar reactions, olefin *dimerization, *metathesis, *telomerization, etc. Especially useful for the formation of C-C bonds are *Friedel-Crafts reactions. Important examples of reactions which form C-C bonds within a ring are pericyclic convs. such as *Diels-Alder reactions and [3+2] *cycloadditions, as well as *cyclooligomerization of alkynes and *Pauson-Khand reactions. In recent years large efforts have been devoted towards new environmentally friendly methods for C-C bond formation: specially the substitution of classical organic reactions producing stoichiometric amounts of byprods. by cat. routes is of interest (*atom economy, *E-factor). This effort is demonstrated clearly by the increased use of metathesis reactions instead of "classical" *Wittig reactions or other methods to generate double bonds. Transition metal cat. C-C coupling procs. are of special value for the generation of bonds between sp and sp^2 carbon atom centers. Here, metal-cat. *cross-coupling reactions (*McMurray,*Stille, *Negishi, *Suzuki) and couplings with unsaturated cpds. (*Heck reaction) are of special interest. **F** formation de C-C bonds; **G** C-C-Bindungsbildung. M. BELLER
2: enzymatic: cf. *aldolases, *isoprenoids, *squalane synthesis.
Ref.: Weissermel/Arpe; Cornils/Herrmann-1, pp. 29, 220, 465, 552; Beller/Bolm; Trost/Fleming.

CCC Criterion Catalyst Company, → catalyst alliances

CCR → continuous catalyst regeneration, → regeneration

CD → cyclodextrin

CDCumene process → CDTech procs.

cddt, CDDT cyclodecatriene as a *ligand or in *complexes

CdF-Chimie Charbonnage de France

CdF Norsorex process first proc. for the cat. *metathesis of norbornene (bicyclo [2.2.1]hept-2-ene) to the *polyalkenamer Norsorex over W cats. **F** procédé Norsorex de CdF, **G** CdF Norsorex-Verfahren. B. CORNILS

CdF/Technip ethylbenzene process manuf. of *ethylbenzene by *alkylation of benzene by means of *Friedel-Crafts cats. The CdF/Technip proc. uses anhydrous ammonia for washing. **F** procédé CdF/Technip pour l'EB, **G** CdF/Technip EB-Verfahren.
B. CORNILS

CDHydro → Lummus procs.

cdt, CDT cyclododeca-1,5,9-triene, inter alia, as a *ligand and in *complex cpds.

CDTech the partnership between ABB Lummus Global and Chemical Research and Licensing, Co. Cf. also *catalyst alliances.
B. CORNILS

CDTech CDCumene process manuf. of high-purity *cumene from propylene and benzene using specially formulated, isothermal *zeolite *alkylation cat. packaged in proprietary structures and another *zeolite *transalkylation cat (Figure, see page 112). **F** procédé CDTech pour le cumène; **G** CDTech Cumolverfahren. B. CORNILS

CDTech ether process cat. manuf. of *MTBE (and other ethers) from methanol and isobutene using a "boiling point reactor" (a fixed-bed downflow adiabatic reactor, where the liquid is heated and restricts the reaction temp.) and a *cat. distillation. **F** pro-

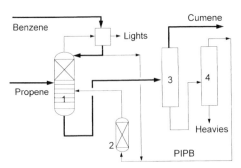

1 catalytic distillation; 2 transalkylator; 3,4 distillation; PIPB = polyisopropyl benzene

CDTech CDCumene process

cédé CDTech pour les éthers; **G** CDTech Ether-Verfahren. B. CORNILS
Ref.: *Hydrocarb.Proc.* **1996**, (11), 113; Ertl/Knözinger/ Weitkamp, Vol. 4, p. 1995.

CDTech ethylbenzene process manuf. of *EB from benzene and ethylene over *zeolite cats. in *catalytic distillation towers. This technique is said to minimize the polyalkylation. **F** procédé CDTech pour l'EB; **G** CDTech EB-Verfahren. B. CORNILS
Ref.: Ertl/Knözinger/Weitkamp, Vol. 5, p. 2127.

CDTech Isomplus process for the skeletal *isomerization of *n*-butenes to *i*-butylene over fixed-bed, *zeolite-based cats. The proc. needs no addition of cat. *activation agents to promote the reaction. **F** procédé Isomplus de CDTech; **G** CDTech Isomplus-Verfahren.
Ref.: *Hydrocarb.Proc.* **1996**, (11), 138. B. CORNILS

Cech Thomas R. (born 1947), professor of molecular biology at Boulder, CO (USA); discovered that not only *enzymes but RNA is catalytically active (*ribozymes). Together with Altman he was awarded the 1989 Nobel prize. B. CORNILS
Ref.: *Angew.Chem.* **1990**, *102*, 735.

Celanese catalysts Celanese manuf. cats. at Oberhausen/Germany (the former Ruhrchemie plant site) since 1936. The product portfolio consists of supported Ni, Co, and Cu process cats. with the trade name CelActiv used in fine and intermediate chemicals processing mainly for *hydrogenation and *amination reactions and the Hoecat product line of cats. for processing fats, oils, and fatty acids. J. KULPE

Celanese hexamethylene diamine (HMDA) process older proc. for the manuf. of *HMDA from cyclohexanone via caprolactone and its *hydrogenation with Raney Cu or *Adkins cat. at 250 °C/28 MPa to hexane-1,6-diol. This diol is *ammonolyzed over Raney Ni cat. with H_2/NH_3 at 200 °C/23 MPa to HMDA. **F** procédé Celanese pour l'HMDA, **G** Celanese HMDA-Verfahren.
B. CORNILS

Celanese Low Water Technology → Hoechst-Celanese Low Water Technology

Celanese *LPO process manuf. of *acetic acid by liquid-phase *oxidation of *n*-butane with oxygen at 175 °C/5 MPa. The cat. is homogeneous Co acetate. Other aldehydes, ketones, and acids are byproducts, e.g. up to 18 % of methyl ethyl ketone. **F** procédé LPO de Celanese; **G** Celanese LPO-Verfahren.
B. CORNILS

***Cellulomonas* sp. glycerol dehydrogenases** → glycerol dehydrogenase

CE Lummus process → Lummus procs.

CEMS → conversion electron Mössbauer spectroscopy

central atom cas. (to which other atoms or atom groups, the *ligands, are attached) are the constituents of *complexes (coordination entities); cf. *coordinastion chemistry theory. Cas. often are charged *transition metal ions. The ligand atoms directly bound to the ca. form the coordination polyhedron. The coordination number of the ca. is the number of σ-bonds between the ligands and the ca. (special rules exist for some organometallic complexes). The oxidation number (state) of a ca. is defined as the charge which results, when all ligands, including their binding electron pairs, were removed from the ca. In cat. active complexes, the ca. together with the characteristics of the ligand sphere determines their basic cat. behavior, selectivity, and efficiency. Cf. also *complex. **F** atom central; **G** Zentralatom.   K. KÖHLER

Ref.: M. Bochmann, *Metallorganische Chemie der Übergangsmetalle*, Wiley-VCH, Weinheim 1997.

cephalosporin An antibiotic which inhibits a transpeptidase involved in the synth. of the bacterial cell wall. See also *ACV synthetase. **E=F=G.**   C.-H. WONG

ceramics in catalysis are used in cat. procs. as cats., high *surface area coatings referred to as *washcoats, and cat. *supports. Ceramics (c.) are defined as all-inorganic non-metallic solids. As cat., rare-earth oxide c. such as La_2O_3, *zeolites such as Cu/ZSM-5, and *transition metal oxides are used, among others. These cats. and their applications are discussed elsewhere. As coatings and supports, c. are critical to the operation of many commercial cats. because they are generally chemically inert, thermally stable, and low in cost. The c. carrier often plays a role in the cat. proc. as well; for example, CeO_2 is added to the washcoat of the *automotive *three-way cat. as an oxygen storage comp.
Cats. are usually placed on high surface area c. material. Some examples are SiO_2, TiO_2, *zeolites, special forms of C such as *activated carbon or graphite, and the crystalline γ-phase of Al_2O_3. This form of *alumina with its advantages of high surface area and low cost can be used as a washcoat placed on a more substantial substrate or it can be formed into beads which are packed randomly in a cat. bed. These high *surface area materials are usually fired at temps. up to about 700 °C and used no higher than approx. 1100 °C.
Substrates are used in cat. procs. when enhanced mechanical durability is required. C. are usually used as the substrate materials onto which high surface area coating and cat. are placed. The substrate can have any of several forms and be composed of any of a number of c. materials, depending of the specific chemical, thermal, and mechanical requirements. Saddles, beads, *honeycombs, foams, and several other structures are available (*forming, *shaping). Saddles and beads are packed randomly in the cat. system and so tend to have considerable mass and press. loss but good *mass tranfer performance. The foam structure gives good mass transfer and has low substrate mass but the channel structure is random, leading to high and variable press. loss. In contrast to the high surface area materials, the supports are processed up to temps. of about 1600 °C.
The honeycomb (*monolithic honeycomb) or cellular structure is uniform in channel shape, both within a single part and from part to part. The mass of this cat. support is moderate, being between that of beads and foam structures. The press. loss is low and the mass transfer is moderate to good depending on the particular application and the design of the substrate. Most automotive cat. supports are cellular c. prods. Although *mullite, *alumina, and porcelain are all available in the cellular structure, the c. substrate material most widely used is *cordierite, an Mg aluminosilicate. The cordierite cellular structure is especially well suited to the rigors of the automotive cat. application due to an especially low thermal expansion coefficient and moderate porosity. The former provides very good thermal shock resistance and the latter allows the high surface area washcoat to be tightly bonded to the underlying substrate. **F** catalyseurs céramiques; **G** keramische Katalysatoren.   J. PAUL DAY

Ref.: A. Cybulski, J.A. Moulijn (Eds.), *Structured Catalysts and Reactors*, Dekker, New York 1998; R.M. Heck, R.J. Farrauto, *Catalytic Air Pollution Contol – Commercial Technology*, van Nostrand Reinhold; New York 1995; C. Misra, *Industrial Alumina Chemicals*, ACS Monograph 184, ACS, Washington DC 1986.

ceria CeO_2, used, inter alia, as a *support for het. cats.

cermets are "metal ceramics" or "ceramic metals", hom. materials consisting of two constituents very different in melting point and hardness. They are manuf. by powder metallurgy or coated by flame spraying. Although in some cases similar in chemical composition, they are high-temp. composites and different from *supported cats. as a consequence of their manuf. procedure. They have been proposed as cats., e.g., for cat. layers in *fuel cells. **E=F=G**.

B. CORNILS

Ref.: Kirk/Othmer-2, p. 596; Ullmann, Vol. A6, p. 74; Winnacker/Küchler, Vol. 4, p. 607; V.I. Trefilov, *Powder Metallurgy and Metal Ceramics*, Kluwer, Dortrecht 1999.

Cetus PO process multi-step biosynthetic proc. for *PO. The first step converts D-glucose by an enzymatic-oxidative reaction to H_2O_2, followed by the *oxidation of propylene over *immobilized *enzymes. The by-prod. is D-fructose through Pd- cat. *hydrogenation of a D-glucose intermediate. **F** procédé Cetus pour PO; **G** Cetus PO-Verfahren.

B. CORNILS

C-F activation The functionalization of polyfluorinated organic cpds. is difficult. First success is based on the stoichiometric reaction of hexafluorobenzene with $L_3Rh\text{-}SiR_3$ the thermodynamics of which clearly relies on the strength of the Si-F bond (L = $P(CH_3)_3$, R = phenyl, Figure).

The Rh-*phosphine *complex was found to allow the cat. *hydrogenolysis of hexafluorobenzene: when this fluorocarbon is used as a solvent, it eliminates F upon heating with the cat. $(PMe_3)_4RhH$ in the presence of a base like NEt_3 and of hydrogen (TON < 114). Het. cat. hydrogenolysis of C-F bonds is known; however, it requires very high temps. and is nonselective. **F** activation C-F; **G** C-F-Aktivierung.

W.A. HERRMANN

Ref.: M. Hudlicky, *Chemistry of Organic Fluorine Compounds*, 2nd Ed., p. 175, Prentice Hall, New York 1992; *Angew.Chem.Int.Ed.Engl.* **1997**, *36*, 1048; *Chem. Rev.* **1994**, *94*, 373.

CFC chlorofluorocarbons. **F** chlorofluorocarbones; **G** Fluorchlorkohlenstoffe (FCK).

CFC destruction method to remove CFCs (and HCFCs, hydrochlorofluorocarbons) from the environment. The proposed techniques include combustion in air, oxygen, ammonia, or water atmospheres, *catalytic combustion, or cat. reaction with ammonia over LaF_3 yielding HCN, F, HCl, and N_2. **F** destruction des chlorofluorocarbones; **F** Zerstörung von FCK (und FCKW). B. CORNILS

Ref.: Ertl/Knözinger/Weitkamp, Vol. 4, p. 1683.

CFR continuous-flow reactor

CFSTR continuous-flow stirred tank reactor, → catalytic reactors

C-H activation In general, any reaction in which a metal *complex weakens ("activates") a C-H bond within a *ligand (usually classified as *agostic interaction). C-Ha. can result in C-H bond cleavage, which is best known from intramolecular decomposition pathways such as *cyclometallation (e.g., *orthometallation in undercoordinated tris (aryl)phosphine complexes of the late *transition metals) or α- and β-hydrogen elimination reactions (e.g., in alkyl *complexes of *Lewis acidic metal centers). Cat. relevant intermolecular C-Ha. can occur acc. to four pathways. 1: A *radical pathway in the presence of very strong oxidants or at high temps. (e.g., the *enzymatic oxidation of alkanes in *cytochrome P-450, the *Gif system, and the mercury-sensitized photochemical *dimerization of alkanes, cf. *photocatalysis). 2: The *oxidative addition of hcs. to a low-valent metal center to form an (organo) (hydrido)metal complex (e.g., the reaction of CpIrH $(PMe_3) \cdot (CH_2Cl_2)$ with methane to give $CpIr(H_2)(PMe_3)CH_3$ or the two-centered oxidative addition of alkanes to the tetramesitylporphyrin-Rh dimer). 3: An electrophilic pathway with a number of late transition metal ions (e.g., the *oxidation of C-H bonds of alkanes in water by a $PtCl_4^{2-}/PtCl_6^{2-}$ system). A related ligand-assisted four-centered electrophilic activation is possible (e.g., in lanthanide and actinide hydride or alkyl complexes). 4: The Pt-cat. nucleophilic attack acc. to the *Shilov reaction (cf. *Pt as cat. metal). Drawbacks of the cat. C-Ha. are low activities due to the high C-H bond energy, the lack of inertness of common ligands and solvents, poor *selectivities due to comparable reactivities of different C-H bonds within the substrate molecules, and the difficulties in suppressing the activation of the usually more reactive C-H bonds of the prod. molecules. **F=E**; **G** C-H Aktivierung. J. EPPINGER

Ref.: *Chem.Rev.* **1995**, *95*, 987; Cornils/Herrmann-1, p. 1081; Hill; *Angew.Chem.Int.Ed.Engl.* **1999**, *38*, 1699.

chain-end control Two mechanisms are responsible for the *stereocontrol of a growing polymer chain in *Ziegler-Natta cat.: the *site-control and the chain-end control mechanisms. In the cec. mechanism the configuration of the last inserted monomer is responsible for the stereocontrol; in the site-control mechanism (also called *catalytic site control or *enantiomorphic site control) the chirality of the cat. site is the responsible factor. The cec. mechanism can be distinguished from site-control by studying the errors of insertion in the polymer chain. While in a site-control mechanism the error affords the formation of a single odd insertion, cec. leads to a polymer, in which the error will be propagated in the growing chain until another misinsertion occurs. The resulting polymer has a stereo block structure. **G** Kettenendkontrolle.

W. KAMINSKY, C. SCHWECKE

Ref.: *J.Polym.Sci Part B* **1965**, *3*, 23; *Macromolecules* **1986**, *19*, 2465; *J.Am.Chem.Soc.* **1984**, *106*, 6355; *Cattech* **1977**, (Dec.), 79.

chain-growth factor The shape of the *Schulz-Flory distribution and the chain length of the *α-olefins are controlled by the geometric chain-growth factor K. This factor is used, e.g., in the *Shell SHOP proc., where the value is usually between 0.75 and 0.80. **F** facteur de croissance de la chaîne; **G** Kettenwachstumsfaktor.

W. KAMINSKY, C. SCHWECKE

Ref.: *J.Am. Oil Chem. Soc.* **1983**, *60*, 653; Cornils/Herrmann-1, p. 254; Mark/Bikales/Overberger/Menges, Vol. 274, p. 288.

chain reaction → radical polymerization

chaperones proteins that aid in the proper folding (and unfolding) of other proteins. Some require an energy source, such as *ATP. They include the foldases, *enzymes that cat. protein *isomerization reactions such as *proline *cis-trans* isomerizations and disulfide exchange reactions. Many (e.g., hsp70) are heat shock proteins, i.e., produced in response to a high-temp. pulse. One family of chaperones, the chaperonins, (e.g., GroEL) are cagelike protein *oligomers with a central *cavity that encompasses the (misfolded) protein substrate, and at least part of the function of such chaperonins is to sequester the misfolded protein to prevent aggregation. **E=F**; **G** Chaperone. P.S. SEARS

Ref.: *Biochem.J.* **1998**, *333*, 233; *FEBS Lett.* **1998**, *425*, 382.

charcoal → activated carbon

Chatt-Dewar-Duncanson model → Dewar-Chatt-Duncanson model

chelate effect results from an increase in entropy by displacing two or more monodentate *ligands by at least one chelate ligand.

W.R. THIEL

chelates, chelating ligands *Multidentate *ligands (ligands bearing two or more donor centers) are called chelating ligands since they can form cyclic structures by coordinating a central metal with more than one of the donor moieties. Generally, chelate *complexes show an enhanced stability toward *ligand exchange compared to complexes with monodentate ligands. There are thermo-dynamic and kinetic reasons for this behavior: the *chelate effect results from an increase in entropy by displacing two or more monoden-tate ligands with at least one chelate ligand. On the other hand, the metal center is shielded against ligand association procs. by the bulky backbone of the chelate ligand.

Owing to the enhanced stability of chelate complexes, multidentate ligands play a domi-nant role in cat. Bidentate *phosphines allow reduction of the P/Rh ratio in Rh-cat. *hydro-formylation (e.g., the ligands *BINAPHOS, *BINAS, *BISBIS). A rapidly increasing field is the application of chelate ligands in *enan-tioselective *transition metal cat. or mediated synth. Here, chelate complexes generally show a better transfer of chirality to the sub-strate, resulting from a more rigid complex geometry and a reduced ligand dissociation tendency compared to complexes with mono-dentate ligands. **F** ligands chélatés; **G** Chelat-liganden. W.R. THIEL

Ref.: Kirk/Othmer, Vol. 5, p. 764; *J.Mol.Catal. A.* **1999**, *143*, 85.

Chem. acronym for *chemisorption

chemical adsorption → chemisorption

chemical intermediate → intermediate

chemical ionization mass spectroscopy (CI-MS) a variant of *mass spectroscopy. It delivers kinetic information and is applied to gas-phase, time resolved, *isotope labeling. The probe needs a sample (reactive auxiliary) gas, the response are ions. The technique is destructive but well suited for the monitoring of cat. procs. (in the gas phase). R. SCHLÖGL

chemically induced dynamic nuclear polarization → CIDNP

Chemical research and Licensing Co. (CR&L) developed cat. procs. for licensing (e.g., MTBE, catalytic distillation, etc.).

chemical sensors (in catalysis) devices which respond with an electric signal to analy-tical information (e.g., press., conc., etc.) which are based on a chemical state. The re-sponses of some sensors are based upon inter-actions between *electrochemistry and cataly-sis. Best known in cat. is stoichiometric (auto-motive) engine operation controlled by an oxygen *sensor (lambda sensor; cf. *TWC). Another connection between sensors and cat. is the use of cat. filters (*guard bed) before entering of the material flow to be measured. **G** chemische Sensoren. B. CORNILS

Ref.: Ullmann, Vol.B6, p. 122; Ertl/Knözinger/Weit-kamp, Vol. 3, pp. 1283, 1335.

chemical vapor deposition (CVD) meth-od for the manuf. of thin films by gas-phase deposition (sputtering) of metals or cpds. (MOCVD is molecular CVD using *single-source *precursors) in ultra-high vacuum and/ or *plasma devices (*metal film cats.). So far, CVD cats. have no commercial importance for cat. **F=E=G**. B. CORNILS

Ref.: Blackborrow/Young; T. Kodas, M. Hampden-Smith (Eds.), *The Chemistry of Metal CVD*, VCH, Weinheim 1994; H.O. Pierson, *Handbook of Chemical Vapor Deposition*, Noyes Publ., New Jersey 1992; Ertl/Knözinger/Weitkamp, Vol. 2, p. 853.

Chemie Linz fumaric acid process manuf. of fumaric acid by *isomerization of aqueous maleic acid to fumaric acid. **F** pro-cédé Chemie Linz pour l'acide fumarique, **G** Chemie-Linz Fumarsäure-Verfahren.

B. CORNILS

Chemische Werke Witten → Witten DMT process

chemisorption *adsorption which involves the formation of a chemical bond between the adsorbed atom/molecule and the *surface. C. is acc. characterized by *adsorption enthalpies comparable to chemical bond energies, i.e., typ-ical adsorption enthalpies for c. are larger than 40 kJ/mol. There is a distinction between *disso-ciative c.* when the adsorption involves the breaking of an intramolecular bond (e.g., Pt/O_2,

Ni/H$_2$) and *associative c.* (e.g., Pt/CO) when the molecule remains intact upon adsorption. The main criterion to distinguish *physisorption and c. is the adsorption enthalpy but spectroscopic techniques are also useful, e.g., vibrational spectroscopy demonstrates the appearance of new vibrations and the modification of existing vibrations when the molecule is bonded to the *surface, or photoelectron spectroscopy shows changes in the electronic structure due to the formation of chemical bonds. **E=F=G**.

R. IMBIHL

Ref.: Somorjai; Ertl/Knözinger/Weitkamp, Vol. 3, p. 911,942; R.P.H. Gasser, *An Introduction to Chemisorption and Catalysis by Metals*, Oxford University Press, Oxford 1985.

chemoselectivity A reaction is called chemoselective when it converts exclusively (chemospecifically) or preferentially (chemoselectively) a functional group (e.g., -CHO in the case of selective *hydrogenation) in the presence of another reactive group (such as, e.g., a C-C double bond). An example is the selective hydrogenation of crotonaldehyde (but-2-enal, CH$_3$-CH=CH-CHO) either to crotyl alcohol (but-2-enol, CH$_3$-CH=CH-CH$_2$OH) or to *n-butyraldehyde (CH$_3$-CH$_2$-CH$_2$-CHO); cf. also *regioselectivity, *selectivity. **F** chémosélectivité; **G** Chemoselektivität.

Ref.: Augustine, p. 315.   B. CORNILS

ChemSystems PO process for the TeO$_2$/I$_2$ cat. *acetoxylation of propylene with acetic acid and oxygen to 1,2-diacetoxypropane. *Hydrolysis yields propylene glycol, *thermolysis PO. **F** procédé ChemSystems pour le PO; **G** ChemSystems PO-Verfahren.   B. CORNILS

Chevron Isocracking (Isomax) process combination of *catcracking and *cat. reforming through conv. of naphtha, *LGO, *HGO, resids, etc. on fixed-bed cats. at 260–400 °C/5–14 MPa. The latter Isomax proc. contains elements of *UOP's Lomax proc. **F** procédé Isocracking de Chevron; **G** Chevron Isocracking-Verfahren.   B. CORNILS

Ref.: *Hydrocarb.Proc.* **1978**, (9), 119; *Hydrocarb.Proc.* **1996**, (11), 124; McKetta-2, p. 883.

Chevron Isodewaxing process *dewaxing and wax *isomerization (for lube oil prod.) on cats. which contain a *hydrogenation comp. (e.g., Pt) on a *SAPO *molecular sieve. **F** procédé Isodewaxing de Chevron; **G** Chevron Isodewaxing-Verfahren.

B. CORNILS

Ref.: Ertl/Knözinger/Weitkamp, Vol. 4, pp. 2014, 2032.

Chevron Isomax process is similar to *Chevron's Isocracking and consists of a two-step proc. 1: the first step comprises *HDS, *HDN, and *hydrocracking; 2: in the *isocracker the cleavage and *isomerization of paraffins and alkylated aromatics, is followed by subsequent *hydrodealkylation (*HDA) of naphthenes. The proc. was lateron licensed also by *UOP. **F** procédé Isomax de Chevron; **G** Chevron Isomax-Verfahren.   B. CORNILS

Chevron RDS and VRDS processes variants of the *hydrotreating proc. for the one-through operation of hydrocarbons from *VGO and *VRDS by cat. treatment. The sulfur content decreases from 3–5 % to 0.5–1 %. **F** procédés RDS et VRDS de Chevron; **G** Chevron RDS und VRDS-Verfahren.

B. CORNILS

Ref.: *Hydrocarb.Proc.* **1978**, (9), 148; *Hydrocarb.Proc.* **1996**, (11), 132.

Chevron Rheniforming process cat. *reforming proc. to convert low-octane naphthas to high yields of high-octane gasoline (RON clear 103) blendstock or aromatics plant chargestock by fixed-bed operation on *bimetallic cats. (Pt-Re; cf. *Re as catalyst metal). **F** procédé Rheniforming de Chevron; **G** Chevron Rheniforming-Verfahren.

B. CORNILS

Ref.: *Hydrocarb.Proc.* **1978**, (9), 165; Winnacker/Küchler, Vol. 5, p. 59.

Chevron VGO and DAO process *hydrotreating proc. to *desulfurize high-boiling, high sulfur-content gas oils by one-through operation on fixed-bed cats. **F** procédé VGO

et DAO de Chevron; **G** Chevron VGO und
DAO-Verfahren. B. CORNILS
Ref.: *Hydrocarb.Proc.* **1972**, (9), 184.

Chevron VRDS process → Chevron RDS
and VRDS processes

Chichibabin reaction direct *amination
of pyridine in the 2-position using Na or Li
amide as cats. in hot toluene or dimethylani-
line. A second amino group may be introduced in the 6-position. When these positions
are occupied, the 4-position is aminated. 2-
Amino-3-picoline and 2-amino-4-picoline are
obtained in 70–80 % yield from picoline.
Quinoline and its derivatives may be converted analogously. Together with $KMnO_4$,
Sodium amide in liqu. NH_3 also allows the
amination of di-, tri-, and tetrazines. **F** réaction
de Tchitchibabin; **G** Tschitschibabin-Reaktion.
H. BÖNNEMANN
Ref.: *Adv.Heterocycl.Chem.* **1988**, *44*, 1.

chi (χ) value is an electronic parameter for
*phosphine ligands as defined by *Tolman (cf.
*Tolman concept). χ is the difference in the
IR frequencies of the symmetric CO stretches
of the vibrational spectrum of $Ni(CO)_3P(t-Bu)_3$ as the reference as compared to another
ligand to be measured. Per definitionem the
χ-value of $Ni (CO)_3P(t-Bu)_3$ is 0, that of PPh_3
is 13, that for $P(CF_3)_3$ is 59. B. CORNILS
Ref.: Moulijn/van Leeuwen/van Santen, p. 204.

chiral auxiliaries → auxiliary compounds

chiral catalysts Most of the synth. *chiral
cats. are based on *transition metals, modified
by optically active *additives, or *transition
metal *complexes, containing optically active
*ligands. Ccs. may be applied het. or homoge-
neously. Also, organic cpds. such as *amino
alcohols can act as ccs. The majority of the
*enzymes are enantiospecific chiral cats. **F** catalyseur chiral; **G** chiraler Katalysator.
H. BRUNNER

chiral compounds *asymmetric cpds.
which can be converted via substitution into
diastereomers.

chiral methyl groups The methyl group
becomes chiral when two of the hydrogens
are replaced with deuterium and tritium (cf.
*isotope labeling). Acetate and pyruvate with
chiral methyl groups have been used to study
the *stereochemistry of biochemical methyl
transfer reactions. **F** groupes méthyles chirals;
G chirale Methylgruppen. C.-H. WONG
Ref.: *Nature (London)* **1969**, *221*, 1212, 1213; *Methods
Enzymol.* **1982**, *87*, 126.

chiral pool includes all *chiral natural
products such as carbohydrates, hydroxycar-
boxylic acids, amino acids, terpenes, alkaloids,
etc. Particularly important as starting material
for the synth. of chiral *ligands and cats. for
*asymmetric catalysis. The Figure shows se-
lected chiral ligands (*chelating ligands) derived from the chiral pool (trivial name/cata-
lytic application/source in the chiral pool).
E=F=G. A. BÖRNER

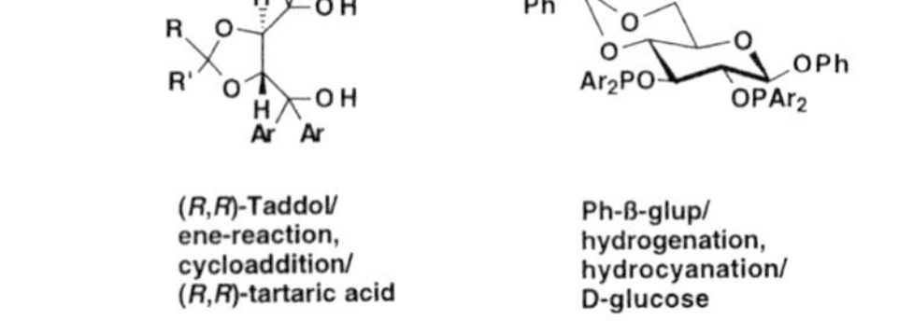

(*R,R*)-DIOP/
hydrogenation/
(*R,R*)-tartaric acid

RoPHOS/
hydrogenation/
D-mannitol

(*S,S*)-pybox/
hydrosilylation,
acyloxylation/
amino acids

(*R,R*)-Taddol/
ene-reaction,
cycloaddition/
(*R,R*)-tartaric acid

Ph-ß-glup/
hydrogenation,
hydrocyanation/
D-glucose

Ref.: R. Scheffold (Ed.), *Modern Synthetic Methods*,
Vol. 2, p. 91, Frankfurt 1980; Beller/Bolm, Vol. 2, p. 3;
J.Org.Chem. **1998**, *63*, 8031; D.C. Walker (Ed.), *Origins
of Optical Activity in Nature*, Elsevier, New York 1979;
Colin/Sheldrake/Crosby.

CHIRAPHOS the well-known optically active *ligand [(2*R*,3*R*)- or (2*S*,3*S*)-bis(diphenylphosphino)butane] for hom. cat., obtained by

tosylation or mesitylation of the enantiopure (2*R*,3*R*)-(–)-butanediol and subsequent reaction of (2*R*,3*R*)-(+)-butanediol-(di-*p*-toluene sulfonate) or dimesitylate with $(C_6H_5)_2PM$ (M = Li, K) (Figure).

H_3C CH_3 CH_3 CH_3

H——⟩—⟨····H   H····—⟩—⟨——H

$(C_6H_5)_2P$ $P(C_6H_5)_2$ $(C_6H_5)_2P$ $P(C_6H_5)_2$

(2R,3R)-CHIRAPHOS   (2S,3S)-CHIRAPHOS

CHIRAPHOS may be sulfonated with fuming sulfuric acid to give the *water-soluble derivative. Rh *complexes of Rh[CHIRAPHOS(cod)]ClO_4 are active in the *hydrogenation of prochiral alkenes; CHIRAPHOS/Pd complexes mediate *C-C coupling reactions and allylic *alkylations. Cu(II) triflate complexes cat. *asymmetric conjugate additions.   O. STELZER

Ref.: *J.Am.Chem.Soc.* **1977**, *99*, 6262; *Inorg.Synth.* **1997**, *31*, 131 and **1998**, *32*, 36; *Tetrahedron Lett.* **1995**, *36*, 2051; *Tetrahedron Asymmetry* **1991**, *2*, 663 and **1997**, *8*, 3987.

chitinase a *glycosidase which cat. the *hydrolysis of chitin, a polymer composed of *N*-acetylglucosamine linked via β-1,4-linkages. **E=F=G**.   C.H. WONG

Chiyoda CT-BISA process manuf. of bisphenol-A reacting phenol and acetone in a synth. reactor which is packed with an *ion (cation)-exchange resin cat. Water of reaction from the *condensation reaction is simultaneously stripped out of the reaction mixture. Reaction effluents flow directly to a primary crystallization; the impurities are removed into the mother liquor. A secondary mother liquor, containing essentially all reaction by-prods., is thermally cracked and cat. rearranged (Figure). **F** procédé bisphénol-A de Chiyoda; **G** Chiyoda Bisphenol-A-Verfahren.
Ref.: *Hydrocarb.Proc.* **1997**, (3), 113.   B. CORNILS

chlorinated (chlorided) alumina an early cat. for the *isomerization of lower hydrocarbons, i.e., light naphthas C_5/C_6, and for

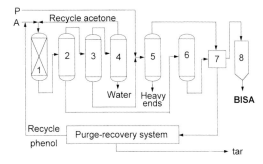

1 reactor; 2–5 distillations; phenol purification; 6 crystallizer; 7 separator; 8 prilling tower; A = Acetone; P = Phenol; BISA = bis-phenol-A.

Chiyoda CT-BISA process

the conv. of *n*-butane to *i*-butane at approx. 100 °C. Chlorination can be achieved with CCl_4, hexachloroethane, thionyl chloride, etc. For practical use, the additional feeding of $AlCl_3$ and/or HCl is possible. For commercial applications, cf. the procs. of *Atlantic Richfield Pentafining, *BP C_4 and C_5/C_6 isom., *CDTech Isomplus, *Houdryforming, *IFP, *Kellogg Isofin, *Lummus, *Mobil LTD, *Monsanto/Lummus, *Standard Isomate, *Shell, *UOP Butamer or Penex, etc. **F** alumine chlorée; **G** chloriertes Aluminiumoxyd.
                                    B. CORNILS

Ref.: *Catal.Letters* **1995**, *31*, 121; Ertl/Knözinger/Weitkamp, Vol. 4, p. 2057; *J.Mol.Catal.A*: **1999**, *148*, 253.

chlorination often proceeds stoichiometrically instead of cat. Cs. by molecular chlorine proceed through *radical or ionic mechanisms. While radical pathways are favored by heat, light, or *radical initiators, ionic mechanisms often involve cats. For the electrophilic *addition to alkenes and the substitution of aromatics *Lewis acids are used.
The major commercial applications of cat. c. are the c. of ethylene, yielding 1,2-dichloroethane (a precursor for *vinyl chloride) and the prod. of chlorotoluenes. The c. of ethylene proceeds in the liquid phase at 60–140 °C over $FeCl_3$ as cat. First group chlorides as *additives increase the *selectivity to 1,2-dichloroethane. The c. of toluene is also carried out in

the liquid phase in the presence of $FeCl_3$, $AlCl_3$, $SbCl_3$, or other *Lewis acids. The position of *substitution is dependent, within wide limits, on the reaction conds. and the cat. The c. of alkenes is cat. by iodine. During the reaction iodine chloride is formed which possibly acts as a nucleophile. Like other *halogenations the *addition of unsaturated cpds. is cat. by Li cations, chlorides (e.g., HCl), or water. For commercial procs. cf. *Bayer (benzene c.) or *Dow (benzene or methane c.). **F** chloration; **G** Chlorierung.   S. TRAUTSCHOLD

Ref.: Ullmann, Vol. A6, p. 266; A. Loupy, B. Tchoubar, *Salt Effects in Organic and Organometallic Chemistry*, VCH, Weinheim 1992; NL 69–01398 (1969).

chlorocarbonylation   *carbonylation of unsaturated halides or of aryl dihalides and diamines using CO yielding Cl-containing or Cl-free intermediates.   B. CORNILS

Ref.: *J. Mol. Catal. A:* **1999**, *143*, 181, 197.

chlorolysis   a mostly *non*-cat. reaction between elemental chlorine and organic substrates with cracking of C-C and C-H bonds; cf. also *halolysis. **F=G** chlorolyse. B. CORNILS

Ref.: McKetta-1, Vol. 8, p. 206; McKetta-2, p. 501

chloroperoxidase   The *enzyme chloroperoxidase (CPO) cat. the asymmetric *epoxidation of alkenes with H_2O_2; cf. also *peroxidase, *oxiranes by hom. cat.   B. CORNILS

chlorosulfonation   → sulfochlorination, *Reed reaction

cholesterol esterase   cat. the *hydrolysis of cholesteryl esters. Bovine pancreatic cholesterol esterase (EC 3.1.1.13) is a serine *esterase. It also cat. the *enantioselective hydrolysis of binaphthol, spirobindanol, and protected myoinositol diesters and secondary alcohol esters. The reactions proceed through a two-step resolution, which can yield prods. with higher *ee than one single-step resolution can. **E=F=G**.   C.-H. WONG

chorismate mutase   (EC 5.4.99.5) This *enzyme cat. the *Claisen rearrangement of

*chorismate to prephenate, the first committed step in the prod. of the hydrophobic amino acids *phenylalanine and tyrosine (and subsequently thyroxine, dopamine, epinephrine, norepinephrine, and others). The enzymes from *Saccharomyces cerevisiae*, *Bacillus subtilis*, and *Escherichia coli* have been crystallized, with and without an *endo*-oxabicyclic *inhibitor.

diequatorial   diaxial   *endo*-oxabicyclic inhibitor

Inhibition by this cpd., which mimics the (pseudo) diaxial conformer of chorismate, suggests that the enzyme binds chorismate in the diaxial conformation. This structure, disfavored over the diequatorial conformation in solution, is the form required for the uncat. pericyclic rearrangement. Enzymatic cat. appears to proceed via a similar mechanism to that in solution, achieving rate enhancement in part by preorganizing the substrate into the correct conformation for cat. Antibodies have been elicited against the *endo*-oxabicyclic inhibitor that show cat. *activity (cf. *pericyclic reactions, *catalytic antibodies). **E=F=G**.

Ref.: *Structure* **1997**, *5*, 1437.   P.S. SEARS

chromia   Cr(III) oxide, a *precursor in cats. for *fluorinations (similar to *alumina; cf. *chromium as catalyst metal)

chromium as catalyst metal   A major cat. application of Cr is for olefin *polymerization. The *Phillips cat. is prepared by *impregnating *silica/*alumina or pure silica *supports with 1–5 wt% CrO_3. After *calcination and *reduction the cat. polymerizes ethylene to *HDPE without any *co-cat. The alternative Union Carbide cat. prepared from Cp_2Cr on silica is highly sensitive toward H_2 transfer for chain-length control and highly selective for ethylene (no *copolymerization with other olefins). Similar but less active cats. with *surface

Cr alkyls can be prepared from tris(allyl)chromium. Even cationic Cr(VI) amido-alkyl *complexes are very active cats. Some hom. cats. prepared from Cr(III) cpds., amines (e.g., pyrroles), and aluminum alkyls selectively *trimerize ethylene to 1-hexene. For chromocenes cf. *ansa-metallocenes.

Cr_2O_3 (*chromia, often combined with Al_2O_3 or Fe_2O_3) can cat. reactions such as *hcs. oxidation, *dehydrogenation, *dehydrocyclization, *fluorination, *hydrotreating, *oxydehydrogenation, (cyclo)dehydroaromatization, or *selective cat. reduction (procs. of *Air Products Catfin, *Atlantic Richfield HR/HDA, *Dow butadiene, *Houdry Catadiene, *Lummus Catofin, *Phillips OXD). Cr_2O_3 (also in mixture with Fe) at 350–400 °C is used as het. cat. for the *WGSR, various Cr carbonyls like $Cr(CO)_5H^-$ are known to play a role in the *WGSR in aqueous solution (and the *Dötz reaction). *Chromia is a constituent of the *Adkins cat. (cf. procs. of *Eastman BHMC, *Celanese HMDA, *Henkel direct hydrogenation), of *methanol synth. cats. (*BASF, *ICI), and a cat. for the *isomerization of allyl alcohol to *propylene oxide (*BASF-Wyandotte). $CrCl_2$ (often *electrocat. generated) has been used for organic reactions.

Metallic Cr *surfaces can activate N_2 similarly to Fe in the *Haber-Bosch proc. Various Cr(III) complexes cat. the *epoxidation of alkenes. *Jacobsen's chiral *salen complexes of Cr(III) cat. the *enantioselective opening of *meso* *epoxides. Cr is an essential metal in biology and may act as the cat. *site in some *enzymes. **F** le chrome comme métal catalytique; **G** Chrom als Katalysatormetall.   R.D. KOHN

Ref.: *Eur.J.Inorg.Chem.* **1998**, 15; Ullmann, Vol. 13, p. 519; *Organic Reactions*, Vol. 53, p. 1; *Methodicum Chimicum*, Vol. XIII/7; *Acc.Chem.Res.* **1996**, 29, 544; *Chem.Rev.* **1999**, 99, 991.

chromogenic assays One of the simplest ways to follow an enzymatic reaction is to monitor the evolution of a colored prod. This is an easy alternative to chromatographic, titrimetric, radioactive, or other methods. A large number of chromogenic assays have been developed for different classes of *enzymes. The *NADH-dependent *oxidoreductases can be followed via the difference in absorption between the oxidized and reduced forms of the nicotinamide *cofactor. For many enzymes, the prod. and substrate do not have very different extinction coefficients at easily monitored wavelengths, but the enzyme reaction may be coupled to another that does. Reactions that consume *adenosine triphosphate, for example, may be coupled to an NADH-utilizing redox reaction that can be monitored directly.

The reactions of various hydrolytic enzymes can be followed via *hydrolysis of artificial substrates in which the leaving group is colored. *p*-Nitroanilides and *p*-nitrophenol esters of amino acids are common substrates for the *proteases, and both classes produce yellow prods. *p*-Nitrophenyl glycosides, likewise, are used to follow the kinetics of glycosidases. **E=F=G**.   P.S. SEARS

chymotrypsin, chymotrypsinogen Chymotrypsin is a mammalian digestive *enzyme secreted into the small intestine by the pancreas. It is a *serine protease with a preference for large hydrophobic residues such as phenylalanine and tyrosine in the S_1 subsite (adjacent to the scissile bond, on the acyl donor side). It is secreted in a virtually inactive *proenzyme form, which has a defective substrate binding pocket until the enzyme becomes activated upon (trypsin-cat.) cleavage between Arg15 and Ile16 (the protein may then become further proteolytically processed to give several other forms, which are also active). **E=F=G**.   P.S. SEARS

Ciamician photodismutation the light-induced photodisproportionation and rearrangement of *o*-nitrobenzaldehyde to *o*-nitrosobenzoic acid. **F=G=E**.   B. CORNILS

CIDNP effect chemically induced dynamic nuclear polarization (CIDNP), affectionately referred to as "kidnap", is a powerful in-situ NMR technique to identify free-*radical procs. (e.g., *radical cat.) via emission and enhanced absorption lines of the associated reaction prods. or unstable *intermediates. Its origin was explained via the radical-pair theory to occur because of the spin selectivity of the combination or *disproportionation procs. of radical pairs – To form a single bond, radical pairs have to encounter as a combined singlet state of the two unpaired radical electrons. Triplet radical pairs may not react. *Electron transfer procs., however, may lead to triplet states of unsaturated prod. molecules. The CIDNP signals are the more intense, the more unstable the radicals are. This is in contrast to *ESR, which favors less reactive radicals. Since only radical pairs yield CIDNP, this technique is less general than ESR. Care has to be taken not to confuse CIDNP with *PHIP, which yields a similar response, but is due to the formation of *parahydrogen upon cooling samples containing H_2 and subsequent breaking of its symmetry via *hydrogenation cats. **E=F=G.**   J. BARGON

Ref.: *Adv.Magn.Res.* **1974**, 7, 157; *Adv.Free Rad.Chem.* **1975**, 5, 318.

CI-MS → chemical ionization mass spectroscopy

cinchona alkaloids The four different cas. of the cinchona tree (quinine, quinidine, cinchonine, cinchonidine) differ in the configurations at C8 and C9 in the crucial β-hydroxyamino moiety and in the presence/absence of a methoxy substituent in the 6'-position of the quinoline system. In *stereochemistry the cas. and their derivatives are employed as bases in the *optical resolution of acids, as phase transfer cats. (*PTC), and as *additives to *chiral cats. such as *supported Pt cats. for the *enantioselective *hydrogenation of α-keto esters or as *ligands in *chiral cats., such as hom. Os cats. for the *dihydroxylation of alkenes. Cf. also *DHQ and *(DHQ)$_2$. **F** cinchona alcaloides; **G** Cinchonaalkaloide.
                                     H. BRUNNER

Ref.: *Tetrahedron:Asymmetry* **1991**, 2, 843; *J.Org. Chem.* **1991**, 56, 4585.

circulating-bed reactor (CBR) → *catalytic reactors, *FCC proc.

CIR-FTIR cylindical internal reflectance FTIR, → Fourier transform infrared spectroscopy

***cis/trans* isomerization** occurs usually as an unwanted side effect in *hydrogenation reactions of alkenes cat. by metals. The extent to which the i. takes place depends on the active metal comp. and likewise on the *reduction conds. as the *lifetime of certain *intermediates in the proc. decides between a branching between i. and hydrogenation. In this way it is possible to deliberately isomerize an alkene almost exclusively when reaction conds. are chosen such that active H is present in only low abundance and if Pd is used as the cat. If active H is present in large excess and the active metal is not Pd, then the i. reaction can almost completely be suppressed. Such a situation occurs with Pt. This behavior is the expression of two different mechanisms occurring with Pd and Pt in *selective hydrogenation. Over Pt the hydrogenation is sequential with one active H atom opening the double bond to the hemi-hydrogenated intermediate. This state can either have one H dissociated and fall back to an (isomerized) alkene or add another active H and react irreversibly to form the alkane. Over Pd the alkene is activated via a cyclic allylic intermediate which causes unavoidable i. before any addition of H. **F** isomérisation cis-trans; **G** Cis-trans-Isomerisierung.

Ref.: Augustine, p. 345 ff.   R. SCHLÖGL

CIS-MS coordination ion spray mass spectrometry

citric acid cycle → Krebs cycle

Claisen condensation self-*condensation of esters containing an α-hydrogen atom in the presence of strong bases to yield α,β-keto-esters acc. to 2 RCH_2COOR^1 → $RCH_2CO-CH(R)COOR^1 + R^1OH$. Cc. can also be carried out with two different esters, one of which must have an α-hydrogen atom. Hence, the reaction is often performed with benzoic acid esters, alkyl carbonates, and alkyl oxalates. The intramolecular Cc. (*Dieckmann condensation) is a convenient method for the synth. of 5–7-membered rings as well as large ring systems (>12 members) via high-dilution techniques. As base cat. for the Cc. the respective *alkoxide is often used because it is generated from the ester. However, the use of stronger bases such as NaH, NaNH$_2$, etc. often increases the yield. **F** condensation de Claisen; **G=E**.   M. BELLER

Ref.: March; Trost/Fleming; *Organic Reactions*, Vol. 2, p. 1.

Claisen rearrangement 1: general: Allyl vinyl ethers undergo a rearrangement to give γ,δ-unsaturated carbonyl cpds.; allyl aryl ethers go to *o*-allylphenols. The mechanism involves *concerted pericyclic [3,3] sigmatropic rearrangements. Cats. for the Ca. are *Lewis acids. The value of the Ca. has been increased recently by developing efficient *domino procs. in which allyl vinyl ethers are generated in situ under the influence of Ti-mediated methylenation of esters. Another procedure is to use diallylic esters which undergo the Ca. when heated with Ru(II) *complexes. **F** réarrangement de Claisen; **G** Claisen-Umlagerung.   M. BELLER

Ref.: March; Trost/Fleming; *J.Am.Chem.Soc.* **1991**, *113*, 2762; *Organic Reactions*, Vol. 22, p. 1.

2: enzymatic: Ca. cat. by *antibodies or cf. *chorismate mutase.

Claus COS conversion removal of COS by reaction with SO$_2$ acc. to COS + 1/2 SO$_2$ ⇌ CO$_2$ + 3/2 S over Co-Mo cats.

Ref.: Ertl/Knözinger/Weitkamp, Vol. 4, p. 1840.

Clausius-Clapeyron equation special variant of the more general Clausius equ.

$$\frac{dp}{dT} = \frac{\Delta S_m}{\Delta V_m}$$

which describes the temp. dependence of the equil. press. of a two-phase system. ΔS_m and ΔV_m are the molar entropy and volume changes, resp., for the phase transition considered. When the vapor in a solid/vapor or liquid/vapor system can be treated as an ideal gas, the Clausius equ. can be rewritten as the Clausius-Clapeyron equ. (C-Ce):

$$\frac{d\ln p}{dT} = \frac{\Delta H_m}{RT^2}$$

with ΔH_m being the molar enthalpy change of the phase transition and R = gas constant. An analogous relationship holds for *adsorption systems and can be applied for the determination of *isosteric heats of adsorption. A set of *adsorption isotherms has to be measured for a given system at different temps., from which the equil. press., p, necessary to obtain the same *surface *coverage at different temps., can be determined. The slope of ln p at constant coverage Θ versus $1/T$ is given by the C-Ce.

$$\left[\frac{d\ln p}{dT}\right]_{\Theta=\text{const.}} = -\frac{\Delta H_{ads}^{iso}}{RT^2}$$

where ΔH^{iso}_{ads} is the isosteric heat of adsorption, which can thus be determined from experimental isotherms. The negative sign results from the convention that the adsorption is treated as an exothermic proc., whereas sublimation and evaporation are treated as endothermic procs. **F** équation de Clausius-Clapeyron; **G** Clausius-Clapeyron-Gleichung.
                          H. KNÖZINGER

Clauspol process → IFP procs.

Claus process removal of S cpds.(*sweetening in wider sense) from gas streams by cat. reaction between H$_2$S and oxygen or SO$_2$ to elementary S acc. to 2 H$_2$S + O$_2$ → S$_2$ + 2 H$_2$O (over Co-Mo on Al$_2$O$_3$, TiO$_2$) and 2 H$_2$S + SO$_2$ → 3/2 S$_2$ + 2 H$_2$O (reversible). The Cp. is used by petroleum refineries to recover S from H$_2$S from their *HDS procs. and by chemical plants if they produce *syngas from S containing heavy fuel oils. The Claus cats. are based on high-purity *alumina, promoted alumina (doped with SiO$_2$, Fe$_2$O$_3$, or group

VIB and VII metals), and TiO_2 cats. The procs. used are licensed by, e.g., *BASF, *BOC, *Comprimo, *IFP Clauspol, *Linde Clinsulf, *Lurgi Sulfreen and Oxy-Claus, *Shell SCOT, *UOP Selectox, etc. For the removal of COS cf. *Claus COS removal. **F** procédé de Claus; **G** Claus-Verfahren.

B. CORNILS

Ref.: Büchner/Schliebs/Winter/Büchel, p.107; Ertl/ Knözinger/Weitkamp, Vol.4, pp.1761, 1840.

clavaminate synthetase

clavaminate synthetase a *monooxygenase which cat. the double oxidative *cyclization of proclavaminic acid to clavaminic acid in the presence of Fe(II), O_2, and α-ketoglutarate. This *enzyme was used in the synth. of a new class of β-lactam analogs. **E=F=G**.

C.-H. WONG

Clavaminic acid

Clavulanic acid　　E : clavaminic acid synthase

Ref.: *Biochemistry* **1990**, *29*, 6499; *J.Am.Chem.Soc.* **1989**, *111*, 7625.

Clay-Kinnear-Perren reaction

Clay-Kinnear-Perren reaction *condensation of alkyl chlorides with PCl_3 and $AlCl_3$ as cats. The intermediate *complexes are *hydrolyzed carefully yielding alkylphosphonyl dichlorides. **F** réaction de Clay-Kinnear-Perren; **G** Clay-Kinnear-Perren Reaktion.

Ref.: Krauch/Kunz.　　　　　　B. CORNILS

Claymet

Claymet metal-doped montmorillonites (*clays) as cats. [e.g., Clayfen = clay-supported Fe(III) nitrate, Claycop = clay-supported $Cu(NO_3)_2$]; cf. *Clayzic.　B. CORNILS

Ref.: M. Balogh, P. Laszlo, *Organic Chemistry Using Clays*, Springer, Berlin 1993; *J.Chem.Soc.Chem.Commun.* **1989**, 1353.

clay (minerals)

clay (minerals) Cs. are aluminosilicate minerals containing *montmorillonites, phyllosilicates (smectites), *micas, *bentonites, etc. Cs. are swellable and can be calcined (*calcination) to mullites (*ceramics). Cs. are *ion exchange active and form with polyoxoaluminum ions extended layers (*pillared clays), the *cavities of which are similar to cat. active *zeolites. Cs. can be used as mildly acidic het. cats. due to the acidity of the *surface OH groups. They are also used as cat. *supports. **F** argile; **F** Ton.　P. BEHRENS

Ref.: M. Balogh, P. Laszlo, *Organic Chemistry Using Clays*, Springer, Berlin 1993; Ertl/Knözinger/Weitkamp, Vol.1, p.387; Ullmann, Vol.A7, p.105; Kirk/Othmer-1, Vol.6, p.381; Kirk/Othmer-2, p.283.

Clayzic

Clayzic a special case of *Claymet: *montmorillonite-supported zinc chloride as a *solid acid cat.

Clemmensen reduction

Clemmensen reduction *deoxygenation of ketones to hydrocarbons with amalgamized Zn and HCl. **F=G=E**.　B. CORNILS

Ref.: *Organic Reactions*, Vol.1, p.155 and Vol.22, p.401.

Clinsulf process

Clinsulf process → Linde procs.

cloning

cloning (expression) The isolation of *enzymes from tissue is only a viable option for the prod. of large amounts of enzyme if the tissue is plentiful and the enzyme is produced to high levels. For other enzymes, cloning, and overexpressing the gene, is a necessity for large-scale prod. Molecular cloning refers to the insertion of a gene into a *DNA *vector that can be replicated independently of the chromosome(s) of a host organism, and introduction of that vector into the host. Under proper growth conditions, the host will express the gene and produce the encoded protein.

One basic approach to c. a gene is to start with a large set of different DNA molecules, which together constitute all of the genetic information of an organism. This DNA can either be partially digested chromosomal DNA (typically used for bacterial genes) or cDNA, made of reverse-transcribed messenger RNAs (used for *eukaryotic genes). The DNA is typically inserted into a bacterial expression vector, and the resulting DNA *library is used to transform bacteria. Ideally, the desired colonies can be found via selection, in which the bacteria that do not prod. the desired protein are not able to survive under the growth conds. used. Alternatively, the resulting colonies may be individually screened for the presence of the desired gene. Screening for an enzymatic activity can be as simple as growing the colonies in the presence of a *chromogenic substrate. Another screening technique is expression cl., in which the desired enzyme creates a new *epitope on the surface of the cell that can be highlighted by fluorescently tagged antibodies; the positive clones may then be selected by fluorescence-activated cell sorting (FACS). **E=F=G.**

P.S. SEARS

clostripain *thiolprotease isolated from *Clostridium histolyticum*. It is highly specific for amide bonds on the C-terminal side of arginine and displays poor activity even toward Lys-Xaa amide bonds. The pH optimum of the *enzyme falls between 7 and 8, depending on the substrate. Like other *thiolproteases, it is inhibited by heavy metals and oxidants, though it requires Ca^{2+} for function and thus EDTA is also an inhibitor. **E=F=G.**

Ref.: *Methods Enzymol.* **1970**, *XIX*,635.   P.S. SEARS

clusters connected groups of at least three atoms, where every atom is connected at least to two others of the group. Cs. can consist of atoms of one type (uni- or mononuclear c.) or two or more elements (heteronuclear cs.). Cs. are usually stabilized by *ligands (L) which saturate the free valences (*vacant sites), but can also be prepared as naked cs., which tend to aggregate and make handling difficult. Cs. are usually ill defined and show a distribution of size, total charge, and number of ligands; however, well defined species can also be isolated and recrystallized. Well defined species are mostly subunits of a hexagonal or

cubic lattice, e.g., closed-shell-cs. possess a *central atom and a number of dense layers with cubically packed atoms, such as $Ru_{13}L_n$ (one layer), $Au_{55}L_m$ (two layers), or $Pd_{561}L_o$ (five layers). Due to the nature of binding (between covalent and metallics), cs. bridge between the chemistry of molecular cpds. and solids (and thus, when applied, between to *hom. or *het. cat.). Compared to the corresponding solids, they show an increased reactivity (cat. *activity), which is partly due to their increased specific *surface, and also to the quantum size effect (the dependence of band position and electrochemical potentials on size). The rapid exchange of ligands makes cs. very efficient in cat., as well as ideal candidates to study *surfaces and surface reactions. Some cs. are observed in enzymatic cat. (cf., e.g., Fe/S clusters in *cofactors, *nitrogen fixation). **E=F=G.**   M. ANTONIETTI

Ref.: Adams/Cotton; M. Moskovits (Ed.), *Metal Clusters*, Wiley, New York 1986; B.F.G. Johnson (Ed.), *Transition Metal Clusters*, Wiley, New York 1980; P. Braunstein et al. (Eds.), *Metal Clusters in Chemistry*, 3 Vols., Wiley-VCH, Weinheim 1999; F.A. Cotton, R.A. Walton, *Multiple Bonds Between Metal Atoms*, Wiley, New York 1982; Ertl/Knözinger/Weitkamp, Vol. 2, p. 793; Gates-1, p. 171; Gates-2; *Chem.Rev.* **1995**, *95*, 511; W. Eckardt (Ed.), *Metal Clusters*, Wiley-VCH, Weinheim 1999; *J.Mol.Catal.A:* **1999**, *145*, 1.

CMC critical micelle concentration, → micelles, *micellar catalysis.

CMP-_N_-acetylneuraminic acid (CMP-Neu-Ac) is synth. from CTP and NeuAc by CMP-NeuAc synthetase (EC 2.7.7.43). An improved synth. procedure which is suitable for multigram-scale synth. has been developed in which CTP is synth. in situ from CMP using *adenylate kinase and *pyruvate kinase. Adenylate kinase cat. the phosphate transfer between CTP and CMP to produce CDP, and the CDP is subsequently phosphorylated by pyruvate kinase to provide CTP. Another procedure has been employed in which the NeuAc used in the synth. of CMP-NeuAc was prepd. in a NeuAc aldolase-cat. reaction of pyruvate with *N*-acetylmannosamine, which was itself generated from *N*-acetylglucosamine by a base-cat. epimerization. A one-pot synth. of CMP-NeuAc which is based on the latter procedure involves the in situ synth. of NeuAc from *N*-acetylmannosamine and pyruvate, cat. by sialic acid aldolase. Chemical synth. of CMP-NeuAc have also been reported (Figure). **F** l'acide CMP-*N*-acetyleneuramine; **G** CMP-*N*-acetylneraminsäure.   C.-H. WONG

E₁: NeuAc aldolase
E₂: CMP-NeuAc synthetase
E₃: Pyruvate kinase
E₄: Adenylate kinase

CMR   catalytic membrane reactor, → membrane reactor

CNG   compressed natural gas, → natural gas

CO   → carbon monoxide, *C$_1$ as a building block

CO$_2$ → carbon dioxide as a building block, *carbon dioxide reforming

CoA   → coenzyme A

coadsorption   simultaneous presence of two chemically different *adsorbates on the same solid *surface. As each surface is typically capable of adsorbing only a few species with high specificity, the degree of interaction between the surface and several adsorbates is usually significantly different. The presence of different chemical species in close proximity to each other often causes strong adsorbate–adsorbate interactions which determine to a large extent the structure of the resulting surface. Typical examples are the simultaneous presence of CO and O atoms on noble metal surfaces, the coa. of alkali metals and small diatomic molecules (CO, NO, N$_2$), or the presence of CO and benzene on noble metals. Such cases have been studied exten-

sively by surface science using *single crystals and electron *diffraction techniques. Surface vibrational spectroscopy proved a valuable tool for studying the adsorbate–adsorbate interaction. The existence of coa. systems is a necessary prerequisite for the occurrence of het. cat. reactions according to the *Langmuir–Hinshelwood mechanism, which requires the simultaneous presence of all reactants on the surface in an adsorbed state prior to reactions. **E=F=G**.   R. SCHLÖGL

Ref.: Somorjai.

coal   → carbon

Coalcon process   → Union Carbide procs.

coal gasification   conv. of *coal to coal gases, preferably *syngas (CO plus hydrogen, water gas, coke oven gas) by special methods (autothermal or allothermal procs.; *partial oxidation). Some cg. procs. are catalytic, either by cat. properties of coal ashes or by specially added cats. such as alkaline or alkaline earth salts (cf. procs. of *Atgas, *Kellogg). **F** gazéification du charbon; **G** Kohlevergasung.   B. CORNILS

Ref.: Falbe-2, p. 114f; Ertl/Knözinger/Weitkamp, Vol. 4, p. 2075; Payne; Ullmann, Vol. A7, p. 197.

coal hydrogenation   procs. for the hydrogenative *liquefaction of coal, based on either the direct or the indirect (solvent refining, hydrogenating extraction) *hydrogenation of coal (cf. the procs. of *Bergius, *Canmet, *Consol, *coal liquefaction, *EDS, *Exxon DS, *German technology, *Hydrocarbon Research H-Oil, *SRC, *Synthoil, etc.). **F** hydrogénation du charbon; **G** Kohlehydrierung.   I. ROMEY

coal liquefaction   1: Term which is used often synonymously with *coal hydrogenation although cl. is more precise since with het. cats. activated hydrogen has a lack of mobility and does not fully penetrate the coal network (cf. *German technology). 2: Term which comprises the four principal sectors of technological developments – 2.1: manuf. of liquid prods. (mainly, but not necessarily hydrocarbons) from coal; 2.2: single- and two-stage direct hydrogenation procs. 2.3: The *coprocessing of coal with heavy *petroleum* fractions. 2.4: Synth. procs. to liquid prods. which can also be manuf. with coal-based raw materials (*Fischer-Tropsch synth., *Mobil's MTG and MTO procs., each from *syngas). **F** liquéfaction du charbon; **G** Kohleverflüssigung.

I. ROMEY

Ref.: Falbe-2, Ullmann, Vol. A7, p. 197, Kirk/Othmer-1, Vol. 6, p. 476; Ertl/Knözinger/Weitkamp, Vol. 4, p. 2078; Payne; McKetta-1, Vol. 9, p. 289.

CoAPO   Co-modified *aluminophosphate

coated catalysts   *egg-shell cats. which are characterized by an only thin outer layer of the carrier loaded with the active comp. (*metal distribution, *precipitation, *impregnation). **F** catalyseurs sur support neutre (coquille d'oeuf); **G** Schalenkatalysatoren (Eierschalenkatalysatoren).

B. CORNILS

cobalamines   derivatives of vitamin B_{12}, cf. *coenzymes, *cofactors, *vitamin B_{12}

cobalt aluminophosphates   (CoALPOs) Replacement of Al(III) framework ions in open-structure *ALPOs by any divalent ion (e.g., Mg, Zn) leads to a *solid acid, for an exchangeable proton (to preserve electroneutrality) must be loosely attached to an adjacent framework oxygen. Co *ion-exchanged ALPOs (like Mn ion-exchanged ones) are particularly important of the so-called metal-ion-exchanged ALPOs (otherwise designated *MAPOs). A good example of Co(II)-substituted ALPO acid cats. is CoALPO-18, the framework structure of which is essentially that of the *zeolite mineral chabazite. With pore apertures of just 3.8 Å and correspondingly small *cavities, the cat. *dehydration of methanol leads to light olefins (cf. *MTO procs. of *Mobil and *UOP/Norsk Hydro). CoALPO-18 is also a remarkably good cat. for the aerobic

*oxidation of linear alkanes or of cyclohexane to cyclohexanol/one. **E=F=G.**   J.M. THOMAS

Ref.: *J.Chem.Soc.Chem.Commun.* **1994**, 603; *J.Phys. Chem.* **1996**, *100*, 8977; *Nature (London)* **1999**, *398*, 227; *Catal.Lett.* **1998**, *55*, 15.

cobalt as catalyst metal   Co is a *transition metal strongly related to Fe and Ni (total prod. in 1998: 33 422 tonnes). It forms numerous cpds. and *complexes/organometal cpds. with a large variety of (even stable) π- or σ-bonded *ligands which can undergo *ligand exchange reactions and are thus often versatile cats. The complexes have varying *coordination numbers and they occur in equil. as more than one stereochemical species. Co exists in the +2 and +3 valence states; its ability to cat. reactions is based on its *redox properties, i.e., the ability to form stable species in multiple oxidation states and the ease of *electron transfer between these states via unstable intermediates or free *radicals.

Over 40 % of nonmetallurgical use of Co is applied in cats., 80 % thereof in three main areas – as a het. cat. in *hydrotreating/*hydrodesulfurization/*hydrodealkylation (in combination with Mo for the oil and gas industry; procs., inter alia, *Atlantic Richfiels HR/HDA, *Exxon EDS, *Howe-Baker, *Houdry-Detol, *ICI, *Rhône-Progil, etc.), as a hom. cat. in *high-press. *oxo syntheses, and for the manuf. of aromatic or aliphatic acids (*acetic, *phthalic, *terephthalic, or *trimellitic acid) in the chemical industry (procs. of *Amoco, *Celanese, *Glitsch, *Mobil, *Hüls, and many others). Secondary alcohols or *KA oils are manuf. via *Bashkirov oxidation. Smaller portions serve for: *amidocarbonylations, *ammonoxidations, *arene-aromatics or aliphatic oxidations (with *Fenton's reagent: *Brichima proc.), *carbonylations (*BASF hydroquinone proc.), *cyclizations, *cyclocarbonylations, *cyclooligomerizations, *fluorinations, *Fischer-Tropsch synth. (cf. *Shell SMDS proc.), *hydrogenations, *isomerizations (e.g. *Atlantic Richfield Xylene-Plus), *Pauson-Khand reactions, and as *polymerization, *cross-linking, or *drier cats. In cat. patent literature, Co often extends the number and scope of claims.

For the *HDS application over Co-Mo cats. during *hydrotreating the sulfidic form is important (cf. *sulfided cats.). The reactions cat. are the *hydrogenation and the *hydrogenolysis of carbon-heteroatom bonds (for *HDN, *deoxygenation, *hydrocracking of higher to lower cpds., too). Different theories describe the action of the *bimetallic and *bifunctional Co-promoted MoS_2 cats. (*pseudointercalation model, *remote control model). In *oxidation reactions (*TPA, *acetic or *adipic acid, *arene-aromatics oxidation, *hc. oxidation, etc.) the Co cats. serve to decompose the hydroperoxides during the *radical chain mechanism. In *hydroformylation or *carbonylation under relatively high press. (>10 MPa) hom. Co carbonyls (possibly *ligand-modified) activate the reactants CO, H_2, and alkenes acc. to the *Heck-Breslow mechanism. The application of *Raney Co is similar to *Raney Ni. Some *enzymes contain Co, such as the enzyme *cofactor B_{12} or *nitrile hydrolyzing enzymes.

In cat. production, the EU classification of Co cpds. has to be taken into account. Both Co sulfate and Co chloride have been classified as highly potent carcinogens. **F** cobalt comme métal catalytique; **G** Kobalt als Katalysatormetall.   B. CORNILS

Ref.: Cornils/Herrmann-1; Falbe-1 and -3; Ertl/Knözinger/Weitkamp; *Cobalt News* (publ. by the Cobalt Dev. Institute) **1998**, (4), 13 and **1999**, (2), 12; Ullmann, Vol. A7, p. 281; *Organic Reactions* Vol. 22, p. 253; *Houben-Weyl/Methodicum Chimicum*, Vol. XIII/9b.

COC → cyclic *copolymers

co-catalyst The term co-catalyst (coc.) is not precisely defined and therefore frequently intersects with *activator or *promoter (or even *modifier, cf. figure under the keyword *additives). In a stringent sense, a coc. may be understood as a substance which adds its own *activity (or *deactivation) for the considered reaction to the cat. itself and increases the overall *rate of reaction by a synergetic contribution to the basic *activity of the main cat. Accepting this definition, the main difference

between an activator and a coc. is ascribed to the fact that an activator alone does not show cat. activity whereas a coc. is at least faintly active in the desired way. A positive influence on the *selectivity is a common feature for both, coc. and activator. For example Pt/Al_2O_3 cats. were replaced by Pt-Ir/Al_2O_3 *cat. reforming of hydrocarbons, resulting in an increase in the yield of aromatics and a stable activity over the time on stream. A cat. with 0.3 wt.% each of Pt and Ir exhibits a markedly superior performance to a cat. with 0.6 wt.% of Pt. Somewhat later the even more beneficial influence of Re to Pt/Al_2O_3 was realized in industrial applicationd. The effect of Ir or Re on Pt is ascribed to *bimetallic *clusters, referring to bimetallic entities highly dispersed on the *surface of a *support (cf. also *bimetallic cats.). A similar principle is claimed for the Ru-Cu system for *hydrogenolysis reactions.

In the *hydroformylation some other materials besides Co and Rh are cat. active. A number of patents and publications claim advantages with respect to activity and *selectivity for mixed-metal cats., e.g., Co/Rh, Co/Pt, Co/Fe, or Co/Ni. Similar to hydrocarbon convs. the improvements in hydroformylation are attributed to a *synergetic effect arising from mixed metal clusters. The Pt/Sn system has been widely investigated for *dehydrogenation reactions indicating the existence of mixed-metal aggregates. In the conv. of *i*-butane to isobutene both activity and selectivity are drastically enhanced over the pure Pt catalyst. Other examples are CH_3I within the cat. cycle of the methanol carbonylation (*Monsanto proc.) which is not consumed in the proc. but the cycle cannot be closed without it, $AlEt_3$ in *Ziegler-Natta cat., $W(OCH_3)_6$ together with $AlEt_2Cl$ in *metathesis reactions, or K (or Cs) salts (or $BaCO_3$) as cocs. for Ag during ethylene oxidation (*EO manuf.).

In other definitions cocs. accelerate reactions when combined with cats. without having any activity themselves (cf. also *cofactors, *coenzymes). Typical for this interpretation of cocs.

(following other definitions: *modifiers) is the addressing of *phosphines or other *ligands as cocs. for the basic *complex cats. with Rh as *central atoms. The differentiation from a *co-promoter is difficult (cf. *promotion). *Cofactors are different in their relation to enzymes. **F** co-catalyseur; **G** Cokatalysator.

C.D. FROHNING

Ref.: J.H Sinfelt, *Bimetallic Catalysts*, Wiley, New York (1983); Gates-1, pp. 99, 110, 392; Falbe-1, p. 44; *J. Catal.* **1998**, *179*, 459; *J. Mol. Catal.* **1999**, *142*, 169.

C-O coupling   → O-C coupling

cocyclization   special case of cat. *cyclization, where two (or more) different unsaturated *monomers are linked together via an intra- or intermolecular ring formation. The cc. between dienes and monoalkenes or alkynes is a domain of zerovalent Ni cats. Depending on the *ligand present at the Ni(0) template, the cc. of butadiene with an alkene leads either to noncyclic 1:1 prods. (i.e., 1,4-hexadienes) or to cyclic 2:1 prods. (1,5-*trans,cis*-cyclodecadienes). The synth. of N-heterocycles can be achieved by cc. of diynes with nitriles. **F** cocyclisation; **G** Cocyclisierung.   H. BÖNNEMANN

Ref.: Cornils/Herrmann-1, Vol. 1, p. 365; *Acc. Chem. Res.* **1977**, *10*, 1.

cod, COD   cyclooctadienyl groups, inter alia as *ligands or in *complex cpds.

codimerization   *dimerization of two structurally different substrates. C. is used commercially to upgrade olefinic streams from various hc. *cracking (*steam or *catalytic) or forming (*Fischer-Tropsch synth. or *Mobil MTG) procs. C. of an odd-numbered (e.g., propylene) and an even- numbered (e.g., butene) *α-olefin is used to prepare higher α-olefins with an odd number of carbon atoms. For this type of c. mainly cationic allyl-Ni complexes and IFP's *Dimersol proc. are used. The c. of ethylene and butadiene is cat. by many *Ziegler-Natta cats., but only the use of $RhCl_3 \cdot H_2O$ yields the commercially important *1,4-*trans*-hexadiene (used as a *termonomer in the prod. of ethylene-propylene-diene [EPDM] elastomers en-

abling *vulcanization of the EPDM by *cross-linking reactions of the pendant double bonds). **F** codimérisation; **G** Codimerisierung.

W. KAMINSKY, M. ARNDT-ROSENAU

Ref.: Cornils/Herrmann-1, p. 258; *J. Am. Chem. Soc.* **1965**, *87*, 5638.

COED process   → FMC procs.

coenzyme   is a relatively small organic molecule that is associated with an *enzyme and participates in the enzymatic reaction. It is chemically altered during the course of the reaction and often must be regenerated in a separate reaction, by that enzyme or another. Some of the common coenzymes are *NADH, *NADPH, *FADH, FMN (which are involved in *oxidation/ *reduction reactions); *coenzyme A, and lipoic acid, involved in acyl transfer; *pyridoxal phosphate, for amine transfer; *thiamine pyrophosphate, tetrahydrofolate, coenzyme B_{12}, and *biotin, which help cat. the formation of *C-C bonds. Cf. *cofactor. **E=F**; **G** Coenzym.   P.S. SEARS

coenzyme A (and analogs)   are useful for the study of CoA-dependent *enzymes. Enzymatic synth. of an easily functionalized thioester analog of CoA from a thioester derivative of pantetheine phosphate provides a new route to different CoA analogs via *aminolysis of the thioester (see also *acetyl CoA). **F** coenzyme A et analogues; **G** Coenzym A und Analoge.   C.-H. WONG

Ref.: *J. Am. Chem. Soc.* **1992**, *114*, 72848.

cofactor   A c. is an additional, non-proteinaceous comp. of an enzymatic reaction which participates in the reaction and in the case of certain cs. (such as *ATP), may end up chemically transformed when the reaction is finished. In such cases, the c. is actually a cosubstrate but, unlike other substrates of the reaction, it is regenerated (*cofactor regen.) and reused with each turnover of reactant. *Regeneration of the c. to complete the *cat. cycle may require a separate *enzyme, as in the case of reactions which *hydrolyze the c.

*ATP, or may occur in a separate step by the same enzyme. Cs. may be transiently (many *coenzymes) or tightly/covalently attached (prosthetic groups). They may be small organic molecules, or they may be metals or metal *clusters. Some examples of cs. are the coenzymes (e.g., *NADH, *ATP, pyridoxal phosphate, etc.); Fe/S clusters and various metal ions; and *porphyrins such as heme and cobalamin (*vitamin B_{12}). Cs. aid in enzymatic reactions but play no similar role as co-cats. in chemical catalysis. **E=F=G**. P.S. SEARS

cofactor regeneration A number of enzymatic reactions require *cofactors. These are too expensive to be used as stoichiometric reagents. Cr. from their reaction prods. is thus required to make procs. using them economical. Cr. can also reduce the cost of synth. by 1: influencing the position of equil.; that is, a thermodynamically unfavorable reaction can be driven toward prods. by coupling it with a favorable cr. reaction; 2: preventing the accumulation of cofactor byprods. that inhibit the forward proc.; 3: eliminating the need for stoichiometric quantities of cofactors and thus simplifying the workup of the reaction; and 4: increasing *enantioselectivity relative to stoichiometric reactions. **E=F=G**. C.-H. WONG

coke → carbon

coking 1: deactivating deposition of carbon and higher molecular cpds. on cat. *surfaces. The *deposits originate from *hcs. during cat. reactions (e.g., *cracking, *deactivation) at higher temps. (*thermolysis); they may be removed by *combustion (burning off, *regeneration). 2: proc. for the thermal degradation (*pyrolysis) of high molecular products, specially residues (e.g., the delayed coker of *Foster-Wheeler) **F** cokéfier; **G** Verkokung.

B. CORNILS

Ref.: Ertl/Knözinger/Weitkamp, Vol. 3, p. 1273; Farrauto/Bartholomew, p. 273.

cold hydrogenation collective term for the *hydrogenation of methylacetylene and propadiene at relatively low temps. to im prove the yield and quality of propylene (c *Bayer procs., *Lummus CDHydro techno ogy and see *low-temperature hydrogen ation).

B. CORNIL

colloids in general are chemical objects of fi nely divided matter, in either the gaseous, liquic or solid state, dispersed in another *phase Therefore cs. are not a distinct class of material: but describe instead a state of matter. Colloid: systems include foams (gas in liquid), *xeroge! (solid *foams, gas in solid), emulsions (liquid i: liquid), *dispersions (solid in liquid), aeroso! (liquid in gas), and smokes (solid in gas). Ther is no clear differentiation between cs. and re lated systems such as *clusters and nanocrysta! (*nanodispersed materials) on the one hand an particulate matter on the other. Typically, th colloidal domain spans over a length scale c 1–1000 nm. In addition, c. can be differentiatec into three categories. 1: Dispersion c., the clas of which is characterized by the *dispersio: state. Usually, the molecules in these entities ar not connected, i.e., they represent classical gase: solids, or liquids. 2: In molecular cs. the constitu ting entities are linked by primary valences: th c. is one molecule. Typical of molecular cs. ar proteins or *polymers. 3: In association cs., th colloidal structure is formed by the assembly o a number of similar subunits, connected eithe by secondary valences or by hydrophobic forces The association c. usually shows high symmetr: and regularity; its formation is reversible anc conc.-dependent.

The most prominent advantage of cs. for cat is the very high specific surface area of 10-3000 m^2/g. Cs. can therefore be efficient cats (if separable, cf. *hom. cat.), e.g., metal *clus ters or nanoparticles, and can also be em ployed to mediate reactions between other wise immiscible comps., e.g., in *phase trans fer cat. Colloidal *Ziegler cats. and *immo bilized cs. are also known. **F** colloides **G** Kolloide. M. ANTONIETT

Ref.: D.F. Evans, H. Wennerström, *The Colloidal Do main*, Wiley-VCH, Weinheim 1999; Ullmann, Vol. A7 p. 341; *J. Mol. Catal.* **1999**, *142*, 201.

column reactor   → bioreactors

combinatorial catalysis   Based on highly parallelized prepn. and testing methods, the discovery of fundamentally new cats. can possibly be accelerated. This approach, including automated methods for cat. synth., testing, analysis, and data handling, is often summarized under the term cc., in analogy to the combinatorial methods used for drug discovery. The major difference of this approach compared to former cat. searches, as for instance demonstrated by *Mittasch in the case of the ammonia cat., is the degree of automation and parallelization possible with modern equipment for control of reaction systems, prod. analysis, data acquisition, and analysis. The field is still in its infancy, and it is not clear whether cc. will be more efficient in discovering novel formulations than the conventional approach. Several startup companies are active in this field, too.

The most important aspect in cc. is the highly parallelized cat. testing which gives activity data related in a meaningful way to the performance of a cat. in the commercial proc. Several techniques for parallel or pseudoparallel activity testing have been reported, including scanning *mass spectrometry, resonance-enhanced multiphoton ionization, fluorescence detection, IR thermography, or the use of multichannel reactors with multiport valves for switching to analytical instruments. Synth. is done by, inter alia, sputtering techniques with adjustable masks, ink jet delivery of cat. precursor solutions, or the use of pipetting robots. Solid handling, however, is still a problem. Cc., even if successful, will probably not replace the conventional cat. development proc., but will allow investigation a wider range of cat. candidates which will then have to be scaled up, shaped, and finetuned (*tailoring) by conventional development methods. The complete proc. from the discovery of an active formulation, through scale-up of cat. synth. and cat. proc. to the implementation of the novel cat. into the commercial proc. on a prod. scale has not been executed,

yet. **F** catalyse combinatorial; **G** kombinatorische Katalyse.   F. SCHÜTH

Ref.: *Angew.Chem.* **1998**, *110*, 2788 and **1999**, *111*, 335, 508; *Angew.Chem.Int.Ed.Engl.* **1998**, *37*, 2333 and **1999**, *38*, 1216, 2495; A. Furka, *Advanced ChemTech Handbook of Combinatorial & Solid Phase Organic Chemistry*, Advanced ChemTech 1999; New journal: Combinatorial & High-Throughput Screening, **1999**, Vol. 2.

combustion   → catalytic combustion, *regeneration

commensurate structure   An *adlayer may form a cs. if the interaction between adsorbate and substrate is strong and specific. Thereby the substrate exhibits special periodically identical *adsorption sites, on which the adsorbate adsorbs preferentially. Thus the substrate induces a structural ordering of the adlayer. To describe such css. the matrix notation is used: the lattice vectors b_1 and b_2 of the adsorbate unit cell are linear combinations of the lattice vectors a_1 and a_2 of the substrate acc. to $b_1 = m_{11}a_1 + m_{12}a_2$ and $b_2 = m_{21}a_1 + m_{22}a_2$, with the matrix

$$M = \begin{vmatrix} m_{11} & m_{12} \\ m_{21} & m_{22} \end{vmatrix}$$

If the coefficients m_{ij} and det M are integers the cs. is called simple. If the coefficients m_{ij} are all rational and det M is a simple fraction, the cs. is called coincident. An alternative notation is the Wood notation. In contrast structures without a common periodicity between the two lattices are called *incommensurate structures. **E=F=G**.   R. SCHLÖGL, Y. JOSEPH

comonomers   → copolymerization

compensation effect   the effect when the *activation energy E and the *pre-exponential factor A change in the same direction with a change in reaction for a given cat. or with a change in cat. for a given reaction. Thus, changes in A compensate for changes in E. **F** effet de compensation; **G** Kompensationseffekt.   B. CORNILS

competitive inhibition Ci. reflects the binding of an *inhibitor (I) to the *enzyme (E) near or at the *active site; this binding prevents the substrate(s) (S) from binding properly or at all (Figure).

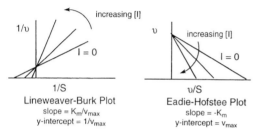

$$EI \underset{}{\overset{K_I}{\rightleftharpoons}} I + E + S \underset{}{\overset{K_S}{\rightleftharpoons}} ES \xrightarrow{k_{cat}} E + P$$

I and S are thus competing for the active site of the enzyme. With this type of inhibition, the values of K_m increase with increasing conc. of the inhibitor, as shown in the *Lineweaver-Burk and *Eadie-Hofstee plots; V_{max} (or k_{cat}) does not change. By increasing the conc. of substrate, eventually V_{max} can be reached.

increasing [I]

$1/\upsilon$

$I = 0$

1/S
Lineweaver-Burk Plot
slope = K_m/V_{max}
y-intercept = $1/V_{max}$

υ

increasing [I]

$I = 0$

υ/S
Eadie-Hofstee Plot
slope = $-K_m$
y-intercept = V_{max}

It is worth noting that for a two-substrate reaction, an inhibitor that shows ci. against one substrate will typically not show ci. toward the other (cf. also *Dixon plot). **F** inhibition compétitive; **G** konkurrierende Inhibierung.

P.S. SEARS

competitive inhibitor → inhibitor

complex catalysis catalyses operated *complexes as (mostly) homogeneous or biocatalysts, cf. *complexes.

complexes 1: The interchangeably used terms complexes and *coordination entities refer to molecules or ions consisting of a *central atom (generally but not necessarily a metal atom) to which other atoms or atom groups, the *ligands, are attached. One ligand saturates a secondary (excess) or a primary (classical or stoichiometric) valence of the *central atom (*coordination chemistry theo-

ry, Werner complexes). The coordination entity can be a cation, an anion, or an uncharged molecule; it is stable in solution and dissociates only partially. The ligand atoms directly bound to the central atom form the coordination polyhedron. The coordination number of the central atom is the number of σ-bonds between the ligands and the central atom (special rules exist for some organometallic cs.). A ligand containing more than one potential coordination atom (donor or ligating atom) is termed a *bi-, *tri-, or *multi-dentate ligand. *Chelating ligands are attached to the central atom through two or more coordinating atoms and form relatively stable cs. Bridging ligands are attached to more than one center of coordination. Acc. to the number of coordination centers in a complex or coordination cpd., "dinuclear" indicates two central atoms, trinuclear three, etc. The oxidation number (state) of a central atom is defined as the charge which results, when all ligands including their binding electron pairs are removed from the central atom. Coordination cpd. and complexation of metal ions are of considerable practical importance.

The bonding situation in cs. can be described by different models (cf. *coordination chemistry theory). Due to their unoccupied valence orbitals the metal ions (central atoms) are *Lewis acids, and the coordinative interaction with occupied orbitals of nonmetal ions as Lewis bases (ligands) can be regarded qualitatively in terms of the *HSAB concept, which gives a rough estimate of the relative c. stability (thermodynamic c. formation or stability constants). Cs. of main group metals having only four stable valence orbitals and two stable oxidation states at maximum behave differently from *transition metal cs., which have (due to the five inner d-orbitals) nine bond orbitals and a variety of oxidation states. Via occupied d-orbitals transition metals can additionally realize an electron donor function (π- or σ-bases), whereas main group metals act as *Lewis acids only. Transition metal cs. are favored for many cat. reactions also due to the variable oxidation state and coordination number. In solution, ligand molecules can dissociate or exchange, or *vacant coordination sites can be occupied by solvent or substrate molecules (cf. *cat. cycle). Metal ions or cs. may act as cats. by bonding to organic reactants, introducing charge (polarizing) into them. The diversity of *central atoms, ligands, and complex structures allows a variety of activation mechanisms.

Involvement of *redox activity in complex cat. increases the number of activation routes. Cat. by cs. can be classified acc. to the structure of the cat. and the corresponding specific interactions and *activation mechanisms (besides *Brønsted and *Lewis acid-base cats.) as i-General complex cat.: the substrate is activated by a coordinative interaction or by a *redox reaction with a metal ion or metal c. (electronic, equil., or stereochemical *ligand effects). Due to the properties of the central atom ion (positive charge, Lewis acidity, polarizing action, spatial arrangement of the ligands by coordination) reactions of ligands with negative, nucleophile reactants are facilitated, *heterolytic bond cleavage becomes possible, and the comb. and correct spatial arrangement of reactants (*template) can be achieved. A stereochemical effect is the asymmetric control of ligand reactions in the coordination sphere of *optically active cs. Fundamental steps of general complex cat. are c. formation with the substrate (substrate activation); the reaction forming the prod. complex; elimination of the prod.; and recovery of the starting complex (or cat.). An optimum of kinetc lability and thermodynamic stability of the metal cs. in the *cat. cycle represents a cond. for effective complex cat. *Product inhibition (not so pronounced with organometallics) should also be mentioned. ii-Organometallic complex cat. (*organometallic cpds. for hom. catalysis): involves the formation or participation of a metal-carbon bond. The key reactions of a typical organometallic cat. cycle are the dissociation step (16-electron complex, cf. *Tolman's 18–16 electron rule), *oxidative addition, *insertion/addition reactions, and *reductive elimination.   K. KÖHLER

2: In biocatalysis: "complex" may imply the bulk of substrate and *biocatalyst/*enzyme or when *enzymes containing an active metal center are involved (coordination via amino acids of peptides, *macrocyclic chelate ligands or nucleobases as comps. of polynucleic acids). **F** complèxes; **G** Komplexe.   P.S. SEARS
Ref.: Ciardelli/Tsuchida/Wöhrle.

Comprimo Superclaus process recovers sulfur from acid gases with H_2S, CO_2, and NH_3 and is thus a variant of the *Claus proc. New propietary cats. are claimed to be very selective. **F** procédé Superclaus de Comprimo; **G** Comprimo Superclaus-Verfahren.
Ref.: *Hydrocarb.Proc.* **1996**, (4), 144.   B. CORNILS

comproportionation   (synproportionation)   *redox reaction in which a cpd. of an element in a medium oxidation state is formed by the reaction of two cpds. with the same element in a higher and a lower oxidation state, e.g., $HClO + HCl \rightarrow H_2O + Cl_2$. The reverse reaction is called *disproportionation.

One of the most important cs. is the decomposition of NH_4NO_3 into either N_2O or N_2. NH_4NO_3 is therefore used as fertilizer and as a gaseous explosive. This decomposition is cat. by many inorganic materials, including chlorides, chromates, hypophosphites, and powdered metals such as Cu, Zn, or Hg, as well as organic substances (oil, paper, sawdust).

The reversible temp.- and press.-dependent c. of CO_2 and C into CO (*Boudouard equilibrium) is an important reaction for technical applications. **F** métamutation; **G** Komproportionierung.   P. ROESKY

computational chemistry all theoretical methods for calcng. molecules, solids, or liquids with the help of computers. Cc. has become a major research discipline in cat., because the postulated *intermediates and *transition states of cat. reactions are difficult to investigate experimentally, and theoretical methods have become more and more accurate in the recent past. The most important cc. methods for calcng. cat. proc. are *density functional theory (DFT), *ab-initio calcns., *semi-empirical methods, and *force-field models. Besides these static methods there are *molecular dynamics (MD) calcns. which simulate the time evolution of particle systems. MD methods are employed to calculate properties such as free enthalpies and entropies, and they have also become important for investigating reaction profiles. *Solvent effects can be calcd. by various methods based on the continuous electric field theory of Onsager, or by *Monte Carlo methods. Some of the methods can be combined, e.g., in *embedding methods where different methods are used to calc. the different parts of a molecule.

There are cc. program packages available which make it possible to calc. in principle every property of a substance. The results of the calcns. may be displayed graphically by methods of *molecular modeling. **G** Computer-Chemie.   G. FRENKING

Ref.: F. Jensen, *Introduction to Computational Chemistry*, Wiley, Chichester 1999; Thomas/Thomas, p. 364; P. von Ragué Schleyer (Ed. in Chief), *The Encyclopedia of Computational Chemistry*, Wiley, Chichester 1998.

computer-aided (assisted) catalyst design

*heuristics on which design of het. cats. is still mainly based. More recently deterministic methods also evolved. Purely deterministic methods are *ab-initio or semi-empirical quantum chemical calcns., which can help build up a model of the cat. system and study effects of modifications. Simulation of the *molecular dynamics of the interaction between reactants or intermediates may describe observed effects in a cat. system and may be able to predict the effect of cat. modifications. The use of chemical reaction engineering related deterministic computational methods for the description of *inter- and intraparticle *diffusion and its interaction with the cat. reaction are also of importance in cat. design. Also, the microkinetic analysis is a supportive means in identifying cat. materials (cf. *microkinetic analysis). **E=F=G**.
                                                    M. BAERNS

Ref.: *Catal. Today* **1991**, *10*, 147, 167 and **1995**, *23*, 32 J.A. Dumesic et al., *The Microkinetics of Heterogeneous Catalysis*, ACS, Washington DC 1993.

computer modeling

(for enzyme inhibition) Computers have become useful too in the design of small molecules to fit and inhibit *enzyme *active sites. The general procedure used by computer programs depend upon whether the enzyme to be inhibited has a known three-dimensional (3D) structure. the enzyme structure is unknown, then or must design an inhibitor that mimics the be substrate, or better still, the postulated *transition state of the reaction. Knowledge of the enzyme structure provides the opportunity to design molecules that take advantage of favorable interactions that are not used by the substrate.

In either case, there are two main approache in *inhibitor design. 1: Library search algorithms, in which large libraries of small molecule structures are compared to the target molecule (for designing substrate/transition state mimics) or are docked into the active site (for enzymes of known 3D structure); and 2: de novo ligand design, in which molecule are built into the enzyme active site. *De novo ligand design is typically accomplishe by positioning atoms or functional groups specific, favorable positions and allowing th computer to fill in the structure to make contiguous molecule. Other approaches ar also in use, such as filling the active site wit hexagonally close-packed atoms and evolvin the system using a *Monte Carlo simulatio in which the atom type, position, etc., are va ied stepwise.

All of these methods require a means of as sessing how well a molecule fits. "Goodnes of fit" refers not only to the enthalpy of bind ing, but also to how stable and how stiff th cpd. is (stiffer molecule = less unfavorable er tropy of binding). Many computer program currently in use have weighting algorithms in corporated which discourage or encourag certain features (such as discouraging het eroatom-heteroatom bonds) within the targe

molecule in order to encourage the selection of a stiff, stable molecule. **E=F=G.**   P.S. SEARS
Ref.: *J.Med.Chem.* **1995**, *38*, 466.

COMS   → surface organometallic chemistry (SOMC)

concerted reaction   a reaction in which more than one bond-breaking and/or bond-forming step occurs almost simultaneously (synchronous reaction, cf. *domino reaction). **F** réaction concertée; **G** Konzertierte Reaktion (Synchronreaktion).   B. CORNILS

condensation   reaction   Condensation means an – often cat. influenced – conv., in which at least two identical or different molecules react with cleavage of smaller molecules such as H_2O, HCl, NH_3, CO_2, or alcohols. Crs. include various inter- or intramolecular transformations, homo- or heterocondensations, or *polycondensations. Examples are *ammondehydrogenations, *esterifications, *Guerbet reactions, *Mannich reactions, *transesterifications, *Ugi four-component reactions, *BISA or *MDI procs., *acylations, *Baekeland-Lederer-Manassen, *Clay-Kinnear-Perren, or *Darzens reactions, and polymer condensations such as polyvinylacetals, polyamides, or polyesters (*DMT, *PET). Crs. are unprofitable on grounds of *atom economy. **F** condensation; **G** Kondensation.   B. CORNILS

Condensol   special series of BASF cats. based on metal salts for the cross-linking of formaldehyde-containing textile auxiliaries. **F=G=E**.   B. CORNILS

conduction band   → band theory

cone angle   → Tolman cone angle

configuration   Many molecules can adopt several forms that have the same connectivities but which cannot interchange without breaking and re-forming bonds. This is the case, for example, in molecules that have un-saturated bonds that cannot rotate freely, or in *chiral molecules. The different forms are said to have different configurations (as opposed to *conformations, forms that can be readily interchanged without disruption of covalent bonds). The distinction between these terms is difficult when talking about rotational *isomers that have extremely hindered rotation. **E=F=G**.   P.S. SEARS

confined catalysts   → *ship-in-the-bottle (or tea bag) catalysts; cf. *constraint chiral cats.

conformation   (conformational change) The cs. of a molecule are the three-dimensional (3D) structures which are able to adopt without altering covalent bonds (cf. *configuration). *Enzymes are large molecules, often made of several hundred amino acids and there are an enormous number of 3D structures that they can adopt. The native (active) c. is usually thought to be the thermodynamically most stable c. (at least at the time of folding), though attaining that c. may require the assistance of protein-folding cats. called *chaperones.
Enzymes are dynamic molecules, for which the side chains and backbones undergo slight (or in some cases, large) conformational shifts. There are many examples of cat. where large shifts in the protein c. are required for *activity. *Hexokinase, for example, protects water-labile intermediates from solvent by folding over the *active site. In other cases, a conformational change is used for regulation of enzyme activity. There are many examples of *feedback inhibition in which the prod. of a *biosynthetic pathway inhibits the enzyme that cat. the first committed step by binding to a site remote from the active site. By binding to that site, it causes a conformational shift in the enzyme that renders it less active. This is called *allosteric inhibition. The conformational shift is frequently very large, involving movement of entire protein domains or subunits with respect to each other. **E=F=G**.   P.S. SEARS

conglomerate → racemate

Consol Synthetic Fuel (CSF) process   a Consolidated Coal Co. variant of *coal liquefaction/extraction with tetralin/cresols or coal-own middle distillates. This technique decreases the press. necessary for cat. *coal hydrogenation and avoids problems with ash removal. **F** procédé CSF de Consol; **G** Consol Synthetic Fuel-Verfahren.   B. CORNILS

constitutive enzymes   *enzymes that are produced at a relatively constant rate. They are typically involved in basic cell upkeep, e.g., the glycolytic enzymes.   P.S. SEARS

constraint chiral catalyst   concept of confinement of large *chiral organometallic cat. entities including chiral *chelate *ligands in various *cavities (cf. *confined cats.).
Ref.: *J. Mol. Catal. A:* **1999**, *141*, 139.   B. CORNILS

contact potential difference (CPD)(measurements)   a method for qualitative and quantitative *chemisorption procs. Surface charges are used as probe; the response is capacitances. The ex-situ method is non-destructive, but the surface potential is extremely sensitive to impurities. The data are local and *surface sensitive.   R. SCHLÖGL

contact process   1-General: proc. for the manuf. of H_2SO_4 by Pt cat. *oxidation of SO_2 to SO_3 (developed by *Knietsch and *Winkler), cf. *Bayer double-contact proc. 2: Next step of proc. development replacing the ancient *lead chamber proc. in the *oxidation of SO_2 to SO_3 under modern industrial conds. acc. to $SO_2 + O_2 \rightarrow SO_3$; $\Delta H_{(298)}$ = –90.1 kJ/mol. Acc. to the thermodynamic equil., low temps. favor the formation of SO_3 requiring an active cat. V oxides are commonly used, either as such, as salts with phosphoric acid (vanadyl pyrophosphates) on an acid-resistant *carrier, or doped with K_2O (*V as cat. metal). Reaction temps. are in the range 450–550 °C, depending on whether the removal of SO_2 or the generation of SO_3 is the target. **F** procédé de contact; **G** Kontaktverfahren.   C.D. FROHNING
Ref.: Ullmann, Vol. A25, p. 635; Büchner/Schliebs/Winter/Büchel, p. 112; Ertl/Knözinger/Weitkamp, Vol. 4, p. 1774.

continuous   catalyst   regeneration (CCR) 1-General: The aging of cats. may be considerable and covers *catalyst lifetimes from min to years, i.e., orders of magnitude between 10^0 and 10^{5-6} (*deactivation). Depending on cat. reaction conds., purity of reactants, content of *poisons, load, etc., the cats. lose *activity and/or *selectivity. To compensate for these losses a total change of cats. or a reactivation (*regeneration) is usual. The regeneration may be practised intermittently (example: coke burn off during special cycles by heating) or continuously (e.g., *FCC cats. or *make-up of the hom. *oxo cat. of Ruhrchemie's former Co process: continuous treatment of a cat. side-stream with thermal [steam] and chemical [precipitation of Co hydroxide] measures). *Make-up steps such as with hom. cats. facilitate the separation of *cat. poisons per pass. A *aqueous-phase cat. is specially suitable for ccr. 2: special UOP technology with stacked radial-flow reactors for the *regeneration of cats. within their BTX aromatics manuf. using the *UOP/BP Cyclar proc. **F** régénération continue des catalyseurs; **G** kontinuierliche Katalysatorregenerierung.   B. CORNILS
Ref.: Falbe-1, p. 80; *Hydrocarb. Proc.* **1997**, (3), 116.

continuous-flow reactor → CFR, *CFSTR *catalytic reactors

continuous membrane reactor   → bioreactors, *catalytic reactors, *membrane reactors

controlled-pore glass (CPG)   special glass *support for cats. (Picture, inter alia for *enzymes), e.g., Trisopor®, or CPG®-240 (with a average pore size of 240 A and 120/200 mesh size).   B. CORNILS

[by courtesy of Schuller GmbH, Germany]

Ref.: *Ann.Chim.Acta* **1994**, *292*, 281; R. Epton (Ed.) *Innovation and Perspectives in Solid Phase Synthesis*, Mayflower Worldwide, Birmingham 1994, p. 131; DE 4413 735 A1.

conversion electron Mössbauer spectroscopy (CEMS)

a Mössbauer variant for the determination of surface chemical constitutions (probes are γ-rays, the response electrons). The ex-situ technique is non-destructive but restricted to some elements (Fe, Sn, Au). The data are integral, *surface sensitive, and have in-situ capabilities.   R. SCHLÖGL

conventional transmission electron microscopy (CTEM)

a variant of electron microscopy, applied to the measurement of local structures, defects, and *nanostructures. The ex-situ, UHV method is non-destructive, but only thin samples can be handled and beam stability can be difficult. The data are local (on an atomic scale) and bulk sensitive.
                                    R. SCHLÖGL

co-oligomerization

of butadiene and ethylene by *codimerization gives *1,4-hexadiene.

cooperative effect

concerted action of two or more active distal site centers (subunits) of a multi-subunit cat. on substrate binding. In a positive ce., the effect at one site increases the affinity for the substrate of the other site. A negative ce. diminishes the affinity for the substrate of the first site. The velocity data for cooperative cats. can be expressed by the *Hill equ.

Ces. can be observed in hom. or in het. cat. as well as in biocat. In hom. and in *enzyme cat., substrate binding at one site frequently induces structural changes in the surrounding unit of the cat. that are transmitted to remote *active sites of the cats. (long-distance communication, *allosteric interactions). Ces. can occur between separate binding sites for the same substrate (homotropic ces.). Cat. may also be effected by binding of a structurally unrelated *ligand at a distant separate site (heterotropic ce.). Examples for ces. in biology are hemoglobin, the *tryptophan suppressor protein, *threonine deaminase, or *aspartate carbamoyltransferase; cf. *cooperativity). An example in het. cat. is the *reduction of NO in the presence of oxygen which proceeds advantageously on mixtures of *alumina-supported Pt. **F** effet coopératif; **G** kooperativer Effect.   D. HELLER, A. BÖRNER

Ref.: *Angew.Chem.Int.Ed.Engl.* **1985**, *24*, 1; R.A. Copeland, *Enzymes*, VCH, Weinheim 1996, p. 279.

cooperativity

Many *enzymes exist as multimers with several *active sites. The *activity of the subunits may be independent, or the binding of substrate by one subunit may affect the activity of the others in a positive or negative fashion. This phenomenon is called c., and was first observed with the binding of oxygen to hemoglobin, a homotetramer. It was found that the binding of one molecule of O facilitated the binding of further molecules, leading to a sigmoidal dependence of the hemoglobin saturation on the partial press. of oxygen. Similar behavior has been observed with a number of *enzymes, including aspartate transcarbamoylase from *E. coli* (see also *allosteric interactions in enzymatic cat.; *Hill equ.). Such *enzymes may display positive or negative c., and the effect of positively (pos) and negatively (neg) cooperative and non-cooperative (no) binding on the appearance of two common plots (rate vs. substrate and *Lineweaver-Burk) is shown below in the Figure. **F** cooperativité; **G** Kooperativität.

P.S. SEARS

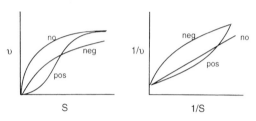

Ref.: I.H. Segel, *Enzyme Kinetics*; Wiley, New York 1975.

coordination chemistry theory   The birth of cct. goes back to A. Werner, who extended the valence theory to account for the fact that certain elements could form cpds. with more ions or molecules than allowed acc. to common concepts at that time. He introduced the term *secondary valence* for the number of groups a *central atom can bind in excess of its classical primary valence. The idea of a *complex cpd. being a central atom (metal) surrounded by a certain number of *ligand atoms or molecules in a definite stereochemical arrangement was in accord with the number of isolated complex isomers. Since there is no basic difference between the two types of valence, these concepts were later replaced by the term *coordination number*, i.e., the number of ligands or ligator atoms directly bound to the central atom, and the *oxidation state* of the central atom. Following the *Lewis acid-base concept, Sidgwick coined the term *coordinative bond*, which implies that the ligand atom provides an electron pair for a covalent bond to the Lewis-acidic central atom. Within *Pauling's valence bond theory, metal s, p, and d acceptor orbitals are hybridized to account for the observed coordination geometries.

Paralleling the octet rule for the main group cpds., a 18-electron rule (*Tolman) for *transition metal complexes is explained. The introduction of high- and low-spin configurations gives reason for the magnetic properties of certain complexes. The justification for these concepts, the capability of interpreting experimental UV/Vis spectra by d-d transitions and the prediction of coordination geometries as a function of d-electron configuration, constitute the main practical achievements of crystal field and ligand field theories which postulate a characteristic energy splitting of metal d-orbitals due to the electrostatic field of a ligand sphere of a certain symmetry. These concepts are now largely replaced and extended by molecular orbital descriptions which help to explain the unusual chemical stability of certain complexes, such as carbonyl complexes, by the introduction of *π-back-donation acc. to the donor-acceptor model (*Chatt-Dewar-Duncanson model), the reactivity of complexes by inspection of frontier orbitals, and other effects. Recently modern *ab-initio calcns. and the *DFT have widely lightened the understanding of the structures, reactivities, and properties of coordination cpds., and of the phenomena of hom. cat. **F** théorie de la chimie coordinative **G** Theorie der Koordinationschemie

W. HIERINGER

coordination entities   → complexes

coordination number   → coordination chemistry theory

coordinatively unsaturated metal centers (cus)   → vacant coordination sites

co-oxidant   → reoxidant

COPE process   Claus Oxygen-based Process Expansion, which exceeds the oxygen content in the combustion air by up to 80–100 %. The overheating of the combustion chamber is avoided by mixing with the recycled gas from the first sulfur condenser, thus reducing the quantity of waste gases by 35 %. **F** procédé COPE; **G** COPE-Verfahren.   B. CORNILS

Ref.: Büchner/Schliebs/Winter/Büchel, p. 108.

Cope rearrangement (catalytic)   the rearrangement of 1,5 dienes under the influence of Pd(II) cpds. The cat. version is 10^{10} times faster than the thermal rearrangement.

Ref.: *Organic Reactions*, Vol. 22, p. 1.   B. CORNILS

copolymerization   *polymerization of two (or more) structurally different monomers A and B (comonomers) into one polymer chain. Based on the distribution of the comonomers in the copolymer formed, four limiting types are differentiated. 1: random copolymers, featuring a statistical distribution of A and B monomer units (-AABABBBAABABABBA-AABA-); 2: alternating copolymers, such as (-ABABABABABABABABABAB-); 3: block copolymers (-AAAAAAAABBBBBB-BBBBAA-); 4: graft copolymers (Figure)

```
    -AAAAAAAAAAAAAAAAAAA-
        B       B       B
        B       B       B
        B       B       B
        |       |       |
```

The commercially most relevant type of copolymers is *LLDPE which is prod. by c. of ethylene with α-olefins using het. *Ziegler-Natta (ZN), *Phillips, or *metallocene cats. Besides their molecular weight distribution cs. are characterized by the homogeneity of comonomer incorporation and the comonomer distribution along the chains. *Single-site cats. usually prod. hom. comonomer incorporation and distribution, while in the case of multi-site cats. (ZN type) the comonomer incorporation and distribution are heterogeneous although each site gives similar chain structures. Cat. by *metallocenes also allows the prod. of ethylene-norbornene copolymers (*COC) and ethylene-styrene copolymers, new materials not available from Ziegler-Natta cats. Olefin-CO c. (*alternating c.) is achieved by hom. single-site cats. based on Pd. These copolymers feature a perfectly alternating microstructure and are therefore semicrystalline polymers (*Carilon). Pd-based cats. also enable the c. of ethylene and acrylates although the acrylate is not incorpo-

rated into the main chain of the polymer. **F** copolymérisation; **G** Copolymerisation.

                        W. KAMINSKY, M. ARNDT-ROSENAU
Ref.: Ullmann, Vol. A21, p. 336; *Adv. Polym. Sci.* **1997**, *127*, 143; Cornils/Herrmann-1, p. 333; *J. Am. Chem. Soc.* **1996**, *118*, 267; Mark/Bikales/Overberger/Menges, Vol. 4, p. 192.

copper amine oxidases   class of *enzymes that oxidatively deaminate organic amines to the respective aldehydes with the generation of NH_3 and H_2O_2. They use a unique *cofactor (6-hydroxydopaquinone or topaquinone [TPQ]), which is covalently bound and formed by the *oxidation of a tyrosine residue in the context of the consensus sequence Asn-Tyr-Asp-Tyr. The cofactor is believed to form a *Schiff base with the substrate. Tautomerization and *hydrolysis gives the aldehyde and the amino form of the cofactor, which is regenerated by Cu-assisted oxidation (with O_2) to the imine followed by hydrolysis. **F** oxidase d'amine de cuivre; **G** Kupferaminoxidase.   P.S. SEARS

Ref.: *Biochemistry* **1998**, *37*, 4946; *Biochemistry* **1998**, *37*, 12513.

copper as catalyst metal   Cu plays a considerable role in cat. as metallic copper, with its valence change Cu(I) → Cu(II), as well as in cats. based on mixed metals. Metallic Cu serves as Raney Cu (in *Celanese HMDA proc. or in *Mitsui's acrylamide proc.) or in various procs. such as the *ethynylation (*Reppe synth., or other convs. with acetylene, e.g., *BASF BDO proc.), *cyclopropanation, or the *Ullmann reaction. *Supported Cu cats. are used for *dehydrogenation reactions. Cu in mixture with other metals is used in the *Simmons-Smith reaction (Cu-Zn), *Rochow sysnth. (Cu-Si), or as *Hopcalit (Cu-Mn or Ag). Cu(I) salts are used in *Dow's phenol synth. (*decarbonylation of benzoic acid), as a *Nieuwland cat., or in the reactions acc. to *Gattermann-Koch, *Glaser, or *Sandmeyer. Cu salts are also employed as *reoxidants in *redox reactions (*Wacker proc., the former *BASF acrylic acid proc., *oxychlorinations).

Cu as intermetallic cpds., *mixed oxides, or in mixture with other *additives (*promoters, etc.) serve as *Adkins cats. (Cu-Cr), for *ammonolysis reactions, *methanol synth. (CuO-ZnO), or *fat hydrogenation (Cu-Zn). Cats. based on Se-CuO have been proposed for *BP/Ugine's ACN proc. Cu is essential for life; several *enzymes such as *copper amine oxidase, *superoxide dismutase, *tyrosinase, or *galactose oxidase contain Cu. **F** le cuivre comme métal catalytique; **G** Kupfer als Katalysemetall.   B. CORNILS

Ref.: C. Kemball, D.A. Dowden (Eds.), *Catalysis*, The Chemical Society, London 1980, Vol. 3, p. 74; *Prog. Inorg.Chem.* **1987**, *35*, 219; *Organic Reactions*, Vol. 19, p. 1 and Vol. 41, p. 135. *Houben-Weyl/Methodicum Chimicum*, Vol. XIII/4; *Chem.Rev.* **1994**, *94*, 737 and **1996**, *96*, 2563; *Acc.Chem.Res.* **1997**, *30*, 227; Murahashi/Davies, p. 303,317; *Chemie in unserer Zeit* **1999**, *33*, 334.

copper chromite   → Adkins catalyst

coprecipitation 1: Different techniques are described for the prepn. of cat. materials, such as *equil. adsorption, *grafting, *impregnation, and coprecipitation (cf. also *precipitation/coprecipitation). The c. technique for cat. prepn. can be used when a very intimate contact is wanted between two or even three comps. During cat. activation, e.g., *calcination, the cat. active *phase is formed from the hom. mixture of coprecipitated *precursors. For this technique the coprecipitated cpds. are required to be insoluble in the same *solvent or in the same pH regime. Under these conds. c. can occur and yields hom. element/precursor distributions. For example, *mixed oxide cats. for *selective (partial) *oxidation can be prepd. by c. of ammonium heptamolybdate, ammonium metavanadate and ammonium metatungstate in acid solutions. Calcination of the mixed precursor finally leads to the active multielement (*bulk) cat.

G. MESTL

2: Although IUPAC defines c. as the simultaneous *precipitation of a normally soluble comp. with a macrocomp. from the same solution by formation of mixed crystals etc., in their daily work specialists understand as c.

also the simultaneous precipitation of cat. active ingredients *on* slurried (or even preformed) carriers, thus yielding *supported cats. (cf. *deposition-precipitation, *sequential precipitation). **F** coprécipitation; **G** Cofällung.   B. CORNILS

Ref.: *Pure Appl.Chem.* **1975**, *37*, 463; Ertl/Knözinger/Weitkamp, Vol. 1, p. 72.

coprocessing refers to the processing of coal- and oil-derived prods. simultaneously, with the objective of liquefying the coal and upgrading the oil (residues). A small fraction of coal mixed with more than 90 % of oil may put the coal in the position of a cat., cat. *support, or *scavenger, supporting the upgrading reactions for the oil. The oil-derived prods. are vacuum distillates, residues from visbreaking or coking, asphaltenes, crude bitumen, or even waste plastics. At higher proc. temps. the thermal fragmentation of the oil into gases and solid residues will be important for the choice of proc. parameters. Taking crude bitumen as an example, the c. was done at 450 °C/7 MPa with cats. consisting of *bimetallic Mo-Fe or Mo-Cr oxides. **E=F=G**.   I. ROMEY

copromoter an *additive which supports the properties of a *promoter in respect of cat. *activity and/or *stability. An example is *n*-butyl phosphonium iodide as *copromoter of the Rh cat./CH$_3$I in *Hoechst's proc. to acetic anhydride. The differentiation from *co-catalysts is difficult. **E=F=G**.   B. CORNILS

cordierite natural *alum(in)osilicate (dichroite) with low thermal expansion and thus best suited for requisites which need to combine heat resistance with resistance to thermal shocks, such as *monolitic honeycombs in *automotive exhaust cats. (*ceramic cats, *catalytic combustion, *diesel engine emissions). For better uniformity the raw materials for monoliths are mainly produced synth. **E=F**; **G** Cordierit.   B. CORNILS

coreductant is a term from *redox reactions in which the metal oxidation (e.g., O$_{2,g}$

2 $Ag_s \rightarrow$ 2 AgO_s for the *EO synth.) is followed by the metal reduction with the co-reductant ethylene (AgO_s + $C_2H_4 \rightarrow$ Ag + EO). Cf. also *co-catalyst.   B. CORNILS

corner atom   → single crystal

corn syrup   manuf. by the *isomerization of fructose to glucose by means of *glucose isomerase

coronands   subclasses of *crown cpds.

correlation energy   The basic idea of the *Hartree-Fock (HF) method for solving the *Schrödinger equ. is the approximation of the many-electron system by a set of one-electron functions, called *molecular orbitals. The calcd. energy which is obtained in this way differs from the true total energy, because each electron is calc. in the average field of the other electrons, i.e., the electron-electron interaction is not treated individually. The difference between the true energy and the HF energy is caused by neglecting the individual correlation of the electrons, and by   neglecting relativistic effects in the HF treatment. Thus, correlation energy is defined as the difference between the nonrelativistic total energy and the HF energy which is obtained with an infinite *basis set. Although the correlation energy is typically only ca. 1–2 % of the total energy, it can become important for accurate calcns. of molecules. There are several methods available for calcng. correlation energy. The most important methods are: Møller-Plesset perturbation theory of nth order (MPn), configuration interaction (CI), coupled-cluster methods (CC), and multiconfigurational SCF (MCSCF). In *ab-initio calcns., the HF energy is calcd. first, followed by calcn. of the correlation energy. In *density functional theory the correlation energy is already included in the total energy which is given as a functional of ρ. The latter can be obtained by solving the Kohn-Sham equ. **E** l'énergie de corrélation; **G** Korrelationsenergie.   G. FRENKING

corundum   crystallographic species of α-Al_2O_3

Cossee (Cossee-Arlman) mechanism   is an early proposal for a *Ziegler-Natta polymerization mechanism.   B. CORNILS
Ref.: *J.Catal.* **1964**, *3*, 80, 89, 99.

cot, COT   *cyclooctatetraene or cyclooctatriene, inter alia, as a *ligand or in *complex cpds.

counter phase transfer catalysis   proc. in which hydrophilic metal *complexes have a double function as *transition metal cat. and as *inverse *phase transfer cat. (inverse PTC) under biphasic conds. (*two-phase cat.), e.g., transition metal complexes of polyethers or sulfonated *phosphines. Though cptc. originally had the same meaning as *inverse PTC, this name is now restricted to distinguish the cat. of hydrophilic metal complexes from the cat. of organic molecules such as pyridine or *cyclodextrin derivatives which have only an inverse phase transfer function. Inverse PTC means transporting hydrophobic molecules from the organic into the aqueous phase and accelerating the reaction with hydrophilic reagents in the aqueous phase. This is in contrast to normal *PTC, which transports anions from the aqueous into the organic phase. Inverse PTC is useful for biphasic reactions using hydrophilic reagents which are not transportable by normal PTC. For the loading of hydrophobic molecules, pyridines or cyclodextrins form (not without steric or electronic limitations) pyridinum salts or inclusion cpds., with the respectively lipophilic *cavities. Hydrophilic transition metal complexes carry the hydrophobic molecules by coordination to the metal center. Since the transition metals are able to react with various kinds of functional groups, a wide range of hydrophobic molecules is transportable in this inverse phase transfer without steric limitation. Pyridine and cyclodextrin derivatives as well as the normal PTCs are not cats. for reaction but cats. for interphase transportation. An additional cat. is normally necessary to achieve a cat. reaction under biphasic conds. A key step of these mixed cat. systems is the reaction between a molecule transported by a phase transfer cat. and another molecule activated by a reaction catalyst. Such reactions sometimes result in low efficiency due to insufficient *activity of one of the two cats., or are accompanied by serious side reactions caused by the excessive reactivity of one cat. Cptc. is useful for such reactions, e.g., for *carboxylations of allyl or benzyl halides, where the hydrophilic transition metal complexes exhibit high efficiency and *selectivity, while those of the mixed cat. system as well as hom. systems using amphiphilic solvents give serious amounts of allyl and benzyl alcohols and ethers. Since in cptc. the carrier molecule itself is a cat. for the reaction, the complexes transport the lipophilic halides from the organic into the aqueous phase, and then change into cats. for the reaction with hydroxyl anions in the aqueous phase. This successive cat. is effective in retarding the unnecessary transportation and the excessive reaction which result in poor efficiency and *selectivity. **E=F=G**.   T. OKANO

Ref.: Cornils/Herrmann-1, p. 221.

covalent catalysis formation of an activated covalent *intermediate between the *substrate and the cat. This is a frequent occurrence in enzymatic reactions. The formation of a covalent adduct provides an efficient means of retaining the substrate within the *active site through several steps and should provide an entropic advantage in the steps following formation of the covalent adduct. Many *enzymes contain an active-site nucleophile which reacts with the substrate to displace a leaving group; the resulting intermediate is then subject to attack by another nucleophile, such as water. This type of mechanism is observed, e.g., in the *serine and cysteine proteases, and many kinases, *cis-trans* *isomerases, and retaining glycosidases (see also *serine proteases; *thiol proteases; *glycosidases).   Electrophilic cat. (cat. via stabili-

zation of a negative charge) also may involv a covalent intermediate with the enzyme, a though the adduct is typically with a *cofacto such as *pyridoxal phosphate or *thiamir pyrophosphate rather than an enzyme sic chain. The pyridoxal phosphate-depender enzymes, which include many transaminase and decarboxylating enzymes, form a *Schi base between the aldehyde of the *cofacto and the amine of the substrate. This iminiu can effectively stabilize negative charge c the C adjacent to the amino group (Figure).

pyridoxal phosphate

iminium

In the case of transaminases, this can lead to a straction of a proton from that position and a α-1,3-tautomeric shift, followed by *hydrolys of the iminium to give pyridoxamine phospha and the aldehyde (see *pyridoxal phosphate). The *thiamine pyrophosphate dependent e zymes, which are responsible for the formatic and cleavage of C-C bonds, form a covalent ac duct between the C2 of the thiazolium and th substrate. For example, pyruvate decarboxylas from yeast converts pyruvate, a three-carbon (ketoacid, to acetaldehyde and carbon dioxid In order to decarboxylate the pyruvate, the re sulting carbanion must be stabilized. The th azolium group of thiamine pyrophosphate ca form a stable carbanion which is readily able t attack the ketone carbonyl of pyruvate, formir a tetrahedral adduct.

tetrahedral adduct

Decarboxylation of the pyruvate forms a carl anion, hydroxyethylthiamine pyrophospha

(HETPP), which is stabilized by delocalization of the electrons to the iminium. After protonation of the carbanion and abstraction of the hydroxyl proton, the thiazolium is released to form acetaldehyde (see *thiamine pyrophosphate, cofactor). **F** catalyse covalent; **G** Covalenzkatalyse.   P.S. SEARS

coverage   the degree Θ_i at which a *surface is covered by adsorbed (e.g., *chemisorbed) atoms or molecules; cf. also *monolayer coverage, *capillary condensation, *Temkin isotherm, *rate of adsorption/desorption. Since the surface of a solid or a liquid is the termination of the bulk state, that is to say, the region where the equs. based on three-dimensionality are no longer sufficient to describe the complete physical state of the system, thus a surface is not necessarily confined to the topmost layer of atoms of the solid or liquid, but may consist of several such layers extending into the bulk. The specific c. which is the ratio between empty and filled sites on the surface is unity. For *chemisorption proc. this is the limiting adsorption capacity. For unspecific *physisorption procs., however, more than one monolayer can be adsorbed on a solid. This multilayer adsorption ends with the condensation of the adsorbate onto the adsorbent forming a film of liquid or of solid "ice". Cf. also *Temkin isotherm. **F** couverture; **G** Belegung.   R. SCHLÖGL, A. SCHEYBAL

cp, Cp   *cyclopentadienyl as a *ligand or in *complexes

CPD   → contact potential difference (measurements)

CPG   → controlled-pore glass

CPI   chemical process industry

CP-MAS   → cross polarization magic angle spinning NMR

CP method   → Car-Parinello method

CPO   → chloroperoxidase

Crabtree catalysts   are superunsaturated Ir or Rh *complexes $L_2Ir(I)$ which are prep. by *reduction of cationic cyclooctadienyl Ir(I) complexes in not-coordinating solvents (e.g. CH_2Cl_2). **F** catalyseurs de Crabtree; **G** Crabtree-Katalysatoren.   B. CORNILS

cracking   → catalytic cracking (catcracking)

CR&L   → Chemical Research and Licensing Co.

Cram   Donald J. (born 1919), professor of chemistry at Los Angeles. Worked on recognition and *host-guest relationships of molecules. Nobel laureate in 1987.   B. CORNILS
Ref.: *Angew.Chem.Int.Ed.Engl.* **1988**, *27*, 1009.

CrAPO   Cr-modified *aluminophosphate

creatine kinase (phosphate)   Creatine phosphate (cph.) is a high-energy phosphate donor commonly used by mammalian cells. Cph. is both created and used by the *enzyme creatine kinase (EC 2.7.3.2) , which cat. the formation of creatinine and *ATP from cph. and *ADP. The "stockpile" of cph. allows for *bursts of high energy use. **E=F=G**.   P.S. SEARS

CRI International   A Royal Dutch/Shell Group subsidiary to operate the Shell cat. business on a global basis. A number of fully or partially owned companies is integrated in the CRI International Group at Houston, Texas. Amongst the fully owned companies CRI Catalysts Co. producing *EO and environmental protection cats. and Catalyst Recovery offering *regeneration of *refinery cats. and *molecular sieves are important. *Criterion Catalyst Co. is one of the companies owned at 50 % (the other partner is Cytec) active in *hydrotreating cats. for refineries, *reforming, and *styrene monomer catalysts.   J. KULPE

Criterion Catalyst Company   → cf. *CRI International and *catalyst alliances

Criterion Lummus Syncat technology
new combination for *hydrotreating procs. consisting of multiple cat. beds within a single reactor shell with intermediate byproduct gas removal and optional counter-current gas flow in the bottom cat. bed. Special applications are deep *desulfurization and deep *hydrogenation of aromatics in diesel fuels. **F** technique Syncat de Criterion Lummus; **G** Criterion Lummus Syncat-Technologie.

B. CORNILS

critical micelle concentration → micelles, *micellar catalysis

cross-coupling (cc.) special coupling reactions of organometallic cpds. with organic halides or related electrophiles cat. by transition metals acc. to $R'-M + R''-X \rightarrow R'-R'' + MX$ with M = $-BR_2$, $-SnR_3$, $-MgR$, $-Li$, $-AlR_3$, etc., X = $-Cl$, $-Br$, $-OTf$, etc. Ni and Pd are suitable cats., both ligand-modified (cf. *Grignard, *McMurray, *Negishi, *Stelzer, *Suzuki, *Stille). **F** couplage croisé; **G** Kreuzkupplung.

B. CORNILS

Ref.: Beller/Bolm, Vol. 1, p. 158, Dietrich/Stang; *Chem. Rev.* **1995**, 95, 2457.

Crossfield catalysts → Synetix catalysts

cross-linking 1: the proc. of forming cross-links between linear polymer molecules. Cross-links have different structures from the primary chains and are relatively short. They may only be single bonds but they are more usually short chains. There are many ways of c-l. preformed linear polymers. Often an easily cross-linkable group is incorporated into the polymer chain to facilitate c-l., as happens in the commercially important cases of sulfur *vulcanization cat. by zinc oxide of polydiene rubbers or *copolymerization of unsaturated polyester resins with styrene. It is also possible to cross-link preformed linear polymers by generating free *radicals, e.g., via *peroxides, high-energy radiation, or UV or visible light. The use of multifunctional *monomers, e.g., divinylbenzene during chain *polymer-

ization, also results in cross-linked polyme Noncovalent bonding forces between chai often have similar effects to c.-l. C.-l. increases the glass transition point a the modulus, and reduces not only elongatic but also cold flow under stress which is t main reason for vulcanizing rubbers.

W. KAMINSKY, C. STRÜB

2: cross-linking of *surface groups on *su ports during prepn. of het. cats. 3: cross-lin ing in *biocatalysis: cf. *enzyme immobiliz tion. **F** réticulation; **G** Vernetzung. B. CORN

Ref.: Adv. Polym. Sci. **1982**, 44, 1; Mark/Bikal Overberger/Menges, Vol. 4, p. 350.

cross metathesis intermolecular olef *metathesis between two different alkenes.

desired product   undesired by-products

Since the m. of acyclic alkenes is essential thermoneutral, a statistical distribution of r actants and prods. may occur. Use of termin alkenes results in the formation of a symme trical internal alkene and of ethylene as vol. tile byprod., which provides the driving forc for the reaction. Therefore, almost all rel vant investigations to develop cm. as a synt method employ reactions between termin. alkenes. The reverse reaction, cm. with eth ene (*ethenolysis) affords linear α-olefir from internal olefins. In a similar way, unsatt rated *polymers (*polyalkenamers) can t degraded by intermolecular cm. with low m lecular weight alkenes. Commercial applic tions of the cm. are, e.g., the W-cat. *triolefi and neohexene procs. of *Phillips or th *Norsorex proc. of *CdF. **F** métathèse crois **G** Kreuzmetathese. T. WESKAN

Ref.: Ivin/Mol; *Angew.Chem.Int.Ed.Engl.* **1997**, 3 2036; *Tetrahedron* **1998**, 54, 4413.

cross polarization magic angle spir ning NMR (CP-MAS) *NMR variant f local bulk and adsorbate structures using hig frequency (MHz) radiation as external magnetic field with mechanical sample rotation (response: variations in high-frequency electrostatic field). The ex-situ technique is non-destructive, applicable for a wide range of nuclei (H, C, P, Al, Si, metals) but only for diamagnetic samples. The in-situ applicability is limited. The data are integral, bulk and *surface sensitive.   R. SCHLÖGL

crotonization   the elimination of water from α-hydroxyaldehydes yielding unsaturated aldehydes which normally follows cat. *aldolization at higher temps. (e.g., procs. of *Degussa acrolein, *Aldox). **F** crotonisation; **G** Crotonisierung.   B. CORNILS

crown compounds   a family of *macrocyclic oligodentate *ligands (ionophores) for the complexation of metal or organic cations; mostly used for complexation in organic *solvents, such as *PTC, and in analytical chemistry. The term encompasses the subclasses *crown ethers, coronands or crowns (similar to crown ethers but with heteroatoms that are not exclusively ether oxygens), *cryptands ("spherical" analogs of crown ethers), and the acyclic *podands. In order to differentiate the ligands from their metal ion complexes, the terms coronate, cryptate, and podate, respectively, are used for the latter. Related neutral organic ligands are *spherands (cyclic oligo *m*-phenylenes with intraanular donor groups) and natural ionophores. Ccs. consist of lipophilic (e.g., alkyl chains) and hydrophilic (e.g., ether O, amine N, functional groups) elements. Cyclic and heterocyclic subunits (benzene, pyridine, etc.) are often incorporated into the molecular framework to increase rigidity and change complexation characteristics. If the molecular backbone is sufficiently flexible, the polar groups are pointing outward in hydrophilic media and the non-polar groups inward, so that solubility is ensured and an endolipophilic *cavity formed. In lipophilic media this arrangement is reversed with the polar groups pointing toward the center of the molecule so

that an electronegative cavity is preorganized for the uptake of cations with formation of a ligand-separated ion pair together with the "naked" counter ion.

1   **2**   **3**   **4**
[12]crown-4   [18]crown-6   pentaglyme   [2.2.2]cryptand

The cation *selectivity and the *stability of the complexes are strongly influenced by the size of the ligand, the number and type of donor atoms, the molecular architecture, and the *solvent. For example, [12]crown-4 (**1**) or [18]crown-6 (**2**) are Li$^+$, Na$^+$, K$^+$, and Cs$^+$ selective. The Figure also shows the podand (**3**) and the [2.2.2]cryptand.

Protonizable or cationic ccs. (e.g., N-containing azacrowns) are also used for the complexation of anions. **F** composés de crown; **G** Kronen-Verbindungen.   C. SEEL, F. VÖGTLE

Ref.: *Top.Curr.Chem.* **1981**, 98; J.D. Atwood, J.E.D. Davies, D.D. MacNicol, F. Vögtle, *Comprehensive Supramolecular Chemistry*, Vol. 1, Pergamon Press, Oxford 1996; F. Vögtle, *Supramolecular Chemistry*, Wiley, Chichester 1991.

crown ethers   macrocyclic oligo(ethylene glycol) ethers, often with various *hydrocarbon building blocks in the molecular framework; a subtype of oligodentate metal *ligands termed *crown cpds. The synth. and the ability to complex metal cations was first reported by *Pedersen. As for the nomenclature: the ring size is given in square brackets, followed by the noun "crown" as family name and the number of donor atoms; additional substituents or fused rings are affixed at the beginning. **E=F; G** Kronenether.

C. SEEL, F. VÖGTLE

Ref.: *Angew.Chem.Int.Ed.Engl.* **1988**, *27*, 1021; *Acc. Chem.Res.* **1997**, *30*, 338.

crude oil   → petroleum

cryptands   neutral, oligodentate metal *ligands of spherical shape with macrooligocyclic framework that usually contain N bridgehead atoms and oligo(ethylene glycol) ether units. Heteroatomic binding sites are often incorporated. Cs. are a subtype of *crown compounds and were introduced in 1970 by *Lehn. **E=F; G** Kryptanden. C. SEEL, F. VÖGTLE

Ref.: *Angew.Chem.Int.Ed.Engl.* **1988**, *27*, 89.

crystallography of surfaces The complete translational invariance of a bulk crystal is destroyed by cleavage. At best, periodicity in only two dimensions is retained. For a strictly two-dimensional (i.e., planar) periodic structure every lattice point can be reached from the origin by translation vectors, $T = ma +nb$, where m and n are integers. The primitive vectors, a and b, define a so-called unit mesh or unit cell. There are five possible unit cells in two dimensions (oblique, hexagonal, rectangular, centered rectangular, and square), called *Bravais lattices*. The specification of an ordered *surface structure requires both the unit mesh and the location of the basis atoms. The latter must be consistent with certain symmetry restrictions. In two dimensions, the only operations consistent with the five Bravais lattices that leave one point unmoved are mirror reflections across a line and rotations about an angle $2\pi/p$ where $p =1, 2, 3, 4,$ or 6. The resulting ten point groups combined with the unit mesh and glide symmetry operations results in a total of 17 two-dimensional space groups. Ideal or simply relaxed surfaces are identified easily by reference to the bulk plane of termination given by Miller indices. The periodicity and orientation of the surface is the same as the underlying bulk lattice. These are called (1 x 1) structures. If the primitive translation vectors of the surface differ from those of the ideal surface, a reconstruction of the surface has occurred.

The most popular method for the investigation of the crystallographic structure of a surface is *low-energy electron diffraction (LEED). Electrons with energies which correspond to de Broglie wavelengths of the order of the interatomic spacings are elastically back-scattered from a crystal surface and will form a diffraction pattern. This pattern is a image of the surface reciprocal lattice. Th energy (voltage) dependence of low-energy electron diffraction beam intensities, the s called $I(V)$ curves, are used in an iterativ procedure to determine the geometrical a rangement of surface atoms within the surfac unit cell. Other methods to determine the su face structure are X-ray scattering with gla cing incidence angles, ion or atom scatterir techniques as well as special microscopi such as *field ion microscopy (FIM) or *sca ning tunneling microscopy (STM), which d rectly provides real space images of the su face topography on the atomic scale. **F** crista lographie des surfaces; **G** Oberflächenkrista lographie. R. SCHLÖGL, Y. JOSEF

CSF process → Consol Synthetic Fu proc. of Consolidated Coal Co.

CSTR continuous stirred tank reactor, - catalytic reactors

CT-BISA process → Chiyoda procs.

CTEM → conventional transmission ele tron microscopy, *electron microscopy

CTSL catalytic two stage liquefaction (c coal), → coal liquefaction

cuen cupriethylene diamine as a *ligand

cumene isopropylbenzene, an importar and cat. manuf. *intermediate for the *Hoc reaction (cf. procs. of *BP/ Hercules, *Do *CDTech, *Mobil/Badger, *Monsanto, *UOP **F** cumène; **G** Cumol. B. CORNII

cumene hydroperoxide intermediate c the *Hock process

Curtin-Hammett principle In a moder context the principle allows conclusions abou reactivity-selectivity relationships in kinet cally controlled reactions. It is applicable t reactions when non-interconvertible prod A' and B' arise from *conformational isomer

A and B which are rapidly interconverting relative to prod. formation. The composition of prods. not only depends on the relative ratio of A and B, [B]/[A] = K, but is controlled by the *Gibbs free energy difference of the representative *transition states. Since the Gibbs free energies of the transition states can be calcd. by proper quantum mechanical methods, the *selectivity S can be derived without knowing the equil. constant K for the two substrate conformers. The principle further provided the basis for understanding why a major prod. can arise from a less stable, possibly not observable, minor conformation. **F** principe de Curtin-Hammett; **G** Curtin-Hammett-Prinzip.   R. TAUBE

Ref.: *J.Chem.Educ.* **1986**, *63*, 42.

cus cordinatively unsaturated metal centers, → vacant coordination sites

CV → cyclovoltammetry

CVD   → chemical vapor deposition

CX process   → Mitsui procs.

cyanoethylation   is the introduction of a -CH$_2$-CH$_2$CN group into organic cpds. In general, the *addition of acrylonitrile in the presence of basic cats. to cpds. with an active hydrogen yields the corresponding cpd. with a cyanoethyl group. The reaction follows the *Michael addition scheme. **F=E**; **G** Cyanoethylierung.   B. DRIESSEN-HÖLSCHER

Ref.: *Angew.Chem.* **1949**, *61*, 229; Kirk/Othmer, 3.Ed., Vol. 7, p. 370; Kirk/Othmer-2, p. 337; *Organic Reactions*, Vol. 5, p. 79.

cyanohydrins (enzymatic synth.)   → oxylonitrilases, *Pseudomonas* sp. Lipases

Cyclar   → UOP procs.

cyclic hydrocarbons from diazoalkanes
Carbene-like *intermediates can be obtained by decomposition of diazoalkanes by means of *transition metal cats. These diazoalkane-derived carbene transfer reactions have

found application in the Rh- and Cu-cat. *stereo- and *enantioselective synth. of cyclopropranes, e.g., pyrethroids, through *asymmetric *cyclopropanation of conjugated alkenes. Intermolecular cyclopropanation of remote double bonds with diazo cpds. causes macrocyclization. Other cats. are Pd, Ru, Os, or Cr. The reaction pathway may be regarded as a *cycloaddition of a carbene fragment to a double bond. It is assumed that a transition metal-carbene intermediate is formed during the *cat. cycle. **F** hydrocarbures cyclique par diazoalkanes; **G** cyclische Kohlenwasserstoffe aus Diazoalkanen.   R. KRATZER

Ref.: Cornils/Herrmann-1, p. 733; *Chem.Rev.* **1998**, *98*, 911.

cyclization the intra- or intermolecular ring formation from unsaturated *monomers giving carbo- or heterocyclic organic cpds. *Chiral *Lewis acid auxiliaries may control the *stereochemistry of *Diels-Alder or 1,3-dipolar reactions.

The selective *cyclo-oligomerization and *co-cyclization of small, unsaturated building blocks (CO, alkenes, alkynes, dienes) yielding 4–14-membered ring systems is the domain of hom. *transition metal cat. (Figure)

Cat. c. is mediated by many transition metals (Fe, Co, Ni, Ti, Au) and has also been applied to a variety of heteroatom monomers, including phospha-alkynes and -alkenes. The prod. distribution (*chemo-, *regio-, and *stereoselectivity) may be controlled by the steric and electronic properties of additional *complex *ligands (*ligand tuning). A *zeolite-mediated c. is included in *Lonza's nicotinamide proc. **F** cyclisation; **G** Cyclisierung.   H. BÖNNEMANN

Ref.: *J.Bras.Chem.Soc.* **1997**, *8(4)*, 289; *Acta Chem.Scand.* **1996**, *50*, 652; Parshall/Ittel; *Angew.Chem.Int.Ed.Engl.* **1973**, *12*, 975; Cornils/Herrmann-1, Vol. 1, p. 358.

cycloaddition a class of pericyclic reactions where two or more molecules with π-bonds combine in a concerted manner to form a new ring cpd. Cs. are characterized by the number of either atoms or electrons involved. The most important cs. are the [4+2] (*Diels-Alder reaction) and the [3+2] cycloaddition, which proceed easily with thermal activation. On the other hand, [2+2] (olefin *dimerization) and [4+4] c. (diene dimerization) do not proceed at all under similar reaction conds. (Woodward-Hoffmann rules, *HOMO-LUMO and *overlap model). The most important c. is the *Diels-Alder reaction. Many c. are thermally or photo initiated; some need *transition metal *complexes as cats. **E=F=G.** M. BELLER

Ref.: March; Mulzer/Waldmann, p. 28; *Angew.Chem.* **1968**, *80*, 329; *Chem.Rev.* **1996**, *96*, 49 and 137; *Organic Reactions*, Vol. 51, p. 331; *Chem.Rev.* **1997**, *97*, 523; K. Rück-Braun, H. Kunz, *Chiral Auxiliaries in Cycloadditions*, Wiley-VCH, Weinheim 1999.

cyclocarbonylation the incorporation of CO as a *C_1 building block into a newly formed ring (carbonylative cyclization). Examples are *domino sequences of *carbonylations of aryl-, vinyl-, benzyl-, and allyl-X (X = Cl, Br, I, OTf, *Heck carbonylations) with various modes of intramolecular trapping reactions. These examples include the carbonylation of 2-halophenols or 2-haloanilines in the presence of alkynes to give annulated O- and N-heterocycles. Similar reactions are performed to yield isocoumarins, quinolinones, indanones, or tetralones. Although Pd *complexes have been mostly applied as cats. other *transition metal cats., such as Ni, Co, or Cu can also been used. The *BASF hydroquinone synth. from acetylene and CO is also a *Reppe cyclocarbonylation.

C. of allylic halogenides containing an additional functional group in a suitable position such as an alkene moiety, OH, or NH_2 group prod. the corresponding cyclpentenone or lactone derivatives. Another important class of c. is the hydroxycarboxylation of alkenes with an appropriate alcohol or amine function. Hence, Co, Rh, or Pd cats. convert allyl and homoallyl alcohols or amines to the corresponding butyrolactones or -lactams, respectively. Lactones have also been synth. by *radical c. from saturated alcohols using stoichiometric amounts of $Pb(OAc)_4$. Another important c. reaction is the *Pauson-Khand reaction. **F** carbonylation cyclique; **G** Cyclocarbonylierung. M. BELLER

Ref.: Cornils/Herrmann-1, p. 148, 187; *Angew.Chem. Int.Ed. Engl.* **1990**, *29*, 1413; *J.Mol.Catal. A:* **1999**, *143*, 137.

cycloco-oligomerization → *cyclo-oligomerization

cyclodextrins (CyD) *macrocyclic, torus-shaped cyclic *oligosaccharides of 6 to 13 glucose units linked by an α-(1–4) glucosidic bond. They are produced through the degradation of starch by the *enzyme *CyD glycosyltransferase. The CyDs. composed of 6, 7, and 8 glucopyranose units are called α-, β-, and γ-CyD. Due to the presence of the hydrophobic host *cavity with precisely positioned functional groups, CyDs. can encapsulate a wide range of organic molecules and cat. their reactions or mimic a step in an enzymatic cat. sequence. Thus, many organic reactions, such as *Diels-Alder, *halogenation, *isomerization, etc., can be promoted by natural CyDs. Attachment of one or several cat. or reactive groups to one of the faces of the toroidal CyD affords *ligands (for hom. cats.) or interesting artificial *enzymes (cf. symbolic Figure with reactive R and inclusion *TPPTS, → *host-guest relations).

These cats. specifically bind the substrate into their host cavity and generally show specificity in the prods. formed, including *stereospecificity. The synth. of artificial enzymes with two or several CyDs. to bind the substrate has also been achieved. These compds. perform selective reactions at particular points in a bound substrate. CyDs. are also efficient *inverse phase transfer cats. as they are able to transport lipophilic molecules from an organic to an aqueous phase via *complexation. In particular, they make possible to increase mass transfer in biphasic systems involving water-soluble organometallic cats. So, native CyDs. are effective *PTCs for a series of chemical reactions. Interestingly, chemically modified CyDs. like dimethylCyD show a better cat. activity than native CyDs. in the *Wacker oxidation, *hydroformylation, *hydrocarboxylation, and the *hydrogenation of aldehydes. Advantageous effects of CyDs. on the *chemoselectivity and the reaction *rate in the presence of het. cats. like *Raney Ni, Pd on *charcoal, or Ir on *alumina have also been reported. **F** cyclodextrine; **G** Cyclodextrin.   E. MONTFLIER

Ref.: *J.Org.Chem.* **1990**, *55*, 1854; *J.Mol.Catal.* **1994**, *91*, L313; *Chem.Rev.* **1998**, *98*, 1977, 1997, 2013, 2035; *Tetrahedron Lett.* **1998**, *39*, 2959; Mulzer/Waldmann, p. 374; Ertl/Knözinger/Weitkamp, Vol. 2, p. 888; J.Szejtli, T.Osa (Eds.), *Cyclodextrins*, Elsevier, Amsterdam **1996**.

cyclodextrin glycosyltransferases

Cyclodextrin (CyD) α-1,4-glucosyltransferase (EC 2.4.1.19) from *Bacillus macerans* cat. the *cyclization of oligomaltose to form α-, β- and δ-CyD, and the transfer of sugars from CyD to an acceptor to form oligosaccharides. This *enzyme is also able to transform α-glucosyl fluoride into a mixture of α- and β-CyDs and malto-oligomers in almost equal amounts.
Unnatural sugar acceptors that are structurally similar to glucose are also substrates for CyD-glucosyl transferase. **E=F=G**. C.-H. WONG

cyclodimerization special case of cyclic *addition with ring closure during reaction, e.g., the c. of butadiene to 1,5-cyclooctadiene with Ni cats. These cats. are prep. by reacting Ni bis(acac) with AlEt₂(OEt) in the presence of a bulky *ligand. Mechanistically, the reaction is assumed to occur by coupling of two butadiene molecules to give a bis(allyl)Ni *complex, whereby the *ligand hinders the coordination of a third butadiene molecule. By tuning the c. reaction conds. and the cats., butadiene can also cyclodimerize to give other cyclic prods. such as 1,2-divinylcyclobutane, 4-vinylcyclohexene (which can be hydrogenated to ethylbenzene), and 1-vinylidene-2-vinylcyclopentane (Figure). **F** cyclodimérisation; **G** Cyclodimerisierung.

W. KAMINSKY, I. ALBERS, O. PYRLIK

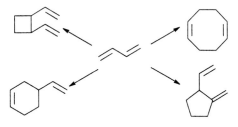

Ref.: *Angew.Chem.Int.Ed.Engl.* **1988**, *27*, 186.

cyclometallation a reaction sequence of a metal *complex in which a C-H bond undergoes a reaction, e.g., an *oxidative addition to the metal center to yield a cyclic complex (*metallacycle, for example *palladacycles). Sometimes M-C and M-H bonds are formed when the hydride can be transferred to an organic group which is then eliminated as an alkane, etc. (Figure).

There are three main classes of reactions that lead to the formation of three-, four-, five-, or six-membered rings. 1: The sp^3 hybridized C is attached to a *phosphine, *cp, or other *ligand. 2: Metal alkyls can also undergo *cyclization. 3: The most common type is usually termed *orthometallation, i.e., the *ortho* hydrogen of an aryl group can be transferred or lost. Examples are aryl phosphines or *phosphites, azobenzenes, arylketones, etc. In c. the initial step is sometimes assumed to involve *agostic M · · · H-C interactions. **F=E**; **G** Cyclometallierung. F.E. KÜHN

Ref.: *Coord.Chem.Rev.* **1980**, *32*, 235; *Chem.Rev.* **1981**, *81*, 229; *Polyhedron* **1984**; *3*, 1073; *J.Organomet.Chem.* **1980**, *200*, 307.

cyclooctadiene (COD) specially *cis,cis*-1,5-COD is readily available by the *transition-metal-cat. *cyclodimerization of butadiene. This isomer is ideally preorganized to bind transition metals via the alkene at a distance of approx. 200 pm with a *bite angle of nearly 90° to give very stable *chelate *complexes (Figure).

Many cod complexes serve as cats. with two *vacant *coordination sites for cat. by replacement of cod. **F** cyclooctadiène; **G** Cyclooctadien. R.D. KOHN

cyclooctatetraene (COT, cot) an eight-membered tub-shaped ring system with four individual double bonds. C. is obtained by a cat. *cyclization of acetylene at 70–120 °C/1.5–2.5

MPa in 70 % yield (*Reppe synth.). Ni(II) salts such as cyanides, acac, or *salen cpds. are used as cats. Reaction of acetylene and 1-alkynes gives monosubstituted cs. in 15–25 % yield while disubstituted cs. are obtained from internal alkynes in modest yield. **F** cyclooctatetraène; **G** Cyclooctatetraen. P. ROESKY

Ref.: Schröder, *Cyclooctatetraen*, VCH, Weinheim 1965; *Ann.* **1965**, *560*, 1.

cycloolefin coplymer (COC) is prepared by *transition-metal- cat. *polymerization of aliphatic and cyclic alkenes, most commonly ethylene with norbornene or tetracyclododecene. The *polymer has interesting properties; COCs are mechanically very stable up to the glass temp. Cats. applied are *Ziegler-Natta types like TiCl$_4$, VCl$_4$, or *metallocenes, the latter representing current state-of-the-art technology. Required *co-cats. are alkylaluminum halides such as (C$_2$H$_5$)$_2$AlCl or *methylalumoxane (MAO) at 50–100 °C/0.5–5 MPa. Depending on the metallocene structure, both a strictly alternating (semicrystalline, transparent) or a randomly distributed (amorphous polymer) sequence of monomers is accessible. The *activities of the cats. are up to 10^5 kg of polymer per mole of cat. *Solvents like toluene, cyclohexane, or decalin are recommended. COCs are applied in optoelectronics or as substitutes for polycarbonate. Tradenames are Apel, Elmit, or Topas. **E=F=G**. C.W. KOHLPAINTNER

Ref.: *Angew.Makromol.Chem.* **1994**, *223*, 121; *Kunststoffe* **1995**, *85*, 1038; *Polimery* **1997**, *42*, 587; *Nachr. Chem.Tech.Lab.* **1999**, *47*, 338.

cyclooligomerization special case of *oligomerization in which monomers such as alkenes, dienes, or alkynes (cf. the following entry) give rings instead of chains. Examples (of some commercial relevance) are the preparation of benzene from three molecules of acetylene or *cyclooctatetraene from four molecules, respectively. Cyclooctadienes and cyclododecatrienes can be obtained through cyclo*di*- or *tri*merization of butadiene. Common cats. are *transition metals like Ti, Fe,

Pd, etc., with Ni-*ligand systems representing the most versatile and active cats. for cs.

Cyclo-*co*-oligomerization of 1,3-dienes with strained olefins or alkynes are also possible and are almost exclusively cat. by zerovalent "naked" (*bare) Ni complexes. With alkenes cyclic 2:1 (diene:olefin) prods. are usually formed whereas with alkynes incorporation of more than two alkyne molecules is possible and thus yields larger rings. **F** oligomérisation cyclique; **G** Cyclooligomerisierung.   A. ECKERLE

Ref.: *Angew.Chem.* **1963**, *75*, 10; Abel/Stone/Wilkinson.

cyclooligomerization of alkynes is

mediated by various hom. or het. *transition metal cats., e.g., Fe, Co, and Ni carbonyls, further Cr, Ni, Pd, and predominantly Co organometallics. The organo Co(I) cat. c. allows the total synth. of benzene derivatives; substrates with sensitive functional groups may also be reacted under mild conds., e.g., in the *aqueous phase. *Supercritical conds. have been applied, too. The broadly applicable c. of alkynes and nitriles relies preferentially on cpCo(I) cats. and leads to a large variety of pyridine derivatives (Figure).

A photoassisted modification of the Co(I)-cat. [2+2+2] *cycloaddition of alkynes and nitriles runs efficiently under ambient conds. and makes it possible to substantially enlarge the pattern of substituents. **F** oligomérisation cyclique des alcynes; **G** Alkin-Cyclooligomerisierung.   H. BÖNNEMANN

Ref.: *Angew.Chem.Int.Ed.Engl.* **1984**, *23*, 539; Cornils/Herrmann-1, Vol. 2, p. 1102; *Nachr.Chem.Tech.Lab.* **1999**, *47*, 9.

cyclooxygenase (COX) cat. the *oxida-

tion of arachidonic acid to prostaglandin G$_2$. Two *isozymes have been found. COX-1 is re-sponsible for prostaglandin and thromboxin biosynthesis in the gastrointestinal tract and blood platelets. COX-2 plays a major role in prostaglandin biosynthesis in inflammatory cells and is thus a target for the development of antiinflammatory agents. **E=F=G**.

            C.-H. WONG

Cyclopol process *oxidation of cyclohex-

ane to cyclohexanone (inter alia, *KA oil) by means of a *bimetallic cat. Developed by the Industrial Chemistry Research Institute in Warsaw. **F** procédé Cyclopol; **G** Cyclopol-Verfahren.   B. CORNILS

cyclopolymerization *Polymerization of

multifunctional molecules may proceed not only intermolecularly to produce branched and *cross-linked molecules but also intra/ intermolecularly with the formation of rings in chains. Thus, c. is defined as any type of polymerization that leads to formation of cyclic structures in the main chain of the polymer. C. occurs mainly with monomers having C-C double bonds in the 1,5- or 1,6-position, such as 1,5-hexadiene. Suitable monomers also include diallyl cpds., dialdehydes, diepoxides, diisocyanates, and diisonitriles. Certain monomers (e.g., 2,6-diphenyl-1,6-heptadiene) undergo c. via all of the well-known methods of initiation of polymerization such as *radical, cationic, and *Ziegler-Natta cat. to yield essentially the same cyclopolymer. C. with *metallocene cats. is also known. **F** polymérisation cyclique; **G** Cyclopolymerisation.

    W. KAMINSKY, I. ALBERS, O. PYRLIK

Ref.: Mark/Bikales/Overberger/Menges, Vol. 4, pp. 423 and 543; *Comprehensive Polym.Sci.* **1989**, *4*, 423; *Polymer Bull.* **1997**, *38*, 141.

cyclopropanation The synth. of cyclopro-

pyl moieties has attracted considerable interest because of the unique properties of the cyclopropane ring and its importance in naturally occurring molecules. Central to all synth. efforts is the ability to control the *regio-, *stereo-, and *enantioselectivity of the reactions, potentially up to three *chiral centres being formed in one step. Among the most useful and versatile methods employed for the synth. of cyclopropane rings are the *Simmons-Smith reaction and the cat. c. of alkenes with diazo cpds. which can be regarded as [2+1] *cycloaddition of a carbene fragment to a double bond. The most frequently employed cats. are based on Cu, Rh, Ru, and Pd. The cats. usually provide a mixture of diastereomers with the *anti (trans)* prod. predominating. Diazomethane is mostly employed with Pd cats. which are on the other hand often more efficient for the c. of electron-poor alkenes and strained olefins, whereas Ru-, Rh-, and Cu-based cats. are better suited to the c. of electron-rich olefins with α-diazocarbonyl cpds. Elusive metal-carbene *complexes ("carbenoids") are supposed to be short-lived intermediates in *carbene transfer reactions (cf. *Fischer carbenes). Although some metallacyclobutanes have been shown to deliver cyclopropanes by *reductive elimination, their intermediacy remains unproven in many c. reactions (*metathesis). Asymmetric c., cat. by a *chiral transition metal complex, is a very expedient route to *enantiomerically pure cyclopropanes. In this respect, Cu(I) complexes associated with chiral *oxazoline ligands and dirhodium(II) complexes of *optically active pyrrolidone-5-carboxylic acid are to date among the most efficient ones for delivering high enantiocontrol (e.g., for the synth. of the adjuvant Cilastatin). Alternative methods resort to chiral auxiliaries (e.g., *Oppolzer's sultam) to induce desired stereochemistries. The *Kulinkovich reaction provides direct acces to hydroxycyclopropanes. **F**=E; **G** Cyclopropanierung.

A.F. NOELS

Ref.: *Russ.Chem.Rev.* **1993**, *62*, 799; *J.Chem.Soc. Chem. Commun.* **1997**, 211; Cornils/Herrmann-1, p. 733.

cyclotetramerization special case of *cyclo-oligomerization with four unsaturated monomers being linked to a ring system. The most prominent example is the c. of acetylene with Ni(CN)$_2$ or Ni (CO)$_2$[(PPh$_3$)$_2$] as cats. at

65 °C/ 1.5–2.5 MPa to give 1,3,5,7-cyclooctatetraene (*COT, Figure).

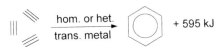

Monosubstituted alkynes may also be included in the c., giving 1,2,4,7-, 1,2,4,6-, and 1,3,5,7-tetrasubstituted derivatives. **F** tetramérisation cyclique; **G** Cyclotetramerisierung.

H. BÖNNEMANN

Ref.: Ullmann, Vol. 8, p. 103; *Angew.Chem.Int.Ed. Engl.* **1988**, *27*, 185; *Chem.Ber.* **1997**, 823.

cyclotrimerization cat. *cyclization of three unsaturated synthons, a special case of the *cyclooligomerization via *cycloaddition. The very exothermic cat. c. of acetylene giving a mixture of aromatic cpds. including benzene and naphthalene (first observed by Berthelot in 1866) is quoted as one of the earliest examples of *transition metal cat. Monosubstituted alkynes, cat. by Co, lead to 1,2,4- or 1,3,5-trisubstituted benzene derivatives (Figure).

Hom. *Ziegler-type cats. prepared by *reduction of Ti, Cr, or Ni salts with organoaluminum cpds. are used in the commercial proc. for the c. of butadiene to give 1,5,9-cyclodododecatriene (*CDT). The distribution of the isomers depends on the metal salt used. **F** trimérisation cyclique; **G** Cyclotrimerisation.

H. BÖNNEMANN

Ref.: *Liebigs Ann.Chem.* **1866**, *141*, 173; *Angew.Chem. Int.Ed.Engl.* **1984**, *23*, 539; Cornils/Herrmann-1, Vol. 1, p. 358; Ullmann, Vol. 8, p. 205.

cyclovoltammetry (CV) method of recognizing the preferential locations of, e.g., metals by probing electrons at varying potentials. The response is a current. The nondestructive method is used for the determination of redox properties, reaction kinetics, and elementary reaction steps, and for *electrocatalysis. The data are integral, *surface sensitive, and express in-situ capabilities.   R. SCHLÖGL

CyD → cyclodextrin

cytochrome P 450 system *heme-containing *monooxygenase that cat. the activation of molecular O to form an oxo-iron species for the *oxidation of alkanes, alkenes, and heteroatoms. The reactions may go through a *radical or a cation radical *intermediate. Using bicyclo[2.1.0]pentane as a probe for microsome P450, a radical pair has been determined to form in the *hydroxylation step that collapses with stereochemical specificity at a rate $1.5 \cdot 10^9$ s^{-1}. Oxidation of alkyl sulfides by microsome P450 indicates that both S-oxidation and S-dealkylation occur; the S-*dealkylation takes place more rap-

idly with substrates bearing a more acidic α-hydrogen, indicating the cation radical mechanism. In any case, chiral sulfoxides can be prepared via monooxygenase reactions. Oxidation at Se also occurs similarly followed by a [2,3] sigmatropic rearrangement. The alcohols obtained were achiral; perhaps *racemization of selenoxide is too fast. Cyt P450 also cat. oxidative O-dealkylation and N-dealkylation. **E=F=G**.   C.H. WONG

Ref.: P.R. Ortiz de Montellano, *Cytochrome P450*, Plenum Press, New York 1986.

cytosine monophosphate (CMP) → sugar nucleotide synthesis, *sugar nucleotide regeneration

χ → chi

D

DABCO   a bicyclic, tertiary amine, used as a cat., e.g., in *hydroxyalkylations.   H. GEISSLER

DAHP synthetase → 3-deoxy-D-arabino-2-heptulosonic acid 7-phosphate synthetase

(+)-DAIB   a bicyclic and *bidentate *ligand (Figure), specially recommended for *enantioselective *alkylation at aromatic aldehydes (cf. *chelating ligands).   W.R. THIEL

Daicel glycerol process   starting from allylic alcohol (from *propylene oxide via *isomerization) which is epoxidized to glycidol with peracetic acid (*PAA) in a combined PAA/glycerol plant. The glycidol is *hydrolyzed to glycerol. **F** procédé Daicel pour le glycérol; **G** Daicel Glyzerin-Verfahren.
                                B. CORNILS

Daicel picoline process   manuf. of 3-picoline (3-methylpyridine) by cat. *condensation of acrolein with ammonia in the presence of *bimetallic cats. on Al_2O_3/SiO_2 *supports at 350–400 °C in the gaseous phase. **F** procédé Daicel pour le picoline; **G** Daicel Picolin-Verfahren.   B. CORNILS

Daicel propylene oxide process   manuf. of *propylene oxide (*PO) by *epoxidation of propylene with separately produced peracetic acid from acetaldehyde and H_2O_2 with

acid cat. (*Daicel glycerol process). The ratio between the byprod. acetic acid and PO is 1.3:1. **F** procédé Daicel pour PO; **G** Daicel PO-Verfahren.   B. CORNILS

Ref.: *Hydrocarb.Proc.* **1977**, *56*(11), 221; Winnacker/Küchler, Vol. 6, p. 58.

Damköhler number   The Damköhler number I (*Da*I) is used to characterize the conv. behavior of ideal, isothermal reactors and represents the ratio of the space time or hydrodynamic residence time τ (reaction volume V_R/volumetric flow rate V_0) to the time constant t_R of the reaction (equ. 1)

$$Da\mathrm{I} = \frac{\tau}{t_R} = \frac{|\nu_A| r_A \tau}{c_{A,0}} \qquad (1)$$

For educt A, ν_A is the stoichiometric coefficient and $c_{A,0}$ the initial conc., and r_A the reaction rate. Under consideration of the design equs. for a batch reactor (BR) and (*PFR) on the one hand and a *CSTR on the other hand and by using the corresponding rate laws for chemical reactions, different relations between the Damköhler number I and the degree of conversion (X_A) can be obtained.

To characterize the influence of *mass transport on chemical reactions in the gas phase, the Damköhler number II (*Da*II) is used which is defined as ratio of reaction rate and transport by diffusion (equ. 2)

$$Da\mathrm{II} = \frac{kc_{A,g}^{n}}{k_g a c_{A,g}} = \frac{kc_{A,g}^{n-1}}{k_g a} \qquad (1)$$

where k ($m^{3(n-1)}/mol^{(n-1)} \cdot s$) is the rate constant of a reaction of order n, $c_{A,g}$ the gas-phase conc. of reactant A (mol/m^3), k_g the mass transfer coefficient (m/s) and a the interfacial area per unit volume (m^2/m^3). Moreover, in het. cat. reactions further dimensionless groups DaIII and DaIV exist: for checking interphase or intraparticle heat transfer in cases where the temp. throughout the cat. particle cannot be regarded as isothermal because of strongly exothermic or endothermic reactions, DaIII and DaIV are used as diagnostic criteria. **F** nombre de Damköhler; **G** Damköhler-Zahl.   P. CLAUS, D. HÖNICKE

Danishefsky's diene is *trans*-1-methoxy-3-(trimethylsilyloxy)-1,3-butadiene, an activated, nucleophilic diene which is intensively applied in *Lewis acid-cat. hetero *Diels-Alder *cycloadditions, reacting with electron-deficient hetero dienophiles such as benzaldehyde to form, e.g., dihydropyrans. Thus, total synths. of highly functionalized target molecules can be accomplished. Following a *chelation-controlled mechanism, in these Diels-Alder reactions a pronounced *endo* *selectivity is observed with, e.g., Eu(fod)$_3$ or with Er(hfc)$_3$ as cats. **F** diène de Danishefsky; **G** Danishefkys Dien.   G. GERSTBERGER
Ref.: *J.Am.Chem.Soc.* **1982**, *104*, 358 and **1983**, *105*, 3716.

DAO   deasphalted oil

dark field imaging (DF)   a special version of electron microscopy for contrast formation using selected diffracted beams which are slightly off- angle from the transmitted beam. The technique is applied with *AEM, *CTEM, *HRTEM, and *TEM and is non-destructive.   R. SCHLÖGL

Darzens reactions   1: Darzens-Erlenmeyer-Claisen *condensation (glycidyl ester condensation, mechanistically related to the *Knoevenagel condensation followed by nucleophilic substitution) comprises the reaction of aldehydes or ketones with α-halogen carboxylic acid esters in presence of a base to give α,β-epoxy carboxylic acid esters which react further to the homologous aldehydes via *decarboxylation. 2: Darzens-Kondakoff acylation of cyclo-

olefins (also Darzens-Nenitzescu or *Nenitzescu acylation) is the *Lewis acid cat. conv. of cycloalkenes with acid chlorides yielding acyl-substituted hydrochlorinated alkanes which eliminate HCl during heating or in presence of a base cat. to give the corresponding conjugated acyl-substituted cycloalkenes. **F** réactions de Darzens; **G** Darzens-Reaktionen. R.W. FISCHER
Ref.: *Organic Reactions*, Vol. 5, p. 413.

Davison Chemical   → Grace catalysts

Davy, Sir Humphrey (1778–1829), professor of chemistry at the Royal Institution in London, important scientist and discoverer of K, Na, Al, Mg, chlorides etc., and inventor of the *safety lamp.   B. CORNILS
Ref.: *Phil.Trans.Roy.Soc.(London)* **1808**/I, p. 1; Neufeldt.

Davy-Powergas oxo process → LPO process of Union Carbide, *Union Carbide procs.

Davy Process Technology   licenses a proc. for the manuf. of saturated fatty acids by liquid-phase, cat. *hydrogenation (*hardening) of unsaturated fatty acids over fixed-bed Ni cats. at 90–230 °C/0.7–7 MPa. The space velocity is between 0.5 and 10 h. **F** procédé de Davy; **G** Davyverfahren.   B. CORNILS
Ref.: Ertl/Knözinger/Weitkamp, Vol. 5, p. 2227; S.L. Taylor (Ed.), *The Chemistry and Technology of Edible Oils and their High Fat Products*, Academic Press, London 1989.

DBNi catalyst   diborane nickel cat., prepared from Ni salts and diboranes; cf. *nickel boride cats.

DBP   dibutyl phthalate, a *plasticizer based on *n*-butanol and phthalic anhydride. **F** phthalate de butyle, **G** Dibutylphthalat.

DCC process   deep catalytic cracking, → Stone & Webster procs.

DCD   → Dewar-Chatt-Duncanson model

DCP 1: dicyclopentadiene; 2: dicyclohexyl phthalate (dihexyl phthalate), a *plasticizer based on cyclohexanol and *phthalic anhydride

DDDA dodecanedioic acid

DDP dodecyl phthalate, a *plasticizer based on dodecanol (C_{12}-alcohol) and phthalic anhydride

de diastereomeric excess, analogous to *ee

DEA diethanolamine

Deacon, Henry (1822–1876), chemist and company owner, developed the *Deacon process

Deacon process formerly used to generate chlorine by *oxidation of HCl in the presence of *supported CuCl$_2$ as cat. at approx. 430 °C. The Dp. has been obsolete for over hundred years (invented 1868), but the reaction has gained considerable importance in another field. Today it is applied to form 1,2-dichloroethane (EDC), a precursor for *vinyl chloride monomer (VCM). The *oxychlorination combines the generation of chlorine with the chlorination of ethylene:

$$H_2C=CH_2 + 2\ CuCl_2 \rightarrow Cu_2Cl_2 + ClCH_2CH_2Cl$$
$$O_2 + Cu_2Cl_2 \rightarrow CuOCuCl_2$$
$$2\ HCl + CuOCuCl_2 \rightarrow 2\ CuCl_2 + H_2O$$

The reaction is carried out in the vapor phase in a fixed- or fluid-bed proc. in the presence of a Cu cat. supported predominantly on *silica at 220–300 °C/0.2–1.5 MPa, resulting in about 95 % selectivity to EDC. Cf. the *Kellogg Kelchlor proc. **F** procédé de Deacon; **G** Deacon-Verfahren.   C.D. FROHNING

Ref.: Kirk/Othmer-1, 3rd Ed., Vol. 23, p. 872.

deactivation In contrast to *fouling (by proc. constituents) d. originates from proc. impurities. It is the collection of effects which reduce the intrinsic activity of a cat. system. By definition the cat. is recovered from a *cat. cycle without changes in its structure and composition. This would imply infinite life of the cat. at steady activity (*catalyst lifetime). As cats. are subject to chemical transformations during each turnover cycle there are numerous possibilities that the molecular structure may change during operation (*aging). In addition, all feedstocks contain impurity comps. which can undergo chemical reactions with the cat. causing irreversible changes in its structure. Such cpds. are called *poisons. Finally, the cat. can undergo slow transformations initiated by reaction conds. other than the presence of the reactant molecules. Temp., solvents, high press., or the mechanical forces during stirring or motion of the cat. bed can cause structural changes, fragmentation of the solid, formation of stable complexes, or loss of active phase by friction, which all lead to d. A further group of reactions which can endanger the life of a cat. are caused by *deposition of material from the reactants or formation of impurities on the active *surface (*fouling).

In hom. cat. the addition of above-stoichiometric amounts of *ligands is a known measure to retain the activity of the metal atoms as long as possible (cf. *P-C bond cleavage). In het. cat. *co-cats. or structural *promoters are added with the feed to counterbalance irreversible structural transformations or the loss of *selectivity. Most measures against d. have, however, to be taken during prepn. and choice of materials for het. cats. To retain lifetime and to avoid unwanted side reactions cat. by decomposition prods. of the initial cat., are major concerns of optimization strategies. In supported cats. the *support material is optimized to retain the *stability and *particle size distribution (dispersion) of the active phase. Often minute impurities such as alkali ions can drastically alter the adhesion properties of an active comp. leading to a critical dependence of high-performance systems on the continuous maintenance of the quality of the support.

D. is often the result of cooperation of several of the above-mentioned factors. *Directed poisoning (directed d.) can be successful for the reduction of *activity while accelerating the *selectivity (e.g., *Lindlar's cat., *Raney cats., *Rosenmund reaction, cf. *sulfided cats.). The control of d. behavior is thus only possible in a reaction environment identical to that of the technical reaction. Short-term d. tests or the application of laboratory feed-stocks can be grossly misleading for any pre-diction of the lifetime of a given cat. batch. As every batch of a commercial cat. is slightly differently manuf., the lifetime can vary over an order of magnitude between batches. To minimize prod. problems it is very desirable to understand mechanisms of stabilization or at least of d. for cats. Thus d. is still a poorly controllable phenomenon in all branches of catalysis. Cf. also *modifiers. **F** déactivation; **G** Deaktivierung. R. SCHLÖGL

Ref.: Ertl/Knözinger/Weitkamp, Vol. 3, p. 1263; Anderson/Boudart, Vol. 6; Butt/Petersen; Farrauto/Bartholomew, p. 265; Trim; Delmon/Froment; J. Oudar, H. Wise, *Deactivation and Poisoning of Catalysts*, Dekker, New York 1985; Thomas/Thomas, p. 417; *Studies in Surface Science and Catalysis*, Vols. 6, 34, 68, 88; *Chem. Rev.* **1994**, *94*, 1021.

dealkylation → hydrodealkylation (HDA)

deamination reactions play an important role in biochemistry. D. is an elimination of the amino group from amines or amino acids, where the nitrogen function is replaced by H. The reaction is initiated by pentyl nitrite or can be carried out via isocyanides. The de-struction of proteins is very important and corresponds to the *enzyme-cat. reaction of amino acids yielding nitrogen-free cpds. and ammonia. Mammalian RNA editing events, often represented by citidine-to-uridine and adenosine-to-inosine conversions, are pre-dominantly mediated by base deamination. **F=E**; **G** Entaminierung. B. DRIESSEN-HÖLSCHER
Ref.: *Cell. Mol. Life Sci.* **1998**, *54*, 946.

dearomatization is the proc. for the removal of aromatics in (preferably) cyclo-alkanes (e.g., procs. of *Atlantic Richfield Duotreat, *Criterion SynCat, *IFP PDH, *Lummus Arosat, *Toray Hytory, *Union Oil

Unisar, *UOP BenSat and Hydrar). D. is also a part of *hydrotreating. **G** Entaromatisie-rung. B. CORNILS

De Boer-Hamaker forces Three types of forces that exist between atoms and mole-cules are recognized: 1: dipole-dipole force between molecules possessing permanent di-pole moments (orientational interaction); 2: dipole-induced dipole (or induction inter-action) where a fixed dipole in one molecule induces a dipole in another; and 3: induced dipole-induced dipole force (or dispersion in-teraction) where in molecules lacking a per-manent dipole moment there are fluctuating and interacting dipoles caused by very rapid electron movements within the molecules (or atoms). All these forces vary as r^{-6} where r is the distance between the interacting species: collectively they are known as *van der Waals-London forces. Forces of this type are the cause of *physical adsorption, a phenome-non which is widely employed to measure the total *surface area of a solid: the physical ad-sorption isotherm for N_2 or Xe is interpreted using the *Brunauer-Emmett-Teller equation. The development of the theory, refined by De Boer and Hamaker, naturally presupposes an understanding of the forces at work. The de BHfs. are thus the subset of van der Waals or dispersion forces that apply to non-bonding interactions between atoms and solids or between two solids. **F** forces de De Boer-Hamaker; **G** De Boer-Hamaker-Kräfte.

G.C. BOND

Ref.: Moulijn/van Leeuwen/van Santen; G.K. Vemula-palli, *Physical Chemistry*, Prentice-Hall, Engelwood Cliffs, NJ 1993; Ch.32; P.W. Atkins, *Physical Chemistry*, 4th Ed., Oxford U.P., Oxford 1990, Chapter 22; A.W. Adamson, *Physical Chemistry of Surfaces*, 2nd Ed., Wiley Interscience, New York 1967.

DECADE (Design Expert for CAtalyst DE-velopment) uses knowledge processing methods (i.e., an *expert system shell) for CO-*hydrogenation cat. M. BAERNS
Ref.: *Comput. Chem. Engng.* **1987**, *11*, 265 and **1988**, *12*, 923.

decarbonylation   a reaction (also called hydro*de*formylation or deformylation) in which CO is liberated from organic carbonyl cpds. In general, aliphatic and aromatic aldehydes are decarbonylated in the presence of *transition metal cats. such as *Wilkinson's cat. or Pd-*phosphine *complexes. The prods. of the d. are alkanes or aromatic cpds. acc. to R-CHO → R-H + CO. In addition to the cats. mentioned, aliphatic aldehydes can be decarbonylated *radically by *peroxides. This reaction occurs (even without cats.) at very high temps. (500 °C) or *photocatalytically. The d. of acyl halides in the presence of Rh or Pd complexes yields the corresponding organic halides, acc. to RCO-X → R-X + CO with X being halogen. D. of benzoic acid chlorides and recently benzoic anhydride with Pd cats. and alkenes in the presence of a base cat. has been used to prepare aromatic alkenes (*Blaser-Heck reaction). The reaction proceeds via *oxidative addition of the Pd(0) complex into the CO-X bond, subsequent d., and olefin insertion. D. of benzoic acid with Cu cats. (*Dow phenol proc.) is a commercial proc. **F=E**; **G** Decarbonylierung.   M. BELLER

Ref.: March; *J.Am.Chem.Soc.* **1977**, *99*, 5664; S. Patai (Ed.), *The Chemistry of Functional Groups*, Suppl.B, Wiley, New York 1979, pt.2, p. 825.

decarboxylase   Enzymatic *decarboxylation involves different mechanisms. Tyrosine d. and glutamate d., for example, are *pyridoxal phosphate-dependent. α-Ketoacid ds., such as *pyruvate d., are *thiamine pyrophosphate-dependent.

A d. from *Alcaligenes bronchisepticus* was used in the asymmetric *decarboxylation of a number of arylmethylmalonic acids to (*R*)-arylpropionic acids. Pyruvate d. is the main *enzyme involved in *acyloin condensation. Other enzymes that cat. decarboxylations include, for example, gluconate 6-phosphate *dehydrogenase and isocitrate dehydrogenase (a Mn^{2+}- dependent enzyme). **E=F=G**.

C.-H. WONG

Ref.: *Biocatalysis* **1991**, *5*, 49; *J.Am.Chem.Soc.* **1990**, *112*, 4077 and **1991**, *113*, 8402.

decarboxylation   thermal or cat. scission of CO_2 from carboxylic acids or from their salts. In contrast to simple aliphatic carboxylic acids, carboxylic functionalized acids in the α- or β-position are often easily decarboxylated under mild conds. In the case of β-ketoacids a six-membered *transition state to form an enol and consecutively free CO_2 beneficially contributes to the reaction *rate. Aromatic carboxylic acids are decarboxylated by heating the acids in quinoline in the presence of cat. amounts of Cu, forming Cu salts of the aromatic acids as *intermediates (e.g., *Dow phenol proc.). Aromatic carboxylic acids with electron-donating substituents in *o*- and *p*-positions decarboxylate upon heating in the presence of sulfuric acid. Cat. (e.g., ThO_2) or non-cat. pyrolysis of aliphatic carboxylic acid salts yields ketones (e.g., by acetone from Ca acetate, cf. *ketonization). *Radical reaction mechanisms occur under electrolytic d. conds. (Kolbe synthesis) or in the d. of silver carboxylates with bromine (Hunsdiecker-Borodin reaction). **E=F**; **G** Decarboxylierung.

R.W. FISCHER

Ref.: H.-G. Frank, J. W. Stadelhofer, *Industrial Aromatic Chemistry*, Springer, Berlin 1988, p. 154.

decobalting   the decisive step of Co removal during *workup of *oxo crudes and other Co-containing reaction prods. The reaction mixture contains soluble hydridocarbonyls with Co^{-1} as long as a stabilizing CO and H_2 partial pressure is maintained (cf. Figure).

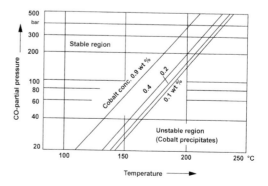

For many reasons (deposition of Co precipitates in downstream apparatuses and procs., initiation of considerable side reactions by traces of metal, coloring of final products, best possibility for the removal of *poisons prior to cat. make-up) these metal carbonyls ought to be removed *before* further processing. After depressurizing, the two methods of decobalting are the *thermal* or hydrothermal decomposition of $Co^{-1} \rightarrow Co^0$ (*Ruhrchemie oxo proc.) or the *chemical* treatment under valence change Co^{-1} to Co^{2+} (*BASF oxo proc.). Without reduction of press. the removal and recycling as Co^{-1} (*Kuhlmann proc.) is preferred. Besides oxo prods., Co traces may also cause problems with coloring of various other prods. (e.g., *DMT). **F** décobaltation; **G** Entkobaltung.   B. CORNILS

Ref.: *Chem.-Ztg.* **1973**, 97, 374; Kirk/Othmer-2, p. 827

deep catcracking (DCC) fluidized proc. with traditional *FCC cats. including two different modes: maximum propylene (*riser and bed cracking in severe cond.) and max. *i*-alkenes (only riser cracking in milder conds.) operation; cf. *catcracking, *severity.   B. CORNILS

defect structure Each solid matter is fully described only when its ideal structure (is.) and the intrinsic deviations from it are known (cf. also *ideal crystal structure). The existence of structural deviations, which can be chemical substitutions or geometric displacements of atoms relative to the is., is unavoidable as for a perfect crystal the destabilizing effect of the entropy term in the heat of formation becomes dominant. Typical defect densities in real crystals are one defect of any kind per about 10 unit cells ultrahigh-quality crystals exhibit about six orders of magnitude fewer defects. The defects can be atomically dispersed and are then statistically distributed throughout the volume of the solid. Such point defects are difficult to determine and to characterize. When they order to line defects, to internal *surfaces (grain boundaries) or to three-dimensional *clusters (ex-solution phases) they are easier to analyze and can be better studied. All defect types affect directly the physical and chemical properties of a sol-

id. The location and description of their nature are thus indispensable prerequisites to understanding and an important piece of information for the characterization of a solid. Usually the defects are statistically distributed on all scales of dimensions. As defects are dynamic objects they can order and form defect superstructures which are detected in diffraction experiments. The dynamic nature of defects couples their existence and nature to the environment (chemical, thermal, mechanical) of the solid under investigation and this is one fundamental reason why het. cats. undergo the transformation from a precatalyst (*precursor) into an *active phase under working conds. The presence of defects can be beneficial or detrimental for a het. cat. as they can affect the kinetics of the cat. surface reaction (beneficial), the kinetics of the transformation into the active phase (beneficial), and the kinetics of the transformation of the active phase into other phases (detrimental). The detection and analysis of defects are usually only possible indirectly. Surface analytical probes either are not sensitive enough for detection of many types of defects or their sensitivity relies on translational symmetry which is usually not characteristic for defective structures (with the important exception of the surface of a solid). Cf. also *single crystal.
                                                  R. SCHLÖGL

DeFine → UOP procs.

deformylation → decarbonylation

DEG diethylene glycol

Deguphos → Degussa procs.

Degussa acrolein process an older proc. via cat. (Na silicate/ SiO_2) *aldolization of acetaldehyde and formaldehyde, followed by *crotonization at 300–320 °C in the vapor phase. The newer version uses the het. cat. *oxidation of propylene over *bifunctional cats. **F** procédé Degussa pour l'acroléine; **G** Degussa Acrolein-Verfahren.   B. CORNILS

Degussa allylic alcohol process manuf. of allylic alcohol via *selective *hydrogenation of acrolein on het. Cd-Zn cats. in the vapor phase. **F** procédé de Degussa pour l'alcool allylique; **G** Degussa Allyalkohol-Verfahren.

B. CORNILS

Degussa BMA (HCN) process manuf. of hydrogen cyanide by *ammondehydrogenation of methane and ammonia at 1300 °C in Pt- or Ru-coated Al_2O_3 tubes or with *gauze cats. At complete ammonia convs. the *selectivity to HCN exceeds 90 %. **F** procédé BMA de Degussa; **G** Degussa BMA-Verfahren.

B. CORNILS

Degussa catalysts Founded in 1843 in Frankfurt (Germany) as a gold and silver refinery, precious metals have been a core business within Degussa since then. Today Degussa's cat. activities are in *automotive exhaust cats., stationary emission control cats., precious metal cats., and activated base metal cats. for both powder and fixed-bed application, as well as *silicas and *zeolites.   J. KULPE

Degussa Deguphos process manuf. of optically active fine chemicals (e.g., phenylalanine or *Dopa) by *asymmetric *hydrogenation of suitable *precursors in various solvents. Generally, Rh *complexes with *chiral *phosphines are used as cats. at 20–50 °C/1–10 MPa. **F** procédé Deguphos de Degussa; **G** Degussa Deguphos-Verfahren.   B. CORNILS
Ref.: *Chem.Ber.* **1986**, *119*, 3326; *Appl.Catalysis* **1994**, *115*, 190, 203.

Degussa Deloxan method This pilot plant development uses *immobilized hom. cats. for the manuf. of different fine chemicals on Deloxan (organofunctional polysiloxane) *carriers, e.g., for selective *hydrogenations. **F** méthode Deloxan de Degussa; **G** Degussa Deloxan-Methode.   B. CORNILS
Ref.: DE 4.035.032 (1990); *Appl.Catal.* **1994**, *115*, 190.

Degussa Desonox process treatment of power plant effluents comprising the selective

cat. *reduction (*SCR) of NO_x by NH_3 on *aluminosilicate *zeolites and *oxidation of SO_2 to H_2SO_4. The proc. was jointly developed by Degussa, Lurgi, and Lentjes. **F** procédé Desonox de Degussa; **G** Degussa Desonoxverfahren.   B. CORNILS
Ref.: *Proc.Eng.* **1988**, 69(2), 37.

Degussa glycidol process manuf. of glycidol from allyl alcohol and H_2O_2 with *Milas reagent (WO_3/H_2O_2). **F** procédé Degussa pour le glycidol; **G** Degussa Glycid-Verfahren.

B. CORNILS

Degussa L-*tert*-leucine process reaction of 2-oxo-3,3-dimethylbutyric acid with the *enzymes leucine dehydrogenase and *formate dehydrogenase.   B. CORNILS
Ref.: *Tetrahedron: Asymmetry* **1995**, *6*, 2851; *Bioproc. Engng.* **1996**, *14*, 291.

Degussa nicotinamide process manuf. of nicotinic acid amide via cat. *ammonoxidation of 3-picoline to 3-cyanopyridine and subsequent *hydrolysis to nicotinamide. **F** procédé Degussa pour nicotinamide; **G** Degussa Nicotinamid-Verfahren.   B. CORNILS

dehalogenases cat. the *hydrolysis of haloalkanes to alcohols. Haloalkane dehalogenase from *Xanthobacter autotrophicus* contains Asp124, His289, and Asp260 as a *cat. triad. Asp124 first attacks the C-containing halogen group to form a covalent intermediate. His289 is then involved as a *general base cat. to hydrolyze the intermediate and form the alcohol. Application of the dehalogenase from *Alcaligenes sp.* and *Pseudomonas sp.* to the synthesis of (S)- and (R)-glycidol from (R,S)-3-chloro-1,2-propanediol has been reported. **E=F**; **G** Dehalogenase.   C.-H. WONG

Ref.: *Nature (London)* **1993**, *363*, 693; *Bioorg.Med. Chem.Lett.* **1991**, *1*, 343.

dehalogenation → dehydrohalogenation, β-*elimination reactions

DEHP → DOP

dehydration 1-chemical: D. reactions belong to the class of *elimination reactions in which water is formed as a prod. via C_α-OH and C_β-H bond cleavage yielding alkenes. Different mechanisms are known (β- or E1-elimination). In a concerted E2 mechanism the two leaving groups/atoms are typically eliminated from a *trans*-configuration even on solid cats. such as *alumina.

Ds. are cat. in hom. reaction by proton acids, basic, and aprotic cpds. in solution and by hydrogen halides in the gas phase. Gas-phase ds. cat. by *solid acids, *solid bases, or amphoteric materials (typically crystalline or amorphous oxides) have frequently been reported. The reaction mechanism chosen is determined by the acid-base properties of the cat., the C_α-OH and C_β-H bond strengths, stereochemical factors, and the temp. In het. cat. by oxides, *dehydrogenation may occur as a competing reaction. Prod. distributions are often used for the qualitative characterization of acid-base properties of solid cats. (cf. *acid-base cat.). Procs. of interest are *BASF THF, *GAF THF, *IFP isoprene, *Jefferson morpholine, *Knapsack ACN, etc. Cf. also *oxydehydration. **F=E; G** Dehyratisierung.

H. KNÖZINGER

Ref.: Ertl/Knözinger/Weitkamp, Vol. 5, p. 2370; S. Patai (Ed.), *The Chemistry of the Hydroxyl Group*, Interscience 1971, p. 641.

2-thermal: D. also describes the proc. of water removal from solid materials by thermal treatment (*drying) and the thermal transformation of hydroxides into *oxides. H. KNÖZINGER

3-enzymes, organic solvents: The use of hydrolytic *enzymes in organic *solvents with minimal water present allows the shifting of the hydrolytic equil. toward synth. It has been noted by many groups that enzymes are typically most stable and active in the solvents that are least polar. Evidence has mounted to demonstrate that the hydrophilic solvents de-

activate the enzyme by stripping water away from them. Studies with *lysozyme have shown that very dry enzyme is inactive, and that the addition of water will reactivate the enzyme. Enzyme activity requires less than monolayer coverage by water (see also *enzyme cat., *solvents, and *equil.-controlled synth.). **F=E, G** Dehydratisierung. P.S. SEARS

dehydrocyclization formally the simultaneous *dehydrogenation and *cyclization of aliphatic hydrocarbons. More precisely, d. means the conv. of $>C_{6+}$ aliphatics into aromatics and is often synonymous with *aromatization (of C_{6+} paraffins). D. is one of the most important reactions of cat. naphtha *reforming, which is commercially mostly realized on metallic/acid *bifunctional cats., especially on *supported Pt on *alumina or *zeolites. Monofunctional group VIII metal cats., oxide catalysts such as Cr_2O_3/Al_2O_3, and acid zeolites are also used. Depending on the cat. type, reactants, and conds., several mechanisms of C_{6+} alkane d. have been proposed. D. on bifunctional cats. has been explained in terms of a cooperation of dehydrogenating (but cyclizing as well) metallic sites and skeletal *isomerization (specially cyclizing, cf. *isomerization) acid sites. As general reaction routes, C_6 ring closures as well as the *dehydroisomerization of C_5 cyclic intermediates into aromatics have been suggested to occur in parallel on such cats. Examples are given under the entries *BASF GBL proc. and *Phillips STAR proc. **F** déhydrocyclisation; **G** Dehydrocyclisierung. H. LIESKE

Ref.: Gates-1, p. 297; B.C. Gates et al., *Chemistry of Catalytic Processes*, McGraw-Hill, New York 1979, p. 184; *Adv. Catal.* **1983**, *29*, 273; *Catal. Rev.-Sci. Eng.* **1997**, *39*(1/2), 5.

dehydrodimerization *dimerization of cpds. after preceding *dehydrogenation, e.g., by metal-photosensitized cat. (with Hg, cf. Figure), with *dehydrating cats., or by *Raney-Ni cats. (e.g., *Lebedew proc.). Cf. *aromatization. **F** déhydrodimérisation; **G** Dehydrodimerisierung. B. CORNILS

$$2 \ H_3C-\underset{\underset{H_3C}{|}}{\overset{\overset{H_3C}{|}}{C}}-H \ \xrightarrow[hv]{Hg(^3P_3)} \ \underset{\underset{H_3C}{|}}{\overset{\overset{H_3C}{|}}{C}}-\underset{\underset{CH_3}{|}}{\overset{\overset{CH_3}{|}}{C}}-CH_3$$

Ref.: Cornils/Herrmann-1, p. 941, 1108.

dehydroesterification

new proc. for the *C-H activation of alkanes. The cat. d. of alkanes with methyl formate yields methyl esters of carboxylic acids in the presence of sulfuric acid and *tert*-butyl cation *precursor (e.g., *tert*-butyl chloride or *MTBE) acc. to $RH + HCOOMe + Me_3CX \rightarrow RCOOMe + Me_3CH + HX$. The yields are moderate (e.g., 60 % methyl-2,2,3-trimethylbutanoate).

Ref.: *J. Mol. Catal. A:* **1997**, *125*, 33.   B. CORNILS

dehydroformylation

$\rightarrow$ decarbonylation

dehydrogenases

$\rightarrow$ oxidoreductases

dehydrogenation

abstraction of hydrogen, yields alkenes (from alkanes), dienes (from alkenes), or carbonyl cpds. (aldehydes, ketones) from primary or secondary alcohols. Representing the reverse of *hydrogenation, d. reactions are endothermic (110–130 kJ/mol for alkanes/ alkenes; 65–80 kJ/mol for alcohols). Conv. therefore is limited by the thermodynamic equil. and is favored by high temps. (250–600 °C), which demand thermally stable and robust cats. Thermodynamics are favorably influenced by the addition of oxygen (*oxidative d.), turning the overall reaction enthalpy negative by the formation of water (–286 kJ/mol). The addition of steam, H, or N as diluents and carrier gases increases the *selectivity. A number of different cats. have been in use acc. to the requirements of proc. and feedstock. For thermodynamic reasons, atmospheric or reduced press. is advantageous.

D. reactions are generally classified as *structure insensitive, i.e., the active comps. should be finely divided without special requirements with respect to crystallinity. A high specific *surface of metals is advantageous but difficult to stabilize at high reaction temps. Pt is the most active and versatile cat. metal for the d. of alkanes, finely distributed onto *silica, *alumina or silica-alumina, in a conc. range from 0.01 to 1.5 wt.%. *Bimetallic combinations are common, e.g., Pt-Sn (3:1 mol/mol), being selective to the corresponding alkenes and more stable toward oxidative regeneration than Pt alone. Primary or secondary alcohols (except methanol) readily undergo d. in the presence of Cu cats. (Zn, Cr, Mg, Ca, Ba as *activators; SiO_2 as support). Depending on the molecular weight (boiling point) of the alcohol, gas-phase or liquid-phase operation is applied with temps. of 250–320 °C. For examples of commercial procs. cf. *Air Products (propylene), *BASF (*GBL), *BP (butenes), *Dow (*styrene), *Lummus (*styrene), *Phillips STAR, *Union Carbide (ethylene), *UOP Pacol and Oleflex, etc. **F** déhydrogénation; **G** Dehydrierung.

C.D. FROHNING

Ref.: *J. Mol. Catal.* A:Chemical **1998**, *133*, 267; *J. Catal.* **1998**, *179*, 459; DE-OS 19 626 587 (1996); DE-OS 19 609 954 (1996); *J. Catal.* **1986**, *102*, 160; Ullmann, Vol. A13, p. 487; *J. Catal.* **1973**, *30*, 128; Kirk-Othmer-1, Vol. 13, p. 848; Ertl/Knözinger/ Weitkamp, Vol. 5, pp. 2140, 2151 and 2159; Farrauto/Bartholomew, p. 450; McKetta-1, Vol. 14, p. 276.

dehydrohalogenation

elimination of hydrogen halides from an organic halide (also: hydro*halo*elimination, and often simply called dehalogenation, a β-*elimination reaction). The reaction is very general and is accomplished with alkyl, alkenyl, and even aryl halides acc. to $RCH_2\text{-}CH(X)\text{-}R^1 + OH^- \rightarrow R\text{-}CH{=}CH\text{-}R^1 + H_2O + X$. As halide ions X, chlorides, fluorides, bromides, or iodides can be used. To induce d., the addition of stoichiometric amounts of base (NaOH, KOH, NaH, R_3N, pyridine, NaOAc, etc.) is necessary. A convenient method to perform d. is the *phase transfer cat. The orientation of the double bond is not simply predictable and is influenced by both steric and electronic effects. Most ds. require stoichiometric rather than cat. conditions. Commercially important is the thermally induced d. of *EDC to *vinyl chloride (e.g., by the

*EVC proc.). Dehydrochlorinations under *hydrotreatment conds. are called *hydrodechlorinations. **F** déhydrohalogénation; **G** Dehydrohalogenierung.   M. BELLER

Ref.: March; *Organic Reactions*, Vol. 29, p. 163; *J.Mol. Catal.A*: **1999,** *148*, 1.

dehydroisomerization   simultaneous *dehydrogenation and *isomerization of hydrocarbons. Commercially it is the conv. of C$_5$ cyclic hcs. and of C$_{6+}$ aliphatics, specially C$_{6+}$ alkanes, into corresponding aromatics (*reforming, *dehydrocyclization, *aromatization). Alkylcyclopentanes as native naphtha comps. or as intermediates undergo, after dehydrogenation, a C$_5$/C$_6$ ring enlargement into a cyclohexane derivative, followed by dehydrogenation into the corresponding aromatic, e.g., methylcyclopentane → methylcyclopentene → cyclohexene → benzene. D. proceeds with high efficiency on metallic/acid *bifunctional cats. such as Pt supported on acid *alumina via cooperation of metallic sites (for *dehydrogenation) and acid sites (ring enlargement). With less efficiency d. is also possible on monofunctional metal cats. such as Pt and other group VIII noble metals, unsupported or on non-acid carriers. **F=E**; **G** Dehydroisomerisierung.   H. LIESKE

dehydrosulfurization → hydrodesulfurization (HDS)

dehydroxidation   proc. for simultaneous cat. *oxidation and *dehydrogenation, e.g., the conv. of butane to maleic anhydride (procs. of *Lummus ALMA, *Alusuisse, *Lurgi Geminox) or the *MMA synthesis from isobutene (*Escambia proc.); cf. also *dehydration, *oxydehydrogenation, *deoxygenation. **F=E**; **G** Dehydroxidation.
                                         B. CORNILS

delayed coking processes → Foster-Wheeler procs.

Deloxan   an organofunctional polysiloxane *carrier, proposed for the *immobilization of hom. cats (cf. *Degussa procs.).

demetall(iz)ation 1: important step during handling of reaction prods. of hom. or het. cat. because metals may influence further processing of prod. streams, can be deposited in apparatus or on cats., or may cause the quality to deteriorate (e.g., through inadmissible color values). Since even metal contents in the ppm range are too high, special methods have to be applied, amongst them are chemical and/or thermal treatments (e.g., for Co and Rh, *decobalting) or maintenance of appropriate partial press. (e.g., CO, CO$_2$, H$_2$, NH$_3$ for Co, Ni, Fe, Rh, or Pd). For the d. of prods. for het. cat. procs. cf. *Demex procs. or *guard beds. The metal content of prods. of hom. cat. can be elegantly controlled by *complex binding or by aqueous-phase cat.
2: *hydrodemetallation (HDM) is the removal of traces of metals from crude oil fractions (*petroleum processing industry) under *hydrotreating conds.; cf. procs. of *Howe-Baker, *IFP Hyval, *Lummus, *Shell Hycon, etc. **F** démétallisation; **G** Entmetallisierung.
                                         B. CORNILS

Ref.: McKetta-1, Vol. 14, p. 302; Vol. 26, p. 474.

demethanation   cat. cleavage of methane as an example of hom. gas/gas cat. (e.g., demethanation of 2-methyl-2-pentene by HBr; *Goodyear/SD isoprene proc.).   B. CORNILS

DEMEX process UOP proc. for the *demetall(iz)ation of petroleum prods. by extraction.

denaturation (enzyme inactivation) the unfolding from the native (natural, active) state of a protein. D. can in general be induced by high concs. of denaturing agents such as urea or guanidinium hydrochloride, detergents, and high temps. Many proteins will also unfold with extremes of pH or ionic strength. Not all proteins are capable of refolding after denaturation; some require the presence of folding cat. called *chaperones. *Ribonuclease A, for example, was the first protein that was demonstrated to fold and unfold reversibly, and it was concluded that proteins in general adopt the lowest-energy (thermodynamically optimal) *conformation, although it is certainly conceivable that an *enzyme could be kinetically trapped in a metastable state, particularly in the cases where the enzyme is modified (e.g., via proteolysis) after folding. **E=F=G**.   P.S. SEARS

dendrimers highly symmetrical, three-dimensional molecules with a defined structure, grown exponentially acc. to the Mandelbrot principle in fractal geometry (cascade polymers, Figure).

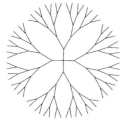

There are divergent and convergent formations of ds. Divergent dendritic construction results from the addition of sequential *monomers beginning at the core and proceeding to the outward macromolecular surface, building up a symmetrical, hom. d. Acc. to the convergent mode of dendritic construction, the branched polymeric arms (dendrons) are synth. from the "outside-in", often resulting in a het. bifunctional backbone. Branching depends on building block valences. The application of d. in cat. results from the functionalization of the macromolecule surface. Various reactions are used to attach metal *complexes at the ends of the dendritic arms, which are able to generate a cat. effect in the periphery of the d. (d.-supported cats.) and which are thus *ligands. Ds. are used inter alia in *membrane reactors. **F** dendrimères; **G** Dendrimere.   J.P. ZOLLER

Ref.: G.R. Newkome, C.N. Moorfield, F. Vögtle, *Dendritic Molecules: Concepts, Syntheses, Perspectives*, VCH, Weinheim 1996; *Angew.Chem.Int.Ed.Engl.* **1999**, *38*, 885; *Chem.Rev.* **1999**, *99*, 1665.

denitration → hydrodenitrogenation (hydrodenitration, *HDN), *cat. nitrate reduction, *denitrification reactions, *DeNOx.

denitrification reactions → DeNOx reactions

denitrogenation   → hydrodenitrogenation (HDN)

de-novo ligand design method, in which molecules are built into the *enzyme active site. De-novo *ligand design is typically accomplished by positioning atoms or functional groups at specific, favorable positions and allowing computers to fill in the structure to make a contiguous molecule (cf. *computer modeling for enzyme inhibition).   P.S. SEARS

Denox (DeNOx, DeNO$_x$) processes →proc. of *Didier, *La Grande Paroisse; *Rhône-Poulenc; cf. also *nitrogen oxide removal.

DeNOx reactions (denitrification) reactions to remove (i.e., convert into innocuous N_2) nitrogen oxides (NO_x), mostly from flue/exhaust gases of mobile and stationary sources, such as cars (*automotive exhaust cat.), power or chemical plants (*selective cat. reduction, SCR). NO_x removal is of increasing importance with changes in legislation reducing limits on NO_x emissions which have harmful effects on the environment (smog, acid rain, ozone layer depletion; cf. *green chemistry, *environmental issues). NO_x are formed either from reaction of N-containing cpds. in a chemical/combustion (*catalytic combustion) proc. or through *oxidation of nitrogen from air at high temps. NO_x control is divided into *primary* (prevention of NO_x formation) and *secondary* measures (DeNO$_x$, DeNOx). Procs. to remove NO_x can be classified into wet or dry, reaction or absorption, cat. or non-cat., selective or non-selective. The DeNOx proc. most frequently used to clean effluents in power and nitric acid plants is *selective catalytic reduction (SCR) with ammonia (alternatively urea) as a reducing agent to give N_2 and H_2O. The NH_3/NO ratio determines the degree of NO reduction, but is limited by excess ammonia leaving the plant (*ammonia slip). SCR is performed at temps. between ca. 473 and 773 K with TiO_2-supported V_2O_5 monolayer cats., often with admixtures of W and Mo oxides (*titania). Other cat. systems include Cr- or Fe-containing oxides supported on Al_2O_3, or Cu on various *supports. Typical cat. formulations are plates, *honeycombs, and *pellets. SCR units can be placed early in the flue gas treatment scheme to minimize the need to reheat the gases, or later, after dust and SO_x removal, to minimize cat. *deactivation. SCR can be combined with other flue gas treatments to simultaneously remove NO_x and SO_x (DeNOx/DeSOx).

A currently used cat. system in automotive pollution control is $Rh-Pt/CeO_2/Al_2O_3$ (*three-way converter), reducing NO with CO and H_2 from the exhaust at temps. between ca. 523 and 773 K. There is an ongoing search for new cats. in automotive exhaust control for several reasons such as cost (expensive Rh), susceptibility of the cats. toward *poisons (sulfur), and changes in engine technology (e.g., lean-burn engines with higher air excess in the exhaust, *zero-emission automobiles). Noble metals supported on *zeolites and *perovskites are among the materials tested for NO_x reduction with CO/H_2. Recent efforts involve hcs. as reducing agents and cat. systems under investigation include mixed oxides (e.g., Cu-Zr-O) and zeolite (mostly MFI)-supported Cu or Co (also Ga, Mn, Ni) but these systems are negatively affected by sulfur cpds. and water. Attempts to decompose NO directly (into N_2 and O_2) were performed mainly using Cu supported on zeolites, with limited applicability because of low H_2O resistance of these cats. For the simultaneous removal of NO_x and S cpds. cf. *Haldor Topsøe's SNOX proc. **F** réactions DeNOx; **G** DeNOx-Reaktionen.   F. JENTOFT

Ref.: *Catal. Today* **1998**, *46*, 233; Ertl/Knözinger/Weitkamp, Vol. 4, p. 1633.

density functional theory (DFT) an alternative method to *ab-initio calcns. for obtaining geometries, bond energies, and other physicochemical properties of molecules and solids. The basic idea of DFT is to replace the complicate N-electron wavefunction Ψ and the associated *Schrödinger equ. by the much simpler electron density $\rho(r)$ and the associated calcn. scheme. The Hohenberg-Kohn theorem states that the ground-state energy of an atom or molecule is completely determined by the electron density. Unfortunately, the exact functional dependence of the energy on the electron density is not known. Many approximate functionals based on the Kohn-Sham (KS) formalism have been developed, which are a set of one-electron equs. that are formally related to the *Hartree-Fock (HF) method. Unlike the HF equs., which aim at giving one-electron energies and *MOs for constructing the wavefunction, the KS equs. give orbitals whose significance is to construct the electron density $\rho(r)$. However, it has been shown that the KS orbitals and the HF orbitals usually have a similar shape. Both types of orbitals can be expressed through atomic basis sets which are Gaussian or Slater functions.

Early versions of DFT are the multiple-scattering and X_α methods, which are in general not accurate enough to compete with ab-initio methods for calcng. molecules. They became important, however, in solid-state physics. A breakthrough in the application of DFT to molecules came with the development of gradient-corrected functionals, in which the energy is given as a functional of the density *and* its first derivative $E[\rho, \nabla\rho]$.

This is also called non-local density functional theory (NL-DFT). The results of NL-DFT calcns. showed accuracies which were similar or even better than standard ab-initio calcns. This makes DFT very attractive for calcns. of molecules, because DFT shows only an N^3 scaling of the computational costs, where N is the size of the one-electron basis set in the KS equs. In contrast, ab-initio calcns. already have N^4 dependence for HF calcns., and become even more costly when *correlation energy is to be calcd. Correlation energy is already included in DFT calcns., which strive at giving correct total energies. The disadvantage of DFT is that the results obtained from a given functional can not be improved at a higher level of theory, while ab-initio calcns. form a hierarchy of methods which approach the correct result, albeit at increasing costs. The local density functional (LDF) method is an elder variant of DFT in which the functional depends only on the density. **E=F=G.**

Ref.: Thomas/Thomas, p. 364.   G. FRENKING

density of polyethylenes → *polyethylene, *high-, *low-, *linear-low density, and *ultrahigh molecular weight polyethylenes.

Polyethylenes	Pressure applied	Degree of branching*	Density [kg/m³]
HDPE	Low	None	>940
UHMWPE	Low	Very low	930–935
LLDPE	Low	Medium	915–925
LDPE	High	High	910–940

* cf. polyethylene

density of states (DOS)   In the band structure of a solid the number of electronic states $N(E)$ which exist in an energy interval of width dE, i.e., the DOS, is given by $N(E)dE$. For *surfaces it is more convenient to consider the DOS for a single layer of atoms, called the local DOS or LDOS. **F** densité des états; **G** Zustandsdichte.   R. IMBIHL

Ref.: Thomas/Thomas, p. 403.

deoxygenation   a general term for removal of oxygen and/or hydrogen, supported by cats., e.g., by *oxydehydrogenation, *dehydroxidation, *oxidative coupling, *Clemmensen reduction, etc. Direct d. of organic cpds. (elimination of carbonyl groups of aldehydes or ketones) occurs while coupling aldehydes or ketones to form alkenes in the presence of activated Ti cpds. (oxophilic titanium) which act as O acceptors (*McMurray coupling). Recent improvements for d. are hydridic Ti(II) species resulting from treatment of $TiCl_3$ with activated MgH_2. Cat. d. has been reported applying TMSCl as oxygen acceptor and 5–10 mol% $TiCl_3$. **F** désoxygenation; **G** De(s)oxygenierung.   R.W. FISCHER

Ref.: Cornils/Herrmann-1, Vol. 2, p. 964.

3-deoxy-D-*arabino*-2-heptulosonic acid 7-phosphate (DAHP) synthetase
(EC 4.1.2.15)   DAHP, also known as phospho-2-keto-3-deoxyheptanoate, is a key intermediate in the shikimate pathway for the biosynth. of aromatic amino acids in plants. In vivo, DAHP synthetase cat. the synth. of DAHP from PEP and D-erythrose 4-phosphate. **E=F=G.**   C.-H. WONG

Ref.: *J. Am. Chem. Soc.* **1986**, *108*, 8010.

3-deoxy-D-*manno*-2-octulosonate aldolase (DMOA, EC 4.1.2.23)   also known as 2-keto-3-deoxyoctanoate (KDO) aldolase, cat. the reversible *addition of pyruvate with D-arabinose to form KDO. KDO and its nucleotide-activated form, CMP-KDO, are key *intermediates in the synth. of the outer membrane lipopolysaccharide (LPS) of gram-negative bacteria.

R_1 = OH; R_2 = H (S); kinetically favored
R_1 = H; R_2 = OH (R); thermodynamically favored

Investigations of this *enzyme showed high specificity for KDO in the direction of cleavage, whereas the addition reaction proceeds with some flexibility; several unnatural, weak substrates, including D-ribose, D-xylose, D-arabinose 5-phosphate, and N-acetylmannosamine have been reported. The KDO aldolase from *Aureobacterium barkerei*, strain KDO-37-2, accepted an even wider variety of substrates, which included trioses, tetroses, pentoses, and hexoses as substrates. The *enzyme is specific for substrates having an R configuration at C-3, but the stereochemical requirements at C-2 are less stringent. Under kinetic control, the C-2 S configuration is favored while the C-2 R configuration is thermodynamically favored. **E=F=G**.   C.-H. WONG

Ref.: *J.Am.Chem.Soc.* **1993**, *115*, 413.

3-deoxy-D-*manno*-2-octulosonate 8-phosphate synthetase (EC 4.1.2.16) also
known as phospho-2-keto-3-deoxyoctanoate (KDO 8-P) synthetase, cat. the irreversible *aldol reaction of PEP and D-arabinose 5-phosphate to give KDO 8-P. KDO 8-P itself has been synth. using KDO 8-P synthetase where the starting material, D-arabinose 5-phosphate, was generated either by hexokinase-cat. phosphorylation of arabinose or by *isomerase-cat. reaction of D-ribose 5-phosphate (cf. *3-deoxy-D-*manno*-2-octulosonate aldolase). **E=F=G**.

Ref.: *Tetrahedron Lett.* **1988**, *29*, 427.   C.-H. WONG

D-arabinose 5-P

KDO 8-P

2-deoxyribose 5-phosphate aldolase
(DERA, EC 4.1.2.4) The *enzyme DERA is unique among the *aldolases in that the donor of the *aldol reaction is an aldehyde. In vivo, this enzyme cat. the reversible *condensation of acetaldehyde and glyceraldehyde-3-phosphate (G3P) to form D-2-deoxyribose 5-phosphate. DERA is a type I aldolase and has been isolated from animal tissues and several microorganisms.

acetaldehyde   G3P   D-2-deoxyribose 5-P

A number of unnatural substrates are accepted by DERA, and the newly generated

*chiral center always has the S-configuration. Specificity studies have been conducted on DERA from *Lactobacillus plantarum* (specific activity of ca. 6000 U/mg for condensation) and *E. coli* (V_{max} = 21 U/mg for aldol cleavage). The enzyme from *L. plantarum* accepts various acceptor substrates, including L-G3P, but not D-ribose or glyeraldehyde. Only propionaldehyde could weakly replace acetaldehyde as the donor. The *E. coli* enzyme, on the other hand, is able to utilize several other donor substrates, including acetaldehyde, propionaldehyde, etc. A number of aliphatic aldehydes, sugars, and sugar phosphates are acceptor substrates. **E=F=G**.   C.-H. WONG

Ref.: *J.Am.Chem.Soc.* **1990**, *112*, 2013 and **1992**, *114*, 741.

deparaffination   → dewaxing

deposition   1: In the synth. of het. cats. d. is the proc. of laying down a solid from a liquid or a gaseous *precursor (cf. *precipitation). In this way thin films of metallic or oxidic cpds. are obtained. On oxidic *supports a small abundance of deposited material can be used to form small particles when the initially present thin film aggregates as a consequence of a chemical reaction. Such a 8reaction involving *oxidation or *ligand abstraction is carried out during a heat treatment (*calcining). There are numerous procs. where this technique is used for manuf. of commercial and *model cats.

The critical conds. of d. (concs., *residence time over the substrate and temp.) have to be maintained with minimum gradients over the substrate in order to reach a uniform quality. It is often advantageous to utilize organometallic precursors (cf. *organometallic cpds.) or inorganic *complexes with sufficient chemical stability. It is essential to prevent the nucleation of the material to be deposited outside the substrate *surface. Therefore, the stability should be high enough to ensure a long lifetime of the precursor in the transport phase (gas or liquid). The growth of the phase to be deposited can be initiated either by crystallization (condensation) procs. or by a direct chemical reaction between precursor and substrate.

2: D. is the formation of an unwanted solid material on the *surface of an active het. cat. during operation. The formation of a deposit blocks the access to *active sites and is thus a serious *poisoning effect. The most common proc. is the formation of a carbon *overlayer during processing of organic molecules in gas-phase reactions. *Polymerization and *dehydrogenation of activated small organic molecules can lead to the formation of carbonaceous deposits. This material is hc. polymer with a C/H ratio below 0.4. It can be burnt off with oxygen which leads, however, to a secondary deposit of graphitic carbon that cannot be removed. Procs. which are strongly affected by deposition deactivation include petrochemical procs. and *selective hydrogenations. In the cat. literature deposition from the proc. constituents is referred to as *fouling whereas d. from proc. impurities is designated *deactivation. Deposits other than carbon arise from aerosol or colloidal impurities in feed streams. They can form strongly adhering passivating layers, destroying cat. systems progressively from the reactor inlet. Cf. also *electrochemical d. **F** déposition; **G** 1: abscheiden, 2: Ablagerungen. R. SCHLÖGL Ref.: Ertl/Knözinger/Weitkamp, Vol. 2, p. 626.

deposition-precipitation special method for the prepn. of supported cats. by 1: adding, e.g., urea to the slurry of metal salt solution and the *support, heating to 90 °C with decomposition of the urea to NH_3 and CO_2, and 2: hom. *precipitation of the metals on the *surface of the support.

D-p. permits thus the prep. of supported cats. if precipitation from solution of an active *precursor is carried out in the presence of a finely divided suspended *support. Provided the interaction between the nuclei of the insoluble active precursor and the *surface of the suspended *support is strong enough, precipitation will occur exclusively at the support *surface. For a hom. distribution of the active precursor to be achieved within the *pore system of the support, the pH and the valence state of the precursor have to be controlled and *complexing agents may be added, which keep the active precursor in solution under conds. which would otherwise lead to precipitation away from the support surface. Hom. distributions of active species on preshaped support *pellets or granules can also be achieved by d-p. unter certain conds. **F** déposition-précipitation; **G** Fällung. H. KNÖZINGER Ref.: Ertl/Knözinger/Weitkamp, Vol. 1, p. 240.

deposits organic materials laid down (lay-downs) on cat. *surfaces, e.g., by *coking, *cracking, *polymerization, etc., or as by-prods. of reactions (*fouling, e.g., *WGSR). As a consequence, *active sites within het. cats. are blocked and this prevents accesss of reactants to active cat. surfaces (*coking, *deactivation, *deposition). R. SCHLÖGL

deprotection (enzymatic) → enzymatic deprotection

Desonox process → Degussa procs.

desorption In general use, the term is used differently. 1: To indicate the direction from which equil. has been approached, e.g., d. curve or d. point; 2: in d. (or stripping), a comp. of a solution leaves the liquid and enters the gas phase (reverse proc. of *absorption); 3: under gas-solid conds., a liquid leaves the solid and diffuses into a carrier gas (i.e., drying, the reverse proc. of adsorption). In cat. terminology among the other elementary steps, i.e., *diffusion, adsorption, and reactions of adsorbed species, the d. is the final step in any cat. reaction. An elementary step is called the d. proc., when molecules leave the *surface layer of the adsorbent. D. has adsorption as its counterpart, denoting the reverse proc. The term molecular d. is applied to the reverse of non-dissociative adsorption, while associative desorption is the reverse of dissociative adsorption, where a molecule is adsorbed with dissociation into two or more fragments. The existence of multiple d. procs. are an indication either that different kinds of adsorbed species coexist on the surface or that the d. induces an interconv. from one species to another as *coverage is depleted. D. of an adsorbate can be induced thermally, or it can be stimulated by means of irradiation by light (photodesorption), or it can be induced by electrons (electron-stimulated d.). Usually, the proc. of d. is analyzed by means of *temp.-programmed desorption (TPD; synonymous with *TDS and *FDS), i.e., heating the adsorbate-covered solid in an UHV or in an inert carrier gas. D. rates are often described by the *Polanyi-Wigner equ. The kinetic order of d. suggests the nature of the elementary steps involved. D. from a multilayer can often be described by zero-order kinetics where the rate of d. is independent of *coverage. The d. of a single surface species is usually of first-order while the recombination of adsorbate atoms leading to the formation of a diatomic molecule, which is then released, follows second-order kinetics. Furthermore, two adsorbed species may recombine to an adsorbed molecule which in turn desorbs leading to precursor-mediated d. kinetics. *Transition state theory (TST) yields further insights into a microscopic description of the d. pathways via an activated *complex and provides the link between macroscopic reaction *rates and molecular properties. The assumption of TST is that equil. is established between the reactants and an activated complex, which is a reactive chemical species that is in transition between reactants and prods. The terms *immobile* and *mobile* are used to describe the freedom of the molecules of adsorbate to move on the surface. Depending how many degrees of freedom are permitted in the activated *complex and in the chemisorbed state the preexponential factor ranges from 10^{12} to 10^{18} s^{-1} (first-order d.) or 10^{-5} to 10 cm^{2}s^{-1} (second-order d.). Compared to TST, *MD calcns. (based on potential-energy surfaces) may yield a better understanding of surface reactions in the future. Cf. also *TPD. **E=F=G**. M. MUHLER

desulfurization the proc. through whi sulfur S and cpds. (*poisons in most cat. rea tions) are removed from inorganic or orgar substrates by various chemical or physic measures (e.g., purification of *syngas, *Cla procs., *sweetening).

When applied to S cpds. in fossil fuels, the proc. is commonly referred to as *hydrodest furization (HDS) and requires H$_2$ and he cats. under high press. In the *absence* of H$_2$ of fused-ring thiophenes in naphthas and cc can be achieved using microorganisms und either aerobic or anaerobic conds., the d. *hc solutions with alkyl bisbenzothiophen through π-acceptor molecules which form i soluble and thus easily separable charg transfer *complexes with thiophenic cpc The d. of thiols from gases and light oil fra tions can be obtained with the *UOP Merc proc. Metal-cat. d. in *biphasic systems usin *phase transfer cats. are known for thiobenz phenones and alkyl or aryl mercaptans. *Ox desulfurization comprises the treatment pyritic and organic S in coal with water and in the presence of alkaline salts. The cat. d. SO$_x$ emissions is still without solution thoug the hom. cat. *hydrogenation of SO$_2$ to and H$_2$O has been reported. Commerci procs. for the d. of flue gas are availab (*Claus proc.); also for waste gases (e.g. *Ha dor Topsøe's SNOX proc.).

Chemical and electrochemical *oxidation sulfide ions is used for the conv. of absorbe H$_2$S to S and polysulfides. The hom. d. of c ganosulfur cpds. (thiols, sulfides, dithioaceta sulfoxides, sulfones, thiophenes) transformin C-S bonds into C-H or C-C bonds is we known in organic synth. Common cat. or sto chiometric methods involve the use of *met carbonyls, low-valent *transition metal con plexes, metal hydrides, *Grignard reagents, other transition metal cpds. Certain Mo e zymes undergo d. of Mo=S groups with fc mation of Mo=O moieties to desulfo deriv tives with distinct *enzymatic activity. **F** d sulfurisation; **G** Entschwefelung. C. BIANCHI

Ref.: Topsoe; Cornils/Herrmann-1, p. 969; Weisserm Arpe, p. 65; J.H. Clark, *Chemistry of Waste Minimiz tion*, Blackie Academic, London 1995; *Synthesis* **1990**, 89; *J.Chem.Soc.Dalton Trans.* **1996**, 801; *Hydrocarb. Proc.* **1975**, *54*, 93; McKetta-1, Vol. 14, p. 302, Vol. 15, p. 145, Vol. 55, p. 355.

Detol process   → Lummus/Houdry procs.

Deutsche Texaco isopropyl alcohol process   for the manuf. of isopropyl alcohol by direct *hydration of propylene with cocurrent water over acid *ion exchangers in trickle-bed operation. Another proc. dehydrogenates isopropanol to acetone. **F** procédé de Deutsche Texaco pour l'isopropanol; **G** I-sopropanol-Verfahren der Deutschen Texaco.

B. CORNILS

Ref.: *Hydrocarb.Proc.* **1972**, (9), 113 and **1977**, *56*(11), 121.

Deutsche Texaco MIBK process   Manuf. of methyl isobutyl ketone (MIBK) from acetone via one-step synth. (*aldol condensation and *hydrogenation) on metal-doped *ion exchangers or *zeolites. **F** procédé de Deutsche Texaco pour MIBK; **G** MIBK-Verfahren der Deutschen Texaco.   B. CORNILS

Devarda's alloy   Cu-Al-Zn alloy for lab-scale *hydrogenations of substrates in alkaline solutions. **F** alliage de Devarda; **G** Devardasche Legierung.   B. CORNILS

Dewar-Chatt-Duncanson (DCD) model. In 1952 two research groups suggested a model for the chemical bonding between a *transition metal and ethylene which soon became very useful for the understanding of chemical bonds in transition metal *complexes. The DCD model considers the bonding in a complex to arise from donor-acceptor interactions between a metal and the *ligands. Two major comps. are suggested to dominate the binding interactions. One comp. is the donation from filled σ-orbitals of the ligand into empty d-type orbitals of the metal. The second comp. is the *back-donation from filled d(π)-orbitals of the metal into empty ligand π-orbitals. The DCD model usually considers the energetically highest occupied and lowest unoccupied

*MOs of the ligand, because the most important contributions to the orbital interaction energy should come from the *frontier molecular orbitals. In case of ethylene, these are the C=C occupied π- and unoccupied π^*-orbitals. The former orbital has σ-symmetry in the complex.

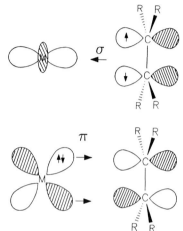

The advantage of the DCD model is that only a few (usually two) MOs of the ligand need to be considered for the binding interactions. The properties of many transition metal complexes can be understood and even predicted in terms of orbital interactions in the framework of the DCD model. The analysis of *ab-initio calcns. showed that the DCD model is a valid description of the binding interactions in transition metal complexes.   G. FRENKING
Ref.: Moulijn/van Leeuwen/van Santen.

dewaxing   removal of waxes (deparaffination) from petroleum- based feedstocks either physically by *adsorption or chemically/cat. by *wax cracking (w/o hydrogen; procs. of *BP, *Chevron Isodewaxing, *Mobil MDDW and LDW); cf. also *hydrodewaxing. **F** élimination de ciré; **G** Entwachsung.   B. CORNILS
Ref.: McKetta-1, Vol. 15, p. 346; Ertl/Knözinger/Weitkamp, Vol. 4, p. 2031.

DF   → dark field imaging

DFT   → density functional theory

DHD process a special version of *coal hydrogenation proc. for the *dehydrogenation treatment of coal-based gasoline using process-own middle distillates and Mo-Al$_2$O$_3$ cats. **F** procédé DHD; **G** DHD-Verfahren.

B. CORNILS

DHQ (dihydroquinine) a *chiral *ligand, used as a *chelating ligand for, e.g., *enantioselective *cis*-*hydroxylation of alkenes, cf. *cinchona alkaloids, *(DHQ)$_2$-PHAL.

B. CORNILS

(DHQ)$_2$- and (DHQD)$_2$-PHAL derivatives of bis-cinchona alkaloid *ligands with a 1,4-phthalazine *spacer (PHAL) (Figure).

They are prepared from dihydroquinine or dihydroquinidine with 1,4-dichlorophthalazine in toluene using K$_2$CO$_3$ as base. They have been employed in Os cat. asymmetric *dihydroxylation of alkenes. Cf. also *cinchona alkaloids, *DHQ. **E=F=G**. O. STELZER

Ref.: *J.Org.Chem.* **1992**, *57*, 2768 and **1995**, *60*, 3940; *Angew.Chem.Int.Ed.Engl.* **1996**, *35*, 448 and **1996**, *35*, 451.

Diaden process (two-tower process) an early version of the commercial *hydroformylation procs. consisting of a system of "two towers" for 1: a tower of *supported, fixed-bed cobalt *oxo cat. for reaction, and 2: an adsorption tower with fixed-bed support for *decobalting. Periodically the order of the oxo and adsorption reactors was changed. This operation at *Ruhrchemie (1942–1947) was not convenient. **F** procédé des Diades; **G** Diaden-Verfahren. B. CORNILS

Ref.: Falbe-3, p. 63.

Dian → bisphenol-A

diars phenylenebis(dimethylarsine), used as a *ligand.

diastereomers *stereoisomers which do not form image and mirror image. They originate from substitution of *chiral substrates. **F=E**; **G** Diastereomere. H. BRUNNER

diastereoselective synthesis In ds. one or several *diastereomers are formed preferentially out of a larger number of possible diastereomers. H. BRUNNER

diatomaceous earth, diatomeous earth, diatomite → kieselguhr

DIBP diisobutyl phthalate, a *plasticizer based on *iso*-butanol and phthalic anhydride. **F** phthalate d'isobutyl; **G** Diisobutylphthalat.

dicarboxylation the simultaneous introduction of two carboxylic groups into an organic molecule. D. of alkenes to 1,2-dicarboxylic acids can be achieved in moderate conds. (rt/0.1–0.3 MPa) in alcohol in the presence of metal *complexes, especially with Pd as mediating metal, e.g., C$_6$H$_5$-CH=CH$_2$ + CO + CH$_3$OH → C$_6$H$_5$-CH(CO$_2$CH$_3$)-CH$_2$CO$_2$CH$_3$. In principle, d. can be performed with a Pd cat. and the use of an re-oxidant for Pd(0) to Pd(II). Typically stoichiometric or (in the presence of air) even cat. amounts of Cu chloride have been used, but with low *TONs (cf. *redox reaction, *reoxidant, *Wacker proc.). Thus, stoichiometric amounts of Pd(II) salts are still used for the d. of aliphatic as well as aromatic alkenes. To avoid side reaction caused by water, the latter has to be removed from the reaction mixture. Depending on the oxidative reaction conds., besides dicarboxylated prods. unsaturated carboxylic acid derivatives can be formed. Thus, linear 1-alkenes yield β-alkoxy esters under neutral oxidative conds., but in sodium

butyrate buffer) 1,2-diesters are formed. **E=F**;
G Dicarboxylierung.   R.W. FISCHER
Ref.: Cornils/Herrmann-1, Vol. 1, p. 193; *J.Organo-met.Chem.* **1979**, *181*, C14.

Didier Werke DeNox process is a proc.
for the treatment of effluents from nitric acids
plants by selective cat. reduction by NH_3 over
K_2CrO_4/Fe_2O_3.   B. CORNILS
Ref.: *Appl.Catal.* **1994**, *115*, 179.

Dieckmann condensation an intramolecular cyclization of esters (cf. *Claisen condensation).

Diels-Alderases antibodies active in the
*Diels-Alder reaction.
Ref.: *Science* **1998**, *279*, 1929 and 1934.

Diels-Alder reaction In its most general
form a double or triple bond adds in 1,4-mode
to a conjugated diene ([4+2] *cycloaddition)
(Figure).

Although the reaction is easily performed and
of very broad scope, sometimes it is advantageous to add a cat. to increase the reaction
rate and *regio- as well as *stereoselectivity.
As cats. typically *Lewis acids such as $ZnCl_2$,
$FeCl_3$, BX_3 (X = Hal), AlX_3 (X = Hal, OR),
etc. are employed. DArs. are of special importance in *asymmetric synth. since up to four
stereogenic centers can be formed in a single
reaction step. The DAr. is usually reversible
(retro-DAr.) and has been used to protect
double bonds in organic synths. From a commercial standpoint DArs. of cyclopentadiene
with ethylene and terminal alkenes are important for the synth. of norbornene (*cycloolefin copolymers) and substituted norbornenes.
DArs. cat. by *antibodies are also known
(*pericyclic reactions). For commercial procs.
cf., e.g., *Bayer anthraquinone proc. **F** réaction de Diels-Alder; **G** Diels-Alder-Reaktion.
                              D. SCHICHL, M. BELLER

Ref.: J. Mulzer et al. (Eds.), *Organic Synthesis High-lights*, VCH, Weinheim (Germany) 1992; I. Fleming, *Frontier Orbitals and Organic Chemical Reactions*, Wiley, London 1976; M.J. Taschner, *Asymmetric Diels-Alder Reactions*, JAI Press, London 1989; *Chem.Rev.* **1992**, *92*, 1007; *Organic Reactions*, Vol. 4, p. 1,60, Vol. 32, p. 1, Vol. 52, p. 1, Vol. 53, p. 223.

dien diethynetriamine

diesel engine emissions (catalytic abatement) The emissions of diesel engines are
more complex than emissions from Otto engines (*automotive exhaust catalysts) and
consist of solids (respirable aerosols with high
polynuclear aromatic content, dry carbon
soot together with adsorbed sulfuric acid, liquids such as soluble organic fraction [un-
burned diesel fuel, SOF], and hydrocarbons,
CO, NO_x, and SO_2). The way to control the
emissions of diesel engines is via diesel oxidation catalysts (DOC), basing on *cordierite
monolithic *honeycombs (*ceramic or metal)
layered with *washcoats with Pd or Pt, supported on γ-*alumina (for details cf. *automotive exhaust cats., *nitrogen oxide removal.
F émissions   des   moteurs   Diesel;
G Dieselmotoremissionen.   B. CORNILS
Ref.: Farrauto/Bartholomew, p. 602; Ertl/Knözinger/
Weitkamp, Vol. 4, p. 1568.

Difasol process → IFP procs.

differential reactor → Berty reactor

differential   scanning   calorimetry
(DSC) is applied for solid- state structures
by specific reactions and for phase transformations (by linear variation of temp.). The
samples are mostly destroyed; polymorph
analysis requires solid-state samples. The data
obtained are integral and bulk-sensitive.
                                        R. SCHLÖGL

differential thermoanalysis (DTA) →
calorimetry

differential thermogravimetry (DTG)
→ calorimetry

diffuse reflectance (DR) a mode of spectroscopy (for *IR and *UV) with photons for "dark" solid samples. R. SCHLÖGL

diffuse reflectance Fourier transform infrared spectroscopy (DRIFTS) is applied for the vibrational analysis of cats. and adsorbates. The ex-situ method is non-destructive and applicable to a wide range of gas, solid, and liquid samples. The data are local (on an atomic scale), bulk-sensitive and have in-situ capabilities. R. SCHLÖGL

Ref.: *Chem.Ing.Tech.* **1999**, *71*, 861.

diffusion The transport of fluids through a reactor and on the microscopic scale to a reaction site can occur either by viscous flow or by diffusional procs., the latter being mainly the case if the reaction proceeds at an interface or in the *pore volumes of a solid. D. can occur in four independent modes of transport mechanism, i.e., 1: free-molecule or *Knudsen d.; 2: molecular or bulk d. and d. in porous solids; 3: *surface d.; and 4: forced flow (Poiseuille flow). For all d. procs. the flux of the molecules J_i is in accordance with Fick's first law $J_i = -D^e \cdot dc/dx$, where D^e is the effective d. coefficient, which depends on the kind of d. mechanism. The driving force is a conc. gradient along the distance x. For 1: for Knudsen d. the gas density is so low (or the cat. pores are so small) that collisions between molecules are less likely than collisions of the molecules with the walls of a tube (or cat. pore). According to kinetic gas theory, D^e_k ranges from $10 \cdot d_P$ to $100 \cdot d_P$ cm$^2 \cdot$s^{-1}. 2: For higher press., where the mean free path of the molecules is much smaller than the vessel diameter, gaseous collisions will become dominant and under these conds. bulk d. occurs. For binary gas mixtures the effective d. coefficient $D^e_{1,2}$ is of the order of 0.1 to 1 cm$^2 \cdot$s^{-1}. 3: The third mechanism of transport becomes possible when a gas is adsorbed on the inner surface of a porous solid. Transport then occurs by the movement of molecules over the *surface. Typical surface d. coefficients are between 10^{-3} and 10^{-5} cm$^2 \cdot$s^{-1} at rt. 4: If during a reaction in a porous solid the total number of moles changes due to stoichiometry, a press. gradient ΔP may occur between the interior and the exterior, resulting in typical d. coefficients between 0.1 and 10 cm$^2 \cdot$s^{-1}. For d. in porous media, such as the *pores of a solid cat., the effective d. coefficient has to be multiplied by ε/τ to account for the *void volume ε (typically between 0.2 and 0.7) and the tortuosity τ of the pores (typically between 3 and 7). In general, these two parameters are difficult to be calcd. a priori and must be determined experimentally. Reactions which are determined by d. are d. controlled (cf. *d. limitation). **E=F=G**. M. MUHLER

Ref.: J.O. Hirschfelder, C.F. Curtiss, R.B. Bird, *Molecular Theory of Gases and Liquids*, Wiley, New York, 2nd printing, 1964; Wijngaarden/Kronberg/Westerterp, p. 203 .

diffusion control → diffusion, → diffusion limitation

diffusion limitation (in enzymatic catalysis) An enzymatic reaction occurs in (at least) two steps: substrate (S) binding and substrate conv.

$$E + S \underset{k_{-1}}{\overset{k_1}{\rightleftharpoons}} E{:}S \xrightarrow{k_2} E + P$$

If the conv. step (k_2) is extremely rapid (much faster than k_{-1}), then the second-order rate constant is dominated by k_1, the rate of association. The rate of association is dependent on the frequency of collisions (Z); the proportion of collisions in the correct geometry for binding (p); and the proportion of molecules with enough energy to overcome any energy barrier for binding (ΔE_A) acc. to $k_1 = Zp \exp(-\Delta E_A/RT)$. If there is no energy barrier for binding, and if every molecule that collides with enzyme binds ($p = 1$), then $k_1 = Z$, the collision frequency, and the reaction is *diffusion-controlled. Z, in turn, is limited by the temp., viscosity of the solution (η), and the radii of enzyme and substrate (r_E and r_S), acc. to $Z = (2\,RT/3000\,\eta)(r_E + r_S)^2/r_E r_S$. At

25 °C, for a substrate and *enzyme that are almost equal in size, this rate is approx. 10^{10}
$s^{-1}M^{-1}$. Some enzymes approach this limit,
such as β-lactamase, which has a k_{cat}/K_m of
about $10^8 s^{-1}M^{-1}$ (cf. also *mass transfer limitations). **F** limitations de diffusion en catalyse
enzymatique; **G** Diffusionsbeschränkung in
der enzymatischen Katalyse.   C.-H. WONG

diglyme   diethylene glycol dimethyl ether,
a *solvent and *ligand

**2,3-dihydro-2,3-dihydroxybenzoate de-
hydrogenase** cat. the *NAD-dependent
*oxidation   of   2,3-dihydro-2,3-dihydroxy-
benzoate to 2,3-dihydroxybenzoate. It also
cat. the oxidation of 3-hydroxy- cyclohexane-
carboxylate analogs to the corresponding
3-keto species. The *enzyme recognizes
(1R,3R)-dihydro substrates and is potentially
useful for the synth. of a number of *optically
active cyclic ketones and hydroxy compounds.

Another keto acid dehydrogenase is NAD-dependent (R)-mandelate dehydrogenase from
Streptococcus faecalis. **E=F=G.**   C.-H. WONG
Ref.: *Biochemistry* **1990**, *29*, 6789; *Agr.Biol.Chem.*
1986, *50*, 2621.

dihydrofolate reductase (DHFR, EC
1.5.1.3) The *NADP(H)-requiring DHFR
cat. the in-vivo interconv. of folate and tetrahydrofolate. The *enzyme tolerates changes
in the substituents at both the 6- and 7-posi-
tions. The methyl group can be substituted at
the 7-position. The differences at the other
position are much more diverse. Thus, DHFR
provides a means to obtain *chiral derivatives
of dihydrofolate. Preparative synths. with the
*enzyme from *E. coli* proceed with high
*stereoselectivity. **E=F=G.**   C.-H. WONG
Ref.: *Tetrahedron* **1986**, *42*, 117.

dihydroneopterin aldolase (EC 4.1.2.25)
cat. the *aldol reaction of dihydroneopterin.
E=F=G.   C.-H. WONG
Ref.: *J.Biol.Chem.* **1970**, *245*, 3015.

dihydroxyacetone phosphate (DHAP)
Many *aldolases use DHAP as the donor substrate. DHAP can be most conveniently generated enzymatically in situ from FDP using FDP
aldolase and triosephosphate *isomerase (TPI).
FDP *aldolase cat. the retro-*aldol reaction of
FDP to give G3P and DHAP. G3P is rapidly
*isomerized to DHAP with TPI as the cat. One
drawback of this method is that the thermodynamics of the reaction may favor the formation
of FDP rather than the desired prod. Another
enzymatic method for DHAP generation employs *glycerol kinase to cat. the *phosphorylation of dihydroxyacetone (DHA) using *ATP
with in-situ regeneration of ATP. DHAP can
also be synth. non-enzymatically via phosphorylation of dihydroxyacetone. Reductive cleavage of the aromatic groups, or *hydrolysis of the
chlorophosphate (for the third method), yields
a stable dimer *precursor of DHAP that can
easily be converted into DHAP by acid cat. *hydrolysis. A multienzyme system which converts
sucrose to DHAP has also been utilized in aldolase reactions. **E=F=G.**   C.-H. WONG
Ref.: *Tetrahedron* **1991**, *47*, 2643.

dihydroxylation   1: A reaction to introduce
two OH groups into unsaturated organic cpds.
using the oxometals M = Os, Ru, Mn (Figure).

OsO$_4$ is the most reliable reagent used to
form *cis*-diols. After initial work by Markow-
ka and Criegee, Hentges and Sharpless developed an *asymmetric dihydroxylation (AD)
using a variety of chiral *ligands. Inclusion of
a co-oxidant during reaction reoxidizes the

Os(VI) species, thus allowing the use of cat. amounts of the oxometal and a *cat. cycle (*reoxidant). A variety of co-oxidants have been used, e.g., *TBHP, *tert*-amine-*N*-oxide, or K hexacyanoferrate(III). **F**=E; **G** Dihydroxylierung. M. BELLER, C. DÖBLER

Ref.: *Chem.Ber.* **1908**, *45*, 943; *Liebigs Ann.Chem.* **1936**, *522*, 75; *Chem.Rev.* **1994**, *94*, 2483; Cornils/Herrmann-1, p. 1009.

2-enzymatic: → *dioxygenase.

2,5-diketo-D-gluconic acid reductase → metabolic engineering

dimeric acids acids obtained by *dimerization of unsaturated fatty acids over *clay cats. **F** acides dimeriques, **G** Dimer(fett)-säuren. B. CORNILS

Ref.: Kirk/Othmer-1, Vol. 8, p. 223

dimerization formation of a molecule (the dimer) out of two identical chemical cpds. (the *monomer) mostly by a cat. reaction. It can be considered as an *oligomerization, of the second degree. The starting materials usually contain double (alkenes, ketones) or triple (alkynes) bonds, but *carbenes also can dimerize. Ethylene is dimerized to 1-butene by the *Alphabutol proc. of *IFP. The *Shell SHOP proc. – oligomerization of ethylene by Ni cats. – can be optimized to yield 1-butene. Propylene is dimerized to 2-methylpent-1-ene by [NiBr(η^3-allyl)(PCy$_3$)-EtAlCl$_2$. 1-Alkenes are selectively dimerized by *zirconocene/*methylalumoxane cats. Cycloocta-1,5-diene is generated by d. of 1,3-butadiene with NiCl$_2$-Al(alkyl)$_3$ (cf. *cyclodimerization). 4-Hydroxy-4-methylket-2-one (diacetone alcohol) is formed by base-cat. d. of acetone. Many dimers are formed *non*cat. by *Diels-Alder reactions, like dicyclopentadiene, 4-vinylcyclohexene (from 1,3-butadiene) and 3,4-dihydro-2*H*-pyran-2-carboxaldehyde (from acrolein). These prods. are often impurities in the cat. diene *polymerization and could act as an *inhibitor of the cat. For commercial examples of the d. cf. the entries *BASF (anthraquinone), *Goodyear (d. of propylene), *IFP Alphabutol and *IFP Dimersol, *Ube oxamide, etc. Alkanes may de dimerized by photochemical Hg sensitization (cf. *C-H activation). **F** dimérisation; **G** Dimerisierung. W. KAMINSKY, F. FREIDANCK

Ref.: *J.Am.Chem.Soc.* **1996**, *118*, 4715; *Inorg.Chim. Acta* **1998**, *270*, 20.

Dimersol process → IFP procs.

DINP diisononyl phthalate, a *plasticizer based on *iso*-nonyl alcohol and phthalic anhydride. **F** phthalate d'isononanol; **G** Diisononylphthalat.

DIOA diisooctyl adipate, a *plasticizer based on *iso*-octyl alcohol and adipic acid. **F** adipate d'isooctanol; **G** Diisooctyladipat.

DIOP, diop 1: One of the early *ligands for *asymmetric hom. catalysis, derived from tartaric acid. Polymer-attached and soluble as well as analogous cpds. containing different alkyl and aryl groups at P instead of phenyl are known (Figure).

(S,S)-DIOP

DIOP is synth. by reaction of 1,4-di-O-*p*-toluenesulfonyl-2,3-O-isopropylidene threitol with Na diphenylphosphide or addition of Na diphenylphosphide to (–)-1,2 : 3,4-diepoxybutane, followed by ketalization with (CH$_3$)$_3$C (OCH$_3$)$_2$. DIOP is extensively applied as a *ligand for Rh-cat. asymmetric *hydrogenation of enamides, *hydroformylation of prochiral alkenes, and *hydrosilylation. Pd(0) *complexes of (*R,R*)- or (*S,S*)-DIOP affect the addition of HCN to the *exo*-face of norbornene. **E**=F=**G**. O. STELZER

Ref.: *J.Am.Chem.Soc.* **1972**, *94*, 6429 and **1973**, *95*, 8295; *Tetrahedron: Asymmetry* **1990**, *1*, 693, 913; and **1991**, *2*, 173; *Tetrahedron Lett.* **1988**, *29*, 4755 and **1987**, *28*, 3675; *Organometallics* **1988**, *7*, 1761; *Pure Appl. Chem.* **1975**, *43*, 401; Cornils/Herrmann-1, Vol. 1, p. 207.

2: diisooctyl phthalate, a *plasticizer based on *iso*-octyl alcohol and phthalic anhydride. **F** phthalate d'isooctanol; **G** Diisooctylphthalat.

dioxygenases   *Enzymes such as the non-heme iron(III) enzymes *catechol 1,2-d., and *protocatechuate 3,4-d. cat. *dihydroxylation reactions using both atoms of O_2. Most of the ds. require *NADH or *NADPH as an indirect electron donor. The prosthetic group can be heme or non-heme Fe or Cu. See also *arene dioxygenase and *lipoxygenase. **E=F=G**.

C.-H. WONG

DIPA   diisopropanolamine

DIPAMP, dipamp (R,R)-bis[(o-methyloxyphenyl) phenylphosphino]ethane, a well-known *ligand for *asymmetric conv. (Figure).

DIPAMP may be applied for asymmetric Rh-mediated *hydrogenations of prochiral alkenes, e.g., for the synth. of L-Dopa (*Monsanto proc.) and for *hydrosilylations.

O. STELZER

Ref.: *J.Am.Chem.Soc.* **1977**, *99*, 5946; *Acc.Chem.Res.* **1983**, *16*, 106; *Organometallics* **1992**, *11*, 3588; Cornils/Herrmann-1, Vol. 1, p. 207; Collins/Sheldrake/Crosby, p. 37.

DIPHOS, diphos   1,2-bis(diphenylphosphino ethane (also DPPE), one of the first *bidentate *phosphine *ligands for stable *complexes with *transition metals. *Sulfonation yields the 3-sulfonatophenyl (Figure) analog.

The synth. proceeds via reaction of $(C_6H_5)_2$PNa with 1,2-dichloroethane in liquid

ammonia. Pd(II) and Pt(II) complexes may be used in the presence of $AgPF_6$ as cats. for the *addition of methanol to alkyne carboxylic esters to give vinyl ethers. Pd/DIPHOS complexes cat. the *cyclization of propargylic cpds. in *domino reactions. For *hydroformylations and *cross-coupling reactions Rh- or Ni-mediated DIPHOS complexes are used.

O. STELZER

Ref.: *J.Chem.Soc.(A)* **1962**, 1490; Patai-2; *Inorg.Chem.* **1994**, *33*, 164; *Chem.Lett.* **1994**, 1283; Cornils/Herrmann-1, Vol. 1, p. 276; *J.Am.Chem.Soc.* **1993**, *115*, 5865; *J.Chem.Soc.,Chem.Commun.* **1995**, 2031; *J.Org. Chem.* **1984**, *49*, 478.

DIPPP, dippp   a special electron-rich, bulky and *chelating *phosphine as a *ligand for Ni, Pd, Fe, Rh, and Ir *complexes. DIPPP [1,3-bis(diisopropylphosphino)propane] is prep. by lithiation of $(i$-$C_3H_7)_2$PH with n-BuLi and reaction of the Li phosphide with 1,3-dichloropropane. DIPPP complexes with Pd for alkene *arylations, *carbonylations, or *alternating copolymerizations.

O. STELZER

Ref.: *J.Organomet.Chem.* **1985**, *279*, 87; *Organometallics* **1993**, *12*, 4734; *J.Am.Chem.Soc.* **1989**, *111*, 8742; *Macromolecules* **1996**, *29*, 6377.

dipy   → bipy

directed evolution (in biocatalysis)   Evolution refers to the accumulation of favorable traits through successive rounds of *mutagenesis and selection. By selecting organisms with certain desired characteristics, evolution can be guided in a certain direction. The discovery of chemical mutagens has allowed us to increase the pace of evolution, since the mutation rate was no longer determined by the natural error rate of DNA polymerase. The invention of *DNA manipulation techniques has allowed specific genes or specific regions within genes to be targeted for mutagenesis, thus avoiding the lethality of rapid mutagenesis of the entire chromosome.

One goal of mutagenesis is to produce *enzymes with altered cat. *activities. The number of mutants one would like to sample, however, can get impossibly large very quickly.

With an excellent *selection technique, in which survival is linked to the desired enzymatic activity, this is feasible. However, with triple mutants, one becomes limited by bacterial transformation efficiencies (the best transformations yield $\sim 10^{10}$ colonies). In-vitro evolution of cat. nucleic acids (*ribozymes, *DNAzymes) does not suffer from problems of low transformation yields, since nucleic acids can be amplified via *PCR (and therefore do not require a living system for amplification). Even with such a system, however, the number of sequences can very rapidly get too large to handle. A randomized DNA 35mer will have $4^{35} \simeq 10^{21}$ members, and the average molecular weight will be $\sim 11\,200$. A *library containing a single representative of each sequence will require the synth. of 18 g of nucleic acid. These limitations put a cap on the mutagenesis rate. To improve this only those mutants with improved properties are selected for further mutagenesis. In this way, a large number of favorable mutations can be gradually accumulated (cf. *in vivo evaluation of biocats.). **F** évolution dirigée; **G** gerichtete Evolution.   P.S. SEARS

Ref.: *Angew.Chem.Int.Ed.Engl.* **1998**, *37*, 3105.

directed poisoning (*deactivation) the addition of special ingredients to cats. to decrease *activity and to increase *selectivity (e.g., with *Lindlar's cat., *Rosenmund reduction).
Other examples are the dosing of Pt cats. with Fe^{2+}/Zn^{2+} which enables the Pt cat. to selectively *hydrogenate unsaturated aldehydes to unsaturated alcohols, or the sulfidation of Pt cats. (cf. *sulfided cats.), which may be used to reduce aromatic nitro to amino groups without totally hydrogenating the aromatic system. Cf. also *sulfided cats., *dectivation, *inhibitor, *modifier. **G** gezielte Vergiftung.
                                         B. CORNILS

Ref.: Augustine, p. 221; *Proc.Chem.Soc.* **1964**, 398.

direct hydrogenation process → Henkel procs.

direct saturation sites → single turnover

dismutation → disproportionation

dispersion In het. cat., the *rate of a cat. reaction should be proportional to the *surface area of the active comp., provided the reaction is not limited by *mass transfer either within or outside the cat. particles. It is therefore usually desirable to have the active phase in the form of the smallest possible particles, where the quotient of *surface to total atoms is maximal, viz. in the highest possible degree of d. (sometimes called dispersity). Only very occasionally is it necessary to conduct a reaction under *diffusion-limited conds., when (as for example with *fat hardening) the yield of an intermediate prod. has to be maximized (see *selective hydrogenation).
The science of forming active metals into particles less than about 2 nm in size, e.g., corresponding with Pt to a d. of about 50 %, is now very well developed, but such small particles will quickly sinter, with heavy loss of surface area, unless they are separated from one another by being mounted on a *support. As particle size (*cat. particle size) is lowered, especially below 5 nm, the proportion of surface atoms having unusually low coordination number, by reason of their occupying *edge or *corner sites (cf. *single crystals), begins to rise, so that it is not only the area of the *surface that increases but also the type of atom composing it. As size decreases and surface atoms make up a progressively larger fraction of the whole, energy levels that formerly merged into bands begin to separate, metallic character is lost, melting temp. decreases, and a host of other related changes become apparent. The connection of morphology, geometry, and electronic character with cat. activity continues to be debated, although there are well-documented instances of *structure sensitivity where rate or selectivity changes markedly with d.
The conception of d. is less thoroughly developed in the case of oxides and sulfides, as here the active phase is often not formed as particularly small pieces, and with few excep-

tions the active phase is not (as with metals) stabilized by a chemical glue to the support (cf. *binder). The few exceptions include the Co-Mo/Al_2O_3 *HDS cat. and the V_2O_5-TiO_2 *selective oxidation cat.
The degree of d. of supported metals is estimated either by selective gas chemisorption or by *TEM: *EXAFS, *XPS and magnetic measurements may also lead to information on d. **E=F=G**.   G.C. BOND

dispersity → dispersion

disposal of catalysts → final disposal

disproportionation (dismutation) is a *redox reaction in which a cpd. in a medium oxidation state is transferred into two cpds. with a higher and a lower oxidation state respectively, e.g., $2 H_2O_2 \rightarrow H_2O + O_2$. The reverse reaction is called *com- or synproportionation. Usually, d. only takes place under the influence of cats., e.g., H_2O_2 is decomposed by het. cats. such as Ag, Au, Pt, or MnO_2, or by hom. cats. such as OH^-, Fe^{3+}, or Cu^{2+}. In nature H_2O_2 is removed from cells by the tetrameric *heme *enzyme *catalase. The reversible d. of CO into CO_2 and C (*Boudouard reaction) is an important economical proc. Other important ds. include the *Cannizzaro reaction in which aldehydes are converted to the corresponding carboxylic acids and alcohols using strong bases. Similar is the *Tishchenko reaction yielding the corresponding esters. Crossed Cannizzaro and Tishchenko reactions and intramolecular Cannizzaro reactions are also possible. Even *metathesis, converting alkenes to lower and higher molecular weight olefins, is called a d. For examples for commercial applications cf. entries *Atlantic Richfield (xylenes), *Henkel (*TPA), *Kvaerner; *Lummus Detol, *Mobil LTD and LTI; *Toray Tatoray, etc. **F** dismutation; **G** Disproportionierung.
                                    P. ROESKY

dissociation constants (enzymatic) Acc. to *transition state theory, *enzymes cat. re-

actions by stabilizing the transition state of a reactant with respect to the ground state. The enzyme should therefore bind the transition state more tightly than the substrate(s) (or prods.). This can be exemplified by a thermodynamic cycle. The ratio of the cat. rate to the uncat. rate will be equal to the ratio of the dissociation constants of enzyme-substrate and enzyme-transition state complexes. **E=F=G**.
                                    P.S. SEARS

$$E + S \underset{K^{\neq}}{\rightleftharpoons} E + S^{\neq} \longrightarrow E + P$$

$$K_S \updownarrow \qquad \qquad \updownarrow K_T$$

$$ES \underset{K^{\neq}_{cat}}{\rightleftharpoons} [ES]^{\neq} \longrightarrow E + P$$

Transition State Theory: $\dfrac{k_{cat}}{k} = \dfrac{K^{\neq}_{cat}}{K^{\neq}}$

Thermodynamic Cycle: $\dfrac{K^{\neq}_{cat}}{K^{\neq}} = \dfrac{K_S}{K_T}$

$$\therefore \quad \frac{k_{cat}}{k} = \frac{K_S}{K_T}$$

dissociative chemisorption the *adsorption proc. which involves the breaking of intramolecular bonds in the adsorbing molecule for the formation of *chemisorption bonds. Well-known examples are the dc. of homonuclear diatomic molecules such as O_2, H_2, N_2, etc. leading to the formation of the atomic adsorbates on the *surface. A gas like CO can be adsorbed as molecules or dissociatively, depending on the metal substrate. Dc. is usually explained by means of a 1D potential energy diagram where dissociation takes place at the intersection of the potential curves of the molecular adsorbate and the atomic adsorbate. Dc. is an essential mechanistic step in nearly all important cat. reactions. **F** chemisorption dissociative; **G** dissoziative Chemisorption.   R. IMBIHL

Distillate HDS process → IFP procs.

Distillate Dewaxing process → Mobil procs.

distribution of metals → metal distribution

Dixon plot The Dixon plot is a useful means of plotting *enzyme inhibition data to determine both the mode of inhibition and the inhibition constant. A plot of $1/\upsilon$ vs. [I] (inhibitor conc.) at several values of [S] (substrate conc.) will have a characteristic appearance depending on the mode of inhibition. A competitive inhibitor will appear as a collection of lines that intersect at a point to the left of the vertical axis, at $[I] = -K_i$ and $1/\upsilon = 1/V_{max}$. A replot of the slopes of those lines vs. $1/[S]$ will give a line that passes through the origin and has slope $K_m/V_{max}K_i$.

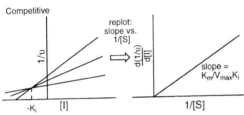

Noncompetitive inhibitors will give Dixon plots in which the lines intersect on the horizontal axis, at $[I] = -K_i$. A replot of slope vs. $1/[S]$ will give a line that intersects the horizontal axis at $-1/K_m$, again with slope $K_m/V_{max}K_i$. Uncompetitive inhibitors have very characteristic Dixon plots: all of the lines are parallel. **E=F**; **G** Dixon-Diagramm. P.S. SEARS

Ref.: I.H. Segel, *Enzyme Kinetics*, Wiley, New York 1975.

DMC dimethyl carbonate

DMD process → IFP procs.

DME dimethyl ether

DMF dimethylformamide

DMPO 5,5-dimethylpyrroline-*N*-oxide, a *radical scavenger

DMSO dimethyl sulfoxide

DMT dimethyl terephthalate

DMT from terephthalic acid Various proprietary procs. for DMT manuf. are based on the *Amoco/Mid-Century terephthalic acid (PTA) proc. *p*-Xylene is *oxidized to crude PTA with air in acetic acid using hom. Co-Mn cats. and a bromine *promoter (*Amoco procs.). PTA is then *esterified with methanol. In some instances acetaldehyde, paraldehyde, and methyl ethyl ketone have been used as promoters in place of bromine. **F** DMT de l'acide terephthalique; **G** DMT aus Terephthalsäure. B.L. SMITH

DMT synthesis DMT is a raw material for the worldwide prod. of poly(ethylene terephthalate), PET. Historically, DMT was made via nitric acid oxidation of *p*-xylene to *terephthalic acid (or PTA = purified terephthalic acid) followed by methanol *esterification. This proc. suffered from undesirable nitrated byprods. and was replaced by *autoxidation methods which utilized O or air and Co salts as cats. Most of the world's DMT is currently made via the *Witten proc. DMT competes commercially with PTA, which is also used for PET synth. DMT's main advantage over PTA is that it can be easily purified by distillation and/or crystallization whereas PTA manuf. requires exotic metallurgy. However, the *Witten proc. is more complicated and suffers from lower yields (87–89 % vs. 95 %), and slower reaction rates. DMT producers must also recycle methanol recovered from the PET proc. On a new plant design, PTA is considered more economical than DMT as a raw mat. for PET. In order to compensate for differences in yield and rate, a few DMT producers use proprietary variations of the *Amoco/Mid-Century PTA proc. followed by methanol esterification of the resultant PTA. Other Witten DMT producers sell valuable byprods., including methyl benzoate, *AA, and formic acid. Residues from the Witten proc. are also useful as raw materials in the polyurethane industry. **F** synthèse de DMT; **G** DMT-Synthese. B.L. SMITH

Ref.: McKetta-1, Vol. 16, p. 38.

DNA desoxyribonucleic acid

DNA and RNA oligomers Recombinant DNA technology depends on a number of enzymes that make it possible to introduce the desired genes into an organism. Synth. of recombinant DNA involves chemical and/or enzymatic synth. of primers, *polymerase chain reactions (PCRs) to extend these primers, restriction enzyme and DNA ligase reactions. These transformations cannot be accomplished by classical organic synth. techniques. Although the enzymes used in recombinant DNA technology are very expensive, large-scale work is unnecessary because DNA can be amplified in cells. For the prepn. of small oligonucleotides in large quantities for use in antisense and genetic engineering technology, enzymatic synth. may ultimately prove useful, especially for the synth. of small *RNA oligomers. The formation of phosphodiester links often requires *ATP (e.g., DNA and RNA ligases) and regeneration of ATP will probably be required. The enzyme *T$_4$ RNA ligase, for example, cat. the synth. of single-stranded oligonucleotides of different lengths from a 3'-terminal hydroxyl acceptor and a 5'-terminal phosphate donor through the formation of a 3'→ 5' phosphodiester bond, with *hydrolysis of ATP to AMP and pyrophosphate. The 5'-phosphate donor can be mono- or polynucleotides, and the 3'-terminal hydroxyl acceptor can be a trimer or oligomer. The enzyme also accepts single-stranded DNA as a substrate. This enzyme has been used to couple short, chemically synthesized oligonucleotides to longer oligomers, to introduce radioactive probes, and to modify RNA. Similarly, T$_4$ phage DNA *ligase is an ATP-requiring enzyme that cat. the formation of phophodiester bonds between nucleotide chains; in vivo it acts to repair and replicate DNA strands (Figure).

T$_4$ DNA ligase may be used in the synth. of certain oligoribonucleotides. Since the chemical methods for the synth. of RNA are not as well developed as those for DNA, RNA ligase could play an important role in RNA engineering. Two other examples of polynucleotide synth. deserving mention are the polynucleotide phosphorylase-cat. synth. of polynucleotides from nucleoside diphosphates and ribonuclease-cat. synth. of trinucleotide codons. These two enzymes have a broad substrate specificity and are thus particularly useful for the synth. of oligomers. **F** oligomères de DNA et RNA; **G** DNA- und RNA-Oligomere.   C.-H. WONG

Ref.: *Methods Enzymol.* **1979**, 65; *Biochemistry* **1980**, *19*, 6138; *Biochemistry* **1978**, *17*, 2069.

DNA and RNA with unnatural bases

Incorporation of modified or unnatural nucleosides into *DNA or *RNA provides new opportunities for the study of DNA and RNA recognition, for *cross-linking, and for expanding the genetic code. A series of modified and fluorescent nucleosides and their triphosphates have been prepared and exploited as biological probes or as alternative substrates for *enzymes that utilize nucleoside phosphates. Etheno-bridged *ATP, for example, is a good substrate for several ATP-dependent *kinases and *RNA *polymerase.

N^7-Methyl-2'-deoxyguanosine has been incorporated into DNA via DNA polymerase and

DNA ligase. A dGTP analog containing an aziridine group was incorporated into DNA by *E. coli* DNA polymerase and avian myeloma virus *reverse transcriptase in the place of dGTP, and the aziridine group was rapidly opened by the N_4 of the complementary cytosine group to give cross-linked DNA. The DNA *polymerase from *E. coli* also accepted O^4-methylthymidine to replace thymidine. The 2'-deoxy derivative of N^4-(6-aminohexyl)cytidine, 8-azido-dATP, and several UTP and 2'-deoxy UTP derivatives with a substituent at the 5-position are substrates for the polymerase reactions. Site-specific enzymatic incorporation of an unnatural base has also been reported.

Other base pairs have also been enzymatically incorporated into DNA or RNA. In the presence of non-standard tRNA containing the anticodon CUisodG and charged with iodotyrosine, there was a high degree (~90 %) of read-through of the isoCAG codon, compared to ~35 % for the iodotyrosine-tRNA-CUA- and the UAG non-sense codon system. **F** DNA/RNA avec des bases non naturel; **G** DNA/RNA mit unnatürlichen Basen.

C.-H. WONG

Ref.: *CRC Crit.Rev.Biochem.* **1984**, *15*, 125; *J.Am. Chem.Soc.* **1993**, *115*, 4461; *Nature (London)* **1992**, *356*, 537.

DNA ligase → DNA and RNA oligomers

DNA shuffling a combinatorial *mutagenesis method which mimics homologous recombination in vivo. In short, fragments of similar genes are combined together randomly, and the resulting chimeric genes are then selected for the desired property. For example, it is possible to begin with a *library of genes encoding mutant *enzymes that have been preselected for some function, and use DNA shuffling as a means of finding the best comb. of mutations. Alternatively, it is possible to start with homologous enzymes from several organisms which, through evolution, have diverged; thus the homologs provide, in essence, a library of "acceptable substitutions." After shuffling, one

might select for improved thermal stability, or increased cat. *activity.

Shuffling genes in vitro has been made feasible by the invention of the *polymerase chain reaction (PCR). The genes are mixed, then partially digested with deoxyribonuclease I. Melting the (double-stranded) pieces and then cooling allows homologous regions of different genes to anneal together; extension with a DNA polymerase will fill in the gaps, and carrying out the PCR with primers corresponding to the front and back of the full gene will amplify the "mixed and matched" mutants. Using error-prone PCR can introduce additional diversity into the mutant library.

1. Make single mutants
2. Select/Screen for best resulting proteins

3. Fragment with DNase

4. Melt/reanneal

5. PCR

This type of method has been used to evolve, e.g., a β-galactosidase into a β-fucosidase. Mutants with improved activity toward a colorimetric substrate (X-Fuc) were shuffled, and the resulting mutants selected for improved cat. One of the resulting mutants had a 1000-fold improved specificity for fucose *vs.* galactose. (cf. *in vitro evolution of biocats.). **E=F=G**.

P.S. SEARS

Ref.: *Proc.Natl.Acad.Sci.USA* **1997**, *94*, 4504; *Nature (London)* **1998**, *391*, 288.

DNAzymes are cat. deoxyribonucleic acids. The first DNAzyme isolated had ribonuclease activity, and cat. the Pb^{2+}-dependent cleavage of RNA. Since then, DNAzymes have been selected with other cat. activities, such as DNA *hydrolysis and *porphyrin *metallation. Like the RNA *enzymes (*ribozymes), most but not all DNA enzymes isolated to date require metal ions for cat. and/or folding. **E=F**; **G** DNAzyme.   P.S. SEARS

Ref.: *Chem.Biol.* **1994**, *1*, 223 and **1997**, *4*, 579.

DNP dinonyl phthalate, a *plasticizer based on *n*-nonyl alcohol and phthalic anhydride. **F** phthalate de n-nonanol; **G** Dinonylphthalat.

DOA dioctyl adipate, a *plasticizer based on *n*-octyl alcohol and adipic acid. **F** adipate d'octanol; **G** Dioctyladipat.

DOC diesel oxidation catalyst, → diesel engine emissions

Döbereiner Johann Wolfgang (1780–1849), professor of chemistry and pharmacology in Jena (Germany). Active in catalysis (prep. of formic acid, acetaldehyde, acetic acid, etc., *Döbereiner's lighter).   B. CORNILS

Ref.: *Nachr.Chem.Tech.Lab.* **1999**, *47*, 326.

Döbereiner's lighter the first lighter which ignites hydrogen (prepared from Zn and sulfuric acid) on finely dispersed Pt. **F** briquet de Döbereiner; **G** Döbereinersches Feuerzeug.

Ref.: Neufeldt.   B. CORNILS

Doebner reaction → Knoevenagel condensation

Dötz reaction α-Aryl substituted *Fischer-type *carbene *complexes react with alkynes under mild conds. to give $Cr(CO)_3$-coordinated annulated aromatic cpds. This Dr. is formally a [3+2+1] *cycloaddition. The benzannulation is compatible with a wide range of substituents and allows the synth. of densely functionalized arenes. The metal of choice is Cr; carbene complexes containing Mo or W react less cleanly.

The mechanism of the Dr. involves a reversible *decarbonylation of the metal center to give an alkyne-carbene-carbonyl complex. Insertion of the alkyne into the metal-carbene bond occurs *regioselectively.

The Dr. has been used for the synth. of indenes, indenopyrroles, and carbazole derivatives. **F** réaction de Dötz; **G** Dötz-Reaktion.

                M. BELLER

Ref.: Beller/Bolm, Vol. 1, p. 335; *Angew.Chem.Int.Ed.Engl.* **1975**, *14*, 644.

domain swapping Many proteins are built in a modular fashion: the protein is arranged as a set of discrete folded units or domains (which may be formed from continuous stretches of the polypeptide chain or from separate polypeptide segments that come into close proximity during folding), connected by polypeptide tethers. In a number of multifunctional *enzymes, the different enzymatic activities are cat. by different domains. Many of the *polyketide (macrolactone) synthases are arranged in this fashion; the *acyltransferase, reductase, dehydratase, and thioesterase functions are carried out by different "modules."

For enzymes in which the domains are contiguous polypeptide segments, the domains can often be exchanged with domains from other enzymes to produce a new enzyme with novel activity. Such a procedure is called ds. The modular polyketide synthases, for example, have been the subjects of extensive ds. experiments. The resulting macrolactones produced have modified stereochemistry, functional groups, and ring sizes. Ds. is also frequently used to investigate the purpose, the boundaries, and the specificity determining regions of different domains, particularly in cases where the 3-D structure is unknown. Domain swapping also refers to the ability of some multidomain menomeric proteins (eg., diptheria toxin) to dimerize or mulitmerize by exchanging equivalent domains. **E=F=G**.

                P.S. SEARS

Ref.: *Proc.Natl.Acad.Sci.USA* **1996**, *93*, 6841; *Chem.-Biol.* **1997**, *4*, 667.

domino reactions (tandem, cascade, concerted reactions) sequences of reactions in which several bonds are formed consecutively in one reaction. Among the classification of dr. are *transition metal-cat. examples such as *hydroformylations/ Wittig reactions (*concerted reactions). B. CORNILS

Ref.: *Nachr.Chem.Tech.Lab.* **1997**, *45*, 1181; L.F. Tietze, U. Beifuß, *Domino Reactions in Organic Synthesis*, Wiley-VCH, Weinheim 1998; *J. Org.Chem.* **1998**, *63*, 428; *Angew.Chem.* **1999**, *111*, 1022; *Chem. Rev.* **1996**, *96*, 115.

DOP (DEHP) dioctyl phthalate or di(2-ethylhexyl) phthalate, the most produced *plasticizer, based on cat. *esterification of 2-ethylhexanol and phthalic anhydride. **F** phthalate d'octyle; **G** Dioctylphthalat. B. CORNILS

dopamine monooxygenase (dopamine β-hydroxylase, dopamine β-monooxygenase, EC 1.14.17.1) cat. the *hydroxylation of dopamine to norepinephrine with the concomitant *oxidation of ascorbate to dehydroascorbate. It is a Cu-dependent *enzyme, and the ascorbate substrate binds first and reduces the Cu^{2+} to Cu^+. Dopamine binds next and, in the presence of oxygen, is hydroxylated. **E=F=G**. P.S. SEARS

Dopamine — Norepinephrine

dopant → *doping, *promoter, *co-catalyst, *additive

dopaquinone, 6-hydroxy → copper amine oxidases

doping the addition of a small abundance of a foreign material to a cat. in order to control or improve its function or *stability. Doping levels range from ppm to a few percent in a cat. matrix. The dopant is usually added during cat. manuf. The addition of dopants with the reaction feed (e.g., the addition of Cl-con-taining cpds. during *isomerization over *alumina; cf. *chlorinated alumina) is usually termed modification and the material used is referred to as a *modifier. Dopants which increase the cat. *activity are termed *promoters. **E=F=G**. R. SCHLÖGL

DOS → density of states

double absorption process → *Bayer double contact process

double carbonylation comprises (as a special case of *carbonylation) the introduction of *two* CO molecules in organic cpds. in a one-pot reaction. The dc. is cat. by late *transition metal *complexes, especially their carbonyl complexes (e.g., Ni, Co, Fe, Ru, Rh, Pd, etc.). Organic halides (aryl-X, Bn-X, alkyl-X; X = hal) yield α-keto carboxylic acid derivatives; dc. of alkenes under oxidative conds. (in presence of $CuCl_2$) yields succinic acid derivatives (Figure).

$$R\text{-}X + 2\,CO \longrightarrow R\text{-}CO\text{-}CO\text{-}X$$

$$H_2C = CH_2 + 2\,CO + 2\,ROH \longrightarrow ROOCH_2CH_2COOR$$

While Pd cats. have been mainly used for dc. of aryl bromides or iodides, Co cats. such as $HCo\,(CO)_4$ were applied for reactions of alkyl and benzyl halides. In the case of Pd the mechanism proceeds via *oxidative addition of a Pd(0) complex in the C-X bond. Subsequent insertion of CO into the Pd-C bond and attack of a nucleophile (e.g., RO^-) on coordinated CO gives a Pd-acyl alkoxycarbonyl or -amidocarbonyl complex. *Reductive elimination yields the prod. and an HPdX complex which regenerates the active cat. in the presence of a base. **F** carbonylation double; **G** Zweifach(Doppel)carbonylierung. M. BELLER

Ref.: *J.Am.Chem.Soc.* **1985**, *107*, 3235 and **1976**, *98*, 1806; Cornils/Herrmann-1.

double Hock the application of a twofold *Hock reaction, e.g., manufg. hydroquinone

from *p*-diisopropylbenzene (*Mitsui proc.).
F double réaction d'Hock; **G** Doppel-Hock-
Reaktion.　　　　　　　　　　B. CORNILS

Dow benzene chlorination manuf. of
chlorobenzene by FeCl₃-cat. *chlorination of
benzene in the liquid phase (cf. *Bayer benzene chlorination). **F** Dow chloration de benzène; **G** Dow Benzolchlorierung. B. CORNILS

Dow butadiene process manuf. of butadiene by gas-phase *dehydrogenation of butene on Ca-Ni phosphate/Cr₂O₃ cats. in the
presence of steam as diluent. **F** procédé Dow
pour le butadiène, **G** Dow Butadien-Verfah-
ren.　　　　　　　　　　　　B. CORNILS

Dow cumene process manuf. of cumene
from propylene and benzene over modified
(highly dealuminated) *mordenite *zeolites
at 130 °C. **F** procédé de Dow pour le cumène;
G Dow Cumol-Verfahren. B. CORNILS

Dow ethyleneimine process manuf. of
ethyleneimine by conv. of 1,2-dichloroethane
with ammonia and CaO. **F** procédé Dow pour
l'éthylèneimine; **G** Dow Ethylenimin-Verfah-
ren.　　　　　　　　　　　　B. CORNILS

Dowex *ion exchange resins from Dow,
mostly with a polystyrene backbone cross-
linked through *copolymerization with divinylbenzene. Beads are typically 0.3–1.2 mm.
The resins are functionalized with amines for
anionic or sulfonic acid for cationic exchange,
with a wide range of basicity/acidity available.
Also used as cats., e.g., in acidic form for *esterification; or as cat. *supports, e.g., to *immobilize metal *clusters for *hydrogenation,
*hydroformylation, *isomerization, *hydra-
tion of alkynes, etc. **E=F=G.** F. JENTOFT
Ref.: *J.Mol.Catal.A:* **1999**, *144*, 159.

Dow methane chlorination manuf. of
methane *chlorination prods. by UV-induced
chlorination and special workup of the reaction

prods. **F** procédé Dow pour la chloration de
méthane, **G** Dow Methanchlorierung.
　　　　　　　　　　　　　　B. CORNILS

Dow phenol process manuf. of phenol by
two-step conv. of toluene. 1: Cat. liquid-phase
*oxidation of toluene with Co salts to benzoic
acid; 2: conv. of benzoic acid to phenol in molten
solvents by cat. *decarbonylation in the presence of promoted Cu salts (*molten salt media).
F procédé Dow pour le phénole, **G** Dow Phenol-Verfahren.　　　　　　　B. CORNILS

Dow styrene process manuf. of styrene by
cat. *dehydrogenation of ethylbenzene in fixed-
bed shaft furnaces over propietary cats. The endothermic heat of reaction is transferred by
superheated steam. **F** procédé Dow pour le styrène, **G** Dow Styrol-Verfahren. B. CORNILS

DPG process　→ Lummus procs.

DPPB, dppb (1,4-bis(diphenylphosphinobu-
tane), prep. by *alkylation of alkali metal phos-
phides with X-(CH₂)₄-X (X = Cl, Br) or by
*electrochemical coupling. DPPB may be ap-
plied as *ligands or co-ligands in *hydroformylations (Rh, Pt), *hydrocarboxylations (Pd),
*reduction of CO₂ to formic acid (Rh), *Suzuki
cross-coupling (Pd), and *silylations.
　　　　　　　　　　　　　　O. STELZER
Ref.: *Can.J.Chem.* **1979**, *57*, 180; Patai-2; *J.Organo-
met.Chem.* **1993**, *457*, 273 and **1994**, *475*, 257; *J.Mol.
Catal.* **1992**, *77*, 7; *J.Chem. Soc.,Chem.Commun.* **1995**,
1479; *Tetrahedron Lett.* **1991**, *32*, 2273; *Tetrahedron*
1994, *50*, 335.

DPPE → DIPHOS

DPPF, dppf prep. from *ferrocene by di-
lithiation and reaction with chlorodiphenylphosphine (Figure).

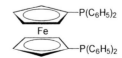

DPPF forms *complexes with Ni(II) and
Pd(II) for *Grignard cross-couplings. Differ-

ent Pd complexes have been engaged in *hydrocarboxylations, *carbonylations, and interior intramolecular *Suzuki cross-couplings.

O. STELZER

Ref.: *J.Organomet.Chem.* **1971**, *27*, 241; *J.Am.Chem.Soc.* **1984**, *106*, 158 and **1989**, *111*, 314; Beller/Bolm, Vol.1, p.52; *J.Mol. Catal.* **1994**, *93*, 1; *Tetrahedron Lett.* **1994**, *35*, 5697.

DPPO, dppo (diphenylphosphinophenyloxazoline, Figure) is a *bidentate *chelating *ligand, preferably used for *enantioselective *allylic substitution or *cyclopropanation. **E=F=G**. W.R. THIEL

DR → diffuse reflectance

driers *Transition metal salts (specially metallic soaps, such as naphthenates, resinates, 2-ethylhexanoates of Co, Pb, Mn) are *additives for drying oils in paints, effective as accelerators for cat. *autoxidation and thus drying, cross-linking, etc. of the oils. **F** siccatif; **G** Sikkativ, Trockenstoff. B. CORNILS

DRIFTS → diffuse reflectance Fourier transform infrared spectroscopy

drop-coagulation from sols is an alternative method for the manuf. of *pellets.

H. KNÖZINGER

drop-in ligands In general hom. *Lewis acidic *transition metal centers form with basic dils. (mainly donor bases like amines, amino alcohols, sulfides, *phosphines, etc.) coordinative *complexes which show different cat. behavior from the noncoordinated, *ligand-free cats. The complexes may be prepd. in situ easily by "dropping" the ligand (solution) into the dissolved cat. or cat. *precursor (cf. *preformation). Often a considerable excess of the dils. has to be used. Due to the presence of the coor-

dinating dils., the cat. *activity and *selectivity can be changed. Prominent examples are *enantioselective *epoxidation of allylic alcohols with a Ti/tartaric acid cat. system (*Sharpless reagent) and *aminohydroxylation with Os/*DHQ cats. Dils. are widely used in the area of *asymmetric synth. Applying dils., sometimes a *ligand acceleration effect can be observed. **E=F=G**. R.W. FISCHER

Ref.: *Angew.Chem.* **1996**, *108*, 1406; *J.Organomet. Chem.* **1995**, *500*, 149; *Acc.Chem.Res.* **1997**, *30*, 169.

dry impregnation → incipient wetness method, *impregnation

drying in catalyst preparation Drying removes moisture from a solid by evaporation to produce a (relatively) dry substance. The evaporation requires energy which can be provided directly or indirectly. With direct (adiabatic) d. the energy for the evaporation of the liquid is introduced by the sensible heat of a flowing transfer medium (carrier gas) which also removes the vapors generated. Flash, spray, and fluid beds are typical dryers for suspended cat. particles, whereas tray, chamber, belt, tunnel, and rotary dryers can be classified as bed dryers. With indirect (nonadiabatic) d. the heat transfer medium is separated from the prod. to be dried by a (metallic) wall. Heat transfer fluids may either be of the condensing type (e.g., steam) or liquids (hot water, oil). Vacuum shelf, freeze, plate, tube, paddle, or rotary dryers are typical equipment.

The d. proc. may be subdivided into several phases. During the heat-up period surface moisture is removed. Mechanically bound moisture from the interstices of the solid moves to the surface by diffusion or press. gradients during the period at constant temp. Chemically bound moisture emerges from the structural reorientation of hydrates leading to a change of the morphology of the solid during prolonged drying at high temp. The last step can be critical with regard to the cat. properties of the material. **F** séchage; **G** Trocknung. C.D. FROHNING

Ref.: Ullmann, Vol.B2, p.4–1 and Vol.A5, p.351; Kirk/Othmer-2, p.373.

DSC → differential scanning calorimetry

DSM/Stamicarbon HPO cyclohexanone oxime process *oximation of cyclohexanone in phosphoric acid/hydroxylamine buffers ("hydroxyl-phosphate-oxime"). Hydroxylamine is generated by hydrogenative conv. (*hydrogenation with Pt cats.) of ammonium nitrate in presence of H_3PO_4: NH_4NO_3 + $2 H_3PO_4$ + $3 H_2$ → $(NH_3OH)H_2PO_4$ + $NH_4H_2PO_4$ + $2 H_2O$. After separation of the cat. the solution reacts directly with cyclohexanone. **F** procédé DSM/Stamicarbon HPO pour l'oxime de cyclohexanone; **G** DSM/Stamicarbon HPO-Cyclohexanon-Verfahren.

B. CORNILS

Ref.: *Hydrocarb.Proc.* **1972**, (9), 92; Büchner/Schliebs/Winter/Büchel, p. 54.

DSM pyrrolidone process three-step manuf. of pyrrolidone via 1: the *hydrocyanation of acrylnitrile to succinic acid dinitrile, 2: its partial *hydrogenation on Ni cat. to 3-aminosuccinic acid nitrile, and 3: its *cyclization under press. to pyrrolidone. **F** procédé DSM pour le pyrrolidone; **G** DSM Pyrrolidon-Verfahren.

B. CORNILS

DSM/Toyo aspartame process manuf. of the sweetener *aspartame (Asp-Phe-OMe). The *N*-protected *precursor is made *enzymatically from the protected amino acid with the *protease thermolysin (2 000 tpy). **F** procédé DSM/Toyo pour l'aspartame; **G** DSM/Toyo Aspartame-Verfahren.

B. CORNILS

Ref.: R.A. Sheldon, *Chirotechnology*, Dekker, New York 1993.

DTA → differential thermal analysis

DTG → differential thermogravimetry

dual-bed converter forerunner of the *TWC, cf. *automotive exhaust cat.

dual-function catalyst → bifunctional catalyst

Duotreat process → Atlantic Richfield procs.

DuPHOS, duphos a class of chiral C_2-symmetric *bidentate *ligands (Figure).

Their prep. is achieved by *metallation of 1,2-bis (phosphinobenzene) with *n*-BuLi and subsequent treatment with $(2R,5R)$-2,5-hexanediol cyclic sulfate. Rh *complexes with different DUPHOS derivatives are used for enantioselective *hydrogenations (inter alia in scCO₂, cf. *supercritical phase cat.) or for *reductive aminations.

O. STELZER

Ref.: *J.Am.Chem.Soc.* **1993**, *115*, 10125 and **1995**, *117*, 8277; Beller/Bolm, Vol. 2, p. 20; *Pure Appl.Chem.* **1996**, *68*, 37; *Tetrahedron* **1994**, *50*, 4399.

DuPont acrylonitrile process former proc. for the manuf. of *acrylonitrile by *nitrosation of propylene over cats., based on Ag_2O/SiO_2, Tl, or Pb cpds. **F** procédé DuPont pour l'acrylonitrile; **G** DuPont Acrylnitril-Verfahren.

B. CORNILS

DuPont aniline process manuf. of aniline by cat. *ammondehydrogenation of benzene and NH_3 over Ni-NiO at 350 °C/30 MPa. The benzene conv. is 15 % per pass; the byprod. hydrogen reduces the cat. which has to be regenerated continuously. **F** procédé DuPont pour l'aniline, **G** DuPont Anilin-Verfahren.

B. CORNILS

DuPont glycol process manuf. of glycolic acid and ethylene glycol by 1: *hydroxycarbonylation of formaldehyde under acid cat. to glycolic acid and 2: the *esterification of the glycolic acid with methanol and the *hydrogenation of the methyl ester to ethylene glycol. The proc. is now obsolete. Newer developments start from *syngas with Rh cats. under extemely high press. (up to 340 MPa). **F** procédé DuPont pour l'éthylène glycol; **G** DuPont Ethylenglykolverfahren.

Ref.: Parshall/Ittel. B. CORNILS

DuPont hydrocyanation process replaces the former *adiponitrile proc. using butadiene, HCl, and NaCN as starting materials. It covers about 75 % of the world demand for *adiponitrile. The overall proc. is described as the *addition of two equivalents of HCN to butadiene in the liquid phase.
The cat. consists of tetrakis(triarylphosphite)-Ni(0), an excess of phosphite *ligand and a *Lewis acid *promoter. The h. of butadiene proceeds via a sequence of distinct reactions. Initially pent-3-ene nitrile and a minor amount of the branched regiomer 2-methyl-but-3-ene nitrile are formed by monoaddition of HCN to a mixture of 3- and 4-pentene nitriles (4-PN). In the final step the second equivalent of HCN is added to 4-PN to give adiponitrile besides minor amounts of 2-methylglutaronitrile. **F** procédé d'hydrocya-

nation de DuPont; **G** DuPont Hydrocyanie-rungsverfahren. S. KRILL
Ref.: Cornils/Herrmann-1, p. 465; Cornils/Herrmann-2, p. 39; Ullmann, Vol. A1.

DuPont Nixan process *oximation, in the two-step manuf. of cyclohexanone oxime, by 1: liquid-phase *nitration of cyclohexane with nitric acid and 2: cat. *hydrogenation of nitro-cyclohexane. The term Nixan means NItro cycloheXANone proc.). **F** procédé Nixan de DuPont; **G** DuPont Nixanverfahren.
B. CORNILS

Dutch Staatsmijnen Comp. → DSM

DVB divinylbenzene

dynamics of surface reactions deals with the reactivity of surface reactivity from the point of view of macroscopic kinetics and microscopic phenomena. The topic is highly speculative. B. CORNILS
Ref.: Ertl/Knözinger/Weitkamp, Vol. 3, p. 972,991.

Dynamit Nobel DMT process → Hüls DMT process

E

Eadie-Hofstee plot a graphical tool used for determining *enzyme kinetic parameters. The *Michaelis-Menten equ. for enzyme kinetics, $\upsilon = V_{max}[S]/([S] + K_m)$ where $[S]$ is the substrate conc., υ is the reaction velocity, V_{max} is the maximal reaction velocity at saturation and K_m is the substrate conc. at which $\upsilon = V_{max}/2$ can be rearranged to give $\upsilon = -K_m\upsilon/[S] + V_{max}$.

A plot of υ vs. $\upsilon/[S]$ will give a line with a slope of $-K_m$ and a y-intercept of V_{max}.

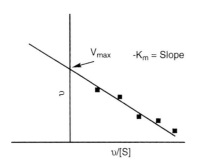

Unlike the *Lineweaver-Burk plot, this plot does not overweight on points taken at low substrate concs. (cf. *competitive inhibition). **E=F; G** Eddie-Hofstee-Diagramm.   P.S. SEARS

Eastman acetic anhydride process a liquid-phase *carbonylation of CH_3COOCH_3, hom. cat. by Rh cpds., promoted by CH_3I, *co-promoted by LiI/CH_3COOLi at 190 °C/5 MPa CO/H_2 (yield based on MeOH is approx. 96 %). The highly integrated proc. starts with coal, producing firstly *syngas (2:1 CO/ H_2) via *coal gasification. Recycle acetic acid (*AA) (from cellulose acetate prod.) is esterified with MeOH to give CH_3COOCH_3, which carbonylates with syngas to *acetic anhydride (Ac_2O, AAA). The overall carbonylation rate is strongly dependent on the Li cation conc.

At high LiI concs., the rate-determining step is *oxidative addition of CH_3I to $[Rh(CO)_2I_2]^-$ to form $[Rh(CH_3)(CO)_2I_3]^-$ followed by facile CO *insertion (cf. the mechanism of *Monsanto's AA procs.). At low LiI concs. (<10:1 mol ratio LiI/Rh) the rate determing step becomes the iodolysis of CH_3COOCH_3 with LiI to form CH_3I and CH_3COOLi. Thus, the reaction mechanism is very similar to Monsanto's proc. except for the final step, which involves reaction of CH_3COI with CH_3COOLi to form Ac_2O instead of reaction with H_2O to AA. MeOH can be co-fed with CH_3COOCH_3 to produce AA also. Some H_2 in CO feed is essential as a reducing agent to minimize the formation of inactive Rh(III). Major byprods. are ethylidene diacetate from *hydrocarbonylation of CH_3COOCH_3 and "tars" derived from thermal decomposition of Ac_2O. The commercial carbonylation proc. involves formation of Ac_2O in an essentially anhydrous AA/AAA reaction medium containing a Rh salt, LiI, CH_3I, and CH_3COOCH_3. The crude prod. and light ends are separated from the reactor cat. via flash distillation and returned to the reactor with light ends recycle streams from the purification section. Crude prod. is purified in a series of distillation towers and finished prod. is obtained as a sidedraw from the final column. **F** procédé Eastman pour l'acide acétique; **G** Eastman Essigsäure-Verfahren.   P. TORRENCE

Ref.: V.H. Agreda, J.R. Zoeller (Eds.), *Acetic Acid and Its Derivatives*, Marcel Dekker, New York 1993, p. 145; *J.Chem. Educ.* **1986**, *63*, 206; Cornils/Herrmann-1, Vol. 1, p. 116.

Eastman BHMC process two-step manuf. of *BHMC (1,4-dimethylol cyclohexane) by 1: *hydrogenation of *DMT to *trans*-1,4-cyclohexanedicarboxylic acid dimethyl ester over Pd; and 2: hydrogenation of the dimethyl ester to the diol over a Cu-Cr *Adkins cat. **F** procédé Eastman pour BHMC; **G** Eastman BHMC-Verfahren. B. CORNILS

Eastman hydroquinone process a (now obsolete) proc. based on the *oxidation of aniline to quinone by MnO_2 in the presence of sulfuric acid. The quinone was reduced to hydroquinone by Fe in the presence of water. **F** procédé d'Eastman pour l'hydroquinone; **G** Eastman Hydrochinonverfahren. B. CORNILS
Ref.: US 1.880.534 (1932), US 2.716.138 (1955).

Eastman Kodak methyl methacrylate process two-step proc. proposal starting from isobutyric acid by 1: *oxydehydrogenation with the hom. cat. HBr at 160–175 °C to methacrylic acid; and 2: *esterification of the acid to methyl methacrylate. **F** procédé d'Eastman-Kodak pour le méthyle méthacrylate; **G** Eastman Kodak Methylmethacrylatverfahren. B. CORNILS
Ref.: Weissermel-Arpe, p. 284.

Eastman terephthalic acid process for the manuf. of purified *terephthalic acid (PTA) by co-*oxidation of p-xylene and acetaldehyde, cat. by Co salts in acetic acid solution. **F** procédé Eastman pour l'acide terephthalique; **G** Eastman Terephthalsäure-Verfahren. B. CORNILS

EB → ethylbenzene

Ebemax catalyst → Südchemie catalysts

ebullated-bed reactor → slurry-bed reactor

EC abbr. for Enzyme Commission of the International Union of Biochemistry and Molecular Biology. The EC number is a series of four numbers. The first number gives the general class: (1) *oxidoreductase (reactions CH → COH or CH-CH → -C=C-); (2) *transferase (cat. the conv. of acyl- or phosphoryl groups to other molecules); (3) *hydrolase

(*hydrolysis of esters or amides); (4) *lyase (cat. additions to π-bonds); (5) *isomerase (cat. isomerizations); (6) *ligase (cat. the formation of C-O, C-S, C-N, etc. bonds). The second number indicates the sub-class: for *oxidoreductases, this indicates the type of electron donor (1 = alcohol, 2 = ketone or aldehyde, etc.); transferases, the group transferred (1 = one carbon unit; 2 = aldehyde or ketone unit, etc.); hydrolases, the type of bond hydrolyzed (1 = ester, 2 = glycosidic, etc.); lyases, the link broken or formed (1 = C-C, 2 = C-O, etc.); isomerases, the type of isomerization; ligases, the type of bond formed (1 = C-O; 2 = C-S; etc.). The third number gives a more precise description: for oxidoreductases, it determines the electron acceptor (1 = *NAD(P), 2 = *cytochrome, etc.), while for the other classes it gives more specific information about the groups transferred or the bonds made/broken. The last number is the serial number. EC 1.1.1.1, for example, is *alcohol dehydrogenase. **E=F=G**. P.S. SEARS
Ref.: E.C. Webb, *Enzyme Nomenclature*, Academic Press, San Diego 1992; D. Schomburg, D. Stephan, *Enzyme Handbook*, Springer, Berlin 1996.

ED electron diffraction

EDA ethylene diamine

EDAX, EDX → energy-dispersive X-ray emission analysis; *X-ray emission spectroscopy

EDC 1,2-dichloroethane, cf. *vinyl chloride, *Kellogg EDC process

edge atoms → single crystal

Edman degradation the stepwise removal of amino acids from the N-terminal end of a protein with phenyl isothiocyanate followed by HF, yielding the thiazolinone which can be converted to the phenylthiohydantoin by treatment with aqueous acid. A great advantage of this technique is that it can readily be automated. **F** dégradation d'Edman; **G** Edman-Abbau. P.S. SEARS

EDS process → Exxon procs.

EDTA ethylenediamine tetraacetic acid, a *chelating agent

EDX → energy-dispersive X-ray emission analysis, *X-ray emission spectroscopy

ee (EE) enantiomeric excess, the surplus of one enantiomer (*R* or *S*) in a mixture of the racemate (*R* plus *S*) over the other:

$$\% \ ee = \frac{R-S}{R+S} \cdot 100 = [\%R - \%S] = \text{optical yield (oy)}$$

Cf. also *enantioselective syntheses, *resolution, optical; *optical yield. **E=F=G.**

                        M. BELLER

EELS electron energy loss spectroscopy, → high-resolution electron loss spectroscopy

E factor (environmental factor) an alternative concept to *atom economy. Defining waste as everything except the desired product, the *E* factor is the ratio (kg/kg) of byprods. to prod. and indicates the environmental acceptability. From oil refining (~0.1) through bulk chemicals (<1–5) to pharmaceuticals the *E* factor increases to >100. Regarding the nature of wastes, the *E* factor can be refined to *EQ* (the environmental quotient) which is obtained by multiplying *E* by an arbitrarily assigned "unfriendliness quotient" *Q*. The *E* factor is very helpful in evaluating environmentally new routes to chemicals and in this respect superior to the term *atom economy. Catalysis plays a decisive role in improving traditional routes to bulk and fine chemicals with respect to environmental, *selectivity, and yield aspects. For the first time, BASF announced in 1997 for its Antwerp plant (and thus for bulk prod.) 9926 tonnes of waste which is in relation to the total prods. an *E* factor of less than 1. **F=G=E.**   B. CORNILS
Ref.: *CHEMTECH* **1994**, (3), 38; *ChemPress* (NE, ISSN 0009–3173) **1998**, *32*(17), 5; *Green Chem.* **1999**, *1*(1), G3.

effectiveness factor The *rate of a cat. reaction at an interface or in *pores may be impeded by *mass transport limitations. This is due to a decrease in reactant conc. at the *active site if the reaction is faster than the transport of the reactants. The ef. η is then defined as the ratio of the observable *rate of reaction to the intrinsic rate of reaction without transport limitations. For the relation of η and the Φ cf. the *Thiele modulus. Usually η is 1 or <1 but, for negative reaction orders where the reaction rate is limited by a reactant, η exceeds 1 under diffusional limitation conds. due to a decrease in the reactant conc. In strongly endo- or exothermic reactions temp. gradients in the cat. affect the reaction rate and therefore the ef. For $\beta < 0$ (β being the Prater number) η becomes smaller while for $\beta > 0$ it increases and may even exceed 1. **E=F=G.**   M. MUHLER
Ref.: Wijngaarden/Kronberg/Westerterp, p. 113,221,222; Gates-1, p. 228; Ullmann, Vol. A5, p. 324.

effectors cpds. which bond to enzymes and induce structural changes and which thus influence the proc. of regulation.

EG ethylene glycol

egg-shell, egg-white, egg-yolk → metal distribution

Eglington method → Glaser coupling reaction

2-EH 2-ethylhexanol, the most important *plasticizer alcohol, manuf. by hom. cat. *oxo synthesis via *n*-butyraldehyde → aldolization to 2-ethylhexenal → *hydrogenation to 2-EH. **F** 2-éthylhexanol; **G** 2-Ethylhexanol.

EHD process → Monsanto procs.

elastase Elastase (pancreas, EC 3.4.21.36; leukocyte, EC 3.4.21.37) is a *serine endoprotease related to *trypsin and *chymotrypsin. It is usually isolated from bovine pancreas and was named for its ability to *hydrolyze the extracellular matrix protein elastin. It is also produced by leukocytes as part of the in-

flammatory proc. It is relatively nonspecific, but prefers small neutral residues (Ala, Gly, Val, Ser) in the S_1 subsite (acyl donor side of the scissile bond) and does not accept proline in the S_1' site (amine side of scissile bond). **E=F=G**. P.S. SEARS
Ref.: Schomberg/Salzmann, Vol. 5.

electric fields (in catalysis) are on the one hand relevant for several investigation methods (such as *field emission and *field ion microscopy) for detailed in-situ information about cat. surface reactions. On the other hand, high-energy electric fields can influence *surface properties of cats. such as field desorption, field ionization, electrochromy, electrostriction, or dipole orientation. The topic is different from *electrocatalysis; so far, there is no application for commercially relevant cats. **F** champ èlectrique; **G** elektrisches Feld. B. CORNILS
Ref.: Ertl/Knözinger/Weitkamp, Vol. 3, p. 1104

electrocatalysis → electron transfer chain (ETC) catalysis

electrochemical deposition special methods for the prep. of *supported metal cats. **F** déposition électrochimique; **G** elektrochemische Fällung. H. KNÖZINGER
Ref.: Ertl/Knözinger/Weitkamp, Vol. 1, pp. 240,257; *Chem. Rev.* **1995**, *95*, 477.

electron diffraction (ED) yields information (local, i.e., on an atomic scale and bulk-sensitive) about local crystallographic analysis of solid particles and thin films (cf. *SAD). R. SCHLÖGL

electron donor ligands *Ligand/metal bonding can be described simply as an interaction between *Lewis bases (ligands) and a Lewis acid (as the metal center, *central atom). Thus, electron density is transferred from the donor (the ligand) to the acceptor (the metal), which can be confirmed, e.g., by *NMR or *IR spectroscopy or by electrochemical methods. Lewis basicity of a certain ligand depends mainly on the nature (element, hybridization) but also on the chemical environment (e.g. the substituents) of its actual donor center. **F** ligands doneurs d'électrons; **G** Elektronendonor-Liganden.
W.R. THIEL

electron energy loss spectroscopy
(EELS) cf. high-resolution electron energy loss spectroscopy. EELS yields information about chemical bonding of light elements and lateral distribution analyses with atomic resolution (solid-solid interfaces). It is suitable for *TEM. R. SCHLÖGL
Ref.: Ibach/Mills.

electronic effects (in homogeneous catalysis) chemical effects which determine the electronic structure of the cat.-substrate entity (i.e., the bonding behavior of the reactant towards the cat. and vice versa), and affect the cat. reactivity through the free energy of activation. Ees. are difficult to distinguish from *steric effects, and it is often the synergism between the two effects which governs the cat. performance in terms of efficiency and *selectivities. Ees. arise from the cats. or precats. (*precursors), the substrate (e.g., by *tailoring with electron-donating or -withdrawing groups), the *solvent (stabilization or destabilization of polar/non-polar reactant molecules or cat. intermediates), and *salt effects (e.g., the *lithium effect). In particular, *complexes and *organometallic cpds. display an electronic variability suitable for hom. cat. Both metal- and *ligand-based ees. have to be considered. Cat. sequences based on *oxidative addition and *reductive elimination involving d-transition metals depend on *Tolman's 18-electron rule and the *coordination chemistry theory. The intrinsic electron configuration and charge density of the metal or metal ion place the reactivity standard in terms of *redox behavior, electronegativity, electron affinity, ionicity, oxophilicity, electrophilicity, and *Lewis acidity. Ligands tune these intrinsic properties not only by changing or polarizing the electron density through interaction with the metal center (cf. *ligand effects, *lig-and-accelerated cat., *ancillary or *acceptor ligands, *hemilabile ligands, *ligand tuning, etc.) via the generalized *acid-base approach (*acid-base cat., *HSAB principle, *backdonation), but also by their coordination mode. Coordinated ligands themselves are excellent probes for ees. of approaching ligands (e.g., stretching frequencies of carbonyl ligands, *Blyholder model). Electronic procs. in hom. cat. are also treated by quantum chemical approaches (*frontier molecular orbitals, *ab-initio calcns, *DFT). Ees. are crucial in all cat. steps including cat.-substrate *preformation, *Lewis acid-base interaction, *oxidative addition, *reductive elimination, *insertion, *transmetallation, *ligand exchange in *metathesis, etc. **F** effets électroniques en catalyse homogène; **F** elektronische Effekte in der hom. Katalyse.   R. ANWANDER

Ref.: Yoshida/Sasaki/Kobayashi.

electronic factors   All *surface procs., including cat. reactions, involve some movement of electrons between the solid and the reacting molecules. The phenomenon is most clearly detected with metals and *alloys, where the number and class of electrons at the surface of these electron-rich materials can differ very greatly, with important consequences for their abilities in *chemisorption and cat. Differences in behavior have for long been attributed to the completion of the d-electron shell or (equivalently) to the filling of the d-band, *vacancies in which were therefore held to be necessary for *chemisorption. This abrupt change in cat. performance is observed with many types of reaction, and is the observational basis for the electronic effects in cat. Besides a collective electron model there are more factors affecting the cat. properties of alloys than was previously thought, especially when they are in the form of *supported *bimetallic particles. 1: Surface segregation commonly occurs, the comps. of lower surface energy coming preferentially to the surface; this effect can however be counteracted if the adsorbate interacts strongly with the comp. of higher surface energy. 2: Segregation

of the lower surface energy comp. to the various types of site present on small particles takes place preferentially where it can have the greatest impact on the energy of the whole particle, i.e., first at corners, then at edges, and only finally on terraces (cf. *single crystal). In interpreting the cat. trends shown by alloys and bimetallic particles, much more emphasis is now placed on the population of the groupings (*ensembles) having at least the minimum number of the active atoms needed to secure reaction.

This change of emphasis does not of course eliminate the electronic factor, but places it in a new perspective. Trends in *cat. activity, derived from changes in *adsorption enthalpies, on traversing the periodic table find their origin in the availability of electrons to form the necessary chemisorptive bonds: thus the elements in groups 9 and 10 are unable to synth. NH_3 because they cannot chemisorb nitrogen dissociatively. It needs to be stressed, however, that it is impossible in a general way to disassociate electronic from geometric factors, e.g., lengths of metal-metal bonds are a periodic function of the number of valence electrons. It nevertheless remains certain that the ability of atoms and molecules to form essentially covalent bonds utilizing electrons drawn from the solid is a critical factor in het. cat. For ef. in hom. cat., cf. *χ-value and *ligand effects. **F** facteurs électroniques; **G** elektronische Faktoren.   G.C. BOND

Ref.: Yoshida/Sasaki/Kobayashi.

electronic promoter   a substance which is added to the active comp. and whose effect is to strongly enhance the cat. *activity without, however, exhibiting substantial cat. activity itself (cf. *promoter, *promotion). A well-known example is the cat. for commercial *ammonia synth., consisting of Fe_3O_4, K_2O, CaO, and Al_2O_3 before *reduction. The active comp. is Fe and the other substances are promoters. While the *structural* promoters (Al_2O_3) prevent the sintering of Fe particles, *electronic* promoters (K_2O) modify the cat. properties of the metal. The promotion effect of K in ammonia synth. has been thoroughly investigated. The main effect of K is to increase the *adsorption strength of the molecular N_2 species, thus increasing the *sticking coefficient for dissociative *chemisorption, and to reduce the adsorption energy of ammonia (cf. *product inhibition). **F** promoteur électronique; **G** elektronischer Promoter. R. IMBIHL

Ref.: *J.Catal.* **1988**, *109*, 51; *Catal.Rev.Sci.Eng.* **1980**, *21*, 201.

electronic structure the distribution of the electrons of a molecule or solid among various energy states. The es. of a solid is given by the *band structure in which the individual orbitals of atoms merge into energy bands and the electrons are delocalized over the whole solid. The band structure is described by the energy vs. momentum (wave vector) relationship; the *density of states (DOS) and the position of the *Fermi level determine which states are occupied and which are empty. The es. of a *surface is different from that of the bulk because on surfaces new states may appear (*surface states*) which are forbidden in the bulk, and/or existing states of the bulk can be enhanced at surfaces (*surface resonances*). The es. of a solid is essential for the understanding of cat. because the *adsorption properties as well as cat. effects such as *poisoning or *promotion are discussed in terms of the es. The formation of chemical bonds between a surface and an adsorbate will depend on the availability of filled and empty energy states of the surface in a very similar way to that in which the formation of a chemical bond in a molecule takes place via the interaction of filled and empty orbitals as described by MO theory. Static adsorption properties such as *adsorption enthalpy and geometry depend on the es. as well as dynamic properties like the *sticking of a molecule on a metal surface. The translational energy can be very effectively transferred from the molecule to electrons which are excited into states above the Fermi level. Theoretically a large number of quantum chemical methods are available with which the es. of surfaces and their influence on the adsorption properties can be calculated. The main method to determine experimentally the es. of a solid surface is *photoelectron spectroscopy (PS); the valence band region is probed by *ultraviolet photoelectron spectroscopy (UPS), the core level range by *X-ray photoelectron spectroscopy (XPS). For single crystal surfaces angle resolved UPS (= ARUPS) can be applied to obtain the es. wave vector resolution (band mapping). On surfaces the es. can be studied locally on an atomic scale by spectroscopic variants of *scanning tunneling microscopy (STM). **F** structure électronique; **G** elektronische Struktur. R. IMBIHL

Ref.: R. Hoffmann, *Solids and Surfaces: A Chemist's View of Bonding in Extended Surfaces*, VCH, Weinheim 1988; R.A. van Santen, *Theoretical Heterogeneous Catalysis*, World Scientific, Singapore 1991.

electron microprobe analysis (EMA), **electron probe microanalysis** (EPMA) X-ray techniques for determining compositions on the submicron scale. EPMA is a non-destructive UHV technique and needs flat samples. R. SCHLÖGL

electron microscopy (EM) uses electrons and because of their wavelength of less than 1 A comes close to seeing atomic details. The figure explains the interactions between the

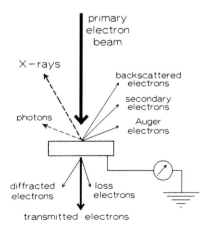

primary electron beam and the sample in EMs.

Especially relevant for cat. research are the EM variants conventional transmission EM (CTEM, operating in the 100–200 kV range of electron energies), high-resolution EM (HREM, operating at electron energies of 0.5–1 MeV), *transmission EM (TEM), *scanning EM (SEM), and *scanning transmission EM (STEM).   R. SCHLÖGL

Ref.: Anderson/Boudart, Vol. 7; Ertl/Knözinger/Weitkamp, Vol. 2, p. 493; Niemantsverdriet, p. 165; Ullmann, Vol. B6, p. 229.

electron paramagnetic resonance (EPR)
→ electron spin resonance spectroscopy

electron probe microanalysis (EPMA)
→ electron microscopy, *electron microprobe analysis

electron spectroscopy for chemical analysis (ESCA) nowadays more commonly referred to as *X-ray photoelectron spectroscopy (XPS).

electron spin resonance spectroscopy
(ESR)   is based on magnetic resonance and uses high-frequency radiation in the GHz range as probe (response: high-frequency radiation). It is applied for paramagnetic centers and *radicals in and on cats. The ex-situ, non-destructive, very sensitive technique can use solid or liquid samples, but has only limited in-situ capabilities. The data are integral and bulk-sensitive. Identical with EPR.   R. SCHLÖGL

Ref.: Ullmann, Vol. B5, p. 471; Ertl/Knözinger/Weitkamp, Vol. 2, p. 641.

electron transfer chain (ETC) catalysis
belongs to the family of chain reactions and is also called electrocatalysis – the cat. of reactions by electrons. The reaction is cat. by an electron or by an electron hole. If the uncat. reaction is too slow, the electron or electron hole transforms at least one of the reactants, for instance a stable diamagnetic molecule, into a very labile *radical anion or radical cation whose reactivity is of the order of

10^6 to 10^{10} times larger than that of the starting stable cpd. Thus, the kinetic gain is enormous. The reactions are usually non-*redox reactions such as nucleophilic substituition, *ligand exchange, *isomerization, *insertion, and extrusion. Occasionally, however, the ETC-cat. reaction can be a *disproportionation of a binuclear *complex with a metal-metal bond to an anion and a cation.

Like all chain reactions, the ETC-cat. conv. has initiation, chain, and *termination comps. Initiation can be achieved using electricity (electrochemistry), light (photochemistry, *photocatalysis), or a redox reagent. It is important whether the initiation must be done by *oxidation or by *reduction. The side reactions of the *propagation cycle (which are fast since they are based on radicals) must be considered. In most cases this problem is solved by comparing the electron richness of the starting and the final cpds., respectively. If the prod. is more electron-rich than the reactant, the reaction is usually initiated by reduction. If the prod. is more electron-poor than the reactant, the conv. is usually initiated by oxidation. In both situations, the initiation is thus chosen in such a way that the redox reaction of the propagation cycle is exergonic (favorable). Since an exergonic electron transfer is fast and an endergonic one is slow, a side termination reaction would be faster than a slow redox propagation step. The *propagation cycle contains the redox step and one or several "chemical" (non-redox) step(s). The chemical propagation step is also important, but it is mostly very difficult to estimate its exergonicity. Thus, it is best to do the molecular engineering at the level of the redox propagation step, as indicated above. In one case, it has been possible to compare reductive and oxidative initiations in order to test the above hypothesis.

Electrochemistry (cyclic voltametry) is excellent for checking the efficiency of an ETC-cat. reaction. Indeed, the return wave of the reactant must disappear and let a new wave appear for the prod. (at more negative potential in the case of a reductive scan, at more positive potential in the case of an oxidative scan). The change of scan rate gives a qualitative picture of the *kinetics and more precise analysis gives the kinetics of the ETC-cat. reaction. Pioneering work in this respect has been done by Kochi for *ligand substitution of acetonitrile by *phosphine in Mn complexes.

Such ligand substitution reactions of a solvent ligand by a variety of other better π-acceptor ligands work well in the chemistry of monometallic complexes when they are initiated by oxidants. On the other hand, the exchange of a carbonyl ligand by a phosphine using reductive initiation is marred by side reactions in monometallic complexes, but works well in *cluster chemistry. These ETC-cat. reactions have been studied mechanistically in detail using $[Fe(\eta^5\text{-}C_5H_5)(\eta^6\text{-}$ arene$)]^+$ as a model. For ETC-cat. reactions initiated by oxidation, ferricinium is useful and, if a stronger oxidant is needed, the "super ferricinium", a 17-electron complex $[Fe(\eta^5\text{-}C_5Me_5)(\eta^6\text{-}C_6Me_6)][SbCl_6]_2$ can be used.

ETC cat. of ligand substitution in a pre-cat. can be coupled with hom. cat. to increase the *selectivity and efficiency of cat. procs. For instance, the *polymerization of terminal alkynes cat. by W(0) complexes is very slow at $20\,^\circ C$ and must be carried out at $100\,^\circ C$. If cat. amounts of ferricinium are added to W(0), the reaction becomes fast at $20\,^\circ C$ with high yields and selectivities. This example opens a promising future for applications of ETC-cat. reactions. For ETCs in enzymatic chemistry, cf. *electron transport. **E=F=G.**   D. ASTRUC

Ref.: Astruc; Ertl/Knözinger/Weitkamp, Vol. 3, p. 1325. *Acc.Chem.Res.* **1980**, *13*, 323, **1993**, *26*, 455 and **1987**, *20*, 214; *J.Organomet.Chem.* **1986**, *302*, 389; *Coord. Chem.Rev.* **1985**, *63*, 217 and **1987**, *16*, 1; Sullivan/ Krist/Guard; Yoshida/Sasaki/ Kobayashi.

electron transfer mechanisms   *redox

reactions in het. cat. in which electrons of the cat. take part in reactions of the *adsorbate (cf. also *ETC). This group of reactions includes *hydrogenation, *dehydrogenation,

*oxidation, etc. Typical redox cats. are metals and *metal oxides with varying oxidation number. The latter group typically represents a *semiconductor. Acceptor reactions are reactions in which electrons are transferred from the metal to the adsorbate, e.g., the adsorption of O_2 on an n-type semiconductor surface like ZnO. Donor reactions are reactions in which electrons from the adsorbate are transferred to the substrate. Such reactions are favored by p-type semiconductors. For a cat. metal *supported on a semiconductor, an additional pathway for electron transfer has to be taken into account in which electrons are transferred via the metal/semiconductor interface. **F** mécanisme de transfert d'électrons; **G** Elektronentransfermechanismus.   R. IMBIHL

Ref.: I. M. Campbell, *Catalysis of Surfaces*, Chapman and Hall, London 1988.

electron transport (enzymatic)   There are

several instances in nature in which cpds. with strong reducing power are used to drive a transmembrane *proton pump* via the gradual transfer of electrons from molecules of low to those of high reduction potential. In the respiratory electron transport chain, electrons are provided by reduced *nicotinamide (NADH) and *flavin (FADH$_2$) *cofactors to an *electron transport chain (ETC) which ultimately reduces molecular oxygen to water. The proton gradient created across the mitochondrial membrane (or plasma membrane, in the case of prokaryotes) is then coupled to the synth. of *ATP via a proton-translating ATP synthase (F_oF_1- ATPase).

ETCs are also involved in the *light reactions* of *photosynthesis, in which light is captured by photosystems, generating cpds. with very low standard reduction potentials. These cpds. release electrons to electron carriers (or hydride carriers, in the "proton pumping" steps) of successively higher *redox potential. These ETCs not only generate a proton gradient capable of driving a proton-translocating ATPase, but also generate (via a pair of electron transport chains) a reductant strong enough to reduce NADP to NADPH. **F** transport des électrons; **G** Elektronentransport.   P.S. SEARS

electrophilic catalysis (enzymatic)   refers to reaction acceleration via the stabilization of a negative charge on the substrate. The stabilization may occur via noncovalent associations, as in the polarization of amide carbonyls by Zn^{2+} in the *metalloproteases, or the stabilization of the oxyanion *transition state of amide and ester *hydrolysis via the "oxyanion hole" of *serine and *thiol proteases, which provides several H-bond donors. There are also many examples of covalent enzymatic electrophilic cat. The *cofactors *thiamine pyrophosphate and *pyridoxal phosphate form covalent adducts with substrates. Both have excellent electron sinks which can effectively stabilize *carbanion formation in the substrate (see also *covalent cat.). **F** catalyse électrophile; **G** elektrophile Katalyse.
P.S. SEARS

Eley-Rideal mechanism   In het. cat., reactions between two molecules A and B on a solid *surface may involve three kinds of general mechanisms: a *Langmuir-Hinshelwood mechanism, an ERm., or a *precursor mechanism. In an ERm. the reactant A is chemisorbed first. A then reacts directly with an incoming molecule B to form an adsorbed product P, without B becoming adsorbed prior to reaction. There is not always a clear distinction between the ERm. and a precursor mechanism in which B collides with the surface first and enters a weakly bound and thus highly mobile precursor state B′. This B′ rebounds to the surface until it reacts with adsorbed A to form the adsorbed prod. P. In both cases the prod. P is then desorbed. All these reaction steps may also occur in reverse (reverse ERm.). There are only very few examples of reactions such as H_2-D_2 exchange reactions and reactions in semiconductor film growth that follow a true ERm. rather than a precursor mechanism. The reason for this is that a gas/surface collision lasts for only about a picosecond while the lifetime of a precursor

state can be in the order of several microseconds. Hence, the probability of a molecule B reacting in an ERm. is up to 10^6 times lower than that of a reaction acc. to a precursor mechanism. The kinetics of many het. cat. reactions seems to follow that of an ERm. in a certain range of reaction conds., i.e., partial press. of reactants, temp., etc. Under such conds. their reaction rate is proportional to the conc. of A in the adsorbed state multiplied by the gas-phase conc. of B, similarly to the relationship derived for a true ERm. This expression, on the other hand, can also be derived for a Langmuir-Hinshelwwod mechanism if the adsorption constant for B or the conc. of B is very low. **F** mécanisme d'Eley-Rideal; **G** Eley-Rideal-Mechanismus.
M. MUHLER

ELF/IFP process for TAME   → IFP procs.

elimination reactions   a class of reactions in which atoms or molecules are eliminated from C-C cpds. yielding double or triple bonds (β-elimination, e.g., *hydration, *dehalogenation, *dehydrohalogenation, *ester splitting), *carbenes (α-e.), or *cyclopropanes (γ-e.). In some respect *addition reactions are the reverse of ers. **G** Eliminierungsreaktionen.   B. CORNILS
Ref.: Ertl/Knözinger/Weitkamp, Vol. 5, p. 2370.

ellip.   → ellipsometry

ellipsometry (ellip.)   uses visible light for the determination of thin film properties. The method is non-destructive and suitable for solid samples or flat model cats. The data are integral, *surface-sensitive, and express in-situ capabilities.   R. SCHLÖGL

Elmit   trade-name of a *COC polymer from Mitsui

Elovich equation   describes the *adsorption of gases, e.g., the *chemisorption of H onto *mixed oxides which show discontinuities (which in turn indicate distinct rate procs.).

The equ. was originally proposed to describe the *kinetics of cat. *oxidation. **F** l'équation d'Elovich; **G** Elovich-Gleichung. B. CORNILS
Ref.: *Chem.Phys.Lett.* **1973**, *18*, 423; Thomas/Thomas, p. 89.

Eluxyl process → IFP procs.

EMA → electron microscopy

Embden-Meyer-Parnas pathway → glycolysis

embedding methods Quantum chemical methods aim in most cases at calcng. the properties of isolated molecules. There are two problems associated with this approach. One concerns the neglect of the medium effect in a condensed phase (solution or solid state). The second is that the computational costs of *ab-initio, *DFT, and even *semiempirical calcns. sharply increase with the size of the molecule. One solution to the problems is to combine the expensive quantum chemical (QC) methods with computationally cheap *force-field (molecular mechanics, MM) methods. The basic idea is to treat the electronically demanding part of the system with a QC method, and then "embedding" the rest of the system with force-field calcns. It is also possible to combine different QC methods, such as DFT and semiempirical procedures, in embedding calcns. A critical point of the em. is the coupling of the theoretical techniques when different parts of a single molecule are to be calcd. with different methods. For example, bulky substituents of a molecule may be calculated with MM methods while a QC method is used for the core structure. Several coupling techniques have been developed.
The QM/MM methods may have mechanical embedding, where the QM region is calcd. independently and has no interaction at all with the MM region. More advanced programs use electronic embedding schemes which allow for interactions between the QM and MM regions, for example via electrostatic interac-

tions. Ems. are promising tools for accurate theoretical calcns. of realistic cats.
Ref.: Thomas/Thomas, p. 397. G. FRENKING

emim 1-ethyl-3-methylimidazolium, a prototype of *non-aqueous liquids (Figure):

Bmim is the 1-butyl derivative. B. CORNILS
Ref.: *Chem.Eng.News* **1998**, March 30, 32.

Ems-Inventa polycaproamide process *polymerization of ε-caprolactam (LC) monomer. **F** procédé Ems- Inventa pour le polycaproamide, **G** Ems-Inventa Polycaproamidverfahren. B. CORNILS
Ref.: *Hydrocarb.Proc.* **1997**, (3), 147.

en ethylene diamine group in *complex cpds. or as a *ligand

enantiomeric excess → *ee*

enantiomers Two *stereoisomers, the image and mirror image of which are non-superimposable, are called es. H. BRUNNER

enantiomorphic site control Besides *chain-end control, esc. is one of two mechanisms that give rise to stereocontrol in the *metallocene-cat. *polymerization of α-olefins. Since a chiral environment is required around the cat. center, the esc. is exerted by bridged, i.e., stereorigid metallocenes (such as *ansa*-metallocenes). The *chiral structure of the cat. site and the growing polymer chain force the incoming *monomer to coordinate in a certain way, thus generating a regular *configuration on the tertiary C atom. The *insertion mode of the α-olefin is determined by the structural symmetry of the cat. site. C_2-symmetric cats. generate mainly *isotactic* polymer chains whereas C_s-symmetric cats. result in *syndiotactic* polymers (cf. *tacticity). The esc. mechanism can be distinguished from *chain-end control by analyzing the type of stereoerrors which occur during *polymerization. **E=F=G**.   W. KAMINSKY

Ref.: *J.Am.Chem.Soc.* **1955**, *77* ,1708 and **1988**, *110*, 6255; *Angew.Chem.Int.Ed.Engl.* **1995**, *34*, 1143.

enantioselective acylation (hydrolysis) → proteases, *lyases

enantioselective catalysis → asymmetric catalysis

enantioselective catalysts → chiral catalyst

enantioselective opening of *epoxides by means of nucleophiles is promoted by *chiral *transition metal *complexes. **F** ouverture des composés cyclique; **G** enantioselektive Ringöffnung.   B. CORNILS

Ref.: Mulzer/Waldmann, p. 62.

enantioselective synthesis In an es. a prochiral *precursor is transformed into a *chiral prod. In the presence of an optically active auxiliary, such as *solvents, *additives, reactants, or cats. (even circularly polarized light), one of the two enantiomers is formed preferentially. Additionally, *kinetic resolutions belong to the group of es. **F** synthèse énantioselective; **G** enantioselektive Synthese.   H. BRUNNER

Ref.: Mulzer/Waldmann; Noyori; Ojima.

enantioselectivity (in enzymatic cat.) In an enantioselective enzymatic transformation, two *enantiomeric substrates or two enantiotopic faces or groups compete for the *active site of the *enzyme. Using the *steady-state or *Michaelis-Menten assumptions, the two competing reaction rates are $\upsilon_A = (k_{cat}/K_m)_A[E][A]$ and $\upsilon_B = (k_{cat}/K_m)_B [E][B]$. The ratio of these two reaction rates is therefore $\upsilon_A/\upsilon_B = (k_{cat}/K_m)_A[A]/(k_{cat}/K_m)_B[B]$.

This analysis shows that the ratio of specificity constants $[(k_{cat}/K_m)_A/(k_{cat}/K_m)_B]$ determines the e. of the reaction. Since these specificity constants are related to free-energy terms (that is, $\Delta G^{\neq}_A = -RT\ln(k_{cat}/K_m)_A$ and $\Delta G^{\neq}_B =$

$-RT\ln(k_{cat}/K_m)_B$), the e. of the reaction is related to the difference in energy of the diastereomeric transition states.

$$\Delta\Delta G^{\neq} = (\Delta G_A^{\neq} - \Delta G_B^{\neq}) =$$
$$-RT \ln (k_{cat}/K_m)_A/(k_{cat}/K_m)_B$$

In an enzyme-cat. *kinetic resolution which proceeds irreversibly, the ratio of specificity constants (or the e. value, E) can be further related to the extent of conv. (c) and the *enantiomeric excess (ee). The parameter E is commonly used in characterizing the e. of a reaction. **F** enantiosélectivité; **G** Enantioselektivität.   C.-H. WONG

Ref.: *J.Am.Chem.Soc.* **1982**, *104*, 7294.

enantiotopic differentiation A *prochiral substrate with two identical functional groups attached to the prochiral centre, or a *meso* substrate with identical functional groups attached to chiral centers of opposite *configuration, may be selectively derivatized at one of those functional groups to produce a *chiral molecule.

prochiral diamine (serinol)   *oxidoreductase* [O]   L-serine

A chiral cat. such as an *enzyme may be able to selectively create a single enantiomer by positioning cat. groups and/or cosubstrates at a particular position; this is called enantiotopic differentiation. **E=F=G**.   P.S. SEARS

encapsulation of catalysts → final disposal of catalysts

endoglycosidase → glycosidase

endo H, endo F, endo M → glycosidases

energy bands are bands of energy describing the interaction of orbitals on neighboring atoms and their *overlap.

Ref.: Thomas/Thomas, p. 378.

energy-dispersive X-ray emission analysis (EDX, EDAX) has been used for the determination of the composition of individual particles in bimetallic cats., i.e., for local elemental compositions and the analysis of *nanostructures, supported metal particles, etc. The electron micrograph technique (ex situ) is non-destructive but has problems with light elements. The data are local and bulk-sensitive. R. SCHLÖGL

Engelhard catalysts Engelhard Corp. at Iselin, NJ, has a special business unit for *additives and petroleum cats. Engelhard has probably the most complete portfolio of het. precious and base metal cats. of any single manufacturer. The cats. are used in the *automotive, *petroleum, chemical, food-processing, and pharmaceutical industries. J. KULPE

Engelhard fluid cracking (FCC) process for an efficient carbon rejection from feedstocks containing asphaltenes using the Asphalt Residual Treating (ART) technology over a rare earth/ *zeolite cat. The proc. delivers a stable *syncrude containing no bottoms. **F** procédé FCC d'Engelhard; **G** Engelhard FCC-Verfahren. B. CORNILS
Ref.: *Hydrocarb. Proc.* **1996**, (11), 121.

Engelhard HPN process Cat. *hydrogenation of pyrolysis naphtha (HPN) to produce stable gasoline blend stocks. **F** procédé HPN d'Engelhard, **G** Engelhard HPN-Verfahren. B. CORNILS
Ref.: *Hydrocarb. Proc.* **1978**, (9), 134.

Engelhard Magnaforming process → ARCO processes.

enhancement factor In the case of fluid-fluid systems where the *rate of reaction is high compared to the transport rate, i.e., for high values of the *Hatta number (Ha >3), the *mass transport is enhanced by the reaction, resulting in an extra decrease in the conc. of educt A. The ratio of the molar flux J_A (kmol/m^{2}s) with chemical reaction to that without chemical reaction (purely physical absorption) is called the ef., defined as E_A, which depends on the Hatta number and the order of the chemical reaction. Thus, the three regimes of the reaction rate differ on the basis of different values of the Hatta number and are characterized by different values of E_A. 1: In the case of slow reactions (Ha <0.3) where the reaction (first-order) does not have any influence on the rate of mass transfer, it follows that E_A = 1. 2: For fast reactions (Ha >3) where the reaction takes place near the interface in a film, the ef. is given by $E_A \sim Ha$. 3: In the case of an instantaneous reaction (Ha >>3) where the reaction (irreversible, second-order) between two educts A and B proceeds in the boundary layer, the ef. depends on the *diffusion coefficients. If both coefficients are similar, E_A is mainly determined by the conc. ratio $c_{B,1}/c^*_A$. In the special case of c^*_A = 0, the reaction surface moves into the interface between the two fluid phases. For a hyperbolic form of the intrinsic kinetic, as frequently observed in hom. cat. (gas-liquid) reactions, a theoretical analysis is possible to evaluate quantitatively the mass transfer effects. If the reaction is assumed to occur in bulk liquid, a generalized Hatta number is used to obtain an approximate analytical solution for E_A. If the reaction occurs completely in the film, a transition in the regimes of absorption with change in Ha is indicated which reflects a change in the conc. of the cat. This approach covers all the regimes of hom. cat. gas-liquid reaction, and allows a quantitative prediction of mass transfer effects on the kinetics of this operation mode. **F** facteur d'amplification; **G** Verstärkungsfaktor. P. CLAUS, D. HÖNICKE
Ref.: L.K. Doraiswamy (Ed.), *Frontiers in Chemical Reaction Engineering*, Proc.Int.Chem.React.Eng. Conf., New Delhi, Wiley, New York 1984, *1*, p. 291.

Enichem ammoximation process conv. of cyclohenanone to its oxime via *ammoximation with ammonia/H$_2$O$_2$ over *Ti silicalite cats. without any NO$_x$ byprod. formation.
B. CORNILS

Enichem LDPE-EVA process a high-press. *autoclave or tubular reactor *copolymerization proc. (ethylene and vinyl acetate). **F** procédé Enichem pour LDPE-EVA; **G** Enichem LDPE-EVA-Verfahren.   B. CORNILS

Ref.: *Hydrocarb.Proc.* **1997**, (3), 154.

Enichem dimethyl carbonate process proceeds via *oxidative carbonylation of alcohols with CO and oxygen with Pd cats. in the liquid phase at 120 °C/2.5 MPa acc. to 2 ROH + CO + 0.5 O_2 → RO-CO-OR + H_2O; the basic flowscheme is given with the Figure. **F** procédé d'Enichem pour le carbonate de diméthyle; **G** Enichem Dimethylcarbonatverfahren.   R.W. FISCHER

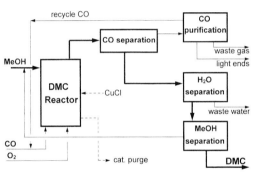

Ref.: *Appl.Catal.* **1994**, *115*, 182; *Chem.Eng.News* **1987**, *44*(11), 65.

Enichem (Eniricerche) propylene oxide process manuf. of *PO by *epoxidation of propylene with H_2O_2 over *Ti silicalite TS-1. **F** procédé d'Enichem pour l'oxyde de propylène; **G** Enichem Propylenoxidverfahren.

Ref.: *Appl.Catal.* **1994**, *115*, 188.   B. CORNILS

Enichem hydroquinone process manuf. of hydroquinone (and catechol as byprod.) by *hydroxylation of phenol with H_2O_2 in the presence of Ti silicalite TS-1 (cf. *titania, *titanium silicalite, *Ti as catalyst metal) in *SBRs at 100 °C. **F** procédé d'Enichem pour l'hydroquinone; **G** Enichem Hydrochinonverfahren.   B. CORNILS

Ref.: *Appl. Catal.* **1994**, *115*, 173; Ertl/Knözinger/Weitkamp, Vol. 5, p. 2332.

enoate reductase (EC 1.3.1.31) isolated from *Clostridia* sp., cat. stereospecific reduction of a number of α,β-unsaturated carboxylates, aldehydes, and ketones. Mechanistic studies indicate that the *enzyme contains an iron-sulfur-flavin *cluster that accepts and transfers electrons to substrates. Proton exchange between the reduced cluster and water was observed. The Fe/S cluster can be reduced with *NADH, with reduced methyl- and benzyl*viologen generated electrochemically, with H_2 cat. by hydrogenase present in the cell, or with Pd, Pt, or *Raney Ni modified with fluorine-containing surfactants. In practice, whole cells instead of free *enzyme are used for reactions, presumably due to the instability of the free enzyme. The productivity numbers of the system are often 10–100 times higher than those found for reduction with yeasts. The system has also been used to prepare isotopically labeled chiral δ-aminolevulinic acid (*radioactive labeling).

In addition to α,β-unsaturated systems, many ketones and 3-oxocarboxylic acid esters can also be reduced stereoselectively with *Clostridia* in the presence of H_2 with very high *ee.

Many α-keto mono- or dicarboxylates can also be reduced by *Proteus vulgaris* in the presence of *methyl- or benzylviologen to the corresponding (R)-hydroxy species with very high ee. The viologen mediator can be regenerated either electrochemically or by H_2 and a hydrogenase present in the microorganism, or by formate and a *formate dehydrogenase, also found in the cell. As mediator carbamoyl methylviologen (CAV) can be used; the oxidation of CAV^+ to CAV^{2+} may be carried out electrochemically or with air. It is, however, sensitive to oxygen when used as a reductase.

Similar reduction of α,β-unsaturated systems was carried out with Baker's yeast or fungi. **E=F=G.** C.-H. WONG

Ref.: *Angew.Chem.Int.Ed.Engl.* **1985**, *24*, 539, **1986**, *25*, 462, and **1987**, *26*, 128.

ensembles Experiments had shown that small amounts of CO had a disproportionate effect on the *activity of a Cu cat. Thus it was proposed that for certain reactions only a small fraction of the *surface was cat. active, and that once this was *poisoned activity would disappear. The active fraction was composed of *active sites (or active centers). This concept was later extended by Kobozev and Balandin, who proposed that some reactions needed a group of like atoms to form an active center. Kobozev used the term *ensemble* to describe this group, while Balandin wrote of *multiplet theory. This caught the scientific imagination most strongly when applied to the *dehydrogenation of cyclohexane, which was supposed to need a group of seven metal atoms, i.e., a hexagon with one central atom. One of the most firmly established examples of a geometrically defined center is the high activity shown by the Fe (111) surface for NH_3 synthesis.

The ideas underlying active centers and es. are now widely used in interpreting cat. phenomena. That the smallest (contiguous) group of cat. active surface atoms show size effects is thus useful for the understanding of the action of mono- and *multimetallic cats. (cf. *alloys, *supports, *metal-support interactions, *clusters, etc.). **E=F**; **G** Ensemble.

G.C. BOND

Ref.: Moulijn/vanLeeuwen/van Santen; Ertl/Knözinger/Weitkamp, Vol. 2, p. 821 and Vol. 3, p. 1077; *Surf. Sci.* **1969**, *18*, 62; *J.Catal.* **1977**, *50*, 77.

ensemble size effects → ensembles

enthalpy (entropy) changes (enzyme catalysis) → activation energy (enzyme catalysis)

Entner-Douderoff pathway → glycolysis

entrapment, entrapping, encapsulation 1: enclosed cat. active metals (or metal *complexes), cf. *ship-in-the-bottle cats.; 2: → enzyme immobilization, *immobilization

entropy of adsorption entropy difference of the adsorbent before and after adsorption. If the adsorbate is immobile on the *surface then the three translation degrees of freedom of the gas molecule have to be converted into vibrational degrees of freedom. For this localized adsorption the contribution of configurational entropy has also to be taken into account. If the adsorbate is mobile and can be treated as a 2D gas then two translational degrees of freedom are retained in the *adsorption proc. The eoa. can be determined from adsorption/*desorption equil. measurements. **F** entropie d'adsorption; **G** Entropie der Adsorption. R. IMBIHL

Ref.: V. Ponec et al., *Adsorption on Solids*, Butterworth, London 1974.

environmental catalysis Protection of the environment against risky or toxic emissions is a large application of het. cat. The precise definition includes all applications which remove unwanted prods. from procs. In a wider definition are included all synth. applications of het. cat. in which environmentally useful prods. (MTBE) as fuel additives or substitution prods. for *CFCs are produced and even cat. procs. in which higher *atom economy variants are applied. In a political definition all cat. procs. which help to save resources and reduce emissions are considered as ec. In the present context only the precise definition will be considered.

About 40 % of the cat. sales belong to ec. (*catalyst market). The single largest application with about 50 % of this value is the utilization of *automotive exhaust (*three way) cats. which made the field of cat. science and technology known to the general public. The development of ec. is driven not by technology but by legislation. The state of California takes a leading role in setting environmental standards (cf. example under the keyword *hydrotreating, *zero-emission automobiles) in many technologies which must be met usually by an average of the related prods. (cars, power stations, etc.). This policy drives the evolution of technology in ever more complex fields with a wide gap in the application of these technologies worldwide (lower standards in Europe, different standards in Japan, no standards in other areas of the world).

Despite the enormous efforts in the development of ec., only a few applications are at present technologically feasible. The respective cat. systems are *supported precious metal particles (Pt, Rh, Pd, Au) or very disperse V oxides on Ti-containing oxides. The applications include: removal of CO, NOx and hydrocarbons from *automotive exhausts (*TWC) and *diesel engine emissions; removal of NOx from *stationary power stations (*DeNOx) by *selective catalytic reduction (SCR); removal of SO_2 emissions by cat. oxidation and precipitation (*SNOX, DeSOx); removal of polyaromatic chlorinated hydrocarbons from incineration plants (dioxin removal), and removal of gaseous organic cpds. from a wide range of applications by cat. total oxidation (*VOC combustion). In the near future the legislation of CO_2 limitation will cause the very severe problem of removing NOx from automotive sources with high-efficiency internal combustion engines under oxidizing conds. (lean-burn exhausts). To this problem no technologically sound solution is available yet.

The development of stable cat. procs. in this area of application is severely hampered by the following factors: the operation under uncontrolled conds. by inexperienced users for all small-scale applications (cars, VOC combustion); operation under frequent changes of gas composition and of cat. temps. (changes of 500 K within a minute); operation without control of function by the user for *automotive application; operation in the unprecedented presence of cat. poisons in power and incineration plants; operation under minimum press. drop conds. (gas flow in smoke-stacks or in auto exhaust pipes); operation in the presence of extreme excesses of water vapor and CO_2; and operation under mechanical shock conds. for automotive sources.

These boundary conds. preclude in almost all cases the transfer of existing technology for feed gas purification to environmental applications. Additional complications are the price limitations and the requirement of minimum cat. loss during its lifetime (uncontrolled *deposition of chemically active species in the environment). All these requirements represent massive barriers for the development of improved or new environmental cleanup technologies and cause excessive additional efforts over the already demanding solution of the purely chemical problems.

A different aspect of ec. is the vast and complex chemistry occurring in the Earth's atmosphere. Here a selection of small molecules emitted from the biological and geological procs. react with each other and with O or N from the atmosphere in the presence of dust particles or in hom. cat. procs. These molecules occurring naturally in concs. of 10–10 000 ppb are: CO, CH_4, HCHO, NOx, NH_3, H_2S, SO_2, CS_2, COS, CH_3Cl, HCl, and others. In concentrations of a few ppb a large number of complex organic molecules (aldehydes, ketones, aromatics, terpenes) are emitted by biological procs. which also react in mostly hom. cat. reactions with the small-molecule population in the atmosphere. Ice particles and included metal oxide traces contained in clouds are a major source of het. procs., many of them occurring as photocatalytic oxidation reactions. The phenomena of ozone depletion and of summer smog are well-known examples of complex sequences of ec. procs. These reactions would also occur without the interference of mankind, whose anthropogenic emissions cause, however, changes and imbalances in the naturally occurring reaction networks. Their extreme complexity and political bias have up to now precluded a clear and complete picture of the proc. occurring in the chemical reactor "Earth". Particularly the following keywords extend the topic: the removal of sulfur cpds., NO_x or NO_x plus SO_2 from gases and flue gases – *Claus proc., *sweetening, *selective cat. reduction, *DeNOx, *SNOX proc.; the decrease of pollutants – *automotive exhaust gases; *hydrotreating of petroleum fractions; purification of waste water – *catalytic waste water treatment, *catalytic nitrate reduction, or catalytic *CFC destruction; cf. also *green chemistry, *environmental issues. **F** catalyse d'environnement; **G** Umweltkatalyse.

R. SCHLÖGL

Ref.: Ertl/Knözinger/Weitkamp, Vol. 4, pp. 1559; Farrauto/Bartholomew, p. 580; Ullmann 6th ed.; F.I.I.G. Janssen, R.A. van Santen, *Environmental Catalysis*, Imperial College Press, London **1999**; Heck/Farrauto.

environmental issues   and cat. technologies of the future will concern acid rain (NO_x, SO_x, corrective cat. technology: *HDS, *HDN, *DeNOx), ozone layer, global warming and minimization of greenhouse gases (CFCs, CO_2, CH_4), green fuels, photochemical smog (NO_x and *hc. in urban areas, *combustion), fresh and waste water (*cat. water treatment), soil, *recycling, environmentally friendly procs. (zero-waste procs., reduction in volume of byprods., *E factor, *CO_2 as a building block, replacement of corrosive liquid acid cats.), *automotive exhaust cats., and introduction of *fuel cells ("cat. automobiles").

B. CORNILS

Ref.: *Cattech* **1997**, *1*, 15; Thomas/Thomas, p. 55.

environmental scanning electron microscopy (ESEM)   a variant of electron microscopy for the determination of morphologies in reactive atmospheres. The method is non-destructive and excellently suited for insulating samples; it works in water and in air at approx. 50 mbar press. The data are local (atomic scale) and *surface-sensitive.

R. SCHLÖGL

enzymatic deprotection (in peptide synth.)   *Enzymes can be used in the selective deprotection and liberation of the α-amino group, the carboxy group, and the various side-chain functionalities in peptide synth. The phenylacetyl group can be removed from

dipeptides in reactions cat. by *penicillin acylase without affecting other protecting groups. The C-terminal ester can be hydrolyzed by *carboxypeptidase Y and *thermitase. Thermitase also cat. the deprotection of C-terminal *tert*-butyl ester. The lipase from *Mucor javanicus* has been used in the deprotection of C-terminal glycopeptide heptyl ester. The C-terminal amide can be deprotected with a peptide amidase from the flavedo of oranges without affecting the peptide bonds and N-protecting groups. This enzyme accepts a variety of substrates with L-configuration at the C-terminal residue. Enzymes are also known for the deprotection of both Cbz- and Boc-groups, although the specificities are quite limited. **E=F=G**.

C.-H. WONG

enzymatic hydrolysis   → Nitto process for acrylamide; *Lonza nicotinamide process

enzyme catalysis (in organic solvents) Many enzymatic transformations can be performed in organic *solvents containing minimum amounts of water.   Further studies suggest that *enzymes (es.) only need a thin layer of water on the *surface of the protein to retain their cat. active *conformation. The most useful nonaqueous media are hydrophobic solvents that do not displace these essential molecules of water from es. Water-immiscible solvents containing water below the solubility limit permit certain dry es. (crystalline or lyophilized powder) to be cat. active.   Lyophilization of es. dissolved in optimal-pH solution provides the most active forms to be used in organic solvents. Within this range of water content, the enzymatic activity in an appropriate organic solvent can be optimized and, in some cases, is comparable to that in aqueous solution, and the cat. follows *Michaelis-Menten kinetics. Mechanistic investigations of serine protease-cat. reactions in organic solvents suggest that the *transition state structure in nonpolar organic solvents is nearly the same as that in aqueous solution, indicating that the microenvironment of the e. active site in nonpolar organic solvents is the same as that in water. Higher thermostability of some es. in organic solvents than in water has also been reported, presumably because enzymes are conformationally less flexible in nonaqueous media. Changes of stereoselectivity in going from water to organic solvent was also observed. The change in *stereoselectivity comes mainly from the different importance of the release of water during the binding of isomeric substrates to the e. Many reactions that are sensitive to water, or are thermodynamically impossible to perform in water, become possible in organic media. Enzyme-cat. *dehydrations, *transesterifications, *aminolyses, and *oxidoreductions in organic solvents are now common. Novel enzymatic reactions in gases and *supercritical fluids have also been exploited. *Product or *substrate inhibition can be lessened in certain enzymatic reactions in which prods. or substrates partition preferentially into the nonaqueous phase. In most cases, es. are insoluble in organic media; they can therefore be recovered by centrifugation or filtration and used repeatedly. For an example cf. *oxylonitrilase. **F** catalyse enzymatique en solvants organiques; **G** Enzymkatalyse in organischen Lösemitteln.   C.-H. WONG

Ref.: *Acc.Chem.Res.* **1990**, *23*, 114; Wong.

enzyme-catalyzed reactions (rate acceleration) *Enzymes (es.) are remarkable for their dramatic enhancements of reaction rates, often 10^8–10^{11}-fold over background. Their ability to provide such tremendous rate accelerations comes from their special 3D structures, which allows them often to fully encompass the substrate(s) and provide stabilization of the *transition state (ts.). Much of the cat. power of es. comes from proximity effects: in bimolecular reactions, the substrates are held close together, achieving a higher local conc. than would be found in solution. In addition, the 20 amino acids commonly found in proteins provide a number of acidic and basic side chains which the e. can maintain in close proximity to the substrate for general *acid-base cat., as well as nucleophilic resi-

dues for nucleophilic catalysis. H-bonding, electrostatic, hydrophobic interactions are also used to stabilize the ts., and because of the enzyme's well-defined tertiary structure, the groups responsible for such interactions can be held in the ideal geometry to maximize their stabilizing power.

Conformational restriction of both stabilizing groups and substrate(s) is a key factor in cat. Reactions in solution are slow due to the rotational and translational freedom of the substrate and *solvent (which often contributes to transition state stabilization), some of which is lost upon reaching the ts. Achieving this high-energy state thus requires not only an enthalpic cost, but a large entropic one as well. The entropic loss required to align the ts. stabilizing groups within an enzyme, however, was lost when the e. folded, and thus the entropic loss of aligning these groups does not factor into the reaction rate. Much of the conformational restriction of the substrate, which also contributes to the (negative) entropy of activation, is achieved during the initial binding step to the e., and is offset by favorable interactions with the enzyme (later studies showed this mech. to be unlikely). **F** réactions catalysés par des enzymes; **G** enzymkatalysierte Reaktionen.   P.S. SEARS

Ref.: K.H. Drauz, H. Waldmann, *Enzyme Catalysis in Organic Synthesis*, VCH, Weinheim (Germany) 1995; Sinnott; Wong; R.L.Ornstein, *Improving Enzyme Catalysis*, Dekker, New York 1999.

enzyme-catalyzed reactions, transition state theory → transition state theory

enzyme immobilization (entrapment, encapsulation, cross-linked enzyme crystals)
For enzymatic reactions, the most expensive reagent is typically the *enzyme itself. Economical large-scale enzymatic reactions require either enzyme that has been *immobilized, or a downstream enzyme *recovery step. The enzyme may be covalently linked to a solid *support (cf. immobilization). Alternatively, the enzyme may be linked to a *membrane, as in a *membrane reactor or *hollow-

fiber reactor, and the substrate solution(s) can pass them. An additional advantage of ei. is that multipoint attachment of the enzyme (attachment at several positions on the enzyme) typically stabilizes the enzyme toward *denaturation. Ei. can affect the kinetics and substrate preferences of the enzyme, however, due to direct effects upon the enzyme (distortion of the enzyme *active site during i., for example, or steric hindrance of the active site) and to indirect effects, such as *mass transfer limitations, which can be severe for i. to microporous supports. Enzymes have a variety of functional groups that can be derivatized for covalent i. Along a similar vein, enzymes are sometimes not linked to a support, but are simply cross-linked to each other. Glutaraldehyde cross-linked crystals of thermolysin, for example, showed a greatly increased stability toward protease digestion and thermal and cosolvent-induced denaturation. If the enzyme loses activity upon covalent modification, it can simply be embedded in a resin such as agarose, gelatin, or polyacrylamide to achieve a similar effect. Some resins, such as *ion exchange resins, can be used to simply adsorb the enzyme, although under some conds. (such as high ionic strength) these will allow the enzyme to bleed off (cf. *leaching) into solution.

Another alternative to covalent i. is to entrap soluble enzyme behind a membrane that provides little impedance to the flow of small molecules such as the substrate and prod. but restrains the enzyme. The enzymes may be enclosed in dialysis membrane (membrane-enclosed enzyme cat., MEEC). **F**=E; **G** Enzym-Immobilisierung.   P.S. SEARS

Ref.: *Methods Enzymol..* **1987**, *135*, 30; *J.Am.Chem.-Soc.* **1992**, *114*, 7314; *Methods Enzymol.* **1987**, *44*, 19; A. Wiseman (Ed.), *Handbook of Enzyme Biotechnology*, Ellis Horwood, Chichester 1985.

enzyme inactivation → stability of enzymes, *thermostability

enzyme inhibition the decrease in cat. *activity of an *enzyme as a result of a change

of reaction conds. or the addition of certain cpds. to the solution. These conds. can cause *conformational changes within the enzyme, blocking of *active sites, or covalent or non-covalent modification of the enzyme. Certain cpds. may inhibit enzymes. *Metalloenzymes can often be inhibited by chelators, by small *ligands that can fit in the active site and *poison the metal, or by metals other than those naturally bound to the enzyme's active site, if they can compete with the natural metal for the binding site. *Inhibition can be caused by the substrates and/or prods., or analogs of these cpds. (see *substrate and *product inhibition). Cpds. that mimic the *transition state of an enzymatic reaction will be good inhibitors of that enzyme, and so transition state analog design is a common approach for designing specific eis. though it requires detailed knowledge of the enzymatic reaction mechanism.

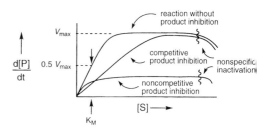

Inhibition may be reversible or irreversible. There are several modes of reversible inhibition: competitive, *noncompetitive, *uncompetitive, and *mixed. These types of inhibition can be distinguished experimentally and are usually characterized by graphing the kinetic data taken at several different inhibitor concs. on *Dixon, *Lineweaver-Burk, or *Eadie-Hofstee plots. Irreversible inhibition (*inactivation) can be caused by enzyme unfolding or by covalent modification of key groups within the enzyme, either through the action of nonspecific reagents, such as iodoacetamide, or through specific active-site directed irreversible inhibitors. **F** inhibition d'enzymes; **G** Enzyminhibierung.   P.S. SEARS

Ref.: H. Zollner, *Handbook of Enzyme Inhibitors*, 2 Vols., Wiley-VCH, Weinheim 1999.

enzyme mechanism (kinetic) The kinetic mechanism (as opposed to the chemical mechanism) of an *enzyme is the sequence of the binding and reaction events. For a reaction with more than one substrate, the kinetic mechanism describes whether the substrates bind separately, in a set order (ordered sequential) or in a random order (random sequential); or whether one substrate binds, reacts with the enzyme, and ejects the first prod. before the second substrate enters (ping-pong). The number of reactants and prods. are described as *uni* (single), *bi* (two), or *ter* (three). Thus, an ordered sequential bi-bi reaction would be one in which there are two substrates which bind in a certain order and react to form two prods. which exit in a set order.

Kinetic reaction mechanisms are derived from a large variety of data. **F** méchanisme enzymatique; **G** enzymatischer Mechanismus.

P.S. SEARS

Ref.: I.H. Segel, *Enzyme Kinetics*, Wiley, New York 1975.

enzyme membrane reactor → membrane reactor

enzyme model → active site model

enzyme selectivity (solvent polarity) → enzyme catalysis

enzymes are biological *cats. composed primarily of protein, often with *cofactors such as *NADH and *ATP, or metals/metal *clusters. Like other cats., they merely accelerate the *rate of reaction, rather than changing the reaction equil. They are noteworthy for their exquisite acceleration of reaction *rates (often 10^8–10^{11}-fold increases over background), their extreme *stereo- and *regioselectivities, and typically very high substrate specificities. As biological cats., they usually can work under very mild conds. of temp. and pH, and are in fact are usually deactivated under harsh conds.

Es. are classified according to the type of reactions they catalyze (cf. *EC). E.g., *transferases cat. the transfer of any of a variety of groups, such as phosphates, acyl groups, or sugars, etc., from a donor to an acceptor. Transferases for which the acceptor is water are given a class of their own: the *hydrolases, which include the *proteases, *lipases, glycosidases, *phosphatases, and so on. *Oxidoreductases cat. *oxidation/*reduction reactions, and nearly always require organic *cofactors, metals, and/or metal *clusters to act as electron carriers (though some use unusual amino acids, such as selenocysteine). *Lyases cat. the addition of some molecule across a double-bond (or in the reverse direction, cat. *elimination of some group to form a double bond). *Isomerases cat. a variety of isomerization reactions, such as the conv. of glucose to fructose (*glucose isomerase). *Ligases cause the joining of two molecules with the concomitant *hydrolysis of a nucleotide triphosphate (usually *ATP). **E=F**; **G** Enzyme. C.-H. WONG

Ref.: Schomberg/Salzmann; Wong; W. Gerhartz (Ed.), *Enzymes in Industry*, VCH, Weinheim (Germany) 1990.

enzyme-substrate complexes The first step in enzymatic cat. is the binding of the substrate(s) to the *enzyme. The complex formed is called the enzyme-substrate complex. Because of the exquisite specificity many enzymes display for their natural substrates, it was initially proposed that the enzyme had an active site that fitted the substrate perfectly, like a *lock and key. Later it was realized that enzymes must stabilize the *transition state of the reaction with respect to the ground state of the substrate in order to cat. the chemical transformation.

There is a fine balance, then, between binding the substrate well enough to allow for a sufficient degree of association between the enzyme and substrate, and binding the substrate so tightly that the *transition state cannot be further stabilized. There clearly must be an optimal association constant (cf. *Scatchard plot) for a given reaction which balances the need for binding so well that, with respect to in-vivo concs. of substrate, the enzyme can bind enough substrate to function, with the need for binding so weak that the transition state can be further stabilized. **F** complèxes enzyme-substrate; **G** Enzym-Substrat-Komplexe.

P.S. SEARS

EO ethylene oxide, manuf. by heterogeneous vapor-phase oxidation over Ag cats.

Epal process → Ethyl procs.

EPDM an elastomeric *terpolymer containing ethylene, propene, and a small amount (<12 wt.%) of a nonconjugating diene made by vanadium-based *Ziegler-Natta- or *metallocene-cats.

W. KAMINSKY

epimerase → mutarotation

epitopes Different *antibodies will bind to different regions of the surface of macromolecules. The particular area to which a given antibody binds is called its binding epitope.

P.M. SEARS

EPMA electron probe microanalysis, → electron microscopy

epoxidation The conv. of an alkene to the corresponding *oxirane (epoxide) is referred to as epoxidation. The conv. is achieved with a variety of oxidants and metal cats. Alkenes not containing reactive allylic C-H bonds (e.g., ethylene, butylene, etc.) are epoxidized in the gaseous phase with molecular O_2 or air over *supported Ag cats. E. with alkyl *hydroperoxides proceeds with early *transition elements in high oxidation states, e.g., Mo(VI), W(VI), V(V), and Ti(IV), and is broadly applicable. A het. Ti(IV)-SiO_2 cat. is also very effective (cf. *ARCO and *Shell PO procs.). V is the cat. of choice for e. of allylic alcohols with alkyl hydroperoxides. E. with (aqueous) H_2O_2 can be performed with a tungstate/phosphate cat. under *PTC conds., *methyltrioxorhenium in CH_2Cl_2/H_2O or a

het. *Ti silicalite cat. Asymmetric e. of allylic alcohols with alkyl hydroperoxides is cat. by Ti(IV) complexes of tartrate esters. Asymmetric e. of unfunctionalized prochiral alkenes with NaOCl is cat. by Mn(III) *complexes of *chiral *Schiff's bases. Examples of commercial procs. are *Butachem, *Daicel, *Laporte, *Lummus.

The *enzyme chloroperoxidase (CPO) cat. the asymmetric e. of olefins with H_2O_2. For enzymatic epoxidation cf. *monooxygenase. **F=E**; **G** Epoxidierung.

R.A. SHELDON

Ref.: Cornils/Herrmann-1, p. 411; McKetta-2, p. 431; *J.Mol.Catal.* **1980**, 7, 107; *J.Org.Chem.* **1996**, 61, 8310; Ojima, pp. 101 and 159; Ertl/Knözinger/Weitkamp, Vol. 5, p. 2244.

epoxide hydrolases have been used in selective *hydrolysis of epoxides. Synth. of D-, L-, and *meso*-tartaric acid from the precursor epoxides cat. by microbial ehs. still represent the most useful reactions of epoxide hydrolysis. The microsomal ehs. prefer cyclic *cis*-meso-epoxides as substrates, attacking the S-center to give the (R,R)-diol. The microsomal eh. cat. involves an ester intermediate, formed by the reaction of a carboxylate group of the *enzyme and the epoxide. **E=F=G**.

Ref.: *Tetrahedron* **1997**, 53, 15617.

C.-H. WONG

epoxy resins → *CAT-resins

EPR electron paramagnetic resonance spectroscopy, → electron spin resonance spectroscopy

equilibrium adsorption → grafting, *coprecipitation

equilibrium-controlled synthesis (enzyme-cat.) Since *enzymes, like all cats., merely increase the *rate of reaction without changing the equil., it is possible to use enzymes in the reverse of their natural direction through a variety of techniques to shift the reaction equil. (e.g., by raising the conc. of one of the reactants). For hydrolytic enzymes such as *proteases, the water conc. to shift the equil. toward synth. can be reduced to (cf.

*enzyme catalysis in organic solvents). Alternatively, substrates may be selected such that the reaction prod. is insoluble in water and precipitates, driving the reaction to completion. This is the case, for example, in the *thermolysin-cat. synth. of the artificial sweetener *aspartame. **F** synthèse par équilibre controlé; **G** gleichgewichtskontrollierte Synthesen.   P.S. SEARS

erionite a naturally occurring *zeolite with the simplified formula $(Na_2,K_2,Ca)_4$ [Al_8-$Si_{28}O_{72}$]·28 H_2O. It belongs to the group of small-pore zeolites. Its channels, which build up a 3D-channel system, are lined by elliptical eight-membered ring windows having an aperture of 3.6 × 5.1 Å. The *zeolite structural code for the tetrahedral framework structure of erionite topology is ERI. Other materials possessing this topology are LZ-220, Linde T and the → aluminophosphate AlPO-17. **E=F=G**.   P. BEHRENS

ESCA electron spectroscopy for chemical analysis, → X-ray photoelectron spectroscopy.

Escambia methacrylic acid process

proc. proposed for the manuf. of methacrylic acid (MA) by *oxidation of *i*-butene with N_2O_4 to γ-hydroxyisobutyric acid, followed by *dehydration to MA. **F** procédé Escambia pour l'acide méthacrylique, **G** Escambia Methycrylsäure-Verfahren.   B. CORNILS
Ref.: Weissermel-Arpe, p. 283.

Eschenmoser salt a condensation agent, → *Mannich reaction

Eschweiler-Clarke reaction → Leuckart reaction, *transfer hydrogenation

ESEM → environmental scanning electron microscopy

ESKA (Experten System für KAtalysatoren = expert system for cats.) was designed at BASF, especially for *hydrogenation reactions. The main comp. of a cat. is proposed on the basis of activity patterns of cat. for different types of hydrogenations. Secondary cat. comps. and *support materials may also be suggested.   M. BAERNS
Ref.: *DECHEMA Monographs* **1989**, *116*, 46.

ESR → electron spin resonance spectroscopy

esterases are *enzymes which cat. the *hydrolysis of esters and also *transesterification, *esterification, and *aminolysis reactions in high concs. of organic solvents. These enzymes often contain a serine group in the *active site and the reaction proceeds through an acyl-enzyme intermediate. Many *proteases and *lipases also exhibit esterase activities. **E=F=G**.   C.-H. WONG
Ref.: U. Bornscheurer, R. Kazlauskas, *Hydrolases in Organic Synthesis*, Wiley-VCH, Weinheim 1999.

esterification a *condensation reaction and formation of esters (carboxylic acid esters) by reaction of acids (organic or inorganic) with alcohols involving elimination of water (*Taft equ.) acc. to R-COOH + R'OH → R-COOR' + H_2O. The reverse reaction is *hydrolysis (*saponification). Esters are also formed by other reactions, including convs. with acid anhydrides, acid chlorides (Schotten-Baumann reaction), amides, nitriles, unsaturated hcs., etc., including ester interchange (*transesterification). Es. are cat. by mineral acids, metal salts such as Sn or Ti alkoxides, BF_3, or *ion exchange resins. Esters play an important role either in organic and biochemistry (*triglycerides, *fats, *plasticizers) or with *polymers (*DMT, *PET). **F** estérification, **G** Veresterung.   B. CORNILS
Ref.: Kirk/Othmer-1, Vol. 9, p. 755; Kirk/Othmer-2, p. 433; McKetta-1, Vol. 19, p. 381; McKetta-2, p. 637; Ullmann, Vol. A9, p. 561.

ester interchange → transesterification

ester splitting (cracking) Cat. proc. for the treatment of the ester fraction of *hydroformylation prods. (Co-based version). The butyl esters and the residua of the aldehyde workup are cracked, e.g., over special *aluminas increasing

the yield of *n*- and *iso*-butyraldehydes and alcohols. **F** cracking des esters; **G** Esterspaltung.
Ref.: DE 1.817.051, US 3.462.500.   B. CORNILS

ESYCAD (Expert SYstem for CAtalyst Design) has been designed for selecting cat. materials for different reaction classes and individual reactions. The system possesses an extensive knowledge base and an acquisition facility which allows the user to enter his own knowledge (expertise) and to apply the program to cat. systems acc. to his interest.
Ref.: *Chem. Ing. Tech.* **1990**, *62*, 365.   M. BAERNS

et, Et ethyl group

ETC → electron transfer chain catalysis

ethanol (ethyl alcohol, EtOH) manuf. by *fermentation procs. and industrially by *hydration of ethylene (e.g., *Shell or *VEBA ethyl alcohol procs.). EtOH is, like *MTBE, used as an oxygenate (antiknock) additive in gasoline (*oxyfuel) and is favored over MTBE acc. to the new US legislation.
                                                 B. CORNILS

ethenolysis cat. *cross-metathesis of linear alkenes and ethylene yielding α-olefins (*Shell's SHOP proc.).

etherification, ethers is in cat. terms a proc. for the manuf. of *MTBE or *TAME by *addition reaction of *tert.* alkenes (such as *i*-butene, amylene) and alcohols (procs. of *CDTech., *Hüls/UOP, *IFP, *Neste NEx, *Phillips, *Snamprogetti). **F** formation d'éthers; **G** Veretherung.   B. CORNILS
Ref.: Kirk/Othmer-1, Vol. 9, p. 860; Kirk/Othmer-2, p. 437; Ullmann, Vol. A10, p. 23; Ertl/Knözinger/Weitkamp, Vol. 4, p. 1986.

Ethermax process → Hüls/UOP procs.

Etherol process → BP/Erdoelchemie process

ethoxylation (oxethylation) *insertion reaction of -CH$_2$-CH$_2$-O- groups into cpds. with

acid H atoms such as alcohols, alkylphenols, fatty amines, etc. Typical are reactions of *EO or *PO (propoxylation) at higher temp. and press. yielding commercially valuable ethoxylates (propoxylates) under base cat. (NaOAc, NaOH). Other cats. are alkaline-earth metal salts. **F** ethoxylation; **G** Ethoxylierung.
                                                 B. CORNILS

Ethyl Alfen process two-step manuf. of linear α-olefins (LAO) by ethylene *oligomerization with AlEt$_3$ cats. (*Ziegler cats.). **F** procédé Alfen d'Ethyl; **G** Ethyl Alfen-Verfahren.
                                                 B. CORNILS

Ethyl Alfol process multi-step proc. for the manuf. of primary linear alcohols by 1: ethylene *oligomerization, with AlEt$_3$ cats., 2: *alkoxide formation with air, and 3: their *hydrolysis (cf. *aufbau reaction). **F** procédé Alfol d'Ethyl; **G** Ethyl Alfol-Verfahren.
                                                 B. CORNILS

ethylbenzene (EB) important, cat. manuf. intermediate for *styrene synth. via *dehydrogenation (e.g., procs. of *ALBENE, *BASF/Badger, *CdF/Technip, *CDTech, *Lummus/UOP, *Mobil/Badger, *Monsanto/Lummus, *Mobil, *UOP Alkar, etc.) or from methylphenylcarbinol as a byprod. of propylene oxide plants (procs. of *ARCO, *Halcon, *Oxirane, *Shell SMPO).   B. CORNILS
Ref.: Ullmann, Vol. A10, p. 35; McKetta-1, Vol. 20, p. 77.

Ethyl Epal process a variant of the Ethyl Alfol proc. including a *transalkylation step (narrow chain length) (Figure).   B. CORNILS

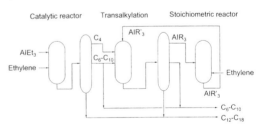

ethynylation 1,2-addition of terminal alkynes to carbonyl cpds. to form the correspond-ng propargyl alcohols (*Reppe, 1928). The most important es. are those with acetylene. E. of aldehydes proceeds best when cat. with Cu. However, the reactivity of higher aldehydes decreases rapidly, so the most important prod. is butynediol from HCHO and acetylene. Cu cpds. on inert *carriers and/or unsupported cats. such as Cu malachite are used. The cat. active comps. are the intermediately formed Cu acetylides (cf. *copper as cat. metal).

A variant of e. is the *Mannich reaction, which is also cat. by Cu. Propargylamines are manuf. from aldehydes by reaction with alkynes and secondary amines. Ketones do not react with Cu but need basic cats. (e.g., het. with *ion exchangers, hom. with KOH/NH₃). F=E; **G** Ethinylierung.   J. HENKELMANN

eukaryotes organisms with true membrane-bound *organelles* (enclosed capsules which sequester different functions of the cell). The term actually means *true nucleus*, and in eu. the DNA is enclosed in the membrane-bound nucleus. In the prokaryotes (eubacteria and archaea), by comparison, the DNA is not separate from the cytoplasm. Eukaryotic cells are different from the *prokaryotes in many fundamental ways. They are structurally very distinct, with a variety of very specialized organelles that enclose small volumes with often very different environments from that found in the cytoplasm. Examples of organelles are mitochondria (respiration, *Krebs cycle), *peroxisomes (oxidative reactions, destroying peroxides), or lysosomes (degradation of macromolecules). **E=F=G**.   P.S. SEARS

Euler-Chelpin Hans von (1873–1964), professor of organic chemistry at Stockholm (Sweden). Worked on fermentation *enzymes and confirmed the structure of *NAD⁺. Nobel laureate in 1929 (together with A. Harden).
Ref.: *Chem.-Ztg.* **1964**, *88*, 933.   B. CORNILS

Eupergit C polyacrylamide beads, activated with epoxide groups resin, commonly used for the *immobilization of *enzymes.
Ref.: *Biotechnol.Bioeng.* **1980**, *2*, 157.   P.S. SEARS

Eurecat S.A. a joint venture of *Akzo Nobel catalysts and is headquartered in La Voulte sur Rhône (France). Eurecat offers cat. *regeneration and cat. pretreatment services and has operations in the US, Japan, and Saudi Arabia.   J. KULPE

EVA (E/VAC) are ethylene-vinyl acetate *copolymers

EVC International ethylene dichloride process further development of Stauffer's *EDC proc. by *oxychlorination

EVC vinylchloride process manuf. of vinyl chloride monomer (VCM) by thermally induced dehydrochlorination (cf. *dehydrohalogenation) of 1,2-dichloroethane which is produced from ethylene via chlorination (*halogenation) or *oxychlorination (Figure). **F** procédé EVC pour le chlorure de vinyle; **G** EVC-Vinylchlorid-Verfahren.
                                   W.A. HERRMANN

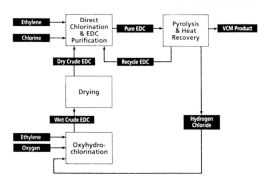

Ref.: B.E. Leach (Ed.), *Applied Industrial Catalysis*, Academic Press, New York 1983; *Hydrocarb.Proc.* **1997**,(3),161.

evolution → directed evolution, *in-vivo evolution of biocats.

EXAFS → extended X-ray absorption fine structure spectroscopy

EXCAT LLDPE from Exxon's *Exxpol process

EXELFS → electron microscopy

exon → intron

expert systems in catalysis → knowledge-based systems in catalysis

extended X-ray absorption fine structure spectroscopy (EXAFS, XANES) a variant of the X-ray absorption technique (probe and response: X-ray photons). The focus of investigation is local atomic structure (cf. *NEXAFS). The ex-situ technique is non-destructive to amorphous or crystalline samples and good for *in situ* studies. The data are local and bulk-sensitive. R. SCHLÖGL
Ref.: Niemantsverdriet, p. 150; Ertl/Knözinger/Weitkamp, Vol. 2, p. 475.

external mass transport → interparticle mass transport

extraction procedure to separate, concentrate, and purify the different metals of a reworked, spent cat., especially by a combination of *precipitation and liquid/liquid extraction (liquid ion exchange). *Extraction is also used for the workup of deposits on loaded cats. by solvents or *sc. fluids (cf. *regeneration, e.g., *Fischer-Tropsch cats.) and for the separation of reaction prods. from the catalyst/prod. mixture from hom. cat. reactions. **F=E**; **G** Extraktion. C.D. FROHNING

extraparticle (interphase) mass transport → mass transport

extremophile an organism that lives in an extreme environment; commonly "extreme" in terms of salinity, temp., or pH. The thermostable DNA *polymerases from such organisms, e.g., have proven very useful in the *polymerase chain reaction (PCR), since they remain stable at temps. where double-stranded DNA melts, thus permitting the successive melting, reannealing, and extension reactions required for PCR (cf. also *thermostability, enzymes). **E=F=G**. P.S. SEARS

Exxon alkylation process for the manuf. of high-octane comps. by *alkylation of C_3-C_5 alkenes and *i*-butane in the presence of liquid H_2SO_4 cat. in stirred, autorefrigerated reactors (Figure). **F** procédé Exxon d'alkylation **G** Exxon Alkylierung. B. CORNILS

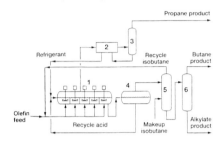

1 reactor; 2 compressor; 3 depropanizer; 4 settler; 5 deisobutanizer; 6 debutanizer.
Ref.: *Hydrocarb.Proc.* **1996**, (11), 91.

Exxon catalytic reforming *catalytic reforming proc. over Pt-Ir on *alumina cats.

Exxon DODD (diesel oil deep desulfurization) process → Exxon Hydrofining process

Exxon Donor Solvent (EDS) process for the hydrogenative extraction of coal with a hydrogen-generating solvent (donor solvent) at >425 °C/17.5 MPa. The coal/solvent slurry moves upward through the reactor together with H_2. A portion of the distillate stream is used as the recycle solvent. Before recycling, the H_2-poor solvent is regenerated over a fixed-bed loop, using conventional Co-Mo or Ni-Mo cats. (e.g., Ni-MoO$_3$). The prods. are gases (C_1/ C_2 as *SNG, C_3/C_4 as *LPG), naphtha, middle distillates and residue, which may pass a *hydrotreating step or be coked in a *flexicoker. Approx. 50–70 % of the coal feed is converted to gaseous and liquid prods., depending on the type of coal and operating conds. A pilot plant in Baytown (TX) operated between 1976 and 1985 with max. 250 t/d. **F** procédé EDS d'Exxon; **G** Exxon EDS-Verfahren. I. ROMEY

Exxon Flexicoking process an integrated fluid coking and coke gasification of vacuum

residuum. **F** procédé Flexicoking d'Exxon;
G Exxon Flexicoking-Verfahren.   B. CORNILS
Ref.: *Hydrocarb.Proc.* **1978**, (9), 104; McKetta-2, p. 882.

Exxon Flexicracking process   for the
fluidized cat. (*FCC) conv. of virgin, cracked,
or deasphalted oils to lower prods. such as alkenes, gasoline, middle distillates, etc. **F** procédé Flexicracking d'Exxon; **G** Exxon Flexi-
cracking-Verfahren.   B. CORNILS
Ref.: *Hydrocarb.Proc.* **1978**, (9), 107; *Hydrocarb.Proc.*
1996, (11), 121.

Exxon GO Fining process   fixed-bed *desulfurization (*HDS) of virgin vacuum gas
oils, thermal and cat. cycle oils, or coker gas
oils.; also used to pretreat *catcracker feeds
to improve crackability and reduce SO_2 pollution. *Residfining is a variant with an incorporated reactor plugging protection system.
F procédé GO Fining d'Exxon; **G** Exxon GO
Fining-Verfahren.   B. CORNILS
Ref.: *Hydrocarb.Proc.* **1978**, (9), 130.

Exxon Hycracking process   for *hydrocracking virgin and cracked gas oils (atmospheric and vacuum gas oil) in a fixed-bed
cat. operation. **F** procédé Hycracking d'Exxon; **G** Exxon Hycracking-Verfahren.
Ref.: *Hydrocarb.Proc.* **1978**, (9), 116.   B. CORNILS

Exxon hydrofining process   for the cat.
(Ni-Mo, Co-Mo) fixed-bed treatment of various feedstocks, including *sweetening and
improvement of color, sulfur content
(*DODD), stability, and burning qualities of
kerosene and jet fuels. Other versions include
trickle-bed reactors. **F** procédé Hydrofining
d'Exxon; **G** Exxon Hydrofining-Verfahren.
                                     B. CORNILS
Ref.: *Hydrocarb.Proc.* **1978**, (9), 139; *Hydrocarb.Proc.*
1996, (11), 132.

Exxon LAO process   for the manuf. of ethylene oligomers (*LAOs) with Al alkyl chlor-
ide/$TiCl_4$ cats. **F** procédé Exxon pour LAOs;
G Exxon LAO-Verfahren.   B. CORNILS

Exxon LDPE process   proceeds under high
press. in free-*radical *polymerization in tubular reactors (or stirred tank reactors)   for
the manuf. of LDPE homopolymers or *EVA
copolymers. **F** procédé Exxon pour LDPE;
G Exxon-LDPE-Verfahren.   B. CORNILS

Exxon LLDPE process   Besides all their
other advantages, the use of *metallocene
cats. only causes a minimal change in processing technology for polymer prods. So many
plants used for free-*radical ethylene *polymerization have been converted for the cat.
synth. of *LLDPE. Production is possible
with both stirred and tubular *autoclaves operating at 30–200 MPa and 170–350 °C, i.e., a
mixture of *sc. ethylene and molten polymer
being the reaction medium (*catalytic reactors). The *residence times in these reactors
are short, ranging from 1 to 5 min.
Exxon is using such a *polymerization proc.
(*Exxpol) with metallocenes to produce ethylene *copolymers with 1-butene and 1-hex-
ene, called *EXCAT resins. The comb. of advantageous properties results in a lower
amount of extractables, a lower tendency of
such films to block, and a better impact
strength compared to conventional LLDPEs.
F procédé Exxon pour LLDPE; **G** Exxon
LLDPE-Verfahren.

                    W. KAMINSKY, U. WEINGARTEN

Exxon Neo acid process   for the manuf. of
branched acids by *hydroxycarboxylation of
alkenes under the influence of acid cats. like
H_2SO_4, HF, BF_3/H_3PO_4, etc. (similar to the
*Shell Versatic proc.) (Figure see p. 214).
F procédé Neo acid d'Exxon; **G** Exxon Neo
Acid-Verfahren.   B. CORNILS
Ref.: *Hydrocarb.Proc.* **1977**, 56(11), 186.

Exxon oxo process   is the former *Kuhlmann proc.

Exxon Powerforming process   a cyclic
or semi-regenerative cat. *reforming proc. for
the conv. of low-octane naphthas to gasoline
blending stocks with up to 105 RON clear. P.

also produces large quantities of H. The cats. are *bimetallic on acid *supports. **F** procédé Powerforming d'Exxon; **G** Exxon Powerforming-Verfahren. B. CORNILS

Ref.: *Hydrocarb.Proc.* **1972**, (9), 118 and **1978**, (9), 164.

Exxon Residfining process is the *Exxon GO Fining proc. without a prereactor

Exxpol catalyst Exxon's *single-site cats. (*metallocene cats.) together with the appropriate proc. for the prod. of polyolefins.

Eyring equation (in enzyme catalysis). The Ee. relates the rate constant of a reaction to the difference in free energies of the *transition state and ground states. It is fre-

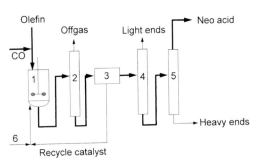

1 reactor; 2 stabilizer; 3 catalyst separation; 4, 5 distillation columns.

Exxon Neo acid process

quently used in calculations of this "free energy of activation", and has been applied to enzymatic reactions as well. In the limit of low substrate concs. [S], these reactions follow second-order kinetics $v = (k_{cat}/K_m)$ [E][S]. The second-order rate constant, k_{cat}/K_m, can be used to calc. the free energy of activation of the bimolecular reaction (assuming the *enzyme has a simple kinetic mechanism) via the Ee. Applying a similar calcn. to the first-order rate constant k_{cat} gives the *activation energy required to reach the *transition state from the *enzyme-substrate complex. This approach has proven useful in understanding enzyme cat. It underscores the importance of the idea that an enzyme must stabilize the transition state of the reaction, not the substrate. Efforts to increase cat. rates must focus on reducing the energy barrier between reactants and the transition state. *Protein engineering to reduce K_m by increasing the affinity of the enzyme for substrate without a corresponding (or greater) increase in the affinity for the transition state will decrease k_{cat}. The energetic advantages of alterations in the enzyme or substrate can be calcd. from the rate constants. Efforts to design *inhibitors of the enzyme are better focused on cpds. that resemble the transition state, not the substrates or prods. (see also *activation energy). **F** équation d'Eyring; **G** Eyring-Gleichung.

P.S. SEARS

F

Fab → antigen binding fragment

face (planar) atoms → single crystals

FAD → flavine adenine dinucleotide

farnesyl diphosphate synthetase → ju-veline hormone

fat hardening an important application of *selective (partial) *hydrogenation, developed in 1902 by *Normann. Nature provides numerous oils which contain glycerides which are the glycerol esters of long-chain fatty acids containing up to three C=C bonds, mostly in the *cis*-configuration. They are oxidatively unstable and their use is limited to cooking: to provide a butter substitute, i.e., margarine, it is necessary to hydrogenate some of the C=C bonds, leaving only one *cis*-configured bond. The prod. is then stable and spreadable. The C=C bonds are initially non-conjugated, but are brought into conjugation by double-bond migration, after which the selective *reduction of those which are not desired becomes possible. This is achieved by a $Ni-SiO_2$ cat. (or Ni-kieselguhr, doped with Zr, or Ni-Ag cats.) working under conds. of partial *diffusion-limitation. **F** solidification des graisses et huiles; **G** Fetthärtung.   G.C. BOND

Ref.: Ertl/Knözinger/Weitkamp, Vol. 5, p. 2221; Farrauto/Bartholomew, p. 441.

fat hydrogenation the saturation of fat's naturally occurring double bonds; a special case is *fat hardening. Fatty alcohols are obtained by high-press. *hydrogenation of fatty acids or fatty acid esters over Cu chromite or other (e.g., Ba-promoted) *Adkins cats. For environmental reasons there is a trend to substitute Adkins by Cr-free Cu-Zn cats. (cf. procs. of *Henkel direct hydrogenation, *Lur-gi). **F** hydrogénation de graisse; **G** Fetthydrierung.   B. CORNILS

Ref.: Ullmann, Vol. A10, p. 245; McKetta-1, Vol. 21, p. 207.

fats and oils General term for renewable materials which are of solid, soft, or liquid consistency (e.g., butter, palm kernel, soybean, castor or fish oil, tallow, peanut butter) from bioorganic origin (plants, animals); chemically the mixed glycerol esters of fatty acids (saturated, mono-, or multiply unsaturated) of even C numbers. Cat. procs. such as *hydrolysis, *hydrogenation (e.g., *Henkel Direct Hydrogenation), *fat hardening, or *Twitchell cleavage convert fs. to glycerol and fatty acids. **F** corps gras et huiles; **G** Fette und Öle.   B. CORNILS

Ref.: Ullmann, Vol. 10, p. 173, 245, 277; Kirk/Othmer, Vol. 10, p. 252.

fatty acid biosynthesis Fatty acids are synth. in the endoplasmic reticulum by a multienzyme system called fatty acid synthase. The pathway starts with *acetyl CoA, which is converted to malonyl CoA via a *carboxylation reaction cat. by acetyl CoA carboxylase. These two building blocks are then transferred from the thiol of the *coenzyme A (CoA) to a free thiol of the acyl carrier protein (ACP), a reaction cat. by the enzyme ACP transacetylase. The *condensation begins with acetyl-SACP as an electrophile and malonyl-SACP as a nucleophile under the cat. of β-ketoacyl ACP synthase.
Further reactions yield fatty acyl CoA via C_4-β-ketoacyl-SACP, butyryl-SACP, and transacylation. Unsaturation may then occur using molecular O and a *nicotinamide cofactor. **F** biosynthèse des acides gras; **G** Fettsäure-Biosynthese.   C.-H. WONG

fatty acid biosynthesis

fatty acid degradation   occurs via β-oxidation. A fatty acid is first converted to the thioester, fatty acyl *coenzyme A (CoA) in the presence of *ATP and CoA-SH. The fatty acyl CoA is *oxidized to the *trans*-α,β-enoyl-CoA by acyl-CoA dehydrogenase, a flavoenzyme. Water is added across the double bond of the α,β-unsaturated thioester by enoyl-CoA hydratase, and the resulting β-hydroxyacyl-CoA is oxidized further by an *NAD$^+$-dependent dehydrogenase to give the β-ketoacyl-CoA. Further steps involve β-ketoacyl thiolase for cleavage of the ketoester. The *enzyme intermediate will then react with another CoA-SH molecule to give another fatty acyl-CoA for another round of β-oxidation. **F** dégradation des acides gras; **G** Fettsäureabbau.   C.-H. WONG

fatty acid synthase   → fatty acid biosynthesis

FAU   *zeolite structural code for *faujasites

faujasite   Fs. (*zeolite structural code: FAU) represent the prototype group of *alum(in)osilicate *zeolites (cf. Figure under keyword *zeolite). They are built from sodalite-type cages (β cages with a free inner diameter of ca. 7 Å) which are linked together by double six-membered rings. By this way of linking, a large cage, the so-called *super-cage with a diameter of ca. 13 Å and an access win-

dows of 7.4 Å, is formed. The cages in this very open structure are accessible by a three-dimensional channel system. The name-giving mineral faujasite was formed hydrothermally in basaltic rocks. Synth. fs. are also prepared hydrothermally with Si:Al ratios up to 2.5. For many cat. purposes, even higher Si:Al framework ratios are desirable, but these cannot be obtained by direct synth., but only by post-synthetic dealumination. Designations of strongly or fully dealuminated faujasites are US-Y (ultrastable zeolite Y) or DA-Y (dealuminated zeolite Y), resp. (*zeolite structural code).

Fs. possess many very important industrial applications in het. cat., especially in *petroleum processing. Mostly the intrinsic acidic properties of their protonated forms are exploited, but cat. *activity may also be traced back to metal ions that have been incorporated via *ion exchange. Many new ideas in the field of zeolite cat. are tested first using fs., for example zeolites for basic cat. (generated by introducing cesium ions or cesium oxide *clusters into the voids) or *enzyme mimicking by zeolites (based on the deposition of metal complexes with *phthalocyanines as *ligands in the super-cages). **E=F**; **G** Faujasit.   P. BEHRENS

Favorsky-Babayan reaction   synth. of secondary and tertiary acetylenic alcohols or glycols from terminal alkynes and ketones or aldehydes. The reaction is cat. by strong bases and can be carried out in liquid ammonia. **F** réaction de Favorsky-Babayan; **G** Favorsky-Babayan-Reaktion.   A.F. NOELS

FCC process   fluid catalytic cracking, a special variant of the *cat. cracking procs. of *petrochemistry (*petroleum processing industry, *refinery) for the cat. cleavage of higher to lower molecules at higher temps. and press. in *fluidized-bed or *riser reactors (cf. also circulating-bed reactors, CBR, *catalytic eactors). The target of FCC are higher yields light olefins, gasoline, jet fuels) and better qualities (octane number, color value).

The spent cats. are collected in stand-pipes, purged with an air/oxygen stream to the bottom of a *regenerator and blown through the regenerator, where the coke is burned off (Figure; cracking under hydrogen atmosphere, *hydrocracking). The regenerated cat. is collected and purged with the gas oil feed in *risers to the reactor. The cats. consist of *clays (acid-treated), synth. *silica-*alumina, silica-*magnesia, or *zeolites. Examples: procs. of *Amoco, *CdF, *Engelhard, *Exxon, *Gulf, *Kellogg, *Lumnus, *Stone & Webster, *Texaco, *UOP. The global value of FCC cats. is approx. 10 % of the total *catalyst market. **E=F=G.**   B. CORNILS

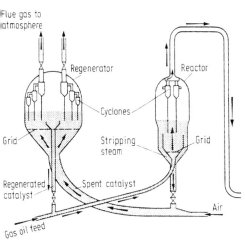

Ref.: Kirk/Othmer-1, Vol. 5, p. 419; Ertl/Knözinger/ Weitkamp, Vol. 4, p. 1955; M.L. Occelli, P. O'Connor, *Fluid Cracking Catalysts*, Dekker, New York 1998; Ullmann, Vol. A18, p. 62.

FDP → D-fructose 1,6-diphosphate

FDS flash desorption spectroscopy, → thermal desorption spectroscopy

FEAST process   → Shell procs.

feedback inhibition A common and logical method of controlling biochemical pathways in cells is for the prod. of a biosynth. pathway inhibit the *enzymes that cat. the

first committed step in its synth. This is called fi. and is found, for example, in the *biosynthesis of the aromatic amino acids tyrosine (Tyr), phenylalanine (Phe), and tryptophan (Trp). These are produced from *chorismate, a molecule made in several steps from *phosphoenolpyruvate and erythrose-4-phosphate. Chorismate can be used in the synth. of Tyr and Phe via the action of *chorismate mutase which isomerizes it to prephenate, or can be converted to tryptophan (Trp) in several steps beginning with conv. to anthranilate. In *eukaryotes, high concs. of Tyr or Phe turn off chorismate mutase, which shunts the chorismate into Trp synthesis. Moreover, they inhibit enzymes much further upstream in the reaction pathway, including the first step, the conv. of erythrose-4-phosphate and phosphoenolpyruvate to 2-keto-3-deoxyarabinoheptulosonate, a precursor (by several steps) of chorismate. **E=F=G.**   P.S. SEARS

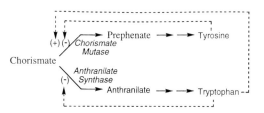

Felkin-Anh model predicts the stereochemical outcome of the kinetically controlled addition of nucleophiles to aldehydes or ketones with a stereogenic center α to the C=O group. Acc. to Felkin (steric control) the preferred *transition state is based on a *conformation where the bulkiest of the α *ligands (L) adopts a perpendicular relationship to the plane of the carbonyl group *anti* to the incoming nucleophile. The sterically next most demanding substituent (M) is placed *gauche* to the carbonyl function (Figure).

Anh stressed the importance of stabilizing orbital interactions (stereoelectronic control). Thus, the stabilization of the HOMO derived from the interaction of the HOMO of the nucleophile with the π^* orbital (LUMO; cf. *frontier molecular orbital) of the CO group is best stabilized by overlap with the the the σ^* antibonding orbital of the antiperiplanar α substituent. In order to avoid disfavored intermolecular orbital *overlap between the nucleophile and the non-symmetric π^* orbital an angle of attack of greater than $90°$ occurs (Bürbi-Dunitz trajectory). **F** modèle de Felkin-Anh; **G** Felkin-Anh-Modell.

D. HELLER, A. BÖRNER

Ref.: *Nouv.J.Chim.* **1977**, *1*, 61.

FEM → field emission microscopy

Fenton's reagent is a mixture of H_2O_2 and Fe(II) salts for the prepn. of carbonyl cpds. by *oxidation of hydroxy cpds. (e.g., *Brichima proc.). Special variants are the method of Ruff-Fenton for sugar degradation, the *hydroxylation of aromatic cpds., the photo-Fenton degradation, and the provision for radicals (*radical inhibitors). **F** réactif de Fenton; **G** Fentons Reagens. B. CORNILS

Ref.: Krauch/Kunz; *J.Mol.Catal.A:* **1999**, *144*, 77; *Acc. Chem.Res.* **1998**, *31*, 155, 159 and **1999**, *32*, 547.

FER abbr. for *ferrierite

fermentation *biocat. by means of microorganisms or *enzymes (e.g. ethanol, compost, silage, sauerkraut, etc.). Examples for industrial proc. were the *protein proc. of *BP and *ICI. **E=F=G**. B. CORNILS

Fermi level (Fermi energy, E_F) separates occupied from unoccupied electronic states in solids. At T = 0 K filled and unfilled states are separated sharply at E_F by a step-like function. For T > 0 K the distribution of electrons on the available energy levels obeys the Fermi-Dirac statistics. The Fl. represents the chemical potential of electrons. A metal is characterized by a substantial population of electronic states at E_F. **F** niveau de Fermi; **G** Fermi- Niveau. R. IMBIHL

Ref.: Thomas/Thomas, p. 389.

ferrierite a naturally occurring *zeolite with the simplified formula $(Na_2,K_2,Mg,Ca)_3$ $[Al_6Si_{30}O_{72}] \cdot 18\ H_2O$. It belongs to the group of medium-pore zeolites, containing two types of channels, a larger one which is lined by a ten-membered ring with an aperture of 4.2×5.4 Å and a smaller eight-membered ring channel (3.5×4.8 Å). These two channels build up a two-dimensional channel system. The *zeolite structural code for the tetrahedral framework structure of ferrierite is FER. Other, synthetic, materials possessing this topology are NU-23, ZSM-35 and FU-9. Pure-silica ferrierite can also be synthesized. Ferrierites with low aluminum contents and in their protonated form (H-ferrierite) are of interest as acidic cats. and have been studied, for example, with regard to butene *isomerization. **E=F**; **G** Ferrierit. P. BEHRENS

ferrocene(s) archetypal representatives of the *metallocenes. They are obtained by reaction of cyclopentadienyl anions with Fe(II) salts. Chemically they are very robust, behaving much like organic aromatic cpds. They can be reversibly oxidized to isolable 17-electron ferricium ions. This behavior contributes to the great value of their use in electrochemical devices. In cat., fs. act as versatile *ligands. Since f. itself has only a very weak donor ability, donor substituents have to be introduced into the f. molecule, the most important donor substituent being the phosphine moiety, R_2P (Figure). There are two types of ferrocenyl ligands which differ in that in type **A** the donor residues are fixed to both cp ligands (e.g., *dppf), whereas in type **B** both donor substituents are bound to only one ligand species (*ppfa; thus the planar *chiral molecule introduces chirality into the ferrocenyl ligand). A great variety of chiral fs. can be synth. and used efficiently in *enantioselective hom. cat. (e.g., with Pd and Zn cats.). Fs. can be constituents of *poly-ferrocenes. **E=F=G**.   P. HÄRTER

A
(R = H, dppf)

B
(ppfa)

Ref.: Togni-1, Vol. 2, p. 685; Togni-2; Mulzer/Wald-mann, p. 73.

ferrodoxins   *clusters containing Fe and thiolato *ligands which play a role in *bio-synthesis (cf. *sulfur in catalysis).   P. KALCK

Ferrofining process   → BP procs.

FFM   → force-field model

ficin (EC 3.4.22.3)   a *thiol protease from *Fi-cus* sp. that *hydrolyzes polypeptide amides and esters with broad specificity but with a prefer-ence for *hydroxylated or basic residues and cer-tain hydrophobic amino acids in the S_1 subsite (acyl donor side of scissile bond). Activity is en-hanced in the presence of reducing agents and *chelators such as *EDTA. **E=F=G**.   P.S. SEARS

Ref.: Schomberg/Salzmann, Vol. 5.; *Methods Enzymol.* **1970**, *19*, 261.

Fick's law   → diffusion, *Knudsen diffusion

field emission microscopy (FEM)   works at ultrahigh voltages with electrostatic fields. The ex-situ, non-destructive method detects single atoms on metals but is very limited in materials and samples (tip). The data are lo-cal (on atomic scale) and *surface-sensitive.

Ref.: Niemantsverdriet, p. 186.   R. SCHLÖGL

field ion microscopy (FIM)   cf. *field emission microscopy, the response comprises ions instead of electrons.   R. SCHLÖGL

filtration in catalyst preparation   is the intermediate step between *precipitation or *impregnation and *drying of solids. The tar-get is the removal of the mother liquor and of undesired ions from the *surface and from the *pore system of the solid by the washing liquid (usually water). Depending on the amount of material to be handled and on the required intensity of the washing procedure, different devices are applied, such as Büchner funnels (vacuum or pressure, discontinuous), centrifuges, decanters, or filter presses (plate or frame, all semicont. when automatized), or rotary or bed filters (cont.).

On washing with a liquid flowing through a filter cake, three regimes may be distin-guished: displacement, intermediate, and dif-fusional. The fitration time depends on var-ious variables such as the *diffusion constant, the flow resistance of the filter cake, or the fractional amount of washing liquid. As a rule of thumb the volume of the washing liquid should equal at least three times the volume of the cake. The application of press. intensi-fies the washing but may cause a reduction of the *pore volume and an increase of the *bulk density of the solid. **E=F=G**.

C.D. FROHNING

Ref.: *Chem.Ing.Tech.* **1983**, 55, 823 and **1983**, 55, 8; Ull-mann, Vol.B2, p. 10–1.

FIM   → field ion microscopy

final disposal of catalysts   For final dis-posal (decommissioning) spent cats. can be uti-lized as landfill (where permitted). Acc. to en-vironmental legislation it may be necessary to reduce the leaching: this is possible by stabiliza-tion (chemically by converting the contami-nants to an insoluble form) or by encapsulation (by sealing with bitumen, polyethylene, or inor-ganic coatings like cement, Pozzolan(ic) ce-ment, or incorporation into glassy materials, si-milarly to the *AVM proc.). Wherever possible, *makeup and workup (e.g., by hydro- or pyro-metallurgical procs.) are more advantageous: "post mortem" solutions may be expensive. **F** stockage final des catalyseurs ; **G** Katalysatorendlagerung.   B. CORNILS

Ref.: Trim, p. 41; Twigg, p. 188; R.B. Pojasek (Ed.), *Toxic and Hazardous Waste Disposal*, Ann Arbor Science Publ., Michigan 1970, Vol. 1; Ertl/Knözinger/Weitkamp, Vol. 3, p. 1278 and Vol. 4, p. 1815.

Fischer Emil (1852–1919), professor of organic chemistry at Würzburg (Germany) and Berlin. With his work on natural prods. (sugars, purines) and the *lock-and-key principle of biocats. he is one of the founders of *biocatalysis. Nobel laureate in 1902.   B. CORNILS

Ref.: Neufeldt; *Z.Chem.* **1970**, *10*, 41.

Fischer Ernst Otto (born 1918). German organometallic chemist, professor at Technische Universität Munich. He elucidated the structure of *ferrocene in competition with G. *Wilkinson, and later dibenzenechromium, the first authentic *sandwich complex of benzene. Nobel laureate in 1973, together with G. Wilkinson.   W.A. HERRMANN

Fischer Franz (1877–1947), professor of chemistry in Berlin, head of Kaiser-Wilhelm-Institut (now Max-Planck-Institut) für Kohlenforschung at Mülheim/Ruhr (Germany). Developed the *Fischer-Tropsch-synthesis (FTS) together with Hans *Tropsch.
                            B. CORNILS

Ref.: Neufeldt; M. Rasch, *Geschichte des KWI für Kohlenforschung 1913–1943*, VCH, Weinheim 1989; F. Fischer, *Leben und Forschung*, MPI Mülheim, 1957.

Fischer carbenes of the late *transition metals, e.g., $(CO)_5W[=C(OMe)C_6H_5]$, are normally electrophilic *ligands which can often be transferred to alkenes (cf. *cyclopropanation). **F**=**E**; **G** Fischer-Carbene.

Ref.: *Angew.Chem.* **1974**, *86*, 651.   W.A. HERRMANN

Fischer indole synthesis was discovered by *E. Fischer and comprises the reaction of phenylhydrazines and carbonyl cpds. to phenylhydrazones which cyclize via a [3,3] rear-rangement eliminating NH_3. Indolization takes place in the presence of acidic cats. or under thermal treatment (>180 °C). A variety of cats. is known such as HCl, pyridine hydrochloride, *zeolite Y, as well as *Lewis acids. The most employed cats. are $ZnCl_2$ and PCl_3. **F** synthèse d'indoles de Fischer; **G** Fischer'sche Indolsynthese.   C. BREINDL, M. BELLER

Ref.: B. Robinson, *The Fischer Indole Synthesis*, Wiley, New York 1982.

Fischer projection   a standard method for drawing a 3D tetrahedral molecule in two dimensions. Tetrahedral molecules are drawn such that the four bonds are at right angles to each other, and those bonds that are horizontal represent bonds projecting out of the page, while vertical bonds are projecting into the page. Rotation of the projection by 180° gives a molecule of the same stereochemistry, while rotation by 90° or 270° results in the opposite *enantiomer. The Fp. is helpful for determining which sugars are D or L. **F** projection de Fischer; **G** Fischer-Projektion.   P.S. SEARS

Fischer-Tropsch synthesis (FTS)   In 1922 *Fischer and *Tropsch discovered the het. cat. reaction between CO and H_2 (*syngas) which yields mixtures of mainly linear alkanes and alkenes with few oxygenates as byprods. The reaction was considered as an alternative to *coal liquefaction and served as a source for hydrocarbons (*hcs.) specially during World War II. The reaction proceeds via a series of complicated parallel and consecutive reactions whose relative velocity depends on cat. type and reaction conds. The stoichiometry of hc. formation can be derived from the two basic reactions, the formation of hcs. (on Co cats.)

$$CO + 2 H_2 \rightarrow \text{-(-CH}_2\text{-)-} + H_2O \quad \Delta H_{227\,°C} = -165 \text{ kJ}.$$

and the water gas shift reaction (*WGSR) on iron cats.

$$CO + H_2O \rightarrow H_2 + CO_2 \quad \Delta H_{227\,°C} = -39.8 \text{ kJ}.$$

The overall stoichiometry is $2\ CO + H_2 \rightarrow$ -(-CH$_2$-)- + CO_2 with $\Delta H_{227\,°C} = -204.8$ kJ.

The maximum attainable yield is 208.5 g of alkenes C_nH_{2n} per Nm^3 of syngas for complete conv. The workup of the reaction mixtures is complicated. Via different cats., reaction conds. (temp., CO/H_2 ratio, etc.) the prod. spectrum can be shifted more or less toward alkenes, gasolines, diesel oils, waxes, or oxygenates.

Several mechanisms have been developed and discussed during FTS history (*Anderson-Emmett-Kölbel, Pichler-Schulz, and variants thereof). Today the broadly accepted interpretation follows the suggestion of the inventor Fischer: the primary step consists of the formation of a carbide-like structure on the metallic iron *surface of the cat., followed by alternating *hydrogenation and *chain growth steps. Different requirements enforced various technical solutions of the FT synthesis (cf. procs. of *ARGE, *Kellogg Synthol, *Sasol, *Hydrocol). Today FT synthesis is carried out commercially in South Africa by *Sasol on the basis of syngas from coal and in Malaysia by Shell (*Shell SMDS proc.) on the basis of natural gas. **F** synthèse de Fischer-Tropsch; **G** Fischer-Tropsch-Synthese. C.D. FROHNING

Ref.: Storch/Golumbic/Anderson; Anderson/Boudart, Vol. 1; Falbe-1; Falbe-2; *Technikgeschichte* **1997**, *64*(3), 205; *Appl.Catal.A:* **1999**, *186*,3 *CHEMTECH* **1999**, (10), 32.

fixation　→ immobilization

fixed-basket reactor　→ catalytic reactors

fixed-bed reactor　→ catalytic reactors

flash desorption spectroscopy (TDS) → thermal desorption spectroscopy

flame hydrolysis　proc. for the prepn. of cat. *supports in which a mixture of the cat. or the support *precursor, hydrogen, and air are fed into a flame. The *precursors (e.g., $SiCl_4$ to *silica, $TiCl_4$ to *titania, $SnCl_4$) are hydrolyzed by steam and oxides are precipitated. Producers are, inter alia, Cabot, Degussa, Wacker; well-known prods. are Cabosil or

Aerosil (cf. also *thermal decomposition). If prods. of fh. are cat. active they would be *bulk cats., e.g., the prods. of the *synthane proc. which produces sprayed Raney Ni cats. for the *methanation of *syngas, using tube-wall reactors (TWR). **F** hydrolyse en flammes; **G** Flammenhydrolyse.　B. CORNILS

Ref.: Ertl/Knözinger/Weitkamp, Vol. 1, p. 94.

flavine adenine dinucleotide (FAD) a *cofactor used in many *redox reactions, e.g., in the α,β-*dehydrogenation of fatty acyl CoA molecules by acyl-CoA dehydrogenase, the first step toward the *catabolism of the acyl-CoA by a two-carbon unit. **E=F=G**.

P.S. SEARS

flavine mononucleotide (FMN) a *cofactor used in *redox reactions. Unlike the *nicotinamide redox cofactors, the flavin cofactors can undergo either a one- or two-electron *reduction, and are therefore useful as adapters between two-electron donors and one-electron acceptors in, e.g., *ETCs. **E=F=G**.　P.S. SEARS

flavoenzymes　→ flavine adenine nucleotides; *FAD, *FMN

Flexi processes　→ Exxon procs.

fluid catalytic cracking (FCC)　→ FCC process

fluidized-bed reactor　→ catalytic reactors

fluorenyl　Although flu may be regarded as a dibenzo-annelated cp *ligand it possesses a somewhat different frontier orbital structure and coordination chemistry. It may be mono-, tri-, or penta*hapto*-bonded at the metal and hexa-*hapto* through the six-membered ring or in an *exo*cyclic trihapto benzallyl mode. There is still a limited number of flu *transition metal *complexes. Cats. based on group 4 fluorenyl *metallocenes are known. Ti and Zr complexes containing the linked amidofluorenyl ligand show cat. activity toward living propylene *polymerization and ethylene-styrene *copolymerization. **E=F=G**.   J. OKUDA

fluorination The most important f. involves the replacement of C-Cl bonds by C-F using anhydrous HF, either under het. conds. at moderate temps. in the presence of Cr(III)- or Al-based cats. or under hom. conds. with Sb(V)-based cats. F. is widely applied (pre-Montreal Protocol) for industrial synths. of CFCs; now used for CFC replacements, hydrofluorocarbons (HCFs) and the transitional cpds., hydrochlorofluorcarbons (HCFCs). An important example is the thermodynamically limited conv. of CF_3CH_2Cl to CF_3CH_2F by HF at ≥ 623 K. The most widely used cat. precursors for this and related reactions are amorphous *chromia and *γ-alumina, which may be *promoted by other metal cations. Precursor activation by prefluorination with F, HFCs, or HCFCs is required to promote surface (*Lewis) acidity. *XPS data indicate that *surface f. is slow and, for chromia, incomplete. Surface metal(III) centers with disordered O/F environments have been postulated as *active sites. The compact structures of α-MF_3 (M = Cr, Al) phases give rise to low (Cr) or negligible (Al) cat. activity; other, more open AlF_3 phases ($\beta, \eta, \Theta, \kappa$) have significant activity, no pretreatment being required. Doping β-MF_3 (Al, Cr) with metal(II or III) fluorides can be beneficial. Radiotracer studies involving HF-prefluorinated chromia indicate that not all surface fluoride is cat. active and that cat. *chlorination can occur even in the presence of a large excess of surface fluoride. Hence cat. f. of $C_2Cl_3F_3$ and related CFCs is more properly described by a halogen exchange model comprising non-concerted F-for-Cl and Cl-for-F surface procs. that involve transfer of halogen species between the surface and the organic cpd. Dehydrochlorination/hydrofluorination pathways are also likely in cat. f. of some C_2 hydrochlorocarbons or CFCs, but dehydrofluorination of CF_3CH_2F is inhibited by a large conc. of surface fluoride. Cat. f. of CCl_4 under het. conds. has been variously formulated as sequential fluorination steps or as concurrent mono- and difluorination pathways. **F=E**; **G** Fluorierung.

J.M. WINFIELD

Ref.: *J.Catal.* **1997**, *169*, 307; **1996**, *159*, 270; **1995**, *152*, 70; **1993**, *140*, 103 and **1998**, *174*, 219; *J.Fluorine Chem.* **1992**, *59*, 33; *J.Chem.Soc.Chem.Commun.* **1996**, 1947, *Appl. Catal.* **1991**, *79*, 89.

fluorous biphase catalysis is a special variant of the *two-phase cat. and is based on the limited miscibility of various perfluoro derivatives with common organic *solvents. A fluorous (fl.) biphase cat. consists of a fl. phase containing a preferentially fl.-soluble cat. and a second prod. phase, which may be any organic or non-organic solvent with limited solubility in the fl. phase. Cats. can be made fl.-soluble by attaching fluorocarbon moieties to *ligands in appropriate size and number. The most effective moieties are linear or branched perfluoroalkyl chains with high C number that may contain other heteroatoms (*fluorous ponytails*). Because of the well-known electron-withdrawing properties of the F atom, the attachment of fl. ponytails changes significantly the electronic properties and consequently the reactivity of fl. cats. Therefore, the insertion of electronically insulating groups before the fl. ponytail may be necessary to decrease the strong electron-withdrawing effects. Fl. biphase systems are well suited for converting apolar reactants to prods. of higher polarity, as the partition coefficients of the reactants and prods. will be higher and lower, respectively, in the fl. phase. The net results are no or little solubility limitation on the reactants and easy separation of the prods. Furthermore, as the conv. level increases the amount of polar prods. increases further enhancing separation (Figure).

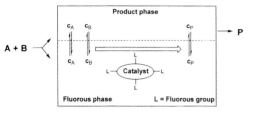

It should be emphasized that some fl. biphase systems can become a single phase at increaed temps. Although most molecular cats. could be made fl.-soluble, only *transition metal *complexes have been converted through ligand modification. Fl.-soluble *phosphines, *phosphites, *porphyrins, phthalocyanines, etc., have been prepared und successfully used in *hydroformylation, *hydrogenation, *hydroboration, *hydrosilylation, *oligomerization, etc. **F** catalyse fluoro-biphasique; **G** fluorous Zweiphasenkatalyse.
Ref.: *Acc.Chem.Res.* **1998**, *31*, 641.   I.T. HORVÁTH

fluxional ligands ligands which exist in various structural forms, such as π-allyl groups.

FMC COED process Modified proc. for the stepwise fluidized-bed, low-temp. carbonization and *gasification of coal yielding 25 % gas, 25 % tars, and 50 % coke. Cat. *methanation of the gas yields *SNG.   B. CORNILS
Ref.: Falbe-2; Kirk/Othmer-2, p. 540.

FMN → flavine adenine mononucleotide

foam catalysts cats. on *supports such as pumice, perlite (volcanic rock), expandable clays, or synth. "foam glass" or foamed carbon, sometimes applicated in foam-bed reactors. Cf. *ceramics in catalysis, *colloids.
          B. CORNILS
Ref.: Büchner/Schliebs/Winter/Büchel, pp. 400,404,486; K.D.P. Nigam, A. Schumpe, *Three-Phase Sparged Reactors*, Gordon and Breach, London 1996.

fod, FOD 1,1,1,2,2,3,3-heptafluoro-7,7-dimethyl-4,6-octanedionate in *complex cpds. or as a *ligand.

foldases → chaperones

force-field model (FFM) considers a molecule as a set of atoms which are points in space whose location is a function of parametrized forces. Another name for FFM is molecular mechanics, because the mathematical expressions for describing a molecule in a FFM. are the same as in classical mechanics without quantum theoretical methods being involved. Because molecules are treated as classical particles, electronic effects can not be described. A FFM is determined by 1: the types of forces which are considered, such as stretching, bending, Coulomb interactions, *van der Waals forces, etc.; 2: by the mathematical equs. which are chosen for the forces; and 3: by the parameters which determine the strength of the forces as a function of interatomic distances. The most imporant forces which are found in many FFMs. are bond stretching, angle bending, torsional rotation, and charge interactions. The parameters are either derived from experimental values or from high-level *ab-initio calcns. FFMs. are the fastest methods in *computational chemistry. They are particularly common for calcng. large organic cpds., because there are many accurate experimental and quantum chemical reference data for the development of reliable FFMs. of cpds. containing C, H, N, O. Ffms. of heavier atoms and less common structural types may not be very accurate because of the lack of reference data.

There are methods and programs available for calcng. large molecules which combine FFMs. with quantum chemical methods such as ab-initio calcns., *density functional theory or *semi-empirical methods. Such combinations are called *embedding methods. FFMs. have also been developed for *transition states in order to study the effect of substituents on activation barriers.   G. FRENKING

formaldehyde important *intermediate, manuf. cat. acc. to the procs. of *Haldor-Topsøe, *Perstorp-Reichhold; cf. *formaldehyde synth.

formaldehyde synthesis   F. is manuf. by cat. *oxidation of methanol with O_2 or air in the presence of Ag or metal oxide cats. in a gas-phase reaction. The overall reaction is highly exothermic. Ag cats. are applied in the form of *gauze cats. or of granular silver crystallites deposited on an inert *carrier in the form of a fixed bed which is passed by a mixture of methanol vapor and air. MeOH is fed in stochiometic surplus in order to minimize the total oxidation to CO_2 and to be in the non-flammable regime. F. prod. proceeds at reasonable rates above 500 °C, but for acceptable methanol conv. temps. in the range 550–700 °C are common. At these temps. the equil. predicts decomposition of MeOH as well as of f., which is circumvented by extremely high *space velocities (*GHSV $>10^4$–10^5) or short residence times, respectively (*gauze reactor). The gaseous reactor effluent is quenched with water and the ingredients are separated and concentrated by distillation.

Fe-Mo oxides (ferric molybdate $Fe_2(MoO_4)_3$) or W-Mo oxides may be used instead of Ag as cats. and are claimed to permit higher methanol convs. (up to 97–98 %) leading to higher-conc. aqueous solutions of f. The procs. for Ag and oxide cats. are almost equal in importance with some advantages for the silver process. Commercial procs. are described under the keywords *Haldor Topsoe and *Perstorp Formox. **F** synthèse de formaldehyde; **G** Formaldehydsynthese.   C.D. FROHNING

Ref.: B.L.Shapiro (Ed.), *Heterogeneous Catalysis*, Texas A&M University, 1988; Ullmann, Vol. A11, p. 619; Kirk/Othmer-1, Vol. 11, p. 929; Kirk/Othmer-2, p. 527; McKetta-1, Vol. 23, p. 350.

formate dehydrogenase   → nicotinamide cofactor regeneration

formation of final catalyst   → forming of catalysts

forming of catalysts   (*shaping) comprises extrusion, tableting, pelletizing, and grinding. Extrusion is widely used as a relatively simple and cheap method. Moreover, most *precursors retain high *porosity, and the accessibility to the outer *surface is improved by applying special profiles (e.g., trilobes, quadrolobes; cf. *catalyst forms). The precursor, usually a powder, is kneaded to a paste by addition of (predominantly) water and extrusion aids (mineral acids, *clays, cements, methylcellulose, starch, polyglycols, *graphite), and the paste is fed to a screw or ring-roll extruder with a forming nozzle. The cutted extrudates are dried and calcined to yield the shaped cat.

Tableting, i.e., the mechanical compression of powders or granules into a predetermined shape, renders uniform cat. particles of high mechanical strength and high density. It is carried out on tableting machines which require good mechanical flow of the precursor and a minimum particle size. Powders (from previous *drying or *calcining) generally have to pass a densification step (granulator) prior to tableting. Additives and lubricants such as Mg stearate, polyglycols, or graphite are necessary to ensure the flow of the material and to reduce the abrasion of the tableting tools.

Pellitizing is the agglomeration of fines to coarser particles by controlled addition of *binders, mostly a liquid w/o additional binders. Commonly the liquid is sprayed onto the fines in a rotating vessel (pelletizer), thus generating agglomerates which are dried later. The mechanical strength of pellets is generally lower than that of tablets. Grinding of particles yields powders with an average particle diameter of 10–100 μm which are suitable for filtration (cf. *shaping, *catalyst forming). **F** mise en forme des catalyseurs; **G** Katalysatorformgebung.   C.D. FROHNING

Ref.: *Catal. Today* **1997**, *34*, 535; Wijngaarden/Kronberg/Westerterp; Ertl/Knözinger/Weitkamp, Vol. 1, pp. 264,413; Farrauto/ Bartholomew, pp. 88,102; Ullmann, Vol. A5, p. 352.

form of pores   → pore models

Formox process   →   Perstorp-Reichhold process

formylation introduces the formyl group HCO- into other organic cpds. by reaction of a formyl donor such as formic acid or formic acid esters with hydrogen active cpds. such as alcohols, amines, amides, thiols, etc., acc. to HCO-X + R-OH → HCOOR + HX (with X = OCCH$_3$, H). A formylation reagent of special reactivity is a mixture of formic acid and Ac$_2$O which gicves in situ the mixed anhydride HCOOCOMe. F. of alkenes can be accomplished with N-disubstituted formamides and POCl$_3$. The similar f. of aromatic rings can be achieved by the *Vilsmeier or the *Gattermann reaction. Cats. are HCl or Zn(CN)$_2$. **F=E; G** Formylierung.   M. BELLER
Ref.: March.

Foster-Wheeler delayed coking process manuf. of petroleum coke and to upgrade residues to lighter hydrocarbons. The charge feed (crudes, resids, shale oils, tar-sands liquids, cat. cycle oils, etc.) is fractionated into light prods. and bottoms, which are coked in drums. The coke drum overhead enters the fractionator, too. Each unit has at least two drums, one in service, the other being decoked with high-press. water jets. **F** procédé delayed-coking de Foster-Wheeler; **G** Foster-Wheeler Delayed Coking-Verfahren.
                                B. CORNILS

fouling physical inactivation of the *surfaces of het. cats. by reaction-own solids or *poisoning cpds., e.g., by *deposits of *coke, resids, or tars (cf. the f. by higher oxygenates in *BASF's high-press. methanol synth. by deposition of Fe from the cat. Fe carbonyl). Some of these deactivating impurities may be burned off. Cf. *deactivation. **F=G=E**.
                                B. CORNILS
Ref.: Ertl/Knözinger/Weitkamp, Vol. 3, p. 1267; Farrauto/Bartholomew, p. 273.

Fourier transform infrared spectroscopy (FT-IR) is applied for the vibrational analysis of cats. and adsorbates. The ex-situ method, non-destructive to the sample, is extremely versatile, excellent in interpretation and chemical and structural information, and has good in-situ capabilities. The data are integral and bulk/*surface-sensitive.
Ref.: Niemantsverdriet, p. 200.   R. SCHLÖGL

Fourier transform Raman spectroscopy (FT-R) similar to FT-IR, using Raman spectroscopy.

Fp abbr. for FeCp(CO)$_2$; see also *cp

Frankland Edward (1825–1899), professor of chemistry in London, discovered diethylzinc (mobile fluid) and ethylzinc iodide (white mass of crystals), the first authentic *transition metal alkyl complexes.
                                W.A. HERRMANN

free energy of activation → *Eyring equation

free-radical polymerization → radical polymerization

frequency-modulated Raman spectroscopy → Raman spectroscopy

Freundlich isotherm The Freundlich isotherm $\theta = c_1 p^{1/c}{}_2$ with c_1 and c_2 being constants describes the relation between the surface coverage θ and the pressure p of the adsorbing gas assuming that the adsorption enthalpy depends logarithmically on the pressure. Other adsorption isotherms are the *Langmuir isotherm, the *BET isotherm, and the *Temkin isotherm. **F** isotherme de Freundlich; **G** Freundlich-Isotherme.   R. SCHLÖGL, A. SCHEYBAL

Friedel-Crafts reactions belong to the class of reactions in which *Lewis acids cat. or promote *substitutions, *additions, *isomerizations, or *polymerizations. The most important examples are FC *alkylation and FC *acylation acc. to

$$Ar\text{-}H + R\text{-}X \rightarrow Ar\text{-}R + HX$$
$$(Ar = aryl; X = F > Cl > Br > I)$$
$$Ar\text{-}H + RCOX \rightarrow Ar\text{-}(CO)\text{-}R + HX$$
$$(X = Cl, OOCR)$$

The FC *alkylation describes the reaction be-tween aromatic rings with alkyl halides, al-kenes, alcohols, ethers, mercaptans, etc.; the FC *acylation is the most important method for the prep. of aromatic ketones using mostly acid chlorides and anhydrides.

Commercially relevant FC alkylations are the manuf. of ethylbenzene (benzene + ethyl-ene; procs. of *BASF/Badger, *CdF/Technip, *Lummus/UOP, *Mobil/Badger, *Monsanto/ Lummus, *UOP Alkar, etc.) or of cumene (benzene + propylene; cf. the procs. of *CdTech/CDCumene, *Mobil/Badger, *UOP). For FC *isomerizations cf. the procs. of *Mitsu-bishi or *IFP and Lummus/ABB (over chlo-rinated *alumina).

While FC alkylations proceed in the presence of cat. amounts of *Lewis acids such as $AlBr_3$, $AlCl_3$, BF_3, $FeCl_3$, $ZnCl_2$, $SbCl_5$, etc., FC acyl-ations require more than stoichiometric amounts of cats. due to the consumption of the Lewis acids by coordination with the aro-matic ketone produced. Newly developed acylation cats. are activated Fe sulfates or ox-ides, *heteropoly acids, lanthanide triflates (*lanthanides as cat. metals), diphenylboryl hexachloroantimonate, CF_3SO_3H, *zeolites, and liquid HF. The last two examples have also been commercially used for the synth. of ibuprofen (*profens).

For reactions similar to FC, cf. the *Balsohn olefin addition and the *Fries rearrangement. **F** réaction de Friedel-Crafts; **G** Friedel-Crafts-Reaktion.   M. BELLER

Ref.: Oláh-1 and –2; Trost/Fleming, Vol. 2, p. 733 and Vol. 3, p. 293; March; Kirk/Othmer, Vol. 11, p. 1042; *Organic Reactions*, Vol. 3, p. 1 and Vol. 5, p. 229 .

Fries arrangement   of phenol esters to hy-droxyphenyl ketones with *Friedel-Crafts cats. The *o-* and *p-*isomers accrue depending of the reaction conds. A variant is the *photo Fries reaction **F** réarrangement de Fries, **G** Fries'sche Umlagerung.   B. CORNILS

Ref.: *Organic Reactions*, Vol. 1, p. 342.

frontier molecular orbitals (FMOs) Electrons in molecules have discrete energy values. In the framework of the *Hartree-Fock theory, the electrons are assigned pair-wise to *molecular orbitals, which are func-tions of the space coordinates and the spin. The solution of the diagonalized Hartree-Fock equs. leads to canonical orbitals which also have discrete energy values. The energe-tically lowest lying orbitals are occupied and the higher MOs are unoccupied in the elec-tronic ground state of a molecule. The fron-tier orbitals are the highest occupied MO (HOMO) and the lowest unoccupied MO (LUMO). HOMO and LUMO play a central role in the *MO theory of chemical reactivity. The FMO theory states that the most impor-tant electronic interactions between two re-acting molecules occur between the HOMO and the LUMO of the reactants. The interac-tions which determine the reaction course arise between the HOMO of one reactant and the LUMO of the other reactant. Since the strength of the interactions is inversely propor-tional to the orbital energy difference, it follows that the HOMO-LUMO interactions are the strongest, because the energy difference is small. Furthermore, the *regioselectivity and *stereoselectivity of a chemical reaction can also be predicted by the FMO model. This is be-cause MOs are usually expressed as a linear combination of atomic orbitals, and the contri-bution of the different atoms and the spatial ex-tension indicate the orientation and stereose-lection of the reaction.

The FMO model has been very successful in explaining the reaction course of organic re-actions, particularly for pericyclic reactions. Cf. also *Felkin-Anh model.   G. FRENKING

Ref.: I. Fleming, *Frontier Orbitals and Organic Chemi-cal Reactions*, Wiley, Chichester 1976.

fructose 1,6-diphosphate aldolase (FDP, EC 4.1.2.13). FDP aldolase cat. the reversible *aldol addition of dihydroxyace-tone phosphate (DHAP) and *D-glyceralde-hyde 3-phosphate (G3P) to form D-fructose 1,6-diphosphate (FDP). The equil. constant, $K_{eq} \approx 10^4$ M^{-1}, favors FDP formation. Both type I and II *enzymes have been isolated from a variety of eukaryotic and prokaryotic sources (see also *aldolase).

DHAP

(PO = phosphate)   D-Gly 3-P

D-FDP

Most of the mechanistic studies have been carried out on FDP aldolases from rabbit muscle or yeast. Generally, the type I FDP aldolases exist as tetramers (mol.weight ca. 160 kDa), while the type II enzymes are dimers (ca. 80 k Da). The sequences of the type I enzymes have a high degree of homology (>50 %) with the active site sequence being conserved throughout evolution. Significant differences in the C-terminal regions have, however, been identified that may be important in mediating substrate specificity. FDP aldolase is the most widely used aldolase in organic synth. This enzyme accepts a wide range of aldehyde acceptor substrates with DHAP as the donor to generate ($3S,4R$)- vicinal diols, stereospecifically, with D-*threo* *stereochemistry. Suitable acceptors include unhindered aliphatic aldehydes, α-heteroatom substituted aldehydes, and monosaccharides and their derivatives. Phosphorylated aldehydes react more rapidly than do their unphosphorylated analogs, but aromatic aldehydes, sterically hindered aliphatic and α,β-unsaturated aldehydes are generally not substrates. The specificity for the donor substrate is much more stringent. Only three DHAP analogs have been found to be substrates for RAMA and are all poor (ca. 10 % of the activity of DHAP). The diastereoselectivity exhibited by FDP aldolase was found to be dependent on the reaction conditions. **E=F=G**.   C.-H. WONG

Ref.: Wong/Whitesides, Vol. 12, p. 195.

fructose-6-phosphate phosphoketolase (EC 4.1.2.22) cat. the synth. of fructose-6-phosphate from glyceraldehyde and erythrose-4-phosphate. **E=F=G**.   C.-H. WONG
Ref.: *Methods Enzymol.* **1962**, 5, 276.

FT synthesis → Fischer-Tropsch-synthesis

FT Fourier transformation

FT-IR (R) → Fourier transform infrared (Raman) spectroscopy

FTS → Fischer-Tropsch synthesis

fucose 1-phosphate aldolase → fuculose 1-phosphate aldolase

fucosyltransferase (FucT) cat. the transfer of fucose from GDP-fucose to an acceptor with an α-linkage. Fs. are involved in the *biosynthesis of many *oligosaccharide structures such as blood-group substances and cell-surface and tumor-associated antigens. Fucosylation is one of the last modifications of oligosaccharides *in vivo*. Several fs. have been isolated and used for *in vitro* synth., including FucTs III, IV, V, VI, and VII. The Lewis A α-1,4-fucosyltransferase has been shown to transfer unnatural fucose derivatives from their GDP esters. Furthermore, this *enzyme will transfer a fucose residue that is substituted on C 6 by a very large sterically demanding structure. **E=F=G**.   C.-H. WONG

fuculose 1-phosphate (Fuc 1-P) aldolase (EC 4.1.2.17), rhamnulose 1-phosphate (Rha 1-P) aldolase (EC 4.1.2.19), and tagatose 1,6-diphosphate aldolase. Fuc 1-P aldolase cat. the reversible *condensation of *DHAP and L-lactaldehyde to give L-Fuc 1-P, and with the same substrates, Rha 1-P aldolase produces L-Rha 1-P. Both of these *enzymes are type II *aldolases and are found in several microorganisms. Fuc 1-P aldolase and Rha 1-P aldolase have been cloned and over-expressed and subsequently purified. Tagatose 1,6-diphosphate (TDP) aldolase, a type I

E₁ = pyruvate kinase
E₂ = rhamnulose kinase

E_1 = pyruvate kinase
E_2 = rhamnulose kinase

aldolase involved in the galactose metabolism of *cocci*, cat. the reversible condensation of *G3P with DHAP to give D-TDP.

Both Fuc 1-P and Rha 1-P aldolase accept a variety of aldehydes, and generate vicinal diols with D-*erythro* and L-*threo* configurations, respectively. While the enzymes yield products with the ($3R$)-configuration, stereospecifically, the *stereoselectivity at C 4 is somewhat diminished for a few substrates. Sterically unhindered 2-hydroxyaldehydes normally give very high diastereoselectivities. These two aldolases also show significant kinetic preference for the L-enantiomer of 2-hydroxyaldehydes, so they facilitate the kinetic resolution of a racemic mixture of these compounds. Both enzymes have been used in the synthesis of rare ketose 1-phos-

phates and several iminocyclitols. Fuc 1-P and Rha 1-P aldolases have also been utilized in whole cell systems with DHA and cat. inorganic arsenate.

An alternative method for preparing the same ketose 1-phosphates available from the Rha 1-P and Fuc 1-P aldolases is to use the rhamnose and fucose *isomerases (Rha I and Fuc I). The *isomerase only partially converts the aldose to the ketose, so a second step which phosphorylates the ketose drives the reaction to completion. **E=F=G**.   C.-H. WONG
Ref.: *Angew.Chem.Int.Ed.Engl.* **1991**, *30*, 555.

fuel cell catalysis   occurs in two parts of the *fuel cell system: in the processing of the fuel before feeding it into the fc. stack and in the cat. at the electrodes of the fc.; both are cat. heterogeneously. The latter phenomenon is called *electrocatalysis because the fuel oxidation consists of electrochemical reactions, involving the transfer of one or more electrons from the reactants to the electrodes or vice versa. At the negative electrode (anode) of the fc. the anodic oxidation of the fuel proceeds which delivers electrons to the anode. On the other side of the fc. the cathodic reduction of oxygen takes place at the positive electrode (cathode) while taking up electrons, acc. to $H_2 \rightarrow 2\,H^+ + 2\,e^-$ and $\frac{1}{2}\,O_2 + 2\,H^+ + 2\,e^- \rightarrow H_2O$. Depending on the nature of the electrolyte between the electrodes, ions are formed which travel through the electrolyte as long as the overall reaction continues, i.e., electrical current flows which is drawn from the cell. The passage over into the electrolyte of the ions formed from the gaseous (or liquid or dissolved) reactants at the electrodes also has to be accelerated by electrocatalysts, which have to be electronically conducting.

fuel cell catalysis

	AFC	PEFC	PAFC	MCFC	SOFC	DMFC
Electrolyte	KOH	Polymer	H_3PO_4	Molten carbonate	Ceramic	Polymer
Temp.[°C]	<120	<80	<200	<650	<1000	<80
Fuel	H_2	H_2, ref.CH_3OH	H_2, ref.CH_4	H_2,CO, CH_4	H_2,CO, CH_4	CH_3OH

Abbrs.: ref. = reformed; for other abbrs. see text.

Depending of the fuel applied and the mode of electrolyte, different fcs. with characteristic cats. can be distinguished (Table).

The cat. active materials are: AFCs (alkaline fcs.): Ni, Ni-Ti, Pt-Pd for the anode and Pt on carbon, Ag, or different *perovskites or *spinels for the cathode. PEFCs (proton exchange fcs.; PEMFC = proton exchange membrane fc.) use at the anode Pt (which may be *poisoned by CO), carbon-supported Pt, Pt-Ru and on the cathode Pt alloys or Pt-C. PAFCs (phosphorus acid fcs.), anode: same as PEFCs; cathode: same as PEFCs or Pt plus Cr or Co. MCFCs (molten carbonate fcs.), anode: pure Ni or Ni-Cr; cathode: lithiated NiO. SOFCs (solid oxide fcs.), anode: Ni-ZrO_2(Y)-*cermet layers; cathode: *perovskites, e.g., $LaMnO_3$, Sr-doped. DMFCs (direct methanol fcs.), anode: Pt-Ru, Pt-Sn, Pt-WO_3; cathode: Pt (affected by methanol crossover), metal chelates, thiospinels. DHFCs (direct hydrocarbon fcs.) operate at 100–400 °C and are still in the research stage. **F** catalyseur pour piles à combustion; **G** Brennstoffzellen-Katalysatoren. G. SANDSTEDE

Ref.: G. Sandstede, *From Electrocatalysis to Fuel Cells*, University of Washington Press, Seattle 1972; A.J. Appleby, F.R. Foulkes, *Fuel Cell Handbook*, Van Nostrand-Reinhold, New York 1989; K. Kordesch, G. Simader, *Fuel Cells and Their Applications*, VCH, Weinheim (Germany) 1996; Ertl/Knözinger/Weitkamp, Vol. 4, p. 2090.

fuel cell system The system comprises the devices necessary for successful and economic operation. In cat. respects the fcs. includes, besides the *fuel cell catalysis itself, gas processing (e.g., *reforming of hcs. like methane or methanol) and the cat. burner for the off-gases (Figure). **F** système des piles à combustible; **G** Brennstoffzellensystem. B. CORNILS

Ref.: Stimming, presentation at Dechema, Jan. 1999.

fuel methanol mixtures of methanol and higher (homologous) alcohols as gasoline additives or for fuel cells; *homologation, *Lurgi Octamix proc., *gasohol. B. CORNILS

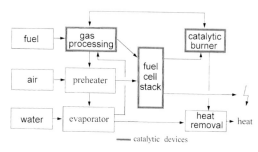

fuel cell system

fumarase (EC 4.2.1.2) cat. the stereospecific *hydration of fumarate to L-malate. This reaction has been used in a microbial whole cell proc. for the commercial synth. of L-malic acid. The *enzyme from pig heart has been examined for synth. utility, and found to have narrow substrate specificity. No other nucleophiles can replace water for the *addition reaction, and only a few fumarate analogs are acceptable. Chlorofumarate and difluorofumarate have been used as substrates and converted to L-*threo*-chloromalic acid and 2,3-difluoromalate respectively. **E=F=G**.

C.-H. WONG

Ref.: *J.Org.Chem.* **1987**, *52*, 2838.

fumed silica → silica

Furukawa modification → Simmons-Smith reaction

fused catalysts *bulk pre-melted cats. prepared by fusion of various *precursors. Examples are *alloy cats. (e.g., Pt-Rh gauze cats. for the *oxidation of ammonia, cf. *gauze cat., *metallic glasses), oxidic cats. (e.g., fused Fe oxide for *ammonia synth. or for the *Fischer-Tropsch synth., *Synthol proc. of *Kellogg), and the Adams cat. **F** catalyseurs de fusion; **G** Schmelzkatalysatoren. B. CORNILS

Ref.: Twigg, p. 35; Ertl/Knözinger/Weitkamp, Vol. 1, p. 54.

G

GAF General Aniline & Film Corp., a pre-war subsidiary of I.G. Farbenindustrie AG of Germany

GAF butanediol process manuf. of 1,4-butanediol (BDO) by *Reppe reaction of acetylene and aqueous HCHO solution cat. by Cu acetylide/Bi on *supports. The intermediate 2-butyne-1,4-diol is *hydrogenated on het. Cu-Cr or Ni cat. in trickle-bed operation (*BASF BDO proc.). **F** procédé de GAF pour le butanediol; **G** GAF Butandiol-Verfahren.   B. CORNILS

GAF tetrahydrofuran process manuf. of THF by acid-cat. (H_3PO_4) *elimination (*dehydration) of water from *BDO; cf.*BASF THF proc. **F** procédé GAF pour THF; **G** GAF THF-Verfahren.   B. CORNILS

galactokinase cat. the synth. of α-galactosyl phosphate using ATP

galactose oxidase (EC 1.1.3.9) Go. from *Dactylium dendroides* contains one atom of Cu(II) per molecule as the *cofactor and cat. the *oxidation of D-galactose at the C 6 position in the presence of oxygen to the corresponding aldehyde and H_2O_2.

The *enzyme also cat. the *stereospecific oxidation of glycerol and 3-halo-1,2-propanediols to the corresponding L-aldehydes. Several unusual L-sugars have been prepared from polyols using this enzyme. The byproduct H_2O_2 must usually be destroyed with *peroxidase to avoid inactivation of the enzyme. **E=F=G**.   C.-H. WONG

Ref.: *Biochem.Biophys.Res.Commun.* **1982**, *108*, 804; *J.Am. Chem.Soc.* **1985**, *107*, 2997; *Nature (London)* **1991**, *350*, 87.

galactose-1-phosphate uridyltransferase → UDP-glucose (UDP-Glc) and UDP-galactose (UDP-Gal)

galactosidases cat. the *hydrolysis of galactosides. Both types of *enzymes (α and β) have been used in the synth. of galactosides (cf. also *glycosidase). **E=F=G**.   C.-H. WONG

galactosyltransferase (β-1,4-Galactosyltransferase [UDP-Gal: N-acetylglucosamine β-1,4-galactosyltransferase, EC 2.4.1.22]) cat. the transfer of galactose from UDP-Gal to the 4-position of β-linked GlcNAc residues to produce the Galβ1,4GlcNAc substructure. In the presence of lactalbumin, however, glucose is the preferred acceptor, resulting in the formation of lactose, Galβ1,4Glc. The *enzyme has been employed in the in- vitro synth. of N-acetyllactosamine and glycosides thereof, as well as other galactosides. With regard to the donor substrate, the β-galactosyltransferase also transfers glucose, 4-deoxygalactose, arabinose, glucosamine, galactosamine, N-acetylgalactosamine, 2-deoxygalactose, and 2-deoxyglucose from their respective UDP-derivatives, providing an enzymatic route to oligosaccharides which terminate in β-1,4-linked residues other than galactose. **E=F=G**.

C.-H. WONG

gallium in catalysis Since the chemistry of Ga resembles that of Al only some minor applications have been proposed, mainly Ga or Ga/Al incorporation in *zeolites (e.g., *aromatization of propane/butanes over Ga-doped zeolite, *UOP/BP Cyclar proc.; for De-NOx reactions).   B. CORNILS

Ref.: *Houben-Weyl/Methodicum Chimicum*, Vol. XIII/6.

gas chromatography (GC) a chromato-graphic separation for mixtures of substances which are gaseous or which can be evaporated without decomposition. A mobile phase is used as carrier gas. The cpds. to be separated are distributed between a solid and the mo-bile phase. The non-destructive method (de-tection via conductivity, ionization, electron capture, etc.) is used for the determination of quantitative composition of liquid or gaseous samples, reaction sequences, kinetics, etc. GC can be combined with other methods such as *MS (*GC-MS). Using chiral stationary phases (CSPs) even racemic mixtures can be separated.

High-press. (or perfomance) liquid chroma-tography (*HPLC) works without evapora-tion and is universally suitable for liquid sam-ples. **F** chromatographie en phase gazeuse; **G** Gaschromatographie.   R. SCHLÖGL

Ref.: Ullmann, Vol.B5, pp. 182,239.

gas chromatography coupled mass spectroscopy (GC-MS) is ideal for the identification of molecules after GC separa-tion and for *isotope labeling studies. In con-trast to GC, GC-MS is destructive to the sam-ple.   R. SCHLÖGL

gaseous/gaseous-phase homogeneous catalysis → homogeneous catalysis

gaseous hourly space velocity (GHSV) The GHSV indicates the volumetric gas flow per unit time (hour) through a unit volume of cat. as $L_{gas}/L_{cat} \cdot h$ $[h^{-1}]$. The volume of cat. corresponds to the space filled with cat. in the reactor and may thus deviate from its *bulk volume. In a few cases the GHSV is based on the reaction volume instead of the cat. vol-ume to describe, e.g., the efficiency of a tank reactor volume. **F** vitesse spatiale horaire ga-zeuse; **G** Gas-Raumgeschwindigkeit.
                                          C.D. FROHNING

gasohol is the generic US term for *gasoline* and alc*ohol* (ethyl alcohol) mixtures as motor fuels. The ethanol (10–20 %) is based on agri-cultural feedstocks and on *fermentation procs., *not* on synth. routes (similar develop-ments in Brazil). In Germany methanol is used as a diluent for gasoline. **F=G=E**.
                                          B. CORNILS

Ref.: Kirk/Othmer-1, Vol. 1, p. 826; Kirk/Othmer-2, pp. 50, 552.

gasol early term for liquid gases from *Fischer-Tropsch plants.

Ref.: *Technikgeschichte* **1997**, *64*, 205.

gatsch slack wax, → Fischer-Tropsch syn-thesis; *waxes

Gattermann-Koch synthesis   for the prep. of aromatic aldehydes by *formylation of activated arenes with CO (and HCl) in the presence of cat. amounts of Cu(I) chloride and AlCl₃ acc. to Ar-H + CO + HCl → Ar-CHO. The reaction mixture may be prepared in situ by dropping chlorosulfuric acid into the solution of formic acid to produce CO, Cl, and H₂SO₄. Based on the GKs. other formyla-tion procedures have been developed, e.g., with a mixture of HCN and HCl (Gattermann variant in situ with Zn(CN)₂). In general the GKs. is performed in chloro- or 1,2-dichloro-benzene as solvent in presence of the cats. The mechanism has not been investigated in detail. **F** synthèse de Gattermann-Koch; **G** Gattermann-Koch-Synthese.
                                W.A. MORADI, M. BELLER

Ref.: *Org.React.* **1957**, *9*, 37; *J.Am.Chem.Soc.* **1923**, *45*, 2373; *Organic Reactions*, Vol. 5, p. 290.

gauze catalysts *bulk cats. made of wo-ven metal or *alloy wires, firstly applied in *Davy's *safety lamp. Reactors filled with gauze cats. (*metal gauze reactors, shallow-bed reactors) express extremely low press. losses and high heat transfer values; they are very suitable for highly exothermal reactions such as *oxidation of ammonia, *EO manuf. by ethylene oxidation, *formaldehyde synth., or the synth. of HCN by *ammondehydrogenation from MeOH or from methane. Cf. also *grid cats. **F** catalyseurs de gaze (de tissu métallique); **G** Netzgewebekatalysatoren.

B. CORNILS

Ref.: Ertl/Knözinger/Weitkamp, Vol. 4, pp. 1750,1757.

GBL   γ-butyrolactone, a solvent and precursor for other chemicals such as *N*-methylpyrrolidone (e.g., procs. of *BASF, *GAF, *Lurgi Geminox, etc.).

GC   → gas chromatography

GC-MS   → gas chromatography coupled with mass spectroscopy

GDP-mannose and GDP-fucose (GDP-Man, GDP-Fuc) GDP-Man is prepared from Man-1-P and GTP using GDP-Man pyrophosphorylase (Figure).

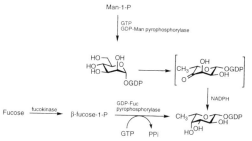

GDP-Fuc is biosynth. in vivo from GDP-Man by an *NADPH-dependent *oxidoreductase *enzyme system. Such systems have also been utilized for in-vitro synth. of GDP-Fuc. Using a similar procedure, GDP-Fuc has been generated in situ for use in a glycosylation reaction. *Enzymes from a minor biosynth. pathway that synth. GDP-Fuc from L-fucose-1-phosphate or L-fucose have also been exploited for synth. Fucose was phosphorylated by fucokinase (EC 2.7.1.52) to produce Fuc-1-P, which subsequently underwent a GDP-fucose pyrophosphorylase-cat. reaction with GTP to provide GDP-Fuc. **F=F=G**.

C.-H. WONG

Geminox process   → Lurgi procs.

general acid-base catalysis   1: *acid-base catalysis; 2-in enzyme-cat. reactions: Proteins have several side chains with pK_as relatively close to neutrality (e.g., aspartic acid) which can function as general acids and bases in cat., and indeed much of *enzyme cat. is *acid-base cat. A large number of reactions involve nucleophilic attack on a molecule, and this is often facilitated by the abstraction of a proton by a base held in close proximity by the enzyme. Other reactions proceed via *elimination, where protonation of the leaving group by a nearby acid accelerates the rate. Typically, both will be observed: a base to facilitate nucleophilic attack and an acid to donate a proton to the leaving group.

As a result, enzymes typically have well-defined pH ranges in which they work. If an enzyme requires a certain group to act as a general base, it will not function at a pH where that base is protonated. The pK_as of general acids and bases in the active site may be estimated from the pH-rate profile, although this approach is not without potential pitfalls. (A change in rate-limiting step with pH, for example, will give a false pK_a). **F** catalyse général d'acide-base; **G** allgemeine Säure-Basen-Katalyse.

P.S. SEARS

Ref.: Gates-1, pp. 27,35.

General Aniline & Film Corp.   → GAF procs.

generator gas   a gas mixture of CO, H_2, and N_2, originating from the autothermal *coal gasification with air (as oxidant instead of oxygen); used as fuel gas or as *syngas for *ammonia. **F** gaz de gazéificateur; **G** Generatorgas.

B. CORNILS

Ref.: Ullmann, Vol. 12, p. 169.

genetic engineering   refers to the modification of organisms at the genetic level. It has been a useful technique in the prod. of organisms with improved properties concerning hardiness, disease resistance, or prod. of a desired molecule. See also *metabolic engineering and *in-vivo evolution of biocats. **E=F=G**.

P.S. SEARS

German technology the I.G. Farbenindustrie version of *coal hydrogenation, an advanced development of the *Bergius-Pier proc. as realized between 1928 and 1945 in 12 plants (with a capacity of approx. $4 \cdot 10^6$ t/a) in Germany. The pulverized and dried coal was fed, together with cat. (based on *bulk sulfides of Fe, W, or Mo; later, red mud – Fe_2O_3) and proc.-derived solvent, into a mixing vessel forming a slurry with solid contents of approx. 42 %. After pumping the slurry up to 30 MPa, H_2 is added and the slurry enters the hydrogenation reactor. The temp. is controlled by quenching cold H_2 into the reactor. Part of the hydrogenative work was done in the slurry or in the gaseous phase. The reaction prods. are 7 % of gases (H_2S, NH_3, CO, CO_2, etc.), 23 % of C_1-C_4 gaseous prods.(*SNG and *LPG), 15 % naphtha (ranging from C_5 up to 200 °C), middle distillates (200–325 °C), and residues (solids like ash included). A special comb. of two hydrogenative steps (including a second fixed-bed) and a workup is called the IGOR proc. The principal prods. can be upgraded to naphtha, diesel, heating oil, etc. Based on these experiences a 200 t/d pilot plant of VEBA Oel/Ruhrkohle AG was run at Bottrop. **F** technologie allemande; **G** Deutsche Technologie.   I. ROMEY
Ref.: Falbe-2.

GHSV → gaseous hourly space velocity

Gibbs free energy (in enzyme cat.) Gfe. affects enzymatic cat., as it does non-enzymatic cat., by setting the equil. position of the reaction. It is important to recognize that *enzymes cannot change the equil. position of a reaction, although they very frequently couple an unfavorable reaction with a very favorable one (such as *hydrolysis of ATP to AMP + 2 P_i). Gfe. is not an absolute quantity, but is quantified by comparison with a *standard state, usually a slightly modified standard state from the one chemists use. Because most enzymatic reactions take place in water near neutral pH (and many cpds. are unstable at very low or high pH), the biochemical standard state is a 1 M ideal solution at 25 °C, normal press., and pH 7.0. **F** ènergie libre de Gibbs; **G** Gibb'sche freie Energie.   P.S. SEARS

Giese-Stork method The high toxicity of Sn cpds., which are difficult to separate, is a disadvantage in *radical initiated- procs. To overcome this, the G-Sm. was developed where the produced Sn halides are reduced in situ by Na cyanoborohydride to Bu_3SnH. Thus, only cat. amounts of toxic Sn cpds. are involved in the reaction cycle. **F** méthode de Giese-Stork; **G** Giese-Stork-Methode.
Ref.: Giese.   T. LINKER

Glaser coupling reaction the *oxidative coupling of terminal alkynes in the presence of Cu(I) salts. The reaction is carried out at pH 4–5 with a mixture of CuCl/ammonium acetate/ HCl. The reaction is well suited to the prod. of symmetrical diynes. The reactivity is reduced with decreasing electron density at the C-C triple bond. As a result, conjugated alkynes react much more slowly. During the reaction Cu(I) is oxidized to Cu(II), thus this reaction is not purely cat. A range of variants have been developed, such as the Eglington method (with Cu(II) acetate/ pyridine). The most important variant is the Hay variant in which cat. quantities of Cu(I) salts are *complexed with amines. The Cu(II) formed during reaction is reduced in situ. **F** couplage de Glaser; **G** Glaser Kupplungsreaktion.   J. HENKELMANN

Glitsch BTX process (formerly GT-Aromex) manuf. of *BTX by extractive distillation from refinery or petrochemical aromatics streams, such as *cat. reformate or pyrolysis gasoline. **F** procédé Aromex de Glitsch pour BTX; **G** Glitsch GT-Aromex-Verfahren.
Ref.: *Hydrocarb.Proc.* **1997**, (3), 114.   B. CORNILS

Glitsch DMT process manuf. of *dimethyl terephthalate (DMT) from *p*-xylene via *oxidation with air (cat. by Co-Mn) in a specially designed oxidation section with efficient cat. recovery, *esterification of the intermediate acid with methanol, and purification of the DMT via fractionation/crystallization (Figure). **F** procédé Glitsch pour le DMT; **G** Glitsch DMT-Verfahren.   B. CORNILS

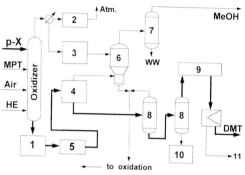

1 cat. recovery; 2 p-X recovery; 3 methanol/ acid recovery; 4 improved reactor design; 5 p-X/MeBz removal; 6–8 distillation towers; 9 crystallizer; 10 residue treatment. p-X: p-xylene.

Ref.: *Hydrocarb.Proc.* **1997**, (3), 122.

glucan phosphorylase → glycosyl phos- phorylase

glucoamylase (EC 3.1.2.3) is an exo*glyco- sidase, usually isolated from *Aspergillus* or *Rhizopus*, that cat. the sequential removal of glucose from the non-reducing end of starch. It is used commercially to degrade corn syrup amylopectin to glucose, which may be crystal- lized, used as syrup, or (partially) isomerized to fructose with glucose (xylose) *isomerase to give high-fructose corn syrup. **E=F=G.**

P.S. SEARS
Ref.: A. Wiseman (Ed.), *Handbook of Enzyme Bio- technology*, Ellis Horwood, Chichester 1985.

glucoisomerase → glucose isomerase

gluconate 6-phosphate dehydrogenase → decarboxylase

glucose dehydrogenase → nicotinamide cofactor regeneration

glucose isomerase (GI, or xylose isomer- ase, EC 5.3.1.5) cat. the *isomerization of

fructose to glucose and is used in the food in- dustry for the prod. of high-fructose corn syr- up. GI also accepts fructose analogs that are modified at the 3,5- and 6- positions. Various *FDP aldolase prods. can be isomerized to a mixture of the ketose and aldose, and subse- quently, the two isomers can be separated using Ca^{2+}- or Ba^{2+}-treated *cation exchange resins. Aldose analogs including 6-deoxy, 6- fluoro, 6-*O*-methyl and 6-azidoglucose have been synth. using this FDP aldolase/GI meth- odology. **E=F=G.**   C.-H. WONG
Ref.: *J.Am.Chem.Soc.* **1986**, *108*, 7812.

glucose oxidase cat. the *oxidation of glucose using molecular oxygen to form glu- conolactone and H_2O_2. **E=F=G.**   C.-H. WONG

glucose 6-phosphate and analogs → hexokinase

glucose 6-phosphate dehydrogenase → nicotinamide cofactor regeneration

glutamate dehydrogenase (EC 1.4.1.3) The natural substrate for gd. in the oxidative direction is glutamic acid. The *oxidation of racemic amino acids destroys a *chiral center of the L-enantiomer, and the D-enantiomer may be prepared in this manner. In the *re- ductive amination direction, though 2-oxoglu- taric acid is the best substrate, the *enzyme accepts other mono- and dicarboxylic acids.

$$R-C(=O)-CO_2^- + NH_4^+ \xrightarrow[\text{NADH}]{\text{GluDH}} R-\overset{NH_3^+}{\underset{}{C}}H-CO_2^-$$

Gd. can therefore be used to synth. several unnatural amino acids such as L-α-aminoadi- pic acid or L-β-fluoroglutamic acid. The use- fulness of the *enzyme as a synth. cat. can be coupled with its ability to regenerate *NADP. Gd. remains useful in coupled systems for *cofactor regeneration and has been used in *biphasic systems for this purpose. This com- mercially available enzyme can be easily *im- mobilized on solid *support to increase its sta- bility. **E=F=G.**   C.-H. WONG

Ref.: *J.Org.Chem.* **1989**, *54*, 498; *J.Chem.Soc.Perkin Trans.* **1992**, *1*, 3253.

glutamyl proteases Gps. (e.g., EC 3.4.21.9, the *V8 protease from *Staphylococcus aureus*) specifically cleave Glu-Xaa bonds. They have been useful not only in the cleavage of proteins for characterization, but also in the synth. of peptides. Highly specific *proteases can be advantageous in the *condensation of large polypeptide fragments since their specificity precludes them from hydrolyzing many of the amide bonds of the substrates. **E=F=G**.

P.S. SEARS

Ref.: *Tetrahedron Lett.* **1991**, *32*, 3421; *J.Biol.Chem.* **1992**, *247*, 6720.

glutardialdehyde for enzyme immobilization G. readily cross-links aminated cpds. It has long been used as a tissue fixative for this reason, and has been used for *enzyme immobilization on solid *supports as well. Enzymes may be attached to a solid *support derivatized with amines (e.g., aminopropyl glass). Enzymes immobilized in this fashion are typically linked via several lysines (attachment is multipoint), and so although any single imine may be *hydrolyzed, the enzyme as a whole is immobilized in a reasonably stable fashion. The stability may be further increased by reducing the imine to the secondary amine with sodium cyanoborohydride. Glutaraldehyde has also been used to stabilize enzymes or enzyme crystals by cross-linking. **E=F=G**.

P.S. SEARS

glutathione (reductase, peroxidase, transferase, coenzyme) G. is a tripeptide *cofactor, γ-L-glutamyl-L-cysteinyl-L-glycine. It is formed from the N-to-C terminus by a series of *enzymes that activate the amino acids in an *ATP-dependent reaction to form an acyl phosphate, then couple the activated acyl donor with the amine of the next amino acid (cf. *non-ribosomal peptide synth.). It is used by many enzymes as a *cofactor for *oxidation-reduction and *isomerization reactions. The dimeric oxidized form of the cofactor, GSSG, may be reduced in a *nicotinamide adenine dinucleotide phos phate (NADPH) requiring reaction cat. by the enzyme *glutathione reductase.

G. peroxidase, an enzyme prevalent in red blood cells, is a *selenoenzyme that causes the reduction of *peroxides with concomitant oxidation of glutathione. A GSH-dependent hy droperoxidase activity is associated with the enzyme prostaglandin endoperoxide synthase which cat. the addition of two equivalents of di oxygen to arachidonic acid to form one cyclic peroxide and one hydroperoxide, then reduces the hydroperoxide to an alcohol (prostaglandin H_2) with oxidation of glutathione.

A number of *isomerases which cat. *cis-trans* *isomerization of alkenes without migration of the double-bond are gl.-dependent. For example, the enzyme that cat. the *cis-trans* isomerization of maleate derivatives to the corre sponding fumarate derivatives proceeds through a covalent GSH-substrate adduct.

G. is also a substrate, consumed stoichiometri cally and not regenerated, of the glutathione S-epoxide transferases, enzymes that destroy *epoxides via nucleophilic addition of the thiol group of the g. The glutamate and gly cine are hydrolyzed and the only remaining part of the original GSH, the cysteine, is acetylated. In the case of arene epoxides, the mercapturic acid prod. may be excreted **F=G** glutathion.

P.S. SEARS

glyceraldehyde 3-phosphate (G3P) de-hydrogenase an *enzyme reponsible for the coupled *oxidation and phosphorylation of glyceraldehyde 3-phosphate (G3P) to 1,3 phosphoglyceric acid, a step in the glycolytic degradation of glucose to pyruvate. It reduces *NAD$^+$ to *NADH and consumes inorganic phosphate, and releases the first "high-energy" intermediate of glycolysis (the first intermedi ate with a phosphate of a high enough free energy to phosphorylate ADP). The enzyme is proposed to work by formation of a thiohemia cetal, which becomes oxidized by NAD$^+$ to the thioester. Phosphate then attacks the thioester to regenerate the thiolenzyme and generate the phosphorylated product (Figure).

P.S. SEARS

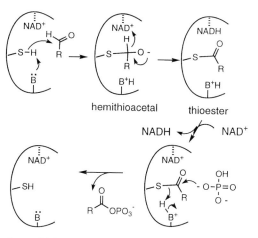

hemithioacetal   thioester

glyceraldehyde 3-phosphate dehydrogenase

D-glycero-D-galacto-2-nonulosonic acid, *N*-acetyl-5-amino-3,5-dideoxy → *N*-acetylneuraminate aldolase.

glycerol dehydrogenase (EC 1.1.1.6) Gd.

reduces α-hydroxy ketones to chiral (*R*)-1,2-diols. The chiral hydroxy ketones are obtained by *oxidation of 1,2-diols or via the *kinetic resolution of racemic hydroxy ketones. The *in vivo* role of the *NAD(H) requiring Gd. is the interconversion of dihydroxyacetone and glycerol. The purified *enzyme is commercially available, it is stable and can be *immobilized but must be used under an inert atmosphere. Many acyclic or cyclic substrates have been found for this enzyme. Hydroxy aldehydes are not substrates and hydroxy esters are substrates of one strain, *Geotricum candidum*. For synth. applications, the enzyme from *Cellulomonas sp.* is superior because of its higher specific activity and lower cost.

Reactions proceeding in the direction of *oxidation usually suffer from slow rates and incomplete conversions due to *product inhibition. **E=F=G.**   C.-H. WONG

Ref.: *J.Org.Chem.* **1986**, *51*, 25; *Tetrahedron Lett.* **1988**, *29*, 2453.

glycerol kinase (EC 2.7.1.30, ATP: glycerol

3-phosphotransferase) cat. the synth. of *sn*-glycerol 3-phosphate from ATP and glycerol. It was also used in the prepn. of dihydroxyacetone phosphate, a useful phosphate in aldolase-cat. reactions. The preparation of *sn*-glycerol-3-phosphate and several *chiral phosphate derivatives of glycerol – potential precursors to interesting phospholipid derivatives – has been demonstrated. Many analogs of glycerol function as substrates for the kinase reaction. The phosphorylation produced chiral phosphate derivatives from racemic alcohols (90 % *ee*) in good to excellent chemical yields. As with most examples, the ATP-dependent reaction was accomplished using in- situ regeneration of the ATP. **E=F=G.**   C.-H. WONG

Ref.: *J.Am.Chem.Soc.* **1985**, *107*, 7008, 7019.

glycerol 3-phosphate → glycerol kinase

glycolate oxidase (EC 1.1.3.15) from spi-

nach was used in the *oxidation of glycolate to glycoxylate. The reaction was performed in the presence of *catalase from *Aspergillus niger* and ethylenediamine. The reaction is known to proceed via a two-electron transfer from glycolate to *FMN and reoxidation of the reduced FMN by O_2 to produce H_2O_2, which is decomposed with catalase. **E=F=G.**
Ref.: *J.Org.Chem.* **1993**, *58*, 2253.   C.-H. WONG

glycolysis refers to the series of reactions by which an organism breaks glucose down into smaller units, providing energy for the generation of *ATP in the proc. Other sugars feed into the glycolytic scheme via conv. to one of the *intermediates: galactose, for example, can be epimerized to glucose, while mannose can be phosphorylated by *hexokinase, and then *isomerized to fructose 6-phosphate by phosphomannose *isomerase. Most organisms use the same pathway, called the Embden-Meyer-Parnas pathway. Some bacteria use an alternate route, the Entner-Douderoff pathway. The pyruvate formed from gl. may undergo oxidative degradation in the *Krebs cycle, or in the absence of oxygen (or an equivalent electron acceptor) it may be converted to any of a number of small *fermentation prods.

(ethanol, lactate, etc.) resulting in oxidation of the reduced *nicotinamide *cofactors generated during gl. Many of the intermediates of gl. feed directly into the synth. of other biomolecules. Pyruvate, for example, may be reductively aminated to give the amino acid alanine, while 3-phosphoglycerate is the precursor of serine biosynthesis.

Many of the enzymes of gl. are useful in organic synth. *Pyruvate kinase, in particular, has been widely used as an integral member of cofactor regeneration schemes to replenish nucleoside triphosphates from nucleoside diphosphates and phosphoenolpyruvate.

F=G glycolyse. P.S. SEARS

glycoprotein remodeling the enzymatic modification of the oligosaccharide structures of glycoproteins by removing and adding sugar units (*remodeling*) and making new types of protein-oligosaccharide conjugates.

C.-H. WONG

glycoproteins Many proteins, particularly those produced in *eukaryotic cells but also many in the prokaryotes, are glycosylated through a variety of linkages. The most common linkages are to the amide groups of asparagine and the hydroxyl groups of the hydroxylated amino acids, serine and threonine, but a large variety of other linkages have been observed (Figure).

In eukaryotic cells, glycosylation is most commonly performed within the endoplasmic reticulum (ER) and the *Golgi apparatus, membrane-bound organelles responsible for the prod. of proteins destined for the plasma membrane, for secretion, or for other organelles such as the lysosomes. O-glycosylation occurs via the stepwise addition of sugars to the glycoprotein. N-glycosylation is somewhat more complex.

Protein glycosylation has many effects on proteins. The presence of sugars change the physicochemical properties of the protein, including its solubility and the viscosity of the protein solution. The glycans often stabilize the protein toward proteolytic degradation,

and help it fold via both conformational restriction and binding to *chaperones. The activity of many proteins is affected by glycosylation. Finally, glycosylation is often responsible for the localization, targeting, and clearance of proteins. **E=F**; **G** Glycoproteine.

P.S. SEARS

Ref.: *Bioorg.Med.Chem.* **1995**, *3*, 1565; *Glycobiology* **1993**, *3*, 97; *Angew.Chem.Int.Ed.Engl.* **1999**, *38*, 2301.

glycosidases, glycosides (*endo-* or *exo*-glycosidases, *inhibitors*) Gs. are *enzymes that cleave a glycosidic bond (acetal) between two saccharides to form the alcohol and the hemiacetal. They may cleave with inversion or retention of the anomeric configuration, and may cleave specifically at the ends of or within the polymer.

An endoglycosidase cleaves glycosidic bonds within the middle of an oligosaccharide. Some important examples of endoglycosi-dases are the endoglycosidases D, M, H, and F (EC 3.2.1.96). These enzymes cleave between the N-acetylglucosamine units of the chitobiose core of N-linked oligosaccharides on glycoproteins, but differ in the types of glycans they accept. They are commonly used to remove the saccharides of glycoproteins for characterization of either the protein or saccharide moiety. Exoglycosidases, on the other hand, cleave sugars away from the nonreducing end.

Studies of the mechanism of glycosidases have indicated that *hydrolysis proceeds in an S_N1-like fashion, with the glycosidic bond mostly breaking before attack of the nucleophile, generating an oxocarbonium species that adopts a flattened half-chair *conformation. This is cat. by a pair of carboxylic acids which act as a *general acid and base. The attacking nucleophile may be water, or in the case of many retaining glycosidases, it may be one of the *active-site carboxylic acids. In the latter case, the covalent intermediate is then hydrolyzed to regenerate the enzyme; thus retaining glycosidases proceed through a double displacement mechanism.

Glycosidases have also been used in the synth. of glycosides by the inclusion of large amounts of the acceptor nucleophile and addition of organic cosolvent to suppress hydrolysis. Some excellent glycosidase inhibitors have been designed and synth. Glucals have also been used in the induction of *antibodies that cat. glycosidase reactions. **E=F=G.**

P.S. SEARS

Ref.: *J.Am.Chem.Soc.* **1988**, *110*, 8551; *Science* **1993**, *262*, 2030; *Bioorg.Med.Chem.Lett.* **1998**, *8*, 1145.

N-glycosides, N-glycopeptides (*N*-glycosylation)   *N*-Glycosides are saccharides conjugated to another molecule via nitrogen. In particular, in *N*-glycopeptides or *N*-glycoproteins, the anomeric carbon of the reducing end sugar is linked to the side-chain amide nitrogen of asparagine. In *eukaryotes, the sugar is *N*-acetylglucosamine, and this linkage is created by *oligosaccharyltransferase which transfers a polysaccharide from a specialized lipid, dolichyl pyrophosphate, to the asparagine acceptor. Lysosomal degradation of glycosylasparagine occurs via *hydrolysis of the side-chain amide, rather than the glycosidic bond (cf. also *glycoprotein). **E=F=G.**   P.S. SEARS

glycosylation (inhibition)   Many gl. *enzymes release nucleotide-activated sugars as saccharide donors, and much of the binding energy for these comes from the nucleotide portion of the donor substrate. As a result,

the enzymes typically suffer from extreme *product inhibition by the released nucleotides. This has important ramifications. When using *glycosyltransferases for synth., yields are increased if the inhibition is alleviated by substrate regeneration or nucleotide *hydrolysis. This characteristic has important implications for the design of glycosyltransferase inhibitors, too. Clearly, some sort of nucleotide analog must be included in the design for effective inhibition. Indeed, most of the best glycosyltransferase inhibitors have a nucleotide attached. **E=F**; **G** Glycosylierung.   P.S. SEARS

glycosyl donors   The common gds. in living systems are nucleotide sugars, sugar phosphates, lipid-linked sugars, and sometimes di- or polysaccharides. *Enzymes responsible for protein glycosylation in *eukaryotes typically use lipid-linked sugars (dolichyl phosphosugars) and nucleotide-activated sugars. In mammals, these are cytidine 5′-phospho-β-NeuAc and others, though plants, eubacteria, and the archaea are different in their nucleotide usage. **E=F=G.**   P.S. SEARS

glycosyl phosphorylases   cat. the *hydrolysis and phosphorolysis of *glycosyl phosphates. Examples are the synth. of sucrose and trehalose, cat. by sucrose phosphorylase (ph.) and trehalose ph., respectively. Examples of other *enzymes of this class are those involved in synth. of dextrans, levans, and starch-like polyglucose.

Potato ph. (EC 2.4.1.1), for example, has been used in the presence of primers to synth. polysaccharides. Improvement of this system is accomplished with the use of a coupled enzyme system where glucose 1-phosphate is generated in situ from sucrose and inorganic phosphate, cat. by sucrose ph. The inorganic phosphate liberated by potato ph. is used by sucrose ph. to drive the formation of polymer, thereby increasing the yield. This coupled-enzyme system also allows for regulation of the molecular weight of the polysaccharide prod. by control of the conc. of the primer. Unnatural primers bearing functional groups can also be used to prepare tailormade polysaccharides for further manipulation, e.g., attachment to proteins or other cpds. **E=F=G.**

C.-H. WONG

glycosyltransferases Gts. are *enzymes that transfer sugars from a donor substrate to an acceptor, typically a lipid, protein, or another sugar. Dolichyl phosphosugars, sugar phosphates, disaccharides, and nucleotide sugars are used as sugar donors. Most enzymes are quite specific for the donor substrate, though some of them do not show absolute specificity. As a rule, gts. show nearly absolute *regio- and *stereoselectivity for the linkage they create, and as a result they provide an excellent alternative to chemical techniques for synth. glycoconjugates. The gts. which transfer sugars from nucleotides to proteins are typically type II membrane proteins, though the transmembrane domain can generally be proteolytically cleaved or removed via genetic manipulation to generate the soluble enzyme without loss of activity. The optimal pH values for the *glycosyltransferases are similar, falling between 6 and 7. Most gts. require Mn^{2+}, but many will also work with Mg^{2+}, Ca^{2+}, Co^{2+}, or other divalent cations. As a rule, they also suffer from extreme *product inhibition by the released nucleotides. For this reason, cells either destroy or sequester the prods. The nucleotides may be hydrolyzed, and in fact yeasts are dependent on the activity of a *Golgi resident guanosine diphosphatase for *glycosylation. Likewise, acid *phosphatase, a *trans*-Golgi resident enzyme, hydrolyzes CMP. **E=F**; **G** Glycosyltransferase.

P.S. SEARS

glyme ethyleneglycol methyl ether

GO-Fining process → Exxon procs.

gold as catalyst metal Het. gold cats. involve metallic Au which can be highly dispersed on a variety of *surfaces such as TiO_2, α-Fe_2O_3, Co_3O_4, NiO, $Be(OH)_2$, or $Mg(OH)_2$ as well as supported gold halides. The former cats. are active in the low- temp. oxidation of CO in the presence of O_2 and the *reduction of CO_2 with H_2, the combustion and *partial oxidation of *hcs., and in the *WGSR. *Immobilized Au halides can be used for the hydrohalogenation of ethylene. Hom. Au cats. comprise species with gold in the oxidation state +I and +III. Au(I) and Au(III) *alkoxides have been shown to cat. the *condensation of benzaldehyde with cpds. containing an active methylene group. Au(III) species are cat. active in the *cyclization of alkynyl amines and in other reactions (e.g., *aldol reactions with Au(I)/*BPPFA *ligands). Au(I) cats. can be applied for the *addition of alcohols to alkynes and for the *carbonylation of alkenes. Cats. made from Au/Ag alloys have been proposed for the manuf. of ethylene oxide. *Chiral ferrocenylphosphine-Au(I) *complexes are cat. active in the synth. of asymmetric heterocycles. **F** l'or comme métal catalytique; **G** Gold als Katalysatormetall.

H. SCHMIDBAUR, A. SCHIER

Ref.: *Catal.Lett.* **1997**, *44*, 83; *J.Phys.Chem.* **1996**, *100*, 9929; *Gold Bull.* **1996**, *29*, 123 and **1996**, *29*, 131; *Synthesis* **1991**, 975; *Angew.Chem.Int.Ed.Engl.* **1998**, *37*, 1415; *J.Org.Chem.* **1997**, *62*, 1594; *Tetrahedron Asymm.* **1994**, *6*,1091 and **1995**, *6*, 2503; Adams/ Cotton; *Coord. Chem.Rev.* **1981**, *35*, 259; *Chem.-Ing. Tech.* **1999**, *71*, 869; H. Schmidbaur, *Gold*, Wiley, Chichester 1998.

Golgi apparatus The Golgi apparatus ("Golgi") is an organelle in eukaryotic cells that is the location for much of the processing of proteins and lipids destined for the outside of the cell or for other organelles. Many posttranslational modifications of proteins occur here. Proteins enter the Golgi via vesicular transport from the endoplasmic reticulum, and are shuttled through the Golgi in a directional fashion, *cis* → medial → *trans*. The processing *enzymes are localized in specific compartments within the Golgi. For example, *sulfotransferases are typically found only in the *trans*-Golgi, while *N*-*acetylglucosaminyl-transferase I, an enzyme involved in the elaboration of N-linked oligosaccharides on proteins, is found in the *cis*-Golgi. **E=F=G.**

P.S. SEARS

Gomberg Moses (1866–1947), professor of chemistry in Ann Arbor (USA), discovered in 1900 the first *free-radical triphenylmethyl.

B. CORNILS

Goodyear/Scientific Design (SD) dipropylene process for the manuf. of dipropylene (mainly 2-methyl-1-pentene as an octane booster) by *dimerization of propylene via tripropylaluminum as cat. under press. **F** procédé de Goodyear/SD pour le dipropylène; **G** Goodyear/SD-Verfahren zur Propylendimerisierung. B. CORNILS

Goodyear/SD isoprene process manuf. of isoprene from 2-methyl-1-pentene through 1: *isomerization over *supported acid cats. to 2-methyl-2-pentene, and 2: *demethanation to isoprene under the influence of cat. amounts of HBr (hom. gas/gas cat.; cf. *hom. cat.). **F** procédé Goodyear/SD pour l'isoprène; **G** Goodyear/SD Isopren-Verfahren.
Ref.: Weissermel/Arpe, p. 117.   B. CORNILS

G3P → D-glyceraldehyde 3-phosphate

Grace & Co. (W.R. Grace) the parent company of *Grace Davison Catalysts

Grace Davison Catalysts a subsidiary of W. R. Grace & Co., manuf. cats., *silica prods., and *zeolite adsorbents. The cat. portfolio consists of *FCC, *hydroprocessing, *polyolefin, and chemical processing cats. In 1924 Grace Davison introduced *Raney nickel cats. It still owns the tradename and produces *promoted and unpromoted Ni, Co, and Cu Raney cats. Grace Davison is considered as a leader in the FCC cats. business with 45–50 % market share.

J. KULPE

grafting in cat. terms is a proc. of equil. *adsorption of an active cat. species (metal or *precursor) by covalent bonding to a solid *support from a solution in which the support is suspended. Thus, *supported cats. are manuf. (*immobilization; cf. also *impregnation).

B. CORNILS
Ref.: Ertl/Knözinger/Weitkamp, Vol. 1, p. 207,

graphite → carbon

GRAS abbr. for Generally Regarded As Safe, indicating the status of enzymatic work (e.g. *yeasts)

grazing incidence X-ray powder diffraction (GR-XPD) This method (*ex situ*, non-destructive to the sample) uses X-ray photons for the analyses of phases, film thickness and roughness of thin- film solid samples. Morphology analysis is difficult for complex film compositions. The data are *surface-sensitive. R. SCHLÖGL

green chemistry is dedicated to environmentally benign chemical syntheses and processing and includes sustainable development as a necessary goal for achieving societal and economic objectives. Among other targets (Figure) catalysis (which is not mentioned as underlying technique) will be of decisive importance. **F** chimie verte pour l'environnement; **G** Grüne Chemie, Umweltchemie. B. CORNILS

Ref.: P. Tundo, P.T. Anastas, *Green Chemistry: Challenging Perspectives*, Oxford Science, Oxford 1999; *Green Chem.* **1999**, *1*, 3.

green fluorescent protein (GFP) a protein produced by the jellyfish *Aequorea vic-*

toria. It contains a fluorophore that is created in an *autocatalytic fashion from three consecutive amino acid residues: Ser65, Tyr66, and Gly67. Substitutions have been made to several positions to produce mutants with shifted excitation and emission spectra.

The fluorophore, which is not fluorescent when isolated, is found at the center of a cylinder made of 11 β-strands (a "β-can"). **E=F=G.**

P.S. SEARS

Ref.: *Curr.Opin.Struct.Biol.* **1997**, *7*, 821.

grid catalyst crystallization of cat. active constituents (e.g., *zeolites) on the (inactive) network of *gauzes yields gcs. Gcs. express extremely low press. losses and high heat transfer values; they are very suitable for highly exothermal reactions; *particle size, composition, or cat. density may be controlled during crystallization. (Cf. *metal distribution). **F** catalyseur en treillis; **G** Gitterkatalysatoren.

B. CORNILS

Ref.: *Chem.Ing.Tech.* **1999**, *71*, 388.

Grignard Victor A.F. (1871–1935), professor of chemistry in Nancy and Lyon (France), discovered the Grignard reagents RMgX and the *Grignard reaction, a broad chemistry concerning nucleophilic organyl-transfer reagents. Nobel prize laureate 1912 (together with *Sabatier).

W.A. HERRMANN

Grignard cross-coupling is the transition metal cat. coupling of Grignard reagents with organic electrophiles. Ni *complexes are the most common cats. for this reaction, but complexes of Fe, Pd, Cu, and others are also known. Pd is mostly limited to reactions of aryl Grignard or alkyl Grignard reagents lacking β-hydrogens. In the Ni-cat. reaction β-hydrogens may be present in the Grignard

reagent; hence primary alkyl halides can be applied. Secondary and tertiary Grignard reagents tend to rearrangements and reduction of the electrophile. The order of reactivity of aryl halides with Grignard reagents correlates with the ability of *oxidative addition of Ar-X to the low- valent *transition metal center (ArI > ArBr > ArCl > ArF). In the presence of Ni cats. even fluorobenzene can be coupled with Grignard reagents. Coupling reactions of aryl or alkenyl Mg reagents with alkyl halides and related cpds. are almost limited to primary alkyl electrophiles and Cu cats. Few examples of different cat. are known (e.g., Ag). **F** couplage croisé de Grignard; **G** Grignard-Kreuzkupplung.

A. ZAPF, M. BELLER

Ref.: Beller/Bolm, Vol. 1, p. 158; Diederich/Stang; Trost/Fleming, Vol. 3, p. 435.

Grignard reaction the addition of Grignard reagents (i.e., organo Mg derivatives) to aldehydes or ketones and the subsequent *hydrolysis (Figure).

Formaldehyde gives primary alcohols, other aldehydes yield secondary alcohols, and ketones give tertiary alcohols. Hetero-analog carbonyl cpds., carboxylic acid derivatives, nitriles, oxiranes, and alkyl halides also react. The most important side reactions are enolization of the carbonyl cpds. and *reduction. The mechanism of the Gr. is not known in detail; four-centered cyclic *transition states or *single electron transfer procs. are discussed. The presence of traces of impurities in Mg seems to have considerable (catalytic?) effect on the kinetics of the reaction. **F** réaction de Grignard; **G** Grignardreaktion.

A. TILLACK, M. BELLER

Ref.: March; Ullmann, Vol. A15, p. 62; Kirk/Othmer-1, Vol. 12, p. 768; Kirk/Othmer-2, p. 571; H.G. Richey, *Grignard Reagents*, Wiley, Chichester 1999.

group transfer polymerization (GTP) A polymerization (p.). which in contrast to free- *radical p. delivers low and narrow molecular weight distributions; is mostly obtained with anionic p. such as Li alkyls at low temps. B. CORNILS

Ref.: Mijs (Ed.), *New Methods of Polymer Synthesis*, Plenum, New York 1991.

GR-XPD → grazing incidence X-ray powder diffraction

GT-Aromex process → Glitsch procs.

GT-DMT process → Glitsch procs.

GTP 1: *guanosine triphosphate; 2: *group transfer polymerization.

guanosine triphosphate (GTP) a nucleotide *cofactor. Many GTP-binding proteins are interconverted between active and inactive forms by the binding of GTP, its *hydrolysis to GDP + P_i, and exchange of GDP for a new molecule of GTP. GTP hydrolysis is also coupled to protein translation, and GTP is used as an activating group for several glycosyl donors such as *GDP-fucose and *GDP-mannose.

GTP can be made from *ATP and GDP with the enzyme nucleoside disphosphate kinase. E=F; G Guanosintriphosphat. P.S. SEARS

guanylate kinase → nucleoside diphosphate synthesis

guard bed an additional cat. bed installed upstream to remove *poisons or particulate matter and thus to protect the main cat. bed. A guard bed is helpful when classical purification methods are not possible or too expensive (e.g., procs. of *Aluminum Co, *Selexsorb, *Houdryforming, *Union Oil Unicracking). A comb. of adsorption/desorption towers and a cat. reactor is applied in *UOP's *SafeCat technology. F lit catalytique de sécurité; G Sicherheitsbett. B. CORNILS

Guerbet reaction procedure for the manuf. of *Guerbitols, i.e., higher alcohols, via the pressurized *condensation reaction of lower alcohols in presence of Cu and Na. The overall Guerbet reaction is a sequence of *dehydration of the starting alcohols, *aldolization of the resulting aldehydes and their subsequent *hydrogenation to the higher alcohols. F=G=E. B. CORNILS

Guerbitols tradename for alcohols from the Guerbet reaction

guests → host-guest relationships

Guinier plot a method of evaluating the data obtained in small-angle scattering experiments using either (laser) light, X-ray, or neutron radiation. It allows the determination of the average size of a collection of small particles. The Gp. technique works in the size range below 100 nm, when the wavelength λ of the scattered radiation is adapted to the particle size. In the simplified equ. $\log(I) = k_1 - k_2 R_g^2 \lambda^2 (2\angle)^2$, I is the intensity distribution observed at small scattering angles $2\angle$ ($2\angle$ is the angle between the primary beam and the scattered beam and is typically <1°, k_1 and k_2 are constants and R_g is the so-called radius of gyration, i.e., the mean square distance of the surface of a particle from its center of gravity. The Gp. is a plot of $\log(I)$ against $(2\angle)^2$. Acc. to the equ. above, this plot should give a straight line, from the slope of which the radius of gyration can be determined. E=F, G Guinier-Diagramm. P. BEHRENS

Gulf α-olefin process for the manuf. of linear α-olefins (LAOs) by cat. one-step *oligomerization of ethylene with $AlEt_3$ under high press. (25 MPa). **F** procédé Gulf pour LAO; **G** Gulf LAO-Verfahren. B. CORNILS

Gulfining, Gulfinishing process
→ Gulf procs.

Gulf FCC process
for high conv. of different feedstocks including *desulfurized residues to maximize gasoline or furnace oil yields with low coke formation by a proprietary *FCC technique and fluid cats. **F** procédé FCC de Gulf; **G** Gulf FCC-Verfahren.
Ref.: *Hydrocarb.Proc.* **1972**, (9), 133.   B. CORNILS

Gulf Gulfining process
fixed-bed cat. *HDS proc. for the hydrodesulfurization of heavy distillate gas oils in special Gulfining reactors and a special arrangement of high-press. separator, scrubber, and stripper. **F** procédé Gulfining de Gulf; **G** Gulf Gulfining-Verfahren.   B. CORNILS
Ref.: *Hydrocarb.Proc.* **1978**, (9), 131.

Gulf Gulfinishing process
cat. *hydrotreatment proc. for color improvement, neutralization, and impurity reduction in lube oils, waxes, and specialty oils of solvent-extracted raw paraffinic neutrals, bright stocks, naphthenic distilllates, etc. **F** procédé Gulfinishing de Gulf; **G** Gulf Gulfinishing-Verfahren.   B. CORNILS
Ref.: *Hydrocarb.Proc.* **1978**, (9), 132.

Gulf-Houdry (H-G) *hydrocracking process
a one- or two-stage cat. proc. for the conv. of light and heavy gas oils into lower- boiling prods. To minimize cat. *aging the recycle hydrogen may be treated for removal of extraneous materials. **F** procédé hydrocracking de Gulf-Houdry; **G** Gulf-Houdry-Hydrocracking-Verfahren.   B. CORNILS
Ref.: *Hydrocarb.Proc.* **1978**, (9), 114.

Gulf lube oil hydrotreating
for the manuf. of single or multi grade lube oils from almost any crude source by simultaneous desulfurization (*HDS), denitrogenation (*HDN), and demetallization (*HDM) over *hydrotreating cats. at medium or high severity. **F** hydrotreating BP pour l'huile de lubrification; **G** BP Schmierölhydrotreating-Verfahren.   B. CORNILS
Ref.: *Oil&Gas J.* **1972**, 70(6), 94; *Hydrocarb.Proc.* **1972**, (9), 169.

Gulf phenol process
proposal for the manuf. of phenol by *oxychlorination of benzene with aqueous HCl in the liquid phase (het. cat. with diluted HNO_3, hom. with Pd) and the subsequent *hydrolysis (cat. rare earth phosphates) of the chlorobenzene(s) to phenol. **F** procédé Gulf pour le phénol; **G** Gulf Phenolverfahren.   B. CORNILS

Gulf Resid HDS process
high-press. cat. proc. for the *hydrodesulfurization of high-sulfur atmospheric or vacuum residua for fuel oil prod. and for upgrading of crudes or bitumen at 350–450 °C and *LHSVs of 0.2–2. The reactor section is usually duplicated in two trains. **F** procédé Resid HDS de Gulf; **G** Gulf Resid HDS-Verfahren.   B. CORNILS
Ref.: *Hydrocarb.Proc.* **1972**, (9), 157 and **1978**, (9), 149.

Gulf THD process
a thermal proc. for the *hydrodealkylation of aromatic feedstocks, e.g., toluene → benzene, at 650–820 °C/3.5–7 MPa (cf. *HDA proc. of *Atlantic Richfield). **F** procédé THD de Gulf; **G** Gulf THD-Verfahren.   B. CORNILS

gyrase → topoisomerase

H

HA-84 process → Sinclair procs.

Haber, Fritz (1868–1934), professor of chemistry in Karlsruhe (Germany), head of Kaiser-Wilhelm-Institut in Berlin. Famous for his work on electrochemistry, *ammonia synthesis (*Haber-Bosch process), chemical warfare and combat gases, isolation of gold from sea water. Nobel laureate in 1918/1919.

B. CORNILS

Ref.: Ch. Haber, *Mein Leben mit Fritz Haber*, Econ, Düsseldorf 1970; D. Stoltzenberg, *Fritz Haber*, VCH, Weinheim 1994; M. Szöllösi-Janze, *Fritz Haber*, Beck, München 1998; *Nobel Lectures, Chemistry 1922–1941*, Elsevier, Amsterdam 1966.

Haber-Bosch process the large-scale industrial proc. for the synth. of ammonia out of the elements nitrogen (N_2) and hydrogen (H_2), named after its inventors *Haber and *Bosch. The HBp. consists of *syngas generation and purification, *ammonia synth. prod. separation, and the recycling of the unconverted reaction gas. The ammonia synthesis reaction $N_2 + 3H_2 \rightarrow 2NH_3$; $\Delta H = 92.4$ kJ is performed at a press. of 20 MPa and at temps. between 475 and 600 °C over an iron catalyst promoted with the oxides of potassium, calcium, and aluminium; cf. procs. of *Haldor Topsøe, *ICI, *Casale, *Kellogg, *Linde, etc.

R. SCHLÖGL, A. SCHEYBAL

Hafnium as catalyst metal Owing to the lanthanoid contraction, the atomic parameters of the group IV elements Hf and *Zr differ only marginally. Therefore, Hf cpds. generally show analogous cat. properties to the homologous Zr derivatives. In *metallocene application Hf is said to exhibit special advantages. W.R. THIEL

Ref.: *Houben-Weyl/Methodicum Chimicum*, Vol. XIII/7.

Hägg carbide → Kellogg Synthol synthesis

HAI/LPI process → Mobil procs.

Halcon acetic acid anhydride process manuf. of *acetic acid anhydride (AAA) by ligand-modified Rh-cat. *carbonylation of dimethyl ether or *methyl acetate. Realized in the plant of *Tennessee Eastman. **F** procédé Halcon pour anhydride de l'acide acétique; **G** Halcon Acetanhydrid-Verfahren.

B. CORNILS

Halcon/Oxirane ethylene glycol process the two-step manuf. of ethylene glycol by 1: *acetoxylation of ethylene (cat. by TeO_2/HBr), and 2: by *hydrolysis of the diacetates. **F** procédé Halcon/Oxirane pour éthylèneglycol; **G** Halcon/Oxirane Ethylenglykolverfahren. B. CORNILS

Halcon/Oxirane propylene oxide process → ARCO PO process

Halcon/SD formic acid process manuf. of formic acid by Na *alcoholate (methylate) cat. *carbonylation of methanol under press. to *methyl formate and its *hydrolysis to formic acid and recycling of MeOH. **F** procédé Halcon/SD pour acide formique; **G** Halcon Ameisensäureverfahren. B. CORNILS

Halcon vinyl acetate monomer process a multi-step proc. for the manuf. of *vinyl acetate monomer (VAM) from *syngas by 1: synth. of *methyl acetate, 2: press. *carbonylation of methyl acetate (cat. iodine/picoline-modified Rh), 3: *hydrogenation of the *intermediate to 1,1-diacetoxyethane, and 4: the *acid-based cat. cleavage and elimination of acetic acid to VAM. **F** procédé Halcon

pour l'acétate de vinyle; **G** Halcon Vinylacetat-Verfahren.   B. CORNILS

Haldor Topsøe catalysts Topsøe (50 %
owned by Snamprogetti) develops and offers
procs. and cats. for the petrochemical and
chemical industry. Cats. are manuf. in Houston (TX) and in Frederiksund (Denmark).
The main fields of activity include *steam reforming, *hydrotreating, sulfur removal,
*methanation, and synth. of *methanol or
*ammonia. Procs. and cats. are either offered
in combination or separately. The company is
running intensive research in procs. and related cats.   C.D. FROHNING

Haldor Topsøe formaldehyde process
for the manuf. of formaldehyde by *oxidation
of methanol in tubular reactors. The cat. consists of Mo oxide (80 %) and Fe oxide (20 %),
promoted with Cr oxide. In order to maintain
optimum temp. control and to reduce byprod.
formation the cat. in the upper part of the
tubes is diluted with inert material. By changing the amount of process water fed to the absorber, the conc. of the aqueous HCHO solution can be varied between 37 and 55 wt.%.
Alternatively, an aqueous urea solution can
be used instead of process water for prod. of
urea formaldehyde (UF) precondensates (cf.
Figure). **F** procédé Haldor Topsøe pour le
formaldéhyde; **G** Haldor Topsøe Formalde-
hyd-Verfahren.   B. CORNILS

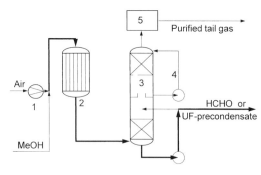

1 blower; 2 reactor; 3 absorber; 4 water or urea
solution; 5 incinerator or catalytic afterburner.

Haldor Topsøe hydrogen process for
the manuf. of H_2 by decomposition of methanol in the presence of water and *shift conv.
cats. such as CuZnAl at 250 °C/1 MPa. **F** procédé de Haldor Topsøe pour l'hydrogène;
G Haldor Topsøe Wasserstoffverfahren.
                    B. CORNILS
Ref.: US 4.316.880 (1982); *Chem.Eng.News* **1985**, (1).

**Haldor Topsøe low-energy ammonia
process** for the manuf. of NH_3 from natural gas, including *desulfurization, primary and
secondary *reforming, two-step *shift conversion, CO_2 removal, *methanation, compression, *ammonia synth., and prod. recovery. A
co-production of methanol is also possible.
F synthèse d'ammoniaque de Haldor Topsøe;
G Haldor Topsøe Ammoniaksynthese.
                    B. CORNILS

Haldor Topsøe methanol process a special version of the medium- press. *methanol
synth. over CuO-ZnO_2/Al_2O_3 cats. at 230–
260 °C/10–15 MPa. Coproduct. of *ammonia
is also offered. **F** procédé Haldor Topsøe pour
le méthanol; **G** Haldor Topsøe Methanol-Ver-
fahren.   B. CORNILS

Haldor Topsøe SNOX process method
for the removal of sulfur *and* NO_x from flue
gases; cf. *SNOX.

**Haldor Topsøe syngas (ATR) genera-
tion** Prod. of CO-rich *syngas by a combi-
nation of *partial oxidation and adiabatic
steam *reforming using a fixed-bed Ni cat.
F procédé gaz de synthèse (ATR) de Haldor
Topsøe; **G** Haldor Topsøe ATR-Synthesegas-
Verfahren.   B. CORNILS
Ref.: *Hydrocarb.Proc.* **1996**, (4), 148.

Haldor Topsøe syngas generation Prod.
of hydrogen- and/or CO-rich gas using advanced steam *reforming of *hcs. (up to heavy naphtha) over Ni cats. in tubular reactors.
F procédé gaz de synthèse de Haldor Topsøe;
G Haldor Topsøe Synthesegas-Verfahren.
Ref.: *Hydrocarb.Proc.* **1996**, (4), 146.   B. CORNILS

halogenation   the conv. of an element or a cpd. into the corresponding halide or the introduction of a halogen atom into organic cpds. Many h. reagents are used in stoichiometric amounts and not catalytically. For the substitution of aromatic cpds. as well as *addition to unsaturated organic cpds., *Lewis acids support the electrophilic attack of the halogen molecule by polarizing the halogen-halogen bond. The major commercial application of this cat. is the *chlorination of ethylene. For the introduction of halogens into aromatic cpds. the *Sandmeyer reaction uses Cu(I) salts as cats. The addition of halogens to double bonds is known to be cat., by both anionic and cationic cats. Addition to the corresponding halide (e.g., HX or tetrabutyl-ammonium halides) during the h. increases the reaction rate.

Li ions or protons are examples of cationic cats. Similarly to the above-mentioned Lewis acids, cations polarize the halogen-halogen bond. The h. of ketones is cat. by acids or bases. Without cats., this reaction takes place *autocat. under the influence of the hydrogen halide formed. Depending on the different halogens, other cats. are known. For the *chlorination of double bonds, e.g., iodine has a cat. effect. The *Hell-Volhard-Zelinsky reaction uses PBr_3 as cat. for the *regioselective bromination of carboxylic acids. **E=F**; **G** Halogenierung.   S. TRAUTSCHOLD

Ref.: A. Loupy, B. Tchoubar, *Salt Effects in Organic and Organometallic Chemistry*, VCH, Weinheim 1992; Ertl/Knözinger/ Weitkamp, Vol. 5, p. 2348.

halolysis (halogenolysis)   the cleavage of a bond by halogens. The most important application is the chlorolysis of C_1-C_3 hydrocarbons and their chlorinated derivatives for the prod. of CCl_4 or tetrachloroethylene. This uses waste gases from other chlorination procs., e.g., from methane chlorination or *vinyl chloride prod. The reaction (at 600 °C/ 10–15 MPa) is governed by the equil. 2 CCl_4 $\rightarrow$ C_2Cl_4 + 2 Cl_2, which is temp.- and press.-dependent. The chlorolysis is noncatalyzed

and mainly used for the prod. of tetrachloroethylene. **F=G** halolyse.   S. TRAUTSCHOLD

Ref.: Ullmann, Vol. A6, p. 266.

haloperoxidases   *enzymes which cat. the *oxidation of halogenide ions to hypochlorous acid, acc. to $H_2O_2 + X^- + H^+ \rightarrow HOX + H_2O$. Hs. may contain Fe or V; some are metal free. **E=F=G**.   B. CORNILS

Ref.: *Angew.Chem.Int.Ed.Engl.* **1999**, *38*, 977.

Hammett $\rightarrow$ Curtin-Hammett principle

Hammett constants (ρ,σ) σ is defined by log $(K_a/K_a^0)(=pK_a-pK_a^0)$, for K_a etc., cf. *Hammett equ.) and describes the relative basicity (nucleophilicity) of aromatic bases in water. The Hammett reaction constant ρ is the slope of the logarithmic plot of the rate constant for a particular series of reactions versus the Hammett substituent constant σ according to the relationship log $(k/k^0)=\rho\sigma$ (for k cf. *Hammett equ.). ρ describes the dependence of the reaction on the basicity of the aromatic cpd. **F** constant de Hammett; **G** Hammett-Konstante.   R. VAN ELDIK

Ref.: Wilkins; Stumm; Connors; Martell/Hancock; Atwood.

Hammett equation   correlates the *rates of reaction of a series of *meta-* and *para*-substituted aromatic cpds. with a common substrate log $(k/k^0) = \rho$ log $(K_a/K_a^0) = \rho\sigma$ (k, k^0 are the reaction rate constants and K_a, K_a^0 the acid dissociation constants for the X-substituted and unsubstituted aromatic cpds.). This is a very useful linear free energy relationship (LFER) in which usually log k or log (k/k^0) is plotted against the Hammett substituent constants σ; the shape is the *Hammett reaction constant ρ. The standard reaction is the aqueous ionization equil. of *m-* and *p*-substituted benzoic acid, for which the reference cpd. is benzoic acid. These reactions led to the *Hammett constants σ_m and σ_p. The constant σ depends only on the substituent and is independent of the reaction series, whereas ρ depends only on the nature of the reaction examined. The Hammett substituent constant σ has for instance been effectively correlated with the rate of base *hydrolysis of a series of *complexes of the type $Co(en)_2(X-C_6H_4COO)_2$, etc. The predictive importance of the Hammett relationship applies also to substitution and oxidation reactions of aromatic cpds., even to *enzyme cat. reactions. **F** équation de Hammett; **G** Hammett-Gleichung.

R. VAN ELDIK

Ref.: Wilkins; Stumm; Connors; Martell/Hancock; Atwood; *Chem.Rev* **1991**, *91*, 165; *J.Am.Chem.Soc.* **1989**, *111*, 6864.

Hammond postulate This postulate states: "If two states ... occur consecutively during a reaction proc. and have nearly the same energy content, their interconversion will involve only a small reorganization of the molecular structure." Although reactions have been found that violate this principle, it is often correct, and has several consequences. If there is a relatively unstable intermediate adjacent to the *transition state on the reaction pathway, then the transition state looks more like the intermediate than like the prod. or substrate. For a reaction involving no intermediates, but a large change in energy from reactant to prod., the postulate has been used to model the transition state on the species (reactant or prod.) to which it is more similar in energy. This postulate applies to enthalpic differences, rather than entropic ones, and so it does not apply well to bimolecular procs. It does, however, apply to enzymatic cat. after the initial binding step. **F** postulat d'Hammond; **G** Hammond-Postulat. P.S. SEARS

Ref.: *J.Am.Chem.Soc.* **1953**, *77*, 334.

Hansch equation Many substituents on a molecule change the partitioning of the cpd. between water and an organic phase in a predictable and additive fashion. If P_0 represents the partitioning of the parent cpd. between the aqueous and organic phases (i.e., [solubility in water]/[solubility in organic phase]), then the "hydrophobicity constant" π of a substituent group can be used to predict ▮ the partitioning of the derivatized cpd., wit▮ the Hansch equation $\pi = \log_{10}(P/P_0)$. The hy▮ drophobicity constants of many substituent▮ are tabulated. The approach works best (tha▮ is, π is most constant) for non-polar group▮ and poorly for strongly electron-donating o▮ -withdrawing groups. For these cpds., π varie▮ with the parent cpd.

For *enzymes with hydrophobic bindin▮ pockets, the Hansch equ. has been used to re▮ late the hydrophobicity of different substrate▮ with the log of the observed k_{cat}/K_m. A plo▮ of log (k_{cat}/K_m) for the *hydrolysis of differ▮ ent amino acid ester substrates vs. the Hansch▮ π values for the side chains was found to b▮ linear, with substrates as widely different a▮ alanine and phenylalanine. **F** équation d▮ Hansch; **G** Hansch-Gleichung. P.S. SEAR▮

Ref.: *J.Pharm.Sci.* **1970**, *59*, 731; *FEBS Lett.* **1972**, *2▮ 122.

hapten a small molecule used to includ▮ *antibodies that is conjugated to a large *car▮ rier. If the h. is a *transition state analog of ▮ reaction, then some of the antibodies induce▮ are expected to be cats. of that reaction.

P.S. SEAR▮

hapticity (η^n) gives the number n of dono▮ atoms by which a *ligand is bound to the met▮ al. **F** hapticité; **G** Haptizität. R.D. KOHM▮

haptotropic shift When a *cp *ligand i▮ coordinated to a metal center, any distortion o▮ the central pentahapto bonding mode can resul▮ in better access to the metal center and ulti▮ mately to nucleophilic displacement (η^0). The▮ first crystallographically authenticated case of ▮ trihapto-bonded cp ligand was observed by▮ Brintzinger in $cp_2W(CO)_2$, which is formally ▮ 20-electron *complex (*Tolman's 18–16 elec▮ tron rule). Monohapto-bonded cp ligands were▮ known, e.g., in $cpFe(CO)_2(\eta^1$-cp) which under▮ goes ring whizzing, i.e., 1,2-migration of the Fe▮ fragment on the periphery of the cp ring. Basol▮ postulated that nucleophilic *substitution reac▮ tions at coordinatively saturated metal centers▮

where the incoming nucleophile is a small electron donor such as PMe_3, are aided by hs. **F** déplacement haptotropique; **G** haptotrope Verschiebung.   J. OKUDA

Harden Arthur (1865–1940), professor in London and head of the biochemical department of the Jenner Institute of Preventive Medicine. Worked on the degradation and the *fermentation of sugars, *enzymes, and *coenzymes. Nobel laureate in 1929 (together with Euler-Chelpin).   B. CORNILS
Ref.: *J.Chem.Soc.* **1943**, 334.

hardening of fats Fats are esters of glycerol with randomly distibuted linear fatty acids of approx. 12 to 22 C atoms (triglycerides; cf. *fat hardening). The fatty acids have even C-numbers and contain one to five non-conjugated C-C double bonds in *cis*-configuration. Consequences of the *cis*-unsaturation are low melting points (mp) and limited stability. These disadvantages are greatly reduced by partial *hydrogenation (hy.) of multiply unsaturated fatty acids to the mono-unsaturated level, thereby increasing the mp and the stability. For nutritional reasons *cis*-configurated mono-insaturates are preferred. Typical example is the h. of linolenic acid (all-*cis* 9,12,15-octadecatrienoic acid, mp –11 °C) – main constituent of soybean oil – to oleic (*cis*-9-octadecenic acid, mp 13.2 °C).
The raw feedstock (vegetable or animal fats) is upgraded before h. to remove impurities by bleaching, degumming, deacidification, etc. The hy. step is carried out in the liquid phase, mostly in stirred-tank reactors equipped with heating/ cooling devices without gas recycle ("dead-end hy."). Continuous hy. is seldom applied due to the unavoidable changes in feedstock quality. The progress of the hy. is controlled by the gas uptake, the change in iodine value, and the increase in mp.
Ni cats. are applied mostly, Cu cats. only seldom to slightly reduce nnsaturation. Today's Ni cats. are *precipitated onto a *carrier (*kieselguhr, *silica, *alumosilicate, *alumina), reduced, and enveloped in hardened fat

(flakes, prills, granules) to prevent the access of O_2. When added to the hot feedstock, the protective medium melts and the cat. powder is set free. Depending on the quality of the feedstock, the initial degree of unsaturation, and the *activity of the cat., as less than 150 ppm of Ni are sufficient to reduce the iodine value in 2 h at 180 °C/0.25 MPa. The used cat. (still retaining some activity) is filtered off and can be re-used or *regenerated. Cf. procs. of *Buss, *Davy, *Henkel, *Lurgi. **F** solidification des matières grasses et huiles; **G** Fetthärtung.   C.D. FROHNING
Ref.: D. Swern, *Bailey's Industrial Oil and Fat Products*, Wiley, New York 1986; Ertl/Knözinger/Weitkamp, Vol. 5, p. 2221.

hard-soft acid-base (HSAB) concept
→ Pearson acid-base concept

Hartree-Fock (HF) method is the most important approximation to the *Schrödinger equ. in *ab-initio calcns. In the HF approximation it is assumed that the electrons in an atom or molecule move independently of each other. This approximation, which is also called the independent particle model, leads to the set of one-electron HF equs. Unlike the *Schrödinger equ., the HF equs. can be solved, because the molecule is treated like a two-particle system, where one electron moves in the average field of all the other particles (nuclei and electrons). The one-electron function which describes the electron in a molecule is called a *molecular orbital (MO). The total wavefunction of the system is then given by the prods. of the MOs which are arranged in a determinant, called the Slater determinant. The form of a determinant rather than a simple prod., which was originally suggested by Hartree, is necessary to fulfill the Pauli principle. MOs are usually expressed by a linear comb. of atomic arbitals (LCAOs). The advantage of the LCAO approximation is that the HF equs. become linear, while the original HF equs. are integro-differential equs., which are difficult to solve. Since the solution of the one-electron function in a molecule requires knowledge about the other electrons, the HF equs. can only be solved iteratively. One starts with a guess for the solution, solves the equs., and then repeatedly uses the solutions for new equs., until convergence of the solution is achieved. This is called the self-consistent field (SCF) approach.

The quality of a HF calcn. is determined by the number of basis functions which are used for the LCAO-SCF equs. Even with an infinite number of basis functions there would still be a difference between the HF energy and the correct non-relativistic total energy. This difference is called the correlation energy. The error arises from the assumption that the electrons move independently of each other in an average field of the other particles. Although the correlation energy is typically only 1 – 2 % of the total energy, correlation contributions can become important for chemical problems. **F** méthode de H-F; **G** H-F-Methode. G. FRENKING

Ref.: Thomas/Thomas, p. 364.

Hatta number In two-phase systems in which the cat. reaction between a gaseous reactant (A_g) and a liquid reactant (B_l) takes place in the liquid phase, A_g has to be transferred over the gas/liquid boundary layer into the liquid phase. Concerning the macrokinetics of such a reaction, the ratio of the maximum conv. rate in a film of thickness δ per unit area interface to the maximum diffusional transport through a film without any chemical reaction is characterized by a dimensionless number, the Hatta number (Ha, Equation).

$$Ha = \frac{1}{k_L}\sqrt{\frac{2}{n+1}\,k_{n,m}\,D_A\,c_{Ai}^{n-1}\,\bar{c}_B^{m}}$$

where k_L is the liquid-site mass transfer coefficient (m/s), $k_{n,m}$ is the reaction *rate constant $[m^{3(n+m-1)}/(kmol^{(n+m-1)}\,s)]$ of reaction of order n, m as in the rate law $r_A = kc_A^n c_B^m$, and D_A is the diffuision coeficient of A. For first-order reactions (n = 1, m = 0) the rela-

tionship can be simplified to an equ. which is closely analogous to the *Thiele modulus for a porous cat. Thus, the Hatta number is often called the reaction modulus for fluid phases. For analysis of coupled fluid-fluid systems it is useful to distinguish between at least three regimes of the reaction rate which are characterized by different values of Ha: (a) $Ha < 0.3$: *slow reactions*, controlled by the kinetics (no steep conc. gradients, reaction occurs mainly in the bulk of the reaction phase); (b) $Ha > 3$: *fast reactions* (steep conc. gradients, reaction exclusively takes place in the boundary layer); (c) $Ha \gg 3$: *instantaneous reactions* (conc. drops to zero within the boundary layer, resulting in a reaction plane). For the three regimes described, different expressions for the *enhancement factor are obtained. **F** nombre de Hatta; **G** Hatta-Zahl. P. CLAUS, D. HÖNICKE

Hay variant a catalytic variant of the *Glaser coupling reaction

hc(s). *hydrocarbon(s)

H-coal process → Hydrocarbon Research process

HC-SCR process an alternative technology to the *selective cat. reduction (SCR) with hydrocarbons instead of ammonia or urea (which may cause secondary ammonia emissions, *ammonia slip). The *DeNOx proc. for *stationary exhaust sources is under development; possible cats. are *ZSM-5 supported Cu cats. B. CORNILS

Ref.: *Appl.Catal.B* **1991**, *69*, L15 and *70*, L1; Ertl/Knözinger/Weitkamp, Vol. 4, p. 1633.

HDA → hydrodealkylation

HDC → hydrodechlorination

HDC Unibon process → UOP procs.

HDI → hexamethylene diisocyanate (1,6-hexylene diisocyanate), formerly HMDI

HDM → hydrodemetallation

HDN → hydrodenitrogenation

HDO → hydrodeoxygenation

HDPE → high-density polyethylene

HDS → hydrodesulfurization

HDS residual process → Shell procs.

heat of adsorption   When a single atom or molecule strikes a *surface and forms a bond with it, heat evolves because the *adsorption proc. is always exothermic (for historical reasons the heat of adsorption ΔH_{ads} is always denoted as having a positive sign – unlike the enthalpy ΔH, which for an exothermic proc. would be negative according to the usual thermodynamic convention). This hoa. reflects the strength of the adsorbate-substrate bond. As more atoms or molecules are adsorbed, each one contributes its hoa. (q_{ads}) to the total (integral) measured value (ΔH_{ads}). As more atoms and molecules are adsorbed, an attractive or repulsive adsorbate-adsorbate interaction may occur, which is usual for most *chemisorption systems, and q_{ads} may become dependent on the coverage. Thus determining the hoa. as a function of the coverage can provide information about the adsorbate-substrate and adsorbate-adsorbate interactions. If ΔH_{ads} is differentiated with respect to the coverage, the differential hoa. is obtained, which is commonly called the *isothermal or *isosteric heat of adsorption. The integral and differential hoas. are determined by measuring the *adsorption isotherms or *adsorption isobars for a given system at different press. or temps. From the slope of the plots in an *Arrhenius plot ($\ln P$ vs. $1/T$) the isosteric heats

of adsorption are determined by the use of the *Clausius-Clapeyron equ.

If adsorption is not activated, the hoa. is equal to the desorption energy and can be determined by *temp.-programmed desorption (TPD). If the adsorption proc. induces structural changes in the substrate, adsorbate reactions, or decomposition, it has to be noticed that the heat needed for restructuring or bond breaking is part of the measured heat of adsorption. **F** chaleur d'adsorption; **G** Adsorptionswärme.   R. SCHLÖGL, Y. JOSEPH

heat of reaction   The heat of reaction q is the heat which is released or consumed in the course of a chemical reaction. If heat is produced, the reaction is called exothermic with q being negative; if on the other hand heat is consumed, then the reaction is called endothermic and q is positive. The amount of transformed heat depends on the conds. under which the reaction is performed, thus distinguishing between the heat of reaction at constant volume q_v and reactions carried out at a constant press. (q_p). When no other work than pV work (volume work) interferes, then it follows from the first law of thermodynamics that $\Delta_r U = q_v$ and $\Delta_r H = q_p$, where $\Delta_r U$ and $\Delta_r H$ are the energy and the enthalpy of reaction respectively. Between $\Delta_r U$ and $\Delta_r H$ holds then the relation $\Delta_r H = \Delta_r U + \Delta(pV)$. The heat of reaction can be determined experimentally by *calorimetry. Simultaneous heat and *mass transfer id difficult to handle. **F** chaleur de réaction; **G** Reaktionswärme.   R. SCHLÖGL, A. SCHEYBAL
Ref.: Ertl/Knözinger/Weitkamp, Vol. 3, p. 1209.

heat transfer in catalysis   Efficient ht. is one of the substantial prerequisites for performing cat. reactions in a limited temp. range and is essential to attain high prod. yields. For *hom. cat. reactions the ht. correlations used for standard heat exchangers are relevant; in the case of a two- or multiphase system the ht. proc. depends on the single transitions between the comps. or *phases (cf. *cat. reactors). The ht. in such systems shows complex behavior and is strongly affected by the intensities and directions of the liquid and gas flow as well as by the *diffusion coefficients of the individual comps. For *het. cat. reactions of comps. in one or more fluid phases which occur at the *surface of solid cats., the ht. depends on the thermal properties of the cats. and the velocity of the fluid phase. Significant intrapellet temp. gradients can exist (*intraparticle mass and heat transfer). In most of the het. cat. reactions fixed-bed cats. and gaseous reactants are used and heat and mass are transferred between solid and fluid by similar mechanisms. Data on ht. in fixed beds are correlated in the same way as those on mass transfer (*mass transfer in cat.):

$$J_{heat} = \frac{h}{c^p \cdot m} \cdot Pr^{2/3}$$

with h being the ht. coefficient, c_p the specific heat capacity, m the mass velocity, and Pr the Prandtl number. J_{heat} is apporoximately equal to the Chilton-Colburn factor J_{mass} [$= Sh/(Re \cdot Sc^{1/3})$], and this forms the basis for estimating mass transfer coefficients from ht. data. **F** transfert de chaleur en catalyse; **G** Wärmetransport in der Katalyse.   P. CLAUS, D. HÖNICKE

heavy gas oil (HGO) liquid fraction of petroleum refining, boiling range 316–420 °C

Heck-Breslow mechanism Heck and Breslow in 1960 presented a proposal for the mechanistic sequence of steps in the hydroformylation of alkenes which has been generally accepted over the years. Formulated originally for unmodified Co cats., the mechanism has later been transferred to today's most used complex Rh-phosphine cats. The sequence starts with HRh(CO)(PR$_3$)$_3$, the *precursor for the active *complex. One mole of *phosphine is split off forming a *vacant site, and the starting alkene is bound to the Rh via a π-complex, which rearranges to an alkyl and – after insertion of CO – to an acyl complex. The subsequent *oxidative addition of H$_2$ yields the aldehyde and the cat. precursor is *regenerated by addition of one mole of

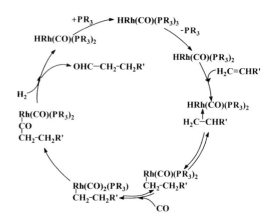

Heck-Breslow mechanism

phosphine (Figure). **F** mécanisme de Heck-Breslow; **G** Heck-Breslow-Mechanismus.
Ref.: Falbe-1, p. 4.   C.D. FROHNING

Heck reaction *transition metal-cat. C-C coupling reaction for the *olefination of aryl-X, vinyl-X, or benzyl-X cpds. (X = Cl, Br, I, OMs, OTf, N$_2^+$, COCl, SO$_2$Cl) (Figure).

RX + ⤳R′ →[cat / base]→ R⤳R′ + HX

The reaction is usually cat. by hom. Pd cats. More recently cats. based on Co, Ni, Ir, Pt, and het. Pd cats. have been reported. The mechanism involves *oxidative addition of R-X to the metal cat., coordination, and *insertion of the olefin into the M-R bond, β-hydride elimination, and release of the product alkene. If there is no possibility of a β-hydride elimination toward the C-C bond created, other reactions can occur and a new stereogenic center (for *enantioselective Hr.) is formed. The Hr. has a broad tolerance for functional groups. Het. Hrs. are described, too. The H. reaction was independently from each other discovered by Heck and by *Mizoroki. **F** réaction de Heck; **G** Heck-Reaktion.   T. RIERMEIER, M. BELLER

Ref.: Heck; J. Tsuji, *Palladium Reagents and Catalysts*, Wiley, Chichester 1995; *Tetrahedron* **1997**, *53*, 7371; Beller/Bolm, *J.Mol. Catal.* **1999**, *142*, 275; Diederich/Stang.

H-elimination → hydride shift

Hell-Volhard-Zelinsky reaction a meth-od for the prepn. of α-halocarboxylic acids by *chlorination or bromination of the acids in presence of phosphorous halides as cats. **F** réaction de Hell-Volhard-Zelinsky; **G** Hell-Volhard-Zelinsky-Reaktion.   B. CORNILS

heme The *chelate *complex of *porphyrins with Fe(II) is called heme; the Mg complex is chlorophyll.   P. ROESKY

Ref.: G.C. Ferreira et al. (Eds.), *Iron Metabolism*, Wi-ley-VCH, Weinheim 1999.

hemilabile ligands polyfunctional *chelate *ligands in which one part binds strongly, the other part weakly, to a metal center. Often a "soft-hard" combination like P~O, P~N, or η-C_5R_5~N is used, where two or more ligands are connected by a suitable *spacer. This spacer al-lows the labile group to dissociate, but it always keeps the ligand close to the metal center. The labile part may be regarded as an intramolecu-lar solvent molecule, able to reversibly create or protect a "*vacant" coordination site (Figure; S = substrate).

Depending on the strength of the metal-ligand interaction, reactive intermediates may be isolated. In other cases it was possible to enhance the *activity and *selectivity in cat. procs. such as *polyolefin synth., alkyne *car-bonylation, stereoselective *hydrogenation, *hydrosilylation, or *hydroformylation. **F** li-gands hémilabiles; **G** hemilabile Liganden.

Ref.: *Coord.Chem.Rev.* **1991**, *108*, 27.   M. ENDERS

Henderson-Hasselbach equation relates the pH of a solution to the relative conc. of an acid/conjugate base or base/conjugate acid pair and the pK_a of that pair:

$$pH = pK_a + \ln [A^-]/[HA]$$

where [A^-] and [HA] are the concs. of the protonated and deprotonated forms of the acid (or base), respectively. Many *enzymes, such as most *hydrolases and transferases, rely on *acid-base cat. during key steps in the reaction pathway, and so the HHe. provides a useful tool for determining the effect of pH on the ionization state of the cat. residues in-volved (and thus on the reaction *rate itself). See also *pH dependence of enzymatic cat. **F** équation d'Henderson-Hesselbach; **G** Hen-derson-Hesselbach-Gleichung.   P.S. SEARS

Ref.: A. Fersht, *Enzyme Structure and Mechanism*; W.H. Freeman and Co., New York 1985, p. 155.

Henkel direct hydrogenation proc. for the direct manuf. of fat alcohols by *hydro-genation of fats (i.e., triglycerides) on Cu-Cr (*Adkins) cats. at 200–300 °C/20–30 MPa (fixed-bed mode) without prior *transesterifi-cation and thus with glycerol as byprod. **F** procédé Henkel de hydrogénation directe; **G** Henkel-Direkthydrierung.   B. CORNILS

Ref.: Ertl/Knözinger/Weitkamp, Vol. 5, p. 2227.

Henkel terephthalic acid processes I and II now obsolete procs. for the manuf. of *terephthalic acid (TPA) by 1: *isomerization of dipotassium phthalate over Zn-Cd cats. at > 400 °C/CO_2 press., or 2: by *disproportiona-tion of K benzoate on Zn-Cd benzoate cats. under CO_2 press. to a mixture of benzene and TPA. **F** procédés Henkel pour TPA; **G** Hen-kel TPA-Verfahren.   B. CORNILS

Ref.: Weissermel/Arpe, p. 397.

Henry aldol addition describes the reac-tion of aldehydes or ketones with nitroalk-anes acc. to $R^1CHO + RCH_2NO_2 \rightarrow R^1CH(OH)C(R)NO_2$. The resulting *aldols can readily be converted into 1,2-aminoalco-hols and/or 2-hydroxycarbonyl cpds. The Haa. is thus a powerful synth. tool to natural prods. A cat. variant was reported using heterome-tallic lanthanoids (*BINOL). **F** aldolisation de Henry; **G** Henry'sche Aldoladdition.

          M. BELLER

Ref.: March; *J.Am.Chem.Soc.* **1992**, *114*, 4418; *Tetrahedron* **1994**, *50*, 12313.

Henry isotherm → adsorption isotherm

Henry's law applies to ideal dilute solutions of a soluble gas A in a liquid L. It states that the mole fraction of the solute is directly proportional to its partial press. in the gaseous phase. Hl. constant is a distribution coefficient which usually depends strongly on temp. Strong deviations from Hl. occur for gases which interact chemically with the solvent, such as CO_2 with H_2O. In chemical reaction engineering, Hl. is applied to gas-liquid reactions within the two-film theory which assumes that there is no resistance to *mass transfer at the interface between the gas film and the liquid film. Thus, Hl. relates the partial press. of the gas-phase species to their concs. in the liquid phase. Hl. is the thermodynamic basis for *gas chromatography. **F** loi d'Henry; **G** Henry'sches Gesetz. M. MUHLER

heparin a highly sulfated linear polysaccharide comprising of repeating 1,4-linked uronic acid and glucosamine residues. It has been used clinically as an inhibitor of the serine protease thrombin. A pentasaccharide repeat of heparin has shown high anticoagulant activity.

Heparin sulfate is similar to *heparin in structure, but with different ratios of N-acetylation to O-sulfation. **E=F=G**. C.-H. WONG

Heraeus catalysts The family-owned company is active in refining and reclaiming of precious metals. Main fields of activity in cat. are prod. of *organometallic cpds. (Pd, Pt, Rh, Ir, Ru, Os) as *precursors or cats. for hom. cat. Heraeus also produces het. precious metal cats. (Pd, Pt Rh, Ru) on different *supports, for *hydrogenation of fine chemicals, *selective hydrogenation in petrochemistry, and cat. gas purification as main applications. J. KULPE

Herrmann carbenes metal *carbenes, specially metal-metal bridged sytems. Typical examples are the Mn or Rh dimetallacyclopropanes of formulae [(C_5H_5)Mn(CO)$_2$]$_2$ (μ-CH$_2$) and [(C_5Me_5)Rh(CO)]$_2$(μ-CH$_2$), resp., made from diazoalkanes. The hcs. are molecular models for chain-growing procs. such as the *Fischer-Tropsch synth. **F=E; G** Hermann-Carbene. D. MIHALIOS

Ref.: *Adv.Organometal.Chem.* **1982**, *20*, 159; Cornils/Herrmann-1, p. 747.

Herrmann catalysts are the Re cpd. **1** and the Ru. cpd. **2**, resp. They are applied in the chemistry of alkenes (**1** for *epoxidation, *metathesis; **2** for *ROMP) (Figure). D. MIHALIOS

1   2

Cy = cyclo-hexyl

Ref.: Cornils/Herrmann-1, p. 430; *Acc.Chem.Res.* **1997**, *30*, 169; *Chem.Rev.* **1997**, *97*, 3197; *Angew.Chem.Int. Ed.Engl.* **1998**, *37*, 2490.

Hertz-Knudsen formula When an adsorbable gas is brought in contact with an adsorbing *surface, acc. to the kinetic theory of gases the particle flux J is defined as the number of molecules which collide with unit area of surface in unit time (molecules cm^{-2} s^{-1}). At a gas press. p one arrives at the HKf. by $J = N_A p/(2\pi MRT)$, where N_A is the Avogadro number and M the molecular weight. If the *sticking probability approaches unity, then the rate of adsorption at most reaches a value corresponding to the HKf., for example for oxygen (assuming 10^{15} adsorption sites to exist on 1 cm^2) the surface will be completely covered within 1 s at a press. of 10^{-6} torr (which is equal to dosing a gas for one Langmuir (L); 1 L = 10^{-6} torr s) and at a temp. of 300 K. **F** formule d'Hertz-Knudsen; **G** Hertz-Knudsen-Formel.   M. MUHLER

Hesorb Isom process   → IFP isomerization proc.

het.   heterogeneous, heterogeneously

heterobimetallic catalysis   → bimetallic catalysts, *bifunctional cat.

heterogeneous catalysis (particulate, surface catalysis) is one of three different kinds of *catalysis; the others are *homogeneous and *biocatalysis. Het. catalysis (hc.) is defined as 1: catalysis with the cat. in a different phase than the reactants, and 2: with the cat. on solid, porous material or impregnated in such material. (The term "solid" does not exclude mobile portions such as potassium cpds. in *steam reforming cats. or Cl in *isomerization cats; cf. *chlorinated alumina). Whereas the cat. is solid, the reactant and the prod. phase may be liquid or gaseous, thus determining the chemical engineering of the cat. reactions (cf. *catalytic reactors).

Het. cat. reactions take place on the *surface of the solid cats. The following sequences of phenomena are significant: *adsorption of the reactants (*physisorption and eventually *chemisorption), mono- or bimolecular reaction of the chemisorbed reactants with other adsorbed species or with reactants in the gas phase, and the *desorption of the prods. Adsorbed molecules are *activated by interaction with the cat. surface (which ought normally to be as great as possible) and thus reducing the energy barrier or *activation energy (cf. diagram under entry *activation energy; compare also *Langmuir-Hinshel-

wood kinetics, *thermodynamics in cat., *Arrhenius plot, *isotherms, *mass and *heat transfer, *single crystals, *active sites, etc.). After desorption the active sites on the surface regenerate and are ready for a new *catalytic cycle.

Acc. to the *shape of the het. cats. (e.g., granules, *gauzes, *pellets, extrudates, rings, etc.) various chemical engineering solutions of the problem of transporting reactants to the cat. particles and the prods. away from it have been realized (e.g., *fixed-bed, *slurry, *gauze, entrained, tubular *fluidized-bed, bubble columns, etc.) in *autoclaves, *CSTR, *batch, or other reactors. The interaction of gases, liquids, and surfaces on solids is extremely complex and thus difficult to treat with chemical kinetics and reaction engineering. The understanding of het. cat. is fragmentary (cf., e.g., *pressure gap, *material gap) and needs experienced specialists. This is one of the important disadvantages of het. cat. The others are that the variation of electronic properties of the cats. (*tailoring) is more or less empirical and that the mechanistc understanding of the cat. procs. on the surfaces is largely hypothetical.

Nevertheless, the use of het. cats. is essential: their recycling (i.e., the start of a new cat. cycle) proceeds without any problems because the separation of products and cat. is simple (in contrast to *hom. cat.) and the catalyst *lifetimes are high. The handling of het. cats. is normally simple also: a bunch of decisive reasons which explain why commercially approx. 80 % of all cat. procs. are run with het. cats. (cf. *catalysis). **F** catalyse hétérogène; **G** heterogene Katalyse.

B. CORNILS, W.A. HERRMANN, R. SCHLÖGL

Ref.: Anderson/Boudart; Augustine; Bond-1; Butt/Petersen; Dines/Rochester/Thomson; Ertl/Knözinger/Weitkamp; Falbe-4; Farrauto/Bartholomew; Fürstner; Gates-1; Guisnet et al.; Hegedus; Iwasawa; Kirk-Othmer-1, McKetta-1; Moulijn/van Leeuwen/van Santen; Rylander; Schwab; Somorjai; Stiles; Thomas/Thomas; Twigg; van Santen; Ullmann; Wijngaarden/Kronberg/Westerterp; Yermakov/Zakharov/Kuznetsov; J.F. LePage, *Applied Heterogeneous Catalysis – Design, Manufacture*, Technip, Paris 1987.

heterogeneous equilibrium   chemical equilibria with more than one *phase involved. **F** équilibre hétérogène; **G** heterogenes Gleichgewicht.

heterogenization   → immobilization

heteroleptic complexes   constitute neutral or charged cpds. of the type $(MX_nY_mZ_o)^{\pm}$ with the metal center M coordinated by non-identical neutral or charged *ligands X, Y, and Z. Hcs. applied in cat. typically comprise an *ancillary ligand X – such as a neutral *bisphosphine or a divalently linked bis*indenyl ligand, and a participating ligand such as a hydride or an alkyl group which can easily be cleaved or exchanged. **F** complèxes hétéroleptiques; **G** heteroleptische Komplexe.

Ref.: *Acc.Chem.Res.* **1974**, *7*, 209.   R. ANWANDER

heterolysis   the splitting of a molecule into two ions $AB \rightarrow A^+ + B^-$ (heterolytic cleavage), indicating that the binding electron pair remains with one of the two molecule as fragments. Because of a solvation of ions h. takes place preferably in polar solvents. H. plays a role in nucleophilic *substitutions, *eliminations, *addition reactions, or fragmentations. **F** hétérolyse; **G** Heterolyse.   B. CORNILS

heterolytic cleavage   → heterolysis

heteropolyacids   The amphoteric metals of groups VB (V, Nb, Ta) and VIB (Cr, Mo, W) in the +5 and +6 oxidation states, respectively, form weak acids that readily condense (polymerize) to form anions containing several molecules of the acid anhydride. If these condensed acids contain only one type of acid anhydride, they are called *isopolyacids*, and their salts are called isopoly-salts. The acid anhydrides also can condense with other acids (e.g., phosphoric or silicic acids) to form *heteropolyacids*, which can form heteropoly-salts. The *condensation reactions, which occur reversibly in dilute aqueous solution, involve the formation of oxobridges by elimination of water from two molecules of the weak acid.

About 70 elements can act as *central (hetero) atoms in heteropolyanions. In general, free heteropolyacids and salts, of which the heteropolymolybdates and heteropolytungstates are the best known, have very high molecular weights (some above 4000) as compared with other inorganic electrolytes; are very soluble in water and organic solvents; are almost always highly hydrated, with several hydrates existing; and are highly colored. The most common structural feature is the Keggin anion build up by 12 MO_6 octahedra and one central XO_4 tetrahedron.

Has. are used as *bulk cats. in many different reactions. They can be applied in various *phases, as hom. liquid cat., as well as in liquid/solid and solid/gas combs. Because of their acidic and *redox properties they cat. a wide range of reactions. *Hydration and *dehydration, *condensation reactions as well as *reduction, *oxidation, and *carbonylation chemistry (with Keggin type hpas. with V, Mo) can be performed with has. Other commercially important reactions are the *selective *oxidation of methacrolein by a Cs salt of $H_4PVMo_{11}O_{40}$, *hydration of isobutylene, or the *polymerization of tetrahydrofuran. Cf. also *polyoxometallates. **F** hétéropolyacides; **G** Heteropolysäuren.

R. SCHLÖGL, M. DIETERLE

Ref.: Ertl/Knözinger/Weitkamp, Vol. 1 p. 118; *J.Mol. Catal. A:* **1997**, *127*, 33.

heterotropic cooperative effect   → cooperative effect

heuristics (in cat. development)   are qualitative relationships between variables. They may be supported by theoretical concepts, or just rules-of-thumb learned from experience. Since up to now it has been impossible to predict the cat. behavior of a solid by deterministic methods alone, the application of heuristics is crucial in the design of het. cats. H. describe methods by which knowledge is acquired and used in a logical but non-mathematical manner, based on a knowledge of 1: cat. chemistry; 2: materials, e.g., their synth.

characterization, texture, and structure; 3: chemical reaction engineering principles, i.e., thermodynamic equil., kinetics of the cat. reaction, and transport procs. as well as their interplay. All these qualitative aspects affect heuristic searches for an optimum cat. (cf. also *knowledge-based systems). In the broadest sense, trial-and-error procedures may also be considered to be of a heuristic nature when based on some rationale. **F** heuristiques; **G** Heuristik.   M. BAERNS

1,4-hexadiene manuf. by hom. *codimerization (*co-oligimerization) of ethylene and butadiene with cats. based on $NiCl_2(PBu_3)_2$ and $Al_2Cl_2(i\text{-}Bu)_4$ or $[RhCl(C_2H_4)_2]_2$. The reaction prods. (mainly *trans*-hexadiene) are separated by distillation. 1,4-H. is used as a comonomer in *EPDM elastomers. **E=F=G**.
                                    B. CORNILS

hexamethyleneimine (HMI) intermediate, manuf. by selective *hydrogenation of ε-caprolactam (*Mitsubishi HMI proc.).

hexokinase cat. the synth. of *glucose 6-phosphate from *ATP and glucose. Although h. is involved biosynthetically in the prod. of glucose 6-phosphate, its breadth of specificity has allowed its use in the synth. of other phosphate sugars. Its use as a cat. in the prod. of arabinose 5-phosphate from arabinose and ATP provided the key step in the synth. of *3-

deoxy-D-manno-2-octulosonate-8-phosphate (KDO-8-P). H. has also been used in the synth. of several uncommon sugar phosphates, including fluoro derivatives. **E=F=G**.
                                    C.-H. WONG

Ref.: *Tetrahedron Lett.* **1988**, *29*, 427; *J.Org.Chem.* **1985**, *50*, 5912; *Adv.Catal.* **1979**, *28*, 323.

hexosaminidases (β-*N*-acetylhexosaminidases) glycosidases that *hydrolyze β-linked *N*-acetylated hexosamines (*N*-acetylglucosamine, GlcNAc, and *N*-acetylgalactosamine, GalNAc) from the non-reducing end of a saccharide. The *enzymes have been isolated from a variety of sources, and differ with respect to the aglycone preferred. **E=F=G**.   P.S. SEARS

Hexorb Isom process → IFP isomerization process

3-hexulose phosphate synthase
(EC 3.7.1.3) cat. the formation of 3-hexulose phosphate from formaldehyde and 2-pentulose phosphate. **E=F=G**.   C.-H. WONG

Ref.: *Tetrahedron Lett.* **1991**, *32*, 3159.; *Appl.Microbiol.Biotechnol.* **1991**, *34*, 604.

hfac hexafluoroacetyl acetonate in *complex cpds.

HF alkylation → alkylation, e.g., processes of *UOP or *Phillips

hfc, HFC (+/-)-3-perfluorobutyryl camphor in *complex cpds. or as *ligands

HF method → Hartree-Fock method

HG hydrocracking process → Gulf (Gulf-Houdry) procs.

HGO → heavy gas oil

HGR process Hot-Gas Recycle process for the cat. *methanation of *syngas with Raney Ni. **F** procédé HGR; **G** HGR-Verfahren.

HICAPDH → hydroxyisocaproate dehydrogenase

Hieber Walter (1895–1976), professor of chemistry in Munich and one of the pioneers in metal carbonyl chemistry. He discovered $HCo(CO)_4$, which is relevant for cat. hydrogen transfer reactions (e.g., *hydroformylation). The nucleophilic *addition to metal carbonyls is known as the *Hieber base reaction.

W.A. HERRMANN

Hieber base reaction nucleophilic attack at metal-attached CO to yield functionalized derivatives acc. to the net equ. $M\text{-}CO + Nu^-$ → $[M\text{-}C=ONu]^-$. The reaction was discovered with $Nu^- = OH^-$ by *Hieber for the example of $Fe(CO)_5$. In more general terms, nucleophilic *addition reactions such as

$$L_xM\text{-}CO + LiCH_3 \rightarrow Li^+[L_xM\text{-}COCH_3]^-$$
$$L_xM\text{-}CO + NaN_3 \rightarrow Na^+[L_xM\text{-}CON_3]^- \rightarrow$$
$$N_2 + Na^+[L_xM\text{-}N=C=O]$$

with $L_x = x$ *ligands are also called "base reactions". They led to a great variety of metal carbonyl derivatives and thus belong to the much applied organometallic transformations. Cationic metal carbonyls react more readily with nucleophiles. Hbrs. are part of the mechanistic cycle of the cat. *WGSR acc. to the Figure. **F** réaction de base de Hieber; **G** Hieber'sche Basenreaktion.   W.A. HERRMANN

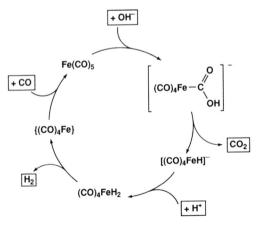

Ref.: *J.Organomet.Chem.* **1990**, *383*, 21; *Acc.Chem.Res.* **1981**, *14*, 31.

high-density polyethylene (HDPE)
During the 1950s three research groups working independently discovered three different cats. which allowed the prod. of essentially linear polyethylene at low press. and temp. These polymers had densities of approx. 960 kg/m³, and became known as high-density polyethylenes (HDPE), in contrast to the polymers produced by the extensively commercialized high-press. proc., which were named *low-density polyethylenes (LDPE).
These discoveries laid the basis for the *coordination cat. of ethylene *polymerization, which has continued to diversify. Of the three discoveries at Standard Oil (MoO_3 on Al_2O_3), at *Phillips Petroleum (CrO_3 on *silica), and by Karl Ziegler at the MPI Mülheim (Germany) ($TiCl_4/(C_2H_5)_3Al$), the latter two have been extensively commercialized. HDPE is a white opaque solid that is rigid and forms films which have a turbid appearance and a crisp feel. HDPE does not dissolve in any solvent at room temp., but dissolves readily in aromatic and chlorinated hydrocarbons above its melting point. In contrast to LDPE and LLDPE it is essentially free of both long and short branching. The molecular weight distributions depend on the cat. type but are typically of medium width; the applications are widespread.
The heat of polymerization of ethylene is 93.6 kJ/mol (3.34 kJ/g). Since the specific heat of ethylene is 2.08 J °C⁻¹ g⁻¹, the temp. rise in the gaseous phase is ca. 16 °C for each 1 % conv. to polymer. Heat removal is thus a key factor in commercial polymerization proc. HDPE is mainly produced via a *suspension (slurry) proc. in various types of reactors and polymerization procedures. The procs. use a *supported Ziegler-Natta cat. containing Ti as active metal center or chromium (*Phillips) catalysts. **F** polyéthylène à haute densité; **G** Polyethylen hoher Dichte.

W. KAMINSKY, O. PYRLIK

Ref.: Ullmann, Vol. A21, p. 487; Kirk/Othmer-1, Vol. 17, pp. 702, 724; Mark/Bikales/Overberger/Menges, Vol. 6, p. 454.

high-energy intermediates   As the name implies, heis. are labile chemical species on the reaction pathway, and are different from *transition states in that the intermediate is at the bottom of a local energy well, whereas the transition state is at the peak of an energy barrier. The *Hammond postulate states, however, that the transition state resembles an adjacent high-energy intermediate more than it does either the substrate or the prod. Some examples of heis. found in *enzyme cat. are free-*radical species, *carbocations, and carbanions (cf. *intermediates). **F** intermédiaire à haute énergie; **G** hochenergetisches Zwischenprodukt.

P.S. SEARS

high-fructose corn syrup   Fructose is a more efficient sweetener than sugar, and so the conv. of corn syrups, which contain mostly glucose polymers (amylopectins), to syrups high in fructose is a valuable transformation in the US food industry.
This commercially important cat. transformation is usually performed with a pair of *enzymes: a *glycosidase, such as amyloglucosidase, which degrades amylopectin to glucose; and glucose (xylose) *isomerase, which converts glucose to fructose. **E=F=G**.   P.S. SEARS

highly dispersed metals   are metals with a high ratio of *surface atoms, N_S to the total number of atoms, N_T. This case is realized when a metal is dispersed into small particles. The limiting case, $N_S/N_T = 1$, is reached for single atoms but for a metal *cluster size of 100 Å the dispersion, N_S/N_T, is already as low as 10^{-3}. Hdms. are the metal particles in *supported cats. where it is important for cost effectiveness to keep the dispersion ratio for noble metals as close as possible to one. **F** méteaux hautement dispersé; **G** hochverteilte Metalle.   R. IMBIHL

high-mileage catalysts   economic cats. with low cat. consumption, e.g., for *HDPE manuf.

high-pressure catalysis   is carried out in liquid phases at press. between 100 and 1500 MPa. Due to the low compressibility of liquids, such press. can safely set up with commercially available equipment ranging from small- (10 cm³) to large-scale (1000 L) reactors. Typical reaction temps. that can be applied in high-press. vessels extend from 0 to 100 °C, but care has to be taken in the choice of the *solvents due to the substantial decrease in their melting points under press. Cat. procs. having a negative *activation volume, i.e., the volume of the transition state of a reaction is smaller than the volume of the reactants, will be accelerated by press. For reactions that would be particularly difficult to carry out otherwise (as demonstrated for certain *Diels-Alder reactions) a combination of press. and a *Lewis acid has been found to be advantageous. Moreover, *Pd-cat. coupling reactions such as the *Heck reaction can be accelerated not only by press. but also forcing unreactive substrates such as aryl chlorides to undergo the desired reaction. It has also been discovered that for the latter reaction the lifetime of the cat., reflected in *TOF and cycles, can be dramatically increased by press. Also by using press., considerably lower temps. may often be used, allowing a cleaner course of reaction. Besides *reactivity, *selectivity of cat. procs. is also influenced by press. While *diastereo- and *enantioselectivity decrease with very few exceptions, *regio- and *chemoselectivity can be favorably altered through press. This offers the possibility of designing new cat. reactions or combining reactions to give new *domino procs.   O. REISER

Ref.: Beller/Bolm, Vol. 2, p. 442; *Topics in Catal.* **1998**, 5, 105; *Rev.High Press.Sci.Technol.* **1998**, 7, 1241 and **1998**, 8, 111; van Eldik/Hubbard.

high-pressure liquid chromatography (HPLC) → gas chromatography

high-resolution electron energy loss spectroscopy (HREELS) a method for high-resolution EELS on single crystals, based on the vibrational analysis of adsorbates on flat surfaces. The probes are low-energy, extremely monochromatic electrons, the responses are electrons. The data are integral, surface-sensitive and ex-situ. The UHV technique is non-destructive, and the interpretation is similar to *FT-IR.   R. SCHLÖGL

Ref.: Ullmann, Vol.B6, p. 94.

high-resolution electron microscopy

(HREM, HRTEM)   This variant of electron microscopy is applied for the projection of atomic structures into two dimensions, the visualization of lattice fringes, and the analysis of solid interfaces and defect structures. The technique is non-destructive and needs (under high vacuum) solid samples, transparent for electrons. The data are on an atomic scale and bulk-sensitive; their interpretation is possible only with model calcns. and crystallographic models.
R. SCHLÖGL

high-temperature shift (HTS) → water-gas shift reaction

high-throughput screening   The rapid identification and measurement of relative *activity of suitable cats. from a large pool of potential candidates requires reliable and robust screening techniques. Methods currently developed (*combinatorial work, robotics, data management and handling, new analytical methods, e.g., infrared thermography, *REMPI, scannning mass spectrometry) make possible the identification of specific cat. transformations of interest, promising lead cats., compositions, and interesting combs. of *central atoms and *ligands, etc. worthy of further investigation and follow-up. **E=F=G**.   B. CORNILS

Ref.: Angew.Chem.Int.Ed. **1999**, 38, 2522.

Hill constant, Hill equation   In an effort to model the sigmoidal binding curve of the homotetrameric oxygen-binding molecule hemoglobin (Hb), it was assumed that the four sites bound in an all-or-none fashion; that is, all of the sites on a tetramer were either completely bound or unbound, with no mixed

states. Taking the logarithm of both sides and plotting [log $(Y_S/(1-Y_S))$] vs. (log $[p_{O2}]$), the Hill plot, gives a line with slope α_H, the Hill constant. This type of plot is also useful for assessing whether a multimeric *enzyme displays cooperative binding of substrate or *inhibitor (cf. also *cooperative effects). **F** constant (équation) de Hill; **G** Hill-Konstante (Gleichung).   P.S. SEARS

Ref.: I.H. Segel, Enzyme Kinetics, Wiley, New York 1975.

Hill photoreduction   acc. to Hill the *photolysis of water over chloroplasts without catalyst. *NADP$^+$ and CO_2 takes place even in the presence of reducible cpds. such as Fe(III) oxalate or benzoquinone. **F** photoréduction de Hill; **G** Hill'sche Photoreduktion.

Ref.: Nature (London) **1937**, 139, 881.   B. CORNILS

Hinshelwood   Cyril (1897–1967), professor of physical chemistry in Oxford. Worked on thermodynamics, kinetics, gaseous phase and chain reactions. Nobel laureate in 1956 (together with Semenov).   B. CORNILS

Ref.: Neufeldt; Chem.Brit. **1967**, (3).

Hinshelwood kinetics   The *Hinshelwood kinetics arises from a general mechanistic model (in contrast to the *Eley-Rideal mechanism) in order to derive *surface-related reaction rates. Within this model it is assumed that for a het. reaction two gaseous reactants A and B have to be adsorbed on the cat. surface. The surface-related reaction rate neglects all transport procs. and assums a *steady state of *chemisorption. Assuming that species B is in excess, two border cases depending on the partial pressure of B can be distinguished. For small partial press. of A, the reaction *rate is proportional to p_A, whereas for high partial press. of A the reaction rate is in reciprocal relation to its partial pressure. The formal reaction order is –1 (cf. also *Langmuir-Hinshelwood kinetics). **F** cinétique de Hinshelwood; **G** Hinshelwood-Kinetik.   R. SCHLÖGL, M.M. GÜNTER

histidine ammonia lyase (EC 4.3.1.6)   cat. the deamination of histidine to urocanic acid and ammonia. **E=F=G**.   P.S. SEARS

history of catalysis The science of cat. is driven by inventions and technology. The most important discoveries are given in the following Table.   B. CORNILS

Year	Discovery or event
1746	*Lead chamber process
1781	*Oxidation $SO_2 \rightarrow SO_3$
1781	Parmentier: acids for the conv. of starch to sugar
1806	Desormes, Clément: hom. NO action in lead chamber
1823	*Döbereiner's lighter and other Pt catalyses
1831	*Philips: first patent for Pt-cat. SO_2 oxidation
1832	*Davy's safety lamp
1833	Payen, Persoz: "diastase" from malt discovered
1836	*Berzelius coined the term "catalysis"
1838	*Kuhlmann: *oxidation of ammonia to nitric acid
1849	*Frankland: first organometallic cpd. $Zn(C_2H_5)_2$
1853	*Cannizzaro dismutation
1858	*Pasteur's first kinetic resolution using microorganisms
1875	*"Kontakt" process for H_2SO_4 by *Winkler, *Knietsch
1876	*Markovnikov addition
1876	Pasteur effect: anaerobic glycolysis
1877	*Friedel-Crafts reaction
1894	*Ostwald: precising the term "catalysis"
1894	*Fischer's *lock-and-key principle in *biocatalysis
1897	*Buchner: zymase-based *fermentation
1899	*Sabatier and Senderens: *hydrogenations over Ni
1902	*Fat hardening by *Normann
1907	*Baekeland's formaldehyde/phenol resins, "bakelite"
1909	*Haber-*Bosch ammonia process
1912	*Klatte/Zacharias: hom. cat. with Hg salts
1913	*Bergius: *coal liquefaction
1918	*Langmuir: *adsorption of gases on *surfaces
1923	*Mittasch, *Pier: *methanol synth. on $ZnO-Cr_2O_3$ cats.
1923	*Brønsted: *acid-base concept
1923	*Lewis: *acid-base concept
1925	*Fischer-Tropsch synthesis

1931	Development of *catcracking by *Houdry
1933	*Sulfochlorination of paraffins by Reed
1936	*Ipatieff: cat. *alkylation and upgrading of gasolines
1936	Fawcett et al.: *ethylene polymerization at ICI (*LDPE)
1937	*Ethynylation discovered by *Reppe
1938	*BET method of Brunauer, Emmett, Teller
1938	Discovery of *hydroformylation by *Roelen
1938	*Birch: *alkylation of hydrocarbons with sulfuric acid
1940	*W.K. Lewis: development of the *FCC process
1944	*Hock's process for phenol via cumene hydroperoxide
1953	*Ethylene polymerization with Al/Ti cpds. by *Ziegler
1954	Propylene polymerization by *Natta
1955	*Kutepow: high-press. synthesis of *acetic acid (BASF) with Co cats.
1957	Hafner: *Wacker process for acetaldehyde from ethylene
1957	*Idol: development of the *ammonoxidation process
1962	Solid-phase peptide synthesis (*Merrifield technique)
1964	Plank, *zeolites for the petrochemical industry
1965	*Wilkinson: ligand-modified Rh complexes for hom. cat.
1971	Roth: low-press. synthesis of *AA with Rh cat., (*Monsanto)
1974	*Methanol-to-gasoline (MTG) proc. on *zeolites (Chang)
1974	Introduction of the *three-way cat. by General-Motors and Ford
1975	*L-Dopa synth. by *Knowles

Ref.: Neufeldt; Ullmann (5th), Vol. A18, p. 216; Kirk/Othmer-1, Vol. 5, p. 323; Anderson/Boudart, Vol. 1 and Vol. 3; A. Mittasch, *Kurze Geschichte der Katalyse in Praxis und Theorie*, Springer, Berlin 1939; A. Mittasch/Theis, *Von Davy und Döbereiner bis Deacon*, Berlin, 1932; Ertl/Knözinger/Weitkamp, Vol. 1, p. 13.

HIV protease, aspartyl protease The human immunodeficiency virus (HIV) produces many of its structural proteins as a long polyprotein which must be cleaved into individual units. Inhibition of this processing reaction prevents viral reproduction. The cleavage is cat. by a specific aspartyl *protease, HIV protease, which has a very unusual substrate preference: it cleaves the peptide bond between tyrosine and proline. The *enzyme is a homodimer with the active site at the interface between the two subunits. **E=F=G**.

                                         P.S. SEARS

HLADH $\rightarrow$ horse liver alcohol dehydrogenase

HMDA hexamethylene diamine, an important diamine, manuf. by cat. procs. (e.g., *Celanese procs.).

HMDI → HDI

Hock Heinrich (1887–1971), professor in Clausthal-Zellerfeld (Germany), discovered the *Hock reaction (Hock cleavage) for the manuf. of phenol and acetone from benzene and propylene via *cumene hydroperoxide.
B. CORNILS

Hock reaction the cleavage of organic *hydroperoxides to form carbonyl cpds. and alcohols as reaction prods. via hemiacetals, described by *Hock and Lang in 1944, acc. to R^1R^2CH-O-O-H + H^+ → $R^1CH(=O)$ + R^2-OH. The major application of the Hock reaction is the prod. of phenol and acetone from isopropylbenzene (cumene) via cumene hydroperoxide (cf. procs. of *BP, *Hüls, *Kellogg/Hercules/BP, *Lummus, *Mitsui, *Rhône-Poulenc, *Sumitomo, *UOP Cumox, etc.). Besides phenol dihydroxylated benzene derivatives also can be synth. from p-diisopropylbenzenes such as hydroquinone (*double Hock). Performing *oxidations of alkyl-substituted aromatic cpds., the Hock cleavage can cause *selectivity problems due to oxidation of the isopropyl group and formation of the corresponding alcohol after transfer of the peroxidic oxygen in the course of the oxidation pathways. Easy elimination of water forms highly active aromatic isopropenyl derivatives which can form heavy ends via *oligo- and *polymerization. To avoid this, special reaction conds. have to be provided: comparable mild reaction temps., tuned cats., or co-oxidation conds. to decrease *radical attack at the isopropyl group. **F** réaction de Hock; **G** Hock-Reaktion. R.W. FISCHER

Ref.: H.-G. Frank, J. W. Stadelhofer, *Industrial Aromatic Chemistry*, Springer 1988, p. 148.

Hoechst acetic acid process a hom. liquid-phase *oxidation of CH_3CHO (*AcH) at 60–80 °C/0.3–1 MPa in bubble columns, cat. by metal acetate salts, particularly Mn, Co,

and Cu. Conv. of AcH is >90 % and *selectivity to *acetic acid (AA) is >94 %. The reaction path involves free-*radical chains that produce peracetic acid as the key *intermediate. From AcH, acetaldehyde monoperacetate is formed which decomposes to AcH and AA. The metal salts increase significantly the reaction of peracetic acid and AcH to CH_3COOH. With a Mn salt, the reaction is first order in peracetic acid, AcH, and Mn. Depending on the metal cat. used, particularly Co and Cu acetates, acetic anhydride can also be produced (*Hoechst-Knapsack AAA proc.). **F** procédé Hoechst pour l'acide acétique; **G** Hoechst AAA-Verfahren.
P. TORRENCE

Ref.: Cornils/Herrmann-1, p. 424; Weissermel/Arpe, p. 171.

Hoechst acetic acid/anhydride process the co-prod. of acetic acid/anhydride (AA/AAA) by Rh-cat. *carbonylation of methyl acetate or dimethyl ether (cat. doped with CH_3I and heterocyclic aromatic nitrogen cpds. or *phosphonium salts as *copromoters). The ratio AA/AAA varies between 10/90 and 90/10. The proc. (although ready for operation) did not go on stream (cf. Figure). **F** procédé Hoechst pour AA/AAA; **G** Hoechst AA/AAA-Verfahren. B. CORNILS

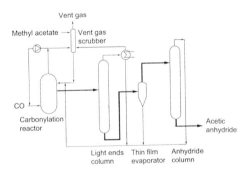

Ref.: Cornils/Herrmann-1, p. 104.

Hoechst-Knapsack acetic acid anhydride process for the manuf. of *acetic anhydride (AAA) by liquid-phase *oxidation of *acetaldehyde with oxygen (Cu-cat. reaction of the acetyl cation with AA), cat. with heavy metal acetates such as Cu, Co, Mn at 50 °C/ 0.4 MPa. When run as a co-production proc. AAA/AA ratios of 56:44 can be obtained. **F** procédé de Hoechst-Knapsack pour AAA; **G** Hoechst-Knapsack AAA-Verfahren.

B. CORNILS

Ref.: Weissermel/Arpe, p. 180; Winnacker-Küchler, Vol. 6, p. 91.

Hoechst-Celanese low-water acetic acid technology

significant improvement to *Monsanto's AA process. Low-water tech affords considerable reduction in operating costs and low-cost expansion potential. Proc. enables higher cat. activity at lower reaction water concs. with improved CO and MeOH efficiencies and improved cat. *stability via incorporation of an inorganic or organic iodide as a rate-enhancing *co-promoter and *stabilizer. Enhanced reaction rates due in part to a synergistic cat. effect of methyl acetate and iodine salt to reduce HI and presumably form a strong nucleophilic dianionic cat. species $[RhI_2(CO)_2-(L)]^{2-}$ (L = I^- or CH_3COO^-). *WGSR is suppressed with reduction of HI to account for improved raw material efficiencies (cf. also *BP AA proc.). **F** technologie "low water" de Hoechst-Celanese; **G** Hoechst-Celanese Low-Water-Technologie.

P. TORRENCE

Ref.: Cornils/Herrmann-1, Vol. 1, p. 104; *J. Mol. Catal.* **1987**, *39*, 115.

Hoechst chlorolysis process

is a non-cat. proc. for the manuf. of CCl_4 by *chlorolysis of chlorine-containing residues at high press. and temps. and short *residence times. **F** procédé chlorolyse de Hoechst; **G** Hoechster Chlorolyseverfahren.

B. CORNILS

Ref.: Weissermel/Arpe, p. 54.

Hoechst resorcinol process

an abandoned proc. for the manuf. of resorcinol by *sulfonation of benzene with SO_3 and subsequent direct neutralization of the melt. **F** procédé Hoechst pour le résorcine; **G** Hoechster Resorcin-Verfahren.

B. CORNILS

Hoechst sulfoxidation process

for the manuf. of alkylsulfonates by UV light cat. *sulfoxidation with SO_2/O_2 mixtures in bubble-column reactors. **F** procédé sulfoxidation de Hoechst; **G** Hoechster Sulfoxidations-Verfahren.

B. CORNILS

Hoechst vinyl acetate (VAM) process
→ Bayer-Hoechst VAM proc.

Hoechst Wacker acetaldehyde process
→ Wacker process

Hoffmann-La Roche process for vitamin A

is similar to *BASF's vitamin A proc. and includes for the C_5 building block of vitamin A the *hydroformylation of *trans*-1,4-diacetoxy-2-butene with a *ligand-modified Rh cat. From the 1,4-diacetoxy-2-formylbutane obtained the AcOH is removed, and *isomerization of the *exo* double bond yields 2-formyl-4-acetoxybutene (γ-acetoxytiglic aldehyde). **F** procédé de Hoffmann-La Roche pour vitamin A; **G** Hoffmann-La Roche-Verfahren für Vitamin A.

R. KRATZER

Ref.: Cornils/Herrmann-1, p. 8; US 4.124.619.

Hofmann-Löffler-Freytag reaction

is the presumably *radical-cat. cyclization of *N*-haloamines. **F** réaction de Hofmann- Löffler-Freytag; **G** Hofmann-Löffler-Freytag-Reaktion.

B. CORNILS

Hofmeister series

At very low ionic stregths, addition of salt provides counterions for the ionic groups of a protein and thus increases the solubility of the protein; this is called "salting in." At very high ionic strengths, most salts cause proteins to precipitate out, presumably due to competition for solvating water molecules. This is called "salting out," and is very commonly used in protein purification. The order of effectiveness of certain ions for stabilizing a protein is called the Hofmeister series. **E=F=G**.

P.S. SEARS

H-oil process
→ Hydrocarbon Res. process

hollow-fiber reactor reactors (preferably for *enzyme reactions) with fibers that have an absorbent fiber lined with a macromolecule-impermeable "skin" which can be saturated with, e.g., enzyme solution.

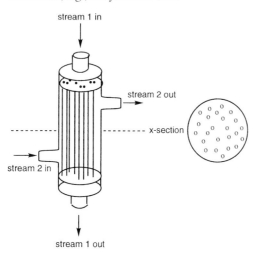

stream 1 in

stream 2 out

x-section

stream 2 in

stream 1 out

Organic *solvent on the outside of the fibers will not dissolve the enzyme, and the enzyme is kept out of the inner aqueous stream by the impermeable skin (cf. also *bioreactor and *membrane reactor). **F** réacteur à fibres creuse; **G** Hohlfaserreaktor.   P.S. SEARS

holoenzyme refers the entire *enzyme in its active state: that is, both the polypeptide chain and any *cofactors necessary for its activity. This is to be compared to the *apoenzyme, which refers solely to the polypeptide portion of the enzyme without the cofactors. **E=F=G**.   P.S. SEARS

hom. homogeneous, homogneously

HOMO highest occupied molecular orbital, → frontier molecular orbitals

homogeneous catalysis (solution catalysis) one of three different kinds of *catalysis; the others are *heterogeneous (particulate) and *biocatalysis. Homogeneous catalysis (hc.) is defined as catalysis with the catalyst in the same phase as the reactants (see Table).

Phases/catalyst reactants	Type of cat.		Examples of cat. reaction
Homophasic			
Liquid	*Acids, protons		*Esterification
	*Lewis acids		*Isomerization
			*Alkylation
	*Bases		*Aldolization
			*Condensation
	*Organometals		*C-C bond formation
Gaseous	NO/NO$_2$ [NOHSO$_4$]		*Lead chamber proc.
	NO$_2$/HCl [NOCl]		*Kellogg Kelchlor proc.
	Oxygen		*LDPE synthesis
	Halogens		*Chlorinations
	Halides		*Eastman MMA
			*Goodyear/SD isoprene
	Radicals (*initiators)		*Radical reactions
	H-Hal		*Demethanation
Heterophasic			
Liquid/liquid	*Organometals		two-phase operation, *Hydroformylation *Shell SHOP

For catalysis to be homogeneous, all demands of the definition must be fulfilled. Thus, hom. cat. include cats. which 1: are molecularly dispersed "in the same *phase"; 2: are unequivocally characterized chemically and spectroscopically; 3: permit the reaction *kinetics unequivocally; 4: pass a detectable *cat. cycle, and 5: can be tailormade for special purposes (this is especially true for organometallic *complexes as cats.; cf. *immobilization). Nevertheless and in the case of *heterophasic* operation the cat. action is definitely *homogeneous* although the cat. is *heterogeneously* dispersed in relation to the phases of the reactants.

Contrary to many publications, hc. is mature and started with the action of nitric oxides in the *lead chamber proc. and thus prior to the recognition of the effect of het. cats. (*history of catalysis). The ratio of applications of het. versus hom. cat. is approx. 80:20. The reaction conds. of hc. are relatively mild, typically 0.1–25 MPa and 50–150 °C, often depending on the stability of the active cat. and cat. *precursors (cf. stability plot under keyword *decobaltation). The cats. may be used as ready species or be formed under cat. conds. from *precursors. An example of a hom. *cat. cycle is given by the Ni-cat. *carbonylation of ethyl-ene to propionic acid (Figure 1):

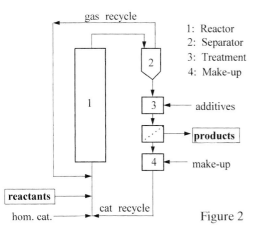

Regarding *asymmetric reactions, which are a special domain of hc., the different energy profiles of the pathways with normal and with *chiral *ligands have to be observed. Hom. cats. are usually gaseous reactants (e.g., NO/NO$_2$ in the lead chamber), acids (protons, *Lewis acids), bases, *free radicals (although not *real* cats. since they may disappear during cat.), and organometallic cpds. Metals which readily undergo valence changes are especially suitable (*transition metals, Al, Bi, Sn, Se). The active species of the hom. cat. depend on their structure and the reaction conds.; the exchange of *ligands before or during operation or within the cat. cycle is typical (*oxidative addition, *reductive elimination). Hom. cats. may be extremely active and selective (*selectivity, *turnover number,

*turnover frequency). The general industrial potential of hc. is shown in Figure 2.

The *disadvantage* of hc. is the immense difficulty of cat. separation and recycling because the cat. is molecularly dissolved. Therefore, the application phase of hc. was also changed, by *immobilization by anchored cats. or heterogenization through *two-phase or *aqueous-phase hc. In both cases the ease of separation and recycling is similar to het. cat., thus allowing the benefit of the tremendous advantages of hc. **F** catalyse homogène; **G** Homogenkatalyse.   B. CORNILS

Ref.: Basset; Beller/Bolm; Bond-2; Burgess; Clark; Ugo; Cornils/Herrmann-1 and -2; Falbe-1 and 4; Fink et al.; Fürstner; Gates-2: Togni-1; Hartley; Long; Noyori; Parshall/Ittel; Pignolet; Sheldon/Kochi; Morokuma/Lenthe/Leeuwen; Mortreux/Petit; Moulijn/van Leeuwen/van Santen; Mulzer/Waldmann; *Organic Reactions*, Vol. 24, p. 1.

homoleptic (isoleptic) complexes

neutral or charged cpds.(MX$_n$) or $[M^1X_n]^{z-}$ $[M^2(L)]^{z+}$ (*ate* *complex, L = solvent *ligand), comprising a central metal M coordinated by n identical neutral or charged ligands X (Figure).

e-homoleptic   d-homoleptic

ate-e-homoleptic   ate-d-homoleptic

Hcs. can be further classified acc. to whether the ligands are equally (*e*-homoleptic) or differently (*d*-hom.) coordinated to the metal center. Hcs. are widely used in *hom. cat., e.g., *ate-e*-hcs. ($M^1{\neq}M^2$) are prominent heterobimetallic cats. **F** complèxes homoleptiques; **G** homoleptische Komplexe.   R. ANWANDER

Ref.: *Acc.Chem.Res.* **1974**, *7*, 209; *Top.Curr.Chem.* **1996**, *179*, 1.

gas recycle

1: Reactor
2: Separator
3: Treatment
4: Make-up

additives

products

make-up

reactants

cat recycle
hom. cat.   Figure 2

homologation 1–generally: the method for the incremental extension of chemical individuals, e.g., the chain extension of aliphatic hydrocarbons by a -CH$_2$- group yielding the next hc. 2–special: h. is the cat. chain extension of aliphatic alcohols by means of hydrogen rich *syngas

$$\text{R-OH} + \text{CO} + 2\,\text{H}_2 \xrightarrow{\text{cat.}} \text{R-CH}_2\text{-OH} + \text{H}_2\text{O}$$

Carbonyls of, e.g., Co, Ru, Rh, Ir are cat. active. The h. of other structures (formaldehyde, butanone, methyl acetate, carboxylic acids) is also possible. H. was important for the conv. MeOH → EtOH, and thus after *dehydration of EtOH a connection between syngas and ethylene or as a basis for *fuel methanol prod. H. of acetic acid to higher acids with Ru cats. is also known. Diazomethane carboxylic acids are subject to h. acc. to the *Arndt-Eistert reaction. The opposite of h. is chain shortening with degradation reactions. **F** homologation; **G** Homologisierung. B. CORNILS
Ref.: Falbe-1, p. 226

HOMO-LUMO model → frontier molecular orbitals

homolysis splitting of a molecule into two *radicals, indicating that the covalent bond is separated symmetrically acc. to AB → A· + B· (homolytic cleavage). H. plays a role in thermal or photochemical dissociations, *radical-cat. additions, fragmentations, *substitutions, or other *radical reactions. **F** homolyse; **G** Homolyse. B. CORNILS
Ref.: Giese.

homolytic splitting (cleavage) → homolysis

homopolymers → polymerization

homotropic cooperative effect → cooperative effect

honeycombs special arrangement of catalysts or of *ceramic cat. *supports, especially in *automotive exhaust cats.; cf. Figure (g) on page 105.

Hooker bisphenol (BISA) process manuf. of *bisphenol-A by *condensation of acetone and phenol with dry HCl as cat. and methylthiols as cat. *promoter. **F** procédé Hooker pour le bisphénol-A; **G** Hooker Bisphenol A-Verfahren. B. CORNILS

Hopcalit cat. for the oxidation CO → CO$_2$, consisting of Mn-Cu oxides or Mn-Cu-Co/Ag oxides on *supports.

Horiuti-Polanyi mechanism a four-step mechanism originally suggested by Horiuti and Polanyi in 1934 to explain the kinetics of the *hydrogenation of cyclohexene on Pt in the liquid phase. The mechanism assumes *dissociative *chemisorption of H$_2$ and the formation of an RH intermediate on the *surface (R representing the alkene) before the fully hydrogenated product RH$_2$ is released into the gaseous phase. **F** mécanisme de Horiuti-Polanyi; **G** Horiuti-Polanyi-Mechanismus. R. IMBIHL
Ref.: M. Boudart and G. Djega-Mariadassou, *Kinetics of Heterogeneous Catalytic Reactions*, Princeton University Press 1984.

Horner Leopold (born 1911), prof. of chemistry in Mainz (Germany), worked with special *phosphines and hom. cat., discovered independently from *Knowles that prochiral olefins may be hydrogenated *enantioselectively by means of *Wilkinson cats., modified by optically actice *ligands such as *DIPAMP or *DIOP. B. CORNILS

Horner phosphines optically pure *phosphines in which the *chiral center is at P and/or in a P-bonded organic group (e.g., (–)-menthyl) (Figure).
The prep. is achieved by classical routes; the synth. of optically avtive phs. requires either a *resolution step or the transformation of a chiral precursor. H.phs. were first used by *Horner and *Knowles in 1968 as co-cats. for *enantioselective Rh-cat. *hydrogenation of prochiral alkenes. The H.phs. have been replaced in enantioselective cat. by the more

selective and active C_1 and C_2 ligands, such as *BPPFA, *BINAP, or *DUPHOS. O. STELZER

Methyl-
n-propylphenyl-
phosphine
Horner, 1961

Methylcyclohexyl-
o-methoxy-
phenylphosphine
Knowles, 1974

Ref.: Cornils/Herrmann-1, Vol. 2, p. 1027; *Tetrahedron Lett.* **1961**, 161 and **1979**, 1069; *Pure Appl. Chem.* **1964**, *9*, 225; Patai-2, Vol. 1, p. 51; *Phosphorus and Sulfur* **1983**, *14*, 189.

horse liver alcohol dehydrogenase

(HLADH; EC 1.1.1.1) cat. the *oxidation of ethanol to acetaldehyde using *NAD as the *cofactor. It also accepts a number of other alcohols. The *enzyme can differentiate between prochiral groups or faces in symmetrical or *meso* cpds. and make distinctions between *enantiotopic groups or faces and geometric isomers.

HLADH contains bound zinc ions acting as cofactors and thus is susceptible to *deactivation by metal-*chelating reagents. The Zn ion can be replaced by other metals. The cubic space section model for prediction and explanation of enzymatic activity has proven to be widely applicable and accurate, particularly for cyclic cpds. The cofactor nicotinamide ring is firmly in place with transfer of the pro-*R* hydride to the *re* face of the carbonyl group and similarly with abstraction of the substrate pro-*R* hydrogen to the *coenzyme *re* face during oxidation. The *active site Zn atom must also be coordinated to the carbonyl or hydroxyl oxygen atom.

In addition the enzyme will reduce a significant number of aldehydes. Each of these reactions is potentially useful since only the pro-*R* H of the alcohol substrate is removed in the oxidation and hydride transfer occurs to the *re* face of the substrate in the reduction. Primary alcohols labeled at either the pro-*R* or pro-*S* position can be obtained easily.

HLADH's inability to oxidize methanol is attributed to a lack of productive binding in the active site. For substrates insoluble in aqueous solution, biphasic systems, reverse *micelles, or organic *solvents can be used. The *enantioselective oxidation of primary alcohols cat. by HLADH provides a route to obtain *chiral aldehydes but the yield is relatively low due to the problem of *product inhibition by the aldehyde.

The oxidation of an enantiotopic group of prochiral *meso-* diols yields lactones with a significant degree of enantioselectivity. The reaction presumably proceeds via the hemiacetal after the initial primary alcohol oxidation. The oxidations occur with high pro-*S* *selectivity. These reactions provide a method to obtain valuable five- and six-membered chiral lactones, which are useful synth. *intermediates.

Monocyclic *meso*-diols can similarly provide bicyclic chiral lactones. Oxidation occurs with the same absolute stereospecificity in each case, with the pro-*S* hydroxyl preferentially oxidized with all carbocyclic and the previously mentioned acyclic substrates. The heterocyclic cpds. exhibit pro-*R* enantiotopic selection. A number of bicyclic *meso*-diols can be oxidized to lactones by HLADH. The pro-*S* hydroxyl groups of the carbocyclic substrates and pro-*R* hydroxyl groups of the heterocyclic cpds. are preferentially oxidized. These chiral lactones can be used as synth. *precursors for many molecules.

Acyclic ketone substrates are typically poor substrates for HLADH and few have been reported. Cyclic ketones are good substrates for HLADH. The enzyme can also tolerate heterocyclic rings containing O or S without a change in stereospecificity. N apparently coordinates the active site zinc atom and cannot be used in the cyclic substrates. For mono- and bicyclic ketones, the *hydride transfer from *NADH is to the *re* face of the carbonyl and primarily in the equatorial direction. In addition, *desulfurization with lithium in diethylamine results in the breakage of one or both S bonds, depending upon reaction conds., to yield an acyclic chiral alcohol.

HLADH cat. the reduction of many different *cis-* and *trans-*decalindiones. These highly symmetrical cpds. are potentially useful chiral synthons. HLADH also accepts a variety of cage-shaped molecules as substrates. These cpds. – including *meso* diketones – are reduced with high enantioselectivity. **E=F=G.**

C.-H. WONG

Ref.: *J.Am.Chem.Soc.* **1988**, *110*, 577; *Tetrahedron* **1986**, *42*, 3351.

horseradish peroxidase → peroxidase

host-guest relationships Hosts and guests are the two or more molecular or ionic comps. which comprise supramolecular *complexes. The host is usually concave and its binding sites converge in the *complex in analogy to biological receptors; the (convex) guest has binding sites which diverge, similarly to substrates, *inhibitors, or *cofactors. Binding forces are *hydrogen bonding, ion pairing, metal-to-*ligand binding, *π-stacking, cation-π attractions, electron donor-acceptor, dipole-dipole, and *van der Waals interactions. Since these forces are weak, a structural well defined pattern of multiple interactions is usually required for strong and selective binding. In addition to size, number, and shape (conformation), the binding sites are characterized by their electronic properties (charge, polarity, polarizability), lipo- or hydrophilicity, and arrangement in the molecular framework as well as their structural dynamics.

Interactional and geometrical complementarity of the binding sites – as in *Fischer's *lock-and-key priciple – leads to *molecular recognition, which in addition to mere binding implies and determines the binding selectivity. Structural preorganization of the binding sites is an important factor for the binding strength and consequently of great impor-

tance for the design of selective binding agents. Macrocyclic hosts often exhibit a higher degree of molecular recognition as compared to open-chain analogs, as was shown for *crown ethers and *podands; this is referred to as the *macrocyclic effect.* Likewise the balance between rigidity and flexibility of the host's molecular framework determines the binding, since loss of conformational degrees of freedom upon binding causes an entropic penalty, whereas an unfavorable rigidity hinders induced-fit adjustments and reduces the enthalpic contribution to the free energy of binding. **F** relations host-guest **G** Wirt-Gast-Beziehungen. C. SEEL, F. VÖGTLE

Ref.: F. Vögtle, *Supramolecular Chemistry*, Wiley, Chichester 1991; J.-M. Lehn, *Supramolecular Chemistry*, VCH, Weinheim 1995; *Comprehensive Supramolecular Chemistry*, Pergamon Press, Oxford 1996; Ertl/Knözinger/Weitkamp, Vol. 2, p. 888.

hot bands → Boudouard equilibrium

HOT process → UOP procs.

hot spots In exothermic het. cat. it is difficult to remove the heat of reaction from the parts of the *cat. reactor where the *activity is highest. In order to transport the reaction heat through cat. beds which are often poorly thermally conducting (metal oxides in particulate form) reactor designs are chosen with many thin tubes embedded in a thermally well-conducting medium. Along the tubes however, heat transport is more difficult. In particular at the location of the most active part of the cat. bed, local temp. rises can occur which can destroy the cat. and even endanger the whole reactor. As the cat. is being deactivated during its lifetime and the amount of deactivated cat. increases from the inlet slowly through the reactor tube, the hot spot is also moving slowly through the reactor tube. To minimize the local temp. rise the cat. beds are either diluted or even divided into zones, allowing the feed of additional cold gas. It is a major goal in chemical engineering for exothermic reactions to minimize the temp. rise of the hot spot by many different measures. The extreme solution is the design of the fluidized-bed reactor in which a very uniform distribution of mass and energy over the entire reactor is ensured.

Hot spots can occur not only in the macroscopic structures of a reactor but also in the pore system or in the *defect structure of an individual cat. particle. Such microscopic hss. are caused by an uneven distribution of reaction gases in the cat. or by the unwanted dissolution of reactants in the cat. and their reaction within the solid cat. The consequence of such micro-hss. is the mechanical disruption of massive cat. particles leading to press. drops or loss of cat. and hence to early *deactivation. The temp. rise in the hs. is a major factor in the deactivation scenario of the cat. In exothermic *oxidation reactions the temp. rise can be as high as several hundred degrees over the nominal operation temp. This rise can further be increased by improper or accidental changes in the gas composition of the feed or in flow of the gas through the reactor tube.

Hss. also occur in *automotive exhaust cleaning systems where they can seriously damage the *ceramic *monolith system and even lead to the evaporation of the precious metal active phase. Malfunction of the oxygen sensor or the fuel injection system even for short times can cause hss. in car systems. **E=F=G.**

R. SCHLÖGL

Houben-Fischer nitrile synthesis

a *condensation reaction for the manuf. of nitriles by reaction of aromatics or phenols with trichloroacetonitrile under the influence of AlCl$_3$. The *intermediate ketimino cpds. are treated with KOH. **F** synthèse de Houben et Fischer pour nitriles; **G** Houben-Fischer'sche Nitrilsynthese.   B. CORNILS
Ref.: Krauch/Kunz.

Houben-Hoesch synthesis

*acylation of reactive aromatics (e.g., phenols) with nitriles under the influence of ZnCl$_2$. **F** synthèse de Houben-Hoesch; **G** Houben-Hoeschsynthese.
Ref.: *Organic Reactions*, Vol. 5, p. 387,   B. CORNILS

Houdresid → Houdry proc.

Houdry Eugene (1892–1962) French-born inventor and founder of Houdry Process Corp., who developed in 1931 the *catcracking technology for the cracking of low-grade crude oil in the gaseous phase over Al silicate cats. This cat. procedure substituted the thermal cracking and improved the *petroleum refinery considerably. He was also active in the research for cat. mufflers.   B. CORNILS

Houdry Catadiene process for the one-step, fixed-bed manuf. of butene by cat. *dehydrogenation (*oxydehydrogenation) of butane at 600 °C/reduced press. using Cr-Al$_2$O$_3$ cats. (possibly doped with K) which must often be regenerated by air blowing due to severe coke formation. Similarly isoprene is manuf. from isoamylene. **F** procédé Catadiene de Houdry; **G** Houdry Catadiene-Verfahren.   B. CORNILS
Ref.: Weissermel/Arpe, p. 110.

Houdry Catofin process similar to *Houdry Catadiene proc. with a cat. based on Cr$_2$O$_3$ on *alumina, doped with K or ZrO$_2$. The proc. is now licensed by Lummus (cf. *Lummus proc.)   B. CORNILS

Houdry Detol process *hydrodealkylation (HDA) proc. for the degradation of toluene and thus adjusting the prod. of benzene acc. to demand. Cats. are Co, Mo, Cr, or Rh on supports at 540–650 °C/3.5–10 MPa. The S content of the benzene is low. **F** procédé Detol de Houdry; **G** Houdry Detol-Verfahren.
B. CORNILS

Houdryflow process → Houdry Houdresid process

Houdryformate → Houdry Houdriforming process

Houdry-Gulf H-G Hydrocracking process → Gulf procs.

Houdry Houdresid (Houdriflow) process cat. *cracking proc. for petroleum residues; the Houdryflow version with entrained-bed cats. based on *zeolites. **F** procédé Houdresid (Houdryflow) de Houdry; **G** Houdry Houdresid (Houdryflow)-Verfahren. B. CORNILS

Ref.: *Hydrocarb.Proc.* **1972**, (9), 137; Ullmann, Vol. A18, p. 53,61.

Houdry Houdriforming process *reforming proc. for the manuf. of gasoline and aviation blending stocks by conv. of straight run or cracked naphthas using Pt-Al$_2$O$_3$ (with traces of hydrogen halide or hydrogen halide precursor to the feed) or *bimetallic cats. at approx. 500 °C/2.7–4 MPa. High-sulfur feeds may be pretreated in *guard beds; the Houdriformate requires no further treatment. **F** procédé Houdryforming de Houdry; **G** Houdry Houdriforming-Verfahren. B. CORNILS

Ref.: *Hydrocarb.Proc.* **1972**, (9), 115,116 and **1978**, (9), 162.

Houdry Pyrotol process for the cat. manuf. of high-purity benzenes by *hydrodealkylation of pyrolysis gasoline or impure *BTX streams with simultaneous *desulfurization and *hydrocracking of non-aromatics. **F** procédé Pyrotol de Houdry; **G** Houdry Pyrotol-Verfahren. B. CORNILS

Ref.: *Hydrocarb.Proc.* **1978**, (9), 146.

house of cards structure → *pillared clays

Howe-Baker Arofining process removal or reduction of the aromatics in petroleum distillates by cat. *selective *hydrogenation to naphthenes at 180–320 °C/6 MPa and *LHSVs of 6. S, N, or O cpds. are removed almost quantitatively. **F** procédé Arofining de Howe-Baker; **G** Howe-Baker Arofining-Verfahren. B. CORNILS

Ref.: *Hydrocarb.Proc.* **1972**, (9), 150.

Howe-Baker catalytic reforming proc. for the increase of the octane number of straight run or cracked naphthas for gasoline by cat. *reforming over Pt or *bimetallic cat. **F** procédé catalytic reforming de Howe-Baker; **G** Howe-Baker Catalytic Reforming-Verfahren. B. CORNILS

Ref.: *Hydrocarb.Proc.* **1996**, (11), 96.

Howe-Baker hydrotreating process for the reduction of the S, N, and metals content of various feedstocks (*HDS, *HDN, and *HDM procs.) by cat. treatment over Co-Mo, Ni-Mo, or Ni-W cats. at 280–400 °C/2.8–10 MPa (Figure). **F** procédé hydrotreating de Howe-Baker; **G** Howe-Baker Hydrotreating-Verfahren. B. CORNILS

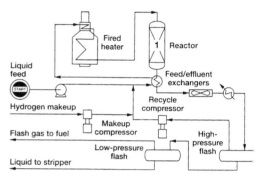

Ref.: *Hydrocarb. Proc.* **1996**, (11), 134.

Howe-Baker Mercapfining process cat. *sweetening of hydrocarbons in a liquid-phase, fixed-bed operation, converting mercaptans with oxygen to disulfides, which remain dissolved in the *hc. stream. Periodically the cat. is regenerated with a readily available liquid regenerant. **F** procédé Mercapfining de Howe-Baker; **G** Howe- Baker Mercapfining-Verfahren. B. CORNILS

Ref.: *Hydrocarb.Proc.* **1972**, (9), 212.

HPLC high pressure liquid chromatography, → gas chromatography

HPN process hydrogenation of pyrolysis naphtha process, → Engelhard procs.

HPO process hydroxyl-phosphate-oxime, process → DSM/Stamicarbon procs.

HPS heavy paraffin synthesis, cf. *Shell SMDS process

HR → *Hydrocarbon Research procs.

HREELS → high-resolution electron energy loss spectroscopy

HREM high-resolution electron microscopy, identical with HRTEM

HRTEM → high-resolution electron microscopy

HSAB concept Pearson's "principle of hard and soft acids and bases" qualitatively describes the extra stability of *complexes formed by the interaction of generalized acids (A) and bases (B). The equil. constant K adequately estimates the stability of complexes A-B by the consideration of both intrinsic strengths (S) and softness parameters (σ) (Equ.).

$$A + IB \rightleftharpoons A{-}B$$

$$\log K = S_A S_B + \sigma_A \sigma_B$$

The HSAB c. states that hard (soft) acids preferentially associate with hard (soft) bases. Hard acids comprise small acceptor atoms of high positive charge and low polarizability, containing no unshared pairs of electrons in their valence shells (e.g., H^+, Ln^{3+}, BF_3, SO_3). Soft acids comprise large acceptor atoms of low positive charge and high polarizability, containing unshared pairs of electrons (p or d) (e.g., Cu^+, I_2, CH_2). Hard bases display donor atoms of low polarizabilty which hold their valence electrons tightly and are hard to oxidize (H_2O, F^-, NO_3). Soft bases are, e.g., R_2S, I^-, CO).
Pearson proposed a scheme for quantifying the hardness concept by defining hardness operationally as half the difference between ionization potential and electron affinity of a substrate molecule S, which turns out to be half the energy change for the disproportionation reaction acc. to $S + S \rightarrow S^+ + S^-$. The HSAB c. contributes to the understanding of a wide variety of chemical phenomena in both hom. and het. media and the prediction and correlation of reactivity in basic procs. and in catalysis. **F** HSAB concept; **G** HSAB-Prinzip. R. ANWANDER

Ref.: *J.Am.Chem.Soc.* **1963**, *85*, 3533 and **1983**, *105*, 7512; *Science* **1966**, *151*, 172; *Chem.Rev.* **1975**, *75*, 1; *Nov.J.Chim.* **1983**, *7*, 499.

HSDH → hydroxysteroid dehydrogenase

HTS high temperature shift, → water-gas shift reaction

Hüls acetic acid process for the manuf. af AA by liquid-phase butene *oxidation in the presence of steam using Ti-V or Sn-V cats. **F** procédé Hüls pour AA; **G** Hüls Essigsäure-Verfahren. B. CORNILS

Hüls cyclododecatriene process
→ Hüls dodecanedioic acid process

Hüls dimethylterephthalate (DMT) process manuf. of fiber-grade *DMT from p-xylene and p-methyl toluate by oxidation with air at ~160 °C/0.4–0.8 MPa in the presence of heavy metal (Co, Mn) cats. yielding p-toluic acid and monomethyl terephthalate. The acids are converted to p-methyl toluate and DMT and the oxidate is *esterified with methanol (Figure). The proc. was developed

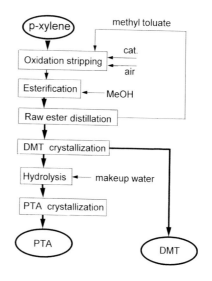

by Dynamit Nobel. **F** procédé Hüls pour le DMT; **G** Hüls DMT-Verfahren. B. CORNILS

Ref.: *Hydrocarb.Proc.* **1977**, *56*(11), 1147 and **1997**, (3), 122.

Hüls dodecanedioic acid process four-step proc. for the manuf. of 1,12-dodecane-dioic acid by 1: Ni(0) cat. (or *Ziegler cats.) *trimerization of butadiene to cyclododeca-triene (*CDT), 2: Ni-cat. *hydrogenation of CDT to cyclododecane, 3: boric acid-cat. *oxidation of the hc. yielding a mixture of cyclo-dodecanol and cyclododecanone, which is 4: converted by ring-opening oxidation with nitric acid to dodecanedioic acid. **F** procédé Hüls pour dodécane diacide; **G** Hüls Dode-candisäure-Verfahren. B. CORNILS

Ref.: Weissermel/Arpe, p. 243.

Hüls MTBE process manuf. of *MTBE from isobutene and methanol over *ion ex-change cats. This proc. is the forerunner of the *Hüls/UOP Ethermax proc. B. CORNILS

Ref.: *Hydrocarb.Proc.* **1977**, *56*(11), 185.

Hüls Octol process manuf. of C$_8$ alkenes by C$_4$ *oligomerization in the presence of *clay-supported Ni cats. in the liquid hase. **F** procédé Octol de Hüls; **G** Hüls Octol-Verfahren.

Ref.. *Hydrocarb.Proc.* **1992**, *61*(2), 45. B. CORNILS

Hüls/Phenolchemie phenol process
a variant of the *Hock process

Hüls SHP process for the reduction of re-sidual butadiene in cracker butene fractions (*raffinates) by adiabatic liquid-phase *selec-tive *hydrogenation over Pd cats. at 30 °C. The butadiene level is below < 10 ppm; cf. *low-temp. hydrogenation. **F** procédé SHP d'Hüls; **G** Hüls SHP-Verfahren. B. CORNILS

Ref.: M. Baerns, J. Weitkamp (Eds.), Proceedings of the DGMK conference *Selective Hydrogenations and Dehydrogenations*, Kassel (Germany) 1993.

Hüls terephthalic acid process manuf. of *PTA by *hydrolysis of *DMT. **F** procédé Hüls pour PTA; **G** Hüls PTA-Verfahren.

B. CORNILS

Hüls THF process → BASF tetrahydrofur-an process

Hüls/UOP Ethermax process manuf. of *MTBE and other ethers using reactive distil-lation technology (*UOP Ethermax proc.). The reactive alkene (e.g., *i*-butene) is com-bined with the alcohol over an acidic *ion ex-change resin at mild temp. and moderate press. in a fixed-bed reactor. In the subse-quent distillation the simultaneous reaction of the remaining *i*-butene and distillation (*cata-lytic disitillation, cf. figure there) occur. **F** procédé Hüls/UOP pour les éthers; **G** Hüls/ UOP Ether-Verfahren. B. CORNILS

Ref.: *Hydrocarb.Proc.* **1996**, (11), 113; Ertl/Knözinger/ Weitkamp, Vol. 4, p. 1996.

Hüls Vestenamer process manuf. of a polyoctenamer (Vestenamer) by *ring-open-ing metathesis of cyclooctene with W cats. **F** procédé Vestenamer de Hüls; **G** Hüls Ves-tenamer-Verfahren. B. CORNILS

Hugen-Watson kinetics → Langmuir-Hinshelwood kinetics

hybrid catalyst is a cat. which combines hom. and het. catalyses, see *immobilization

hybridoma technology A hybridoma is an *antibody-producing cell line produced by fusing a lymphocyte with a myeloma cell, a can-cerous form of the antibody-secreting plasma cell. Although lymphocytes cannot be continu-ously propagated in cell culture, the hybrid is "immortal": it can be grown indefinitely. Thus, hybridoma technology provides stable mono-clonal antibody-producing cell lines. (cf. also *monoclonal antibodies). **F** technique d'hybri-dome; **G** Hybridoma-Technik. P.S. SEARS

Hycon process → Shell procs.

Hycracking → Exxon procs.

HYD → hydrodearylation

hydantoinase Both D- and L-amino acids can be prepared from the corresponding 5-substituted DL-hydantoins via hydantoinase-cat. *hydrolysis. The advantage of this system is that the unreacted hydantoin racemizes spontaneously, and *enantioselective *hydrolysis of the interconverting enantiomers leads to complete conv. of a racemic mixture to one enantiomer. **E=F=G**. C.-H. WONG

Hydeal process *HYdroDEALkylation (*HDA), cf. the proc. of *Atlantic Richfield, *Gulf, *Houdry, *Shell Bertol, *Union Oil, *UOP Hydeal, etc.

Hydrar process → UOP procs.

hydration *addition of water to unsaturated cpds. yielding oxo or hydroxy derivatives. Economically most important is the h. of alkenes to alcohols (cf., e.g., the procs. of *Deutsche Texaco, *Nippon Oil, *Shell EtOH, *Tokuyama, *VEBA); h. of unfunctionalized alkenes needs to be cat. H. of acceptor-substituted alkenes gives the corresponding linear alcohols. Here, the reaction proceeds via a nucleophilic mechanism. Hence, the addition of base is known to cat. the reaction. Alkenes are *indirectly* hydrated by the use of oxymercuration with $Hg(OAc)_2$ and subsequent *reduction of the organomercury derivative with $NaBH_4$. The h. of alkynes is still carried out with Hg salts as cats. Since the addition follows *Markovnikov's rule, only acetylene yields an aldehyde; all other alkynes give ketones. For the *enzymatic h. cf. *fumarase. **F=E**; **G** Hydratisierung, Hydratation. M. BELLER
Ref.: March; *Organic Reactions*, Vol. 13, p. 1.

hydrazine decomposition is applied as the motive power for a monopropellant in jet-propulsion devices (rockets, gas generators yielding H_2 and N_2 for turbines), in *fuel cells, or as a consequence of the corrosion protective properties of hydrazine. Apart from a variety of metals or *supported metals, the best suited commercial cat. is an Ir cat. supported on *alumina (Shell 405). Ru-based cats. show also sufficient activity. **F** décomposition de hydrazine; **G** Hydrazinzersetzung. B. CORNILS
Ref.: Ertl/Knözinger/Weitkamp, Vol. 4, p. 1795.

hydrazine synthesis by reaction between NaOCl (*Bayer proc.) or H_2O_2 (*PCUK Kuhlmann proc.) and NH_3 and in the presence of ethyl methyl ketone and a mixture of acetamide/NaH_2PO_4 as *activators. The initially obtained methyl ethyl ketoneazine is *hydrolyzed to hydrazine. **F** synthèse de hydrazine; **G** Hydrazinsynthese. B. CORNILS
Ref.: Büchner/Schliebs/Winter/Büchel, p. 45; *Chem. Week* **1982**, Jan., 28; Büchel/Moretto/Woditsch.

hydride shift the migration of a negatively charged or polarized H substituent or *ligand in a closed-shell proc. The terminology is frequently applied in a more general sense to describe the net transfer of a proton and a pair of electrons without any mechanistic implications. The motion can occur either intra- or intermolecularly. The two possibilities are sometimes (but not here) distinguished as *hydride shift* and *hydride transfer*, respectively. Cat. procs. involving hss. are frequently encountered with noble metal cats., metal complexes, or *enzymes. Technically important examples include *hydrogenation, *hydroformylation, *isomerization, *hc. *cracking, *methanol synth., or the *Fischer-Tropsch proc. Hydride migration also plays an important role in metal-cat. *transfer *hydrogenations and in biological *redox couples.
In organometallic cats. the intramolecular hs. of a metal-bound hydrogen substituent to another unsaturated ligand such as alkene, diene, CO, ketones, etc. is equivalent to the migratory insertion into the metal hydride bond. The reaction generally (but not universally) involves reversible migration of a *cis*-hydride from the same face as the metal via a four- or three-centered *transition state. The reverse proc. is known as *H-elimination. **F** le "schift" du hydrure; **G** Hydridverschiebung. W. LEITNER

Ref.: Abel/Stone/Wilkinson, Vol. 12, Chapter 1, p. 4; A. Dedieu (Ed.), *Transition Metal Hydrides*, VCH, Weinheim 1992, p. 185; *Adv. Phys. Org. Chem.* **1988**, *24*, 57.

hydride transfer, enzymatic → oxidoreductases, *regeneration of NAD

hydride transfer (reactions) → hydride shift

Hydrisom process → Phillips procs.

hydroacylation the hydrogenating introduction of acyl groups (*acylation). The intramolecular h. yields cyclic cpds., e.g., the Rh cat. cyclization of unsaturated aldehydes such as 4-pentenal achieves cyclopentanone.

M. BELLER

Ref.: Beller/Bolm, Vol. 1, p. 136; *Angew. Chem.* **1998**, *110*, 150.

hydroamination the formal addition of an N-H moiety to a C-C multiple bond. As the key step for amine synth. without any by-prod., the reaction is of general interest. The direct *addition of ammonia or simple amines to alkenes is thermodynamically allowed but kinetically strongly inhibited. Since then reaction is also entropically unfavored, a cat. course of the reaction is indispensable. Either the amine or/and the alkene can be activated. The activation of the amine is realized by formation of the amide while the alkene activation proceeds via π-complex formation. Different cat. systems like amides of strongly electropositive metals such as alkali metals, alkaline earth metals, lanthanides, or suitable *transition metal *complexes are described in the literature, but until now none of them has proven to be usable for commercial purposes. **F=E; G** Hydroaminierung.

R. TAUBE

Ref.: Cornils/Herrmann-1, p. 507; *Chem. Rev.* **1998**, *98*, 675; Chem. in uns. Zeit **1999**, *33*, 296.

hydroaminomethylation cat. transformation of alkenes to aminomethylated cpds. with one C atom more under the conds. of *hydroformylation and *reductive amination, e.g., *n*-hexylamine from 1-pentene. **E=F; G** Hydroaminomethylierung.

M. BELLER

Ref.: *Angew. Chem. Int. Ed. Engl.* **1999**, *38*, 2372.

hydroboration reaction of alkenes with boranes or organoboranes (R_2BH, RBH_2) to yield the corresponding alkylborane cpds. following the *Markovnikov rule, acc. to $R\text{-}CH=CH_2 + R^1_2BH \rightarrow RCH_2CH_2BR^1_2$. With disubstituted alkenes high *regioselectivities are observed using sterically hindered boranes such as 9-borabicyclo[3.3.1]nonane or disiamylborane. Although the h. takes place upon simple heating, catalytic variants are known, e.g., with the *Wilkinson cat. Except for that, electrophilic organolanthanide *complexes can also be used as cats. **E=F; G** Hydroborierung.

M. BELLER

Ref.: H.C. Brown, *Boranes in Organic Chemistry*, Cornell University Press, New York 1972; H.C. Brown, *Organic Synthesis Via Boranes*, Wiley, New York 1975; Beller/Bolm; *Angew. Chem. Int. Ed. Engl.* **1985**, *24*, 878; *J. Am. Chem. Soc.* **1988**, *110*, 6917; Kirk/Othmer-2, p. 612; Murahashi/Davies, p. 483.

hydrocarbon resins These resins (M_n approx. 500 – >1000) are produced by cat. *polymerization of olefinic *hcs., their derivatives, and mixtures. Raw materials are predominantly petrochemical *steamcracker fractions (aromatic resins from C_9/C_{10}, aliphatic resins from C_5) and coal-tar light oil (indene-coumarone resins), and also pure olefinic monomers and natural terpene monomers. The most often applied cats. are BF_3 and its adducts, $AlCl_3$, H_2SO_4, or H_3PO_4. Prod. capacity is ~10^6 tpy worldwide. Main applications are in adhesives, coatings, printing inks, and rubber formulations. Cf. also *CAT-resins. **G** Kohlenwasserstoffharze.

G. COLLIN

Ref.: R. Mildenberg, M. Zander, G. Collin, *Hydrocarbon Resins*, Wiley-VCH, Weinheim 1997; Mark/Bikales/Overberger/Menges, Vol. 7, p. 758.

hydrocarbons abbr. as *hc., boiling ranges (°C) of the major hydrocabon classes obtained from crude oil (different values due to differing sources) **F** hydrocarbures; **G** Kohlenwasserstoffe.   B. CORNILS

Gases C_1–C_4	up to 38
Light naphtha	up to 150
Gasoline	up to 180
Heavy naphtha	150–205
Kerosene	205–260
Stove oil	205–290
Light gas oil	260–315
Heavy gas oil	315–425
Lube fractions	420–538
Vacuum gas oil	425–600
Asphaltenes	> 550
Residuum	> 600

hydrocarbon oxidation → *oxidation of hydrocarbons, e.g., procs. of *BP, *Celanese LPO, *Hüls

Hydrocarbon Research H-coal (H-oil) process
the *hydrogenation of coal with Co-Mo oxidic cats. (Co-MoO$_3$, support: *alumina, kaolin, or Al silicate) in an ebullated-bed reactor at 400–450 °C/1.4–2.8 MPa. Similar: H-oil proc. at 370–470 °C/4–14 MPa.
F procédé H-coal de Hydrocarbon Res.;
G Hydrocarbon Res. H-coal-Verfahren.

I. ROMEY

Ref.: *Hydrocarb.Proc.* **1972**, (9), 142,158; Winnacker/Küchler, Vol.5, pp. 420f,460; Ullmann, Vol.A18, p. 75.

Hydrocarbon Research M-B-E process
manuf. of methane, benzene, and ethane from liquid feedstocks by a proprietary cat. *hydrocracking proc. with simultaneous *HDA step.
F procédé de M-B-E de Hydrocarbon Res.;
G Hydrocarbon Res. M-B-E-Verfahren.
Ref.: *Hydrocarb.Proc.* **1977**, 56(11), 150.   B. CORNILS

hydrocarbonylation
1: *Carbonylation procs. with water as reactant (hydroxycarbonylation, to be distinguished from *hydrocarboxylation), one of the developments of *Reppe.

$$CH_2=CH\text{-}CH_3 + 3\ CO + 2\ H_2O \xrightarrow{\text{cat.}}$$
$$CH_3\text{-}CH_2\text{-}CH_2\text{-}CH_2OH + 2\ CO_2$$

The h. is cat. by Fe(CO)$_5$ and modifiers such as tert. amines or *N*-butyl pyrrolidine. Propylene as feed yields butanols with an *n/iso* ratio of 85:15. It is said that the cat. can be decanted as a separate phase from the reaction mixture, thus allowing a *two-phase cat. Despite the high *n/iso* ratio the reaction is inferior to the *hydroformylation of propylene because only alcohols and no C$_4$ aldehydes are obtained (e.g., *BASF butanol proc.). DuPont's glycol proc. is another example. Alkenes, water, and CO are also reacted under the influence of acid or *Lewis cats. (*Koch synthesis, *hydrocarboxylation). 2: A seldom used expression for *hydroformylation. **F=E**;
G Hydrocarbonylierung.   B. CORNILS
Ref.: Falbe-1, p. 250; *J.Mol.Catal.A:* **1999**, *143*, 11, 23.

hydrocarboxylation
(hydroxycarboxylation)   the *addition of CO and water to an alkene or alkyne. H. is cat. by *transition metal cpds. based on Co, Ni, Pd, Pt, Rh, and Ru, as well as by strong acids via *carbenium ions (*Koch synth.). Depending on the feedstocks the carboxylic acids obtained are *n*- and/or *iso*-configurated (*n/iso* ratio). The h. was developed by *Reppe. Today's most important commercial application is the manuf. of propionic acid. As to the mechanism, cf. the proc. of *BASF. Improved *regioselectivities have been obtained by Co carbonyl/pyridine cats. With Pd-, Rh-, and Pt-based cats. high regioselectivities can sometimes be achieved since these metals permit milder reaction conds. Asymmetric h. with *enantioselectivities up to 84 % *ee* have been achived by reacting the prochiral alkene 6-methoxy-2-vinylnaphthalene in the presence of PdCl$_2$/CuCl$_2$ and a chiral phosphate *ligand (e.g., *BNPPA). The conversion of acetylene to acrylic acid was a proc. run by BASF (Equation).

By using Co cats. it is possible to obtain succinic acid at higher temps. (100 °C) and press.

(10–20 MPa), as result of a second h. step.
F=E; **G** Hydrocarboxylierung.

W. MÄGERLEIN, M. BELLER

$$\equiv\!\!\equiv \;+\; CO \;+\; H_2O \;\xrightarrow{\;NiBr_2/CuI\;}\; \diagdown\!\!\diagup\!\!\diagdown COOH$$

Ref.: Heck; *J.Mol.Catal.* **1995**, *104*, 17; H.M. Colquhoun, D.J. Thompson, M.V. Twigg, *Carbonylation*, Plenum Press, New York 1991.

Hydrocol process fluidized-bed variant of the *Fischer-Tropsch synth., realized in Brownsville (TX) with a capacity of 360 000 tpy. The alkalized Fe cat. of the proc. was originally prepared from *Alan-Wood ore, later from millscale. Due to the high-press. (3MPa), high temp. (350 °C) operation the prods. contained highly branched *hcs. rich in alkenes. The plant was shut down in 1957 due to uneconomic operation.   B. CORNILS

Ref.: Falbe-2, p. 294; Winnacker/Küchler, Vol. 5, p. 533.

hydrocracking H. is an important step in the petroleum manuf. chain and a special case of *hydrotreating reactions where the quality of petrochemical prods. is upgraded by *hydrogenation of unwanted structures. H. removes excessive fractions of aromatic molecules from usually low-value oil feeds. It converts them into alkane chain molecules. It is important to avoid formation of small molecules below the C_4 fraction as these are difficult to reuse for petroleum prods *Dual-function cats. with a solid *acid comp. for *isomerisation cat. and a hydrogenation comp. are used. Pt particles on *alumosilicates or *zeolites (*mordenite) are technical realizations of such cats. The cats. have to withstand very aggressive conds. of 600–900 K operating temp. and press. between 0.4 and 5 MPa.

The operation of h. procs. is an essential prerequisite to supply sufficient amounts and the right mix of light oil prods. The proc. is involved in the prod. of gasoline, kerosene, and diesel fuels. In order to use the cats. in h. plants effectively it is essential to remove *poisoning heteroatoms such as S, N, and certain metals from the oil feed by hydrotreating

(*HDS, *HDN) reactions usually performed in combined plants with dual reactors for hydrotreating and h. The hydrotreating is performed over *alumina-supported Co-Mo composite cats. in similar conds. (at higher press.) to the h. proc.

The mechanism of the h. reaction is very complex as many reaction channels are open and mixtures of molecules enter the proc. *Hydrogenolysis of larger molecules and *carbocation-driven isomerizations are key steps in the reaction scenario. In order to achieve large yields of isomerized prod. the initially formed saturated alkanes have to be *dehydrogenated over the precious metal comp. to alkenes which isomerize under the influence of protons passing from and to the acidic comps. to secondary and tertiary alkenes.

The application of modern stable varieties of zeolitic cats. allowing of control of the product spectrum by shape-selective reactions has greatly added to the value of this cat. *refinery process. Procs. of interest are *BASF, *BP, *Exxon, *Gulf, *Houdry, *IFP, *Kellogg MAK, *Lummus LC, *Texaco T-Star, *UOP Unibon and Unocat, *VEBA, etc. **E=F=G**.

R. SCHLÖGL

Ref.: Ertl/Knözinger/Weitkamp, Vol. 4, p. 2017; Topsøe/Clausen/Massoth; Farrauto/Bartholomew, p. 550.

hydrocyanation general method for the synth. of nitriles by addition of HCN to π-bonded systems. The concept includes the formation of cyanohydrins by reaction of HCN on aldehydes or ketones as well as its addition to alkenes, acetylenic cpds., and the corresponding α,β-unsaturated derivatives (Scheme).

Nitriles are valuable intermediates for industrial key procs. (e.g., *adiponitrile) and the prod. of amino acids. H. of carbonyl cpds. is usually cat. by bases as well as by alkyl Al cpds. and the cyanide anion itself. Non-activated alkenes and acetylenes are hydrocyanated in the presence of hom. cats. consisting of *transition metals and an excess of appropriate *ligands. Complexes of metals of the first and second rows of group VIII, VI, and

$(R_1\text{-}R_4 = H$, alkyl, aryl, vinyl, carbonyl, NC, R_2N, OR, Hal, etc.; $X = O$, NR$)$

Ib exhibit the best cat. performance in the presence of *phosphines, *phosphites, and phosphinites as ligands. Commonly NiL_4 (L = phosphite) is used as the cat. for hom. h. H. of alkenes is further promoted by the presence of *Lewis acids as *co-cats. The reaction proceeds via C-C coupling of the cyanide carbon to a C atom of the substrate. Mechanistic studies suggest HCN activation by oxidative addition of the metal into the C-H bond. Regiocontrol as well as *enantioselectivity of the addition to a given olefinic substrate is effectively influenced by adjusting steric and electronic properties of ligands and the co-cat. For the h. of aldehydes the chirality is induced by employing Ti(IV) cpds. in the presence of *bi- or *tridentate ligands derived from *Schiff bases of, e.g., acyclic dipeptides. The h. of butadiene with Rh(0)/phosphine-phosphite cats. yields *adiponitrile (*Butachimie process). For the h. of acrylonitrile cf. *DSM pyrrolidone proc. **E=F**; **G** Hydrocyanierung.

S. KRILL

Ref.: Cornils/Herrmann-1, p. 465; Cornils/Herrmann-2, p. 393.

hydrocyclization (hydrogenative cyclization) see under *hydrogenation of γ- or δ-nitro carbonic acids or -esters.   B. CORNILS

Ref.: *J.Chem.Soc.* **1951**, 671.

hydrodealkylation (HDA)   comprises the removal of lighter alkanes from petroleum fractions or of alkyl groups from cpds. such as, e.g., methylnaphthalenes (to naphthalene, or toluene to benzene), cf. *hydrotreating. Among procs. of interest are, i.a., *Atlantic Richfield HDA, *Gulf THD, *Houdry Detol and Pyrotol, *Mitsubishi MHD, *Shell Bextol, *Union Oil Unidak, *UOP Hydeal, etc.

B. CORNILS

hydrodearylation (HYD)   comprises as a subreaction of *hydrotreating the removal of aromatics or polyaromatics acc. to legal limits (as for example the environmental limits in California); cf. *dearomatization; procs. of *Toray Hytoray, *UOP BenSat.   B. CORNILS

hydrodechlorination (HDC)   a subreaction of *hydrotreating which comprises the removal of chlorine from petroleum fractions (as HCl over Ni-Mo cats. on *alumina) for environmental and corrosion reasons; cf. *dehydrohalogenation.   B. CORNILS

hydrodeformylation   → decarbonylation

hydrodemetallization (HDM, also dehydrometallation)   comprises the removal of metal in order to avoid metal deposits on cats. and apparatus; cf. *demetallization, *hydrotreating. Among procs. of interest are *Howe-Baker hydrotreating, *IFP Hyval, *Lummus ABB, *Shell residual HDS and Hycon, *Union Oil Unicracking.   B. CORNILS

Ref.: Ertl/Knözinger/Weitkamp, Vol. 4, p. 1928.

hydrodenitrification   → catalytic nitrate reduction

hydrodenitrogenation (HDN, hydrodenitration)   comprises the removal of N cpds. as ammonia from *FCC or *hydrocracking feeds (on Ni-Mo, Co-Mo, or Ni-W cats. at higher press. compared to *HDS) which may contain polycyclic aromatics (quinoline, indole, etc.). N cpds. tend to build up NO when combusted in *stationary or *automotive power sources;

cf. *hydrotreating. Among procs. of interest are *BASF hydrocracking, *Howe-Baker hydrotreating, *Lummus PGO, *Union Oil Unioncracking and Unifining, etc.   B. CORNILS
Ref.: Ertl/Knözinger/Weitkamp, Vol. 4, p. 1910.

hydrodeoxygenation (HDO) a subreaction of *hydrotreating which comprises the removal of O (as H_2O) from (mainly) coal-derived fuel fractions (which may contain phenols, aryl ethers, benzofurans, etc.). Oxide cats. are applied. Cf. *hydrotreating and the procs. of *BP Ferrofining or *Lummus PGO. **E=F=G**.   B. CORNILS
Ref.: Ertl/Knözinger/Weitkamp, Vol. 4, p. 1912; *J. Catal.* **1983**, *80*, 56.

hydrodesulfurization (HDS) the commercial proc. for the removal of S from fossil feedstocks (0.1–5 wt% S) in refineries. The HDS proc. (33×10^6 bpd worldwide, 1997) involves the conv. of the organo S cpds. (thiols, sulfides, disulfides, thiophenes) to H_2S and hcs. by *hydrogenation at high temps. (300–425 °C) and high press. (3.5–17 MPa) acc. to [RS] + H_2 → R-H + H_2S. H_2S may later be disposed of as elemental S via the *Claus proc. In addition to preventing *poisoning of the *cracking and *reforming cats. the HDS proc. reduces the emission of S oxides into the atmosphere when combusting fuels. Most conventional cats. contain Mo or W sulfides *supported on *γ-alumina. Other *transition metals such as Co, Ni, Ru, Pt, and Pd may be incorporated as *promoters to increase the cat. *activity; Pt and Pd are most effective. Deep HDS is required to meet current legislation in the USA and Europe which sets a max. S content of 0.001 wt% in urban diesel (cf. *hydrotreating).

The HDS mechanisms are still rather obscure. It is generally agreed that the main role of the *promoter is to favor the activation of the S cpds., especially the thiophenes which are most difficult to degrade, while the MoS_2 or WS_2 units activate H_2. The *RDS step is generally related to the rate of creation of S vacancies over the cat. surface. In reality the sit-

uation is more complex since the reactions are not completely independent of each other. Model studies applying *organometallic *complexes (*hydrogenation or *hydrogenolysis of thiophenes in various liquid-phase and aqueous-biphase systems) have provided valuable mechanistic information on the elementary steps of the HDS proc. (substrate adsorption, C-S insertion, C-S cleavage, H_2S formation). The *immobilization of hom. cats. on various *supports is pursued to mimic the HDS of thiophene over *single-site metal species. Het. cat. HDS with *Raney Ni is one of the most important methods of reducing organosulfur cpds. in organic synth. (*desulfurization). Examples are procs. of *BASF hydrocracking and HDS, *BP Autofining, Hydrofining, *BP residue desulfurization, *Chevron RDS, VRDS, VGO and DAO, *Criterion Lummus Syncat, *Exxon GO Fining, *Gulf Gulfining, HDS, lube oil hydrotreating, and Resid HDS, *Houdry Pyrotol, *Howe-Baker hydrotreating, *Hydrocarbon Res. H-Oil, *IFP Hyval, distillate HDS, lube hydrotreating and hydrofinishing, *Kellogg MAK, *Lummus LC Fining, *Mobil Octagain, *Shell HDS and Hycon, *Standard Oil hydroprocessing and Ultrafining, *Union Oil Unicracking, *UOP Unifining, RCD Unibon, and Hydrobon, *VEBA Combi Cracking, etc. **F=G=E**.   C. BIANCHINI
Ref.: Topsøe,; J. Scherzer et al., *Hydrocarbon Science and Technology*, Marcel Dekker, New York 1996; Cornils/Herrmann-1, p. 969; *Acc.Chem.Res.* **1998**, *31*, 109; *Synthesis* **1990**, 89.

hydrodewaxing a proc. for the cat. *dewaxing of petroleum- derived feedstocks in the presence of H with selective cracking of normal and branched paraffins (e.g., *BP dewaxing, *Chevron Isodewaxing, *Mobil MDDW, *Mobil LDW procs.). **F=G=E**.
B. CORNILS

hydrodimerization a special case of *telomerization. In h. *taxogens such as butadiene and *telogens such as water in the ratio 2:1 react in the presence of special Pd cats.

which give the linear dimer 2,7-octadien-1-ol in high *selectivity. Not only zerovalent Pd complexes such as Pd(PPh$_3$)$_4$ but also bivalent cpds. such as Pd(OAc)$_2$ can be used in comb. with excess PPh$_3$.

$$2 \diagdown\diagup + H_2O \longrightarrow \diagup\diagdown\diagup\diagdown\diagup\diagdown_{OH}$$

Efficient h. requires the presence of *additives such as CO$_2$, trialkylammonium hydrogencarbonate, soda, and an aprotic *solvent (e.g., acetonitrile, sulfolane) which dissolve both butadiene and water. Thus, the reaction in sulfolane with [NEt$_3$H][HCO$_3$] at 70 °C yields 92 % selectivity of the octadienol, which is the precursor for the commercial prod. of 1-octanol (by *hydrogenation), 1,9-nonanediol (by *hydration), and 1,9-nonanediamine (*Kuraray telomerization proc.). **F** hydrodimerisation; **G** Hydrodimerisierung.   N. YOSHIMURA

Ref.: Tsuji; *Adv.Organomet.Chem.* **1979**, *17*, 141; Cornils/Herrmann-1, p. 351.

hydroesterification   the *addition of CO and alcohols to alkenes or alkynes to produce monocarboxylic esters. The reaction is closely related to *hydrocarboxylation. Consequently, the same cats. (carbonyl *complexes of Co, Ni, Pd, Pt, Rh, Ru) are effective. The decisive step in h. is the *alcoholysis of the acyl complex (for the mechanism cf. the propionic acid proc. of *BASF). In the case of neutral or basic conds. with Pd cats. another mechanism proposes an *insertion of CO into the Pd-O bond of a Pd *alkoxide complex. Insertion of the alkene is followed by protonolysis to yield the ester and the Pd *alkoxide. In h. of terminal alkenes high selectivities for the linear *n*-isomer (*n/iso* ratio) can be achieved by using *ligand-stabilized *bimetallic cats. based on SnCl$_2$ and Pt or Pd cpds. Conjugated dienes undergo h. under specific conds., where the competing *telomerization or *oxidative carbonylation is suppressed. Of special interest is the three-stage synth. of *adipic acid from butadiene (cf. *BASF procs.). **F=E**; **G** Hydroesterifizierung.

W. MÄGERLEIN, M. BELLER

Ref.: *J.Mol.Catal.* **1995**, *104*, 17; H.M. Colquhoun, D.J. Thompson, M.V. Twigg, *Carbonylation*, Plenum Press, New York 1991; Falbe-1.

hydrofining   → hydrotreating

Hydrofining process   → Exxon procs.

hydrofinishing   a low-*severity *hydrotreating variant. The term is somewhat obsolescent and may also include procs. under normal hydrotreating conds. **E=F=G**.

B. CORNILS

Hydrofinishing process   → BP procs.

hydroformylation (oxo synthesis, Roelen reaction)   H. is the cat. *addition of CO and H$_2$ – the constituents of *syngas – to the double bond of alkenes or their functionalized derivatives:

$$\overset{\text{cat}}{R\text{-}CH_2\text{=}CH_2 + CO + H_2 \longrightarrow R\text{-}CH_2\text{-}CH_2\text{-}CHO}$$
$$+ R\text{-}CH\text{-}CH_2\text{CHO}$$
$$\underset{CH_3}{|}$$

n-aldehyde   *iso*-aldehyde

The first prods. of h. are thus aldehydes. Generally the h. of alkenes which are not asymmetric and/or do not form isomers via double-bond migration leads to the formation of isomeric aldehyde mixtures (unbranched *n*- and branched *iso*-aldehydes, the ratio of which is an important criterion for each oxo proc. and cat. system, *n/iso* ratio). The aldehydes may be used as *intermediates (via *aldolization, *hydrogenation, *oxidation, *amination, etc.); the bandwidth of cpds. accessible through h. is considerable. The main outlets are oxo alcohols which are used as *solvents, as *plasticizers, or as detergent alcohols. The annual manuf. of oxo products sums up to approx. 6.5×10^6 tpy, of which 5×10^6 tpy are C$_4$ aldehyde being converted to *2-EH. Economically important as oxo feedstocks are alkenes from C$_2$ (ethylene) to approx. C$_{16}$. The virtue of h. lies in its applicability to a broad variety of substrates with C-C and some C-heteroatom bonds. For the reactivity of different olefinic structures, especially to fine chemicals (e.g., the h. of allyl alcohol → *BDO, h. in the series of *vitamin A synth.) cf. special literature.

H. cats. $H_xM_y(CO)_zL$ typically consist of a *transition metal M as the *central atom which enables the formation of metal-carbonyl hydride species. Optionally these *complexes are modified by additional *ligands L. The oxo active species may be formed by *precursors of different compositions converted under syngas press. and at higher temps. Today's h. focuses almost exclusively on the metals Co and Rh. The generally accepted order of h. activity with regard to the central atom is Rh >> Co >> Ir,Ru > Os > Pt > Pd > Fe > Ni. Commercial h. plants are run exclusively with cats. like $HCo(CO)_4$ (ligand-free), $HCo(CO)_3PR_3$, or $HRh(CO)(PR_3)_3$ (with phosphine ligands PR_3 = unsubstituted or substituted triphenylphosphine). Polymetallic h. cats. Co-Rh or Pt-Sn systems have also been studied as well as *cluster and *colloid cats., but gained no commercial relevance.

The mechanism of h. has been clarified by *Heck and Breslow in one of the first *cat. cycles investigated, consisting of the pathway π-complex → alkyl Co carbonyl → acyl Co carbonyl and the eventual associative formation of aldehyde and recovery of the starting cat. $HCo(CO)_3$ (Figure 1).

Natta et al.'s kinetic relationship best represents all effects observed in the applied pres. and temp. range.

$$\frac{d(\text{aldehyde})}{dt} = k \, [\text{olefin}][\text{Co}] \frac{P_{H2}}{P_{CO}}$$

Other useful relationships have been published for Rh or ligand-modified cats. Acc. to the cat., operation (process), hoped-for selectivity, etc., the proc. parameters of h. vary considerably, e.g., 90–180 °C, 2–30 MPa total press., 0.001–1 % cat.conc., and surplus of lig-

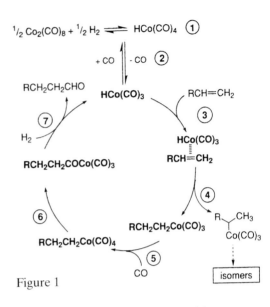

Figure 1

ands 1–200. The h. is exothermal by approx. 30 kcal mol^{-1}.

Besides the *central atom (see above) the cat. and operation of h. may be influenced by a modification of the ligand sphere (*ligand-modified cats.) and a variation of the application phase. A comparative study of Ph_3E (E = main group V element) as ligands for h. showed the order of reactivity Ph_3P >> Ph_3N > Ph_3As, Ph_3Sb. Of considerable interest are *phosphites $(RO)_3P$ and water-soluble ligands such as triphenylphosphine trisulfonate (*TPPTS). Especially with chiral diphosphines (*DIOP, *DIPAMP), *asymmetric h. with high *enantioselectivities (*ee values) are feasible. So far only phosphines and phosphites are used in large-scale operations. Variation of the application phase of the h. cats. has been tried with anchored cats. (*immobilization), by *SLPC cats., or by *aqueous-phase operation. Only the latter is used in a commercial scale (*Ruhrchemie/Rhône-Poulenc proc.).

The industrial versions of h. date back to two basically different techniques (Figure 2): the *first-generation* procs. with unmodified $HCo(CO)_4$ cats. operate under high press. in continuous-flow tank reactors with simultaneous *decobaltation and cat. formation during each pass of the cat., thus demonstrating the enormous expenditure to overcome the problem of hom. cat., the separation between cat. and prod. The *second-generation* technique for the Rh-based, ligand-modified proc. uses *CFSTRs with special devices for this separation (flashing with *LPO procs., phase separation with *aqueous-phase procs.) (Figure 2).

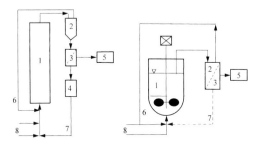

Figure 2

1 reactor; 2 separator; 3 cat-separator;
4 make-up; 5 further processing; 6 gas recycle;
7 cat recycle; 8 reactants

For the details of fundamental reaction engineering (*heat and *mass transfer, fluid dynamics), basic engineering (mechanical engineering, automation, corrosivity), and technical solutions (choice of cat., *phase behavior, *demetallization, further processing of the byproducts, etc.) cf. specialist literature.

The various industrially relevant solutions are compiled in the Table on p. 282. The Table indicates the characteristic differences and the special suitability of certain procs. for the manuf. of various building blocks. Although it is a mature technique the research on h. is extremely active in applying new results from the organometal progress. The research work concentrates on ligand syntheses, especially of *chiral ones for enantiomeric conversions.
E=F; G Hydroformylierung.   B. CORNILS

Ref.: DE 849.548; *Angew.Chem.Int.Ed.* **1994**, *33*, 2144; Falbe-1,3; Cornils/Herrmann-1,2; *J.Mol.Catal.* **1995**, *104*, 17; Kirk/Othmer-1, Vol. 17, p. 902; Kirk/Othmer-2, p. 826; *Chem.Rev.* **1995**, *95*, 2485.

hydrogenation   *addition of hydrogen to organic molecules, preferably catalytic h. This is almost always performed with molecular

H_2, although there is limited scope for using other H donors (e.g., formic acid). The H molecule is very stable; its dissociation energy is 434 kJ mol^{-1}, and H atoms are only formed by thermal decomposition at such high temps. as to be useless in any practical sense. By far the best method to generate H_2 is to allow the molecule to undergo *dissociative *chemisorption at the *surface of a suitable solid, and to employ the atoms thus formed in cat. h. The H_2 is chemisorbed as atoms on many clean metal *surfaces, especially those of the later *transition elements, on Ni, and the noble metals of groups 8 to 10, where the atoms are most reactive because at high *coverage they are only weakly bonded. The *adsorption occurs with almost no *activation energy and may therefore be observed at very low temp. H_2 is also adsorbed with dissociation on certain *oxides (ZnO, Cr_2O_3) and *mixed oxides (*mixed oxide cats.), which have limited practical use, and of course with a number of inorganic *complexes such as *Wilkinson's or *Vaska's complex, which may therefore be used for hom. cat. h.

Almost any unsaturated function in an organic molecule is susceptible to *reduction by h. The most easily reduced is the C–C double bond, which has been hydrogenated as low as 213K; many other groups are reducible, given the appropriate catalyst, at or close to ambient temp. (e.g., C≡C, -NO$_2$, the aromatic ring), while ketones and aldehydes need somewhat more forcing conds. The hardest to reduce is the carboxyl group (-COOH) and the corresponding anion (-COO$^-$). The ease of reduction of a chosen function may depend on the presence of other functional groups, which although unreactive can serve to anchor the molecule to the surface. Inorganic ions in solution can also be reduced cat. (e.g., Fe^{3+} to Fe^{2+}). It is often necessary to bring about a partial reduction or the reduction of one function in the presence of another (*selective h.).

A particularly important aspect of *selectivity is proved by *chiral h., where it is desired to form one optical isomer exclusively: this is

Industrially relevant oxo processes (to hydroformylation)

Variant	Catalyst metal			
	Cobalt		Rhodium	
	Unmodified [1,2,3]	Modified [4]	Modified [5]	Modified [6]
Cat. species	$HCo(CO)_4$	$HCo(CO)_3L$	$HRh(CO)L_3$	$HRh(CO)L_3$
Temp. [°C]	150–180	160–200	60–120	110–130
Press. [MPa]	20–30	5–15	1–5	4–6
Cat. conc.[%][7]	0.1–1	0.6	0.01–0.1	0.001–0.1
LHSV[8]	0.5–2	0.1–0.2	0.1–0.2	> 0.2
Products	Aldehydes	Alcohols	Aldehydes	Aldehydes
Byproduct formation	High	High	Low	Low
n/iso ratio[9]	80:20	88:12	92:8	>95:5
Restrictions of feed olefins	No	No	Yes	Yes

[1] *BASF; [2] *Ruhrchemie; [3] Exxon; [4] *Shell; [5] *LPO; [6] *RCH/RP [7] % relative to olefin; [8] throughput in *LHSV; [9] for propylene

achieved when two H atoms add preferentially to one specific side of the reactant molecule, and this in turn follows if that molecule is constrained by geometry to be chemisorbed only with that one particular side downwards. Simultaneous adsorption of a suitable optically active *template (e.g., an alkaloid) can secure this, and may also increase the rate.

H. may be performed either in the gas phase or in a *three-phase system where the organic molecule is in liquid form or in solution (*cat. reactors). The reaction is exothermic, and especially when used in the gas phase *supported metal cats. quickly deactivate due to formation of coke, etc. The applied press. may range according to the equipment to tens of MPa. Efficient agitation minimizes the distance the dissolved H_2 molecule has to travel from the gas-liquid to the liquid-solid interface, and therefore helps to overcome *diffusion limitation.

Metallic cats. are used in a variety of forms. The most active are the noble metals of groups 8 to 10, of which Pd and Pt are most often used, *supported on *alumina or *silica for gas-phase applications, but often on *activated carbon for liquid-phase use. Ni on SiO_2

is used for *fat hardening and it is also widely used as *Raney Ni. The noble metal cats. have to be carefully crafted so as to avoid loss of metal during use, as proc. economics depend critically on the successful recovery and re-use (*regeneration) of all the metal deployed.

Many detailed studies have been performed to elucidate the mechanism by which cat. h. occurs. Of importance are, e.g., procs. of *adsorption, double-bond migration, cis,trans *isomerization, and the *Horiuti-Polanyi mechanism.

H. of alkenes is widely regarded as being a *structure-insensitive reaction as *rates are not greatly dependent upon metal particle size, and on many metals show the same *activation energy (ca. 40–45 kJ mol^{-1}) which may be regarded as a true parameter (see *Arrhenius plot and *apparent activation energy) because of the near constancy of surface concs. It is however possible sometimes to see an inversion of the *Arrhenius plot at high temp., due to the desorption of one or other of the reactants. The reason for the structure insensitivity has been much debated: it is now thought that especially in gas-phase reactions only a small part of the surface remains active, the fraction of which is not responsible to particle size. It has even been suggested that reaction occurs on top of a "coke *deposit", although this does not explain the different characteristics that metals show. The more strongly adsorbed alkynes can draw metal atoms from their normal lattice sites, and perhaps alkenes can also do this to some extent: this effect would also help to explain the lack of particle-size dependence.

A clear distinction needs to be drawn between h., where reduction is due to addition of H atoms to a bond having π-character, and *hydrogenolysis, where a covalent bond is broken by the action of a hydrogen molecule: this may also be seen as reduction if it results in the removal of an electronegative atom (O, Cl, etc.).

H. is well investigated in *biocatalysis. As an example 2-heptanone may readily be hydrogenated with formic acid to S-(+)-2-heptanol with the *enzyme *alcohol dehydrogenase (ADH). As *cofactor *NADH is oxidized to NAD^+, which is regenerated to NAD by a second enzyme, *formate dehydrogenase, and closes the *cat. cycle. **F** hydrogénation; **G** Hydrierung.   G.C. Bond

Ref.: Rylander; Cerveny; B. James, *Homogeneous Hydrogenation*, Wiley, New York 1973; F.J. McQuillen, *Homogeneous Hydrogenation in Organic Chemistry*, Reidel, Dordrecht 1976; Ertl/Knözinger/Weitkamp, Vol. 5, p. 2165; Farrauto/Bartholomew, p. 411; Z. Paál, P.G. Menon, *Hydrogen Effects in Catalysis*, Dekker, New York 1988; Guisnet/Barrault et al.; Ullmann, Vol. A13, p. 487; McKetta-1, Vol. 27, p. 53; Werner/Schreier; *Houben-Weyl/Methodicum Chimicum*, Vol. IV.

hydrogen bonding   in enzyme catalysis *Enzymes cat. reactions by stabilizing the reaction *transition state. In many cases, the transition state is very polar or charged. The charged and polar groups are often stabilized by H bonds to enzyme side chains/backbone amides. The basicity and acidity of enzymatic groups involved in general *acid-base cat. can be increased by hb.

There is heated controversy regarding the strength of this hb. In the gaseous phase, very short (less than the sum of the van der Waals radii of donor and acceptor), strong hbs. have been observed that display a single energy "well" for the proton positioned between the donor and acceptor, rather than the two wells observed in a "normal" hb. It has been postulated that perhaps a low-barrier H bond (LBHB; barrier to proton transfer is less than the zero-point energy) might form in the serine *protease transition state between the active-site aspartic acid and the active-site histidine. Such a bond, found in the *transition state but not the ground state, might be responsible for the incredible rate accelerations observed by this class of proteases. Calculational studies, however, indicate that the presence of an LBHB, even if present, would not be nearly as strong as LBHBs present in the gas phase. Moreover, the formation of a LBHB is promoted in non-polar media, but most cat. takes place by stabilization of polar intermediates. **F** liaison d'hydrogène; **G** Wassserstoff-Bindung.   P.S. Sears

Ref.: *Proc. Natl. Acad. Sci. USA* **1996**, *93*, 13665; *Science* **1994**, *264*, 1927, 2887; *J. Am. Chem. Soc.* **1995**, *117*, 6970; *Ann. Rev. Phys. Chem.* **1997**, *48*, 511.

hydrogen cyanide   manuf. as byprod. of *acrylonitrile plants or by oxidative reaction of methane and ammonia (*Andrussov reaction, *ammoxidation) over Pt-Rh *gauze cats.
                                                      B. Cornils

hydrogen donors   In *hydrogenation procs. not only H_2 is used but also other H-containing cpds. called "hydrogen donors", especially in large-scale commercial hydrogenations where explosive hydrogen has to be avoided. Typically, the *transfer hydrogenation (of alkenes) used H sources such as primary or secondary alcohols (which are converted into the corresponding aldehydes and ketones) or formic acid, which is converted into CO_2. A convenient H source is the commercially available 5:2 azeotrope of formic acid and NEt_3. Also, CO in combination with water giving CO_2 and H_2 can be used as an H_2 source using the *water-gas shift reaction (WGSR). An example, typically enantioselective, is the transfer hydrogenation of itaconic acid, which is reduced to methyl-succinic acid (Figure).

The most convenient H sources for making metal hydrides, however, are NaH (ionic) and $K[HBR_3]$ (covalent, R = alkyl). **F** donors d'hydrogène; **G** Wasserstoffdonoren.

W.A. HERRMANN

Ref.: Cornils/Herrmann-1, p. 201; W.A. Herrmann (Ed.), *Synthetic Methods in Organometallic and Inorganic Chemistry*, Thieme, Stuttgart 1995–1998.

hydrogen finishing process → Texaco procs.

hydrogen manufacture H_2, being essential for *hydrogenations, is manuf. via *syngas (from *steam reforming of *natural gas or naphthas, partial *oxidation of heavy oil or resids, *coal gasification, *Haldor Topsøe proc.), via water electrolysis, from byprods. in *refineries or petrochemical plants (*aromatization, *dehydrogenation, *reforming, methane convs. such as the *UOP Hypro proc.), or via other procs. such as thermal scission of water or thermochemical cyclic procs. In special cases, H_2 can be stored and released from hydrides (e.g., LiH, $TiFe_2$) or *hydrogen donors such as HCOOH or *liquid syngas can be used (cf. *transfer hydrog., *Meerwein-Ponndorf-Verley reaction). **F** production d'hydrogène; **G** Wasserstoffherstellung.

B. CORNILS

Ref.: Farrauto/Bartholomew, p. 341; Büchel/Moretto/ Woditsch.

hydrogenolysis cleavage of a bond by reaction with molecular hydrogen in the presence of cats. acc. to A-B + H_2 → A-H + B-H. Although originally used for bonds such as *transition metal alkyls, today the term mainly refers to organic substrates. Some industrial procs. such as *dehalogenation, *HDS, or *desulfurization of oil are h. reactions. Another well-known h. which has a great importance in the synth. of peptides is the cleavage of heteroatom-benzylic bonds. The benzoyloxycarbonyl group (-C(=O) -O-CH_2-C_6H_5), introduced as a protecting group, is finally eliminated by h. **E=F**; **G** Hydrogenolyse.

B. DRIESSEN-HÖLSCHER

Ref.: Freifelder, *Catalytic Hydrogenolysis in Organic Synthesis*, Wiley, New York, 1978; Augustine, p. 511; Gates-1, p. 388.

hydrogen-oxygen titration Reactive oxygen in cat. materials can be titrated using the *temp.-programmed reduction (TPR). In this method the cat. under study is placed in a tubular reactor and the O-releasing cat. is reduced in a flow of inert gas, usually Ar or N_2 containing a few percent of H_2. At the same time, the temp. is linearly increased. The off-gases are continuously monitored and the consumption of H_2 is recorded as a function of the reaction temp. While increasing the temp. linearly, the material under study will begin to be reduced at certain temps. The rate of reduction will increase as the temp. is further increased and at a maximum reduction rate, H consumption will be recorded. Finally, the amount of reducible material in the reactor, and thus the H consumption, will drop to zero as the temp. is further increased. Integration of the H_2 consumption signal allows the determination of the total amount of H_2 used to titrate the reactive O_2 in the cat. material. The H_2 consumption thus can be used to calc. the average degree of *reduction reached after the TPR experiment. Variation of the heating rate results in shifts of the temp. of maximum H_2 consumption. This rate-dependent shift can be used to determine the *activation energies of the reduction proc. The TPR experiment must be carried out under differential conds., in order to avoid *mass and *heat transport problems during reduction. H_2 conc. and temp. gradients in the reactor and across the cat. bed lead to erroneous consumption measurements and thus to wrong activation energies. In order to avoid these problems, the experimental parameters must be set at optimum (which is a function of the heating rate, the amount of reducible material, the flow rate of the reducing gas mixture, and the H_2 conc.). *Temp.-programmed oxidation (TPO) is an equally valid technique to titrate the amount of reduced species in a cat. material. The experimental setup of a TPO experiment is identical to that of a TPR. Therefore, both techniques can easily be combined, and valuable information is obtained. Oxidation and reduction kinetics of cats. can thus be studied and important information on oxygen release or storage properties of cat. materials is obtained.

Such combined techniques allow the determination of the reactivity of cats. toward *reduction and *reoxidation. It can be determined whether the cat. under study releases or takes up oxygen in a single-step or a multiple-step proc. *Metal-support interactions may be identified, if this interaction leads to changes in the reducibility of the active species or the cat. *precursor. The effects of *promoters or *additives on the redox properties of cats. can be detected using TPR/TPO experiments. The combination of such a TPR/TPO setup with CO *temp.-programmed desorption (TPD), moreover, allows the titration of coordinatively unsaturated metal centers on the cat. *surface as a function of the TPR/TPO pretreatment. **F** dosage de l'hydrogène et de l'oxygène; **G** Wasserstoff-Sauerstoff-Titration.   G. MESTL

Ref.: R.B. Anderson, P.T. Dawson, *Experimental Methods in Catalytic Research*, Academic Press, New York 1976, Vol. 2, p. 43; Ertl/Knözinger/Weitkamp, Vol. 2, pp. 445,676; Farrauto/Bartholomew, p. 149.

hydrogen peroxide in catalysis H_2O_2 is a strong and clean synth. oxidant ($H_2O_2 \rightarrow H_2O$ + "O"). It is manuf. via ethylanthraquinone

(worldwide 5×10^5 t/a): EA + $H_2 \rightarrow$ EAH$_2$ + O_2 $\rightarrow$ EA + H_2O_2. H. decomposition is cat. by Fe-dependent *enzyme (*catalase), and by *transition metal ions capable of one-electron *redox reactions, e.g., Fe(II/III) or Co (II/III). In the Fe(II/III)/ h. system, alkanes and arylarenes undergo *hydroxylations. V(V) cpds. in AcOH solution and MoO_4^{2-}, WO_4^{2-}, or $Ca(OH)_2$ at pH > 7 cat. $^1O_2$ formation. V(V)/CF$_3$COOH cat. ozone synth. acc. to 3 $H_2O_2 \rightarrow$ 3 H_2O + O_3. *Epoxidation of activated alkenes XCH=CH$_2$ (X = CN, COOR, NO$_2$, SO$_2$R, etc.), *oxidations of aromatic aldehydes (*aromatic oxidation) to phenols are cat. with bases. *Alkylation of h. with alcohols and alkenes to alkyl *hydroperoxides or dialkyl *peroxides, conv. of aliphatic and aromatic acids to peracids or diacyl peroxides, *oxidation of ketones to esters are H$^+$-cat. reactions. A number of cpds. (HCOOH, *AA, CF$_3$COOH, (CF$_3$)$_2$CO, Mo(VI), W(VI),Ti(IV) cpds., Fe *porphyrins, and *MTO/py) are used to epoxidize alkenes and oxidize amines, *phosphines, mercaptans, sulfides, and *sulfoxides. On a large scale, allylic alcohol is converted into glycerol ester and glycerol over WO_3/AcOH and phenol to catechol and hyroquinone over Ti silicalite (e.g., *Shell glycerol or acrolein procs.) (cf. also *Fenton's reagent, *Milas reagent, *Prilezhaev reaction, *Sharpless reaction; procs. of *Daicel PO, *Degussa glycidol, *Bayer-Degussa PO, and *Cetus PO). **F** peroxyde d'hydrogène; **G** Wasserstoffperoxid.

                                                              I.I. MOISEEV

Ref.: W. Weigert (Ed.), *Wasserstoffperoxid und seine Derivate*, Hüthig, Heidelberg 1978; Schirrmann, Delavarenne, *Hydrogen Peroxide in Organic Chemistry*, Soc.Exp.Techn.Econ., Paris 1980; G. Strukul, *Catalytic Oxidations with H_2O_2 as Oxidant*, Kluwer, Dordrecht 1992.

1,2-(1,3)-hydrogen shift $\rightarrow$ hydride shift

hydrohaloelimination $\rightarrow$ dehydrohalogenation

hydroisomerization   the simultaneous *hydrogenation and *isomerization of, e.g., straight run C_5/C_6 paraffins ("light ends") from refinery distillation fractions to gasoline comps. with improved octane numbers over Pt on *mordenite cats. In most cases this total isomerization proc. (e.g., *Shell Hysomer, *UCC TIP) is combined with the separation of *n*- and isoparaffins. Other examples are the procs. of *ARCO Octafining (*EB to xylenes), *IFP propylene (h. of crude C_4 streams), *Union Carbide TIP, *VEB Leuna Aris (h. of C_8 aromatics to xylenes). In the *UOP Butamer proc. the isomerization in the presence of H suppresses the *polymerization of alkenes. **F** hydroisomérisation; **G** Hydroisomerisierung. B. CORNILS

Ref.: Moulijn/van Leeuwen/van Santen, p. 33; Ertl/ Knözinger/Weitkamp, Vol. 4, pp. 1998, 2035.

hydrolases a class of *enzymes that cat. the nucleophilic displacement of a leaving group with water. The bond broken is typically C-N (as in *amidases), C-O, or C-S (as in esterases), O-P, or S-P (*phosphatases). Many proceed via nucleophilic cat., in which an enzyme nucleophile attacks the bond to be broken and forms a covalent enzyme adduct with the substrate, which is then hydrolyzed. All hydrolases have *EC numbers that begin with 3.

Hs. are the most commonly used enzymes in organic synth. Of particular interest among the classes of hydrolytic enzymes are amidases, proteases, esterases, and *lipases; these enzymes cat. the *hydrolysis and formation of ester and amide bonds.These reactions may or may not proceed through covalent intermediates. In general, there are four types of proteases and the mechanism of each is known: *serine proteases, *thiol proteases, *metalloproteases and *aspartyl proteases. Examples of serine proteases include *trypsin, *chymotrypsin, *subtilisin, α-lytic enzymes, *elastase, and *V8 protease. The serine-type esterases include *pig liver esterase, *cholesterol esterase, *acetylcholine esterase, *lipase, *carboxypeptidase Y, and *penicillin acylase.

Thioproteases are similar to the serine-type proteases. Examples of thioproteases are *papain, *clostripain, and *cathepsin. Metalloproteases often utilize a Zn^{2+} ion as a *Lewis acid that is coordinated in the cat. step to the nucleophilic water (to increase its acidity) and to the carbonyl group of the scissile bond. A carboxylate group often serves as a base to abstract a proton from the nucleophilic water molecule. No covalent intermediate is formed. Thermolysin, acylases, and carboxypeptidase A represent this type of enzyme. Aspartyl proteases utilize a carboxylate group as a general base and a carboxylic acid as a general acid for hydrolysis. Examples of aspartyl proteases include *pepsin and *HIV protease. **E=F=G**. C.-H. WONG, P.S. SEARS

Ref.: U. Bornscheurer, R. Kazlauskas, *Hydrolases in Organic Synthesis*, Wiley-VCH, Weinheim 1999.

hydrolysis the cleavage of cpds. by water acc. to the general equation A-B + H_2O → A-H + B-OH. Thus, h. is a reverse neutralization. The extent of the h. reaction can be calcd. using the mass-action law and the relevant equil. constants. Often the sole presence of water is not enough to start or maintain h. In this case h. has to be cat. by *Brønsted acids (H_3O^+), strong bases like OH^- or *Lewis acids and bases. The latter are often applied for selective (also *enantioselective, e.g., by *salen complexes) h. of organic cpds., e.g., the ring opening of *oxiranes to form *vicinal* glycols. H. reactions are essential for organic chemistry, e.g., h. (or *saponification) of esters, ethers, acetals, and ketals, the h. of nitriles to amides or carboxylic acids (e.g., acrylamide procs. such as *Mitsui, *Nitto), or the hydrolytic cleavage of acid chlorides and acid anhydrides (similar the *Raschig phenol proc.). Commercially important hs. are the cleavage of triglycerides to obtain gylcerol and fatty acids (*fats), h. of saccharose to obtain glucose and fructose, h. of starch and cellulose to glucose, the hydrolytic cleavage of proteins to yield amino acids or the h. of side-chain-chlorinated arenes to obtain selectively benzylic alcohols, aldehydes, or acid chlorides. To avoid incomplete conv. due to the equilibration of the reaction, h. reactions are often performed in basic medium. In this cases – in contrast to most acidic h. – irreversible h. takes place. Acid-free h. under special activated reaction conds. (fast heating and subsequent fast cooling) is achieved by the application of *microwaves. However, h. reactions are often not planned but rather are unfavorable side reactions which ought to be avoided by working under strict exclusion of water or moisture. Working under an inert gas atmosphere in absolutely dry solvents is often an important prerequisite for the handling of metal organic cpds. Hydrolytically active *enzymes (*hydrolases) are applied as highly selective cats. For other commercial procs. involving hs. cf. *Atochem (to C_{11} acid), *Butakem (tartaric acid), *ChemSystems (*PO), *Daicel (glycerol), *Gulf (phenol), *Halcon (formic acid), *Nitto (acrylamide), Toyo Soda (*BDO), *Toray (lysine), Raschig-Hooker (phenol), etc. **F=G** Hydrolyse.

R.W. FISCHER

hydroperoxides cpds. of the general formula R-O-OH which are important in *aut(o)oxidation reactions (e.g., *anthrahydroquinone), the drying of oils (in lacquers; cf. *driers), and specially for the manuf. of *propylene oxide, phenol (*Hock proc.), or hydroquinone, where the hydroperoxides of isobutane, ethylbenzene, cumene, or diisopropylbenzene play the decisive role as *reaction carriers (*epoxidation, *peroxides).

B. CORNILS

Ref.: Sheldon/Kochi; Emanuel et al., *Oxidation of Organic Compounds*, Pergamon, Oxford 1984; Kirk/Othmer-1,Vol. 18, p. 202; S. Patai, *The Chemistry of Peroxides*, Wiley, Chichester 1983; Ullmann, Vol. 19, p. 210.

hydroperoxides (enantioselective acylation) → *Pseudomonas* sp. lipase

hydrophobicity (hydrophobic packing in enzyme structure)   H. refers to the water-excluding behavior of nonpolar cpds. such as alkanes. The tendency of hydrophobic groups to group together and form a separate *phase provides much of the driving force for protein

folding (cf. *chaperones). Proteins fold in such a way as to keep the nonpolar side chains on the inside, away from water, and the polar/charged residues on the outside, where they may be solvated. The interior of a protein is characterized, therefore, by a low dielectric constant; the electrostatic interactions (e.g., *salt bridges) found on the inside are expected to be stronger than those near the surface. **F** hydrophobicité; **G** Hydrophobizität.

P.S. SEARS

hydrophosphinylation is an *addition reaction of H-P bonds of *sec.*-phosphine oxides, $HP(O)R_2$, to alkynes (Equ. 1). Although *radical initiators promote the reaction, the *regio- and *stereoselectivities are better controlled by using Pd-phosphine *complex cats. Pd-cat. hs. of terminal acetylenes normally end up with anti-*Markovnikov adducts, but the reaction run in the presence of Pd complex and acidic additives such as phosphinic and phosphoric acids affords Markovnikov adducts.

Pd complex cats. promote the closely related addition of hydrogen phosphonates $HP(O)(OR)_2$, namely hydrophosphorylation, to alkynes to give Markovnikov prods (Equ. 2). The prods. of h. and hydrophosphorylation can be useful building blocks for flame retardants, *ligands for *hom. cat., and agrochemicals. **E=F=G**.

M. TANAKA

Ref.: *J.Am.Chem.Soc.* 1996, *118*, 1571; *Organometallics* 1996, *15*, 3256; *Angew.Chem.Int.Ed.Engl.* 1998, *37*, 94.

hydrophosphorylation → hydrophinylation

hydroprocessing comprises cat. hydrogenative procs. for conv. and upgrading of petroleum fractions; cf. the general term *hydrotreating or hydrorefining and the special procs. such as *HDA, *HDO, *HDN, or *HDS. B. CORNILS

hydroquinone important intermediate, manuf. by various cat. procs. such as *Brichima, *BASF, *Mitsui, Rhône-Poulenc, *Ube.

hydrorefining is a low-*severity *hydrotreatment proc.

hydrosilylation (hydrosilation) *addition of organic or inorganic Si hydrides to multiple C-C, C-O, or C-N bonds yielding organosilicon cpds. It is also a unique and efficient method for selective *reduction of the C-heteroatom (O,N) bond, including *asymmetric synth. Although the reaction can occur by a free *radical mechanism generated in the reaction mixture, most cats. (*tert.* amines, *Lewis acids, *supported metals, *transition metal *complexes of Pt, Pd, Rh, and predominantly Ni. *Ziegler systems, metal carbonyls) accomplish the proc. through a *heterolytic mechanism. Chloroplatinic acid (*Speier cat.) is the best *precursor for hom. cats. for all h. procs. Cats. of alkene h. involve activation of the *precursor, coordination of the unsaturated cpd. competing with oxidative addition of the hydrosilane to the metal center, *cis*-*insertion of the *ligand (π/σ-rearrangement) into the M-H bond, and *reductive elimination of the prods. Acc. to newer postulations alkene insertion into an M-Si bond is followed by C-H *reductive elimination. Under some conds. the dehydrogenative silylation of alkenes by hydrosilanes especially with Fe and Co triad complexes, becomes a separate method for the direct synth. of alkenylsilanes.

The cat. addition of substituted silanes to alkenes occurs predominantly acc. the *anti-*Markovnikov rule, resulting in alkylsilanes with terminal silyl groups, but under some conds. (e.g., with cats. such as Pd, Ni complexes) this prod. is accompanied by an α-adduct with an *internal* silyl group. Organosilicon derivatives from the h. of a carbonyl group or a C-N bond (imines, oximes, isocyanates) are *intermediates which usually lead,

when reacted via *solvolysis, to the desired organic prods. (complexes of Rh, e.g., the *Wilkinson cat., Ru, and Pt, are cats. for selective and *asymmetric h. of carbonyl cpds.). H. of unsaturated organosilicon cpds. has also found several applications in molecular and polymer organosilicon chemistry (e.g., addition of polyfunctional Si hydride to poly(vinyl)organosiloxane with *Karstedt cat., cure for silicon rubber, etc.). **E=F**; **G** Hydrosilylierung. B. MARCINIEC

Ref.: B. Marciniec (Ed.), *Comprehensive Handbook on Hydrosilylation*, Pergamon, Oxford 1992; Patai-1, p. 1479.

hydrothermal synthesis prepn. method for *zeolites, *molecular sieves, etc. by a sequence of water-based steps, supersaturation, nucleation, and crystal growth. Hydrothermal reactions also play a role in various steps of *decobaltation and *makeup of *hydroformylation cats. **F** synthèse hydrothermale; **G** Hydrothermalsynthese. B. CORNILS

Ref.: Ertel/Knözinger/Weitkamp, Vol. 1, p. 311; Kirk/Othmer-1, Vol. 13, p. 1014.

hydrotreating (hydroprocessing, hydro[re]-fining especially if coal-derived prods. are concerned; hydrofinishing for low-severity h.) is the collective term for reductive treatment of oil fractions and comprises the removal of S (*hydrodesulfurization, HDS), N (*hydronitrogenation, HDN), O (*hydrodeoxygenation, HDO), metals (*hydrodemetallization or hydrodemetallation, HDM), aromatics or alkyls (*hydrodealkylation, HDA, *hydrodearylaton, HYD), and halogens, mainly chlorine (*hydrodechlorination, HDC). *Hydrocracking also belongs to the h. procs. This important part of *petroleum processing purifies oil prods. to decrease air-polluting emissions such as sulfur or nitrogen oxides, to improve further processing of fractions by removal of halogens and metals, and to meet the increased quality demands of the market. Table 1 (average figs., legislation for the year 2000) exemplifies the discrepancies between legislative prod. specifications and typical

Table 1

Product	S(wt.%)	aromatics(vol.%)	benzenes(vol.%)	N(ppm)
Californian gasoline	0.003	22	0.8	No limit
European gasoline	0.01	No limit	1	No limit
Diesel (worldwide)	0.05	10	No limit	No limit
Gas oil	1.5	25–40	Included	100
FCC oil	0.7	60–90	Included	1000

prod. streams from conventional *petroleum processing without h. The problems of N and metal contents are severe in the heavy residues from the initial distillations which are used as feeds for *catalytic cracking. H. is thus mandatory in order to sustain the *lifetime of the downstream cat. procs. for generating light oils.

The relevant procs. are conducted at temps. of about 573 to 723 K and at press. between 3 and 10 MPa hydrogen. As only low *liquid hourly space velocities of about 0.2 can be realized the relevant instrumentation is for large volumes and requires very large loads of cat. Typical cat. *shapes are cylinders, rings or trilobe extrudates. They have to be of high *mechanical strength. The large cat. volumes needed require a cheap manuf. and the application of non-precious metal comps. For this reason the function of the Co/Ni-MoS system has to optimized to the maximum level. Typical parameters for the *support are *surface areas of 100 m^2/g and typical loadings are 1–4 % for Co and Ni and 8–16 % for Mo. The oxidic cat. *precursors need to be *presulfided before use. To this end H is loaded in the start-up phase with alkyl sulfides and/or with H$_2$S. The proc. of sulfidation is extremely critical to the later function of the cat. and many proprietary knowhow deals with the de-

Table 2

Acronym	Process	Feed	Target
HDS	Hydrodesulfurization	Diesel	Environmental limits
		Gasolines (straight run)	Deodorization
		Distillates	Environmental limits
		FCC/HCR feed	Avoid cat. poisoning
HDC	Hydrodechlorination	All feeds	Remove chlorine
HCR	Hydrocracking	FCC/HCR feed	Conversion to light fractions
HDA	Hydrodealkylation	Diesel	Dewaxing, Remove alkenes, stabilization
HDN	Hydrodenitrogenation	FCC/HCR feed	Avoid cat. poisoning; environmental limits
HDM	Hydrodemetallization	All heavy fuels	Avoid metal deposits
HDO	Hydrodeoxygenation	Only for coal-derived prods.	
HYD	Hydrodearomatization	Diesel	Reduce aromatics
		Kerosene	Reduce aromatics
		FCC	feed Remove polyaromatics
		Cracked feeds	Stabilization

tails of this step in the cat. *activation (cf. also *sulfided cats.). Table 2 lists the characteristic procs. of h. and selected petrochemical targets for each reaction.

The result of all these h. procs. is removal of heteroatomic species, reduction of the content in aromatic molecules, and under severe conds. reduction in chain length and branching of the alkane fractions. These modifications are needed to meet the modern requirements of gasoline, to protect the automotive cat. and the environment.

H. cats. contain Co, Ni, or W sulfides as *bulk or as *supported cats. Mixtures are also usual (e.g., Co-Mo or Ni-Mo on *alumina, in which case Co or Ni are considered to act as *promoters of Mo's activity; cf. *Mo as cat. metal). Bulk h. cats. are mainly used for the *hydrogenation of aromatics or of coal (*coal liquefaction, *coal hydrogenation). A low-*severity h. is sometimes called a *hydrofinishing (hydrofining) proc.

Except for special *HDS, *HDM, or other procs., typical hydrotreating procs. are described inter alia by *Atlantic Richfield Pentafining, *BP (Autofining, Ferrofining procs.), *Chevron RDS and VGO, *Criterion Lummus Syncat technology, *Exxon GO fining and hydrofining; *Gulf Gulfinishino or Gulfining procs., *Lummus ABB hydrotreating, LC-Fining, or PGO hydrotreating procs., *Howe-Baker, *IFP Hyval, *Standard Oil Ultrafining, *Texaco T-Star, *Union Oil Unionfining, *UOP Hydrobon or Unibon, etc. H. procs. operate in diverse technological variants. The details of *reaction mechanisms and of the mode of operation of the cats. are still subject to scientific debate. A rich selection of models and scenarios can be found in the literature. Cf. also *hydrocracking. **E=F=G.** R. SCHLÖGL

Ref.: Anderson/Boudart, Vol. 11; B.C. Gates, J.R. Katzer, G.C.A. Schuit, *Chemistry of Catalytic Processes*, McGraw-Hill, New York 1979; D.D. Eley, P.H. Selwood, P.B. Weisz (Eds.), *Advances in Catalysis*, Vol. 27, Academic Press, New York 1978; Topsøe/Clausen/Massoth; Thomas/Thomas, p. 636; Farrauto/Bartholomew, p. 519; Trimm/Akashah et al; Ertl/Knözinger/Weitkamp, Vol. 4, p. 1908.

hydrovinylation Acc. to *hydroformylation (and the *addition of H/HCO to double bonds) the term h. was coined to describe the addition of H/CH=CH$_2$ to the neighboring C-atoms of a second alkene molecule.

The most active h. cats. contain Ni or Pd, but Ru, Rh, or Co have been described, too. Accompanying side reactions such as *isomerizations can be suppressed by adding suitable *activators such as BF$_3$, SbF$_5$, or ClO$_4^-$ or P-donor *ligands. Hydrovinylating styrene to 3-phenyl-1-butene, the activity of ligand-modified PdL$_x$BF$_4$ cats. is found to increase in the order P(i-Pr)$_3$ < P(O-menthyl)$_2$Ph < PPh$_2$(O-menthyl) < P(O-menthyl)$_3$ < PPh$_3$ < P(OPh)$_3$. In the presence of *bidentate ligands, either the conv. is low or no reaction occurs.

Acc. to the Figure, in h. a new asymmetric center is generated, thus yielding up to 95 % of the R,R-isomer in the case of styrene. In this h. with P(menthyl)$_2$ (i-Pr) modified cats. the optical activity decreased in the order SbF$_6^-$ > PF$_6^-$ > Et$_3$Al$_2$Cl$_3$ ~ BF$_4^-$ > CF$_3$SO$_3^-$ ~ ClO$_4^-$. **E=F**; **G** Hydrovinylierung. B. CORNILS

Ref.: Cornils/Herrmann-1, p. 1024.

hydroxyalkylation *condensation of aldehydes or ketones with aromatic rings or activated alkenes. H. is cat. by acids or bases depending on the substituents and heteroatoms of the aromatic moiety (Equ.).

The reaction of phenols with aldehydes is cat. by alkali hydroxides and yields *o*- and *p*-derivatives. The aldehydes used ought to have no α-hydrogen atom, to avoid *aldol reactions. The reaction of formaldehyde (*Baekeland-Lederer-Manasse or Lederer-Manasse reaction) is a special case (*hydroxymethylation). H_2SO_4, HCl, or strong *Lewis acids ($AlCl_3$, BF_3, $ZnCl_2$) are used as acid cats.; the resulting alcohol can react with another aromatic ring in a *Friedel-Crafts reaction yieding diarylalkanes. This can be avoided by using aldehydes having electron-withdrawing substituents in the α-position to the carbonyl moiety.

The reaction of alkenes (Morita-Baylis-Hilman reaction) is cat. by tertiary amines (e.g. *DABCO) or *phosphines as well as *transition metal hydrides. Attempts to achieve *enantioselectivity by using *chiral tertiary amines have been reported only with low *ees. **E=F**; **G** Hydroxy-alkylierung.   H. GEISSLER

Ref.: E. Ciganek, *Org. Reactions*, **1997**, *51*, 201; Olàh, Vol. 2, p. 597.

hydroxyamination (oxyamination, *aminohydroxylation) introduction of an amine and an alcohol in a single step to a carbon skeleton. The simplest and most important access to 1,2-aminoalcohol derivatives is the direct *addition of the heteroatoms to alkenes. Acc. to Sharpless OsO_4 and *tert*-butylamine react to generate the *tert*-butyl imido Os species, which reacts later with alkenes to yield 1,2-aminoalcohols in a *regio- and *stereoselective way (Equ.).

The reaction was later made catalytic by using chloramine-T (TsNClNa) or *N*-chloroargentocarbamates as N sources. Asymmetric h. was developed which utilized the optical active alkaloids *(DHQ)$_2$-PHAL and *(DHQD)$_2$-PHAL as *ligands. Styrene- type cpds. react with enantioselectivities of >99 % *ee*. **F=E**; **G** Hydroxyaminierung.   C. DÖBLER, M. BELLER

Ref.: *J.Am.Chem.Soc.* **1975**, *97*, 2305 and **1978**, *100*, 3596; *Angew.Chem.Int.Ed.Engl.* **1996**, *108*, 449 and **1997**, *109*, 2751.

2'-hydroxybenzalpyruvate aldolase cat. the synth. of 2'-hydroxybenzalpyruvate from pyruvate and 2-hydroxybenzaldehyde. Ref.: *J.Org.Chem.* **1993**, *268*, 9484.

hydroxycarbonylation   → hydrocarbonylation

hydroxycarboxylation   → hydrocarboxylation

hydroxyisocaproate dehydrogenases (HICAPDH) cat. oxidoreductions similar to those of LDH. D-HICAPDH from *Lactobacillus casei* reduces 2-oxoacids to the corresponding *R*-hydroxyacids and oxidizes solely the *R*-hydroxyacids to the ketoacids. Likewise, L-HICAPDH from *Lactobacillus confusus* yields *S*-α-hydroxyacids as reduction prods. and oxidizes the *S*-α-hydroxyacids preferentially. Both *enzymes use *NAD(H) as *cofactor and show no activity with NADP(H), and also reduce and oxidize a variety of similar substrates. Compared to *lactate dehydrogenases, the HICAPDHs are better able to utilize substrates with longer side chains. Comparing D- and L-HICAPDH, the D enzyme appears to have a somewhat broader substrate specificity and a greater synthetic value. **E=F=G**.   C.-H. WONG

Ref.: *Appl.Microb.Biotechnol* **1984**, *19*, 167; *Enzyme Microb. Technol.* **1992**, *14*, 28.

4-hydroxy-2-ketovalerate aldolase cat. the formation of 4-hydroxy-2-ketovalerate from pyruvate and acetaldehyde. **E=F=G**. Ref.: *J.Bacteriol.* **1993**, *175*, 377.   C.-H. WONG

hydroxylamine important intermediate manuf. either by NO reduction over supported Pt (Bayer) or Pd (Inventa AG) cats. acc. to $2\ NO + 3\ H_2 + H_2SO_4 \rightarrow (NH_3OH)_2SO_4$ (both cats. may be selectively poisoned to enhance selectivity), or by ammonium nitrate reduction in phosphorous acid solution with Pt or Pd on charcoal (cf. procs. of *BASF, *DSM, *Raschig). **E=F=G**.   B. CORNILS

Ref.: Büchner/Schliebs/Winter/Büchel, p. 52; Büchel/Moretto/Woditsch.

hydroxylase → monooxygenase

hydroxylation introduction of hydroxy groups into organic cpds. H. can be performed either by substitution of functional groups or hydrogen atoms (e.g., *oxygenation), by *addition of water or hydroxylating agents to the double bond of unsaturated systems such as alkenes, or by *isomerization (cf. *BASF-Wyandotte proc. with *PO to allylic alcohol). Alkanes with *tert*-C-H groups are converted highly selectively to the corresponding alcohols via stable *radicals as reaction *intermediates using O_3, $K_2Cr_2O_7$, $KMnO_4$ (basic conds.), or organic peracids as oxidizing reagent. Allylic alcohols are formed by *oxidation of alkenes with SeO_2 as selective oxidant. The formation of glycols (*vic*-diols) from alkenes is achieved by oxidation of simple or functionalized alkenes with $KMnO_4$ (Baeyer's test) under mild reaction conds. to avoid the formation of α-hydroxy ketones. Such oxidations can also be performed efficiently under water-free conds. in organic solvents if *crown ethers are applied to dissolve the $KMnO_4$. For better selectivities OsO_4 is used cat. to generate vicinal *cis*-glycols. OsO_4 is a highly selective and reactive oxidant. The Os(VIII) species is regenerated by oxidants like H_2O_2, *TBHP, organic N-oxides (R_3N^+-O^-), $K_3Fe(CN)_6$, or inorganic chlorate salts. *Enantioselective *dihydroxylations are achieved by applying OsO_4 as oxidation cat. in the presence of asymmetric *drop-in ligands such as quinine derivatives (cf. DHQ). To generate *vic,trans*-glycols OsO_4 or $KMnO_4$ cannot be applied. In this case *epoxidations are often used as the oxidation step followed by acid- or base-cat. ring opening of the *oxiranes. For this purpose either organic peracids (*Prilezhaev reaction) or epoxidation cats. such as Mo or W (acidic or basic conds.) are used. A highly selective alternative for the synth. of pure *vic,trans*-glycols is the application of $CH_3Re(=O)(O_2)_2 \cdot H_2O$ derived from CH_3ReO_3 and H_2O_2. The Re peroxo complex acts as a highly active epoxidation cat. The Re-coordinated water molecule shows signifi-cant acidic behavior and cleaves epoxides under very mild reaction conds. The rate of this cleavage can be controlled by temp. and basic co-ligands. With this system arenes also can be hydroxylated, but due to the high oxidation potential further oxidation to the corresponding quinones often occurs. Alternative h. of aromatic cpds. is a possible application of *Fenton's reagent. For anthraquinone hydroxylation, cf. the *Bohn-Schmidt reaction. **E=F**; **G** Hydroxylierung.   R.W. FISCHER

Ref.: Cornils/Herrmann-1, Vol. 2, p. 848; *Acc.Chem.Res.* **1997**, *30*, 169; Ertl/Knözinger/Weitkamp, Vol. 5, p. 2329.

hydroxymethylation Use of formaldehyde is a special case of *hydroxyalkylation (*Baekeland- Lederer-Manasse or Lederer-Manasse reaction), yielding mostly Bakelite-type condensates instead of the monohydroxymethyl aromatics obtained when reacting HCHO with phenols. **E=F**; **G** Hydroxymethylierung.   H. GEISSLER

hydroxymethylglutaryl-CoA reductase (inhibitor) → isopentenyl pyrophosphate

hydroxynitration the simultaneous oxidation and nitration of aromatics with nitric acid with Hg salts as cat. (*Wolffenstein-Böters reaction). **E=F**; **G** Hydroxynitrierung.
B. CORNILS

17-α-hydroxyprogesterone aldolase (EC 4.1.2.30) cat. the aldol reaction of 17-α-hydroxyprogesterone. **E=F=G**.   C.-H. WONG

Ref.: *Biochem.Pharm.* **1981**, *30*, 1827.

hydroxysteroid dehydrogenase (HSDH) cat. the *regiospecific *oxidation of a hydroxy group or reduction of a ketone functionality of a *steroid. The natural specificity of each particular *enzyme is designated by its name. For example, 7α- HSDH cat. the oxidation and reduction of steroids at position-7. The enzymes selective for positions 3, 7, 12, and 20 have been reported. Due to high regio- and *stereoselectivity, this group of enzymes shows great potential in the synth. of steroids, bile acids, and other steroid derivatives. Not only can these enzymes tolerate different side chains on the steroid skeleton, but some can cat. transformations of unrelated molecules. In particular, 3α-HSDH appears to be especially useful, oxidizing a range of aromatic *trans* diols with high enantioselectivity. Additionally, 3α,20β-HSDH also accepts non-steroid substrates. The coupling of different HSDH such as 3α-HSDH followed by 3β-HSDH or vice versa can invert the hydroxy center selectively at the 3-position. Inversion at C 7 has also been accomplished similarly.

Some steroid derivatives poorly soluble in aqueous solution have been used in biphasic systems (cf. *two-phase cat.) in reactions coupled with *cofactor regeneration. **E=F=G**.

C.-H. WONG

Ref.: *J.Org.Chem.* **1986**, *51*, 2902 and **1993**, *58*, 499.

Hygas process the cat. *methanation of *syngas in fixed-bed adiabatic reactors at 300–500 °C, developed by the Institute of Gas Technology. **F** procédé de Hygas; **G** Hygas-Verfahren. B. CORNILS

hypervalent compounds are those of the 13th-18th groups of the periodic table, for which the *Lewis structures demand the presence of more than eight valence electrons (inert gas electron configuration) at the *central atom. While such deviations from the octet rule are rare among period 2 elements (exceptions are polyborane *clusters), the phenomenon of hypervalence is abundant among period 3 and subsequent elements (e.g., $[AlF_6]^{3-}$, $[SiF_6]^{2-}$, PF_5, SF_6, SO_4^{2-}, IF_7, XeF_6).

In the case of period 2 elements, hypervalence is normally restricted to cpds. with two-electron multicenter bonds. In contrast, elements of higher periods may also employ *vacant d-orbitals to accommodate excess electrons. Recent quantum chemical calcns. provide subtle theoretical insight into hypervalent cpds. without making use of d-orbitals. **F** composés hypervalents; **G** hypervalente Verbindungen. M. WAGNER

Ref.: *Chem.Rev.* **1993**, *93*, 1371.

Hypro process → UOP procs.

Hysomer process → Shell procs.

hysteresis the delay between the prod. of a physical effect and the event producing it. This phenomenon is well known for magnetic materials since the degree of magnetization lags behind the applied magnetic field when this field is varied through a cycle of values. *Isotherms originating from physical *adsorption on *mesoporous solids display a h. loop between the lower branch obtained by progressive addition of adsorbate to the system and the upper branch obtained by progressive withdrawal of the adsorbate. The h. loop is due to the occurrence of *capillary condensation which lowers the vapor press. over a concave meniscus (i.e., angle of contact <90°). Different shapes of the h. loop make it possible to identify porous solids made up of spherical particles of fairly uniform size (type A, H1), solids composed of plate-like particles (type B, H3), and solids with ink-bottle pores consisting of a large *cavity with a small neck (type E, H2). Cf. also *catalyst surface. **E=F**; **G** Hysterese. M. MUHLER

Hytex process → Texaco procs.

Hytoray process → Toray procs.

HZSM → zeolites

I

i abbr. for *iso*, used for branched isomers or for isomers differing in ring position; important for *isomerization or *hydroformylation reactions

IBA *i*-butyraldehyde, isobutyraldehyde

i-bu, i-Bu *iso*-butyl

ICI AMV process manuf. of *ammonia from hydrocarbon feedstocks by a comb. of a two-step *reformer stage (700–800 and 900–950 °C/2,8–3,5 MPa), CO_2 removal, and *ammonia conv. at 380 °C/8 MPa. The AMV proc. (AMmonia-5) claims excellent energy efficiency together with simplcity and reduced capital cost. The variant for low capacities is called LCA (Leading Concept Ammonia). **F** procédé AMV d'ICI; **G** ICI AMV-Verfahren.   B. CORNILS
Ref.: *Chem.Eng.Progr.* **1983**, *79*(3), 62; *Appl.Catal.* **1994**, *115*, 193.

ICI catalysts ex *ICI Katalco*, now *Synetix, headquartered at Billingham (UK) and in operation since 1963, a division of ICI PLC and a producer of cats. and supplier of cat. procs. for the chemical and the petrochemical industry. The main field of activity covers cats. for the prod. of H_2, *methanol, and *ammonia. Typically ICI cat. types are described by a combination of numbers connected by a hyphen (e.g., 53–3 for a Cu-based methanol cat.).
                                              J. KULPE

ICI chloropropanoic acid process manuf. of *S*-2-chloropropanoic acid by *resolution from the *racemic acid with dihalogenase *enzymes.   B. CORNILS
Ref.: *Appl.Catal.* **1994**, *115*, 184.

ICI Katalco → Synetix catalysts

ICI LCA process → ICI AMV process

ICI methanol process In 1966 ICI introduced a new methanol synth. cat., based on Cu-Zn oxide/*alumina, with much higher activity than the $ZnO-Cr_2O_3$ based cat. used so far in high-press. operation. The increased activity permitted operation at 5–10 MPa and at temps. less than 300 °C. The low-press. proc. is more efficient, has lower capital costs, and is cheaper to operate than the early high-press. procs., which consequently became obsolete (*BASF proc.). However, the Cu cat. is sensitive to *poisons, especially to S, and therefore requires a carefully purified *syngas, whereas the old $ZnO-Cr_2O_3$ based cat. is virtually resistant to sulfur.
The cat. is prepared by continuous *precipitation at constant pH around 7. The aim is to achieve a high Cu *surface without generating γ-alumina with acidic surface properties (formation of dimethyl ether). Part of the ZnO-alumina moiety can also be added in the form of zinc *spinel, increasing the thermal stability of the cat. (refractory support). The properly reduced cat. is run with 200/240 °C inlet/outlet temp. at 10 MPa in reactors similar in construction to ammonia converters. **F** procédé d'ICI pour méthanol; **G** ICI Methanolverfahren.   C.D. FROHNING
Ref.: M. Twigg (Ed.), *Catalyst Handbook*, 2nd ed., Wolfe Publishing, London 1989; *Appl.Catal.* **1988**, *38*, 1. *J.Catal.* **1996**, *161*, 1; EP 217 513 (1986); *Appl.Catal.* **1988**, *36*, 1; *J.Catal.* **1987**, *103*, 79.

ICI protein process *biocat. manuf. of high-protein feeding stuff (trademark Pruteen) from methanol by *fermentation in *bioreactors. The proc. is run continuously consuming air and ammonia, mineral nutrients, and growth factors and withdrawing the circulating culture. The proc. (and the prod.) are now obsolete. **F** procédé d'ICI pour protéine; **G** ICI-Proteinverfahren. B. CORNILS

Ref.: *Hydrocarb.Proc.* **1977**, 56(11), 223.

ICI syngas generation *steam reforming proc. for the manuf. of *syngas from hcs. (up to naphtha) by three steps. 1: *hydrodesulfurization of *hcs. with CoO-MoO$_3$/Al$_2$O$_3$ cat. at 350–450 °C; 2: cat. *cracking in tube reactors (primary *reformer) at 700–800 °C/1.4–4 MPa on Ni-K$_2$O/Al$_2$O$_3$ cat.; 3: burning of parts of the gas over Ni cats., air, and steam in shaft furnaces (*secondary reformer*). No carbon black is formed, therefore no *regeneration of the cat. is necessary. **F** procédé d'ICI pour le gaz de synthèse; **G** ICI Synthesesgas-Verfahren.

B. CORNILS

Ref.: *Hydrocarb.Proc.* **1997**, (3), 108.

ICP-AES inductively coupled plasma-atomic emission spectroscopy

ICP-MS → inductively coupled plasma-mass spectrometry

ideal crystal structures The characteristic property of a crystalline solid is the transitional periodicity of its unit cell content. This requires complete identity of all unit cells and does not account for extra (or missing) atoms located in the unit cells or even in the voids of the structure. The periodic arrangement accounts for the long-range ordering and for all properties of a crystalline solid which are not confined to one structural motif, i.e., one hypothetical molecule of the solid material. These long-range properties such as conductivity of heat and electrons or the enhanced stability of a solid to disintegration by "cooperative force stabilization", are sensitive functions of the degree of perfection. The term ideal structure (is.) negates the existence of any defect and is hypothetical as the end of the crystal, i.e., its surface, is an unavoidable property of any solid which is not accounted for by the ideal structure. Thus the *surface properties of a solid are rarely described correctly in terms of its ideal (bulk) structure. As

the surface structure is very rarely known, the practice exists of intersecting the ideal structure and taking the resulting surface terminations as models for the description of cat. properties. This is only a very crude approximation, as is known from the few cases (element metal crystals, a few binary oxides, a few semiconductors) where the surface structures are known.

The proc. of intersecting the is. is referred to as hypothetical cleavage. Certain rules apply to the selection of allowed cleavage planes from the infinite number of lattice planes of the ideal structure which are, a priori, all possible cleavage planes. These rules say that the planes which intersect a minimum of chemical bonds in the structural motif are most probable. In addition, those planes which allow the preservation of electroneutrality, i.e., which contain not just one atomic species, are more likely as they allow easier minimization of the surface energy. In reality the solids rarely behave as rigid bodies within the ideal structural model but tend to change the near-surfacer structure to accommodate for the surface effect. This process is called *reconstruction* and often prevents the prediction of cleavage planes. R. SCHLÖGL

iditol dehydrogenase The NADH-dependent IDH from *Candida utilis* (also known as sorbitol or polyol dehydrogenase, EC 1.1.14) cat. the *reduction of fructose to sorbitol and the carbonyl group of other ketoses to the *S*-alcohol. The corresponding *R*-alcohol can be obtained by non-stereoselective reduction of a ketone with NaBH(OAc)$_3$ from which the *S*-epimer was selectively removed by *oxidation cat. by IDH. **E=F=G**.

Ref.: *J.Am.Chem.Soc.* **1989**, *111*, 9275. C.-H. WONG

Idol, James D. (born 1928), inventor, who devised in 1957 SOHIO's (Standard Oil of Ohio) proc. for the *ammonoxidation of propylene/air/ammonia mixtures over Bi-P cats. to acrylonitrile. B. CORNILS

IETS inelastic electron tunneling spectroscopy, → electron microscopy

IFP Alphabutol process   *dimerization of ethylene to 1-butene over a hom. *Ziegler-Natta type Ti cat., containing *co-cats. (AlEt₃). At partial convs. at 50–55 °C/1–2.5 MPa selectivities of 95 % are obtained. There is no cat. cycle, the cat. is destroyed by addition of amines (Figure). **F** procédé Alphabutol d'IFP; **G** IFP Alphabutol-Verfahren.

B. CORNILS

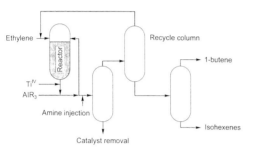

Ref.: *Hydrocarb.Proc.* **1984**, *63*, 118; *Information Chimie* **1987**, *281*, 149; *Hydrocarb.Process.* **1984**, *63*(11), 118.

IFP aromatization BTX process   cat. *aromatization of naphtha to *BTX in moving-bed operation, combined with continuous cat. *regeneration. **F** procédé d'IFP pour BTX; **G** IFP BTX-Verfahren.   B. CORNILS
Ref.: *Hydrocarb.Proc.* **1997**, (3), 114.

IFP butene isomerization process   liquid-phase *isomerization of 1-butene to 2-butenes in C₄ cuts. **F** procédé d'IFP pour butène isomérisation; **G** IFP Butenisomerisierungs-Verfahren.   B. CORNILS

IFP catalytic reforming process   Proc. for the upgrading of naphtha to high-octane reformate, *BTX, and *LPG in moving-bed reactors using two options, 1: through cat. *reforming at 1.2–2.5 MPa and with cyclic (intermittent) cat. *regeneration periods, or 2: with continuous cat. regeneration at 1.8 MPa (Octanizing proc.). The cats. are Pt-based, *bi- or multimetallic. **F** procédé "catalytic reforming" d'IFP; **G** IFP Catalytic Reforming-Verfahren.   B. CORNILS
Ref.: *Hydrocarb.Proc.* **1978**, (9), 161 and **1996**, (11), 98.

IFP Clauspol process   Claus tail-gas treatment proc. for the removal of S by mutual absorption of H₂S and SO₂ in a solvent which contains a dissolved cat. that promotes the *Claus reaction in the liquid phase, producing elementary S. **F** procédé Clauspol d'IFP; **G** IFP Clauspol-Verfahren.   B. CORNILS
Ref.: *Hydrocarb.Proc.* **1996**, (4), 110.

IFP Difasol process   conv. of IFP's *Dimersol-X proc. to a NAIL-based cat. (*non-aqueous ionic liquids) converting butenes to isooctenes.   B. CORNILS
Ref.: *Chem.Eng.News* **1998**, March 30, 32.

IFP Dimersol process   a hom. *dimerization and *oligomerization proc. which upgrades the light olefins from *steam cracking or *FCC procs. There are three variants. Dimersol-E and -G convert ethylene and propylene into gasoline with high octane numbers. Dimersol-X dimerizes butenes (or mixtures of propylene and butene) into lowly branched dimers (or *codimers) as feedstocks for the oxo reaction. The proc. takes place in the liquid phase at 40–60 °C without any solvent by means of soluble *Ziegler-Natta cat. systems based on Ni salts, activated by Al derivatives (Figure). 35 plants are in operation. Dimersol can be fitted with biphasic systems using *NAILs (* IFP Difasol proc.). **F** procédé Dimersol d'IFP; **G** IFP Dimersol-Verfahren.   H. OLIVIER-BOURBIGOU

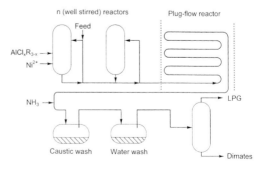

Ref.: *Hydrocarb.Proc.* **1977**, *56*(11), 170; Cornils/Herrmann-1, p. 258; *Appl.Catal.* **1994**, *115*, 178.

IFP DMD process → IFP isoprene (DMD) process

IFP Eluxyl-Octafining process manuf. of *p*-xylene from mixed xylenes originating from *steam reforming or *steam cracking. The proc. comprises an Eluxyl unit for a *p*-xylene separation proc. (based on countercurrent adsorption) and an Octafining unit for the cat. C_8 aromatics *isomerization. **F** procédé Eluxyl-Octafing d'IFP; **G** IFP Eluxyl-Octafining-Verfahren. B. CORNILS

Ref.: *Hydrocarb.Proc.* **1997**, (3), 145.

IFP ether process manuf. of ethers like *MTBE from appropriate alcohols and *tert*-alkenes over acid *ion exchange resins in fixed-bed, adiabatic (1st step) and expanded-bed (2nd step) reactors at ~70–85 °C/0.5–1.5 MPa. The distillation contains a cat. zone (*cat. distillation, *Catacol technology). The proc. for *TAME is commercialized jointly with Elf (Figure). **F** procédé d'IFP pour éthers; **G** IFP Ether-Verfahren. B. CORNILS

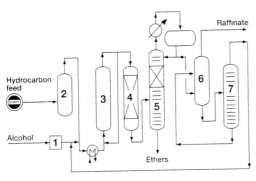

Raffinate

Hydrocarbon feed

Alcohol

Ethers

1,2 purification; 3,4 reaction section; 4 fixed-bed reactor; 5 Catacol column (*cat. distillation); 6 waterwash; 7 distillation.

Ref.: *Hydrocarb.Proc.* **1996**, (11), 114; *Appl.Catal.* **1994**, *115*, 178.

IFP HBL (Houillères du Bassin Lorraine) process Older proc. for the purification of benzene by *hydrogenation over suspended Raney Ni at 200 °C/1.5 MPa. **F** procédé HBL d'IFP; **G** IFP HBL-Verfahren. B. CORNILS

IFP HDS process for the improvement of distillates ranging from light gasoline to heavy vacuum gas oil by removal of S, N, and metals by *hydrodesulfurization at 340–400 °C. The feed passes a fixed-bed reactor with a suitable HDS cat. **F** procédé d'IFP pour HDS; **G** IFP HDS-Verfahren. B. CORNILS

Ref.: *Hydrocarb.Proc.* **1972**, (9), 152.

IFP hydrocracking process for the upgrading of vacuum gas oil to middle distillates etc. over a *dual cat. system consisting of a *hydrotreatment cat. (amorphous type) and a *hydrocracking cat. (*zeolite). For the earlier BASF/IFP proc., cf. *BASF procs. **F** procédé hydrocracking d'IFP; **G** IFP Hydrocracking-Verfahren. B. CORNILS

Ref.: *Hydrocarb.Proc.* **1996**, (11), 124, 126.

IFP Hyvahl hydrotreating process for upgrading of high-metal atmospheric or vacuum resids to cleaner fuel and/or cracker feeds by hydrogenative treatment over *dual cats.: a first-stage cat. for *demetallation and conv. and a second-stage cat. for *hydrodesulfurization and *hydrogenation (Figure). **F** procédé Hyvahl d'IFP; **G** IFP Hyvahl-Verfahren.

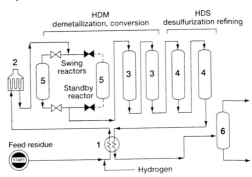

HDM demetallization, conversion | HDS desulfurization refining

Swing reactors
Standby reactor

Feed residue

Hydrogen

1 heat exchanger; 2 preheater; 3 demetallation reactors; 4 desulfurization/denitrification reactors; 5 guard bed; 6 distillation. B. CORNILS

Ref.: *Hydrocarb.Proc.* **1996**, (11), 130.

IFP isomerization process for the *isomerization of C_5/C_6 paraffin-rich *hc. streams to high RON and MON prods. for the gaso-

line pool in two variations (Ipsorb Isom and Hesorb Isom) over *zeolite or over *chlorinated *alumina. Both cats. can be regenerated. The Ipsorb part contains *molecular sieves for paraffin separation. **F** procédé isomérisation d'IFP; **G** IFP Isomerisierungs-Verfahren.   B. CORNILS
Ref.: *Hydrocarb.Proc.* **1996**, (11), 138.

IFP isoprene (DMD) process is an elder proc. for the manuf. of isoprene via 4,4-di-methyl-*m*-dioxane (DMD), which is produced by acid. cat. *addition of formaldehyde to *i*-butene at 100 °C (*Prins reaction) and its *dehydration over phosphoric acid, sulfuric acid, or cationic *ion exchange resins. **F** procédé d'IFP pour isoprène; **G** IFP Isopren-Verfahren.   B. CORNILS

IFP lube oil hydrofinishing for the improvement of color and oxidation stability, impurity reduction, neutralization, and for V.I. improvement in lube oils by cat. *hydrogenation in fixed-bed operation and a subsequent four-column distillation. **F** IFP hydrofinishing de l'huile de lubrification; **G** IFP-Schmierölhydrofinishing.   B. CORNILS
Ref.: *Hydrocarb.Proc.* **1972**, (9), 167.

IFP lube oil hydrotreating prod. of high quality lube oils and waxes by hydrogenative treatment (*hydrotreating) in a single stage operation using a selective cat. in order to maintain initial viscosity and high yields. **F** IFP hydrotreating de l'huile de lubrification; **G** IFP Schmierölhydrotreating.
Ref.: *Hydrocarb.Proc.* **1972**, (9), 168.   B. CORNILS

IFP Meta-4 process a *metathesis proc. for the conv. of 2-butene with ethylene to propylene. The proc., jointly developed by IFP and Chinese Petroleum Corp., uses a cat., continuous moving-bed *regeneration technology at 35° C/6 MPa (similar to that used in *reformers); the cat. is Re$_2$O$_7$ on *alumina. The Meta-4 proc. is advantageously combined with *MTBE plants. **F** procédé Meta-4 d'IFP; **G** IFP Meta-4-Verfahren.   B. CORNILS

Ref.: *Hydrocarb.Proc.* **1997**, (3), 156; Ertl/Knözinger/Weitkamp, Vol. 5, p. 2398.

IFP Octanizing process → IFP catalytic reforming process

IFP PDH process for the selective *hydrogenation of aromatics to cycloalkanes with soluble *Ziegler-type cats. **F** procédé PDH d'IFP; **G** IFP PDH-Verfahren.   B. CORNILS
Ref.: Parshall/Ittel, p. 180; Winnacker/Küchler, Vol. 5, p. 134.

IFP propylene process a combination of two IFP developments. 1: the manuf. of *MTBE from isobutene (obtained by *hydro-isomerization of crude C$_4$ streams), 2: the *metathesis of the remaining 2-butene and ethylene to polymer grade propylene (*IFP Meta-4 proc. (Figure). **F** procédé propylène d'IFP; **G** IFP Propylen-Verfahren.   B. CORNILS

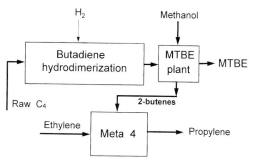

Ref.: *Hydrocarb.Proc.* **1997**, (3), 156.

IFP residue hydrodesulfurization proc.
→ BASF/IFP process

IFP terephthalic acid process manuf. of terephthalic acid (TPA) by liquid-phase *oxidation of *p*-xylene in acetic acid over Co/Br cats. at 180 °C/1 MPa. **F** procédé d'IFP pour l'acide terephthalique; **G** IFP Terephthalsäure-Verfahren.   B. CORNILS

IFP Selectopol process prod. of polymer alkenes as the base for additional gasolines if refinery olefins are available. The alkenes are cat. *oligomerized (over H$_3$PO$_4$). **F** procédé

Selectopol d'IFP; **G** IFP Selectopol-Verfahren.   B. CORNILS

Ref.: Ullmann, Vol. A18, p. 83.

IFP variant special variant of the *oxidation of cyclohexane to *adipic acid with addition of boric acid (*Bashkirov oxidation).

B. CORNILS

I.G. Farbenindustrie process of *coal hydrogenation, cf. German technology

IGOR process is a variant of the coal hydrogenation, cf. *German Technology

Imhausen process → Witten process

iminocyclitols are *inhibitors of glycosidases because they mimic the *transition state of glycosidase reactions. Two potent glycosidase inhibitors, deoxynojirimycin (DNJ) and deoxymannojirimycin (DMJ), were readily prepared in three steps with rabbit muscle fructose bisphosphate aldolase (RAMA) being used in the key C-C bond-forming step. DNJ and DMJ can also be prepared exclusively by utilizing the respective optically pure azidoaldehydes.

azidoacetaldehyde

1,4-dideoxy-1,4-imino-D-arabinitol

RAMA has also been used in the synth. of iminocyclitols corresponding to N-acetylglucosamine and N-acetylmannosamine, based on the analogous RAMA-cat. *aldol reaction/ *reductive amination procedure from (S)- and (R)-3-azido-2-acetamidopropanal, respectively. Synth. of the *precursor aldehydes necessary for the *aldolase reaction started from lipase-resolved (R)- and (S)-3-azido-2-hydroxypropanal diethyl acetal. When 2-azidoaldehydes are used as substrates for the RAMA-cat. aldol reaction/reductive amination, a number of polyhydroxylated pyrrolidines were synth.

6-Deoxyiminocyclitols and their analogs can also be prepared by direct *reductive amination of the aldol prods. prior to removal of the phosphate group. **E=F=G**.   C.-H. WONG

Ref.: Acc.Chem.Res. **1993**, 26, 182; Angew.Chem.Int. Ed.Engl. **1994**, 34, 412, 521.

immobilization stands in general for the permanent fixation (anchoring) of cat. active comps. on solid or *liquid *supports. In particular *enzyme i. is involved in *biocatalysis, and in *hom. cat. the vision of combining the positive aspects of a hom. cat. (with reference to high *activity, high *selectivity, the variability of steric and electronic properties; cf. Table under the entry *catalysis) with the advantages of a het. cat. (i.e., long lifetime, ease of separation). In this respect especially the possibility of a heterogenization of hom. cats. offers all the advantages (i.e., chemical reaction engineering simplifications). As a consequence of the *heterogenization* via anchoring, the cat. is immobilized ("hybrid") and thus no longer homogeneously distributed and no more "volatile".

In general six different methods have been applied in hom. cat. (*supported hom. cats.). 1: Anchoring of the cat. active species (i.e., soluble metal *complexes) via covalent bonds on suitable inorganic or organic supports, such as inorganic polymers, *silicons, oxides, *zeolites, polystyrenes, styrene-divinylbenzene copolymers, etc. The *polymerization or *copolymerization of suitably functionalized monomeric metal complexes is also known. 2: Chemical fixation via ionic bonding, especially for cationic or anionic metal complexes, using *ion exchange. Through *impregnation and *ion exchange cat. active metals may be incorporated in cages of, e.g., *zeolites. 3: Anchoring via *chemi- or *physisorption, *CVD or *MVS and entrapping in porous materials (e.g., metal carbonyls on activated carbon,

Al_2O_3, *zeolites). The *ship-in-the-bottle principle belongs to this concept, too. 4: With cat. active species that are dissolved in *ligands supported on porous solids *supported liquid-phase cats. (*SLPCs) are accessible (*SAPC). 5: Besides *solid* supports *liquid* supports are possible and used (*two-phase cat., *aqueous-phase cat., *fluorous phase cat.). In this case the cat. is also heterogenized and thus immobilized, but the cat. reaction takes place in the monophase and thus there are no underlying *mass transfer limitations. 6: Via molecularly defined *surface organometallic chemistry (SOMC) a transition from hom. to het. cat. is also possible.

Except from immobilized *metallocenes (which are not recycled but stay in the product) no heterogenized metalorganic cats. have been used on a commmercial scale so far. This is mainly due to *leaching of the active species from the support and to mass transport difficulties. Polymer-bound substrates can act to distinguish between hom. and het. catalysis. For the *biocatalytic entrapment cf. *enzyme immobilization. **F** immobilisation; **G** Immobilisierung.   B. CORNILS

Ref.: Stiles; Yermakov; Cornils/Herrmann-1, p. 605f.; Hartley; Pignolet; Marks; Clark; *Acc.Chem.Res.* **1992**, *25*, 57; *J.Am. Chem. Soc.* **1984**, *106*, 2569; Mark/Bikales/Overberger/Menges, Vol. 2, p. 708; Ertl/Knözinger/Weitkamp, Vol. 5, p. 2436.

IMP → ion microprobe

impregnated catalysts For a number of reasons *impregnation techniques may show advantages over *precipitation techniques in special applications of het. cat.; e.g., *supported precious metal cats. have good metal dispersions and high specific metal *surfaces; for high-temp. applications, ics. are stable and resistant *carriers for the active cat. phase. Additionally, ic. have advantages when a non-homogeneous distribution of the active phase is required (*egg-shell, *egg-yolk, cf. *metal distribution) or when soluble comps. which are *leached during precipitation (e.g., alkali metal cpds.) must be added. As far as *zeolites are concerned a metal content is easier

to introduce by *ion exchange rather than by incorporation during zeolite synth.

For impregnated cats. the conc. of the reactive ingredients in the cat. is principally restricted (*pore volume impregnation) and barely exceeds 25 wt%. The impregnation may be carried out by dipping the carrier into an excess of solution, leading to an equil. absorption of the solution constituents on the surface of the carrier pores. Depending on the conc. to be achieved, the procedure may be repeated several times. The absorbed material may be fixed on the carrier surface, if necessary, by subsequent impregnation with a precipitating agent, e.g., caustic. Uneven distribution of the acvtive phase can be advantageous. If the reaction is limited by pore diffusion effects, a *shell-type distribution of the active phase is desirable which may be generated by presoaking of the carrier with an inert solvent, thus limiting the access to the inner pore structure of the support. An *egg-yolk distribution results after leaching active cpds. from the outer sphere of an evenly impregnated carrier particle.

Interactions between the support material and the cat. active comps. are known for nearly all types of cats. **F** catalyseur impregné; **G** Tränkungskatalysator.   C.D. FROHNING

impregnation of a *support material by a solution (typically aqueous) containing an appropriate *precursor cpd. (e.g., a metal salt) is a convenient route toward prep. of precursors of supported cats. For large-scale manuf. the so-called *incipient wetness impregnation (also called *pore volume, or *dry or *capillary i.) is the most advantageous method. In this approach, the support is brought into contact with a solution the volume of which corresponds to the total pore volume of the solid and which contains the appropriate amount of precursor cpd. If cats. with high loadings of the active comp. are to be made, limited solubility of the precursor cpd. may cause problems, and multiple is. may have to be applied. With *incipient wetness i., even precursor cpds. which do not interact with the support surface can be deposited when the solvent is removed during a subsequent *drying procedure. In contrast, specific interactions are required between precursor cpd. and support surface during equil. adsorption from a large excess of solution (cf. also *coated cats., *grafting, *ion exchange, *metal distribution). **E=F**; **G** Imprägnierung. H. KNÖZINGER

Ref.: Ertl/Knözinger/Weitkamp, Vol. 1, pp. 191,243; Chem.Rev. **1995**, 95, 477; Ullmann, Vol. A5, p. 349.

imprinted nanostructural materials
porous solids that contain chemical functionalities that are organized into precise geometrical positions on a molecular-sized length scale by the use of an imprinting molecule (im.) during their prepn. The im. organizes precursors (binding monomers) to the solid materials via covalent bonds and/or non-covalent interactions after which these *precursors are reacted (e.g., polymerized) to form the solid product (cf. Figure).

From Inert Cross-linking Monomer
From Binding Monomer
Imprint

Removal of the imprint creates a functional, nanostructured material. Organic polymers and inorganic metal oxide solids have been prepared in this manner. Typical solids of this type contain a distribution of functional group positionings, indicating that the imprinting proc. remains imperfect. Cat. by these solids is emerging. **E=F=G**. M.E. DAVIS

Ref.: *ACS Symp.Ser.* **1986**, 308, 186; *Trends Polym.* **1994**, 2, 166; *Angew.Chem.Int.Ed.Engl.* **1995**, 34, 1812; *Bio/Technology* **1996**, 14, 163; *Chem.Mater.* **1996**, 8, 1820.

IMR inorganic membrane reactor, → membrane reactor

IMR-MS ion molecule reaction mass spectroscopy, → nuclear magnetic resonance

inactivators → active site-directed irreversible inhibitors, *enzyme inhibition

INCAP (Integration of Catalyst Activity Patterns) an *expert system which rates the applicability of cat. comps. for the desired reaction based on known activity patterns for different cat. properties. It was successfully applied for the selection of *promoter comps. for the *oxidative *dehydrogenation of *ethylbenzene to *styrene. **E=F=G**. M. BAERNS

Ref.: *Chem.Lett.* **1988**, 1269; *Appl. Catal.* **1989**, 48, 107 and **1989**, 50, L11.

INCAP-MUSE (INCAP for MUlti-Component catalyst SElection) an improved version of INCAP and selects as many cat. comps. as necessary until all required cat. properties are present. M. BAERNS

Ref.: *Chem.Eng.Sci.* **1990**, 45, 2661; S. Yoshida, N. Takezawa, *Catalytic Science and Technology*, TCH, New York 1991, Vol. 1, p. 285; *Stud.Surf.Sci.Catal.* **1993**, 75, 489.

incipient wetness method (dry impregnation) is an alternative impregantion method and is successfully applied when conc. profiles of the active comp. are to be prepared within preshaped cat. *supports (beads, *pellets, extrudates). By careful choice of the *precursor cpd. and of the pH of the solution, and eventual addition of competitive co-adsorbates, various distributions of the active comp. such as hom. (uniform), egg-shell, egg-white, and egg-yolk type can be predetermined (cf. *metal distribution and also *pore volume impregnation, *capillary impregnation). **E=F=G**.

H. KNÖZINGER

Ref.: Ertl/Knözinger/Weitkamp, Vol. 1, pp. 191,243; Chem.Rev. **1995**, 95, 477.

incommensurate structures When two solids form an intimate interface it is required that the structural motifs of both solids get into a regular arrangement with respect to each other. The more complicated these motifs are and the more different the chemical characters of the (atomic) heterointerfaces, the less likely it will be that such an arrangement is possible. In fortunate cases there exists a translational superlattice in which the structures of both cpds. coincide in translational symmetry after not one, but several, unit cell repeat. The number of unit cell repeats can be different in different directions of the interface. This case is called a *commensurate* superlattice structure. If the number of repeat distances becomes large (above 20) or if an infinite number of repeats would be required for a match between the two structures, then the two structures are *incommensurate* and form no interface with a translational interface. To minimize the interface energy such iss. tend to form *islands or to build three-dimensional microcrystals of one phase supported on the other phase. The interfaces can exist between two bulk solids or between a solid and an adsorbate. Iss. tend to be reactive and to form complex three-dimensional aggregates instead of flat and extended interfaces.   R. SCHLÖGL

incorporation of unnatural amino acids into proteins, → unnatural amino acids

indenyl (ind) This benzo-annelated *cp *ligand is one of the most frequently used substitutes of the cp ligand. I. ligands are generally more electron donating and less kinetically inert. Permethylated and substituted i. ligands are also known. The i. ligand has a higher propensity for "ring-slippage", or *haptotropic shift from pentahapto to trihapto bonding, compared to, e.g., unsubstituted cp. This so-called "indenyl effect" is promoted by the aromaticity gained by the benzene ring upon trihapto bonding of the five-membered moiety and was first observed in indCo(CO)$_2$.

This proc. has important consequences for reactions at the coordinated metal center, e.g., in assisting associative ligand *substitution reactions at saturated metal centers by an effect creating an extra coordination site by decreasing the i. *hapticity. This effect may also play a role during cat., including *polymerization. Linked i. ligands make up a substantial proportion of the ligands in use for *ansa*-metallocenes. **E=F=G**.   J. OKUDA

indicator ligands special (mostly sterically hindered) phosphines which signal, in mixtures with highly efficient *phosphites as *ligands in Rh-based *oxo procs., the exhaustion of the (more expensive) phosphites.
Ref.: US 5.741.943 (1998).   B. CORNILS

induced fit theory (in enzyme catalysis) The induced fit model of enzymatic cat. proposes that the binding of substrate to an *enzyme induces a conformational change in the enzyme (cf. *lock-and-key principle, *host-guest relationships, *molecular recognition). This conformational change may serve a variety of purposes: it may improve the enzyme's binding affinity for substrate (or more appropriately, for the *transition state); it may swing cat. residues into the proper *conformation for cat.; and it may serve to allosterically regulate the activities of other subunits or domains. A number of proteins have been demonstrated to behave in this manner. *Hexokinase, for example, undergoes an enormous rearrangement of its domains to fold over and bury the active site, bringing the substrates into close proximity and excluding water from the active site. **E=F=G**.   P.S. SEARS

inductively coupled plasma-mass spectrometry (ICP-MS) is used for elemental chemical and trace analysis (probe: atoms in solution; response: ions). The ex-situ technique (destructive to the sample) requires extensive chemical prepn. of solids. The data obtained are integral and bulk-sensitive.
                                         R. SCHLÖGL

industrial catalytic reactors → catalytic reactors

inelastic mean free path (IMFP, λ) plays an important role in *photoelectron spectroscopy (pes.). The IMFP is a measure of the mean distance traveled by the electron without energy loss, thus also containing information regarding the depth that can be effectively sampled by pes. The IMFP is also called the escape depth, and relates the intensity I_d of the photoelectrons obtained from a layer of material of thickness d, with I_∞, the photoelectron intensity from an infinitely thick layer, by the equ. $I_d = I_\infty(1-e^{-d/\lambda})$. By successively putting d = λ, d = 2, d = 3, it can be seen that this equ. implies that 63 % of the total signal intensity is due to electrons originating from a layer of thickness λ, 87 % from 2λ, and 95 % from 3λ. This means that most photoelectrons originate within a surface thickness of about 3λ. λ is a function of photoelectron kinetic energy (universal curve of electron mean free path) and depends on the material. Chosing different core levels of an element, data from different depths of a sample can be obtained by pes. Variation of the mean free path is also possible by variation of the photon energy (synchrotron radiation). **F** non-élastique longeur moyenne de chemin libre; **G** unelastische mittlere freie Weglänge.

R. SCHLÖGL, I. BÖTTGER

inert complexes complexes with kinetically stable ligands. **F** complexe inerte; **G** inerter Komplex. R.D. KOEHN

infrared spectroscopy (IR) The principle of IRs. for cat. is well known from conventional IR spectroscopy. For the study of cat. materials, thin, self-supporting wavers are usually used to obtain IR spectra in the transmission mode. Recently, IRs. in diffuse reflectance (*DRIFTS) has attracted increasing interest. Transmission IRs. in principle is a bulk rather than a *surface-sensitive method. It is determined by the optical properties of the solid powder. A weak bulk absorption and an average particle size d smaller than the wavelength of the IR radiation is necessary for optimum IR transmission. The particle size condition d $<< \lambda_{IR}$ usually leads to strong scattering in the near-IR regime, while strong bulk absorption occurs below 1100 cm^{-1} for most oxidic cats. Thus, the optimum frequency window for transmission IRs. of cat. materials is the mid-IR regime between 5000 and 1000 cm^{-1}.

Because IRs. in principle is a bulk-sensitive method, it is necessary to prove that the detected vibrational bands belong to surface species. This can be done for example by adsorbing probe molecules like NH_3 or CO and following the characteristic band shifts upon adsorption. The extinction coefficient (coeff.) of the surface groups or adsorbates determines the sensitivity of the technique. Thus, the extinction coeffs. may vary from about 5×10^{-18} cm^2 molecule^{-1} for the stretching mode of CO to about 10^{-20} cm^2 molecule^{-1} for CH stretching vibrations of saturated hcs. Exact absorption coeffs. of adsorbed probe molecules, however, are unknown. Only estimates from gas-phase absorption coeffs. can be used under the assumption that they are unchanged by the interaction with the surface. The small values of the extinction coeffs. requires a high surface-to-volume ratio for spectra with good signal-to-noise ratios. A rough estimate of the lower detection limit of adsorbed probe molecules of about 0.02 surface coverage can be calcd. from typical cat. *surface areas of 100 m^{2}g^{-1}, a sample weight of ca. 20 mg cm^{-2}, an absorption coeff. of 10^{-19} cm^2 molecule^{-1}, and 5 % absorption. This simple calcn. shows that the IR technique is able to detect even one-tenth of a monolayer of adsorbed probe molecules.

Typically, surface OH groups of oxidic cats. and their interaction with usually basic probe molecules like NH_3, pyridine, CO, or even N_2, are studied. Surface OH groups, besides coordinatively unsaturated (*cus) metal centers, are one of the most important functional surface groups due to their acidic or basic properties. The OH group density and their basic or acidic properties, e.g., H bond donor or acceptor strength, can be determined by quantitative IRs. of the interaction with the above-mentioned probe molecules, provided the Lambert-Beer law is applicable.

OH stretching vibrations occur in the range between 3200 and 3800 cm^{-1}. The bending mode of adsorbed water is located at 1600–1650 cm^{-1}, whereas the bending mode of M-OH groups is found at about 870 cm^{-1}. The combination of the two deformation modes $\delta_2 + \delta_3$ of water is detected at about 5200 cm^{-1}, as a function of the degree of *hydration. The combination of the stretching and the bending mode $\nu_{OH} + \delta_{OH}$ of OH groups occurs at 4450 cm^{-1}.

OH stretching frequencies in the range between 3200 and 3800 cm^{-1} can be assigned to different OH surface species, terminal OH groups, hydroxyls bridging between two or three metal centers, and OH groups interacting via H-bridges. Coordination of OH to metal centers generally shifts the stretching frequency to higher energies relative to the gas-phase frequency of 3735.2 cm^{-1}. Terminal OH groups adsorb in a small frequency range at about 3750 cm^{-1}. The stretching frequency shifts to lower wavenumbers with increasing coordination of the OH group to metal centers. The magnitude of the shift depends on the local electric field polarizing the OH group. Thus, differently coordinated bridging hydroxyls adsorb in the frequency regime between 3745 to 3590 cm^{-1}. A further shift to lower energies of the stretching vibration results from the polarization of the OH groups by H bridging to neighboring OH groups and leads to a broad band between 3650 and 3200 cm^{-1}.

IRs. of adsorbed CO is a pure surface-sensitive method and allows the characterization of *cus metal centers. Thus, for example CO adsorbed at Mo^{5+} centers has an absorption at 2205 cm^{-1}, while CO at Mo^{4+} centers exhibits a band at 2174 cm^{-1}, and CO at Mo^{2+} has a band at 2110 – 2100 or 2050 cm^{-1}, depending on their location at surface steps or kinks. CO interacting with acid OH groups leads to an absorption at about 2150 cm^{-1}, whereas physisorbed CO gives rise to a band at 2130 cm^{-1}. **F** spectroscopie IR; **G** IR-Spektroskopie.

G. MESTL

Ref.: Niemansverdriet, pp. 193,230; Ullmann, Vol. 5, p. 429.

inhibition → enzyme inhibition, *product inhibition, *substrate inhibition, *competitive inhibition.

inhibitor Acc. to Gates an i. is a substance that reduces the rate of a cat. reaction, often as a result of bonding chemically to the cat. (for the genesis, cf. *additives). Examples are Pt cats. on CaCO$_3$, poisoned by Pb (*Lindlar's cat., *directed poisoning), the action of acids in aprotic solutions (forming H bonds with cat. molecules in competition with other reactant molecules), the action of *scavengers in radical reactions, the action of large surpluses of *ligands in *ligand-modified Rh-cat. *hydroformylation or *Wilkinson-type *hydrogenation, or structural *effectors in *enzyme cat. which bond to the active sites in competition with the reactant.

Competitive inhibitors which slow down the reaction by competing with the reactants in bonding to the cat. are also known (e.g., *substrate is.). A strong i. is a cat. poison; e.g., sulfur for Ni *hydrogenation cats. which bonds with Ni and excludes the reactants from bonding with the cat. **F** inhibiteur; **G=E**. B. CORNILS

Ref.: Augustine, p. 303; Gates-1, pp. 3,431.

initial rate The ir. of an enzymatic reaction is dependent on the substrate conc., which changes as the reaction proceeds. Furthermore, as a significant conc. of prod. builds up, there may be a noticeable back-reaction, and so the net rate of conv. of substrate to prod. will be affected by the prod. conc. as well. For the purpose of extrapolating enzyme kinetic parameters, therefore, it is best to measure the initial rate of an enzymatic reaction, the reaction velocity (*rate of reaction) early in the reaction when very little of the substrate has been converted to product (and thus the substrate conc. is well known and the prod. conc. is minimal).

P.S. SEARS

initiator 1: The discovery that the cat. properties of pure cpds. may be excessively influenced by *additives, which by themselves were inactive in the reaction under consideration, was the starting point for the enormous development which cat. procs. have undergone. However, some confusion about the "real" influence of *promoters (cf. also *activator) has also been generated, and far from all of their effects have been explained. Several classes of activators may be discerned.

Structural promoters influence the texture of the *surface of het. cats., thus generating preferred *chemisorption sites for desired molecules (*structure sensitive reactions; *ensemble effect), which leads to an increase in *selectivity; *electronic promoters* influence the electronic character of the active phase of the cat. and therefore the chemisorption of reactants; *textural promoters* stabilize the structure of the cat. by inhibition of crystallite growth and increase the long-term stability of the cat. under harsh reaction conds. Although some general rules have been identified (*acidity/basicity of the cat. surface; atomic distances in *alloys; deficiency/surplus of electronic charge), the choice of a suitable activator mainly depends on trial-and-error methods. The search for suitable substances increasing the performance of a cat. is accelerated by modern methods for *high-throughput parallel testing of candidates (*combinatorial cat.). **E=G**; **F** initiateur. C.D. FROHNING

2: cf. → *radical initiator

Innovene process → BP procs.

insertion 1-organometallic: Insertion reactions taking place at organometallic *complexes play an important role in metal mediated transformations of a broad variety of substrates. In the mechanistic description of transition metal cat. reactions like *hydroformylation, *hydrogenation, or the *Heck reaction, the insertion of

small molecules into an M-C bond usually follows an *oxidative addition step and leads to bond formation between two substrates. The resulting metal-bound fragment is released after *reductive elimination. *Alkene polymerization can also be described as an (infinite) sequence of insertions; cf. *insertion reaction (polymer). **F** insertion organometallique; **G** organometallische Insertion. W.R. THIEL

Ref.: *Angew.Chem.Int.Ed.Engl.* **1999**, *38*, 871.

2-polymer: Classical anionic, cationic, or free-*radical *polymerizations proceed by addition of monomers to active chain ends that are separated from covalently bound *initiator fragments by many monomeric units (*chain-end control). In the i. reaction, the new monomer molecule (e.g., ethylene) is inserted between the initiator complex (catalyst M) and the growing chain (P) via a coordinative bond:

$$M\text{-}CH_2\text{-}CH_2\text{-}P + \mathbf{H_2C\text{=}CH_2} \rightarrow$$
$$M\text{-}CH_2\text{-}CH_2\text{-}\mathbf{CH_2\text{-}CH_2}\text{-}P$$

I. plays a major role in the *Ziegler-Natta polymerization, *metathesis polymerizations, and probably also in the so-called group transfer polymerization. Monomers which undergo insertion reaction include olefinic double and triple bonds as well as *carbenes and cpds. which exhibit carbene-like structures (e.g., diazo, isonitriles, carbon monoxide, cf. *alternating copolymers). The major importance of the i. reaction is in the homo- and *copolymerization of mono- and diolefins. **F** réaction d'intercaler; **G** Einschubreaktion.

W. KAMINSKY, M. GOSMANN

Ref.: Mark/Bikales/Overberger/Menge, Vol. 4, p. 1 and Vol. 8, 147.

Insite catalysts Dow's *single-site cats. (*metallocene cats.) for the prod. of polyolefins.

in-situ catalysts Hom. *transition metal cats. mostly contain special *ligands, e.g., *phosphines. Sometimes, these cats. can be assembled in situ, i.e., a suitable metal cpd. and he ligand (both preferably commercially vailable, air stable, storable) are added separately to the reaction mixture. In solution hey combine to give the active cat. To prepare an in situ cat. does not require any additional synth. steps. A typical example is the preformation of the Rh cat. of *RCH/RP's hydroformylation proc. In some in-situ cats. heir exact nature and structure of the active pecies are unknown, e.g., in the case of the co-cat. *methylalumoxane in *metallocene hemistry. **F** catalyseur in-situ; **G** in-situ Katalysator.   H. BRUNNER

n-situ polymerization   generates polymer blends by polymerization of a monomer A in the presence of a smaller proportion of polymer B. In the most important industrial blends, polymer B is grafted, and *cross-linkble diene rubber and monomer A leads to a hermoplastic. The blends are thermodynamically immiscible. There are various cases hom. or het. blends, materials with microphases, etc.). The most important technical case of isp. leads to rubber-modified, impactmodified thermoplastics. A grafted or crossinked polymer B, e.g., a diene rubber, is disolved in monomer, e.g., styrene. The *polymerization leads to the thermoplastic. Usually a multiphase system is obtained, in which the rubber phase is predominant in the disperse phase. Depending on the portion, size, size listribution, and consistency of the disperse phase, impact-modified polymers with different characteristics are obtained. **F** polymérization in-situ; **G** in-situ Polymerisation.

W. KAMINSKY, M. GOSMANN

Ref.: Elias, Vol. 2, p. 624, 650.

n-situ reaction monitoring   techniques o examine a chemical reaction under the specific set of reaction conds. without disturbing he reaction system (i.e., press., temp., stirring, etc.). As opposed to het. cat. (*pressure gap, *material gap), hom. cat. provides various echniques for obtained data from pressurzed, corrosive, diluted, or concentrated reacion systems under real reaction conds. (e.g.,

high-press. *infrared spectroscopy, CIR = cylindrical internal reflectance; OF CIR = optical fiber CIR; *NMR, *CIDINP = chemically induced dynamic polarization; flah pyrolysis, etc.). Pressurized devices and reactors for hom. cat. investigations need windows such as sapphire, ZnS, CaF_2, or NaCl which withstand press. from 1000 MPa (sapphire) to 20 MPa (NaCl) and temp. from >250 °C (sapphire) to 100 °C (NaCl). The level of information that can be extracted from the real-time (and real-conds.) spectra is excellent

Most techniques for examination of surface states and het. catalytic *intermediates require ultra high vacuum (which is far from the real reaction conds.) and thus permit only indirect data. Additionally, the techniques are often destructive to the samples (e.g., *electron spectroscopy (sp.), *atom scattering, *CP-MAS, *calorimetry, *X-ray sp., *ion microprobe sp., *ion scattering sp., often *mass sp. variants, *molecular beam scattering, *porosimetry, *SIMS, *SNMS, *TDS, *TGA, *TPD, etc. The level of information from these spectra is not real-time and not real-conds. **F** in-situ moniteur de la réaction; **G** in-situ Reaktionskontrolle.   B. CORNILS

Ref.: W.R. Moser, D.W. Slocum, *Homogeneous Transition Metal Catalyzed Reactions*, ACS, Washington D.C., 1992.

in-situ studies   For many reasons in-situ studies of working cats. are essential but difficult to carry out in practice (*press. gap, *material gap, *in-situ reaction monitoring). Only when appropriate techniques required to obtain such information are available can one hope to fashion high-performance, new cats.

Determining the precise structure of an *active site on inorganic solid cats. is not easy. As compared to *hom. cat. or *biocat. the advent of synchrotron sources has radically changed the prospects for attacking in-situ problems in het. cat. The first combined use of X-ray absorption spectrocopy (XAS, which monitors directly and in quantitative detail the immediate environment of the active site) and X-ray diffraction (XRD, which probes the long-range structural integrity and crystallographic phase of the cat.) was reported in 1991. Since then many refinements have been registered. A specific example is *EXAFS measurements of the Ti-SiO$_2$ system for alkene *epoxidation (cf. also *titanium silicalite). The results prompt the design of a comparable active site which has been "engineered" in such a manner as to replace one of the Si atoms of the tripodally bound titanol group by Ge, thereby giving rise to enhanced cat. performance. A set-up for combined *XAS and *XRD has also been published. An even more striking illustration of the power of combined XAS-XRD is given in the Figure (Fourier transform of EXAFS taken during thermal activation), which tracks the emergence of a metal *cluster (Cu$_4$C$_2$Ru$_{12}$) *hydrogenation cat. (on the *mesoporous silica *support MCM-41) from mixed metal (Ru-Cu) carbonylates (Cu K-edge).

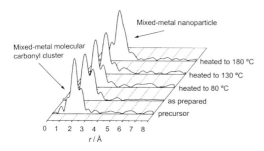

Mixed-metal nanoparticle

Mixed-metal molecular carbonyl cluster

heated to 180 ºC
heated to 130 ºC
heated to 80 ºC
as prepared
precursor

0 1 2 3 4 5 6 7 8
r / Å

Other techniques may also be adapted. Sum frequency generation (SFG) and *scanning transmission microscopy (STM) are another pair of techniques, and *Raman spectroscopy as well as *NMR in a variety of different modes, *ellipsometry, *Mössbauer spectroscopy, *neutron scattering, or positron emission tomography (PET) have also to be mentioned.
There is now available an exceptional method (that of judicious isotopic substitution) of determining with great precision the radial distribution around an atom of key cat. importance. It has yet to be fully exploited, but judging by the success in determining the *hydration properties of Ni^{2+} ions in solution, one

can expect significant advances in the unde: standing of any Ni-containing cats. **F** étude in-situ; **G** in-situ Studien. J.M. THOMA

Ref.: Thomas/Thomas; *Topics in Catal.* **1999**, , 23,35,45,81, 133; *Nature (London)* **1991**, *354*, 465 an **1995**, *378*, 159; *Science* **1964**, *265*, 1675; *Chem.Eur. **1997**, *3*, 1557 and **1998**, *4*, 1214; *J.Phys.Chem.B* **199** *102*, 1849; *Angew.Chem.Int.Ed.Engl.* **1999**, *38*, Dec. i sue; *Chem.Soc.Rev.* **1995**, *25*, 159.

inteins Some proteins are spliced posttran: lationally to remove polypeptide segment the inteins. The sequences flanking the intein are called exteins. I. excision is initiated whe a cysteine residue at the N-terminal end the intein attacks the peptide bond joinin the intein N-terminus with the C-terminus the first extein, forming a thioester. A cy steine residue at the N-terminus of the secon extein attacks the thioester, displacing the cy steine of the i. The reaction is complete when an asparagine residue at the C-terminu of the intein cyclizes to the aspartimide, re leasing the N-terminus of the second extei This amine attacks the thioester to form th peptide product. **E=F=G**.

P.S. SEAR

Ref.: *Chem.Biol.* **1997**, *4*, 187.

intercalation compounds molecule which occupy the spaces between the sheet of layered materials such as, e.g., *clays; c also *cavities. They are cpds. in which interac tions exist between the foreign and the gues atoms (cf. also *interstitial phases). Ics. ar not common in het. cat. as they tend to b very labile in their structure. Even in case where the guest species are bonded by *io exchange to the host species, such as in (*pi lared) clay systems, the stability of the com posite is a very limiting factor to its applica tion. R. SCHLÖG

Ref.: Gates-1, p. 300.

interceptor → scavenger

interesterification (in enzyme catalysis) refers to the reaction between an ester and a alcohol or acid, in which the alcohol or acid moiety of the ester is transferred to the added alcohol or acid. These reactions occur in organic solvents or at an oil-water interface. A *regioselective i. of a triglyceride in hexane with an *immobilized *lipase was used for the prepn. of cocoa butter-like fat from olive oil and steric acid. The oleic acid at the 1- and 3-positions of the lipid was replaced with steric acid, cat. by the lipase from *Rhizopus delemar* (*transesterification). **F** transestérification; **G** Umesterung.   C.-H. WONG

Ref.: *Eur.J.Appl.Microbiol.Biotechnol.* **1982**, *14*, 1.

intermediate   any chemical substance produced during the conv. of some reactant to a prod. Apart from substances that can be recovered as prods. if the reaction is finished at the point of generation of the i., unstable molecules are either only known or hypothesized to be intermediate, even if they have not yet been isolated. Chemical i. in gas-solid or liquid-solid het. catalyses are adsorbed on the cat. *surface. Absorbed molecules can be observed by various surface techniques such as *HREEL or *IR reflection adsorption spectroscopy. Nevertheless the species observed on a solid are not always the reaction is. *Isotope labeling experiments enable to distinction between reaction intermediates or stable adsorbed molecules on the surface. In the case of cat. transformation of alkenes it is generally believed that the reactive intermediates are represented by absorbed *carbenium ions, resulting from the protonation of double bonds in the initial molecule. Subsequent reactions (skeletal *isomerization, *hydride transfer, etc.) of this i. lead to the different reaction prods. Knowledge of intermediates in hom. cat. is much broader. Cf. *high-energy intermediates. **F** intermédiaire chimique; **F** chemisches Zwischenprodukt.

R. SCHLÖGL, M. DIETERLE

intermetallic compounds   cpds. of two or more metals which crystallize in a lattice very different from those of the constituting elements (intermetallic *alloys). The binding is predominantly metallic, but also contains ionic as well as covalent comps., resulting in both metallic and *semiconductor properties. Stoichiometric ics. possess common bond order numbers, whereas the more frequent non-stoichiometric ics. often show unconventional binding situations and can be either hom. or het. (mixed crystals). The stability of ics. is driven by the tendency toward maximal packing density and a maximal coordination number, while the metallic character increases with the coordination number. Ics. with high ionic or covalent character crystallize in lattices with low coordination numbers (e.g., the Zintl phases) and have pronounced semiconductor properties. Other important subgroups of ics. are the Laves phases (with AB_2 composition and hexagonal or cubic symmetry) and the Hume-Rothery phases. Metals of the same subgroup of the periodic table do not form ics. (*Tamman rule). Typical properties of ics. are their very high brittleness as well as the frequent occurence of crystal defects. Melting points of ic. are usually well above those of the pure comps., whereas their conductivity can range between that of pure *transition metals and semiconductors. Thus, ics. are interesting as semiconductors rather than as cats. (cf. *alloy cats.). **F** composés intermétalliques; **G** Intermetallische Verbindungen.   M. ANTONIETTI

Ref.: G. Sauthoff, *Intermetallics*, VCH, Weinheim 1995.

internal mass transport   → intraparticle mass transport

Interox process   → Solvay/Laporte/Carbochimique process

interparticle mass and heat transfer
Gradients which persist throughout the gross confines of the reactor are termed interparticulate. In fixed-bed reactors (cf. *cat. reactors) for *het. cat. reactions, axial and radial gradients of concs., and gas-solid temps. may exist. The related *diffusion coefficients

(more precisely: dispersion coefficients; cf. *dispersion) govern the axial and radial diffusional mass transport in the reactor. However, they are generally not molecular in Nature, being determined in most cases by turbulence within the reactor and by variations of the gas velocity. Therefore, the gross transport coefficients describe the axial and radial dispersion of mass and heat as well as the heat transport coefficients at the reactor wall on the reaction and coolant side. For defining the interparticle gradients throughout the reactor, the following dimensionless parameters established from the continuity equs. are inportant: axial Peclet number Pe_a (axial convection/axial dispersion), radial Peclet number Pe_r (axial convection/radial dispersion), axial aspect (bed length/particle diameter), radial aspect number (bed radius/particle diameter), overall reactor aspect number (bed length/bed radius), dimensionless time (real time/contact time), and wall Biot number (temp. in core of bed/temp. gradient in wall film) as well as the *Damköhler number DaI, Arrhenius number, and the dimensionless adiabatic temp. change β.

Useful criteria for the absence of mass and heat transfer effects are given in the literature. Cf. also *mass transport in cat., *Bodenstein number, *Damköhler number. F interparticle transfert de matière et de chaleur; G Interpartikel-Stoff- und -Wärmetransport.

D. HÖNICKE, P. CLAUS

Ref.: L.K. Doraiswamy, M.M. Sharma, *Heterogeneous Reactions: Analysis, Examples, and Reactor Design*, Vol. 1: Gas-Solid and Solid-Solid Reactions, Wiley, 1984, p. 178; Wijngaarden/Kronberg/Westerterp, p. 141; Thomas/Thomas, p. 425.

interphase mass transport → mass transport in catalysis

interrupted Pauson-Khand reaction → Pauson-Khand reaction

interstitial phases Solids contain regular voids in their structure. These can be the octahedral and tetrahedral sites in close-packed structures or they can be much larger and represent two-dimensional infinite planes in *van der Waals bonded crystals of layered cpds. The properties of such solids can be altered drastically when additional foreign atoms (typically non-metal atoms in metal crystals or alkali atoms in layered cpds.) are intercalated in these voids. I. metal atoms in metal hosts are designated as *alloys. In a strict definition i. cpds. are characterized by structures in which the foreign atoms are exclusively surrounded from matrix atoms whereas cpds. in which interactions exist between the foreign or guest atoms are referred to as intercalation cpds.

I. hydrides, nitrides, carbides, or oxides are very common as impurity phases in cat. active metals. In the Fe cat. for *ammonia synth., N atoms are incorporated in the octahedral voids of the metal structure. *Hydrogenation cats. often contain dissolved C atoms (sometimes called *carbides*) which distort the electronic structure of the noble metal and hence modify positively the *selectivity of the cat. reaction. In most cases the existence and function of interstitial cpds. in solid cats. is poorly documented or even unknown. Cf. also *intercalation cpds.   R. SCHLÖGL

intramolecular catalysis refers to cat. of a reaction within a molecule by groups (e.g., base, acid, etc.) within the same molecule. The constrained proximity of the cat. group(s) increases the speed of the reaction over what would be observed by inclusion of those cat. groups in the *solvent (due to a reduced entropy of activation). F catalyse intramoléculaire; G intramolekulare Katalyse.   P.S. SEARS

intraparticle (internal) mass transport If diffusion of an educt into a porous cat. particle is relatively slow so that the chemical reaction proceeds only at the outer region of the cat. *surface, i.e., before the educt has diffused far into the particle, the intraparticle pore resistance for *mass transport is strong and reduces the overall reaction rate. In such a scenario, an nth-order reaction behaves like a reaction of order $(n + 1)/2$, and the observed *activation energy corresponds to approximately one-half of the true activation energy. For a single, irreversible reaction which is described by a power-rate law, an experimental diagnostic criterion for the absence of imt. can be applied via the *Thiele modulus Φ. **F** transport de matière intraparticulaire; **G** Intrapartikel-Stofftransport.

P. CLAUS, D. HÖNICKE

Ref.: Thomas/Thomas, p. 418; Gates-1, p. 224.

intrazeolites   → ship-in-the-bottle cats.

intrinsic binding energy   The ibe. of a small molecule relative to an *enzyme is the maximum binding energy achievable, given perfect complementarity of the enzyme and the substrate. The interaction of certain *aminoacyl-tRNA synthetases, such as tyrosyl-tRNA synthetase, with their cognate amino acids are thought to tend to this value, due to their need to discriminate between closely related amino acids (Tyr vs. Phe). **F** énergie de liaison intrinsèque; **G** intrinsische Bindungsenergie.

Ref.: *Biochemistry* **1980**, *19*, 5520.   P.S. SEARS

intron   In *eukaryotes, the *DNA sequence encoding a single protein may not be contiguous in the chromosome: it may be separated by sequences that are transcribed into *RNA, but are removed prior to translation. The noncoding segments are called introns, while the segments of DNA that encode proteins are called exons. Is. must be spliced out at the mRNA level before translation. Some of these is. are actually self-splicing; no protein cat. is required. Cf. also *ribozymes. **E=F=G**.

P.S. SEARS

Inventa process to hydroxylamine   → hydroxylamine

inverse phase transfer catalysis   one type of phase transfer catalysis (*PTC). PTCs. such as *crown ethers or quaternary ammonium cpds. transport anions in a *biphasic

system from the aqueous into the organic *solvent *phase, whereas the iPTCs. transport hydrophobic organic molecules from the organic into the aqueous phase. The iPTCs accelerate the biphasic reactions of hydrophobic molecules with hydrophilic reagents or metal *complexes which cannot be transported by the normal PTC. Typical iPTCs. are 4-dimethylaminopyridine, pyridine oxide, *cyclodextrins, and hydrophilic *transition metal complexes. The pyridine derivatives carry carboxylic acid halides from the organic into the aqueous phase by forming the pyridinium salts. This inverse phase transfer is limited by the functional groups of the hydrophobic molecules. The cyclodextrins carry the hydrophobic molecules by forming the inclusion cpds. with their lipophilic *cavities. There is no restriction as to functional groups but there is a *steric* limitation due to the cavity size. The cyclodextrins can accelerate biphasic reactions using water soluble transition metal cats. These metal complexes transport the lipophilic substrates/metal center cpd. without steric limitation. The hydrophilic complexes differ essentially from the other iPTCs. in that these have the functions of not only the phase transfer cat. but also the transition metal cat. (*counter-phase transfer catalysts). **E=F=G**.

T. OKANO

Ref.: Cornils/Herrmann-2, p. 221; *J.Am.Chem.Soc.* **1986**, *108*, 1093; *Chem.Lett.* **1986**, 1463.

inverse substrates   In the kinetically controlled synth. of polypeptides from peptide esters, one factor that has restricted the choice of both ligation site and protease cat. has been the substrate *specificity of the protease. A protease with very broad specificity can ligate at many sites, but since *hydrolysis of the substrates and prods. is a competing side reaction in kinetically controlled peptide ligations, broadly specific proteases may also hydrolyze the peptide substrates (and products) at various positions to give multiple side products. Conversely, a highly specific protease such as trypsin will not hydrolyze the peptides at many sites, but there may be few or no sites within the target polypeptide for ligation. One way around this problem is to use a very specific protease with an "inverse substrate," a substrate in which the leaving group is able to bend around and satisfy the requirements of the S_1 site (S_1–S_4 are the acyl donor residue-binding subsites of the enzyme active site, while S_1'-S_4' are the leaving group subsites). Iss. for trypsin, which prefers basic groups in the S_1 site, have p-hydroxybenzamidine leaving groups. Because the leaving group satisfies the specificity of the S_1 site, trypsin is acylated by such substrates even in the absence of a basic S_1 residue. **F** substrates inverses; **G** inverse Substrate.   P.S. SEARS

Ref.: *J.Am.Chem.Soc.* **1977**, *99*, 4485; *Angew.Chem. Int.Ed.Engl.* **1997**, *36*, 2473.

in-vitro evolution   → directed evolution

in-vitro evolution of biocatalysts Although many different *enzymes can be used to cat. reactions in the field of organic chemistry, detergents, and procs. of ecological interest, the naturally occuring form (wild-type) often shows insufficient *activity, *stability, and/or *selectivity. In-vitro evolution (directed e.) is a new type of protein engineering acc. to which superior enzymes can be created on the basis of random mutagenesis in combination with a proper gene expression system and an efficient method for selection or screening. Accordingly, the gene encoding for the wild-type enzyme is subjected to mutagenesis by the error-prone *polymerase chain reaction or by other methods of molecular biology such as *DNA shuffling. Following expression in suitable microbial hosts, *combinatorial *libraries of mutant enzymes are created which are then selected or screened for the desired cat. properties. The optimal enzyme is identified, and mutagenesis is performed once more on the corresponding gene, a proc. that is repeated as often as necessary. This strategy of evolution in the test tube can be used not only to increase activity and selectivity, but also to create highly *enantioselective enzymes for use in synth. or-

ganic chemistry without any knowledge of the three-dimensional structure of the enzyme nor of the enzyme mechanism. **F** évolution de biocatalyseurs in vitro; **G** in vitro-Evolution von Biokatalysatoren.   M.T. REETZ

Ref.: *Acc.Chem.Res.* **1998**, *31*, 125; *Nature (London)* **1994**, *370*, 389; *Angew.Chem.Int.Ed.Engl.* **1997**, *36*, 2830; *Top.Curr.Chem.* **1999**, *200*, 31.

ion exchange   consists of replacing an ion A on a solid *surface by another ion B via electrostatic interactions, when the solid is brought into contact with a large excess volume (compared to the pore volume of the solid) of a solution containing ion B. Typical solid ies. are *zeolites and *clays, materials in which the framework bears excess negative charge which has to be compensated by exchangeable cations. In this case the number of exchangeable cations is predetermined by the chemical composition of the framework and independent of the solution pH. In contrast the overall surface charge of oxides can be positive or negative depending on the pH value of the contacting solution. The pH value at which the ζ-potential is zero (equal number density of positive and negative charges) is called the *isoelectric point. The pH dependence of the surface charge of oxides is related to the interface equilibria (where S symbolizes surface): S-OH $\rightleftharpoons$ SO$^-$ + H$^+$ and S-OH + H$^+$ $\rightleftharpoons$ S-OH, which describe the deprotonation and protonation of the surface. For the prep. of *precursors of *supported cats. it is therefore essential to choose cationic or anionic precursors, depending on the excess surface charge, which can be controlled by the solution pH. The solution pH also determines the saturation value of adsorbed ionic species under equil. conds. (cf. also *impregnation, *adsorption). Ion exchange in zeolites can also be achieved by solid-state ion exchange. Examples for cat. prepns. with ie. resins are the procs. of *BP/Erdölchemie Etherol, *Deutsche Texaco MIBK, *IFP isoprene, *Neste NEx ether, *Phillips etherification. **F** échange d'ions; **G** Ionenaustausch.

H. KNÖZINGER

Ref.: Ertl/Knözinger/Weitkamp, Vol. 1, p. 191; *J.Chem. Soc., Faraday Trans.* **1992**, *88*, 1345; Kirk/Othmer-1, Vol. 14, p. 737; Ullmann, Vol. A14, p. 393.

ionic catalysis   another term for *aqueous acid-base catalysis

ionic liquids   → non-aqueous ionic liquids

ion microprobe (IMP)   applied for surface elemental mapping (cf. *SIMS) and uses high-energy imaging ions as the probe (response: sample ions). The data obtained are local and *surface-sensitive. The extremely sensitive technique is ex situ and destructive to the sample.   R. SCHLÖGL

ion molecular reaction mass spectro-scopy   → mass spectroscopy

ionophores   → crown compounds

ion pair extraction   → phase transfer catalysis

ion scattering spectroscopy (ISS, LEIS) is based on the elastic scattering of a beam of ions. Probe: ions of noble gases; the responses are also ions. The data are integral and *surface-sensitive, and relate to the elemental analysis of the topmost atomic layer; the UHV technique is ex situ and destructive to the sample. The quantification is difficult.   R. SCHLÖGL
Ref.. Ullmann, Vol.B6, p. 71; Ertl/Knözinger/Weitkamp, Vol. 2, p. 620.

IPA   abbr. for *i*-propanol (isopropanol, isopropyl alcohol).

Ipatieff, Vladimir N. (1867–1952), Russian chemist (Petersburg, from 1931 in Chicago). Developed the *alkylation (and upgrading) of gasolines.   B. CORNILS

Ipsorb Isom   → IFP isomerization process

IR(S)   → infrared spectroscopy

iridium as catalyst metal   Besides its function as a companion of Pt, Pd, Rh, and other precious metals Ir plays its own role in *BP's Cativa process to *acetic acid (anhydride) and in some proposals for *metathesis reactions. The occupation with *hydroamination is without commercial consequence so far.   B. CORNILS
Ref.: Farrauto/Bartholomew, p. 413; *Platinum Met.Rev.* **1987**, *31*(1), 32; *Houben-Weyl/Methodicum Chimicum*, Vol. XIII/9b; *Coord. Chem.Rev.* **1982**, *35*, 113.

iron as catalyst metal   Fe shows valency changes between –2 and +3 and thus plays an important role in the essential biochemical *redox systems and *enzymes (cf. *catalase, *catechol dioxygenase, *cytochrome P-450, *haloperoxidase, *heme, *lipoxygenase, *nitrogenase, *peroxidases, *reductases, and others). In these reactions Fe is often bonded to sulfur in mononuclear form or as an Fe-S *cluster. Iron is the basis for some economical large-scale cat. procs. such as the *ammonia synth. or the *Fischer-Tropsch synth. (cf. *ARGE synth., *Kölbel-Engelhardt synth., millscale cat./*Kellogg Synthol, *Oxyl synth.). The cats. are bulk or *supported. Iron oxides serve as *coal hydrogenation (*red mud) or as *dehydrogenation cats. (cf. *Phillips isoprene proc., *UOP Smart proc.); mixed metal cat. with Fe are used for the *water-gas shift reaction (Fe-Cr), manuf. *formaldehyde (Fe-Mo), or *ammoxidation (Fe-Mo). In hom. cat. Fe is involved in *Reppe reactions (*BASF BDO or hydroquinone, as carbonyls), *dimerizations and *cyclizations (cf. *BINOL) or with $FeCl_3$ in *Diels-Alder reactions, *chlorinations (e.g., *Dow benzene chlorination), *Friedel-Crafts reactions, etc. *Fenton's reagent serves for *oxidations and *hydroxylations, some Fe chelates in *Wheelabrator's proc. for the H_2S removal. **F** le fer comme métal catalytique; **G** Eisen als Katalysatormetall.   B. CORNILS
Ref.: *Appl.Catal.* **1986**, *25*, 313; G.C. Ferreira et al. (Eds.), *Iron Metabolism*, Wiley-VCH, Weinheim 1999; *Houben-Weyl/Methodicum Chimicum*, Vol. XIII/9a; *Chem.Rev.* **1996**, *96*, 2335; *J.Mol.Catal.A* **1999**, *148*, 9.

iron sulfur clusters   *enzyme prosthetic groups used in *redox reactions; for example, they are key members of *ETCs. Iscs. consist of a variable number of iron atoms coordinated to sulfur (sulfide and cysteine), and are capable of performing *SETs. The three most common types differ in the number of irons present: Fe-S, 2Fe-2S, and 4Fe-4S.

$$Cys-S \diagdown \underset{Fe}{} \diagup S-Cys \qquad Cys-S \diagdown \underset{Fe}{} S \underset{Fe}{} \diagup S-Cys$$
$$Cys-S \diagup {} \diagdown S-Cys \qquad Cys-S \diagup {} S {} \diagdown S-Cys$$
$$[Fe-S] \qquad\qquad [2Fe-2S]$$

Cys-S—Fe—S
Fe—S-Cys
S—Fe—S-Cys
Fe—S
/
Cys-S [4Fe-4S]

In each case, the irons are each coordinated to four sulfurs, which may be from sulfide or from a cysteine side chain. The *redox potential (Fe^{3+}-Fe^{2+}) of the iron within an iron-sulfur cluster (ca. –0.6 V) is quite different from that of free metals (+0.8 V). An important example of iron-sulfur enzymes is *nitrogenase, the enzyme responsible for the reduction of N_2 to ammonia. **F** clusters fer-soufre; **G** Eisen-Schwefel-Cluster.   P.S. SEARS

islands   less organized parts of *overlayers

iso (Iso, *iso*)   a prefix to characterize an isomer of a cpd. from others with the same empirical formula. Acc. to the IUPAC rules this prefix ought to be avoided, but many common names incorporating it are still widespread, especially in the fine chemicals industry. **F=G=E**.   B. CORNILS

ISO   International Organization for Standardization, responsible for the standardization and for the improvement of of intellectual, scientific, technological, and economic cooperation.   B. CORNILS

isobutyl oil synthesis   variant of the *Fischer-Tropsch synth. and a proc. for the manuf. of mixtures of methanol and isobutylalcohol by reacting *syngas over ZnO-Cr oxide cats. at 425 °C/32 MPa. **F** synthèse de l'huile isobutyrique; **G** Isobutylöl-Synthese.
Ref.: Falbe-2, p. 323.   B. CORNILS

isocitrate dehydrogenase   → decarboxylase

Isocracking (Isomax) process   → Chevron procs.

Isodewaxing process   → Chevron procs.

isoelectric point   The iep. characterizes the properties of an electrical double layer on a *surface; specifically the conds. (conc. of charge-determining ions, e.g., pH) in which no charge is measured by electrokinetic measurements, e.g., when the electrophoretic mobility of a *colloid is zero. The iep. measures the situation when the ζ-potential equals zero, i.e., essentially when the diffuse charge is zero. Under pristine conds. (absence of specific adsorption), the iep., the point of zero charge (pzc, representing an uncharged surface), and the pristine point of zero charge (ppzc., pzc of an uncontaminated surface) become identical. In the presence of specific *adsorption of ions, i.e., adsorption by nonelectrostatic forces, the iep. shifts in the opposite direction to the pzc, i.e., toward lower cation conc., e.g., increasing pH, upon adsorption of cations, and toward higher cation conc., e.g., decreasing pH, upon adsorption of anions. Specific adsorption and surface charges are important in the prepn. of *supported cats. from solution, when one or more species (e.g., metal *complexes) are to be anchored on the surface (e.g., by *precipitation). **F** point iso-électrique; **G** isoelektrischer Punkt.
Ref.. Ullmann, Vol. B5, p. 320.   F. JENTOFT

isoenzymes   → isozymes

IsoFin process   → Kellogg procs., *Mobil procs.

Isolen process   → Toray procs.

isoleptic complexes → homoleptic complexes

Isomar process → UOP procs.

Isomate process → Standard Oil procs.

Isomax (Isocracking) process → Chevron procs.

isomerases *enzymes that cat. *isomerization reactions (which are often coupled to an energy-providing reaction, such as the *hydrolysis of *ATP, as in the case of the *prokaryotic *topoisomerases). Some examples are the *cis-trans* isomerases, protein disulfide isomerases, topoisomerases, racemases, various aldose-ketose isomerases, and *mutases. **E=F**; **G** Isomerasen.   P.S. SEARS

isomerization a rearrangement reaction that occurs when cpds. with the same formula exhibit different structures, e.g., double-bond i. (1- and 2-butene), structural i. (propylene oxide (PO) ⇌ allylalcohol), substitution i. (*o-*, *m-* or *p*-xylenes), skeletal i. (methylcyclopentane ⇌ cyclohexane), *cis/trans* i., conformational i., valence i., etc. Most is. require cats. (or UV radiation, protons, etc.) or *isomerases. Of technical importance are the *Beckmann rearrangement, *catalytic reforming, etc. and the butane/butene i. (via the procs. of *BP, *CDTech Isomplus, *IFP, *Kellogg Isofin, *Mobil IsoFin, *Phillips, *Shell, *Snamprogetti, *UOP Butamer); the i. of cyclic hydrocarbons (via the procs. of *ARCO Octafining, *Atlantic Pentafining, *BP, *IFP Eluxyl, *Lummus ABB, *Mitsubishi Xylene Plus, *Mobil HAI/LPI and LTI, *Phillips Hydrisom, *Shell Hysomer, *Toray Isolen, *UCC TIP, *UOP Isomax, Isomar, Penex, and HOT, *VEB Aris); the i. of *PO to allyl alcohol (*BASF/Wyandotte, *Olin, *Rhône-Progil), and others (*Chevron Dewaxing, *Chemie Linz fumaric acid, *Shell SHOP, *Henkel TPA); cf. *isomerization cats. **F** isomérisation; **G** Isomerisierung.   B. CORNILS

Ref.: Ertl/Knözinger/Weitkamp, Vol. 3, p. 1123, Vol. 4, p. 1998, and Vol. 5, p. 2137; Farrauto/Bartholomew, p. 562; *Organic Reactions*, Vol. 29, p. 345; McKetta-1, Vol. 27, p. 435.

isomerization catalysts Examples of commercially relevant isomerization procs. are 1: Skeletal i. of C_4–C_8 hydrocarbons to obtain more valuable branched hydrocarbons with higher octane numbers. It is desirable for cats. to be active at low temps. at which branched prods. are favored by thermodynamics. Examples are metal-solid acid *bifunctional cats. such as Pt *supported on Y-*zeolite (active at about 520 K) or Pt supported on chlorided Al_2O_3 (active at 423–673 K, *chlorinated alumina). Recent (and less corrosive) developments include cats. based on sulfated metal oxides (e.g., *sulfated zirconia) whose high activity in comparison to the zeolitic cats. corresponds to about an 80 K lower reaction temp. 2: I. of alkylbenzenes to obtain *p*-xylene or *ethylbenzene. These reactions can be performed in the presence of solid acids, with the side reactions (*cracking, *disproportionation) being best suppressed by use of a *microporous cat. such as the *zeolite HZSM-5. Technically important isomerizations including heteroatoms are 3: Aliphatic/aromatic *epoxides to unsaturated alcohols, e.g., the manuf. of allylalcohol from *propene oxide which can be cat. by Li phosphate. 4: *Beckmann rearrangement of cyclohexanone oxime to give ε-caprolactam, with attempts to replace the sulfuric acid cat. by *mixed oxides or zeolites. **F** catalyseurs pour l'isomérisation; **G** Isomerisierungskatalysatoren.   F. JENTOFT

Ref.: *Stud.Surf.Sci.Catal.* **1989**, *51*; Ertl/Knözinger/Weitkamp, Vol. 4, p. 2010.

Isomplus process → CDTech procs.

isopenicillin N → *CV synthetase

isopenicillin-N synthetase a non-heme monooxygenase that cat. the *cyclization of tripeptide   δ-(L-α-aminoadipyl)-L-cysteinyl-

D-valine (ACV) to isopenicillin-N in the presence of oxygen, Fe(II), and ascorbate. This step is a key to the *biosynthesis of all penicillin and cephalosporin antibiotics. The enzyme possesses a relaxed requirement for variation of the Val residues (Figure).

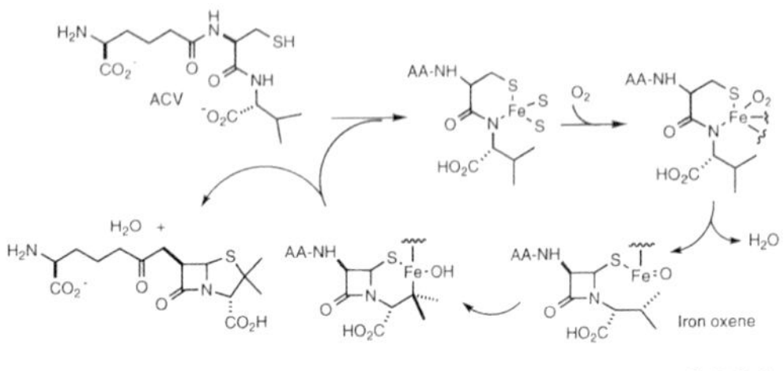

When tripeptides containing unsaturated Val analogs were used as substrates, prods. from both desaturative and hydroxylative pathways were obtained. Expansion of penicillins to cephalosporins is cat. by the expansion enzyme from the same species. Substrate specificity studies indicate that the enzyme also has a broad side chain specificity. See also *ACV synthetase. **E=F=G**.   C.-H. WONG

Ref.: *Nat.Prod.Rep.* **1988**, 5, 129; *Nature (London)* **1997**, *387*, 827.

isopentenyl (dimethylallyl) pyrophosphate
Isopentenyl and dimethylallyl pyrophosphates (iPP and dPP) are the five carbon precursors to steroids, terpenes, ubiquinone, and other isoprenoid cpds.

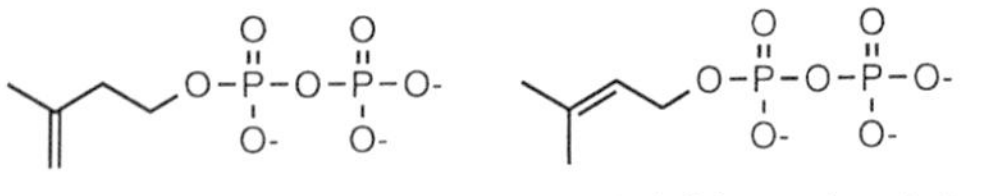

Isopentenyl pyrophosphate   Dimethylallyl pyrophosphate

IPP is synth. ultimately from three molecules of *acetyl coenzyme A (acetyl-CoA): two combine to give *acetoacetyl-CoA (cat. by the *enzyme thiolase), and a third is added in a *Claisen condensation to give hydroxyl-methylglutaryl-CoA (HMG-CoA). HMG-CoA is reduced and *hydrolyzed to mevalonate by HMG-CoA reductase, which is phosphorylated (twice) and *decarboxylated to give the unsaturated iPP. IPP isomerase in-

terconverts iPP and dPP (cf. also *isoprenoids). **E=F=G**.   P.S. SEARS

isopolyacids → heteropolyacids

isoprenoids a very large class of cpds. synth. from five- carbon precursors, *dimethylallyl pyrophosphate and *isopentenyl pyrophosphate, including the polyprenols, dolichols, sterols, carotenoids, terpenoids, and others. Chain elongation can be accomplished by head-to-tail joining of an isoprenyl-PP (PP, pyrophosphate) and dimethylallyl-PP, which is believed to be initiated by cleavage of the PP from the dimethylallylic substrate, forming a resonance-stabilized *carbocation. The carbocation is attacked by the π-bond of the isopentenyl-PP, and abstraction of a proton by a base within the *enzyme generates geranyl-PP. "Head-to-head" joining also occurs, and has been observed, for example, in *squalene synthase.

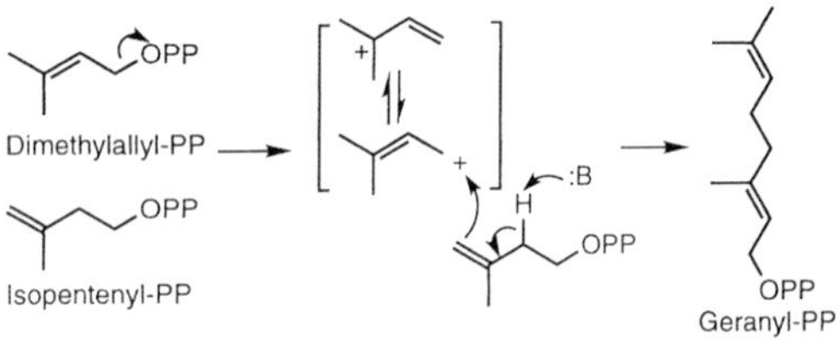

Dimethylallyl-PP

Isopentenyl-PP

OPP
Geranyl-PP

The diversity of the isoprenoids (>24 000 known, to date) is generated in large part by different *cyclization reactions. Cyclizations occur by one of three mechanisms: carbocation formation by elimination of PP; protonation of a C-C double bond; or *epoxidation of a double bond, protonation of the *epoxide to generate the carbocation, and attack by another π-bond (cf. also *squalene synthase). **F** isoprénoides; **G** Isoprenoide.   P.S. SEARS

Ref.: *Science* **1997**, *277*, 1788; C. Walsh, *Enzyme Reaction Mechanisms*, W.H. Freeman and Co., New York 1979.

IsoSiv process → Union Carbide procs.

isosteric heat → Clausius-Clapeyron equation

isotactic polymers are poly-1-alkenes and vinylic polymers (ps.) with the same relative confiation at their tertiary C-atoms (cf. *atactic ps., *stereoselectivity, *tacticity). Successive monomer units are linked in *meso* form (m.). The degree of isotacticity (i.) is measured by ^{13}C-NMR spectroscopy and quoted as pentad i. or %mmmm. The notion i.was introduced by *Natta and indicated for isotactic *PP by the fraction of insoluble PP in boiling heptane. The most important isotactic polymer is i-PP (a crystalline p. with a melting point of 165 °C), prod. with *Ziegler-Natta or *metallocene cats. Many Ziegler-Natta cats. contain $MgCl_2$ and $TiCl_4$ in combination with internal or external donors (ethyl benzoate, silanes, ethers to deactivate atactically working centers) and $AlEt_3$ as *co-catalyst. **F** polymères isotactique; **G** isotaktische Polymere.   W. KAMINSKY

Ref.: *Makromol.Chem.* **1955**, *16*, 213; *Angew.Chem.Int. Ed.Engl.* **1985**, *24*, 507; *Angew.Chem.* **1980**, *92*, 869.

isotherms → adsorption, *Langmuir-Hinshelwood kinetics, *Langmuir isotherm

isotope effects 1: Despite the mostly minute differences between the masses of regular cpds. relative to their isotope-labeled counterparts, small differences in their physical properties and their chemical behavior may occur stemming from the different zero-point energies or from the small deviations of their intermolecular interactions. The different thermodymamic properties of isotopomers provide the basis of various isotope separation techniques. Small differences in the kinetic constants of the isotope-labeled cpds. are responsible for the kinetic ie. The kinetic ie. is most pronounced if during a chemical reaction a bond to an isotope is broken or formed; therefore, the observation of a kinetic ie. proves that a bond to an isotope is broken or formed as the slowest step of the reaction; i.e., it reveals what the nature and the timing of bonding changes during the reaction. Several concepts exist to model kinetic ies. Ies. are smaller, the smaller the differences in

masses. Ies. are taken advantage of in synth. and are important to study the kinetics of certain reactions (e.g., cat. reactions), to investigate equlibria, to elucidate *enzymatic reactions, etc. Isotopes typically have a spin different from that of the nucleus they replace, which gives rise to pronounced differences in the spectra from various types of spectroscopy and in the rates of spin-dependent reactions.   J. BARGON

Ref.: Willi, *Isotopeneffekte bei chemischen Reaktionen*, Thieme, Stuttgart 1983; Cleland et al., *Isotope Effects of Enzyme Catalyzed Reactions*, University Park Press, Baltimore 1977; *Acc.Chem.Res.* **1978**, *11*, 218.

2-in enzyme cat.: Secondary isotope effects, where the isotopic substitution occurs at a position adjacent to the scissile bond, give information regarding changes in orbital hybridization at that site. S_N2 reactions show no secondary isotope effect, while S_N1 reactions show a small but detectable effect. This has been useful for dissecting the mechanism of *glycosidase and *glycosyltransferase reactions, for example; the presence of a detectable secondary isotope effect is considered good evidence that these reactions proceed through an S_N1-like mechanism. **F** effets isotopiques; **G** Isotypieeffekte.   P.S. SEARS

Ref.: C. Walsh, *Enzyme Reaction Mechanisms*, Freeman and Co., New York 1979; *Chem.Rev.* **1990**, *90*, 1171.

isotope labeling Unique isotopes of many elements (e.g., D = ^2H or T = ^3H for ^1H or ^{13}C or ^{14}C for ^{12}C) are used to elucidate reaction mechanisms or to trace the fate of chemicals. Il. cpds. make it possible to distinguish alternative reaction pathways. To qualify as radioactive tracers, the radioisotopes should have half-lives ($t_{1/2}$) which are neither be too short not too long while they still emit sufficiently strong radiation. In addition to ^3H and ^{14}C up to 40 other radioisotopes are in use, among them cat. active metals like ^{60}Co, ^{106}Ru, or ^{59}Fe. Especially useful is ^{32}P. Isotope-labeled cpds. react essentially identically to unlabeled ones, unless an isotope effect differentiates their behavior. Thanks to the sensitivity of the analytical methods used for their detection, only low concs. of radioactive tracers are used. Even multiple tracers can be incorporated into one and the same molecule, since ^{3}H, ^{14}C, and ^{32}P can be monitored simultaneously and independently. Unfortunately, some of the most significant elements of organic chemistry or biology, namely N or O, possess no radioactive isotopes. In these cases stable isotopes are used as tracers. If they are employed, a relatively high degree of their enrichment is desirable. Special isotopes can be incorporated via exchange reactions (scrambling) or by means of special synth. starting from suitable isotope-marker *precursors. The analytcal method of choice to detect their subsequent whereabouts depends on the properties of the individual isotope. Isotopes with a magnetically active nucleus, e.g., ^{13}C or ^{17}O, can be identified and monitored using

*NMR spectroscopy; ^{18}O and most other types can be discriminated using *mass spectroscopy. **F** marquage isotopique; **G** Isotopenmarkierung.   J. BARGON

Ref.: Duncan, Susan, *Synthesis and Applications of Isotopically Labeled Compounds*, Elsevier, Amsterdam 1983; Ertl/ Knözinger/Weitkamp, Vol. 3, p. 1005.

isozymes (isoenzymes) molecular species of an *enzyme that are genetically differentiated. With similar activities, their properties (pH optimum, *Michalis constant, ionic strength, etc.) may be different.   B. CORNILS

Ref.: Z.-I. Ogita, C.L. Markert, *Isozymes*, Wiley, New York 1990.

ISS → ion scattering spectroscopy; identical with LEIS

J

Jacobsen complex (R,R)-$(-)$- or (S,S)-$(+)$-N,N'-bis(3,5-di-*tert*-butylsalicylidene)-1,2-cyclohexane diamino-Mn chloride, a *ligand for *asymmetric *epoxidations (Figure cf. *Jacobsen epoxidation). B. CORNILS

Ref.: *J.Am.Chem.Soc.* **1990**, *112*, 2801; *Angew.Chem. Int.Ed.Engl.* **1997**, *36*, 2060.

Jacobsen epoxidation a (salen)Mn(III)-cat. *asymmetric *enantioselective reaction of alkenes with NaOCl as oxidizing agent. Values of 97 % *ee have been achieved using *cis*-disubstituded or trisubstituted alkenes. However, *trans*-disubstituted alkenes are epoxidized with only low to moderate enantioselectivity. **F** époxidation de Jacobsen; **G** Jacobsen-Epoxidierung.

W. MORADI, M. BELLER

Ref.: Ojima, Chapter 4.4; *J.Am.Chem.Soc.* **1994**, *116*, 6937; *J.Org.Chem.* **1993**, *58*, 6939.

Japan Synthetic Rubber produces a *copolymer (Arton), basing upon *ROMP *Diels-Alder adducts of dicyclopentadiene with acrylates.

Jefferson morpholine process manuf. of morpholine by *dehydration of diethanolamine or the *hydrogenating *amination of diethylene glycol with NH_3/H_2 over Ni, Co, or Cu cats. at 150–250 °C/3–20 MPa. **F** procédé Jefferson pour le morpholine; **G** Jefferson Morpholin-Verfahren. B. CORNILS

Johnson Matthey was founded 1817 and is still active in manuf., refining, and *reclaiming of precious metals. In 1874 Johnson Matthey produced the Pt/Ir *alloy for the standard meter and kilogram for the International Metric Commission. Today JM. is selling a huge number of hom. cats. (Pt, Pd, Rh, Ru, Ir, Os; for, e.g., *hydroformylations) and het. cats. (Pt, Pd, Rh, Ru, Ir) for the main use in chemical procs. for *hydrogenation, *carbonylation, *dehydrogenation, *oxidation, and air purification. JM. is the leading supplier in *automotive exhaust cats. with around 30 % market share. J. KULPE

Johnson Matthey process → LPO process of *Union Carbide

juvenile hormone is synth. by farnesyl diphosphate synthetase (EC 2.5.1.10) using methyldihomofarnesol, of which the *S*-enantiomer was used as a *precursor for the synth. of a juvenile hormone. **F** hormone juvénile; **G** Juvenilhormon. C.-H. WONG

Lit: *J.Am.Chem.Soc.* **1987**, *109*, 2853.

K

KAAP process   Kellogg advanced ammonia process, → Kellogg procs.

Kagan reagent   is samarium(II)iodide, cat. the coupling of organic halides with carbonyl cpds. in THF acc. to R-X + R$_1$-C(=O)-R$_2$ → R$_1$(R$_2$)-C-H(OH) + HX. **F** réactif de Kagan; **G** Kagan-Reagenz.   B. CORNILS
Ref.: *Chem.Rev.* **1996**, *96*, 307 and **1999**, *99*, 745.

KA oil   ketone-alcohol oil, anone-anol, the mixture of cyclohexanone and cyclohexanol, *intermediate for the manuf. of *adipic acid by hom. *LPO *oxidation; manuf. from cyclohexane (via *Bashkirov oxidation). B. CORNILS

Karstedt's catalyst   is sometimes used in place of *Speier's cat. for the *hydrosilylation of alkenes. Kc. is obtained by treating H$_2$[PtCl$_6$] with (H$_2$C=CH-)(Me$_2$)SiOSi(Me$_2$)-(CH=CH$_2$).   W.A. HERRMANN
Ref.: US 3.775.452 (1973); *Angew.Chem.Int.Ed.Engl.* **1991**, *30*, 458; *Organometallics* **1987**, *6*, 191.

Katalco   the former catalyst business of *ICI; now *Synetix.

KataLeuna catalysts is the former cat. production of Leuna Werke (East Germany). The first cat. was produced in 1921 for the manuf. of *ammonia. Main areas today are selective hydrogenation (inter alia in *petroleum processing), gas production, *isomerization, and environmental protection. Ultimate parent company: Royal Dutch/Shell Group.
                                   J. KULPE

Katalyse-Institut   a   (disadvantageously) politically influenced institute in Darmstadt (Germany) for environmental resources and for ecological problems, i.e. – in spite of the designation – *not* a scientific institution for catalysis.   B. CORNILS

Katapak structures   cat. *supports and cat. structures (metallic or ceramic) with open crossflow derived from the SMV motionless mixing priciple (static mixers, crossed corrugated-plate packing). The continuous splitting and relocating of the flowing fluids and gases lead to homogenization of conc., temp., and velocity profiles over the cross-section of the reactor; inter alia, suited for reactive distillations (*catalytic distillation). **F** structures de Katapak; **G** Katapak-Strukturen.   B. CORNILS
Ref.: *Chem.Ing.Tech.* **1999**, *71*, 131, 388.

Katzschmann process   → Witten DMT process

KDN, L-synthesis   → *N*-acetylneuraminate (NeuAc) aldolase

KDO aldolase   → 3-deoxy-D-manno-2-octulosonate aldolase

KDO-8-phosphate synthetase
→ 3-deoxy-D-manno-2-octulosonate 8-phosphate

Keggin structures   → heteropolyacids

Kel-chlorine   → Kellogg procs.

Kellogg advanced ammonia process (KAAP)   Developed by Kellogg and BP, the KAAP uses a new Ru-based cat. for *ammonia synth. which is supported on graphite-containing carbon and copromoted with Cs and Ba. The KAAP operates at milder conds. and at high ammonia concs. while maintaining higher conv. than a conventional Fe-based cat. A retrofit downstream of an existing ammonia plant consists of a two-bed radial converter with interbed cooling, whereas a grassroots plant employs a vertically oriented four-bed vessel and a single-stage synth. gas compressor. **F** procédé KAAP de Kellogg; **G** Kellogg KAAP-Verfahren. M. MUHLER

Kellogg cascade sulfuric acid alkylation converts refinery propylene, butylenes, and amylenes to high octane gasoline by selective *alkylation with isobutane in the presence of concentrated sulfuric acid cat. The special *alkylation reactor is a cascade, autorefrigerated device with an acid recycle. **F** procédé Kellogg de l'alkylation en cascade avec l'acide sulfurique; **G** Kelloggs Kaskadenalkylierung mit Schwefelsäure. B. CORNILS
Ref.: *Hydrocarb.Proc.* **1972**, (9), 127.

Kellogg coal gasification The Kcg. (formerly called the molten salt gasification proc., MSGP) employs a (highly corrosive) molten bath (cf. *molten salt media) from Na or Ca carbonate. Oxygen and steam are introduced into the bottom of the gasifier, coal is fed beneath the surface. The cat. *partial *oxidation of the coal yields CO which with steam is partially converted to hydrogen. The *WGSR occurs in the gas phase, thus yielding *syngas. The Rockgas proc. is a press. variant (1000 °C/2 MPa). **F** gaséification Kellogg pour charbon; **G** Kellogg-Kohlevergasungsverfahren.
B. CORNILS

Ref.: Ertl/Knözinger/Weitkamp, Vol. 4, p. 2077.

Kellogg 1,2-dichloroethane process
manuf. of 1,2-dichloroethane (EDC) by liquid-phase *oxychlorination of ethylene at 180 °C/1.2–1.8 MPa over hydrochloric acid $CuCl_2$ solutions. **F** procédé Kellog pour l'EDC; **G** Kellog EDC-Verfahren.
B. CORNILS

Kellogg/Hercules/BP phenol process a variant of *Hock's phenol synth. via *cumene hydroperoxide and its cracking in hom. solution with cat. amounts of H_2SO_4 at 60 °C (Figure). **F** procédé de Kellog/Hercules/BP pour

le phénol; **G** Kellog/Hercules/BP Phenol-Verfahren.
B. CORNILS

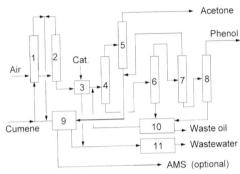

1, 2 oxidation reactors; 3 cleavage; 4-8 fractionation; 9 hydrocarbon recovery; 10 phenol recovery; 11 effluent treatment. AMS = α-methylstyrene.

Kellogg Isofin process cat. proc. for the skeletal *isomerization of *n*-butenes to isobutenes (Figure). **F** procédé Isofin de Kellogg; **G** Kellogg Isofin-Verfahren. B. CORNILS

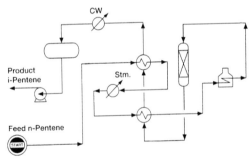

Kellogg Kelchlor process for the workup and recov. of waste HCl to Cl_2 with oxygen with $NO-NO_2/H_2SO_4$ acc. to 2 HCl + O_2 → Cl_2 + H_2O (cf. *Deacon process; cat. CuCl). **F** procédé Kelchlor de Kellog; **G** Kellog KelChlor-Verfahren. B. CORNILS
Ref.: McKetta-1, Vol. 8, p. 95; *Hydrocarb.Proc.* **1977**, 56(11), 139.

Kellogg KRES (Kellogg reforming exchange system) process for the manuf. of raw *syngas by cat. steam hydrocarbon *reforming in a pressurized heat exchange based system. **F** système KRES de Kellogg; **G** Kellogg KRES-System.   B. CORNILS

Kellogg MAK (Mobil/Akzo/Kellogg) hydrocracking process

a cat. proc. for the conv. of various feedstocks (vacuum gas oil, FCC cycle oils) into low-sulfur distillate fuels, LPG, high-octane gasoline, etc., by *hydrocracking over multiple cat. systems in multi-bed reactors. **F** procédé MAK hydrocracking de Kellogg; **G** Kellogg MAK-Hydrocracking-Verfahren.   B. CORNILS

Ref.: *Hydrocarb.Proc.* **1996**, (11), 126.

Kellogg Synthol process

is one of three commercial variants of the *Fischer-Tropsch synthesis, based on the former *Hydrocol proc., developed by the M.W. Kellogg Co. but refined by *Sasol. Similarly to *FCC procs., the backbone of the Synthol technology is an entrained-bed technology.

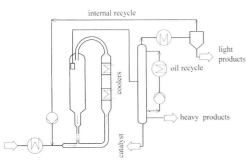

internal recycle

light products

oil recycle

coolers

heavy products

catalyst

*Syngas entrains the cat. powder from bottom to top through the synth. reactor, passing internal heat exchangers to limit the increase in temp. The cat. powder is separated from the prod. vapors in a cyclone-like collector; the non-converted syngas and the gaseous prods. are withdrawn via filters. Some cat. fines are swept out and reclaimed by an oil wash. The *hcs. are condensed and separated, the syngas is recycled. The reaction conds. are 280–350 °C/2.2–2.5 MPa, and H_2:CO ratios of 6:1 at the reactor entrance; the syngas conv. is approx. 85 %. The prod. composition, optimized to generate a max. yield of motor fuels, is [% of carbon converted]: C_1 10, C_2–C_4 33, C_5–C_{12} 39, C_{13}–C_{30} 11, and oxygenates 7.

The cat. is prepd. by fusing iron ore or *millscale, doped with K_2O, MgO, CaO, or Al_2O_3, leading to magnetite Fe_3O_4 (*fused cats.). The milled and classified *precursor is reduced in a fluid-bed reactor to metallic iron with porous structure. After some time under reaction conds. the cat. consists of nearly equivalent parts of magnetite and Fe_2C (Hägg carbide). The specific cat. consumption is in the range of 5–10 kg per t of prods. The latest version of the proc. is in use at Secunda (South Africa) by Sasol with an estimated capacity of approx. 4×10^6 tons of *hcs. per year. **F** procédé Synthol de Kellogg; **G** Kellogg Synthol-Verfahren.

C.D. FROHNING

Kellogg Ultra-Orthoflow process

a special variant of the *FCC technology and a combination of the cat. *regeneration of *Amoco UltraCat and the *riser and converter configuration of Kellogg. **F** procédé Ultra-Orthoflow de Kellogg; **G** Kellogg Ultra-Orthoflow-Verfahren.   B. CORNILS

Ref.: *Hydrocarb.Proc.* **1978**, (9), 111 and **1996**, (11), 122.

Kelvin equation   The Ke. offers on the basis of basic thermodynamic principles an explanation for the hydrostatic effect leading to *capillary condensation observed in *mesoporous adsorbents. This equ. relates the curvature of the liquid meniscus in a pore to the relative press. at which condensation occurs. The lowering of vapor press. is dependent on the pore radius. The Ke. is applicable over the mesopore range and assumes that the meniscus curvature is controlled by *pore size and shape. An estimation of the radius of the pore is possible under the assumption of a wetting angle of 0° and an adsorbed film on walls of a cylindrical pore. **F** équation de Kelvin; **G** Kelvin-Gleichung.

R. SCHLÖGL, M.M. GÜNTER

2-keto-3-deoxy-L-arabinate aldolase

(EC 4.1.2.18)   cat. the formation of 2-keto-3-deoxy-L-arabinate from pyruvate and glycoaldehyde. **E=F=G**.   C.-H. WONG

Ref.: *J.Biol.Chem.* **1972**, *247*, 2238 .

2-keto-3-deoxy-D-glucarate (KDG) aldolase (EC 4.1.2.20) cat. the reversible reaction of pyruvate and tartronic acid semialdehyde to form KDG. Various bacteria have been the source of this *enzyme. Several non-natural aldehyde acceptors act as substrates for KDG aldolase, including glycoaldehyde, glyoxylate, and D- and L-glyceraldehyde. **E=F=G**.

C.-H. WONG

pyruvate · tartronic acid semialdehyde · KDG

Ref.: *Methods Enzymol.* **1966**, *9*, 529.

2-keto-3-deoxyoctanoate aldolase → 3-deoxy-D-manno-2-octulosonate aldolase 2-keto-3-deoxy-6-phosphogluconate (KDPG) aldolase (EC 4.1.2.14) KDPG *aldolase from *Pseudomonas putida* cat. the reversible *condensation of pyruvate with G3P to form KDPG which, in vivo, favors the condensation prod. A number of unnatural aldehydes are accepted, albeit at rates much lower than the natural substrate. As long as there is a polar functionality at C 2 or C 3, however, there appears to be no other structural requirement for the acceptor aldehyde. The KDPG aldolase reaction *stereospecifically generates a new stereocenter at C 4 with S-configuration. Other related enzymes also showed narrow substrate *specificity. **E=F=G**. C.-H. WONG

Ref.: *J.Org.Chem.* **1992**, *57*, 426.

2-keto-3-deoxy-D-xylonate aldolase (EC 4.1.2.28) cat. the formation of 2-keto-3-deoxy-D-xylonate from pyruvate and glycoaldehyde. **E=F=G**. C.-H. WONG

Ref.: W.A. Wood, *The Enzymes*, Boyer, P.D. (Ed.) Academic Press, New York 1970, Vol.VII, p. 281.

ketoglutarate/glutamate dehydrogenase → regeneration of NAD(P)H

2-keto-4-hydroxyglutarate (KHG) aldolase (EC 4.1.2.31) KHG aldolase cat. the reversible *condensation of pyruvate and glyoxylate to form KHG. Studies of substrate *specificity on KHG aldolase from bovine liver indicate that it accepts both *enantiomers of KHG equally well, and also cleaves 2-keto-3-deoxyglucarate, 2-keto-4,5-dihydroxyvalerate, and oxaloacetate. In the condensation direction, this *enzyme is relatively specific for glyoxylate, but some pyruvate derivatives are accepted. In the condensation reaction, glyoxylate can be replaced with glyoxaldehyde, formaldehyde, acetaldehyde, and formic acid, while pyruvate can be substituted by α-ketobutyrate and bromopyruvate. **E=F=G**.

C.-H. WONG

pyruvate · glyoxylate · KHG

Ref.: *J.Chem.Soc.Perkin Trans.* **1992**, *1*, 1085.

2-keto-4-hydroxy-4-methylglutarate aldolase (EC 4.1.3.17) cat. the formation of 2-keto-4-hydroxy-4-methylglutarate from two molecules of pyruvate. **E=F=G**. C.-H. WONG

Ref.: *J.Chem.Soc.Perkin Trans.* **1992**, *1*, 1085.

ketonization manuf. of ketones via pyrolytic decomposition of metal carboxylates by simultaneous vapor-phase *decarboxylation and *dehydration of carboxylic acids, e.g., AcOH → acetone. The reaction is cat. by oxides of Th, Ce, Zr, Ti, or rare earth metals (cf. *decarboxylation). **F** cétonisation; **G** Ketonisierung. B. CORNILS

Ref.: *Stud.Surf.Sci.Catal.* **1993**, *78*, 527; *Appl.Catal.A* **1995**, *128*, 209 and **1998**, *166*, 201; *J.Mol.Catal.A:* **1999**, *139*, 73.

ketopantoaldolase (EC 4.1.2.12) cat. the *aldol condensation between formaldehyde and α-keto-β-methylbutyric acid. **E=F=G**.

Ref.: *J.Biol.Chem.* **1957**, *228*, 499. C.-H. WONG

ketotetrose phosphate aldolase (EC 4.1.2.2) cat. the condensation between dihydroxyacetone phosphate and formaldehyde to form ketotetrose phosphate. **E=F=G**.

Ref.: *Methods Enzymol.* **1962**, *5*, 283. C.-H. WONG

Keulemans' rule   As an irregularity with branched alkenes Keulemans et al. found that no quaternary carbon atoms are formed by alkene *hydroformylation under standard conds. Accordingly *i*-butene forms almost exclusively 3-methylbutyraldehyde. **F** règle de Keulemans; **G** Keulemans-Regel.   B. CORNILS

Ref.: *Recl.Trav.Chim. Pays-Bas* **1948**, *67*, 298; *J.Mol. Catal. A:* **1999**, *143*, 123.

Kharasch addition   comprises the *addition of polyhalogenoalkanes to alkenes in presence of *radical initiators such as *peroxides or UV light acc. to R-CH=CH$_2$ + CX$_3$Br → R-CHBr-CH$_2$-CX$_3$. As side reactions *telomerizations and *polymerizations can occur. Because of the lower bond energy of the C-Br bond compared to the C-Cl bond, the polybromoalkanes react more easily than the poly*chloro*alkanes. **F** réaction de Kharasch; **G** Kharasch-Reaktion.

A. TILLACK, M. BELLER

Ref.: S. Patai, *The Chemistry of Alkenes*, Wiley, London 1964.

Kharasch-Sosnovsky reaction   → *acetoxylation

Kieselguhr   Kieselguhr (diatomeous or diatomaceous earth, diatomite, kieselgur) is a siliceous sediment composed of the skeletal remains of microscopic algae (diatoms). These aquatic plants had the unique ability to absorb silica from the surrounding water and to form complex skeletal frameworks of intricate, meshlike structures with a variety of shapes. Main ore locations are the USA, Mexico, France, Spain, and Germany. The raw ore from open-pit mining is crushed, classified, and *calcined in order to remove organic residues and to improve the color. The chemical comp. in [wt%] may vary broadly: SiO$_2$ 80–92, Al$_2$O$_3$ 3–15, Fe$_2$O$_3$ 0.5–3, CaO and MgO 0.2–1, and Na$_2$O/K$_2$O 0.2–10. Density ranges from 2.0 to 2.5 kg/L, and the loose weight (powder) is approx. 0.1 kg/L. The specific surface may reach 200 m^2/g. K. finds broad application as filter aids, fillers for many materials, and cat. *supports. **F=G=E**.   C.D. FROHNING

Ref.: Kirk/Othmer, Vol. 8, p. 108; McKetta-1, Vol. 15, p. 436.

kinases   *enzymes that transfer phosphate from a high-energy donor (esp. *ATP, also other nucleotides) to an acceptor, which may be a protein, sugar, lipid, or a variety of other molecules. *protein ks. phosphorylate amino acid residues of proteins, particularly at Ser, Thr, Tyr, His, and Asp, and these modifications are used to regulate the activity of the proteins. There are many ks. in primary metabolic pathways, too. **E=F=G**.   C.-H. WONG

kinetically controlled synthesis   synth. of a cpd. which, while not the thermodynamically favored prod., is the prod. formed faster. Termination of the reaction before it has reached equil. provides substantial amounts of the desired cpd.

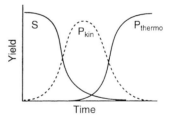

For example, in the kcs. of a peptide from an ester and an amine with a serine protease, the peptide may be synth. in aqueous solution by the addition of a large amount of amine nucleophile and conv. of the *active-site serine of the *enzyme to a cysteine (which favors *aminolysis vs. *hydrolysis). Although the hydrolytic prod. is thermodynamically favored, the amide is quicker to form, and so terminating the reaction before equil. has been reached can result in good yields of the peptide. **F** synthèse controllé par cinétique; **G** kinetisch kontrollierte Synthese.   P.S. SEARS

Ref.: *Angew.Chem.Int.Ed.Engl.* **1991**, *30*, 1437.

kinetic isotope effect   → isotope effect

kinetic rate factor (synonymous with *ki-netic rate limitation*, opposite to *transport limitation*) The rate of any reaction depends on the conc. of the reactants, the resistance of the system to *mass transport, and the microkinetics of the proc. If, in a kinetic experiment, the concs. of the reactants are known by setting the educt concs. and by measuring all prod. concs. then it is still *not* possible to determine the kinetic *rate constant of the reaction from a single experiment. Only when the reaction parameters are systematically varied, then conds. may be found in which either the microkinetics or the transport kinetics are determining the *overall rate. In normal laboratory conds. a mixed situation is typical with combined influences of transport and chemical kinetic constants. This is a major source of discrepancy in kinetic parameters quoted from literature studies of the same reaction and the same cat. The problem of measuring transport kinetic data which characterize only the lab reactor instrumentation and not the chemical proc. is the greater, the faster the chemical reaction. Such cases are typically *oxidation reactions of small molecules. Very low *apparent activation energies (0–15 kJ/mol) may be an indication for not measuring the microkinetic parameter. This *kinetic rate factor* is a convolution of the kinetic laws of all rate-determining elementary step reactions which may not necessarily be only one process in a complex reaction sequence.

R. SCHLÖGL

Ref.: J. Dumesic et al., *The Microkinetics of Heterogeneous Catalysis*, ACS, Washington 1993.

kinetic rate limitation → kinetic rate factor

kinetic resolution 1-general: With *chiral cpds. two and only two *stereoisomers (*enantiomers) exist. They rotate the plane of polarized light in opposite directions, but in equal amounts, and they react with other chiral cpds. at different rates. Equimolar mixtures of enantiomers are called *racemic mixtures, which are optically inactive, since the rotations cancel.

*Diastereomers are stereoisomers which are not enantiomers, with consequently differing physical and chemical properties. A pair of enantiomers is usually separated by conv. to diastereomers with an optically pure resolving agent (auxiliary) followed by fractional crystallization or chromatography. Commonly, carboxylic acids such as tartaric acid and camphorsulfonic acid, and amines such as brucine and *cinchona derivatives, are used. The *kinetic resolution of a racemate relies on the relative *rates of reaction of a pair of enantiomers. If one enantiomer reacts faster than the other, then an optical enrichment is achieved (e.g., *asymmetric *epoxidation of racemic mixtures of allylic alcohols using optically pure diethyl tartrate in addition to $(CH_3)_3CO_2H$ and $Ti(OiPr)_4$ (Sharpless), or the asymmetric *hydrogenation of racemic mixtures of alkenes with an adjacent polar group using Ru complexed with the *bulky chiral *ligand *BINAP [Noyori]).

M. MUHLER

2-enzymatic: an enzymatic proc. which cat. the differentiation of enantiomers through selective reaction with one or the other, such as the selective *hydrolysis of racemic esters, amides, phosphates (or their reverse reactions), or through oxidoreductions. **F** résolution cinétique; **G** kinetische Isomerentrennung.

C.-H. WONG

kinetics of catalyzed reactions The phenomenon of cat. designates the modification of the ks. of a given chemical reaction. This modification can be an acceleration (mostly wanted; by cats.) but also an inhibition (by *inhibitors) of a proc. Cats. and inhibitors can only be effective when the reaction is kinetically limited. This means that the yield and the *selectivity observed are not determined by a *thermodynamic equil. situation. The functional materials act upon the reaction by changing the *activation energy for the *rate-determining step (rds.) of the proc. Each chemical reaction is composed of a large number of elementary step reactions comprising *adsorption, dissociation, *oxidation, *reduction, association and *desorption steps.

Each of these steps requires the exchange of chemical and motional energy with the surrounding gas and solid *phases. Thus, even the chemically very simple reactions require the coordinated interaction of a large number of elementary procs. which all exhibit their characteristic temporal evolutions. One of these steps will be the slowest in the overall sequence. This step is determining the overall *rate. A cat. is only active if it affects exactly this rds. which is in almost no cases known in its chemical nature.

The determination of the macroscopic reaction ks. of a cat. or non-cat. reaction never gives insight into the nature of the rds. as it is not sensitive to elementary reactions. The very nature of a cat. action thus alters the rds. and thus affects the course of the chemical reaction. For this reason it is to be expected that the prod. distribution will change, when a given non-cat. reaction is carried out over a cat. If the interaction of a species with the cat. becomes the rds. (which is often the case) then the nature and type of the cat. affects directly the ks. and prod. distribution of a cat. reaction, leading to a dependence of the result of a reaction on the nature of the cat.

A k. analysis is only useful when the parameters measured are not dependent on time. In this case a *steady state of the system is achieved. If then the *mass and energy fluxes into and out of the reactor are exactly known, a macrokinetic model can be constructed allowing the prediction of the result of a cat. reaction under varying external reaction conds. Such a model does not require and thus does not contain any information about the physical nature of the elementary steps or about the rds. This can only be derived if a large number of elementary steps are characterized in their ks. in isolated model experiments requiring *surface science techniques. These elementary steps are then combines into a microkinetic model which uses kinetic constants derived under isolated (non-reactive) conds. to describe the macrokinetic behavior. Only for very simple chemical reactions (e.g., CO *oxidation to CO_2, *ammonia synth.) does such a complete description of the ks. of cat.

reaction exist. The very large number of macrokinetic models known for technically relevant procs. are often controversial and in most cases valid only for very limited ranges of reaction parameters. One of the reasons for this very serious problem in cat. research is the neglect of the time invariance of the data measured. Cats. tend to *deactivate and some of them do this so quickly that at no time is a stationary state of the system reached. Then all assumptions about kinetic constants and models are invalid and hence so are the macrokinetic models.

Another serious problem for cat. reactions is the determination of the observed ks. by transport phenomena and not by chemical reactions. The *heat transport to the cat. for endothermic reactions or the removal of heat from the cat. for exothermic reactions is often hampered by the nature of the cat. (poor heat conductor, loose particles) and by the design of the reactor (*catalytic reactors). Then the true cat. surface temp. deviates from the average measured temp. Such deviations can be as large as several hundred K and thus seriously affect the cat. reaction. The deviation further depends on the local conv. and is thus different for different *activity levels and for different gas-phase comps., e.g., along a reactor tube. Isothermal reaction conds. as assumed in all cat. models are very rarely even only crudely approximated in reality. A similar situation holds for the transport of chemicals (*mass transport); the homogeneity of the gas flow in solid or liquid media, the problem of *diffusion into cat. *pores of very wide ranges of *shapes and sizes (*intraparticle transport), and the transport through the macroscopic cat. bed (*interparticle transport) are rarely hom. and free of gradients. As consequence k. parameters are often obtained which describe the resistance of the experimental setup to the flows of mass and/or of energy. The ks. depends then on the apparatus used and on the details of the external reaction parameters and not on the catalyst/substrate chemical system. The proper application of chemical engineering not only in industrial but also in research reactors is of importance to ensure the relevance of cat. reaction data. Reports in which no indications about the validation of the data against transport artifacts are reported, should be considered with caution. **F** cinétique des réactions catalytiques; **G** Kinetik katalysierter Reaktionen.   R. SCHLÖGL

Ref.: Connors; Laidler; Schwab; Somorjai; J.R. Anderson, K.C. Pratt, *Introduction to Characterisation and Testing of Catalysts*, Academic Press, London 1985; J.A. Dumesic, D.F. Rudd et al., *The Microkinetics of Heterogeneous Catalysis*, ACS, Washington 1993; J. Hagen, *Industrial Catalysis*, VCH, Weinheim 1999; Wilkins; M. Boudart, G. Djega-Mariadassou, *Kinetics of Heterogeneous Catalytic Reactions*, Princeton University Press, 1984; Ertl/Knözinger/Weitkamp, Vol. 3, p. 1189; *Chem.Rev.* **1995**, *95*, 667; I.H. Segel, *Enzyme Kinetics*, Wiley, New York 1975.

KIST   Korean Institute of Science and Technology

Klatte, Fritz (1880–1934), industrial chemist with Griesheim-Elektron (Frankfurt, Germany), active with polymerizations of vinyl chloride, vinyl acetate, etc.   B. CORNILS

Knapsack-Griesheim acrylonitrile process

Obsolete proc. for the manuf. of *AN by a two-step synthesis, 1: the base-cat. *addition of hydrogen cyanide to acetaldehyde, and 2: the H_3PO_4-cat. *dehydration of the lactonitrile to ACN. at 650 °C. **F** procédé Knapsack-Griesheim pour ACN; **G** Knapsack-Griesheim ACN-Verfahren.   B. CORNILS

Ref.: Weissermel/Arpe, p. 303

Knietsch Rudolf (1854–1906), chemist with BASF, developed the "Kontakt" (*contact) process for sulfuric acid (together with *Winkler) and observed the necessity of ultrapurification of the reaction gases because of the poisoning by arsenic.   B. CORNILS

Knoevenagel condensation

*condensation of aldehydes or ketones with cpds. with activated methylene groups acc. to R^1-CH_2-R^2 + RCHO → R-CH=C(R^1R^2). The Kc. is cat. by both acids and bases. With strong bases even alkynes undergo the Kc. (*Nef reaction). When pyridine is used the reaction is known as the Doebner modification. **F** condensation de Knoevenagel; **G** Knoevenagel-Kondensation.   M. BELLER

Ref.: March; Trost/Fleming; *Chem.Rev.* **1996**, *96*, 115; *J.Mol. Catal.A:* **1999**, *142*, 361; *Organic Reactions*, Vol. 15, p. 204.

Knoop synthesis   converts α-oxocarboxylic acids to α-amino acids via the intermediate imino acids through *hydrogenation over Pt or Pd cats. or *Raney Ni. **F** synthèse de Knoop; **G** Knoop-Synthese.   B. DRIESSEN-HÖLSCHER

knowledge-based (expert) systems in catalysis

Kbs. are computer programs which apply knowledge about a specific domain in order to derive new conclusions. These are on the level of a human expert in this field, but constrained by the field of expertise. Applied to cat., the knowledge base contains *heuristics about relationships between chemical and physicochemical properties of solids and their cat. properties, as well as known properties for such solids. An expert system is able to combine this knowledge about different cat. properties which may be necessary to catalyze the required reaction steps, or which should be avoided because they cat. side reactions. Proposals for the selection of a cat. material and reaction conds. are reached on different levels of fundamental knowledge: on the most simple level, known cats. are selected for the desired reaction; cats. are selected which are known to cat. a reaction which belongs to the same reaction class as the desired reaction; on the most detailed level of knowledge, reaction steps are postulated by formal analysis of necessary breaking or formation of bonds, and cat. comps. are selected on the basis of knowledge about the ability of different cat. cpds. in the steps required in a specific reaction. **F** système expert; **G** wissensbasierendes System der Katalyse.   M. BAERNS

Ref.: *CHEMTECH* **1994**, *24*(8), 23; M. Doyama et al. (Eds.), *Computer Aided Innovation of New Materials II, Part 1*, Elsevier, Amsterdam 1993.

Knowles William S. (born 1917), developed with Monsanto's L-Dopa proc. (independently of *Horner) the *enantioselective *hydrogenation of 3,4-dihydroxy-*N*-acetylamino cinnamic acid with a cationic Rh-biphosphine *complex, in which the *enantioselectivity is induced by the *chiral biphosphine.

B. CORNILS

Ref.: *Acc.Chem.Res.* **1983**, *16*, 106; US 4.008.281 (1977)

Knudsen diffusion   *Diffusion is the proc. of molecular transport due to a conc. gradient associated with the stochastic movement of the individual diffusants. Diffusion coefficients D (diffusivities) in porous media may be defined on the basis of the generalized Fick's first law. They depend on the given porous medium and the temp., as well as on the diffusants involved and their concs.

At low gas press. and/or small pore sizes the diffusing molecules interact much more strongly with the walls of the pores than with each other. This type of diffusion is named *Knudsen diffusion* in contrast to molecular diffusion. This is the case if the mean free path length λ of the molecules is bigger than the pore diameter d_p. The parameter for Knudsen diffusion (gas press. p and pore diameter d_p) can be estimated.

d_p (nm)	<1000	<100	<10	<2
p (10^5 Pa)	0.1	1	10	50

λ, the Knudsen diffusion flux J_i through a cylindrical pore, and D^K_i (the Knudsen diffusion coefficient) can also be calc. in analogy to Fick's first law.

If not only a cylindrical pore but a whole porous medium is considered, an effective relative pore volume ε_p (typical values for ε_p are 0.2–0.7) and a tortuosity factor τ (typical values for τ are 3–7) have to be introduced. ε_p describes the fact that the pores are not cylindrical and τ takes into account that they are connected with each other. The expression $\varepsilon_p d_p/3\tau$ is called structure factor K_0 for Knudsen diffusion in porous media. **F** diffu-

sion de Knudsen; **G** Knudsen-Diffusion.

R. SCHLÖGL, M. HAEVECKER

Koch carboxylic acids   The reaction between (branched) alkenes, CO, and water in the presence of protons yields Koch's carboxylic acids (cf. *Koch or Koch-Haaf synth.). The reaction proceeds at <100 °C/<10 MPa in the presence of mineral or *Lewis acids. Olefinic feedstocks of medium chain length (C_7–C_{11}) are converted to C_8–C_{12} carboxylic acids which are commercially available (Versatic Acids by *Shell, Neo-Acids by *Exxon), useful for lubricants, resins, and paints. **F** l'acide Versatic (Neo); **G** Versatic(Neo)-Säuren.

C.D. FROHNING

Koch (Koch-Haaf) synthesis   In the presence of mineral or *Lewis acids, alkenes are *carboxylated to yield carboxylic acids:
$H_2C=CR\text{-}CH_3 + CO + H_2O \rightarrow H_3C\text{-}CR$ $(CH_3)COOH$. H_2SO_4, H_3PO_4, HF, or Lewis acids such as BF_3 are used as cats., which either deliver protons or stabilize the intermediate *carbenium ion. At first the alkene reacts with the acid, forming a carbenium ion, which adds CO. The intermediate complex with the acid is then *hydrolyzed, leading to the free carboxylic acid. Alcohols can be used instead of water, yielding esters directly. The reaction occurs at mild temps. (–20 to 100 °C) and press. up to 100 MPa. The initially developed high-press. proc. (DuPont 1933) was followed by a medium-press. proc. (Koch-Haaf 1955) at 40–60 °C/6–10 MPa in the presence of $H_3PO_4/BF_3 \cdot H_2O$ (*Shell proc.) or $BF_3/$ H_2O (*Exxon proc.). **F** synthèse de Koch (Koch-Haaf); **G** Koch (Koch-Haaf)-Synthese.

Ref.: Falbe-1, p. 372.   C.D. FROHNING

Kogasin process   Hydrocarbons emerging from *Fischer-Tropsch procs. were split in the first step by fractionating condensation. Besides motor fuels, two fractions received special attention: C_{11}–C_{14} (boiling range 190–260 °C; Kogasin I) and C_{15}–C_{18} (boiling range 260–320 °C; Kogasin II). Both fractions consisting of liquid (Kogasin I) and semi-liquid

(Kogasin II) *hcs. were rich in linear alkenes, predominantly *α-olefins (up to 50 wt%). Accordingly, they found application as raw materials for anionic detergents (alkane and alkylbenzene sulfonates resp.) with excellent tenside and ecological properties. After shutdown of the Fischer-Tropsch plants in Europe they were replaced by the cheap but less degradable tripropylene consecutive prods. **F** procédé Kogasin; **G** Kogasin-Verfahren.

Ref.: Falbe-2, p. 391.   C.D. FROHNING

Kölbel-Engelhardt synthesis special variant of the Fischer-Tropsch synthesis, starting from CO and water:

$$3 CO + H_2O \rightarrow \text{-(-CH}_2\text{-)-} + 2 CO_2$$
$$\Delta H_{227\,°C} = -244.5 \text{ kJ}$$

The max. attainable yield amounts to 208.5g of alkenes per m^3 of gas at complete conv. The reaction was discovered, making direct use of low-value gases rich in CO (coke oven, carbide prod.), combining the *WGSR and the *hc. synth. Cats. are based on Fe, Co, Ru, or Ni, suspended in long-chain hcs. generated in the proc. The temp. is 240–280 °C, press. approx. 2 MPa; the reaction takes place in simple tank reactors. The proc. has been demonstrated on pilot plant scale. **F** synthèse de Kölbel-Engelhardt; **G** Kölbel-Engelhardt-Synthese.

C.D. FROHNING

Kontakt Historical term for a catalytically active substance, coined by Mitscherlich in 1833. So far, only a little experience had been gathered concerning the acceleration of chemical reactions by metals. *Davy had discovered the positive effect of a Pt wire in the *oxidation of H by air (1816), presumably the first example of cat. *Berzelius created the term "catalysis" in 1835, *Ostwald collected and extended in 1896 the former understanding and formulated the hitherto valid definition of cat. By choosing the German expression "Kontakt" (contact) for a cat. active substance, Mitscherlich wanted to indicate that such a substance was able to influence the

rate of a chemical reaction only by "contact" with the reactants . **E=F=G**.   C.D. FROHNING

Kontakt process → contact process

Krebs cycle Also called the tricarboxylic acid cycle or the *citric acid cycle, this cat. cycle of reactions results in the *oxidation of acetate groups (fed into the cycle in the form of *acetyl coenzyme A) to CO_2. Each round of the cycle generates two CO_2 molecules and results in the prod. of 3 equiv. of *NADH (and H^+) from NAD^+, one equiv. of reduced FADH$_2$ from FAD (flavine adenine dinucleotide) and one equiv. of GTP from GDP and P_i. The Kc. produces the precursors of a variety of other biomolecules. Many of the amino acids are formed from Kc. intermediates. Glutamate and glutamine are formed directly from α-ketoglutarate, for example, while aspartate and asparagine are synth. from oxaloacetate. Acetyl-CoA, an intermediate formed from the pyruvate formed during glycolysis, is also used in the synth. of lipids. **F** cycle de Krebs; **G** Krebs-Cyclus.   P.S. SEARS

KRES process Kellogg reforming exchange system, → Kellogg procs.

Kriwitz-Prins olefin-formaldehyde addition acid-cat. *addition of alkenes (preferably tertiary and asymmetric) and formaldehyde to 1,3-glycols or cyclic acetals of formaldehyde.   B. CORNILS

Ref.: Krauch/Kunz.

Kuhlmann Karl Friedrich (1803–1881), professor of chemistry in Lille (France) and company owner, worked on the *oxidation of ammonia to nitric acid.   B. CORNILS

Kuhlmann (PCUK) oxo process special variant of Co-based hydroformylation with its own cat. cycle technology, involving two steps, 1: the recovery of Co from the crude oxo reaction prods. as sodium cobalt carbonylate, and 2: its regenerative conv. to HCo (CO)$_4$, which is extracted by the feed alkene. The reaction takes

place at 160–180 °C/30 MPa total press. in special loop reactors which ensure circulation and
mixing of liquid and gaseous reactants by the
mammoth pump principle (Figure 1).

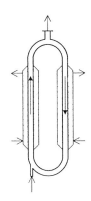

Originally designed for propylene hydroformylation, the technology was later transposed
to higher olefins and taken over by Exxon
(now "Exxon proc.") (Figure 2). **F** procédé
oxo de Kuhlmann; **G** Kuhlmann Oxoverfah-
ren.   B. CORNILS

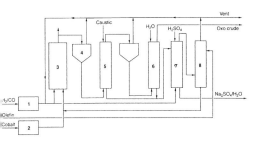

3 reactor; 4 separator; 5–8 catalyst recycle

Ref: *Hydrocarb.Proc.* **1966**, *45*(2), 148; Falbe-1, Cor-
nils/Herrmann-1, p. 71.

Kulinkovich hydrocyclopropanation

The reaction of a *Grignard reagent with a
carboxylic methyl ester in the presence of up
to one equiv. of Ti isopropoxide produces 1,2-
trans-disubstituted hydroxycyclopropane in
good yields (*cyclopropanation).

$$RCH_2\text{-}CO_2Me + 2\,R_1CH_2CH_2MgBr + Ti(O\text{-}iPr)4 \longrightarrow$$

Titanacyclopropanes are key intermediates in
the reaction, which is diastereoselective and

has been extended to the synth. of cyclopro-
pylamines from acid dialkylamides. **F=G=E**.
                                                 A.F. NOELS

Ref.: *Synthesis* **1991**, 234; *J.Am.Chem.Soc.* **1994**, *116*, 9345.

Kuntz-Cornils catalyst   the Rh(I) *complex of the trisulfonated triphenylphosphine
(*TPPTS), the most active cat. in the *aqueous-phase *hydroformylation (*Ruhrchemie/
Rhône-Poulenc proc.) of lower alkenes (Figure).

This cat. landmarks the breakthrough development of commercial biphasic catalysis and con-
tradicted the prejudice that organometallic
cpds. could not be applied in the aqueous phase.
F catalyseur Kuntz-Cornils; **G** Kuntz-Cornils-
Katalysator.   W.A. HERRMANN

Ref.: *Angew.Chem.Int.Ed.Engl.* **1993**, *32*, 1524; *Adv.
Organometal.Chem.* **1992**, *34*, 219; *J.Organometal.
Chem.* **1995**, *502*, 177.

Kuraray telomerization process   manuf.
of 1-octanol (*n*-octanol) and 1,9-nonanediol
through a multistep series of reactions. 1: the
*telomerization/*hydrodimerization of butadiene and water over Pd/*TPPMS/*phosphonium salt cats. to 2,7-octadien-1-ol (Figure); 2:
the cat. *hydrogenation of this dienol to *n*-oc-
tanol; or 3: the *hydroformylation of the dienol to 7-octenal, which is 4: hydrogenated to

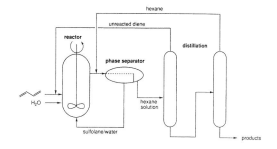

1,9-nonanediol. The prods. are extracted with hexane; the Pd cat. in the residual aqueous phase is reused. **F** procédé telomérisation de Kuraray; **G** Kuraray Telomerisations-Verfahren.   N. YOSHIMURA

Ref.: Cornils/Herrmann-1, p. 315; Cornils/Herrrmann-2, p. 408.

Kureha vacuum residue (VRC) process for upgrading residues by *steam cracking and simultaneous steam stripping with superheated steam at atmospheric press. **F** procédé VRC de Kureha; **G** Kureha VRC-Verfahren.   B. CORNILS

Kutepow Nikolaus von (1910–1986), industrial chemist with BASF, developed the BASF high-press. acetic acid proc. (*BASF AA proc.).   B. CORNILS

Ref.: *Chem. Ing. Tech.* **1965**, *37*, 383.

Kutscherov acetylene hydration synth. of carbonyl cpds. through *hydration of acetylenes cat. by Hg salts or BF₃. Vinyl alcohols are formed at an intermediate stage. Acc. to *Markownikov's rule 1-alkynes and diarylacetylenes yield ketones. **F** Kutscherov hydratation d'acétylène; **G** Kutscheroff Acetylenhydratisierung.   B. CORNILS

Ref.: Krauch-Kunz.

Kvaerner process for the manuf. of ethyl acetate (acetic acid ethyl ester) by *Tishchenko reaction of acetaldehyde. The special proc. starts with an impure ethanol feedstock (e.g., from *Fischer-Tropsch plants) which is firstly *dehydrogenated on a promoted Cu cat. at 200–250 °C. **F** procédé Kvaerner pour l'acé-

tate d'éthyle; **G** Kvaerner Ethylacetatverfahren.   B. CORNILS

Ref.: *ECN* **1999**, Feb.1–7, 26.

L-kynurenine hydrolase (kynureninase; EC 3.7.1.3) a pyridoxal phosphate dependent *enzyme that cat. the synth. of L-kynurenine from L-alanine and *o*-aminobenzoic acid. **E=F=G**.   C.-H. WONG

Ref.: *J. Am. Chem. Soc.* **1991**, *113*, 7385.

Kyowa Hakko L-lysine process for the manuf. of L-lysine monohydrochloride from carbohydrate materials (or acetic acid) by *biocat. *fermentation with the help of *Corynebacterium* gluamicum at a controlled pH by feeding the *bioreactor with air, ammonia, and other nutrients at 30 °C (Figure). **F** procédé de Kyowa Hakko pour le L-lysine; **G** Kyowa Hakko L-Lysinverfahren.

B. CORNILS

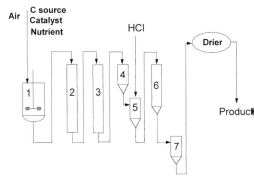

1 fermentation; 2, 3 adsorption, elution to and from ion exchange resin; 4, 5 concentration, neutralization; 6, 7 crystallization

Ref.: *Hydrocarb. Proc.* **1977**, *56*(11), 178.

L

L *ligand

LAB linear alkylbenzenes, manuf. by cat. *alkylation of benzene, e.g., proc. of *UOP

labile complexes complexes with quickly exchanging *ligands. **F** complexe labile; **G** labiler Komplex. R.D. KOEHN

laboratory catalytic reactors → autoclaves, *Berty reactor, *Carberry type reactor, *Parr hydrogenator.
Ref.: Ertl/Knözinger/Weitkamp, Vol. 3, p. 1377; Augustine.

LABS linear alkylbenzene sulfonate

LAC → ligand-accelerated catalysis

LAC process Linde ammonia concept, → Linde processes

lac-repressor fusion Nucleic acid can be readily amplified, either by replication in a living system or via the *polymerase chain reaction (PCR). Polypeptides, however, cannot be amplified unless they are linked in some fashion to their cognate nucleic acid and passed through a living system. A variety of biological techniques exist for producing *libraries of polypeptides linked to their cognate *DNA (or *RNA) in order for affinity selection and peptide replication to be readily coupled. One method involves the prepn. of a fusion gene between a polypeptide of interest and the *lac* repressor gene (*LacI*). **E=F=G**.
P.S. SEARS
Ref.: *Proc.Natl.Acad.Sci.USA* **1992**, 89, 1865; *Methods Enzymol.* **1996**, 267, 171.

lactams An interesting approach to the synth. of the C 13 side chain of the anticancer drug Taxol is based on the lipase-cat. *enantioselective transformation of 3-hydroxy-4-phenyl β-lactam derivatives. *Pseudomonas* *lipase (P-30) was found to be highly selective for the *hydrolysis of the 3-acetoxy derivatives, with the ring nitrogen free or protected with a benzoyl or *p*-methoxyphenyl group. **F** lactames; **G** Lactame. C.-H. WONG

Ref.: *J.Org.Chem.* **1993**, 58, 1068.

lactase (β-galactosidase; EC 3.2.1.23) hydrolyzes β-linked galactose units, in particular that found in lactose (galactose 1,4-glucose). It is used commercially for treatment of milk prods. to *hydrolyze lactose for consumption by individuals with lactose intolerance. Lactase activity has been isolated from a variety of organisms, but the *enzymes from fungal sources (*Saccharomyces lactis, Aspergillus niger*) are most commonly used in food processing. **E=F=G**. P.S. SEARS

lactate dehydrogenase (LDH; EC 1.1.1.27) L-Lactate dehydrogenase (LDH) cat. the *reduction of pyruvate to L-lactate in the presence of *NADH. The commercial sources of this *enzyme are numerous: rabbit muscle, porcine heart, lobster tail, etc., and *Bacillus stearothermophilus*. The equil. of the reaction lies in favor of the reduction.
LDH accepts a variety of 2-oxo acids as substrates. The *S*- hydroxyacids produced are valuable *chiral synthons. Several of the prods. have been used as precursors for useful chiral prods. Due to the cost of the enzyme and ease

of isolation of prods., LDH represents a useful synth. cat.; in addition, this enzyme provides an excellent system for the regeneration of the oxidized *cofactor.

D-LDH cat. the reduction of pyruvate to D-lactate in vivo. The enzyme from *Leuconostoc mesenteroides* reduces a range of 2-oxo acids to the corresponding *R*-hydroxy acids with high enantioselectivity. In a sense, D- and L-LDH are complementary to each other in that the opposite enantiomers of a number of substrates are obtained. The substrate specificity of D-LDH is, however, substantially narrower than that of the L-enzyme. The D-enzyme showed less tolerance for side chains longer than three carbons, as these cpds. were reduced only slowly. **E=F=G**.   C.-H. WONG

Ref.: *J.Am.Chem.Soc.* **1988**, *110*, 2959 and **1989**, *111*, 6800; *Appl.Biochem.Biotech.* **1989**, *22*, 169; *Science* **1988**, *242*, 1541.

lactonization (enzyme-catalyzed) →
Baeyer-Villiger oxidation

lactoperoxidase → peroxidase

La Grande Paroisse DeNOx process   for the treatment of effluents of nitric acid plants by *selective *reduction by NH_3 over a "FeCr (GPPTII)" cat.   B. CORNILS
Ref.: *Appl.Catal.* **1994**, *115*, 179.

lambda (λ) value   is the fuel/air ratio important for the proper operation of *automotive exhaust cats. At λ <1 the *activity for NO reduction is high, but not for the *oxidation of CO and *hcs.; at λ >1 the reverse is true.
    B. CORNILS

LAMMA   identical with *LMMS

Langmuir   Irving (1881–1957), industrial chemist with General Electric Co., worked on surface chemistry, catalysis, the theory of chemical bonds, and the adsorption of monomolecular films. Coined the term "*active sites". Nobel laureate in 1932.   B. CORNILS
Ref.: *Nobel Lectures, Chemistry 1922–1941*, Elsevier, Amsterdam 1966.

Langmuir-Hinshelwood kinetics
describe a method for representing the dependence of the *rate of a cat. reaction on the press. or concs. of the reactant(s) in the fluid phase. The *surface *coverage θ_A of an *adsorbed molecule A with its press. P_A in the gas phase is related acc. to *Langmuir's equ. (often but incorrectly called his isotherm). Equating rates of adsorption and desorption at equil., one obtains the relation $\theta_A = b_A P_A/(1 + b_A P_A)$, where b_A is the adsorption coefficient (i.e., the equil. constant k_a/k_d, these being the rate constants respectively for *adsorption and *desorption). This simple derivation requires a number of assumptions to be met: each adsorption site accommodates only one molecule; molecules are adsorbed without dissociation; all *sites are energetically equal (i.e., the adsorption enthalpy is independent of *coverage). They are not all likely to be valid in practice. This equ. also describes adsorption of A from solution when the solvent is not itself adsorbed. This concept is readily extended to describe the situation where the molecule dissociates into two or more fragments as it is adsorbed. When two molecules A and B compete for the same surface

$$\theta_A = b_A P_A/(1 + b_A P_A + b_B P_B)$$
$$\text{and } \theta_B = b_B P_B/(1 + b_A P_A + b_B P_B)$$

If it is now supposed that the rate r of a unimolecular reaction is proportional to the coverage by the reactant $r = k\theta_A = kb_A P_A/(1 + b_A P_A)$, where k is a constant of proportionality termed the *rate constant. Thus when $b_A P_A \ll 1$, because either b_A or P_A is small, the rate is first order in P_A, but when $b_A P_A \gg 1$, because either b_A or P_A is large, then the rate is independent of P_A, i.e., the reaction is of zero order.

When two molecules A and B, competing for the same surface, react, then the rate r is given by $r = k\theta_A\theta_B = kb_A P_A b_B P_B/(1 + b_A P_A + b_B P_B)^2$. Consider the case where $b_A = 10b_B$. As P_A is increased from a low value, P_B being held constant, the rate at first rises, then reaches a maximum when $\theta_A = \theta_B$ (i.e., when

$P_A = 0.1P_B$), after which it declines continuously (Figure 1). As P_B is increased from a low value, the rate will increase linearly at first (Figure 2), before achieving a maximum when $P_B = 10P_A$.

The further extension of this procedure to account for reactant dissociation, *inhibition by strongly adsorbed prods., etc., is associated with the names of Hugen and Watson. These equs. have very considerable utility in affording empirical descriptions of rate dependencies on reactant press., notwithstanding their insecure theoretical foundation.

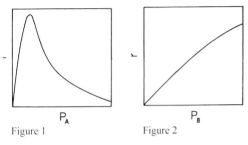

Figure 1   Figure 2

If the temp. is raised, the press. P_A of a single reactant being held constant, its coverage will decrease if (as is usual) its adsorption is exothermic. It is therefore expected to find a progressive *increase* in the order of reaction from close to zero to approaching unity as the range of coverage in which the measurements are made falls from high to low. In the case of the bimolecular reaction, A being again the more strongly adsorbed, raising the temp. will cause the press. of A causing the maximal rate to increase and the order above the maximum to become less negative. Eventually both reactants will be only weakly adsorbed, and so the orders will be first in each. If the effect of raising the temp. is to cause one of the reactants to desorb quickly, so that the product $\theta_A\theta_B$ falls faster than the rate constant rises, then the rate will pass through a maximum, after which the apparent activation energy will become negative (see also *Arrhenius plots, *apparent activation energy, *rate constants).

Since temp. coefficients of surface reactants tend to be greater than those of *mass-transport, a reaction may pass from chemical control to diffusion control when the temp. is raised sufficiently. Deactivation procs. also become more marked and may cause the *apparent activation energy to seem to decrease. **F** cinétique de Langmuir-Hinshelwood; **G** Langmuir-Hinshelwood-Kinetik.   G.C. BOND

Ref.: Laidler; Butt, *Reaction Kinetics and Reactor Design*, Prentice-Hall, London 1980; Thomas/Thomas, p. 65; Moulijn/van Leeuwen/van Santen.

Langmuir isotherm   the elementary and most frequently used *adsorption isotherms, named after I. *Langmuir. The Li. describes an *adsorption/*desorption equil. of a gas in contact with a *surface valid under a number of simplifying assumptions: all sites of the surface are energetically equiv. and no multilayer adsorption occurs. In the Li. the *coverage is proportional to the press. at low press., and at high press., the *coverage asymptotically approaches a monolayer. **F** isotherme de Langmuir; **G** Langmuir-Isotherme.   R. IMBIHL

lanosterol   *2,3-Oxidosqualene lanosterol cyclase cat. the synth. of a number of lanosterol analogs. When using an *enzyme suspension from Baker's yeast containing this cyclase, catalysis was promoted by ultrasonic irradiation. The cyclase was also utilized to synth. a C 30 functionalized lanosterol to form the steroid ring system, while leaving the normally labile vinyl group at C 30 intact. This prod. was subsequently converted to

(+)-30-hydroxylanosterol and the corresponding aldehyde. These cpds. are natural receptor-mediated feedback *inhibitors of HMG-CoA reductase, and, therefore, are of interest in the development of hypocholesteremic drugs. **E=F=G**. C.-H. WONG

lantha La oxide, used in *washcoats for *automotive exhaust cats. (cf. *stabilizers)

lanthanides as catalyst metals → rare earth metals as catalysts

LAO linear α-olefins, prod. by *dehydrogenation of alkanes or *oligomerization of ethylene (procs. of *Ethyl Alfen, *Exxon LAO, *Gulf, *Shell SHOP, etc.). B. CORNILS

Laporte epichlorohydrin process
manuf. of epichlorohydrin by *epoxidation of allyl chloride with perpropionic acid acc. to the *Interox process to yield e. at about 70–80 °C. **F** procédé Laporte pour l'épichlorhydrine; **G** Laporte Epichlorhydrin-Verfahren. B. CORNILS

Laporte Interox process → Solvay/Laporte/Carboquimica process

LAR process → Alusuisse procs.

laser-induced fluorescence spectroscopy (LIF) for kinetic information by in-situ analysis of traces in gaseous phases. The method is non-destructive and very specific (lateral resolution in reaction volume). R. SCHLÖGL

laser microprobe mass spectroscopy (LMMS, LMS, LAMMA) delivers information about local surface and bulk compositions. The *ex-situ* technique (high vacuum) is destructive to the sample, and the analysis of data is complex. R. SCHLÖGL

laser mass spectroscopy (LMS) → laser microprobe MS

laser Raman spectroscopy (LRS) fo the vibrational analysis of cats. with lase light (IR, visible). The in-situ capabilities o the ex-situ method are good; the data are in tegral and bulk-sensitive. R. SCHLÖGL

laser spectroscopy *spectroscopy usin laser light, often coupled with other method such as fluorescence s. (*LIF), *MS (*LMMS), or Raman s. (*LRS). R. SCHLÖGL
Ref.: Ullmann, Vol.B5, p.653; E.R. Menzel, *Lase Spectroscopy*, Dekker, New York 1995.

laydowns → deposits

LCA leading concept ammonia, → IC AMV process

LCAO linear combination of atomic orbit als, → ab-initio calculations

LC-Fining → Lummus procs.

LCT → lower critical temperature

LDA lithium diisopropylamide

LDC ligand-decelerated catalysis, → ligand accelerated catalysis

LDF local density functional method, → density functional theory

L-dopa (L-DOPA, Levodopa)
3-(3,4-dihydroxyphenyl)alanine, a pharmacologically important amino acid; first example of an *enantiomeric reaction via hom. cat synth. (cf. *Monsanto procs.). **F=G=E**.
Ref.: Gates-1, p.85. B. CORNILS

LDPE → low-density polyethylene

LDW process → Mobil procs.

leaching 1: Besides the hydrometallurgical extraction (leaching or bioleaching) of ores in general, *leaching* (bleeding, carry-over), especially in cat. sciences, stands for the loss of cat. active metals, metal *complexes, or sites of hom. cats. from het. or heterogenized (*immobilized) cats. with the reactants or reaction prods. – a severe problem of *anchored cats. The reason is that the metal complexes should not be too strongly bonded to the matrix (the *support), thereby causing *mass transfer problems and a drop in reaction velocity. On the other hand, the linkage is not allowed to be so weak that – despite anchoring – cat. losses occur which would necessitate reprocessing steps. These contradictory requirements lead to leaching, which normally results from the dissociation of the metal from the *support with the anchored ligands, thus liberating the (active) molecular cat. Leaching can also originate from structural changes with concomitant weakening of certain bonds during the cat. cycle, where the coordination spheres of a metal undergo intermittent and severe changes. Plotting the Rh distribution in recovered sectioned reactor beds shows that the total Rh content of the cat. charge decreases and the Rh deficiency starts at the top of the reactor, arising from transport phenomena as a consequence of desorption/absorption effects (Figure).

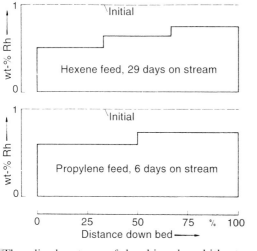

The disadvantage of leaching has hitherto only been discussed with regard to the metal centers of cats.; it is noted, however, that bleeding losses can also involve expensive ligands, especially those necessary in stereoselective cats. So far no commercial proc. with immobilized cat. beds containing metal complexes is operational; in all cases the problem of leaching was underestimated.

2: A procedure for cats. which are prepared from metal alloys or intermetallic cpds. with two or more constituents by selectively dissolving (leaching) more or less all but one metal. The remaining metal has a microscopic spongy network of pores (cf. *Raney cats. *skeletal cats., *Urushibara cats.). **F** lessiver; **G** leaching, Laugung.   B. CORNILS
Ref.: Falbe-1.

lead chamber process   the historic method of producing sulfuric acid by *oxidation of sulfur dioxide, invented in 1746 by Roebuck. SO_2 reacts with air or oxygen in the presence of water and of N oxides. The N oxides (effectively $NOHSO_4$) serve as hom. cats. for the oxidation. Because the reaction is rather slow, sufficient *residence time must be provided for the mixed gases to react. This gaseous mixture is highly corrosive, and the reaction must be carried out in containers made in large, boxlike chambers of sheets of lead. The lcp. has been largely replaced in modern industrial production by the *contact proc. **F** procédé du chambre de plomb; **G** Bleikammerverfahren.   R. SCHLÖGL, M. DIETERLE

lead compounds in catalysis   The first *DNAzyme isolated had ribonuclease activity and cat. the Pb^{2+} dependent cleavage of RNA. Pb cpds. have been proposed for the *remote or cyclocarbonylation (with $[Pb(OAc)_4]$) or as cat. *driers.   B. CORNILS
Ref.: *Organic Reactions* Vol. 19, p. 279; *Houben-Weyl/Methodicum Chimicum*, Vol. XIII/7.

Lebedew process   an older proc. for the synth. of 1,3-butadiene through *dehydrodimerization of ethyl alcohol over $MgO-SiO_2$, $SiO_2-Al_2O_3$, or Al_2O_3-ZnO cats. at 380 °C. This proc. is important for countries without sufficient petrochemical resources. **F** procédé de Lebedew; **G** Lebedew-Verfahren.   B. CORNILS
Ref.: GB 331.482 (1930); *Chem. Ztg.* **1963**, 87, 577.

Lederer-Manasse reaction → hydroxy-alkylation

LEED → low energy electron diffraction

Lehn, Jean-Marie (born 1939), professor of chemistry at Strasbourg (France) and Collège de France (Paris), works on macroyclic cpds. and their complexes (*cryptands) and supramolecular chemistry. Nobel laureate in 1987 (together with Cram and Pedersen). B. CORNILS
Ref.: *Angew.Chem.Int.Ed.Eng.* **1988**, *27*, 90.

LEISS low-energy ion scattering spectroscopy, → ion scattering spectroscopy

Leloir, Luis Federico (1906–1987), professor of physiology in Buenos Aires. Worked on nucleotides, metabolism, etc. Nobel laureate in 1970. B. CORNILS

Leloir pathway refers to the metabolic pathway for galactose utilization (as described by *Leloir), which involves phosphorylation of galactose to galactose 1-phosphate, transfer of UMP to give UDP-galactose, and epimerization to UDP-glucose. **E=F=G**.
P.S. SEARS

Lennard-Jones equation Simple interaction potential between two molecules which is also used to describe the potential energy diagram for *physisorption. The LJe. contains an attractive term A/r^6 and a repulsive term B/r^{12}. The two parameters, A and B, can be calc. from gas-phase polarizabilities and ionization potentials of the atoms involved. **F** équation de Lennard-Jones; **G** Lennard-Jones-Gleichung. R. IMBIHL

Leonard amine process manuf. of methyl amines by *amination (*ammonolysis) of methyl alcohol and NH_3 over Al oxide or Al phosphate cats. at 300–500 °C/2 MPa. Mono-, di-, and trimethylamines are obtained. **F** procédé Leonard pour des amines; **G** Leonard Aminverfahren.
Ref.: Winnacker/Küchler, Vol. 6, p. 127. B. CORNILS

Leonard formic acid process manuf. of formic acid by a two-step proc. 1: synth. of *methyl formate by *carbonylation of methanol and insertion of CO in the O-H bond over *alkoxides (Na methylate) at 70 °C/2–20 MPa; 2: the *hydrolysis of the methyl formate to formic acid. **F** procédé Leonard pour l'acide formique; **G** Leonard Ameisensäure-Verfahren. B. CORNILS

leucine dehydrogenase (EC 1.4.1.9, LeuDH) cat. the *oxidation of L-leucine to the corresponding α-ketoacid using *NAD. LeuDH from *Bacillus cereus* exhibits broad substrate specificity, as does *glutamate dehydrogenase. Although less developed than other ketoacid dehydrogenases, LeuDH has the potential to synth. unnatural amino acids with branched side chains such as *tert*-L-leucine. The *enzyme has been used in a *membrane reactor for the continuous prod. of L-amino acids via *reductive amination of 2-oxoacids. As with the other ketoacid *dehydrogenases the enzyme utilizes *NAD(H) as *cofactor. These *chiral molecules are *precursors to a number of valuable cpds. **E=F=G**. C.-H. WONG
Ref.: *Appl.Microb.Biotechnol.* **1985**, *22*, 306.

Leuckart-Wallach amine alkylation by reaction of primary or secondary amines with carbonyl cpds. and formic acid over *Lewis acids as cats. HCOOH acts as a H source in *transfer hydrogenation as well as in the variant of the Eschweiler-Clarke reaction. **F** Leuckart-Wallach alkylation d'amines; **G** Leuckart-Wallach Aminoalkylierung. B. CORNILS

LEUPHOS, leuphos chiral β-aminoalkyl *phosphines derived from α-amino acids; they are especially suited for *Grignard cross-coupling reactions. LEUPHOS is, e.g., (*S*)-1-dimethylamino-2-diphenylphosphino-4-methylpentane which may be used as ligand for Ni- or Pd-cat. asymmetric *cross-coupling of 1-arylethyl *Grignard reagents (over 80 % **e**e*). O. STELZER
Ref.: *J.Org.Chem.* **1983**, *48*, 2195; *Org.Synth.* **1988**, *66*, 67; Cornils/Herrmann-1, Vol2, p. 764.

Lewis Gilbert Newton (1875–1946), professor of physical chemistry in Berkeley, was active on atomic models, covalency, coordination, extending the acid-base theories of *Brønsted and *Lowry (*Lewis acid-base concept).

B. CORNILS

Ref.: Jensen, *The Lewis Acid-Base Concepts*, Wiley 1980.

Lewis Warren K. (1882–1975), industrial chemist and university teacher, developed together with Edwin R. Gilliland at Standard Oil the *FCC technology for the fluidized-bed cracking of higher to lower molecules of superior quality (e.g., octane numbers of 100).

B. CORNILS

Lewis acid-base concept proposed by G.N. *Lewis as an acid-base theory which is more comprehensive than that of *Brønsted. Lewis acids are neutral or charged electron-deficient species with a vacant orbital which act as electrophilic reagents. Lewis acids include *Brønsted acids, metal ions, metal atoms, neutral *complexes (e.g., BF_3, $AlCl_3$), and so-called *superacids. Lewis bases are also *Brønsted bases, i.e., molecules or ions possessing available electron pairs which act as nucleophilic reagents (e.g. OH^-, R_3N, CO). A Lewis acid-base reaction is in its broadest sense any proc. in which Lewis acid (A,A′) and base (B,B′) molecules or ions associate, dissociate, or exchange partners (Figure).

$$A + |B \rightleftharpoons A—B$$

$$A + A'—B \rightleftharpoons A—B + A'$$

$$|B + A—B' \rightleftharpoons A—B + |B'$$

$$A—B + A'—B' \rightleftharpoons A—B' + A'—B$$

The l.abc. also comprises non-ionic acid-base transformations, including complexation reaction involving different *coordination numbers, e.g., the reaction of Cu^{2+} with four molecules of neutral NH_3 to yield $Cu(NH_3)_4]^{2+}$ (*coordination complex). The bonds formed can be of ionic or covalent nature. When a Lewis acid combines with a base to form a negative ion in which the *central atom has a higher-than-normal valence, the resulting salt is called an *ate complex. Lewis acids and bases are classified acc. to Pearson's *HSAB concept. Lewis acid-base interactions are involved in procs. such as hom. and het. catalyses, CO or oxygen binding to hemoglobin, the hydrogen bonding of base pairs in *DNA, etc. **F** concept acido-basique de Lewis; **G** Lewis Säure-Base-Prinzip.

R. ANWANDER

Ref.: R.S. Drago, N.A. Matwiyoff, *Acids and Bases*, Heath, Boston 1986; Finston/Rychtman; *J.Chem.Educ.* **1974**, *51*, 300.

Lewis acid-catalyzed radical reactions
Compounds like $ZnCl_2$ or $EtAlCl_2$ accelerate the addition rate of *radicals to electron-poor alkenes by complexation with nitrile or ester groups. Such *Lewis acids not only cat. *polymerizations but alter the regioselectivity during *copolymerization. First applications of Lewis acids in diastereoselective radical reactions were found when the Lewis acid complexed the radical precursor or the alkene. The disadvantage of this strategy consists in the necessity for stoichiometric amounts of Lewis acids. In combination with *chiral *ligands, the methodology was extended to enantioselective radical reactions. **F** réactions radicalaires catalysées par acide de Lewis; **G** Lewis-Säure katalysierte Radikalreaktionen.

T. LINKER

Ref.: *Angew.Chem.Int.Ed.Eng.* **1998**, *37*, 2562.

LGO → light gas oil

LHDPE → linear high-density polyethylene, *high-density polyethylene

LHSV → liquid hourly space velocity

library A l. is simply a large collection. It may be a collection of genes, as in a cDNA library or a l. of mutants (which may be inserted into a viral genome, to create a *phage display l.); or it may be a collection of related cpds. derived from a set of chemical transformations in which the precursors are varied, to create a combinatorial library (*combinatorial catalysis). **F** bibliothèque; **G** Bibliothek.   P.S. SEARS

LIF → laser-induced fluorescence (spectroscopy)

lifetime → catalyst lifetime

ligand-accelerated catalysis (LAC) The addition of *ligand increases the reaction *rate of hom. cat. convs. (Equ.). This ligand acceleration and the non-accelerated reactions proceed simultaneously and in competition.

$$A + B \xrightarrow[k_0]{\text{catalyst}} \text{product} \xleftarrow[k_1]{\text{cat. + ligand}} A + B$$

Typical examples are the Ti/tartrate-cat. *asymmetric *epoxidation (cf., e.g., *Butakem tartaric acid proc.) and the OsO_4-cat. asymmetric *dihydroxylation, where the acceleration k_1/k_0 for different alkenes and ligands exceeds a factor of $>10^3$. The extent of the ligand acceleration is a function of the nature of the ligand and the reactant and its conc. LAC is also described with het. cats. (e.g., *hydrogenations employing modified Pd or Pt cats.). The inverse effect is called ligand-decelerated catalysis.   D. HELLER
Ref.: *Angew.Chem.Int.Ed.Engl.* **1995**, *34*, 1059.

ligand-decelerated effect → ligand-accelerated catalysis

ligand effects Ligands (ls.) in *complexes may influence other coordination sites at the metal by les. These can be used to optimize the cat. *activity of a complex by *ligand tuning. Les. can be divided into steric and electronic effects. There have been several concepts to quantify these effects. The steric bulk of *ancillary ls. determines the size and shape of *vacant sites during the cat. *cycle. and can thus influence the *selectivity of cats. Steric effects can also influence the *rate and

*mechanism of l. *substitution procs. (associative vs. dissociative). *Tolman introduced the concept of the *cone angle to quantify the steric factors of *phosphines (cf. *χ- and the *Θ-values). Similar angles in other cat. systems are used, e.g., the *aperture angle in *metallocenes. A quantitative measure of the electronic les. (cf. *electronic factors) may be the gas-phase ionization energies of the donor orbital or pK_a values of the free l., both of which have been used for phosphines and other ls. The electronic influence can often be studied by varying the p-substituents in phenyl groups acc. to the *Hammett σ-values. A sensitive probe for electron les. at the metal are the stretching frequencies of coordinated CO. Electronic les. have a great influence on the *redox potentials and *Lewis acidity-basicity. **F** effets de ligand; **G** Ligandeneffekte.   R.D. KOHN
Ref.: D.R. Salahub, N. Russo (Eds.), *Metal-Ligand Interactions*, Kluwer, Amsterdam 1992; Moulijn/van Leeuwen/van Santen, p. 203.

ligand exchange plays a central role in certain cat. procs., since a variety of substrates can act as a *ligand and/or is activated by coordination to a metal center. Ligand exchange can take place either by a dissociative or by an associative mechanism. In one case, which is observed for most of the hexacoordinated metal centers and some square planar *complexes of the Pt metals, a free site for the coordination of the substrate is generated. However, in most four-coordinate systems of cat. relevance, ligand exchange is induced by coordination of a donor molecule and followed by dissociation of a coordinated ligand. **F** èchange de ligand; **G** Ligandenaustausch.
                                        W.R. THIEL

ligand field theory → coordination chemistry theory

ligand-modified catalysts ligand-derived *complex cats. for the hom. cat. on the basis of e.g., metal carbonyls. Thus, HRh(CO)(TPPTS)

a TPPTS-modified HRh(CO)$_4$ cat. for *hy-
roformylation reactions (cf. also *complex,
ligands, *aqueous-phase cat., *TPPTS).

B. CORNILS

igands 1: Surrounding molecules whose do-
ior atoms are bound to a *central metal atom
r an ion in coordination cpds. A l. can be
ieutral or charged and the number of ligands
s the coordination number. *Complexes with
nly one type of l. are called *homoleptic. A
i- or multidendate l. has two or several
onds to the metal and forms *chelate com-
lexes. The angle between two bonds to the
netal in a bidentate l. is the *bite angle. The
iumber n of donor atoms by which the ligand
s bound to the metal is given by the *hapti-
ity (η^n). Ls. are often classified into *classical*
nd *non-classical*. Classical ls. such as halide
nions, oxygen donors like H$_2$O, ethers, or
etones, amines, or amides, are *Lewis bases
lonating a lone pair of electrons in a single
5-bond to the metal with possible additional
z-donation. Non-classical ls. such as CO,
phosphines, *carbenes, *Cp$^-$, and arenes, are
z-acids with low-lying empty orbitals of π-
ymmetry which allow π-*backbonding from
lectrons in metal orbitals. Since electrons in
l-orbitals are ideally suited for π-bonding,
omplexes with non-classical ls. are mostly
ormed with *transition metals. The empty l.
irbitals are typically low-lying antibonding
irbitals in the l. with a possible contribution
if empty d-orbitals. Other classifications of ls.
ire acc. to their formal number of electrons
lonated, the number of donor atoms, or the
ype of donor atom (e.g., O-donor ls.). An
'ambidentate l. can have different modes of
ionding to the metal, e.g., M-NO$_2$ or MONO.
Acc. to *Pearson's concept of *hard and soft
icids and bases, hard ls. (e.g. H$_2$O, F$^-$, NR$_3$)
vill preferentially bind to hard metal ions
e.g., Ln^{3+}) and soft ls. (e.g., SR$_2$ or I$^-$) to soft
netal atoms or ions (e.g., Au(I). The thermo-
lynamic stability of the binding of the l. L to
i metal M is expressed by the *stability con-
tant K for the reaction ML = M + L, often
neasured in aqueous solution.

Especially stable ls. are multidentate (*chelate effect) and *macrocyclic ls. (*macrocyclic
effects) such as *crown ethers and *porphyrins. Multidentate ls. with one central connecting unit are called *podants. Complexes
with kinetically stable ls. are inert; those with
quickly exchanging ls. are *labile. In reactions
cat. by complexes the conv. usually occurs
when the substrate becomes a l. – often by l.
*substitution reactions. L. which do not take
part directly or are not exchanged during the
cat. cycle are called *ancillary ls. However,
ancillary ls. can greatly influence the reactivity of the complexes by their electron-donating or -withdrawing nature, their steric influence, or their coordination geometry determining the nature (e.g., *bite angle, only fa-
cially or meridionally coordinating)(*ligand
effects). Non-innocent ls. can directly take
part in the cat. cycle by changing their hapti-
city (e.g., ring slip of Cp$^-$ from η^5 to η^3) or the
number of donated electrons or by changing
their formal charge (e.g., neutal bipy or
bipy^{2-}). Acc. to the IUPAC rules for nomenclature abbrs. for ligands should be written in
small letters. **E=F; G** Liganden.   R.D. KOHN

2-enzymatic: cf. *bacterial display.

ligand tuning   The ligands (ls.) in cat. active *complexes can be adjusted to optimize
properties of the cat. such as reactivity, *selectivity, *stability, or *solubility. This can often
be achieved without changing the direct coordination environment of the metal but by
changing l. positions more distant to the metal. This optimization of the cat. using *ligand
effects is called lt. Important factors are the
steric bulk, the electron-donating or -with-
drawing properties or, in multidentate ls., the
*bite angles. The different l. effects are often
probed separately to allow a rational cat. design. The steric bulk of *ancillary ls. determines the size and shape of *vacant coordination sites during the cat. cycle and can thus influence the selectivity of the cat. Steric effects
can also influence the *rate and *mechanism
of l. *substitutions (associative vs. dissociative), e.g., the *Tolman cone angle can suc-

cessfully predict the relative dissociation rates of *phosphines, which is the rate-limiting step in many cat. cycles. Similar angles are defined in other cat. systems, e.g., the *aperture angle in *metallocenes which influences the activity and selectivity of the *polymerization. A typical substitution sequence for probing a steric influence is the substitution of Me by Et, *i* Pr, and *tert*-Bu. Asymmetric substitution with very *bulky substituents on one side and small substitutents on the other – often enhanced by *C_2 symmetry – can lead to highly *enantioselective cats. The electronic properties of a l. can similarly be varied by introducing electron-donating or -withdrawing groups in the l. The electronic l. effects allow a tuning of *redox potentials in the cat. over a wide range. The position of a l. donor atom in the coordination environment of the metal can be tuned by connecting two or more ligands with bridges of different lengths and different rigidity. Bridges between the donor atoms can force them into *cis* or *trans*, *mer* or *fac* positions. Lt. can also influence the properties of the complex that are not directly related to the cat. center, e.g., solubility by introducing hydrophobic groups (for solubility in *hcs.), hydrophilic groups (for solubility in water, cf. *aqueous-phase cat.), or perfluoro groups (cf. *fluorous phase cat.). Cf. also *tailoring. **F** adaptation des ligands; **G** Ligandenanpassung   R.D. KOHN

ligases (synthetases) *enzymes which cat. the connection of two substrates with the energetic help of *NTP (e.g., *ATP) or vice versa.   P.S. SEARS

LIGA technique a special micromechanical technique for the manuf. of *catalytic microreactors   B. CORNILS

ligator atoms → coordination chemistry theory, *ligands

light gas oil (LGO) fraction of petroleum refining, boiling 260–315 °C

Linde ammonia concept (LAC) Proc. fo: the manuf. of ammonia from light hydrocarbon as a combination of a hydrogen plant, a standard N unit, and a high-efficiency *ammonia synth. loop, based on the *Casale axial-radial three-bed converter with internal heat exchangers. **F** Linde concept pour l'ammoniaque **G** Linde Ammoniak-Konzept.   B. CORNILS
Ref.: *Hydrocarb.Proc.* **1997**, (3), 110.

Linde Clinsulf process a variant of the *Claus proc., the cat. reaction of H_2S with oxygen or SO_2 yielding sulfur. **F** procéde Clinsulf de Linde; **G** Linde Clinsulf-Verfahren.   B. CORNIL

Lindemann-Hinshelwood mechanism → Langmuir-Hinshelwood kinetics

Linde-Shell MMA process → Shell-Linde MMA process

Linde type A a synth. molecular sieve with an idealized formula $Na_{12}[Al_{12}Si_{12}O_{48}] \cdot 27$ H_2O and a *pore diameter of 4.1 Å. The *zeolite structural code is LTA; the substance is often called simply "zeolite A".   J.M. THOMA:

Lindlar's catalyst cat. based on Pd-$CaCO_3$· PbO, used for the *selective *hydrogenation o: alkynes to alkenes. Triple bonds are *regioselectively hydrogenated easily without affecting other functionalities like double bonds, etc. The addition of hydrogen is formulated as a stereoselective *syn* *addition. The cat. system can be described as Pd *supported on $CaCO_3$ and *poisoned with Pb salts (*directed poisoning) Disk-shaped Pd particles at the cat. *surface contain the *active sites. The *poisoning of the cat. with lead seems to block active sites and thus to decrease *activity and increase *selectivity. Organic bases such as quinoline or piperidine enhance the selectivity of alkene formation. The additon of sulfur seems to block selectively most of the unselective sites. Commercially available Lindlar cats. are usually gray powders. In a typical het. hydrogenation reaction, Lindlar cat. (usually 5–10 % Pd and <1 %

b by weight) is mixed with alkyne in a solution with H addition. The prod. is then seperated from the cat. Triple bonds can also be hydrogenated selectively with alkali metals (Li, Na) in ammonia and with diisobutylaluminum hydride. **F** catalyseur de Lindlar; **G** Lindlar-Katalysator.   R. SCHLÖGL, A. FISCHER

Ref.: *Helv.Chem.Acta* **1987**, *70*, 627; *Org.Synth.* **1966**, *46*, 89; Augustine.

linear free-energy relationships (in enzymatic reactions) Lfers. are linear equs. relating the logarithm of a *rate constant to the logarithm of an equil. constant. Since the log of a rate constant is proportional to the free energy of activation, and the log of an equilibrium constant is proportional to the *free energy change of the proc. involved, these equations are called free energy relationships. For example, the equil. constant involved may be for a deprotonation reaction (K_a), as in *Hammett and *Brønsted equations, or it may be a partition coefficient, as in the *Hansch equation. **F** relation lineaire d'énergie libre; **G** lineare Beziehung der freien Energie.   P.S. SEARS

linear low density polyethylene (LLDPE) includes only α-olefin copolymers of densities 915–925 kg/m³, usually containing 1-butene, 1-hexene, or 1-octene. Most of these prods. are made with *Ziegler-type cats.; some higher-density polyethylenes are made with the *Phillips cat. The coordination cats. allowed for the first time the *copolymerization of ethylene with other alkenes, which by introducing side branches reduces the crystallinity and allows an *LDPE to be produced at comparatively low press. LLDPE has branching of uniform length which is randomly distributed along a given chain, but there is a spread of average conc. between chains, the highest concs. of branches being generally in the shorter chains. In contrast, LDPE prods. contain many long-chain branches. The number and length of the short-chain branches are directly related to the conc. and molecular weight of the α-olefin comonomer. The selec-

tion of comonomers for LLDPE seems to be based on proc. compatibility, cost, and prod. properties. LLDPE is produced at less than 100 °C/2 MPa and has remarkable properties. **F** polyéthylène lineaire à densitée basse; **G**=E.   W. KAMINSKY, V. SCHOLZ

Ref.: Mark/Bikales/Overberger/Menges, Vol. 6, p. 429; Ullmann, Vol. A21; Kirk/Othmer-1, Vol. 17, p. 756.

Lineweaver-Burk plot Using a rearranged *Michaelis-Menten equ. to plot $1/v$ versus $1/[S]$ should give a line acc. to $1/v = (K_m/V_{max})(1/[S]) + 1/V_{max}$ (Figure).

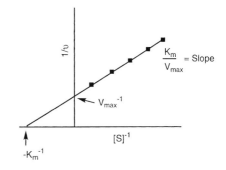

This treatment of the data is often referred to as the Lineweaver-Burk procedure and the plot is called the LB. plot. From this plot, V_{max} (1/y intercept), K_m (–1/x intercept), and V_{max}/K_m (1/gradient) can be obtained. This plot has the disadvantage of compressing the data points at high substrate concs. into a small region. The *Eadie-Hofstee plot, based on a different method of plotting the same data, will not have this problem and is generally considered more accurate, but it is historically the less commonly used in cat. and enzymology (cf. *competitive inhibition). **E=F**; **G** L-B Diagramm.   C.-H. WONG

lipases serine *hydrolases that cat. the *hydrolysis of lipids of fatty acids and glycerol at the lipid-water interface. Unlike *esterases, which show a normal *Michaelis-Menten activity, lipases display little activity in aqueous solutions with soluble substrates. A sharp increase in activity is observed when the substrate conc. is increased beyond its *CMC.

The active center of *Mucor meihei* lipase contains structurally analogous Asp-His-Ser triads, which are buried completely beneath a short helical lid, while the lipase from *Geotrichum candidum* contains a *cat. triad consisting of a Glu-His-Ser sequence, and the *Humicola lanuginosa* lipase requires Asp-His-Tyr for activity. **E=F=G.**   P.S. SEARS

Ref.: *Angew.Chem.Int.Ed.Engl.* **1998**, *37*, 1608.

lipid A synthetase Lipid A is a comp. of the outer membrane lipopolysaccharides of Gram-negative bacteria, and is responsible for many of the pathological effects of endotoxin.

A key step in the *biosynthesis of lipid A is the glycosylation of the 6-position of a 2,3-diacylglucosamine-1-phosphate by the donor *UDP-2,3-diacylglucosamine to produce a lipid A precursor.

This transformation is cat. by the *enzyme lAs. LAs. has been cloned and overexpressed, and used for the in-vitro synth. of the lipid A precursor and analogs thereof. Some examples include *C*-glycoside and phosphate analogs. **E=F=G.**   C.-H. WONG

Ref.: *J.Med.Chem.* **1992**, *35*, 2070; *Annu.Rev.Biochem.* **1990**, *59*, 129.

lipoxygenase a non-heme iron-containing dioxygenase, which cat. the *oxidation of polyunsaturated fatty acids to a *hydroperoxide. In the case of arachidonic acid, the lipoxygenation could occur at carbons 5, 8, 9, 11, 12, or 15 depending on the specific l. used. Arachidonic acid can be converted to the (*S*)-5-hydroperoxide on a synth. scale using potato l., while a soybean l. will convert the fatty acid to the (*S*)-15-hydroperoxide. Linoleic acid was converted to (*S*)-13-hydroperoxide. The (*S*)-5-hydroperoxide is a useful intermediate for the synth. of leukotrienes. The 5-lipoxygenase contains a high-spin Fe(III) center in the active form which specifically abstracts the 7-H_s hydrogen from the substrate. The reaction was proposed to proceed through a free *radical *intermediate that reacts directly with oxygen or a γ-organoiron intermediate.

The detectable *radical intermediate could result from the *homolysis of a C-Fe bond. Recent investigation of the *enzyme regarding its application to the synth. of *chiral diols indicates that a decrease in the hydrophobicity for R and an increase of the chain-length lead to the increase in the prod. with the OH group next to R. **E=F=G.**   C.-H. WONG

Ref.: *Pure Appl.Chem.* **1987**, *59*, 269.

liquefied petroleum gas (LPG) LPGs are byprods. obtained from *natural gas separation plants (natural gas liquids, NGLs) or from petroleum refineries (liquefied refinery gas, LRG). They consist essentially of propane/propylene and butane/butene in various mixtures. **F=G=E.**   B. CORNILS

liquid hourly space velocity (LHSV) The LHSV indicates the volume of liquid fed to a volume of cat. per hour as $V_{liquid}/V_{catalyst} \cdot$ [h^{-1}]. The term "liquid" requires further definition, as it may refer to pure feed or to a dilute solution. The volume of cat. corresponds to the space filled with cat. in the reactor and may thus deviate from its bulk volume (*bulk density). In a few cases the LHSV is based on the *reaction* volume instead of the cat. volume to describe, e.g., the efficiency of a tank reactor. **F** vitesse spatiale horaire de liquide; **G** Flüssig-Raumgeschwindigkeit.   C.D. FROHNING

liquid support A *support – a carrier (matrix) for cat. *active sites or species (e.g., metal *complexes) in a different (heterogeneous) *phase related to the reactants/reaction prods. – enables a simple operation, especially as far as the separation of the cat. is concerned. All of the het. cats. are in a het. phase, most of them supported. The characteristics of a support are not necessarily linked with *solids*. Liquid supports – cats. dissolved in a second, immiscible phase rel. to the reactants/reaction prods. – fulfill the same duties as solid supports; in particular they make possible the elegant separation of cat. and reaction prods. and are thus best suited for hom. cat. These immiscible liquids may be other organic liquids (*two-phase cat.), water (*aqueous-phase cat.), fluorous cpds. (*fluorous-phase cat.), or *molten salt media (e.g., *nonionic liquids). The qualification of the second phase may be ascertained by diagrams.

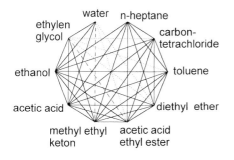

Figure: Miscibility diagram for organic solvents: — miscible in all proportions; – – – limited miscibility; ···· low miscibility; without line: immiscible

The mode (but not the efficiency; cf. *metal support interaction) of cat. (hom. or het.) is independent of the support and the kind and number of phases; e.g., *hydroformylation by the aqueous-phase technique is *homogeneously* cat. although operated in a *heterogeneous* phase. For the partitioning of different *solvents cf. the *Hansch equation. **F** support liquide; **G** Flüssiger Träger.   B. CORNILS

Ref.: Cornils/Herrmann-2; Reichardt.

liquid syngas   liquids (such as formic acid, aqueous formaldehyde, or *methyl formate)

which can readily converted to gaseous *syngas (*carbon monoxide, *syngas).   B. CORNILS

Ref.: *J.Mol.Catal.A:* **1999**, *144*, 295: *Acc.Chem.Res.* **1999**, *32*, 685.

lithium effect   Due to the large polarization power of the small Li$^+$ cation, Li frequently shows specific cat. effects as an *activator (e.g., the special cat. of *Kuraray's *hydrodimerization proc.). The Li effect is normally not paralleled by the higher homologs Na to Cs and is based on an electronic influence which entails structural amendments of the active cat. species. A new example is the stabilizing effect of Li on the stability of cat. systems. **F** effet de lithium; **G** Lithiumeffekt.

Ref.: *J.Mol.Catal.A:* **1999**, *143*, 23.   W.A. HERRMANN

lithium in catalysis   Apart from the *lithium effect, the following Li cpds. have been used/proposed as cats.: Li$_3$PO$_4$ for *dimerizations (*BASF anthraquinone proc.) or for *isomerizations (e.g., *Olin's allyl alcohol proc. from *PO), Li on an MgO *support for *oxidative coupling of methane, CH$_3$COOLi/ LiI in Ac$_2$O synth. acc. to the *Eastman proc., Li cations for the *chlorination of alkenes, or BuLi for *polymerizations. Some other Li cpds. are similar to alkaline earth metals, which explains the insolubility of some salts (such as the carbonate, phosphate, or fluoride) and their use in cat. (cf. *lithium effect). **F** le lithium en catalyse; **G** Lithium in der Katalyse.   B. CORNILS

Ref.: *Acc.Chem.Res.* **1996**, *29*, 552.

living free-radical polymerization   Living polymerization is a chain-growth proc. *without* chain-breaking reactions. It provides end-group control and enables the synth. of *block copolymers by sequential monomer *addition. Lfrps. are carried out in the presence of cat. amounts of, e.g., nitroxide radicals as *radical *scavengers, which terminate the *propagation steps of the radical chain reactions by fast coupling with the chain-end radicals. The prods. obtained are stable, but can regenerate nitroxide and polymer radicals by thermolysis of the N-O bond. This allows the addition of other monomers, resulting in a convenient method of controlling the polydispersity and molecular weight of growing polymer chains (cf. also *polymerization).

E=F=G. T. LINKER

Ref.. K. Matyjaszewski, *Controlled Radical Polymerization*, ACS Symp. Series 685, 1997; *Acc.Chem.Res.* **1997**, *30*, 373.

living polymerization → block polymers

LLDPE → linear-low density polyethylene

LMMS → laser microprobe mass spectroscopy

LMS laser mass spectroscopy

Ln general term for lanthanide metals (→ rare earth metals)

LNG liqified natural gas

local density functional (LDF) method → density functional theory

lock-and-key principle developed by Emil *Fischer as early as 1894, explains the specificity of *complexes of *enzymes and substrates. Fischer proposed that the substrate fits into the enzyme's active site much as a key fits into a lock: the two are perfectly matched. Acc. to this principle the enzyme is a rigid preorganized matrix which fits only into those substrates which are spatially complementary (cf. *host-guest relationships, *induced fit theory). B. CORNILS, P.S. SEARS

Lomax process → UOP procs., *Isomax process of Chevron

long term stability → catalyst lifetime

Lonza nicotinamide process from 2-methylpentane-1,5-diamine through a sequence of *cyclizations over *zeolite, a Pd-cat. *dehydrogenation, the *ammoxidation of 2-methylpyridine, and its *enzymatic *hydrolysis to nicotinamide. B. CORNILS

Ref.: *Chimia* **1996**, *50*, 114.

low-barrier hydrogen bond → hydrogen bonding

low-density polyethylene (LDPE) In 1933 ICI observed by chance that ethylene, in the presence of adventitious oxygen, could be polymerized to a branched polymethylene-like structure. This discovery, a byproduct of the study of chemical reactions at high press., created the basis of the wide range of LDPEs now being made by free-*radical proc. at high press. of up to 276 MPa. Nowadays the group of LDPEs also includes *copolymers containing polar groups (*EVA, vinyl acetate, ethyl acrylate, etc.). The structural features of LDPE can be varied widely by the choice of reaction conds. and type and amount of co-monomer, resulting in a broad spectrum of physical properties and an extremely wide range of applications. The polymers made by free-radical *polymerization are partially crystalline. The densities of completely amorphous and completely crystalline polyethylene would be 880 and 1000 kg/m^3, respectively. Due to side reactions occurring at the high polymerization temp. employed, the polymer chains are branched with densities of 915–925 kg/m^3. LDPE has a random long-branching structure.

LDPE has a unique combination of properties: toughness, high impact strength, low brittleness temp., flexibility, processability, film transparency, chemical resistance, low permeability to water, stability, and outstanding electrical properties. **F** polyéthylène à basse densité; **G** Polyethylen niedriger Dichte. W. KAMINSKY, V. SCHOLZ

Ref.: GB 471.590; Mark/Bikales/Overberger/Menges, Vol. 6, p. 386; Kirk/Othmer-1, Vol. 17, p. 707.

low-energy electron diffraction (LEED) Electrons of low energy (50–200 eV) are not able to penetrate into the bulk (e.g., of solid cats.) due to the strong absorption of the low-energy electrons in matter. However, some of the electrons are scattered at atoms at or near the *surface of a solid material. The scattered electrons will exhibit an interference pattern which can be detected in the region of backscattering. It carries information on the lattice properties and on the structure of the surface and of adsorbate layers.

The geometric evaluation of the diffraction pattern, i.e., the determination of the periodicity of the arrangement of atoms or molecules at or on the surface, is straightforward. LEED is used to verify the structure, orientation, and quality of *single-crystal surfaces as well as to investigate surface reconstructions and the structures of surface adsorbates. More sophisticated applications also allow the determination of the positions of atoms and the nature of defects, e.g., missing atoms or the morphology of surface steps. In het. cat., LEED is mainly applied to the basic study of model systems with regard to surface reconstructions and the *adsorption of molecular species. LEED is not well suited for in-situ studies as the low-energy electrons are easily absorbed in matter and the measurements have therefore to be carried out in a UHV. **E=F=G**.   P. BEHRENS
Ref.: Niemantsverdriet, p. 146; Ullmann, Vol.B6, p. 111.

lower critical temperature the lowest temp. on a curve in a temp./phase diagram of a non-ionic surfactant in water. Above this temp., the transparent *micellar dispersion will change into a two-phase system and tends to become cloudy ("cloud point"). Typical phase diagrams show a miscibility gap with a closed two-phase region, the highest point of which is called the upper consolute temp. (UCT). The two-phase region is important for detergency, *tenside ligands, and *micellar cat.   G. OEHME
Ref.: J.H. Clint, *Surfactant Aggregation*, Blackie, Glasgow 1992, p. 154.

Lowry Thomas Martin (1874–1936), prof. of physical chemistry in London and Cambridge, worked, inter alia, on enantiomeric reactions and acid-base theory (*Brønsted acid-base concept).   B. CORNILS

Lowry acid-base catalysis cat. procs. in mutual presence of acids and bases, e.g., the *mutarotation of tetramethylglucose. Cf. also *Brønsted acid-base concept. **F** catalyse acide-base de Lowry; **G** Lowry Säuren-Basen-Katalyse.   B. CORNILS
Ref.: *J.Chem.Soc.* **1925**, *127*, 1371, 1385.

low-temperature hydrogenation for the removal of methylacetylene and allene (propadiene) from propylene and of ethylacetylene and vinylacetylene from C_4 cuts by *hydrogenation over fixed-bed Pd cats. at 10–20 °C (cf. *Bayer cold hydrogenation proc., *Lummus CDHydro, or *Hüls SHP technology). **F** hydrogénation à basse température; **G** Kalthydrierung.   B. CORNILS

low-temperature shift (LTS) → water-gas shift reaction

LPE process → Phillips procs.

LPG → liquefied petroleum gas

LPO liquid-phase oxidation, e.g., procs. of *Celanese LPO, *SD adipic acid, *Bayer AA, *BP AA.

LPO (low-pressure oxo) processes (general) comprise the *hydroformylation procs. under relatively low press., as compared to the classical Co-cat. oxo synthesis. While the Co variants (*BASF, *Ruhrchemie, *Kuhlmann) require as a consequence of the carbonyl stability (*decobalting) syngas press. of >25 MPa, the modification of the ligand sphere (phosphines/Co, *ligand-modified cats.) or the modification of the *central atom plus the ligand sphere (Rh/*TPP) enables press. as low as 0.5–5 MPa. The name LPO is common with the proc. of *Union Carbide (plus Davy Powergas [now Davy McKee]/ Johnson Matthey & Co), although the term was coined by BP (e.g., GB 1.197.902 [1968]).

Other modern oxo procs. (*BASF, *Mitsubishi, *Ruhrchemie/Rhône-Poulenc) are also LPO procs. **F=G=E.** B. CORNILS

Ref.: Falbe-1, p.158; Cornils/Herrmann-1, p.29; Cornils/Herrmann-2, p.271.

LPO processes (special) 1: *Celanese procs. 2: *Union Carbide procs.

LRG liquefied refinery gas, → liquified petroleum gas

LRS → laser Raman spectroscopy

LTA 1: lead tetraacetate, cf. *acetoxylation; 2: *zeolite structural code for Linde type A zeolites.

LTD process → Mobil procs.

LTI process → Mobil procs.

LTS low-temperature shift, → water-gas shift reaction

Lummus ABB fluid cracking process
for the selective conv. of, e.g., virgin or hydrotreated gas oils to high-octane gasoline etc. The proc. incorporates an advanced *FCC reaction system, a high-efficiency cat. stripper, and a single-stage fluid-bed *regenerator by combustion of coke. **F** procédé FCC de ABB Lummus; **G** ABB Lummus FCC-Verfahren.
Ref.: *Hydrocarb.Proc.* **1996**, (11), 116. B. CORNILS

Lummus ABB hydrocracking process
for *hydrodesulfurization, *hydrodemetallation, and *hydrocracking of atmospheric or vacuum resids using the *Lummus LC-Fining proc. (cf. Figure). **F** procédé de hydrocracking de Lummus ABB; **G** Lummus ABB Hydrocracking-Verfahren. B. CORNILS

Lummus ABB hydrotreating process A low-press. *hydrogenation of distillates over *SynSat cats. in co- and counter-current feed,

contacting in separate cat. beds. **F** procédé hydrotreating de ABB Lummus; **G** ABB Lummus Hydrotreating-Verfahren. B. CORNILS

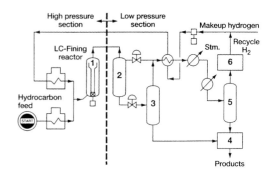

Lummus ABB hydrocracking process
1 expanded catalyst bed; 2 high press separator; 3 low press. separator; 4 product fractionator; 5 recycle hydrogen; 6 hydrogen purification.

Ref.: *Hydrocarb.Proc.* **1996**, (11), 130.

Lummus ABB isomerization process
This proc. for C_5/C_6 *isomerization converts *n*-paraffins to their isomers with higher octane numbers over fixed-bed chlorided-*alumina cats. In addition this technology also achieves *dearomatization via benzene saturation. **F** procédé ABB Lummus pour l'isomérisation; **G** ABB Lummus Isomerisierungs-Verfahren. B. CORNILS
Ref.: *Hydrocarb.Proc.* **1996**, (11), 136.

Lummus ABB SMART process bases on UOP's SMART proc., licensed by Lummus ABB

Lummus ALMA process developed by *Alusuisse, for the manuf. of *maleic acid anhydride (MAA) from *n*-butane by *dehydroxidation with air over doped V_2O_5 cats. ($[VO]_2P_2O_7$, Fe) using a fluid-bed reactor system at 400 °C/0.2 MPa and an organic solvent for continuous anhydrous prod. recovery. **F** procédé ALMA de Lummus; **G** Lummus ALMA-Verfahren. B. CORNILS
Ref.: *Hydrocarb.Proc.* **1997**, (3), 136; *Chem.Rev.* **1988**, *88*, 55; *Appl.Catal.* **1994**, *115*, 195.

Lummus Arosat process for fixed-bed cat. *hydrogenation of aromatics with reformer off-gases yielding low-aromatic white spirits and high-quality jet kerosene. **F** procédé Arosat de Lummus; **G** Lummus Arosat-Verfahren.   B. CORNILS

Ref.: Winnacker/Küchler, Vol. 5, p. 134.

Lummus bisphenol A (BISA) process for manuf. of bisphenol A by *condensation of acetone and phenol with acid *ion exchange resin as cat. Purified BISA is achieved by various crystallization steps and workup of mother liquors. **F** procédé Lummus pour BISA; **G** Lummus BISA-Verfahren.

Ref.: *Hydrocarb.Proc.* **1997**, (3), 113.   B. CORNILS

Lummus Catofin process proc. (derived from the *Houdry Catadiene proc.) for the *dehydrogenation of low alkanes in the presence of Cr oxide on *alumina cats at 600–650 °C/33–50 kPa. Feeding butane yields butadiene in a cyclic, multireactor system. Coke depositions are burned off intermittently; eventually the cat. is regenerated. **F** procédé Catofin de Lummus; **G** Lummus Catofinverfahren.   B. CORNILS

Ref.: Ertl/Knözinger/Weitkamp, Vol. 5, p. 2145f.

Lummus CDHydro technology a variant of *cat. hydrogenation/distillation technology for the depropanation by cat. *hydrogenation of methylacetylenes and propadienes as part of a SRT (short residence time) cracker (cf. *Bayer cold hydrogenation proc., *low-temp. hydrogenation). **F** technologie CDHydro de Lummus; **G** Lummus CDHydro-Technologie.

Ref.: *Hydrocarb.Proc.* **1997**, (3), 127.   B. CORNILS

Lummus Detol process manuf. of high-purity benzene and xylenes from toluene and other aromatics by cat. treatment with hydrogen (*disproportionation). Toluene and C_{9+} aromatics are recycled internally (Figure). **F** procédé Detol de Lummus; **G** Lummus Detol-Verfahren.   B. CORNILS

Ref.: *Hydrocarb.Proc.* **1977**, 56(11), 132 and **1997**, (3), 112.

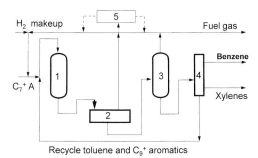

Lummus Detol process
1 catalytic reactor; 2 condensor; 3 stabilizer; 4 distillation

Lummus DPG process cat. fixed-bed *hydrogenation of pyrolysis naphtha to improve the stability of blending feedstocks. **F** procédé DPG de Lummus; **G** Lummus DPG-Verfahren.   B. CORNILS

Ref.: *Hydrocarb.Proc.* **1978**, (9), 126.

Lummus LC-Fining process cat. *hydrogenation of heavy *hcs. such as gas oils, residues, coal liquids, or *syncrudes, tars, and shale oils. *Severity and cat. *selectivity of the suspended cats. (Co-Mo/Ni-Mo on Al_2O_3) in expanded-bed, turbulent operation can be varied for either *hydrocracking or *hydrodesulfurization. **F** procédé LC-Fining de Lummus; **G** Lummus LC-Fining-Verfahren.

B. CORNILS

Ref.: *Hydrocarb.Proc.* **1978**, (9), 141 and **1996**, (11), 123; McKetta-1, Vol. 47, p. 497.

Lummus PGO hydrotreating process for the cat. upgrading of highly unstable pyrolysis gas oils (PGO) by a fixed-bed low-*severity *hydrotreating, accompanied by *hydrodesulfurization and *hydrogenation of oxygenates and nitrogen cpds. **F** procédé PGO hydrotreating de Lummus; **G** Lummus PGO Hydrotreating-Verfahren.   B. CORNILS

Ref.: *Hydrocarb.Proc.* **1978**, (9), 144.

Lummus phenol process an improved variant of the *Hock process from *cumene via *cumene hydroperoxide, which is cleaved

in a two-stage Advanced Cleavage Technology (ACT) proc. into acetone and phenol. Refined α-methylstyrene (AMS) prod. is optional. **F** procédé Lummus pour le phénol; **G** Lummus Phenol-Verfahren. B. CORNILS
Ref.: *Hydrocarb.Proc.* **1997**, (3), 145.

Lummus PO process manuf. of *propylene oxide via the older chlorohydrin route by 1: *chlorination of *tert*-butanol to *tert*-butyl hypochlorite; 2: hypochlorination of propylene; and 3: *saponification of the chlorohydrin to *PO. **F** procédé Lummus pour PO; **G** Lummus PO-Verfahren. B. CORNILS

Lummus terephthalic acid process proc. proposal for the manuf. of pure *terephthalic acid by *ammonoxidation of *p*-xylene over V_2O_5-Al_2O_3 cats. at 400–450 °C to terephthalonitrile. Subsequently three steps convert the nitrile to purified terephthalic acid (*PTA). **F** procédé Lummus pour PTA; **G** Lummus PTA-Verfahren. B. CORNILS

Lummus/UOP classic styrene process manuf. of polymer-grade *styrene monomer by first *alkylating benzene with ethylene to form *ethylbenzene (over *zeolite, liquid-phase operation; cf. *Lummus/UOP EB proc.) and then *dehydrogenating the EB (in a multi-stage cat. reactor with steam) to form styrene. **F** procédé styrène classic de Lummus/UOP; **G** Lummus/UOP Classic Styrol-Verfahren. B. CORNILS
Ref.: *Hydrocarb.Proc.* **1997**, (3), 162.

Lummus/UOP ethylbenzene process for the manuf. of EB from benzene and ethylene using a fixed-bed, liquid-phase, *zeolite cat. technology. The more highly ethylated benzenes are *transalkylated with recycle benzene to produce additional EB (Figure); **F** procédé de Lummus/UOP pour le styrène; **G** Lummus/UOP Ethylbenzol-Verfahren. B. CORNILS
Ref.: *Hydrocarb.Proc.* **1997**, (3), 126.

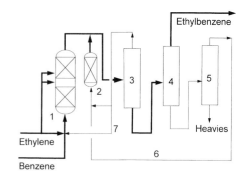

1 alkylation reactor; 2 transalkylation reactor; 3-5 distillation (3 benzene column; 4 EB column; 5 PEB column); 6 polyethylbenzenes; 7 recycle benzene

Lummus Transcat process for the manuf. of *vinyl chloride from ethane and chlorine with molten *oxychlorination cats. ($CuO \cdot CuCl_2$) at 450–500 °C. The reduced cat. is oxidized from $Cu(I) \rightarrow Cu(II)$ in a special *regeneration step (cf. *reoxidant). **F** procédé Transcat de Lummus; **G** Lummus Transcat-Verfahren. B. CORNILS

Lummus visbreaking process a non-cat. proc. for upgrading atmospheric or vacuum-reduced crude by thermal convection. **F** procédé visbreaking de Lummus; **G** Lummus Visbreaking-Verfahren. B. CORNILS

LUMO lowest unoccupied molecular orbital, → frontier molecular orbitals

Lurgi Bamag Sulfreen process a variant of the *Claus process for the purification of Claus tail gas or lean H_2S waste gas. The gas-phase proc. operates with fixed beds, consisting of impregnated activated *alumina. **F** procédé Sulfreen de Lurgi Bamag; **G** Lurgi Bamag Sulfreen-Verfahren. B. CORNILS
Ref.: *Hydrocarb.Proc.* **1996**, (43), 142.

Lurgi/BP butanediol (Geminox BDO) process manuf. of BDO (or *tetrahydrofuran, or *γ-butyrolactone) from *n*-butane;

developed jointly by Lurgi and BP (Figure). *n*-Butane and air are *dehydroxidized in a fluid-bed cat. reactor to maleic anhydride. The resulting MAA is sent directly to the fixed-bed cat. *hydrogenation reactor, which delivers BDO (yield up to 94 %) or (by adjustments to the reactor, cat. section, and the recovery/purification section) mixtures of BDO with THF and/or *GBL. **F** procédé Geminox de Lurgi/BP; **G** Lurgi/BP Geminox-Verfahren.   B. CORNILS

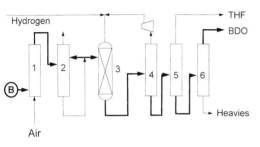

B butane feed; 1 catalytic reactor; 2 scrubber; 3 cat. hydrogenation; 4–6 distillation

Ref.: *Hydrocarb.Proc.* **1997**, (3), 118.

Lurgi high-pressure fatty alcohols process

for the conv. of fatty acids to fatty alcohols at 300 °C/30 MPa in the presence of suspended Cu chromite (*Adkins) cat. slurries.
Ref.: *J.Am.Oil Chem.Soc.* **1984**, *61*, 350.   B. CORNILS

Lurgi methanol process

In order to remove the exothermic heat of reaction of the methanol synth. conveniently and to limit the peak temp. in the cat. bed, several types of reactors have been designed and used. The low-press. Lurgi methanol synth. reactor has been designed according to experience gathered during the development of the *Fischer-Tropsch fixed-bed synth. (*ARGE reactor), where the problem of heat removal is of comparable importance. The cat. (CuO-ZnO at 250 °C/5–8MPa; cf. *methanol synth.) is arranged in tubes which are surrounded by boiling water. One reactor shell contains several thousand tubes. The temp. in the water shell is at a constant level according to the press., whereas the tubes are dimensioned in such a manner that the temp. rise in the cat. is limited to 8–10 °C (30–50 mm inner diameter). As a consequence of the good temp. control, the reaction equil. is shifted in favor of methanol formation and less cat. is required per unit prod. capacity. About 50–60 m^3 of cat. is necessary for a 1000 t/d unit at 5 MPa. The heat of reaction is recovered as steam (4–6 MPa) which can be utilized for the recycle loop compressor and the methanol purification by distillation. **F** procédé Lurgi pour méthanol; **G** Lurgi-Methanolverfahren.

     C.D. FROHNING

Ref.: *Appl.Catal.* **1994**, *115*, 180; *Hydrocarb.Proc.* **1984**, *63*(7), 34C and **1985**, *64*(11), 145.

Lurgi Octamix process

manuf. of gasoline additives (fuel methanol) by a modified methanol proc. Over modified Cu cats., variable CO/H$_2$ ratios, and 300 °C/5–10 MPa a mixture of approx. 50 % methanol and 50 % higher alcohols is obtained (cf. *fuel methanol, *gasohol). **F** procédé Octamix de Lurgi; **G** Lurgi Octamix-Verfahren.   B. CORNILS

Lurgi OxyClaus process

a supplement to existing *Claus units for the *desulfurization of gases by cat. reaction of H$_2$S with oxygen or air in a multipurpose burner at 1300 °C. **F** procédé OxyClaus de Lurgi; **G** Lurgi Oxy-Claus-Verfahren.   B. CORNILS

Ref.: *Hydrocarb.Proc.* **1996**, (4), 130.

lyases

*enzymes that cat. additions across, or formation of, double bonds. **E=F**; **G** Lyasen.   P.S. SEARS

lyonium ion

→ Brønsted acid-base concept

lyophilization of enzymes

→ enzyme catalysis in organic solvents

lysine-arginine transaminase

→ site-directed mutagenesis

L-lysine process

→ Kyowa Hakko proc.

lysozyme An *enzyme usually isolated from (chicken) egg-white, l. cleaves between the *N*-acetylmuramic acid (MurNAc) and the *N*-acetylglucosamine (GlcNAc) moieties of the poly[GlcNAcβ1,4-MurNAcβ1,4] backbone of bacterial peptidoglycan. It can also *hydrolyze chitin, poly [GlcNAcβ1,4], though polymers with less than five residues are hydrolyzed slowly. The X-ray crystal structure of l. complexed with various GlcNAc polymers together with model building has provided useful information regarding the cat. behavior of the enzyme. L. has binding subsites for six sugars. **E=F**; G Lysozym. P.S. SEARS

lysyl protease (*Achromobacter* protease I, EC 3.4.21.50) is a serine endoprotease that cat. the *hydrolysis of amide bonds following lysine residues. It is quite unusual in that it can accept proline in the P_1' position (immediately following the scissile bond). **E=F=G**. P.S. SEARS

α-lytic protease (serine protease) α-Lp. is a thermostable serine *protease isolated from *Sorangium* sp., named for its ability to lyse other bacteria. It has regions of homology to the mammalian *enzyme *chymotrypsin, but very different specificity: it is more like *elastase, preferring small residues such as Ala, Gly, Asn in the P_1 position (ahead of the scissile bond). **E=F=G**. P.S. SEARS

Ref.: *Methods Enzymol.* **1970**, *19*, 599.

M

MA 1: → methacrylic acid; 2: → maleic acid

MAA → maleic acid anhydride

macrocyclic compounds cyclic cpds. with large rings (>C$_{13}$), often synth. by cat. reactions (*cyclization, *cycloaddition, *cyclooligomerization, *metathesis, *RCM, etc.). For special cat. purposes *crown cpds., *cryptands, *cyclodextrins, etc., are used. **F** composés macrocycliques; **G** Makrocyclische Verbindungen.

B. CORNILS

Ref.: Dieterich/Viout/Lehn; N.V. Gerbeleu, V.B. Arion, J. Burgess, *Template Synthesis of Macrocyclic Compounds*, Wiley-VCH, Weinheim 1999; J.P. Sauvage, C. Dietrich-Buchecker (Eds.), *Molecular Catenanes, Rotaxanes and Knots*, Wiley-VCH, Weinheim 1999.

macrocyclic effect the difference in the degree of *molecular recognition between macrocyclic hosts such as *crown cpds. (cf. *host-guest relationships) and open-chain analogs such as *podands. **F** effet macrocyclique; **G** makrocyclischer Effekt. B. CORNILS

macropores Most cats. of practical importance are highly porous and possess large specific *surface areas. Although the cat. *activity may be only indirectly related to this total surface, the determination of surface area is generally considered to be an important requirement in cat. characterization. A distinction is made between *micropores (pore of width less than 2 nm), *mesopores (pore of width between 2 and 50 nm) and macropores with a width of greater than 50 nm. **E=F**; **G** Makroporen. R. SCHLÖGL, C. KUHRS

Madelung energy (constant, sum) represents the electrostatic energy of an ion in an ionic solid. Successive shells of ions surrounding a central ion have to be taken into account (Madelung sum). The Me. (E_M) can be expressed as $E_M = M \cdot z \cdot e/r$ with M being the M. constant whose value depends solely on the type of ionic structure, r being the interionic distance, and z being the charge of the ion. The surface Me. is reduced from the bulk value due to a lower coordination number of the *surface atoms. **F** l'énergie de Madelung; **G** Madelungenergie. R. IMBIHL

magic acid acc. to Oláh, the super acid $FSO_3H \cdot SbF_3$, which forms *carbocations and which converts *hcs. like CH_4 to CH_5^+.

magic angle spinning → cross-polarization magic angle spinning NMR

Magn. → magnetic susceptibility measurements

Magnaforming process → ARCO procs.

Magnéli phases understoichiometric oxides Ti$_n$O$_{2n-1}$ (4<n<10), interesting as cat. carrier. **F=G=E**. B. CORNILS

magnesium and Mg compounds in catalysis Apart from the *Grignard reaction, MgO (magnesia) and *spinel play a certain role as *supports for het. cats. and in some special reactions (e.g., *dehydrodimerization, *Lebedev proc.); Mg ethylate is a support for *Ziegler-Natta cats., too. Na or K supported on magnesia are important as *isomerization cats. in *Shell's SHOP proc. *Polyribonucleotide phosphorylase is an *enzyme that, in the presence of Mg^{2+}, *polymerizes ribonucleoside diphosphates into *RNA polymers; Mg is bonded with *porphyrin as chlorophyll. MgO (together with ZnO, CaO, NiO, CoO, and other *transition metal oxides) is well investigated in oxide solid solutions and in some applications such as *dehydration, *dehydrogenation, and *selective hc. *oxidation. MgEt₂ or MgCl₂ serve as bisupport (for *metallocene or *Ziegler-type cats.) in *polymerization.   B. CORNILS

Ref.: Ertl/Knözinger/Weitkamp, Vol. 2, p. 845; *J.Mol. Catal.A:* **1999**, *144*, 61; *Houben-Weyl/Methodicum Chimicum*, Vol. E 18, p. 664,816; H.G. Richey (Ed.), *New Developments*, Wiley-VCH, Weinheim 1999; *J. Mol.Catal.A:* **1999**, *145*, 265; *Adv.Polym.Sci.* **1987**, *1*, 81.

magnesium stearate   acts as a lubricant during tableting of het. cats. (→ forming)

magnetic susceptibility measurements (Magn.) for the determination of bulk electronic structures and of *redox states are made under magnetic field conds. with a magnetic balance. Data are integral and bulk-sensitives.   R. SCHLÖGL

Ref.: P.W. Selwood, *Chemisorption and Magnetization*, Academic Press, New York 1975; Imelik/Védrine, p. 585.

magnetite catalyst   Fe₃O₄, a *fused cat. for the *Kellogg Synthol proc. and *ammonia synthesis. Cf. also *spinel.

MAK   Mobil/Akzo/Kellogg   hydrocracking proc., → Kellogg procs.

makeup steps   As a first stage of *regeneration, makeup means a procedure for moving catalysts (usually homogeneous) which is between *recycling and a drastic regeneration (total workup, e.g., with recovery of the active metals, e.g., precious metals). The aims of the m. of cats. (cf. Figure under the entry *hom. cat.) are the measurement and adjustment of concs. (e.g., metal content, content of active species, *ligands, and ligand surplus) and of activity, securing of special properties (e.g., by different acidities), etc. Typical examples are the m. of Co cats. in *Kuhlmann's oxo proc. which takes place without any valence change, and the completion of organic chlorides during isomerization procs. (*BP C₅/C₆

isomerization proc., *chlorided alumina) In some cases the possibility of excluding cat. *poisons is offered during makeup (recycle of the older *Ruhrchemie oxo proc.) although the borderlines between m. and regeneration are moving. M. procedures for het. cats. are also often called *regeneration. **F=G=E.**

Ref.: Farrauto/Bartholomew, p. 58.   B. CORNILS

malachite   the mineral copper hydroxide carbonate, used, e.g., to cat. the *ethynylation reaction.   B. CORNILS

maleic (acid) anhydride (MA, MAA)   an important *intermediate, manuf. by het. *VPO *oxidation of benzene (procs. of *SD, *UCB/Alusuisse), butenes (*Mitsubishi, *Bayer), butane *dehydroxidation (procs. of *Alusuisse, *Lummus ALMA, *Lurgi Gemonox), or from wastewaters from phthalic acid production (*UCB or *BASF procs.). The het. cats. are based on doped V oxides; cf. *alkane oxidation. **F** acide (anhydride) maléique; **G** Maleinsäure (anhydrid).   B. CORNILS

Mallinckrodt catalysts   Mallinckrodt's cat. business, including Catalysts Resources and Calsicat, was acquired by *Engelhard Corp.
                                                  J. KULPE

Manassen's principle (principle of Manassen/Whitehurst) Compared to the other "classical" homogeneously cat. procs. (classical in respect of the organic phase usually employed), the use of a secondary *phase is a revolutionary development acc. to a paper by Manassen/Whitehurst, expressing the vision of a *two-phase catalysis for the first time. This quotation indicates that the extraordinarily important advantage (in terms of proc. technology) of a cat. phase which is immiscible with the reaction prods. was the actual impetus for the *two-phase cat. proc. outlined by Manassen (cf. also *aqueous-phase cat.). **F** principe de Manassen; **G** Manassens Prinzip.   B. CORNILS

Ref.: J. Manassen/D.D. Whitehurst, in Barolo/Burwell (Eds.), *Catalysis Progress in Research*, Plenum, London 1973, p. 177.

mandelonitrile lyase → oxylonitrilases

manganese as catalyst metal Since Mn forms stable cpds. in the formal oxidation states from -III to +VII, its cpds. are widely used as *redox cats. Additionally, Mn(III) salts are applied as redox promoters in the Co cat. oxidation of alkyl-substituted aromatic cpds. leading to the corresponding aromatic acids (e.g., *terephthalic acid, *DMT, and the procs. of *Amoco TPA, *Hüls DMT, *Glitsch DMT). Mn cpds. are also used for oxidations to sec. alcohols or *KA oil (*adipic acid; corresponding to the *Bashkirov oxidation), phthalic acid anhydride (*Rhône-Poulenc) or quinone (*Eastman hydroquinone proc.); cf. also *aliphatic aldehyde oxidation, *aromatic oxidation. A problem of Mn- mediated redox reactions is the *autoxidation of *ligands coordinated to the active metal. This has been partially overcome by introducing stable (aromatic) N-donor ligands such as halogenated *porphyrin or *Schiff bases. Mn-porphyrin *complexes have been used in the *oxidation of C-H bonds leading to C-X and C=X fragments. Chiral Mn-Schiff base complexes are at present the best cats. for the *enantioselective *epoxidation of unfunctionalized alkenes and are active in the alkene epoxidation with different aldehydes as co-reagents (e.g. *Merck & Co. proc.). Zr-Mg-Mn cats. serve in *Petrotex's Oxo-D proc. for the *oxydehydration (*oxidative dehydrogenation) of butenes to butadiene. *Hopcalit (Mn(Cu) cat. the conv. of CO to CO_2.
*Decarboxylase, *isocitrate dehydrogenase, or *superoxide dismutase are Mn containing *enzymes (cf. *metalloenzymes). **F** le manganèse comme métal catalytique; **G** Mangan als Katalysatormetall.   W.R. THIEL

Ref.: *J.Mol.Catal.A:* **1999**, *138*, 221; *Organic Reactions*, Vol. 49, p. 427; *Houben-Weyl/Methodicum Chimicum*, Vol. XIII/9a; *Chem.Rev.* **1994**, *94*, 807, **1996**, *96*, 2841, 2927.

Mannich reaction the *condensation reaction between formaldehyde, an ammonium derivative (or primary or secondary amine), and a cpd. which contains an active hydrogen (C-H acid/electron-withdrawing group [EWG] as in ketones, esters, nitro derivatives, nitriles) to so-called "Mannich bases".

$$HCHO + NH_4Cl + R\text{-}CH_2\text{-}EWG \rightarrow$$
$$H_2N\text{-}CH_2CH(R)\text{-}EWG + H_2O$$

Moreover, HCN, alkynes, indenes, etc. and heteronucleophiles (e.g., alcohols) may be applied. The Mr. is cat. by either acid or base. However, using the classical protocols the Mannich base often undergoes further condensation reactions with the starting aldehyde and/or the active hydrogen cpd. In the acid cat. reaction iminium ions are produced as intermediates. New modifications using preformed iminium salts (e.g., the Eschenmoser salt, $H_2C=N(Me)_2{}^+I^-$ or imines) prove higher *chemo- and *regioselectivities. The reaction of imines with silyl enolates in the presence of *Lewis acids provides an especially promising method for Mannich bases. Apart from classical *Lewis acids, lanthanide triflates (*lanthanides as cat. metals) have also been employed as recoverable hom. cats. (cf. also variations of *ethynylation). **F** réaction de Mannich; **G** Mannich-Reaktion.   M. BELLER

Ref.: March; Trost/Fleming, Vol. 2, p. 893; *Tetrahedron Lett.* **1995**, *36*, 5773; *Tetrahedron* **1994**, *50*, 2785; *Organic Reactions*, Vol. 1, p. 303.

mannosidases cat. the *hydrolysis of mannosides.

mannosyltransferases cat. the transfer of mannose from a donor such as *GDP-mannose or mannosyl phosphoryldolichol to an acceptor. Various ms. have been shown to transfer mannose and 4-deoxymannose from their respective GDP adducts to acceptors. α1,2-M. was employed to transfer mannose to the 2-position of various derivatized α-mannosides and α-mannosyl peptides to produce the Manα1,2Man structural unit. **E=F**; **G** Mannosyltransferase.   P.S. SEARS

MAO → methylalumoxane

MAPO (MeALPO) metal *ion-exchanged aluminum phosphates (*ALPOs), e.g., *CoALPO. In MAPOs the Al(III) framework in open-structure ALPOs is replaced by divalent ions such as Mg(II) or Zn(II). These *solid acids contain an exchangeable proton to preserve electroneutrality which is loosely attached to an adjacent framework oxygen (cf. *metal alum(in)ophosphate). MeALPOs can also function as powerful Redox cats., e.g., for the selective (terminal) Oxidation of n-alkanes.   J.M. THOMAS

Ref.: Thomas/Thomas; *Nature* **1999**, *398*, 227.

Markovnikov addition The regiochemistry of the *addition of nucleophiles (H-Nu; Nu = NR₂, OR, SR, etc.) to an asymmetrically substituted alkene is classified into Markovnikov (Ma.) and *anti*-Markovnikov (aMa.) additions. The Ma. which is usually favored leads to branched prods., i.e., when hydrogen halides are added onto a terminal alkene, the more electrophilic part (typically H) is added to the C-atom bearing the most H atoms. Common electrophiles are water, alcohols, amines, or HCN. Acids, metal salts, or *transition metal *complexes are used as cats. for this reaction. Whereas acid-cat. Mas. proceed via *carbenium ions, base cat. Mas. run via nucleophilic attack on the olefin bond. In the case of transition metal cats. the insertion of the alkene into the M-X bond (X = H, C, heteroatom) determines the regiochemistry, which is influenced by steric and electronic factors. The acid-cat. Ma. *hydration of alkenes such as propylene and butenes is commercially relevant for the prod. of isopropanol and *tert*-butanol (cf. the procs. as mentioned under *hydration).

Exceptions to the Ma. are obtained with activated alkenes, e.g., butadiene. In general a successful aMa. on neutral alkenes would be of great commercial interest; first examples are known. **F** addition de Markovnikov; **G** Markownikoff-Addition.

                 H. TRAUTHWEIN, M. BELLER

Ref.: Weissermel/Arpe; March; *Chem.Rev.* **1998**, *98*, 675; *Science* **1986**, *233*, 1069; *Angew.Chem.Int.Ed.Engl.* **1997**, *36*, 2225.

Markovnikov rule an empirical rule about the regiochemistry of electrophilic *addition on asymmetric alkenes or alkynes. When hydrogen halides, water, alcohols, amines, etc. are added onto a terminal alkene, the more electrophilic part of the reagent (typically H) is added to the C atom bearing the most hydrogen atoms. Mechanistically the Mr. is explained by the intermediate formation of *carbenium ions. The rule does not apply if the alkene is substituted with strong electron-withdrawing groups like -CN, -CHO, -COR, -COOR, etc. In the case of a *radical mechanism the Mr. is not applicable (*Kharasch reaction; *peroxide effect). The Mr. is of importance for nearly all kinds of cat. alkene functionalization (cf. *n/iso* ratio, *regioselectivity) **F** règle de Markovnikov; **G** Markownikoff'sche Regel.

                 H. TRAUTHWEIN, M. BELLER

Ref.: March; *J.Chem.Educ.* **1969**, *46*, 601; *Acc.Chem. Res.* **1976**, 106.

Mars-van Krevelen mechanism This mechanism and kinetic model were developed in studies on the oxidation of SO_2 and naphthalene using vanadia cats. The key observation was that in various reactions, lattice oxygen from the oxide is brought into reacting molecules and the lattice defect is subsequently removed by a reaction with gaseous dissociative *chemisorbed oxygen. This mechanism has been confirmed experimentally either by analyzing the interaction of the cats. with *hydrocarbons (hcs.) in the absence of gaseous O_2 or by using labeled $^{18}O_2$ or ^{18}O-labeled oxides, or other chemicals, and analyzing the transient reactivity.

The oxidation of the hc. is thus mediated by the cat., but the key feature is that the lattice oxygen of the oxide has different characteristics from molecular oxygen or from *chemisorbed O_2 species, having a nucleophilic-type character. Thus, the role of the oxide is not only to activate O_2 and the *hc. (as on metal *surfaces), but to be the direct oxidizing agent able to make types of *selective oxidation reactions not possible with a *Langmuir-type mechanism operating.

Many industrial cats. for selective oxidation reactions operate through a MvKm. of which both steps (substrate oxidation by the oxide and reoxidation of the reduced oxide by gaseous O_2) occur simultaneously in the reactor. In this case, the key feature of the oxide is rapid incorporation of adsorbed oxygen forms which may give rise to unselective reactions. It is possible, however, to separate the two steps of the reaction physically, with the interaction of the hc. with the cat. in one reactor and reoxidation of the cat. in a second reactor. The continuous transfer of the cat. between the two zones maintains the proc. productivity constant. This solution has been implemented recently. The reactor configuration is analogous to modern *FCC units. This reactor configuration opens new possibilities for selective oxidation reactions, because it may be possible either to carry out reactions with oxides in a partially reduced state (not stable when the hc. and O_2 are co-feeds) or to use cats. which are unselective in the presence of both hc. and O_2. The limitation of this technology for implementing the MvK. concept, however, is the low productivity, which is of the order of about 1:1000 based on substrate produced and recirculating cat.

The MvKm. also may be easily transformed to a kinetic model. It is possible then to assume the rate-determining step of the reaction (for example oxidation of the substrate by lattice oxygen) and use these expressions to derive the kinetic rate equation. **F** mécanisme de Mars-van Krevelen; **G** Mars-van Krevelen-Mechanismus.

G. CENTI

Ref.: *Chem.Eng.Sci.* **1954**, *3*, 41; *Stud.Surf.Sci.Catal.*, **1997**, *110*, 1; *Catal. Today* **1988**, *3*, 175 and **1997**, *34*, 401.

Maruzen XIS (xylene isomerization)

process   manuf. of *p*-xylene by *isomerization of *m*-xylene over $Al_2O_3 - SiO_2$ cats. at 400–500 °C. **F** procédé XIS de Maruzen; **G** Maruzen XIS-Verfahren.   B. CORNILS

MAS   → Mössbauer absorption spectroscopy

MAS-NMR magic angle spinning nuclear magnetic resonance, → cross-polarization magic angle spinning NMR (CP-MAS)

mass spectroscopy (MS) an analytical (physical) method for producing ions from neutral species, their separation and identification acc. to their mass/charge ratio. For cat. work, the interpretation of mass spectra reveals valuable information about mechanisms, *intermediates and their stability, and the scope of reactions. Special variants are *chemical ionization MS (via charge exchange with reactant gases for sensitive and/ or non-volatile cpds.), *TOF, *secondary ion MS or *secondary neutral MS, spark source MS (SSMS), etc. MS can be coupled with *gas chromatography (*GC-MS), laser microprobe analysis (LMMS), etc. **F** spectroscopie de masse; **G** Massenspektroskopie.   B. CORNILS
Ref.: Ullmann, Vol. 5, p. 515.

mass transfer limitations (in enzymatic catalysis)   Few *enzymes are so fast that the rate of substrate diffusion to the enzyme, when both are dissolved in solution, affects the overall rate of reaction (cf. *diffusion limitations in enzymatic cat.). When the enzyme is restrained on or within a solid *support, the situation is quite different. Mtls., both to and within the solid support, can have a significant effect on the overall *rate of reaction. The conc. of substrate in the local region near the enzymes can become depleted with respect to the bulk substrate conc., and thus the enzymatic rate will be diminished (unless the local substrate conc. is still much greater than the *Michaelis constant.) Mass transfer to the particle (external mass transfer) can be improved by increasing the flow rate of solution past the support, which will minimize the thickness of the boundary layer through which the substrate must diffuse. Mtls. through the support (*internal mass transfer) can be improved by reducing the particle size and enzyme loading. **F** limitations de transfert de matière en catalyse enzymatique; **G** Stoff-transportbegrenzungen in der enzymatischen Katalyse. P.S. SEARS

Ref.: H.S. Fogler, *Elements of Chemical Reaction Engineering*, Prentice-Hall, Englewood Cliffs, NJ 1986.

mass transport in catalysis Mt. in het. cat. may exist in two forms: between fluid *phases (gaseous, liquid) and cat. particles (*extraparticle mt. or interphase mass transfer), or inside a cat. particle (*intraparticle or internal mt., or *pore diffusion regime). If the overall *rate of the reaction is controlled by interphase mass transfer, the reactant conc. has already fallen off at the external cat. *surface. Thus, the difference in conc. between bulk fluid and external surface becomes significant and depends on the rate constant of the chemical reaction and the mass transfer coefficient can be correlated in terms of dimensional groups representing fluid characteristics (*Sherwood, *Schmidt, and Reynolds numbers). The reaction appears to be first order with apparent *activation energies between 4 and 12 kJ/mol. The effect of internal mt. is to decrease the educt conc. within the cat. particle. The average rate for the whole cat. particle at *steady state will be equal to the overall rate at the location of the cat. in the reactor, but the conc. of the bulk fluid at this location may not be equal to the values at the external surface of the cat. particle. For evaluating the influence of intraparticle mt. on a cat. reaction it is necessary to estimate the effective diffusivity in a porous cat. This can be predicted by combining equations for the molar flux in a single pore with geometric models of the pore structure of the cat. (e.g., parallel-pore model, random pore model).

Three cases of internal mt. have to be considered; 1: During bulk diffusion in large pores of the cat., the walls of the pores are not involved in this proc., which can be described by the laws of fluid *diffusion. 2: If the *pore radius is much lower than the mean free path, the probability of collisions between fluid reactants and the pore walls increases in relation to the frequency of collisions among the molecules themselves (*Knudsen diffusion). 3: In the transition region where both bulk and Knudsen diffusion contribute to the diffusion resistance, the effective diffusion coefficient can be calcd. from the law of additive resistances. With the diffusion coefficient D_{eff} calcn. of the *Thiele modulus is possible and, thus, by using experimental diagnostic criteria, the evaluation of mt. effects (*intraparticle mt.). Whether bulk or Knudsen diffusion predominates depends on the effects of press., temp., and pore size on both types of mt. in a porous cat.

The contribution of surface diffusion (Vollmer diffusion) to the total mt., where the adsorbed educt may be transported by migration to an adjacent *active site on the cat. surface, is comparatively small. The effects of internal and external transport resistance can be combined to give the overall rate in terms of bulk fluid properties. It is important that under specific experimental conds. internal and external mt. can affect not only the overall rate and apparent activation energy, but also the *selectivity to a desired prod. in consecutive and parallel reactions, provided that the reaction orders of individual reactions are different.

Besides external and internal conc. gradients which can build up in the fluid boundary layer around the cat. particles and inside their pores, other factors such as axial dispersion (*Bodenstein number) and radial gradients may cause conc. gradients within a bed of cat. particles; they are called reactor gradients. In hom. cat. reactions which are often conducted in gas/liquid or liquid/liquid mode, mt. occurs, too, because conc. gradients may exist near the fluid/fluid interface (*Hatta number, *enhancement factor). **F** transfert de matière en catalyse; **G** Stofftransport in der Katalyse.

P. CLAUS, D. HÖNICKE

Ref.: Ertl/Knözinger/Weitkamp, Vol. 3, p. 1209; Wijngaarden/Kronberg/Westerterp, p. 61,141.

MAT → micro activity test

material gap a term introduced to describe the difference in materials between commercial het. cat. and the *single-crystal experiments typically used in surface science. Besides the *pressure gap* the mg. is the second fundamental problem in cat. research with the so-called surface science approach. Commercial cats. are mostly complicated mixtures of various comps. which, in addition, may undergo chemical transformations before the actual cat. active state forms (*activation, *preformation). The composition of commercial cats. is typically optimized empirically in screening methods. Various strategies have been suggested to overcome the mg. problem among, which the design of *model cats. is the most important one. **E=F=G**.   R. IMBIHL
Ref.: Somorjai.

Mayo-Lewis parameters define the relationship of *rate constants during a *copolymerization. They can be evaluated experimentally from the ratios of the two *monomers and the composition of the resulting polymer. This method is based on the work of Mayo and Lewis in which they presented a new possibility of the comparison of the different behavior of monomers in the *radical *polymerization of *styrene and *methyl methacrylate. **F** paramètres de Mayo et Lewis; **G** Mayo-Lewis-Parameter.
                    W. KAMINSKY, M. VATHAUER
Ref.: *J.Am.Chem.Soc.* **1944**, 66, 1594.

MBS → molecular beam scattering

MC methylcellulose

MCM catalysts 1: multicomponent metal oxides, → multicomponent cats., cf. e.g. *Asahi MA proc. 2: New mesoporous siliceous solids with channel apertures from 25 to 100 Å (e.g., MCM-41).
Ref.: Thomas/Thomas, p. 361,618,623.

McMurry coupling a reductive coupling of aldehydes or ketones to alkenes *promoted by lower-valent Ti. The various procedures described differ essentially in the source and the actual nature of [Ti], which is prep. from $TiCl_4$ or $TiCl_3$ and reducing agents such as K, C_8K, Mg, $LiAlH_4$, Zn, or Li. The reaction can be stopped at the intermediate pinacol stage by lowering the temp. An important extension pertains to intramolecular *cross-coupling reactions of aldehydes or ketones with esters or amides, respectively, providing inter alia furan, benzofuran, pyrrole, and indole derivatives. The McMc. in its classical form is a stoichiometric proc. driven by the formation of highly stable Ti oxides as inorganic byprods. For intramolecular cases, however, a cat. version can be achieved via a *multicomponent *redox system comprising $TiCl_3$ cat., Zn, and R_3SiCl. In this case, the admixed R_3SiCl converts the Ti oxides primarily formed into Ti (oxy)chlorides, which are then reduced to [Ti] in the presence of the substrate by means of Zn as stoichiometric reducing agent (Figure).

The basic principle can be extended to many other reactions. **F** couplage de McMurry; **G** McMurry-Kupplung.   A. FÜRSTNER
Ref.: *Angew.Chem.Int.Ed.Engl.* **1996**, 35, 2442; *Chem. Rev.* **1989**, 89, 1513; *J.Org.Chem.* **1994**, 59, 5215; *J.Am. Chem.Soc.* **1995**, 117, 4468 and **1996**, 118, 12349; *Chem.Eur.J.* **1998**, 4, 567.

MCSCF multi-configuration SCF, → ab-initio calculations

MD → molecular dynamics

MDDW process → Mobil procs.

MDI 4,4′-diphenylmethane diisocyanate (methanediphenyl diisocyanate), manuf. cat. (e.g., *Asahi, *Atlantic Richfield procs.).

MDPE medium-density polyethylene, → polyethylene

me, Me methyl group

MEA monoethanolamine

MeA(L)PO (*MAPO) general term for metal ion-exchanged alum(in)ophosphate, cf. *metal alum(in)ophosphates, *VAPO, CrAPO, etc.

mechanical properties (of catalysts) After production, shipping, and loading into *catalytic reactors, het. cats. underlie many mechanical strength-reducing stresses (temp., press., abrasion by gaseous reactants, *attrition, chemical *poisons, crushing, deposition by byprods. (such as *coke formation, etc.) which need to be observed because these influences co-determine the cat. *lifetime, too. *Monoliths, for example, frequently do not provide high *surface areas or keep the active phase in a highly dispersed state, but are merely mechanical supports (*ceramics in catalysis). Such ceramic monoliths are used primarily in *automotive exhaust treatment systems. Besides the proper cat. action, these mechanical properties influence the economy of cat. procs. considerably, e.g. in the case of *hydrotreating. **F** qualité mécanique; **G** mechanische Eigenschaften.   B. CORNILS

Ref.: Farrauto/Bartholomew, p. 136; Ullmann, Vol. A5, p. 355.

mechanisms → reaction mechanisms

medium metal-support interaction → metal-support interaction

MEEC membrane-enclosed-enzyme cat., a reaction system in which the *enzymes are enclosed in dialysis membranes (cf. *bioreactors, *cat. reactors, *enzyme immobiliza-tion, *membrane reactors, *hollow fibre reactors).   P.S. SEARS

Meerwein-Ponndorf-Verley reaction the *reduction of aldehydes or ketones to primary or secondary alcohols with Al tri-2-(propanolate) as cat. in *iso*-propanol as reagent and hydride source (cf. *transfer hydrogenation) acc. to:

$$R^1C(=O)R^2 + CH_3CH(OH)CH_3 \rightarrow$$
$$R^1C(OH)HR^2 + CH_3C(=O)CH_3$$

This reaction is reversible (cf. *Oppenauer oxidation). Alkene bonds will not be reduced. The mechanism of the reaction is assumed to run via a six-membered *transition state with Al^+ and H^- as bridging ions between the carbonyl function and the alcohol group (R_2C-O-Al) of the isopropylate unit. The *hydride transfer occurs from the sterically less hindered position. The reaction can be applied widely. Simple aldehydes and ketones, cyclic ketones, aromatic carbonyl cpds. (benzaldehyde and acetophenone derivatives), or even phenyl hydrazine derivatives are converted to the corresponding alcohols (or amines). Besides aluminum *tris*-alcoholates lanthanoid alkoxides [$Ln_5O(i\text{-}Pr)_{13}$, $LnI_2(OR)$], and $IrCl_4$ also cat. the MPVr. As an alternative formic acid can also act as the hydride transfer agent. The driving force of the reaction is the formation of CO_2. **F** réaction de MPV; **G** MPV-Reaktion.   R.W. FISCHER

Ref.: B.P. Mundy, M. G. Ellerd, *Name Reactions and Reagents in Organic Synthesis*, Wiley, New York 1988, p. 144; *Chem.Lett.* **1987**, 181; *Synthesis* **1994**, 1007; *Organic Reactions*, Vol. 2, p. 178.

Meerwein reaction (arylation) the substitution of α-hydrogen by aryls when reacting α,β-unsaturated alkenes with aryldiazonium halides under the influence of Cu^{2+}. **F** réaction de Meerwein; **G** Meerwein-Reaktion.

Ref.: Krauch/Kunz   B. CORNILS

MEK methyl etyl ketone

membrane-enclosed enzyme catalysis
→ enzyme immobilization

membrane reactor Acc. to IUPAC, a mr. is a "device for simultaneously carrying out a reaction and membrane-based separation in the same physical enclosure". Within the reactor the *membrane can function in different ways, as shown schematically in the Figure.

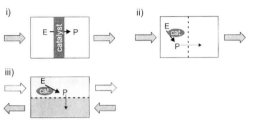

In case i) the membrane material may be cat. active (e.g., with Pd membranes) or the cat. active *complexes may be deposited within the pores or on the surface of the membrane by adsorption, covalent bonding, etc. (*immobilization). This type of reactor is called a catalytic membrane reactor (CMR) or an inorganic membrane reactor (IMR). Examples are the *reduction of NOX with V_2O_5-impregnated Al_2O_3 membranes or the cleavage of racemates by a *lipase, immobilized in the pores of a membrane. In reactor ii) the membrane may be used to retain a suspended or hom. soluble cat. while educts and prods. can pass through the membrane. Examples are the retention of enzymes by *ultrafiltration membranes or of *dendrimer bound cats. with *nanofiltration membranes. In case iii) the membrane can be used to stabilize the interface between two immiscible solvents, allowing removal of prods. or delivery of substrates. Examples are H_2 delivery or removal through Pd membranes (also type i), the removal of H_2O in *esterification reactions using pervaporation (*membranes) or of *fermentation prods. by different membrane procs. Acc. to the application and conds. various types of membranes can be used. *Hollow-fiber reac-

tors are reactors with membranes specially configured as hollow fibers. **F** réacteur à membrane; **G** Membranreaktor.   U. KRAGL

Ref.: Cornils/Herrmann-1, Vol. 2, p. 832; Ertl/Knözinger/Weitkamp, Vol. 3, p. 1387; H.P. Hsieh, *Inorganic Membranes for Separation and Reaction*, Elsevier, Amsterdam 1996; Ullmann, Vol. A16, p. 187; *Cattech* **1997**, *1*, 67.

membranes are thin sheets or surface films manuf. from synth. or modified natural polymers as well as inorganic or metallic materials that separate comps. on the basis of their properties such as size, charge, hydrophobocity, etc. Membrane procs. can be classified acc. to the size of particles separated or the driving force (Table).

Membrane process	Driving force	Mol.weight range
Microfiltration	Pressure	$> 0.1\ \mu m$
*Ultrafiltration	Pressure	$10^3–10^6$
*Nanofiltration	Pressure	500–5000
*Reverse osmosis	Pressure	<500
Electrodialysis	Electric field	10–500
Dialysis	Conc. gradient	$10–10^6$
Diafiltration	Pressure	$10–10^6$
Pervaporation	Pressure/vacuum	Volatile cpds.
Perstraction	Conc. gradient	–

In many cases there is no sharp distinction between the procs. The membranes are often thin-fim composites having a thin active layer on an inert support. There are porous ms. (micro- and ultrafiltration, perstraction) or nonporous ms. In the latter case the transport is due to solubility and diffusion of the comps. within the membrane material. **F**=E; **G** Membranen.   U. KRAGL

Ref.: M. Mulder, *Basic Principles of Membrane Technology*, 2nd ed., Kluwer, Dordrecht 1996; Ullmann, Vol. A16,p. 187; J. H. Fuhrhop, J. König *Membranes and Molecular Assemblies* RSC, London 1998.

Merc.   → mercury porosimetry

Mercapfining   → Howe-Baker procs.

Merck & Co. process for indenepoxide
manuf. of a precursor of the anti-HIV drug Crivixan by *epoxidation of indene with NaOCl

under the influence of an (S,S)-Mn salen (*Jacobsen) *complex (TOF ~140).

Ref.: *Chem.Ind.* **1996**, (June), 412.   B. CORNILS

mercury as catalyst metal   Because of its toxicity Hg is not a common cat. metal (*toxicity in catalysis). However, it has been applied in the classical acetaldehyde synth. (Equ.), in *C-H activation, *dehydrodimerization, alkene *hydration, alkyne oxidation, methane *oxidation, and vinyl ester synth. (*VAM).

$$HC \equiv CH + H_2O \xrightarrow{Hg^{2+}} \left[\begin{array}{c} H \\ C \\ H_2C \quad OH \end{array} \right] \rightarrow \begin{array}{c} O \\ \parallel \\ C \\ H_3C \quad H \end{array}$$

AcH synth. is effectively cat. by Hg_2^{2+} and Hg^{2+} salts. However, Hg^{2+} partially oxidizes the AcH to AcOH, and is itself reduced to metallic Hg. In a former *Wacker proc. the metallic mercury was reoxidized to Hg^{2+} by Fe(III) sulfate. Hg-photosensitized ($^3P_1$ excited state) *dehydrodimerization of *hcn. has been developed into a useful organic synth. method in which the *radical reaction prods. are protected from further transformation simply by condensation. In methane oxidation to methyl sulfate, Hg^{2+} was used as a cat. by *Catalytica workers. Pure sulfuric acid was used as a solvent and as the reoxidant of the metal. For dihalogencarbenes cf. *Seyferth's reagent. **F** le mercure comme métal catalytique; **G** Quecksilber als Katalysatormetall.   W.A. HERRMANN

Ref.: Parshall/Ittel, p. 199; Cornils/Herrmann-1, pp. 274, 940, 1081; Larock, *Organomercury Compounds in Organic Synthesis*, Springer, Berlin 198; F.K. Steinberger, *Die Acetylen-Chemie in Höchst*, Farbwerke Hoechst AG, Frankfurt 1946; *Houben-Weyl/ Methodicum Chimicum*, Vol. E 18, p. 827.

mercury porosimetry (Merc.)   for the determination of *pore size distributions and pore shapes in solids (cats.) by filling the void space with mercury. This destructive routine

method complements *BET measurements. The data are integral and bulk-sensitive.
          R. SCHLÖGL

Ref.: Ertl/Knözinger/Weitkamp, Vol. 2, p. 437; van Santen/van Leeuwen/Moulijn/Averill.

Merrifield Robert B. (born 1921), professor of biochemistry at the Rockefeller University, New York. Developed the *solid-phase technique for the synth. of peptides. Nobel laureate in 1984.   B. CORNILS

Ref.: *Angew.Chem.Int.Ed.Eng.* **1985**, *24*, 799; Gates-1, p. 144.

Merrifield technique   → solid-phase technique

MES   Mössbauer emission spectroscopy, → Mössbauer spectroscopy

mesophilic enzymes (temperature stability)   → thermostability of enzymes

mesoporous pores   → mesoporous solids

mesoporous solids   Acc. to the IUPAC definition, the diameter of mesopores is between 2 and 50 nm. The pores can be structural pores, as for instance in the ordered mesoporous oxides of the M41S type, or interparticle pores formed by agglomeration of particles (*pore structure). Important characteristics of an ms. are the average *pore size, the *pore size distribution, and the *pore volume. These parameters can most easily be determined by N_2 adsorption at the temp. of liquid N_2, where the presence of mesopores is typically indicated by a capillary condensation step. The basis for the determination of the pore size and the pore size distribution is the Kelvin equ. Although many materials can be prepared as mss. the most important ones are either oxides, such as *silica or *alumina, or *activated carbons. Cat. relevant mesoporous oxides are usually prepared by precipitation under carefully controlled conds., *sol-gel routes, or *precipitation in the presence of liquid-crystal forming molecules which are later removed from the oxide by *calcination, leaving behind

the mesopores. The latter synth. pathway is especially interesting in that it allows formation of a highly ordered pore arrangement and results in a narrow pore size distribution.
With respect to cat., the importance of the mss. lies in their high *surface area, which is typically in the range of several hundred m^2/g. This makes them ideally suited for use as cat. *supports. Solids with smaller pore sizes usually strongly restrict diffusion of reagents and prods. in the pore system so that the cat. reaction easily becomes mass-transfer limited. **F** matière solide mesoporeux; **G** mesoporöse Feststoffe.   F. SCHÜTH
Ref.: *Chem.Rev.* **1997**, *97*, 2373; Thomas/Thomas, p. 608; van Santen/van Leeuwen/Moulijn/Averill.

Meta-4 process → IFP procs.

metabolic engineering   Because most *enzymes work under similar conds., it is possible to conduct multienzyme reactions in a single pot (*multienzyme systems). One approach to multienzyme reactions is based on whole-cell or fermentation procs. Cells containing the desired multienzyme systems can be constructed through metabolic engineering via recombinant *DNA methods or via selective pressure. In principle, genes encoding the enzymes responsible for the synth. of a target molecule can be cloned and localized in one species or in a single plasmid ("plasmid-based biocatalysis"). By localizing the genes for 3-*deoxy-D-arabino-heptulosonate phosphate (DAHP) synthetase and *transketolase or the same plasmid, a new *E. coli* strain is able to produce high levels of DAHP. Further manipulation of the cells has led to the high-level microbial prod. of intermediates used in the shikimate pathway. "Interspecies" cloning of antibiotic *biosynthesis genes has also been used to express in the same cell two biosynthetic pathways to make hybrid antibiotics structurally different from those produced by the parent organisms. **E.=F=G.**   C.-H. WONG
Ref.: *J.Am.Chem.Soc.* **1990**, *112*, 1657.

metal alum(in)ophosphate   (metallo alum(in)ophosphate, MeAPO, *MAPO) MeAPOs (MeALPOs) are crystalline *microporous solids belonging to the family of *zeolite-type solids. They are related to *alum-(in)ophosphates (ALPOs) by formal substitution of Al for Me ions (Me = Be, Mg, Ti, V, Cr, Fe, Co, Mn, Zn, Ga). Usually, the metal atom is in tetrahedral coordination, so that the structures of MeAPOs consist of tetrahedral frameworks, as is the case for zeolites. The MeAPO *molecular sieves span a wide range of compositions within the general formula $R_a[(Me_x Al_y P_z)O_2]$, where R is a *templating molecule or water and $x+y+z = 1$. The fraction of Al substituted for Me ranges from $x = 0.01$ to 0.25. Like zeolites and ALPOs, MeAPOs are usually generated in template-directed hydrothermal synth. In some cases, Me^{2+} or Me^{3+} or Al^{3+} may be in addition coordinated to additional, non-framework ligands, e.g., water, resulting in a higher coordination number than four. The (unofficial) nomenclature for MeAPOs uses the element symbol (or its first letter) and the abbreviation "APO" (ALPO) for the alumophosphate part, e.g., MAPOs for *magnesium alumophosphates, VAPOs for vanadium aluminophosphates or CoAPOs for cobalt ALPOs.
The formal substitution of Al^{3+} for divalent Me^{2+} results in a negative framework charge. This has to be balanced by non-framework cations. In contrast to ALPOs, MeAPOs therefore exhibit *ion exchange properties and, when the counter cations are protons, acidic properties. Their acid strength varies from weakly to strongly acidic, depending upon the framework metal Me and the structure. MeAPOs with transition metal atoms Me show *redox properties based upon the corresponding redox pairs, such as, e.g., Mn^{2+}/Mn^{3+}, Fe^{2+}/Fe^{3+}, and Co^{2+}/Co^{3+}. Oxidation reactions are also in a realm where these cpds. have been tested extensively for cat. applications. Compared to zeolites and ALPOs, the stability of MeAPOs is limited. Leaching of the metal atoms out of the framework presents a particularly severe problem. No com-

mercial procs. using MeAPOs are currently being carried out. **E=F=G**.   P. BEHRENS

metal catalysts The metallic character of a cat. depends strongly on the reaction atmosphere. Under oxidizing conds., only the precious metals (Au, Ag, and the Pt group metals) remain in metallic form because they form unstable oxides. Under reducing conds. only the metals that form easily reducible oxides can be used as metallic cats. (not Cr, Mn, and Ge).

Furthermore, the cat. *activity depends on the *chemisorption properties of the metal. In general the *activity decreases from left to right in periods 4–6 of the periodic system to the iron group metals. Group VIIIA metals are used for *hydrogenation reactions. The ability to chemisorb H is facilitated by an electron gap in the d-band of the metal; it decreases with the filling of the d-band (Ni-Co-Fe). Group VA and VIA metals develop too strong adsorption bonds to render fast reactions. Group IB metals show little chemisorption of hydrogen. Well-known examples of this reaction type are *NH$_3$ synth. with Fe cats. or the *Fischer-Tropsch synth. with Co cats. In general, every hydrogenating metal could also be used for *dehydrogenation reactions. But because of the high temp. of this reaction the metal oxides are the preferred cats. It is possible to *alloy a cat. active metal with another metal to change the activity and *selectivity or suppress an undesired reaction and accelerate a desired one. The addition of oxidic *activators or *inhibitors is yet another way to improve the performance of the cat. The addition of group IB metals to Ni, Pt, or Pd decreases the hydrogenation activity of these metals. A Pt/Rh alloy is the favored cat. for the *oxidation of ammonia. Metal cats. show high selectivity in *partial oxidation of alkenes and in preferential hydrogenation of C≡C, C=C, C=O, or C≡N bonds in the presence of other functionalities.

Metals as commercial cats. are mostly *dispersed or *supported; only in a few cases are they used in the form of particles, *skeletal cats., or wire gauzes (*gauze cats.), to ensure more economic utilization and higher activity per unit mass of the active metal, and better resistance to *deactivation by temp. and *poisons. **F** catalyseurs métallique; **G** Metall-katalysatoren.   R. SCHLÖGL, H.J. WÖLK

Ref.: Fürstner.

metal complex effect The addition of metal *complex cpds. to *cluster-type cats. can considerably modulate both the *activity and the *selectivity for, e.g., the *hydrogenation of α,β-unsaturated aldehydes or *o*-chloronitrobenzene. This phenomenon is called the mce.; it is related to the kind of metal *central atom ions and the *ligands, as well as to the number of ligands coordinated to the central metal ion. However, the mce. is not the sum of the effects of the metal central ions and the ligands. It is generally believed that the interaction between metal complexes and the metal clusters may change the electronic density of the metal active site, which in turn influences the cat. performance.   HANFAN LIU

Ref.: *J.Mol.Catal.* **1997**, *126*, L5.

metal distribution The md. may strongly influence the cat. properties; e.g., small metal particles, finely dispersed on a *support, are wanted for high yields. In the case of highly exothermic reactions, however, low metal *dispersions and large *particle sizes are optimal. Often unsupported metals or metal grids (cf. *grid cats.) are then used. To control the md. on the *surface of the cat., the cat. prepn. procedure, the cat. *activation, and the type of *support are very important. In order to obtain high mds., prepn. procedures are used which lead to a rather strong fixation of the metal *precursor complexes on the support *surface. Thus, for example, *ligand exchange reactions of the precursor complex with surface OH groups lead to direct bonding and strong fixation. During cat. activation the complexes decompose and small metal *clusters are formed depending on temp. and activation atmosphere. Usually, the average metal *particle size increases with increasing activation temp. and in a reducing atmosphere. Oxidizing conds. often lead to metal oxide formation during activation, which may spread over the support surface, thus leading to high metal dispersions when subsequently reduced under mild conds.

The *metal support interaction also strongly influences the md. Thus, more or less inert supports which hardly interact with the metal lead to pronounced particle *sintering and therefore low metal dispersions during activation or cat. action. Noble metals cannot easily be supported on C in high dispersions, because of the weak support interaction. Usually large particle sizes are generated under thermal load. High noble metal dispersions can often be stabilized on oxidic supports. However, the *SMSI with the oxidic support may induce a change in the electronic properties of the small metal clusters, which may appear to be in a $\delta+$ state of oxidation.

The most direct way to determine the md. in cats. is to use *SEM or *STEM. Angle-dependent measurements allow the generation of 3D information on metal particle sizes and clustering. The average location of metal particles can also easily be determined their location at pore mouths, in the interior of pores or at the outer support surface (Figure; A: uniform distribution, B: egg-shell [pellicular], C: egg-white, D: egg-yolk).

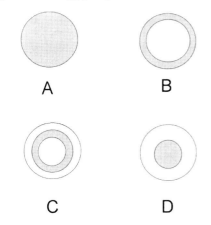

A   B

C   D

Very small metal particles can be further characterized by electron microdiffraction and thus direct information is obtained on their *crystal structure which can be compared with the known bulk values.

Mds. on *model cat. surfaces are characterized usually by STM when the model support is electrically conductive, as for highly oriented pyrolytic graphite, or by *AFM when the model surface is nonconducting, as for quartz or sapphire single crystals. **F** distribution de métal; **G** Metallverteilung.

G. MESTL

Ref.: Augustine, p. 301; Ertl/Knözinger/Weitkamp, Vol. 1, p. 205.

metal film catalysts are manuf. by galvanizing, *CVD, *plasma techniques, etc. and are used exclusively as model cats. for the study of single cat. steps, layers, *surfaces, adsorbates, adsorbents, etc. (surface sciences). Newly developed procs. combine UHV and surface-sensitive methods for cat. research. Films consist of small crystallites. Special prepns. allow *alloy films, ultrathin films, metal overlayers, etc. B. CORNILS

Ref.: Ertl/Knözinger/Weitkamp, Vol. 2, p. 786.

metal gauze reactors (shallow-bed reactors) for reactions with high exothermicity at high temps. and low press. drops, dominated by *mass transfer from the gaseous phase to the cat. Commercial applications include pads consisting of layers of finely interwoven Pt or Pt/Rh gauze for *ammonia oxidation to nitric oxides, for the *Andrussov proc., or Ag gauzes for *formaldehyde synth. (cf. *gauze cats., *grid catalysts, *cat. reactors). **F** réacteurs avec des catalyseurs de gaze (tissu de grillage métallique); **G** Reaktoren mit Metallnetzgewebekatalysatoren.

B. CORNILS

Ref.: Ertl/Knözinger/Weitkamp, Vol. 4, p. 1750.

metallacyclus cyclic organometallic cpd., containing C-metal-C bonds (via *cyclometallation); cf., e.g., *palladacycles.

metallic glasses → alloys

metallic oxides → oxides, *transition metal oxides, *mixed oxide catalysts

metal load effect The performance of het. cats. made from metal particles on a *support does not scale in a linear fashion with the amount of metal per unit mass support (metal load or metal loading). This effect, which includes not only variations in specific *rate but also in the *activation energy and even in the *selectivity of given procs., has many different origins. It is usually very difficult to trace back to a structural understanding of the cat. and hence is subject to extensive speculations in each case. For this reason almost all studies on supported metal cats. measure their effects as a function of metal loading.

Often a maximum in the plot of catalytic rate vs. metal loading is identified. This *volcano plot indicates that the *shape and size of the particles are different in different cat. preparations. When the metal loading is changed the chemistry of the *precursor and its interaction with the substrate is also modified. Synth. is usually initiated by an *impregnation technique from solution. As a function of ion conc., *precipitation of different solution complexes and *ion exchange reactions with functional groups of the support will occur simultaneously. In the following *calcination step a different surface mobility of intermediates and different nucleation phenomena will lead to different particles of the active phase although apparently only the conc. of the precursor solution was changed.

The development of a generalized understanding of the dependence of a cat. function on *particle size for *nanosized particles is a research objective. Little undisputed information other than the existence of very strong particle size effects (e.g. phenomena of *clusters) is available at present. **E=F=G**. R. Schlögl

metalloalum(in)ophosphate → metal alum(in)ophosphate

metallocenes complexes of s-,p-,d-, and f-block elements that contain two, normally pentahapto-bonded *cp *ligands. They are often defined more narrowly to contain two parallel cp ligands, as found in the prototypical *ferrocene. The two planar ligands may be substituted cp, ind, or fluorenyl ligands, five-membered heterocycles, or sometimes other ligands that show some kind of similarity to the cp ligand, such as benzene, cycloheptatrienyl, or cyclooctatetraene (such as actinocenes $M(\eta^8-C_8H_8)_2$). The term m. is often used synonymously for sandwich *complexes. Ms. of formula $M(C_5H_5)_2$, analogous to *ferrocene, are known for a number of divalent metals, but only a few possess the ferrocene-type *sandwich structure. Except for Mg, V, Cr, Ru, Os, Co, and Ni, other structures with cp ligands not pentahapto-bonded are encountered. For Ti and all 4d and 5d metals except Ru and Os the ms. are of fleeting existence. Electropositive metals such as the heavier alkaline earths or Zn favor ionic structures; Mn is intermediate. With the cp* ligands titanocene and silicocene became accessible.

The most important ms. are the bent ms. of the type $M(C_5R_5)_2L_nX_m$ where M may be a group 3–6 metal, and where the sum of ligands n+m does not exceed three. Although titanocene dichloride had long been in use in hom. *Ziegler-Natta cat. group 4 ms. only gained prominence as *polymerization cats. for ethylene, 1-alkene, and other monomers after the development of *methylalumoxane and subsequently *Lewis acid *co-cats. such as $B(C_6F_5)_3$. In the presence of co-cats. the m. dichloride is transformed into a 14-electron alkyl cation of the type $[(C_5R_5)_2MR^+]$ that is stabilized by a weakly coordinating anion. The exceedingly fast insertion of a C-C double bond into the M-C bond occurs via a four-membered ring *transition state. The term *m. cats.* usually refers to these hom. polymerization cats. The most significant ms. of this class are *ansa*-ms. **E=F**; **G** Metallocene. J. Okuda

Ref.: Long; Togni-1; W. Kaminsky (Ed.), *Metalorganic Catalysts*. Springer, Berlin 1999.

ansa-metallocenes Metal *complexes containing a linked *cp *ligand system w/o additional ring substituents may have significantly different structures and reactivity compared to a corresponding non-bridged *metallocene (m.). When a linked cp, containing at least one ring substituent, is coordinated to a metal center, the planar *chirality gives rise to a racemic mixture of *complexes which in principle can be resolved into its optically active forms. The bridge may be alkanediyl -CH₂- or -CH₂CH₂-), silanediyl (SiMe₂), or incorporated in more elaborate structures such as helicene, or be part of a heterocycle, including optically pure ones such as 1,1′-binaphthyldioxy. Thus, by introducing a -CH₂CH₂- bridge in *ferrocene, its parallel sandwich structure is distorted to a bent m. structure with the non-bonding electron pair more exposed, which explains the higher basicity. The ring strain caused by shorter bridges can be utilized in *ring-opening polymerizations. Increased *Lewis basicity and the ability to add additional two-electron ligands such as CO was found for *ansa*-chromocene, whereas the unbridged chromocenes interact only weakly with CO. Group 3, 4, and 5 *ansa*-metallocenes are referred to as *Brintzinger-type m. and are important as so-called *stereorigid mc. cats. in *stereoselective reactions, most notably in *polymerization of propylene. *Chiral and optically active group 3 and 4 *ansa*-ms. have remarkable properties, e.g., in fractional crystallization of optically active ligands (e.g., mandelate), enantioselective *hydrogenations, carbomagnesiation or alumination, *hydroamination, or 1-alkene *polymerization when activated by appropriate *co-cats. The structure of the bridged bis(cp) ligands of *ansa*-zirconocene cats. is intimately connected to the microstructure of the resulting polymer and was pivotal in the understanding of stereospecific propene polymerization (cf. *tacticity). **E=F=G**. J. OKUDA

metalloenzymes Metals such as Ca, Mg, Mn, Fe, Mo, Cu, Co, and Zn are found as *cofactors in *enzymes, where they serve a variety of functions. They may not participate in cat., but simply act to stabilize the enzyme's tertiary structure (such as the function of Ca in several *serine proteases such as *subtilisin). *Transition metals are frequently used as *Lewis acids, stabilizing a negative charge on the substrate/ transition state. For example, in the metalloproteases, zinc coordinates both with the nucleophilic water, lowering its pK_a, and with the carbonyl of the scissile amide bond, stabilizing formation of the oxyanion in the *transition state. Another common function of transition metal cofactors is to act as electron donors/acceptors during *redox reactions. In many oxidases, O_2 is the oxidant. O_2 can form a complex with the metal which increases the reactivity of the oxygen. **E=F**; **G** Metalloenzyme.   P.S. SEARS

metalloproteases comprise one of the four main classes of protein-hydrolyzing enzymes. They often utilize a Zn^{2+} ion as a *Lewis acid that is coordinated in the cat. step to the nucleophilic water (to increase its acidity) and to the carbonyl group of the scissile bond. A carboxylate group often serves as a base to abstract a proton from the nucleophilic water molecule.

No covalent intermediate is formed. Thermolysin, acylases, and *carboxypeptidase A are members of this class of enzyme. **E=F**; **G** Metalloproteasen.   C.-H. WONG

metal macrocycles large molecules analogous to macrocyclic cpds. containing *transition metal atoms (Figure) as building blocks for *crown ethers, *sandwich cpds., etc. **E=F**; **G** Metallomakrozyclen. B. CORNILS

Ref.: *Chem.Commun.* **1998**, 413 and 2521; *Angew. Chem.* **1998**, *110*, 2340.

metalorganic compounds → organometallic compounds

metal single crystal → single crystal

metal-support interactions (MSIs)

When metal particles are supported over an oxide surface in order to increase the active *surface and reduce the rate of *sintering of the metal *nanoparticles, the interaction occurring between the lattice oxygen of the *support and the metal particles (the reason for their stabilization against their thermodynamic tendency to reduce the free surface energy) also influences the cat. behavior of the active metal comp.

This support effect may take place in two distinct ways – 1: the support modifies the electronic character of the metal particles, influencing its *adsorption and *reactivity properties, or 2: the bond between the metal particle and the support influences the *shape or geometry of the metal particles, i.e., the type of metallic *clusters or crystalline faces present. These effects take place especially for very small metal particles, but have different influences on the cat. behavior. The electronic effect could change the activity of the sites on the metal surface, i.e., their *TON, whereas the geometric effect would modify the number and type of *active sites present.

Despite many publications concerning metal-support interactions there is still no clear identification of the general rules determining the predominant effect on the cat. *activity, although experimental data have definitively confirmed that metal-support interactions affect the activity. Metals supported on TiO_2 were found to have much higher activities than when supported on other oxide supports such as *alumina or *silica, and metals supported on *zeolites exhibited higher activities

than when supported on amorphous metal oxides. Altering the acidity of zeolites was also shown to have a considerable effect on cat performances of supported Pt and Pd. For the latter effect, several models have been proposed. Recent data have shown that the MS is similar for zeolites and amorphous support and indicate that a model based on a direct influence of the *Madelung potential of the oxide support on the electronic structure of the metal particles could explain the various results. The Madelung potential is the result of electrostatic interactions in an oxide lattice and is determined by all the elements in the material.

MSIs can be defined as being weak, medium or strong. Non-reducible metal oxides such as *silica, *alumina, and magnesia as well as *carbon or *graphite have only weak MSIs Zeolites exert a medium metal-support interaction, while metals supported on reducible oxides (*titania, *ceria) when reduced at high temps. exhibit a strong metal-support interaction (SMSI). The MSI depends not only on the support, but also on the metal. Sintering supported Ag cats. in vacuum indicated that the effect decreases in the $SiO_2 > Al_2O_3 > C$ sequence, while for Pt cats. the sequence is $Al_2O_3 > SiO_2 > C$.

The SMSI effect is the better known example of MSI. When metals supported on titania are heated in hydrogen at relatively high temps. a dramatic decrease in H_2 and CO *chemisorption occurs. Data have shown that small Pt particles on reduced titania become partially encapsulated and migration of reduced TiO_x *ensembles over the *surface of the metal occurs The latter effect enhances the formation of direct metal-support (M-S) bonds with respect to the sample before SMSI. The tendency to give rise to SMSI effects also depends on the nature of the metal. Pt enters the SMSI state more easily than does Rh, and Pt is also more stable in the SMSI state than Rh.

Even though the *chemisorption of both H_2 and CO is reduced on cats. in the SMSI state these cats. show enhanced activity for the *hydrogenation of CO. This is related to the fact

hat in these cats., due to the presence of M-S
direct bonds, cooperative *adsorption of the
CO at the interface between the Pt and the
TiO$_x$ overlayer is possible. The C atom in CO
is coordinated to Pt ions and the oxygen on
the oxygen-deficient TiO$_x$. This effect has
been detected for a variety of other substrates
containing carbonyl groups, and used to enhance the *selectivity in *hydrogenation.
The SMSI effect indicates the role of the
*contact interface between oxide and metal
particles in determining the cat. behavior. Recent data on supported gold nanoparticles indicate that the perimeter junction contains
the sites for the activation of oxygen for CO
oxidation and for the *epoxidation of propene. The metal-support interaction is thus
not only a way to disperse and stabilize metal
nanoparticles (*nanodisperse materials), but
also a way to control their reactivity, by electronic or geometric effects, or by creating new
cat. functionalities at the metal-oxide interface. A better understanding of these effects
will be the challange to design new improved
metal-supported cats. (cf. *oxide cats.). **F** interaction métal-supports; **G** Metall-Träger
Wechselwirkungen.   G. CENTI

Ref.: *Angew.Chem.* **1959**, *71*, 101; T.M. Tri, J. Massar-
dier, P. Gallezot, B. Imelik, *Metal-support and Metal-
additive Effects in Catalysis*, Elsevier, Amsterdam
1982; *Stud.Surf.Sci.Catal* **1996**, *101*, 1165 and **1997**, *110*,
123; *Catalysis (London)* **1983**, *6*, 27; *Acc.Chem.Res.*
1987, *20*, 389; Ertl/Knözinger/Weitkamp, Vol.2, p. 752;
Augustine, p. 169.

metal vapor syntheses (MVS) metal-
mediated synth. to produce labile *complexes
that are adsorbed on porous *supports and
form cat. active metal *cluster cpds. May also
be important as *precursors for hom. cats.; cf.
*CVD. **F=G=E**.   B. CORNILS

Ref.: Blackborrow/Young, *Metal Vapor Synthesis in
Organometallic Chemistry*, Springer, Berlin 1979.

metathesis a cat. reaction in which alkenes are converted into new prods. via the
rupture and reformation of C-C double
bonds. An incomplete survey of the broad

scope and their interrelations is given in the
Scheme.

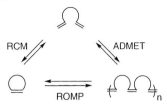

Depending on the starting material (cyclic or
acyclic alkenes) and the reaction parameters,
*ring-closing metathesis (RCM), acyclic diene
metathesis (ADMET), or *ring-opening metathesis (ROMP) proceed. *Cross-metathesis
of internal and terminal alkenes (e.g., *ethenolysis), metathesis of alkynes, and combinations of the reactions mentioned (e.g., *cross-
m. between cyclic and acyclic alkenes or
between alkenes and alkynes) complement
the intriguing possibilities of m. The reaction
mechanism proceeds via a metal carbene
chain mechanism. The propagation reaction
involves a *transition metal *carbene as the
active species with a *vacant coordination
site. The alkene coordinates at this vacant site
and subsequently a metallacyclobutane intermediate is formed, which cleaves to form a
new metal-carbene *complex and a new
alkene. For cyclic alkenes this mechanism
results in unsaturated polymer chains.
The m. can be cat. by both het. and hom. systems. These include cats. containing *transition metals in high as well as in low oxidation
states, based on W, Mo, Re, Ru, Os, Ir, Ta, or
Ti. Whereas het. systems generally consist of
a transition metal oxide or an organometallic
*precursor deposited on a high-*surface *support, hom. systems mainly consist of a comb.
of a transition metal cpd. (usually an oxo
chloride), an organometallic cpd. as a *co-catalyst, and sometimes a *promoter cpd. In
some cases cats. for m. are well-defined carbene complexes of transition metals. For commercial examples, cf. the entries *CdF, *Hüls,
*IFP Meta-4 or propylene, *Phillips POCT or
Triolefin, *Shell FEAST and SHOP, etc.
F metathèse; **G** Metathese.   T. WESKAMP

Ref.: Ivin/Mol; Mark/Bikales/Overberger/Menges, Vol. 9, p. 634; Ertl/Knözinger/Weitkamp, Vol. 5, pp. 2380,2387; *J.Mol.Catal.A:* **1999**, *140*, 287; *Acc. Chem. Res.* **1995**, *28*, 446; *Angew.Chem.Int.Ed.Engl.* **1999**, *38*, 2416.

metathesis of alkanes metathesis under the conds. of *surface organometallic chemistry (Figure).   W.A. HERMANN

$$2 \ CH_3-CH_3 \xrightarrow{\text{cat.}} CH_4 \ + \ CH_3CH_2CH_3$$

$$\left.\begin{array}{c}\textbf{polyethylene} \\ or \\ \textbf{polypropylene}\end{array}\right\} \xrightarrow[H_2]{\text{cat.}} 2 \ C_nH_{2n+2}$$

$$n \geq 1 \ \textbf{(mainly } CH_4 \ \& \ C_2H_6)$$

Ref.: *Angew.Chem.Int.Ed.Engl.* **1998**, *37*, 806.

methacrylic acid an important intermediate, manuf. as the acid or as an ester (MMA = methyl methacrylate), either by 1: the non-cat. acetone cyanhydrin proc., 2: via cat. procs. starting from isobutene (*Escambia), *iso*-butyraldehyde (*Eastman), *iso*-butyric acid (*Kodak), *Mitsubishi), *tert*-butanol (*Asahi), or propionaldehyde (*BASF proc.). F acide méthacrylique; G Methacrylsäure.
                    B. CORNILS

methanation the highly exothermic *hydrogenation of CO to CH_4 is cat. by group VIII metals as well as by Mo and Ag; Ni is the most commonly reported cat. M. is being applied as a well-proven means for conv. of traces of CO in hydrogen used for, e.g., *ammonia synth. In the 1960s and 1970s m. was believed to be an alternative to naturally occurring methane, which was to be realized by *gasification of coal and subsequent conv. to methane (*substituted natural gas, SNG). Various types of reactors were considered for the reaction to handle the severe adiabatic temp. increase due to the high exothermicity (cf. *catalytic reactors; procs. of *FMC COED, *Hygas, *Parsons RM). Since the 1980s, however, an abundance of natural gas exists worlwide and demonstration plant work on m. was terminated. F méthanation; G Methanisierung.   M. BAERNS

Ref.: Falbe-1; Falbe-2; Twigg, p. 340.

methane coupling → oxidative couplin(of methane

methane monooxygenase from metha notrophic bacteria cat. the *oxidation o(methane to *methanol acc. to $CH_4 + NADH + H^+ + O_2 \rightarrow CH_3OH + NAD^+ + H_2O$. It als(cat. the oxidation of other alkanes containin(two to eight carbons to the corresponding pri mary and secondary alcohols, alkenes to *ep oxides, thioethers to sulfoxides, and aromatic to the monohydroxy aromatics. The *enzyme from *Methylococcus capsulatus* and *methylo sinus trichosporium* are *complexes of three proteins: Protein A, a non-heme iron comp. that contains a (μ-oxo) diiron center and act: as an oxygenase; Protein C, a 2Fe-2S/FAD protein (cf. *iron-sulfur clusters) that accept: electrons from *NADH and acts as a reduc tase; and protein B, a small regulatory protei■ that regulates the interaction of substrate: with protein A. A *steady-state kinetic analy sis indicates that methane binds to the en zyme complex first, followed by NADH t(form a ternary complex. This complex ther binds oxygen to form a secondary ternary complex which gives rise to methanol anc water. **E=F=G**.   C.-H. WONC

Ref.: *J.Am.Chem.Soc.* **1993**, *115*, 939; *Nature (London* **1993**, *366*, 537.

methanol synthesis is governed by three reactions, all of which are exothermic:

$$CO + 2\,H_2 \;\rightarrow\; CH_3OH \quad \Delta H_{298\,K} = -90.64\;kJ/mol$$

$$CO_2 + 3\,H_2 \rightarrow CH_3OH + H_2O \quad \Delta H_{298\,K} = -49.47\;kJ/mol$$

$$CO + H_2O \leftrightarrow CO_2 + H_2 \quad \Delta H_{298\,K} = -41.17\;kJ/mol$$

The decrease in volume together with the limited *activity of the cats. available initially led to the development of high-press. (25–35 MPa) and high-temp. (300–450 °C) procs. (cf. *BASF methanol proc.). The development of cats. with increased activity allowed a reduction of synth. temp. as well as press. and led to the modern technology of low-press. methanol procs. ICI was in the lead to develop the Cu-ZnO-Al$_2$O$_3$ (series 51; type 51–1) cat. with much higher activity than the ZnO-Cr$_2$O$_3$-based cat. used so far in high-press. operation. The increased activity permitted operation at 5 MPa and at temps. around 260 °C initially. Later, the press. was increased to 10 MPa, accompanied by the development of cats. with increased activity. The low-press. proc. is more efficient, has a lower capital cost, and is cheaper to operate than the early high-press. proc., which consequently became obsolete. However, the low- press. Cu-containing cat. is sensitive to *poisons, especially to sulfur, and therefore requires a carefully purified *syngas (cf. procs. of *ICI, *Lurgi, *Haldor Topsoe). **F** synthèse de méthanol; **G** Methanolsynthese.   C.D. FROHNING

Ref.: Farrauto/Bartholomew, p. 370; Twigg, p. 441; Kirk/Othmer-1, Vol. 16, p. 537; Kirk/Othmer-2, p. 757.

methanol-to-gasoline process (MTG)
→ Mobil procs.

methanol-to-olefin process (MTO) →
Mobil procs.

methoxycarbonylation
an unusual term for the hom. cat. *carbonylation of alkynes in the presence of methanol, yielding, e.g., *methyl methacrylate from propyne, CO, and MeOH. Formally CO is inserted between the alkyne and the O-atom of methanol.
                    B. CORNILS

methyl acetate
(acetic acid methyl ester) an important *intermediate for *acetic acid (*anhydride) and *vinyl acetate, manuf. by *carbonylation of methanol and thus based on *syngas only. **F** acétate de méthyle; **G** Methylacetat.   B. CORNILS

Ref.: Cornils/Herrmann-1; p. 104.

methylalumoxane
(MAO) the prod. of the partial *hydrolysis of trimethylaluminum (TMA) and water. Known since the end of the 1970s by Sinn and Kaminsky for the ability to promote the metallocene cat. as a *co-cat., MAO is prepared by an incomplete hydrolysis of TMA acc. to different technologies (Equ.):

$$n\;Me_3Al + n\;H_2O \longrightarrow \left[\begin{array}{c} CH_3 \\ | \\ Al-O \\ \end{array} \right]_n + 2n\;CH_4$$

MAO is an amorphous white solid which is usually readily soluble in aliphatic and aromatic *hcs. It contains differing quantities of TMA, undergoes self-ignition and reacts vigorously with water. Although MAO is widely used in industry its structure and function in cat. are not yet completely understood. MAO seems to be a complex mixture of linear, cyclic- and cagelike elements. The most important use of MAO is as a co-cat. in *metallocene cat. alkene *polymerization. It increases the efficiency of the *transition metal cats. over several magnitudes and gives the possibility of producing high molecular weight prods. The function of MAO during the *polymerization is three-fold- 1: *alkylation of the transition metal *complex; 2: *scavenging cat. *poisons; 3: serving as a weakly coordinating anion which stabilizes the polymerization active cationic cat. species. Acc. to first reports, MAO can serve as a cat. for the *epoxidation of unsaturated alcohols. **E=F=G**.

            W. KAMINSKY, R. WERNER

Ref.: Ullmann, Vol. A18; *New Sci.* **1993**, (8),28; *Nachr.Chem. Tech.Lab.* **1993**, *41*, 1341; B. Tieke, *Makromolekulare Chemie*, Wiley-VCH, Weinheim 1997, p. 140; *Green Chem.* **1999**, *1*, 27; *J.Mol.Catal.A*: **1999**, *148*, 29.

methyl *tert*-butyl ether (MTBE) an important antiknock additive (*oxyfuel), manuf. by cat. (acids, *zeolites) *addition of *i*-butene and methanol acc. to various procs. (*CDTech, *Hüls/UOP, *IFP); cf. *etherification. The importance of MTBE will diminish because of the new US legislation which will favor ethanol at the expense of MTBE. **F** tertiobutyl méthyl éther; **G=E**.   B. CORNILS

Ref.: Ullmann, Vol. A16, p. 543 US-EPA No. 510-F-97-016; *Chem.Eng.* **1994**, *101*, 61.

methylchymotrypsin   a chemically modified form of the mammalian *serine protease, *chymotrypsin, in which the ε2 nitrogen of the *active-site histidine (h. 57) is methylated. In order for the histidine to act as a general base, the ring must flip so that the ε2 nitrogen faces the active-site aspartic acid, and the δ1 nitrogen is adjacent to the active-site serine. This disrupts the *catalytic triad, and prevents stabilization of the protonated histidine by the active-site aspartic acid. With respect to the wild-type *enzyme, this enzyme has been shown to have an increased aminolysis/hydrolysis ratio for the synth. of peptides from peptide or amino acid esters. It has a severely impaired active site, however, and rates of reaction (*hydrolysis) for both amide and ester substrates that are diminished by several orders of magnitude. For this reason, peptide synth. with this enzyme works best with activated ester substrates such as cyanomethyl esters. **E=F=G**.   P.S. SEARS

Ref.: *J.Am.Chem.Soc.* **1990**, *112*, 5313; *Biochemistry* **1970**, *124*, 13.

methylenation   → olefination

methylene blue   used as an electron transfer mediator in cat., cf. *regeneration of NAD(P).

methyl formate   an important building block, manuf. via base-cat. methanol *carbonylation. Acc. to its cat. *decarbonylation (RO$^-$ or metal cats.) it can be considered as a C_1 building block and as "*liquid *syngas".

Thus, most of cat. reactions involving CO + MeOH have been transposed to mf. The most relevant include its formal *isomerization to *acetic acid, its carbonylation to *acetaldehyde or methyl acetate, and its reaction with halide derivatives or alkenes to give the corresponding methyl esters. *Isomerization has attracted the most attention. The reaction is cat. by various *transition metals (Co, Pd, Ni, Ru, Ir, Rh) with an iodine *promoter. The mechanism includes formation of the mixed formic-acetic anhydride, which finally decomposes into acetic acid and CO. The other reactions generally proceed via mf. decarbonylation; however, direct activation of the C-H bond of mf. by the metal center has also been proposed. Mf. is used commercially for the manuf. of formic acid (*Leonard proc.). **F** formiate de méthyle; **G** Methylformiat, Ameisensäuremethylester.   Y. CASTANET

Ref.: *Appl.Catal.A:* **1995**, *21*, 25.

methyl methacrylate (MMA)   methacrylic acid methyl ester; cf. *methacrylic acid

methyl transfer reactions   → *S*-adenosylmethionine

methyltrioxorhenium (MTO)
CH_3ReO_3, the simplest of the alkylrhenium trioxides, can be synth. by various methods, e.g., by reaction of dirhenium heptaoxide Re_2O_7 with tetramethyltin $Sn(CH_3)_4$ as alkylating agent in dry tetrahydrofuran under reflux conds. Recent developments reveal an efficient Re recycle to cat. active alkylrhenium species. Higher and functionalized homologs of MTO are synth. either by the anhydride route using tin organyls or by *alkylation of Re_2O_7 with zinc organyls at low temps. MTO and its homologs are soluble in any organic solvent as well as in water, where their water adducts (hydrates) react as rather strong *Lewis acids ($pK_a^{22\,°C}$ ca. 3.82). Due to their expressed Lewis acidity, MTO has high activity as *oxidation cat. for the oxidation of double- and triple-bond systems, of alkanes, alcohols, and of hetero atoms. With H_2O_2

MTO (and its congeners) form mono- and bisperoxo complexes $CH_3ReO(O_2)_2 \cdot H_2O$ or $RReO(O_2)_2 \cdot H_2O$ which are highly active in a broad variety of oxidation reactions (e.g., alkene *epoxidation, alkene *trans-*hydroxylation, alkyne *oxidation, *Baeyer-Villiger oxidation, oxidation of aromatic compounds to quinones, etc. MTO-cat. carbon transfer procs. are alkene *metathesis (inter alia with dissolved or surface-*supported MTO), *olefination of aldehydes, synth. of cyclopropanes, aziridines, epoxides, etc. **E=F=G**. R.W. FISCHER
Ref.: *Acc.Chem.Res.* **1997**, *30*, 169; *J.Organomet.Chem.* **1995**, *500*, 149.

methylviologen common name for 1,1'-dimethyl-4,4'-bipyridinium dichloride, used as an electron transfer mediator (e.g., in *enzyme cat.; cf. *enoate reductase, *regeneration of NAD(P)). **F** viologène de méthyle; **G** Methylviologen. B. CORNILS

metolachlor reaction the key step in a novel *enantioselective proc. for the grass herbicide metolachlor. The commercial prod. exists as a mixture of four stereoisomers, but approx. 95 % of the activity is due to the two S-diastereomers. As a consequence, a chiral switch based on an Ir-mediated *asymmetric *hydrogenation of 2-methyl-5-ethylaniline (MEA) imine to N-alkylated aniline (S-NAA, Equ.) gives selective access to these stereoisomers and leads to approx. 35 % less environmental load (50 °C/8 MPa hydrogen). The *ligand is *xyliphos. T. WESKAMP

xyliphos

Ref.: *Chem.Ind.(New York)* **1996**, *68*, 153; *Chimia* **1999**, *53*, 275.

Metton a poly dicyclpentadiene, manuf. by Hercules in a *ROMP reaction, cat. by W salts

mevalonic acid a six-carbon carboxylic acid *precursor to *isopentenyl and dimethylallyl pyrophosphates, which are used in the synth. of steroids and various other polyisoprenoid cpds. such as dolichol and ubiquinone. It is produced by the *reduction of hydroxymethylglutaryl CoA (HMG-CoA), cat. by HMG-CoA reductase. HMG-CoA reductase is a highly regulated *enzyme, and is a target for many cholesterol-reducing drugs such as Lovastatin. **F** acide mévalonique; **G** Mevalonsäure. P.S. SEARS

Meyer-Schuster rearrangement Substituted alkynols undergo proton-cat. rearrangement to α,β-unsaturated ketones; cf. *Rupe reaction. **F** réarrangement de Meyer-Schuster; **G** Meyer-Schuster-Umlagerung. B. CORNILS
Ref.: Krauch-Kunz.

MFI *zeolite structural code for *ZSM-5 (called Mobil five)

MHD process → Mitsubishi procs.

MIBK methyl isobutyl ketone

mica a silicate (laminated talc) which can be cleaved into thin, flexible sheets. M. is proposed to be a *support for het. cats. **E=F**; **G** Glimmer. B. CORNILS

microreactors → catalytic microreactors

micellar catalysis the enhancement of reaction rates in aqueous media by addition of surfactants above the critical micelle conc. (CMC, between 10^{-4} and 10^{-2} mol·dm^{-3}). *Micelles are association *colloids of surfactants. In aqueous dispersion they have at low concs. a spherical structure with a polar surface (interface) and a hydrophobic core, at higher concs. a rod-like shape. Aqueous micelles can solubilize reactants from the surrounding water phase. In the case of ionic educts and ionic micelles the incorporation depends on the charges. The kinetic treatment is similar to a *Michaelis-Menten equation in *enzyme cat. There are the following areas of mc.- 1: reactions in which the micelles are reagents; 2: reactions in which interaction between micelles and embedded reactants affect the conv.; 3: reactions in which the micelles contain cat. active groups. Most reactions influenced by micelles belong to type 2. The rate enhancement of chemical reactions in micelles can be a comb. of different effects: i. there is a medium effect because of the lower dielectric constant in comparison to water; ii. the *transition state can be stabilized by interaction with the headgroup; iii. reactants are concentrated by interaction with the micelle surface or due to incorporation. Thus, the rate of bimolecular reactions should be increased. Besides this acceleration of reactions *chemoselectivities and *stereoselectivities can also be influenced by micelles; e.g., in comparison to water the *hydrolysis of p-nitrophenyl heptanoate is increased by 10^5-fold at pH 8.5 in micelles of octadecyldimethylhydroxyethylammonium bromide. Micelle formation in water also accelerates the acid-cat. *hydrolysis of hexadecyl sulfate to 46 times faster than that of the non-micelle-forming methyl sulfate. A variety of cat. reactions can be promoted by surfactants, e.g., *oxidations, *reductions, *asymmetric *hydrogenations, and *C-C bond formation.

Other surfactant aggregations such as rod-like micelles, vesicles, or giant micelles are also able to promote chemical reactions. Noteworthy are "reverse micelles" which are formed by association of polar headgroups of amphiphiles with colloidal drops of water. Like aqueous micelles, reverse micelles exist in highly diluted organic systems; they are so far of preparative interest.   G. OEHME

Ref.: J.H. Fendler, E.J. Fendler, *Catalysis in Micellar and Macromolecular Systems*, Academic Press, New York 1975; J.H. Fendler, *Membrane Mimetic Chemistry*, Wiley, New York 1982; J.H. Clint, *Surfactant Aggregation*, Blackie, Glasgow 1992; Cornils/Herrmann-1, p. 193.

micelles assemblies of water-soluble surfactants (amphiphiles) containing a hydrophi-

lic headgroup and a hydrophobic (lipophilic) tail. Ms. are formed above a particular conc., the critical micelle conc. (CMC; cf. *micellar cat.) and above a characteristic temp. (Krafft's temp.). The surface of the spherical ms. is formed by headgroups and the core contains the hydrophobic chain. The shape of the ms. can change at higher surfactant conc. to a rod-like structure. The CMC can be investigated through the conc. dependence on the osmotic press., surface tension, solubilization of polar cpds., turbidity and sometimes by calorimetric titrations. In the case of ionic surfactants, the measurement of conductivity is also useful. The CMC depends on the structure and the purity of the surfactant. Examples of typical structures (anionic, ionic, zwitterionic, nonionic) and their CMCs [$mol \cdot dm^{-3}$] are Na dodecyl sulfate (SDS, 8.1×10^{-3}), cetyltrimethylammonium bromide (CTABr, 9.2×10^{-4}), or polyoxyethylene(20) hexadecyl ester (Brij 58, 7.7×10^{-5}). Ms. appear to be relatively small with aggregation numbers <100 and radii between 1.2 and 3 nm. Between the micelle-water interface and core is an extremely strong polarity gradient, and it is a typical property of a m. to solubilize polar and nonpolar cpds. Reaction rates in the surrounding water phase can be influenced by ms. This acceleration is known as *micellar catalysis. Stepwise enhancement of the water content and addition of a co-surfactant can lead to microemulsions. **E=F; G** Micellen.   G. OEHME

Ref.: Y. Moroi, *Micelles*, Plenum Press, New York 1992; J.H. Clint, *Surfactant Aggregation*, Blackie, Glasgow 1992; M.J. Rosen, *Surfactant and Interfacial Phenomena*, Wiley, New York 1982.

Michael addition 1: describes the *addition of active methylene groups to electron-poor alkenes in the presence of sufficiently strong bases acc. to $H_2C=CH-R + R^1-CH_2-R^2 \rightarrow R^1R^2CH.CH_2CH_2-R$. C nucleophiles in general are malonates, cyanoacetates, acetoacetates, β-keto esters and less C-acidic cpds. such as esters, ketones, aldehydes, etc. Even indenes, fluorenes, and *carbene *complexes can be applied. In some cases *Lewis acids ($TiCl_4$, $ZnCl_2$) have been proven to accelerate the reaction. In recent years the Ma. has been performed diastereo- as well as *enantioselectively. Here a number of *chiral *Lewis acids and diphosphine/Rh cpds. (e.g., *TRAP) are effective as *asymmetric cats.   M. BELLER

Ref.: March; Trost/Fleming; *Top.Stereochem.* **1989**, *19*, 227; *J.Mol.Catal.* **1999**, *142*, 7; *Organic Reactions*, Vol. 10, p. 179 and Vol. 47, p. 315.

2: Michael and related additions, are found in a number of *enzymatic reaction mechanisms, too. For example, *cis-trans* *isomerases that act upon α,β-unsaturated carbonyl cpds. work via the Michael-type addition of glutathione to the β-carbon, free rotation about the α-β bond, and elimination of glutathione. In addition, many metabolic pathways, such as those of lipid *catabolism, involve the addition of water to an α,β-unsaturated carbonyl. **F** addition de Michael; **G** Michael-Addition.

P.S. SEARS

Michaelis complex   a term for the noncovalently associated enzyme-substrate complex.   P.S. SEARS

Michaelis constant (K_m), Michaelis-Menten equation, Michaelis-Menten kinetics

The multistage reaction proc. in *enzyme cat. requires that the substrate(s) initially bind noncovalently to the enzyme at a special site on its surface called a *specificity pocket. The collection of specificity pockets for all the reactants is called the *active site of the enzyme. The complex of substrates and enzyme is called the *Michaelis complex* and provides the proper alignment of reactants and cat. groups in the active site. It is this active site where, after formation of the Michaelis complex, the chemical steps take place. Because each molecule of enzyme has only a limited number of active sites (usually one), the number of substrate molecules that can be processed per unit of time is limited. After an enzyme is mixed with substrate, an initial transient occurs in which the concentration of the reaction *intermedites are in rapid

flux. This is called the *pre-steady state condition. After a short period of time, the intermediates reach relatively stable concentrations that change more slowly. This is called the *steady state condition. Since there is a slow depletion of substrate, the steady-state assumption (that the rate of change of intermediates is small) is of course not always valid; however, restriction of rate measurements to this time interval is a good approximation to conds. used in synth. Steady-state rates are measured because these data are easier to collect (as compared to most pre-steady state rates) and generate the most reliable and relevant enzymatic rate constants.

Many reactions of enzymes follow a pattern of kinetic behavior known as *Michaelis-Menten kinetics*. By applying MM kinetics, the measured reaction rates or velocities (v) can be transformed into rate constants that describe the enzymatic mode of action. Useful constants such as k_{cat}, K_m, and k_{cat}/K_m (below) can be determined. In most systems, the rate of reaction at low conc. of substrates is directly proportional to the conc. of enzyme $[E]_0$ and substrate $[S]$. As the conc. of substrate increases, a point will be reached where further increase in substrate conc. does not further increase v. This phenomenon is called substrate saturation. The reaction velocity that is obtained under saturating concs. of substrate is called V_{max}.

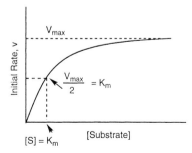

The *Michaelis-Menten equ.*, which relates the reaction velocity to the substrate and enzyme concentrations, is $v = [E]_0 k_{cat}[S]/(K_m + [S])$. In this equ. $k_{cat}[E]_0 = V_{max}$, [S] is the substrate conc., K_m represents the conc. of substrate at which $v = V_{max}/2$, and k_{cat} is the apparent first-order enzyme rate constant for conv. of the enzyme-substrate complex to product; k_{cat} is also called the *turnover number. At high concs. of substrate, the equ. simplifies to $V_{max} = k_{cat}[E]$. Correspondingly, at low concs. of substrate, the equ. simplifies to $v = (k_{cat}/K_m)[E]_0[S]$. In the last equ., k_{cat}/K_m represents the apparent second-order rate constant for enzyme action.

Although not all enzyme systems follow the same mechanistic pathway, most of them can be reduced at least approximately to the above relationships, and they are widely used in considering applications of enzymes in synthesis.

The upper limit of k_{cat}/K_m is k_1, the diffusive rate of substrate binding (10^8–10^9 M^{-1}s^{-1}). At this upper limit in rate, k_{cat} is no longer rate limiting and the Michaelis-Menten kinetics changes to *Briggs-Haldane kinetics. **F** constant de Michaelis, équation et cinétique de Michaelis-Menten; **G** Michaelis-Konstante; Michaelis-Menten-Gleichung und -Kinetik.

C.-H. WONG

Ref.: A. Fersht, *Enzyme Structure and Mechanism*, 2nd. Ed., W.H. Freeman and Co., New York 1985.

Michaelis-Menten mechanism For a simple one-substrate *enzymatic reaction which proceeds through a single binding and reaction step, the MMm. assumes that the initial binding step is very rapid with respect to the chemical transformation step.

$$E + S \underset{k_{-1}}{\overset{k_1}{\rightleftarrows}} ES \xrightarrow{k_2} E + P$$

Thus, the binding step is essentially at equil. at any given time, and so the conc. of ES can be approximated by $[ES] = (k_1/k_{-1})[E][S] = [E][S]/K_S$ where K_S is the dissociation constant of the ES complex. A mass balance on the enzyme, $[E] = [E_o] - [ES]$ (E_o = total enzyme added) can be inserted, and the equation rearranged to give $[ES] = [E_o][S]/(K_S +$ [S]). Since the velocity of the overall reaction is $v = dP/dt = k_2[ES]$, the reaction rate is: $v = k_2[E_o][S]/(K_S + [S])$.

This is the Michaelis-Menten equ. In this special case, the constant in the denominator (the *Michaelis constant, usually written K_m) is equal to the dissociation constant of the enzyme-substrate complex. **F** mécanisme de Michaelis-Menten; **G** Micheaelis-Menten-Mechanismus. P.S. SEARS

micro activity test (MAT) bench scale (screening) test for the evaluation of cats. under research or development with regard to *activity, *selectivity, *regenerability, or lifetime. B. CORNILS

microcalorimetry is used for the measurement of differential heats of *adsorption of probe molecules in heat-flow *calorimeters and differential scanning calorimeters. The data provide information about the acid/base strength distribution. B. CORNILS

microchannel reactor a modern variant of fixed-bed reactors designed with a stack of microstructured, (typically) metallic wafers (cf. Figure 2,h under the entry *catalytic reactors). By microstructuring, microchannels are formed having equivalent diameters of ≤500 μm. This short pathway from the bulk of the gas to the channel wall leads to very rapid mass and heat transfer procs. In addition, by a crosswise stacking of the wafers a cross-flow reactor is build, which allows heat removal directly from wafer to wafer, by which isothermal conds. are attainable. Mrs. are in the development stage. **F** réacteur à canal micro; **G** Mikrokanalreaktoren. P. CLAUS, D. HÖNICKE

Ref.: W. Ehrfeld, V. Hessel, *Microreactors*, Wiley-VCH, Weinheim 1999.

microkinetic analysis supportive technique for the synth. of cat. materials. The derivation of the microkinetics is not necessarily based on detailed kinetic experimentation but, by analogy, on similarities with other known cat. procs.

A microkinetic simulation is performed using rate constants of the postulated elementary reactions. The mechanistic and kinetic models applied have to be consistent with experimentally identified gas-phase prods. and *surface intermediates observed. By using these models and kinetic parameters having physical and chemical meanings the general trend of the kinetic relationships is expected to predict the cat. performance of a solid material. **F** analyse microcinétique; **G** mikrokinetische Analyse.   M. BAERNS

Ref.: J.A. Dumesic, D.F. Rudd et al., *The Microkinetics of Heterogeneous Catalysis*, ACS, Washington 1993; E.R. Becker, C.J. Perreira (Eds.), *Computer Aided Design of Catalysts*, Dekker, New York 1993, Ch.2; *Catal. Today* **1991**, *10*, 147.

micropores acc. to IUPAC, pores with a size smaller than 2.0 nm. Additionally, ms. are characterized through their sorption behavior ("micropore filling", occurring in principle whenever adsorptive and pores are of similar size). The nature of *adsorption is described as an overlap of adsorption forces across the pore, affecting the isotherm shape. Adsorption isotherms are typically determined with N_2, Ar, or alkanes. "M. volumes" are calcd. from m. capacities (moles of a particular gas adsorbed at a stated temp.) using the bulk liquid density, i.e., making the questionable assumption of liquid adsorbate in the pores, and should therefore be termed "apparent". Ms. are sometimes subdivided into ultramicropores (<0.6 nm) and supermicropores (0.6–1.6 nm). **E=F**; **G** Mikroporen.   F. JENTOFT

microporous solids materials with *micropores, which may additionally have *meso- or *macropores or an appreciable external *surface. Typical representatives are *zeolites, amorphous *silica and *titania, modified *clays, *heteropoly cpds., phosphates, niobates, germanates, *activated carbons, carbogenic *molecular sieves, and *polymers. Micropores can be 1: the void spaces in a crystal structure (e.g., zeolite framework), or 2: the *interstitial spaces which are created when primary particles agglomerate (e.g., void

spaces in amorphous silica). The *pores are characterized by their sizes (*pore size distribution), *shape (e.g., cylindrical vs. slits), and the apparent micropore volume. These parameters are measurable through sorption experiments. Surface areas, which are often reported, may reach a few thousand m^2/g, but reflect the sorption capacity rather than the accessible area because they are based on the *BET model, which is invalid in the presence of micropores.

There are a variety of commercial applications for mss. (*molecular sieve effect, adsorbents, for separation purposes, *ion exchangers). Advantages of m. materials as cats. (predominantly zeolites) lie in the steric hindrance imposed by the pore system, i.e., restriction of molecules entering (*reactant shape selectivity, cf. *selectivity) or exiting (*product shape selectivity, cf. *zeolites) the pore system, and restriction of the formation of certain molecules in the pores (*transition-state shape selectivity). Mss. also assist in the stabilization of small metal particles, such as highly dispersed noble metals, which have many applications as cats. Side reactions such as the formation of *coke can be suppressed in the confined space of the pore system; however, reactions can always take place on the external surface, and pores can be plugged, leading to *deactivation of the cat. Restriction of pore mouth size can also be advantageous in order to tune the shape selectivity and can be accomplished by, e.g., silanization (zeolites), or carbon *deposition (carbogenic molecular sieves). Disadvantages of microporous materials as cats. include transport limitations. **F** solides microporeuses; **G** mikroporöser Feststoff.   F. JENTOFT

Ref.: Thomas/Thomas, p. 608; Beyer.

microscopic reversibility The principle of microscopic reversibility states that every reaction is reversible, though the rate of the reverse reaction may be extremely slow. Thus, macroscopically a steady state results which is described by the mass action law. Mr. has implications for *enzymatic (or any) catalysis: the cat. cannot affect the position of the reaction equil. If it increases the reaction *rate in one direction tenfold, it must increase it tenfold in the opposite direction, too. This principle is most easily visualized on a reaction energy diagram (Figure).

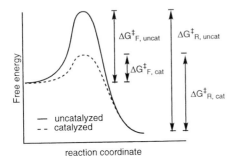

$$\Delta G^{\ddagger}_{F, uncat} - \Delta G^{\ddagger}_{F, cat} = \Delta G^{\ddagger}_{R, uncat} - \Delta G^{\ddagger}_{R, cat}$$

since rate $\propto \exp(-\Delta G^{\ddagger}/RT)$

$$k_{F, uncat}/k_{R, uncat} = k_{F, cat}/k_{R, cat}$$

It can be seen that even a reaction that is often called "irreversible" does have a reverse reaction ($k_R \neq 0$); furthermore, a cat. that lowers the energy barrier to the reaction (*activation energy) in one direction must necessarily lower the barrier in the opposite direction by the same amount. **F** réversibilitée microscopique; **G** mikroskopische Reversibilität.
P.S. SEARS

microsomes are small, membrane-bound vesicle derived from the endoplasmic reticulum produced upon disruption of the cell. Microsomes are rich in ribosomes and contain the membrane-bound machinery for protein translocation and glycosylation. As a result, they are frequently used for the in-vitro study of these procs. **E=F**; **G** Microsomen. P.S. SEARS

micro units version of *miniplant work within *cat. pilot plants (pps.), as a step between lab work and pp. Since each step of the

sequence laboratory → micro unit (benchtop) → miniplant → pp. → demonstration plant involves an exponential increase in resources, time, and money, the least expensive simulation of the planned reaction is required. Mps. can be constructed from glass or steel and any other special material; they are fully equipped with process control devices. For the tasks of micro units cf. *catalytic pilot plant, *catalytic microreactors. **F=G=E**.
B. CORNILS

microwaves in catalysis The application of microwaves in cat. includes microwaves for heating (*drying) in the prepn. of cats., catalysis via microwave irradation (mostly in solventless organic synth. as a low-waste route), and microwave plasma cat. (PACT, *plasma catalysis technique). So far, it is a point for debate whether microwaves only function via superheating or via an electromagnetic effect; cf. *hydrolysis. **F** micro-ondes en catalyse; **G** Mikrowellen-Katalyse. B. CORNILS
Ref.: Ertl/Knözinger/Weitkamp, Vol. 3, p. 1347; *Cattech* **1998**, (6), 75; Beller/Bolm, Vol. 2, p. 436; *Green Chem.* **1999**, *1*, 43.

Midrex process → CO_2 reforming

Milas reagent consists of WO_3/H_2O_2 for the *epoxidation of unsaturated systems, e.g., *Degussa Glycidol process. **F** réactif de Milas; **G** Milas-Reagenz. B. CORNILS

millscale catalyst → Kellogg Synthol process, *Hydrocol process

Miller-Bravais indices → crystallography of surfaces

miniplants → catalytic pilot plant, *micro units

mirror electron microscopy (MEM) a variant of electron microscopy, specially suitable for the measurement of local *work functions. R. SCHLÖGL

Mitsubishi acrylic acid process *oxidation of acrolein over Mo-V oxide cats.

B. CORNILS

Mitsubishi butanediol process manuf. of *1,4-butanediol (BDO) from 1,3-butadiene via *acetoxylation on Te-doped Pd cats. on carbon as *support. **F** procédé Mitsubishi pour le BDO; **G** Mitsubishi BDO-Verfahren.

B. CORNILS

Ref.: Cornils/Herrmann-1, p. 402.

Mitsubishi Gas terephthalic acid process two-step proc. for the manuf. of *terephthalic acid (TPA) by *carbonylation of toluene at 30–40 °C with a HF · BF$_3$ cat., followed by cat. *oxidation of the resulting p-tolylaldehyde. **F** procédé Mitsubishi Gas pour le TPA; **G** Mitsubishi Gas TPA-Verfahren.

B. CORNILS

Mitsubishi hexamethyleneimine process for the manuf. of hexamethyleneimine (HMI) by selective *hydrogenation of ε-caprolactam over a Co cat. **F** procédé de Mitsubishi pour le hexaméthylène imine; **G** Mitsubishi Hexamethylenimin-Verfahren.

B. CORNILS

Mitsubishi maleic anhydide process manuf. of *MA/MAA by fluid-bed *oxidation of butenes over a V$_2$O$_5$/H$_3$PO$_4$ cat. **F** procédé Mitsubishi pour l'anhydride de l'acide maléique; **G** Mitsubishi Maleinsäureanhydrid-Verfahren.

B. CORNILS

Mitsubishi methacrylic acid process manuf. of *MA from iso-butyric acid by *oxydehydrogenation over a Mo-V-P (*VPO) cat. at 250–350 °C. **F** procédé Mitsubishi pour MA; **G** Mitsubishi MA-Verfahren. B. CORNILS

Mitsubishi MHD process a thermal variant of the *hydrodealkylation. **F** procédé MHD de Mitsubishi; **G** Mitsubishi MHD process. B. CORNILS

Mitsubishi oxo process *LPO variant of the *hydroformylation of propylene with an Rh cat., modified with *triphenylphosphine at 100 °C/1.5–1.8 MPa. The separation of the cat. solution and the butyraldehydes takes place via a comb. of gas and liquid recycle with a strip column (Figure). **F** procédé oxo de Mitsubishi; **G** Mitsubishi Oxoverfahren. B. CORNILS

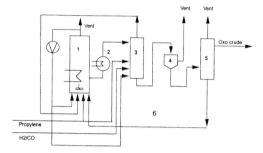

1 reactor; 2 cooler/heater; 3 stripper; 4 gas-liquid separator; 5 distillation; 6 catalyst recycle.

Ref.: Cornils/Herrmann-1, p. 84.

Mitsubishi trimellitic acid process a two-step manuf. of *trimellitic anhydride by 1: cat. *carbonylation of m-xylene with CO (cat. HF · BF$_3$) and 2: *oxidation of the *intermediate 2,4-dimethylbenzaldehyde (with MnBr$_2$/HBr). B. CORNILS

Ref.: Weissermel/Arpe, p. 316

Mitsubishi xylene process *isomerization of m-xylenes under *Friedel-Crafts conds. with HF · BF$_3$ at 100 °C. **F** procédé Mitsubishi pour des xylènes; **G** Mitsubishi Xylol-Verfahren. B. CORNILS

Mitsui CX process for the cat. manuf. of bimodally molecular-weight-distributed *HDPE and *MDPE under low-press. slurry conds. **F** procédé CX de Mitsui; **G** Mitsui CX-Verfahren. B. CORNILS

Ref.: $Hydrocarb.Proc.$ **1997**, (3), 148.

Mitsui hydroquinone process for the manuf. of hydroquinone from p-diisopropylbenzene via a variant of the *Hock proc. (*double Hock). **F** procédé de Mitsui pour l'hydroquinone; **G** Mitsui Hydrochinon-Verfahren. B. CORNILS

Mitsui-Toatsu acrylamide process
manuf. of *acrylamide by cat. *hydrolysis of *acrylonitrile over *Raney Cu at 80–120 °C. **F** procédé Mitsui-Toatsu pour l'acrylamide; **G** Mitsui-Toatsu Acrylamid-Verfahren.
B. CORNILS

Mittasch Alwin (1869–1953), worked with BASF on catalysis and catalysts, and developed various cats. for *ammonia and *methanol synthesis. B. CORNILS

mixed metals → multicomponent catalysts, *mixed oxide catalysts

mixed oxide catalysts cats. containing several metal oxides, the active phases of which are *bimetallic. The presence of various metal oxides permits a proper adjustment of local electronic properties, ensures defined coordination, limits the extent of *redox phenomena, and may be helpful in stabilizing the cat. by smooth *sintering. Examples are Cu chromite (CuCr$_2$O$_4$-CuO, *Adkins cat.), Cu-Zn chromite (Cu$_x$Zn$_{1-x}$Cr$_2$O$_4$-CuO, *methanol low-press. synth.), or Fe molydate [Fe$_2$(MoO$_4$)$_3$, *Topsøe formaldehyde synth.]. Mocs. include also *vanadia-*molybdena (maleic anhydride from benzene or butene), *chromia-*alumina (*dehydrogenation), *perovskites, or systems with MgO, ZnO, etc.
A famous example of moc. is the *ammoxidation (cf. *bismuth molydbates, *antimonates, *Aurivillius phases). **F** catalyseurs d'oxydes des métaux mixte; **G** gemischte Metalloxidkatalysatoren (Mischoxide). B. CORNILS
Ref.: Ertl/Knözinger/Weitkamp, Vol. 1, p. 101; Gates-1, p. 320; *Ind.Eng.Chem.,Prod.Res.Dev.* **1986**, 25, 171; *Catal.Rev.* **1998**, 40, 175.

mixed-type inhibition (in enzyme systems) A mixed-type *inhibitor is one which affects both the k_{cat} and the K_m values of an *enzyme reaction. This can occur when the inhibitor

binds both the free enzyme and the enzyme-substrate *complex, but with different affinities.

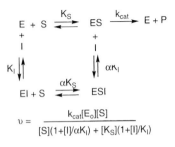

$$\upsilon = \frac{k_{cat}[E_o][S]}{[S](1+[I]/\alpha K_I) + K_S(1+[I]/K_I)}$$

This is not the only scheme in which mixed inhibition is observed. Many different scenarios can lead to mixed inhibition patterns. For example, if the inhibitor binds to two independent sites (with different affinities), and binding to one site gives purely competitive inhibition while the other gives purely non-competitive or uncompetitive inhibition, the overall inhibition pattern will be mixed. **F** inhibition mixte; **G** gemischte Inhibition.
P.S. SEARS
Ref.: I.H. Segel, *Enzyme Kinetics*, Wiley, New York 1975.

Mizoroki-Heck reaction This term is used synonymously for the Pd-cat. *Heck reaction taking into account its independent discovery by Mizoroki and Heck. **F** réaction de Mizoroki-Heck; **G** Mizoroki-Heck-Reaktion.
V. BÖHM
Ref.: *Bull.Chem.Soc.Jpn.* **1971**, 44, 581; *J.Org.Chem.* **1972**, 37, 2320.

MM molecular mechanics, → embedding methods

MMA 1: methyl methacrylate (*methycrylic acid); 2: monomethylamine

MMO catalysts multimetal-multiphase oxide catalysts, → *transition metals (het. catalysis)

MnAPO Mn-modifieed *aluminophosphate

MO molecular orbital, → molecular dynamics, *molecular modeling, *molecular orbital theory.

Mobil Airlift TCC process was primarily used to produce high quality gasoline and to reduce the yield of residual fuels by *thermo-for cracking (TCC) of virgin or previously processed gas oils (in some cases the whole crude). The charge to the reactor (at 440–510 °C) is a vapor/liquid mixture. The effluent vapors from the cat. reactor flow to prod. fractionation and gas recovery. The cat. flows by gravity through the reaction bed, passes vapor collecting grids, and through a purge zone. The purged spent cat. then flows through the kiln where *coke is completely burned from the cat. After the cat. is regenerated it passes through coolers and into the lift pot, is lifted by air to the surge vessel and starts cat. cycle again. **F** procédé Airlift TCC de Mobil; **G** Mobil Airlift TCC-Verfahren.

Ref.: *Hydrocarb.Proc.* **1972**, (9), 131.   B. CORNILS

Mobil/Badger cumene process fixed-bed *alkylation proc. for the manuf. of *cumene (isopropylbenzene, a *precursor for phenol acc. to the *Hock proc.) by reaction of benzene and chemical-grade propylene or propylene/propane mixtures over Mobil MCM-22 *zeolites. Polyalkylates have to be recycled to the separate *transalkylation unit. **F** procédé Mobil/Badger pour le cumène; **G** Mobil/Badger Cumol-Verfahren. B. CORNILS

Ref.: *Hydrocarb.Proc.* **1997**, (3), 121.

Mobil/Badger EBMax process liquid-phase proc. for the manuf. of *ethylbenzene (EB) from benzene and ethylene over cat. fixed beds of Mobil MCM-22 *zeolite/Trans-4 molecular sieve. Polyethylbenzenes are *transalkylated with benzene to form EB. **F** procédé EBMax de Mobil/Badger pour l'éthylbenzène; **G** Mobil/Badger EBMax-Verfahren.   B. CORNILS

Ref.: *Hydrocarb.Proc.* **1997**, (3), 127.

Mobil/Badger 3GEB process Vapor-phase (third generation) proc. for the manuf. of *ethylbenzene from benzene and ethylene over fixed beds of ZSM-5 (*zeolite) cats. (cf. *Mobil/Badger EBMax proc.) at 380–450 °C/2–3 MPa and *WHSV >100 (Figure). Byprod. polyethylbenzenes are recycled and *transalkylated. A narrow optimal temp. is secured by interbed quench by the reactants. **F** procédé 3GEB de Mobil/Badger pour l'éthylbenzène: **G** Mobil/Badger 3GEB process.

B. CORNILS

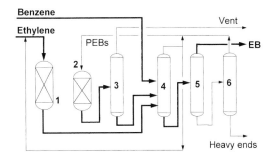

1 alkylation reactor; 2 transalkylation reactor; 3 stabilizer; 4 benzene column; 5 EB column; 6 polyethylbenzene column.

Ref.: Ullmann, Vol. A10, p. 39.

Mobil benzene reduction (MBR) process The fluid-bed MBR proc. with a *zeolite cat. reduces the benzene content of reformate streams, reducing pool benzene content to below 1 % while boosting pool octane by one number. The cat. *alkylates benzene with

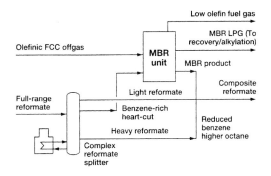

light olefins such as ethylene or propylene (Figure). **F** procédé BR de Mobil; **G** Mobil BR-Verfahren. B. CORNILS

Ref.: *Hydrocarb.Proc.* **1996**, (11), 94.

Mobil Distillate Dewaxing (MDDW) process for the cat. *hydrodewaxing of high pour-point liquids to diesel and heating oils over fixed-bed cats. (ZSM-5 *zeolite) which selectively crack the normal and slightly branched paraffins to gasoline. **F** procédé Destillate Dewaxing de Mobil; **G** Mobil Destillate Dewaxing-Verfahren. B. CORNILS

Ref.: *Hydrocarb.Proc.* **1978**, (9), 124; *Oil Gas J.* **1977**, *15*, 165 and **1990**, Aug.13, 51.

mobile support → liquid support

Mobil *p*-ethyltoluene process for the manuf. of PET (which is the basis for *p*-methylstyrene through *dehydrogenation) by reacting ethylene with toluene over modified *zeolite ZSM-5 cats. **F** procédé de Mobil pour PET; **G** PET-Verfahren von Mobil. B. CORNILS

Ref.: Ertl/Knözinger/Weitkamp, Vol. 5, p. 2129.

Mobil HAI/LPI (high-activity isomerization/low-pressure isomerization) process for the manuf. of *p*-xylene and/or *o*-xylene from C_8 aromatics-rich feedstocks (mainly *ethylbenzene) by ZSM-5-based *isomerization. HAI coprocesses hydrogen; LPI uses a dual-bed cat. system. **F** procédé HAI/LPI de Mobil; **G** Mobil HAI/LPI-Verfahren. B. CORNILS

Ref.: Ertl/Knözinger/Weitkamp, Vol. 5, p. 2137.

Mobil IsoFin (isomerization of olefins) process a *zeolite based proc. for the *isomerization of *n*-butene and *n*-pentene from C_4 and C_5 streams, which are the intermediates for *MTBE or *TAME. **F** procédé IsoFin de Mobil; **G** Mobil IsoFin-Verfahren. B. CORNILS

mobility of surfaces → restructuring

Mobil LDW (lube oil dewaxing) process cat. fixed-bed proc. for the cat. *hy-drodewaxing of solvent-refined raffinates, *hydrocracked feedstocks, etc. (over *zeolites) to yield low pour-point lubricating base oils. The hydrodewaxing reactor outlet is fed to a second *hydrotreating reactor to stabilize the dewaxed oil. **F** procédé LDW de Mobil; **G** Mobil LDW-Verfahren. B. CORNILS

Ref.: *Oil Gas J.* **1990**, Aug.13, 51.

Mobil LTD (low-temperature disproportionation) process for the manuf. of xylenes by *disproportionation and *transalkylation of polymethylbenzenes and toluene at 260–320 °C/4.5 MPa over zeolite cats. **F** procédé LTD de Mobil; **G** Mobil LTD-Verfahren. B. CORNILS

Mobil LTI (low-temperature isomerization) process manuf. of xylenes by *disproportionation and *transalkylation of polymethylbenzenes and toluene at 80–125 °C/3.5–7 MPa over AlCl$_3$ or BF$_3$ cats., doped with HCl, HF, or 1,2-dichloroethane. **F** procédé LTI de Mobil; **G** Mobil LTI-Verfahren. B. CORNILS

Mobil M2 Forming process manuf. of petrochemical-grade benzene, toluene, and xylenes (*BTX) by *aromatization of propane and butanes over *zeolites (ZSM-5). **F** procédé M2 Forming de Mobil; **G** Mobil M2 Forming-Verfahren. B. CORNILS

Ref.: Ertl/Knözinger/Weitkamp, Vol. 4, p. 2069.

Mobil MOGD process (methanol-to-gasoline-and-distillate) a combination of Mobil's *MTG proc. (operated at low temps. and high press.) together with a *hydrogenation of the received oligomers, producing premium-quality jet and distillate fuels. **F** procédé MOGD de Mobil; **G** Mobil MOGD-Verfahren. B. CORNILS

Ref.: *Hydrocarb.Proc.* **1985**, *64*, 72; *Am.Inst.Chem. Eng.J.* **1986**, *32*, 9.

Mobil MTG process (methanol-to-gasoline) Upon investigating the cat. properties of the strongly acidic *zeolite H-ZSM 5 that was in-house developed, its ability to convert methanol to a mixture of mainly aromatic hydrocarbons was discovered. Further experiments confirmed a reaction sequence starting from a methanol-dimethyl ether equil.acc. to $[2\ CH_3OH \leftrightarrow CH_3OCH_3 + H_2O] \rightarrow$ light alkenes $\rightarrow$ alkenes + aromatics + *iso*-paraffins + 2 H_2O. At 370 °C/0.1 MPa and 100 % methanol conv., about 55 % C_1–C_6 hydrocarbons (hcs.) and 45 % C_6–C_{10} aromatics were formed. About 40 % of the *hcs. were *iso*- and *n*-butane, and the aromatic fraction was rich in mono- to tetramethylbenzenes, a result of the *alkylating properties of H-ZSM 5. In the range from 350–450 °C the *selectivity (= product pattern) was nearly constant. However, decreasing partial press. of methanol favors the formation of alkenes instead of aromatics (cf. Mobil *MTO proc.). With about 1600 kJ/kg methanol converted the reaction is highly exothermic, leading to deactivation of the zeolite with time by *coke deposition and to changes in prod. composition. Oxidative *regeneration restores *activity and selectivity. Subsequently, Mobil developed a two-stage proc. for commercialization. In the first stage the equil. between MeOH and dimethyl ether was established by a conventional *alumosilicate. Approx. 20 % of the reaction heat is removed in this stage. In the second stage multiple fixed-bed reactors at different levels of *deactivation are used to maintain constant overall prod. composition. In 1986 the first and so far only commercial plant faced start-up at Motinui (New Zealand) with a nameplate capacity of some 15000 bbl/d. A 83:13 (w/w) MeOH/water mixture is fed with 2–3 *LHSV at 360/420 °C inlet/outlet temp./2.2 MPa press. and yields 44 % of hcs. and 56 % of water, accompanied by small amounts of CO_2 (and coke). The application of the fluid-bed technology has been successfully demonstrated in a 100 bbl/d pilot plant at Wesseling (Germany). **F** procédé MTG de Mobil; **G** Mobil MTG-Verfahren.

C.D. FROHNING

Ref.: *CHEMTECH*, **1978**, Oct., 624 and **1988**, Jan., 32; *Erdöl-Kohle-Erdgas-Petrochem.* **1984**, *37*, 558; *Catalysis Today* **1992**, *13*, 103; Ertl/Knözinger/Weitkamp, Vol. 4, pp. 1894,1905.

Mobil MTO process (methanol-to-olefins) Methanol can be converted to a mixture of hydrocarbons (hcs.) and water in the presence of acidic cats. (cf. *Mobil MTG proc.). On investigating the bandwidth of the reaction it turned out that the prod. spectrum could be shifted from predominantly C_6–C_{10} aromatics to C_2–C_4 alkenes by alteration of the *zeolite, by reducing the partial press. of MeOH, and by increasing the reaction temp. Besides Mobil a number of other companies and research groups have been active in increasing *selectivity (to ethylene/propylene) and shelf-life of the cats. Depending on the type of cat. and the reaction conds. a typical prod. composition of 2.5 % methane, 7.1 % ethylene, 31.3 % propylene, 18 % butenes (besides 6 % of saturated C_2–C_4 *hcs. and 36 % of C_{5+} *hcs.), can be attained.

A variety of acidic *zeolites have been investigated in a broad band of reaction conds. If the formation of aromatics is suppressed (high temp., high water content in feed), methane formation is favored, whereas high conv. of MeOH and high space-time yields depress the selectivity to short-chained alkenes. So far the reaction has not been applied on an commercial scale. **F** procédé MTO de Mobil; **G** Mobil MTO-Verfahren.

C.D. FROHNING

Ref.: *CHEMTECH* **1987**, Oct., 624; DE-OS 3.118.954 (1981); EP 74.075 (1982); SRI PEP Report 79–3–2 (Stanford Research Institute, Menlo Park, USA); *Hydrocarb.Proc.* **1982**, (11), 117; Ertl/Knözinger/Weitkamp, Vol. 3, p. 1137 and Vol. 4, p. 1894.

Mobil Octagain process cat., fixed-bed *hydrofinishing proc. to reduce sulfur and olefins in *FCC gasoline without loss of octane. **F** procédé Octagain de Mobil; **G** Mobil Octagain-Verfahren.   B. CORNILS

Mobil PTA (pure terephthalic acid) process older proc. for the manuf. of PTA by co-*oxidation of *p*-xylene and methyl ethyl ketone over Co salts at 130 °C/1.5 MPa. **F** procédé PTA de Mobil; **G** Mobil PTA-Verfahren.   B. CORNILS

Mobil Selectoforming process a variant of the *hydrocracking of reformate or stabilized reformate in a one-step, fixed-bed operation at 315–530 °C/1.5–4 MPa over special ZSM-5 *zeolites. This proc. permits simultaneous *alkylation of aromatics with alkene fragments from paraffin cracking and an increase of the *ON. **F** procédé Selectoforming de Mobil; **G** Mobil Selectoforming-Verfahren. B. CORNILS

Ref.: *Hydrocarb.Proc.* **1972**, (9), 120.

Mobil STDP process a further development of *Mobil's TDP-3 proc. with co-fed hydrogen. The *p*-selectivity is 87 %. The *coking can be tailored acc. to the level of *selectivity desired; the cat. is regenerated with air. **F** procédé STDP de Mobil; **G** Mobil STDP-Verfahren. B. CORNILS

Ref.: *Oil Gas J.* **1992**, Oct.12, 60.

Mobil TDP-3 process (toluene disproportionation, 3rd generation) Proc. for the cat. manuf. of mixed xylenes and benzene by *disproportionation and *transalkylation of toluene (and limited amounts of C_9 aromatics) over ZSM-5 *zeolite in the vapor phase at >400 °C (Figure). **F** procédé MTDP-3 de Mobil; **G** Mobil MTDP-3-Verfahren. B. CORNILS

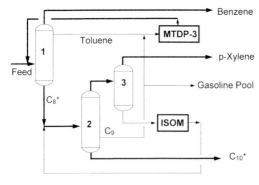

1 fractionation section; 2 C_8^+ fractionation section; 3 p-Xylene recovery. MTDP-3 reactor, ISOM = isomerization.

Ref.: Ertl/Knözinger/Weitkamp, Vol. 5, p. 2138.

Mobil TransPlus process for the cat *transalkylation of C_{9+} heavy aromatics and toluenes into higher-value mixed xylenes and high-purity benzene in fixed-bed operation (cf. *Mobil LTD and LTI procs.). **F** procédé TransPlus de Mobil; **G** Mobil Trans Plus-Verfahren. B. CORNILS

Mobil VPI process vapor-phase isomerization proc., a forerunner of the MHAI process

MOCVD molecular *CVD

model catalysts are used in a strategy to overcome the *materials gap problem in het. cat. The aim is to design a cat. whose complexity is greatly reduced to that of a commercial cat. (*real catalyst) but which still retains some of the essential properties of the mc. *Single-crystal *surfaces and, in particular, stepped single-crystal surfaces are entitled to be called mcs. but the term is commonly used to describe systems which are built up from at least two comps. For example, Al_2O_3 evaporated on a Fe(110) single crystal has been used as an mc. for the commercial cat. employed for *ammonia synth. It was shown that the more open and more reactive Fe(111) and Fe(211) orientations develop when the mc. is subjected to reaction conds. at high press. (cf. *pressure gap). **F** catalyseurs de modèle; **G** Modellkatalysatoren. R. IMBIHL

Ref.: *J.Catal.* **1987**, *103*, 213.

models (of enzymes) → active sites of enzymes

model surfaces of supported catalysts molecular approaches to "oxo *surfaces" binding metal/*ligand functionalities of cat. relevance. Ms. were developed to mimic oxide-*supported species involved in het. cat., in order to investigate the mechanism of both their formation (cf. *SOMC) and cat. behavior by additional access to various methods. Monoanionic siloxide ligands like OSiPh_3 or OSi(O*t*-Bu)_3 are used to model isolated silanol groups on *silica; incompletely condensed polyhedral oligosilasesquioxanes (POSS, $T_7(OH)_3$) exhibit a realistic electronic and steric situation for a set of surface silanol groups (Figure).

cyclopentyl-$T_7(OH)_3$

The extensive Si-O framework of the trisilanol $T_7(OH)_3$ displays a close-range geometric similarity to known SiO_2 morphologies such as the idealized trisilanol sites available on the (111) octahedral face of cristobalite and the (OO1) rhombohedral face of tridymite. $T_7(OH)_3$ form PO*MSS* (*m*etal-substituted POSS) via corner-capping reactions with metal *complexes. POMSS were shown to model *active sites for, e.g., *epoxidations with [PO(TiR)SS] (cf. *Enichem PO proc.), *Diels-Alder cyclizations with POAlSS, or alkene *metathesis with [PO(MoR$_4$)SS]. Other applications use *calixarenes or *polyoxometallates. **F** modèles pour les surfaces des catalyseurs supportés; **G** Oberflächenmodelle für Trägerkatalysatoren. R. ANWANDER

Ref.: *Polyhedron* **1995**, *14*, 3239; *Inorg.Chem.* **1995**, *34*, 1413; *J.Am.Chem.Soc.* **1985**, *107*, 8261; *J.Phys.Chem.* **1994**, *98*, 2817; *Chem.Eur.J.* **1999**, *5*, 19.

modifiers 1-in het. cat.: additives (dopants, promoting agents) to the feed which control the cat. performance of het. reactions. In *selective *hydrogenation ms. are, e.g., used routinely to achieve and maintain the *selectivity of the proc. Typically, organic sulfides are used to selectively *poison the *surface of Pd or noble metal cats. This selective poisoning reduces (*deactivates) the rate of reaction but creates the *regio- and *chemoselectivity of the reduction reaction. As the m. is slowly de-

composed by the reactant hydrogen, it has to be added with the feed (e.g., Cl-containing cpds. for the re-activation of *alumina *isomerization cats.; cf. *chlorinated alumina and the procs. of *BP, *Houdry Houdriforming, *Lummus/ABB, *Mobil LTD, *Monsanto/Lummus, *Standard Oil of Indiana Isomate, or *IFP isomerization). Other modifiers are organophosphates added to *selective *oxidation cats. in order to maintain high levels of selectivity, or organochloro cpds. added to the ethylene *epoxidation reaction. The mode of action of these additives is unclear and difficult to evaluate. *Chemisorption studies of ms. show effects on structure and *surface properties of the cat. Ms. are strong adsorbents to the cat. surface and often poison the activity completely in low-temp. operation. As their proper function is coupled to technical operation conds. where the fate of the modifier is extremely difficult to track down by in-situ experiments, there is still room for many speculations about the multiple role of ms. Under reaction conds. the additives are not stable and can act as volatile delivery agents for heteroatoms such as Cl, S, or P which are not only adsorbed but undergo reactions to metastable cpds. with the cat. base material. The consequence would be a local modification of the electronic (e.g., electronegative m.) and geometric structures resulting in the desired change in cat. reactivity.

R. SCHLÖGL

Ref.: M.P. Kiskinova, *Poisoning and Promotion is Catalysis*, Studies in Surface Science and Catalysis Vol. 70, Elsevier, Amsterdam 1992; Farrauto/Bartholomew, p. 58.

2-in hom. cat.: cpds. or subgroups (e.g., *ligands) which modify the character of cats. or cat. systems in different directions as far as *activity and *selectivity are concerned. The best known example is the ligand modification of Rh cats. for *hydrogenation, *epoxidation, or *hydroformylation. Acc. to the *ligand-accelerated (or -decelerated) cat., the addition of *ligands influences the reaction *rate of hom. cat. convs. This ligand acceleration/deceleration proceeds simultaneously and in competition. Typical examples are the Ti/tartrate-cat. *asymmetric *epoxidation (cf., e.g., *Butakem tartaric acid proc.) and OsO₄-cat. *asymmetric *dihydroxylation where the acceleration for different alkenes and ligands exceeds a factor of 10^3. In *Wilkinson-type *hydrogenations and in Rh-cat. propene *hydroformylation, a growing surplus of *phosphines increases slightly the *selectivity* (expressed as the *n/iso* ratio of the butyraldehydes) whereas the *activity* decreases considerably. Ms. may also be *inhibitors, so that a *directed poisoning of the cat.'s activity may occur (e.g., the addition of thiourea or quinoline-sulfur while reducing heterocyclic acyl halides acc. to *Rosenmund).

The extent of the modifying effect is a function of the nature of the ligand and the reactant and its conc. *Ligand effects may also influence other *coordination sites at the metal. These can be used to optimize the cat. *activity of a complex by *ligand tuning. **F** agents modifiants; **G** Modifikatoren, Modifyer.

B. CORNILS

Ref.: Gates-1, p. 79; Falbe-1, p. 45; *Angew.Chem.Int. Ed.Engl.* **1995**, *34*, 1050; Moulijn/van Leeuwen/van Santen, p. 203.

MOGD process → Mobil procs.

Mössbauer spectroscopies → see entry "Mos"

Moiseev reaction synth. of vinyl acetate by *acetoxylation of ethylene with acetic acid/oxygen in presence of Pd-Cu cats. in the liquid phase (*VAM synthesis). **F=G=E.**

B. CORNILS

Ref.: *Dokl.Akad.Nauk.SSSR* **1960**, *133*, 377.

molecular beam scattering (MBS) is essential for the determination of the kinetics of elementary steps and of surface structures. A molecular beam is used; the response comprises scattered or reacted molecules. The UHV method needs *single crystals.

R. SCHLÖGL

molecular catalysts are cats. on a molecular, defined, and reproducible basis. These *hom. cats. are normally highly active and selective, they operate under relatively mild conds., because of the molecular distribution there are no *mass transfer or diffusion problems, and their steric and/or electronic properties are highly variable. The mechanistic understanding of their operation is excellent and is the base for their successful *tailoring (cf. *hom. cat., *immobilization). **F** catalyseur moléculaire; **G** Molekularer Katalysator.

Ref.: *Topicin Cat.* **1996**, (3), 1. B. CORNILS

molecular CVD with *single-source *precursors, deposits that are cpds. or *alloys already in the right ratio of constituents and with the demanded ratio of application.

B. CORNILS

molecular dynamics (MD) MD calcns. simulate the time evolution of a many-particle system. There are two major pieces of information which can be obtained from the results of MD techniques. One concerns properties which depend on ensembles of particles rather than individual species, such as *free energies and entropies. The other is the time evolution of reaction procs. Classical MD calcns. involve the solution of Newton's equ. $F = m \times a$ for the ensemble of particles. The positions in space and the forces of the particles (given by the energy derivatives) are taken from either *ab-initio calcns., *semi-empirical methods, *density functional theory, or *force-field methods. MD calcns. can also be used for exploring the potential energy surface of a single (large) molecule, because changes in the positions of the atoms may lead to different *conformations. MD is therefore also used for conformational searching of macromolecules. Since the algorithms for MD simulations involve calcns. of the forces at subsequent time steps, where the maximum time step must be significantly smaller than the rate of the fastest process, MD calcns. must be carried out for time steps of the order of femtoseconds (10^{-15} s) or less.

It follows that the simulation of procs. which take place in one nanosecond (10^{-9} s) requires the calcn. of the energies and the forces at 10^6 time steps. Therefore, MD simulations of very large systems can only be carried out in conjunction with force-field methods. A breakthrough in the use of quantum chemical methods for MD simulations was made with the *Car-Parrinello method, which treats the wavefunction like a particle with a fictive mass. Car-Parrinello simulations of cat. procs. are very helpful to gain insight into the detailed mechanism of the reaction. **E=F=G**.　　G. FRENKING

Ref.: Ertl/Knözinger/Weitkamp, Vol. 3, p. 1173; Thomas/Thomas, p. 365.

molecular imprinting Synthetic cats., e.g., cat. *antibodies, imprinted amorphous organic polymers, *silicas, or *zeolites, may be prepd. by designing them via the use of rationally constructed organic *templates that mimic reaction *transition states. **F=G=E**.

Ref.: *Cattech.* **1997**, (3), 19.　　B. CORNILS

molecular mechanics → force-field model

molecular modeling a very loosely defined term which is used when the structure and properties of a molecule are taken from theoretical calcns. of any method, with particular emphasis on the graphical display of the results. The graphical representation of molecular structures and properties with the help of *computational chemistry methods and modeling software may be called mm. It is very helpful to visualize 3D properties which are otherwise difficult to recognize, such as the structure of large molecules like proteins or peptides. It is important to recognize that the information which is given by the pictures is determined by the theoretical methods which are used to calc. the properties, and that the use of computer graphics is solely a tool to visualize the property. Mm. is widely used in computational chemistry, particularly for research in pharmaceutical chemistry. Mm. uses all the techniques of computational chemistry which are helpful for calculating molecular properties. *Force-field methods are particularly important for mm., because computer graphics are helpful mainly for large molecules.　　G. FRENKING

Ref.: K.E. Gubbins, N. Quirke, *Molecular Simulation and Industrial Application*, Gordon and Breach, London 1996.

molecular orbital (MO) theory describes the electronic structure of a molecule in terms of the one-electron function ϕ called a molecular orbital. A spin-free molecular orbital ϕ is a mathematical function which gives the spatial extension of an electron in a molecule. Empty MOs give the region in a molecule where excited electrons are located. The square of the function $\phi(x)$ (more precisely, the product of ϕ with the complex conjugate function ϕ^*) gives the electron density at point x. Two electrons with different spin may occupy the same orbital. The total wavefunction Ψ is constructed from the set of MOs by way of the Slater determinant, which fulfills the Pauli principle. The MOs are usually formed by a linear combination of atom-centered functions ϕ called atomic orbitals (*LCAOs). The optimal mathematical functions for ϕ are Slater functions $\exp(-\zeta r)$, where ζ is the exponent of the function, and r is the radius. For practical purposes it is better to use Gaussian functions $\exp(-\zeta r^2)$, because the integrals over Gaussian functions are much easier to calc. than those over Slater functions. MOs evolved originally from the approximation of the wave function via the *Hartree-Fock method. It was later found that MOs can also be used for chemical models which are helpful for the understanding of the structure and reactivity of molecules. In particular *frontier molecular orbitals were shown to be useful for predicting the *stereochemistry of pericyclic reactions. **G** MO-Theorie.　　G. FRENKING

Ref.: M. Ladd, *Symmetry and Group Theory in Chemistry*, Horwood Publishing, Aldingbourne 1998.

molecular recognition describes the proc. of selective binding between a substrate and a receptor through immanent interactional and geometrical (steric or electronic) factors as a consequence of the rigidity and flexibility of both binding partners (cf. *host-guest relationships, *lock-and-key principle). **F** identification moléculaire; **G** molekulare Erkennung.

C. SEEL, F. VÖGTLE

molecular sieves The descriptive term signifies a class of solid materials which – without regard to their composition and exact atomic structure – have the common property of selectively sorbing molecules. The *selectivity is based on the size and the *shape of the *pores in a solid material, which compare to the size and the shape of molecules to be separated. In addition to *size and *shape, chemical effects such as hydrophobocity/-philicity may also influence the separation proc.
In order to exert a high selectivity in the sorption proc., the porous material must have a narrow pore width distribution. Within the *microporous regime (i.e., for pores smaller than 20 A), *zeolite and zeolite-type solids fulfill the requirement and are the archetypical mss. Their pore size can vary from 3 to 7 A. Their chemical character extends from strongly hydrophilic (and hygroscopic) for *alumosilicate zeolites to hydrophobic for porous *aluminophophates and pure silica *zeosils. Mss. are widely used in lab. and industrial work. Typical commercial applications are the separation of O and N, removal of sulfur cpds. (*sweetening) from natural gas and petroleum, and separation of isomers, e.g., the separation of linear alkanes from branched or cyclic ones or of the *o-*, *m-* and *p-* isomers of xylene. Hydrophilic zeolites in their desiccated forms are used as water-scavenging agents in the drying of gases or organic solvents or for removing water formed in reaction (e.g., *esterification). These procs. are also used in the lab. Zeolites and other mss. (such as *ALPO, *SAPO) are also important cats. and cat. *supports. Their cat. reactivity can be combined favorably with their molecular sieving properties, giving rise to *shape-selective cat. procs. Examples of

prepns. of cats. with molecular sieves are the procs. of *Chevron Isodewaxing, *UCC Iso-Siv, *UCC TIP, *Union Oil Unicracking, *UOP Pacol-Olex, *UOP Parex, etc. **F** tamis moléculaires; **G** Molekularsiebe. P. BEHRENS

Ref.: Ertl/Knözinger/Weitkamp, Vol. 1, p. 286; R. Szostak, *Handbook of Molecular Sieves*, Van Nostrand Reinhold, New York 1992; *Chem.Rev.* **1997**, 97, 2373.

molecular sieve catalysts The term generally refers to inorganic solids, usually *zeolites, *mordenite, or *faujasite, which are able to separate molecules by virtue of the latter's cross-sectional areas. *Linde type A ms. has an idealized formula $Na_{12}[Al_{12}Si_{12}O_{48}] \cdot 27$ H_2O and a *pore diameter of 4.1 Å. When however, the Na^+ ions are replaced by Ca^{2+} the pore dimension effectively increases to close to 5 Å. Methane, with an effective molecular diameter of 3.8 Å, or ethylene (3.9 Å) readily diffuse through zeolite A; isobutane (5.0 Å), on the other hand, cannot penetrate into the interior of Na^+-exchanged zeolite, nor can benzene (5.5 Å). Thus, ms. cats. are the prime examples where *shape selectivity and sometimes *regioselectivity in cat. occurs. **F** catalyseurs des tamis moléculaires; **G** Molekularsiebkatalysatoren. J.M. THOMAS

Ref.: *J.Amer.Chem.Soc.* **1956**, 78, 5972; *Z.Kristogr.* **1969**, 128, 352.

molecular traffic control (MTC) a concept which presupposes preferential *diffusion of *reactants* into one channel of *zeolites and diffusion of *products* out of another channel. Thus, counter diffusion is minimized and product *selectivity maximized. **E=F=G**.

B. CORNILS

Ref.: Farrauto/Bartolomew, p. 73; *J.Mol.Catal.A:* **1997**, 125, L87.

molten salt media are another possibility of conducting *two-phase catalyses and thus separating the reaction prods. from the cat. without decomposition of the cat. Suitable molten salts can be stable, nonvolatile *solvents for reactants and cats. from which organic prods. are readily separated. Three different types of media are known. 1: High-melting inorganic salts with high thermal stability, low vapor press., excellent thermal and electrolytic conductivity, low viscosity, and good solubility for oxides, hydrides, metals, carbides, etc. Despite corrosivity at the necessarily high application temp., inorganic molten salts are applied for the *oxychlorination of methane or ethane in molten $CuO \cdot CuCl_2$ or $KCl \cdot CuCl_2$ (*Lummus TransCat proc.) or for molten carbonate cells (MCFC *fuel cells). *Dow's phenol proc. use a molten Cu salt medium for the *decarbonylation of benzoic acid. 2: Special variants of *coal hydrogenation (or *gasification) use $ZnCl_2 \cdot KCl$ melts, cat. active molten iron bath (*Atgas proc.), or molten carbonates (*Kellogg procs.). 3: *Nonionic organic liquids (NAILs) such as $[NR_4]GeCl_3$, $[NR_4]SnCl_3$, $R[AlEtCl_3]$, or *emim, which are stable, low-melting ($<150\,°C$) cpds. are suitable solvents for substrates (such as alkenes) and cat. active hom. *complexes. They may be used for *hydroformylations, *hydrogenations, and *oligomerizations. **F** sels fondus média; **G** Salzschmelzen.   B. CORNILS

Ref.: *J.Am.Chem.Soc.* **1972**, *94*, 8716; *CHEMTECH* **1995**, (9), 26; *Catal.Rev.Sci.Eng.* **1975**, *11*, 197.

molybdena molydenum dioxide, MoO_2, inter alia a base for cats. or *supports (cf. *molydenum oxides).

molybdenum as catalyst metal

Mo is commonly used in multifunctional cats. Mo-containing cats. are important as: 1: Bi-Mo mixed oxides (promoted with Fe, P, K, and other metals) for propylene *oxidation and *ammoxidation (e.g., *SOHIO proc.). 2: sulfided Co-Mo or Ni-Mo on *alumina (promoted with B, P, K and rare earth oxides) for *hydrogenations (e.g., *hydrotreating, *HDS, *sulfided cats.) in the refinery industry; 3: Co-Mo oxide and Mo oxide on acid *supports for *hydrocracking procs.; 4: Mo-Fe cats. for methanol *oxidation to *formaldehyde; 5: hom. Mo cats. for propylene *epoxidation with *hydroperoxides (*ARCO, *oxirane proc.); 6: supported *Mo oxides and other Mo

cpds. for alkene metathesis; 7: Mo-V cats. for acrolein oxidation to *acrylic acid. Further cat. applications are given by Mo-containing *heteropolyacids (acid and *redox function) and by use of Mo in *photooxidation reactions. Mo-Fe- containing proteins are essential for the *nitrogen fixation. **F** le molybdène comme métal catalytique; **G** Mo als Katalysatormetall.   B. LÜCKE

Ref.: Adams/Cotton; Coughlan, *Molybdenum-Containing Enzymes*, Pergamon Press, Oxford 1980; *Science* **1996**, *272*, 1599, 1615; *Houben-Weyl/ Methodicum Chimicum*, Vol. XIII/7; *Chem.Rev.* **1996**, *96*, 2983.

molybdenum oxide Mo-oxygen cpds. are of technical relevance. MoO_3 and polymolybdate are important in cat., both as cats. and cat. *precursors; unsupported (*bulk cats.) and *supported. Mo cats. both play an important role in *HDS, *HDN, and *HDM (*Mo as cat. metal). Furthermore, Mo-based cats. are used in *selective (partial) *oxidation and the cat. *reduction of NO_x. Mo-containing cats. are used too for *oxidative coupling reactions and are active for *photocatalytic oxidations. The cat. properties of MoO_3 are determined by its morphology or crystallinity. The monoclinic β-phase of MoO_3 is reported to have a higher cat. *activity in the partial oxidation of alcohol than the usual, orthorhombic α-phase. The *selective cat. reduction of NO_x was also reported to be influenced by the MoO_3 particle morphology. Early investigations interpreted the complex layered structure of MoO_3 as built up by strongly distorted MoO_6 octahedra, which are bound by a common edge and corner in one direction (c-axis) to form zigzag rows. The octahedra are bound in the perpendicular direction (a-axis) by common corners. An alternative structural description of MoO_3 is based on the tendency of Mo atoms to fivefold coordination.

The most precise structure determination of MoO_3 led to a different structural description. The MoO_3 crystal structure is better described as built up from distorted MoO_4 tetrahedra which are connected via common corners along the c-axis to form chains. These chains are moderately condensed to form the MoO_3 half-layers. The *binding energy between the chains is about 2–6 times higher than that between the half-layers and about 4–8 times smaller than that within the chains of tetrahedra. **F** oxide de molybdène; **G** Molybdänoxid. G. MESTL

Mond, Ludwig (1839–1909), German-English inventor and company owner (Widnes, UK), described the first binary metal carbonyl, $Ni(CO)_4$. This discovery initiated *Sabatier' work on the cat. *hydrogenation of C-C double bonds. W.A. HERRMANN

monoclonal antibodies *Antibodies are proteins produced by B-lymphocytes whose purpose is to act as an "adaptor" between an foreign antigen (such as a foreign protein or saccharide) and various elements of the immune system (such as macrophages). Diversity is generated through several recombination and mutagenesis mechanisms that operate at the *DNA level, and thus the changes are permanent and consistent within a cell line. The (homogeneous) antibodies isolated from a single cell line are called "monoclonal." Mas. may be generated against *haptens that resemble the *transition state of a reaction in order to produce cats. of that reaction. Cf. *catalytic antibodies. **F** anticorps monoclonaux; **G** monoklonale Antikörper. P.S. SEARS

Ref.: C.A. Janeway; P. Travers, *Immunobiology*, 3rd Ed., Garland Publishing, New York 1997.

monolayer coverage → capillary condensation

monolytic honeycomb cylindrical structures, traversed by a multitude of usually straight channels. Depending on the application,they are made out of cat. *inactive* material (*ceramics, metal) as support for washcoated cats., or from cat. *active* material when the whole honeycomb structure serves as cat. For both categories, a broad variety of monolith (m.) designs (differing in shape, dimen-

sions, length, shape and size of the channels, thickness and porosity, etc.) is used. The shapes can be square, round, oval, or of "racetrack" (flattened ellipses) design. The size of the frontal area expressed on the basis of an equiv. diameter ranges from about 2 to 50 cm for bigger monoliths. *Ceramic* ms. have typically round, hexagonal, triangular, or square channels; the channel shape of *metallic* ms. resembles a trigonometrical sinus curve. The width of the m. channel and the thickness of the wall fix the so-called cell density (the number of channels [cells] per unit of frontier surface area), which is between 7 and 240 cells/cm^2. The monolith are used for *automotive emission control cat. They are loaded with a *washcoat (cf. Figure g on p. 105). The exhaust gas flow through the channels is laminar, thus assuring a low press. drop.

Ceramic ms. are made of porous *cordierite. The porosity of the m. walls ranges from 15 up to 40 %, caused by *macropores with an average diameter of a few μm. The bulk density is approx. 0.5 kg/L. These ms. are made by an extrusion proc., followed by a *drying and a firing step. These ceramic ms. have an extremely low thermal expansion coefficient and thus show a sufficient thermal shock resistance. They are placed in a muffler-like converter housing made of high-quality corrosion-resistant steel. Special precautions ensure the proper, homogeneous exhaust gas flow. Metallic ms. consist of a metallic outer shell, in which a honeycomb-like metallic matrix structure is fixed, the foils of which are non-porous and fixed to each other and to the outer shell by a brazing technique.

The major application of washcoated mhs. is the cat. aftertreatment of exhaust gases from engines in on-road transportation. The honeycombs made entirely out of cat. active inorganic oxides are mainly used in the cat. aftertreatment of exhaust gases from stationary emission sources, especially in *SCR systems for the elimination of NO_x emissions from fossil fuel power plants (cf. also *cat. combustion). **E=F=G**. E.S.J. LOX

Ref.: Ertl/Knözinger/Weitkamp, Vol. 3, p. 1402.

monomers are the constituent units of *polymers, e.g., the m. ethylene yields *polyethylene by cat. with *radicals, cations, or anions (*polymerization). Ms. contain one or more functional groups, such as C-C double bonds, cyclic ethers or amides, or H-acid functions. M. may be single (*homopolymers) or different cpds. (*copolymers). **F** momomères; **G** Monomere.   B. CORNILS

monooxygenase *enzymes capable of activating molecular oxygen for *oxidation of a number of inert organic cpds., including alkanes, alkenes, aromatics, and heteroatoms. The non-heme m. from *Pseudomonaoleovoran* is a *NADH/O_2 dependent enzymatic system that contains three comps.: reductase, *rubredoxin (r.), and *hydroxylase. The reductase is a flavoprotein that transfers two electrons from NADH to the Fe-centre of r. via its bound *FAD. The r., a non-heme iron protein, accepts and transfers two electrons from reduced r. one by one to the iron-containing hydroxylase, which then accepts a substrate and a molecular O, activates the O, and inserts the oxygen to the substrate.

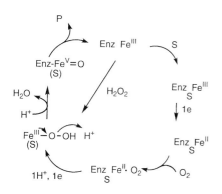

Mechanistic investigations of the hydroxylase regarding the activation of oxygen indicates similarity to the other type of m., i.e., heme-containing P450 system, where the active oxygen species has been established to be a high-valent iron oxo species, Fe(V)=0 or Fe(IV)=0 coordinated with a cation radical ligand. The mechanism of O transfer from the iron-oxo

species to substrates is still not well established and three possible mechanisms have been proposed. These include *radical, cation radical, and *concerted pathways. Also cf. *Baeyer-Villiger oxidation, *isopenicillin N synthase, and *methane monooxygenase. **E=F=G.**   C.-H. WONG

Ref.: *J.Am.Chem.Soc.* **1990**, *112*, 3993 and **1991**, *113*, 5878; P.R. Ortiz de Montellano, *Cytochrome P450*, Plenum Press, NY 1986, p. 217; *Chem.Rev.* **1996**, *96*, 2841.

Monsanto acetic acid process is a low-press., hom., liquid-phase *carbonylation of MeOH, cat. by Rh cpds. and *promoted by iodine at 150–200 °C/3–4 MPa with MeOH efficiencies of 98 %. This technolgy accounts for >50 % annual *AA world capacity and >90 % new capacity worldwide. The reaction is first order in Rh and CH_3I and zero order in MeOH and in CO >0.2 MPa. The mechanism is well studied and includes a nucleophilic attack by the active Rh(I) cat. ($[Rh(CO)_2I_2]^-$) on CH_3I to form $[Rh(CH_3)(CO)_2I_3]^-$. Additional reaction steps are rapid CO *insertion to form an acyl *intermediate, $[Rh(CH_3CO)(CO)_2I_3]^-$, followed by *reductive *elimination of CH_3COI with its subsequent *hydrolysis to form CH_3COOH and regeneration of $[Rh(CO)_2I_2]^-$. Numerous Rh cpds. can be used as *precursors. The overall *reaction rate depends on the reaction medium, but the overall *kinetics are unaffected by the media. H_2O provides rate enhancement in most reaction systems. Water serves to reduce inactive Rh(III) carbonyl species via the *WGSR. AA/H_2O is the preferred medium. Reaction *rate and cat. *stability decline with reduction of reaction water. CO_2 and H_2 are major gas byprods. from the competitive WGSR. Propionic acid is the main liquid byprod. Basic commercial procs. consist of reactor and flasher sections for carbonylation, the separation of cats. from crude prod., and a purification section. **F** procédé Monsanto pour l'acide acétique; **G** Monsanto-Essigsäureverfahren.

                                                                 P. TORRENCE

Ref.: Cornils/Herrmann-1, Vol.1, p. 104; *Appl.Ind.Catal.* **1983**, *1*, 275.

Monsanto adipic acid process manuf. of *adipic acid (AA) by Pd-cat. *carbonylation of 1,4-dimethoxy-2-butene. **F** procédé de Monsanto pour l'acide adipique; **G** Monsanto Adipinsäure-Verfahren. B. CORNILS

Monsanto cumene process obsolete proc. in which propylene and benzene were converted to *cumene over AlCl₃/HCl as cat. The low energy offered a 99 % yield (based on benzene). **F** procédé de Monsanto pour le cumène; **G** Monsanto Comol-Verfahren. B. CORNILS

Monsanto/Lummus ethylbenzene process manuf. of *EB from benzene and ethylene with AlCl₃ as cat., doped with ethylchloride, at 140–200 °C/0.3–1 MPa. Higher alkylbenzenes are *dealkylated or *transalkylated externally. **F** procédé de Monsanto/Lummus pour l'EB; **G** Monsanto/Lummus EB-Verfahren. B. CORNILS

Monsanto EHD process for the manuf. of *adiponitrile via (noncatalytic) electro*hydrodimerization of acryl nitrile. B. CORNILS

Monsanto L-dopa process L-Dopa, (S)-3,4-dihydroxyphenylalanine, is a drug for Parkinson's disease, needed in several hundred tpy. The commercial prod. of L-dopa by a Rh-cat. *hydrogenation of the corresponding dehydroamino acid was introduced by Monsanto (*Knowles) as the first *transition metal *enantioselective cat. The cat. is a Rh(I) *complex of the *chelate *ligand *DIPAMP, containing chiral phosphorus atoms. The hydrogenation proceeds quantitatively with approx. 94 % *ee. Recrystallization gives the optically pure L-dopa separated from the cat., which can be recycled (cf. *VEB Isis-Chemie L-dopa proc.). **F** procédé Monsanto pour L-dopa; **G** Monsanto L-Dopa-Verfahren.
H. BRUNNER

Ref.: *Acc.Chem.Res.* **1983**, *16*, 106; *J.Am.Chem.Soc.* **1975**, *97*, 2567.

Monte Carlo (MC) method MC methods are the second major technique besides *molecular dynamics simulations which are used to calc. properties that depend on an ensemble of particles rather than on the property of a single species such as *free energy and entropy. MC simulations start from an initial geometry and generate new points in phase space by random variations of the coordinates of one or more particles. If the energy of the new geometry is lower than the previous energy, the new structure is taken as starting point for the next step. Otherwise, the Boltzmann factor $\exp(-\Delta E/kT)$ is calcd. and compared to a random number x with $0 \le x \le 1$. If the Boltzmann factor is smaller than x, the new structure is accepted. Otherwise the old structure is used again. This Metropolis procedure ensures that the ensemble of points in the phase space obeys a Boltzmann distribution. Since higher-energy configurations are accepted, it is possible to climb uphill and find other conformations in the vicinity of the starting structure. MC calcs. are therefore also used for the conformational analysis of macromolecules. Since millions of energy calcns. are necessary to collect a canonical ensemble, most MC calcns. are done in conjunction with *force-field methods for calcng. the energies. **G** Mc-Methode. G. FRENKING

Ref.: Ertl/Knözinger/Weitkamp, Vol. 3, pp. 986,1173; Thomas/Thomas, p. 365.

Montecatini oxychlorination for the manuf. of 1,2-dichloroethane from ethylene, HCl, and O₂ over *supported Cu on *alumina (plus *promotors) at 225–240 °C/0.3–0.5 MPa. **F** procédé de Montecatini d'oxychloration; **G** Montecatini Oxychlorierungsverfahren. B. CORNILS

Ref.: US 4.587.230 and EP 0.176.432; *Appl.Catal.* **1994**, *115*, 180.

Montedison acrylonitrile process manuf. of *acrylonitrile via *ammonoxidation of propylene in fluidized-bed reactors at 420–460 °C with Te-Ce molybdates (Bi-free) on SiO₂ cats. **F** procédé Montedison pour

'ACN; **G** Montedison ACN-Verfahren.
B. CORNILS

Ref.: *Hydrocarb.Proc.* **1977**, 56(11), 125.

Montell Spheripol process is the former Himont proc. for the manuf. of propylene-based *polymers including homopolymers and random and heterophasic impact *co-polymers in liquid propylene within a loop tubular reactor and using *supported cats. (even with chemical-grade propylene). **F** procédé Spheripol de Montell; **G** Montell Spheripol-Verfahren.
B. CORNILS

Ref.: *Hydrocarb.Proc.* **1997**, (3), 155.

montmorillonite an alumohydroxysilicate and the main constituent of *clay minerals. Ms. are 2:1 clays with one octahedral AlO_6 layer being sandwiched between two tetrahedral SiO_4 layers. M. are reversibly swellable and *ion-exchange active. Ms. are used as cat. *supports. They are well suited for the generation of *pillared clays (materials with a bimodal *micro-/*mesoporous *pore size distribution) and for the prod. of mesoporous materials by leaching with inorganic acids, yielding in both cases materials with high *surface areas.
P. BEHRENS

Moore Stanford (1913–1982), biochemist at the Rockefeller Inst. for Medical Research, New York. Worked on the structure of *ribonuclease. Nobel laureate in 1972 (together with Anfinsen and Stein).
B. CORNILS

Ref.: *Angew.Chem.* **1973**, 85, 1074.

MOR *zeolite structural code for *mordenite zeolites

mordenite a naturally occurring *zeolite with the simplified formula (Na_2,K_2,Ca) $[Al_2Si_{10}O_{24}] \cdot 7\ H_2O$. It belongs to the group of large-pore zeolites, containing two types of channels. The larger one, which runs straight, is lined by a 12-membered ring with an aperture of 7.0×6.5 A. These channels are connected by short alternating eight-membered ring channels (with an aperture of 3 A). These

two channels build up a two-dimensional channel system (cf. *molecular traffic control). The *zeolite structural code for the tetrahedral framework structure of mordenite is MOR. Other, synthetic, materials possessing this topology are Na-D and zeolon. Ms., especially synth. species with low framework Al contents and in their acidic form or *transition-metal ion-exchanged materials, are important industrially used cats. **E=F**; **G** Mordenit.
P. BEHRENS

Morita-Baylis-Hilman reaction → hydroxyalkylation

Mössbauer absorption spectroscopy (MAS) is applied to geometric and electronic structures in cat. Probe and response are γ-rays. The data are integral, bulk-sensitive, and have in-situ capability. The method is non-destructive to solid crystalline or amorphous samples.
R. SCHLÖGL

Mössbauer spectroscopy is a nuclear technique and restricted to those elements that exhibit the Mössbauer effect (Fe, Sn, Ir, Ru, Co indirectly, Pt, etc.); it provides information about oxidation states, phases, lattice symmetries, and lattice vibrations. Mössbauer emission spectroscopy (MES) and *Mössbauer absorption spectroscopy (MAS) are the usual variants.
R. SCHLÖGL

Ref.: Niemantsverdriet; Ertl/Knözinger/Weitkamp, Vol. 2, p. 512; T.E. Cranshaw et al., *Mössbauer Spectroscopy and its Applications*, Cambridge University Press, Cambrigdge 1985.

motor fuel alkylate → alkylate

MS 1: → mass spectroscopy; 2: → molecular sieves

MSGP molten salt gasification process, → Kellogg procs.

MTBE → methyl *tert*-butyl ether

MTC → molecular traffic control

MTDP process → TDP-3 process, *Mobil procs.

MTGD process methanol-to-gasoline and distillate process, → Mobil procs.

MTG process methanol-to-gasoline proc., → Mobil procs.

MTO process 1: methanol-to-olefin process, → Mobil procs.; 2: *methyltrioxorhenium.

MTPD metric tons per day, the measure for production figures of high-tonnage chemical plants such as *ammonia or *methanol units.

Mukaiyama aldol reaction *condensation of aldehydes or ketones with trimethylsilyl enol ethers to form the corresponding β-hydroxycarbonyl cpds. (Figure).

$$R^1R^2C{=}CR^3[OSi(CH_3)_3] \; + \; R^4R^5CO \xrightarrow[\text{solv.}]{\text{Lewis acid}}$$

The reaction is promoted by *Lewis acids such as TiCl$_4$, SnCl$_2$, BF$_3$ etherate, or ZnCl$_2$. *Promoters inhibit the retro *aldol reaction by, e.g., the formation of Ti *chelates. Recently lanthanide triflates have been also used as cats. in aqueous solution (*rare eath metals as cat. metals). In general, the Mar. gives a mixture of stereoisomers (*threo* and *erythro*) of the aldol in comparable amounts. The use of *chiral Lewis acids or of *bimetallic *complexes allow diastereo- and in some cases even *enantioselective aldol reaction under mild conds. **F** aldol réactions de Mukaiyama; **G** Mukaiyama Aldolreaktion.

D. SCHICHL, M. BELLER

Ref.: Beller/Bolm, Vol. 1, p. 285; *Organic Reactions* **1982**, Vol. 28, 203; *J.Am.Chem.Soc.* **1974**, 96, 7503 and **1991**, *113*, 1041.

Müller-Rochow synthesis → Rochow re action

mullite an alumosilicate constituent o *ceramic catalysts

multicomponent catalysts systems com posed of two or more monofunctional cats. hav ing only loose contact between them. An mc. i. easy to prepare by simply mixing cats. each con taining one *active site. The intimacy of the dif ferent sites in this mixture as described by the *Weisz criterion is thus not as high as in multi functional cats. Nevertheless, the decoupling o the cats. in the system can have some advan tages, e.g., it can avoid negative interaction be tween two given active sites.
In the literature, the distinction mentioned be tween mcs. and multifunctional cats. is not ac cepted in general. In addition, many authors have used the notion "mcs" for characterizing cats. containing not only the active species but also some *additives in order to monitor the cat properties. Examples are the different multime tallic cats. used to optimize *hydroformylations by facilitating the *hydrogen transfer proc. with the help of a second metallic comp. This broad definition, however, soon leds to the statement that nearly all cats. of practical importance are mcs. Mo, V, W, Cr, Bi, and other metals are often constituents of mcs. (*ammoxidation, *MCM, *SOHIO proc.). **F** catalyseurs multi components; **G** Multikomponenten-Kataly satoren.

D. HESSE

multidentate ligands *ligands bearing two or more donor centers (cf. *chelate lig ands, *bi- or *tridentate ligands). They can form cyclic structures by coordinating a cen tral metal with more than one of the donor moieties (cf. *chelates). **F** ligands multiden taire; **G** vielzähnige Liganden. W.R. THIEL

multienzyme systems (multifunctional en zymes, multienzyme complexes) Most *en zymes generally function under the same or similar conds. (aqueous, pH ~ 7, rt.), and so several reactions can be carried out in one

pot. Mss. can be used not only to simplify reaction procs. but also to shift an unfavorable
equil. toward the desired prod. by removing
coprods., and to avoid enzyme *product inhibition. Many mss. have been developed; multienzyme *cofactor regeneration schemes
have been used in large-scale synth. In oligosaccharide synth., it has been demonstrated
that more than six enzymes can be used in
one pot to produce oligosaccharides effectively with concurrent regeneration of sugar
nucleotides. All enzymes used in the one-pot
synth. can be co-*immobilized to a solid *support. The efficiency of the mss. can be further
improved by *cross-linking or by gene fusion.
F système multienzymatique; **G** Multienzym-
systeme.   C.-H. WONG

Ref.: *Nature Biotechnology* **1998**, *16*, 769.

multifunctional catalysts   cats. containing two or more different cat. functions acting
synergistically to facilitate a sequence of cat.
reactions. To produce a desired chemical species in such a one-pot reaction system, the
polyfunctional cat. used must fulfill a number
of requirements: for it to have synergistic action, and for the time constants of the single
reaction steps as well as the distance between
the single *active sites to obey the *Weisz criterion. In addition, each monofunctional cat.
must maintain its activity even in the presence
of a relatively large number of cpds. Moreover, the *selectivity of each of the active
sites must be high to avoid a decrease on the
conc. of the *intermediate by dilution procs.
Otherwise, the rates of the single reaction
steps will decrease and, additionally, the separation of the target prod. from the complex
reaction medium will become extremely expensive. These demands are best obeyed by
preparing an mc. by fixing the monofunctional part to a common *support, which itself
can have cat. properties, e.g., an acidic function. Although these restrictions are serious,
mcs. play an important role in the petrochemical industry. Examples are the cats. used in
the manuf. of gasoline of high antiknock quality, e.g., Chevron's *Rheniforming proc. In

this case the het. cat. contains an optimized
*dehydrogenation/*hydrogenation   function
through a combination of Pt, Re, and an
acidic function on the surface of the *alumina
support to facilitate the complicated reaction
network during the *dehydrocyclization reaction of hcs. (*metal-support interactions, *zeolite supported cats.).

While in this example certain byprods. are tolerable, the situation is much more severe for
cases where only one target prod. formed by a
sequence of reaction steps is desired. An example is the manuf. of *2-ethylhexanol from
propylene and syngas via the *Aldox proc., a
variant of *hydroformylation. The reaction
sequence is *n*-butyraldehyde → 2-ethylhexa-
nal → 2-ethylhexanol. Rh cats. on a polymer
matrix have been developed to cat. the *oxo
reaction as well as the *hydrogenation steps.
Amino groups attached to the polymer cat.
the *aldol reaction. Other examples are the
synth. of *bisphenol A or the *alkylation of
benzene, *Michael additions with *BINOL
complexes of La-Na, etc. **F** catalyseurs multi-
fonctionels; **G** multifunktionelle Katalysato-
ren.   D. HESSE

Ref.: Gates-1, p. 219; *Proc.Int.Congr.Catal.*, 6th meeting, **1976**, *9*, 499; *CHEMTECH* **1983**, Sept., 556.

multiple steady states   Het. catalytic reactions may in a certain parameter range exhibit mss., i.e., for one parameter value more
than one (typically not more than two) *steady-state (ss.) solutions exist for the corresponding differential equs. which describe the
kinetics. Experimentally, a *hysteresis is observed if two stable ss. solutions exist. Typical
examples are ignition/extinction phenomena
or the coexistence of a high- and a low-rate
branch in het. cat. reactions such as cat. CO
*oxidation on Pt. The mapping of the ss. solu-
tions in parameter space, the testing of their
stability, and their classification are the subject of bifurcation theory. In chemical engineering science mss. are an important subject
because the stability of reactor is essential for
commercial use. Today easy-to-use computer
algorithms exist which allow to carry out a bi-

furcation analysis without having to delve too deeply into mathematical theory. **E=F=G**.

R. IMBIHL

Ref.: *Chem.Rev.* **1995**, *95*, 697; L. Lapidus, N. R. Amundsen, *Chemical Reactor Theory*, Prentice-Hall, Englewood Cliff, 1977.

multiplet theory postulates "that the *activity of a cat. depends to a large degree on the presence on the *surface of correctly spaced groups (or multiplets) of atoms to accommodate the various reactant molecules" (cf. also *ensemble, *Weisz criterion).

Ref.: Thomas/Thomas, p. 32. B. CORNILS

multiplication factor of stereochemical syntheses, cf. *asymmetric synth.

Ref.: *Chem.Soc.Rev.* **1989**, *18*, 187.

Murahashi synthesis → alkoxylation

mutagenesis → random mutagenesis. *site-directed mutagenesis, *directed evolution, *DNA shuffling

mutarotation (tautomeric catalysis) conv. of sugars or optical active substances into their anomers (epimers) upon dissolving. The m. may be accelerated by *acid-base cat. (cf *Lowry) or by *enzymes such as mutarotases or *epimerases. **E=F=G**. B. CORNILS

mutase *isomerase that transfers a functional group (such as a phosphate or an acetate) from one position to another within a molecule; mutases are intramolecular transferases. **E=F=G**. P.S. SEARS

MV → metal vapor synthesis

N

n   *normal*, referring the linear (unbranched) isomers, important for *isomerization or *hydroformylation reactions; cf. *iso.

N-acetyl   → see entry "acetyl"

NAD, NADP   → nicotinamide adenine dinucleotide (NAD) and the analogous 2′-phosphate (NADP)

NADH, NADPH   The nicotinamide ring system is *redox active, accepting a hydride or two electrons and a proton to form the 1,4-dihydronicotinamide derivatives (NADH or NADPH)

NAD hydrolase and glycohydrolase → nicotinamide adenine dinucleotide (NAD) hydrolase

NAD kinase   → nicotinamide adenine dinucleotide (NAD)

NAD pyrophosphorylase   → nicotinamide adenine dinucleotide (NAD)

Nafion   perfluorinated ionomer belonging to the class of *superacids containing hydrophobic -CF_2-CF_2- and hydrophilic -SO_3 or -COOH regions in its polymeric structure. Nafion can be used as a solid cat. for various reactions (e.g., *alkylation, *isomerization, *acylation). By *copolymerization with other *monomers, *ion exchange *membranes can be obtained to be used in *fuel cells, electrolysis, or electrodialysis (*membranes).

U. KRAGL

Ref.: G.A. Olàh, A. Molnar, *Hydrocarbon Chemistry*, Wiley, New York 1995; Ullmann, Vol. A11, p. 373 and Vol. A16, p. 211.

NAIL   → nonaqueous ionic liquids, *molten salt media

nanodispersed materials   materials which are stabilized in particles with dimensions of about 1 nm or less (also aggregates or *clusters). Particles can consist of *transition metals, *metal oxides, or mixtures of different metals or metal oxides and are often stabilized on high *surface area *supports (typically about 1–5 %w/w of active material) or in *colloids. Highly dispersed particles give the most efficient use of cat. active comps. with a greater fraction of the atoms available to reactants for smaller particles. For Pt metal clusters a 1.5 nm diameter cluster (~100 atoms) has a dispersion (surface atoms/total atoms) of 0.5. Synth. can usually be divided into two steps: *deposition of a *precursor (or in some cases the active material itself) followed by *calcination and/or *reduction. *Deposition methods include *impregnation, *vapor deposition, *ion exchange, and *anchoring and *grafting. Commonly used supports are SiO_2, Al_2O_3, TiO_2, and *zeolites. Support surface properties and geometric structure are critical for maintaining high dispersion. N. metals often have properties between those of the bulk material and of individual atoms, and the material properties can be influenced by the support (*SMSI). *Particle size can influence cat. *activity (*structure-sensitive cat.). Measurement of dispersion for nanosized clusters can be complicated by a distribution of *particle sizes in the same sample, and by support-covered particles. It can be difficult to determine both cluster size and shape. Techniques for size measurement include chemisorption, *EXAFS spectroscopy, *TEM, *SAXS, and

*NMR. **F** matériel nanodisperse; **G** nanodisperse Materialien. R. JENTOFT
Ref.: Ertl/Knözinger/Weitkamp, Vol. 2, p. 752; *Chem. Rev.* **1999**, *99*, 1641; J.H. Fendler (Ed.), *Nanoparticles and Nanostructured Films*, Wiley-VCH, Weineim 1998; *J.Mol.Catal.A:* **1999**, *145*, 1.

nanofiltration the pressure-driven *membrane-based separation proc. in which particles and dissolved molecules smaller than approx. 2 nm (corresponding to a molecular weight range between 500 and 5000 g mol^{-1}) are retained while smaller molecules can pass the membrane. Multivalent ions are retained as well, while monovalent ions will pass. The membranes are often made from organic *polymers such as polyamide. In principle, the membranes are nonporous solution-diffusion membranes. In cat., n. is used for the separation of amino acids and peptides, the recovery of *dendrimers, etc. **E=F=G**. U. KRAGL
Ref.: Ullmann, Vol. A28, p. 74; *J.Carbohydr.Chem.* **1999**, *18*, 41.

NAPHOS, naphos the precursor of *BINAS, obtained in 97 % *ee by SiHCl$_3$/NEt$_3$ *reduction of (*S*)-(–)NAPHOS dioxide, prep. by Michaelis-Arbusov reaction between (*S*)-(–)-2,2′-bis(bromomethyl)-1,1′-binaphthyl and (C$_6$H$_5$)$_2$POCH$_3$ (Figure).

NAPHOS is used in Rh-cat. *hydrogenations and *hydroformylations and in Ni-cat. *Grignard cross-couplings. O. STELZER
Ref.: *Tetrahedron Lett.* **1977**, 1389; *Synthesis* **1992**, 503; Patai-2, Vol. 1, p. 52; *J.Organomet.Chem.* **1997**, *532*, 243; *Organometallics* **1995**, *14*, 1961.

naphtha reforming → catalytic reforming

natrolite sub-group of fibre *zeolites. **E=F**; **G** Natrolith.

Natta Giulio (1903–1979), professor of industrial chemistry in Milano (Italy). Active especially in *polymerizations (*polypropylene, *tacticity) and industrially relevant procs., e.g., the *oxo synthesis (*Natta equation). Nobel laureate in 1963 (together with *Ziegler). W.A. HERRMANN

Natta (Natta-Ercoli) equation Kinetic relationship for the high-press. *hydroformylation. C.D. FROHNING
Ref.: *Chim.Ind.* **1955**, *37*, 6; *Chem. Ber.* **1968**, *101*, 2209; Falbe-1, p. 16.

natural bite angle The *Tolman cone angle of a *bidentate *phosphine is affected not only by the M-P bond length but also by the P-M-P bite angle. Acc. to Casey's definition, the nba. is the "preferred chelation angle determined only by *ligand backbone constraints and not by metal valence angles"; it "is independent of any electronic preference in bite angle imposed by the metal center and is based solely on steric considerations." The nba. of a bidentate phosphine with respect to a fixed M-P bond length and a specific metal center can be calcd. by *molecular mechanics. Thereby, the energy of the diphosphine metal fragment is minimized without consideration of the P-M-P bending force (force constant zero) as well as the remaining coordination sphere. The resulting value depends on the specific *ligand backbone conformation and on the *force-field model used. Both *electronic and *steric effects of *phosphine ligands have a strong influence on the cat. *activity, and differences between *complexes with two monophosphine ligands and those containing only *chelating ligands are frequently observed. Some correlations between nbas. and *regioselectivities in Rh-cat. *hydroformylation have been postulated, but a generalization seems to be deceptive. Principally, the nba. can be determined for all bidentate *chelating ligands. **F** l'angle de la morsure naturel; **G** natürlicher Bißwinkel. D. GLEICH
Ref.: *Isr.J.Chem.* **1990**, *30*, 299; *J.Am.Chem.Soc.* **1998**, *120*, 11616.

natural gas naturally occurring gas consisting mainly of methane, together with other saturated *hcs. such as ethane, propane, etc. (*liquefied petroleum gas, *NGL). **F** gaz naturel; **G** Erdgas. B. CORNILS

Ref.: Ullmann, Vol. A17, p. 73; Kirk/Othmer-1, Vol. 12, p. 318.

Nazarov cyclization comprises the acid (*Lewis acid)-cat. formation of 2-cyclopentenones from allyl vinyl ketones. **F** cyclisation de Nazarov; **G** Nazarow-Cyclsierung.

Ref.: *Organic Reactions*, Vol. 45, p. 1.   B. CORNILS

NBA *n*-butyraldehyde, → hydroformylation

NBD 2,5-norbornadiene in *complex cpds. or as a *ligand

Nboc *N-tert*-butoxycarbonyl as a protective group

n-bu, n-Bu *n*-butyl group

NC-AFM non-contact AFM, → atomic force microscopy

N$_n$ chelates are *chelates between metals (mostly *transition metals) and cpds. with *n*-coordinating N atoms, e.g., porphyrins or phthalocyanines (both *n* = 4, i.e., N$_4$ chelates). N$_4$ chelates, deposited on *carbon black (e.g., Co phthalocyanine), are suggested as cat. active electrode material for *fuel cells. **E=F**; **G** Chelate. B. CORNILS

Ref.: *J.Electrochem.Soc.* **1993**, *140*, No.7; *J.Appl.Electrochem.* **1997**, *27*, 77.

NDP nucleoside diphosphate, → nucleoside diphosphate synthesis

near-edge X-ray absorption fine structure spectroscopy (NEXAFS, also called XANES) is similar to *EXAFS and is applied for the investigation of local electronic structures of solids. X-rays with variable wavelengths are used as probes. The ex-situ method is non-destructive, good for in-situ studies and suitable for fingerprint techniques. The data obtained are local and surface- and bulk-sensitive.

R. SCHLÖGL

Neber rearrangement synth. of aminoketones from ketoxime tosylates with strong bases such as *alkoxides or pyridine. **F=G=E**. B. CORNILS

Nef reaction the base-cat. manuf. of carbonyl cpds. from nitroalkanes; cf. *Knoevenagel condensation. **F=G=E**. B. CORNILS

Ref.: *Organic Reactions*, Vol. 38, p. 655.

Neftekhimia/Leuna process an elderly variant of the *hydroformylation technique with Co carbonyls, developed jointly by N/L. **F** procédé N/L; **G** N/L-Verfahren. B. CORNILS

negative catalysis → anticatalysis

Negishi reaction the Pd- or Ni-cat. *cross-coupling reaction of organozinc cpds. with aryl or vinyl halides. Many functional groups are tolerated. **F** réaction de Negishi; **G** Negishi-Reaktion. V. BÖHM

Ref.: *Acc.Chem.Res.* **1982**, *15*, 340; *Tetrahedron* **1992**, *48*, 9577.

Negishi zirconocene coupling a non-catalytic coupling of Si substituted bisalkynes with cp$_2$Zr (Negishi reagent; cf. *carbometallation). B. CORNILS

Ref.: *J.Am.Chem.Soc.* **1998**, *120*, 964.

Nenitzescu acylation (*Darzens-Nenitzescu) acylation of cycloalkenes (e.g., cyclohexene) with acetyl chloride over AlCl$_3$. **F** acylation de N.; **G** Nenitzescu-Acylierung.

B. CORNILS

Neo acids → Exxon procs.

Neste NEx ether processes manuf. of ethers (*MTBE, *TAME, ETBE, and others) from C$_4$–C$_7$ tertiary alkenes by reaction with methanol or ethanol over *ion exchange resin cats., followed by distillative purification.

F procédé NEx de Neste; **G** Neste NEx-Ether-Verfahren.   B. CORNILS

Ref.: *Hydrocarb.Proc.* **1996**, (11), 110.

neural network for catalyst development

a simplified model of the human brain for correlating output data (results) with input data (factors). Nns. have been applied, e.g., to estimate or predict structure-*activity/*selectivity relationships for the *acid-cat. oxidative *dehydrogenation of *EB to *styrene by promoting SnO_2 with various inorganic cpds.

Input data are properties of cat. comps., whereas output data are the observed cat. performance. Commonly, a so-called intermediate hidden layer of neurons is included between the input layer and the output layer, and the artificial nn. is set by calcng. the activity of a neuron as a linear combination of the activity of the neurons of the preceding layer. The artificial nn. is trained with a given set of experimental data, by adjusting the weights for the linear combinations so that it describes the experimental data with a minimum error. **F** réseau neural; **G** neuronale Netzwerke.   M. BAERNS

Ref.: J. Zupan, G. Gasteiger, *Neural Network for Chemists*, VCH, Weinheim 1993; *CHEMTECH* **1995**, *25*(2), 18; *CICSJ Bulletin* **1991**, *9*, 15 and *11*(5), 2; *Anal.Sci.* **1991**, *7*, 761; *Ind. Eng.Chem.Res.* **1992**, *31*, 979; *Catal. Today* **1995**, *23*, 347.

neuraminidase

an *enzyme which cat. the *hydrolysis of sialosides

neutron scattering (NS)

technique based on neutrons (probe and response). The insensitive, ex-situ method is applied for the structure of solids and liquids (magnetic ordering, hydrogen surfaces). The data gained are integral and bulk-sensitive and versatile for many structural aspects.   R. SCHLÖGL

Ref.: *Chem.Eng.Technol.* **1999**, *22*, 135.

NEx process → Neste process

NEXAFS → near-edge X-ray absorption fine structure spectroscopy

NGL   natural gas liquid, see also LPG

***N*-glycosides** → see entry "glycosides"

nickel as catalyst metal

Ni is essential for life; several *enzymes contain Ni (e.g. *methane monooxygenase or *urease). For the cancer risk of Ni cats., cf. *toxicology in cat. Since *Sabatier's day Ni has been one of the earliest cat. metals. Metallic Ni as *Raney or *Urushibara Ni and Ni on suitable *supports are used for *hydrogenations (including *fat hardening, *dehydrogenations, *ammon-dehydrogenations [*Degussa BMA proc.]), or for special reactions (e.g., *Exxon EDS proc. on Ni-Mo cats.). Some reactions with Ni cats. include an intermediate dehydrogenation such as *ammonolysis or ethylene diamine synth. (*Berol proc.) or *reductive amination. The formation of $Ni(CO)_4$ is easy; *carbonylation reactions are thus widespread, such as the *hydrocarboxylation of acetylene (with the *redox system $NiBr_2$/CuI, *BASF acrylic acid proc.), *cyclocarbonylations, or *Reppe reactions. *Dimerizations (e.g., *IFP Dimersol proc.), *oligomerizations (*Shell SHOP proc.) *cyclizations, *cyclodi-, tri-, and tetramerizations (*Hüls dodecandioic acid proc.), or *cyclooligomerizations via hom. Ni cats. also proceed. Het. cats. (Ni on *clay) fulfill the same purpose (*Hüls Octol proc.). Ni catalyzes *addition reactions such as *hydrocyanations and *hydrovinylations.

Other het. Ni cats. give considerable service in the *petroleum processing industry for, e.g., *reforming/*steam reforming, *cracking, or *hydrotreating procs. Ni is part of newer developments for *fuel cells. **F** le nickel comme métal catalytique; **G** Nickel als Katalysatormetall.

B. CORNILS

Ref.: Jolly/Wilke; *Chem.Ind.* (Düsseldorf) **1984**, *36*, 380; *Angew.Chem.* **1990**, *102*, 251; *Houben-Weyl/Methodicum Chimicum*, Vol. XIII/9b; *Organic Reactions* Vol. 19, p. 115; *Chem. Rev.* **1994**, *94*, 2421 and **1996**, *96*, 2515; Murahashi/Davies, p. 159.

nickel boride catalysts

Prepared from nickel salts with Na or K borohydride, Ni borides are active cats. for the *hydrogenation of

, number of functional groups. **F** catalyseurs
de borure de nickel; **G** Nickelborid-Katalysa-
toren.   B. CORNILS
Ref.: Augustine, p. 233.

nicotinamide adenine dinucleotide
(NAD) hydrolase cat. the *hydrolysis of
NAD to nicotinamide and ADP-ribose, but can
be used to cat. the exchange between nicotinamide analogs and the nicotinamide moiety of
NAD. The *enzyme accepts nicotinamide analogs with modifications at the amide functional-
ty as substrates. Depending on the structures of
the nicotinamide analogs used, the reaction
may be either reversible or irreversible.
E=F=G.   C.-H. WONG

nicotinamide adenine dinucleotide
(NAD) phosphate (NADP) synthesis
NAD and NADP are both *cofactors for *enzymatic oxidoreductions. Combined chemical
and enzymatic techniques have been used to
produce NAD and NADP. Although it is not
as efficient as fermentative procedures, enzymatic synth. does provide a route to NAD
analogs for research purposes. Ribose 5-phos-
phate is converted to nicotinamide mononucleotide (NMN) in two chemical steps. This
intermediate is coupled with AMP by the
*ATP-dependent NAD pyrophosphorylase
(EC 2.7.7.1). The NAD can be converted into
NADP by the action of NAD kinase. **E=F=G**.
                                          C.-H. WONG

nicotinamide adenine dinucleotide (NAD) phosphate (NADP) "synthesis"

nicotinamide cofactor regeneration
→ regeneration of NAD(P) and NAD(P)H

nicotinamide cofactors NAD and NADP
are involved in many two-electron *oxidations
cat. by *dehydrogenases. The nicotinamide ring
system is *redox active, accepting a hydride or
two electrons and a proton to form the 1,4-di-
hydronicotinamide derivatives (NADH or
NADPH).

The reversible hydride transfer from a reduced
substrate to NAD(P), and that from NAD(P)H
to an oxidized substrate, is *stereoselective and
characteristic of individual *enzymes. Each enzyme is able to stereospecifically transfer one of
the diastereotopic methylene hydrogens at C 4
of NAD(P)H to a substrate carbonyl group or
an equivalent sp^2 center (C=C or C=N) with
high enantio- or diastereofacial selectivity. **E=F**;
G N. Cofaktoren.   C.-H. WONG

Nieuwland acetylene hydration comprises the *addition of water to acetylenes,
cat. by HgO/BF_3 in *AAA, with simultaneous

*esterification of the hydroxyl group. **F** hydration d'acetylène de Nieuwland; **G** Nieuwland Acetylenhydratisierung. B. CORNILS

Nieuwland catalyst mixture of CuCl and NH₄Cl for the *ammonolysis of chloroaromatics (with aqueous ammonia, 180–220 °C/6–7.5 MPa) or the *hydrocarbonylatin of acetylene with HCN at 80–90 °C, formerly the preferred route to *acrylonitrile. **F** catalyseur de Nieuwland; **G** Nieuwland-Katalysator.
B. CORNILS

niobium and tantalum as catalyst metals are not very well recognized as cats. in hom. or het. cat. In fact they play a role in the elucidation of the mechanistic understanding of important cat. procs. A convincing example for modeling the *Fischer-Tropsch synth. is the reaction cascade (*domino reaction) of [(SiO)₂Ta(H)₂]₂ (Si=Si(*t*-Bu)₃) with CO which results in the cleavage of two CO molecules and the formation of an μ-ethane-diyl *ligand as well as two μ-oxo ligands. Complex Ta[CH₂ (*t*-Bu)]Cl₄ is an interesting cpd. to model alkene *polymerization and *metathesis. Under reductive conds. and in the presence of trimethylphosphine it polymerizes ethylene. The mechanism differs from the *Cossee mechanism, since hydride-*carbene *complexes are involved in the reaction. Low-valent Nb and Ta complexes cat. the trimerization and *cyclotrimerization of alkynes. Notably, the *Pedersen reagent, NbCl₃(dme), and (arylO)₂ TaCl₃(ether)/Na/Hg are often used in such reactions. The former is also used in stoichiometric *cross-couplings of imines with ketones leading to ethanolamines as well as 1,2-ethylenediamines. Incorporated in *zeolites, Nb together with V has been proposed as a cat. for *oxidative *dehydrogenation. **F** Nb et Ta comme métaux catalytiques; **G** Nb und Ta als Katalysatormetalle. P. HÄRTER
Ref.: J.A. Labinger, M.J.Winter, *Comp.Organomet. Chem.*, Pergamon, Oxford 1995, Vol. 5, p. 57; *Houben-Weyl/Methodicum Chimicum*, Vol. XIII/7.

Nippon Oil *t*-butyl alcohol process for the manuf. of *t*-butyl alcohol (TBA) by *hydration of isobutene with aqueous HCl and in presence of metal salts. **F** procédé de Nippon Oil pour le TBA; **G** Nippon Oil TBA-Verfahren. B. CORNILS

Nippon Shokubai acrylic acid process Proc. (today mainly two-step) for the manuf. of *acrylic acid by partial *oxidation of propylene over *multicomponent cats. containing Mo, Te, etc. and various *promoters at 260–350 °C/1 MPa. **F** procédé de Nippon Shokubai pour le AA; **G** Nippon Shokubai AA-Verfahren. B. CORNILS
Ref.: *Hydrocarb.Proc.* **1972**, (9), 85 and **1977**, 56(11) 123.

Nippon Shokubai paraffin oxidation manuf. of secondary alcohols by boric acid cat. *Bashkirov oxidation of paraffins at 140–190 °C. **F** procédé Nippon Shokubai pour l'oxidation des paraffines; **G** Nippon Shokubai Paraffinoxidation. B. CORNILS

Nippon Steel picoline process for the manuf. of 2-picoline and 2-methyl-5-ethylpyridine from ethylene and NH₃ at 100–300 °C/3–10 MPa over a hom. system consisting of Pd salts/CuCl. **F** procédé Nippon Steel pour le picoline; **G** Nippon Steel Picolinverfahren. B. CORNILS

Nippon Zeon Zeonex a *ROMP based norbornene polymer

***n/iso* (*n/i*) ratio** is the ratio between linear and branched isomers. This value is important for the reaction prods. of *isomerizations and propylene *hydroformylation, since *n*-butyraldehyde (*NBA) is more valued than *i*-butyraldehyde (*IBA) (only NBA is converted to the most important downstream product, *2-EH). The demand for IBA (for *i*-butanol, isobutyric acid, or neopentyl glycol) is quite low. In higher oxo reaction prods. the *n/iso* ratio plays a role with detergent alcohols because their biodegradability (and that of derivatives) depends strongly on their content of unbranched isomers. The *n/iso* ratio of plasticizer alcohols may disadvantageously influence their compatability with, e.g., PVC. The *n/iso* ratio is influenced by the reaction parameters (temp., press.) as well as by the CO/H_2 ratio, the *central atom of the *complex cat., or the nature of the *ligands. **F** rapport n/i; **G** n/i-Verhältnis.   B. CORNILS

Ref.: Falbe-1.

Nissan Girdler catalysts → Südchemie catalysts

nitration   the introduction of nitro groups ($-NO_2$) into organic molecules by means of nitric acid, alkyl nitrates, N_2O_4, etc. The n. proceeds mainly through *radical mechanisms at higher temps. without cats. (cf. *aromatic substitution). **E=F; G** Nitrierung.   B. CORNILS

nitric acid synthesis   → ammonia oxidation

nitric oxide synthase (NOS) cat. the *oxidation of arginine to citrulline and NO. The *enzyme is produced as cytokine-inducible (iNOS), endothelial (eNOS), and neuronal (nNOS) *isozymes. The iNOS isozyme is critical for the immune response, but is also implicated in most diseases involving NO overproduction. Constitutive eNOS and nNOS generate NO that is related to blood circulation and signal transmission in the nervous system. Much of the biological activity of NO is due to its activating effect on the enzyme guanylate cyclase through its coordination to enzyme-bound ferrous heme. This enzyme cat. the generation of 3′,5′-cyclic GMP (from GTP), a second messenger that signals vascular smooth muscle relaxation and inhibition of platelet aggregation and adhesion. The NO synthase reaction involves two steps. The first step in the heme-based oxidation of L-arginine to N^ω-hydroxy-L-arginine (NOH-L-Arg), while the second step is the further *oxidation of NOH-L-Arg. **E=F=G.**

Ref.: *Science* **1997**, *278*, 425.   C.-H. WONG

nitrilases   → nitrile-hydrolyzing enzymes

nitrile hydratase   →   nitrile-hydrolyzing enzymes

nitrile-hydrolyzing enzymes Microbial *hydrolysis of nitriles may be cat. by nitrilase to give the corresponding acids, or by nitrile hydratase to give amides. The amides may be hydrolyzed to acids in reactions cat. by amidases. An enzymatic synth. of acrylamide from acrylnitrile has become an industrial proc. for prod. of acrylamide based on the resting cells of a *Pseudomonas* sp. (cf. *Nitto proc.). The enzyme contains a ferric ion and a *cofactor pyrroloquinoline quinone (PQQ). Both are involved in the hydration of nitriles. Several other nitrile hydratases containing cobalt have also been isolated. **E=F; G** Nitrilhydrolysierende Enzyme.   C.-H. WONG

Ref.: *Pure Appl.Chem.* **1991**, *62*, 1441.

nitro aldol reaction   → Henry aldol addition

nitrogenase   → nitrogen fixation

nitrogen fixation Nitrogen *reduction is an energetically expensive proc. As a result, very few organisms are able to achieve it. Among them are members of the genus *Rhizobium*, which are symbiotic bacteria that live in root nodules of leguminous plants. The *enzyme that accomplishes this reaction, nitrogenase, is composed of two protein complexes: an Mo-Fe protein, a 220 kDa $\alpha_2\beta_2$ heterotetramer containing Fe and Mo, and an Fe protein, a 64 kDa homodimer containing iron. Most of the iron in the enzymes is contained in Fe-S *clusters: the Fe protein has one per subunit, while the MoFe protein has four Fe-S clusters and an unusual Mo-Fe-S cluster. The reaction carried out is

$$N_2 + 12\ ATP + 6e^- + 6H^+ \rightarrow$$
$$2NH_3 + 12\ ADP + 12P_i,$$

although there is a futile cycle in which H_2 is formed. The actual *reduction occurs on the

Mo-Fe protein in three steps: 1: reduction of nitrogen to the diimine: $N \equiv N + 2H^+ + 2e^- \rightarrow$ H-N=N-H; 2: reduction of the diimine to *hydrazine: $H-N=N-H + 2H^+ + 2e^- \rightarrow H_2N-N_2H$; 3: reduction of hydrazine to 2 equiv. of *ammonia: $H_2N-N_2H + 2H^+ + 2e^- \rightarrow 2NH_3$. The reducing equivs. are contributed from *ETCs. The electrons are fed to the MoFe protein by the Fe protein, while the equivalents of ATP (2 ATP per electron passed) are bound to and hydrolyzed by the Fe protein. It is believed that the ATP *hydrolysis caused a conformational change in the Fe protein that alters its *redox potential.

The enzyme is extremely oxygen sensitive, and *Rhizobium* sp. and their plant hosts work together to protect it by producing an oxygen *scavenger, leghemoglobin: the protein chain is provided by the plant while the *porphyrin is provided by the bacterium. **F** fixation de l'azote; **G** Stickstoff-Fixierung.   P.S. SEARS

Ref.: *Nature* **1997**, *387*, 352, 394; Graham, *Symbiotic Nitrogen Fixation*, Norwell-Kluwer, Dordrecht 1994.

nitrogen oxide removal (DeNOx) Various techniques are used to remove N oxide pollutants from vent streams of stationary and mobile sources. The approaches used depend strongly on the O content of the effluent gas stream. The cat. decomposition of NO is at present far from technologically significant. The best available cats. provide insufficient *activity and are easily *poisoned, e.g., by oxygen. N oxides are, therefore, reduced by *hcs. or NH_3 over selective or non-selective cats. In non-selective NOx *reduction (usually performed with noble-metal cats.), the (hc.) reductant converts both O and NOx present in the effluent gas. It is therefore used mainly for streams of limited velocity and low O content, e.g., nitric oxide plant tail gas. *Three-way cats. (*automobile exhaust cats.) are also non-selective, but the stoichiometric control of the air/fuel ratio prevents unacceptable fuel losses for oxygen reduction.

In *selective cat. reduction (SCR), the N oxides are reduced by NH_3 or hcs. over selective cats. SCR with ammonia is an established technology for flue-gas purification with stationary sources, e.g., power plants (cats.: V_2O_5-WO_3-TiO_2, V_2O_5-TiO_2) and is now also being applied with heavy diesel engines (urea injection). SCR with hcs. as a potential future technology for NOx abatement with diesel and lean-burn gasoline engines is currently subject to intensive research efforts (cf. procs. of *Degussa Desonox, *Didier DeNOx, etc.). **F** suppression des oxydes de l'azote; **G** Stick-oxidentfernung.   M. MUHLER

Ref.: *Catal. Today* **1998**, *46*, 233.

nitrosation the introduction of nitroso groups into organic molecules by means of nitrous acid, nitrogen oxides, or NOCl. Primary, secondary, or tertiary amines react via nitrosation to yield different prods: primary amines yield the corresponding primary alcohol and alkene, secondary amines give the corresponding nitrosamine, and tertiary amines react only at higher temps. at pH 3–6 to yield complex prod. mixtures leaving the nitrogen untouched. N. is an important reaction for *DuPont's AN proc. or *Toray's PNC proc. **F**=**E**; **G** Nitrosierung.

B. DRIESSEN-HÖLSCHER

Nitto Chemical acrylamide process manuf. of acrylamide (30 000 tpy) by *hydrolysis (*saponification) of acrylonitrile by anchored *enzymes (*nitrile hydrolyzing enzymes) on polyacrylamide. **F** procédé Nitto Chemical pour l'acrylamide; **G** Nitto Chemical Acrylamidverfahren.   B. CORNILS

Ref.: *Trends Biotechnol.* **1989**, *7*, 153.

Nixan process *Ni*tro Cyclohe*xan*e pr., → DuPont processes

NLDA nonlocal density approximation, → local density approximation

NMP nucleoside monophosphate, → NDP synthesis

NMR → nuclear magnetic resonance

N,O chelates *chelates between metals (mainly *transition metals) and cpds. with co-ordinating N and O atoms, e.g., with *salen ligands (cf. *Jacobsen's complex).

B. CORNILS

NOE → nuclear Overhauser effect

NOESY nuclear Overhauser effect spectro-copy, → nuclear Overhauser effect

non-aqueous ionic liquids (NAILs) var-ants of *molten salt media and constituted of ions. Most of them melt below rt and exhibit densities of >1. They are easily prepared and stable and have no measurable vapor press. They are generally based on large organic cat-ons such as N-alkylpyridinium, N,N'-dialkyl-imidazolium (e.g., 1-ethyl-3-methylimidazo-ium, abbr. *emin), alkylsulfonium, tetraalkyl-ammonium, or tetraalkylphosphonium and organic or inorganic anions. Examples of the most frequently described anions are $AlCl_4^-$, $Al_2Cl_7^-$, $CuCl_2^-$, $CF_3CO_2^-$, $N(CF_3SO_2)_2^-$, BF_4^-, SbF_6^-, or PF_6^-. NAILs have been used in many applications. Because of their large spectrum of miscibility with organic sub-strates, their tunable coordinating ability and *Lewis acidity (depending on the nature of the anions), they can be considered as a new family of *solvents for cat. and non-cat. or-ganic *biphasic liquid/liquid reactions. They have found applications as solvents of *transi-tion metal complexes for *hydrogenation, *isomerization, *hydroformylation, *telomer-ization, *dimerization, and *codimerization of alkenes, and as acidic cats. and solvents for *alkylations. Their *solvent effect in *Diels-Alder reactions and nucleophilic substitution is remarkable. *Dimersol procs. can be real-ized with the biphasic system using NAILs (*IFP Difasol proc.). The performances of the proc. are improved and the cat. disposal can be reduced. NAILs promise significant envi-ronmental benefits. **E=F=G**.

H. OLIVIER-BOURBIGOU

Ref.: Cornils/Herrmann-2, p.554; *C&EN* **1998**, March 30, 32; *Green Chem.* **1999**, *1*, 23; R.E. Malz, *Catalysis of Organic Reactions*, Dekker, New York 1996; *Chem. Rev.* **1999**, 99, 2071; *J.Mol.Catal.A*: **1999**, *148*, 43.

noncompetitive inhibition A noncom-petitive inhibitor affects k_{cat} but not K_m (cf. *competitive i.). This type of *inhibition can be observed if the inhibitor and the substrate are not at the same site, so that the binding of inhibitor does not affect the binding of the substrate, only its reaction rate:

$$E + S \xrightleftharpoons{K_S} ES \xrightarrow{k_{cat}} E + P$$

$$EI + S \xrightleftharpoons{K_S} ESI$$

Noncompetitive Inhibition

$$v = \frac{[E_0][S]k_{cat}/(1+ [I]/K_I)}{[S] + [K_S]}$$

This pattern can be observed, for example, if the inhibitor binds at a site removed from the active site, but causes a change in the position of one of the cat. residues. An interpretation of non-competitive inhibition is that the inhi-bitor binds equally well to the *enzyme and the enzyme-substrate (*Michaelis) complex. The inhibitor will bind to the enzyme with or without the substrate present (cf. *Dixon plot). With nci., there is a common intersec-tion on the x-axis of the *Lineweaver-Burk plot, while the lines are parallel for the *Eadie-Hofstee plot, indicating an effect on V_{max} rather than an effect on K_m. **E=F=G**.

P.S. SEARS

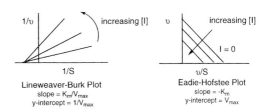

Lineweaver-Burk Plot
slope = K_m/V_{max}
y-intercept = $1/V_{max}$

Eadie-Hofstee Plot
slope = $-K_m$
y-intercept = V_{max}

non-contact atomic force microscopy
(NC-AFM) → atomic force microscopy

nonribosomal peptide bond formation

In the *biosynthesis of nonribosomal peptides (e.g., peptide antibiotics and glutathione), an amino acid is activated enzymatically to aminoacyladenylate; this species then reacts with the amino group of another aminoacyl thioester to form a peptide bond.

The peptide chain is thus extended from N- to C-terminus on a multienzyme template. A typical example is the biosynth. of δ-(L-α-aminoadipyl)-L-cysteinyl-D-valine (ACV) catalyzed by ACV synthetase. The *enzyme accepts many non-protein amino acids as well as hydroxyacids. Since ACV is a precursor of penicillins and cephalosporin, the enzymatically formed ACV analogs may be converted to the corresponding β-lactam derivatives. Another class of non-protease *enzymes potentially useful for peptide synth. is aminoacyl-tRNA synthetases. These enzymes cat. the synth. of aminoacyl-tRNA in the presence of ATP with the formation of an aminoacyl-AMP intermediate.

C.-H. WONG

norb 1-norborneyl in *ligands or *complex cpds.

Norit carbon special activated carbon, → carbon

Normann, Wilhelm (1870–1939), industrial chemist, worked on high-press. *hydrogenations, *hardening of fats, manuf. of fat alcohols from fatty acids, etc. B. CORNILS

NORPHOS, norphos is an optical active bicyclic *ligand, prep. by a multistep synth. including a *Diels-Alder addition (Figure).

Neutral or cationic Rh *complexes of NORPHOS may be used in *asymmetrical *hydrogenations and *hydrosilylations. NiCl$_2$/NORPHOS complexes cat. *Grignard cross-coupling. O. STELZER

Ref.: Angew.Chem.Int.Ed.Engl. **1979**, *18*, 620; Chem. Ber. **1981**, *114*, 1137 and **1992**, *125*, 2085; Organimetallics **1986**, *5*, 739; J.Organomet.Chem. **1981**, *209*, C1.

Norsorex → CdF procs.

Northrop John H. (1891–1987), biochemist at Princeton and Berkeley (USA). Worked on kinetics of enzyme-cat. reactions and succeeded in the crystallization of biochemical cats., the *enzymes, and *antibodies. Nobel laureate in 1946 (together with Sumner and Stanley). B. CORNILS

Ref.: J.Gen.Physiol. **1930**, *13*, 739 and **1981**, *64*, 597.

NOXSO → SNOX process

NQR nuclear quadrupole resonance (spectroscopy), → nuclear magnetic resonance

NS → neutron scattering

NTA nitrilotriacetic acid, a *chelating agent.

NTP 1: nucleoside triphosphate, → NTP synthesis; 2: normal temperature and pressure (as reaction conds.)

N-transribosylase see under keyword "T"

nuclear magnetic resonance (NMR)

This form of spectroscopy, based on magnetic resonance, is used for a wide range of cat. topics, e.g., the local structure of cats., model cpds., adsorbates, dynamic motion of molecules, and for in-situ imaging. Probe and response: magnetic field, high-frequency (MHz) radiation; ex-situ technique, non-destructive. The informations are integral, bulk-sensitive, and have in-situ capability. Cf. also *CP-MAS, *MAS-NMR. R. SCHLÖGL

Ref.: Ullmann, Vol.B5, p.471; Anderson/Boudart, Vol.10.

nuclear Overhauser effect (NOE) a

phenomenon in *NMR spectroscopy, based on magnetic dipole interactions between nuclear spins. It is used both to enhance the signal intensities of like or unlike nuclei and to determine the 3D structure of simple or of complex macromolecules, e.g., of proteins. To do this, spin relaxation procs. are exploited to transfer the higher population difference of one type of nuclei to another, magnetically weaker or less abundant spin systems thereby enhancing the intensity of the NMR signal associated with the latter. The signal enhancement of the nucleus A that is coupled to another nucleus X which is saturated by a strong irradiation at its resonance frequency depends on the magnetogyric ratios of the nuclei. The signal enhancement is most pronounced for nuclei in the immediate spatial vicinity, even if they are separated by many chemical bonds. The NOE(SY) is routinely applied in NMR studies of heteronuclei, in particular in NMR of rare or magnetically weak nuclei to boost their sensitivity, e.g., in ^{13}C or ^{29}Si. **E=F=G**.

J. BARGON

Ref.: J.P. Hore, *Nuclear Magnetic Resonance*, Oxford University Press, Oxford 1995.

nucleases *phosphodiesterases in which the substrate is nucleic acid. Ns. are distinguished by the class of nucleic acid they accept (*DNA, *RNA, or both), by their preference for external or internal phosphodiester bonds (exo- vs. endonucleases), and their degree of *specificity. Restriction endonucleases, for example, are highly specific endodeoxyribonucleases that accept typically a specific four, six, or eight-base, C_2-symmetric sequence and in general have virtually no activity for even closely related sequences. *Staphylcoccal* nuclease, at the other extreme, will accept both DNA and RNA substrates. **E=F**; **G** Nukleasen.

P.S. SEARS

nucleic acid a linear *polymer of nucleotides C containing either ribose or 2'-deoxyribose (B, in Nature; cf. Figure) which are linked via a phosphodiester that links the 3'-hydroxyl of one with the 5'-hydroxyl of the following one. Nucleic acid analogs have been synth. chemically which have been modified at the sugar, the base A, or the linkage, such

as the "peptide nucleic acids" (PNAs) which are linked through peptide bonds instead of phosphodiesters; or the *phosphorothioate-linked nucleic acids. **F** acides nucléiques; **G** Nukleinsäuren.

P.S. SEARS

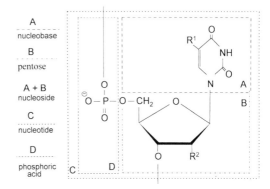

Ref.: Blackburn, Gait, *Nucleic Acids in Chemistry and Biology*, 2nd. Ed., Oxford University Press, Oxford 1996; Z.A. Shabarova, A.A. Bogdanov, *Advanced Organic Chemistry of Nucleic Acids*, VCH, Weinheim 1994.

nucleophilic catalysis (in enzymatic reactions) Nucleophilic cat. is a special case of *covalent cat. that is frequently used by *hydrolases and *transferases. Many such *enzymes proceed via a double displacement mechanism, in which a nucleophilic group within the enzyme *active site (e.g., an active site residue such as serine or cysteine) attacks the substrate, expelling a leaving group and forming a covalent intermediate, which is then attacked by another substrate or by water to generate the product and regenerate the enzyme. Another mechanism (*addition-elimination) is observed in many *cis-trans* *isomerases. An active-site nucleophile (typically a thiol) attacks the β-carbon of an α,β-unsaturated ketone to give the enolate, which can freely rotate about the α,β-bond. The thiol is then eliminated. **F** catalyse nucléophile; **G** nucleophile Katalyse.

P. SEARS

Ref.: C. Walsh, *Enzyme Reaction Mechanisms*, W.H. Freeman, New York 1979.

nucleoside diphosphate synthesis The preparation of *NDPs from *NMPs requires different *enzymes for each NMP. Adenylate *kinase (EC 2.7.4.3) phosphorylates AMP and CMP, and also slowly phosphorylates UMP. Guanylate kinase (EC 2.7.4.8) cat. the phosphorylation of GMP. Nucleoside monophosphate kinase (EC 2.7.4.4) uses *ATP to phosphorylate *AMP, *CMP, GMP, and UMP. For those kinases requiring ATP as a phosphorylating agent, ATP is usually used in cat. amounts and recycled from ADP using pyruvate kinase/PEP or acetate kinase/ acetyl phosphate. **F** synthèse de NDP; **G** NDP-Synthese. C.-H. WONG

nucleoside phosphorylase cat. the reversible formation of a purine or pyrimidine nucleoside and inorganic phosphate from ribose-1-phosphate (R-1-P) and a purine or pyrimidine base, with the equil. lying well in favor of nucleoside formation. Nucleoside synth. has relied on the transfer of the ribose moiety of a readily available nucleoside to a different purine or pyrimidine base or analogs through the inter mediacy of R-1-P. This work has been done primarily with isolated *enzymes but whole cells have also been employed in a few cases. Direct purine-to-purine exchange reactions have been conducted without isolation of R-1-P using activated purine derivatives as the ribosyl donors.
The *nucleoside phosphorylases have been found to accept a wide range of nucleoside analogs, with modifications in both the base and glycosyl components, as substrates. **E=F=G.** C.-H. WONG
Ref.: *TIBTECH* **1990**, *8*, 348.

nucleoside triphosphate (NTP) regeneration NTPs other than ATP are most often used in reactions involving the transfer of a nucleoside phosphate (or pyrophosphate) moiety, and create inorganic phosphate or pyrophosphate as a product.

Useful regenerative synths. of NTPs such as CTP and UTP have been accomplished using the relatively readily available monophosphates as the starting reagent. Regeneration of CTP has been demonstrated and used in a multienzymatic synth. of N-acetylneuraminic acid derivatives. Similarly, acetate and pyruvate *kinase have been used in the in-situ regeneration of UTP from *UDP for use in *enzyme cat. synth. of carbohydrate derivatives. Regeneration of NTPs from the corresponding diphosphates is not a problem because both acetate and pyruvate kinase have broad substrate *specificities. Preparation of NTPs from unphosphorylated nucleosides is possible using nucleoside kinases. **F** régénération de NTP; **G** NTP-Regenerierung. C.-H. WONG

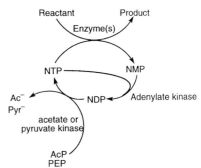

Regeneration of NTP from NMP

Ref.: *Appl.Biochem.Biotech.* **1990**, *23*, 205; *J.Org. Chem.* **1982**, *47*, 5416.

nucleoside triphosphate (NTP) synthesis Enzymatic synth. of NTPs involves the sequential use of two *kinases to transform the monophosphates to NDPs and then to NTPs. Any of three kinases may be used to synth. NTPs from the corresponding NDPs, each of which uses a different phosphoryl donor: *pyruvate kinase uses phosphoenolpyruvate (PEP) as a phosphoryl donor; *acetate kinase uses acetyl phosphate; and nucleoside diphosphate kinase uses ATP. **F** synthèse de NTP; **G** NTP-Synthese. C.-H. WONG

O

9-O-acetylneuraminic acid → *N*-acetyl-neuraminate aldolase

OAE optoacoustic effect, → photoacoustic spectroscopy

OAS → optoacoustic spectroscopy

O-C coupling 1: phenol coupling reactions can proceed in different *regioselective directions, either to the formation of quinones (C-C coupling) or to aromatic oligo- or polyethers (O-C coupling). Cu(I) salts or their *complexes are used as cats. Commercially relevant is the *oxidative coupling of 2,6-xylenol to give polymeric *p*-phenylene oxide, (PPOs), used as high-melting thermoplastics.

$$2,6\text{-}(CH_3)_2C_6H_3\text{-}OH + 0.5\,O_2 \rightarrow$$
$$^1/_n\,[\text{-}2,6\text{-}(CH_3)_2C_6H_3\text{-}O\text{-}]_n$$

For such phenol oxidations Cu(I) chloride in the presence of dibutylamine, NaBr, and a quarternary ammonium salt are typically used. Copper(I) phenolates are the reaction intermediates, reacting further to the polyaryl ether via *radical reaction pathways. 2: *Telomerization reactions applying water as solvent/reactant (*hydrodimerization) can also be regarded as O-C coupling reactions. Telomerization reactions of dienes are commercially important. **F** couplage O-C; **G** O-C-Kupplung. R.W. FISCHER
Ref.: Parshall/Ittel, p. 174; Cornils/Herrmann-1, Vol. 1, p. 351.

Octacat common name for a Y-type zeolite

Octafining process → Atlantic Richfield and *IFP procs.

Octagain process → Mobil procs.

Octamix process → Lurgi procs.

Octanizing process → IFP cat. reforming process

Octol process → Hüls procs.

oil processing industry, oil refining → petroleum processing industry; *refinery.

Okazaki fragments → telomerase

Oláh George A. (born 1927), professor of organic chemistry at UCLA in Los Angeles (USA), works on reaction mechanisms, *Friedel-Crafts reactions, *carbocations. Nobel laureate in 1994. B. CORNILS

olefination a class of reactions in which a new C-C double bond is introduced into an organic molecule. Often os. (also called vinylations or methylenations) are carried out with aldehydes or ketones. However os. of carbon-X bonds (X = halogens, OTf, N_2^+, COX) or even with double bonds themselves are known. Examples of stoichiometric organic o. reactions of aldehydes or ketones include the *Wittig o. as used in *BASF's vitamin A synth. Non-cat. *transition metal-mediated os. of aldehydes and carboxylic acid derivatives have been achieved using the *Tebbe-Grubbs reagent. Cat. o. are possible when certain diazoalkanes are used as alkylidene group transfer reagents in the presence of *phosphines and cat. amounts of Mo or Re *complexes acc. to $RCHO + R^1R^2CN_2 + PPh_3 \rightarrow R\text{-}CH=CR^1R^2 + Ph_3P=O$. Other important cat. o. reactions are the *Heck reaction and indirectly the commercially important *Balsohn reaction, which yields araliphatic alkenes after *dehydrogenation of the initial addition prods. (e.g., benzene and ethylene yield *EB, which is converted to styrene). O. of heteroatoms is also known; thus the first vinyl esters were prepared by the reaction of acetic acid and acetylene by hom. cat. with $HgSO_4$. Recent literature describes the o. of carboxylic acids in the presence of Pd, Pt, or Ru *complexes. **F=E**; **G** Olefinierung.   M. BELLER

Ref.: Weissermel/Arpe; Cornils/Herrmann-1; *Organic Reactions*, Vol. 14, p. 270 and Vol. 46, p. 105.

olefin metathesis → metathesis

Oleflex process → UOP procs.

oligodentate ligands *ligands which complex *central atoms in hom. cats. via N *and* O atoms, e.g., *Schiff bases or *salen *complex ligands (e.g., the *Jacobsen complex, *N,O chelates).   B. CORNILS

oligomerization the synth. of linear, branched, or cyclic prods. which repeats only a few different or the same constitutional *monomer units. Oligomers are produced from monomers or mixtures of monomers or by degradation of *polymer prods. Cats. for polyreactions are widely used as oligomerization *initiators. For cat. alkene o., *transition metals like Pd, Co, or Ni are of importance. The alkene oligomers manuf. this way are important for detergents, *plasticizers, and fine chemicals. Of greater commercial significance, however, are the Ni hydride *complexes of phosphorus/oxygen *chelates that are used in the *Shell SHOP proc. for the o. of ethylene to *α-olefins. The o. of alkenes with functional groups (e.g., *hydrodimerization of *acrylonitrile to form *adiponitrile) and the *co-oligomerization reactions of dienes and alkenes (for example the synthesis of *trans*-1,4-*hexadiene, which is used for prod. of *EPDM elastomers) are also of comercial interest. An attractive perspective of *enantioselective cat. may be the ability to form *asymmetric co-oligomerization reactions for the prod. of *optically active prods. For examples of commercial interest, cf.,

procs. of *Bayer Isobutene, *Ethyl Alfol, *Exxon LAO, *Gulf LAO, *IFP Selectopol, *Shell SHOP, *UOP InALK, etc. **F** oligomérisation; **G** Oligomerisierung.

W. KAMINSKY, R. WERNER

Ref.: Mark/Bikales/Overberger/Menges, Vol. 10, p. 432; Entelis et al., *Reactive Oligomers*, VSP, Utrecht 1989; Uglea, Negulescu, *Synthesis and Characterization of Oligomers*, CRC Press, Boca Raton 1991.

oligosaccharides linear or branched *oligomers composed of 2 to approx. 20 monosaccharides and/or monosaccharide derivatives that are linked by means of glycosidic bonds. The most prevalent monosaccharide unit in os. is D-glucose, followed by D-mannose, D-fructose, D- and L-galactose, D-xylose, etc. Cyclic os. composed of glucose units are called *cyclodextrins and constitute a particular class of os. which find numerous applications in cat. Os. can be manuf. by means of classical carbohydrate chemistry or by *enzymatic pathways. The latter involve generally the utilization of *glycosyl transferases and/or of *glycosidases. The enzymatic synth. of os. on solid supports has also been successfully achieved by using matrix materials of which the interior is accessible to the enzyme (*Merrifield [*solid-phase] technique). Numerous chiral chelating *ligands can be easily obtained from disaccharides. These *ligand-exhibited high *enantioselectivity and activity in hom. cat. reactions (*asymmetric *hydrogenations, *hydroformylations, etc.). The selective transformation of os. can be performed efficiently with het. or hom. cats. For instance, hydrogenation of di- and trisaccharides using Ru/C or hom. Ru *triphenylphosphine *complexes has been reported. The *telomerization of butadiene with disaccharides as nucleophiles or the *decarbonylation of disaccharides can be cat. by Rh complexes in hom. or organic-aqueous *two-phase systems. **E=F**; **G** Oligosaccharide.   E. MONTFLIER

Ref.: US 2.868.847 (1959); US 3.935.284 (1976); *J.Org. Chem.* **1989**, *54*, 5257; *Tetrahedron* **1989**, *45*, 5365; *Tetrahedron Lett.* **1993**, *34*, 3091; *Synthesis* **1991**, 499; FP 2.693.188 (1992).

oligosaccharides (enzymatic synthesis)
→ sugar nucleotide synthesis, *sugar nucleotide regeneration

oligosaccharyltransferase   cat. the transfer of an oligosaccharide consisting of two N-acylglucosamine, nine mannose, and three glucose units from a dolichyl pyrophosphate intermediate to an Asn residue of a nascent peptide or protein. The *enzyme also transfers the minimal structure GlcNAcβ1,4Glc-NAc from the corresponding dolichyl pyrophosphate donor or from a derivative in which the lipid comp. is truncated or simplified. The minimal peptide structure which will serve as an acceptor is the tripeptide Asn-Xaa-Ser/Thr (Xaa = any amino acid save Pro or Asp). O. has been utilized for the in-vitro synth. of several peptides containing glycosylated Asn residues. **E=F=G**.   C.-H. WONG

Olin allyl alcohol process   *isomerization of *propylene oxide over Li phosphate in the vapor phase at 250–350 °C. The *selectivity is 97 %. **F** procédé d'Olin pour l'alcool allylique; **G** Olin Allylalkohol-Verfahren.
                                                              B. CORNILS

OM   → optical microscopy

ON   octane number (RON: research octane number, MON: motor ON)

operon   In prokaryotic organisms, genes encoding *enzymes that all function in a particular *biosynthetic pathway are frequently found clustered together on the chromosome. The transcriptional regulation of the entire gene cluster is coupled: the genes are turned on and off together. Such a gene cluster is called an operon. **F** opéron; **G=E** .   P.S. SEARS
Ref.: Lewin, Benjamin, *Genes*, Vol. 5, Oxford University Press, Oxford 1990.

Oppenauer oxidation   the oxidative transformation of secondary alcohols to the corresponding carbonyl cpds. under mild, acid-free conds. An excess of simple ketones such as acetone or cyclohexanone serves as the source of the oxidizing agent (hydrogen acceptor; cf. *transfer hydrogenation). In the presence of cat. amounts of aluminum tris (isopropylate) the oxidizing ketones are converted to the alcohols while forming ketones from the alcohol substrates. Thus the Oppenauer *oxidation can be understood as the reverse of the *Meerwein-Pondorf-Verley *reduction. Under acid-free conds. the Oo. is a favorable method for the oxidation of acid-labile substances. Related to the Oo. are oxidations (hydride transfer reactions) applying trichloroacetaldehyde (Cl_3CCHO) in the presence of Al_2O_3. This reaction works highly selectively for the oxidation of secondary alcohols (in the presence of primary alcohols which do not undergo this oxidative transformation) as well as substrates which carry other oxidative labile functional groups. **F** oxidation d'Oppenauer; **G** Oppenauer-Oxidation.
                                                              R.W. FISCHER
Ref.: *Chem.Lett.* **1987**, 181; *Synthesis* **1994**, 1007; *Organic Reactions*, Vol. 6, p. 207.

Oppolzer's sultam   First 1984, O.s. (bornane-10,2-sultam) is one of the most useful *chiral *auxiliaries. Both *enantiomers are commercially available. They are effective in a wide variety of metal-promoted reactions and, more uniquely, also in several important classes of thermal reactions, including *radical reactions where metals are absent. **F** sultam d'Oppolzer; **G=E**.   A.F. NOELS

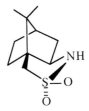

Ref.: *Tetrahedron* **1993**, *49*, 293.

optical microscopy (OM)   OM (visible light) has a certain importance for the determination of morphologies, phase transformations, or in-situ reaction monitoring. The ex-situ method is non- destructive. Combination with other methods (e.g., polarization analysis) is possible. The data are integral, bulk- and *surface-sensitive.   R. SCHLÖGL

optical resolution   → resolution

optical yield (oy, *ee*)   In a given chemical re- action of *chiral starting materials the *enantio- meric excess (*ee*) of the prods. may differ from that of the feed. In order to define the degree of change of enantiomeric excess during a reaction the term *optical yield* was introduced as the ra- tio of the enantiomeric excess of the prods. to the enantiomeric excess of the starting materi- als (optical rotation of prods. divided by the op- tical rotation of pure material). If the reaction is stereospecific, the optical yield is 100 %. On the other hand, if *racemization of the chiral start- ing materials occurs the oy. is 0 %. **F** rendement optique; **G** optische Ausbeute.   M. BELLER

Ref.: E.L. Eliel, S.. Wilen, *Stereochemistry of Organic Compounds*, Wiley-Interscience, New York 1994; Au- gustine, p. 333.

optimum strength of bonding   → Sabatier's principle

optoacoustic spectroscopy (OAS)   ope- rates with laser pulses in the range of 10^{-11} to 10^{-5} s, generating ultrasound pulses, and is used for highly sensitive absorption measurements (cf. *photoacoustic spectroscopy). **F** spectro- scopie optoacoustique;   **G** Optoakustische Spektroskopie.   A. HEUMANN

Optimized Inc. Deoxy process   for the removal of oxygen from lean *NGL streams by cat. oxidation (with consumption of O) to ppm levels. **F** procédé Deoxy de Optimized Inc.; **G** Optimized Inc. Deoxy-Verfahren.

B. CORNILS

Ref.: *Hydrocarb.Proc.* **1996**, (4), 114.

organic container compounds   → host- guest relationships, *cyclodextrins, *macro- cyclic cpds.

organoaluminum compounds   cpds. con- taining at least one Al-C bond, e.g., trialkyl- aluminum (R_3Al; R = alkyl), dialkylaluminum hydrides (R_2AlH), dialkylaluminumchlo- rides (R_2Al-Cl) alkylaluminumdichlorides (R-AlCl$_2$), or alkylaluminumsesquichlorides (R_2Al-Cl · R-AlCl$_2$). They were first synth. in 1859. The industrially used aluminum al- kyls are highly air- and moisture-sensitive colorless liquids. The short-chain alkyls un- dergo self-ignition and react explosively with water. They form dimers and have good solubility in saturated and aromatic hc. solvents. They are prepared on an industrial scale. An important method is the *Ziegler direct synth. Activated aluminum, hydrogen and an α-olefin react under increased press. and temp. according to Al + 3 C_nH_{2n} +H$_2$ → $(C_nH_{2n-1})_3Al$.

An important application of oc. is in the Zieg- ler growth reaction (*"Aufbau"reaction) for the synth. of long-chain primary alcohols for the detergent industry. Another important use of aluminum alkyls is in *Ziegler-Natta cats. They are composed of a *transition met- al cpd. (e.g., TiCl$_4$, VCl$_3$ or Cp$_2$TiCl$_2$) and an oc., which may also contain halide substitu- ents (e.g., Et$_3$Al, Et$_2$AlCl or EtAlCl$_2$). Further applications of aluminum alkyls are as alkylating agents in the synth. of organo- Zn or -Sn cpds. and for the dimerization of α- alkenes.   **F** composés   organo-aluminium; **G** Organoaluminiumverbindungen.

W. KAMINSKY, R. WERNER

Ref.: Ullmann, Vol. 1, p. 543; McKetta-1, Vol. 3, p. 1; Winnacker-Küchler, Vol. 3, p. 341, Vol. 4, pp. 38, 62, 116, and Vol. 5, p. 180; Kirk/Othmer-1, 3rd. Ed., Vol. 2, p. 195; Abel/Stone/Wilkinson, 2nd. Ed., Vol. 1, p. 432.

organometallic compounds for homo- geneous catalysis   Due to their analytical, structural, and electronic variability, ocs. are best suited as molecular cats. of high *activity and *selectivity for hom. cat. A certain struc- tural principle can be "tuned" to become even more efficient by substitution within the *ligand sphere and by changing the metal it- self or its oxidation state (*tailoring of cats.). Organometallic cats. must meet the following conds.: 1: high thermal *stability toward deg- radation to the bulk metal or other less active species; 2: easy accessibility or means of in-situ generation from easily available *precursor cpds.; 3: a high degree of stability in a given cat. cycle, which normally involves many reaction steps changing the composition and structure of the organometallic species. Once these prerequisites are fulfilled, the basic working principle is as follows. 1: Certain ligands stabilize the metal from being precipitated as (inactive) bulk metal (e.g., as metal black), thus keeping the metal in a molecular, structurally precisely defined environment; 2: other – or the same – ligands provide active *coordination sites (*vacant sites) at the metal center to start and continue the cat. reactions (Figure, examples **1** to **3**).

1

R = alkyl, aryl,
     alkoxy,
     aryloxy

n = 2, 3, or 4

2

R = alkyl, aryl,
     alkoxy, aryloxy

3

R = alkyl
R$'$ = alkyl, aryl,
     alkylaryl

The ubiquitous *phosphine *ligands can both stabilize and activate the cat. active metal through dissociation while the N-heterocyclic *carbene of example 3 only stabilizes the metal but does not dissociate from it; therefore, the main activation path comes in this case from auxiliary ligands of the (dissociating) phosphines.

The position of the respective metal in the periodic table gives some orientation regarding its organometallic chemistry and thus the applicability in certain types of cat. convs. Thus, the electron-poor oxophilic metals such as Ti, Hf, or Zr are not suitable for CO reactions while electron-rich metals such as Co, Rh, and Pd are excellent *carbonylation metals. Re, W, Mo, and Ru are successful in alkene *metathesis; Pd has recently been found suitable in *alternating copolymerizations. Sterically strained, *chiral organometallics (e.g., *ansa-metallocenes) yield *stereoselec-

tivity in reactions like propylene *polymerization. Metal-attached ligands may or may not participate in cat. transformations. In cats. **2** and **3**, the H-, CO-, and =C(H)C$_6$H$_5$ ligands initiate the reaction sequence and become part of the prod. In the case of *polymerization (cpd. **3**), the initiating group ends up as the terminus of the growing C-C chain. In many if not most cases, organometallic cats. undergo the basic steps of *oxidative addition, *reductive elimination, and frequently intramolecular migration. **F** composés organometallique; **G** Organometallverbindungen.

W.A. HERRMANN

Ref.: W.A. Herrmann (Ed.), *Synthtic Methods in Organometallic and Inorganic Chemistry*, Thieme, Stuttgart 1995–1998; Abel/Stone/ Wilkinson; J. Buckingham, *Dictionary of Organometallic Compounds*, Chapman and Hall; London 1984; Parshall/Ittel; *Angew.Chem.Int.Ed.Engl.* **1999**, *38*, 1194; cf. also *organotransition metal cpds.

organoperoxides → peroxides

organosilanes
are important for the preparation of silylated synthons as reactive *intermediates. Mono- up to tetra*alkyl*silanes are manuf. through the reaction between the respective halosilanes and lithiated alkyl cpds., with alkylmagnesium halides (*Grignard reaction), alkyl halides acc. to Wurtz-Fittig, or by the *addition of silanes to alkenes (*hydrosilylation). *Vinyl*silanes are accessible by direct procs. from vinyl chloride, metallation reactions, *hydrosilylation of alkynes, or the *hydrogenation of alkynylsilanes over *Raney Ni. *Allyl*silanes are mainly manuf. by reductive *silylation or *hydrosilylation of dienes. *Alkynyl*silanes can be obtained by silylation of metallated alkynes or by elimination reactions of halogenated alkanes or alkenes in the presence of silylating agents. **E=F**; **G** Organosilane.   B. MARCINIEC

Ref.: K.M. Lewis, D.G. Rethwish (Eds.), *Catalyzed Direct Reactions*, Elsevier, Amsterdam 1993, p. 1; *J.Catal* **1994**, *147*, 15; E.W. Colvin, *Silicon Reagents in Organic Synthesis*, Academic Press, London 1988.

organotransition metal compounds
a huge class of inorganic-organic cpds. with ap-

prox. 120 000 derivatives known by now. The first example was the *Zeise salt. Later, metal alkyls (*Frankland), metal carbonyls (*Mond), and *complexes containing alkynes, *aromatics, *carbenes, carbynes, nitrosyl, *phosphines, etc. followed. *Transition metals are compatible with organic *ligands (usually classified by the electron count, e.g., 1e, 2e, 6e ligands) in various oxidation states, for example $[Re(CO)_4]^{3-}$, $Re_2(CO)_{10}$, CH_3ReO_3, depending on the electronic situation: high oxidation states stabilize bonds with π/σ-donating ligands (e.g., alkyl in addition to oxo), low oxidation states favor π-acceptor ligands like CNR, CO, carbenes, etc. **F** composés organique des métaux de transition; **G** organische Übergangsmetallverbindungen.

W.A. HERRMANN

Ref.: J.P. Collman, L.S. Hegedus, J.R. Norton, R.G. Finke, *Principles and Applications of Organotransition Metal Chemistry*, University Science Books, Mill Valley 1987; C.H. Elschenbroich, A. Salzer, *Organometallics*, VCH, Weinheim, 1989; R.H. Crabtree, *The Organometallic Chemistry of the Transition Metals*, Wiley, New York 1988; C.M. Lukehart, *Fundamental Transition Metal Organometallic Chemistry*, Brooks/Cole, Monterey 1985; A. Yamamoto, *Organotransition Metal Chemistry*, Wiley, New York 1986; Parshall/Ittel; Gates; Cornils/Herrmann-1.

ortho-hydrogen (o-H_2) is one of two modifications of H_2 with parallel nuclear spin ($I = 1$) with odd rotational quantum numbers (p-H_2: antiparallel nuclear spin, $I = 0$, cf. *parahydrogen). At and above rt, H_2 is a mixture consisting of essentially three parts of o-H_2 and one part of p-H_2; the fraction of the latter can be enriched via equilibration in the presence of a cat. (such as *activated carbon) at low temps. Orthohydrogen labeling is also possible and equivalent to *parahydrogen labeling, but it yields a lower enhancement of the *NMR signals, which occur with an inverted phase relative to p-H_2 labeling. **E=F=G**. J. BARGON

Ref.: A. Farkas, *Orthohydrogen, Parahydrogen, and Heavy Hydrogen*, Cambridge University Press, Cambridge 1935; P.W. Atkins, *Concepts in Physical Chemistry*, Freeman & Co., New York 1995.

orthometallation The o. reaction is an intramolecular oxidative *addition of an aryl or aliphatic C-H bond to a metal forming a *chelate ring with an M-C bond. There are numerous examples of o. of *ligands containing P, N, and O atoms to a variety of transition metals.

$$Pd(O_2CCH_3)_2 \;+\; P\!\!\left(\!\!\text{\Large\textcircled{}}\!\!\right)_3 \xrightarrow[-HO_2CCH_3]{\text{toluene}} \left[\text{\Large\textcircled{}}\!\!-\!\!Pd\!\!<\!\!^{OAc}_{P(o\text{-tolyl})_2}\right]_2$$

The o. of Pd *complexes is particularly important due to its utility in organic synth. For example, *palladacycles are efficient cats. for the *Heck vinylation of aryl halides. Benzylamines react with Li_2PdCl_4 under mild conds. to form stable orthopalladated complexes in high yields. The resulting aryl Pd(II) complexes undergo a variety of *ortho* functionalization reactions in which the aryl group is attached to a number of functional groups; e.g., orthopalladated benzylamines insert into conjugated enones. Orthopalladated species also undergo electrophilic cleavage with acyl halides, in which the Pd is replaced by the corresponding acyl or aryl group. The result is a regiospecific *acylation or *alkylation of aromatic systems. Synth. applications of *cyclometallation usually involve aromatic systems. **E=F**; **G** Orthometallierung.

G. STARK, M. BELLER

Ref.: *Eur.J.Inorg.Chem.* **1998**, 29; J.P. Collmann et al., *Principles and Applications of Organotransition Metal Chemistry*, University Science Books, Mill Valley 1987.

oscillations Os. in catalyzed reactions or biological systems are rhythmic changes of colors (e.g., the hom. cat. *Belousov-Zhabotinsky reaction), temps. (gas reactions in flames, combustion), or currents (het. reactions of electrochemistry). They are important in *autocatalytic procs. with *back-donation in multicomp. and multivariable coupled systems. B. CORNILS

Ref.: Kuramoto, *Chemical Oscillations, Waves, and Turbulence*, Springer, Berlin 1984; *Angew.Chem.Int. Ed.Engl.* **1978**, *17*, 1; *J.Am.Chem.Soc.* **1976**, *98*, 4345; Robertson, *Biological Oscillations*, Halsted, New York 1977.

osmium as catalyst metal Os is best known for the cat. asymmetric reactions developed by *Sharpless. Asymmetric *dihydroxylation is a typical case of *ligand-accelerated cat. In the presence of chiral *cinchona ligands the *oxidation of olefins by OsO_4 proceeds in a biphasic reaction mixture (*two-phase cat.). The osmylation takes place in the organic layer giving rise to an Os(VI) glycolate. *Hydrolysis leads to a 1,2-diol and an Os(VI) species, which is reoxidized in the aqueous layer to OsO_4, generally by $K_3Fe(CN)_6$. Most alkenes, even sterically hindered or electronically deactivated ones, can be converted to diols with high *ees. *Aminohydroxylation (hydroxyamination) also proceeds over Os cpds. Os carbonyl *clusters such as $Os_3(CO)_{12}$ are known as cat. *precursors for a variety of organic reactions such as alkene *isomerization, alkyne *cyclotrimerization, C-N bond activation, *hydroformylation, and CO *hydrogenation reactions. OsO_4 cats. the *ROMP reaction of norbornene. The *epoxidation of alkenes proceeds with a variant of *Milas' reagent (with OsO_4 instead of WO_3). Alkanes, alkenes, or alcohols are photooxidized by Os (VI) complexes. Using Os, care has to be taken (*toxicity in cat.). **F** Osmium comme métal catalytique; **G** Osmium als Katalysatormetall.   T. STRASSNER

Ref.: *J.Mol.Catal.A:* **1995**, *96*, 231; *Chem.Rev.* **1994**, *94*, 2483 and 993; Cornils/Herrmann-1; *Houben-Weyl/ Methodicum Chimicum*, Vol. XIII/9a; *Coord.Chem. Rev.* **1982**, *35*, 41.

Ostwald Wilhelm (1853–1932), professor of physical chemistry in Leipzig (Germany). Worked on thermodynamics and cat. phenomena. One of the founders of scientific catalysis. Nobel laureate in 1909.   B. CORNILS

Ref.: G. Ostwald, *Wilhelm Ostwald, mein Vater*, 1953.

Ostwald ripening   → ripening

overall rate of reaction   the rate at which the reactant(s) are transformed into prods. of whatever kind. Thus the prods. may be either desired or unwanted, and may embrace the formation of byprods. that are toxic to the catalyst (*autogenic toxins). If the reactants comprise three kinds of molecule, A, B, and C, such that A + C → X and B + C → Y (X and Y being prods.), the overall rate of reaction is given by the combined rates of disappearance of A and B. In this case, the two rates may not be the same, their relative values depending upon their adsorption coefficients, i.e., the amounts of *surface that each covers.

Where the molecule C can react with A to form either X or Y by the procs. A + C → X and A + 2C → Y, the rates of removal of A and of C will not be the same. Very often a third reaction is possible, viz., X + C → Y where X is the desired prod. and Y is undesired, the degree of *selectivity with which X is formed depends upon the relative adsorption coefficients of A and X (see also *rate of reaction, *selective hydrogenation). **F** vitesse totale de la réaction; **G** Gesamtreaktionsgeschwindigkeit.   G.C. BOND

overlap   Quantum theory describes an electron with a mathematical function called an orbital which depends on four quantum numbers. Three of them represent the extension in 3D space, while the fourth determines the spin. The square of the orbital gives the probability of finding the electron in the given space. Although an orbital decays exponentially with its distance from the nucleus, it has a finite value even far away from it. The spatial integral over the product of two orbitals is called the overlap. The numerical value can be exactly unity, if the two orbitals are at the same nucleus, and if they differ only by the spin.

The value can be zero if positive and negative values of the two orbitals cancel by symmetry, e.g., an s orbital and a p orbital at the same nucleus. For all other cases the overlap has values between zero and one. Note that the integration over the spatial and the spin coordinates of two electrons having different spins and occupying the same orbitals is zero, because it violates the Pauli principle. Cf. *semi-empirical methods. **E=F=G**.   G. FRENKING

O/W   oil in water (emulsions)

oxazolines (bis[oxazolines]) belong to the C_2-symmetric *bidentate *ligands, having a one-carbon spacer between the oxazoline rings. Os. are readily available from *amino alcohols and 1,3-diketo cpds. These ligands form six-membered metal *chelates where the substituents on the ring carrying the *chiral information are close to the metal center (e.g., Cu, Fe, Pd, Ru, Zn). Their application comprises asymmetric *cyclopropanation, *Diels-Alder reactions, nucleophilic *addition reactions, *Mukaiyama aldol reactions, *allylic substitutions, and aziridination of alkenes and imines. The Rh complexes of the *tridentate pyridyl-bridged 2,6-bis (oxazolin-2-yl)pyridines (pybox, cf. *chiral pool) serves in *hydrosilylation of aldehydes. **E=F; G** Oxazoline.   H.W. GÖRLITZER

Ref.: *Tetrahedron Asymmetry* **1998**, *9*, 1; *Angew.Chem. Int. Ed.Engl.* **1994**, *33*, 497.

OXD process   → Phillips procs.

oxethylation   → ethoxylation

oxidation 1-heterogeneously catalyzed: Het. *oxidations are cat. by 1: metals (*oxidation of NH_3 over a Pt) or 2: *metal oxides (H_2SO_4 by oxidation of SO_2 to SO_3 over V oxide, *contact proc.). Under experimental conds. all oxidation cats. are able to change their valence state at the *surface easily. The need for variable valence states of the cat. material makes some of *transition metals into ideal oxidation cats. For example, all the oxides that are active in oxidation cat. are *semiconductor oxides with metals like Mo or V. Substoichiometric *phases are believed to be cat. active. Cat. oxidation provides an easy route for the functionalization of *hydrocarbons. More than 20 % of industrial organic materials are obtained by cat. oxidation. Most oxides used for the oxidation of organic cpds.

are transition metal oxides or mixtures of these. For example propene is oxidized over an Mo-Bi-Fe oxide to acrolein, while acrolein is oxidized to acrylic acid over an Mo-V oxide (similarly to the procs. of *Asahi, *BASF, *Nippon). Commercial procs. comprise, inter alia those of *Alusuisse, *Amoco, *BASF, *Hüls, *Mitsubishi (*phthalic, *terephthalic, or *trimellitic acids), *Nippon (and other *Bashkirov oxidations), *BP or *Celanese (oxidations of hydrocarbons), *Mitsubishi, *Sumitomo, *UCC sorbic acid (oxidation of aldehydes), and various *ammoxidation procs. Cf. also *catalytic oxidation.   R. SCHLÖGL, M. DIETERLE

Ref.: Bielanski/Haber; Centi/Triforo; Strakul.

2-homogeneously catalyzed: While *het.* cat. *oxidation is commercialized widely (e.g., manuf. of *sulfuric or *nitric acid, *phthalic anhydride, *EO, etc.) *hom.* cat. oxidations are usually not of pronounced economic importance. However, there are some hom. cat. procs., mostly for the introduction of functionalities into organic cpds. Different oxidizing agents, such as oxygen, *hydrogen peroxide, alkyl *hydroperoxides, iodosobenzene, *hypochlorite, and even chlorine, can be activated by *transition or main group element cats. for reaction with C-C double bonds, C-H bonds, aromatic cpds., hydrogen, or heteroatoms. These procs. are key steps in the manuf. of *epoxides (procs. of *Shell or *ARCO), *cis-* or *trans-*diols (*dihydroxylation), carboxylic acids and the corresponding anhydrides (e.g., *AA and *AAA, *terephthalic acid), phenols (*Hock reaction), quinones, aliphatic alcohols (*Bashkirov reaction, procs. of *Nippon Shokubai, *Hüls C_{12} acid), aldehydes (*Wacker proc.), ketones, vinyl chloride, or *hydrogen peroxide (*anthraquinone proc.).

Additionally to this, cat. lab scale procs. with enantioselective transformations are interesting, e.g., the enantioselective *epoxidation of allylic alcohols using Ti alkoxides/*TBHP/ dialkyl or aryl tartrates (cf. *Sharpless epoxidation) or enantioselective *cis-*hydroxylation in the presence of cinchona alkaloids.

          W.R. THIEL

Ref.: Barton/Martell/Sawyer; Bielanski/Haber; Siman-di; Sheldon/Kochi; Strakul.
F oxidation avec catalyseur homogène (hé-térogène); **G** Oxidation, homogen (hetero-gen) katalysiert.

oxidation of ammonia   The cat. *oxidation of ammonia by atmospheric oxygen to give NO (*Ostwald proc.) is one of the most effective cat. reactions and is utilized for the prod. of nitric acid. It consists of three consecutive steps. 1: The oxidation of ammonia by atmospheric air to give NO acc. to $4 NH_3 + 5 O_2 \rightarrow 4 NO + 6 H_2O$ ($\Delta H = -940$ kJ). 2: The oxidation of NO by air to give nitrogen dioxide acc. to $2 NO + O_2 \rightarrow 2 NO_2$. 3: The reaction of NO_2 with water to give nitric acid acc. to $3 NO_2 + H_2O \rightarrow 2 HNO_3 + NO$. In addition, ammonia is oxidized on a small scale using pure oxygen and steam for the prod. of pure *hydroxylamine.
The oxidation of NH_3 is carried out by means of a cat. at 800–950 °C at atmospheric press. or increased press. of up to 1.2 MPa. Besides the desired reaction acc. to step 1, several side and consecutive reactions can occur:

$$4 NH_3 + 4 O_2 \rightarrow 2 N_2O + 6 H_2O \quad \Delta H = -1140 \text{ kJ}$$
$$4 NH_3 + 3 O_2 \rightarrow 2 N_2 + 6 H_2O \quad \Delta H = -1268 \text{ kJ}$$
$$4 NH_3 + 6 NO \rightarrow 5 N_2 + 6 H_2O \quad \Delta H = -1808 \text{ kJ}$$
$$2 NO \rightarrow N_2 + O_2 \quad \Delta H = -90 \text{ kJ}$$

It is therefore important that the *residence time of the reacting gases should be very low. This requirement is reached by the use of Pt or Pt-Rh (3–10 % Rh) *gauze cats. having about 1000 meshes per cm^2 resulting in a contact time of approximately 10^{-3} s (*gauze reactors). **F** oxidation de l'ammoniaque; **G** Ammoniakverbrennung. R. SCHLÖGL, A. SCHEYBAL
Ref.: Twigg, p.470; Ertl/Knözinger/Weitkamp, Vol.4, p.1748; *Ind.Eng.Chem.Res.* **1990**, *29*, 1125; Farrauto/Bartholomwe, p.480.

oxidation of hydrocarbons   with air or oxygen is a very broad field. These reactions may produce utilizable energy (combustion) and/or targeted prods. which are commercially important. Many oxidations proceed satisfactorily without added cats., particularly those that produce peroxides as final prods. or *intermediates. Others may require a cat. or give a better prod. distribution when cats. are used. O. may be classified as vapor- or liquid-phase (VPO or LPO), cat. or non-cat., and homogeneous or heterogeneous. Non-cat. o. proceeds primarily by *homolytic mechanisms (one-electron change) while some cat. os. may include *heterolytic mechanisms (two-electron changes) as well.
Cat. hom. VPO has not proven viable for commercial prod. of chemicals. Examples of major prods. from het. cat. VPO include *ethylene oxide (from ethylene; Ag cat.), *maleic anhydride (*n*-butane; *mixed oxide cats.), acrolein/*acrylic acid (propylene; mixed oxide cats.), butadiene (*n*-butane; Fe_2O_3, Zn, Mn, Mg, or Bi/Mo oxide cat.), *vinyl acetate (ethylene and acetic acid; Pd, BiO_2/NaOAc cat.), or *phthalic anhydride (*o*-xylene; mixed oxide cat.). Examples of major prods. from cat. hom. LPO include *terephthalic acid (*p*-xylene; Co, Mn, Br^- cats.), cyclohexanol/cyclohexanone (*KA oil from cyclohexane; Co, Mn, [HBO_2] cat.; cf. *Bashkirov reaction), acetaldehyde (ethylene; $PdCl_2$/$CuCl_2$ cat.), *acetic acid (*n*-butane/naphtha; Co, Mn, Cr cat.), or benzoic acid (toluene; Co cat.). Cat. het. LPO is ambiguous since the effective cat. may be material dissolved from the heterogeneous cat. **F** oxidation des hydrocarbures; **G** Kohlenwasserstoffoxidation.

C.C. HOBBS
Ref.: Ullmann, Vol.A18, p.261; Kirk/Othmer-1, Vol.13, p.682; Cornils/Herrmann-1, p.512; Ertl/Knözinger/Weitkamp, Vol.5, p.2253; McKetta-1, Vol.26, p.352.

oxidation of sulfur dioxide   → sulfuric acid synthesis

oxidation-reduction   → redox reactions

oxidative addition   Oa. is part of *ligand exchange: a metal or a metal *complex (with the metal as *central atom) reacts with a dissociable substrate A-B to form a new *complex in which in the formal oxidation state of the metal is increased. Both two-electron and one-electron oa. are known. While the oas. increases the *coordination number of the metal centre, the reverse reaction, *reductive elimination, decreases the coordination number (Equ., cf. *arene coupling).

Depending on the substrates the oa. is the rate-determining step of the *cat. cycle. Oa. to H-H, C-H, Si-H, and C-X bonds (X = Cl, Br) is important for organic synth. as well as for commercial operation. The oa. of hydrogen to complexes of Rh, Ni, Pd, Ru, etc. is an important step in alkene, alkyne, and arene *hydrogenations, but also for *hydroformylations or *Fischer-Tropsch reactions. While oa. to H-H or C-M bonds (*transmetallations) is easily achieved, the selective activation and subsequent cat. functionalization of C-H and C-C bonds is extremely difficult. **F** addition oxidative; **G** oxidative Addition.   M. BELLER

oxidative ammonolysis → ammoxidation

oxidative carbonylation Under the conds. of oc. (CO, presence of alcohols, oxygen, cat. $PdCl_2$, $CuCl_2$ or other *transition metal cats.; Cu as *co-catalyst for the reoxidation of the Pd(II) species from the Pd(0) formed, esters of α,β-unsaturated carboxylic acids, β-alkoxy esters, and 2-substituted dialkyl succinates are formed from esters:

$$CH_2=CH-R^1 + CO + R_2OH \rightarrow R^2O-CHR^1-CH_2-COOR^2 + R^2OOC-CHR^1-CH_2-COOR^2$$

The reaction of ethylene with CO and oxygen using alcohols as solvents and $PdCl_2/FeCl_3$ as the cat. system gives acrylates or dialkyl succinates in fair yields. The water formed is removed by water-trapping agents such as orthoformates or *AAA. Intramolecular ocs.

are observed if an olefinic double bond and a hydroxyl group are in appropriate positions relative to each other. *Cyclization reactions occur preferentially to substituted pyran ring systems. The oc. of alkenes leading to α,β-unsaturated carboxylic acids may be substituted for industrial procs. such as the carbonylation of acetylene or the *oxidation of propene, e.g., the synth. of crotonic acid could be substituted by a single-step oc. of propene. Propionic acid can be synth. by analogous oc. of ethylene. Considerable interest has been paid to the oc. of styrene cinnamic acid derivatives. The oc. of alcohols is used for the manuf. of dialkyl carbonates (*Enichem, *Ube procs.). Oc. of aniline yields *MDI (*Asahi proc.). Cf. also *alkoxycarbonylation. **F** carbonylation oxidative; **G** oxidative Carbonylierung.

R.W. FISCHER

Ref.: Cornils/Herrmann-1, Vol. 1, p. 169; Falbe-1; *Cat. Surveys from Japan*, **1997**, *1*, 77; Bayer AG, US 5.821.377; *J.Mol.Catal.A*: **1999**, *148*, 289.

oxidative carboxylation → carboxylation

oxidative coupling of methane Coupling of methane mainly to ethylene and partly to ethane and some higher hcs. has been applied industrially at temps. of at least 1600 K for thermodynamic reasons. At the former Hüls AG the reaction was carried out in an electric arc. This type of operation has, however, been terminated for reasons of economics and operability. To overcome the thermodynamic limitations the oc. of methane has been explored extensively. As cats. numerous *mixed metal oxides have been applied which are mainly based on alkali-doped alkaline-earth oxides and alkaline-earth-doped rare earth oxides; a somehow different cat., i.e., $Mn-Na_2WO_4-SiO_2$, has been used successfully. Best *selectivities to C_2 hydrocarbons achieved amount to approx. 85–90 mol % at approx. 10–15 % methane conv.; with increasing methane conv. requiring a higher oxygen partial press., selectivity decreases significantly. Thus, the maximum yields obtained so far amount to about 25 mol%. These data refer to a one-through operation of the cat. reactor. Improvements appear possible when a recycle operation is applied in which low convs. are accepted and then after separation of the desired prods. unconverted methane is returned. This operation is at present not economic. **F** couplage oxidatif de methane; **G** oxidative Kupplung von Methan.

M. BAERNS

Ref.: *J.Catal.* **1982**, *73*, 9; *Chem.Ztg.* **1983**, *107*, 223; *Nature (London)* **1985**, 721; *Angew.Chem.Int.Ed.Engl.* **1995**, *34*, 970; Ertl/Knözinger/Weitkamp, Vol. 4, p. 1843.

oxidative coupling of phenols is cat. by

Cu-amine *complexes, acc. to the work of Hay et al. 2,6-Disubstituted phenols yield poly(1,4-phenylene ether) besides small amounts of substituted diphenoquinone. **F** couplage oxidative de phénol; **G** oxidative Kupplung von Phenol.   B. CORNILS

Ref.: *J.Mol.Catal.A:* **1999**, *140*, 241 and *148*, 289; Mark/ Bikales/Overberger/Menges, Vol. 13, p. 1.

oxidative dehydrogenation a variant of

the general *dehydrogenation reaction (*oxydehydrogenation) and includes the addition of O, which facilitates the abstraction of H. Od. is exothermic and can thus be realized at lower temps. There are different types of ods., depending on the substrate and the cat. Common to all mechanisms is a multistep reaction path, on which C-H bonds are broken consecutively. Od. differentiates itself from dehydrogenation reactions by the formation of water from the abstracted H atoms. This H abstraction *per se* is already an *oxidation reaction of the organic substrate. Of course, in od. reactions, oxygen may also be inserted into the organic substrate during a reaction step after the H abstraction.

Different types of reactions are described. 1: *Inter*molecular od. in which the two H atoms originate from two different organic molecules, which dimerize during the course of the reaction (e.g. *oxidative coupling). 2: *Intra*molecular od. with the two H atoms originating from the same organic molecule and forming a C=C double bond (e.g., *Union

Carbide's Ethoxene proc., butene from butanes). 3: An O atom may be added to the substrate after the first H-atom abstraction. After the oxygen addition, the second H atom is abstracted to give water and the oxygenated substrate (*acrylic acid synth., *Lummus ALMA proc. to *MAA).

The od. of alkanes is a commercially important reaction for the prod. of alkenes. Because the direct dehydrogenation is thermodynamically limited, the coupling of this reaction with the exothermic formation of water in the od. shifts the thermodynamic equil. far to the right and the overall reaction enthalpy becomes exothermic. The od. of alkenes to dienes and organic acids, such as butadiene and acrylic acids, is a commercial proc. In the first step of the reaction, the α-H atom is abstracted to form an allylic *intermediate, which is also the slow rate-determining step. It is suggested that this H abstraction occurs via a *surface *acid-base reaction, in which an oxide ion has basic properties and a coordinatively unsaturated metal (*cus) center acts as a *Lewis acid (procs. of *Dow or *Snamprogetti to butadiene). Ods. of alkylaromatics are also important, e.g., the dehydrogenation of ethylbenzene to styrene (procs. of *BASF, *Lummus/UOP). The mechanism suggested describes this reaction as a direct transfer of two hydrogen atoms to an activated oxygen molecule at a Lewis site. Od. of alcohols is another important class. The synths. of *formaldehyde from methanol, glyoxal from glycol, and acetone from propanol are commercial and are carried out over metallic Ag or Cu cat. **F** déhydrogénation oxidative; **G** oxidative Dehydrierung.   G. MESTL

Ref.: Ertl/Knözinger/Weitkamp, Vol. 5, pp. 2140, 2274.

oxidative Glaser coupling → Glaser

coupling reaction

β-oxidation → fatty acid degradation

oxide catalysts Many cats. contain oxides,

oxidic *supports such as γ-Al$_2$O3, SiO$_2$, or

TiO$_2$, or oxidic active comps., e.g., *hetero-poly acids or V pyrophosphate, etc., or are active themselves such as *zeolites in *acid (-base) cat. The role of the support oxide is, first, to stabilize a high *dispersion of the active cats. This guarantees maximum cat. *activity. Therefore, oxides with high specific *surface areas are usually used as supports, such as *aerogels, amorphous *silicas, *titania, or *γ-alumina. Secondly, the support oxide may also directly affect the cat. activity and/or *selectivity, either by playing a direct role in cat., for example in *bifunctional cat. due to its acid or basic surface OH groups, or by its direct interaction with the active cat., as for example in the ease of *strong metal-support interaction (SMSI). In bifunctional cat. the reactant, for example, may be activated by the acidic properties of the support oxide in the first reaction step and selectively oxidized on the active comp. in the second. In SMSI, the reduced support oxide, e.g., a reduced titania species, spreads onto the active noble metal cat., thereby modifying its electronic properties and possibly the geometric constraints of the active metal centers. Thirdly, the molecular structure of active cats. supported on oxidic carriers is also affected by the chemical properties of the support. Thus, the surface pH value of the oxide support defines, e.g., the structure of Mo cats. The surface of amphoteric *alumina, for example, is positively charged under typical impregnation conds., therefore, the interaction with negatively charged *polyoxymetallate anions leads to strong adsorption. Finally, the mechanical properties of support oxides are of high technical relevance, e.g., for long-term stability operation in fluidized-bed reactors.

Active comps. in cats. may also be oxidic cpds. *Selective (partial) *oxidation cats. usually contain a multitude of oxidic cpds., which are necessary to fine-tune the selectivity to the wanted prod. and increase the reactant conv. as much as possible. The distinction between active oxide cpds. and *promotor oxides can often not be drawn for this class of cat. materials, especially not in the case of multiele-ment cats. with comparable amounts of different oxides. **F** catalyseurs d'oxide; **G** Oxidkata-lysatoren.

G. MESTL

Ref.: C.N.R. Rao, B. Raveau, *Transition Metal Oxides*, VCH, Weinheim 1995.

oxides cpds. of oxygen wherein O represents the more strongly electronegative comp. It is to differentiate between os. with O in the oxidation state –2 and between other special classes, namely *hyperoxides* (formal oxidation state –1/2, e.g., in KO$_2$), *peroxides* (oxidation state –1, e.g. in H$_2$O$_2$) and *subox-ides* with low oxygen contents where bonds are usually present between the atoms of the other constituent (e.g., Cs$_7$O). Os. are further differentiated according to their reaction with water or in aqueous solutions. Acidic os. dissolve with formation of acids or anions (most oxides of non-metals, e.g., CO$_2$, SiO$_2$, NO$_2$, P$_2$O$_5$, SO$_3$, and *transition metal oxides with their highest oxidation state, e.g. V$_2$O$_5$, CrO$_3$, Mn$_2$O$_7$); basic oxides form hydroxides or dissolve by forming bases or cations (oxides of most metals, especially electropositive ones, in their low and medium oxidation states, e.g., Na$_2$O, MgO, lanthanide oxides); *amphoteric* oxides form cations in acidic and anions in basic milieu (e.g. ZnO, Al$_2$O$_3$, As$_2$O$_3$); *indif-ferent* (neutral) oxides do not give acidic or basic reactions in water (e.g., CO, N$_2$O). Ternary and higher oxides are usually more stable when composed of acidic and basic comps. rather than from constituents from the same class.

Oxidic cpds. span a wide range of properties. Under ambient conds., oxides of non-metals are mostly gaseous, whereas those of metals are usually solid. Liquid oxides are exceptions (e.g., H$_2$O, Mn$_2$O$_7$). Partial metallic bonding can be observed in transition metal oxides in low oxidation states. With regard to electronic properties, oxides cover the range from the best insulating materials (e.g., SiO$_2$) through *semiconductors (TiO$_2$, MnO, CoO, NiO, ZnO) to metallic conductors (TiO, NbO, ReO$_3$, *tungsten bronzes), and superconductors (e.g., $BaPb_{1-x}Bi_xO_3$ and high-T_c super-conductors as $YBa_2Cu_3O_{7-x}$).

Oxides have many applications. The *adsorption properties of oxides are strongly influenced by their *surface chemistry. Besides metals, oxides form the most important class of het. cats. They function as *acid-base and as *redox cats. In hom. cat., molecular transition metal oxides with the transition metal in a high oxidation state (e.g., *methyltrioxo-Re) can be used as cats. in olefin *metathesis and *oxidations. **F=E**; **G** Oxide.   P. BEHRENS

oxidoreductases

oxidoreductases cat. oxidoreduction of a wide variety of molecules. Many os. have synth. applicability by virtue of their substrate *specificity, but remain to be developed for large- scale synth. Among those *enzymes are: alanine dehydrogenase (EC 1.4.1.1) from *Bacillus* for *reductive amination; glycerol 3-phosphate dehydrogenase (EC 1.1.1.8) for the *oxidation of phosphonate and difluoromethylene phosphonate; Sepiapterin reductase (EC 1.1.1.153) from *Drosophila* and carbonyl reductase from *Mucor ambiguus*, both for *reduction of diketones. The latter reduces cyclic conjugated and cyclic ketones to diols. The former enzyme reduces cyclic substrates in addition to linear diketones to the diols. Dihydroxyacetone reductase transfers the hydride from the nicotinamide *cofactor to the *si* face of a variety of acyclic and cyclic substrates with high *enantioselectivity. **E=F=G**.   C.-H. WONG

Ref.: *J.Chem.Soc.Chem.Commun.* **1988**, 1169; *Eur.J. Biochem.* **1988**, *174*, 37.

oxidosqualene lanosterol synthetase
→ juveline hormone, *lanosterol

oximation

oximation formation of oximes, i.e., cpds. with the R_1R_2=NOH group by, e.g., oximation of ketones with hydroxylamine (*DSM/ Stamicarbon HPO process), UV-induced reaction between *hcs. and NOCl (*Toray PNC proc.), or selective *hydrogenation of nitro cpds. (*DuPont Nixan proc.).   B. CORNILS

oxiranes

oxiranes 1-by het. cat.: The prepn. of *epoxides (oxiranes) by direct *oxidation of alkenes would open a convenient route to a class of very useful organic intermediates. With the help of het. cats. only *ethylene oxide can be prep. in large-scale quantities. All higher oxiranes, in particular the next-larger molecule *propylene oxide, are until now inaccessible by het. cat. in any commercially feasible way. Ethylene epoxidation occurs over *alumina-*supported Ag with dioxygen as oxidant. *Promoters such as Cs, chloride (fed continuously as an organo chloro cpd.) and recently Re have improved the *selectivity of the proc. to values well over 90 %. The general explanation for the extreme difficulty of producing higher oxiranes is the unwanted function of all cats. used so far, of isomerizing the reactant into an allylic *intermediate (cf. *isomerization) which is easily highly oxidized by the aggressive atomic O species. The use of mild reaction conds. and the application of H_2O_2 as oxidant or the application of electrocatalysis (*electron transfer chain cat.) are current trends in this area of research.   R. SCHLÖGL

2-by hom. cat.: Os. as three-membered ring heterocycles are produced by *epoxidation of alkenes. "Oxirane" (ethylene oxide, *EO) is manuf. by gas-phase epoxidation of ethylene with O_2 or air over a *supported Ag cat. Methyloxirane (*propylene oxide, PO) is produced by epoxidation of propylene with an alkyl *hydroperoxide (*tert*-butyl or ethylbenzene hydroperoxide) in the presence of a hom. Mo cat. (*ARCO PO proc.) or a het. Ti(IV)/SiO$_2$ cat. (*Shell SMPO proc.). A wide variety of os. can be prepared by Mo(VI)-cat. epoxidation with *tert*-butyl hydroperoxide (*TBHP). Os. from allylic alcohols are best prepared using V(V) complexes as hom. cats. Many oxiranes can also be prepared by epoxidation with H_2O_2 in the presence of tungstate/phosphate or *ethyltrioxorhenium. Optically active os. are prepared by asymmetric epoxidation of prochiral alkenes, e.g., Ti(IV) tartrate ester complexes cat. the asymmetric epoxidation of (pro)chiral alkyl alcohols with TBHP. Mn(III) complexes of chiral *Schiff base complexes cat. the formation of optically active os. by epoxidation of unfunctionalized prochiral alkenes with NaOCl. The *enzyme *chloroperoxidase (CPO) cat. the asymmetric epoxidation of alkenes with H_2O_2. Optically active os. are also prod. by asymmetric ring opening of racemic or prochiral epoxides with water in the presence of a Co(III) complex of a chiral Schiff base or the *enzyme *epoxide hydrolase. **F** oxiranes via catalyse homogène; **G** Oxirane durch homogene Katalyse. R.A. SHELDON

Ref.: Ugo, Vol. 4, p. 3; G. Wilkinson et al. (Eds.), *Comprehensive Coordination Chemistry*, Pergamon, Oxforf 1987, Vol. 6, p. 317; Cornils/Herrmann-1, p. 411; *J. Am. Chem. Soc.* **1980**, *102*, 5974 and **1990**, *112*, 2801; *Science* **1997**, *277*, 936; *J. Mol. Catal. B: Enzymatic* **1998**, 5, 79.

Oxirane process manuf. of *propylene oxide by indirect oxidation/co-oxidation, developed jointly by Halcon and Atlantic Richfield (*ARCO). The two-step proc. consists of 1: the *peroxidation of an auxiliary cpd. (*reaction carrier, such as acetaldehyde, isobutane, or *ethylbenzene which is peroxidized) with air or oxygen at approx. 100 °C and with partial convs.; and 2: the transfer of half of the peroxide oxygen selectively to propylene with formation of propylene oxide at approx. 100–130 °C/2–6 MPa over Mo- or W-based cats. in horizontal two-stage reactors, having partitional reaction zones. The cats. are mainly hom.; fixed-bed cat. have been described, too (*Shell SMPO). The other half of the oxygen remains as secondary product O (as alcohol, ketone, or acid) with the reaction carrier and can be either recycled or converted by additional steps to a coproduct. The ratio between coproduct and PO is between 3:1 and 4:1 (in the case of *tert*-butanol) and 2.4:1 (styrene). The various proc. developments use different reaction carriers (Table). **F** procédé Oxirane; **G** Oxirane-Verfahren. B. CORNILS

Ref.: S. Patai, *The Chemistry of Peroxides*, Wiley, Chichester 1983; Kirk/Othmer-1, Vol. 20, p. 271; Ullmann, Vol. A22, p. 239.

Reaction carrier	Peroxidizing agent	Co-oxidized reactant or co-product
Acetaldehyde[1]	AcOOH	Acetic acid or anhydride
Isobutane[2]	$(CH_3)_3COOH$	*t*-Butanol, isobutene, MIBK, MA
Iopentane	C_5H_{11}-OOH	Isopentanol, *i*-pentene, isoprene
Cyclohexane	C_6H_{11}-OOH	Cyclohexanol/one (*KA oil)
Ethylbenzene	$C_6H_5CH(CH_3)$-OOH	Phenylethanol, styrene
Cumene	$C_6H_5C(CH_3)_2$-OOH	α-Methylstyrene

[1] Daicel development; [2] Texaco

Oxirane PO process → ARCO processes

oxo synonym for *hydroformylation.
Ketone formation, e.g., diethyl ketone from ethylene, was noticed by *Roelen when he discovered the *hydroformylation reaction, giving rise to the term "oxo". Ketone formation may occur with other starting alkenes but is negligible – so the term "oxo" is a misinterpretation. **F=G=E**. B. CORNILS

Oxo-D process → Petrotex process

oxo products prods. manuf. via direct oxo synthesis (*hydroformylation), such as aldehydes or alcohols – depending of the special variant of an oxo process. In the extended meaning prod. such as carboxylic acids, esters (e.g., *DOP from *2-EH and phthalic acid), or diols (e.g., neopentylglycol) are also called oxo prods. **F** produits d'oxo; **G** Oxoprodukte. B. CORNILS

oxo synthesis → hydroformylation

oxyacetylation → acetoxylation

oxyamination → hydroxyamination

oxychlorination a special proc. for the *chlorination of organic cpds. by O_2 and HCl (from waste gases). The major commercial application is the o. of ethylene yielding 1,2-dichloroethane (*EDC, the precursor or *vinyl chloride) at 200–300 °C/0.1–1 MPa with Cu(II) chloride as cat. in fixed or fluid-bed reactors. $CuCl_2$ chlorinates and CuCl is reoxi-dized by oxygen and HCl.

$$C_2H_4 + 2\,CuCl_2 \quad\quad\quad \rightarrow C_2H_4Cl_2 + Cu_2Cl_2$$

$$Cu_2Cl_2 + 2\,HCl + 0.5\,O_2 \quad \rightarrow 2\,CuCl_2 + H_2O$$

$$C_2H_4 + 2\,HCl + 0.5\,O_2 \quad \rightarrow C_2H_4Cl_2 + H_2O$$

A proc. in the liquid phase has been devel-oped by *Kellogg (180 °C); other variants are described by *Gulf (o. of benzene), *Lummus Transcat (*molten media), *PPG, *Raschig-Hooker (o. of benzene), and *Rhône-Pou-lenc. **F=E**; **G** Oxychlorierung.

S. TRAUTSCHOLD

Ref.: Ullmann, Vol. A6, p. 266 and Vol. A19, p. 306f; US 3.197.515 (1962).

Oxyclaus process   → Lurgi procs.

oxydehydration   formation of keto cpds. by the *dehydration (elimination of one mol-ecule of water) of *vic* glycols. The acid- cat. dehydration of 1,2-glycols is often followed by a rearrangement, e.g., the elimination of water from 2,3-dimethylbutane-2,3-diol (pina-col) with H_2SO_4 or *Lewis acids as cats. yields *tert*-butyl methyl ketone (pinacolone) in high yields (*pinacol rearrangement). The analo-gous dehydration of isobutylene glycol yields selectively *iso*-butyraldehyde. Dehydration of 1,4- or 1,5-diols gives cyclic ethers, especially if *tert*-hydroxy groups are involved.   **E=F**; **G** Oxydehydratisierung.   R.W. FISCHER

oxydehydrogenation   a special variant of the cat. *dehydrogenation in the presence of oxygen for, e.g., the manuf. of butadiene from butenes or butanes at high temp. (>600 °C, *Houdry Catadiene proc.), ethylene from ethane (*UCC Ethoxene), or *MMA from isobutyric acid (procs. of *Eastman, *Mitsu-bishi). In addition, oxygen facilitates the oxi-dative regeneration of the cat. (e.g., *Petrotex Oxo-D proc.; *Phillips OXD proc., *Standard Oil [NJ] butadiene). Cf. the general terms *dehydroxidation and *oxidative dehydro-genation. **F** oxydéshydrogénation; **G** Oxyde-hydrierung.   B. CORNILS

oxydesulfurization   a proc. for the non-cat. treatment of pyritic and organic S in coal with water and oxygen in the presence of al-kaline salts. **F** oxydésulfurisation; **G** Oxyde-sulfurierung, oxidierende Entschwefelung.

C. BIANCHINI

oxyfuel   gasoline with up to 15 % of *MTBE or 7.3 % of *ethanol.

oxygenases   → monooxygenase, *dioxygen-ase

oxygenation   the controlled introduction of oxygen (*oxidation), of OH groups (*hy-droxylation), or of -O-O- groups (peroxida-tion, formation of *peroxides or *hydroperox-ides) in molecules. All basic reactions can be catalyzed (by *transition metals, e.g., oxo *complexes) or *biocatalyzed (*oxygenases, *di- or *monooxygenases). **F** oxygénation; **G** Oxygenierung.   B. CORNILS

oxylonitrilases   (cyanohydrin synthetases) the class of *enzymes that cat. the *addition of HCN to aldehydes to form cyanohydrins (cys.). The substrate *specificity of these en-zymes is relatively broad, as illustrated by the several different aromatic aldehydes used as acceptors for   (R)-oxylonitrilase (or mandel-onitrile lyase) (EC 4.1.2.10). (R)-Cys. are the predominant prods. in these reactions; in aqueous solutions, however, cys. racemize and competing non-enzymatic cyanide addition results in prods. with low optical purity. To cir-cumvent this problem, the enzyme has been immobilized on cellulose or the reaction has been carried out in ethyl acetate. The result is products with higher yields (77–99 %) and im-proved optical purity (73–99 % *ee). Another method of generating cys. uses acetone cyano-hydrin as the HCN donor in the enzymatic synth. of several aliphatic cys., thus avoiding the use of the highly toxic HCN gas. In addi-tion to aromatic cys., aliphatic (R)-cyanohy-drins were prepared by this method.
(S)-Oxynitrilase (EC 4.1.2.11) was recently used in the synth. of several (S)-cy. deriva-

tives. The enzyme from *Sorghum bicolor* is active in ethyl acetate at a pH of 3–4, where racemization and non-enzymatic cy. formation can be minimized. **E=F**; **G** Oxylonitrilasen.

C.-H. WONG

Ref.: *Angew.Chem.Int.Ed.Engl.* **1987**, *26*, 458; *Tetrahedron Lett.* **1991**, *32*, 2605.

Oxyl synthesis manuf. of synth. alcohols by *hydrogenation of CO in the gaseous phase with fixed-bed cat. at 180–220 °C/1–3 MPa. *Syngas at a ratio of approx. 2:1 H_2/CO is introduced with 100–300 *GHSV onto precipitated Fe cats., doped with Cu, K, Ce, and/ or V; a gas recycle is advantageous. Alcohols and *hc. are formed in about 55:40 m/m ratio.

Main alcohol comps. are (in wt% of primary alcohol) C_1 0.1, C_2 9.9, C_3 6.9, C_4 5.8, C_5 4.9, C_6 4.3, C_7 23. Predominantly primary linear aliphatic alcohols are formed together with some ketones, esters, and carboxylic acids. The proc. has not been commercialized. **F** synthèse d'Oxyl; **G** Oxyl-Synthese.

C.D. FROHNING

oxymercuration the indirect *hydration of alkenes to alcohols with $Hg(OAc)_2$ and subsequent *reduction of the organomercury derivative with $NaBH_4$ (cf. *hydration).

M. BELLER

oy → optical yield, *ee (enantiomeric excess)

P

(P)-   as an abbr. in formulae or schemes, 1: a growing *polymer chain; 2: polymeric *support or a polymer backbone for the *immobilization of *hom. cats.

PA   1: → phthalic (acid) anhydride; 2: → polyacrylate; 3: polyamide

PAA   1: peracetic acid; 2: → phthalic acid anhydride

Pacol-Olex process   → UOP procs.

PACT   plasma and catalysis technoloy, → plasma catalysis techniques; *microwaves in catalysis

palladacycles *metallacycles of Pd. Ps. with C-Pd-C bonds are key intermediates in the mechanism of reactions via *C-H activation, intramolecular *Stille reactions, and *Heck-type *cyclizations, if β-hydrogen elimination is inhibited during the formation of tertiary or quaternary C centers. Generally, intramolecular Pd cat. reactions involving sequences of *oxidative addition and *transmetallation proceed via ps. Six-membered ps. are subject to facile *reductive elimination, whereas five-membered ps. are stable and can react further in a *domino proc. (Figure).
Heteroatom *ligands on Pd can undergo *cyclometallation and form ps. with an X-Pd-C structure via *C-H activation. Complexes of this type are thermally stable and with X = P exhibit unique cat. properties for Heck-type reactions. For ps. of the type X-Pd-X with two heteroatoms X such as N,P,O, or S attached to the Pd center, cf. *chelating ligands. **F=E; G** Palladacyclen.   V. BÖHM

Ref.: *Chem.Ber.Recueil* **1997**, *130*, 1567; *J.Organomet.Chem.* **1999**, *576*, 23.

palladium as catalyst metal   1-hom. cat.: Pd is the only element with a d^{10} electronic ground state and thus it is a unique cat. metal. Pd *complexes are usually stable, easy to handle, and non-toxic. Pd is less expensive than Rh, Ir, or Pt. Hom. Pd cat. reactions can be classified into oxidative reactions of Pd(II) and Pd(0)-mediated coupling reactions. In the case of Pd(II) it is reduced to Pd(0) in the course of the reaction and has to be reoxidized in situ; in the latter case the *cat. cycle starts and ends with Pd(0).
Pd cats. are used commercially in the *Wacker proc. Aromatic cpds. can also be oxidized, yielding either symmetric biaryls by *oxidative coupling or acetoxyaryls by *acetoxylation. $PdCl_2$ or Pd-C as cats. and butyl nitrite as a *reoxidant allow amines and alcohols to be oxidatively *carbonylated to oxamides or oxalate esters (cf. *Ube procs. to DMC, DMO). Pd(0) cats. have to be stabilized by *ligands. Pd(0) cpds. add aryl and alkenyl halides, triflates, and diazonium salts oxidatively, thus activating them for the reaction with alkenes in the presence of bases (*Heck reaction), and alkyl, aryl, and alkenyl organometallic cpds. (Sn, B, Zn. Mg; *couplings of *Stille, *Suzuki, *Negishi, *Grignard). Terminal alkynes (*Sonogashira reaction), disilanes,

sec-*phosphines and phosphine oxides, sec-amines (Buchwald-Hartwig reaction; cf. *amination), and *alkoxides can be coupled with aryl halides by similar procs.

Allylic cpds. such as allyl acetate and conjugated dienes form π-allyl-Pd(II) complexes from Pd(0). They can react with nucleophiles (*Tsuji-Trost reaction). Dienes can be selectively *dimerized in the presence of nucleophiles (*telomerization, *hydrodimerization, *Kuraray proc.). The *carbonylation of alkenes with Pd(0) cats. in the presence of alcohols gives saturated esters, and in the presence of H_2O carboxylic acids. The carbonylation of ethylene using $Pd(OAc)_2$/*dppp *ligand in the presence of acid in methanol results in the formation of *polyketones (*alternating copolymerization). V. BÖHM

2-heterogeneous: Pd shares properties with Pt as well as with Ni, it is cheaper than Pt and is one of the more important metals in the catalyst family. Pd metal solutes H considerably and cleaves the H-H bond with formation of Pd hydrides, which makes Pd a valuable *hydrogenation cat. Pd cat. the hydrogenation of multiple bonds of alkenes, acetylenes, and dienes (hydrogenation of the ethylene stream from steam crackers) as well as the *hydrogenolysis of C-C, C-O, C-X, and C-N bonds (commercially, e.g., in the *anthraquinone proc. to H_2O_2) or N-O to *hydroxylamines (cf. also *catalytic nitrate reduction). The *activity of pure Pd may be influenced by alloying with Sn. Supported Pd cats. are used for *acetoxylations (*VAM synth.), and as hom. cats. for the *Wacker proc. Pd is engaged in *fuel cells as well as in the *cat. combustion of chlorinated hcs. (Pd-Pt alloy on $LaCeCoO_3$). Named cats. include *Lindlar's and *Pearlman's cats. **F** le palladium comme métal catalytique; **G** Palladium als Katalysatormetall.

B. CORNILS

Ref.: Heck; Adams/Cotton; J. Tsuji, *Organic Synthesis with Palladium Compounds*, Springer, Berlin 1980; J. Tsuji, *Palladium Reagents and Catalysts*, Wiley, Chichester 1995; Farrauto/Bartholomew, p. 413; *Organic Synthesis* Vol. 46, p. 89; *J.Organomet.Chem.* **1998**, *555*, 141; *Acc.Chem.Res.* **1980**, *13*, 385; *Houben-Weyl/Methodicum Chimicum*, Vol. XIII/9b; *Coord.Chem. Rev.* **1982**, *35*, 143; Murahashi/Davies; Trost/Fleming,

Vol. 4, p. 833; Rylander; J.L. Malleron, J.-C. Fiaud, J.-Y. Legros, *Handbook of Palladium-Calalyzed Organic Reactions*, Academic Press, San Diego 1997.

palladium black serves mainly as a het. cat. for *hydrogenations. Hom. cat. with Pd(0) cpds. suffer from decomposition to aggregates of elemental Pd (*deactivation). The colloidal particles obtained are usually less reactive and can exhibit different *chemo- and *regioselectivities. Pd black is the aggregated form that is observed in most cases, although sometimes defined *colloids or even Pd particles with metallic luster can be obtained. Preventing early Pd black precipitation is possible by an excess of *ligands (which in turn reduces the reactivity) or by employing more stable cats. like *palladacycles. **F** noir de Pd; **G** Pd-Mohr (Pd-Schwarz). V. BÖHM

PAMP, pamp a *Horner-type *phosphine (a monophosphine related to DIPAMP), prep. from (–)-ephedrine by a multi-step synth. via the *chiral BH_3-protected *intermediate. PAMP has been proposed as Rh *ligand in *hydrogenations. O. STELZER

Ref.: *Tetrahedron Lett.* **1990**, *31*, 6357; Patai-2; *Acc. Chem.Res.* **1983**, *16*, 106.

PAN polyacrylonitrile

papain (EC 3.4.22.2) a *thiolprotease from papaya latex that prefers basic residues (Lys, Arg) in the S_1 site (carbonyl side of scissile bond) but has extremely broad *specificity. It shows poor activity until activation (reduction of the *active-site thiol) with mild reducing agents such as β-mercaptoethanol and metal *chelation with, e.g., *EDTA. It is quite stable to thermally induced denaturation and can withstand very high concs. of organic *solvents and denaturants, but unfolds rapidly in acid solution. It has been known for many years to function as an acyltransferase, and is used in peptide synth. reactions. **E=F=G**.

P.S. SEARS

Ref.: *Methods Enzymol.* **1970**, *XIX*, 226; *J.Org.Chem.* **1982**, *47*, 5300.

PAPS → 3'-phosphoadenosine 5'-phospho-sulfate

parahydrogen labeling a technique based upon parahydrogen- induced nuclear polarization (PHIP), the consequence of the PASADENA effect, which in turn is due to breaking the high symmetry of a nuclear spin isomer of H_2, typically of p-H_2, via a chemical reaction. When starting from pure p-H_2, the resulting *hydrogenation prods. are formed initially with selectively populated nuclear spin states, from which transitions to unoccupied levels, which were initially forbidden in the highly symmetric p-H_2, are now allowed in these less symmetric prods. Accordingly, PHIP causes strongly enhanced *NMR absorptions and emissions during in-situ investigations of hydrogenations with p-H_2. In principle, both para- and ortho-H_2 qualify, giving rise to complementary sequences of absorption and emission lines (phases) in the NMR multiplets of the hydrogenation prods., respectively. Since p-H_2 can be enriched more readily and since this isomer yields more intense PHIP signals anyway, it is used preferentially. The observation of PHIP during *in situ* NMR proves that in such systems the two atoms of H_2 are transferred synchronously (i.e., pair-wise). Due to the associated high signal enhancement (typically up to 10^4 fold), PHIP is a powerful tool to investigate not only hydrogenations but also any chemical reaction, provided that at least one of the reactants is instantly generated from p-H_2, a suitable precursor, and cat. Therefore, the technique of pl. represents an attractive way to investigate both cat. mechanisms and stoichiometric reactions of organometallic *complexes. PHIP is capable of identifying intermediates and yielding kinetic rates and is also attractive for screening cats. **F** marquage par para-hydrogène, **G** para-Wasserstoff-Markie-rung.   J. BARGON
Ref.: Cornils/Herrmann-1, p. 672.

Parex process → UOP procs.

Parr hydrogenator a lab-scale (up to 2 L volume) batch reactor for low-press. (approx. 0.3 MPa) *hydrogenation experiments. The Ph. is connected with a reservoir for some liters of hydrogen. Gas-liquid mass transport limitations are avoided by shaking agitation by means of motor-driven eccentrics. The progress of the reaction is monitored by the press. drop in the system which allow to be calcd. easily the gas consumed.   B. CORNILS
Ref.: R.L. Augustine, *Catalytic Hydrogenation*, Dekker, New York 1965, p. 8.

Parsons RM process a *methanation proc. for the conv. of H_2-enriched *syngas over Ni cats. in six succesive adiabatic fixed-bed reactors at 2.8 MPa. **F** procédé RM de Parsons; **G** Parsons RM-Verfahren.   B. CORNILS

partial hydrogenation → selective hydrogenation

partial oxidation (selective oxidation) 1-general: the formation or transformation of oxygen functions in organic molecules by cat. *oxidation, commonly with pure O_2, air, *hydroperoxides, or N_2O. For oxidation reactions with oxygen the O_2 molecule may be cat. activated to species such as singlet O_2, O_2^- (peroxidic), O•, which are sufficiently highly electrophilic to attack organic bonds with higher electron density (alkenes, aromatics), e.g., as in the o. of alkenes to *epoxides or of aromatics to acid anhydrides like *PA or *PAA. O_2^- species as nucleophilic reactants give the opportunity for oxygen *insertion reactions after activation of the organic molecule (e.g., by H abstraction). The o. itself is realized by redox steps of transition metal cations of the *redox cats. (*Mars-van Krevelen mechanism).
The o. *selectivity is often limited by the possibility of consecutive and parallel reactions and sometimes by the fact that splitting of C-H bonds in the educts is more difficult than in the derived prods. Important gas-phase selective os. are i: ethylene → *ethylene oxide over *supported Ag cats.(e.g., *Shell proc.); ii: propylene → *acrolein over Bi-Mo cats.

(*Degussa or *Shell proc.); iii: acrolein → *acrylic acid with Mo-V- containing cats. (e.g., *Nippon Shokubai proc.); iv: *iso*-butene (or *t*-butanol) → methacrolein over Bi-Mo cats. (proc. of *Asahi, *Escambia); v: methacrolein → *methacrylic acid on Mo-V-P-O *heteropolyacid cats.; vi: *n*-butane → maleic anhydride over $(VO)_2P_2O_7$ cats. (procs. of *Alusuisse, *Lummus ALMA, Mitsubishi); vii: *o*-xylene → *phthalic anhydride with V-Ti cats. (procs. of *BASF, *Rhône-Progil, *Wacker); viii: ethylene and acetic acid → *vinyl acetate (Pd cats.); ix: methanol → formaldehyde over Ag, Fe, Mo, or Cu cats. (procs. of *Haldor or *Perstorp Formox); x: ethylene → acetaldehyde (over hom. Pd-Cu systems, *Wacker proc.); xi: propylene → *propylene oxide with hydroperoxides over Ti or Mo cats. (*ARCO, *Chemsystems, *Shell SMPO, *Texaco); xii: aliphatic aldehydes → acids with metal cats. (*aliphatic aldehyde o.); xiii: aromatics → quinones over Re cats. (*Re as catalyst metal); and xiv: alkylaromatics → arylcarboxylic acids such as *terephthalic acid over Co-Mn cats. New and interesting developments are the introduction of *Ti silicalite cats. at moderate temps. (*Eniricerche) and the *epoxidation of butadiene over Ag cats. (Eastman). Of future interest are the po. of methane, the *hydroxylation of benzene, and the direct *epoxidation of propylene. The increased application of oxidation cats. in the formation of fine chemicals suffers today from the relatively poor yields. B. LÜCKE

2-special: a proc. for *syngas generation by partial oxidation of part of the feed (*natural gas or coal, cf. *Haldor ATR syngas proc., *coal gasification). **F** l'oxidation partielle; **G** Partialoxidation. B. CORNILS

Ref.: Ertl/Knözinger/Weitkamp, Vol. 5, p. 2253; *Chem. Eng.Sci.Special Suppl.* **1954**, *3*, 41; Centi/Trifiro; Farrauto/Bartholomew, p. 487; Werner/Schreier; van Santen/van Leeuwen/Moulijn/Averill.

particle size → catalyst particle size

particulate catalyst are insoluble cats., → heterogeneous catalysts

PAS → photoacoustic spectroscopy

Paternò-Büchi reaction a [2+2] photocycloaddition of a carbonyl cpd. and an alkene, yielding oxetanes. The reaction is initiated by $n\pi^*$-excitation of the carbonyl cpd. and proceeds diabatically via 1,4-biradicals as *intermediates, the stability of which determines the *regioselectivity.

The reaction is of broad scope (R = alkyl, aryl; R^1,R^3,R^5 = aryl, alkyl, H; R^2,R^4 = alkyl, H, heteroatom substituent) and is applied to the prepn. of fine chemicals and of special intermediates. The stereochemical outcome is often predictable on the basis of empirical data and theoretical considerations. **F** réaction de Paternó-Büchi; **G** Paternò- Büchi-Reaktion. T. BACH

Ref.: Houben-Weyl, 4th Ed., Thieme, Stuttgart 1995; *Synthesis* **1998**, 683.

Pasteur Louis (1822–1895), professor in Dijon, Strasbourg, Lille, and Paris (all in France) and founder of the Institute Pasteur (Paris). Worked on *fermentation, microbiology, sterilization, *racemization, etc. He was the first to carry out a *kinetic resolution of tartaric acid using a microorganism. B. CORNILS

Ref.: Neufeldt.

Pauling Linus C. (1901–1994), chemist and professor in Pasadena, USA. Worked in diverse fields, from the valence bond theory and *MO theory to the proposal of α-helix and β-sheet structures in proteins. He was the winner of *two* Nobel prizes, one in Chemistry in 1954 and the Peace prize in 1963 for his opposition to nuclear bomb testing. P.S. SEARS

Pauson-Khand reaction a *transition metal-mediated [2+2+1] carbocyclization reaction employing alkynes, ethylenes, and CO to form cyclopentenones (Equ.)

The PKr. was originally conducted with stoichiometric amounts of $Co_2(CO)_8$ at 60–120 °C under a CO atmosphere. The yields vary and are improved when the preformed *complexes (alkyne)$Co_2(CO)_6$ are used. The *stereochemistry of the *cyclization requires that the larger residue of the alkyne is positioned *vic* to the carbonyl function of the cyclopentenone. Addition of N-oxides of tertiary amines give better yields; and the reaction time necessary is shortened if amines (also aqueous NH_3) are used as solvents. Under CO press. the PKr. becomes catalytic. Also known are pressurized or photochemical versions of the PKr., even under *supercritical conds. *Transition metal *clusters efficiently cat. the PKr. Asymmetric induction was also achieved by modifications of the PKr. High *enantioselectivities were found when the carbonyl *ligands were partially substituted by *chiral *phosphines. Chiral amine oxides were also used. Variable yields of cyclopentenenone are partly due to the high sensitivity of these amine systems toward oxygen, which causes the reaction to cease. This interrupted PKr. is a new approach to cyclopentenes. **F** réaction de Pauson-Khand; **G** Pauson-Khand-Reaktion.   P. HÄRTER

Ref.: A. de Meijere, H. tom Diek (Eds.), *Organometallics in Organic Synthesis*, Springer, Berlin 1987, p. 233; Cornils/Herrmann-1, Vol. 2, p. 1092; *Chem. Rev.* **1997**, *97*, 523; *Angew. Chem. Int. Ed. Engl.* **1998**, *37*, 911; *Organic Reactions*, Vol. 40, p. 1.

PBT, PBTP   → polybutylene terephthalate

pc, Pc, PC   1: phthalocyanine configurations in *complex cpds. or as *ligands; 2: polycarbonate

P-C cleavage   a severe side-reaction of hom. cat. with metal-*phosphine *complexes. In most cases this reaction is the source of *deactivation of phosphine-modified cats. It occurs most often in alkaline media and under conds. of *phase transfer cat. Moreover, the cleavage of the P-C bond is often cat. by the *transition metals themselves through *oxidative insertion of the metal into the P-C bond of the phosphine. In the case of triarylphosphines, aryl group migration leads to the formation of phosphido-bridged dimers and *clusters, which act as cat. *poisons. Further reactions of the aryl and phosphido groups result in formation of a large number of further byprods. and poisons. Although *ortho-metallation is also discussed as a reason for P-C bond cleavage, analysis of "exhausted" cat. solutions confirms the above-mentioned mechanism. **F** scission P-C; **G** P-C-Spaltung.

                                                    R. ECKL

Ref.: *Organometallics* **1984**, *3*, 649 and **1990**, *9*, 530; *J. Chem. Soc., Chem. Commun.* **1987**, 235

PCO   polycarbonate

PCR   → polymerase chain reaction

PCUK   Produits Chimiques Ugine Kuhlmann, subsidiary of Péchiney Ugine Kuhlmann (PUK), → Kuhlmann processes.

PDH process   → IFP procs.

PDMS   plasma desorption mass spectroscopy, → mass spectroscopy

PE   → polyethylene

Pearlman's catalyst   nonpyrophoric Pd $(OH)_2$ on charcoal; specially suitable for the hydrogenative *elimination of N-benzyl groups. P's cats. with Rh or Ru are also used. **F** catalyseur de Pearlman; **G** Pearlman-Katalysator.   B. CORNILS

Ref.: Augustine, p. 275.

Pearson acid-base concept   → HSAB concept

Pechiney Ugine Kuhlmann hydrazine process proceeds between H_2O_2 and NH_3 and in the presence of ethyl methyl ketone and a mixture of acetamide/NaH_2PO_4 as *activators. The initially obtained methyl ethyl ketoneazine is *hydrolyzed to hydrazine. **F** procédé PUK pour le hydrazine; **G** PUK-Hydrazinverfahren.   B. CORNILS
Ref.: Büchner/Schliebs/Winter/Büchel, p. 45; *Chem. Week* **1982**, Jan., 28.

Pechmann reaction synth. of coumarins by *condensation of phenols and β-ketocarboxylic acid esters. Cats. are *Lewis acids or *zeolites. Mechanistically the formation of an ester of the phenol by *transesterification of the β-ketoester is followed by an intramolecular *hydroxyalkylation and finally a *dehydration. Specially suitable are *m*-substituted phenols (-OH, -NH$_2$, NHR, OCH$_3$, -R). **F** réaction de Pechmann; **G** Pechmann-Reaktion.   B. DRIESSEN-HÖLSCHER

pectinase, pectin Pectin is a polysaccharide (polygalacturonic acid) found in fruit. It is responsible for gelation of fruit pulp following heating and cooling, and thus is useful in the prepn. of jellies. The gelation is problematic in the prod. of fruit juices, however, and for this reason pectinase (polygalacturonidase; EC 3.2.1.15), which digests the α1,4-galacturonic acid linkage, is used to *hydrolyze this polysaccharide during juice preparation. **E=F**; **G** Pektin.   P. SEARS
Ref.: A. Wiseman (Ed.), *Handbook of Enzyme Biotechnology*, Ellis Horwood, Chichester 1985.

Pedersen Charles John (1904–1989), chemist with DuPont. Worked on special cpds. such as *crown ethers or cyclic *oligomers and coined the term *host-guest relationships for structure specific interactions. Nobel laureate in 1987 together with Cram and Lehn.
                    B. CORNILS
Ref.. *Angew.Chem.Int.Ed.Eng.* **1988**, 27, 1021.

Pedersen reagent NbCl$_3$(dme), used for the *trimerization of alkynes (*Nb and Ta as cat. metals).   B. CORNILS

PEEM photoemission electron microscopy

PEG polyethylene glycol

PE-HD (HDPE) → high-density polyethylene

PEK poly(ether ketones) with the structural elements x and y (Figure). x = y = 1: PEK; x = 2, y = 1: PEEK = poly(ether ether ketones); x = 1, y = 2: PEKK = poly(ether keton ketones); x = y = 2: PEEKK poly(ether ether ketone ketones).

PEKs are manuf. by cat. poly*condensation of diphenyl ether and 4-phenoxybenzoyl chloride under *Friedel-Crafts conds. (*alternating copolymerizatin). **E=F=G**.   B. CORNILS
Ref.: Kirk/Othmer-2, p. 928.

PE-LD (LDPE) → low-density polyethylene

PE-LLD (LLDPE) → linear low-density polyethylene

pelletizing manuf. of catalyst *pellets from cat. masses or *precursors; cf. *forming.
                    B. CORNILS

pellets In het. cat. procs. in fixed-bed reactors, pressure drops are created across the cat. bed. These press. drops must be high enough to guarantee an even distribution of the reactant fluid across the cat. bed, but they must not be too high because the proc. would then require cost-intensive compression and recycling of reactant gases. The press. drop across a fixed cat. bed is strongly dependent on the geometrical shape (*shaping) and size of the cat. particles. The cat. material must therefore be shaped so as to provide optimal conds. as egards the press. drop. Depending on the proc. the cat. is therefore applied in the form of pellets or granules having typically spherical or cylindrical shapes (beads or extrudates, respectively) of varying dimensions (manuf. by *pelletizing). Rings and carriage-wheels are also used. Beads in the size range of a few millimeters can be produced by drop-coagulation from sols. The cat. raw-material is often a powder from which larger particles are formed by pelletizing, granulation, and extrusion procs (cf. *forming, *shaping). *Binders are often added to provide sufficient mechanical stability of the final shaped cat. pellet. Pastes have to be prepared from the cat. powder for the extrusion proc. **E=F=G**.

H. KNÖZINGER

Ref.: Ertl/Knözinger/Weitkamp, Vol. 1, pp. 50, 412; Wijngaarden/Kronberg/Westerterp, p. 177.

pellicular an egg-shell distribution of cat. active metals on the outer shell of the support (*metal distribution). B. CORNILS

Penex process → UOP procs.

penicillin an antibiotic which inhibits a transpeptidase involved in bacterial cell wall *biosynthesis. Also see *ACV synthetase, *Isopenicillin N synthetase. C.-H. WONG

penicillin acylase (EC 3.5.1.11) from *E. coli* (ATCC 9637) is a serine type of *hydrolase, which cat. the *hydrolysis of benzoyl penicillin to 6-aminopenicillanic acid (6-APA). 6-APA can be converted to different acyl penicillin derivatives. The enzyme also cat. the cleavage of the N- or O-phenylacetyl protecting group from α-, β-, or γ-amino acids, peptides, amines, alcohols, or sugars (cf. *amidases). In addition to the phenylacetyl group, o-hydroxyphenylacetyl and other groups can also be used as enzymatically removable protecting groups. In general, penicillin acylase accepts substrates with stereostructures related to L-amino acids. The enzyme was also used in the acylation of amines with methyl phenoxyacetate. **E=F=G**. C.-H. WONG

PennPhos a *bidentate, bicyclic diphosphine *ligand (Figure).

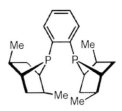

P. is useful for the *enantioselective *hydrogenation of ketones to alcohols (96 % *ee). **E=F=G**. B. CORNILS

Ref.: *Angew. Chem.* **1998**, *110*, 1203.

Pentafining → Atlantic Richfield procs.

pentose phosphate pathway an alternative pathway to glycolysis for the *oxidation of glucose. The *dehydrogenases (DHases) of this pathway generate reducing power in the form of *NADPH, whereas those of glycolysis produce NADH. P.S. SEARS

PEP → phosphoenolpyruvate

pepsin is an *aspartyl protease* from the mammalian digestive tract. It is moderately sized (327 amino acids), and prefers hydrophobic amino acids (esp. Phe, Tyr) on either side of the scissile bond. It has a pH optimum between 2 and 3, determined by the pK_a values of its two cat. aspartic acids: one (the acid) at 1.1 and the other (the base) at 4.5. Such a pH optimum is consistent with its function at the low pH of the stomach. It does not hydrolyze most esters well, unlike many other proteases. **E=F=G**. P.S. SEARS

pepsinogen the inactive *proenzyme form of the mammalian acid protease pepsin, which contains an extra 44 amino acids on the N-terminus. It is stable near neutral pH, but upon exposure to acid pH the first 44 amino acids are cleaved away. This activation step does not appear to be intermolecular, as *immobilized pepsinogen is still capable of activating itself. **E=F=G**. P.S. SEARS

peptide amide synthesis Carboxyl terminal amidation is an important post-translational modification in the synth. of polypeptide amides; the reaction requires two *enzymes acting in sequence. The first enzyme is *peptidylglycine α-monooxygenase (EC 1.14.17.3) which cat. the formation of the α-hydroxyglycine derivative of the C-terminal glycine using a proc. dependent on ascorbic acid, copper, and molecular oxygen. The second enzyme is *peptidylamidoglycolate lyase (EC 4.3.2.5) which cat. the breakdown of the C-terminal α-hydroxyglycine derivative to produce the amidated peptide and glyoxylate. The oxygenase abstracts the Pro-S hydrogen from the Gly residue to form (S)-hydroxy-Gly; this cpd. is the substrate for the second enzyme. The enzymes from horse serum are commercially available. Application of the enzymes in the synth. of human growth hormone releasing factor analog GRF(1–44)-NH$_2$ has been reported.

An alternative route to peptide amides is enzymatic transpeptidation with amino acid amide. Another method is *papain- or *subtilisin-cat. *aminolysis of peptide esters with trimethoxy-benzyl amine, followed by deprotection with trifluoroacetic acid to give N-protected peptide amides. **F** synthèse de peptide d'amide; **G** Peptidamidsynthese.   C.-H. WONG

peptidylamidoglycolate lyase
(EC 4.3.2.5) cat. the breakdown of the C-terminal α-hydroxyglycine derivative to produce the amidated peptide and glyoxylate.
                                      C.-H. WONG

peptidylglycine α-monooxygenase (EC
1.14.17.3) cat. the formation of the α-hydroxyglycine derivative of the C-terminal gly-

cine using a proc. dependent on ascorbic acid, copper, and molecular oxygen.   C.-H. WONG

pepzyme was claimed cat. activity of smal synth. peptides modeled to mimic the *active-site structures of trypsin and chymotrypsin **E=F=G**.   B. CORNILS
Ref.: *Proc.Natl.Acad.Sci.USA* **1994**, *91*, 4106, 4110 Mulzer/Waldmann, p. 185.

pericyclic reactions A number of antibodies (*cat. antibodies) have been elicited to *transition-state analogs of pericyclic reactions such as the *Diels-Alder reaction and Cope rearrangement. Some of these antibodies have proven to be catalytic. The first antibody-cat pericyclic reaction was the *Claisen rearrangement of chorismate to prephenate, a reaction cat. by *chorismate mutase in vivo.

Several Diels-Alder antibodies have been elicited now, as well as antibodies which cat. an oxy-Cope rearrangement and selenoxide *syn* elimination. **F** réactions pericycliques; **G** pericyclische Reaktionen.   P.S. SEARS
Ref.: *J.Am.Chem.Soc.* **1997**, *119*, 3623 and **1994**, *116* 2211; *Proc.Natl.Acad.Sci.USA* **1993**, *90*, 8663.

periodic operation → unsteady-state processes

Perkin reaction the base-cat. reaction between aromatic aldehydes and carboxylic anhydrides yielding α,β-unsaturated carboxylic acids. **F=G=E**.   B. CORNILS

perovskite a naturl mineral with formula CaTiO$_3$ which has given its name to a class of mainly oxidic, mostly synth. cpds. of general formula ABX$_3$ that share certain common structural characteristics. The most important common feature among ps. is the simultaneous presence of a small, often highly charged, cation B and a large cation A, usually of low charge. Ps. of lower symmetry than cubic offer various interesting physical properties (magnetism, ferroelectricity, piezoelectricity, superconductivity) and are used as electroceramics. Other applications of perovskites are as cats. in *fuel cells, for *combustion, and for *DeNOx reactions; cf. also *Aurivillius phases. **F**=**E**; **G** Perowskit. P. BEHRENS

peroxidases such as chloroperoxidase (CPO, EC 1.11.1.10) cat. oxidative reactions using *peroxides. The *enzyme from *Caldariomyces fumago* has recently been used in the presence of TBHP and other peroxides for the *enantioselective *oxidation of aromatic sulfides to (R)-sulfoxides with very high enantioselectivity. Chiral *hydroperoxides such as 1-phenylethyl hydroperoxides are also acceptable as oxidants and the R-enantiomers are selectively accepted as substrates.

The reactions with horseradish p., however, showed very low enantioselectivity. The chloroperoxidase from *Caldariomyces fumago* cat. the oxidation of all halide ions except fluoride, lactoperoxidase oxidizes bromide and iodide, and horseradish p. only oxidizes iodide ion. All reactions require H_2O_2 as the natural oxidation reagent. In the presence of an unsaturated acceptor, the prods. formed by haloperoxidase-cat. reactions are consistent with the reaction with hypohalous acid.
In addition to monooxygenases, horseradish p., cytochrome C p., haloperoxidases and lactoperoxidase all utilize ferryl-oxo [Fe(IV)=O] species

for reaction with substrates. This reactive species perhaps forms a bound hypohalite which could undergo *regio- and *stereoselective bromohydration of glycals. Similarly to monooxygenases, the chloroperoxidase from *Caldariomyces fumago* also cat. the *epoxidation of alkenes in the presence of H_2O_2. Horseradish p. and lactoperoxidase were, however, inactive toward alkenes. Of several alkenes tested, *cis*-olefins are the best, giving epoxides with very high *ee. P. was also used in the polymerization of ethylphenol in reverse micelles. **E**=**F**; **G** Peroxidasen. C.-H. WONG

Ref.: *J.Org.Chem.* **1992**, *57*, 7265; *J.Biol.Chem.* **1970**, *245*, 3135.

peroxide effect describes the influence of *radical initiators such as *peroxides on the *regiochemistry of the addition of nucleophiles to asymmetrically substituted alkenes acc. to *Markovnikov or *anti*-Markovnikov rules. B. CORNILS

peroxides cpds. containing O-O bonds, sometimes referred to as *peroxo (inorganic) or peroxy (organic) cpds. H_2O_2 and alkyl *hydroperoxides (ROOH) are widely used in comb. with metal cats. for the selective *oxidation of various functional groups. Often the intermediate formation of peroxo-metal *complexes (M-OOH or M-OOR) is involved. Alkyl hydroperoxides (e.g., ethylbenzene hydroperoxide; cf. procs. of *Arco PO, *Oxirane) are produced by the free-*radical *autoxidation of (cyclic) alkanes and aralkanes. Autoxidation of alkenes with reactive allylic C-H bonds affords the corresponding allylic hydroperoxides. Peroxycarboxylic acids, R(CO)OOH, are prod. by autoxidation of aldehydes (cf. *aliphatic aldehyde ox.) or by the acid-cat. reaction of H_2O_2 with carboxylic acids. They are widely used, in the absence or presence of metal cats., as oxidants. The same is true for inorganic peroxides such as peroxoborate, peroxomonosulfate (oxone), and peroxodisulfate. Dialkyl peroxides (ROOR), diacyl peroxides, R(CO)OO(CO)R), and peroxycarboxylic esters, R(CO)OOR, are not active oxidants and are used as free-radical *initiatiors. Dioxiranes are three-membered ring cyclic peroxides that are prod. by reaction of peroxosulfate with ketones. They selectively oxidize a variety of functional groups, e.g., as in alkene *epoxidation. **F=G** Peroxyde.

R.A. SHELDON

Ref.: S. Patai (Ed.), *The Chemistry of Peroxides*, Wiley, New York 1992; R.A. Sheldon, *Metal-catalyzed Oxidation of Organic Compounds*, Academic, New York 1981; Strakul.

peroxisomes organelles (subcellular, membrane-bound structures) in which peroxide-dependent reactions and peroxide-degrading *enzymes are sequestered. P.S. SEARS

Perstorp-Reichhold Formox process
manuf. of formaldehyde by *selective oxidation of methanol with air over MoO_3-Fe_2O_3 cats. in tubular reactors. **F** procédé Formox de Perstorp-Reichhold; **G** Perstorp-Reichhold Formox-Verfahren. B. CORNILS

PES photoelecton spectroscopy

PET 1: poly(ethylene terephthalate), manuf. from bis(hydroxyethyl) terephthalate (BHET) by *polycondensation at 270–305 °C/vacuum in the presence of *Lewis acid metal cpds. such as Ti alkoxides, dialkyltin oxide, etc. 2: *p*-ethyltoluene (*Mobil proc.). B. CORNILS

Ref.: Cornils/Herrmann-1, p. 545; Ertl/Knözinger/Weitkamp, Vol. 5, p. 2129.

Petro Canada process → Canmet process

petrochemistry involves the manuf. of chemical prods. (bulk and fine chemicals) and of fuels from natural gas or petroleum. Thus p. is closely related to *petroleum processing (refinery, refining) and so on the one hand to the manuf. of fuels, gasoline, diesel fuel, kerosene, light and heavy fuel oils, etc. On the other hand, petrochemical key prods. are lower alkanes (C_1–C_4) and alkenes (particularly ethylene, propylene, butylenes), and *BTX aromatics as building blocks for organic chemistry and for polymers. Cat. procs. are common such as 1: *catcracking (specially *FCC) on ZSM-5 or other *zeolites; 2: *reforming of naphtha as primary source of BTX on modified Pt-*alumina cats.; 3: *isomerization of xylenes on ZSM-5 zeolites or related systems with *shape selectivity (*petroleum processing). Commercially important petro*chemical* reactions involve the ethylene processing with: i: *oligomerization or *polymerization over supported Cr or with Al-Ti cats. acc. to * Ziegler-Natta (*Phillips, *HDPE, *LDPE); ii: oxidations to ethylene oxide (supported Ag cats., e.g., *Shell proc.) or to acetaldehyde (hom. $PdCl_2$-$CuCl_2$ cats.; *Wacker proc.); iii: *alkylation of benzene (to *ethylbenzene, e.g., over zeolites, such as the procs. of *BASF/Badger, *CdF, *Lummus/UOP, *Mobil/Badger, *UOP, etc.), iv: *oxidative *acetoxylation to *vinyl acetate (Pd-Cu cats., procs. of *Bayer/Hoechst or *Halcon); v: *addition of water to ethyl alcohol over acid cats. (e.g., procs. of *Shell or *VEBA); vi: *addition of chlorine to ethylene, followed by abstraction of HCl forming *vinyl chloride.

The processing of propylene includes: 1: *hydroformylation (over Co or Rh cats.; procs. of *Union Carbide, *Ruhrchemie/ Rhône-Poulenc, or *Mitsubishi) to butyraldehydes and *2-EH; 2: *ammoxidation to *acrylonitrile (*SOHIO proc. with promoted Bi-Mo cats.); 3: oligomerization or polymerization to *PP with Ziegler-Natta or *metallocene cats.; 4: formation of *PO (*ARCO proc.); 5: oxidation to *acrolein or *acrylic acid; 6: the alkylation of benzene to cumene (procs. of *BP, *CDTech, *Mobil/Badger, or *UOP, further *Hock reaction to phenol).

Further important petrochemical reactions are *partial oxidations in the gaseous phase to maleic anhydride from *n*-butane or benzene (over V/P/O cats., procs. of *Alusuisse, *Lurgi Geminox, *Lummus ALMA, *Mitsubishi), phthalic anhydride from *o*-xylene (V/Ti/O cats., procs. of *BASF, *Rhône-Progil, *Wacker), or to methacrylonitrile from isobutene (ammoxidation on Bi/Mo cats.). *Partial oxidations in the liquid phase include acetic acid from *n*-butane (or butene, procs. of

Bayer, *Celanese, or *Hüls), or *terephthalic acid from *p*-xylene (procs. of *Glitsch,
Henkel, *Hüls, *Lummus, *Mobil). *Substitution procs. starting from alkanes (*chlorination, *sulfoxidation, *sulfochlorination, *nitration, etc.) are mainly *radical reactions
partly supported by cats.).

The future development will be influenced by
higher environmental demands (*environmental issues and future demands) for fuel prods.
e.g., increasing demand of separation of aromatic cpds. for petrochemical use) and by increasing consumption of natural gas. In general,
the overlap between refinery and petrochem-
cal industries will increase and cats. will play an
ever-increasing role in this interaction. **F** pétro-
himie; **G** Petrochemie.   B. LÜCKE

petroleum (crude oil)   is a naturally occuring liquid, consting mainly of hydrocarbons,
but also cpds. with heteroatoms such as S, O,
N, etc.   B. CORNILS

petroleum processing (refinery, refining) Pp. and refining are synonyms. They
include the generation and recovery of usable
and salable prods. from crude oil. This is done
either by physical fractionation or by chemical treatment of the crude oil constituents or
of fractions thereof in the presence of heat,
press., or cats. A refinery is the sum of corre-
ponding plants. Its purpose is to convert
crude oil to prods. acc. to suitability and demand. Acc. to the Figure this is done by the
following sequence of procs.: 1: Distillation
(atmospheric and/or under vacuum), the
point where refining begins. The normal fractions are (approx. boiling range, °C): light
naphtha (1–150), gasoline (1–180), heavy
naphtha (150–205), kerosene (205–260), stove
oil (205–290), light gas oil (260–315), heavy
gas oil (315–425), lubricating oil (>400), vacuum gas oil (425–600) and residuum (resid,
>600; cf. *hydrocarbons). Without further
processing they are *straight run. 2: Thermal
cracking (TCC), a proc. to increase the
amount of lower-boiling prods. *Visbreaking
is a mild cracking proc. for prods. lower in

viscosity and pour point. The highest-boiling
outlets are tars. Delayed *cokers for the conv.
of heavy feedstocks deliver coke besides gas
oil and lighter prods. As a main refinery proc.
thermal cracking is older and obsolete. 3:
*Catalytic cracking (catcracking) the thermal
decomposition of petroleum constituents under the influence of cats.; *hydrocracking involves cracking in presence of hydrogen. Various configurations of the catcracking (*FCC,
TCC) make different *regenerations of the
cats. (which are precipitated with coke) possible. 4: Different *hydroprocessing steps (*hydrorefining, *hydrotreatment, *hydrofining)
which serve for the removal of aromatics
(*dearylation), chlorine (*hydrodechlorination), sulfur (*desulfurization, *hydrodesulfurization), nitrogen (*hydrodenitrogenation),
metals (*demetall[iz]ation), *n*-paraffins or alkyls (*hydrodealkylation, *dewaxing), O-containing cpds. (*hydrodeoxygenation), or olefins.
5: *Reforming, to improve the *ON of fractions
boiling in the naphtha range (e.g., naphthenes)
and convert them to high-octane gasoline. 6:
*Isomerization procs., to deliver specially *i*-al-
kenes (isobutene, isoamylenes) for *alkylation
and as intermediates for *etherification procs.
7: The combined reaction of alkenes and paraffins to higher isoparaffins with convenient high
*ONs (or the alkylation of aromatics) is termed
*alkylation. 8: The conv. of lower alkenes (olefi-
nic gases, i.e., ethylene, propylene, butylenes) to
higher molecular weight prods. which are suit-

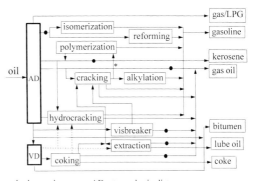

• hydrotreating   AD atmospheric dist.

VD vacuum dist.   * olefins

able for gasolines or to other liquid fuels (*alkylates by *polymerization). 9: Treatment steps (S, acids removal by *clay, solvent extraction, etc.). 10: Gas processing for gases, *LPG.

The combination and configaration of these different procs. vary with the refinery involved (e.g., the availability of hydrogen), the feed crude oil (aromatic, naphthenic, paraffinic), the market demand, etc. **F** rafinage du pétrole; **G** Mineralölverarbeitung.   B. CORNILS

Ref.: Kirk/Othmer-1, Vol. 4, p. 590; Vol. 18, pp. 342, 433; Kirk/Othmer-2, pp. 187, 851; McKetta-1, Vol. 13, p. 238; Vol. 22, p. 232; Vol. 26, pp. 159, 424; Vol. 27, pp. 97, 435, 460; Vol. 35, pp. 86, 115; Vol. 47, pp. 71, 468; Vol. 48, p. 1; McKetta-2, p. 851; Trimm et al.; Ertel/ Knözinger/Weitkamp, Vol. 4, pp. 1801, 1908, 1928, 1939, 1955, 2017; G.W. Dyroff, *Petroleum Products,* 6th Ed., American Technical Publ., Hitchin, Herts, UK; Farrauto/Bartholomew, p. 521; Anonymous, *Modern Petroleum Technlogy*; D.S.J. Jones, *Elements of Petroleum Processing*, Wiley-VCH, Weinheim 1996; Ullmann, Vol. A18, p. 51; W.F. Bland, R.L. Davidson (Eds.), *Petroleum Processing Handbook*, McGraw Hill, New York 1976; van Santen/van Leeuwen/Moulijn/Averill.

Petrolite Corp. Bender sweetening

process for the *sweetening of distillates from light naphtha to jet No.2 by converting mercaptans to disulfides in the presence of solid cats. H_2S is removed with caustic. **F** procédé Bender sweetening de la Soc. Petrolite; **G** Petrolite Bender sweetening-Verfahren.

Ref.: *Hydrocarb.Proc.* **1978**, (9), 212.   B. CORNILS

Petrotex Oxo-D process for the manuf. of butadiene by *oxydehydrogenation of butenes with air or oxygen over a cat. consisting of Zr-Mn-Mg oxides. The *selectivity is controlled via steam (cf. also *Phillips OXD proc.). **F** procédé Oxo-D de la Soc. Petrotex; **G** Petrotex Oxo-D-Verfahren.   B. CORNILS

PE-UHMW (UHMWPE) → ultrahigh molecular weight polyethylene

PF resins phenol-formaldehyde resins, manuf. by acid cat. *Baekeland-Lederer-Manasse reaction.

PFR → plug flow reactor, *catalytic reacₜtors

PGO pyrolysis gas oil

PGO hydrotreating → Lummus procs.

Ph phenyl (C_6H_5-) group in *complex cpd or *ligands

phage display Starting with a *library of bacteria, each of which contains a gene for a single protein mutant, isolation of a single member producing a protein with desirable properties can be challenging. Screening is not feasible for libraries greater than perhaps 10^4 members. For larger libraries, selection techniques are necessary. One such technique is phage display, in which the gene for the protein (or peptide) of interest is placed in front of a gene encoding viral coat protein, and the fused gene is placed either within the phage genome or in a *vector that contains a packaging signal allowing it to be incorporated into a phage particle. The size of the library is limited by the bacterial transformation yield, generally about 10^9 members. When the phage are produced, the coat protein will display the exogenous peptide/protein, and importantly, the *DNA encoding that protein/ peptide will be inside. Affinity chromatography of the phage library will select those phage displaying protein species that have the desired binding properties, and the selected phage can be easily reamplified by inoculation into a culture of their bacterial host (Figure).

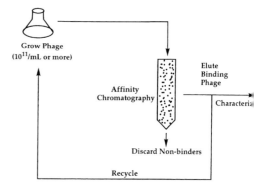

Grow Phage
(10^{11}/mL or more)

Affinity Chromatography

Elute Binding Phage

Characteriz

Discard Non-binders

Recycle

Selection with *transition state analogs or mechanism-based inactivators can be used to select for catalysis. For example, penicillin sulfones, which are mechanism-based inactivators of β-lactamase, have been used to select for active β-lactamase *enzyme from libraries of active-site mutants. Alternatively, customized techniques in which the desired enzyme cat. either binding to or release from a solid *support can be used. A third technique involves linking the cat. to the infectivity of the phage.

Many types of phage have been used for phage display, but the most commonly used species is still the filamentous *coliphage M13*, which has several coat proteins that can be modified (esp. coat proteins III, VIII, and VI). This phage is very robust, surviving for short periods of time at temps. up to 80 °C, and can handle large inserts into the genome: the phage merely gets longer (some phage have a fixed coat geometry and are very limited in the size of the genome that can be inserted). **E=F=G**.   P.S. SEARS

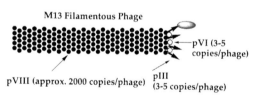

M13 Filamentous Phage

pVI (3-5 copies/phage)

pVIII (approx. 2000 copies/phage)

pIII (3-5 copies/phage)

Ref.: *Proc.Natl.Acad.Sci.USA* **1998**, *95*, 10523 and **1997**, *94*, 11777; *Appl.Biochem.Biotech.* **1994**, *47*, 175.

phases 1: In thermodynamic systems phases are uniform states of matters of substances which are separated (and separable) from each other by unequivocal phase boundaries, e.g., for water-ice – liquid water – steam. 2: Different states of aggregation such as solid/liquid, liquid/gaseous or oil/water (O/W) or water/organic liquids which determine cat. procs. (*hom. cat., *het. cat., *phase-transfer cat.). In this respect interphases as well as phase boundaries are of importance. **E=F**; **G** Phasen.   B. CORNILS

phase transfer catalysis (PTC) This technique allows efficient convs. between cpds. which remain in different *phases and therefore do not react easily as a rule. Typical is the *substitution reaction of an alkyl halide, R-X, and a salt, NaY, to give R-Y in a water/non-polar solvent mixture (liqu./liqu. PTC). This reaction proceeds by the presence of a few mol% of the PT cat., mostly quaternary ammonium or phosphonium salts (abbr. Quat or Q⁺). QX exchanges counter-ions at the interphase and extracts sparingly solvated anions in the form of ion pairs (Q⁺Y⁻) into the non-polar organic phase, where they react rapidly, and cat. QX is regenerated. Instead of a Q⁺, *complex *ligands for inorganic cations can be cats.: *crown ethers, *cryptands, *PEGs, and their ethers.

Important parameters determining the feasibility of such "ion pair extraction" are the relative lipophilicities of all the anions within the sytem. A very organophilic leaving group of a displacement reaction will pair up with the cat. permanently and thus block the cat. proc. (*poisoning). Sometimes PTC is executed with solid salts/organic solvent ("solid/liqu. PTC"), and the solvent itself might either be substituted by excess liqu. organic reagent or be dispensed with altogether ("PTC without solvent"). Most commonly used org. solvents in liqu./liqu. PTC are toluene and other *hcs., chlorobenzene, and – in the lab – chlorinated solvents such as CH_2Cl_2 and $CHCl_3$. For solid/liqu. PTC the more polar acetonitrile and even *DMF are employed, too.

PTC reactions under neutral conds. include *substitutions with many ions, but also *reductions and *oxidations (e.g., with $[MnO_4]^-$, $[CrO_4]^-$, $[OCl]^-$, etc.). Even more widely applicable are base-catalyzed PT reactions using aqueous concentrated or solid NaOH, KOH, K_2CO_3, NaH, etc. These include *alkylations, *isomerizations, *addition reactions, *condensations, eliminations, *hydrolyses, nucleophilic aromatic *substitutions, dihalocarbene generation, etc. PTC alkylations in the presence of 50 % aqueous NaOH are possible for substrates with pK_as up to 23 or more. Cats. used more frequently are NBu_4Br, NBu_4HSO_4, and $MeNOct_3Cl$ ("Aliquat 336") for neutral reactions and $PhCH_2NEt_3Cl$ ("TEBA, TEBAC") or cetyltrimethylammonium bromide ("cetrimide") for basic PTC. Newer variants of PTC are *triphase cats. (in which the cat. is anchored to a polymer for ease of removal); *gas/liqu. PTC* (in which one of the reagents is a gas); *inverse PTC*; *extraction of cations* for electrophilic reactions by large lipophilic cat. anions ("*reverse PTC*"); and extraction of uncharged species into org. media by onium salts QX. These include *transition metal salts (complex formation with, e.g., CuX, $PdCl_2$), acids, hydrogen peroxide, and amines (H_2O_2, HX, HNR_2), e.g., form weakly hydrogen-bonded complexes with QX.

Besides its utilization in preparative chemistry, further extensions of PTC are being pursued both in industry and labs for *polymer manuf. and modification, and in organometallic and analytical chemistry. **F** catalyse par transfer des phases; **G** Phasentransfer-Katalyse.   E.V. DEHMLOW

Ref.: Dehmlow/Dehmlow, *Phase Transfer Catalysis*, 3rd. Ed., VCH, Weinheim 1993; Starks/Liotta/Halpern, *PTC: Fundamentals, Applications, and Industrial Perspectives*, Chapman & Hall, New York 1994; Sasson/Neumann, *Handbook of PTC*, Blackie Academic & Professional, London 1997; Keller, *PTC Reactions*, Fluka-Compendium, 3 Vols., Thieme, Stuttgart 1992; Kirk/Othmer-1, Vol. 18, p. 662; Ullmann, Vol. A19, p. 293.

ph-β-glup   chiral ligand, cf. *chiral pool.

pH dependence of enzyme catalysis   By varying the pH of an enzymatic reaction the conformation and/or the ionization status of the *enzyme and reactants can be changed. The new conformation and charge distribution may or may not correspond to an active enzyme, and it may or may not alter substrate *selectivity, depending upon the particular protein. Another consideration involving pH is the effect it will have on possible side-reactions. It has been shown with many reactions

of *pig liver esterase (PLE) that, although the cat. rate of *hydrolysis is much faster at pH 8.0, it is beneficial to lower the pH so that the contribution of the non-enzymatic hydroxide-induced hydrolysis is small.   By reducing the rate of uncat. hydrolysis, the *ee can be increased. Optimization of pH can be extremely helpful in optimizing enzyme-assisted synth. reactions. **F** dépendance de pH; **G** Abhängigkeit vom pH-Wert.   C.-H. WONG

Ref.: A. Fersht, *Enzyme Structure and Mechanism*, W.H. Freeman, New York 1985.

PHEN, phen   1,10-phenanthroline group in *complex cpds. or *ligands. Phen is a *bidentate ligand for *transition metal-cat. reactions (e.g., for *reductive carbonylation of nitroaromatics).   W.R. THIEL

PHENAP, phenap   a *phosphine- and naphthalene-based *ligand, prep. by a Michaelis-Arbusov reaction and subsequent *reduction. It has been used for *hydroformylations (Rh) and aryl couplings (Ni).   O. STELZER

Ref.: WO 87/07.600; WO 90/06.295; Cornils/Herrmann-1, Vol. 1, p. 86.

phenol   important bulk chemical, cat. manuf. by *hydrolysis of chlorobenzenes (*Gulf proc.), *hydrogenation of *KA oil (*SD proc.), oxidation of toluene (*Dow proc.), or the *Hock process (procs. of *BP, *Kellogg, Hercules,   *Lummus,   *Rhône-Poulenc). **F=G=E**.   B. CORNILS

phenylalanine ammonia lyase (EC 4.3.1.5)   cat. the (reversible) *elimination of ammonia from L-phenylalanine or L-tyrosine (poorly) to give *trans*-cinnamate (a precursor to alkaloids and lignin in plants) or *trans*-coumarate, respectively. Other substrates include *m*-Tyr, which is a much better substrate than the normal *p*-Tyr, and *p*-nitrophenylalanine. The *enzyme has an unusual dehydroalanine prosthetic group, made by *dehydration of an essential serine residue. **E=F=G**.   P.S. SEARS

Ref.: *Proc. Natl. Acad. Sci. USA* **1995**, *92*, 8433; *Arch. Biochem. Biophys.* **1970**, *141*, 1.

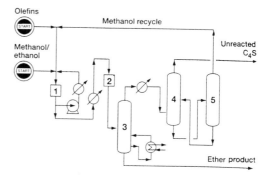

R = H, OH

phenylalanine ammonia lyase

phenylalanine dehydrogenase

cat. the *oxidation of L-phenylalanine to the corresponding α-ketoacid using *NAD. The *enzyme from *Rhodococcus* accepts a range of aromatic 2-oxoacids as substrates, and produces unnatural amino acids including *p*-halogenated phenylalanine and homophenylalanine. The in-vivo role of phenylalanine dehydrogenase is the *reductive amination of 2-oxo-3-phenylpropanoic acid (phenylpyruvate) to yield phenylalanine.   P.S. SEARS

Phenylalanine dehydrogenase from *Bacillus sphaericus*

SCRC-R79a, coupled with *formate/formate dehydrogenase for NADH regeneration, was used for the synth. of a number of natural and unnatural L-amino acids with the enzyme enclosed in a dialysis tube. **E=F=G**.   C.-H. WONG

Ref.: *J.Org.Chem.* **1990**, *55*, 5567; *Appl.Microbiol.Biotechnol.* **1987**, *26*, 409.

phenylalanine hydroxylase → phenylalanine monooxygenase, *tyrosine hydroxylase

phenylalanine monooxygenase

(phenylalanine hydroxylase; EC 1.14.16.1) a pterin-dependent *enzyme that oxidizes phenylalanine to tyrosine (see also *tyrosine hydroxylase). **E=F=G**.   P.S. SEARS

phenylation → arylation

Philips P.

in 1831 filed the first patent on SO_2 oxidation with air over Pt cats. (GB 6096)
                                                    B. CORNILS

Phillips catalyst

class of cats. based on Cr oxides supported on SiO_2 or SiO_2/Al_2O_3. They are manuf. by *impregnation of the *supports with CrO_3 solutions, *drying/heating to 500–800 °C and *reduction with H_2, CO, or hydrides. Phillips cats. are very effective for the *polymerization of ethylene (*Phillips LDPE proc.), but not for propylene. **F** catalyseur de Phillips; **G** Phillips-Katalysator.   B. CORNILS

Phillips catalytic isomerization process

for the *isomerization of *n*-butane to *i*-butane by treatment over cats. consisting of $AlCl_3$ on *bauxite (fixed-bed, gas-phase operation at 100–140 °C/1–2 MPa). **F** procédé Phillips pour l'isomérisation catalytique; **G** Phillips katalytisches Isomerisierungsverfahren.   B. CORNILS

Ref.: W.F. Bland, R.L. Davidson (Eds.), *Petroleum Processing Handbook*, McGraw Hill, New York 1976, p. 3.

Phillips etherification process

manuf. of *MTBE, *TAME, or other ethers from corresponding alkenes (*i*-butene, amylenes) and methanol over acidic *ion exchange resins in fixed-bed, liquid-phase operation. Ethanol

1,2 reactor section; 3 fractionator for MTBE; 4 methanol extractor (to the *Phillips STAR proc.); 5 methanol fractionator.

delivers ETBE and TAEE (Figure). **F** procédé Phillips pour des éthers; **G** Phillips Veretherungs-Verfahren.   B. CORNILS

Ref.: *Hydrocarb.Proc.* **1996**, (11), 114.

Phillips HDPE process   involves the manuf. of *high-density polyethylene with CrO₃ supported on *silica (*Phillips cat.). It is assumed that CrO₃ reacts with the silanol groups of the *support, forming oxo bonds with Si atoms of the silica *surface. Initiation is thought to take place during the reaction between an ethylene unit and chromium oxide. This reaction leads to the formation of a σ-Cr-C bond, into which further ethylene units are inserted. The Phillips cat. can be used in solution, slurry, or gas-phase procs. About 60 % of all high-density polyethylene produced worldwide is made with the Phillips cat. **F** procédé de Phillips pour HDPE; **G** Phillips-HDPE-Verfahren.

W. KAMINSKY, D. ARROWSMITH

Ref.: US 2.825.721 (1958); *Ind.Eng.Chem.* **1956**, *48*, 1152; Ertl/Knözinger/Weitkamp, Vol. 5, p. 2400; Farrauto/Bartholomew, p. 716; *Ind.Eng.Chem.Res.* **1988**, 27, 1559.

Phillips HF alkylation process   manuf. of high-octane motor fuels from lower alkenes and *i*-butane (*alkylation) by means of a circulating HF cat. stream (in a reactor-settler system) (Figure 1) which serves as a large heat sink for the mildly exothermic reaction (Figure 2). **F** procédé HF alkylation de Phillips; **G** Phillips HF Alkylierungs-Verfahren.

B. CORNILS

Ref.: McKetta-2, p. 886, *Hydrocarb.Proc.* **1996**, (11), 91.

Phillips Hydrisom process   for the cat. selective *hydrogenation (removal of diolefins) and simultaneous *isomerization of alkenes prior to their *etherification or *alkylation. The proc. takes place in a once-through operation the liquid-phase/fixed-bed reactor (Figure). **F** procédé Hydrisom de Phillips; **G** Phillips Hydrisom-Verfahren.   B. CORNILS

Ref.: *Hydrocarb.Proc.* **1996**, (11), 140.

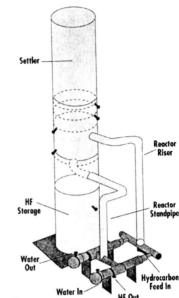

Figure 1

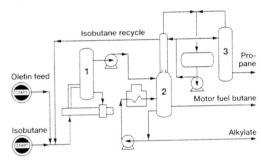

Figure 2
1 reactor-settler; 2 main fractionator; 3 HF stripper.

Phillips HF alkylation process

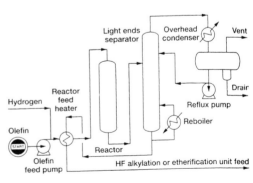

Phillips Hydrisom process

Phillips isoprene process   manuf. of isoprene by a two-step proc. consisting of 1: the
*metathesis of propylene and *i*-butene to a
mixture of ethylene and 2-methyl-2-butene at
450 °C/2.1 MPa over a WO_3/SiO_2 cat., 2: the
*dehydrogenation of the methylbutene to iso-
prene over Fe-oxides. **F** procédé Phillips pour
l'isoprène; **G** Phillips Isoprenverfahren.

B. CORNILS

**Phillips olefin conversion technology
(POCT) process**   manuf. of propylene from
butylenes and ethylene by fixed-bed cat. *me-
tathesis (cf. *Phillips Triolefin proc.) (Figure).
F procédé POCT de Phillips;   **G** Phillips
POCT-Verfahren.   B. CORNILS

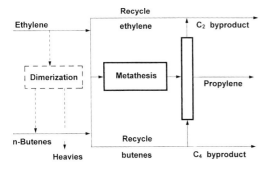

Ref.: *Hydrocarb.Proc.* **1997**, (3), 158.

Phillips OXD process   for the two-step
manuf. of butadiene from butane (butane →
butene → butadiene) via *oxydehydrogenation at 480–600 °C over $Cr-Al_2O_3$ cats. **F** procédé OXD de Phillips; **G** Phillips OXD-Ver-
fahren.   B. CORNILS

**Phillips STAR (steam active reforming)
process**   a *reforming proc. for the manuf.
of either alkenes (*i*-buten, isoamylenes) from
$<C_5$ paraffins (by *dehydrogenation), or aromatics from $>C_6$ paraffins (by *dehydrocyclization). The reaction takes place in the gaseous phase in tube reactors over a promoted
noble-metal cat. ($Pt-ZnAl_2O_4$, doped with
Sn). The combination of various Phillips
procs. is shown in the Figure. **F** procédé
STAR de Phillips; **G** Phillips STAR-Verfah-
ren.   B. CORNILS

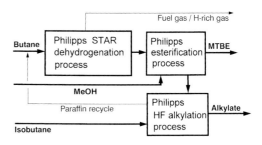

Ref.: *Hydrocarb.Proc.* **1997**, (3), 142.

Phillips Triolefin process   for the manuf.
of ethylene and butylenes (mainly 2-butene)
by fixed-bed *metathesis over a WO_3-SiO_2
cat. at 370–450 °C and 0.7 MPa (cf. *Phillips
POCT and *Phillips isoprene procs.). **F** procédé Triolefin de Phillips; **G** Phillips Triolefin-
Verfahren.   B. CORNILS

Ref.: Ertel/Knözinger/Weitkamp, Vol. 5, p. 2400.

PHIP parahydrogen induced nuclear polarization, → parahydrogen labeling

phosphanes   → correct IUPAC nomenclature (Rule R-2.0) for *phosphines. **E** phosphines; **F=G** phosphanes.

phosphatases   (EC 3.1.3.) hydrolyze phosphate esters of a variety of alcohols. Substrates include proteins (phosphorylated at
hydroxylated amino acids), sugar phosphates,
and nucleoside phosphates, among others.
This class of *enzymes has been useful in
synth. Phosphatases (ph.) have been used as
deprotecting agents in the *hydrolysis of
phosphates to alcohols. A particularly useful
ph. in this regard is the commercially available potato acid ph. 5′-Ribonucleotide ph. (5′-
ribonucleotidase) has been used in the *enantioselective *hydrolysis of racemic carbocyclic
ribonucleotides to prepare (−)-aristeromycin
and a 2′-fluoroguanosine derivative. Wheat
germ acid ph. cat. the selective hydrolysis of
L-*O*-phospho-threonine from the racemic
mixture. Calf intestine alkaline ph. cat. the
transfer of a phosphoryl group from pyro-

phosphate to various alcohols to form mono-alkyl phosphates and to decompose nucleo-side phosphates selectively in the presence of sugar nucleotides. **E=F**; **G** Phosphatasen.

C.-H. WONG

phosphines common name for P,H cpds. P_nH_{n+2} (e.g., PH_3, P_2H_4) and their organic derivatives. In analogy with amine derivatives with one, two, or three organic substituents they are called primary, secondary, and tertiary phs., e.g., *triphenylphosphine $[(C_6H_5)_3P = PPh_3]$. They can be obtained, e.g., by *alkylation of PH_3, by reaction of PCl_3 with Li organics or *Grignard cpds., or by Pd-cat. P-C coupling reactions of PH_3, primary, or secondary phs. with aryl halides in presence of a base.

In contrast to the widely used tertiary phs., PH_3 and primary/secondary phs. have not been employed as cat. comps. PPh_3 is well known as a *ligand and constituent of different cats. such as the *Vaska cpd., *Wilkinson cat., and the sulfonated species (*TPPTS) as the standard ligand for Rh *complexes in *aqueous phase cat. Special phs. serve as ligands in *Shell's oxo proc.(*ph. ligands). Primary and secondary phenylphosphines are used extensively as starting materials for the synth. of achiral and chiral ph. ligands employed in cat. **G** Phososphane.

O. STELZER

Ref.: *Catal. Today* **1998**, *42*, 413; Cornils/Herrmann-1, Vol. 1; Cornils/Herrmann-2; Beller/Bolm; Patai-2, Vol. 1; Pignolet; G.M. Koslapoff, L. Maier (Eds.), *Organic Phosphorus Compounds*, Vol. 1 Wiley-Interscience, New York 1972; K.B. Dillon, F. Mathey, J.F. Nixon, *Phosphorus: The Carbon Copy*, Wiley-VCH, Weinheim 1997; Kirk/Othmer-1, Vol. 18, p. 656; *Acc. Chem.Res.* **1999**, *32*, 9.

phosphine ligands are the most versatile group of *ligands in *coordination chemistry. For the non-bonding electron pair and the empty 3d orbitals of the P atom they show both σ-donor and π-acceptor properties. This stabilizes high and low oxidation states of *transition metals and makes it possible to fine-tune the electronic properties of the cat. active center. Their steric demand can control the *selectivities of cat. reactions (e.g., in *hydroformylation, *Tolman cone angle; for *bidentate phs. the *natural bite angle). Both their electronic and their steric properties can be varied synthetically over a wide range: PF_3 is an excellent π-acceptor and a poor σ-donor ligand, whereas triscyclohexylphosphine is a predominantly σ-donating *ligand.

Pls. are prepared from PX_3 (X = Br, Cl) by nucleophilic attack of *Grignard reagents or lithiated substrates. Alternatively, PH_3 can be reacted with electrophiles, or secondary pls. (HPR_2) may be cat. coupled with organic triflates by Pd cpds. The slow inversion of the pyramidal structure of P(III) allows the synth. of centrally chiral phs., bearing three different substituents. In solution most phosphine ligands are very susceptible to oxidation to P(V).

Monodentate pls. (PR_3), preferably triphenylphosphine PPh_3 for its ease of prep. and low price, are used extensively in hom. cat. Important commercial procs., in special Rh-cat. hydroformylations, use high PPh_3/Rh ratios to stabilize the active cat. species under the reaction conds.

Due to the *chelate effect *bidentate ligands are less inclined to dissociation from the metal center and therefore find wide application in *asymmetric cat. Chirality can be introduced at the P(III) center of the pl. or in its backbone. Both concepts have found commercial application: the synth. of *L-dopa with *DIPAMP, and the prep. of the herbicide *Metolachlor with diphosphine ligands with a chiral and *ferrocene backbone. Other important chiral diphosphines are *BINAP, *CHIRAPHOS, *DIOP, etc.

Much work has been done on the *immobilization of cat. *complexes via pls., for problems of *solubility in *fluorous- phase cat., or in *membrane applications. **F** ligands phosphiniques; **G** Phosphin-Liganden. F. RAMPF

Ref.: Tolman; *Chem.Rev.* **1977**, *77*, 313; *Angew. Chem.Int.Ed.Engl.* **1993**, *32*, 1524; G.M. Kosolapoff, L. Maier (Eds.), *Organic Phosphorus Compounds*, Wiley-Interscience, New York 1972; Hartley; Pignolet.

phosphinylation → aminophosphine phosphinites

phosphites triesters $P(OR)_3$ of phosphorous acid H_3PO_3. They are prep. by reaction of phenols or alcohols with PCl_3 in the presence of bases or by *transesterification with low molecular weight phosphites. $P(OC_6H_5)_3$ is used as a co-ligand in a variety of hom. cat. procs. such as Pd-mediated *hydrovinylation *Pauson-Khand reactions or Ni-mediated *oligomerizations. Although phs. do not support *isomerization of the feed alkenes, as *ligands in Rh-cat. *hydroformylations phosphines are cheaper and more resistant against moisture. Phosphine-phosphite type ligands such as *BINAPHOS are employed in the asymmetric Rh- cat. version. Some phs. are very effective ligands for the hydroformylation of higher alkenes.   O. STELZER

Ref.: *J.Prakt.Chem.* **1993**, *355*, 75; Cornils/Herrmann-1; *J. Organomet.Chem.* **1991**, *421*, 121; *Organometallics* **1997**, *16*, 2929 and 5681; *J.Chem.Soc., Dalton Trans.* **1995**, 409.

3′-phosphoadenosine 5′-phosphosulfate

(PAPS) a *cofactor for enzymatic *sulfation of carbohydrates, proteins, and other molecules.

PAPS

APS

Two *enzymes generate PAPS from *ATP and sulfate: ATP- sulfurylase (EC 2.7.7.4) cat. the reaction of ATP and sulfate to form *adenosine 5′-phosphosulfate (APS) and APS-kinase (EC 2.7.1.25) cat. the phosphorylation of APS with ATP to form PAPS. Coupling of the ATP-sulfurylase reaction with pyrophosphatase drives the reaction toward APS formation. The *sulfotransferases use PAPS in sulfation; 5′-phosphoadenosine 5′-phosphate (PAP) is generated as a byproduct. A possible *regeneration of PAPS from PAP is via AMP. For small-scale sulfation reactions, PAPS can be used stoichiometrically, and a number of chemical and enzymatic methods have been reported for the synth. of PAPS. **E=F=G**.

C.-H. WONG

Ref.: *Methods Enzymol.* **1987**, *143*, 329; *Biochem.Biophys. Acta* **1977**, *480*, 376.

phosphodiesterases *enzymes that *hydrolyze phosphodiesters to the phosphomonoester and alcohol. One important example is the 3′,5′-cyclic AMP ph. which hydrolyzes the second messenger, cAMP, to the adenosine 5′-phosphate.

3′,5′-cyclic AMP phosphodiesterase

Deoxyribonucleases (DNases) and ribonucleases (RNases) are also ps., and are commonly used reagents in molecular biology. **E=F**; **G** Phosphodiesterase.   P.S. SEARS

phosphoenolpyruvate (PEP; pyruvate kinase PK) This system, using PEP as the phosphate donor in a pyruvate kinase (PK; EC 2.7.1.40)-cat. reaction, is the best enzymatic system for ATP regeneration.

PEP has a high phosphate donor capability with a $\Delta G^{\circ\prime}$ (hydrolysis) = -12.8 kcal/mol. The stability of PEP in solution is excellent. Although the synth. of PEP is slightly more difficult and expensive than those of other phosphate donors, the higher stability and donor capability offset this minor disadvantage. A more serious disadvantage of the PEP/PK system is the problem of *product inhibition. Pyruvate is a *competitive inhibitor of the PK reaction. In order to minimize this problem, the reaction must be run in dilute solutions, high concs. of PEP must be used, or pyruvate must be removed from the system. Pyruvate is also reactive toward NAD(P) in some circumstances, and may interfere with some nicotinamide *cofactor-requiring reactions (see also *adenosine triphosphate regeneration). **E=F**; **G** Phosphoenolpyruvat.   C.-H. WONG

Ref.: *J.Org.Chem.* **1982**, *47*, 3765.

phosphofructokinase cat. the phosphorylation of β-D-fructose-6-phosphate to form β-D-fructose-1,6-diphosphate using *ATP.

P.S. SEARS

phosphoglucomutase cat. the reversible conv. between α-glucose-1-phosphate and glucose-6-phosphate.

P.S. SEARS

phosphoglycerate mutase cat. the reversible *isomerization of D-3-phosphoglycerate to D-2-phosphoglycerate.

P.S. SEARS

phospho-5-keto-2-deoxygluconate aldolase (EC 4.1.2.29) cat. the formation of phospho-5-keto-2-deoxygluconate from dihydroxyacetone phosphate and 3-oxopropionic acid. **E=F=G.**

C.-H. WONG

Ref.: *J.Biol.Chem.* **1971**, *246*, 5662.

phospholipases cat. the *hydrolysis of phospholipids. Four different types of *enzymes with different *regio- and *stereoselectivity have been identified; their cleavage sites are indicated in the Figure. The enzymatic cat. generally occurs at the interface of the aggregated substrates.

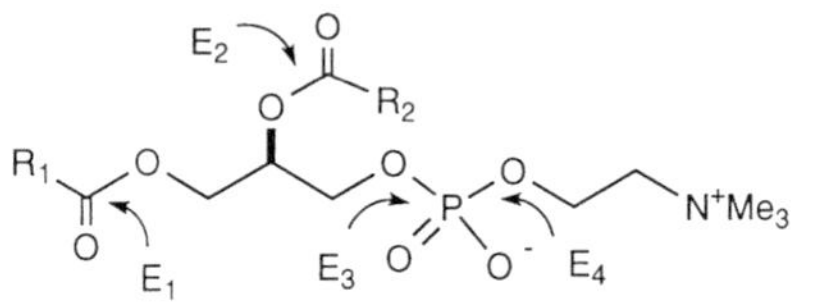

E$_1$: phospholipase A$_1$ E$_2$: phospholipase A$_2$
E$_3$: phospholipase C   E$_4$: phospholipase D

Of many ps. known, phospholipase A$_2$ from Cobra venom is the best studied. The enzyme-substrate interaction facilitates the diffusion of the substrate from the interfacial binding surface to the cat. site, rather than facilitating a conformational change in the enzyme (as suggested in the case of lipase).
A number of phospholipids chiral at phosphorus have been prepared to study the *stereospecificity of ps. Phosphatidyl-specific p. C reaction proceeds through the formation of inositol 1,2-cyclic phosphate via a general

base-cat. intramolecular attack of a neighboring hydroxyl group, followed by *hydrolysis of the cyclic phosphate to a phosphomonoester. The proc. is similar to that cat. by some *ribonucleases (e.g., ribonuclease A.).
P. D from *Streptomyces sp.* was used to cat. the exchange of choline from a phospholipid with other primary alcohols such as nucleosides. **E=F; G** Phospholipase.

C.-H. WONG

Ref.: *J.Am.Chem.Soc.* **1997**, *119*, 9933; *Science* **1996**, *380*, 595.

phosphonium salts quaternary onium cpds. of the type $[PY_4]^+$ X with Y = halogens or organic substituents and X = anions of low basicity. Simple ps. (e.g. $[CH_3P(C_6H_5)_3]^+I^-$) function as methyl group transfer agents in methanol *homologation procs. Long chain derivatives act as effective *promoters in Rh-cat. *hydroformylation of higher alkenes. *Carbonylations are described; Hoechst's *AAA proc. used *n*-butyl phosphonium iodide as *copromoter of the Rh cat. In *Kuraray's *hydrodimerization proc. ps. stabilize the cat. system. Amphiphilic *phosphines with a phosphonium end group form Rh complexes which are very active *hydrogenation cats. in *two-phasic systems.

O. STELZER

Ref.: Cornils/Herrmann-1, Vol. 1,2, p. 353, 911; Cornils/Herrmann-2, p. 166; EP 602.463 (1994); DE 4.301.310 (1994); *J.Mol.Catal.* **1993**, *84*, 157; *J. Organomet.Chem.* **1991**, *419*, 403.

phosphoranes P organic cpds. with coordination number 5, derived from hypothetical PH$_5$ (acc. to IUPAC rules: λ^5-phosphanes, but the name is also used for ylides, e.g., $(C_6H_5)_3P=CH_2$. On the other hand, phosphanes are "phosphines" in American colloquial use). The stereochemical arrangement of the *ligands at P is preferably that of a trigonal bipyramid, the alternative square pyramidal structure being found only rarely in stable cpds. (Figure).
It is adopted by reaction intermediates and is involved in the pseudorotation proc. (Berry pseudorotation) that interchanges *ligands in axial and equatorial positions of the trigonal bipyramidal structure.

O. STELZER

phosphoranes

Ref.: G.M. Koslapoff, L. Maier, *Organic Phosphorus Compounds*, Wiley-Interscience, New York 1972, Vol. 3, p. 185f.; Cornils/Hermann-1, Vol. 1,2, p. 254, 663.

phosphorolysis, polysaccharides → *glycosyl phosphorylase

phosphorothioate-containing DNA and RNA (DNA-S and RNA-S) Both polymers can be prepared from the monomers (S_P-NTPαS) enzymatically.

$$S_P\text{-dNTP}\alpha S \xrightarrow{\text{DNA polymerase}} R_P\text{-DNA-S}$$

$$S_P\text{-dNTP}\alpha S \xrightarrow{\text{RNA polymerase}} R_P\text{-RNA-S}$$

These polymers containing R_P-phosphorothioate are resistant to nuclease and are recognized by *kinases, *ligases and restriction *enzymes. **F** phosphothioate avec DNA et RNA; **G** DNA und RNA enthaltendes Phosphothioat.   C.-H. WONG
Ref.: *TIBS* **1989**, 97.

phosphorous oxy- or sulfochloride
POCl$_3$ is manuf. by *radically initiated reaction of PCl$_3$ and oxygen at 50–60 °C. Fe, Cu, and S act as *inhibitors. PSCl$_3$, an intermediate for pesticide building blocks, originates from the sulfonation with AlCl$_3$ as cat. **G** Phosphoroxy- oder Sulfochloride.   B. CORNILS
Ref.: Büchner/Schliebs/Winter/Büchel, pp. 92, 93; Büchel/Moretto/Woditsch.

phosphorylation
the introduction of the phosphoryl group -P(O)(OH)$_2$. P. is important in enzymatic cat. It is a common motif for –1: driving unfavorable reactions forward by coupling them with highly favorable ones; 2: activation of functional groups for, e.g.,

elimination or *substitution reactions; 3: regulation of *enzyme activities (see *protein kinase). **G** Phosphorylierung.   P.S. SEARS

phosphotransacetylase → acetyl CoA regeneration

photoacoustic spectroscopy (PAS) a
physical, non-destructive method for the analysis of electronic absorption spectra of crystalline solids, amorphous substances, semisolids, gels, and liquids. The method is based on the optoacoustic effect (OAE, Bell), more precisely the transformation of optic to acoustic energy. Periodic interrution of irradiation with sunlight (more generally *UV-vis) of a substance in a "listening" tube generates sound waves of the same frequency. When light is absorbed the excitation energy can be transformed via photochemistry into vibrational and translational energy. The resulting periodic warming/cooling express themselves as press. pulses (sound). Applications: quantitative determination of impurities (ppm scale) in gases, solids, and liquids; analysis of opaque materials; *surface chemistry with materials having profiles between 0.5 and 50 µm; color-pigment powders; corrosion of metals; *semiconductors; determination of quantum yields, etc. mechanistic studies of photochemical procs. and cat. (cf. also *optoacoustic sp.). **F** spectroscopie photoacoustique; **G** Photoakustische Spektroskopie.   A. HEUMANN
Ref.: *Angew.Chem.Int.Ed.Eng.* **1978**, *17*, 238; *Acc. Chem.Res.* **1973**, 6, 329.

photoactivated (photoassisted, photoenhanced, photoinduced reactions) → photocatalysis

photoassisted reactions → photocatalysis

photocatalysis
an important domain in organic synth. that combines the advantages of hom. or het. cat. with simple and mild working conds. Reactions are performed at rt. and normal press. and the cat. handling does

not require special precautions since the active species is formed in situ. IUPAC defines p. as "a cat. reaction involving light absorption by a cat. or by a substrate". Acc. to the pragmatic definition p. concerns reactions mediated with consumption of photons and in the presence of an added substance (metal *complex in metal- cat. p.). More simply, two combinations may be distinguished, one operating with a cat. quantity of both photons and added substances, generally called *photoinduced* cat. In the other possibility, the *photoassisted* reaction, the number of photons consumed corresponds to the amount of cpds. produced, and the light is necessary throughout the reaction time. Photoinduced reactions are characterized by possible quantum yields greater than 1 ($\Phi > 1$), a possible induction period, and the continuation of the cat. reaction without light. The action of the photons (light) is limited to the generation of the active cat. from the pro-cat., e.g., from light-sensitive coordination *complexes (Figure, scheme A). An illustrative example is the activation of $Fe(CO)_5$ by loss of one CO ligand in the *photoisomerization of pentene. The *activation of the *ligand, photochemically dissociated or associated with the metal, may produce interesting photoinduced *chain procs. such as the *photooxidation of alcohols.

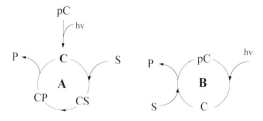

pC: pro-catalyst; P: product; S: substrate; C: catalyst; CS: complex between C and S; CP: complex between C and P.

In *stoichiometric* photoinduction the photon acts inside the cat. cycle. For every S → P transformation a photon is consumed and the overall quantum yield for the production of P is smaller than 1 ($\Phi < 1$, scheme B). Several situations should be mentioned, such as reaction of the substrate in an excited state with the cat., photochemical acceleration of different steps in the *cat. cycle, or photosensitization with cat. quantities of photosensitizer (with energy or *electron transfer).

The most interesting p. reaction is the natural *photosynthesis in plants for the storage and transformation of sunlight. A more commercial version of extraterrestrial energy storage would be photocat. splitting of water and prod. of H_2 (*photoinduced oxidation/reduction of water). In organic synth. many of the metal-cat. reactions are suitable for p. Examples are activation of unsaturated bonds, activation of small molecules, *metathesis, electron transfer procs., *insertion reactions, *cyclization, *cycloaddition, wastewater treatment, or even *C-H activation. Drawbacks are the restrictions encountered in ordinary photochemical reactions such as low quantum yields, rapid recombination procs., and possible photosensitivity of the reaction prods. Cf. also *titania photocatalysis. **F** photocatalyse; **G** Photokatalyse. A. HEUMANN

Ref.: M. Chanon (Ed.), *Homogeneous Photocatalysis*, Wiley, Chichester 1998; M. Schiavello (Ed.), *Heterogeneous Photocatalysis*, Wiley, Chichester 1997; Cornils/Herrmann-1, p. 929; N. Serpone, E. Pelizzetti (Eds.), *Transition Metal Complexex in Photocatalysis*, Wiley, New York 1989, pp. 1, 489; Beller/Bolm, p. 412; *Chem. Rev.* **1995**, 95, 735.

photocleavage → photolysis

photocyclization access route to various cyclic and polycyclic carbo- and heterocycles via a free-*radical mechanism (e.g., arene radicals) and heteroatom cycloisomerization via a *domino reaction. These photoreactions involve olefinic, aromatic, or acetylenic unsaturated groups and proceed as photodimerization, cross-cycloaddition with simple Cu salts such as CuOTf. An example is the [2+2] cycloaddition to strained four-membered rings in polycyclic systems. **F** photocyclisation; **G** Photocyclisierung. A. HEUMANN

Ref.: Trost/Fleming, Vol. 5, p. 147; *Angew.Chem.Int. Ed.Engl.* **1993**, *32*, 131; L.A. Paquette (Ed.), *Encyclopedia of Reagents for Organic Synthesis*, Wiley, Chichester 1995, Vol. 2, p. 1370; *Tetrahedron* **1983**, *39*, 485.

photodehydrogenation → photooxidation

photodeposition of metals on *single crystals occurs when an electron that has been excited into the *conduction band is transferred to a metal ion in solution. For *deposition of metal to occur, the energy of the conduction-band electron must be more negative than the standard *redox potential (E_0) of the metal^{n+}/metal couple. Photodeposited metals are often used as *photocatalysts. An improvement of the cat. activity can be achieved when the deposited metal can decrease the recombination of electron-hole pairs in the photocatalyst. **E**=**F**; **G** Photoabscheidung.   R. SCHLÖGL, B. BEMS

photodimerization → *photocyclization

photoelectrochemistry   To use solar energy requires the photochemical activation of molecules. As very few receptor materials exist which can produce the necessary charge carriers with sufficient energy to break chemical bonds (above 3 eV) and as the solar energy spectrum on Earth is very weak in this energy range, efforts have started to combine photochemistry with electrochemistry; cf. also *photoinduced oxidation/reduction of water. **F** photoélectrochimie; **G** Photoelektrochemie.   R. SCHLÖGL

photoelectron diffraction → electron microscopy

photoelectron emission microscopy (PEEM) a variant of *photoelectron microscopy, suitable for, e.g., the measurement of local *work functions.   R. SCHLÖGL

photoelectron spectroscopy cf. *UPS and *XPS.

photoinduced reactions → photocatalysis

photoinduced oxidation/reduction of water   Decomposition of water into O_2 and H_2 is of interest to convert photon (solar) energy into chemical energy, with water and sunlight being almost unlimited resources and H_2 being a high-energy, environmentally sound fuel and a valuable commodity (cf. *hydrogen manufacture). Photocatalytic water decomposition can be achieved using het. catalysis, with the cats. being highly dispersed *semiconductors, i.e., TiO_2, Ta_2O_5, titanates, or niobates which are promoted with, e.g., Pt, Rh, RuO_x, or NiO_x. The decomposition can also be performed in an electrochemical cell (photoelectrolysis), e.g., with a metal (Pt) cathode and and a semiconductor ($SrTiO_3$; in early investigations, TiO_2) photoanode, which in this case is possible without external bias and even yields stoichiometric amounts of O_2 and H_2. Hom. cat. has also been applied to cleave water, with, e.g., heteropolyanions giving O_2 evolution, and, e.g., Fe, Ru, Co *complexes giving H_2 evolution. **F** photo oxidation/réduction de l'eau; **G** Photooxidation/Reduktion von Wasser.   F. JENTOFT

Ref.: M. Grätzel (Ed.), *Energy Resources through Photochemistry and Catalysis*, Academic Press, New York 1983; *Catal. Today* **1998**, *44*, 17; *J.Am.Chem.Soc.* **1976**, *98*, 2774; *Nature (London)* **1972**, *37*, 238; Thomas/Thomas, p. 594.

photoisomerization the *transition-metal cat. transposition of unsaturated (mainly olefinic) systems such as double-bond migration of alkenes (e.g. 1-pentene over Fe(CO)$_5$ as prototype), *N*-allylamines and amides, allylic *hydrogen shifts, formation of vinyl ethers from unsaturated ethers, ketones from unsaturated alcohols, or the *cis/trans* *isomerization of 1,5,9-cyclododecatriene over Cu cats. P. sometimes accompanies *dimerization of alkenes, e.g., norbornadiene, resulting in the formation of different highly complex polycyclic cpds. P. of multiple types of dienes leads with CuOTf to functionalized polycyclic cpds. Valence isomerization of norbornadiene over CuCl yields quadricyclane. **F** photo-isomérisation; **G** Photoisomerisierung.

A. HEUMANN

Ref.: see refs. under keyword *photocatalysis

photolysis the photochemical decomposition or splitting (cleavage) of bonds and the activating step of the metal cats. in photoinduced reactions. The photolytic step liberates a *ligand (e.g., cp-CoCOD → cp-Co + COD) or provokes cleavage of the metal-metal bond in dimeric complexes (e.g.,$(CO)_5$Re-Re$(CO)_5$ → 2 Re$(CO)_5$) creating a *vacant site on the metal. Metal-cat. photolytic water splitting to H_2 and O_2 may be the solution to solar conversion and storage (cf. *photoinduced oxidation/reduction of water, *thermocatalysis). **F** photolyse; **G** Photolyse. A. HEUMANN

Ref.: M. Chanon (Ed.), *Photocatalysis*, Wiley, Chichester 1998.

photonitrosation → Toray PNC process

photooxidation the oxidation with participation of light, which proceeds with (*photooxygenation*) or without (*photodehydrogenation*, *decarboxylation) incorporation of O atoms into the reaction prods. (cf. *Hill reaction). Alcohols are transformed to aldehydes with Pt cats. and hcs., to alkenes with Cu, heteropolytungsten acids/Pt, or Rh *complexes. Cyclooctane is oxidized to cyclooctene with quantum yields of 0.10 and *TONs up to 5000. The active cat. species Rh$(PMe_3)_2$Cl is formed by photoextrusion of CO from Rh$(PMe_3)_2$(CO)Cl delivering the energy for the thermodymamically unfavored *C-H activation proc. With other transition metals (Ru, Os) oxygen is incorporated, e.g., in the transformation of cyclohexane to *KA oil. Phenols or benzylic oxidation prods. are formed from arenes over Cu, Co, or Fe cats. P. of alkenes generally proceeds with incorporation of dioxygen. *Photooxygenation* prods. are *peroxides, hydroperoxides, ketones, *epoxides, and even epoxy alcohols via *ene*-reaction over Ti, V, Mo. The photosensitized metal transfers energy to the ground state triplet oxygen which changes to singlet O_2 and reacts to the peroxo cpd., allowing the *regio- or *stereo-selective introduction of the oxygen-containing function. **E=F=G**. A. HEUMANN

Ref.: *J.Am.Chem.Soc.* **1989**, *111*, 7088 and **1994**, *116*, 2400; *J.Org.Chem.* **1994**, *59*, 3341.

photopolymerization *polymerization, initiated by light

photosynthesis refers to the set of reactions by which plants, algae, and certain bacteria and protozoans use light to synth. high-energy molecules (*ATP and a strong reductant, *NADPH) which may then be used to *reduce CO_2 to sugars. It falls naturally into two sets of reactions: the "light reactions," in which light is actually harvested and the ATP and NADPH are generated; and the "dark reactions," in which CO_2 is reduced. See also *Calvin cycle, *electron transport. **F** photosynthèse; **G** Photosynthese. P.S. SEARS

phthalic (acid) anhydride (PA, PAA) one of the important, cat. manufd. bulk chemicals via *VPO *oxidation of *o*-xylene or naphthalene (e.g., procs. of *BASF, *Rhône-Progil, *Wacker). **F** anhydride phthalique; **G** Phthalsäureanhydrid. B. CORNILS

physical adsorption → physisorption

Physisorp. abbr. for *physisorption

physisorption is *adsorption in which the molecule/atom is held by weak intermolecular forces characterized by adsorption energies between –10 and –40 kJ/mol, i.e., within the range of condensation enthalpies. The intermolecular forces can be the attraction between permanent or induced dipoles or quadrupoles, respectively, and/or the attraction due to *van der Waals-London forces. In contrast to *chemisorption, which involves overlap between orbitals and therefore requires short bonding distances from the *surface in the physisorbed state, the bond distance to the surface is greater. The potential energy diagram for p. is usually described by a diagram of the *Lennard-Jones type.

Due to the low *adsorption energy physi-sorbed molecules are highly mobile on the *surface. Physisorbed states of a molecule often play the role of a precursor state for chemisorption. **E=F=G.**   R. IMBIHL
Ref.: Somorjai; Gates-1, p. 327.

Pichia sp.   → yeast

π-donation   → backdonation

pig insulin   differs from human insulin by a single amino acid: the β-chain C-terminal residue is Ala for pig insulin and Thr for human. Carboxylase A or the protease from *A. lyticus* cat. the exchange of the C-terminal residues with Thr *tert*-butyl ester. The prod. can then be converted to human insulin after deprotection of the ester group. **F** insuline de porc; **G** Schweineinsulin.   C.-H. WONG
Ref.: *TIBTECH* **1987**, 5, 164.

pig liver esterase (EC 3.1.1.1, PLE)   a serine type of *esterase that cat. the *stereoselective *hydrolysis of a wide variety of esters. Of several models developed for interpreting and predicting the specificity of PLE, the cubic-space *active-site model has been applied to a number of prochiral, cyclic, and acyclic *meso* esters, and racemic esters.

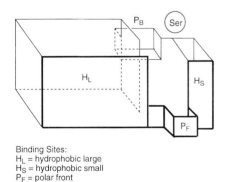

Binding Sites:
H$_L$ = hydrophobic large
H$_S$ = hydrophobic small
P$_F$ = polar front
P$_B$ = polar back

The *enzyme also cat. the hydrolysis of lactams, lactones, cyclic carbonates, phosphonates, and dimethyl aziridine 2,3-dicarboxylates. The enzymatic hydrolysis of γ-phenyl-γ-

butyrolactone is selective for the *S*-enantiomer (>94 % *ee*), but no *selectivity was observed in the hydrolysis of the corresponding acyclic ester (e.g., γ-phenyl-γ-hydroxy substituted methyl butanoate), suggesting that the *acylation step shows the selectvity. A high *enantioselectivity (92 % *ee*) was observed for the *meso*-aziridine substrate without protection of the N-group. A low enantioselectivity was, however, observed in the enzymatic hydrolysis of racemic or N-protected *meso*-aziridine derivatives. PLE was also used to cat. enantioselective *transesterifications of chiral alcohols in a *biphasic organic-aqueous system having high volume fractions of the organic phase. **F** esterase de foie de porc; **G** Schweineleberesterase.   C.-H. WONG
Ref.: *J. Am. Chem. Soc.* **1990**, *112*, 4946; *Organic Reactions* **1989**, Vol. 37, p. 1.

Pier Matthias (1882–1965), industrial chemist with BASF and professor of physical chemistry at Heidelberg (Germany). Worked on catalysis and especially on *methanol synthesis and *coal hydrogenation.   B. CORNILS
Ref.: *Erdöl und Kohle-Erdgas-Petrochem.* **1982**, 35(6), 283.

PILCs   pillared clay catalysts, cf. *pillared clay

pillared clay   Clay minerals which contain negatively charged layers due to exchange of Mg^{2+} for Al^{3+} in the sublayer with octahedral coordination can in many cases exchange the simple charge-compensating interlayer cations (as Na^+, K^+, Ca^{2+}) against other cations. Also, they can intercalate neutral molecules like water, a proc. whereby the distance between the *alumosilicate layers can increase drastically ("swelling"). If in an *ion-exchange proc. large oligomeric inorganic cations are used, the layer distance also shows a large increase. When upon *calcination of the composite material at higher temps. the guest species form strong covalent bonds with the silicate layers, then they act as pillars and prop up the interlayer space, thus creating *porosity. Such a material is termed a "pillared clay". Typical pillaring agents are $[Al_{13}O_4(OH)_{24} (H_2O)_{12}]^{7+}$ with a *Keggin ion structure and the "zirconium tetramer" $[Zr(OH)_2 \cdot 4\ H_2O]_4^{8+}$. Pcs. usually exhibit a bimodal distribution of *pore diameters, with *microporosity stemming from the pillared interlayer regions, and *mesoporosity from the irregular arrangement of the layers after the ion-exchange proc. (so-called "house-of-cards" structure). Pcs. with different compositions and different *surface areas have been tested in various cat. applications. Other materials that can be pillared include zirconium phosphates and layered double hydroxides. **E=F=G**.   P. BEHRENS

Ref.: Ertl/Knözinger/Weitkamp, Vol. 1, p. 387.

pilot plant → catalytic pilot plant

pinacol rearrangement rearrangement of vicinal diols initiated by the *addition of acids (H_2SO_4, HNO_3, *Lewis acids) to yield aldehydes or ketones acc. to R^1-$CR^2(OH)$-$CR^3(OH)$-R^4 → R^1-CO-$CR^2R^3R^4$. Although the reaction is mostly carried out with tri- or tetra-substituted glycols, the pr. has also been accomplished with terminal and disubstituted dialcohols. If at least one substituent is hydrogen, aldehyde formation is favored at lower temps. and with weaker acids as cat. The mechanism of the pr. involves protonation of a hydroxy group and *dehydration followed by a simple *1,2-hydride shift or the shift of the R group. The driving force is the stabilization of the *carbenium ion by an O atom. It has been demonstrated that epoxides are sometimes intermediates in the pr., which also occurs with epoxides when treated with acidic agents. **F** pinacol réarrangement; **G** Pinakol-Umlagerung.   M. BELLER

Ref.: March; Mulzer/Waldmann, p. 113.

π-stacking (arene stacking) describes the association of aromatic rings in solution and solid state driven by a combination of *van der Waals interactions, electrostatic, dispersive, and polarization forces. In the case of

*complexes between π-donors and π-acceptors electron donor-acceptor interactions also emerge. The strength of the effect is determined by the polarity and polarizability of the solvent and the cohesive molecular interaction: it is especially strong in weakly polarizable but polar media such as water (hydrophobic effect), but weak in non-polar but easily polarizable organic solvents. Edge-to-face arrangements (often with T-shaped geometry) in which a positively polarized H atom of one aromatic ring is directed toward the π-electron cloud of another are frequently observed. A common feature of donor-acceptor complexes is an offset face-to-face arrangement of the chromophores where the orientation is controlled by the distribution of the partial charges. A further example is the stabilization of the *DNA double helix by vertical intrastrand base stacking. **E=F=G**.

C. SEEL, F. VÖGTLE

Ref.: J.Am.Chem.Soc. **1990**, 112, 5525.

Pittsburgh Plate and Glass → PPG procs.

PIXE → proton-induced X-ray emission (spectroscopy)

PK (pyruvate kinase) → phosphoenolpyruvate

plasma catalysis techniques Gas-phase plasma procs. generate activated species (*radicals, electrons, ions), and the ions cause sputtering of the electrode material (cf. *chemical vapor deposition). Pcts. can therefore be used to support cat. reactions as well as to *regenerate spent cats., and to prepare het. cats.

In high temp. plasma chemistry, thermal plasmas szuch as arcs are used technically for the manuf. of C_2H_2, NO_x, or HCN. Non-equil. plasms at lower temps. offer potential advantages if combined with the use of cats. This combination is relatively unexplored; the cat. *surface probably provides sites for radical combination. The corrsponding cat. can be placed in or outside the plasma zone. Due to selective *activation and stabilization (lifetime) of the reactants by non-equil. pct., in some cases the chemical equil. is affected advantegeously.

The prepn. of cats. by a non-equil. pct. takes place far from the thermodynamic equil. By *deposition and surface etching, unusual solid-state structures and defect sites are created which can serve as *active centers for the cat. reaction. Besides sputtering, *metal oxides are deposited from organometal cpds. in oxygen plasma. So far, only a few applications for pct. exist in catalysis. In the oxidative cleaning of exhaust gases by the plasmacat proc., volatile organic cpds. (*VOCs) are totally oxidized when passing the cat. bed. **F** techniques catalytiques plasma; **G** Plasma-Katalyse-Techniken.   J. CARO, J.P. MÜLLER
Ref.: *Ind. Eng. Chem. Res.* **1997**, *36*, 632.

plasmid an extrachromosomal, circular piece of *DNA which includes an origin of replication for the host organism(s). Artificially constructed plasmids are used for gene transfer into an organism, and typically encode a protein conferring antibiotic resistance, as a means of selection. Many plasmids are commercially available. These differ with regard to 1: the type of promoter; 2: the copy number; 3: the selection method; 4: the multiple cloning site; and 5: sequences adjacent to the multiple cloning site which will be transcribed and translated as fusions to the protein of interest. **E=F=G**.   P.S. SEARS

plasmid-based catalysis → metabolic engineering; multienzyme systems

plasticizer cpds. incorporated in and compatible with a material to increase its workability, flexibility, or distensibility. Organic p. are normally high molecular weight and high-boiling liquids. Chemically they are mostly esters of carboxylic or phosphoric acids (cf. *2-EH, *DOP, *phthalic anhydride, *n/i ratio). **F** plastifiant; **G** Weichmacher.
                                        B. CORNILS

Ref.: Kirk/Othmer-1, Vol. 19, p. 258; Kirk/Othmer-2, p. 902; Ullmann, Vol. A20, p. 439.

platforming a proc. for the upgrading of straight-run gasolines by treatment (*cyclizations, degradation, and *isomerization of paraffins, *dehydrogenation of naphthenes, increase of *ON, etc.) over Pt cats. at 450 °C/5 MPa. Procs. of interest are *Arco Magnaforming, *Exxon Powerforming, *Houdry Houdriforming, *Howe-Baker, *IFP, *UOP proc.). **F=G=E**.   B. CORNILS

platinum as catalyst metal 1-heterogeneous: Pt is the "historic" metal in cat. (*Davy, *Döbereiner, *history in catalysis); its properties are similar to those of Pd. Metallic Pt is used as Pt black or as *gauze cat.; *Adams' cat. has its own name. Pt is famous for its *hydrogenative (and *dehydrogenative) capabilities for nearly all functional groups at relative low temps. and press. (*UOP Oleflex or Hydrar procs.; *hydroxylamine) including *dehydrocyclization, *hydroxylamine, or the synth. of HCN from methane and *ammonia (*Degussa BMA proc.). *Isomerization (*dehydroisomerization) and *reforming reactions are cat. by Pt cats. (e.g. *ARCO Octafining, *BP butane isom; *Houdriforming, *Howe-Baker reforming, *UOP reforming, *Phillips STAR, *platforming procs., etc.). Pt cats. serve for *oxidation reactions as well (*contact proc. for SO_2 oxid.; *oxidation of ammonia, *automotive exhaust cats.). Other Pt cats. are used for *hydrosilylation, *aromatization (Pt supported on *zeolites), *hydroformylation (Sn-doped Pt), or the manuf. of H_2O_2 (*anthraquinone proc.).
                                        B. CORNILS

2-homogeneous: As one of the most frequently encountered cat. metals, Pt effects many reactions such as *asymmetric *hydroformylations (Pt-Sn), *C-H activation (*Shilov reaction), the enol-ether synth. (sodium hexachloroplatinate), *hydrosilylation (*Speier cat.), *hydrocarboxylation (*ligand modified Pt *complexes), *water-gas shift reaction (Pt complexes), and *isomerizations. Commercially, the largest amount of Pt in hom. cat. is used for the versatile silylation procs. In some Pt cat. procs., *redox steps $Pt^0 \rightleftharpoons Pt^{2+} \rightleftharpoons Pt^{4+}$ are involved, while Pd reactions do not appear to contain Pd^{4+} intermediates (cf. *Heck reaction). **F** le platine comme métal catalytique; **G** Platin als Katalysatormetall. W.A. HERRMANN

Ref.: *Platinum Metals Review* **1992**, *20*(1), 12; Farrauto/Bartholomew, p. 413; Rylander; R.N. Rylander, *Organic Synthesis with Noble Metal Catalysts*, Academic Press, New York 1973; *Appl. Catal.* **1983**, *8*, 167; *Houben-Weyl/Methodicum Chimicum*, Vol. XIII/9a; *Coord. Chem. Rev.* **1982**, *35*, 143; B. Lippert (Ed.), *30 Years of Cisplatin*, Wiley-VCH, Weinheim 1999; Cornils/Herrmann-1 and -2; Kirk/Othmer-1, Vol. 19, p. 347; McKetta-1, Vol. 38, p. 401 and Vol. 41, p. 218.

platinum black → platinum as catalyst metal

platinum oxide Highly dispersed Pt incorporates considerable amounts of oxygen, which results in its cat. activity in *oxidation procs. For example, SO_2 is oxidized to SO_3 even at 400 °C over Pt (*sulfuric acid synth.), and ammonia is oxidized to NO in 98 % yield over Pt (or Pt-Rh) *gauze cats. at about 600 – 700 °C during very short contact times (*oxidation of ammonia).
Pt oxides are strong oxidants. Reddish-brownh PtO_3 can be synth. by anodic oxidation. Blackish-brown $PtO_2 \cdot n\, H_2O$ (n = 1, 2, 3, 4) is formed from hexahydroxyplatinic acid during its dehydration. It decomposes into PtO under loss of O at 400 °C, the latter decomposes into Pt and oxygen at 560 °C. G. MESTL

plug-flow reactor (PFR) Plug flow is an idealized picture of the fluid motion whereby all the fluid elements move with uniform velocity in parallel streamlines. This is an ideal in the *steady state and can be approximated by tubular reactors. From that, it is assumed that there is no axial mixing (back mixing) and no radial gradients of conc. or fluid velocity. Therefore, the fluid elements pass through the cat. bed in the plug flow; all molecules have a same residence time τ. Thus, the ideal tubular reactor is a called plug-flow or piston-flow reactor. **F** réacteur au régime bouchon; **G** Strömungsrohrreaktor.

P. CLAUS, D. HÖNICKE

PMA poly methyl acrylate.

PNC process → Toray procs.

P-N ligands contain both trivalent P and trivalent N atoms interconnected by a chain of carbon or heteroatoms. The combination of *hard and *soft donor atoms makes P-N *ligands valuable for hom. cats. and reactions such as *carbonylation, *hydroformylation, or *hydrovinylation. An example is the Pd-cat. *methoxycarbonylation of propyne to methyl methacrylate (MMA). Replacement of *triphenylphosphine by the P-N *ligand 2-pyridyldiphenylphosphine results in an increase in activity by some three orders of magnitude and selectivity (*carbonylation of alkynes). P-N ligands were also successfully used in some *stereoselective reactions such as *Grignard cross-couplings, *hydroboration, or *hydroformylation. Misleadingly, the term P-N ligands is sometimes utilized to describe phosphines with direct P-N bonds, such as aminophosphines, *aminophosphinites, etc. **E=F**; **G** P-N-Liganden. R. ECKL

Ref.: *Tetrahedron Asymmetry* **1993**, *4*, 743 and **1995**, *6*, 2593; *Angew. Chem. Int. Ed. Engl.* **1994**, *33*, 504; *Chem. Rev.* **1996**, *96*, 663.

PO → propylene oxide

POCT process → Phillips procs.

podand(t) multidentate *ligand with one central connecting unit. **E=F=G**. R.D. KOHN

point of zero charge (pzc) → titania photocatalysis

poison A *catalyst poison is a substance which interacts strongly with a catalyst or cat. ingredients (e.g., by chemical or *surface reaction) and thus causes a substantial loss in activity (*deactivation, *regeneration). A typical example are S cpds. which form cat. inactive sulfides with metals (*sulfur poisoning). Solid bases are subject to *poisoning by H_2O and CO_2, cf. also *superbases. **F=E; G** Gift.

B. CORNILS

Ref.: J. Oudar, H. Wise, *Deactivation and Poisoning of Catalysts*, Dekker, New York 1985; Twigg, p. 76; Ertl/ Knözinger/ Weitkamp, Vol. 3, p. 1084; Butt/Petersen; Thomas/Thomas, p. 417; Farrauto/Bartholomew, p. 266.

poisoning → deactivation, *directed poisoning

Poisson distribution A Pd. is formed if all the chains of a polymer are initiated at the same time and therefore grow simultaneously, independently of previous additions, presuming the initiation reaction is faster than the growth reaction and no termination reactions take place (*living polymerization). The ratio of number-average molecular weight M_w to number- average molecular weight M_n approaches 1 at infinitely high molecular weights ($P_n \rightarrow \infty$). In contrast to the *Schulz-Flory distribution (where M_w/M_n equals 2), the Pd. is thus a very narrow distribution. **F** distribution de Poisson; **G** Poisson-Verteilung.

W. KAMINSKY, I. BEULICH

Ref.: G. Henrici-Olivé, S. Olivé, *Polymerisation Katalyse-Kinetik-Mechanismen*, VCH, Weinheim 1969; Ullmann, 4. Ed., Vol. 1, p. 556.

Polanyi-Wigner equation Thermal *desorption is often described in terms of the Polanyi-Wigner equ. $r_{des} = -dN_A/dt = A \cdot N_A^x \exp(-E_{des}/RT)$, r_{des} being the rate of *desorption, N_A the *surface conc. of adsorbed species, t the time, E_{des} the *activation energy of desorption, A the *preexponential factor of the desorption rate constant, x the kinetic order of desorption, R the gas constant, and T the temp. E_{des} and A can both be a function of coverage. The PWe. serves as the basis of the evaluation of the desorption parameters in the so-called complete methods (proposed by King and Chan, Aris and Weinberg). **F** équation de Polanyi-Wigner; **G** Polanyi-Wigner-Gleichung.

M. MUHLER

pollution abatement → environmental catalysis

Polonovki reaction the base-cat. demethylation of amino-N-oxides with *AAA or acetyl choride. **F** réaction de Polonovki; **G** Polonovki-Reaktion.

B. CORNILS

polyacrylates -(CH$_2$-CH(COOR))$_n$ can be obtained from a variety of acrylic monomers, such as *acrylic and *methacrylic acids, their amides, esters, and salts. Commercially most important are poly(methyl methacrylate), poly(ethyl acrylate), and poly(sodium acrylate) (which is used as a polyelectrolyte).
*Polymerization to atactic (cf. *tacticity) polymers can be initiated by free-*radical cats. such as *peroxides, or by organometallic cpds. such as BuLi. Isotactic ps. are obtained using *metallocene/*methylaluminoxane cats. Isotactic polymers are crystallizable and insoluble in most solvents. **E=F=G**.   W. KAMINSKY

Ref.: Kirk/Othmer-1, Vol. 1, p. 386 and Vol. 8, p. 459; *Macromolecules* **1996**, *29*, 1847.

polyalkenamers → ring-opening metathesis

polyazobenzenes (polyazophenylenes) molecules bearing the -N=N- fragment in the polymer main chain. Convenient access to these cpds. is provided by the *oxidative coupling of aromatic diamines in pyridine solution using Cu(II) as cat. (Figure).

Ps. belong to the general class of photoaddressable polymers wherein reversible light-induced *trans-cis* *isomerization of the -N=N-chromophor can be used to induce conformational changes of the macromolecule. Ps. are thermostable up to 350 °C; in some cases they show liquid-crystalline behavior or (semi)conducting properties. **F**=E; **G** Polyazobenzole.

Ref.: *Encycl.Polym.Sci.Engng.* **2**, 166.   M. WAGNER

polybutadiene the most used synth. rubber (BR, butadiene rubber, general structure -(-CH$_2$-CH=CH-CH$_2$-)-). Different microstructures of BR such as 1,4-*cis*, 1,4-*trans*, 1,2-atactic, or 1,2-syndiotactic (cf. *tacticity) are known, e.g., in mixtures with styrene (SBR), polymerized in bulk, solution, or in aqueous emulsions via free-*radical cats. Metal alkyls such as alkali metals are used as *initiators in anionic *polymerization. Li alkyls give a polymer with higher *trans*-1,4 portion (55 %), *cis*-1,4 (35 %), and 1,2-atactic (10 %). A fraction of 1,2-structures is needed for *vulcanization with sulfur. *cis*-1,4-Polybutadiene is produced preferentially with mixed cats. such as Co(O$_2$CR)$_2$/H$_2$O/AlEt$_2$Cl (1:10:200) or TiCl$_4$/I$_2$/Al(*i*-Bu)$_3$ (1:1.5:8). Allyluranium *complexes are very active cats., too. To obtain highly *trans*-1,4-polybutadiene, *Ziegler-Natta cats. (TiCl$_4$/AlR$_3$) can be used. The synth. of crystalline 1,2-polybutadiene is possible with cpds. of Ti, V, and Cr (e.g., {(allyl)$_3$Cr}). **F** polybutadiène; **G** Polybutadien.

W. KAMINSKY

Ref.: Elias; *Acc.Chem.Res.* **1980**, *13*, 1; *Rubber Chem. Technol.* **1972**, *45*, 1252; *Prog.Polym.Sci.* **1991**, *16*, 405; *J.Polym.Sci.Polym.Chem.Ed.* **1983**, *21*, 1853.

polybutylene terephthalate (PBT) a polymer manuf. from 1,4-butanediol and terephthalic acid.

polycarboxylates water-soluble linear *polymers or *copolymers containing carboxylic groups; they are, as well as *zeolites, constituents of modern detergents. Most ps. contain polymerisates of *acrylic or *maleic acid. **F**=E; **G** Polycaroxylate.   B. CORNILS

polycondensation → condensation reaction

poly(ether ketone) → PEK

polyethylene (PE) belongs to the class of polyolefins with the simplest structure element -[-CH$_2$-CH$_2$-]-. It is manuf. by two major methods, the high-press. proc. which leads to branched polymer with low density (*LDPE), and the low-press. proc. which generates a linear, unbranched PE with high density and crystallinity (*HDPE). By *copolymerization with higher α-olefins (e.g., butene, hexene, or octene) the degree of branching can be controlled at low press. giving linear polymers with low density (*LLDPE)(Figure). Worldwide PE production is in the region of 45 × 10^6 t/a; about the relationship between polymerization conds., density, and degree of branching cf. *density of polyethylenes.

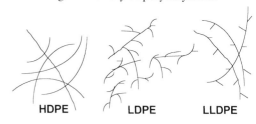

HDPE   **LDPE**   **LLDPE**

The polymerization is cat. by *radicals (*peroxides, oxygen, low-density PE) or by *transition metal cpds. such as Cr-Si *surface *complexes (*Phillips cat., *Phillips HDPE proc.), MoO$_3$ on an *alumina *support (*Standard Oil HDPE proc.), or Ti/Mg/Al organo cpds. (*Ziegler-Natta catalysis). **F** polyéthylène; **G** Polyethylen.   W. KAMINSKY, I. BEULICH

Ref.: Ullmann, Vol. A21, p. 487; Kirk/Othmer-1, Vol. 17, p. 702.

polyferrocenes a class of *ferrocene-containing macromolecules. There are three different subgroups; 1: polymer chains bearing pendant ferrocenyl substituents, e.g., polyvinylferrocenes; 2: macromolecules [-(C$_5$H$_4$)Fe(C$_5$H$_4$)-]$_n$, in which 1,1′-ferrocenylene fragments are directly attached to each other. Their prep. is possible by homocoupling of 1,1′-diiodoferrocene (DIF) with stoichiometric amounts of Mg or by the Pd-cat. *cross-coupling of DIF and 1,1′-bis(dihydroxyboryl)ferrocene; 3: macromolecules consisting of 1,1'-fragments in the main chain that are connected by *spacer units like -CH₂-CH₂-, -P(R)-, or -Si(R)₂- (Equ.).

These polymers are synth. by the thermal or cat. *ring-opening polymerization of strained, ring-tilted *ansa*-ferrocenes (*ansa*-metallocenes). **E=F=G**.   M. WAGNER

Ref.: *Angew.Chem.Int.Ed.Engl.* **1996**, *35*, 1602.

poly(β-hydroxybutyrates)

synth. is via a *biosynthetic sequence that utilizes *acetyl CoA. Many whole cell systems have been used to synth. this polymer and related materials. *Copolymers consisting of (R)-3-hydroxybutyl and (R)-3-hydroxyvaleryl units were prepared by feeding propionate to whole cells of *A. eutrophus*.

The first *enzyme in the synth., acetoacetyl-CoA thiolase (EC 2.3.1.9), cat. the C-C bond-forming step. The *active site of acetoacetyl-CoA thiolase contains a cysteine, which attacks acetyl-CoA to form a acetyl thioester enzyme intermediate. This species then reacts with the enolate derived from enzymatic deprotonation of the other acetyl-CoA. Mechanistic studies that were performed on this enzyme from *Zooglea ramigera* showed that the thiolase forms acyl enzyme *intermediates with a number of acyl-CoA derivatives, but only accepts acetyl-CoA as the nucleophile. After reduction of the ketone by acetoacetyl-CoA reductase (EC 1.1.1.36), the resulting β-hydroxy thioester is *polymerized by polyhydroxybutyrate synthetase. These polymers are synthetically useful as a source of (R)-β-hydroxyacids. **E=F=G**.   C.-H. WONG

Ref.: *J.Biol.Chem.* **1987**, *262*, 97; *J.Am.Chem.Soc.* **1989**, *111*, 1879; *Tetrahedron Lett.* **1984**, *25*, 2747.

poly(β-hydroxybutyrate) synthetase →
poly (β-hydroxybutyrates)

polyketide synthase

Polyketides are assembled from *acetyl- CoA and malonyl CoA through a multienzyme-cat. proc. The initial proc. is similar to *fatty acid biosynthesis, i.e., *condensation between the two- and three-carbon building blocks to form β-ketoacyl-SACP which is then reacted with a certain number of the three-carbon building blocks to form polyketides. In addition to malonyl-CoA, methylmalonyl-CoA and ethylmalonyl-CoA can also serve as chain extenders (the nucleophiles) to produce methyl or ethyl side chains. Reduced polyketides are also formed via *reduction of the keto group (with ketoreductase), *dehydration of the reduced prod. (via dehydratase), and saturation (enoyl reductase). The polyketides generated can undergo an intramolecular *aldol reaction followed by *reduction, *dehydration, etc. to form aromatic polyketides. The final step in the biosynth. of polyketides is the release of the prod. via an intramolecular reaction with the thioester linkage to the *enzyme. **E=F=G**.
Ref.: *Chem.Rev.* **1997**, *97*, 2465.   C.-H. WONG

polyketones → alternating copolymerization, cf. also *PEK

polymerase chain reaction

(PCR, gene amplification) The PCR provides a rapid and simple means of amplifying a sequence of *DNA, given that the sequence of at least 15–18

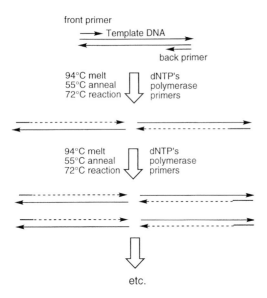

etc.

bases at the 5′ and 3′ ends are known. Two oligonucleotides are synth. chemically which correspond to the first 15–18 bases of each end of the sequence, in the 5′-3′ direction. (This means that one will correspond to the sense strand, while the other will correspond to the antisense strand). The primers are mixed together with the gene of interest (to act as a template), and a thermostable DNA polymerase is added, along with deoxyribonucleotide triphosphates (dNTPs). The mixture is then cycled repeatedly through a high temp. (melts the DNA apart)/low temp. (allows the primers to anneal to the template)/moderate temp. (usually the ideal reaction temp. for the *enzyme) sequence. For each cycle, the amount of DNA is roughly doubled, until the dNTPs are depleted. **F** réaction de chaine avec polymérase; **G** Polymerase-Kettenreaktion.   P.S. SEARS

polymeric support   → support

polymerization   the conv. of low-molecular cpds. (*monomers) to high-molecular prods. (*polymers). P. reactions can be classified as either step-growth or chain-growth reactions. In p. proc. three kinds of reactions are defined: *initiation, *propagation, and *termination. If this termination step does not oc-

cur the reaction is called *"living" polymerization because the active center in the polymer remains and p. will continue as long as monomer is added. A p. can be started either by *initiators (e.g., *radicals, anionic or cationic cats.) or by heat, radiation, or light (photopolymerization).
If the p. is carried out with only one kind of monomer it is called *homopolymerization, if two different monomers are involved, *copolymerization. To describe the geometrical arrangements of the monomer base units a special terminology has been developed (e.g., *tacticity, *stereoselectivity of polymers). Since the p. is residue-free, the atom economy is high. **F** polymérisation; **G** Polymerisation.

W. KAMINSKY, M. VATHAUER

Ref.: Mark/Bikales/Overberger/Menges, Vol. 3, p. 549; A.D. Schlüter (Ed.), *Synthesis of Polymers*, Wiley-VCH, Weinheim 1999; Ciardelli/Tsuchida/Wöhrle; Elias; Fink/Mühlhaupt/Brintzinger; Kaminsky/Sinn; Quirk.

polymers   result from *polymerization of *monomers and prods. of the plastic industry which is sold in amounts up to 100 million tpy (up to 25 MM tpy per single p.; PE: 45 MM tpy). P. are economically very important materials (Figure). **F** polymères; **G** Polymere.

B. CORNILS

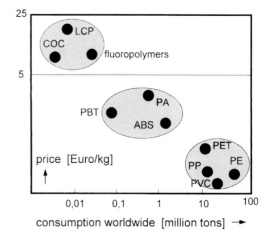

polyol dehydrogenase   cat. the *oxidation of one of the several hydroxyl groups of an alditol or aminoalditol (cf. *iditol dehydrogenase). **E=F; G** Polymere. C.-H. WONG

Ref.: *Angew.Chem.Int.Ed.Engl.* **1993**, *32*, 1197; *J.Org. Chem.* **1992**, *57*, 5899.

polyolefins → LDPE, *LLDPE, *HDPE, *PP, *UHMW-PE

polyoxometallates (POM) and *heteropolyacids (HPA) are used as both oxidation and acid cats. HPA, are more acidic than H_2SO_4 by 2 pK_a units and acid cat. can be 10–1000 times more effective with HPAs. HPAs are used in het. and hom. cat., e.g., *hydration and *etherification of alkenes, *hydrolysis, *polymerization, *alkylation of aromatics and alkanes with alkenes and alcohols, *esterification, and hydrocarbon formation from MeOH (*MTG proc. of *Mobil). Isopropanol is manuf. from propene (e.g., *Tokuyama proc.), *tert*-butanol from a C_4H_8 mixture by selective *hydration of isobutene, polyTHF from *THF or 1,4-butanediol, and *MTBE from MeOH. POM can be used as oxidation cats. in two mechanistic modes; 1: as *dehydrogenation cats. in *Mars-van Krevelen or *redox-type reactions; reoxidation of the reduced cat. is with O_2. This concept is being used for methacrolein → methacrylic acid (e.g., *Asahi, *Mitsubishi, or *Nippon Shokubai procs.). *Wacker-type *oxidation of alkenes, *dehydrogenation of alcohols and amines to aldehydes, alkenes, and dienes to aromatics, and *dimerization of activated phenols are known. 2: POM and *transition-metal substituted POM can activate oxygen donors, e.g., PhIO, *hydrogen peroxide, *TBHP, ozone, and O_2 to form activated intermediate oxo and *peroxo species. Oxidation of alkenes, alkanes, and sulfides is possible. **E=F; G** Polyoxometallate. R. NEUMANN

Ref.: *Chem.Rev.* **1998**, *98*, 77, 239, 273.

polyphosphate kinase from *E. coli* cat. the *phosphorylation of *ADP and other NDPs to NTPs using polyphosphate, a useful proc. for the regeneration of nucleoside triphosphates. **E=F=G.** C.-H. WONG

polypropylene (PP)   a thermoplastic polyolefin with the structural element -[-CH(CH₃)-CH₂-]-. Propylene can be polymerized stereospecifically to yield the highly crystalline *isotactic* PP, the less crystalline *syndiotactic* PP, and the amorphous *atactic* PP (cf. *stereoselectivity of polymers).

PP is produced in *suspension, in bulk, or in the gas phase (*catalytic reactors). The cats. are *supported *complexes such as MgCl₂/TiCl₄/(C₂H₅)₃Al (*organoaluminum cpds.); the cat. is thus heterogeneous. *Surface geometry and various *additives such as *Lewis acids (external and internal donors), ethyl benzoate, or silane ethers control the stereoregularity of PP. New cats. such as *metallocene/*methylaluminoxane lead to precise control and definite *active sites and thus to narrow molecular weight distributions. While C_2-symmetric metallocenes produce isotactic PP, C_s-symmetric cats. give syndiotactic PP (*enantiomorphic site control). **F** polypropylène; **G** Polypropylen. W. KAMINSKY, I. BEULICH

Ref.: *Polymer* **1987**, *28*, 683; W. Kaminsky, H.-J. Sinn (Eds.), *Transition Metals and Organometallics as Catalysts in Olefin Polymerization*, Springer, Berlin 1988; E.P. Moore (Ed.), *Polypropylene Handbook*, Hanser, München 1996.

polyribonucleotide phosphorylase (EC 2.7.7.8)   an *enzyme that, in the presence of Mg^{2+}, *polymerizes ribonucleoside diphosphates into *RNA polymers with the release of inorganic phosphate. In the presence of Fe^{3+}, it can also accept deoxyribonucleoside diphosphates as substrates. It does not require a *template, and has been used in the synth. of a variety of homo- and heteropolymers. **E=F=G.** P.S. SEARS

Ref.: *J.Mol.Evol.* **1996**, *42*, 493.

polysilanes   cpds. with Si-Si bonds which are used in photophysical applications and in ceramics. Dialkylpolysilanes with 40 Si atoms in a ring and linear *polymers with more than 40 000 Si atoms have been isolated. The usual synth. is by Wurtz-type *condensation of diorganodichlorosilanes with Na. A *transition metal cat. *polycondensation (dehydrocou-

pling) of hydrosilanes (*silane coupling) offers an alternative. **E=F**; **G** Polysilane. B. MARCINIEC
Ref.: Patai/Rappoport, Chapters 19 and 24.

polysiloxanes ("silicones") mainly polymethylsilicones which have widespread use because of their valuable properties. The starting materials are diorganodichlorosilanes (mainly dimethyldichlorosilane, *silanes). The *hydrolysis of diorganodichlorosilanes give silandiols which under the influence of cats. (e.g., alkali hydroxides) lead either directly or via cyclosiloxanes to ps. Cat. with *Lewis or Brønsted acids is also known. **E=F**; **G** Polysiloxane. B. MARCINIEC
Ref.: Patai/Rappoport, Chapter 21, p. 1289.

polysome display a method for attaching a *library of peptides to their cognate nucleic acids in order for affinity selection of the peptides to be coupled to their replication. As a ribosome proceeds down a messenger *RNA (mRNA) molecule, translating the RNA into protein, another ribosome can bind to the start of the message before the first ribosome has finished translation. A third can bind after the second has moved a short way down the length of the mRNA, and so forth. The resulting assembly is called a polysome (polyribosome). Polysomes made from a library of mRNA molecules and "frozen" in the middle of translation with the antibiotic chloramphenicol can be subjected to an affinity selection scheme to select those polypeptide species that have a desired binding property. Since the polypeptides are attached to the RNA encoding them, their characterization is relatively straightforward: the RNA is reverse-transcribed into DNA, and the DNA can be amplified via PCR or transformation into a bacterial host (after insertion into a plasmid *vector). The amplified DNA can then be sequenced or transcribed in vitro to produce RNA for another round of panning. This procedure is called "pd.", and has an advantage over other display-selection techniques such as *phage and *bacterial display and *lac repressor fusions in that the library size is not limited by the transformation efficiency of the bacterial host. **E=F=G**. P.S. SEARS
Ref.: *Methods Enzymol.* **1996**, *267*, 195.

polystyrene (PS) polymer with the structure -[-CH$_2$-CH(C$_6$H$_5$)-]$_n$. The molecular masses are between 10 000 and 1 000 000. Styrene is produced by the dehydrogenation of *ethylbenzene. Different microstructures of PS such as atactic, isotactic, syndiotactic (*tacticity) are possible. Free-*radical cats. such as *peroxides are often used for *polymerization in bulk, solution, and in aqueous emulsion and suspension.
Crystallizable, isotactic PS with a melting point of 240 °C has been formed in the presence of alkylK and alkylNa or *Ziegler cats. Compared to syndiotactic PS, it has a low crystallization rate which makes industrial use more difficult. Syndiotactic PS, synth. with cyclopentadienyl-TiCl$_4$ or triethoxy/*methylaluminoxane cats., shows a melting point of 275 °C. **F** polystyrène; **G** Polystyrol.

W. KAMINSKY

Ref.: Brighton et al., *Styrene Polymers*, Applied Science Publ., Barking 1979; *Pure Appl.Chem.* **1976**, *45*, 39; *Macromolecules* **1986**, *19*, 2464.

POM 1: polyoxymethylene (polyformaldehyde); 2: *polyoxymetallates

porcine pancreatic lipase (PPL) a serine-type *lipase cat. for the *hydrolysis of esters and *esterification and interesterification reactions. The *enzyme is selective for esters of primary alcohols, not secondary alcohols. The *enantioselective hydrolysis of *meso*-diacetates by PPL is complementary to the *pig liver esterase-cat. hydrolysis of the corresponding *meso*-1,2-dicarboxylates, and the enantioselectivity can be improved by modification of the substrate. **E=F=G**. C.-H. WONG
Ref.: *Tetrahedron Lett.* **1992**, *22*, 1399.

pore diffusion For porous cat. pellets, it is often difficult to identify the kinetic regimes when all the *heat and *mass transfer effects can intrude. Often pore diffusion resistance is the only physical factor which affects the *rate. The slowdown in rate caused by diffusional resistance is dependent on the *Thiele modulus (TM). Factors affecting the TM. are particle size, temp., and press. In the strong pore resistance regime the observed *activation energy E_{obs} can be expressed as $E_{obs} = (E_{reaction} + E_{diffusion})/2 \approx \frac{1}{2} E_{reaction}$ with $E_{diffusion}$ usually being very small. The pd. regime is entered at high temps. when the rate speeds up so much that the reactant cannot penetrate the particle. By lowering the temp., the pd. regime can be left. The influence of pd. increases with growing particle size.

In the strong pd. regime, the observed reaction order n_{obs} can be expressed as $n_{obs} = (n_{true} + 1)/2$. Therefore a change in reaction order occurs for all reactions with a *reaction order different from $n = 1$. The existence of strong pd. can be determined by calcng. the TM., by comparing the rate for different *pellet sizes, and by noting the drop in the observed *activation energy with rise in temp., coupled with a possible change in reaction order. To find the rate equ. from experimental data, at least three good data points are necessary, two at different temps. in one regime, one in the other. From these three points two lines (one slope double the other) must be drawn in an *Arrhenius diagram. Their intersection represents the transition from diffusion-free conds. to the strong pore diffusion regime. **F** diffusion poreuse; **G** Porendiffusion.   R. SCHLÖGL, C. KUHRS

pore disruption on drying   The drying of cat. *precursor pellets normally proceeds at elevated temp. in a stream of gas with controlled humidity of the solvent (mostly water). Overall, low temps. and high humidity produce the best cat. However, as this procedure is time-consuming and costly, a compromise has to be found. As a rule of thumb, the drying temp. should be kept 10 °C below the boiling temp. of the solvent. Drying first evaporates the *surface liquid, then the liquid in the pores. The rate of drying in both *phases is different: the first stage proceeds at (nearly) constant velocity (removal of superficial moisture), in the second stage (pore drying) the drying rate slows down, and in the third stage (capillary drying) the drying rate proceeds asymptotically. In the second stage, vapor generated inside the pores forces the evaporating liquid out. If the drying rate is excessive, vapor will be generated within the pellets faster than moisture can be forced out of the pores and the pellet structure will be damaged. As drying proceeds, the free liquid in the pores disappears and only adsorbed humidity will remain. Pellets damaged internally will crush easily in subsequent operations. **F** écroulement des pores en séchage; **G** Porenbruch beim Trocknen.   C.D. FROHNING

pore filling method   → pore volume impregnation

pore models   In a macroscopic view, pore models give a rather subjective impression to the viewer. More scientific aspects arise in comb. with methods for the determination of the *surface area such as *BET and *mercury porosimetry. It is possible to distinguish between an open and a closed *pore. The former is defined as a *cavity or channel with access to the *surface, the latter is not connected to the surface. Both of them could be connected with a so-called interconnected pore, open or closed. The pore size is defined by IUPAC as: *micropores (<2 nm), *mesopores (2–50 nm), and *macropores (>50 nm). The *adsorption or *desorption of a gas depends on the pore shape with reference to the BET method. Furthermore, there are cylindrical pores (open at both ends or at only one), bottle-, wedge-, and slit-shaped pores. *Hysteresis in BET curves (*adsorption, *desorption) makes it possible to distinguish between these shapes. In a closed cylindrical pore the condensation starts at the end of the pore wirth formation of a spherical meniscus; desorption happens in the reverse order (no hysteresis). The open cylindrical pore condensation and *desorption start at both ends (hysteresis is possible). The behavior in the case of the bottle-shaped pore is like the closed cylindrical pore and depends on the neck ratio of the bottle. In the case of the wedge-shaped pore adsorption and desorption are reversible (no hysteresis). The slit-shaped pore shows a stepwise condensation of layers and *capillary evaporation (hysteresis). Cf. also *pore volume. **F** modèles des pores; **G** Porenmodelle.

R. SCHLÖGL, H.J. WÖLK

Ref.: van Santen/van Leeuwen/Moulijn/Averill.

pore size distribution   A detailed characterization of the *pore geometry of a cat. give the psd. function $f(r)$. This function usually is defined in such a way that $f(r)\mathrm{d}r$ represents the volume $\mathrm{d}V_g$ of pores per unit weight of the cat. which have a radius between r and $\mathrm{d}r$. The method used for determination of the psd. depends on the pore size (*micro-, *meso- or *macropores). *Mercury porosimetry and *capillary condensation techniques are used. The manufng. procedure for the cat. determines the psd. and the average pore size. **F** distribution de la taille des pores; **G** Porengrößeverteilung.   R. SCHLÖGL, B. BEMS

pore structure   The term ps. refers mainly to *surface area, *pore volume, pore shape (cf. *pore models), *pore size distribution, and pore connectivity. Texture is another expression often used in connection with ps.
Cats. of high surface area have exceedingly complex pore structures. The practical consequences of the ps. are of importance, e.g., they are concerned with the *adsorption properties and their dependence on molecular size, the reversibility of adsorption, the kinetics of adsorption and *desorption, permeability and flow, wetting and dewetting, mechanical strength, and the cat. properties such as conv., *selectivity, and *deactivation behavior. In het. cat. reactions gradients of composition and/or press. may appear within the ps. These gradients give rise to diffusional (*pore diffusion) and/or permeation *mass transport which may considerably affect the process *kinetics.

Pore sizes roughly divide into the groups *micro-, *meso-, and *macropores. Only when the ps. is sufficiently coarse can its characteristics be investigated by direct observation using optical or electron microscopy. In other cases scattering techniques (with either X rays or neutrons) may give some information. The sensitivity of cat. procs. to the cat.'s structure has long been recognized and is known as *structure sensitivity. A variety of methods have been developed to modify the ps., especially of *zeolites and related materials, with the aim of *tailoring the cat. properties in the sense that selectivity toward certain prods. could be maximized. A classic example is varying the ps. by using different parameters in the *sol-gel proc. **F** structure des pores; **G** Porenstruktur.   R. SCHLÖGL, M. HAEVECKER

Ref.: van Santen/van Leeuwen/Moulijn/Averill.

pore volume   The pv. in porous material can vary over a wide range. Pore sizes roughly divide into the following groups *micropores with dimensions <2.0 nm, *mesopores in the range of 2.0 – 50 nm, and *macropores with dimensions >50 nm (acc. to IUPAC). Micropores are sometimes subdivided into ultramicropores (< 0.6 nm) and supermicropores (0.6 nm – 1.6 nm). The basis of this classification is that each of the size ranges corresponds to characteristic *adsorption effects.
In the case of micropores the proc. of *adsorption can be described as micropore filling. *Adsorption effects are dominated by strong interaction with the walls of the pore. Important materials which possess pores of molecular dimensions are, e.g., *zeolites. The volume is usually determined by gas adsorption experiments and interpretation of the resulting *isotherms (e.g. *BET). In the mesopore range intermolecular interactions become dominant. The volume of macropores is usually examined by mercury penetration (mercury picnometry).
Nevertheless no experimental method can be expected to provide an evaluation of the "absolute" pv. and the interpretation of the derived values becomes more difficult as the pore width is reduced to molecular dimen

sions. **F** volume des pores; **G** Porenvolumen.

R. SCHLÖGL, M. HAEVECKER

pore volume impregnation   In order to disperse the cat.-active *phase thoroughly onto the accessible *surface of a carrier, pvi. is frequently applied with the additional advantage that only a minimal volume of solvent has to be removed. The absolute amount of the impregnating (impr.) compds. is chosen with regard to the surface of the carrier to achieve a monolayer or – if desirable – a higher extent of *coverage. The volume of impr. solution exactly corresponds to the pore volume of the *support; the conc. of the impr. solution is thus determined. This so-called *dry impr. (*incipient wetness method, pore filling method) is preferentially applied in industrial operation. The impr. of pores with narrow diameters is facilitated by previous vacuum degassing of the *carrier. The impr. solution is either sprayed onto the carrier powder or added in bulk as shaped particles under intensive mixing. Several sequences of impr. can be used to achieve a sufficiently high loading (about 25 % w/w; cf. *capillary impregnation, application: *SLPC technique). The impr. support is dried and usually *calcined to remove the solvent and undesired ions or other constituents of the impr. cpd. (e.g., complexing agents). Dry impr. differs from the equil. impr. method, which makes it possible to influence several parameters such as the nature and pH of solution, conc., temp., additives, time to reach equil., etc. **F** imprégnation; **G** Imprägnierung.   C.D. FROHNING

porosity   the state of a material of having pores, the quality of being permeable. The p. ε is defined by the ratio of *pore volume to the apparent volume of the particle or the granule acc. to $\varepsilon = 1 - \rho_a/\rho_t$, ρ_a being the *apparent density (obtained from the volume which is accessible to a probe substance; cf. *bulk density) and ρ_t the total external density. This definition does not include those pores that are closed (saccate) or otherwise inaccessible. If the density of the solid matrix ρ_s is known,

the total porosity ε_{tot} and hence also the closed porosity ε_{clos} can be calculated.

While pores in surfaces, i.e., metals, are considered as faults in the material, they often have positive effects in the performance of cats. They expand the active *surface of the cat. and modify strongly the transport of reactants in the cat. Therefore, they strongly influence the prod. yield and the long term performance of commercial cats.

Characteristic parameters for the p. of a material are also the *pore size distribution, the *pore structure, and the inner and outer *surface areas. **F** porosité; **G** Porösität.

R. SCHLÖGL, M. HAEVECKER

Ref.: S.J. Gregg, K.S.W. Sing, *Adsorption, Surface Area and Porosity*, 2nd Ed., Academic Press, London 1982.

porous glass   → controlled-pore glass

porphyrin as a ligand   Ps. are cyclic tetrapyrroles which are derivatives of porphin. The eight β-H atoms of the parent porphin are completely or partly substituted by side chains. Many metal ions are complexed by the *N_4 *chelate p., forming metallo-ps. which are active cats. in vivo and in vitro. The *chelate *complex of p. with Fe(II) is called *heme; the Mg complex chlorophyll. Hemoproteins control fundamental biochemical and biocat. procs., including the *reduction of oxygen to water (cytochrome *c* oxidase; cf. *photoinduced oxidation/reduction), the biosynth. of steroids and the detoxification of organisms (cytochrome P 450-dependent monooxygenases), and the destruction or utilization of H_2O_2 (*catalases, *peroxidases). In synth. chemistry, studies have been performed on the reactions of alkanes with oxidants (e.g., H_2O_2, amine *N*-oxides, peracids) in the presence of Fe, Mn, Co, and Ru-p. cats. to give the corresponding alcohols, ketones, or peroxides. Asymmetric *hydroxylation of alkenes is cat. by metallo-ps. vaulted with a chiratropic bridge. **F** porphyrine; **G=E**.   P. ROESKY

Ref.: Montanari/Casella; Sheldon; *J.Mol.Catal.* **1999**, *138*, 145; *J.Mol.Catal.A:* **1999**, *139*, 11.

positional isotope exchange (PIX) a useful technique for acquiring mechanistic information about a reaction, such as determining the point in a reaction at which a bond is cleaved.   P.S. SEARS
Ref.: *Biochem.Biophys.* **1988**, *267*, 54.

POSS polyhedral oligosilasesquioxanes, → model sufaces of supported catalysts

potential energy diagrams describe the energetics of molecular/atomic interactions, in particular the interaction of an atom/molecule with a *surface (cf. Figure under keyword *activation energy). In a ped. the potential energy is shown as function of the interatomic distances. For the description of *adsorption peds. of the *Lennard-Jones type are typically used, showing the variation of the potential energy as a function of the distance from the surface. Such 1D diagrams are greatly simplified because the internal degrees of freedom and the variation of the lateral position parallel to the surface are neglected. A high-dimensional energy surface would therefore be adequate for the correct description of adsorption. Such energy surfaces can in principle be calc. with *QC methods. The computational effort is, however, exceedingly large. For *dissociative *chemisorption of diatomic molecules 2D peds. have been calc. in which the intramolecular bond length of the molecule and the distance from the surface are the independent variables. With this type of ped. the different roles of translational energy and vibrational energy for the crossing of the activation barrier could be demonstrated. **F** diagramme de l'énergie potentielle; **G** Potentialenergiediagramme.
                                        R. IMBIHL
Ref.: P.H. Emmett (Ed.), *Catalysis*, Reinhold, New York 1954, Vol. 1, p. 236; R.D. Levine and R.B. Bernstein, *Molecular Reaction Dynamics and Chemical Reactivity*, Oxford University Press 1987.

Powerforming process → Exxon procs.

power plant exhaust gases → stationary exhaust catalysis

PP → polypropylene

ppfa, PPFA a *ferrocene-derived *ligand manuf. by ortholithiation of a ferrocene with a substituent $CH(Me)NMe_2$, followed by reaction with diorganophosphane (cf. Figure under keyword *ferrocene). With a comb. of the procedures toward *dppf and ppfa, synth routes to planar *chiral derivatives (e.g., Figure) are obtained.   P. HÄRTEI

PPG chloroethylene process manuf. of chloroethylenes, specially tri- and tetrachloroethylene, by *oxychlorination of ethylene with Cl_2/O_2 over a $CuCl_2$ cat. in fluidized-bed operation at 420–450 °C and at slightly increased press. **F** procédé PPG pour des chloroethylènes; **G** PPG Chlorethylenverfahren.   B. CORNILS

PPh₃ *triphenylphosphine (TPP)

PQQ → quinone proteins

Pr the C_3H_7-group in *complex cpds. of *ligands

Prater number → effectiveness factor

precatalyst → precusor

precipitation/coprecipitation Solid materials can be produced by *precipitation from hom. solutions, provided the overall change in free energy, including the interface free energy decreases. Kinetically, the precipitation proc. is initiated by the formation of a nucleus followed by growth procs. Precipitates can be produced as *precursor materials of cat. *supports or of active cats. The properties of the precipitate are determined by several critical parameters, including the nature of the raw material, its conc. and composition *solvent effects, temp., pH, and *additives. P. can be carried out batchwise at variable or constant pH or in continuous operation. Materials containing more than one comp. can be prep. by coprecipitation, which permits the generation of a hom. distribution of the cat. comps. and the synth. of *precursor materials with a definite and well-defined stoichiometry (cf. *coprecipitation, *metal distribution). **F** précipitation/coprécipitation; **G** Fällung/Cofällung.   H. KNÖZINGER

Ref.: Ertl/Knözinger/Weitkamp, Vol. 1, pp. 7,106; Chem.Rev. **1995**, *95*, 477.

precipitation in catalyst preparation is

a method frequently applied to generate a cat. *precursor (and eventually a *bulk cat.). Most common is the p. of cations from aqueous solutions by adding caustic or anions forming an insoluble cpd. of the cation (oxide, hydroxide, salt; cf. *isoelectric point). The precipitate is separated from the supernatant solution, freed from unwanted ions by washing, and dried before further processing.

During p. the atomically dispersed ions in the solution are transferred into a solid with properties (crystallite chemical and physical comp., size, density, solubility, etc.) depending on numerous variables. Temp. and conc. of the metal solution and the precipitant are the main tools to control the size of the crystallites; low temp. and high concs. favor the generation of small crystallites (which may present difficulties during filtration). Intensive mixing is generally applied in order to reduce the supersaturation of the solution, thus leading to more hom. crystals with respect to composition and size.

Historically p. comprises the combination of (at least) two solutions of controlled conc. to reach a predetermined final value, e.g., the pH. In this procedure the concs. in solution change with the progress of the p. and in parallel the composition of the precipitate is influenced. If high homogeneity of the precipitate is required two solutions are added simultaneously to the p. vessel in a ratio that the pH (or another conc. value) is kept constant.

Modern systems use proc. control of this step, as it determines most of the properties of the final cat. Subsequent steps after p. are *filtration, washing, *drying, *shaping, *reduction. P. delivers *bulk cats.; if p. is conducted in the presence of *supports (*coprecipitation) *supported cats. are manuf. **F** précipitation; **G** Fällung.   C.D. FROHNING

Ref.: *Catal. Today* **1997**, *34*, 281; Ullmann, Vol.B2, Chapter 3–1 and Vol. A5, p. 348; Ertl/Knözinger/Weitkamp, Vol. 1, p. 72.

precursors in cat. procs. are agreed to be

the immediate storable predecessors of the cat. active species. P. is thus synonymous with *precatalyst (cf. also *in-situ catalysts, *preformation). Examples are V phosphates (*VPOs) as ps. for het. V cats. for *ammoxidations, or Co acetate or $RhCl_3$ as ps. for Co or Rh carbonyl cats. in *hydroformylations. For an explanation of ps. in enzymatic cat., cf. *cathepsin/pro-cathepsin. **F** précurseurs; **G** Präkursoren.   W.A. HERRMANN

precursor states intermediate stages in an

*adsorption proc. when an adsorbing molecule, before reaching the final stable state with the highest *adsorption enthalpy, becomes adsorbed in more weakly held states. Such states can be isolated and characterized spectroscopically at low temp. when the activation barrier (cf. *activation energy) for reaching the more stable state is too great to be surmounted. In adsorption experiments, evidence for a ps. is typically given when the *sticking coefficient does not decrease with rising *coverage as expected from the limited availability of *vacant sites, but remains constant (e.g., Pt/CO, W/N_2). It is assumed that the molecule when hitting an occupied site, is trapped in the second layer, where it is highly mobile: thus it has a chance to find a vacant site in the first layer before *desorption occurs. **F** l'état précurseur; **G** Präkursorzustand.

Ref.: Somorjai.   R. IMBIHL

pre-enzyme → proenzyme

pre-exponential factor In order to locate any point on an *Arrhenius plot it is necessary to define a temp.-dependent quantity, i.e., the slope, which equals $-E/R$, E being the *activation energy, and a temp.-independent quantity, i.e., the intercept. Thus either $k = A_t \exp(-E_t/RT)$ or rate $= A_a \exp(-E_a/RT)$, where the subscripts t and a respectively denote *true and *apparent (see *Arrhenius plot), and A is a pef. The *transition state theory gives further insights. A simplified treatment leads to the general expression $k = (k_B T/h)\exp(\Delta G^{\neq}/RT)$ where $-\Delta G^{\neq}$ is the free energy of activation, whence the pef. is given by $\ln A = \ln(k_B T/h) + \Delta S^{\neq}/R$, $\Delta S^{\neq}$ being the entropy of activation. Analysis of the factors determining the rates of cat. reactions is therefore best conducted in terms of E_t (which is almost the same as the enthalpy of activation $\Delta H^{\neq}$) and $\Delta S^{\neq}$.

A linear correlation is frequently observed between E_a and $\ln A_a$, the significance of which has been much debated: the term *compensation (effect) is applied to this phenomenon, as a high value of E_a [which means a small value of $\exp(-E_a/RT)$] is compensated by a high value of $\ln A_a$. **E=F=G.** G.C. BOND

preformation (of catalysts) the formation of the active cat. species from *precursors through a special formation step. Precursors may be two different ingredients (such as Rh 2-ethylhexanoate and the *ligand modifier *TPPTS) which react under suitable conds. and in the presence of the reactant *syngas spontaneously to form the well-defined cat. species HRh(CO)(TPPTS)$_3$. In other cases two defined constituents form the inexactly known cat. species during p., as in the case of *MAO and *metallocene cats. for *PP *polymerization. Even dissociations may be a consequence of p. An illustrative example is the

Ni(PR$_3$)$_4$ ⇌ Ni(PR$_3$)$_3$ + PR$_3$
  1   **2**

Ni(PR$_3$)$_3$ ⇌ Ni(PR$_3$)$_2$ + PR$_3$
  2   **3**

sterically and electronically influenced dissociation of *phosphine ligands from the inactive precursor complex **1** to give the less crowded and thus active species **2** and **3** (Equation). (cf. *drop-in cats.; *in-situ cats., *precursor). **F** préformation; **G** Präformierung. W.A. HERRMANN

Prelog's rule describes the stereochemical course of an alcohol dehydrogenase reaction. In most cases, the *enzyme cat. the transfer of pro-R hydride from *NADH or *NADPH to the re face of the carbonyl substrate. **F** règle de Prelog; **G** Prelogs Regel. C.-H. WONG

Ref.: Pure Appl.Chem. **1968**, 9, 119.

preparation (of catalysts) Chemical composition is the major factor determining the properties of a cat. With a constant chemical composition, the cat. characteristics may vary over a wide range depending on the conds. and methods of cat. prepn., owing to the changes in the nature of interaction of cat. comps., *dispersion, *pore structure, crystallochemical changes, *surface properties, and other factors which may greatly influence cat. reactions. The main hurdles in solid- state chemistry are the hidden variables that control nucleation and growth of the desired solid *phase or phases. The prepn. determine the *activity, *selectivity, and reproducibility of the cat. formed.

A wide variety of materials are used in the prepn. of solid-state cats. Cats. prepd. from natural sources, such as *bauxite or silicates, are used for *cat. cracking. Usually these cats. need a following *activation step. Today these cat. have lost importance and are substituted by more active or selective cats. One of the most commonly used methods for prep. cats. is *precipitation or *coprecipitation. By this method not only cats. but also *supports are produced. The wet solid is converted to the finished cat. by *filtration, washing, *drying, *calcination, and *activation. The other widely used prepn. method for cats. is *impregnation of the porous support with a solution of the active comp. Further known methods include *sol-gel procs., solid-state reactions, and *chemical vapor deposition for prepn. of *fused, *metallic film, or *skeletal metal cats. (e.g. *Raney cat.). **F** préparation; **G** Präparierung.   R. SCHLÖGL, B. BERNS

Ref.: Augustine; Ertl/Knözinger/Weitkamp, Vol. 1, p. 412; G. Poncelet et al., *Preparation of Catalysts V: Scientific Bases for Preparation of Heterogeneous Catalysts*, Elsevier, Amsterdam 1991; M. Sittig, *Handbook of Catalyst Manufacture*, Noyes Data, New Jersey 1978; A.B. Stiles, T.A. Koch, *Catalyst Manufacture*, Dekker, New York 1995; *Studies in Surface Science and Catalysis*, Vols. 1, 3, 16, 31, 63, 88.

pre-proenzymes Secreted *enzymes are usually translocated across a membrane to the outside of the cell (or a compartment destined for the outside) during translation in a proc. where a special N-terminal segment of the protein (the "signal peptide") acts as a signal targeting the ribosome-mRNA-nascent protein *complex to the endoplasmic reticulum (*eukaryotes) or the cell membrane (*prokaryotes). The ribosome docks to the membrane and, with the help of many accessory proteins, the enzyme is extruded through the membrane as it is created. The signal peptide is typically cleaved off the enzyme by *signal peptidases, proteases which are specific for certain types of sequences. The enzyme, complete with signal sequence, is called a pre-enzyme or p-p. (depending on whether it is secreted as an inactive *proenzyme or as the fully active enzyme). **E=F=G**.   P.S. SEARS

pressure gap the term introduced to describe the different reaction conds. in commercial het. cat. ($p \approx 1{-}30$ MPa) and in surface science studies ($p \approx 10^{-10}{-}10^{-6}$ mbar), which differ over many orders of magnitude in press. The pg. together with the *material gap represents a fundamental problem in cat. research because an extrapolation of the results of surface science studies to high press. is required. Such an extrapolation may in some cases be justified but not in general, because the mechanistic pathways of the reaction may be different for a number of reasons. At high press. certain surface or bulk phases may

form which are not allowed at low press. due to thermodynamic constraints. High *coverages at high press. will lead to strongly non-ideal behavior of adsorbates, resulting in a change of the *kinetics. An example where the extrapolation of low press. data to the kinetics of commercial cat. has been successfully demonstrated is *ammonia synth. **E=F=G**.   R. IMBIHL

Ref.: *Phys.Rev.Lett.* **1985**, *55*, 2502.

pre-steady state (enzyme-catalyzed reactions) When a substrate and *enzyme are first mixed, there is a very short period of time before the intermediates of the reaction reach (pseudo) *steady-state concs. This initial transient, when the concs. of all of the reactive intermediates are rising to their steady-state levels, is called the pre-steady state period. Kinetic studies of this transient can give information on individual rate constants of a multistep reaction, but such studies are difficult to perform. The pre-steady state reaction can be followed in some cases with very rapid monitoring techniques such as continuous or *stopped-flow methods.   P. SEARS

Ref.: A. Fersht, *Enzyme Structure and Mechanism*, W.H. Freeman, New York 1985.

presulfidation   → directed poisoning, *sulfide cats.

prices of catalyst metals The prices [US dollars per mole] were 1997: **Ti** 1.44; **V** 1.01; **Cr** 0.78; **Mn** 0.08; **Fe** 0.07; **Co** 2.83; **Ni** 0.39; **Cu** 0.16; **Zr** 9.87; **Nb** 17; **Mo** 2.88; **Ru** 173; **Rh** 897; **Pd** 508; **Ag** 19; **Hf** 28; **Ta** 52; **W** 2.76; **Os** 3161; **Ir** 1554; **Pt** 2369; **Au** 2398. The prices may in some cases (precious metals, Rh, Ir, Os) be subject to speculation.   B. CORNILS

Ref.: A. Behr (Dortmund/Germany), private communication.

Prilezhaev reaction the reaction of alkenes with peracids to *epoxides. As peracids mainly *m*-chloroperbenzoic acid, perbenzoic acid, performic acid, trifluoroperacetic acid, and monoperterephthalic acid are used. From a practical standpoint it is important that the epoxidation proceeds easily with acids generated in situ. The addition of the epoxide oxygen is stereospecific, i.e., a *trans*-alkene gives a *trans*-epoxide and *cis* forms *cis*. Conjugated dienes and allenes can also be epoxidized. **F** réaction de Prilezhaev; **G** Prileschajew-Reaktion.   M. BELLER

Ref.: D. Swern (Ed.), *Organic Peroxides*, Vol. 2, p. 355, Wiley-Interscience, New York 1971; March.

primary structure (of proteins) → secondary structure

principle of least structure variation
Acc. to this principle those elementary reactions will be favored that involve the least change in atomic position and electron configuration. The pronciple can be used to predict the energetically most favored reaction course. E.g., in the case of the allyl-metal *complex cat. butadiene *polymerization, it can be concluded that the *insertion reaction with the η^4- or η^2-coordinated butadiene in the single *cis* configuration must lead to an *anti* structure of the new butenyl end group, while the butadiene coordinated in the single *trans* configuratuion always results in a *syn* structure (*anti/syn* insertion). Furthermore, the *anti* or the *syn* structure of the reacting butenyl end group correlates with the *cis* or *trans* configuration of the double bond generated the growing chain. **G** Prinzip der geringsten Strukturänderung.   R. TAUBE

Ref.: *Adv.Phys.Org.Chem.* **1977**, *15*, 1; *Angew.Chem.Int. Ed.Engl.* **1980**, *19*, 495; Cornils/Herrmann-1, p. 283.

Prins reaction   the *addition of formaldehyde to alkenes in the presence of acid cats. Depending on the reaction conds. as well as the substrates, the main prods. are 1,3-diols, allylic alcohols, or substituted dioxanes. Although the cats. are generally acids or *Lewis acids (e.g., SnCl$_4$), the Pr. has also been carried out with basic cats. or with *peroxides as cats; cf. also the *Kriwitz-Prins reaction. Apart from HCHO the reaction has been performed with other alde-

hydes and even with ketones. Aldehydes or ketones with electron-withdrawing substitutents (e.g., chloral, acetoacetic esters) react even without cats. Due to the reversibility of the reaction β-hydroxyalkenes can be cleaved thermally to yield the corresponding carbonyl cpds. and alkenes. **F** réaction de Prins; **G** Prins-Reaktion.   M. BELLER

Ref.: *Synthesis* **1977**, 661; *Acc.Chem.Res.* **1980**, *13*, 426; D. Schinzer (Ed.), *Selectivities in Lewis Acid Promoted Reactions*, NATO ASI Series, Kluwer, Dordrecht 1989, p. 147.

probe molecules   The interaction of pm. with a cat. *surface is analyzed to obtain information on surface species believed relevant to a reaction, or to obtain information on surface properties. The pms. *adsorption isotherm, *heat of adsorption, and *activation energy of *desorption can be measured by volumetric, *calorimetric, and thermal desorption experiments and these data allow identification of the type and strength of interaction. Spectroscopic methods, e.g., *IR, *NMR, *XPS, help to characterize the nature of the adsorbed state and of the *sites interacting with the pms. Criteria for the selection of pms. are: selective interaction with sites of interest, small size (pore access, limited interaction with neighboring groups), no or little reactivity at the selected temp., and for *spectroscopy a measurable response upon adsorption. Typical applications are 1: analysis of properties of *supported metals, using H$_2$, CO; and 2: characterization of acidic and basic surface sites, including the distinction between *Brønsted and *Lewis sites. Frequently used probes for acid sites are: NH$_3$, pyridine, alkylamines, CO, PPh$_3$, N$_2$, H$_2$. Efforts to find good probes for basic sites have been made, but, they are often amphoteric in nature: CO$_2$, SO$_2$, pyrrole, CHCl$_3$, C$_2$H$_2$. **E=F=G**.   F. JENTOFT

Ref.: Ertl/Knözinger/Weitkamp, Vol. 2, p. 698; Thomas/Thomas, p. 147.

procatalyst   precatalyst, → photocatalysis, *precursor

prochiral selectivity (in enzymatic reactions) → enantioselectivity in enzymatic catalysis.

product inhibition 1-general: Acc. to the principles of cat. reaction engineering, *activity can be described as a function of the substrate conc. for an irreversible one-substrate reaction without inhibition. For a reversible reaction A ⇌ B, the pi. is considered as a competition of A and B for the *active sites. This can include substrate surplus inhibition as well as multisubstrate reactions and may end up in a decrease of *rate of reaction (inhibition by the product as opposed to *substrate inhibition). An example of *template-directed reactions is described under the entry *self-replication. B. CORNILS

2-in enzymatic reactions: The prods. of many enzymatic reactions bind with relatively high affinities to the *enzyme active site, and thus inhibit the enzyme by blocking substrate from binding. This is true, for example, in the case of the *glycosyltransferases that use nucleotide-activated sugars as substrates. The released nucleotides strongly inhibit the enzyme by occupying the glycosyl donor binding site. Product inhibition patterns can give useful information regarding the kinetic mechanism of a reaction (see *enzyme mechanism), but can be problematic for achieving high yields in enzyme-cat. synth. reactions. In such cases, product removal (e.g., by extraction, precipitation, or further reaction) is usually necessary. **F** inhibition par le produit; **G** Produktinhibierung. P.S. SEARS

production-integrated environmental protection comprises the integration of plants and procs. for the improvement of environmental protection at an early stage of development to meet today's high standards of low emissions, workplace protection, operational safety, etc. All national decrees and regulations must be obeyed. *Catalysis and *catalytic procs. are the focal point to meet these reqirements, as well as *atom economy, *E factor, *environmental catalysis, etc.

F protection de l'environnement par "production intégrée"; **G** Produktintegrierter Umweltschutz. B. CORNILS
Ref.: Ullmann, Vol. B8, p. 213; C. Christ, *Production-Integrated Environmental Protection and Waste Management*, Wiley-VCH, Weinheim 1999.

product (shape) selectivity a term derived from the *shape selectivity. Ps. is high when among all cpds. formed within the *pores of, e.g., *zeolites only those with small enough dimensions can diffuse out of the pores as products. An example is the methylation of toluene. Although the substitution reaction initially leads mainly to *o*- and *p*-xylene, intra-zeolitic *isomerization of the xylenes yields a mixture of all isomers, including the *m*-isomer. This mixture may reach a stationary equil. As the *p*-isomer has the smallest kinetic diameter, it can most easily diffuse out of the zeolite framework. Correspondingly, the quasi equil. of the xylenes in the zeolite pores is disturbed, and the more slowly diffusing *o*- and *m*-xylene molecules isomerize inside the zeolite to *p*-xylene. Consequently, *p*-xylene is the main (or only) product, in strong difference to the prod. distribution which is obtained when the reaction is carried without a shape-selective cat. (cf. *microporous solids. **F** sélectivité de produit; **G** Produktselektivität. P. BEHRENS
Ref.: Thomas/Thomas, p. 614.

proenzyme inactive *precursor of an *enzyme which can be converted to active enzyme by proteolysis. Many proteases (e.g., *chymotrypsin and *subtilisin) are produced in proenzyme (or "zymogen") form; cleavage of specific bonds generates the active form. The prosequence, which may remain attached to the enzyme following cleavage, has been shown in some cases to act as a "molecular chaperone," aiding in the proper folding of the protein. **E=F**; **G** Proenzym. P.S. SEARS
Ref.: *Mol.Microb.* **1991**, 5, 1507.

profens international name for analgesic pharmaceuticals which have the 2-arylpropionic acid strucural unit in common. Most important ps. are naproxen (2-[6-methoxy-naphthyl]propionic acid) and ibuprofen (2-[4-iso-butylphenyl]propionic acid). While naproxen is sold as a single enantiomer, ibuprofen is commercialized as a racemic mixture. Other commercial ps. are flurbiprofen, fenoprofen, ketoprofen, etc. In cat. respects the ps. are interesting for their alternative synth. (e.g., naproxen is still manuf. via *Grignard coupling and resolution as key steps, while ibuprofen is prod. by a Pd-cat. *carbonylation of 2-(4-iso-butylphenyl)-2-propanol). Apart from enantioselective *hydrogenations of the corresponding α,β-unsaturated acids, asymmetric *carbonylations (*hydroformylations, *hydrocarboxylations; cf. *BNPPA) of the corresponding styrenes have been developed. **F=E; G** Profene. M. BELLER

Ref.: Beller/Bolm; A.N. Collins et al. (Eds.), *Chirality in Industry*, Wiley, Chichester 1992, p. 303.

prokaryote A prokaryote, by definition, is an organism that lacks a membrane-enclosed nucleus. Ps. actually lack any membrane-enclosed organelles, although some have extensive invaginations of the plasma membrane. This class includes the eubacteria (true bacteria) and the archaea. These are to be contrasted with the *eukaryotes, which have a variety of membrane-bound organelles in which the different functions of the cell are localized. Many gene expression systems use prokaryotic hosts, as bacteria are typically much easier to grow and manipulate genetically than eukaryotic cells. **E=F=G.** P.S. SEARS

proline *cis-trans* **isomerization** a proc. cat. by the *chaperone peptidyl prolyl *cis-trans* *isomerase during protein folding.
C.-H. WONG

cis ⇌ trans

Ref.: *J.Am.Chem.Soc.* **1994**, *116*, 11931.

promoters 1-general: Ps. are *doping agent (dopants) added to cats. to improve their action (*activity, *selectivity, *stability). Normally, a p. has little or no cat. effect (otherwise it would be a *co-catalyst). Some ps interact with active comps. of the cat. The interaction may cause changes in the electronic or crystal structures of the active solid comp., extend the effective *surface area, or it may influence the *ligand sphere of the cat Commonly used ps. are metallic ions incorporated into metals and *metallic oxide cat. reducing and oxidizing gases or liquids, acid or bases, or ligands (surpluses) added during the reaction or to the cat. before being used cf. *promotion. B. CORNIL 2-in biocatalysis: a p. is an untranslated region of *DNA upstream of a gene which bind *RNA *polymerase and is thus responsible for the initiation of transcription. Specific "consensus" promoter sequences have been identified, which differ between *eukaryotes and *prokaryotes. Cells frequently control the concs. of specific proteins at the level of RNA transcription. A number of regulatory proteins have been characterized which bind to "operator" regions near specific promoters and turn them on (e.g., catabolite activator protein) or off (e.g., *lac* repressor). **F** promoteur; **G** Promotoren. P.S. SEARS

promotion the modification of *activity *selectivity, or *stability of a cat. system by addition of cpds. which would not be active without the base cat. Such cpds. are called *promoters (dopants, *modifiers, in this keyword: ps.). Ps. can be beneficial or essential *Beneficial* substances enhance the desired functional quality, usually with a price to pay for it. The most frequent side effect is a reduction in the rate of selective cat. conv., which is tolerated by means of a well-optimized amount of the promoter. Invariably, there is a sharp optimum conc. at which the beneficial effect is maximal and the side effect is still tolerable.
Essential ps. are often referred to as *co-catalysts. They are usually *ligands to hom. or

*polymerization cat. systems. In hom. cat. the ps. are added with the main comps. to the in-situ synth. In het. systems the ps. are added sometimes to the synth. mixture but frequently also in special post-synth. manufng. steps. In all cases the presence of ps. requires special activation measures to bring them to the desired function. In het. cat. ps. are divided into two classes according to their intended function. *Structural* ps. affect formation and stability of the cat. material, *electronic* ps. directly affect the elementary steps involved in each turnover on the cat. The typical function of structural ps. is the fixation of a given metastable defective structure of a cat. Ps. may segregate into defect sites and inhibit their mobility. In this way they block sintering or *Ostwald ripening of crystals and preserve metastable structures produced during activation of a cat. Typical ps. with this function are Cr oxide, *alumina, CaO, or W oxides. Typical cats. are massive metal and oxide materials such as iron for *ammonia synth. or Fe oxide for *ethylbenzene *dehydrogenation. An example of an essential structural p. are alkali additives to V oxide for *sulfuric acid synth. Here, the active phase is a *supported liquid phase of a ternary alkali vanadate which requires the presence of the p. for its formation.

In *selective *oxidation cat. the addition of cations to Mo oxide massively changes its functionality by the formation of different ternary oxides. It is not clear what the consequence of this cpd. formation really is for the cat. function. The great changes in structure of the base oxide also cause massive differences in the electronic structure rendering the distinction between structural and electronic p. uncertain. This uncertainty is a common feature of the analysis of p. as geometric structural changes inevitably imply electronic structural changes and it remains an open problem what property is more decisive for the function of an *active site.

Electronic ps. affect the local electronic structure of a base cat. Most commonly they add or withdraw electron density near the *Fermi edge in the *valence band. This results in a modification of the *chemisorption properties and hence affects the *surface *coverage with a reactant. In addition the change of electron density near the Fermi edge affects the *redox activity as e.g., electron donation for reductive activation of chemisorbed molecules is a sensitive function of the abundance of weakly bound valence electrons. Minute changes which are barely detectable by analytical techniques can have drastic influences on the chemisorption properties. Thus, amounts sufficient to form sub-monolayer coverage of the p. are sufficient for this mode of operation. Also the formation of bulk or surface *alloys which may only exist under cat. reaction conds. serve as electronic ps. Typical electronic ps. are alkali adsorbates on almost all classes of het. cats. They donate electron density to the base structure. In addition they represent strong point charges at the surface. This affects the electrostatic properties of a surface and can modify the chemisorption of weakly bonding reactants by inducing *van der Waals forces in molecules still in the gas phase before reaching the cat. surface.

*Alloy formation has multiple consequences for base cats., as not only is the electronic structure affected but also the geometric disposition of active centers can change when the p. is inactive and dilutes the active material. As a consequence the local interaction between surface and adsorbate is changed as there are no longer close neighbors of *adsorbates interacting with each other. This has complex effects on the local chemical reactivity of an adsorbed molecule.

One obvious disadvantage of adsorbed promoters is the site blocking, as with the chemisorption of a p. a site of the base cat. material is lost. This immediately reduces the reaction rate per geometric surface area and thus the activity of the reactor. It is of great importance to control the *coverage of the base cat. with a p. which should usually not exceed about 25 % of the available surface area. The enhanced chemisorption activity of promoted

surfaces further makes them highly susceptible to chemisorption of unwanted molecules
from the feed. Promoted cats. are usually
highly sensitive to *poisoning and sorptive
*deactivation. The practical applicability of
electronic promotion is thus limited to cases
where extreme feed purities are available. For
examples cf. *AA procs. of *BP or *Eastman,
*Aldox proc., *Amoco PTA proc., *hydrocy-
anation, *rare earth metals, *WGSR; cf. also
*modifiers. **E=F=G**.   R. SCHLÖGL
Ref.: H.P. Bonzel et al.(Eds.), *Physics and Chemistry
of Alkali Metal Adsorption*, Elsevier, Amsterdam
1989; Ertl/Knözinger/ Weitkamp, Vol.2, p.752 and
Vol.3, p.1084; Gates-1, p.24; Thomas/Thomas, p.417;
Moulijn/van Leeuwen/van Santen, p.337; Farrauto/
Bartholomew, p.59.

propagation reaction   the series of reaction steps in a chain *polymerization in which
many *monomer molecules rapidly add on in
succession to the active center, causing rapid
formation of a polymer molecule once an activated monomer molecule has been formed. In
general a propagation step may be represented
as $I\text{-}(M)_n\text{-}M^* + M \rightarrow I\text{-}(M)_{n+1}\text{-}M^*$ where $I\text{-}M^*$ is
the activated monomer. The proc. continues as
a kinetic chain until the *active center is lost by
termination or is transferred to another molecule. Usually many propagation steps (10^2 –
10^4) follow from a single act of *initiation. For
high molecular weight polymer the propagation
rate must be fast compared to the termination
or transfer rate. **F** réaction de propagation;
G Kettenübertragungsreaktion.

W. KAMINSKY, M. GOSMAN
Ref.: Chien, *Coordination Polymerization*, Academic
Press, New York 1975.

PROPHOS, prophos   *phosphine *ligand,
[(*R*)- or (*S*)-1,2-bis(diphenylphosphino)pro-
pane], for *asymmetric reactions. It is prep. by
a multistep proc. and may be sulfonated to
give the *water-soluble ligand which has been
employed in *aqueous-phase *hydrogenation
of prochiral alkenes. In other hydrogenation
reactions (e.g., for the manuf. of L-amino
acids from (*Z*)-*N*-acylaminoacrylic acids)
PROPHOS acts as an efficient cat. O. STELZER

Ref.: *J.Am.Chem.Soc.* **1978**, *100*, 5491; Patai-2, Vol.1;
Cornils/Herrmann-2; *Tetrahedron* **1986**, *42*, 5157.

propionic acid   → BASF procs.

propoxylation   → ethoxylation

propylene oxide (PO)   an important *intermediate and bulk chemical, manuf. by cat.
*epoxidation or biocat. (e.g., procs. of
*ARCO, *Bayer-Degussa, *Cetus, *ChemSys-
tems, *Daicel, *Shell SMPO, *Solvay/La-
porte/Carbochimique, *Texaco), *Oxirane
proc. The only non-cat. variant is from *Lummus.   B. CORNILS
Ref.: Kirk/Othmer-1, Vol.20, p.271; McKetta-1,
Vol.45, p.88.

protease-catalyzed   peptide   synthesis
→ proteases

protease inhibitors   an important and
quite varied class of cpds. that inhibit proteases. Some are themselves proteins, such as
bovine pancreatic trypsin inhibitor (BPTI),
while others are very small. They may be
quite specific for a particular protease, while
others inhibit entire classes, and can be
reversible or irreversible. The reversible ones
are often substrate analogs (peptide isosteres)
or *transition state analogs. A number of protease inhibitors have found therapeutic use.
F inhibiteurs des protease; **G** Proteaseninhibi-
toren.   P.S. SEARS
Ref.: *Persp.Drug Discov.Des.* **1996**, *6*, 47.

proteases (synthesis)   P. are *enzymes that
*hydrolyze the peptide backbone of proteins.
Many can also accept small peptide or amino
acid amides and esters. P. are split into four
classes, based on their cat. mechanisms: the
*aspartyl proteases, *serine proteases, *thiol
proteases, and *metalloproteases. The strong
*regio- and *stereoselectivity of most ps. has
prompted their use in synth. chemistry. They
are frequently used in the *resolution of
*chiral amino acid esters and amides. For ex-
ample, the aminopeptidases are generally
Zn^{2+}-containing enzymes that hydrolyze an

amino acid from the N-terminus of a peptide, and many will accept simple amino acid amides as substrates. They are commonly used in the prepn. of enantiomerically pure amino acids (Figure). The serine p. *subtilisin has been used in the *regioselective acetylation of sugars at the primary hydroxyl group (with, for example, *vinyl acetate as the acyl donor), and has been used in the preparation of the melanoma associated ganglioside 9-O-acetyl-GD_3 from GD_3.

Ps. have also been used in the synth. of polypeptides. In the thermodynamic approach (see *equilibrium-controlled synthesis), the equil. position for the usual hydrolytic reaction is shifted toward the synth. direction, either by removal of water (the reaction is conducted in dry *solvent) or by, e.g., selective precipitation of the prod. The thermodynamic approach has been used in the thermolysin-cat. synth. of an *aspartame precursor.

An alternative approach is the *kinetically controlled synth. of peptides from amino acid esters, in which an amine nucleophile competes with water for attack of the ester. The ratio of aminolysis to hydrolysis can be shifted in favor of amide synth. by a number of methods, such as lowering the water conc., raising the amine conc., or by *active-site modifications to the enzyme (cf. *thiosubtilisin or *selenosubtilisin). **F** synthèse des proteases; **G** Proteasen-Synthese.   P.S. SEARS

Ref.: U. Bornscheurer, R. Kazlauskas, *Hydrolases in Organic Synthesis*, Wiley-VCH, Weinheim 1999.

protease specificity   refers to the peptide cleavage site of a protease. The *enzyme sub-

sites in which the substrate residues bind are numbered S_1–S_4 (acyl donor side of the scissile bond) and S_1'–S_4' (leaving group), going from lower to higher numbers as one proceeds away from the scissile bond. The residues that fit into these pockets are labeled, accordingly, P_1–P_4 and P_1'–P_4'. For the preferred cleavage sites of common proteases cf. the literature. **F** spécifité de protease; **G** Proteasenspezifität.   P.S. SEARS

protein N-acetylglucosaminylation
Many proteins in the cytoplasm undergo reversible covalent modifications as a means of regulating their activity. One such modification is *phosphorylation, while another modification has been more recently discovered is N-acetylglucosaminylation of serine and threonine residues. This modification is cat. by a heterotetrameric ($\alpha_2\beta$) *enzyme that uses UDP-GlcNAc as a substrate, and has been found to modify the activities of several proteins, including many *protein kinases. **E=F=G**.   P.S. SEARS
Ref.: *Glycobiology* **1996**, *6*, 711.

proteinase inhibitors   → protease inhibitors

protein engineering   refers to the mutagenesis and selection/ screening of proteins with new, desirable properties such as altered catalytic activity or increased thermal stability. **E=F=G**.   P.S. SEARS

protein farnesyltransferase   cat. the farnesylation of C-terminal cysteines in proteins (particularly in the Ras family of signal transduction proteins) ending in CAAX, where A is an aliphatic amino acid and X is Ala, Met, Ser, Gln, or Cys. The reaction proceeds through an electrophilic displacement.
Farnesylation of Ras proteins is often found in cancer development (cf. also *protein prenylation). **E=F=G**.   C.-H. WONG
Ref.: *J.Am.Chem.Soc.* **1996**, *118*, 8761; *Science* **1997**, *275*, 1800.

protein kinase   Kinases are *enzymes that transfer a phosphate from *ATP or another activated phosphate donor to an acceptor. Protein *phosphorylation has been observed at serine, threonine, tyrosine, etc., although not all of these residues are necessarily targets of kinases: phosphocysteine, for example, is found as an enzyme intermediate in phosphatase reactions. O-phosphorylation is a very common means of transiently modulating a protein's function, particularly in *eukaryotes. Most of the kinases have specific consensus sequences: cAMP-dependent protein kinase, an enzyme involved in cAMP signal transduction pathways, recognizes sequences with an Arg to the N-terminal side of Ser, such as Arg-Xaa-Ser, or Arg-Arg-Xaa-Ser, etc. The modification is reversed by a battery of protein phosphatases that *hydrolyze the phosphate esters. Cascades of protein kinase reactions are characteristic of many signal transduction pathways. **E=F=G**.   P.S. SEARS

protein prenylation   a post-translational modification of certain eukaryotic proteins such as the Ras family of GTP binding proteins, nuclear lamins, and the yeast mating factors. A cysteine is linked via a thioether to a farnesyl or geranylgeranyl group; see also *protein farnesyltransferase). **F** prénylation de protéine; **G** Proteinprenylierung. P.S. SEARS
Ref.: *TIBS* **1990**, *15*, 139.

protein processes   continous *biocat. procs. (in *bioreactors) for the prod. of yeasts of high protein content from liquid *n*-alkanes or methanol by selected microorganisms (cf. proc. of *BP Toprina or of *ICI Pruteen). These procs. are now obsolete. **F** procédés pour protéine; **G** Proteinverfahren. B. CORNILS

proteolysis   the *hydrolysis of the amide backbone of a protein. Peptide bonds are quite stable at moderate pH and temp., but can be hydrolyzed rapidly by *proteases. This can be a problem in the isolation of proteins from whole cells, which typically contain endogenous protease activity, particularly if the protein of interest is very protease-sensitive. Protein purification protocols frequently include cocktails of *protease inhibitors to overcome this problem. **F** protéolyse; **G** Proteolyse.   P. SEARS

protocatechuate 3,4-dioxygenase   cat. the *oxidation of protocatechuate to *cis,cis*-β-carboxymuconate. It is an iron-containing *enzyme, capable of activating molecular oxygen for the oxidation reaction. **E=F=G**
     C.-H. WONG

proton catalysis   → acid-base catalysis

proton-induced X-ray emission (spectroscopy) (PIXE)   for the determination of trace elements by means of high-energy protons (accelerator required). The ex-situ method is destructive to the sample; the data are integral and bulk-sensitive.   R. SCHLÖGL

Prototec   → Süd-Chemie AG

proximity effects   → enzyme-catalysed reactions

Pruteen process   → ICI protein proc.

PS   polystyrene

pseudo-Arrhenius   → Arrhenius plot

pseudointercalation model   → Co as catalytic metal

***Pseudomonas* sp. lipases**   The *lipases isolated from different *Pseudomonas* sp. (PSLs) are highly selective, especially for the *hydrolysis of the esters of secondary alcohols, and for the corresponding reverse reactions. Although these lipases seem to accept less bulky substrates than do *Candida* lipases, they possess exceptionally high selectivity on narrow open-chain substrates with *chiral centers located both near and remote from the reaction center. The enzymatic hydrolysis and *transesterification are enantiocomplementary, and allow the preparation of both enantiomers. All *Pseudomonas* sp. lipases possess a stereochemical preference for the *R*-configuration at the reaction center of secondary alcohols (Equation).

> 95% ee

In the resolution of chiral cyanohydrins with PSL-cat. hydrolysis of the esters, the enantioselectivity is very high for aromatic cyanohydrins. The isolation of resolved free cyanohydrins still represents a significant problem. An improvement of this procedure is the use of acetone cyanohydrin for *transhydrocyanation with an aldehyde to form a cyanohydrin, which is subsequently resolved in situ via transesterification using isopropenyl acetate (Equation).

In addition to the enantioselective transformation of many primary and secondary chiral and prochiral alcohols, the acyl transfer capability of PSL has also been used in the enantioselective acylation of hydroperoxides and organometallic alcohols. **E=F=G**.   C.-H. WONG
Ref.: *J.Am.Chem.Soc.* **1991**, *113*, 9360.

pseudopotential   The chemical bonds in a molecule are caused by the interactions of the outermost valence electrons of the atoms. The energetically much lower-lying core electrons are hardly affected by the bonding situation of the atom in a molecule. Since the computational efforts of quantum chemical *ab-initio and *DFT calcns. increase strongly with the

number of electrons, it has been suggested that the chemically nearly inert core electrons may be replaced by a mathematical function which leaves the valence electrons in the same state as in a real atom. Such an approach is called the pp. or effective core potential (ECP) approximation. The parameters for the pp. are derived from all-electron atom calcns. Since the all-electron calcn. may be carried out with inclusion of relativistic effects, which become important for heavier atoms starting from the fourth row of the periodic system of the elements, pp. calcns. are very convenient methods to reduce the computer time required and to include relativistic corrections at the same time. The basic idea of relativistic pp. calcns. is that the core electrons and the relativistic corrections to the valence electrons are not change by alteration of the chemical bonding of the molecule. Systematic comparisons with all-electron calcns. have shown that the error which is introduced by the pp. approximation is in most cases negligible. This is the reason why pps. have become very popular for quantum chemical calcns. of molecules containing heavy atoms.
G Pseudopotentiale.   G. FRENKING

PSO   polysulfone

PTA   purified terephthalic acid, → terephthalic acid (TPA)

PTC   → phase transfer catalysis

PU, PUR   polyurethane

Pullman Kellogg process → Kellogg procs.

purine nucleoside phosphorylase → nucleoside phosphorylase

purine-to-purine exchange reactions → transribosylase

Purofer process   → CO_2 reforming

PVA, PVAL, PVOH   polyvinylalcohol

py pyridine as a group in *complex cpds. or *ligands.

pybox chiral ligand; cf. *chiral pool and *bis(oxazoline)

pyridoxal phosphate a *cofactor found in many transaminases, racemases, and *decarboxylases.

pyridoxal
phosphate

In the transaminases, the cofactor, which contains an aldehyde, forms a *Schiff base (imine) with the primary amine to be transferred. The imine undergoes a tautomeric shift, followed by *hydrolysis of the imine to give the aldehyde and pyridoxamine phosphate. Reversal of the proc. with a new aldehyde substrate results in transfer of the amine.

In the *decarboxylation of α-amino acids, pyridoxal again forms a Schiff base with the α-amino group. The conjugation of the imine with the aromatic ring of the pyridoximine provides an electron sink to stabilize the carbanion formed when the carboxylate is eliminated (as CO_2). **E=F=G**.   P.S. SEARS

pyrimidine nucleoside phosphorylase → nucleoside phosphorylase

pyrolysis thermal degradation of materials and cpds. (*thermochemistry, *thermolysis) yielding lower molecular weight prods. with different properties. P. occurs in chemical procs. known as dry distillation (pyrolyzing) of coal, low-temp. carbonization, cracking, *coking, calcination (*calcining). Cat. variants include *cat-cracking, *TCC or *FCC processes for the prod. of *pyrolysis gasoline or *BTX, *coal liquefaction, etc. **F=G** pyrolyse.
                B. CORNILS

Pyrotol → Houdry procs.

pyrrolines → iminocyclitols

pyruvate dehydrogenase, pyruvate decarboxylase Pyruvate dehydrogenase is a large *multienzyme *complex which converts pyruvate to *acetyl-CoA. The first enzyme of the complex, pyruvate decarboxylase, cat. the addition of thiamine pyrophosphate and decarboxylation of pyruvate to generate hydroxyethylthiamine pyrophosphate (HETPP). The HETPP is converted to acetyl-CoA by a second enzyme within the complex, dihydrolipoyl transacetylase, which contains a lipoic acid prosthetic group. The HETPP carbanion adds to the oxidized form of the lipoic acid to generate the thioester, which is transferred to coenzyme A.

Pyruvate decarboxylase from yeast (EC 4.1.1.1) is an *enzyme which carries out the first reaction of the pyruvate dehydrogenase complex by the same mechanism, but instead of feeding the HETPP to dihydrolipoyl transacetylase, acetaldehyde is released to regenerate the thiamine pyrophosphate (cf. also *thiamine pyrophosphate). **E=F=G**. P.S. SEARS

pyruvate kinase → phosphoenolpyruvate

pzc point of zero charge, representing an uncharged *surface, → isoelectric point, *titania photocatalysis.

Q

QC quantum chemical methods, → embedding methods

Q-Max → UOP procs.

quaternary structure (of proteins) → secondary structure

quinine derivatives are *ligands for hom. cat., cf. *(DHQ)$_2$- and (DHQD)$_2$-PHAL

O. STELZER

quinone proteins *NAD-dependent or -independent quinone proteins (e.g., pyrroloquinoline, PQQ, dependent methanol dehydrogenase) or 6-hydroxydopa containing *enzymes cat. oxidoreductions via *electron transfer reactions. **E=F**; **G** Chinon-Proteine.
Ref.: *Science* **1990**, *248*, 981.   C.-H. WONG

R

R general symbol for alkyl or aryl groups in *complex cpds. or *ligands

racemate A 1:1 mixture of *enantiomers is called a racemic mixture. On crystallization a racemic mixture forms either a *conglomerate of *single crystals of the two *enantiomers or a racemate which contains both enantiomers in a 1:1 ratio within the same crystal (third possibiliy: solid solution). A well-known conglomerate is Na ammonium tartrate, which crystallizes from water if the temp. is below 28 °C (*Pasteur). Interestingly, above 28 °C the salt crystallizes as a racemate (cf. *kinetic resolution). **E=F=G**. H. BRUNNER

racemic temperature is that temp. at which there would be no discrimination betwen *R*- and *S*-enantiomers (cf. *temperature effect). **F** razemic témperature; **G** racemic-Temperatur. C.-H. WONG

racemization In molecules, *chiral elements (central, axial, planar, helical) can be stable or labile. If there is a change in configuration an *optically active cpd. racemizes and under achiral conds., a 1:1 mixture of *enantiomers is obtained. The racemic mixture is favored by an entropy factor, which reduces the free energy of the racemic mixture by approx. 0.4 kcal/mole (at rt.) compared to the free energy of the optically pure components. **F** racémisation; **G** Racemisierung. H. BRUNNER

radial electron distribution (RED) a method for the description of geometric structures in non-crystalline solids; it is synonymous with RDF (radial distribution function). RED can be determined by several methods (cf. *EXAFS, *XRD, *SAX). The data are integral and bulk-sensitive. R. SCHLÖGL

radical addition → special *polymerization

radical catalysis (polymerization) *Low-density polyethylene (LDPE), *polystyrene, and poly (vinyl chloride), three polymers of major economic importance, are obtained by *radical polymerization procs. Other polymers produced by radical polymerization are poly(methyl methacrylate), poly (vinyl acetate), and polyacrylonitrile as well as many *copolymers. *Peroxides such as dibenzoyl peroxides ($C_6H_5COOCOC_6H_5$) or K-peroxodisulfate ($K_2S_2O_8$), *hydroperoxides such as *TBHP, or aliphatic azo cpds. like *AIBN are used as cats. that generate radicals by *homolytic scission when heated or irradiated. The propagating site (*propagation reaction) is a free radical, with an unpaired electron at the last carbon atom of the growing chain. The radical polymerization thus involves many successive steps of growth. In each of them a monomer with a vinyl group is added to the chain, whereby the radical site is re-formed on the newly fixed last unit of the chain:

$$chain\text{-}CH_2\text{-}CHR\cdot + CH_2\text{=}CHR \rightarrow$$
$$chain\text{-}CH_2\text{-}CHR\text{-}CH_2\text{-}CHR\cdot$$

Growing radicals can react with each other in two different ways, either by recombination (a homopolar bond is formed by pairing the single electrons of the free-radical sites of two chains) or by *disproportionation (whereby a H atom is transferred from one chain to the other), to terminate the polymer chain. Two "dead" polymers are formed, one of them bearing a double bond at the chain end. Transfer reactions happen when a free-radical site reacts with a molecule of solvent or monomer, starting a new polymer chain.

Controlled atom transfer radical polymerization (ATRP) is based on reversible formation of growing radicals from dormant macromolecular alkyl halides. The polymerization of styrene, dienes, or methacrylates in the presence of *redox-active *transition metal cpds. (CuBIPY complexes) provides polymers with dispersities of $M_w/M_n = 1{,}05$. (Cf. also *catalytic radical reactions.) **F** catalyse radical pour polymères; **G** Polymer-Radikalkatalyse.

W. KAMINSKY

Ref.: *Comprehensive Chem. Kinetics*, Vol. 14A, p. 15; Remp et al., *Polymer Synthesis*, Hüthig & Wepf, Basel 1986; *Organic Reactions*, Vol. 13, p. 91 and Vol. 48, p. 301.

radical chain reactions Very common in radical chemistry and employed for applications such as *polymerizations or the synth. of complex molecules. A chain mechanism involves three steps: *initiation, *propagation, and *termination (Figure; In = initiator).

Various *radical *initiators are suitable to start rcrs. by homolysis of labile bonds. The propagation step (*propagation reaction) is a succession of elementary reactions in which each radical produced in one reaction is consumed in the next. The chain length is determined by the number of propagation steps (up to 10^6) and can be controlled by the addition of *radical scavengers. Finally, the termination step involves dimerization, *disproportionation, *oxidation, or *reduction of the radicals. **F** réaction radicalaire en chaîne; **G** Radikalkettenreaktion.

T. LINKER

Ref.: Perkin, p. 8.

radical inhibitor → radical scavenger

radical initiator *Radical chain reactions can be started by various initiators which are consumed during reaction and are thus not "pure" cats. (for the definition, cf. *catalysis, radicals). Typically, ris. form radicals by *homolysis of labile C-N or O-O bonds which are cleaved, e.g., *photochemically or by *thermolysis. Examples range from azo cpds. through *peroxides (and oxygen, *LDPE) to peresters, but *azobisisobutyrodinitrile is most commonly used for synth. applications. Triethylborane is an attractive alternative, since radicals are generated at $-78\,°C$ in the presence of molecular oxygen acc. to $Et_3B + O_2 \rightarrow Et_2B\text{-}O\text{-}O\cdot + Et\cdot$ and $Et_3B + Et_2B\text{-}O\text{-}O\cdot \rightarrow Et\cdot + Et_2B\text{-}O\text{-}O\text{-}BEt_2$. Another important class of ris. are *redox initiators, which are mainly employed for *radical polymerizations. Redox is. are a comb. of an oxidative and a reductive species, providing neutral radicals (e.g., *Fenton reaction; see Equs.) or radical anions.

$$Fe^{2+} + H_2O_2 \rightarrow Fe^{3+} + HO^- + HO\cdot$$
$$S_2O_8^{2-} + S_2O_3^{2-} \rightarrow SO_4^{2-} + SO_4^{-}\cdot + S_2O_3^{-}\cdot$$

These radicals are the actual initiators which start the polymerization. **F** initiateur radicalair; **G** Radikalinitiatoren.

T. LINKER

Ref.: Perkin, p. 21.

radical polymerization → radical catalysis (polymerization)

radicals atoms, ions, or molecules with unpaired electrons, e.g., the hydrogen atom H·, ·NO, ·NO$_2$, or ·ClO$_2$. They may be generated via *homolytic reactions or *single-electron transfer. Rs. can act (and accelerate reactions) in small amounts and are in this respect similar to cats. although they do not obey the cat.'s definition as "not themselves being substantially consumed" (*catalysis). They may be regenerated by *chain reactions (*living free-radical polymerization, *catalytic radical reactions). R. are commercially important in some *polymerizations (*radical additions). **F** radicaux; **G** Radikale.

B. CORNILS

Ref.: Giese; Perkins; Trost/Fleming, Vol. 4, p. 779; *Chem. Rev.* **1994**, *94*, 519 and 2549.

radical scavenger *Radical chain reactions are usually terminated by *dimerization or *disproportionation, but the addition of cat. amounts of rss. offers an attractive alternative. Such cpds. are stable aryloxy (e.g., galvinoxyl) or nitroxide *radicals (e.g., 2,2,6,6-tetramethyl-1-piperidinyl oxide, TEMPO), which combine with alkyl radicals under diffusion control (Equation).

Another possibility consists in the trapping of radicals by the addition to nitrones (spin traps, e.g., 5,5-dimethylpyrroline-*N*-oxide, DMPO). An important class of rs. is antioxidants which react rapidly with peroxyl radicals to suppress *autoxidations in polymers, lacquers, foods, etc. (*additives). Most commonly used are sterically hindered phenols. Living organisms are protected from damage by hydroxyl or peroxyl radicals by vitamins A, E, and C, which serve as efficient *antioxidants and rss. (cf. also *inhibitors). **F** capteurs de radicaux; **G** Radikalfänger.   T. LINKER

Ref.: Perkin, p. 132; *Adv.Free-Radical Chem.* **1990**, *1*, 253.

radical starter → radical initiator

radioactive catalyst labeling Cats. are used for the labeling of various biomolecules and pharmacophores with radionuclides (in particular with radiohalogens). The application of cats. for labeling purposes is required for reasons different from the usual ones for other procs. A labeling procedure by isotopic exchange or *substitution must predominantly be fast since most of the radionuclides are short-lived (loss of activity). The labeling yield must be very close to 100 % (no side prods. at all) since no purification can follow the prepn. *Turnover rate is not important and the cats. are usually applied in a similar conc. range to the substrate. Furthermore, in an isotopic exchange reaction (A-I +I* → A-

I* + I) the radionuclide I* without carrier added (isotopically pure) is usually present in extremely low substoichiometric concs. in respect of reactant A-I and cat. The result is a pseudo first-order rate law which is common to all cat. radioactive labeling reactions. The reaction *rate is thus in most cases directly dependent on the amount of cat. In a minor number of cases the introduction of the iodine isotopes ^{123}I, ^{125}I, and ^{131}I (either by isotopic or non-isotopic nucleophilic halogen exchange or by electrophilic/ nucleophilic substitution) have been investigated. The former type of reaction only is strongly cat. by Cu(I) species. Isotopic exchange at activated species occurs much more rapidly (i.e., *m*-iodobenzylguanidin).   R. ALBERTO

Ref.: *Fundamentals of Nuclear Pharmacology*, Heidelberg 1998.

raffinates Raffinate I comprises the C_4 effluents of steam crackers. Separation of butadiene and isobutene yields raffinate II, a mixture of 1- and 2-butenes, as a *hydroformylation feedstock. Raffinate II can be freed from alkynes and butadienes by *selective hydrogenation (by procs. such as *Bayer cold hydrogenation, *Hüls SHP).   B. CORNILS

raft structures highly dispersed monolayer *coverages consisting (Figure see page 482: grey color) of only a few atoms on *supports such as γ-Al_2O_3.
Less organized structures are *overlayers (or *islands).   B. CORNILS

Ref.: Gates, p. 335,381.

RAMA rabbit muscle fructose bisphosphate aldolase, → iminocyclitols

Raman spectroscopy Laser Raman spectroscopy (LRS) is important for the really insitu characterization of cats. It can be carried out at high temps. (1000 K) and press. without interference from the gas phase. The whole spectral range from 100 to 4000 cm^{-1} can be recorded with ms time resolutions for excellent Raman scatterers. Transient temp. or

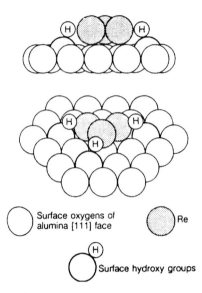

Surface oxygens of alumina [111] face

Re

Surface hydroxy groups

raft structures

press. studies, and pulsed *isotope labeling experiments are thus possible, making it possible to interrelate kinetic and spectroscopic data. Quartz fiber optics render possible spectroscopic access to *cat. reactors of defined and well characterized operation conds. The cat. *activity and *selectivity under *steady-state operation can thus be quantitatively related to the cat. structure as determined by Rs.

Limitations of Rs.: The sample may be sensitive to laser heating which could lead to loss of hydration water, *phase transitions, reduction or even complete decomposition. This effect can be reduced by applying low laser powers (<10 mW), cooling with a thermally conductive, inert gas, cylindrical lens foci, and rotating-sample or focusing-lens techniques. Fluorescence sometimes may overwhelm the Raman spectrum. This may arise from organic impurities, basic surface OH groups, proton superpolarizability, or reduced *transition metal ions. Fluorescence can be overcome by simply burning-off the organic contaminants or dehydroxylating the surface. A change in the excitation frequency may also help in some cases. Frequency- modulation Rs. can also be an alternative to reduce the back-

ground. The inherently low sensitivity of Rs., due to the small Raman scattering cross-sections as compared to IR absorption coefficients, may also be problematic. Raman scattering intensity can be improved by an increase in the excitation frequency due to the v^4 law. A technical solution to this problem was found with the development of high-sensitivity CCD cameras and high-performance, holographic notch filters. These inventions considerably improved the light throughput by a factor of about 10^2. These new-generation Raman spectrometers also allow time-resolved experiments due to their long-term stability. A major drawback of Rs. is the inherently unknown Raman scattering cross-sections. Quantitative Rs. is therefore one of the most difficult tasks, if it is ever possible at an acceptable level of error. Even for pure comps., the Raman scattering cross-section may change as a function of temp., partial press. of the reactive gases, *particle sizes or crystallinity. Furthermore, Raman cross-sections of surface species cannot be compared with those of pure reference cpds. because of the unknown electronic influence of the cat. *support, even when their structure remains unchanged. Thus, the SERS effect observed for Al, Cu, Ag, and Au enhances the Raman scattering efficiency of adsorbates by a factor of 10^6. Moreover, Raman intensities of supported transition metal oxides may change as a function of treatment. Reduction leads to colored or even black samples which have a high absorption coefficient. Self-absorption strongly reduces the intensity of the Raman scattered light. Thus, the intensity of a measured Raman band cannot be directly correlated with the actual conc. of a certain transition metal surface species. Therefore, quantification of Raman spectra, by comparing relative peak intensities, can easily lead to erroneous results because the absolute or even relative Raman scattering efficiencies are not precisely enough known and cannot be determined with the necessary precision. The determination of relative Raman cross-sections of optically hom. samples is already extremely difficult and this task seems to be impossible for het., polydisperse cats. Furthermore, the scattering volume strongly depends on the optical properties of the sample. The scattering volume is considerably reduced for opaque materials due to damping of the incident *and* scattered light and remains undetermined.

Despite all these disadvantages, Rs. is important for the characterization of cats. The advantages of Rs. as compared to other techniques, e.g., simple in-situ Raman reactors, high reaction temps., no interference from gas phase or support, overrule its few disadvantages. Thus, Rs. is perfect for the structural characterization of cat. materials during cat. action, because this technique is sensitive to crystalline, amorphous, glassy, or molecular species. Due to its broad applicability, Rs. finds increasing use to characterize cat. active species, cat. *deactivation procs., and control on-line cat. reactors at their operation maximum. **F** spectroscopie Raman; **G** Raman-Spektroskopie.   G. MESTL

Ref.: Ullmann, Vol. 5, p. 429.

Ramberg-Bäcklund reaction

base-cat. conversion of α-halogen-sulfones to alkenes by SO_2 elimination. **F=G=E.**   B. CORNILS

Ref.: Laue/Plagens.

random mutagenesis

the random introduction of errors into a genetic code. The mutations may be targeted to a specific gene or region of a gene, or they may be targeted to the entire chromosome(s) of an organism. Several cpds. are commonly used for this purpose, which are either *DNA-modifying agents (e.g., deaminating agents such as nitrous acid or sodium bisulfite or methylators such as *N*-methyl-*N'*-nitro-*N*-nitrosoguanidine) or base analogs that are incorporated into DNA but cause poor replication fidelity by DNA polymerase (such as 5-bromouracil). Lately, error-prone *PCR, in which the fidelity of the DNA *polymerase used is reduced by addition of organic cosolvent and/or the addition of excess metals or nucleotide substrate, has been used to target mutagenesis to specific regions of the DNA. Alternatively oligonucleotides with randomized segments may be used to amplify all or a portion of a gene. This mutated piece is then inserted back into the original with *restriction endonuclease digestion + DNA ligase-cat. religation. Rm., coupled with a selection or rapid screening technique, is frequently used for the purpose of isolating a new enzyme with modified properties (cf *screening, *selection). **F** mutagénèse par hazard; **G** Zufallsmutagenese.

                                       P.S. SEARS

Raney catalysts

a group of *skeletal cats., mainly used for *hydrogenation and *dehydrogenation reactions, firstly investigated in 1925 by M. Raney (today the trademark of *Grace Davison cats.). The prepn. is performed by alloying the metals (which achieves excellent mixing of the comps.) and selective dissolution of leachable comps. in acidic or alkaline solution producing *spongy, highly porous cats. with *BET surfaces of 30–120 $m^2 g^{-1}$. *Raney Ni is the most established cat. Raney Cu-Zn cats. are made by leaching of Cu-Zn-Al alloys and are used commercially for *methanol synth. and *water-gas shift reactions. Raney Ag cat. is used for making gas diffusion electrodes for *fuel cells and batteries. Raney Co, Fe, Pd, and Raney Cu-Zn (with Cd modified, *directed poisoning) exhibit a high *selectivity in hydrogenating α,β-unsaturated aldehydes and ketones. Raney-Ru hydrogenates CO and CO_2, whereas Raney Cu-Zr hydrogenates CO_2. Doping of the cat. (e.g., with Cr, Cs, Rb, heteropolyacids, etc.) by *impregnation during the leaching proc. or by soaking in a metal salt solution can improve the cat. performance. **F** catalyseurs Raney; **G** Raney-Katalysatoren.

             R. SCHLÖGL, A. FISCHER

Ref.: Augustine, p. 250; R.E. Malz, *Catalysis of Organic Reactions*, Dekker, New York 1996.

Raney cobalt

prepared from Al-Co alloy (*Raney cats.). It is less active than *Raney Ni but more active than Raney Cu. It is especially useful for the *hydrogenation of nitriles and aldoximes to amines.   R. SCHLÖGL

Ref.: F.E. Herkes, *Catalysis of Organic Reactions*, Dekker, New York 1998; Augustine, p. 248; Ertl/Knözinger/Weitkamp, Vol. 1, p. 67.

Raney nickel widely used *hydrogenation and *dehydrogenation cat. The *skeletal cat. is made by leaching aluminum out of an Ni-Al alloy with an excess of boiling aqu. NaOH. A porous network is created, giving the *spongy metal black, where Ni crystallites (fcc nickel) are supported on Ni-Al cores. Depending on the temp. conds. of preparation, the *BET surface area is of the order of 30 to 120 m^2g^{-1}. Raney Ni is saturated by H during its prepn. ($2\,Al + 2\,OH^- + H_2O \rightarrow 2\,AlO_2^- + 3\,H_2$). Hydrogen is adsorbed in reversible and irreversible forms. Raney Ni is widely used for hydrogenation procs. such as conv. of nitriles to amines, and alkenes to alkanes (predominantly *syn* *addition of H), removal of sulfo groups and halogen from aromatic rings, *desulfurization of thiols and thioethers and *reduction of carbonyl to methylene groups in aldehydes and ketones. Raney Ni acts as a dehydrogenation cat. during the reaction of *methanol to formaldehyde and plays a role in the manuf. of *fuel cell anodes. The *Synthane proc. produces flame- sprayed (*flame hydrolysis) Raney Ni cats. for the *methanation of *syngas, using *tube-wall reactors (TWRs). **E=F=G**.   R. SCHLÖGL, A. FISCHER

Ref.: Augustine, p. 241.

rare earth metals as catalysts In het. cat. the rare earth metals (lanthanides Ln = Sc, Y, La, Ce-Lu) achieved commercial importance as *promoters in Ln-exchanged Y-*zeolites for *FCC procs. creating high electron field gradients and providing high *surface *activity (e.g. *Engelhard FCC proc.). Ce is added to ceramic *afterburner cats. for vehicle emission control (*automotive exhaust cat., *three-way cat., *ceramics). Ce acts as a stabilizer for high-surface *alumina, a *promoter of the *water-gas shift reaction, as an oxygen storage comp., and as an enhancer of NO_x reduction at the metallic comp. Rh. Ce is together with Th the active comp. for *ketonization cats. and is also an *additive to Fe oxide cats. for the *dehydration of *ethylbenzene to styrene and to Mo cats. for the *ammoxidation of propylene. LaF_3 on active carbon has been proposed for the *CFC destruction. Te-Ce cats. serve in *Montecatini's *acrylonitrile proc. or in other *ammoxidation procs.

The cat. efficiency of the lanthanides in hom. cat. is based on their multifaceted nature, the smoothly varying intrinsic radii, electrophilicity, oxophilicity, enhanced hardness acc. to the *HSAB concept, and *redox stability of the trivalent cations. The *precats. range from inorganics to highly moisture-sensitive organometallics; water-stable systems are best represented by Ln triflate *complexes, which are reuseable *Lewis acids. They cat. *aldolizations, *Diels-Alder reactions, *Henry aldoladditions, *Mannich-type and *Michael reactions, *Mukaiyama reaction, as well as *allylations or *cycloadditions effectively. *Heterobimetallic, chiral LnM_3(*BINOL)$_3$ complexes (M = alkali metal) are highly enantioselective cats. for *C-C bond forming reactions. Rare earth *alkoxides are used in functional group transformations, rearrangements, and exchange reactions, *Oppenauer oxidations, or *transesterifications. Rare earth *metallocene *complexes such as Cp_2LnX (X = hydride, amide) have been proposed for *polymerizations or *asymmetric hydrogenation, *hydroamination, *hydrosilylation, or *hydroboration. Multicomponent *Ziegler-Natta cats. containing Nd were used for the manuf. of *cis*-*polybutadiene. Supported (*immobilized) variants of rare earth metal cats. are investigated. Ln complexes with *multidentate ligands are used as artificial *ribonucleases for RNA transesterification and as cats. for *hydrolysis of DNA. **F** les métaux rares comme catalyseurs; **G** Seltenerdmetalle als Katalysatoren.   R. ANWANDER

Ref.: *Catal. Rev. Sci. Eng.* **1977**, *16*, 111; *Prog. Polym. Sci.* **1993**, *18*, 1097; *Top. Curr. Chem.* **1996**, *179*, 247; *Synlett* **1994**, 689; *Angew. Chem.* **1997**, *109*, 1290; Cornils/Herrmann-2, p. 519; Cornils/Herrmann-1, p. 866; Beller/Bolm, Vol. I, p. 285 and 313.

Raschig-Hooker process an older two-step proc. for the manuf. of phenol by 1: *oxychlorination of benzene with HCl/O$_2$, cat. with fixed-bed CuCl$_2$ · FeCl$_3$/Al$_2$O$_3$ at 240 °C, 2: *hydrolysis, cat. with Ca$_3$PO$_4$-SiO$_2$ at 450 °C. **F** procédé Raschig-Hooker; **G** Raschig-Hooker-Verfahren.   B. CORNILS

Raschig hydrazine process by reaction of aqueous NaOCl and NH$_3$ at 0 °C to chloramine and its conv. to hydrazine with a many-fold molar excess of anhydrous ammonia at 130 °C/press. Since the side reaction of chloramine and hydrazine to nitrogen is cat. by Cu, *chelating reagents such as EDTA are used. **F** procédé de Raschig pour le hydrazine; **G** Raschig Hydrazin-Verfahren.   B. CORNILS
Ref.: Büchner/Schliebs/Winter/Büchel, p. 46.

Raschig phenol synthesis *oxychlorination of benzene via chlorobenzene (air and HCl in the presence of a *supported copper cat. at 275 °C). Overchlorination is a problem and is restricted to 5–8 % dichlorobenzenes (refined to *o*- and *p*-dichlorobenzenes) by low conv. (15 %) of the benzene. In the second stage the chlorobenzene formed is converted with steam at 400 to 500 °C on a calcium phosphate apatite cat. (10–12 % conv.; cf. *Raschig-Hooker proc.). A modified *hydrolysis process was developed by *Dow. The chlorobenzene proc. for the prod. of phenol lost its importance during the 1970s. Similarly, the toluene *oxidation process to benzoic acid (applying Co salt cats.) with consecutive oxidative hydrolysis with steam and air at 230–250 °C and 1 MPa is still on stream (Dow). Both procs. are replaced almost completely by the *Hock proc. **F** procédé Raschig pour le phénol; **G** Raschig Phenolverfahren.
R.W. FISCHER
Ref.: H.-G. Frank, J.W. Stadelhofer, *Industrial Aromatic Chemistry*, Springer 1988, p. 148.

rate acceleration, enzyme-cat. → enzyme catalyzed reactions

rate constants With *non*-catalyzed reaction, the term *rate constant means the constant of proportionality which relates the rate to the appropriate functions of the reactant concs. or press., and is usually denoted by k. Thus for the bimolecular reaction A + B, the rate r is expressed as $r = kP_A{}^a P_B{}^b$ or $r/P_A{}^a P_B{}^b = k$, where P is the press. of the subscript reactant and superscript letters denote the orders of reaction. The dimensions of k therefore depend on the values of the orders. Unfortunately with het. cat. reactions the meaning of an rc. is less clear because the concs. of the reacting species are generally unknown. For a bimolecular surface reaction it is usual to write $r = k\theta_A\theta_B$, where the θ terms represent the fractional *surface *coverages by the reactants, i.e., the two-dimensional concs. The problem is then how to relate them to the measurable press. or concs. of the reactants in the fluid phase. This can be done either by assuming that each θ varies with its P according to the *Langmuir adsorption equ., so that rate expressions of the *Langmuir-Hinshelwood type are obtained, or by supposing that the θ-P relationship is adequately modeled by a logarithmic function, which leads to the Power Rate Law formulation $r = kP_A{}^x P_B{}^y$.
A rate constant, if properly derived from the experimental measurements, ought to vary with temp. according to the *Arrhenius equ. $k = A\exp(-E_t/RT)$, where A is the *pre-exponential factor and E_t the *true activation energy (see also *Arrhenius plots, *apparent activation energy). This implies that the experiments are performed under conds. where the slow step occurs on the surface, i.e., there is no *diffusion limitation (cf. *rate of reaction). **F** constante de vitesse de réaction; **G** Reaktionsgeschwindigkeits-Konstante.
G.C. BOND

rates of adsorption/desorption The term *adsorption embraces both physical adsorption (which happens immeasurably quickly) and *chemisorption, for which there is always an *activation energy. The dissociative *chemisorption of H on many clean metal surfaces is almost non-activated, while that of N on iron is clearly activated. The term *desorption always relates to chemisorbed species, and the proc. is inevitably activated because the energy released when adsorption occurs (i.e., the enthalpy of adsorption) has to be supplied to reverse it.

Rates of adsorption may be measured on almost any solid surface, but best results are obtained with metal surfaces that can be thoroughly cleaned (e.g., *single crystals). Possible techniques include change in the press. or volume of the adsorbate, or in some property of the adsorbent (e.g., temp. rise due to adsorption enthalpy, *work function, or magnetic character). Desorption is best characterized by relatively slow or rapid heating. Each combination of adsorbate and adsorbent may exhibit a number of chemisorbed states, which complicate the interpretation of the observed rates. During thermal desorption, new states may be generated which did not pre-exist.

A general expression for rate of desorption v is $v = \sigma P \ (2 \ \pi \ mk_BT) \ f(\theta)\exp(-E_{ads}/RT)$, where the term $\sigma P \ (2\pi mk_BT)$ gives the probability of a molecule of mass m at press. P adsorbing at temp. T, σ being the condensation coefficient. The rate is a function of coverage θ, and it may decrease quickly with coverage because of an increase in E_{ads}. From thermal desorption it is possible to derive an activation energy for desorption E_{des} which is given by $E_{des} = -\Delta H + E_{ads}$, where $-\Delta H$ represents the change of enthalpy due to the adsorption. **F** vitesse d'adsorption/désorption; **G** Adsorptions/Desorptions-Geschwindigkeit. G.C. BOND

Ref.: D.P. Woodruff, T.A. Delchar, *Modern Techniques of Surface Science*, Cambridge University Press, Cambridge 1986; D.O. Hayward, B.M.W. Trapnell, *Chemisorption*, 2nd Ed., Butterworths, London 1964; van Santen/van Leeuwen.

rates of reaction Definition of this term in the context of het. catalysis is more difficult than for reactions proceeding hom. in the gas or liquid phase. 1: The observed rate may not be that of the *surface reaction, but may be partly or wholly limited by the rate of transport of reactants to the surface (specifically, to the active centers) or of prods. away from the surface: reactions under this cond. are termed *diffusion limited or *mass-transport limited (as for the effect of mixing, cf. *catalytic reactors). 2: With thermoneutral reactions, the temp. at the surface of the cat. may be either lower or higher than that recorded outside the cat.; this temp. therefore varies with conv., extent of *deactivation, etc., and this affects diffusion rates within the cat. bed or particle. The observed rate is not therefore that characteristic of the measured temp., except at very low convs. 3: Cat. reactions frequently suffer deactivation due to a number of causes: it may be rapid or slow, but whichever it is, it is necessary to specify whether the rate is being measured early in the cat's. life, or at a steady (partially deactivated) state.

Numerous artifices have been developed to measure rates when deactivation is severe. In a pulsed-flow reactor, short pulses of reactant are passed over the cat. bed, reactivation being effected between pulses. Very useful information is obtained by imposing sudden changes on the reactant conc. (i.e., observing *transient kinetics; cf. *temporal analysis of products (TAP).

For a reaction proceeding in an open system (e.g., a flow reactor), an instantaneous or a *steady-state rate may be defined first as the mass of reactant transformed per unit volume of cat. in unit time. Thus for example the rate r_A of the unimolecular transformation of a reactant A is $r_A = d\alpha/d \ (V/F_A^0) = d\alpha/d\tau$, where α is the conv., V the cat. bed volume, F_A^0 the flow-rate of A at the beginning of the bed, and τ the apparent contact time. It is, however, more profitable to express the rate in terms of either cat. weight or its *surface area (i.e., specific or areal rate): if the number of active centers is known, the rate expressed as molecules per active center in unit time is the *turnover frequency (TOF). Kinetic information is obtained by changing the flow rate and the reactant conc(s).

For reactions proceeding in a closed system (e.g., an *autoclave), the rate decreases as the reaction continues (except in the case of a zero-order reaction) due to depletion of the reactants. Kinetic information comes from analysis of this time-dependent rate and by changing the initial press. or conc. of the reactant(s). **F** vitesse de réaction; **G** Reaktionsgeschwindigkeit.   G.C. BOND

Ref.: Ertl/Knözinger/Weitkamp, Vol. 3, p. 958.

RBS → Rutherford backscattering spectroscopy

RCC process → Ashland procs.

RCD Unibon process → UOP procs.

RCH/RP process → Ruhrchemie procs.

RCM → ring-closing metathesis

RDF radial distribution function, synonymous with *radial electron distribution (RED)

RDS desulfurisation of atmospheric residuum (*petroleum processing)

reactant inhibition → substrate inhibition

reactant (shape) selectivity bases on the principle of size exclusion of certain prods. An example for rs. is the *dehydration of a mixture of *n*-butanol/*iso*-butanol over a *zeolite A cat. As a consequence of the larger kinetic diameter of the branched alcohol, it cannot enter the zeolitic pore system and is excluded from the cat. conv. to the alkene, whereas the linear alcohol is dehydrated (cf. also *microporous solids). **F** sélectivité des réactifs; **G** Reaktandenselektivität.   P. BEHRENS

reacting intermediates → intermediates

reaction carrier special cpd. which is able to transfer reactive groups such as -O- or -O-O- from one molecule to another and which can be recycled, e.g., *hydroperoxides of hydrocarbons (*ARCO or Oxirane procs.) or of *anthrahydroquinone for the manuf. of H_2O_2.
                                                    B. CORNILS

reaction coordinate describe the progress of a reaction via the *transition state(s) to the prods. usually by variation of an inter- or intramolecular distance or by a change in a bond angle. Although in general many coordinates are involved in reactions, the high-dimensional problem is often reduced to a 1D potential energy diagram. **F** coordinate de réaction; **G** Reaktionskoordinate.   R. IMBIHL

reaction dynamics show the actual pathway of the reaction with all variations of bond distances and bond angles and all changes in the energy distribution on the various degrees of freedom in the course of a reaction. Experimentally rd. can be investigated in state-resolved molecular beam experiments in which a laser is used to detect the energy distribution of an educt molecule before reacting at a *surface and/or of a prod. molecule released from the surface. On the basis of calc. interaction potentials, *molecular dynamics simulation of reactions can be conducted which show the actual motion of atoms and molecules during a reaction. The trajectories are obtained by integration of the equs. of motion, i.e., basically Newton's laws are used. **F** dynamique de réaction; **G** Reaktionsdynamik.   R. IMBIHL

Ref.: R.D. Levine and R.B. Bernstein, *Molecular Reaction Dynamics and Chemical Reactivity*, Oxford University Press 1987.

reaction kinetics → kinetics

reaction mechanism   1-heterogeneous: The synth. of *ammonia is one of the very few cat. reactions where the whole sequence of elementary steps is known. It is significant that in hom. and het. reactions this sequence is different. In hom. cat. the dinitrogen molecule is coordinatively bound to an active metal and then successively *hydrogenated via diimine and *hydrazine intermediates. Only in the final reaction step is the N-N bond broken after successive weakening by hydrogenation. In het. cat. the di-N molecule is first split into N atoms forming a metal nitride which is successively hydrogenated by *co-adsorbed H. In the final step the metal–N bond is broken. Such knowledge which has been quantified by determining the relative energetic positions of all the reaction *intermediates is extremely rare. It still does not represent a reaction *mechanism. Such a picture can be defined as the projection of the energy hypersurface on the *reaction coordinate. The minima represent the intermediate prod., the maxima contain the information about the *transition state for each elementary step. According to *transition state theory such a picture can only be handled by theory, as transition states cannot be fixed by experiments (except by ultrafast spectroscopy). In the context of cat. procs. the term *reaction mechanism* is often used to describe a hypothesis about the sequence of events in a reaction, of which usually not a single intermediate is known experimentally. The empirical rules on, e.g., *acid-base and *redox reactions, known in chemistry and the vast amount of empirical wisdom are used to depict a plausible reaction pathway.

In het. cat., the methodology of in-situ experiments (*in-situ studies, *in-situ reaction monitoring) allows an approximation to possible reaction intermediates. Unfortunately, the accuracy of structural information which is standard in hom. cat. can hardly ever be reached by in-situ spectroscopy which, in turn, connects the observations direct to the cat. function by measuring the reaction *kinetics simultaneously with the spectroscopic information. The reliable knowledge of reaction intermediates or even the ability to calc. the whole rm. would be the indispensable prerequisites to designing cat. systems. Without knowledge of the kinetically critical steps it is difficult to design even qualitatively a cat. with high specificity in the interaction between substrate and cat. *active site. The collection of theoretical and experimental facts about rms. thus remains a major challenge of cat. science which has been addressed only with moderate success in the past.   R. SCHLÖGL

2-homogeneous: The mechanism of a hom. chemical reaction describes the detailed manner in which the reaction proceeds, i.e., the nature of the different steps involved. It distinguishes between *rate- and non-rate-determining steps, the participation of *precursor and successor *intermediates, parallel and consecutive reactions, acid-base equil., *complex formation and dissociation, and the reversibility of reaction steps.

Mechanistic information can be gained by detection and spectroscopic identification of stable or unstable intermediates, the isolation of side prods., analyses of the reversibility of the proc., and of chemical *kinetics. The latter involves the determination of the empirical rate law by means of a systematic analysis of the influence of all chemical conc. variables on the rate or rate constants describing the time-resolved behavior of the reaction. These experiments result in the empirical rate law that describes the time dependence of the reaction under all the investigated conds. Temp. and press. can reveal further information. The suggested rm. must be in agreement with all kinetic information, and must fit all the chemical knowledge available. Based on the suggested rm., further chemical and kinetic experiments can be designed in order to confirm its validity (on the cat.'s side: cf. *cat. cycle). In this respect, theoretical *ab-initio and *DFT calcns. are employed to confirm the feasibility of the suggested *transition state. **F** mécanisme de la réaction; **G** Reaktionsmechanismus.   R. VAN ELDIK

Ref.: Wilkins; van Eldik/Hubbard; Burgess; Atwood; Ertl/Knözinger/Weitkamp, Vol. 3, p. 1123.

reaction modulus   → Hatta number

reaction parameters in catalysis For the description of a cat. proc., detailed knowledge about *kinetic and *activity parameters and product *selectivity, as well as *cat. lifetime and load of reactor or the cat., is needed. These parameters are determined for a given set of reaction conds. The kinetic parameters include the *reaction rate constant, the *activation energy (*thermodynamics in cat.) and the order of reaction. From them, the resulting reaction rate represents an activity parameter, which is measured for het. cat. reactions in moles converted per unit time referred to either the weight or volume or active *surface of the cat. For hom. cat. reactions the reaction rate (*rate of reaction) is expressed as moles converted per unit time per volume of the reactor (*TON/TOF). In addition, conv., product yield, and *space time yield, the latter expressed by volume or weight of desired product per unit time, are also activity parameters. A very significant parameter is the product *selectivity, defined as moles of desired product related to the moles of converted feed, which in particular control the proc. efficiency with respect to the economic use of raw material and to the proc. design. The reactor or cat. load is expressed as space velocity (*GHSV, *LHSV).

The same definition is used for hom. cat. liquid-phase reactions, but the liquid volume per unit time is related to the reactor volume instead of the cat. volume. The space velocities for liquid feeds can be expressed also as weight of the liquid per unit time per weight of cat. or of the liquid in the reactor. The reciprocal value of space velocities (h^{-1}) leads to the *space time (h), a measure of the hydrodynamic residence time, e.g., in a tubular reactor. A corresponding parameter is the ratio W/F_0 (kgh^{-1}) which is the weight of the cat. per volumetric flow rate of feed and mostly used to characterize the activity dependence on the linear flow of the feed. Furthermore, the cat. lifetime is an important reaction parameter which is often given as time on stream (h) (*cat. lifetime). However, this parameter mirrors the true life time of the cat. only by taking into account the loading of the cat. used. Therefore, the cat. lifetime parameter, expressed as total weight of feed per weight of cat. per unit total time on stream is more meaningful ($kg \cdot h^{-1} \cdot kg^{-1} = h^{-1}$). The time on stream resulted from the amount of feed which is converted within a given limit of product yield. **F** paramètres de réaction en catalyse; **G** Reaktionsparameter der Katalyse. P. CLAUS, D. HÖNICKE

reaction probability The cat. *activity can be specified as the *turnover frequency (*TOF*, units s^{-1}), which is the number of times N that the overall cat. reaction takes place per *active site per unit time under a specified set of reaction conds. such as temp., press., feed gas composition, and degree of conv.: TOF = $1/S \cdot dN/dt$. Thus, the total number of sites S has to be known. The flux of impinging molecules J per unit area is given by the *Hertz-Knudsen formula as a function of temp. and press. The rp. is defined as the ratio of the rate of formation of prod. to the rate of impingement of the reactant. Thus, it can be calcd. by dividing TOF by J, taking the number of *active sites per unit area S/cm^2 into account: rp. = TOF/($J/[S/cm^2]$). Accordingly, the *sticking probability is defined as the rate of *adsorption divided by the rate of impingement. A large set of rps. has been accumulated for different types of hc. reactions cat. by *single crystal *surfaces of *transition metals in the presence of H. The reactions comprise *dehydrogenation, *hydrogenation, *hydrogenolysis, *cracking, *ring opening, *dehydrocyclization, and *isomerization. The initial rps. were found to decrease by two to six orders of magnitude with increasing press. due to the presence of carbonaceous species on the surface. **F** probabilité de la réaction; **G** Reaktionswahrscheinlichkeit. M. MUHLER Ref.: Somorjai.

reaction rate → *rate of reaction

reaction spectroscopy (TPRS; cf. also *temperature-programmed reaction spectroscopy) Temp.-programmed reaction spectroscopy (TPRS, or temp.-programmed surface reaction, TPSR) belongs to the single-cycle transient techniques, whereby the state of the cat. is changed by means of a heating schedule. The prods. leaving the *surface are measured as a function of temp. It is possible to deduce important information about the nature of the *adsorbed species, about the nature of the *bonding between the adsorbates and the adsorbing *surface, about the *reaction mechanism, and about the influence of coadsorbates and of *promoters on the cat. surface chemistry. In *UHV chambers, TPRS can be employed on *single crystal surfaces and on particles on flat model *supports using a temp.-programmed mode either in vacuo or in a reactive environment.

In a flow system on real cats. under working conds., TPSR can be applied to study the *kinetics of cat. reactions, which occur on *active sites at the cat. surface, and to evaluate the role of this active site in the elementary steps of the overall reaction. Usually, two modes of TPSR experiments are applied; 1: Gases are sequentially coadsorbed on the cat. surface and heating is performed in an inert carrier gas. 2: A cat. on which surface species are pre-adsorbed is heated in a reactive carrier gas (such as CO or H_2). Analysis of the desorbed molecules is usually performed by *GC or by *MS. In-situ *reduction or activation of the cat. and *desorption of any unwanted ad-sorbed species are done prior to a TPSR ex-periment. Important parameters in a TPSR experiment are the total number of *active sites, the *space-time throughput, the chosen sieve fraction of the cat. particles, the total press., the feed gas composition, the presence of coadsorbates, the heating rate (usually cho-sen to be linear), and the geometry of the re-actor. The shape of a TPSR profile is mainly influenced by surface heterogeneity (multiple adsorption states), readsorption, limitations caused by *mass and *heat transfer, as well as dissolution or segregation phenomena. The removal of preadsorbed species from cats. by heating in a reactive gas stream can be further employed for the quantitative determination of the *coverage of this adsorbate, which, in the case of saturation and a known adsorption stoichiometry, is equal to the total number of active sites for a particular het. cat. reaction. Possible kinetic studies comprise the determi-nation of the rate-determining step (rds) in an overall reaction and of kinetic parameters such as reaction orders and *activation ener-gies. TPSR has been widely applied for ex-ploring cat. reactions such as *hydrogenation

(TPH), *oxidation (*TPO), *methanation (TPM), sulfidation (*TPS) used for studies on *hydrodesulfurization (HDS), *gasification (TPG), etc. **F** spectroscopie de la réaction; **G** Reaktionsspektroskopie.   M. MUHLER

reaction velocity → rate of reaction

reactivation of catalysts Reactivation is performed after *deactivation, by appropriate means: e.g., by burning off *coke *deposits or dissolving resids with *supercritical solvents. The r. may proceed alternatively with *deacti-vation or simultaneously with the running proc. (cf. *regeneration)   B. CORNILS

reactive distillation → catalytic distilla-tion

reactor gradients → mass transfer, *heat transfer, *reaction parameters

reactors → catalytic reactors

real catalysts commercial cats. as opposed to *model cats. Cats. with *single crystal sites can be regarded as the intermediate state. The use of single crystal *surfaces is at the heart of the so-called surface science ap-proach to het. cat. because this type of ap-proach offers two fundamental advantages: conceptually, the influence of surface struc-ture can be investigated systematically and experimentally, and diffraction methods like *LEED are applicable to study phenomena like reconstruction or the formation of or-dered overlayers of adsorbates. The *material gap and the *pressure gap are the great chal-lenges (and so far unsolved problems) of the development of het. cat.   R. IMBIHL

really ideal mechanism → Rideal-Eley mechanism

recirculation → recycling

reclamation of catalysts A spent cat. has undergone significant structural changes and/or severe *poisoning effects during use.

These changes are normally irreversible and prevent the cat. from being reactivated/*regenerated for further operation. In this case reclamation (or recovery) of some or all of the constituents is a technically and economically reasonable step. Complete mechanical and chemical destruction is necessary in order to recover valuable ingredients, mostly by acid or base leaching and subsequent purification (selective *precipitation, extraction, electrolysis). Reclamation offers an environmentally sound alternative to disposal and therefore is becoming a widespread method. Cf. *final disposal of catalysts. **F** retraitement des catalyseurs; **G** Katalysatoraufarbeitung.

C.D. FROHNING

Ref.: European Catalyst Manufacturers Association (CEFIC), *Guidelines for the Management of Spent Catalysts*, 1995; McKetta-1, Vol. 55, p. 322.

recombinant DNA → cloning, *expression

recombinant yeast The metabolic pathways of yeast can be altered so that the whole cell can be used for biotransformation. A typical example is the assymetric *Baeyer-Villiger oxidation cat. by recombinant Baker's yeast in which cyclohexanone monooxygenase from *Acinetobacter sp.* is produced. **F** levain recombinant; **G** rekombinante Hefe.   C.-H. WONG

Ref.: *J. Am. Chem. Soc.* **1998**, *120*, 3541.

reconstruction (of surfaces) → ideal crystal structure, *restructuring

recovery of catalysts → reclamation of catalysts

recycle reactor → Berty reactor

recycling In recent years technical and economic considerations in reclaiming constituents of used cats. have been supplemented by ecological constraints. Whereas cats. containing precious metals (with few exceptions) have always been reworked, part of the bulk cats. for hydrocarbon processing have been dumped as landfill due to the comparatively

low value of their metal ingredients (*final disposal of cats.). Today not only the price of the metals but also the changes in ecological attitude (sustainability) and legislation are the driving forces for reworking spent cats.

By far the greatest amounts are Ni-Mo and Co-Mo hydroprocessing cats. These cats. operate as sulfides and deactivate as a result of the accumulation of metals and *coke. Reactivation inside the plant is not very successful, so some 10^5 tpy have to be replaced. A number of companies offer a special *regeneration procedure at their site or a complete reworking. The general scheme is oxidative treatment (to remove coke), roasting with caustic (to convert the Mo and V oxides to the salts), and *leaching with water and acids to extract the metals. To separate, concentrate, and purify the different metals a combination of *precipitation and liquid/liquid *extraction (liquid ion exchange) is common, with variation from one company to another. Another mass of spent cat. results from the *hydrogenation of fats and fatty acids. About 6 000 tpy of nickel leave these procs., containing organic materials, filter aids, *supports, and some Ni, in the conc. region of about 10 % w/w. As rejuvenation is impossible, oxidation and subsequent leaching with acid is the method of choice. Most of the important cat. producer offer this recycling to their customers as a service carried out either by themselves or in cooperation with a specialist company. In fact the recycling share in this sector is very high (40–60 %). **E=F=G**.

C.D. FROHNING

Ref.: B.E.Leach (Ed.), *Applied Industrial Catalysis*, Academic Press, New York 1984, p. 40; Ertl/Knözinger/Weitkamp, Vol. 3, p. 1263.

recycling of raw materials → *environmental issues

RED → radial electron distribution

redispersion the opposite to *sintering. It can be interpreted as a decrease in *particle size accompanied by a net increase in *surface area.

Only few catalytic systems are able to undergo redispersion; most of the research work concentrates on the system Pt-Al₂O₃, which is of special industrial interest for *catalytic reforming. Principally redispersion is achieved by a heat treatment (500–550 °C) in an O-containing atmosphere w/o addition of Cl-containing cpds, followed by H *reduction at about 250 °C. Although the reaction is not fully understood, the effects have been demonstrated by an increase in surface area by 15–20 % and an improved *activity after the treatment. Most plausible is the intermediate formation of Pt oxides which causes splitting of the individual particles leading to an increased active surface area after reduction. **E=F=G.**   C.D. FROHNING

Ref.: Anderson/Boudart, Vol. 6; Farrauto/Bartholomew, p. 293.

red mud catalyst   Fe₂O₃ (from the Bayer proc. of *alumina manuf.); cf. *German technology

redox catalysis (electrochemical)   A stoichiometric *redox change can use either a redox reagent or electricity. The cat. can interact with the substrate either in the inner sphere (contact, coordination) or in the outer sphere (without contact through the solvent molecules). Even with mediators (or cats.), electrochemical rc. allows considerable reduction of the kinetic barrier for cathodic reduction or anodic oxidation for redox reactions involving a structural reorganization (bond breaking and formation). Mediators are redox reagents which must be stable in the two oxidation states involved in the mediation (such as electron-reservoir complexes). Their kinetic advantage is due to the fact that they operate in 3D solutions whereas the uncat. redox reaction must deal with only 2D electrode surfaces. The kinetic gain is usually in the order of 0.5 to 1 V. For practical use, the mediator is often attached to the electrode, providing a modified electrode. This is even the case for *enzymes to build up biosensors. Rc. is extensively used in electrochemical synth. as well as in redox reactions which do not involve electrochemistry. Photochemical rc. has been used for energy conv. devices, especially to convert *single-electron transfers into polyelectronic procs. Not only mononuclear transition metal cats. but also *colloids are useful; the latter function as electron reservoirs with multiple metallic sites (e.g., reduction of H⁺ to H₂). Many redox mediators or cats. are combined and used in Nature (e.g., *photosynthesis): cytochromes, mono- up to polynuclear Fe-S proteins (*iron-sulfur clusters), etc. From the pure chemical side, the *Wacker proc. for alkene oxidation using a Pd-Cu redox system is a well-known example. **F** catalyse d'oxydo-réduction; **G** Redoxkatalyse.   D. ASTRUC

Ref.: Astruc; *Coord.Chim.Rev.* **1989**, *93*, 245 and **1990**, *99*, 15; *Chem.Rev.* **1990**, *90*, 1359 and **1992**, *92*, 1411; *Top.Curr.Chem.* **1987**, *142*, 1 and **1979**, *83*, 67; *Acc. Chem.Res.* **1980**, *13*, 155, 323, **1981**, *14*, 154 and **1999**, *32*, 62; *Tetrahedron* **1984**, *40*, 811.

redox reactions   1-heterogeneous: The number of valence electrons attached to one atom in a molecule is changed when it undergoes a rr. This means that addition or removal of H and addition or removal of O are the most prominent families of rrs., which are frequently carried out using het. cats., e.g., the *epoxidation of ethylene to *ethylene oxide (on low-*surface *alumina/Ag cats.), which involves metal oxidation ($O_{2,g} + 2 Ag_s \rightarrow 2 AgO_s$) as a first step and metal reduction with the coreductant ethylene ($AgO_s + C_2H_4 \rightarrow Ag + EO$) as the second. The selectivity of the whole proc. is determined by the occurrence of combustion of ethylene to CO_2, which restricts the selectivity to approx. 86 % (with addition of special *promoters to 90 %, cf. *oxiranes, het.). Other special cases of this type of reaction involve the transformation of alkanes to alkenes or the transformation of an alcohol into a ketone, which are oxidation reactions without adding O atoms to the substrate. Rrs. involve further the addition or removal of halogen atoms from organic molecules or the introduction of N into a hydrocarbon molecule. From the standpoint of terminology the *ammoxidation is a rr., too: the cat. is reduced by ammonia and hcs. and is readily oxidized by dioxygen. The formation of propylene oxide from propylene and *hydroperoxides are cat. by, e.g., *titanium silicalite, which acts as a *Lewis acid and thus is not a pure redox cat.

A considerable fraction of all het. cat. reactions are rrs. Frequently the course of rrs. involves the *activation of C-H bonds. In these cases the het. cat. must not only provide electrons or holes for exchange of charge equivalents but it must also exhibit basic or acidic properties to *support the activation of C-H bonds. The redox activation of these bonds by strong *dehydrogenation cats. such as noble metals is usually extremely difficult to control selectively. The *activity of noble metal active phases in such reactions must often be moderated by the addition of *poisons (called *modifiers; cf. *directed poisoning). Inorganic rrs. such as *ammonia synth. (*reduction), SO_2 oxidation, or HCN synth. (oxidation of ammonia, *hydrogen cyanide) occur under very drastic conds. requiring high temps. and press. Environmentally relevant rrs. are the *DeNOx proc. and the total oxidation of *volatile organic cpds. (VOCs) or the oxidation of CO with NO or oxygen. These reactions are characterized by the application of extremely active cats. in order to drive the reaction to completion in very short contact times and without the application of any press.

Most cats. are *multifunctional and contain sites for anchoring the reactants, sites for generating active oxidants or reductants, and sites for other procs. apart from the exchange of electrons for H or O atoms (acid sites, *isomerization sites, etc.). It is an open problem whether these functions are combined in one location with a multifunctional site or whether several classes of sites coexist and the intermediates diffuse from site to site.   R. SCHLÖGL

2-homogeneous: Many reactions can be regarded as occurring by the loss of electrons from one substance and their gain by another. Redox systems consist of redox partners – reductant and oxidant – which are inevitably connected due to their definition: the oxidant receives electrones from the reductant during the redox proc. Hom. rrs. are equil. reactions, the type and the state of which is determined by the redox potentials of the reactants. Every *oxidation proc. is strictly combined with a *reduction reaction: the electrons which are released from the oxidized cpd. are directly transferred to the oxidizing agent which is reduced at the same time. The nomenclature of the reaction is mainly determined by the prods. desired and the chemicals used to achieve the reaction. Thus the conv. of ethylene to *ethylene oxide will be defined as oxidation of ethylene and not as reduction of oxygen with ethylene.

Important rrs. in cat. chemistry are all types of 1: *hydrogenations (most of them catalytic), such as addition of molecular hydrogen, reduction by (metal) hydrides, *electron transfer by *electrocatalysis followed by proton quenching or addition of H atoms by transfer of H *radicals; and 2: oxidation reactions transferring oxygen or other direct oxidants, and *dehydrogenations or simply reducing the oxidation state by electron transfer to oxidation reagents. For electrocatalytic redox reactions cf. *redox cat. Organo-transition metal *complexes can carry an electron or an electron hole; if they are stable, they serve as electron-reservoir complexes. Redox reactions play an important role in, e.g., *enzymatic hydrogenation, *nitrogen fixation, *photosynthesis, etc. **F** réactions d'oxydo-réduction; **G** Redoxreaktionen.   R.W. FISCHER

Ref.: D. F. Shriver, P.W. Atkins, C. H. Langford, *Inorganic Chemsitry*, Oxford University Press, 1990, p. 229.

reduction   This term is applied to any proc., whether catalyzed hom. or het. or not at all, in which the oxidation state of a reactant is lowered either by addition of H atoms or the removal of electron-rich (electronegative) elements (e.g., O, S, Cl, etc.), or by addition of electrons [e.g., Fe(III) $\rightarrow$ Fe(II)]. In industrial practice these procs. are usually effected by cat. *hydrogenation or *hydrogenolysis using either hom. or het. cats., with molecular hydrogen as the reductant. R. is frequently an important step in the synth. and manuf. of fine chemicals, and here *selective hydrogenation (i.e., partial, *regio- or *chemoselective, or *chiral reduction) is required.

R. is also an important unit operation in the prep. and manuf. of *supported metal cat.: again, H_2 is the usual reductant, but it is advantageously diluted with an inert gas to control the rate, since uncontrolled exotherms may lead to *sintering. There are some advantages in using CO or NH_3 as reductants. Cpds. of the base metals (Fe, Co, Ni, Cu) need somewhat higher temps. for their reduction, but the noble metals of groups 8–10 are formed satisfactorily between 273 and 473 K. The proc. of r. may be analyzed by the use of temp.-programmed reduction (cf. *hydrogen-oxygen titration and *reduction in cat. prep.).

Ref.: Ullmann, Vol. A5, p. 351.   G.C. BOND

reduction-deposition → reduction in catalyst preparation

reduction in catalyst preparation R. of cat. *precursors lowers the valency of the cat. active ingredients. Predominantly metal oxides are transferred to zero-valent metals (e.g., precious metals, Fe, Ni, Co, Cu); sometimes metal cpds. are partially reduced (e.g., MoS_3 to MoS_2). The removal of O from the precursor oxides and the subsequent formation of metal planes and lattices corresponds to a complete change in the morphology of the material (shrinking core model). Therefore, reduction is a critical step in cat. preparation and has to be carried out under carefully controlled conds. R. of precious metal cpds. is favored by thermodynamics and may easily be achieved by reducing agents in solution, such as formates, formic acid, hydrazine, or hydroxylamine at ambient or slightly elevated temp., whereas less noble metals require more severe conds. Commercially, H is the preferred and readily available reducing agent. In the case of Cu oxide and flowing hydrogen, 150–200 °C is the preferred temp.

range, which is easily accessible in most commercial plants. Cu-containing cats. are therefore commonly delivered in the oxidic state and reduced in situ by flowing H_2 diluted with N_2 to control the temp. Ni and Co require temps. of approx. 400–600 °C in H_2 for (nearly) complete conv. Temps. below the *Tammann temps. are advisable in order to avoid sintering of the metal and loss of surface area. The presence of water vapor during the reduction also induces sintering and leads to decreased activity. Instead of H_2, CO can be applied, e.g., for Fe oxide cats. in the *Fischer-Tropsch proc. yielding a mixture of metallic iron and Fe carbides. **F** réduction; **G** Reduktion.   C.D. FROHNING

Ref.: Twigg, pp. 314,400; Farrauto/Bartholomew, p. 101; *Catal.Lett.* **1995**, *35*, 291; *J.Chem.Soc.Faraday Trans.* **1993**, *89*(17), 3313.

reductive alkylation → reductive amination

reductive amination NH_3 and primary and secondary amines add to aldehydes and ketones in the presence of H_2 and *hydrogenation cats. (so-called reductive alkylation of ammonia or amines, *aminating hydrogenation). Concerning the mechanism initially formed α-hydroxy amine is either *hydrogenolized directly or the *intermediate imine is *hydrogenated. Hom. cats. based on Ni or Rh are effective; commercially only *supported cats. with Ni on *alumina or *kieselguhr, Pd-C, or *Raney Ni are applied. In the lab, reducing agents like Zn/HCl, $NaBH_3CN$, $NaBH_4$, or $Fe(CO)_5$ in NaOH/EtOH have been proposed. When formic acid or ammonium salts of formic acid are used, the proc. is called the *Leuckart-Wallach reaction (*Eschweiler-Clarke procedure).

Apart from aliphatic amines anilines also undergo ra. When the amination reagent is ammonia the initial aminated prod. may react further to produce secondary or tertiary amines. This can be avoided by an excess of ammonia. Reductive alkylations have been carried out with nitro, nitoso, and azo cpds.

The ra. has also been used as part of *domino reactions, especially *hydroaminomethylation, which is a comb. of *hydroformylation and subsequent reductive amination. **F** amination réductive; **G** Reduktive Aminierung; aminierende Hydrierung.   M. BELLER

Ref.: Falbe-4, p. 45; March; Ertl/Knözinger/Weitkamp, Vol. 5, p. 2339.

reductive carbonylation comprises the transformation of aromatic nitro cpds. to the corresponding carbamates, e.g., the manuf. of 2,4-toluenediisocyanate (TDI) from 2,4-dinitrotoluene, cat. by *ligand-modified Ru, Rh, and Pd cpds. and in the presence of alcohols and of *co-catalysts like phenanthrolium hexafluorophosphate. Another example is the *ARCO ethylurethane proc. (cf. also *oxidative carbonylation). **F** carbonylation réductive; **G** reduktive Carbonylierung.   B. CORNILS

Ref.: *J.Mol.Catal. A*: **1999**, *144*, 41.

reductive deoxygenation → McMurray coupling

reductive elimination → oxidative addition

Reed reaction → sulfochlorination

refinery is a plant for the raffination of oils, specially mineral oils. The arrangement of the different unit operations and procs. (which are mostly catalyzed) within a refinery is shown in the Figure under the keyword *petroleum processing. Cf. the various keywords such as *hydrotreating, *coking, etc. **F=G** Raffinerie.   B. CORNILS

Ref.: Ullmann, Vol. 18, p. 51; Falbe-4, p. 86.

Reformatsky reaction β-Hydroxyesters are obtained from the reaction of aldehydes or ketones with α-halogen-substituted esters in the presence of zinc in an inert solvent as reaction medium acc. to R^1-C(=O)-R^2 + BrZnCHR^4-COOR^3 → R^1R^2C(OH)CHR^4-COOR^3
      -ZnBr

The reactive intermediate of the Rr. is an organozinc cpd., an anion of an ester which is closely associated with a Zn cation.

Formally, the reaction can be regarded as analogous to the *Grignard reaction with the organozinc cpd. instead of RMgX. The reaction has also been carried out with Sn, activated In, and with a Zn-Cu couple. The aldehyde may be aliphatic, aromatic, or heterocyclic or it may contain various functional groups. The reaction can be run in less time and with higher yields if performed in the presence of *ultrasound. A similar reaction (Blaise reaction) can be carried out on nitriles, to form β-ketoesters from α-haloesters and nitriles. **F** réaction de Reformatsky; **G** Reformatsky-Reaktion.   R.W. FISCHER

Ref.: F. A. Carey, R. J. Sundberg, *Organische Chemie*, VCH, Weinheim (Germany) 1995, p. 1129; *Organic Reactions*, Vol. 1, p. 1 and Vol. 22, p. 423.

reforming   → catalytic reforming

regeneration   1-general: term for the restoration of the former status of *deactivated or otherwise denatured cats. (rejuvenation) which are subjects to a *cat. cycle. A less radical r. is *makeup, especially for hom. cats. The r. procedure depends on the kind of *deactivation (*poisoning, coating of active species, *fouling, *sintering, *alloy formation, forming of volatile or inactive cat. species, etc.), the kind of catalysis (hom. or het.)., and the mode of operation (continuous or discontinuous). The aims of r. of cat. are makeup after measurement and adjustment of concs. (e.g., metal content and degree of *dispersion, content of *active species, conc. of *ligands and ligand surplus, *surface area), and restoration of *activity and *selectivity, securement of special properties like dispersion of metal on the surface, rearrangement of cat. ingredients, adjustment of additives against sintering, screening out the over- or undersize, etc.). Examples of r. steps are: *makeup of hom. oxo cats., treatment of deactivated fluidized- or entrained-bed cat. in separate fluidized beds by burning off *coke *deposits with air or chlorine (cf. *redispersion, e.g., *IFP reforming proc.), restoration of the original valence state of the cat. (*Lummus Transcat proc., *Rhône-Poulenc EDC proc., older *Ruhrchemie oxo proc.), programmed heating for restoration of desired surface areas, *extraction of deposits on loaded cat. by solvents or supercritical fluids (e.g., *Fischer-Tropsch cats.), sieving for removal of undersize, etc. R. procedures such as combustion of coke deposits can be continuous or periodic (intermittent) or can take place in a bypass sidestream (e.g., *riser-tube reactor). The most extreme r. is the total chemical workup of the cat. The mode of r. may influence considerably the cat. lifetime and thus the economics (cf. *continuous cat. regeneration).

2-enzymatic: cf. *ATP and other keywords under regeneration. **F** régénération; **G** Regenerierung.   B. CORNILS

Ref.: Ullmann, Vol. A5, p. 360, Kirk/Othmer, Vol. 5, p. 419; Anderson/Boudart, Vol. 6, p. 1.; Ertl/Knözinger/Weitkamp, Vol. 3, pp. 1263,1279; E.E.Petersen, *Activation, Deactivation and Poisoning of Catalysts*, Academic Press, New York 1988; Trim; Twigg, p. 176; Farrauto/Bartholomew, p. 291 US 5.821.270 (Exxon).

regeneration of acetyl-CoA → acetyl-CoA regeneration

regeneration of ATP → adenosine triphosphate regenaration

regeneration of NAD(P)

The oxidized nicotinamide *cofactors NAD and NADP are used for the synth. of ketones from the racemic mixture of the hydroxy cpds. Regeneration of NAD and NADP from NAD and NADPH is somewhat problematic because of unfavorable thermodynamics and *product inhibition. The *regioselectivity, however, is not a problem in the *oxidation of NAD and NADPH. Both enzymatic and non-enzymatic methods can be used for NAD and NADP. Enzymatic methods seem to be preferred because they are simpler and more compatible with biochemical systems. Electrochemical, chemical and photochemical methods coupled with electron transfer mediators such as *methylene blue, phenazine *methyl viologen, or Ru-tris(bipyridine) have been used for small scale NAD and NADP. A direct *oxidation of NADH with *FMN followed by spontaneous re-oxidation of the reduced *FMN ($FMNH_2$) by O_2 has also been utilized. The reaction is too slow to be practical, however. Other enzymatic methods seem to be the most useful.

The GluDH system (40 U/mg) accepts both NAD and NADP and is thermodynamically favorable. Both ketoglutarate and GluDH are inexpensive, stable, and innocuous to enzymes. The disadvantage of this system is that the glutamate produced may complicate the workup. The LDH system is stable and inexpensive, and has a high specific activity (~1000 U/mg). Although pyruvate tends to polymerize in solution and react with NAD in a proc. cat. by LDH, regeneration of NAD based on LDH has been successfully carried out for enzymatic oxidations of 10–100 mmol of material. The disadvantage of this system is that LDH is specific for NAD. The system based on FMN/FMN reductase accepts both NAD and NADP and has the most favorable thermodynamics due to the ultimate oxidation with molecular O. *Catalase is often added to destroy *H_2O_2. The system is, however, not suitable for enzymes sensitive to molecular O. This problem was also observed in rNAD(P) based on electron-transfer dyes. Of many dyes evaluated for preparative synth., the one based on *methylene blue/O_2 cat. by diaphorase is considered to be the best. Other enzymatic and biological methods were also reported to be useful for nicotinamide cofactor regeneration. **F** régénération de NAD(P); **G** Regenerierung von NAD(P).

C.-H. WONG

Ref.: *J.Org.Chem.* **1985**, *50*, 1992, **1985**, *50*, 5387 and **1982**, *47*, 2816.

regeneration of NADH and NADPH

A useful and practical regeneration system for NADH and NADPH must be highly *regioselective, compatible with the desired enzymatic reduction, and of capable of recycling the *cofactor 10^2 to $>10^5$ times. Enzymatic cat. provides such high selectivity for the reduction of NADH to NADPH. Other non-enzymatic methods are also available. The disadvantages of enzymatic cofactor regeneration are the expense and limited stability of *enzymes.

There have been many enzymatic systems developed for the regeneration of NADH and NADPH. The most convenient and useful systems for NADPH regeneration are formate/formate dehydrogenase (FDH) from *Candida boidinii* sp. (for NADH), isopropanol and the alcohol dehydrogenase from *Thermoanaerobium brockii* (for NADPH), yeast, or *Pseudomonas* sp. (for NAD), and glucose/glucose dehydrogenase (GDH) from *Bacillus* sp. The FDH system is inexpensive and the product isolation is easy. The *enzyme has low activity (3 U/mg), however, and is specific for NAD. The inexpensive GDH system has high specific activity of the enzyme (250 U/mg) and high stability, and is able to regenerate both NADH and NADPH. NADH regeneration based on formate/FDH performed in a membrane-contained reactor has been used for large-scale procs. Up to 600 000 mol of product can be produced per mol of cofactor lost in the proc.

A number of other reactions have been explored for use in reduction of NADP to NADPH, and although they have not so far proved practical in synth., they may provide the basis for future procs. Many *viologen derivatives can be reduced via one-electron procs. by cat. *hydrogenation, *electrochemical or *photochemical reduction, or hydrogenase cat. *reduction. The reduced viologens have been used as mediators for enzymatic regeneration of NADPH. The enzymes used in this proc. often contain a prosthetic group that can be reduced by a viologen radical cation. With an appropriate choice of the mediator, the redox potential can be adjusted over a wide range to perform the desired reaction. A typical example is the use of reduced methylviologen for regeneration of NADH cat. by diaphorase.

When a carbamoylmethylviologen is used with diaphorase, NAD regeneration from NADH becomes favorable. In the photochemical proc., photosensitizers have been used to mediate the reduction of viologen cations to radical cations.

The homogeneous Rh complex was reported to be an effective cat. for reduction of NAD and NADP to NADH and NADPH, respectively, in the presence of formate as hydride donor. The reduced Rh can also be regenerated electrochemically. The Rh complex can be *immobilized without loss of cat. efficiency (TON up to 1000). **F** régénération de NADH et NADPH; **G** Regenerierung von NADH und NADPH.   C.-H. WONG

Ref.: *J.Am.Chem.Soc.* **1981**, *103*, 4890; *Biotech.Appl. Biochem.* **1987**, *9*, 258; *J.Am.Chem.Soc.* **1985**, *107*, 4028.

regeneration of sugar nucleotides → sugar nucleotide regeneration

regioselectivity A reaction is called regioselective when it takes place exclusively (regiospecifically) or preferentially (regioselectively) at only one of two or more possible reaction sites or when two reactants, each possessing a different substitution pattern at its reactive site, give one of the two possible prods. selectively (Figure; cf. *selectivity).

The enhancement of r. plays a fundamental role in the design of new cats. and the improvement of existing cats. or cat. systems (cf. also *tailoring of cats.). One of the outstanding examples describing the influence of parameters on the r. of a cat. proc. is the *n/iso* ratio of the *hydroformylation of alkenes (commercially of extreme importance for the *hydroformylation of propylene; cf. also *Markovnikov rule, *selectivity). **F** régio-sélectivité; **G** Regioselektivität.   W.R. THIEL

$$L_n M - X + A \longrightarrow L_n M - A^{\diagup X}$$

$$L_n M - X + A - B \longrightarrow L_n M - A^{\diagup B - X}$$

Ref.: Ertl/Knözinger/Weitkamp, Vol. 5, p. 2209.

Rehm-Weller equation   → single electron transfer

REHY   rare earth hydrogen Y-type (zeolites for *cracking)

Reimer-Tiemann reaction   synth. of *o*- and *p*-hydroxy-substituted phenols via conv. of phenol (or phenol derivatives) with chloroform (CHCl₃) in the presence of strong bases such as KOH. If the *ortho* position is blocked, the *para* derivative is formed. The reaction goes via *carbenes (dichlorocarbene unit :CCl₂ from CHCl₃) under *phase transfer catalysis. The *carbene reacts with the deprotonated phenolate to form the dichloromethyl phenolate. The latter undergoes *hydrolysis easily to yield the hydroxybenzaldehyde. Instead of phenols, heterocycles such as indoles or pyrroles can also be transformed to their formyl-substituted derivatives. **F** réaction de Reimer-Tiemann; **G** Reimer-Tiemann-Reaktion.   R.W. FISCHER
Ref.: *Organic Reactions*, Vol. 28, p. 1.

re-immobilization counterpart of the *immobilization of hom. cats. containing *sulfonated *ligands with the help of the *liquid support water (*aqueous-phase cat.). R. makes such cats. soluble again in hom. solvents while retaining and using their chemical functionality through reaction with special amines (Figure). The ammonium cations

formed could be characterized as H-ammonium cations, enabling further reactions due to *Brønsted acidity.

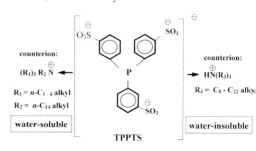

TPPTS

R. opens new opportunities in hom. cat. through simple adjustment of the size of the ligands by reactions with amines (e.g., important for use in *membrane reactors), through selective separation of degradation prods. from *chelating ligands (important for long-term use), through recovery of expensive metals or ligands (by easy neutralization and phase separation), and as a chemical link to *nonionic liquids. **F** réimmobilisation; **G** Re-immobilisierung.   H. BAHRMANN
Ref.: Cornils/Herrmann-2, p. 328.

rejuvenation   → regeneration

relaxation   1– relaxation energy: in photoelectron *spectroscopy, the energy (E_R) which is gained when, after removal of one from N electrons, the remaining (N–1) electrons rearrange into a new state of minimal energy. The energy gained is passed to the photoelectron, which appears at higher kinetic energy than without r.
2– surface relaxation: change in the interplanar spacing of the topmost layer of a *surface to the next layer, to a new value different from the bulk value. Usually a slight contraction by a few percent occurs since the smaller *coordination number of the surface atoms leads to a force directed inwards, into the bulk material. The r. of interlayer spacings continues into the bulk material but with fast decaying amplitude and oscillatory sign. **E=F=G**.   R. IMBIHL

remote carbonylation a very special *carbonylation reaction which takes place not at the initially reactive center but at the more stable remote position. An example is the carbonylation of 1-octanol in the presence of Pb(OAc)$_4$. The one-electron oxidant first generates an oxygen-centered *radical, which in turn reacts with a suitable C-H bond yielding a more stable remote carbon-centered radical. This radical reacts with CO to give an acyl radical which cyclizes after another one-electron oxidation. **F** carbonylation distante; **G=E**. M. BELLER

Ref.: *J.Am.Chem.Soc.* **1994**, *116*, 5473.

remote control model → Co as cat. metal

REMPI → resonance-enhanced multiphoton ionization spectroscopy

reoxidant A r. is the second comp. of a cat. couple, which reoxidizes the active cat. metal, similarly to coreductants in *redox reactions. An example is Cu in *oxidative carbonylation or in *Wacker's proc.: in both cases the proc. relies on the discovery that the Pd(0) formed can be reoxidized by molecular O$_2$ in the presence of CuCl$_2$. The CuCl$_2$ reoxidizes Pd(0) to Pd(II) and the Cu$_2$Cl$_2$ thus formed is readily converted to Cu(II) chloride by air or O$_2$ (e.g., *Lummus Transcat proc., similarly to the *acetoxylation of ethylene acc. to *Wacker; cf. *Nippon Steel picoline proc.). Another variant is the inclusion of a co-oxidant during reaction, which reoxidizes the Os(VI) species of *dihydroxylations, thus making possible both, the use of cat. amounts of the oxometal and a *cat. cycle. **E=F**; **G** Reoxidans(t). B. CORNILS

Reppe Walter (1892–1969), chemist with BASF and professor of applied chemistry in Mainz and Darmstadt (both Germany). Worked on metal carbonyls and economically important syntheses especially with acetylene ("Reppe chemistry", *Reppe syntheses). He

also discovered the metal carbonyl-cat. *cyclooligomerization. W.A. HERRMANN

Reppe syntheses a group of commercially important cat. procs. with acetylene named after *Reppe (*BASF propionic acid proc.). The Rs. can be divided into four categories. 1: vinylation, the reaction of acetylene with protic cpds. such as acids, alcohols, or amides. Important prods. are vinyl chloride (reaction with HCl over HgCl$_2$/C), vinyl esters (with RCOOH over Zn or Hg carboxylates), vinyl ethers (from ROH and alkali metal *alkoxides), acrylonitrile (HCN over CuCl$_2$), and N-vinylpyrrolidone (N-vp; reaction with pyrrolidone over K pyrrolidinate). Only certain vinyl esters apart from N-vp are still produced commercially. 2: *ethynylation, the reaction of acetylene with ketones and aldehydes, e.g., propargyl alcohol and 1,4-butynediol (with HCHO over Cu$_2$C$_2$) and substituted propargyl alcohols (reaction with RC(O)CH$_3$ over alkali metal alkoxides), which are intermediates for *vitamin A, *THF, γ-GBL, and *1,4-butanediol. 3: *carbonylation (*hydrocarbonylation), i.e., the reaction of acetylene with CO (and water). Important prods. are *acrylic acid (NiBr$_2$/CuI cat.) and *hydroquinone (over Fe(CO)$_5$ cat.). 4: *cyclization, i.e., *cyclooligomerization of acetylene, yields benzene (cat. Ni(PPh$_3$)$_2$(CO)$_2$) or *cyclooctatetraene (Ni(CN)$_2$). M. SCHULZ

Ref.: W. Reppe, *Neuere Entwicklungen auf dem Gebiet der Chemie des Acetylens und Kohlenoxyds*, Springer, Berlin 1949; Cornils/ Herrmann-1, p. 269.

residence time The mean rt. τ in a chemical reactor (also called *space-time) is generally defined as the period of time a volume element of an incoming fluid with the volumetric feed rate V remains in the reactor volume V_R (and thus $\tau = V_R/V$). However, only for an ideal plug flow reactor (*PFR) does τ denote the true rt. of all the volume elements, since only in this type of reactor do all volume elements move through the reactor with the same velocity. For all other types of reactors (*cat. reactors) an rt. distribution (rtd.) exists, and thus τ denotes an averaged value. The experimental rtd. can be approximated for real reactors using the tank-in-series model (cascade of *CSTR) or the dispersion model (*PFR with an additional *mass transport term which describes axial diffusion). In *surface science rt. denotes a period of time τ during which the adsorbed molecules diffuse over the *surface of a *single crystal. This rt. can be experimentally determined using modulated molecular beams. For simple molecules it was observed that τ depends on the *activation energy of desorption E_{des} with $\tau = \tau_0 \exp(-E_{des}/RT)$, in accordance with theoretical calcns. (cf. *cat. reactors). **F** temps de séjour; **G** Verweilzeit. M. MUHLER

Residfining process → Exxon procs.

Resid HDS process → Gulf procs.

resols → Baekeland-Lederer-Manasse condensation

resolution separation of *racemates into the *enantiomers. Usually a *racemic mixture of a cpd., e.g., an acid, is resolved by converting it into a pair of *diastereomers with the help of an enantiomerically pure *auxiliary, e.g., a base. Whereas *enantiomers have identical physical properties such as solubility, diastereomers differ in their physical properties and can be separated, e.g., on the basis of solubility differences. After diastereomer separation the *optically active resolving agent is removed, to give the separated enantiomers. Increasingly, chromatography is used to resolve enantiomers analytically (by *GC or *HPLC) and preparatively. In this technique chiral stationary phases (CSPs) cause the diastereomeric relationships with the eluting enantiomers. **F** dédoublement (séparation des racémates); **G** Racemattrennung. H. BRUNNER

resonance-enhanced multiphoton ionization spectroscopy (REMPI) delivers kinetic information by photon-probed in-situ detection of small concs. of prod. species. The technique is non-destructive. *Space-time-re-

solved mechanistic studies or model reactions with *single crystals are also possible. R. SCHLÖGL

restricted transition state selectivity will be observed when the *transition state of a prod. requires more space than available at the *active sites in the *cavities of, e.g., *zeolites. **F** sélectivité de l'état de transition diminué; **G** Selektivität des eingeschränkten Überganszustandes. P. BEHRENS

restriction endonucleases highly specific deoxyribonucleases that accept typically a specific four-, six-, or eight-base C_2-symmetric sequence and in general have essentially no activity for even closely related sequences. The restriction endonucleases are bacterial in origin, and serve a protective function in vivo: they help the bacteria destroy exogenous DNA. Bacteria contain both the endonuclease and a site-specific DNA methyltransferase with the same sequence *specificity. The endonuclease cannot hydrolyze the methylated version, and so the bacterial chromosome, which has been methylated by the endogenous methyltransferase, will be safe. Exogenous DNA, however, will be cleaved. **E=F=G**. P.S. SEARS

restructuring This term designates procs. occurring when one solid is brought into intimate contact with another chemically different solid. Following the principle of minimizing the interfacial energy, the structures of the two solids in contact change. Usually novel and unique structures which are typical for specific interface will result. The proc. of r. requires thermal activation and occurs either during annealing or unintentionally during the initial stage of operation. The limited volume exhibiting the r. proc. and the fact that restructured *surfaces are buried partially at the inner interface between the two solids renders the analysis of the restructured surface difficult. In the surface science literature the term r. is also used for structural change resulting from the presence of a strongly interacting adsorbate such as CO, S, and alkali atoms on metal surfaces.

The process of r. must be distinguished from the phenomenon of *reconstruction, which designates the change in the bulk structure of a solid at its surface. This does not require the presence of a second species but is the consequence of the energy minimization for the terminating periodic crystal structure sometimes referred to as *structural relaxation. Good knowledge of the principles of structural changes at interfaces is only available for metals and for the technologically important *semiconductors. For almost all solid cpds. very little information is available despite the strong effect the structural changes at interfaces exert on the properties of the solid. A common rule of r. is that the structures become more complex at interfaces. Interatomic distances are often reduced at the surface and expanded in the atomic layers below the top surface. Formation of *islands or complicated rotated structures with respect to the principal orientation in the bulk solid are also observed. Methods for the determination of r. are the *scanning probe techniques, the diffraction of electrons and special X-ray diffraction experiments. **E=F**; **G** Restrukturierung.

R. SCHLÖGL

retropinacol rearrangement conv. of an alcohol to a rearranged alkene in the presence of acidic cats. acc. to $R^1R^2R^3C-CH(OH)R^1 \rightarrow R^1R^2C=CRR^3 + H_2O$. The mechanism involves protonation of the hydroxy group and *dehydration. The resulting carbenium ion rearranges toward a more stable *carbocation (tertiary > secondary > primary). Hence, the substituent that is transferred is the one which leads to the most stable carbocation. The rr. is a special case of the *Wagner-Meerwein rearrangement; both rearrangements are often observed with bi- and tricyclic rings, i.e., terpenes. **F** réarrangement retropinacolique; **G** Retropinakolin-Umlagerung.   M. BELLER
Ref.: March.

reverse flow reactor (RFR)   → catalytic reactors

reverse micelles   very small (diameter ~10 nm) spheres of water suspended in a nonmiscible organic solvent, with surfactants to stabilize the interface (cf. *micelles). The surfactants are self-assembled in such a way that the polar group pointing inside and the nonpolar group pointing outside to form a circular assembly. *Enzymes may be enclosed in rms. and suspended in organic solvents. This is particularly useful for reactions in which the substrate and/or product is water-insoluble. The enzyme is surrounded by a small amount of water, and therefore still displays the high enzymatic reaction rates and stability observed in aqueous solution, but the bulk *solvent is organic, and able to dissolve the nonpolar substrates or prods. **F** micelles revers; **G** Reverse Micellen.   P.S. SEARS
Ref.: *Proteins: Struct.Funct.Genet.* **1986**, *1*, 4.

reverse osmosis   a pressure-driven, *membrane based separation proc. in which transmembrane press. causes selective movements of *solvents against its osmotic press. through the membrane. Low molecular weight molecules and ions are retained (the retention of monovalent ions is the most distinct difference from *nanofiltration). Typical values for the press. applied are 1–10 MPa. The membranes are often made from organic polymers such as polyamide or cellulose derivatives. In cat. procs. ro. can be applied to the separation of cat. and prod. and to the conc. of valuable cpds. **F** osmose inversée; **G** Umkehrosmose.
Ref.: Ullmann, Vol. A16, p. 187.   U. KRAGL

reverse PTC   → phase transfer catalysis

reverse transcriptase   a special *DNA polymerase than can use *RNA as a template. Like other DNA *polymerases, it requires an oligonucleotide primer and a *template, and cat. the *polymerization of DNA from deoxynucleoside triphosphates in the $5' \rightarrow 3'$ direction. It is commonly produced as a recombinant *enzyme (originally from retroviral sources), and is used frequently in molecular biology in the synth. of cDNA from messenger RNA for the production of cDNA libraries. **F** transcriptase revers; **G** reverse Transcriptase.   P.S. SEARS

RFR reverse flow reactor, → catalytic reactors

rhamnose isomerase → fuculose-1-phosphate aldolase

rhamnose-1-phosphate aldolase → fuculose-1-phosphate aldolase

Rheniforming R. cats. for *catalytic reforming are a further development (by Chevron, 1967) of *platfomring cats. They serve for the increase of the anti-knock ratings of automotive fuels (straight-run fuel). The proc. is commonly performed at temps. around 500 °C/1–3.5 MPa. The cats. employed in reforming are *bifunctional, possessing both metallic and acidic comps. Pt and Re or Ir are used as metals. These cat. systems exhibit improved performance. Rh. cats. show higher stability and *selectivity, resulting in improved economic properties and longer lifetimes. The ratio of Pt to Re is generally 1:1 (0.3 wt.% of each). For the cat. synth. H_2PtCl_6 and Re_2O_7 are used. The rhenium is present as highly dispersed oxide [Re(IV), formally as ReO_2] as well as Re(0) in typical reforming conds. (formation of bi- or multicomponent clusters of ca. 10 A diameter, due to the extremely small *cluster size). **E=F=G.**
R.W. FISCHER

Ref.: Anderson/Boudart, Vol. 1, p. 16,294; Ertl/Knözinger/Weitkamp, Vol. 4, p. 1947.

rhenium as catalyst metal There are so far three major commercial applications of Re as a cat. metal: together with Pt in *rheniforming procs. (e.g., *Chevron, *Catarol), as a *co-catalytic metal in het. cats. used for the *selective oxidation of ethylene to ethylene oxide (EO) with oxygen, and as Re_2O_7 on

*alumina as a cat. for *metathesis (procs. of *IFP Meta-4 and *Shell FEAST). During the lifetime (ca. five years) of the commonly used Ag-based oxidation cats. *activity and *selectivity drop below economic values. This can be avoided by doping the Ag systems with 5 ppm Re (increase in selectivity from 75 % to 86 %). The Re effect can be explained by the influence of the Re-doped cat. on the electronic properties of the solid-state phase (formation of Ag-O-Re clusters).
Besides these industrial applications Re plays a role in various organic reactions, e.g., as Re metal, salt, or $Re_2(CO)_{10}$ (also in *bimetallic cats. together with Rh(acac)$_2$ or Os) as *hydrogenation cats. converting carboxylic acids to alcohols (>95 % yield). Such cats. show high resistance against cat. *poisons. Of special interest are the properties of Re as the active *central atom in organorhenium oxides in hom. cat. (cf. *MTO). **F** le rhénium comme métal catalytique; **G** Rhenium als Katalysatormetall.   R.W. FISCHER

Ref.: *J.Organomet.Chem.* **1990**, *382* 1 and **1995**, *500*,149; *Chem.-Ing.-Tech.* **1981**, *53*, 850; *J.Am.Oil Chem.Soc.* **1990**, *67*, 21; *Tetrahedron Lett.* **1995**, *36*, 1059; *J.Mol.Catal.* **1999**, *138*, 115; *Chem.Rev.* **1997**, *97*, 3197.

rhenium oxide based structures The structure of ReO_3 is the simplest structural type for AX_3 compositions and is based on a cubic unit cell. The cations A are placed at the corners and the anions X at the centers of the edges of the unit cell so that A is coordinated sixfold octahedrally by X. The octahedra share corners only. When viewed upon a face of the cubic unit cell, the rhenium oxide structure appears as a checkerboard of occupied octahedra and open parts of the structure. At the center of the structure, there remains a void. When this void is filled by a large cation, the perovskite structure is obtained.
There are three classes of structures which are related to the ReO_3 type. The first still has the ReO_3-type linkage of corner-sharing octahedra, but the octahedra are rotated in order to minimize the free space at the center

of the unit cell. The second class is realized when all the octahedra remain connected only via their corners, but with topologies different from the simple cubic one of ReO_3. Typical examples are *tungsten bronzes $A'_x WO_3$; the additional cations A' (Li, Na, K) present in these cpds. occupy voids in the structure. The third class is characterized by the fact that in addition to corner-sharing other types of linkages (edge- or face-sharing) appear between the octahedra. Edge- or face-sharing reduces the relative number of anions in these *shear structures*. They are therefore typical of oxide compounds with a slightly reduced anion content and give rise to continuous solid-solution series or discrete, slightly reduced oxides AO_{3-x}. An example is that of Mo oxides where the slightly reduced derivatives of MoO_3 are formed by the successive introduction of edge-sharing octahedra. Both the W bronzes and the shear structures are redox-active cpds. which can change their *redox state easily by insertion/de-insertion of the metal A' or oxygen. Some of these nonstoichiometric compounds are therefore used in oxidation catalysis. **E=F=G**.   P. BEHRENS

Ref.: Delmon/Froment; *Appl.Catal.* **1983**, *6*, 121; *J.Catal.* **1974**, *39*, 487; *Acc.Chem.Res.* **1997**, *30*, 169.

rhodium as catalyst metal

Rh combines important properties for cat. applications: it changes readily between Rh(I) and Rh(III) and forms square planar d_8-Rh(I) *complexes that are exceptionally prone to oxidative *addition of diatomic molecules (e.g. H_2, MeI). It willingly coordinates alkenes and inserts them into *metal *hydride bonds to yield alkyl intermediates. CO is an excellent *ligand for Rh and is easily inserted into M-C bonds. The prods. of the reactions at the metal center are readily eliminated to yield the final prods. (*reductive elimination). The sum of these qualities make Rh an ideal cat. metal for the *hydrogenation of alkenes and all kinds of *carbonylations. With respect to the high activities achieved with Rh cats., the high price of the metal often becomes of second importance. Nevertheless, cat. *recycling is the

main topic in hom. cat. with Rh. In hom. cat., Rh precursors such as $RhCl_3 \cdot 3 H_2O$, Rh acetate, or $Rh(CO)_2(acac)$, are usually added to the reaction mixture. Under appropriate conds. the active cat. species is formed as an *in-situ cat. The cat. properties of such "unmodified" Rh cats. can be altered dramatically by adding modifying *ligands (*modifiers, e.g., *phosphines).

In het. cat. Rh is usually applied together with other Pt metals. The most important applications are the *oxidation of NH_3 (fine meshes consisting of 90 % Pt and 10 % Rh) and *auto(motive) exhaust cats.

The wide spectrum of reactions involving Rh cats. comprises: *cyclopropanation, *hydroformylation (procs. of *BASF, *Ruhrchemie/Rhône-Poulenc, *Mitsubishi, *Union Carbide), *hydrocyanation (*Butachimie adiponitrile proc.), *hydroacylation, *hydrogenation, *hydrosilylation, *WGSR, carbonylation, *hydroamination, *cyclopropanation, *isomerization, *oxidation of ammonia, *automotive and *stationary exhaust cat., and various procs. of *petroleum refining, in which it is used in mixture with other Pt metals. **F** le rhodium comme métal catalytique; **G** Rhodium als Katalysatormetall.   F. RAMPF

Ref.: Farrauto/Bartholomew, p. 413; Cornils/Herrmann-1; *Houben-Weyl/Methodicum Chimicum*, Vol. XIII/9b; Adams/Cotton; *Coord. Chem. Rev.* **1982**, *35*, 113.

Rhône-Poulenc DeNOx process

low-temp. *SCR proc. for the treatment of the off-gases of chemical plants by cat. reduction with NH_3 over amorphous V_2O_5 on Al_2O_3 at 170–400 °C.   B. CORNILS

Ref.: *Cat.Today* **1989**, *4*, 205; *Appl.Catal.* **1994**, *115*, 179.

Rhône-Poulenc ethylene dichloride (EDC) process

for the manuf. of ethylene dichloride by *oxychlorination of ethylene with O_2/Cl_2 over *supported (and *promoted) $CuCl_2$ cats. in fluidized-bed operation. The cat. is *regenerated via air/HCl treatment. **F** procédé Rhône-Poulenc pour EDC; **G** Rhône-Poulenc EDC-Verfahren.   B. CORNILS

Rhône-Poulenc hydroquinone process

manuf. of hydroquinone by *oxidation (*hydroxylation) of phenol with peroxyformic acid (or mixtures of H_2O_2/carboxylic acids). At low conv. rates (<10 %) the byproducts catechol and resorcinol cannot be avoided. **F** procédé de Rhône-Poulenc pour le hydroquinone; **G** Rhône-Poulenc Hydrochinon-Verfahren.

B. CORNILS

Ref.: Ertl/Knözinger/Weitkamp, Vol. 5, p. 2330.

Rhône-Poulenc/Melle-Bezons acetic acid process

a variant of the manuf. of *AA by cat. *oxidation of acetaldehyde with air (instead of oxygen). **F** procédé Rhône-Poulenc/Melle-Bezons pour l'AA; **G** Rhône-Poulenc/Melle-Bezons AA-Verfahren.

B. CORNILS

Rhône-Poulenc MODOP process

for the treatment of tail gas from sulfur recovery units, including the oxidation of H_2S over TiO_2 based cats. at 180 °C. Developed jointly with Mobil.

B. CORNILS

Ref.: *Oil Gas J.* **1988**, *86*(2), 63; *Appl. Catal.* **1994**, *115*, 179.

Rhône-Poulenc phenol process

a variant of the *Hock proc. with homogeneous cracking of the *cumene peroxide into phenol. **F** procédé de Rhône-Poulenc pour le phénol; **G** Rhône-Poulenc Phenol-Verfahren.

B. CORNILS

Rhône-Poulenc sulfur recovery

a multistep proc. for the treatment of sour natural gas or refinery proc. gases over *Claus-active cats. based on TiO_2 at max. 330 °C. The proc. tolerates up to 1 % of oxygen.

B. CORNILS

Ref.: *Hydrocarb. Proc.* **1982**, (11), 89; *Appl. Catal.* **1994**, *115*, 179.

Rhône-Progil allyl alcohol process

*isomerization of *propylene oxide over Li_3PO_4 cat. (suspended in high-boiling alkylbenzenes) at 275–280 °C. At medium convs. (60 %) the *selectivity is 92 %. **F** procédé Rhône-Progil pour l'alcool allylique; **G** Rhône-Progil Allylalkoholverfahren.

B. CORNILS

Rhône-Progil phthalic anhydride process

manuf. of *PA by liquid-phase *oxidation of *o*-xylene with brominated Co, Mn, or Mo naphthenates or acetates at 150 °C. **F** procédé Rhône-Progil pour le PA; **G** Rhône-Progil PA-Verfahren.

B. CORNILS

ribokinase cat. the *phosphorylation of ribose to ribose-5-phosphate using *ATP. The *enzyme from *Lactobacillus plantarum* (EC 2.7.1.17) has been used to synth. ribose-5-phosphate (r5p), a key intermediate in the synth. of 5-phosphoribosyl-α-1-pyrophosphate and ribulose 1,5-bisphosphate. R5p is a key intermediate in the synth. of 5′-phosphoribosyl-α-1-pyrophosphate (PRPP), which is a precursor to purine, pyrimidine and pyridine nucleotides in *biosynthesis.

R5p has also been used in the synth. of ribulose 1,5-biphosphate (RuBP is a key intermediate in the fixation of CO_2 in plant metabolism). Although the route from AMP via r5p is more direct and convenient, the route from glucose 6-phosphate is also acceptable. **E=F=G**.

C.-H. WONG

Ref.: *J. Am. Chem. Soc.* **1983**, *105*, 7428.

ribonucleases a class of *phosphodiesterases that *hydrolyze *ribonucleic acid. The reaction mechanism of many rs., such as the mammalian pancreatic r. (of which the bovine pancreatic r. A is the best studied), proceeds via two steps. The first is the transfer of the phosphate ester from the 5'-hydroxyl of the leaving RNA chain to the 2'-hydroxyl of the ribose upstream of the scissile bond (to form the 2',3'-cyclic phosphodiester). The second step is the hydrolysis of the cyclic phosphate. The *enzyme cat. the reaction via *acid-base catalysis; in the case of bovine pancreatic r., the acid and base are a pair of histidines. The presence of a cyclic phosphodiester *inter-mediate has allowed the use of cyclic phos-phate substrates in the *kinetically controlled synthesis of nucleic acids from cyclic nucleo-tides. **E=F=G**.   P.S. SEARS

ribonucleic acid $\rightarrow$ DNA and RNA oligo-mers, *RNA

ribonucleotide phosphatase $\rightarrow$ phospha-tases

ribonucleotide reductases cat. the *re-duction of ribonucleotides to the corresponding deoxyribonucleotides with the concomitant *oxidation of a redoxin (thioredoxin, glutare-doxin, or NrdH redoxin) or formate. They are essential *enzymes for deoxyribonucleotide generation. There are three classes of ribonu-cleotide reductases known, all of which work through a *radical mechanism, but which differ in their *cofactor requirements, reductant used, and protein radical species generated, and man-ner of radical generation. **E=F=G**.   P.S. SEARS
Ref.: *Ann.Rev.Biochem.* **1998**, *67*, 71; *Science*, **1999**, *283*, 1499.

ribozyme cat. molecule of *RNA. The dis-covery of self-splicing *introns refuted the idea that nucleic acids were simply passive in-formation carriers. The natural self-splicing introns, self-cleaving RNAs, and ribozymes such as the RNA portion of RNase P acceler-ate reactions involving phosphodiester cleav-age and formation. Selection techniques have been used to identify nucleic acids that can accelerate a number of other reactions, in-cluding carbon ester *hydrolysis, *Diels-Al-der reactions, and amide synth. It is worth noting that in many of the cases reported, the "ribozyme" is "*cis*-acting" only: it is cova-lently attached to one of the reacting groups (e.g. the diene, in the case of the Diels-Alder reaction), and thus there is no turnover. Most, but not all, ribozymes also require metals for activity, either to promote folding or for their direct involvement in the reaction. **E=F**; G Ribozym.   P.S. SEARS
Ref.: *Chem.Rev.* **1997**, *97*, 371; *Cell* **1982**, *31*, 147; Mul-zer/Waldmann, p. 173.

ribulose 1,5-diphosphate $\rightarrow$ Calvin cycle, *pentose phosphate pathway

Rice-Herzfeld mechanism a *radical mechanism for gas-phase reactions at ele-vated temps. (*cracking, *pyrolysis). **F** méca-nisme de Rice-Herzfeld; **G** Rice-Herzfeld-Mechanismus.   B. CORNILS
Ref.: *J.Am.Chem.Soc.* **1934**, *56*, 284.

Rideal mechanism Two simple reactions were used to evaluate metal cat.: The conv. of *parahydrogen, and the equilibration of a mix-ture of hydrogen plus deuterium, viz. $H_2 + D_2 \rightarrow 2 HD$. For the latter reaction the dissociation of the reactants is a necessity; for the former it is permissible but not essential, as in a strong mag-netic field inversion of spin can occur without dissociation of the molecule.

The simple and straightforward mechanism for the H-D reaction is to suppose that both molecules dissociate, forming a pool of H and D atoms, from which by random conjunction of pairs of atoms an equil. mixture is formed. H atoms formed by *chemisorption of *ortho- and *parahydrogen are of course indistin-guishable, so that pairwise recombination leads automatically to an equil. mixture of the spin isomers.

This encountered difficulties, however, be-cause since the reactions are very facile and able to proceed at quite low temp., observed rates of *desorption from monolayers under these conds. are much too slow to explain the occurrence of the reactions. Rideal therefore suggested an alternative mechanism, by which a molecule interacted with an atom near a *vacant site, forming a triatomic species, dissociation of which could lead to prod. (the asterisk in the Figure symbolizes covalent bonds to the *surface).

The same mechanism would clearly explain the parahydrogen conv. Calcn. shows that, on a square array of sites, immobile dissociation chemisorption terminates when 8 % of sites remain unoccupied. This mechanism was also favoured by Eley and is usually known as the *Rideal-Eley (or Eley-Rideal; or really ideal) mechanism.   **F** mécanisme   de   Rideal; **G** Rideal-Mechanismus.   G.C. BOND

Rieche Alfred F. (born 1902), industrial research fellow at Farbenfabrik Wolfen (Germany), professor of technical chemistry at Jena and Berlin and director of the Institute of Organic Chemistry, (East) German Academy of Sciences. Active in *peroxides and ozonides, *autoxidation, and other cat. topics.
                                                         B. LÜCKE

Rietveld analysis The refinement of calc. versus experimental reflection intensities is part of the sequence of methods used in the analysis of crystal structures using diffraction techniques (X-ray or neutron diffraction; cf. *crystallography of surfaces). Refinement sets in when a principal model of the structure is available, which has been obtained either by solving the so-called *phase problem* or by identifying a close relationship to a known structure. Solving the phase problem and subsequently refining the reflection intensities is in most cases no problem when a 3D data set of reflection intensities has been measured on a single crystal of a substance.
Rietveld developed a new method for the refinement of powder diffraction data. Instead of trying to partition the observed intensity profile to single reflection intensities, the whole intensity pattern is refined directly.

This Rietfeld (or *whole pattern fitting*) refinement technique then yields a full simulation of the powder diffraction pattern, i.e., a continuous intensity profile. The results of Rietveld analyses are not as precise as those of crystal structure analyses using *single crystals. The radically new approach of the Rietveld technique and the advent of more and more powerful computers have made the R. analysis a powerful and often used technique for the refinement of the structures of powdered solids from diffraction data. This is important for the characterization of many crystalline het. cats. which are most often used as powders and not as single crystals. **F** analyse de Rietveld; **G** Rietveld-Analyse.   P. BEHRENS

Ref.: A. Young (Ed.), *The Rietveld Method*, Monographs on Crystallography, Vol. 5, Oxford Science Publications, Oxford University Press, Oxford 1993.

ring-closing metathesis (RCM)   the intramolecular alkene metathesis reaction of an acyclic diene (Figure).

Since this reaction is essentially thermoneutral, a statistical distribution of reactants and prods. may occur. In the case of α,ω-alkenes, the reaction can generally be driven to completion by the removal of the volatile byprod. ethylene (*ethenolysis). As a competing pathway, acyclic dienes can undergo acyclic diene metathesis (ADMET). The pathway primarily depends on ring size, dilution, substrate, and cat. Successful applications of RCM have been described for the prep. of five- to eight-membered rings and *macrocycles and thus as a key step in various total synths. Additionally, *stereoselective RCM, RCM of solid supported substrates, and the application of RCM in supramolecular chemistry have been reported. For the cats. see *metathesis.
G Ringschlußmetathese.   T. WESKAMP

Ref.: Ivin/Mol; *Angew.Chem.Int.Ed.Engl.* **1997**, *36*, 2036; *Tetrahedron* **1998**, *54*, 4413.

ring-opening metathesis polymerization (ROMP) the *metathesis of mono- and polycyclic alkenes yielding unsaturated *polymers, the *polyalkenamers (Figure).

ROMP is driven by the release of ring strain in the starting cycloalkene. Compared to other *polymerizations, in ROMP all the C-C double bonds are retained in the polymer. Furthermore, the reaction may be modified to proceed in a stereospecific way, i.e., the double bonds of the resulting polymer can be of the *cis* or *trans* type, depending on the cat. system and the reaction conds. In the presence of certain cats., ROMP can produce living polymers with a very narrow molecular weight distribution or highly *tactic polymers (cf. *tacticity). Cyclic monoenes, dienes, polyenes, bicyclic and polycyclic alkenes, and substituted derivatives of these *"monomers" are used. Several commercial procs. involving hom. ROMP procs. have been brought into practice, e.g., the prods. Norsorex by CdF-Chimie (a polynorbornenamer), Vestenamer by Hüls (a polycyclooctenamer), or various poly(dicyclopentadienes) (*Metton, *Telene), all cat. by tungsten cpds. ROMP of *COT or benzvalene produces conducting polymers of the polyacetylene type. ROMP is the only metathesis variant which has been commercialized so far.   T. WESKAMP

Ref.: Ivin/Mol; Mark/Bikales/Overberger/Menges, Vol. 11, p. 287; Y. Imamoglu (Ed.), *Metathesis Polymerization of Olefins*, Kluwer, Durdrecht 1995.

ripening (Ostwald ripening) the mostly unwanted phenomenon in which large particles of metal or *clusters (e.g., of het. cats.; cf. *precipitation in cat. preparation) tend to grow at the expense of smaller particles. *Sintering is a similar process, although mostly desired. **F** maturation; **G** Reifung.   B. CORNILS

RIS resonance ionization spectroscopy

riser-tube reactor a reactor with vertical tubes in which the reactant(s) transport the cat. upward and in which the reaction occurs. The spent cat. is separated in a cyclone-type device, and carried with an air stream in a standpipe to the *regenerator where the burning of the deposited coke takes place (cf. Figure under keyword *FCC). Then the regenerated cat. returns to the reactant feed. **F=G=E**.   B. CORNILS

Ref.: Ertl/Knözinger/Weitkamp, Vol. 3, p. 1426f.

RM process → Parsons process

RNA (ribonucleic acid) polymer of phosphodiester-linked ribonucleotides. One class of RNAs, the "messenger" RNAs, function as relatively labile intermediates in the in-vivo synth. of proteins encoded by *DNA *templates. Transfer RNAs (tRNAs) are specialized molecules used in protein synth., which function both as "adaptors" to match the correct amino acid with the mRNA code and also function as excellent leaving groups to facilitate the reaction between the growing peptide chain and the free amine of the incoming aminoacyl-tRNA. Ribosomes, the protein synthesis "machines" of the cell, are composed largely of RNA. **E=F=G**. P.S. SEARS

RNA ligase → DNA and RNA oligomers

RNA polymerases *template-dependent *enzymes that polymerize *RNA from ribonucleoside triphosphate substrates. Like *DNA polymerases, they only proceed in the $5' \to 3'$ direction, but unlike their DNA counterparts, they do not require an oligonucleotide primer. T3 and T7 RNA polymerases are enzymes derived from the genomes of the bacteriophages T3 and T7. They are frequently used for the in-vitro synth. of RNA from a DNA *template. **E=F=G**.   P.S. SEARS

Rp-DNA-S → phosphorothioate-containing DNA and RNA

Rochow contact masses → Rochow reaction

Rochow reaction The Rr. (also known as the Müller-Rochow synth.) is the reaction of silicon with organic halides to organohalosilanes in presence of Cu cats. More precisely the Rr. is the synth. of methylchlorosilanes by reaction of Si with methyl chloride acc. to Si_p $+CH_3Cl \rightarrow (CH_3)_xH_y SiCl_z$ (with $x+y+z = 4$) with dimethylchlorosilane being the main prod., obtainable with *selectivities up to >90 %. In the commercial proc., a so-called contact mass, originally consisting of Si powder, Cu powder, or Cu cpds. as cat. *precursors as well as *promoters (e.g. Zn, Sn, or Sb), reacts at 250–320 °C/0.2–0.5 MPa in fluid-bed reactors with CH_3Cl. During an induction period the cat. precursor converts to an active species. Probably not yet defined Cu silicide species are cat. active, with the role of the frequently discussed η-Cu_3Si phase being questionable. The mechanism of the Rr. has not yet been fully elucidated. Recent proposals explain the formation of the individual methylchlorosilanes by insertion of silylene or silylenoid intermediates into the C-Cl bond of adsorbed methyl chloride. **F** réaction de Rochow; **G** Müller-Rochow-Reaktion. H. LIESKE

Ref.: R.J.H. Voorhoeve, *Organohalosilanes*, Elsevier, New York 1967; K.M. Lewis, D.G. Rethwisch (Eds.), *Studies in Organic Chem.*, Vol. 49, *Catalyzed Direct Reactions of Silicon*, Elsevier, New York 1993.

Rockgas process → Kellogg catalytic coal gasification process

Roelen Otto (1897–1993), chemist with the Kaiser-Wilhelm-Institute (later Max-Planck-Institute) in Mülheim and industrial chemist with Ruhrchemie in Oberhausen (both Germany). Worked on general cat., the *Fischer-Tropsch synth., purification of gases, *HDPE. Discovered the *hydroformylation (oxo synthesis, Roelen reaction) in 1938 in the labs of *Ruhrchemie AG. B. CORNILS

Ref.: *Angew.Chem.Int.Ed.Engl.* **1994**, *33*, 2144; Neufeldt.

Roelen reaction → hydroformylation, oxo synthesis

ROMP → ring-opening metathesis polymerization

RON research *octane number

RoPHOS chiral *ligand; cf. *chiral pool

Rosenmund reduction Pd-catalyzed reduction of acid chlorides by hydrogen to give the corresponding aldehydes. Partial *deactivation of the cat. (e.g., by *modifiers such as quinoline/sulfur, *directed poisoning) prevents the subsequent *hydrogenation of the aldehyde to alcohol (Figure).

The organo-Pd species has been suggested as an intermediate in the Rr. During the course of the Rr. moisture has to be excluded to avoid acid chloride *hydrolysis with formation of the anhydride. Keto and nitro groups as well as aliphatic C-halogen bonds are stable under Rosenmund conds. However, hydroxy groups have to be protected. **F** réduction de Rosenmund; **G** Rosenmund-Reduktion. H.J. KREUZFELD, M. BELLER

Ref.: *J.Am.Chem.Soc.* **1986**, *108*, 2608; P.N. Rylander et al., *Catalysis in Organic Reactions*, Dekker, New York 1988, p. 221; *Organic Reactions*, Vol. 4, p. 362.

Rracemic temperature is that temp. at which there would be no discrimination between *R*- and *S*-enantiomers (cf. *temperature effect). **E** rrazemic témperature; **G** Rracemic-Temperatur. C.-H. WONG

R.2.R process → Total/IFP procs.

RRS resonance Raman spectroscopy, → Raman spectroscopy

RS → Raman spectroscopy

rt, RT   room temperature

rubredoxin   → monooxygenase

Ruff-Fenton sugar degradation →
Fenton's reagent

Ruhrchemie GUR process   first proc. for
the manuf. of *ultrahigh molecular weight
polyethylene (UHMPE) by *Ziegler cat. *polymerization of ethylene, developed 1954 by
*Roelen; still market-leading. **F** procédé
GUR de Ruhrchemie; **G** Ruhrchemie GUR-
Verfahren.   B. CORNILS

Ruhrchemie oxo process   now obsolete
proc. variant of the hydroformylation reaction, the Co carbonyl-cat. *addition of H_2 and
CO to the double bond of C_2 up to C_{16} alkenes at 150–170 °C/ 30 MPa total press.
Characteristics of this proc. are the reactor
section and the Co cycle based on a proprietary hydrothermal *decobaltation, and the
use of mixtures of Co metal and Co hydroxide
as cat. *precursors. The proc. was offered in
two versions: using S-free *syngas and syngas
with up to 50 ppm of S. **F** procédé Ruhrchemie d'oxo; **G** Ruhrchemie Oxoverfahren.
Ref.: Falbe-1, p. 162   B. CORNILS

**Ruhrchemie/Rhône-Poulenc (RCH/RP)
process**   a modern proc. variant of *hydroformylation, the cat. addition of CO and H_2
to the double bond of C_3–C_6 alkenes. The
RCH/RP proc. is the first economical application of *aqueous-phase catalysis and thus uses
an Rh carbonyl, *ligand-modified with water-
soluble TPPTS (*triphenylphosphine trisulfonate) as cat. Due to the insufficient solubility
of higher alkenes in water the application of
the proc. is so far restricted to C_3–C_6 alkenes.
The RCH/RP proc. is extremly environmentally sound and economic (Figure). **F** procédé
Ruhrchemie/Rhône-Poulenc; **G** Ruhrchemie/
Rhône-Poulenc-Verfahren.   B. CORNILS
Ref.: *Chem.Ing.Tech.* **1994**, *66*, 916; *CHEM-
TECH* **1995**, (1), 33; *Recl.Trav.Chim.Pays-Bas* **1996**,
115, 211; *Org.Proc.Res. & Dev.* **1998**, *2*, 121.

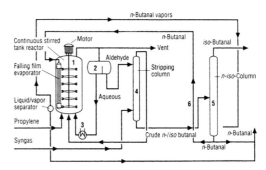

Ruhrchemie/Rhône-Poulenc (RCH/RP) process

runaway of catalysts   Especially with
highly exothermal reactions in fixed-bed operation, cat. runaway is one of the aspects of
operational stability and – ultimately – safety
of procs. when cat. (or even *autocatalyzed)
decompositions can occur (e.g., in combustion
of ethylene while producing ethylene oxide,
or *methanation of CO and H_2 on cat. *deposits during oxo synth.). The best method of
detecting potential runaways is intensive on-
line control of all the relevant parameters, including analyses of prod. streams. Sensitive
procs. have to have rapid shutdown devices
with suitable purging installations. **F** emballe-
ment de catalyseurs; **G** Durchgehen von Katalysatoren.   B. CORNILS
Ref.: Ertl/Knözinger/Weitkamp, Vol. 3, p. 1422; *AIChE
J.* **1988**, *34*, 1663.

Rupe reaction   the acid-cat. rearrangement ot tertiary acetylenic alcohols to α,β-un-
saturated carbonyl cpds.; cf. *Meyer-Schuster
rearrangement. **F** réction de Rupe; **G** Rupe-
Reaktion.   B. CORNILS
Ref.: Augustine, p. 586.

ruthenium as catalyst metal 1-heteroge-
neous: Besides its ability to cat. the *ammonia
synth. (e.g., *Kellogg KAAP), het. Ru cats.
(bulk Ru(IV) oxide) have been used for the
*Pichler synth. of polymethylene from *syn-
gas, and for various *hydrogenations such as
those of aromatic rings without *hydrogenolysis of amino or hydroxy groups, carboxylic

acids, or aldehydes and ketones in aqueous solution (all with Ru on *supports such as *carbon or *alumina). Ru has a considerable potential to cat. *metathesis reactions. Some exotic applications (e.g., *silylcarbonylation) are described, too. **B. CORNILS**

2-homogeneous: Since Ru has a $4d^7 5s^1$ electronic configuration, it has the widest range of oxidation states (from –2 in $Ru(CO)_4^{2-}$ to +8 in RuO_4) of all the elements, and various co-ordination geometries in each electronic configuration, which represents a great potential for the exploitation in cat. reactions. However, until the 1980s the useful methods were limited to a few reactions including proposals for oxidations with RuO_4, hom. *hydrogenations for fine chemicals, *hydroformylations, *Fischer-Tropsch variants (to *waxes), *Reppe synth. of hydroquinone (via *hydrocarbonylation), and *hydrogen transfer reactions. As the coordination chemistry of Ru *complexes has progressed, characteristic features of Ru (e.g., high electron transferability, low redox potentials, stability of reactive metallic species – oxometals, *metallacycles, metal *carbenes) have been opened for a broad variety of cat. transformations. These include *metathesis, *cyclopropanation, *isomerization, or *CO_2 hydrogenation. Additionally, the exceptionally low price of Ru compared to other precious metals (Pd, Rh) makes it an attractive extension for related industrial procs. **F** le ruthénium comme métal catalytique; **G** Ruthenium als Katalysatormetall.

T. WESKAMP

Ref.: *Chem.Rev.* **1998**, *98*, 2599; Falbe-2, p. 334; Adams/Cotton; Farrauto/Bartholomew, p. 413; Cornils/Herrmann-1; *Houben-Weyl/Methodicum Chimicum*, Vol. XIII/9a; *Acc.Chem.Res.* **1997**, *30*, 97; *Coord. Chem.Rev.* **1982**, *35*, 41; *Angew.Chem.Int.Ed.Engl.* **1999**, *38*, 2416; Murahashi/Davies, p. 195,391.

ruthenium black prepared by precipitation of the metal from aqueous Ru chloride, e.g., with HCHO in alkaline medium. Dispersity and therefore *adsorption and cat. properties depend on the composition of the Ru salt and the drying conds. Ru black is mainly used as a cat. for *hydrogenations (especially for aromatics), but application in NH_3 synth., *dehydrogenation, and *Fischer-Tropsch reactions are known as well. **F** noir de Ru; **G** Ruthenium-Mohr (Schwarz). **T. WESKAMP**

Rutherford backscattering spectroscopy

(RBS) based on high-energy ion beam scattering and yields information (integral, bulk-sensitive, ex-situ) about elemental analysis, depth-resolved. The method is non-destructive; solid samples require an accelerator.

Ref.: Ullmann, Vol.B6, p. 77. **R. SCHLÖGL**

rutile structure Named after the important mineral of titanium, rutile (TiO_2, *titania), the rs. is based on a slightly disordered hexagonal closing pack (hcp) of O atoms with half the octahedral interstices being occupied by Ti atoms. The Ti atoms are octahedrally coordinated whereas the O is in a trigonal planar coordination. This structure is commonly adopted by ionic dioxides and difluorides where the relative sizes of the ions are such as to favor six-coordination (i.e., when the ratio of cation/anion lies in the range 0.73–0.41:1). Rutile has found application as a *carrier and *support for cats. Recently, TiO_2 has attracted interest because of its potential in *photocatalysis as a photosensitizer (n-type semiconductor, cf. *titania, *titania photocatalysis). **F** structure de rutile; **G** Rutilstruktur.

R. SCHLÖGL, M.M. GÜNTER

S

Sabatier Paul (1854–1941), professor of chemistry in Toulouse (France). Worked on catalysis, *hydrogenations over Fe, Ni, Co, Cu, and the *mechanism of metal-cat. reactions. Nobel laureate in 1912 (together with Victor *Grignard).   B. CORNILS
Ref.: *Nature (London)* **1954**, 174.

Sabatier-Senderens hydrogenation
cat. *hydrogenation of vaporized unsaturated cpds. over finely divided Ni, prepd. by *reduction of NiO to Ni metal. **F** hydrogénation de Sabatier et Senderens; **G** Sabatier-Senderens-Hydrierung.   B. CORNILS
Ref.: *C.R.Hebd.Séances Acad.Sci.* **1899**, *128*, 1173.

Sabatier's principle
S. emphasized that during the course of het. cat. on the *active site unstable intermediate cpds. are formed with one of the reactants. This intermediate is stable enough to be formed and unstable enough to decompose to yield the final product (optimum strength of bonding). This principle is illustrated by the *volcano plot. **F** principe de Sabatier; **G** Sabatiers Prinzip.
B. CORNILS
Ref.: P. Sabatier, *Catalysis in Organic Chemistry*, van Nostrand, New York 1925; Ertl/Knözinger/Weitkamp, Vol. 1, p. 3; Vol. 3, p. 1082; van Santen/van Leeuwen.

Saccharomyces cerevesiae   → yeast

Sachtler-Fahrenfort plot
a special plot related to *volcano plots.

SAD   → selected area electron diffraction

S-adenosyl-   see under keyword "A"

SafeCat technology   → UOP procs.

safety lamp   invented in 1817 (acc. to other sources, 1832). *Davy discovered that in the presence of platinum, combustible gases were oxidized at low temps. On the basis of this cat. effect he designed a safety lamp for miners in which a Pt net glowed if the flame was extinguished. **F** lampe de sûreté Davy; **G** Davy'sche Sicherheitslampe; Grubenlampe.   B. CORNILS

SAH   → S-adenosylhomocysteine

Sakurai allylation reaction   comprises the 1,4-*addition of allylsilanes to α,β-unsaturated carbonyl cpds. in the presence of *Lewis acids yielding δ,ε-unsaturated aldehydes or ketones acc. to

$$R^1CH=CH\text{-}CO\text{-}R^2 + H_2C=CH\text{-}CH_2\text{-}SiMe_3 \rightarrow$$
$$H_2C=CH\text{-}CH_2CH(R^1)\text{-}CH_2\text{-}CO\text{-}R^2.$$

The Sar. is usually performed in halogenated solvents with stoichiometric amounts of $TiCl_4$ complexation of the resulting enol and Ti. Other Lewis acids such as BF_3, $SnCl_4$, or lanthanide triflates have been used to promote this reaction. Sar. in the presence of *catalytic* amounts of the Lewis acids are possible using electrophilic silicon derivatives, e.g., trimethylsilyl triflate. **F** allylation de Sakurai; **G** Sakurai-Allylierung.   M. BELLER
Ref.: Trost/Fleming.

Salcomine   → Co-salen complexes

salen the bis(salicylaldehyde)ethylenediamine group [H_2salen=N,N'-bis-(salicylidene) ethylenediamine] – a *Schiff base – in *complex cpds. or *ligands; cf., e.g., the *Jacobsen Mn-salen complex. Cr s. complexes were also examined in hetero *Diels-Alder reactions,

Zr s. cpds. in *polymerizations. S. ligands are used in *ship-in-the-bottle cats. Co(II)salen (Salcomine) reacts with O_2 to form labile oxygen complexes which have been used for oxygen storage. **E=F=G**.   R. ANWANDER

salt bridges   (in protein structures)   are electrostatic interactions between groups of opposite charge. In water, and at the surface of proteins, charged groups are well solvated by water and interact weakly with each other. Sbs. at the exterior of a protein thus have little impact on folding stability. In the interior of a protein, where there are *clusters of hydrophobic residues and the local dielectric constant (D) is very low, sbs. can be stronger (that is, the energy of the sb. as compared to isolated charges at the same D is high), since the energy of an sb. varies inversely with the D. For a protein with buried charges, however, folding involves transfer of the charges from the aqueous solution, where they are very well solvated, to the interior, where they are poorly solvated. Thus, the negative free energy of formation for a salt bridge in a low-D environment is balanced by the free energy gain upon desolvation, and the overall effect upon overall folding stability is thought to be small. This does not mean that salt bridges are unimportant, however; sbs. probably add "specificity" to the folded structure: the *conformation that allows most hydrophobic residues to be buried and minimizes the number of unpaired, buried charges will be favored. **F** pont de sel; **G** Salzbrücke.   P.S. SEARS
Ref.: *Protein Sci.* **1994**, *3*, 211.

salt effects   There are two different salt effects. 1: The primary se., due to alteration of the activities of the reactant particles, whether ions or molecules, by salt. 2: The secondary se., in which the effective conc. of a reactant ion coming from a weak electrolyte is decreased by the reduced ionization of the electrolyte due to added salt. Both phenomena play a role in the reaction rates in solution, the activity of acid-base cats., the solubility of non-electrolytes, etc. **F** effets de sel; **G** Salzeffekte.   F.E. KÜHN
Ref.: Reichardt, p. 205; *Solvent Effects on Chemical Phenomena* **1973**, *1*, 227.

salting in/salting out proteins   → Hofmeister series

SAM   1: scanning Auger microscopy, → Auger electron spectroscopy; 2: *S*-adenosylmethionine.

Sandmeyer reaction   an exchange reaction $-NH_2 \rightarrow$ halogen, etc., via diazonium salts in a *radical mechanism. Cats. are Cu(I) halides (*aromatic substitution).   B. CORNILS

sandwich complexes π-complexes (coordination cpds.) between *transition metal atoms and arenes or unsaturated aliphatic molecules in the form of a sandwich structure (*ferrocenes, *metallocenes, *Vaska cpds.).
                B. CORNILS

SAPC   → supported (solid) aqueous-phase catalysis

SAPO   → silicoalum(in)ophosphate

saponification   the opposite reaction to *esterification is a reaction in which esters take up water and are converted to alcohols and carboxylic acids under the influence of cats. such as acids, bases, or *enzymes (e.g. procs. of *Lummus PO, *Shell glycerol, *Twitchell cleavage). In a broader sense any *hydrolysis is called s., e.g., the s. of nitriles to amides (*acrylamide acc. to the procs. of *Mitsui Toatsu or *Nitto). **F** saponification; **G** Verseifung.   B. CORNILS

saponite   a *clay mineral belonging to the group of phyllosilicates with two tetrahedral and one octahedral sublayer. The octahedral layer contains mainly Mg^{2+}, Fe^{2+}, and Fe^{3+}; in the tetrahedral layers, Si^{4+} is partially exchanged for Al^{3+}. The layers are negatively charged. This negative charge is compensated by interlayer cations. A typical formula for a saponite is $Na_x\{(Mg^{2+},Fe^{2+})_3$ $(OH)_2[(Al_xSi_{4-x})$ $O_{10}]\}$. **F=E; G** Saponit.   P. BEHRENS

SASIL zeolite 4A for phosphate substitution in detergents

SASOL South African Coal, Oil, and Gas Corp., Ltd., founded to convert coal to motor fuels. S. operates three coal-based plants (Sasol One to Three) near Sasolburg and at Great Plaines. Their commercial backbone are coal gasifiers (Lurgi type) and subsequent *Fischer-Tropsch syntheses in fixed-bed (*ARGE), entrained-bed (Kellogg's *Synthol), and fixed fluid-bed (Sasol development) operation. Total prod. capacity for hydrocarbons is estimated at $5–6 \times 10^6$ t/year.
                                        C.D. FROHNING

saturation kinetics (in enzyme catalysis) One of the characteristics of a cat. is that it should be saturable: at some conc., all of the cat. sites should be occupied by substrate (unless the *enzyme is so efficient that the overall rate is limited by the speed of substrate diffusion into the enzyme active site). This leads to a saturation curve on a graph of reaction rate vs. substrate conc. The substrate conc. at which the reaction rate reaches half of the maximal rate (V_{max}) is defined as the K_m, the *Michaelis constant.

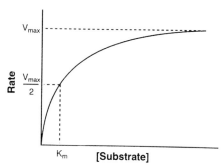

The saturation limit may not be physically attainable if the substrate binds poorly. **F** kinétiques de saturation; **G** Sättigungskinetik.
                                        P.S. SEARS

saturation sites → single turnover (STO)

SAXS → small-angle X-ray scattering

SBA secondary butyl alcohol

SBR styrene-butadiene rubber, → polybutadiene

sc supercritical, → supercritical-phase catalysis

scanning Auger microscopy (SAM) applied to determine the morphology with elemental resolution. Probe/response: electrons; the UHV techniques (non-destructive) need flat samples (cf. *AES).   R. SCHLÖGL
Ref.: Ullmann, Vol.B6, p. 54.

scanning electron microscopy (SEM) is a variant of electron microscopy. It rasters a narrow electronic beam over the surface and detects the yield of either secondary or backscattered electrons as a function of the position of the primary beam. It is applied for measurements of the morphology, *particle size, and *pore size. The ex-situ, UHV method is non-destructive and needs solid, conductive samples. The data gained are integral, surface and bulk-sensitive. SEM is often used in combination with *EDX; cf. *ESEM.
                                        R. SCHLÖGL
Ref.: Niemantsverdriet, p. 168; Ertl/Knözinger/Weitkamp, Vol. 2, p. 493.

scanning near-field optical microscopy (SNOM) can measure the change in chemical compositions as a function of position and time on the nm scale available. Recently used in cat. **G** Optische Nahfeldmikroskopie.
                                        B. CORNILS
Ref.: Appl.Phys.Lett. **1984**, 44, 651; Catal.Lett. **1998**, 56, 1.

scanning probe microscopy → scanning tunneling microscopy (STM) and atomic force microscopy (AFM)

scanning transmission electron microscopy (STEM) This method combines the scanning and the transmission modes of *elec-

tronic microscopy. It is applied for high-resolution structural and compositional analyses
of discontinuous solids (interfaces, supported
particles). Probe: scanning electronic beam;
response: electrons, usually combined with
*EELS. The ex-situ method is non-destructive
and needs flat, thin, solid samples. The data
obtained are integral and bulk-sensitive.

R. SCHLÖGL

Ref.: Niemantsverdriet, p. 168; Ullmann, Vol. B6,
p. 229.

scanning tunneling microscopy (STM)
based on the tunneling of electrons between the
surface and a very sharp tip. It is applied for
*surface imaging with atomic resolution and
limited chemical contrast. The ex-situ method
(UHV, non-destructive, with in-situ capability is
sensitive to regular and defective structures but
very versatile for all kinds of solid conducting
samples. The data obtained are integral and surface-sensitive.   R. SCHLÖGL

Ref.: Niemantsverdriet, p. 182; Ullmann, Vol.B6, p. 82.

scanning tunneling spectroscopy (STS)
a special method for the investigation of local
electronic structures near the *Fermi edge
with atomic lateral resolution (probe: electrostatic potential, gap voltage scanned; response: current; derivative vs. gap voltage).
The ex-situ technique is non-destructive and
ideal for *active-site analysis but exhibits
some problems with sample drift and interpretation due to tip effects. The data are on
an atomic scale and *surface-sensitive.

R. SCHLÖGL

Scatchard plot used to determine the
binding constant between two molecules. For
*enzyme-small molecule binding, the conc. of
the enzyme is held fixed while the small molecule (*ligand) conc. is varied. At each ligand
conc., the amount of free and bound enzyme
is determined. For an enzyme with a single
ligand binding site (or multiple sites with the
same association constants), a plot of [bound
enzyme]/[free enzyme] vs. [bound enzyme]
will give a line with slope = $-K_a$ (association

constant), assuming that the free ligand conc.
is essentially unchanged by the small amount
of enzyme-ligand *complex formed. **E=F; G**
Scatchard-Diagramm(e).   P.S. SEARS

scavenger → radical scavenger

SCD selective catalytic decomposition, cf.
*selective cat. reduction

SCF self consistent field, → ab-initio calculations

scheelite the mineral $CaWO_4$, an important tungsten ore. It is a colorless cpd. with a
hardness of 4–4.5. The tetragonal structure
contains isolated $[WO_4]^{2-}$ tetrahedra and
Ca^{2+} ions which are coordinated octahedrally
by oxygen. Many other tungstates and also
molybdates crystallize in this structure type.
E=F; G Scheelit.   P. BEHRENS

Schiff bases 1-general: as *ligands, are
based on the reaction prods. of the Schiff condensation (imines, azomethines from carbonyl
cpds. and primary amines). Some Sbs. are
N,O-donor (*oligodentate) *ligands in *transition metal *complexes such as Mn (*Jacobsen complex), Ti (for asymmetric epoxidation
or the Me_3SiCN-mediated *hydrocyanation
of aldehydes), or Cu (*cyclopropanation); cf.
also *salen.   B. CORNILS

2-in enzyme cat.: Many enzymatic reactions
involve Sb. intermediates between the *enzyme and substrate. For example, the pyridoxal
phosphate-dependent enzymes (*aminotransferases and many racemases and *decarboxylases) form a Schiff base between the aldehyde
of the *cofactor and an amine on the
substrate. Type I *aldolases contain a lysine in
the active site which forms an Sb. with the donor substrate, promoting attack at the carbonyl of the acceptor substrate. In *copper-
dependent amine oxidases, the topaquinone
cofactor is believed to form an Sb. with the
amine. Tautomerization and *hydrolysis releases the aldehyde product. The cofactor is
regenerated by oxidation to the imine, fol-

lowed by imine hydrolysis. **F** bases de Schiff;
G Schiffsche Basen.   P.S. SEARS
Ref.: Cornils/Herrmann-1, pp. 421,482,563,894,1174.

Schmidt number The Schmidt number
(*Sc*) describes the ratio of kinematic viscosity
to molecular diffusivity. For mixtures of gases,
typical values of *Sc* fall in the range 0.5–3, and
that of liquids between 300–3000. **F** nombre
de Schmidt; **G** Schmidt-Zahl.
                            P. CLAUS, D. HÖNICKE

Scholl reaction   to perylene from naphthalene in *Friedel-Crafts reaction, → arene coupling.

Schrock carbenes form alkylidene *ligands
in *complexes with the early *transition metals,
e.g., $[(\eta^5-C_5H_5)_2Ta(=CH_2)CH_3]^+[PF_6]^-$. Scs.
normally comprise nucleophilic carbenes which
add across C-C double bonds, thus initiating
and cat. propagating the alkene *metathesis.
E=F; **G** Schrock-Carbene.   W.A. HERRMANN
Ref.: *Acc.Chem.Res.* **1979**, *12*, 98.

Schrödinger equation The equ. $H\Psi = E\Psi$ is the quantum theoretical expression for
the energy of microscopic particles such as
atoms or molecules. Unlike macroscopic objects, the energy of microscopic objects is not
continuous, but has discrete values which are
determined by quantum numbers. The energy
expression of classical physics for macroscopic particles is given by the Hamilton function or related expressions, which give the
energy as a continuum. The Sche. is a mathematical expression which regard the energy as
discrete values. The microscopic particle is
described by the wavefunction Ψ, which contains all information about it. Any observable
(measurable) quantity of the particle is given
by action of the appropriate operator on Ψ,
which leads to the corresponding eigenvalue.
The quantum theoretical operator for the
energy is the Hamilton operator H, which
gives the energy of the particle as the eigenvalue E. Since many other properties of atoms
and molecules are calc. as derivatives of the

energy with respect to variables, the calcn. of
the energy is the pivotal part of *QC.
The Sche. cannot be solved exactly for more
than two particles. Thus, energy calcns. for
atoms and molecules with more than one
electron can only be carried out in an approximate way. The most important research area
of method development for *QC *ab-initio
calcns. is the search for good approximations
to the exact solution of the Se. The most important approximation is the *Hartree-Fock
model, which is the starting point for more accurate energy calcns. Although the results of
approximate solutions to the Se. are necessarily not exact, the difference from the correct
results may in practise become smaller than
experimental errors. This is the reason why
*QC ab-initio calcns. and *density functional
theory have become important tools in chemical research, including catalysis.   G. FRENKING

Schulz-Flory distribution (most probable
distribution) A molecular weight distribution which results from several types of *polymerization (or other chain-growing procs.
such as the *Fischer-Tropsch synth.), the distribution function may be derived from the
polymerization kinetics. As so many systems
appear to obey this distribution, more complex distributions are compared to it. The distribution results from 1: chain polymerization
with constant cat. and *monomer concs., with
transfer to solvent but not to monomer and
termination by disproportionation; 2: from
linear *condensation polymerization by assuming equal reactivity of all chain ends; 3: by
linear condensation polymerization allowing
the units to interchange in a random manner;
or 4: from random scission degradation. The
SFd. depends on the *chain-growth factor.
F distribution de Schulz-Flory; **G** Schulz-
Flory-Verteilung.   W. KAMINSKY, C. STRÜBEL
Ref.: *Z.Physik.Chem.* **1935**, *B30*, 379; *J.Am.Chem.Soc.*
1936, *58*, 1877; P.J. Flory, *Principles of Polymer Chemistry*, Cornell University Press, Ithaca 1953.

Schwab Georg-Maria (1899–1984), professor
of physical chemistry in Athens (Greece) and

München (Germany); worked on *kinetics, photochlorination, and catalysis in general. Two so-called *Schwab effects* describe the effect of contact with non-metallic *supports or with metals, respectively on the cat. activity of metals or *semiconductors.   B. CORNILS

SCMCR simulated countercurrent moving-bed chromatographic reactor, → simulated moving-bed reactor

SCOT process   Shell Claus offgas treatment, → Shell procs.

SCR → selective catalytic reduction, *monolithic honeycombs

screening (in biocatalysis) The discovery of *enzyme mutants with novel and desirable properties is generally a two-step proc.: mutagenesis of the enzyme-producing organism (to produce a mutant *library), followed by isolation of colonies with the desired activity (cf. *in-vitro evolution of biocats.). Testing isolated colonies for the desired *activity is known as screening, and can be a time-consuming proc. It is greatly facilitated by the use of chromogenic substrates. For example, a β-galactosidase was recently converted into a β-fucosidase by mutagenesis and screening using the chromogenic substrate X-Fuc (5-bromo-4-chloro-3-indolyl β-D-fucopyranoside). X-Fuc was incorporated into the agar on which the bacterial mutants were grown, and only those colonies with fucosidase activity could hydrolyze the substrate to produce a blue-colored dye. **E=F=G**.   P.S. SEARS
Ref.: *Proc.Natl.Acad.Sci.USA* **1997**, *94*, 4504.

SD   the company Scientific Design

SD adipic acid process manuf. of adipic acid by a two-step procedure, 1: the cat. (by Mn or Co salts) *oxidation of cyclohexane to *KA oil; 2: the liquid-phase oxidation (*LPO) of the KA oil with air to adipic acid over Cu-Mn acetate at 80 °C/0.6 MPa. **F** pro-cédé Scientific Design pour l'acide adipique; **G** Scientific Design Adipinsäure-Verfahren.
B. CORNILS

SD ethylene oxide (EO) process   Older and now obsolete proc. for the manuf. of EO by partial *oxidation of ethylene with air over Ag cats. Today's versions use oxygen. **F** procédé SD pour le EO; **G** SD EO-Verfahren.
B. CORNILS

SD glycol process   for the non-cat. manuf. of ethylene glycols (*EG, *DEG, *TEG) by *hydration of refined ethylene oxide and pure water. **F** procédé SD pour des glycols; **G** SD Glykolverfahren.   B. CORNILS
Ref.: *Hydrocarb.Proc.* **1997**, (3), 133.

SD maleic acid process   for the manuf. of *MAA by *oxidation of benzene with air over doped V_2O_5 cats. in tubular reactors at 400–450 °C/0.2–0.5 MPa. Approx. one quarter of the fed benzene is oxidized totally. **F** pro-cédé SD pour l'anhydride maléique; **G** SD Maleinsäureanhydrid-Verfahren.   B. CORNILS
Ref.: Winnacker/Küchler, Vol. 6, p. 122.

SD phenol process   manuf. of phenol by a two-step procedure, 1: the *oxidation of cy-clohexane to *KA oil (*SD adipic acid process); 2: the *hydrogenation of KA oil at 400 °C over Pt-C or Ni cats. **F** procédé SD pour le phénol; **G** SD Phenolverfahren.
B. CORNILS

secondary ion mass spectrometry (SIMS)   is among the most sensitive surface techniques and thus important for het. cat. research. It is based on the sputtering of atoms, ions and other fragments from the investigated *surface, which are analyzed with a *mass spectrometer. The destructive method is applied for local elemental and chemical compositions in depth profiles. The quantification of the data (integral, surface and bulk-sensitive) is difficult (cf. *ion microprobes, *secondary neutral mass spectroscopy).
R. SCHLÖGL

Ref.: Niemantsverdriet, p. 80; Benninghoven/Rüde-nauer/Werner; Ullmann, Vol.B6, p. 61.; Ertl/Knözin-ger/Weitkamp, Vol. 2, p. 623.

secondary metabolites   cpds. produced by *enzymes that are not part of the essential metabolic pathways (i.e., pathways for energy generation, production of structural comps. such as membrane phospholipids, and organism growth and reproduction). Some examples are: terpenes; *polyketides; β-lactam and other anti-biotics. **F** métabolites secondaires; **G** sekundäre Metaboliten.   P.S. SEARS

secondary neutral mass spectroscopy (SNMS)   a method analogous to *SIMS in which neutral particles are post-ionized be-fore entering the *mass spectrometer. The ad-vantage over *SIMS is that SNMS does not suffer from matrix effects.   R. SCHLÖGL
Ref.: Niemantsverdriet, p. 95.

secondary structure   The ss. of a protein refers to the local *conformation that short, contiguous polypeptide regions adopt. This is to be contrasted with the *primary structure*, which refers to the covalent amino acid se-quence; the *tertiary structure*, which indicates how distant portions of the polypeptide are arranged in space; and the *quaternary struc-ture*, which indicates how different protein subunits or domains interact. Several types of common secondary structural elements are observed. α-Helices are right-handed spirals of amino acids, which average 3.6 amino acids per turn. The side chains project out of the helix, while the amide carbonyl *hydrogen bonds with the amide N-H of the following loop (the [n+4]th residue). β-Sheets are struc-tures in which the polypeptide backbone is extended, with alternate amides and side chains pointing in opposite directions. They are arranged side-by-side in either a parallel or antiparallel fashion with other strands so that the amide carbonyls of one sheet can H-bond with the amide N-Hs of the chain next to it, and vice versa. There are a variety of tight turns that polypeptides can form. One

common class comprise the "β-turns": four-amino-acid bends in the structure that are further classified on the basis of torsion angles between various backbone amides. **F** struc-tures secondaires; **G** Sekundärstrukturen.
                                                      P.S. SEARS

second harmonics generation (SHG)   a *laser spectroscopy variant for the in-situ analysis of gas-solid interfaces. The method is non-destructive and requires flat *surfaces; the data received are integral and *surface-sensitive. Because of non-linear optical effects the interpretation is difficult.   R. SCHLÖGL

D-sedoheptulose-7-phosphate → trans-ketolase

selected area electron diffraction (SAD)   can be used for crystallographic in-formation from *nanostructures in solids.
The UHV electron method (ex situ) is non-destructive and yields local, bulk-sensitive in-formation.   R. SCHLÖGL

selection   The discovery of *enzyme mu-tants with novel and desirable properties is generally a two-step proc.: mutagenesis of the enzyme-producing organism (to produce a mutant *library, *random mutagenesis), fol-lowed by isolation of colonies with the desired activity (*in-vitro evolution of biocats.). Find-ing a desired colony within a large library can be time-consuming if the activity of the colo-nies must be assessed one-by-one. This proc-ess can be accelerated using a selection tech-nique (*screening), in which the survival of the organism is dependent on the production of an enzyme with the desired activity. **E=F**; **G** Selektion.   P.S. SEARS

selective catalytic decomposition
other term for *selective catalytic reduction

selective catalytic reduction (SCR)   SCR with ammonia (alternatively urea) as a reducing agent to give N_2 and H_2O is the *DeNOx proc. most frequently used to clean effluents in power and nitric acid plants (*stationary exhaust cat.). The NH₃/NO ratio determines the degree of NO reduction, but is limited by excess ammonia leaving the plant (*ammonia slip). SCR is performed at temps. between ca. 473 and 773 K with TiO₂-supported V₂O₅ monolayer cats., often with admixtures of W and Mo oxides (*titania). Other cat. systems include Cr- or Fe-containing oxides supported on Al₂O₃, or Cu on various *supports. Typical cat. formulations are plates, *honeycombs, and *pellets. SCR units can be placed early in the flue gas treatment scheme to minimize the need to reheat the gases, or later, after dust and SOx removal, to minimize cat. *deactivation. SCR (sometimes called SCD = selective cat. decomposition) can be combined with other flue gas treatments to simultaneously remove NOx and SOx (De-NOx/DeSOx). Cf. also *titania, *rutile structure. **E=F=G.**   F. JENTOFT

Ref.: Ertl/Knözinger/Weitkamp, Vol. 4, p. 1633; van Santen/van Leeuwen/Moulijn/Averill.

selective hydrogenation   If there are alternative routes in *hydrogenation yielding different prods., the predominant one is said to be formed selectively by a proc. referred to as sh. The degree of *selectivity with which any prod. is formed is expressed as a fraction or percentage of the total, but the term is by custom retained only for the desired product. The sh. of acetylene (C₂H₂) as a byprod. of the *steam cracking of naphtha demands that little or no ethene be reduced to ethane, and that if possible the ethyne should be reduced only to ethene, so as to boost its conc. (*low-temp. hydrogenation). The first requirement is quite easily met, because ethyne is much the more strongly adsorbed, although the ratio of ethene to ethyne is very high. The second requirement is harder to meet, because it is necessary for adsorbed ethene formed from ethyne to leave the *surface before being further reduced. The somewhat complex set of possible reaction steps is illustrated (Figure; the asterisk symbolizes a single covalent bond to the surface).

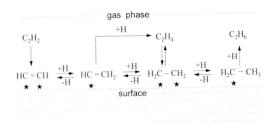

Two criteria for obtaining ethene in high selectivity are that its adsorbed state should be only weakly adsorbed, and that the conc. of adsorbed H atoms should be kept to a minimum.   Both are met by Pd (even better by Pd-group *bimetallic cats.), perhaps partly because of this metal's propensity to dissolve H atoms: its tendency to hold ethene only relatively weakly is mirrored in its effectiveness in double-bond migration and related procs. (see *hydrogenation). Similar considerations arise with gas streams containing propane and C₄ alkenes, where allene (1,2-propadiene), 1,3-butadiene, propyne, and butynes have to be removed, and the methods used to do this are similar. Other metals are capable of reducing alkynes and alkadienes to alkenes with high selectivity (e.g., Cu, Au), but low *activity or other undesirable features limit their use. The aromatic ring is normally reduced to the cyclohexane ring in one step, although useful yields of cyclohexene are obtained from benzene with Ru cats. Phenol is reduced to cyclohexanone by Pd-Al₂O₃ by a combination of partial hydrogenation and *acid-catalyzed *isomerization.

Apart from the major procs. (e.g., *fat hardening) sh. is widely practised in the fine chemicals industry. These applications chiefly demand *regioselectivity or *chiral reduction. One of the classic problems of *chemoselectivity, i.e., the selective reduction of the C=C bond in unsaturated aldehydes, is partly solved by using Ir or Pt cats., the latter being preferably promoted by Fe or Sn. Another problem has been the reduction of chloronitrobenzenes to the corresponding anilines, without effecting *hydrogenolysis of the C-Cl bond: the *p*-substituted reactant can be selectively reduced by Ru cats. and by $Pt-TiO_2$ formed under conds. that result in the *strong metal-support interaction (*SMSI) taking place. Sh. is also important in the *petroleum processing industry (e.g. *BP selective hydrog. proc., *Gulf specialty prods. hydrog., *UOP Hydrobon proc.). **F** hydrogénation sélectif; **G** selektive Hydrierung.   G.C. BOND

Ref.: Augustine; Rylander; Werner/Schreier; *Catal. Rev.* **1998**, *40*, 81.

selective oxidation   → partial oxidation

selective poisoning   → deactivation

selectivity 1-general and het. cat: S. provides comparison between amounts of desired and undesired prods. formed in a complex reaction. The phenomenon may arise in several different ways. If the reaction mixture contains two or more reducible molecules, each containing a single reducible function, one may be reduced faster than, or even to the complete exclusion of, the other, i.e., selectively. The degree of s. depends on how much more of the *surface it occupies than the other, by reason of its stronger *adsorption. If a single reactant contains two or more reducible functions, which may be the same or different, the extent to which one is reduced in preference to another determines the s. with which a particular prod. is formed, and this also depends upon the strength of attachment of each function to the surface. An example of this is the *reduction of crotonaldehyde (but-2-enal, $CH_3-CH=CH-CHO$) either to crotyl alcohol (but-2-enol, $CH_3-CH=CH-CH_2OH$) or to *n-butyraldehyde ($CH_3-CH_2-CH_2-CHO$). The term *chemoselectivity is applied to this type of selective reaction; if however there are two similar reducible functions, the selective reduction of one of them is termed *regioselectivity. The third aspect of s. is when a molecule contains a triple bond, e.g., C≡C, which may be reduced in two stages; the intermediate stage, i.e., reduction to the C=C double bond, may in certain circumstances be achieved with high s. The same

applies to the reduction of conjugated dienes, e.g., buta-1,3-diene. Both of these situations have important practical applications (cf. *asymmetric cat., *stereo- or *regioselectivity).

S. of het. cats. strongly depends on their *chemisorption properties (dissociative/nondissociative, preference in competitive adsorption), since this step activates the reactants for the cat. reaction. Other factors influencing s. are the *redox potential of *surface oxygen in *selective oxidation, or spatial restrictions around a site. Thus, cat. conv. of CO with H_2 strongly depends on the ability of the cat. to dissociate the C-O bond: *hcs. will result if it is cleaved (Fe, Co: *Fischer-Tropsch synth., Ni; *methanation), methanol is obtained if the C-O bond is activated without cleavage (Cu), higher alcohols may be formed on surfaces stabilizing both dissociated and undissociated forms (Rh: oxygenate synth.). In regard to *shape selectivities of *microporous systems or *zeolites cf. *reactant s.; *product s., and *transition state s.

G.C. BOND/M. MUHLER

2-hom. cat.: Since most reaction systems comprise sequential steps, ss. change with reaction progress, and ss. for intermediate prods. decay at higher convs. Therefore, any comparison of ss. is meaningful only at identical conv. Ideal s. may be obtained with cats. exposing to the reaction mixture only *one* suitable active site, as is known from *enzyme reactions or *hom. cat. reactions. This situation is highly unlikely with *het. cats.   High s. is achieved if possible reaction steps combine to reaction paths providing exceptional reaction rates. In cat. design, this may be assisted by selectively poisoning sites that catalyze competing reactions.

M. MUHLER

3-enzyme cat.: *Enzymes are widely recognized for their ability to discriminate between even closely related substances. Such s. results from the ability of the enzyme to provide a binding site that is complementary to the substrate with regard to shape and chemistry (*electrostatic, *hydrogen bonding).   P.S. SEARS

4: differences of reactant s./transition state s./product s.: → zeolites. **F** séléctivité; **G** Selektivität.

Ref.: Davis/Suib; Wijngaarden/Kronberg/Westerterp; Thomas; Augustine, p. 315; P.L. Silverstone, *Composition Modulation of Catalytic Reactors*, Gordon and Breach, London 1998; Ullmann, Vol. 5, p. 322.

Selectoforming process → Mobil procs.

Selectopol process → IFP procs.

selenocysteine → selenosubtilisin, *enzymes

selenoenzymes → glutathione

selenium as catalyst metal Se has been proposed as a cat. metal in *ARCO's ethylurethane (and MDI) proc. (but never applied because of toxic cat. *intermediates), in *BP/Distillers/Ugine proc. to acrylonitrile (over Se-CuO cat.), as *WGSR cats., for the *reduction of nitroarenes, for the *oxidation of aldehydes to carboxylic acids, and for various *hydroxylation variants (*Bohn-Schmidt anthraquinone reaction). Se is a constituent of different *enzymes or *antibodies (*cytochrome P 450, *glutathione peroxidase, *selenosubtilisin). **F** sélénium comme métal catalytique; **G** Selen als Katalysatormetall.

B. CORNILS

Ref.: Parshall/Ittel, p. 114; Cornils/Herrmann-1, pp. 425, 960; *Organic Reactions*, Vol. 44, p. 1.

selenosubtilisin a modified form of the serine *protease *subtilisin in which the active-site nucleophile, serine, has been converted to selenocysteine. The resulting *enzyme can be used for *kinetically controlled peptide synthesis of amino acid (or peptide) esters. It has an excellent *aminolysis/*hydrolysis ratio. The ratio of the second-order rate constants (k_{RNH2}/k_{H2O}) for deacylation of the cinnamoyl enzyme with glycineamide is 27 000 for the seleno-enzyme, as compared to 7400 for the thiol-enzyme and 19 for the serine (wild-type) enzyme. The enzyme is very easily oxidized, however, and requires

active esters to form the acyl intermediate. **E=F=G**.

P.S. SEARS

Ref.: *J.Am.Chem.Soc.* **1989**, *111*, 4513.

self-assembly → self-organization

self-organization This term, often used synonymously with self-assembly, describes the spontaneous non-covalent formation of higher-order structures (often in the *nanoscale range) by molecules or ions that adjust their own positions to reach a thermodynamic minimum by weak, less directional forces such as *hydrogen bonds, ionic bonds, and *van der Waals interactions. The properties of the aggregates differ from those of the comps. but are determined by their nature and positioning. Stability is only ensured if many weak interactions, which are several orders of magnitude weaker than covalent forces, overcome the unfavorable entropy loss caused by the assembly of many individual molecules. Self-organizing synth. is used, e.g., for infinite lattices such as molecular crystals, liquid crystals, *colloids, and *micelles, but also for non-polymeric multicomp. supramolecular complexes (*supramolecular templates for catalysis) of a definite size and molecular composition (self-assembly, e.g., molecular grids, tubes, etc.). Biological precedents for so. are the proc. for protein folding (*chaperones) and the formation of protein aggregates and *enzyme complexes, which result in a higher cat. activity. Furthermore, many organometallic cats. and the preorganization of functional groups in their reactions as well as the *template effect are based on so. phenomena. **F** auto-organisation; **G** Selbstorganisation.

C. SEEL

Ref.: *Science* **1991**, *254*, 1312; J.L. Atwood, J.E.D. Davies, D.D. MacNicol, F. Vögtle, *Comprehensive Supramolecular Chemistry*, Vol. 9, Pergamon Press, Oxford 1996.

self-replication the *template-directed *autocatalytic reproduction of a template molecule upon self-assembly of ternary or higher-order supramolecular *complexes that consist of a self-complementary template C and (at least) two building blocks A and B with reactive groups, which are favorably preorganized by the complexation and react to form a new (identical) template molecule. The generic scheme of sr. is $A + B + C \rightarrow [A \cdot B \cdot C] \rightarrow C + C$. Characteristic of sr. is the dependence of the initial reaction rate of the square root of the initial matrix conc. (square-root law) and a sigmoidal prod. growth. A problem is the occurrence of *product inhibition if the dimer of the template is stable enough to compete with the formation of the complex of template and building blocks. Sr. is of interest in view of the relationship to the origin of life and to prebiotic and living systems in general, e.g., bacterial cells generate new identical cells and are completely replicated. Sr. systems employ the *condensation of trinucleotides in the presence of palindromic hexanucleotide matrices, association via amidinium-carboxylate salt bridges or adenine-imide pairing and *Schiff base or amide condensation. **F** auto-replication; **G** Selbstreplizierung.

C. SEEL, F. VÖGTLE

Ref.: *Angew.Chem.Int.Ed.Engl.* **1986**, *25*, 932 and **1992**, *31*, 1032; *Nature (London)* **1994**, *369*, 184 and **1996**, *382*, 525; *Chem.Eng.News* **1998**, Dec.7, 40.

SEM → scanning electron microscopy

Semenov Nikolai N. (1896–1986), professor of physical chemistry in Moscow (Russia). Worked on *kinetics, *radicals, and *chain reactions. Nobel laureate in 1965 (together with Hinshelwood).   B. CORNILS
Ref.: Neufeldt.

semiconductors are materials with electric conductivity between that of a metal and that of an insulator. In the *band model an *intrinsic* s. results if the *band gap between a fully occupied *valence band and a completely empty *conduction band is of the order of kT so that the thermal energy is high enough to promote some electrons from the valence band into the originally empty conduction band. Thus a small electric conductivity is produced which increases with rising temp. For an *extrinsic* s. the conductivity results

from impurities (*dopants) whose energy states lie in the band gap and which accept or donate electrons, thus also leading to partially filled bands. In contrast to metals, typical elemental ss. like Si or Ge exhibit covalent bonds between atoms. On *surfaces these covalent bonds are broken (dangling bonds) often leading to complex reconstructions, e.g., Si(111)- (7 × 7). If charges are present on an s. surface *band bending occurs. S. surfaces are, with the exception of some metal oxides (cf. *oxidation, het. catalyzed), less important in het. cat. They play a dominant role, however, in *photocatalysis (*photoinduced oxidation/ reduction of water) and photoelectrolysis. **F** semiconducteurs; **G** Halbleiter.   R. IMBIHL

Ref.: H. Lüth, *Surfaces and Interfaces of Solid Materials*, Springer, Berlin 1995; Thomas/Thomas, p. 392; Ullmann, Vol. A5, p. 319.

semi-empirical methods The most important approximation for solving the *Schrödinger equ. in *ab-initio calcns. is the *Hartree-Fock method, which leads to a set of one-electron equs. The computationally most difficult and time-consuming part of solving the Hartree-Fock equs. is the calcn. of the large number of Coulomb and exchange integrals which describe the electron-electron interactions. Sms. avoid the calcn. of the integrals by neglecting integrals between orbitals which have small *overlaps, and by approximating the remaining integrals by experimental values or by multipole expansion of the electronic charge interactions. This can be done at different levels of approximation such as CNDO (complete neglect of differential overlap), INDO (intermediate neglect of differential overlap), or NDDO (neglect of diatomic differential overlap). The basic idea of sms. is to reduce the calcn. of integrals in the *Hartree-Fock calcns. as much as possible by replacing them with empirical values or with simple mathematical expressions. Parameters are then added to the pseudo Hartree-Fock equs., which are fitted to experimental values such as heats of formation, geometries, ionization potentials, and dipole moments. Dewar developed a series of popular programs (cf. MINDO, MNDO, AM1). Later versions with slightly different parametrization are also available. The most widely used program is MNDO, which has parameters for more than 40 atoms. Semi-empirical calcns. of a cpd. are only possible if parameters are available for all atoms in the molecule. Semi-empirical calcns. give also information about the electronic structure of a molecule, because they are still approximate solutions to the Schrödinger equ. Sms. are much faster than ab-initio methods or *DFT, and they can give better results than Hartree-Fock calcns., because the parametrization induces the implicit inclusion of *correlation energy. However, good results can only be expected for classes of cpds. which have been included in the set of reference molecules for parameter optimization. **G** halbempirische Methoden. G. FRENKING

semisynthetic enzymes *enzymes that have been chemically modified at catalytic groups in the *active site in order to modify their behavior. Examples are *seleno- and *thiosubtilisin and *methylchymotrypsin. **F** enzymes semisynthétiques; **G** halbsynthetische Enzyme. P.S. SEARS

sensors → chemical sensors

separation of catalysts 1-het. cat.: The separation between cat. and reaction prods. in het. catalysis depends on the proc. applied. Fixed-bed operation is generally problem-free; in a few cases in the start-up period a polishing filter can be necessary to remove some fines emerging from the filling procedure. In fluid-bed procs. the majority of the cat. remains inside the reactor, but some fines generated by *attrition have to be collected (cyclone, filter bags or candles, sintered metal filters). Entrained-bed procs. use a comb. of cyclones and fines collection to separate the cat. In *slurry procs. the conc. of the cat. and the properties of the product determine the mode of separation (centrifuge, decanter,

Sparkler filters, filter press, sinter metal filter). A special case is Raney Ni, which separates (without mixing) by gravity and frequently is left inside the reactor.
2-hom. cat.: In hom. cat. procs. the separation of the cat. can be the crucial step (exception: *two-phase catalysis). The most important procedures are distillation (*hydroformylation; *carbonylation to *AA), *extraction, or change of *phases (*olefin oligomerization) and destruction (*Ziegler-Natta polymerization; *olefin oligomerization). In some cases the conc. of the cat. is very low and can be tolerated in the product (olefin polymerization with *single-site or *high-mileage catalysts). **F** séparation des catalyseurs; **G** Katalysatorabtrennung. C.D. FROHNING

separation of racemates → resolution, *kinetic resolution

sequential precipitation In forming a cat. *precursor by *precipitation it is desirable to precipitate the single cpds. sequentially, e.g., the *support firstly followed by other constituents which are deposited on the preformed carrier. The procedure makes it possible to start with one solution only for several ingredients, provided the properties of the precursor(s) yield an appreciable cat. Sp. is especially simple if a single anion can be applied, as in the case of some hydroxides. When the solubility constants of the precipitates are known, the sequence of precipitation can be easily determined (Table).

Cation	Precipitate	Solubility constant pK_a
Al^{3+}	$Al(OH)_3$	34.3
Cu^{2+}	$Cu(OH)_2$	19.75
Zn^{2+}	$Zn(OH)_2$	16.75
Ni^{2+}	$Ni(OH)_2$	13.8

In the case of a Cu-Zn-*alumina cat. (for *methanol synth.), upon addition of caustic to an acidic solution of the corresponding nitrates, alumina is precipitated first, followed by Cu and then Zn hydroxides. In the same way Ni can be deposited onto *alumina. The principle can also be applied for different precipitating anions when the solubility constants of the precipitates differ sufficiently. **F** précipitation sequentiel; **G** Sequentielle Fällung.

C.D. FROHNING

serine hydroxymethyltransferase → threonine aldolase

serine protease Serine proteases are characterized by a *cat. triad in the *active site composed of Asp, His, and Ser, with Ser acting as a nucleophile. The reaction proceeds through an initial step, in which serine is acylated; this intermediate is deacylated with water in a second distinct cat. step. In the *hydrolysis of peptide bonds, the acylation is often rate-limiting; in the hydrolysis of esters, deacylation of the acylated *enzyme is rate-limiting. Examples of serine proteases include *trypsin, *chymotrypsin, *subtilisin, *α-lytic enzymes, *elastase, and *V8 protease. The serine-type *esterases include *pig liver esterase, *cholesterol esterase, *acetylcholine esterase, etc. **E=F=G**.

P.S. SEARS

SERS surface enhanced Raman spectroscopy, → laser Raman spectroscopy

SET → single-electron transfer

severity another (industrial) term describing conversion, especially paraphrasing the difficulty of determining conv. of the very different comps. of *cat. cracking. S. is defined, inter alia, as coil outlet temps., propylene to methane ratio, or by a severity index. High s. means high conv. and mainly secondary reactions; low s. means low conv. and primary reactions. S. and cat. are connected by the fact that the expected degree of cracking requires higher temps. and short residence times which, in turn, influence *coking and thus the cat.'s *lifetime. **E=F=G**.

B. CORNILS

Ref.: Kirk/Othmer-1, Vol. 9, p. 884.

SEXAFS → surface-sensitive extended X-ray absorption fine structure

Seyferth's reagent is phenyl(tribromomethyl)mercury for dihalocarbene transfer acc. to the Figure. **F** réagent de Seyferth; **G** Seyferths Reagenz.

B. CORNILS

Ref.: *Acc.Chem.Res.* **1972**, 5, 65.

SFC supercritical fluid, → supercritical phase catalysis

SFG → sum frequency generation

shallow-bed reactor → *metal gauze reactor, *cat. reactors

shape of catalysts → shaping

shape selectivity Shape-selective cat. can occur when het. cat. reactions are influenced by geometric constraints of the solid cat. The archetypical ss. cats. are *zeolite-type materials. These are *microporous solids with a crystalline structure. Their *pore sizes are thus well defined and are in the size range of small organic molecules. The geometric limitations may act either upon the *sorption of reactants, on the *transition state (or the *intermediate) of a reaction to be catalyzed, or on the *desorption of prods. Correspondingly, a distinction is to be made between *reactant selectivity, *restricted transition-state selectivity, and *product selectivity. An example of reactant selectivity is the *dehydration of an n-butanol/iso-butanol mixture over a zeolite A cat. As a consequence of the larger kinetic diameter of the branched alcohol, it cannot enter the zeolitic pore system and is excluded from the cat. conv. to the alkene, whereas the linear alcohol is dehydrated. This principle of size exclusion is also often used in separation procs. As vibrations can influence the effective width of zeolite pores, size exclusion procs. depend upon temp.

An example of ss. controlled by the size requirements of the transition state is the *transalkylation of *m*-xylene. Carried out over an *MFI-type zeolite, this reaction leads exclusively to 1,2,4-trimethylbenzene (this reaction path runs via a "slim" transition state), with no 1,3,5-trimethylbenzene (this reaction path involves a bulkier intermediate) being formed. An example of *product selectivity is the methylation of toluene. Although the substitution reaction initially leads mainly to *o*- and *p*-xylene, intrazeolitic *isomerization of the xylenes yields a mixture of all the isomers, including the *m*-isomer. This mixture may reach a stationary equil. As the *p*-isomer has the smallest kinetic diameter, it can most easily diffuse out of the zeolite framework. Correspondingly, the quasi equil. of the xylenes in the zeolite pores is disturbed, and the more slowly diffusing *o*- and *m*-xylene molecules isomerize inside the zeolite to *p*-xylene. Consequently, *p*-xylene is the main (or only) product, in strong contrast to the prod. distribution which is obtained when the reaction is carried without a shape-selective cat.

In order to obtain strong shape-selective effects, the pore size and geometry have to be adapted carefully to the size of the reactant, the transition state, or the prod. of the reaction to be catalyzed. In this regard, the variety of different zeolite pore structures is helpful. The extent of ss. can depend on the particle size of the het. cat. This is the case when the outer *surface of a uniform cat. is cat. active, too. Larger particles (with smaller total outer surface) then show stronger ss.   P. BEHRENS

Ref.: Ertl/Knözinger/Weitkamp, Vol. 3, p. 1243 and Vol. 4, p. 1963; Thomas/Thomas, p. 614.

shaping   A het. cat. requires optimal contact with its reactants which belong to a different *phase than the cat. Thus transport of gases or liquids to and through solids is of the utmost importance for a stable and efficient operation of any cat. system. To control the *mass transport through a macroscopic cat. bed an accurate design is required of the physical shape of a cat. particle, which is never a loose powder. Considerations of *residence time at given throughput, press. drop and contact between cat. particles for energy transport are important for the choice of the shape of a cat. Well-developed theoretical concepts for the design of the macropore system of a cat. bed exist in chemical engineering sciences. Boundary conds. for the design are the technologies available for forming a given material which are frequently limited by constraints of chemical reactivity of the active mass with auxiliary agents and temp./press. conds. of the shaping proc. Technologies from ceramics and plastic part manuf. are adopted for cat. shaping which often require specific modification, rendering this practically important group of technologies a key know-how factor in cat. manuf. procs. Besides individual particle shapes, complex 3D forms such as monolithic arrays of tubes with complex shapes or platelets with well-defined holes or *cavities, are examples of shaping which greatly surpass the formation of extrudates or split sieved representing the simplest techniques of shaping (cf. *catalyst forms, *forming). **F** mise en forme; **G** Formgebung.

R. SCHLÖGL

Sharpless reagent/Sharpless epoxidation   The S. reagent, a mixture of Ti(IV) isopropoxide and a *chiral tartrate ester (the dimethyl, diethyl, or diisopropyl ester) which generates a Ti(IV) tartrate *complex, is used   for the *asymmetric *epoxidation of allylic alcohols. The Se. is performed by subsequently adding *tert*-butyl hydroperoxide (*TBHP) and the allylic alcohol to the S. reagent at –20 °C. The Se. is widely applicable with good yields (>90 %) and with high enantioselectivities (>90 %). The S. reagent is compatible with many functional groups such as acetal, alkyne, amide, ether, epoxide, ketone, ester, nitro, etc. **F** réagent/époxidation de Sharpless; **G** Sharpless Reagenz/Epoxidation.

H. NEUMANN, M. BELLER

Ref.: Beller/Bolm, Vol. 2, p. 261; *Chem.Commun.* **1990**, *19*, 1364; *J.Am.Chem.Soc.* **1984**, *106*, 8188; *Tetrahedron* **1987**, *43*, 5135; *Organic Reactions*, Vol. 48, p. 1.

Shell acrolein process   First proc. for the manuf. of acrolein by gaseous-phase partial *oxidation of propylene, originally over Cu_2O-SiC cat. and J_2 as promoter at 350–400 °C. Today multicomp. cats. with Bi, Sn, Mo, Sb are usual. **F** procédé Shell pour l'acroléine; **G** Shell Acrolein-Verfahren. B. CORNILS

Shell Bextol process   *hydrodealkylation proc. for the manuf. of benzene from toluene by cat. over supported Cr_2O_3, Mo_2O_3, CoO, or Rh-Al_2O_3 at 500–600 °C/3–5 MPa. The coke *deposition of the cats. requires simultaneous *regeneration. **F** procédé Bextol de Shell; **G** Shell Bextol-Verfahren. B. CORNILS

Shell ethyl alcohol process   for the cat. *hydration of ethylene over acid cats. such as H_3PO_4-SiO_2 at 300 °C/7 MPa. Short residence times minimize the formation of ethylene oligomers and of diethyl ether. **F** procédé Shell pour l'éthanol; **G** Shell Ethanol-Verfahren.
                                                              B. CORNILS

Shell ethylene oxide process   market-leading proc. for the manuf. of *EO by partial *oxidation of ethylene with oxygen over doped Ag cat. at 180–250 °C/1–2 MPa/2000–4000 *GSVH. At low ethylene conv. rates (10 %) *selectivities of >82 % are reached. **F** procédé Shell pour l'EO; **G** Shell EO-Verfahren.   B. CORNILS
Ref.: *Hydrocarb.Proc.* **1977**, *56*(11), 159; Ullmann, Vol. A10, p. 101.

Shell FEAST process   an extension of the *metathesis step of the SHOP proc. in respect of the technology (fixed-bed reactors; regeneration in alternation), the procedure (metathesis of ethylene and cyclolefins such as *COD), and the prod. range (1,5-hexadiene, 1,9-decadiene). The FEAST (Further Exploitation of Advanced Shell Technology) proc. is run on Re_2O_7-Al_2O_3 at <100 °C/<0.2 MPa. **F** procédé FEAST de Shell; **G** Shell FEAST-Verfahren.   B. CORNILS
Ref.: *Chem.Engng.* **1987**, (8), 22; *Appl.Catal.* **1994**, *115*, 182.

Shell glycerol process   a two-step proc. for the manuf. of glycerol from allyl alcohol by *bishydroxylation with H_2O_2 via glycidol (at 60–80 °C over $NaHWO_4$) and its *saponification. **F** procédé Shell pour le glycérol; **G** Shell Glyzerin-Verfahren.   B. CORNILS

Shell Hycon process   for residual hydroconversion in two steps, 1: the *demetallization with SiO_2 cats., 2: *desulfurization with fixed-bed Ni-Mo/Al_2O_3 cats. at 400 °C/15–20 MPa. The degree of S removal is said to be 65–70 %. **F** procédé Hycon de Shell; **G** Shell Hycon-Verfahren.   B. CORNILS
Ref.: *Appl.Catalysis* **1994**, *115*, 178; Ullmann, Vol. A18, p. 75.

Shell Hysomer process   for the upgrading of the octane level of light gasoline fractions by vapor-phase fixed-bed *hydroisomerization of (preferably) C_5/C_6 *hydrotreated feedstocks over *dual-function cats. (Pt on H-*mordenite), and *isomerization of paraffins to branched alkanes. **F** procédé Hysomer de Shell; **G** Shell Hysomer-Verfahren.   B. CORNILS
Ref.: *Hydrocarb.Proc.* **1978**, (9), 171; Ertl/Knözinger/Weitkamp, Vol. 4, p. 2011.

Shell-Linde MMA process   for the manuf. of methyl methacrylate (MMA) by carbonylation of propyne (methylacetylene) with a Pd cat., *ligand-modified by 2-pyridyldiphenylphosphine acc. to

$$H_3C\text{-}C{\equiv}C\text{-}H + CO + CH_3OH \rightarrow$$

$$H_2C{=}C\text{-}COOCH_3$$
$$|$$
$$CH_3$$

The proc. uses byproduct methylacetylene from *cat. cracking sources. **F** procédé de Shell-Linde pour le MMA; **G** Shell-Linde-MMA-Verfahren.   B. CORNILS

Ref.. Cornils/Herrmann-1, p. 1120; *Hydrocarb. Eng.* **1997**, *1*, 41; *Recl. Trav. Chim. Pays-Bas* **1996**, *115*, 248.

Shell liquid-phase *isomerization process

older proc. for the upgrading of light naphtha (the conv. of methylcyclopentane to cyclohexane) in the presence of $AlCl_3/HCl$. **F** procédé d'isomérisation de Shell; **G** Shell Isomerisierungs-Verfahren.   **B. CORNILS**

Ref.: *Hydrocarb. Proc. and Petr. Refiner* **1961**, *40*(9), 171 and **1963**, *42*(7), 125.

Shell oxo process

the first economically realized *hydroformylation proc. with *ligand-modified Co cats. (modifier: *n*-butyl, alkyl or, in particular, special cycloalkylphosphines). Feedstocks are C_7–C_{14} alkenes which react at 180–200 °C/8 MPa under addition of CO and H_2 and under influence of the hydrogenating properties of the cat. system to mainly alcohols. This influence is also responsible for a 20 % portion of hcs. formation from the alkenes. The *n/iso* ratio is up to 90 %. **F** procédé oxo de Shell; **G** Shell Oxoverfahren.   **B. CORNILS**

Ref: Falbe-1, p. 167; Falbe-3, p. 19; Cornils/Herrmann-1, p. 74.

Shell Residual HDS process

for the cat. treatment of distillates (*HDS, asphaltenes, *HDM, and lowering of the viscosity) in trickle-bed operation for their feeding to *FCC units. The run lengths of the cat. are said to be up to 1 year. **F** procédé Residual HDS de Shell; **G** Shell Residual HDS-Verfahren.   **B. CORNILS**

Ref.: *Hydrocarb. Proc.* **1978**, (9), 135 Ullmann, Vol. A18, p. 74.

Shell SCOT process

removal of sulfur cpds. such as CS_2 and COS from S-plant tail gas by reaction cat. under reducing cond. to H_2S. A combination of *Claus and SCOT recovers >99.7 % of saleable sulfur. **F** procédé SCOT de Shell; **G** Shell SCOT-Verfahren.   **B. CORNILS**

Ref.: *Hydrocarb. Proc.* **1996**, (4), 136.

Shell SHOP process

(Shell Higher Olefin Process) The first commercial cat. proc. making use of the *two-phase technology was developed to meet the market needs for linear α-olefins. It is a combination of ethylene *oligomerization, *isomerization, and *metathesis. The ethylene oligomerization is cat. by a Ni complex containing a P-O chelating ligand which is dissolved in 1,4-butanediol, the nonpolar prods., the α-olefins, being almost insoluble in this diol. Thus, the cat. solution can easily be separated from the insoluble prods. in a high-press. separator. In a series of distillation towers, the different α-olefin cuts are isolated. The desired alkenes are those with C_4–C_{18}. The lower and the C_{18+} fractions are combined to be isomerized to internal linear alkenes by a cat. such as Na/K on Al_2O_3 or a MgO in the liquid phase. In this reaction, 90 % of the α-olefins are converted to internal alkenes which are subsequently subject to a *metathesis reaction. Metathesis of the lower and the higher internal alkenes gives a mixture of alkenes with odd and even carbon chain lengths. The mixture comprises about 11–15 % of the desired C_{11}–C_{14} linear internal alkenes, which are separated by distillation (Figure).

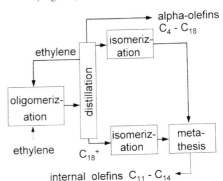

Different possibilities for producing alkenes with the desired chain length and double bond position make the SHOP proc. a very elegant and flexible process. **F** procédé SHOP de Shell; **G** Shell SHOP-Verfahren.

B. DRIESSEN-HÖLSCHER

Ref.: Cornils/Herrmann-1, Vol. 1, p. 245.

Shell SMDS (Middle Distillate Synthesis) process a variant of the *Fischer-Tropsch synthesis (FTS) for countries with large gas deposits instead of mineral oil. Principal prods. are middle distillates (bp 150–400 °C, C_9–C_{25}), i.e., kerosene and gas oil. Shell's first commercial plant (500 000 t/a capacity) was started in Bintulu (Sarawak, Eastern Malaysia). The SMDS proc. is a two-stage proc. In the first stage (*HPS, heavy paraffin synthesis) syngas, generated from natural gas by *partial oxidation with O_2, is converted into paraffins via FTS. In the second stage the liquid *hcs. prod. is *hydrotreated yielding *LPG, naphtha, and predominantly kerosene and gas oil.

The HPS step is carried out in water-cooled multitubular reactors (*ARGE type) in the presence of a Zr-promoted Co cat. on SiO_2 spheres as *support. The cat. has been specially developed for high selectivity to long-chain hcs. at 230 °C/2.6–3 MPa at CO convs. of >95 %. The liquid hcs. are hydrotreated (*hydrocracked) in the second stage in the presence of a proprietary Shell cat. (Zr-promoted Co, 350 °C/4 MPa) leding to LPG (15–25 %), kerosene (25–50 %), and gas oil (60–25 %). **F** procédé SMDS de Shell; **G** Shell SMDS-Verfahren.   C.D. FROHNING

Ref.: ChemSystems PERP Reports 92S12, 85T3; *Catal. Today* **1991**, 8; *Appl.Catal.A:* **1999**, *186*, 27; van Santen/van Leeuwen/Moulijn/Averill.

Shell smoke point improvement converts aromatics into naphthenes for improvement of luminometer number (smoke point) by semi-adiabatic, fixed-bed, trickle operation over noble metal cats. on special *carriers.

Ref.: *Hydrocarb.Proc.* **1972**, (9), 178.   B. CORNILS

Shell Styrene Monomer Propylene Oxide (SMPO) process for the coproduction of *propylene oxide and styrene monomer by combined cat. *epoxidation of propylene and co-oxidation of the auxiliary cpd. (*reaction carrier) *ethylbenzene (*ARCO PO proc., *Oxirane proc.). In the epoxidation step Shell uses a het. system based on V, W,

Mo, or Ti cpds. on SiO_2. **F** procédé SMPO de Shell; **G** Shell SMPO-Verfahren.   B. CORNILS

Shell vapor-phase butane isomerization process an older proc. for the conversion of *n*-butane to isobutane over $AlCl_3$ on *bauxite (similar to *Phillips cat. isomerization proc.) at 100–140 °C/1–2 MPa in a fixed-bed vapor-phase operation. **F** procédé de Shell pour l'isomérisation en phase gazeuse; **G** Shell Gassphasen-Butanisomerisierung.

Ref.: *Petr.Refiner* **1952**, *31*(9), 168.   B. CORNILS

Shell Versatic acid process for the manuf. of highly branched carboxylic acids via *Koch synthesis (*hydroxycarboxylation), reacting alkenes with CO and water at 80–100 °C/2–10 MPa over strong acidic cats. such as sulfuric or phosphoric acid, promoted with HF, BF_3, or SbF_5 (similar: *Exxon Neo Acid proc.). **F** procédé Shell pour l'acide Versatic; **G** Shell Versaticsäure-Verfahren.   B. CORNILS

Ref.: Falbe-1, p. 372; Winnacker/Küchler, Vol. 6, p. 103.

SHG → second harmonics generation

shift conversion → water-gas shift reaction (WGSR)

Shilov reaction a *C-H activation with nucleophilic substitution (Figure).
The Pt^0 is reoxidized to Pt^{2+}. **F** réaction de Shilov; **G** Shilov-Reaktion.   W.A. HERRMANN

$$R\!-\!H + Pt^{2+} \longrightarrow R\!-\!Pt^+ + H^+$$
$$R\!-\!Pt^+ + Nu^- \longrightarrow R\!-\!Nu + Pt^0$$
$$Pt^0 \xrightarrow{\;Ox\;} Pt^{2+}$$

ship-in-the-bottle catalysts discrete metal *complexes encapsulated in the (super) cages or *pores of crystalline microporous solids such as various *zeolite species, formed via intrazeolite assembly (also called "teabag cats."). The physically entrapped metal complex is anticipated to retain many of its solution properties, while displaying "microreactor" confined reactivity behavior synergistically modified by the *shape selectivity, electrostatics, and *acid-base properties of the zeolite. Sitb. species are generated either by 1: penetration of conformationally flexible *ligands (such as *salen ligands) through the restricting windows of zeolites and combination with the metal of ion- exchanged zeolites, or 2: the *template synth., involving the introduction of the metal via ion exchange and subsequent treatment with molecular building blocks such as 1,2-dicyanobenzene or pyrrole/acetaldehyde to produce entrapped metal phthalocyanines or *porphyrin complexes (Figure).

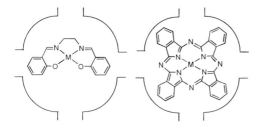

Sitb. species are believed to close the gap between het. and hom. cat. and have been shown to act as *enzyme mimics (zeozymes). Chiral variants are known, e.g., a *Jacobsen sitb. cat. So far, the success of sitb. cats. is limited. **F** Catalyseurs ship-in-the-bottle; **G** Flaschenschiff-Katalysatoren. R. ANWANDER

Ref.: *Inorg.Chim.Acta* **1985**, *100*, 135; *Inorg.Chem.* **1986**, 25, 4714; *CHEMTECH.* 1989, 542; Ertl/Knözinger/Weitkamp, Vol. I, p. 374.

SHOP process → Shell procs.

SHP process → Hüls procs.

shrinking core model → reduction in catalyst preparation

sialic acid → N-acetylneuraminate

sialic acid aldolase → N-acetylneuraminic acid aldolase

sialyltransferases generally transfer N-acetylneuraminic acid to either the 3- or 6-position of terminal Gal or GalNAc residues. Some ss. have been shown to accept CMP-NeuAc analogs that are derivatized at the 5- or 9-position of the sialic acid side chain, such as those in which the hydroxyl group at C 9 is replaced with an amino, fluoro, azido, acetamido, or benzamido group. Analogs of the acceptors Galβ1,4GlcNAc and Galβ1,3GalNAc, in which the acetamido function is replaced by an azide, phthalimide, carbamate, or pivaloyl functionality, are also substrates for the *enzymes. The newly isolated α-2,8-sialyltransferase cat. the *biosynthesis of α-2,8-linked polysialic acids which are comps. of neural cell adhesion molecules. **E=F**; **G** Silyltransferase.

C.-H. WONG

signal peptidase → proenzyme

silica Being relatively inert itself, silica is used as a *support for metals and metal *oxides with a variety of cat. applications. The commercially important ones are: silica-supported Ni, Pd, and Pt as *hydrogenation cats., s.-supported V_2O_5 which is used for SO_2 to SO_3 oxidation in *sulfuric acid synth., s.-supported Cr(II) (*Phillips cat.) for alkene *polymerization, and s.-supported Ti which is used for propene *epoxidation (*titanium silicalite). Silica-*alumina mixtures were early *solid acid cats. and find application for *isomerization reactions. S.-supported materials were among the first cats. used in *Fischer-Tropsch reactions.

S. used for cat. supports is amorphous and can be prepared in two ways; 1: *Sol-gel precipitation giving either xerogels (regular drying) or aerogels. Aerogels can be obtained through drying at high press. (sc. conds.) and require the use of an alternative solvent (usually alcohols). *Surface area, *pore diameter, *pore volume, mechanical stability, and impurities can be controlled by modification of washing (important factor: pH) and drying/hydrothermal procedures. Silica xerogels typically have *surface areas in the range 50–600 m²/g, and *mesopores between 5 and 50 nm in size. 2: *Flame hydrolysis with $SiCl_4$ as a precursor for SiO_2 (*fumed silica, e.g., *Aerosil). Very pure, spherical particles are obtained with surface areas between 25 and 400 m^2/g.

Chemical properties of supports affect the ability to attach and stabilize cat. active species on their surfaces. The point of zero charge of s. is reported to be 1.5–3.0. The surface of silica is characterized by Si-O-Si bridges and by weakly acidic (pK$_a$ ≈ 7) hydroxyl groups (often called siloxane and silanol groups). Isolated, vicinal, and geminal OH groups have been identified on s. using *IR spectroscopy. The density of OH groups on fully hydroxylated surfaces of xerogels is usually ca. 8 μmol/m^2; on fumed silica, the density is approx. 3.3 – 4.0 μmol/m^2. Consistently with these data, only cations may be adsorbed on silica surfaces over most of the pH range, and significant amounts of cations are adsorbed only above pH = 7. Silica xerogels can be dissolved at pH >9. **E=F=G**. F. JENTOFT

Ref.: *Catal.Sci. & Technol.* **1983**, *4*, 40; Farrauto/ Bartholomew, p. 63; McKetta-2, p. 1053.

silane coupling/dehydrocoupling

Homocoupling of substituted silanes and their heterocoupling with substrates such as hydrocarbons, HOR, HSR, or HNR$_2$ occurs in the presence of metal *complexes with elimination of H$_2$, and yielding either Si-Si (coupling of hydrosilanes) or Si-C bonds (silylation of alkenes, *hydrosilylation). Dehydrocoupling of primary or secondary silanes with *metallocenes (mainly of the Ti triad) as cats. ends up acc. to n H-Si(R^1,R^2)-H → H-(SiR^{1}R^2)$_n$-H + (n-1) H$_2$ with *polymers with molecular weights of 7000. **F** couplage/déhydrocouplage de silane; **G** Silankupplung/Dehydrokupplung. B. MARCINIEC

Ref.: G. Larson (Ed.), *Advances in Silicon Chemistry*, JAI Press, Greenwich 1991, Vol. 1, p. 327; Cornils/ Herrmann-1, p. 496.

silanes

(*organosilanes) homologous series based on monosilane, SiH$_4$, with the general formula Si$_n$H$_{2n+2}$. Diorganodichlorosilanes are the starting materials for the manuf. of diorganopolysiloxanes (*polysiloxanes). Dimethyldichlorosilane, the most important one, is made by *Rochow reaction. Other primary prods. formed in the proc. are MeSiCl$_3$, Me$_3$SiCl, and MeHSiCl$_2$. Trichlorosilane (HSiCl$_3$) and trialkoxysilanes (HSi(OR)$_3$, with R = Me, Et) are the basis for the prod. of *silane coupling agents by their addition to alkenes or alkynes (*hydrosilylation). They are produced commercially. **E=F=G**. B. MARCINIEC

silica-alumina strongly acid, amorphous (SiO$_2$)$_n$/(Al$_2$O$_3$)$_m$ support, important for *cracking reactions (*TCC, *FCC). **F=G=E**.

silicalite → titanium silicalite

silicoalum(in)ophosphates (SAPOs) crystalline *microporous solids situated between alumosilicate *zeolites and *alumophosphates (AlPOs). Their framework structures are built from corner-linked tetrahedra [TO$_4$], with T= Al, Si, P. In fact, SAPOs usually are nearer to AlPOs, as their typical composition of R$_a$[(Si$_x$Al$_y$P$_z$)O$_2$] shows: with $x+y+z = 1$, the mole fraction x of Si typically ranges from 0.04 to 0.20. In this formula, R is a *templating molecule or water. Like zeolites and ALPOs, SAPOs are usually produced in template-directed hydrothermal synth., and different templating molecules generate a large variety of different *pore sizes and topologies. Some of these structures also occur in the zeolite family.

The introduction of Si into P sites produces a negative framework charge which results in cation exchange and acidic properties, which are absent in pure AlPOs, but typical of *zeolites. SAPOs exhibit weak to mild acidic cat. *activity. The SAPO frameworks possess high thermal and hydrothermal stability, comparable to that of zeolites and AlPOs. SAPOs are proposed for the *UOP/Norsk Hydro MTO proc. Cf. *alumophosphates. P. BEHRENS

Sillen phases → Aurivillius phases

silox, SILOX siloxide *ligand [OSi(*t*-Bu)$_3$]$^-$

silver as catalyst metal Silver as a cat. metal is essential for two processes: for ethylene *vapor-phase oxidation to *EO (also with Au-Ag *alloys; cf. *epoxidation, *redox reactions) and for the *oxidative dehydrogenation of methanol to formaldehyde. While HCHO is preferably manufd. over bulk Ag crystals or *gauzes, EO prod. is achieved over Ag-*supported cats. (procs. of *SD, *Union Carbide, over specially doped cats.: *Shell proc.) in tubular bundle reactors. Ag_2O on *silica cats. serves in *nitrosation procs. (such as *DuPont ACN or *Toray PNC) or in *Union Carbide's sorbic acid proc. In *fuel cells the cat. active materials are Ni, Ni-Ti, Pt-Pd for the anode, and Pt on carbon, Ag, or various *perovskites or *spinels for the cathode. Ni-Ag cats. have been proposed for fat *hardening (e.g., the manuf. of shortenings). $AgPF_6$ is a *co-cat. for the Pt- or Pd-mediated synth. of vinyl esters (*DIPHOS). **F** l'argent comme métal catalytique; **G** Silber als Katalysatormetall. B. CORNILS

Ref.: *Chem.Rev.* **1994**, *94*, 857; *Coord.Chem.Rev.* **1982**, *35*, 253.

silylation introduction of silyl groups (*hydosilylation, silylcarbonylation, *silylcyanation)

silylcarbonylation describes formally the Si version of *hydroformylation in which hydrogen is replaced by hydrosilane (mainly trialkylsilane, *organosilanes). The s. of alkenes is cat. by $Co_2(CO)_8$, $RhCl(PPh_3)_3$, or by Ru carbonyls to give silyl enol ethers bearing one carbon atom more than the substrate (Equation).

In contrast to hydroformylation, the addition of CO and trialkylsilane to alkynes yields C-centered silanes exclusively, when cat. by Rh or Rh-Co carbonyl *clusters. Rh cationic and zwitterionic *complexes proved to be superior cats. for the hydroformylation of vinylsilanes (*organosilanes) giving either α- or β-silyl aldehydes, depending on the conds. The s.

of dienes is also known; the same is true for Si related *hydrocarboxylations or *hydroesterifications. **E=F**; **G** Silylcarbonylierung.

B. MARCINIEC

Ref.: Patai/Rappoport, Chapter 25, p. 181; Cornils/Herrmann-1, p. 502; *J.Organomet.Chem.* **1997**, *548*, 105.

silylcyanation addition of silyl cyanides (mainly trimethylsilyl cyanide) to aldehydes, ketones, and epoxides as well as to alkenes and alkynes, occurring in the presence of *Lewis acids or *transition metal *complexes. Trimethylsilyl cyanide is more reactive than trimethylsilyl triflate as a silylating agent (*silanes, *silylation). **E=F**; **G** Silylcyanierung. B. MARCINIEC

Ref.: Patai/Rappoport, p. 763.

Simmons-Smith reaction The SSr. is the most important application of organozinc reagents in organic synth. and an efficient method of methylene delivery to a double bond (*cyclopropanation). The classical procedure involves treatment of an alkene with a geminal diiodide and a Zn-Cu couple to produce the corresponding cyclopropane. The high-yielding reaction is *stereospecific with regard to the geometry of the alkene. $IZnCH_2I$ appears to be the intermediate species responsible for the cyclopropanation. An alternative proc. uses Zn iodide and diazomethane. The most notable variation is the Furukawa modification using diethylzinc instead of an insoluble Zn-Cu couple. This methology provides better reproducibility, greater substrate variety, and faster reactions. Metals other than Zn have also been successfully used: Al and Sm alkyls cyclopropanate alkenes in the presence of CH_2I_2; *Seyferth's reagent as well. Recent developments led to *enantioselective versions of the reaction (cf. *sulfonamides). Use of stoichiometric quantites of a chiral additive to modify the reagent has met to date with more success than purely cat. methods. **F=G=E**. A.F. NOELS

Ref.: *Synlett* **1995**, 1197; *J.Org.Chem.* **1997**, *62*, 3390.

SIMS → secondary ion mass spectrometry

simulated moving-bed reactor   reactor configuration for catalytic/chromatographic separations with simulation of counter-current flow, thus allowing rapid product separation and high convs. in both equil.-limited and low per-pass conv. reactions. **F** mobile lit fluidisé simulé: **G=E**.   B. CORNILS
Ref.: *Catal. Today* **1995**, *25*, 159.

simultaneous development   a method for the acceleration of development work by executing the various single steps (*micro, *mini and *pilot plant developent, refinement and transfer to demonstration plant, scaling-up, investigation of specifications, transfer to plant scale, etc.) simultaneously instead of sequentially (cf. *pilot plants; Figure). **F** ingénierie simultane; **G=E**.   B. CORNILS

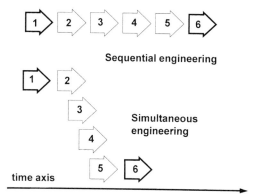

Sequential engineering

Simultaneous engineering

time axis

1 invention; 2 process synthesis; 3 development work; 4 scale-up; 5 planning/comissioning/construction; 6 production.

Sinclair HA-84 process   for the *hydrogenation of petroleum-derived benzene and alkenes in order to increase the yield of chemical raw materials at 150–250 °C/2–5 MPa over Pt or Pd on inert *supports. **F** procédé HA-84 de Sinclair; **G** Sinclair HA-84-Verfahren.
Ref.: Winnacker/Küchler, Vol. 5, p. 224.   B. CORNILS

single-chain antibodies   → antigen bonding fragments

single crystals   are solids with uniform crystallographic orientation. By cutting an sc. rod along a certain crystallographic direction, an sc. *surface with uniform and well defined structure is obtained. The orientation of an sc. surface is defined via the Miller indices (*hkl*); cf. *crystallography of surfaces. Low-index planes (100, 110, 111) which are less rough should be distinguished from high-index planes which may contain atomic steps and kinks as structural elements (cf. Figure for the different types of reactive sites -corner, edge, and face [planar] atoms).

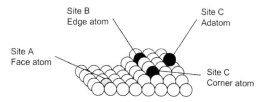

Site B
Edge atom

Site C
Adatom

Site A
Face atom

Site C
Corner atom

[by courtesey reproduced from Augustine, Marcel Dekker, Inc., Figure 3.8]

These high-index planes often may not be structurally stable but may undergo faceting. The use of sc. surfaces is at the heart of the so-called surface science approach to het. catalysis because this type of approach offers two fundamental advantages: conceptually, the influence of surface structure can be investigated systematically and experimentally, and diffraction methods like *LEED are applicable to study phenomena such as reconstruction or the formation of ordered overlayers of adsorbates. **F** monocrystal; **G** Einkristall.   R. IMBIHL
Ref.: G. A. Somorjai; Ertl/Knözinger/Weitkamp, Vol. 2, p. 771; A.M. Kossevich, *The Crystal Lattice*, Wiley-VCH, Weinheim 1999.

single-electron transfer (SET)   is involved in bioprocs.(through σ-bonds), *solid-state phenomena (conductors, *semiconductors, *solid-state reactions, superconductors), solution chemistry, electrochemistry, pulse radiolysis, photochemistry, or cat. (*Wacker proc.). The thermodynamics of electron transfer are given by the Rehm-Weller equ. and depend not only on the standard *redox potentials of two reactants but also on the electrostatic factor. The *kinetics are given by the Marcus equ.: the electron transfer is fast if the thermodynamics are favorable and slow or non-existent if the thermodynamics are not favorable. Chemical reagents can be designed as electron reservoirs in which the redox centre (e.g., Fe) is sterically well protected from the outside. The 19-electron *complexes $[Fe(\eta^5-C_5R_5)(\eta^6-C_6Me_6)]$ (R = H, Me), the most electron-rich neutral molecules, are universal reductants which can effect a variety of single-electron reduction procs. Intramolecular electron transfer between two stable redox centers of a molecule may facilitate charge separation if it is induced photochemically or may define trapped mixed-valence cpds. if it is reversible. SET is discussed for the mechanism of *Grignard reactions. **E=F=G**.

D. ASTRUC

Ref.: Astruc; *Biochem.Biophys.Acta* **1985**, *811*, 265; *Angew.Chem.Int.Ed.Engl.* **1988**, 27, 643; *J.Am.Chem. Soc.* **1990**, *112*, 2420; *Comprehensive Coordination Chemistry*, Pergamon, Oxford 1987, Chapter 7.2.

single-site catalysts cats. which behave in a uniform manner to produce simple statistical distributions of molecular mass. Because the original cats. were bis(cyclopentadienyl) *transition metal cpds., known as *metallocenes, there is a tendency to refer to these recent developements as "metallocene cat.". The more general term is single-site catalysts. The key discovery was the use of *methyl-aluminoxane (MAO) as a *co-catalyst. It is generally accepted that the active species is a transition metal cation associated with an aluminoxane counter-ion. An example is Cp_2ZrCl_2 activated with MAO giving the active species. Sscs. have also been made which have only one cyclopentadienyl *ligand and a fluorenyl, indenyl, or an amino group, e.g., $Me_2Si(Ind)_2ZrCl_2$ or $Me_2Si(Cp)(NR)TiCl_2$. **E=F=G**. W. KAMINSKY, C. SCHWECKE

Ref.: Ullmann, 6th Ed.; van Santen/van Leeuwen.

single turnover (STO) The correlation of cat. *activity with the number and type of

*active sites for het. cats. is difficult because the accuracy of the determination of the active sites (with activity differences) per unit of cat. is contested. To overcome this, het. cats. are used experimentally (by pulsing the reactants) in such a way that a 1:1 site/product ratio is achieved. Thus, a number of "direct saturation sites" can be determined and used to circumvent the determination of *TONs in het. cat. **E=F=G**. B. CORNILS

Ref.: Augustine, p. 41.

sintering welding together of particles by applying heat below the melting point. S. is used in alloy manuf., steel prod., and the processing of high-melting cpds. *Ceramics or *glasses are usually made from powders, are consolidated and densified by sintering, and are finally fixed by firing. The driving force in s. is a decrease in *surface energy; particles coalesce due to viscous flow or diffusion procs. In catalysis, s. procs. are usually observed during prepn. and reaction of *supported oxide or metal cats. S. presents the most important proc. to contribute to the reduction of the active *surface of a cat. in the course of a cat. reaction and can thus be a part of the *deactivation proc. The extent of s. of a given cat. material depends on time of treatment or use, temp. (of special importance are the *Tammann and Hüttig temps.), gasphase composition, presence of additional species on the support surface, and the properties of the support and of the active phase. **F** vitrification, frittage; **G** Sintern.

R. SCHLÖGL, A. FISCHER

Ref.: G.C. Kuczynski et al. (Eds.), *Sintering and Heterogeneous Catalysis*, Plenum Press, New York 1984; Farrauto/Bartholomew, p. 278.

site-control mechanism also called catalytic site control or enantiomorphic site control

site-directed mutagenesis is the targeted replacement of a codon within a gene, which results in the prod. of a protein with a specifically modified amino acid. Sdm. is typically performed via *PCR amplification of the gene of interest using synth. oligonucleotide primers that contain the desired codon substitution. Sdm. provides a useful technique for determining the purpose of a given amino acid with regard to protein folding, stability, or catalysis. It has also been used to modify the stability, cat. activity, and substrate preference of *enzymes. For example, a highly stabilized mutant of the protease *subtilisin was created by specific amino acid replacements at five sites, and an *aspartate aminotransferase was converted to a lysine-arginine transaminase by conversion of an *active-site arginine to an aspartate. P.S. SEARS

Ref.: *J.Am.Chem.Soc.* **1987**, *109*, 2222; *Biochemistry* **1989**, *28*, 7205.

site-reactivity relationship
The amounts of *saturation sites (cf. *single turnover) on the *surface of *supported (e.g. Pt) cats. can be measured. If there is available a series of various cats. with the same metal load but different reactive site densities promoting the same or different reactions, the comparison with Pt *single crystals engaged in the same transformation expresses relationships between the different sites and their reactivity. It is also possible to determine the *reactivity (and *selectivity) of face, edge, and corner atoms (cf. *single crystals). **F** relation entre le site et la réactivitée; **G** Orts-Reaktivitätsbeziehung.

Ref.: Augustine, p. 43. B. CORNILS

site-selective modification
→ semisynthetic enzymes

site time-yield
(STY) in het. cat. is defined as the number of molecules produced per site (measured by *chemisorption) and per second. Similar to *TOF. **F=G=E**.

B. CORNILS

skeletal metal catalysts
*bulk cats. which are prepared from metal alloys or intermetallic cpds. with two or more constituents by selectively dissolving more or less all but

one metal. The remaining metal has a microscopic spongy network of pores. The mode of preparation decisively influences the *activity. *Promoters can be added either to the starting alloy or to the activated cat. A typical example is *Raney Ni, manuf. from an Al-Ni alloy and treatment with NaOH (*Raney cats.). Smcs. are important as unsupported het. cats. (cf. also *Urushiba nickel). **F** catalyseur squelette; **G** Skelettkatalysatoren. B. CORNILS

Ref.: Augustine, p. 241f; Anderson/Boudart, Vol. 6, p. 161; Ertl/Knözinger/Weitkamp, Vol. 1, p. 64.

slippage
a general term from organometal chemistry, indicating that various *ligands (η^3-allyl, η^5-cyclopentadienyl) may be bonded to multiple coordination sites on the *central atom, may move, and may thus be able to create *vacant sites. **E=F=G**. B. CORNILS

slow-binding inhibitors of enzymes
inhibitors that bind slowly with respect to the rate of substrate conv. Determination of the inhibition constant(s) is therefore more challenging, since an initial transient is observed in the rate profile. The slow binding may be caused by a slow on-rate; by a slow *conformational change in the *enzyme upon inhibitor binding which increases the enzyme's affinity for the *inhibitor; or by a chemical step (such as imine formation) following the initial binding step. Determination of the binding constant requires modeling the mode of inhibition (i.e., how many steps are involved; whether the inhibited *complexes are inactive or just display reduced activity; etc.); formulating (time-dependent) kinetic equs. that describe such a model; and curve-fitting the rate data to the model. P.S. SEARS

Ref.: *Methods Enzymol.* **1979**, *63*, 437 and **1995**, *249*, 144.

SLPC
→ supported (or solid) liquid-phase catalysis

slurry-bed reactor
fluidized-bed reactor with cats. and upward-flowing liquids as mov-

ing phase; cf. *catalytic reactors. **F** réacteur de suspension; **G** Suspensionsreaktor. B. CORNILS

Ref.:Ertl/Knözinger/Weitkamp, Vol. 3, p. 1444; McKetta-1, Vol. 46, p. 408; K.D.P. Nigam, A. Schumpe, *Three-Phase Sparged Reactors*, Gordon and Breach, London 1996.

small-angle X-ray scattering (SAXS) technique suitable for the investigation of micromorphologies and *porosities of powder samples of het. cats. by electrons (probe/response). The ex-situ method is non-destructive and complementary to *Merc. and *BET, but the analysis is difficult and requires structural models. The data are integral and bulk-sensitive. R. SCHLÖGL

Ref.: Ertl/Knözinger/Weitkamp, Vol. 2, p. 451.

SMART process → UOP procs.

SMB → simulated moving-bed reactor

SMDS process Shell middle distillate synthesis, → Shell procs.

SMPO Styrene Monomer Propylene oxide, → Shell procs.

SMSI strong metal-support interaction, → metal-support interaction

S$_N$1, S$_N$2 reactions (enzymatic) Nucleophilic *substitution reactions are common in enzymatic reactions, particularly among *hydrolases and transferases. Many *enzymes proceed via *nucleophilic catalysis, in which an *active-site nucleophile displaces a leaving group on the substrate, which is then displaced by a nucleophilic group from a second substrate (or water). This is true, for example, in the case of the retaining *glycosidases, in which an active-site aspartate acts as a nucleophile in an S$_N$1-type mechanism. Other enzymes proceed by accelerating the nucleophilic attack of one substrate by another, as in inverting *glycosidases and *glycosyltransferases (S$_N$1-like) or the methyl-

transferases (S$_N$2-like). **F** réactions de type S$_N$; **G** S$_N$-Reaktionen. P.S. SEARS

Snamprogetti/ANIC acrylonitrile process for the manuf. of *AN by *ammoxidation of propylene with NH_3 and air at 440–470 °C/ 0.2 MPa over a Mo-V, Bi-doped cat. (*bismuth molybdate). Besides AN, HCN and acetonitrile are isolated. **F** procédé Snamprogetti/ANIC pour le nitrile acrylique; **G** Snamprogetti/ANIC Acrylnitril-Verfahren.

B. CORNILS

Snamprogetti butene isomerization process for the *isomerization of *n*-butenes to *i*-butene at 450–490 °C over a cat. which consist of Al_2O_3, surface-modified by SiO_2. **F** procédé Snamprogetti d'isomérisation de butène ; **G** Snamprogetti Butenisomerisierungs-Verfahren. B. CORNILS

Ref.: Moulijn/van Leeuwen/van Santen, p. 59.

Snamprogetti ether process for the manuf. of tertiary methyl or ethyl ethers (e.g., *MTBE) by *addition of alcohols to alkenes over acid *ion-exchange resins under mild cond. **F** procédé Snamgrogetti pour des éthers; **G** Snamprogetti Ether-Verfahren. B. CORNILS

Snamprogetti isoprene process older three-step proc. for the manuf. of isoprene by 1: the cat. (KOH) *addition of acetone and acetylene at 10–40 °C in liquid NH_3, 2: the selective *hydrogenation of the alkynol over Pd cat. to 2-methylbut-3-en-2-ol, and 3: the Al_2O_3-cat. *dehydration to isoprene at 260–300 °C. **F** procédé Snamprogetti pour l'isoprène; **G** Snamprogetti Isopren-Verfahren. B. CORNILS

Snamprogetti Yarsintez FDB process for the *dehydrogenation of lower alkanes to alkenes in a bubbling, staged fluidized-bed reactor at 600–650 °C/atmospheric press. in the presence of doped (K, SiO_2) Cr-*alumina cats. **F** procédé Yarsintez FDB de Snamprogetti; **G** Snamprogetti Yarsintez FDB-Verfahren. B. CORNILS

Ref.: *Chem.Eng.Sci.* **1992**, *47*, 2313.

SNG substitute (synthetic) natural gas

Whereas today the synth. of methane as a substitute for *natural gas is more or less an exception, there have been considerable efforts in the past (1970–1980, and in Germany during World War II) to convert carbon oxides to methane as a shortage of natural gas was envisaged. A number of synth. procs. have been developed, mostly in parallel with *coal gasification procs. in order to make use of cheap and readily available *syngas. Most of these procs. have now disappeared. All the reactions involved in *methanation are exothermal; their equilibria are displaced toward the starting reactants with rising temp. while an increase in press. causes displacement toward the prods. Side reactions are the formation of carbon and CO_2 by decomposition of CO (*Boudouard reaction).

Nickel is the only metal of importance for the methanation. A large number of predominantly oxidic *promoters have been suggested to control the activity, to reduce the sintering and thus to stabilize the cat. structure at the application temp. of 200–400 °C. *Alumina and *alumosilicates are the preferred *supports. Methanation procs. can be characterized by the mode of heat removal and thus the *shape and rearrangement of the cat. The heat of reaction is recovered mostly for steam generation and may be fed back to the gasification step (cf. *Adam and Eve principle). The exit gas is freed from CO_2 by scrubbing and consists of 95–99.5 % CH_4. **E=F=G.**

Ref.: Falbe-1, p. 309.

                                                                        C.D. FROHNING

Snia-Viscosa benzoic acid process

for the manuf. of benzoic acid through *oxidation of toluene with air at 160 °C/ 1 MPa using Co acetate as cat. **F** procédé Snia-Viscosa pour l'acide benzoique; **G** Snia-Viscosa Benzoesäure-Verfahren.   B. CORNILS

Snia-Viscosa caprolactam process

three-step proc. for the manuf. of ε-caprolactam from toluene by 1: *oxidation of toluene to benzoic acid (*Snia-Viscosa benzoic acid proc.),; 2: *hydrogenation of benzoic acid

over Pd-C cat. to cyclohexane carboxylic acid at 170 °C/1.6 MPa in *CSTRs; 3: the conv. of the carboxylic acid with nitrosylsulfuric acid at 80 °C to ε-caprolactam. **F** procédé Snia-Viscosa pour le caprolactame; **G** Snia-Viscosa Caprolactam-Verfahren.   B. CORNILS

Ref.: *Hydrocarb.Proc.* **1997**, (3), 120.

SNMS   secondary neutral mass spectroscopy

SNOM   → scanning near-field optical microscopy

SNOX process

a *Haldor Topsøe proc. for the removal of sulfur *and* NO_x from flue gases. Firstly, NO_x is removed via *SCR cats. After reheating the gases pass an converter which oxidizes SO_2 to SO_3 for sulfuric acid prod. The *ammonia slip from the SCR reactor is oxidized simultaneously (cf. also *Degussa's Desonox proc.). Other terms are NOXSO and SONOx. **F** procédé SNOX; **G** SNOX-Verfahren.   B. CORNILS

Ref.: Ertl/Knözinger/Weitkamp, Vol. 4, p. 1660; Farrauto/Bartolomew, p. 651.

SOD   → superoxide dismutase

SOF   soluble organic fraction, → diesel engine emissions

soft ligands   large, polarizable ligands (e.g., iodine), → *hemilabile ligands

SOHIO acrolein process

for the manuf. of acrolein by fixed-bed partial *oxidation of propylene with air at 300–360 °C/0.2 MPa over cats. consisting of Bi_2O_3-MoO_3 (*bismuth molybdate). For the connection with the *SOHIO AN proc. cf. also *bismuth molybdate. **F** procédé SOHIO pour l'acroléine; **G** SOHIO Acrolein-Verfahren.   W.A. HERRMANN

SOHIO acrylonitrile process

for the manuf. of *AN by *ammonoxidation of propylene with air/ammonia in a fluidized-bed at 450 °C/0.15 MPa. Besides AN, HCN (150 kg

per t of AN) and acetonitrile (30 kg) are isolated. The SOHIO proc. uses *bimetallic cats.
Most efficient *multicomponent cats. use 1: a
multivalent main-group element, preferably
Bi, Sb, or Te; 2: oxidic Mo, and 3: a *redox-active comp. such as $Fe^{2+/3+}$ or $Ce^{3+/4+}$ in a solid-
state matrix. The standard cat. may be formulated as $Bi_2O_3 \cdot nMoO_3$ (*bismuth molybdate).
Acc. to common opinion, both the SOHIO oxidation (*acrolein) and the propene ammonoxidation receive their unexpected *selectivities
from a specific type of crystal-lattice oxygen as
the actual reagent, quite typically exemplifying
the *Mars-van Krevelen mechanism. The cat.
cycle is shown in the Figure.

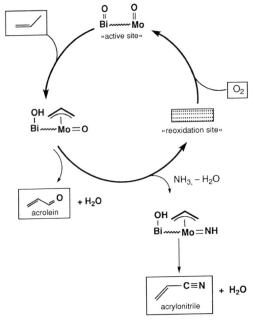

acrolein

acrylonitrile

Bi molybdate is reduced by propylene and
can be reoxidized by air or oxygen yielding
the original valence state (as has been shown
by $^{18}O_2$ labeling experiments). Other procs.
use different cats. **F** procédé SOHIO pour le
nitrile acrylique; **G** SOHIO Acrylnitril-Verfahren.   W.A. HERRMANN

Ref.: *Hydrocarb.Proc.* **1977**, 56(11), 125; Cornils/Herr-
mann-1, p.1156; *Kontakte (Merck)* **1991**, No.3, p.29;
Ertl/ Knözinger/ Weitkamp, Vol.5, p.2302; Winnacker/
Küchler, Vol.6, p.115.

sol-gel catalysts   are amorphous *bulk
cats. prepared by the sol-gel process. The sg.
method is an extremely versatile prep. procedure based on the *hydrolysis/*condensation
reactions of *alkoxides or other hydrolyzable
precursors. It usually provides oxides, but
sgcs. leading to carbides or nitrides are
known. The gelation is accelerated by cats.;
base cats. lead to particle formation and particle gels, while acid cats. usually provide 3D
network gels. Copolycondensation (cf. equ.;
A = builder of porous matrix; B = *active center; C = *surface modifier; D = amorphous
porous oxide, sgc.) during sgc. allows control
of chemical composition and functionality
and the generation of site-isolated active centers.

$$(RO)_4Si + x(iPrO)_4Ti + yR\text{-}Si(OEt)_3 \rightarrow$$
$$\quad A \qquad\qquad B \qquad\qquad C$$

$$(TiO_2)_x(RSiO_{1.5})_ySiO_2$$
$$D$$

Instead of Si as well as Ti many other ions as
well as *enzymes or *antibodies have been introduced. *Modifiers can be any organic
group; alkyl groups have already been used to
control the surface polarity, while suitable ligands *immobilize *transition metals. Depending on the drying and prep. procedure, *pore
size and *porosity can be tailored. *Supercritical drying leads to *aerogels, charaterized by
large surface area and a broad *mesopore distribution (pore size 2–10 nm). Slow drying at
ambient conds. (formation of xerogels) is required to obtain *microporous materials with
a narrow pore size distribution comparable to
that of *zeolites. Such materials, although
amorphous, can show *shape-selective cat.
properties. Sgcs. have been used for many reactions, such as *hydrogenation, *oxidation,
*redox, *hydrocracking, *alkylation, and
*photocatalytic reactions. Of special impor-
tance is the control of surface polarity in sgcs.
Hydrophobic sgcs. have been used to increase
lifetime and activity of immobilized enzymes,
Ti-sgcs. can utilize polar reagents such as

H_2O_2 in water as *selective oxidation re-agents. Sgcs. are a new class of cat. material, whose microstructure can be tailored and controlled to a large extent by the chemistry and the reaction conds. of the mild prep. pro-cedure. **F** catalyseur de sol et gel; **G** Sol-Gel-Katalysatoren.   W.F. MAIER

Ref.: C.J. Brinker, G.W. Scherer, *Sol-Gel Science*, Aca-demic Press, New York 1990; *Catal. Today* **1997**, *35*, 339; *Catal.Rev.Sci.Eng.* **1995**, *37*, 515; *Angew.Chem.Int. Ed.Engl.* **1996**, *35*, 2230; Ertl/Knözinger/Weitkamp, Vol. 1, p. 86; A.C. Pierre, *Introduction to Sol-Gel Pro-cessing*, Kluwer, Dordrecht 1998; Kirk/Othmer-1, Vol. 22, p. 497.

solid acid catalysis Solid acids (sas.) are sought to replace liquid (mineral) acid cats., which are corrosive, difficult to handle, and costly to dispose of. Solid cats. are also easier to separate from reaction prods. Sas. can be classified into *zeolites (e.g., *faujasites, *H-ZSM-5, *mordenite), *clays, metal salts (sul-fates, phosphates, chlorides), cation *ion-ex-change resins (e.g., *Nafion), *mixed oxides (SiO_2-Al_2O_3), mounted *Brønsted acids (H_2SO_4 on SiO_2), mounted *Lewis acids (SbF_5 on SiO_2, $AlCl_3$ on resin), and *hetero-polyacids. Commercially significant reactions are, e.g., *cracking/ hydrocracking, *isomer-ization (skeletal isom. of C_4–C_8 *hydrocar-bons and alkylbenzenes), *alkylation of aro-matics using alcohols or alkenes (synthesis of *p*-xylene, ethylbenzene), *hydration (alkenes to alcohols), *dehydration, *esterification, *hydrolysis, and *polymerization. Some reac-tions seem to require the presence of both acidic and basic sites (*propene oxide to allyl alcohol). Attempts to identify the *active sites and to determine their number and acid strength through sorption and/or thermal as well as spectroscopic methods are sometimes successful. Types of acidic sites are Brønsted (protonic) or Lewis (*cus cations); Lewis sites can be converted into Brønsted sites through (dissociative) adsorption of water, and Brønsted can be converted into *Lewis sites through dehydration/ *dehydroxylation of the *surface. Reaction intermediates are typically of cationic nature, paralleling the solution

chemistry. Side reactions yield *coke, leading to *deactivation of the cat. Strong vs. weak acids and Lewis vs. Brønsted acids favor coke formation, and the presence of alkenes accel-erates (and the addition of hydrogen sup-presses) coke formation. Combustion of the coke can *regenerate the cat. **F** catalyseur d'acide solide; **G** Feste Säure-Katalysatoren.

F. JENTOFT

Ref.: *Stud.Surf.Sci.Catal.* **1989**, *51*; *Catal. Today* **1997**, *28*, 257; *Chem.Rev.* **1995**, *95*, 559; Ertl/Knözinger/Weit-kamp, Vol. 1, p. 404; Anderson/Boudart, Vol. 2; van Santen/van Leeuwen.

solid base catalysis Solid bases are sought to replace liquid cats. which cause cor-rosion and environmental problems. The solid cat. has the additional advantage of easier prod./cat. separation. Sbc. can be classified into *metal oxides (e.g., alkaline earth oxides, alkali metal oxides, rare earth oxides), metal salts (carbonates), *mixed oxides (SiO_2-MgO), *anion-exchange resins (*Dowex), base-modified oxides (Na vapor-treated Al_2O_3, KNH_2-Al_2O_3), hydrotalcites, oxyni-trides, and modified *zeolites. Sbs. can cat. the following reactions: *isomerization, side-chain *alkylation of aromatics, *dehydration of alcohols, *dehydrohalogenation, *polymer-ization of alkenes or epoxides, and *hydro-genation of alkenes or CO. Some reactions seem to require the presence of both basic and acidic sites (e.g., *N*-alkylation of aniline with methanol). In comparison to solid acid catalysis, far fewer investigations have been devoted to sb. cat., and there is not the wide-spread commercial use as with solid acid cat. Properties of basic sites on solids, i.e., site structure, density, and strength, have been less extensively studied. *Brønsted (OH groups) or *Lewis (mostly oxygen anions) types of basic sites exist. Reactions on solid bases involve intermediates of anionic nature. Solid bases are subject to *poisoning by H_2O and CO_2; cf. also *superbases. **F** catalyseur de base solide; **G** Feste Basen-Katalysatoren.

F. JENTOFT

Ref.: *Stud.Surf.Sci.Catal.* **1989**, *51*; *Catal. Today* **1997**, *28*, 321.

solid-phase technique for the stepwise synth. of peptides from amino acids. The first amino acid is covalently bound (*anchored) to a suitable polymer. The whole sequence of amino acids/peptides stays on the resin until it is released by treatment with strong reagents (*Merrifield technique), which is the decisive difference from *immobilized cats. B. CORNILS

solid-solid wetting can occur in multi-comp. cats. consisting of at least two phases, one of which is often the cat. *support. The thermodynamic cond. for wetting of one solid by another solid to occur is a decrease of the overall *surface/ interface free energy, the interface free energy playing a crucial role. Temps. must exceed the *Tammann temp. of the mobile comp. so as to induce sufficiently high mobility. Ssw. phenomena occur in many solid cat. systems and are obviously closely related to the stabilization of the *dispersion of an active *phase on a support. Typical examples are the encapsulation of small metal particles by oxide species of reducible oxide supports, which leads to the phenomenon of *strong metal-support interactions (SMSI), or the *redispersion of small metal particles in their oxide state during oxidative *regeneration. In binary *alloys, wetting of one metal comp. by the second can occur even if the two comps. are immiscible in the bulk. Thus, Ru particles are coated by a Cu overlayer in *bimetallic CuRu-SiO$_2$ cats. although a binary alloy of the two metals does not exist. Finally, disperse oxides (e.g., MoO$_3$ or V$_2$O$_5$) on oxide supports (e.g., Al$_2$O$_3$ or TiO$_2$) can be prep. from the powder mixtures via thermal and/or mechanochemical spreading, which requires the wetting of the support surface by the mobile active oxide phase. **F** imprégnation sec/sec, **G** Trocken/Trocken-Imprägnierung.   H. KNÖZINGER

Ref.: Ertl/Knözinger/Weitkamp, Vol. 1, pp. 100, 216; *Surf. Sci.* **1988**, *201*, 603.

solid-state NMR characterizes the chemical and structural environment of atoms in cats. or in adsorbed species. The application in the gaseous or liquid phase is over a wide range of temps. and press. Nearly all elements can be measured, but such NMR data are often complex (because of dipolar, quadrupolar, or chemical-shift interactions). Information obtainable includes the structure of cats. and their thermal or chemical transformations, adsorbent/adsorbate interactions, and the nature of chemical bond species (for highly resolved NMR data, cf. *cross-polarization magic angle spinning, CP-MAS).   B. CORNILS

Ref.: Ertl/Knözinger/Weitkamp, Vol. 2, p. 525.

solid-state reactions are (topochemical) techniques for the manuf. of *bulk cats. by homogeneous mixing of solid *precursors at the molecular level, mainly for *mixed oxide cats. Such cats. are, e.g., Cu chromites (*Adkins cats.) for various *hydrogenations, *Zn chromite (*methanol synth.), promoted *bismuth molydate (for acrolein or acrylonitrile from propene, *SOHIO proc.), or *perovskite cats. for *cat. combustion or *fuel cell catalysis. Such solid-solid reactions include dry methods (grinding or ball-milling and *calcination, *solid-solid wetting) or wet techniques (*coprecipitation, *spray drying or spray calcination, or chemical complexation methods). For *zeolites solid-state *ion exchange is used. **F** réactions topochimiques; **G** topochemische Reaktionen.   B. CORNILS

Ref.: Ertl/Knözinger/Weitkamp, Vol. 1, p. 100.

solid supports → supports, to be differentiated from *liquid supports

solid (supported) liquid phase catalysis (SLPC) → supported LPC

solubilization application of *solubilizers for the conv. of less soluble compds. into solutions (hydrotropic effect, hydrotropic liquids; *ligand tuning). The mechanisms of s. vary (H-bridges with water/alcohols; associative colloids with water/detergents, microemulsions, *micelle-building amphiphilic cpds., etc.). S. may also be observed via physical means such as *sonochemistry or *microwaves. S. is important for *two-phase cat., specially *aqueous-phase cat., since the initial step of cat. is the solution of the (mostly) organic reactant in water (cf. *aqueous-phase cat.; *two-phase cat.). **F** solubilisation; **G** Solubilisierung.   B. CORNILS

solubilizer is a cpd. which ensures solubilization

solution catalysis → homogeneous catalysis

solvation describes the direct interaction of solvent molecules (e.g., *solvents for cats.) with metal ions. A metal ion in solution has a primary, highly structured s. sheath close to the metal ion (s. or coordination sphere). The solvent molecules associated with the metal ion in this sheath exhibit remarkably different properties than those in the bulk solvent. A second s. sphere (where the solvent molecules have almost bulk properties) surrounds the first one. The solvation number n describes the number of closely associated solvent molecules in the first s. sphere. For instance, n is 6 for Cr^{3+}, Mn^{2+}, or Zn^{2+} in aqueous medium. The number of water molecules in the second s. sphere varies between 12 and 13, i.e., each coordinated water molecule is H-bonded to two water molecules in the second sphere. Depending on the charge of the metal ion and the strength of the metal-solvent bond, coordinated solvent molecules can exchange with bulk solvent molecules at rates between 10^{-10} and 10^9 s^{-1}. Such solvent exchange reactions can proceed acc. to various mechanisms., depending on the electronic structure of the metal ion, its size, and the bulkiness of the solvent. The ability of a solvent to donate an electron pair to a metal cation is of major importance in describing solvation (cf. *solvent). **E=F**; **G** Solvatisierung.   R. VAN ELDIK
Ref.: Wilkins; Dietrich/Viout/Lehn; Connors; Atwood; Burgess; Reichardt; Stumm.

Solvay/Laporte/Carbochimique Interox

process for the manuf. of *propylene oxide

by *oxidation of propylene with perpropionic acid, obtained from recycle propionic acid and H_2O_2. Perpropionic acid is extracted with 1,2-dichloropropane from the hydrogen peroxide solution (cf. *Bayer-Degussa proc.). **F** procédé Interox de Solvay/Laporte/Carbochimique; **G** Solvay/Laporte/Carbochimique Interox-Verfahren.   B. CORNILS

solvent effects account for the influence of solvents on the physical and chemical properties of molecules and ions. Solvents are divided into *protic* (including both proton donors and acceptors), *dipolar aprotic* (solvents with dielectric constants [D] >15 but with H-atoms capable of forming hydrogen bonds), and *aprotic* solvents (having neither acidic nor basic properties). In describing ses., the solvent is regarded as an inert medium that does not participate directly in the chemical proc. The D value of the solvent is the most important parameter, and its effect can be evaluated for ion-ion or ion-dipole reactant interactions, where electrostatic effects dominate the interaction. Another important solvent property is the dipole moment μ, which is defined as the product of charge (usually just a fraction of the electronic charge) and distance between the centers of positive and negative partial charges in the molecule. This parameter controls the interaction of the solvent with polar molecules.

The nature of the solvent will affect the spectroscopic properties of the molecules, especially in those cases where charge transfer bands are monitored. Various solvent parameters were developed to account for these effects, such as the product $D\mu$ (electrostatic factor), Reichardt's parameter E_T, and Kosower's parameter Z. **F** effets des solvants; **G** Lösemitteleffekte.   R. VAN ELDIK
Ref.: Wilkins; Dietrich/Viout/Lehn; Connors; Atwood; van Eldik/Hubbard; Reichardt; Burgess; *J.Mol.Catal. A:* **1999**, *142*, 383; Y. Marcus, *The Properties of Solvents*, Wiley 1998.

solvent power → solvents for catalysis

Solvent Refined Coal process → SRC process

solvents for catalysis are important in many cases to ensure physically homogeneous mixtures of reactants and solvents (including *supercritical fluids, SCFs). A solution is advantageous from a chemical engineering standpoint if the reactant to be subjected to cat. convs. is a solid under reaction conds. and the solvent is recyclable and not reactive. In *two-phase and in *aqueous-phase catalyses the solubility plays a special role since in hom. cat. a one-phase *catalysis* coupled with a biphase cat. *separation* is advantageous. In all cases the miscibility (solvent power) may be estimated following Paracelsus' principle, "*similia similibus solvuntur*" (like dissolves like)(cf. the Figure under keyword *liquid support; *solvation, *solvent effects). For the partitioning of cpds. between two solvents, cf. *Hansch equation. For enzymes cf. *enzyme catalysis in organic solvents. (Cf. *solvation). **F** solvants pour le catalyse; **G** Lösemittel für die Katalyse. B. CORNILS

Ref.: Kirk/Othmer-1, Vol. 22, p. 529; McKetta-1, Vol. 52, p. 458; *Org. Proc. Res. & Dev.* **1998**, *2*, 121; Reichardt; Connors; Martell/Hancock; Y. Marcus, *The Properties of Solvents*, Wiley, Chichester 1998; P. Knochel (Ed.), *Modern Solvents in Organic Synthesis*, Springer, Berlin 1999.

SOMC → surface organometallic chemistry

sonochemistry Ultrasonic waves have been used to accelerate chemical reactions in the liquid phase. Thus, ultrasonic waves (5–50 kHz) are sent out into a liquid via a transducer, the surface amplitude of which is in the range 0.001–0.1 mm, depending on its geometry, type of construction, and the energy available/applicable. The result is an inhomogeneous press. field caused by harmonic vibrations, the spatial intensity of which can be described mathematically by the Helmholtz integral. The cavitation caused by the press. field leads to the formation of bubbles, which on collapsing (imploding) cause momentary and regional changes in temp. (corresponding

to some thousand °C) and in press. (corresponding to some hundred MPa) transferable to molecules in their neighborhood, thus contributing *activation energy to chemical reactions. The interpretation of the effect of ultrasonic irradiation on the rate of a chemical reaction is difficult.

When ultrasonic waves (35 kHz) are applied, e.g., in the *hydroformylation of 1-hexene or diisobutylene in the *RCH/RP proc., the *TOF increases by a factor of 3–5, indicating a positive effect on the *rate of reaction. A two-fold increase in reaction rate is reported for the selective *hydrogenation of cinnamaldehyde to cinnamyl alcohol over Pt/SiO_2. The ultrasonic pretreatment of the cat. increases the *metal-support interaction, thus leading to higher *selectivity for cinnamyl alcohol. **F** sonochimie; **G** Sonochemie, Ultraschallchemie. C.D. FROHNING

Ref.: T.J. Mason, J.P. Lorimer, *Sonochemistry*, Ellis Horwood, Chichester 1988; P.M. Morse, K.U. Ingard, *Theoretical Acoustics*, McGraw-Hill, New York 1968; Ertl/Knözinger/Weitkamp, Vol. 3, p. 1350; EP 173.219 (1985); *Appl. Catal. A:* **1998**, *172*, 225; *Chem. Eng. Technol.* **1998**, *21*, 11; *J. Mol. Catal. A:* **1999**, *149*, 153.

Sonogashira reaction a Pd(0)-cat. coupling of terminal alkynes and aryl halides acc. to the Equ.

$$ArX + R\!\!=\!\!\!=\!\!H \xrightarrow[\substack{base \\ CuI}]{Pd^0L_n} R\!\!=\!\!\!=\!\!Ar + HX$$

In many cases *co-cat. amounts of Cu(I) are necessary (*bimetallic cats.). **F** réaction de Sonogashira; **G** Sonogashira-Reaktion. V. BÖHM

Ref.: Abel/Stone/Wilkinson, Vol. 3, p. 521.

SONOX → SNOX process

sorbitol dehydrogenase → polyol dehydrogenase, *iditol dehydrogenase

sour gas shift the water-gas shift reaction with sour gases on sulfur-tolerant cats.; → water-gas shift reaction. The term includes the COS removal of *syngases on Co-Mo cats. or *aluminas by *hydrolysis acc. to COS

+ H_2O → H_2S + CO_2 (cf. also *Claus COS conversion). **E=F=G.**   B. CORNILS

Ref.: Ertl/Knözinger/Weitkamp, Vol. 4, p. 1838.

SOX, sox is an anionic O,N *chelate *ligand derived from salicylaldoxime (**1**, HSOX) (Figure).

1 (HSOX)   **2** (L = 2 CO, *COD)

The Rh-SOX cpd. **2** combined with P ligands exhibits high *activity for *hydroformylation of alkenes under mild conds. The cat. systems are inert for *hydrogenation under hydroformylation conds.   O. STELZER

Ref.: *J.Mol.Catal.A:* **1998**, *129*, 153 and **1994**, *88*, 277.

SPA solid phosphoric acid, used as an *alkylation catalyst (e.g., for *cumene manuf., *UOP proc.). The cat. was originally developed by *Ipatieff and consists of H_3PO_4 deposited on *kieselguhr or *silica. **E=F=G.**

Ref.: US 2.382.318 (1945).   B. CORNILS

space-charge layer In *band-bending at a *semiconductor *surface the positive/negative charges at the surface are compensated by opposite charges distributed over a certain region inside the bulk which is denoted the scl.

                                                   R. IMBIHL

spacers in cat. sciences and cat. applications, bonding groups between the carrier (*support) and organometallic cat. active species in *anchored cats. (cf. *supported hom. cats.). **F=G=E.**   B. CORNILS

space time The reciprocal value of space velocities (h^{-1}) leads to the *space time (h), a measure for the hydrodynamic residence time, e.g., in a tubular reactor. A corresponding parameter is the ratio W/F_0 [$kg \cdot h \cdot L^{-1}$] which is the weight of the cat. per volumetric flow rate of feed and mostly used to characterize the activity's dependence on the linear flow of the feed (cf. *residence time, *reaction

parameters). **G** Raum/Zeit.

                                   P. CLAUS, D. HÖNICKE

space time velocities   → GHSV, *LHSV, *WHSV

space-time yield The expression space-time yield (STY) is frequently used to characterize the effectiveness of a cat., reactor, or proc. with respect to the amount (volume or mass) of the desired product generated in a distinct volume per unit time (e.g., L/L · h or g/L · h). For different modes of operation the definitions are: STY = (input × conversion × selectivity)/(reactor volume × time of reaction) for discontinuous operation; STY = (feed × conversion × selectivity)/(reactor volume × time) for continuous operation (cf. *reaction parameters). STY is linked to *LHSV or *GHSV respectively by conv. and *selectivity. **F** rendement espace-temps; **G** Raum/Zeitausbeute.   C.D. FROHNING

space velocity The reactor or cat. load is expressed as space velocity (*GHSV, *LHSV), expressed in volume V (liquid or gaseous) per volume V of cat. and per hour ($V \cdot V^{-1} \cdot h^{-1}$).   P. CLAUS, D. HÖNICKE

specific acid-base catalysis   → acid-base catalysis

specific activity The specific activity of an *enzyme preparation is the activity per unit mass or per volume (if the enzyme is in solution). In commerical preparations of enzyme, this is typically given in U/mg or U/µL (where a unit of enzyme will cat. the conv. of 1 µmol of substrate per minute). Cf. *turnover number. **F** activité spécifique; **G** spezifische Aktivität.   P.S. SEARS

specificity constant *Enzymes are remarkable in their ability to discriminate between different substrates (see *selectivity). In a system where two substrates A and B are competing for an enzyme E, the ratio of the rates is given by the ratio of the k_{cat}/K_m val-

ues (multiplied, of course, by the ratio of the substrate concs.) for the two reactions. For this reason, the cat. parameter k_{cat}/K_m is often called the specificity constant.(cf. also *enantio selectivity). **F** constante de spécifitée; **G** Spezifitätskonstante.
P. SEARS

specificity Because *enzymes are large *chiral molecules with unique stereo-structures in the *active site, they can be highly selective for certain types of substrate structures and reactions. Useful types of enzyme-cat. reactions include the chemoselective reaction of one of several different functional groups in a molecule, the *regioselective reaction of one of the same or similar groups in a molecule, the *enantioselective reaction of one enantiomer of a racemic pair or one of the enantiotopic faces or groups, and the *diastereoselective reaction of one or a mixture of diastereomers or one of the diastereomeric faces or groups. All such selective reactions occur because, during a reaction, the prochiral or *chiral reactants form diastereomeric enzyme-*transition state *complexes that differ in transition-state ($\Delta G^{\neq}$) energy. **F** spécifité; **G** Spezifität.
C.-H. WONG

spectator ligands → ancillary ligands

spectroscopy (general) With spectroscopy the interaction of radiation or particles with matter can be investigated. The recorded spectra yield information on the structure, physical nature, chemical composition, or electronic properties of the investigated system by monitoring the response of the system upon variation of the energy or wavelength of the incoming radiation or particles (absorption spectroscopy) or by monitoring the energy or wavelength distribution of radiation or particles emitted upon excitation (emission spectroscopy). Depending on the system and information desired it is necessary to select an appropriate spectroscopic technique.
It is often useful to apply several techniques for a better understanding of a specific system. To characterize *surfaces, surface sensi-

tivity is essential. Therefore either the incoming or outgoing radiation or particles must have a short mean free path (cf. also *IMFP), e.g., electrons, ions, or atoms. The inherent disadvantage of the short mean free path is that measurements need to be carried out in vacuum, which conflicts with the wish to investigate cats. under reaction conds. In contrast, investigations of real cats. under relevant conds. by several in-situ techniques give little information on the catalyst's surface and it is important to guarantee that only the relevant species are detected.
The most common surface spectroscopies in vacuum are *photoelectron spectroscopy (*UPS/*XPS), *Auger electron spectroscopy (AES), and X-ray absorption spectroscopy (e.g., *EXAFS), whereas *Raman spectroscopy, *infrared spectroscopy (IR), or second harmonic generation (SHG) can be applied without vacuum. The Table summarizes this. **F** spéctroscopie (général); **G** Spektroskopie (allgemein).
R. SCHLÖGL, Y. JOSEPH

Reaction conds.	Real cat.	Single crystal
	XRD, TP techniques, IR, Raman, EXAFS, AFM SRM, AFM, Mössbauer, ESR, NMR,	IR, TP techniques
Vacuum	XPS, SIMS, SNMS, LEIS, RBS, TEM, SEM	All techniques

Ref.: Benninghoven/Rüdenauer/Werner; Imelik/Védrine; Niemantsverdriet; Wilson/Decius/Cross; Chastain; Harris/Bertolucci; Ibach/Mills; Ullmann, Vol.B6, p. 23; Ertl/Knözinger/Weitkamp, Vol. 2, p. 539; J.L.G. Fierro (Ed.), *Spectroscopic Characterization of Heterogeneous Catalysts*, Elsevier, Amsterdam 1990; R.J. Clark, R.E. Hester, *Spectroscopy for Surface Science*, Wiley-VCH, Weinheim 1998; van Santen/van Leeuwen.

Speier's catalyst hexachloroplatinic acid in isopropanol, a cat. for the *hydrosilylation of alkenes (Figure).
W.A. HERRMANN

$$R\text{-}CH=CH_2 + R^1_3SiH \xrightarrow{H_2PtCl_6} R\text{-}CH_2\text{-}CH_2\text{-}Si\,R^1_3$$

Ref.: *J.Am.Chem.Soc.* **1957**, *79*, 974; *Adv.Organomet.al.Chem.* **1979**, *17*, 407.

spherand a subclass of *crown cpds., e.g.,
cyclic oligo *m*-phenylenes with intraannular
donor groups

Spheripol process → Montell (Himont)
process

spill out cf. *work function.

spillover In het. catalysis a sequence of different tasks usually has to be fulfilled by the
cat. in order to complete the *cat. cycle. This
is made possible by the multifunctional character of an active *surface. Often the *active
sites for performing different tasks cannot be
chemically identical. Reaction then requires
the transport of activated species on the *surface from one type of sites to another. The
*diffusion of such activated species is referred
to as spillover (Figure). Typically atomic hydrogen or atomic oxygen are s. species
whereas organic moieties usually remain fixed
to their sites of activation.

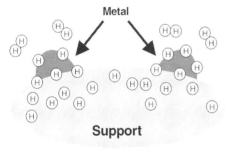

[by courtesey reproduced from Augustine, Marcel
Dekker, Inc., Figure 2.10]

The exact physical nature of the s. proc. has only
very rarely been verified experimentally.
Mostly the term is used in mechanistic argu-
ments when trying to explain the non-linear effects (synergistic effects) of the combination of
chemically different ingredients of a cat. system
on its performance. A characteristic example is
the *addition of Sb oxide to *selective *oxidation cats. Although Sb oxide is absolutely inert
in these reactions, it enhances the cat. *activity
at high levels of *selectivity by a factor of up to

five as compared to the Sb-free system (*antimonates). It is thought that the presence of Sb
enhances the prod. of active O species which
spill over to sorption sites of the organic substrate. Another example is the partial *reduction of *support oxides such as *titania after deposition of noble metal particles (Pt) and addition of H. It is not only the immediate surround-
ings of the oxide that are reduced but the whole
crystal of which only a tiny fraction of its surface
is in contact to the noble metal. **E=F=G**.

R. SCHLÖGL

Ref.: *Chem.Rev.* **1995**, *95*, 759; Ertl/Knözinger/Weit-
kamp, Vol. 3, p. 1064; G.M. Pajonk, S.J. Teichner, I.E.
Germain, *Spillover of Adsorbed Species*, Elsevier, Amsterdam 1983.

spinels The mineral sp. is $MgAl_2O_4$. It has
given its name to the class of spinel materials
which are characterized by certain common
features in composition and structure. The s.
structure is most prominent among oxides but
occurs also in halides and sulfides. Regarding
the general formula AB_2O_4 of a ternary oxide
spinel, there are three different ways to
compensate the −8 charge of oxide ions. Correspondingly, the s. structure is found with
6:1 $(A^{6+}B_2^+O_4)$, 4:2 $(A^{4+}B_2^{2+}O_4)$ and 2:3
$(A^{2+}B_2^{3+}O_4)$ compositions.
Ss. containing transition metal cations often
exhibit interesting magnetic properties, including ferro-, ferri-, and antiferromagnetism.
Especially the spinels containing iron cations
(ferrites) have been thoroughly investigated
for these properties. The binary Fe oxide
*magnetite, $Fe_3O_4/Fe^{2+}Fe^{3+}_2O_4$, also has the
s. structure. It is used for the manuf. of the
cats. for the *Haber-Bosch and *Kellogg
Synthol procs. In defect spinel structures,
some cation sites are *vacant. These cpds. are
usually less stable than stoichiometric ss. and
correspondingly exhibit a higher reactivity.
They can be prepared with larger specific surface areas and are used as cats. and cat. *supports. Typical examples are γ-Al_2O_3 and γ-
Fe_2O_3. **F** spinelle; **G** Spinell.   P. BEHRENS

SPM scanning probe microscopy, → AFM
and STM

sponge metal catalysts → skeletal catalysts, *Raney cats.

spray drying is applied to a slurry of cat. *precursor particles in a *solvent (predominantly water) to generate a dry powder of cat. fines. Sprays are produced by feeding the slurry together with a carrier gas into a swirl chamber followed by a circular orifice outlet (atomizer). By adjustment of gas-to-liquid ratio, flow, and press. extremely fine dispersions are produced which on their further progress through the dryer lose the solvent and accumulate as dry powder. Instead of through a nozzle, the *suspension can be dispersed by being sprayed onto a rotating disk- or cup-shaped device (rotary atomizer), thus avoiding potential clogging of the nozzle. **F** séchage par pulvérisation; **G** Sprühtrocknung.

C.D. FROHNING

Ref.: D.W.Green (Ed.), *Perry's Chemical Engineers' Handbook*, 6th ed., McGraw-Hill, New York 1984, p. 48.

sputtering → chemical vapor deposition

SQS → squalene synthase (EC 2.5.1.21)

squalene Squalene is a C_{30} polyisoprenoid precursor of steroids, formed by the *enzyme squalene synthase (SQS, EC 2.5.1.21) via the head-to-head *condensation of two molecules of farnesyl pyrophosphate (cf. *isoprenoids). The SQS reaction proceeds through two steps. The first step produces an unusual cyclopropylcarbinyl intermediate, presqualene pyrophosphate, which undergoes the second step, a rearrangement and *reduction to squalene. The oxidative *cyclization of squalene to *lanosterol is then carried out by *squalene oxidocyclase. SQS cat. the first committed step in the synth. of steroids, and as such the development of inhibitors is a topic of great interest. **E=F**; **G** Squalen.   P.S. SEARS

Ref.: *J.Am.Chem.Soc.* **1996**, *118*, 13089; *J.Biol.Chem.* **1995**, *270*, 9083.

square root law → self replication

SRC (solvent refined coal) processes for the *hydrogenation of coal at 425 °C/1–1.5 MPa (SRC-I: prods. are distillates and granulated residues) or at 465 °C/>15 MPa (SRC-II: mainly distillates). The proc. is non-catalytic (except for the cat. active coal constituents and except for the use of a downstream *hydrotreater). Coal is slurried with a proc.-derived recycle solvent. The slurry passes through a preheater to the dissolution reactor. Certain amounts of gases and liquids are generated at this stage; the bulk of the coal passes a *hydrotreating step consisting of a stationary or ebullated bed of a highly active cat. similar to those used in *petroleum processing. The prod. from the hydrotreater is fractionated to recover the recycle solvent and to yield the distillates. At the pilot plant in Wilsonville (AL) several configurations of the proc. and numerous cats. were tested. The yields were approx. 15 % of gases, 5 % of *SNG and *LPG, 10 % naphtha, 10 % middle distillates, 40 % of solvent, 5 % of residues and 15 % of undissolved coal. **F=G=E**.

I. ROMEY

Ref.: Kirk/Othmer-2, p. 540; Winnacker/Küchler, Vol. 5, p. 458.

SRT pyrolysis short residence time pyrolysis, → steam reforming

SSITK → steady-state isotope transient kinetics

SSPC → supported solid-phase catalyst

stability 1-general: a measure of the rate of loss of cat. *activity or *selectivity. 2-general: a state of stability (stationary conds.) is reached when within the time of observation the conds. remain constant or return to the state before a disturbance. S. is important for all procs. with excited states, such as cat. *surface conds. or aggregation of fine cat. particles in het. cat., high-energy *intermediates in hom. *cat. cycles, intermediates of *photocat., lifetimes of *radicals, *colloids, or *clusters, etc.

In cat. science, s. is a measure oft the rate of loss of cat. activity or selectivity. S. may be influenced by chemical, thermal, or mechanical degradation or decomposition. Although the definition *catalysis specifies that in cat. cycles the cat. is not consumed, real cats. have a finite lifetime.   B. CORNILS

3-enzyme stabilization: *Enzymes typically prefer an aqueous medium at moderate (<50 °C) temp. and pH, and the free energy required to unfold them is usually quite small (5–10 kcal/ mol). Drastic changes from these conds., such as a large increase in temp., addition of large amounts of organic cosolvents, or addition of other agents (detergents, urea, certain salts) that reduce the free energy of unfolding can cause the protein to denature (unfold) and become inactive. In some cases, the protein can renature to an active form when transferred back to more optimal conds., but in many cases the protein requires a folding cat., or *chaperone, to renature. Covalent modifications may inactivate the enzyme, particularly if the residues involved are near the *active site. Finally, the peptide bonds themselves are subject to *hydrolysis, especially in the presence of *proteases which may contaminate the enzyme prepn. Extremophilic organisms, particularly thermophiles, are a good source of enzymes that are stable toward a broader range of conds. Enzyme *mutagenesis has also provided more robust enzymes. Certain additives, in particular polyols such as glycerol, are excellent stabilizing agents. Finally, covalently restraining the enzyme in its active form, either via multipoint immobilization (cf. *enzyme immobilization) to a support or *cross-linking of the enzyme with a bifunctional "tether" such as glutaraldehyde is frequently used to create a more robust cat. **F** stabilité, l'état stationaire; **G** Stabilität, stationärer Zustand.   P.S. SEARS

Ref.: Butt/Peterson; Delmon/Froment; Ullmann, Vol. A5, p. 323; van Santen/van Leeuwen.

stabilizers reduce phase transformations at the high temps. typical of *oxidation reactions. *Automotive exhaust cats., e.g., contain precious metals (Pt-Rh) distributed in a *washcoat mainly consisting of alumina. Most desirable because of its high specific *surface area is γ-alumina (200–300 m^2/g), however, it converts to α-alumina (2–3 m^2/g) at about 1300 °C. To avoid the drastic loss in surface area (and effectiveness), lanthana (ceria) and barium oxide are homogeneously distributed in the alumina, yielding a residual surface area of 25–35 m^2/g. A similar effect was observed on doping the surface of alumina with 1–3 % of $SiEt_4$. **F** stabilisateurs; **G** Stabilisatoren.   C.D. FROHNING

Stacking   → π-Stacking

standard catalysts In order to ensure comparability (of cats., procs., or apparatus) it is possible to use standard or reference cats. Various cats. have been developed for this purpose under the supervision of the European Association of Catalysis [EUROCATs such as EUROPT-1 (Pt-SiO_2), EURONI-1 (Ni-SiO_2), or EUROTS-1 (Ti silicalite)]. Similar endeavors have been made in the US, Japan, and Russia. **F** catalyseurs de standard; **G** Standardkatalysatoren.   B. CORNILS

Ref.: Ertl/Knözinger/Weitkamp, Vol. 3, p. 1489.

Standard Oil (of Indiana) HDPE process uses standard MoO_3 on *alumina as catalyst; → polyethylene, *high-density polyethylene.

Standard Oil (of Indiana) Isomate process alkane *isomerization for the improvement of the octane quality of light paraffin fractions (mainly consisting of C_5 and C_6 hydrocarbons) by treatment of naphtha (with dissolved HCl as a cat. promoter) with slurried $AlCl_3$. Cat. activity is maintained by periodically injecting a slurry of fresh $AlCl_3$. Effluents from the top of the reactor are cooled and flashed for separation of entrained cat. **F** procédé Isomate de Standard Oil; **G** Standard Oil Isomateverfahren.   B. CORNILS

Ref.: *Petr.Refiner* **1957**, *36*(5), 177.

Standard Oil (of Indiana) Light Naphtha Isomerization process a forerunner of Standard's Isomate proc., using $AlCl_3/HCl$ as a monofunctional cat.

Ref.: *Hydrocarb.Proc.* **1964**, *43*(9), 177. B. CORNILS

Standard Oil (of Indiana) Resid hydroprocessing for the *desulfurization of high-sulfur resids or to improve cat. cracking unit feeds by cat., fixed-bed *hydrodesulfurization.

The proc. includes a low-cost cat. replacement technique for rapid change of cats. **F** procédé Resid hydroprocessing de Standard Oil; **G** Standard Oil Resid Hydroprocessing-Verfahren. B. CORNILS

Ref.: *Hydrocarb.Proc.* **1972**, (9), 174.

Standard Oil (of Indiana) Ultracat cracking converts low value gas oils into more valuable lower-boiling prods. such as high octane gasoline by *cat. cracking in a *riser reactor. The reactor system includes a cat. *regeneration. **F** procédé Ultracat cracking de Standard Oil; **G** Standard Oil Ultracat Cracking-Verfahren. B. CORNILS

Ref.: *Hydrocarb.Proc.* **1972**, (9), 138.

Standard Oil (of Indiana) Ultracracking converts virgin distillates, coker gas oils, etc. to gasoline, naphtha for *reformer charge, etc., by *hydrocracking over "a highly stable, multifunctional cat. that can accomplish *denitrogenation, olefin saturation, *desulfurization, aromatic saturation as well as hydrocracking in a single stage". **F** procédé Ultracracking de Standard Oil; **G** Standard Oil Ultracracking-Verfahren. B. CORNILS

Ref.: *Hydrocarb.Proc.* **1972**, (9), 145.

Standfard Oil (of Indiana) Ultrafining process for *desulfurization, *denitrogenation, and saturation of alkenes and aromatics of virgin feedstocks by *hydrotreating operation over Co-Mo or Ni-Mo cats. at 230–480 °C/1–12 MPa. The space velocity is between 1 and 15 $V \cdot V^{-1} \cdot h^{-1}$. **F** procédé Ultra-

fining de Standard Oil; **G** Standard Oil Ultrafining-Verfahren. B. CORNILS

Ref.: *Hydrocarb.Proc.* **1972**, (9), 189.

Standard Oil (of Indiana) Ultraforming process for the cat. *reforming of low-octane vigin, coker, cat. cracked, or hydrocracked naphthas at up to 550 °C/1 MPa over a Pt fixed-bed cat. including a cyclic cat. *regeneration proc. **F** procédé Ultraforming de Standard Oil; **G** Standard Oil Ultraforming-Verfahren. B. CORNILS

Ref.: *Hydrocarb.Proc.* **1972**, (9), 121; Winnacker-Küchler, Vol. 5, p. 101.

Standard Oil (of Indiana) Ultrasweetening process a dry cat. sweetening of kerosine by removal of mercaptan sulfur via mild *hydrogenation. **F** procédé Ultrasweetening de Standard Oil; **G** Standard Oil Ultrasweetening-Verfahren.

Ref.: *Hydrocarb.Proc.* **1972**, (9), 181. B. CORNILS

Standard Oil (of New Jersey) butadiene process for the two-stage manuf. of butadiene through successive *oxydehydrogenation of butane over a doped $MgO\text{-}Fe_2O_3$ cat. at 670 °C/reduced press. and with large surpluses of steam. **F** procédé de Standard Oil pour le butadiène; **G** Standard Oil Butadienverfahren. B. CORNILS

Standard Oil of Ohio → SOHIO procs.

standard state The (solute) standard state usually used for biochemical applications in energy calculations is 1M ideal solution at 25 °C, 1 atm, and pH 7.0. (The standard state of water is still the pure liquid.) **F** état standard; **G** Standardbedingungen.

P.S. SEARS

STAR process → Phillips procs.

starter radical starter, → radical initiator

stationary exhaust catalysis cf. *DeNOx reactions, *selective catalytic reduction, and also *automotive exhaust catalysts

stationary sources of pollutants (NOx, SO₂, *VOC) are subject to special cat. treatments such as removal of NOx from stationary power stations (*DeNOx) by *selective catalytic reduction (SCR), removal of SO₂ emissions by cat. oxidation and precipitation (*SNOX, DeSOx),removal of polyaromatic chlorinated hydrocarbons from incineration plants (dioxin removal), etc.; cf. also *cat. combustion.   B. CORNILS

Ref.: Heck/Farrauto.

Stauffer (EDC) process precursor of the EDC proc., → EVC International vinylchloride process

steady-state 1-general: a cond. generally characterized by constant properties of any part of a system in the course of a proc., i.e., independent of the time. In chemical reaction engineering one of the important steady-state (stst.) assumptions for continuous operation is expressed by

$$\frac{dc_i}{dt} = 0 \; (i = 1, 2, \ldots, n)$$

at constant reaction conds. over the reaction time. Consequently, the chemical reaction takes place in a reactor at stst. when the rate of prod. removal equals that of introduction of the feed. A chemical stst. can be derived from a proc. where all *transition states are totally subsided since the beginning of the proc. and, therefore, the properties are time-independent and depend only, e.g., on the local conds. From that, reactions carried out continuously in *CSTRs or *PFRs take place in stst. conds., those discontinuous in batch-type reactors, e.g., *autoclaves, in unsteady-state conds. (*catalytic reactors). Concerning reactor characteristics, there is a stst. temp. at which heat generation (exothermic reaction) and heat removal as well as heat consumption (endothermic reaction) and heat supply become balanced. In the case of most applicable exothermic reactions one unstable and two stable stst. may exist. Concerning cat. experiments, the stst. is indicated by a constant de-

gree of conv. over time which is usually observed together with a constant *selectivity. In het. cat., especially in *oxidation reactions, fresh cats. need a period from some hours to several days to attain stst. conds.

P. CLAUS, D. HÖNICKE

Ref.: N.J. Themelios, *Transport and Chemical Rate Phenomena*, Gordon and Breach, London 1995.

2-approximation in enzyme cat. reactions: In order to simplify the rate equations for multi-step reactions (which includes all enzymatic reactions), a few different approximations can be made. In one, the ss. (or pseudo-ss.) approximation, the concs. of all intermediates are assumed to reach a ss. value which does not change appreciably.

$$E + S \underset{k_{-1}}{\overset{k_1}{\rightleftharpoons}} ES \overset{k_2}{\longrightarrow} E + P$$

Thus, for a simple enzymatic reaction in which a single substrate binds to the enzyme, undergoes one transformation, and then leaves, the conc. of [ES], the *enzyme-substrate *complex, would be essentially constant (following an initial transient period). After transformations

$$\upsilon = k_2[ES] = k_2[E]_o[S]/([S] + K_m)$$

is valid. This is still cast in the form of the *Michaelis-Menten equ., but unlike the case of the Michaelis-Menten mechanism, in which the initial binding step is assumed to be at equil. (with an equil. constant K_s), the *Michaelis constant is no longer equal to the dissociation constant for the enzyme-substrate complex. Instead, it is a combination of rate constants. Note that in the limit that $k_2 \ll k_{-1}$, K_m does approach K_s. **F** l'état stabil; **G** stabiler Zustand.   P.S. SEARS

steady-state isotope transient kinetics (SSITK) This *isotope-labeled variant (probe: isotope-labeled educts as pulse in steady-state feed; resonance: scrambled isotope labels in all comps. via *MS analysis) yields kinetic and mechanistic information. The method is non-destructive but needs time-resolved gas-phase analysis.　　　　　　　　　　R. SCHLÖGL

steam cracking　proc. for the non-cat. manuf. of lower alkenes by cracking of suitable distillates. To prevent *polymerization and *condensation reactions the partial press. of the alkenes is lowered by addition of steam. **F=G=E**.　　　　　　　B. CORNILS

steam reforming　converts natural gas with steam into a mixture of hydrogen, CO, and CO_2 acc. to

$$CH_4 + H_2O = CO + 3 H_2$$
$$(\Delta H°_{298} = 206.2 \text{ kJ} \cdot \text{mol}^{-1}) \tag{1}$$
$$CO + H_2O = CO_2 + H_2$$
$$(\Delta H°_{298} = -41.2 \text{ kJ} \cdot \text{mol}^{-1}) \tag{2}$$

This well-known proc. is used for the prod. of *syngas (CO + H_2) and H_2 for other chemical synths. For H_2 prod., CO is removed by the *water-gas shift reaction (equ. 2) and subsequent CO_2 removal. Sr. is also applied to convert naphtha resulting in a gas containing less H_2 acc. to the lower H/C ratio of ~2:1 in naphtha (*naphtha reforming). Several different procs. exist, e.g., *ICI, *Haldor Topsøe, Lurgi/BASF, British Gas, Kellogg, etc.

Endothermic sr. is performed in a two-step het. cat. gas-phase reaction using Ni as cat. *supported on α-Al_2O_3, MgO, or $MgAl_2O_4$ at temps. between 1000 and 1300 K. The feed gas has to be desulfurized in order to avoid cat. *deactivation. In the conventional technology, the natural gas is mixed with steam (H_2O/CH_4 ratio of 3.5:4) and at 3.5 MPa enters the (primary) tubular reformer, in which the gas is heated to ca. 1060 K. The necessary heat is supplied by allothermic heating from the outside in a fired furnace. For NH_3 plants a secondary, autothermic reformer follows in which air is added to the effluent gas of the primary reformer to supply N for *ammonia synth. O_2 and excess methane from the primary reformer are converted into syngas again over an Ni cat. In modern plants, the hot effluent from the autothermic reformer is used to provide the heat for the endothermic primary reformer. Thus, the modern autothermic sr. proc. operates with higher energy efficiency at lower press. and with a decreased steam/carbon ratio. The *activity of the Ni cats. is correlated to the Ni *surface area, which increases with the Ni content. However, the *dispersion of Ni decreases with the Ni content, resulting in an optimal Ni content of about 15 wt.% used in most commercial cats. Alkali cpds. in cats. are known to reduce the activity significantly. Reaction occurs by dissociation of CH_4 and H_2O on the Ni crystallites, releasing H_2 into the gas phase. Cf. also *cat. reforming, *carbon dioxide reforming. **E=F=G**.　　　　　　　M. MUHLER

Ref.: Twigg, p. 225; Ertl/Knözinger/Weitkamp, Vol. 4, p. 1819; Anderson/Boudart, Vol. 5; Farrauto/Bartholomew, p. 341.

Stein William (1911–1980), biochemist at the Rockefeller Institute for Medical Research in New York. Worked on the structure of ribonuclease. Nobel laureate in 1972 (together with Anfinsen and Moore).　　　　　　B. CORNILS

Ref.: Angew. Chem. **1973**, 85, 1074.

Stelzer reaction　the Pd-cat. *cross-coupling of *phosphines HPR_2 (R = H, aryl) with aryl halides or triflates. This method is advantageous for the prepn. of phosphines bearing polar or acidic functional groups which are not tolerated by the traditional synth. via *Grignard reagents. For the prepn. of symmetrical phosphines, PR_3, PH_3 or $P[Si (CH_3)_3]_3$ as starting materials can be used. The reaction also works with $HP(O)R_2$. For related reactions cf. *arene coupling. **F** réaction de Stelzer; **G** Stelzer-Reaktion.　　　V. BÖHM

Ref.: Catal. Today **1998**, 42, 413.

STEM　scanning transmission electron microscopy

stereochemistry　1-general: Most molecules are three-dimensional. All chemical syntheses dealing with spatial relationships contribute to s.　　　　　　H. BRUNNER

2-of NAD(P)-dependent reactions: Both NADH and NADPH contain two diastereotopic hydrogens (*pro-R* or *pro-S* H) that can be transferred as a hydride to an oxidized substrate such as aldehydes, ketones or imines. These substrates also contain two diastereotopic or enantiotopic faces (*re* or *si* face) at the sp^2 carbon center to be reduced. In principle, any NAD(P)H- dependent oxidoreduction should fall into one of these four types of stereospecificity. In practice, most alcohol *dehydrogenases cat. the transfer of *pro-R* H to the *re* face of a carbonyl substrate, a process summarized by *Prelog's rule. This rule applies to enzymes from yeast, horse liver, and *Thermoanaerobium brockii* (E$_3$ in the Figure). The alcohol dehydrogenase from *Mucor javanicus* is specific for the *pro-S* hydrogen of NADH and *si* face of carbonyl substrates (E$_2$) and that from a *Pseudomonas* species is specific for the *pro-R* hydrogen of NADH and *si* face of carbonyl substrates (E$_1$). A *pro-S/re* face-specific alcohol dehydrogenase has not been reported yet. To determine whether the enzyme is *pro-R* or *pro-S* specific for the reduced *cofactor, ^{1}H-NMR is considered to be the most convenient.

E$_4$ (*pro-S/re* face) is unknown. **F** stéréochimie; **G** Stereochemie.   C.-H. WONG

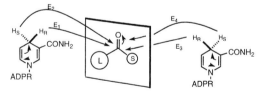

stereocontrol term used in *polymerization, → chain-end control, and *stereoselectivity, → enantiomorphic site control.

stereoelectronic control   → Felkin-Anh model.

stereoisomerism a sub-class of *isomerism, in which the isomers only differ in the arrangement of the atoms in space. Stereo-

isomers can be *enantiomers or *diastereomers.   H. BRUNNER

stereorigid metallocenes bridged, e.g., *ansa*-metallocenes

stereoselective radical reactions
→ Lewis acid-catalyzed radical reactions

stereoselective syntheses Ss. syntheses are subdivided into *enantioselective and *diastereoselective synth.   H. BRUNNER
Ref.: Mander.

stereoselectivity 1-general: S. is the prod. of one steroisomer of the product in preference to another. 2-of polymerization: Prochiral α-olefins can be inserted into the growing polymer chain in two different ways leading to either *S*- or *R*-configuration at the newly formed tertiary C atom. As introduced by *Natta, the resulting relative stereoisomerism of the polymer is called *tacticity. Without stereocontrol the *monomers are inserted randomly and an *atactic* polymner is formed. *Ziegler-Natta cats. and stereorigid *metallocenes offer the possibility of control the stereoregularity. The resulting *isotactic* and *syndiotactic* structures are more regular as shown in the Figure.

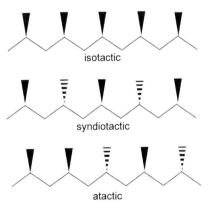

Interaction of the cat., the *monomer, and the polymer chain leads to *diastereotopic* *transition states necessary for the discrimination of the prochiral sites of the α-olefins.

Two mechanisms have been put forward as being responsible for the stereocontrol of the growing polymer chain: *chain-end control and the *enantiomorphic site control mechanism. The polymer properties depend strongly on the type of tacticity induced by the stereoselective polymerization and can be varied considerably. 3-in enzyme cat. reactions: → enantioselectivity in enzymatic cat. **F** stéréosélectivité; **G** Stereoselektivität.

W. KAMINSKY, I. ALBERS, O. PYRLIK

stereospecificity in enzyme catalyzed reactions, → enantioselectivity in enzymatic reactions

steric effects chemical effects caused by the size of a group or a *ligand. Considering an attack of a reagent on a (coordination) cpd., ses. can result in a stabilization of the cpd. or in directing the attack to the least hindered of sterically different sites. The kinetic stabilization gained by attaching *bulky ligands is generally known as steric hindrance and is expressed in the *Taft equ. On the one hand this effect can slow down or suppress desired transformations (e.g., *iso*-aldehyde formation by *hydroformylation) or limit the rotation of substituents around single bonds, being detectable by *NMR spectroscopy. On the other hand, ses. increase the inertness of a cpd. (e.g., $Zn[C(SiMe_3)_3]_2$ can be steam-distilled, $ZnMe_2$ inflames in air) and stabilize multiple element-element bonds or *radicals. Steric repulsion also favors low coordination numbers (e.g., $Pt(PMe_3)_4$ vs. $Pt(PPh_3)_3$ vs. $Pt[P(C_6H_{11})_3]_2$; $M[N(SiMe_3)_2]_3$ with M = Ln, Ti, Cr, Fe) and alternative coordination geometries (e.g., $Sn(\eta^5–C_5H_5)_2$ displays an Cp-Sn-Cp bond angle of $125°$, $Sn(\eta^5–C_5Ph_5)_2$ is linear) via thermodynamic stabilization. Though ses. usually decrease the reaction rates, an emhancement can occur when geometric restrictions (i.e., the orientation of the bond-forming atoms) favor the entropic contribution of the *activation energy. The directing ses. lead to different reactivities of comparable groups or of the two C atoms of

an olefinic double bond (e.g., *anti-Markovnikov *addition in hydrometallations), and to rearrangements of *transition states. In catalysis, the most important ses. are 1: increased cat. *activity by generation of electronically unsaturated species due to steric hindrance of the ligands (e.g., generation of *vacant coordination sites in *transition metal-*phosphine *complexes used as cats. for the *Heck reaction, *hydroformylation, or *hydrogenation reaction; 2: enhanced cat. activity resulting from a decreased activation energy due to a favorable orientation of the substrate molecule (e.g., induced fit in *enzyme cat. or the 10-fold enhanced *polymerization rate for propene by introducing a naphthyl group in the 4-position of *C_2-symmetric bis(indenyl)-zirconocene as compared to *Ziegler-Natta cats.); 3: increased educt-, *product-, and *enantio-/diastereoselectivity of the cat. via the directing effect. A prerequisite to understanding the last two effects is a detailed knowledge of the steric interactions between the cat. and the substrate molecules. *Molecular dynamics calcns. are helpful to understand these effects and facilitate systematic *ligand tuning. In general, ses. are difficult to separate from *electronic effects. **F** effets stérique; **G** sterische Effekte. J. EPPINGER

Ref.: *Adv.Organomet.Chem.* **1994**, *95*, 36; *Organometallics* **1987**, *6*, 650.

steroids → lanosterol, *squalene synthesis

Stevens rearrangement the rearrangement of ammonium ions to tertiary amines, cat. by bases. **F** réarrangement de Stevens: **G** Stevens-Umlagerung. B. CORNILS
Ref: Laue/Plagens.

Stex process → Toray procs.

sticking coefficient (sticking probability) The sc. *s* is the probability that a molecule or atom which hits the *surface sticks, i.e., $0 \leq s \leq 1$ has to be valid. Depending on the later fate of the atom/molecule a distinction can be made between the sc. for dissociative *chemisorption and the reactive sc. where only those atoms/ molecules are considered which adsorb and re- act. The sc. for a given adsorbate/substrate comb. is a function of the energy of the incom- ing particle, the surface temp., and all the adsorbate *coverages. For the same substrate but different *single-crystal orientations the sc. can vary over several orders of magnitude, thus indicating one major reason why a reaction can be *structure-sensitive (e.g., Pt/O_2, Fe/N_2). If the *activation energy for an *adsorption is pos- itive the sc. increases with temp.; if the *activa- tion energy is negative the sc. decreases with temp. The sticking of a molecule is a complex dynamic proc. which requires effective energy transfer from the molecule/atom to the sub- strate. In principle, the sc. depends on the energy distribution over the various vibrational and rotational states of the incoming molecule and on the velocity comps. parallel and vertical to the surface. Experimentally, the sticking of molecules/atoms can be studied state-resolved in molecular beam experiments in comb. with *laser spectroscopy. The sc. can be determined experimentally from uptake curves (coverage vs. exposure) provided the coverage calibration is known, or directly with the molecular beam technique by measuring the intensity of the re- flected beam. Cf. also *reaction probability, *adsorption. **E=F=G**.   R. IMBIHL

Ref.: R.P.H. Gasser, *An Introduction to Chemisorption and Catalysis by Metals*, Oxford University Press, Ox- ford 1985; Ertl/ Knözinger/Weitkamp, Vol. 3, p. 921.

sticking probability → sticking coefficient

Stille cross-coupling the cross-coupling reaction of organotin reagents with electro- philes acc. to R'-M + R''-X → R'-R''+ MX with M = -SnMe₃, -SnBu₃, -SnCl₃. The cat. metal of choice is Pd, *ligand-modified. The reaction has been applied to a number of total synths. of natural prods. **F** couplage croisé de Stille; **G** Stille-Kreuzkupplung.   V. BÖHM

Ref.: Beller/Bolm, Vol. 1, p. 165; V. Farina et al., *The Stille Reaction*, Wiley, Chichester 1998; *Angew.Chem.Int. Ed.Engl.* **1986**, 25, 508; Diederich/Stang; *Organic Reac- tions*, Vol. 50, p. 1.

STM scanning tunneling microscopy, → scanning probe microscopy

STO → single turnover, *turnover number

stoichiometric reactions transformations which follow stoichiometric rules even with those reactants which otherwise act catalyti- cally, e.g., stoichiometric *carbonylations with $Ni(CO)_4$ (Ni carbonyl instead of carbon mon- oxide as CO source) or *McMurray coupling. **F** réactions stoechiométriques; **G** stöchiome- trische Reaktionen.
Ref.: Falbe-3, p. 82.   B. CORNILS

Stone & Webster catalytic cracking (FCC) process for selective conv. of gas-oil feedstocks to high-octane gasolines, distil- lates, and $C_{3/4}$ alkenes. Cat. *regeneration is carried out in a single regenerator equipped with proprietary air rings and a cat. distribu- tion system. **F** procédé FCC de Stone & Web- ster; **G** Stone & Webster FCC-Verfahren.

                                            B. CORNILS
Ref.: *Hydrocarb.Proc.* **1996**, (11), 122.

Stone & Webster DCC technology proc. for the selective conv. of gas-oil feedstocks to C_2–C_5 alkenes, aromatic-rich, high-octane gasoline, and distillates using the deep *cat- cracking (DCC) method. DCC is a fluidized proc. with tradional *FCC cats. including two different modes: maximum propylene (*riser and bed cracking at severe conds.) and max. isoalkenes (only riser cracking at milder conds.) (Figure). **F** technologie DCC de Stone

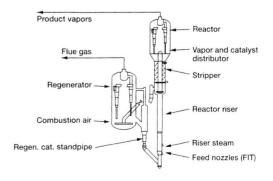

& Webster; **G** Stone & Webster DCC-Technologie. B. CORNILS
Ref.: *Hydrocarb. Proc.* **1996**, (11), 108 and **1997**, (3), 142.

Stone & Webster Resid Catalytic cracking Proc. for the conv. of gas oil and heavy residual feedstocks to high-octane gasoline, distillate, and $C_{3/4}$ alkenes by cat. and selective *catcracking in a special reactor system. The regeneration takes place in two stages. **F** procédé Resid Catalytic cracking de Stone & Webster; **G** Stone & Webster Resid Catalytic Cracking-Verfahren. B. CORNILS
Ref.: *Hydrocarb.Proc.* **1996**, (11), 142.

stop effect in β-*elimination reactions (e.g. the *hydration of alcohols to alkenes in the presence of *alumina), the se. is the time during which the alkene is produced, in amounts ever larger than in the steady state, after stopping the alkene feed. It is presumed that an alkene precursor is kept in a reservoir. **E=F=G**. B. CORNILS
Ref.: *Proc. 7th Int.Congr.Catal.*, Tokyo 1980, p. 853.

stopped-flow kinetics It is sometimes desirable to measure *enzymatic rates at very short times, before the *steady-state reaction rate has been reached. Monitoring *pre-steady state reactions can be accomplished with a very rapid mixing and detection system, such as a stopped-flow apparatus. P.S. SEARS

STR stirred tank reactor, → catalytic reactors

straight run fractions from the distillation of crude oil (*petroleum processing) prior to further processing. For quality reasons straight run prods. need cat. processes (e.g., *reforming) for improvement. **F=G=E**. B. CORNILS

Stratco *alkylation process for the manuf. of high-octane gasolines through reaction of C_3–C_5 alkenes with isobutane with strong liquid sulfuric acid. In the contactor the reactants are contacted at high velocities and ex-

tremely large interfacial areas. The regeneration of the acid may be possible on site (Figure). **F** procédé alkylation de Stratco; **G** Stratco Alkylierungs-Verfahren. B. CORNILS

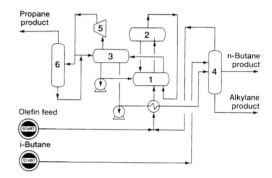

1 contactor (reactor); 2 acid settler; 3 flash drum; 4 deisobutanizer; 5 compressor; 6 depropanizer.

Ref.: *Hydrocarb.Proc.* **1996**, (11), 93.

strong metal-support interaction (SMSI) → metal-support interaction

structural promoter textural promoter, → promotion, *electronic promoter

structural relaxation → restructuring

structure-directed agents *templating molecules, cf. also *zeolites

structure-sensitive/insensitive catalysis The concept of structure sensitivity (ss.) in cat. is most fully developed and clearly apparent in the context of *supported metals. With the advent of techniques and selective gas *chemisorption for estimating the mean size of small metal particles, it became possible to see whether the rate per unit *surface area of a cat. reaction was or was not a function of size. If it was, then it was initially termed *demanding*, but this term was afterwards replaced by *structure-sensitive*. These names implied that the reaction in question required, in order to proceed efficiently, some quite specific grouping of *surface atoms, the population of which varied with particle size. With metals active in *hydrogenolysis of C-C bonds, specific rates increase with decreasing particle size, due to the need for atoms of low coordination number to form multiple C-M bonds. In NH_3 synth., certain crystal faces of Fe, e.g., Fe(*111*), bind N atoms better than others and so are more cat. active. Another aspect of ss. is shown by *alloys comprising active and inactive comps., as increasing the conc. of the latter alters the mean *ensemble size of the active member. Different crystal faces of oxides show different *activities and *selectivities in selective oxidations: this is an important manifestation of structure sensitivity.

It is broadly possible to classify a reaction proceeding on a metal particle as being either structure-sensitive or structure-*in*sensitive, depending upon whether or not the specific rate or *TOF is or is not a function of particle size. In practice the distinction is not always clearcut, and a reaction may be best described as having some degree of structure sensitivity. However, the rates of a number of reactions show zero or very weak dependence on particle size: these include *hydrogenation of alkenes, cyclopropane and alkynes, and isotopic equilibrations. Since change in *particle size is expected to result in significant differences to the probability of finding atoms of a specified coordination number, it is surprising that any reaction should be insensitive to such changes. It has been suggested that "insensitive" reactions in fact proceed on the uniform surface of a film of polymeric or carbonaceous deposit over the metal: more probably the reactants may mobilize surface metal atoms, eliminating differences between them. The mobility of such atoms on very small particles may be quite high, even in the absence of chemisorbed molecules. **E=F=G**. G.C. BOND
Ref.: Gates-1, p. 387; Farrauto/Bartholomew, p. 44.

structures of enzymes → secondary structure

STS → scanning tunneling spectroscopy

STY 1: → space time yield; 2: → site time yield

styrene synth. is from *ethylbenzene (EB) mostly by adiabatic *dehydrogenation in the gaseous phase over unsupported K-*promoted iron oxide at about $600\,°C$ ($\Delta H°_{600}$ = +124.9 kJ/mol). Dilution with heated steam (molar ratio H_2O/EB >10:1) is necessary for thermodynamic reasons, to supply heat, to remove carbonaceous *deposits, and to stabilize the Fe oxide against *reduction to the metallic state. Various procs. are commercial (e.g., *UOP's SMART, *Lummus classic styrene; *Toray Stex). **F** styrène; **G** Styrol. M. MUHLER

Styro-Plus process → UOP SMART process

substitute (synthetic) natural gas → SNG

substitution reactions are reactions in which atoms or groups (or *ligands or *central atoms in *complexes) are substituted by other atoms, groups, ligands, or central atoms and thus release parts of the original molecule. In contrast to *addition reactions, srs. always yield by- or co-products (a typical example is the *ARCO PO proc.) and are thus in this respect by definition disadvantageous in terms of *atom economy. Many srs. are catalyzed. **F** réaction de substitution; **G** Substitutionsreaktion. B. CORNILS

substrate inhibition (also referred to as *reactant inhibition* or *adsorbate inhibition*. 1-general: The *active site of a cat. can only be efficient if any reactant (or product) is only weakly bound to it and thus exhibits a high probability of *desorption. Only then is the *turnover time short and can the cat. be active. Often, however, a reactant, a prod., or an intermediate species is strongly bound to an active site. This will reduce the desorption probability and hence lower the turnover rate. Such a site blocking must be discriminated from *poisoning, where the active site is blocked by a foreign species which does not belong to the reaction system. The determination of reactant site blocking can occur via kinetic tests, and specifically by monitoring the turnover rate as a function of the abundance of each reactant in the system. Typical experiments are systematic variations outside the limits of stoichiometry, or pulse dosing of reactants. Deeper understanding of si. requires *temp.-programmed desorption experiments or *surface spectroscopic investigations. If it is an *intermediate that is inhibiting, the application of theory can help one to understand the effect which is experimentally intangible due to the limited stability of intermediates under nonreacting conds. (see *in-situ studies). Addition of *promoters which affect the binding properties of active sites (such as alkali atoms) or a complete change in the chemical constitution of the cat. are measures to avoid reactant inhibition.   R. SCHLÖGL

Ref.: Gates-1; Augustine; Falbe-1.

2-enzymatic: A substrate can act as an inhibitor of an *enzyme through a number of different mechanisms, e.g., the substrate binds in a non-productive mode to the active site.

$$E + S \rightleftarrows ES \rightarrow E + P$$
$$\Updownarrow$$
$$ES'$$
(nonproductive)

Enzymes suffering from substrate inhibition display nonlinear behavior in *Lineweaver-Burk and *Eadie-Hofstee plots, and the nonlinearity is most pronounced at high substrate concs. **F** inhibition des substrates; **G** Substrat-inhibition.   P.S. SEARS

Ref.: I.H. Segel, *Enzyme Kinetics*, Wiley, New York 1975.

substrates   The term has various meanings in catalytic sciences: reactants (general), *supports (in *het. cat.), or as complementary substrates to *enzymes.

subtilisins   a family of related bacterial serine *proteases secreted by members of the (Gram-positive) genus *Bacillus*. They are secreted as a *pre-proenzyme which remains attached to the membrane of the bacterium

until cleavage of the signal peptide and prosequence, which requires s. activity. Thus s. appears to act as its own "*signal peptidase". The prosequence is required for proper folding of the *enzyme, but it has been shown that the prosequence can be produced as a separate polypeptide chain and still act as a *chaperone for s. folding.

The subtilisins display broad amino acid specificity at most positions, and are stable toward both thermal and detergent-induced denaturation. For this reason, they are frequently used in laundry detergents. S. is capable of *hydrolyzing esters as well as amides, and because of its broad specificity has found diverse uses in synth. chemistry. It has been used in the *regiospecific acylation of sugars. S. and a mutant version, thiolsubtilisin, have proven to be useful in the *kinetically controlled synthesis of peptides, glycopeptides, and peptides containing unusual or unnatural amino acids from the corresponding ester substrates. **E=F**; **G** Subtilisine.   P.S. SEARS

Ref.: *J.Am.Chem.Soc.* **1991**, *113*, 1026 and **1990**, *112*, 945.

sucrose synthetase   cat. the transfer of glucose from *UDP-glucose to fructose to form sucrose. The fructose derivatives 1-azi-do-1-deoxy-, etc. have also been used as glycosyl acceptors in the sucrose synthetase-cat. synth. of sucrose analogs.

6-Deoxy- and 6-deoxy-6-fluorofructose can be generated in situ from the corresponding glucose derivatives with cat. by glucose *isomerase. The enzyme also accepts TDP, *ADP, and *GDP. **E=F=G.**   C.-H. WONG

Ref.: *J.Am.Chem.Soc.* **1984**, *106*, 5348; *Glycobiology* **1993**, *3*, 349.

Süd-Chemie catalysts

Süd-Chemie catalysts with roots back to 1857, comprises three divisions: adsorbents and *additives, catalysts, and environmental technology. Catalysts contributed DM 555 million (1997) to the group sales. Major group companies in the cats. area are: Houdry Inc. (USA), Nissan Girdler Catalysts Co. Ltd. (Japan), Prototec Company (USA), United Catalysts Inc. (USA), Süd-Chemie MT SrI (former Montecatini Techn., Italy), Syncat (South Africa), UCIL (India). Having an almost complete cats. portfolio, Süd-Chemie codes for their cat. prods. start with a letter followed by a number. C-catalysts are produced by United Catalyst Inc., G by Girdler, K are specially treated *montmorillonite *clays, T are products out of R&D (not yet in full-scale industrial use) and K+E catalysts are used exclusively for *hydrogenations.   J. KULPE

sugar nucleotide regeneration

sugar nucleotide regeneration Though analytical- and small-scale synth. using *glycosyltransferases is extremely powerful, the high cost of sugar nucleotides and the *product inhibition caused by the released *nucleoside mono- or diphosphates present major obstacles to large-scale synth. A simple solution to both of these problems is to use a scheme in which the sugar nucleotide is regenerated in situ from the *NDP released. The first example of the use of such a strategy is the *galactosyltransferase-cat. synth. of *N*-acetyllactosamine. A cat. amount of UDP-Gal is used initially to glycosylate GlcNAc; UDP-Gal is regenerated from the product UDP and galactose using an *enzyme-cat. reaction sequence which requires stoichiometric amounts of a *phosphorylating agent. Several oligosaccharides have been prepared using routes based on this concept (Figure). Regeneration of CMP-NeuAc follows the same basic principles as that for the regeneration of *UDP-Gal. The CMP released is converted to CTP and then to CMP-sialic acid for use as a *sialyltransferase substrate. The UDP-Gal and CMP-NeuAc regeneration schemes have been combined in a one-pot reaction and applied to synth. oligosaccharides such as sialyl Lewis X. The development of these regeneration systems, as well as the more recent development of regeneration schemes for UDP-GlcNAc, GDP-Man, GDP-Fuc, and UDP-GlcUA should facilitate the more widespread use of glycosyltransferases for oligosaccharide synth. **E=F=G.**

    C.-H. WONG

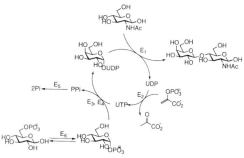

E_1: β 1,4-galactosyltransferase; E_2: pyruvate kinase; E_3: UDP-Glc pyrophosphorylase; E_4: UDP-Glc epimerase; E_5: pyrophosphorylase; E_6: phosphoglucomutase.

Ref.: *Anal.Biochem.* **1992**, *202*, 215.

sugar nucleotide synthesis

sugar nucleotide synthesis The initial step in the synth. of nucleotide-activated sugars is a kinase-mediated *phosphorylation to produce a glycosyl phosphate. This then reacts with an *NTP, cat. by a nucleoside diphosphosugar pyrophosphorylase, to afford an activated nucleoside diphosphosugar. Other sugar nucleoside phosphates, such as GDP-Fuc and UDP-GlcUA, are biosynthesized by further *enzymatic modification of these existing key sugar nucleotide phosphates. **F** synthèse des nucleotides de sucre; **G** Zuckernucleotidsynthese.

    C.-H. WONG

suicide inhibitors

suicide inhibitors → active site-directed inhibitors

sulfated (sulfided) zirconia

sulfated (sulfided) zirconia (SZ) a *solid acid cat. consisting of sulfate-promoted ZrO_2 or $ZrOCl_2$, suitable for, e.g., *alkylation, *acylations, *isomerizations at rather low temps. (cf. *zirconium as catalyst metal).

The structure of SZ is a point of debate. **F** zirconia sulfaté; **G** sulfatiertes Zirkon.

B. CORNILS

Ref.: *Catal.Today* **1989**, *5*, 493 and **1994**, *20*, 219; *J.Catal.* **1997**, *172*, 24; *Green Chem.* **1999**, *1*, 17; Ertl/Knözinger/Weitkamp, Vol. 4, p. 2058.

sulfided catalysts

1: Cats. which are exposed to H_2/H_2S mixtures to attain *directed poisoning. Under precise conds. such as temp., press., amount of S cpds., etc., the presulfiding influences *activity and/or *stability preferentially. Additionally, the resistance against autoignition during shipping to the user is reduced. In some cases, *supports are also presulfided (e.g., *directed poisoning, *sulfided zirconia).

B. CORNILS

2: The most powerful *hydrotreating and *hydrodesulfurization cats. are *supported Mo sulfides, manuf. in situ by reduction of supported Mo *precursor cats. by H_2S or by mixtures of organic sulfides with H_2S. The structural anisotropy of the layered MoS_2 materials as principal comp. of these cats. requires a strict control of the particle shape as only the perimeter of the platelet-shaped sulfide is cat. active. Modern high-efficiency sulfided molybdates carry *cocatalysts of Co, Ni, or mixtures of both. Their location on the sulfided platelets is of key importance for their function. Hence, the conv. of the oxidic to the sulfided form of the cat. is a very important step in the cat. manuf. and requires very strict process control. This is achieved by in-situ sulfidation in the commercial reactor. **F** catalyseurs sulfidés; **G** sulfidierte Katalysatoren.

R. SCHLÖGL

Ref.: Farrauto/Bartholomew, p. 102; *Ann.N.Y.Acad.Sci.* **1967**, *145*, 108; Ullmann, Vol. A5, p. 341; van Santen/van Leeuwen.

sulfochlorination

(chlorosulfonation, *Reed reaction) comprises the introduction of SO_2Cl groups into paraffins or cycloparaffins by reaction with Cl_2/SO_2 mixtures, initiated by UV light or *radical starters. **F** sulfochloration; **G** Sulfochlorierung.

B. CORNILS

sulfodehydrogenation

*dehydrogenation by means of H_2S/S mixtures (*Asahi MA proc.). **F=E**; **G** Sulfodehydrierung.

B. CORNILS

sulfonamides

(bis[sulfonamides], bs.) act as anionic N donor *ligands as hard *Lewis bases acc. to the *HSAB concept. They are deprotonated easily at their acidic NH function and serve as *ligands for highly Lewis-acidic metals such as Al, B, or Ti. In cat. only *complexes of *bidentate bis(sulfonamides) derived from readily available $*C_2$-symmetric diamines are important. They serve as chiral Lewis acid cats. for *Diels-Alder additions, where they are generated in situ from the reaction of $Al(Me)_3$ with a bs. Similarly, cats. for the enantioselective *alkylation of aldehydes are obtained from Ti isopropoxide and the corresponding bs., while dialkyl Zn serves as alkylating agent. *Enantioselective *cyclopropanations of allylic alcohols with diiodomethane (*Simmons-Smith reaction) can be achieved by Zn cats. obtained from the reaction of bs. with $Zn(Et)_2$. **F=E**; **G** Sulfonamide.

H.W. GÖRLITZER

Ref.: *Angew.Chem.Int.Ed.Engl.* **1994**, *33*, 497.

sulfonation

the introduction of sulfo groups $-SO_3H$ by reaction with oleum (H_2SO_4/SO_3). S. is important for the synth. of water-soluble *ligands such as *TPPTS or *BISBIS (cf. *aqueous-phase cat.). Sulfophenylphosphines are superior to carboxy- and amino-substituted *phosphines since they are easy to prepare and dissolve readily (often with a solubility >1 kg/L). Due to the oxophilic properties of trivalent P atoms, precautions have to be undertaken during sulfonation and workup in order to minimize the formation of undesirable phosphine oxides and sulfides. Addition of boric acid to the fuming H_2SO_4 prevents the majority of byprods. (cf. also *aromatic substitution). **F=E**; **G** Sulfonierung.

R. ECKL

Ref.: Cornils/Herrmann-2; *Angew.Chem.Int.Ed.Engl.* **1995**, *34*, 811; Kirk/Othmer-1, Vol. 23, p. 142; McKetta-2, p. 1125; *Organic Reactions,* Vol. 3, p. 141.

sulfotransferase (sulfotransfer reactions) Sts. comprise a family of *enzymes catalyzing sulfotransfer reactions, or the transfer of sulfonate (SO_3) from *3-phosphoadenosine-5'-phosphosulfate (PAPS) to an acceptor molecule. Sts., present in most organisms and in all human tissues, mediate sulfation of different classes of substrates for a variety of biological functions. Cytosolic sts., the largest subset of mammalian sts., are known to cat. the sulfation of aromatic or aliphatic hydroxyls or amines for the purpose of increasing solubility, detoxification, and excretion. Cytosolic sts. have been categorized into four classes. A second mammalian group, the membrane-bound sts., act in the *Golgi to sulfate endogenous molecules such as *glycoproteins, glycosaminoglycans, and protein tyrosines for activation of specialized biological functions. In plants, sts. have been found to regulate signaling and growth, flavonol sts. being the most studied class. **E=F=G**.   C.-H. WONG

sulfoxidation the sulfonation of paraffins by means of SO_2 and oxygen under UV irradiation or in the presence of *radical starters acc. to the overall equation R-H + SO_2 + ½ O_2 → RSO_3H (similarly to *sulfochlorination). The radical chain reaction starts with R• + SO_2 → R-SO_2·; in the presence of water RSO_3H and H_2SO_4 are formed. The reaction is commercially applied (*Hoechst sulfoxidation proc.). The alkylsulfonates produced have many uses. Since the reaction follows a *radical mechanism, alkenes and aromatics act as inhibitors by stabilization of radicals. Except from aliphatic or cycloaliphatic *hc. derivatives (e.g., alkyl chlorides) nitriles, carboxylic acids, alcohols, ethers, and esters are also accessible to s. The sulfonate groups are statistically distributed among the C-atoms of the alkyl chain, whereas sec. C-atoms are preferred over primary ones. **F=G=E**.   J.G.E. KRAUTER

Ref.: *Ann.Chem.* **1952**, *578*, 50; *Chem.Unserer Zeit* **1979**, *13*, 157.

Sulfreen process → Lurgi procs.

sulfur chlorides Sulfur dichloride is manuf. catalytically acc. to S_2Cl_2 + Cl_2 → 2 SCl_2 (cat. I_2). The manuf. of thionyl chloride (from SO_2 + SCl_2) and of sulfuryl chloride (from SO_2 + Cl_2) in both cases is cat. by charcoal. Disulfur dichloride (S_2Cl_2) is made from S and Cl_2, cat. by Fe chloride or Al chloride.

B. CORNILS

Ref.: Büchner/Schliebs/Winter/Büchel, p. 123; Büchel/Moretto/Woditsch.

sulfuric acid synthesis based on the *oxidation of SO_2 from roast gases to SO_3. The *lead chamber proc. uses *N*-oxides as the *hom. cat. Prevailing technology is the het. cat. *contact proc., in which kieselguhr-supported V_2O_5/K_2O is used to oxidize SO_2. The temp. must be chosen to achieve a reasonable reaction *rate and limit thermal decomp. of SO_3, i.e., typically 400–500 °C. Cleaning of the roast gases is necessary to avoid *poisoning of the cat. To form H_2SO_4, SO_3 is absorbed in 98 wt.% H_2SO_4 which is permanently diluted (cf. Bayer procs.). **F** synthèse d'acide sulfurique; **G** Schwefelsäuresynthese.   F. JENTOFT

Ref.: Büchner; Twigg, p. 503; Ullmann, Vol. A25, p. 635; Ertl/Knözinger/Weitkamp, Vol. 4, p. 1774; Farrauto/Bartholomew, p. 474; Büchel/Moretto/Woditsch.

sulfur in catalysis S-containing cpds. are wellknown to exert an *inhibiting or a *poisoning effect in cat. (*sulfur poisoning). In het. cat. S generally forms several ordered structures with increasing *coverage which decreases or even suppresses the reactivity of the metal cat. Although ferrodoxins, which are *clusters containing Fe and thiolato *ligands, play an important role in biosynthesis, S in hom. cat. is most often considered as having negative effects. Het. supported Mo or W sulfides, promoted by late *transition metals, are used to remove S from petroleum feedstocks (*desulfurization, *hydrodesulfurization). However, soluble Rh and Ru complexes are at present explored for *HDS, particularly to transform thiophenic molecules by hom. *hydrogenation or *hydrogenolysis. Recently, promising results have been obtained with *aqueous-phase cats.

S-contaminated CO or CO/H_2 gas contains mainly H_2S and COS. They react with the noble metal cats. to afford inactive complexes, e.g., HRh(CO) (PPh$_3$)$_3$ giving (for instance) Rh(SH)(CO)(PPh$_3$)$_2$. Additionally, in the first-generation procs. of *hydroformylation using HCo(CO)$_4$ cats. and their recycling with the help of H_2SO_4 (*regeneration), S cpds. could poison the Co progressively by incorporation of S, finally giving CoS$_2$.

Thiolato- or thioester-containing ligands have been shown to give very active Rh or Pt cats. in hydrogenation or more specifically in hydroformylation of alkenes. They have a low sensitivity to thiols. Similarly transition metal complexes containing sulfide ligands are active *precursors for various cat. reactions. **F** soufre en catalyse; **G** Schwefel in der Katalyse.   PH. KALCK

Ref.: Cornils/Herrmann-1, p. 969; Cornils/Herrmann-2, p. 477; Weissermel/Arpe; Falbe-1; *Surf.Sci.* **1998**, *395*, 268; *Inorg.Chem.* **1982**, *21*, 2857; *Polyhedron* **1988**, *7*, 2441; *New J.Chem.* **1988**, *12*, 687; *J.Organomet.Chem.* **1993**, *455*, 219 and **1994**, *480*, 177; *J.Mol.Catal.A:* **1998**, *136*, 279.

sulfur poisoning   S is the most prominent *poison for metal-containing cats. The reason is that S is strongly *chemisorbed by most of the cat. active metals and in some cases even converts the metals into sulfides (e.g., Ni$_3$S$_2$), even under reducing conds. (H$_2$ atmosphere in *hydrogenations). But also in cases where no sulfides are formed, the pronounced chemisorption to, e.g., precious metals blocks the active centers and therefore diminishes the cat. *activity. The poisoning effect of S is interpreted as the influence of its electronegativity on the density of electronic states of the metal and is classified as a short range phenomenon, i.e., one sulphur per metal atom is required to poison the cat. center. Based on this assumption, attempts have been made to quantify the influence of S or other poisons on a bulk cat. by the reduction of the available metal *surface area for cat. In contrast to the detrimental effect of S in *hydrogena-

tions, Mo or W sulfides are suitable cats. for *hydroprocessing operations accompanied by hydrogenation (cf. *S in catalysis). **F** empoisonnement par le soufre; **G** Schwefelvergiftung.   C.D. FROHNING

Ref.: Ertl/Knözinger/Weitkamp, Vol. 3; J.B.Butt, E.E.Petersen, *Activation, Deactivation and Poisoning of Catalysts*, Academic Press, New York 1988; *Adv. Catal.* **1951**, *3*, 129; *Catal. Lett.* **1988**, *1*, 1.

sulfur recovery   → Claus process

sum frequency generation (SFG)   a laser spectroscopy variant yielding integral, *surface-sensitive information about in-situ vibrational analysis of small adsorbed molecules. The ex-situ method is non-destructive, the interpretation is like that of *IR. Because of nonlinear optical effects the interpretation may be difficult.   R. SCHLÖGL

Sumitomo cymene/cresol process   an extension of the *Hock proc., converting *m*- and *p*-isopropyltoluene (cymene) to *m*- and *p*-cresol. **F** procédé de Sumitomo pour des crésols; **G** Sumitomo Kresolverfahren.
                B. CORNILS

Sumitomo HCN process   manuf. of hydrogen cyanide by *ammonoxidation of methanol at 460 °C over het. Mo-Bi-P oxides. **F** procédé Sumitomo pour HCN; **G** Sumitomo HCN-Verfahren.   B. CORNILS

Sumitomo isopropyltoluene process for the *alkylation of toluene with propylene over a solid phosphorus acid (*SPA) cat. The reaction mixture contains over 60 % of the *m*-isomer.   B. CORNILS

Sumitomo sebacic acid process   manuf. of sebacic acid through a four-step reaction from naphthalene; 1: naphthalene is cat. *hydrogenated to *cis*-decalene; 2: the decalene is *oxidized to the hydroperoxide, and 3: rearranged to 6-hydroxycyclodecanone and hydrogenated to the corresponding cyclodecanol; 4: the cyclodecanol is *oxidized with nitric acid to sebacic acid. **F** procédé Sumitomo pour l'acide sébacique; **G** Sumitomo Sebacin-säure-Verfahren.   B. CORNILS

Sumner James B. (1887–1955), professor of biochemistry at Cornell (Ithaka/NY). Worked on crystallized enzymes. Nobel laureate in 1946 (together with Northrop).   B. CORNILS
Ref.: *J.Biol.Chem.* **1938**, *125*, 33.

Sun Oil hydrodealkylation process is very similar to *Union Oil's Unidak process

superacids (superacidity) The general definition refers to equil. liquid mixtures which bring about a higher specific conc. of reactive protons than mineral acids. Such systems can be generated with mixtures of, e.g., HF and fluorides and find application in the *activation of C-H bonds via formation of *carbocations. In het. cat. the term is much less well defined and designates oxide solids which bring about the same reactivity as the well-defined liquid superacids. In solids, however, no equil. between an acid (protons) and the conjugate base can be established. Thus the thermodynamic definition of acidity in liquids is changed to a kinetic definition (acidity is set equal to a reaction rate and not to a conc. in equil.) for solids.
A typical example is the application of *sulfated zirconia for the *isomerization of *n*-butane. The term s. is critical in such cases as it is not known whether the reaction which occurs in the liquid phase as an acid-cat. reaction is in fact also acid-cat. as a gas-solid reaction (it could be a *hydroisomerization with intermediate *dehydrogenation). The term remains vague in het. cat., in particular, as no reliable method of determination for superacidity is yet known. The often used *chemisorption of *Hammett indicators is very ambiguous as the type of chemical interaction between the Hammett molecule and an oxidic solid is far from being understood. **F** superacidité; **G** Supersäuren.   R. SCHLÖGL

superbases materials with basic sites stronger than H = 26 (with H corresponding to the pK_a value of the indicator) consisting of alkali metal hydroxides and the alkali metal itself supported on γ-Al_2O_3 ($MOH_x/M_y/$ support; x = 5–15 wt.%; y = 3–8 wt.%). Ss. are cats. (via *carbanions/deprotonation) for *alkylations, *aldolizations, *isomerizations, etc. **F=E**; **G** Superbasen.   B. CORNILS
Ref.: EP 0.211.448 B1; Ertl/Knözinger/Weitkamp, Vol. 5, p. 2131; *J.Mol.Catal.A:* **1999**, *144*, 181.

super-cages By a special way of linking *alumosilicate/ sodalites, a large cage, the so-called super-cage with a diameter of ca. 13 A and access windows of 7.4 A, is formed. This cage may be essential for special cat. action (cf. *zeolites).   P. BEHRENS

supercritical (sc) phase catalysis 1-general: Supercritical fluids (SCFs) are miscible with gases and can dissolve solids and liquids, but appreciable solvent power (*solvents for catalysis) is usually obtained only at densities close to or above the critical density of the pure solvent. The complex phase behavior of mixtures of cpds. requires careful control of the number of phases during reactions involving SCFs. Beneficial effects associated with the specific properties of SCFs have been observed for het., hom., and enzymatic cat. Reaction *rates can be enhanced compared to liquid-phase procs. because of typical gas-like properties, such as high diffusion rates, low viscosity, the absence of surface tension, or the miscibility with other gases. Advantages over gas-phase procs. may arise from better heat transport capacities or the ability to dissolve waxy prods. which would otherwise block access to the *active sites (reduced coking). Changes in *selectivity have been associated with the high compressibility around the critical point, allowing alteration of the density with comparatively small changes in temp. and press. Density changes can be further utilized for effective isolation of prods. and *recycling of hom. cats., or for controlled precipitation of polymers. Other solvent properties of SFCs also vary with press. and temp., thus providing additional potential for reaction tuning. Furthermore, chemical interactions between the SCF and substrates, prods., or cats. are possible.

Commercially important cat. procs. involving supercritical phases include NH_3 synth., *ethylene polymerization (especially *LDPE), and *hydration of light alkenes. Cat. oxidative destruction of organic waste in supercritical water (SCWO) is emerging as another technical application. Sc CO_2, which is commercially well established in extraction of natural prods., is gaining increasing attention as a possible reaction medium for cat. synth. No problems with solvent residues are encountered and its use is environmentally benign compared to that of most organic liquids.   W. LEITNER

2-in enzymatic reactions: One of the most useful SCFs for *enzymatic cat. is supercritical CO_2. Enzyme-cat. reactions in sc CO_2 have potential application in the food and drug industries. Lipases, phosphatases, and other enzymes have successfully been used in this solvent. **F** catalyse en phase supercritique; **G** Katalyse in superkritischen Phasen.   P.S. SEARS

Ref.: P.G. Jessop, W. Leitner (Eds.), *Chemical Synthesis Using Supercritical Fluids*, Wiley-VCH, Weinheim 1999; Ertl/Knözinger/ Weitkamp, Vol. 3, p. 1339; *Chem.Rev.* **1999**, *99*, 353 (and other contributions of this special issue); van Eldik/Hubbard; *Methods Enzymol.* **1997**, *286*, 495; *Science* **1988**, *238*, 387; Murahashi/ Davies, p. 1.

supermicropores → micropores

superoxide dismutase (SOD) a *metalloenzyme (Cu-Zn in mammalian systems, Mn in *E. coli*) that cat. the conv. of the two superoxide radicals (+ two protons) to $H_2O_2 + O_2$, cf. *auto(o)xidation. **E=F=G**.   P.S. SEARS

Ref.: *Cattech* **1997**, *1*, p. 41.

support (catalyst support) 1: The performance of many solid cats. is directly proportional to their *surface area. Increasing the surface area of the active comp(s). of the cat. is one function of the s. (carrier); maintaining a high dispersion of the active comp(s). is the other function. In order to fulfill these functions, a s. material is usually a highly porous and thermostable material. In addition, the s.

might influence the cat. activity of the active component (*metal-support interaction, MSI) or it might be cat. active itself, thus forming a *bifunctional cat. together with the active comp. Ss. are most often used for metallic cats., especially for precious metals such as Pd, Pt, or Rh. Since unsupported metals usually have a very low *dispersion, it would be very uneconomical to use expensive metal cats. without ss. The choice of an appropriate s. can increase the dispersion of the active metal to close to 100 %. There are also examples where nonmetallic cats. are supported, e.g., V oxide on SiO_2 in the sulfuric acid cat. or V oxide, together with other oxides, on TiO_2 in *SCR cats. In this type of cat., the active phase usually spreads as a monolayer on the immobile s. oxide, but can also be present as a liquid phase in the *pore structure of the s., as is the case for the sulfuric acid cat. Typical loadings of ss. are in the range of 1 wt.% for noble metal cats., but can be as high as 20 % for base metal oxide cat. Supported cats. have to be compared with *bulk cats.

The choice of an s. material is usually governed by several factors, the most important of which are: thermostability, stability toward the feed, the chemical nature of the active phase, *mechanical strength, and price. The most frequently used ss. are *alumina, mostly as γ-Al_2O_3, *silica, and *carbon (*activated carbon, *charcoal) in various forms. Oxide supports are usually prepared via *precipitation or *sol-gel procs. Activated carbon supports are typically formed by pyrolysis of carbon-containing natural or synth. polymers; *carbon black is formed by controlled heating of *hydrocarbons in an oxygen-deficient atmosphere. The Table lists several important s. materials together with some important properties. In addition to pure oxides and carbon, more complex s. materials, such as *alumophosphates, silicon carbide, *mullite, zirconium-silicate, *kieselguhr, *bauxite, or calcium aluminate have occasionally been used as s. materials.

Most of the s. materials mentioned above are not purely oxidic unless treated at very high

Support	Crystallographic phases	Properties/applications
Al_2O_3	Mostly α and γ	SA up. to 400; thermally stable three-way cat., *steam reforming and many other cats.
SiO_2	Amorphous	SA up to 1000; thermally stable; *hydrogenation and other cats.
Carbon	Amorphous	SA up to 1000; unstable in oxid. environm., *hydrogenation cats.
TiO_2	Anatase, rutile	SA up to 150; limited thermal stability; *SCR cats.
MgO	fcc	SA up to 200; rehydration may be problematic; *steam reforming cat.
Zeolites	Various (faujasites, ZSM-5)	Highly defined pore system; *shape selective; *bifunctional cats.
Silica/alumina	Amorphous	SA up to 800; medium strong acid sites; *dehydrogenation cats.; bifunctional catalysts.

SA = surface area in $[m^2/g]$

temps. above 1000 K, but contain varying amounts of water in the form of surface hydroxyl groups. In general, the higher the treatment temp. of the s. the lower the water content, but the surface area also usually decreases with the treatment temp. O-containing species are even present in high concs. on carbon surfaces, where phenol, carbonyl, carboxyl, or quinone groups have been identified. Depending on the raw material and the conds. of preparation of carbon-type ss., such materials may contain up to 20 % of oxygen. The surface hydroxyl groups of the oxides and the carbon s. are very important during the loading of the active comp. on the s., since the OH groups are sites at which the *precursor for the active species interacts with the support. A s. can be loaded with the active comp. by various techniques, the most important ones being *impregnation, *ion exchange, *deposition/*precipitation, spreading and *wetting, and *grafting. Depending on the conds., it is possible to produce hom. loaded cats. or cats. with inhom. profiles of the active comp., such as egg-shell or egg-yolk type cats. (*metal distribution). In most cases the form in which the active comp. is loaded onto the s. is not the cat. active form, but a metal salt or *complex which has to be reduced to the metal. In such cases, the loaded s. has to be heat-treated (e.g., *calcined) and/

or *reduced (activated) to yield the active cat. The s. often also provides the reactive surface groups for the *shaping of the cat. into pellets. Surface hydroxyls can condense and eliminate water and can thus *cross-link individual particles. To facilitate this proc. and to strengthen the resulting particle, a *binder with higher reactivity is also used. A special type of shaped s. is the *ceramic *monolith. Such monoliths are frequently not providing high surface area or keeping the active phase in a highly dispersed state, but are merely mechanical supports (*ceramics in catalysis). Such ceramic monoliths are used primarily in automotive exhaust treatment systems. For supported hom. cats.; cf. *immobilization. Manufacturers of ss. are Grace-Davison, Bayer, Alcoa, Condea, Norton, Degussa, Haereus, Cabot, etc.   F. SCHÜTH

2: in enzymatic respect cf. *bacterial display.
F support; **G** Träger.

Ref.: Stiles; Hartley; Yermakov/Zakharov/Kuznetsov; Ullmann, Vol. A5, p. 347; McKetta-1, Vol. 7, p. 1; van Santen/van Leeuwen.

supported (solid) aqueous-phase catalysis (SAPC) the conv. of liquid-phase hydrophobic organic reactants over SAP cats. The cat. consists of a thin film of water that resides on a high-surface-area *support, such as controlled-pore glass or *silica, and contains an active hydrophilic organometallic complex. Reactions take place at the water/ film organic phase interface and are cat. by the phase-*immobilized, organometallic complex (Figure). Other hydrophilic liquids, e.g., glycols, can be used instead of water for the formation of the immobilized liquid layer.

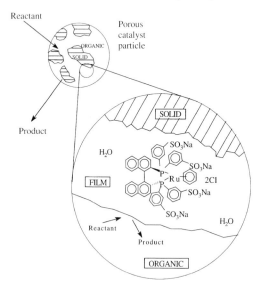

Reactions that have been accomplished successfully with SAPC are *hydroformylation, *Wacker oxidation, allylic *alkylation, alkene *isomerization, and asymmetric *hydrogenation. Extensions to immobilized and cat. *antibodies are possible. **E=F=G**.   M.E. DAVIS

Ref.: *Nature (London)* **1989**, *339*, 454 and **1994**, *370*, 449.

supported catalysts contrast with *bulk cats. which consist only of cat. active ingredients without *support. Scs. are – properly speaking – connected with *het. cats., but supported (*anchored) hom. cats. are also known (*SAPC, *SLPC, *SSPC).   B. CORNILS

supported heterogeneous catalysts As the active sites on a het. cat. are located on its *surface, the preferable way to increase cat. *activity is to have a large cat. active *surface area exposed to the reaction medium, especially when costly active comps. such as noble

metals are used (the opposite of supported cats. are the *bulk cats.). However, the use of cat. *nanoparticles for this purpose may often not be possible, due to the high rate of *sintering of these nanoparticles in the conds. necessary for good cat. *activity. By supporting the cat. particles over an oxide or carbon material (cf. *supports) it may be possible to separate the active particles physically and thus reduce the rate of sintering, but good *stability requires the formation of some form of bonding between the *support and the metal in order to overcome the thermodynamic tendency to reduce the free surface energy of the metal particles. The *support, as a consequence of this interaction, influences the properties and reactivity of the supported metal particles, as well as in a number of cases also directly participating in the mechanism of cat. reactions or playing a role in mitigating active metal deactivation, e.g., by *coke *deposition (cf. *metal-support interaction).

Several widely commercially used cats. are based on supported metal cats., such as noble (Pt, Pd, Rh) or non-noble (Ru, Ni, Fe, Co) metals supported over oxides (*alumina, *silica, *active carbon, and to a lesser extend *titania) for *hydro-dehydrogenation reactions, Ag on alumina for ethene epoxidation, Pt-Rh-Pd on alumina (or modified supports) for NO_x, *hc., and CO removal in *automotive exhaust gases, Pt or Pd on alumina for *cat. combustion of diluted organics in air, supported Pt-based cats. for *cat. reforming, *isomerization of petroleum fractions, etc. Several *additives or *co-catalytic elements are also usually present in these cats., both to improve reactivity and resistance to *deactivation and to enhance mechanical and thermal stability properties of the cats. (cf. *mechanical properties). The *shape of the final cats. may be either *pellets of various forms and dimension or *monoliths.

In the preparation of smcs. it is usually necessary to have a high surface area of the reduced metal deposited on a highly *porous material. Smcs. are generally prepared by reducing a *precursor metal salt which has been added to the support, by either *coprecipitation, *deposition, or selective adsorption/reaction. The choice of the preparation method depends on several factors, but may considerably influence the cat. behavior of the final cat. Also of particular importance, especially for noble metal-based cats., is the conc. profile of the metal along the depth of the support, in order to obtain a better compromise for activity and stability in the presence of deactivation reactions (cf. *metal distribution). **F** catalyseurs hétérogènes supportés; **G** heterogene Trägerkatalysatoren.   G. CENTI

Ref.: Moulijn/van Leeuwen/van Santen; *Chem.Rev.* **1995**, *95*, 523; M.W. Twigg, *Catalyst Handbook*, Wolfe, London 1989; *Catal.Today* **1997**, *34*, 281; Augustine.

supported homogeneous catalysts

This topic includes the use of so-called supported (anchored or *immobilized) metal *complex cats. Anchoring can be achieved by fixation via covalent bonding through donor *ligands anchored to a *support. Organic supports used include polystyrene and styrene/divinylbenzene *copolymer beads, polyvinyls, polyacrylates, and cellulose, the latter being less important. Attached ligands may include diphenylphosphine, tertiary amino, cyanomethyl, thiol, or cyclopentadienyl groups. Generally, a soluble metal complex cat. is reacted with the functionalized polymer in a *ligand displacement reaction, or weakly bridged dimeric metal complex species are split via reaction with the functionalized polymer. Oxidic supports, having a more rigid structure offer a higher stability to temp., solvent, and ageing, and a defined *pore structure that is independent of solvent, temp., and press. Inorganic matrices (such as *silica, *clay, *γ-alumina, *magnesia, *glass, or *ceramics) however provide an upper limit of functional groups of 1–2 meq/g of polymer, whilst organic polymers can carry up to 10 meq/g. The pore sizes of *zeolites are too small. Direct bonding to OH surface groups can also be used; e.g., for the anchoring of metal carbonyl complexes, carbonyl *clustes, or polymerization cats. More common is attachment via a *spacer group which is anchored to the support on one side and acts as a ligand group on the other side. High uniformity of the cat. active surface centers can be guaranteed by reacting a preformed complex bearing silicon-substituted ligands with the support. Alternative concepts involve *polymerization or *polycondensation of suitably functionalized monomeric metal complexes. *Chemoselectivities of immobilized and not immobilized metal complex cats. are similar, whilst due to the diffusional restrictions the *activities are lower than those of the corresponding hom. systems. Fixation to a support by ionic bonds is used also for, e.g., anionic Rh and cationic Pd complexes (*carbonylation) and cationic Ru or Rh complexes (*hydrogenation). The method of ionic fixation is used especially when dealing with *chiral metal complexes to perform *enantioselective hydrogenations. Metal complexes can also be *chemi- or *physisorbed via surface bonding. Simple *impregnation with a liquid containing the metal complex and subsequent removal of the solvent, e.g., taking an inorganic support and dissolved carbonyls, are recommended. Sublimation of metal carbonyls onto the support or metal-atom synth. to produce labile complexes absorbed in porous supports to form cat. active metal cluster cpds. is applied. *SLPC or *SAPC are yielded by impregnation of a solid support with a liquid medium containing a dissolved hom. cat. The *dispersion of the cat.-containing liquid phase onto the high-surface area porous support creates a large gas-liquid transfer or interfacial (liquid-liquid) area. The effectivity of these cats. depends strongly on the degree of pore filling.

Immobilization is not a universally applicable concept, as *leaching of the immobilized metal centre may take place, particularly where strong complexing reactants, e.g., CO in the case of *hydroformylation, are present. **F** catalyseurs homogènes supportés; **G** geträgerte Homogenkatalysatoren.

                                       P. PANSTER, S. WIELAND

Ref.: Hartley; Yermakov/Kuznetsov/Zakharov; Montanari/Casella; Cerveny; Clark; Blackborrow/Young; *J.Catal.* **1966**, *6*, 139 and **1969**, *14*, 142; Mobil, BE 721.686 (1968); *Angew.Chem.Int.Ed.Engl.* **1986**, *25*, 236.

supported (solid) liquid-phase catalysis

(SLPC) SLPC is a technique in which the solution of the hom. cat. in a high- boiling *solvent (e.g., excess *ligand or phthalates; cf. *plasticizers) is distributed in the *pore space of a porous *support under the action of capillary forces (*pore volume impregnation, *capillary impregnation). The reactants pass the cat. in gaseous form. The technique is difficult to control and so far ineffective (cf. *SAPC). **F=G=E.**   B. CORNILS

Ref.: *J.Mol.Catal.* **1980**, *9*, 139,157,241,257,265; Cornils/Herrmann-1, p. 619.

supported solid-phase catalysis (SSPC)

The cats. used for SSPC belong to the group of cats. proposed for heterogenizing hom. cats. The main aim of the research activities is to combine the advantages of het. and hom. cat., but avoiding their disadvantages (*immobilization). In comparison to other types of cats. of this group, e.g., *SLPC, in SSPC the cat. active comp. (mostly a *transition metal *complex) is attached to a solid matrix. In the simplest case the fixation proc. is achieved by adsorption forces using an inorganic solid support such as *silica or *charcoal. For use in liquid-phase reactions, the fixation technique mentioned is ineffective since the active comp. is rapidly washed off the solid support. To overcome this problem the system is heterogenized by chemically bonding the complex to the support. Functionalized inorganic matrixes as well as organic *polymers are used as supports in this case (*supported hom. cats.). Despite all research activities during the last 30 years no commercial use has been noticed so far. One of the main disadvantages with this kind of cats. is the severe *leaching. **E=F=G.**   D. HESSE

Ref.: Iwasaga.

support oxides → oxide catalysts

supramolecular templates for catalysts

different molecular species bind to each other by non-covalent interactions (such as *van der Waals, charge-transfer, electrostatic, hydrogen- bonding, or coordinative) to form supramolecular aggregates. For *enzymes, supramolecular interaction can lead to the binding of a substrate to a receptor and eventually induce a chemical reaction which yields a chemical modification of the substrate. Thus the receptor acts as a *template and cat. the reaction which the substrate undergoes. Dissociation of the modified substrate and attachment to another binding site enables a *cat. cycle to take place. When the substrate itself can act as a cat. for some kind of chemical reaction, the receptor is a supramolecular template for a catalyst (stfc.). An impressive naturally occurring example is double-stranded *DNA. It can act as a template which is transcribed by *enzymes to form *ribozymes or proteins. The RNA-type ribozymes as well as proteins are active as cats. for chemical reactions, e.g., forming C-C bonds; Cf. *C-C bond formation.

Like other examples in the chemistry of cat. *antibodies or imprinted polymers, small molecules (*haptenes) which resemble a model for the overall structure of the *transition state of a reaction are used as sts. The templates are removed after formation of the *polymers or *antibodies, respectively. Both materials possess binding sites for the *transition state of the reaction and are thus able to stabilize the transition state and to catalyze the reaction. Specific reactivities in order to control the *regio- as well as the *stereochemistry of the reaction can be addressed by the choice of appropriate templates. Stfcs. open the way to a rational design of cats. (*tailoring of cats.) in those cases where a specific reaction mechanism and the structure of the transition state are known and can be imitated by simpler molecules. **F** templates supramoléculaires pour catalyseurs; **G** supramolekulare Template als Katalysatoren.

M. ALBRECHT

Ref.: *Nature (London)* **1998**, *395*, 223; *Angew.Chem. Int.Ed.Engl.* **1995**, *34*, 1812; D.N. Reinhoudt (Ed.) *Supramolecular Technology*, Elsevier, Amsterdam 1999.

surface   of particular importance in het. catalysis since it is at the surface that the cat. reaction takes place (cf. *catalyst surface). The s. of a solid or a liquid is the termination of the bulk state, that is to say, the region where the equs. based on three-dimensionality are no longer sufficient to describe the complete physical state of the system. Thus a s. is not necessarily confined to the topmost layer of atoms of the solid or liquid, but may consist of several such layers extending into the bulk. **E=F**; **G** Oberfläche.   R. SCHLÖGL, A. SCHEYBAL

Ref.: Dines/Rochester/Thomson; Somojai; Gates-1, p. 220; Farrauto/Bartholomew, p. 154; I.M. Campbell, *Catalysis of Surfaces*, Chapman and Hall, London 1988; van Santen/van Leeuwen.

surface area   The surface area of a cat. is the real sa. on an atomic scale. It is properly expressed as surface area per mass of catalyst (m^2/g). Often the total sa. of real cats. consists of the *active* surface and the catalytically *inert* sa. Normally the rate per unit area of active sa. (usually per m^2) is used to specify the *activity of a cat. Comparisons between different cats. have to be made on this basis.

It is normally desirable for the cat. to have a high sa. There are two ways to obtaint a high surface: making the particle size very small or making the cat. material porous. Use of very fine powders as cats. is difficult and the ratio of the sa. to cat. mass is much lower than that of a porous material. Formation of a large number of very fine channels or pores through the particles is very efficient. The area of a rough surface is the external sa., whereas the area of the pore walls is the internal sa. To obtain a high surface of the active cat. comp., the active cat. comp. is often dispensed on an (in principle) inert *support, e.g., *zeolites or *alumina. The *Langmuir adsorption isotherm can be used to determine the sa. of a non-porous cat. If the cat. has a porous surface, *BET isotherms are used to

determine the sa. including pores. In some cases selective gas *chemisorption can used to determine the active surface of a cat. For instance, molecules such as H, O, and CO chemisorb selectively on metals at rt. **F** occupation de surface; **G** Oberflächenbelegung.

R. SCHLÖGL, I. BOETTGER

Ref.: S.J. Gregg, K.S.W. Sing, *Adsorption, Surface Area and Porosity*, 2nd Ed., Academic Press, London 1982; Farrauto/ Bartholomew, p. 81.

surface catalysis   → heterogeneous catalysis

surface coverage   is the amount Θ of adsorbed cpds. in relation to the monolayer capacity; cf. *capillary condensation, *overlayers.   B. CORNILS

surface organometallic chemistry

(SOMC) the study of the reactivity of organometallic cpds. with *surfaces, the characterization of the resulting surface organometallic *complexes and their stoichiometric and cat. reactivities. The surface can be either that of metal particles, oxides or *zeolites. The approach is different from that of het. cat., where the knowledge is more empirical, therefore producing very complex and undefined structures. SOMC corresponds to a rational approach towards het. cat. where the cat. *sites are designed using principles of *coordination chemistry and extending them to surface chemistry. This has led to the synth. of many new surface complexes, which can show unusual reactivity that is different from that of their hom. *precursors.

For instance, tetrakis(neopentyl)zirconium reacts with the OH group of *silica dehydroxylated at 500 °C ($\equiv$SiOH) to form a well-defined surface organometallic complex, $\equiv$SiOZrNp$_3$ (Np = neopentyl), which can further react with H$_2$ at 150 °C to form the surface hydride ($\equiv$SiO)$_3$ZrH; both of these were fully characterized and their formation mechanistically understood. These complexes are formally 8e species, which is unusual in classical organometallic chemistry. They also express unexpected reactivities, e.g., the *hydrogenolysis of alkanes and the selective hydrogenolysis of *polyolefins such as *PE and *PP under mild conds. In the case of Ta, which displays a surface chemistry similar to that of Zr, this approach has allowed the discovery of a new catalytic reaction, the *metathesis of alkanes, which transforms a hydrocarbon into its lower and higher homologs. Finally, SOMC shows that the surface cannot be considered as an inert *support as in the case of immobilized hom. cats. Firstly, it allows the formation and the stabilization of highly electrophilic and coordinatively unsaturated species (no dimerization of reactive intermediates). Secondly, the surface and the organometallic complex can play a synergistic role to fine-tune the properties of the cat. obtained by SOMC. F chimie organometallique des surfaces (COMS); G organometallische Oberflächenchemie.

C. COPERET, F. LEFEBVRE, J.-M. BASSET

Ref.: *Science* **1996**, *271*, 966 and **1997**, *276*, 99; *Angew. Chem. Int.Ed.Engl.* **1998**, *37*, 806; *Topics in Catalysis* **1997**, *4*, 211; *Coord.Chem.Rev.* **1998**, *180*,1701; Basset et al.

surface resonance, surface states → electronic structure

surface-sensitive extended X-ray absorption fine structure (SEXAFS) Probing with X-rays with variable wavelengths determines local, *surface-sensitive information about local atomic bonding geometries of adsorbates. The UHV method is non-destructive and uses *single crystals. R. SCHLÖGL

suspended catalysts solid, het. cats. which are suspended (dispersed, slurried) and finely divided in liquids or *molten media. The particle size ranges from mm to *colloids and even *clusters. This "moving-bed" technique offers some advantages: in particular it avoids conc. at temp. gradients and has no restrictions concerning the minimum particle size (as fluidized-bed cats. do). The heat economy in *slurry-bed reactors is excellent; parts of the cat. may easily be removed in a bypass.

*STR, ebullient-bed, or three-phase bubble column reactors are usual. Scs. in the gas phase are entrained-bed cats. (cf. *catalytic reactors). The *Synthol variant of the *Fischer-Tropsch synthesis in an example of an sc. F catalyseurs suspendus (en suspension); G Suspensions (Slurry)-Katalysatoren. Ref.. Thomas/Thomas, p. 483. B. CORNILS

Suzuki coupling the Pd-cat. reaction of aryl or alkenyl halides with organoboron cpds., especially boronic acids and boranes, yielding dienes, styrenes, and biaryls (Equ.).

Sc. proceeds via *oxidative addition of the aryl halide to a low-coordination Pd(0) species, usually a Pd(0) diphosphine *complex. The halide in the σ-aryl Pd(II) species is substituted by the aryl group of the organoboron reagent by *transmetallation. Aryl-aryl reductive elimination leads to the biaryl prod. and the cat. active Pd(0) complex, which can be recycled. A base (e.g., hydroxide, alkoxide, carbonate) or nucleophile (e.g., fluoride) is needed for activation of the organoboron cpd. toward the *transmetallation step.

Arylboronic acids are stable toward air and water, so an inert atmosphere is not needed for reaction and even examples of *aqueous-phase Sc. in water with *TPPTS as *ligand are described. A large variety of functional groups are compatible with the reaction conds. Recently Sc. of aryl chlorides have been described. In these cases Ni can also be used as cat. F réaction de Suzuki; G Suzuki-Reaktion. A. ZAPF, M. BELLER

Ref.: *Acc.Chem.Res.* **1982**, *15*, 178; *Chem.Rev.* **1995**, *95*, 2457; Trost/Fleming, Vol. 3, p. 481; DE-Appl. 195.27.118 and 195.35.528 (1997); *Angew. Chem.Int. Ed.Engl.* **1999**, *38*, 2413; Murahashi/Davies, p. 441.

Suzuki polycondensation   a step-growth *polymerization of aromatic monomers to polyarylenes. It involves Pd-mediated cross-coupling of bifunctional monomers carrying boronic acids (or esters) and preferably iodo-, bromo-, or trifluoromethane sulfonate leaving groups (*Suzuki coupling). Degrees of polym. range between 20 and 100. Spc. proceeds regiospecifically at carbons carrying the functional groups with formation of C-C bonds. It is compatible with many functional groups and tolerates steric complexity near the coupling site. Pd(PPh₃)₄ and Pd(Ptol₃)₃ are used as cat. precursors in most cases (approx. 1 mol%). The cat. *cycle is believed to pass through Pd(0)/Pd(II) intermediates. Mol. weights are limited, e.g., because of stoichiometry problems (water content of boronic acids, their self-condensation, inadvertent deboronification). Occasionally ligand-derived phosphorus incorporation into the backbone is observed. Spc. has been used to synth. soluble, all-hydrocarbon rigid-rod polymers, polyelectrolytes, ladder-type polymers, *dendrimers with a polymeric core, amphiphilic poly(*p*-phenylene)s, etc. **F** polycondensation de Suzuki; **G**=E.   A.D. SCHLÜTER

Ref.: *Acta Polym.* **1993**, *44*, 50; *Chem.Rev.* **1995**, *95*, 2457; *J.Am.Chem.Soc.* **1997**, *119*, 12441.

Swarts fluorination   a fluorination method with SbCl₅/SbF₃; cf. * fluorination.
                                                                                    B. CORNILS

Ref.: Ertl/Knözinger/Weitkamp, Vol. 4, p. 1677 and Vol. 5, p. 2349.

sweetening   removal of evil-smelling sulfur cpds. from petrochemical prods., e.g., procs. of *Beavon, *Exxon Hydrofining, *Howe-Baker Mercapfining, *Petrolite, *Standard Oil Ultrasweetening, *UOP Merox (cf. the removal of S cpds. for chemical reasons: *Claus process, *HDS, *desulfurization,

*sour gas shift). **F** traitement adoucissant; **G** Süßung.   B. CORNILS

SXAPS   soft X-ray appearance potential spectroscopy, a variant of X-ray spectroscopy

Syncat   → Süd-Chemie catalysts

SynC(S)at technology   → Criterion Lummus procs.

synchronous reactions   → concerted reactions

syncrudes   synthetic crudes, either through *coal liquefaction, or by, e.g., *hydrocracking procs. (e.g., *VEBA hydrocracking proc.); cf. also *synfuels. **F=G=E**.   B. CORNILS

syndiotactic polymers   poly(1-alkene)s and vinylic *polymers with a strictly or mostly alternating relative configuration at their tertiary carbon atoms (cf. *stereoselectivity of polymers). Successive monomer units are linked racemically (r). The degree of syndiotacticity (*tacticity) is usually quoted as pentad syndiotacticity or %rrrr, i.e., the probability that four subsequent linkages of monomer units are all racemic. The term *syndiotactic* was introduced by *Natta. Examples of this kind of polymer which demonstrate the decisive influence of the cats. are s. species of *polypropene (by VCl₄/*i*-Bu₃AlCl at low temps. or by the *single site cats. [Me₂C (η⁵-cp)(η⁵-Flu)ZrCl₂/*methylaluminoxane]), *polystyrene (by (η⁵-cp)TiCl₃/methylalumoxane), poly(1,2-butadiene) [by Al(*i*-C₄H₉)/ MoO₂(O-*i*-C₄H₉], poly[vinylchloride), or poly(methyl methacrylate). **F** polymères syndiotactiques; **G** syndiotaktische Polymere.
                                                          W. KAMINSKY, F. FREIDANCK

Ref.: *J.Am.Chem.Soc.* **1962**, *84*, 1488 and **1988**, *110*, 6255; *Macromolecules* **1988**, *21*, 3356; *ACS Symp.Ser.* **1974**, no. 4,15/26; *Makromol.Chem.* **1967**, *100*, 48; *J.Macromol.Sci.Chem.* **1966**, *1*, 61.

synergetic effect   A se. is true if a second substance adds its own activity for a given reaction to the cat. itself and increases the overall rate of reaction by a synergetic contribution to the basic activity of the main catalyst (cf. activators, *bimetallic cats., *co-catalysts, *promoters). Thus it is in principle a positive, nonlinear effect. **F** effet synergétique; **G** synergetischer Effekt.   C.D. FROHNING

synergistic effects → synergetic effects

Synetix Imperial Chemical Industries PLC (UK) has combined its cat. activities into a single business called Synetix. Headquartered at Billingham (UK) the new operation has annual sales of about $200 million (1998). Into Synetix have been merged: 1: *ICI Katalco, with main business areas *methanol, *ammonia, hydrogen and gas processing (the ammonia cats. portfolio include the former *BASF ammonia cat., which had been acquired in 1997); 2: Unichema International, offering cats. for the oleochemical and edible-oil industry; 3: Crossfield's HTC cats. business, offering nickel fixed-bed cats. and procs. for selective *hydrogenations; 4: ICI Vertec, business concerns organotitanium cpds. for use as *polymerization cats; 5: ICI Tracerco, offering a range of diagnostic services using tracers and scanning to the process industry. J. KULPE

synfuels general term for synthetic fuels from cat. reactions such as *Fischer-Tropsch synthesis, *Shell SMDS, *MeOH manuf., *MTG, *alcohols as fuels; cf. also *syncrudes. **F=G=E**.   B. CORNILS
Ref.: Kirk/Othmer-1, Vol. 12, p. 126; Kirk/Othmer-2, p. 538; McKetta-1, Vol. 56, p. 79; McKetta-2, p. 552; Ullmann, Vol. A5, p. 335.

syngas mixture of CO and H_2 in various ratios. S. is an important raw material for cat. chemical syntheses (*syngas reactions, $^*C_1$ as building block). The ratio CO/H_2 may be adjusted by suitable procs. (e.g., *reforming, *shift conversion) acc. to the application, e.g., 1:1 for *oxo synthesis, 1:2 for *methanol prod., 1:3 for *methanation. S. is manuf. from coal (*coal gasification), gaseous or liquid *hcs. (cat. *steam reforming, e.g., *ICI syngas generation, *Haldor Topsøe proc.), or from resids by *partial oxidation. Formaldehyde, formic acid, or methyl formate are each regarded as *liquid syngas. **F** gaz de synthèse; **G** Synthesegas.   B. CORNILS
Ref.: Kirk/Othmer-1, Vol. 2, p. 649; McKetta-1, Vol. 7, p. 251 and Vol. 56, p. 195; Ullmann, Vol. A2, p. 175, Vol. A7, Vol. A12, p. 169, 203, Winnacker/Küchler, Vol. 3, p. 259.

syngas chemistry → C1 as building block, *syngas reactions

syngas reactions → AA and *AAA synthesis, *carbonylations, *Fischer-Tropsch and variants, *homologation, *methanation, *methanol, *oxo, etc.

Synol process a variant of the *Fischer-Tropsch reaction and proc. for the manuf. of primary, aliphatic alcohols from CO/H_2 mixtures at 180–200 °C/2–2.5 MPa over fixed-bed cats., obtained with fused Fe oxides and KOH (*fused cats.). 65 % of the prod. boil between 40–220 °C. **F=G=E**.   B. CORNILS
Ref.: Falbe-2, p. 323; Winnacker/Küchler, Vol. 5, p. 569.

synproportionation → comproportionation

Synsat, SynSat catalyst technology developed by *Criterion and *Lummus for *hydrotreating reactors

Synthane process for the cat. *methanation of *syngas, using tube-wall reactors (TWR) with flame-sprayed Raney Ni cats. (cf. *CVD, *flame hydrolysis). **F** procédé Synthane; **G** Synthane-Verfahren. B. CORNILS

synthetase → ligase

synthetic (substitute) natural gas → SNG

Synthoil process former proc. for the *hydrogenation of coal as developed by the US Bureau of Mines. The ground coal was slurried with recycle oil and passed a fixed bed of pelletized Co-Mo on *alumina in turbulent flow at 450 °C/28 MPa. The yield was said to be 53 %, relative to the fed coal.   I. ROMEY

Synthol process   → Kellogg procs.

synzyme   a synth. *polymer with *enzyme-like properties. The ability of bovine serum albumin to bind a large variety of different molecules, together with the fact that enzymes are able to cat. a wide variety of reactions with a relatively small number of functional groups, prompted the idea that it might be possible to make a cat. by synthesizing a polymer that 1: is "sticky" toward a large number of small molecules, and 2: contains functional groups (such as acids and bases, e.g. carboxylates or imidazoles) capable of catalyzing a number of different reactions. The binding of a small molecule to the polymer would then put it in close proximity to such cat. groups. Such "synzymes" have been synth. that are capable of accelerating, for example, ester *hydrolysis and *decarboxylation reactions. **E=F=G.**   P.S. SEARS

Ref.: *Adv.Chem.Phys.* **1976**, *39*, 109; *Chem.Rev.* **1996**, *96*, 721; T. Scheper (Ed.), *New Enzymes for Organic Synthesis,* Springer, Berlin 1997.

SZ   → sulfated zirconia

T

t, tert *tertiary* in chemical formulae

T4 DNA ligase → DNA and RNA oligomers

TA → thermal analysis methods

tacticity the regularity of monomer units within a polymer, i.e., the stereoisomerization of macromolecules (*stereoselectivity of polymers; *block polymers, *atactic polymers, *isotactic polymers). The t. is influenced by the cats. **F** tacticité; **G** Taktizität. W. KAMINSKY
Ref.: Elias, Vol. 1, p. 136.

Taddol chiral ligand; cf. *chiral pool

Taft equation an equ. with which *acid-base catalytically relevant reactions can be compared with each other via the corresponding reaction velocities (cf. also *steric effects). **F** équation de Taft; **G** Taft-Gleichung.
B. CORNILS

tagatose-1,6-bisphosphate (TDP) aldolase cat. the condensation of dihydroxyacetone phosphate and D-glyceraldehyde 3-phosphate to form *TDP, and has been isolated from several sources. Like the other DHAP *aldolases, TDP aldolase accepts a variety of acceptor substrates for the *aldol reaction, including glycoaldehyde, D- and L-glyceraldehyde, acetaldehyde, and *i*-butyraldehyde. In all cases investigated so far, a diastereomeric mixture of prods. is formed. Also, instead of exhibiting the expected D-tagatose-like *erythro* configuration, >90 % of each product possessed the *threo* configuration similar to D-fructose.

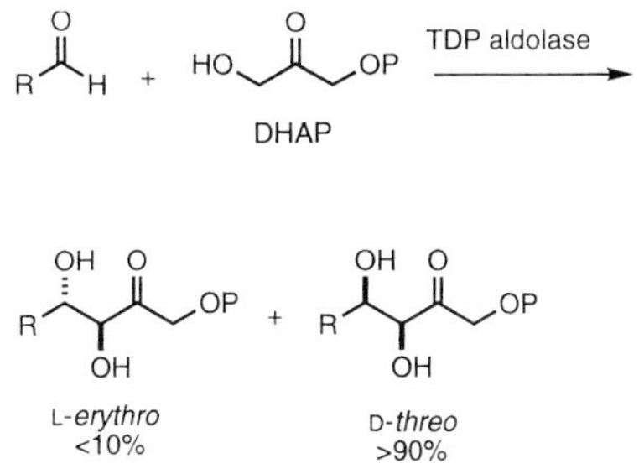

DHAP

L-*erythro*
<10%

D-*threo*
>90%

Only with the natural substrate D-glyceraldehyde does the major prod. (D-TDP) have the tagatose configuration (see also *fuculose 1-phosphate aldolase). **E=F=G.** C.-H. WONG
Ref.: Angew.Chem.Int.Ed.Engl. **1992**, *31*, 56.

tailoring (of catalysts) the guided influence of properties and behavior in design and development of cats. (tailormade cats.). This provides a detailed knowledge of structure-reactivity relationships and is thus more predestined for hom. cats. rather than for het. cats. (cf. Table under keyword *catalysis). Typical examples (among others) are the electronic or steric modification of hom. cats. by *ligands such as *phosphines, the development of *water-soluble ligands for *aqueous-phase cat., the design of *supported cats. as links between *hom. and *het. cat. (cf. *ligand tuning) and the fine-tuning of *oxide cats. by addition of multielement oxides. Tailoring het. (metal) cats. is much more difficult because of the complexity of the solid material "catalyst", many unknown basic *surface effects of that "system" (which are difficult to define and to control), and some ill-defined procedures during their prepn. (such as *precipitation, *impregnation, *shaping, *sol-gel cats., etc.). As an example of t. het. cats. see *V as catalyst metal and *oxide cats. **F** travaillant sur mesure; **G** Maßschneidern. B. CORNILS

Ref.: Becker/Perreira; Dines/Rochester/Thomson; Hegedus; Iwasaga; Yermakov; *Angew.Chem.* **1993**, *105*, 402; *Cattech* **1998**, (6), 87; J.T. Richardson in Twigg/ Spencer, Eds., *Principles of Catalyst Development*, Plenum, New York 1989; D.L. Trimm, *Design of Industrial Catalysts*, Elsevier, Amsterdam 1980.

Takasago process to (–)-menthol
(–)-Menthol, used as a fragrance and in former times isolated from peppermint plants, is now produced on the basis of an Rh-(*S*)-*BINAP *isomerization proc. which converts diethylgeranylamine into citronellal diethyl enamine with >98 % *ee*. The enamine is *hydrolyzed to citronellal, which in an intramolecular ene reaction mediated by $ZnBr_2$ gives isopulegol, the isopropenyl substituent of which is *hydrogenated over *Raney Ni to (–)-menthol. **F** procédé de Takasago pour le menthol; **G** Takasago Mentholverfahren.

H. BRUNNER

Ref.: A.N. Collins, G.N. Sheldrake, J. Crosby, *Chirality in Industry*, Wiley, Chichester 1992, p. 313; *Chem.Ind.* **1996**, 412; Cornils/ Herrmann-1, Vol. 1, p. 552.

TAME   *tert*-amylmethyl ether,   → etherification

Tammann temperature an empirical temp. (approx. half the melting temperature in K) important for various interactions in *solid-state reactions, cat. prepn., or intermetallic cpds. and phases: beyond the Tt. the volume diffusion starts to become important, aggregate consolidation may occur, and atoms in the bulk of the particle may migrate (bulk-to-surface migration). Temps. below the Tt. are advisable for H *reduction of cats. in order to avoid sintering of the metal and loss of *surface area (cf. *intermetallic cpds.). **F** température de Tammann; **G** Tammann-Temperatur.

B. CORNILS

Tanaka-Tamaru plot The cat. *activity can be specified as the temp. necessary to achieve a chosen degree of conv. Thus, the lowest temp. corresponds to the highest activity. In a TTp., this temp. is plotted for a series of metals versus the enthalpy of formation per metal atom of the highest oxide. The maximum of the typically volcano-shaped curve (cf. *volcano plot) reflects the optimum value of the enthalpy: at lower values the reaction intermediate has little propensity to form, whereas at higher values the intermediate is too stable, inhibiting further reaction. **F=G** Tanaka-Tamaru-Diagramm(e).   M. MUHLER

tandem metathesis   → metathesis

tandem oxidation a variant of *Asahi's MA (methacrylic acid) proc. (developed by Japan Methacrylic Monomer Co.) from isobutene via methacrolein and its *oxidation by simultaneous reaction without preceding purification of the methacrolein.   B. CORNILS

tandem reactions   → domino reactions

tantalum as catalyst metal   → niobium as catalyst metal

TAP   → temporal analysis of products

T atoms   → zeolites

Tatoray process   → Toray procs.

tautomeric catalysis   → mutarotation

taxogen is an ethylenic unsaturated monomer M used in *telomerization. An overview of the use of t. in telomerization procs. is given in the literature. **F** taxogène; **G=E**.
W. KAMINSKY, U. WEINGARTEN

Ref.: *Encycl.Polym.Sci.Engng.* **16**,533.

TBA *tert*-butanol

TBHP *tert*-butyl hydroperoxide,   → ARCO PO process

***t*-Bu, *tert*-Bu, *tert* bu**   tertiary butyl group

TCC process   → thermofor process (thermal *catalytic cracking)

TCE   1,1,1-trichloroethane

TCGP   Texaco coal gasification process

TDI   toluene diisocyanate

TDP process   → Mobil procs.

TDS   → thermal desorption spectroscopy

TEA   triethanolamine

teabag catalyst   → ship-in-the-bottle catalysts

Tebbe (Grubbs) reagent   a versatile methylene group transfer reagent which is prepared from titanocene dichloride and hexamethyldialuminum (Equation).

The Tr. allows the *olefination of carbonyl cpds. to terminal alkenes. Compared with the Wittig reagents the Tr. offers several advantages. It also undergoes *homologations, *cycloadditions, and *metathesis reactions with alkenes. **F** réactifs de Tebbe; **G** Tebbe-Reagens.   A.C. FILIPPOU

Ref.: *J.Am.Chem.Soc.* **1978**, *100*, 3611, **1979**, *101*, 5074, and **1983**, *105*, 5490; *Synthesis* **1991,**, 165; *Pure Appl. Chem.* **1983**, *55*, 1733.

TEG   triethylene glycol

telechel   telomer with functional end groups, a prod. of *telomerization

Telene   a polydicyclopentadiene, manuf. by Goodrich in a *ROMP reaction cat. by W salts.

tellurium in catalysis   Te has been used/proposed in hom. cats. for the manuf. of some commercially important bulk chemicals such as *propylene oxide (*acetoxylation of pro-

pene over TeO_2/I_2 in the *ChemSystems PO proc.) or ethylene glycol (acetoxylation of ethylene over TeO_2/HBr in the *Halcon EG proc.). Het. cats. are described for *Montedison's acrylonitrile proc. (via *ammoxidation of propylene over Te-Ce molybdate systems without Bi), for *Nippon's acrylic acid proc. (via *partial oxidation of propylene over Mo-Te cats.), and for the acetoxylation of butadiene on Te-doped Pd cats. on carbon (*Mitsubishi BDO proc.). Te-doped zeolites have been found to be active for *dehydrocyclization reactions. **F** tellurium en catalyse; **G** Tellur in der Katalyse.   B. CORNILS

Ref.: *Houben-Weyl/Methodicum Chimicum*, Vol. XIII/4 and Vol. E12b; *Chem.Rev.* **1994**, *94*, 301.

telogen   is the term for the (solvent) molecule AB that gives the end groups of the *oligomers or low-molecular weight polymers formed during *telomerization. An overview of the ts. being used in telomerization procs. is given in the literature. **F** télogène; **G** Telogen.
                                                         W. KAMINSKY

Ref.: *Encycl.Polym.Sci.Engng.*, **16**, 533.

telomer   is the prod. of *telomerization, i.e., an *oligomer or a low-molecular-weight polymer. In the case of the end groups being functionalized, the telomer is called a *telechel. **F** télomère; **G** Telomere.   W. KAMINSKY

telomerase   in *enzyme cat. *telomers are repeating sequences at the ends of linear chromosomes, important for the mechanism by which linear chromosomes are replicated. All known *DNA *polymerases require a *template and an oligonucleotide primer, and can only replicate in one direction: 5′-3′. This means that replication of one (leading) strand is straightforward, but the opposite strand (lagging) must be replicated in pieces, with transiently created *RNA oligonucleotides (RNA polymerases do not require a primer), called Okazaki fragments, as primers for the DNA polymerase:

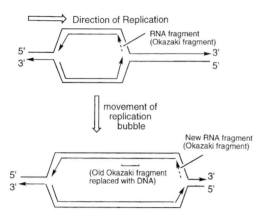

At the extreme ends of a linear chromosome, there will be no space for an RNA primer to prime DNA synth. on the "lagging" strand. The chromosome should get shorter with each round of replication, and important genetic information would be lost if it were not for t., a DNA polymerase which contains its own RNA template (the sequence of which varies between species) which it uses to add a repeating sequence to the ends of linear chromosomes. **E=F=G**.   P.S. SEARS

telomerization   a proc. of polymerization acc. to which a (solvent) molecule AB – the *telogen – and n mol of an ethylenically unsaturated monomer – the *taxogen react to AB + nM → A(M)$_n$B. This *telomer contains the fragments of the telogen as end groups. Telomers with functional end groups are called *telecheles. Normally, the t. is realized (radically initiated by, e.g., *peroxides) as a solution polymerization in solvents with high transfer constants. Halogen cpds., mercaptans, and alcohols, as well as C-H cpds., can be used as telogens. T. can also be initiated by anionic or cationic cats. or under the influence of *transition metal *complexes; cf. also *hydrodimerization. **F** télomérisation; **G** Telomerisierung.   W. KAMINSKY

Ref.: US 2.440.800 (1942); Elias, p. 481; Starks, *Free Radical Telomerization*, Academic Press, Orlando 1974; *Comprehensive Polymer Sci. 3*, 185; Mark/Bikales/Overberger/Menges, Vol. 16, p. 533.

TEM   transmission electron microscopy

Temkin isotherm   A dissociative *adsorption proc. of an adsorbing gas A onto a given *surface follows the *Langmuir adsorption isotherm if a plot of $1/\theta$ versus $1/(P_A)^{1/2}$ (θ = surface coverage; P_A = partial press. of A over the surface) yields a linear result. The Ti. is one of several adsorption isotherms that take into account deviations from the Langmuir adsorption isotherm in cases of dissociative adsorption of gas molecules on *single-crystal surfaces for high or low coverages [e.g., H_2 on Pt(*111*)]. The deviations occur because of 1: different *binding energies of an adsorbate due to independently acting different sites on a given surface (adsorbate/surface interactions) and 2: adsorbate/adsorbate interactions. Besides direct interaction of adjacent molecules, Temkin [1940] considered the effects of indirect adsorbate/adsorbate interactions on adsorption isotherms, where the adsorbate changes the electronic properties of the surface due to donation of electrons affecting the bonding state of all the other adsorbate molecules. He found experimentally that *heats of adsorption would more often decrease than increase with increasing coverage (θ=0.5). Therefore, he developed a model assuming a linear decrease of the heat of adsorption with increasing coverage due to adsorbate/adsorbate interactions, expressed in the following equ.: $\ln (K_{equ}^{A,0}P_A) = \Delta H_{ad}^0 \alpha_T \theta(kT)^{-1}$ with $K_{equ}^{A,0}$ = equil. constant of A for adsorption at $\theta = 0$, P_A partial press. of A over the surface, H_{ad}^0 = equil. enthalpy of adsorption at $\theta = 0$, α_T = fitting parameter, θ = coverage of the surface, k = Boltzmann's constant and T = absolute temp. **E=F=G**.

R. SCHLÖGL, A. FISCHER

Ref.: *Kinetics and Catalysis* **1995**, *36*, N1–2, 1.

Temkin-Pyznev kinetics → ammonia synthesis

temperature effects   Changing reaction temp. is not an obvious approach for optimization of *stereoselectivity, since *enzymes are temp.-labile, but small changes in the reaction temp. can have dramatic effects. E.g., the enzyme *Thermoanaerobium brockii* alcohol *dehydrogenase cat. the reduction of 2-pentanone to (*R*)-2-pentanol at 37 °C, while at 15 °C the prod. is (*S*)-2-pentanol. Similarly, in the *oxidation of 2-butanol cat. by the enzyme *Thermoanaerobacter ethanolicus* alcohol dehydrogenase, the *S*-enantiomer is preferred at <26 °C while the *R*-enantiomer is preferred at >26 °C. A study of the temp.-dependent enantioselectivity of the alcohol dehydrogenase from *Thermoanaerobacter ethanolicus* revealed a linear relation between temp. (K) and the difference in *transition state energies of the two enantiomers ($\Delta\Delta G^{\neq}$) examined. Establishing this linear relationship determined $\Delta\Delta G^{\neq} = \Delta\Delta H^{\neq} - T\Delta\Delta S^{\neq}$ and allowed prediction of *R* or *S*-enantioselectivity at different temps. It also indicated the temp. at which there would be no discrimination between *R* and *S*-enantiomers (the so-called *racemic temperature). **E** effet de la température; **G** Temperatureffekt.   C.-H. WONG

temperature-programmed desorption

(TPD)   kinetic experiments in which the *desorption rate from the *surface is followed while the temp. of the substrate is increased continuously in a controlled way, usually in a linear ramp. TPD is used to characterize *adsorption states and to determine the kinetics of desorption. Qualitatively TPD can be interpreted simply because the higher the desorption temp. the more strongly is an adsorbate bonded to the surface. Since the area under a TPD trace is proportional to the *coverage, TPD spectra allow determination of relative coverages. TPD experiments are frequently carried out in a UHV environment because the analysis of the data is straightforward: provided the pumping rate is high enough no readsorption occurs and the measured partial press. is proportional to the desorption rate. At high press. TPD experiments can be carried out with an inert carrier gas.

Although experimentally simple, a full understanding of TPD spectra requires consider-

able theoretical effort because energetic interactions between the *adparticles require in principle an analysis using the methods of statistical physics. Various approximation methods, however, have been suggested with which kinetic parameters can be extracted from TPD data.

The kinetics of desorption are usually described by the *Polanyi-Wigner equ. in which the following kinetic parameters appear: the order of desorption, the *activation energy for desorption and the *pre-exponential factor. The simplest of these approximation methods originally suggested by Redhead makes it possible to estimate the adsorption energy from the peak maximum of a first-order desorption process (cf. also hydrogen-oxygen titration). **E=F=G.**   R. IMBIHL

Ref.: K. Christmann, *Introduction to Surface Physical Chemistry*, Steinkopff, Darmstadt 1991; van Santen/ van Leeuwen.

temperature-programmed nitridation

(TPN)   temp.-programmed nitridation of a sample with ammonia. The method is destructive and yields integral, *surface-sensitive data about thermodynamics and kinetic parameters of solid-state *HDN reactions or nitride cats. with good in-situ capabilities. **E=F=G.**   R. SCHLÖGL

temperature-programmed oxidation

(TPO)   → hydrogen-oxygen titration

temperature-programmed reaction
spectroscopy (TPRS) This technique is

applied for the determination of temp. profiles of gas-solid reactions and thus makes possible deductions about reaction mechanisms and the influence of, e.g., promoters. The probe consists of a temp. rise with defined time profile; the reaction prods. are detected by *MS. Solid powders or single crystals can be used as samples; the method is non-destructive. The data are integral and *surface-sensitive; cf. *reaction spectroscopy and *TPN, *TPO, *TPR, *TPS). **E=F=G.**

Ref.: Niemantsverdriet, p. 23.   R. SCHLÖGL

temperature-programmed reduction

(TPR)   → hydrogen-oxygen titration

temperature-programmed sulfidation

(TPS)   the sulfidation of samples with a temp. rise with a defined profile (probe) and the measurement of the H_2S consumed (response). The method is destructive and yields integral, *surface-sensitive data about thermodynamic and kinetic parameters of solid-state *HDS reactions. It has good in-situ capabilities. **E=F=G**.   R. SCHLÖGL

Ref.: van Santen/van Leeuwen.

template catalysts   *templates which are able to cat. a chemical reaction. The tc. binds two or more reactants and induces a reaction between them. In principle, this reaction proceeds preferentially for entropic reasons and finally due to an *activation of the reactants by the template. After the reaction the newly formed prod. still binds to the t. and acts as an *inhibitor for its cat. activity. Therefore, it has to be removed to liberate the tc. again and to enable the next *cat. cycle to proceed. This occurs either by dissociation due to low stability of the complex product/template or by substitution of the bound prod. by starting material (Figure).

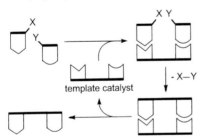

template catalyst

Acc. to this concept, small organic molecules were designed and synth. which are able to act as cats. for the formation of covalent bonds. If the tc. is identical to the prod. which is formed during cat., an *autocatalytic proc. is observed which leads to self-replication of the t. and to an increase in the reaction rate with an increase in time. However, due to self-inhibition of the cat., the reaction rate

drops when a high conc. of the t. is reached. Several systems which show t.-directed autocat. behavior based on small nucleotides have been developed.

*Imprinted molecules or cat. *antibodies are highly sophisticated tcs. Imprinted polymers are prepared by bulk *polymerization of a mixture of the *monomers and a derivative of the monomer which contains a covalently linked guest molecule. After *copolymerization the anchors which connect the guests to the polymer are cleaved and the guest molecules are washed out, leaving a cavity with well-defined shape and functionality. If the guest molecule represents a *transition- state analog of a chemical reaction, the imprinted polymer can stabilize the *transition state of the reaction and can thus be used as a tc. The general principle of *antibody cat. is very similar.

Typical examples of antibody cat. reactions investigated so far are ester cleavage, *Diels-Alder, or *elimination reactions. Most tcs., however, are produced by Nature: the *enzymes. **F** catalyseurs avec templates; **G** Templat-Katalysatoren.   M. ALBRECHT.

template effects   describe the use of metals as reaction centers for the prod. of large organic molecules. The addition of metals to the reaction mixture can result in an increase in yield or modification of the stereochemical nature of the prod., or it may be essential for the buildup of a macrocycle. Alkali and alkaline earth metal cations affect the course of numerous organic reactions in this way. Well-known examples include the synth. of *crown ethers, *macrocycles, and molecular aggregates. Chemical and proximity effects are employed to account for the te. The interaction of reaction centers with metal cations can affect the nucleophilicity/electrophilicty of the reaction center. In addition, complexation by metal cations leads to increased statistical proximity of reaction centers such as chain ends, as well as a loss of conformational entropy through the effect of preorganization.

R. VAN ELDIK

Ref.: Wilkins; Dietrich/Viout/Lehn; Atwood.

templates chemical species which possess one or more binding sites for other chemical species (e.g., substrates X and Y, Figure). Acc. to *Fischer's "*lock-and-key principle" the t. contains the steric as well as the electronic information which is necessary for the selective binding of substrates. Attachment of the substrates to the t. eventually lead to a preorganization which enables a chemical reaction and selectively produces a specific prod. Hereby, in a *stoichiometric proc. the t. can direct the reaction or alternatively it can act as a cat. Thus, a t. can be considered as a mold on the molecular level.

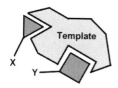

Ts. can be single (metal) atoms, small molecules, *polymers, or biomolecules. Typical examples for t.-directed reactions are the synth. of *crown ethers, catenanes, or naphthoquinones (*Dötz reaction). *Imprinted polymers are *template catalysts for the recognition of substrates. In Nature oligonucleotide derivatives such as *DNA or *RNA act as ts. **E=F**; **G** Template.   M. ALBRECHT

Ref.: N.V. Gerbeleu, V.B. Arion, J. Burgess, *Template Synthesis of Macrocyclic Compounds*, Wiley-VCH, Weinheim (Germany) 1999; J.P. Sauvage, M.W. Hosseini (Eds.) *Templating, Self-Assembly and Self-Organization*, Elsevier, Amsterdam 1999.

templating molecules In *kinetically controlled synth. of solid materials control of the reaction pathway may be executed by molecular species which direct the reaction to a certain prod. structure. Such molecules are designated as tms. or as structure-directing agents. The best-known example for structure direction by molecular species is that of *zeolite-type solids (i.e., *alumosilicate zeolites with a high framework Si/Al ratio, *alumophosphates). In nature, zeolites are formed under mild *hydrothermal conds., and the ar-

tificial synth. of zeolites imitates these conds. (temps. 100–200 °C, applied over several days to basic or fluoride-containing aqueous solutions). Many organic molecules are stable under these conds. Added to the synth. solution, they may become incorporated into the crystallizing material and then influence the type of zeolite framework formed. Typical tms. for the generation of zeolites are amines and alkylammonium cations. According to the degree of control that such molecules exert, one can differentiate between porefillers (which merely prevent the formation of a dense phase), structure-directing agents, and real templates (which under varying synth. conds. form only one type of framework). After the synth., the tms. are burned off by *calcination to make the *pore system of the solid accessible. This process typically involves heating of the material to ca. 500 °C in air.   P. BEHRENS

TEMPO 2,2,6,6-tetramethyl-1-piperidinyl oxide, a *radical scavenger

temporal analysis of products (TAP) This transient method is based upon a pulse technique in vacuum. Pulses containing a small amount of reactants (10^{13}–10^{17} molecules/pulse) are injected via two high-speed valves into the evacuated reactor. The reactor generally contains <1 g of cat. particles placed between two layers of quartz. Gas transport occurs via *Knudsen diffusion. Pulses of different gases can be introduced into the reactor separately or simultaneously. Sequential pulsing with short time intervals between pulses (Δt >0.01 s) is also possible. The reactant (product) molecules leaving the reactor (response signal) are analyzed by *MS with a high time resolution (<100 µs).

In this way *surface procs. (*adsorption, cat. reaction, and *desorption) on solid cats. can be studied. In particular, the reactivity of short-lived surface species can be investigated. Different mechanistic and kinetic models may be discriminated and kinetic parameters of a selected model can be estimated.

The TAP reactor may be also applied at press. up to 200 kPa. In this case the reactor is separated from the vacuum system. **E=F=G**.

O.V. BUYEVSKAYA, M. BAERNS

Ref.: *Catal.Rev.Sci.Eng.* **1988**, *30*(1), 49; *Ind.Eng. Chem.Res.* **1994**, *33*, 2935 and **1996**, *35*, 1556; *J.Catal.* **1995**, *154*, 151 and **1994**, *150*(1), 71; *Catal.Lett.* **1995**, *33*, 291.

tenside ligands ligands that exhibit the characteristics of a surfactant, i.e., they aggregate in solution to form *micelles, they partition to liquid/solid or liquid/liquid interfaces, or they may act as *solubilizing agents. Water-dispersible tensides are characterized by the presence of a strongly hydrophilic region of the molecule and a large hydrophobic region. Some tensides may be dispersed in a non-aqueous phase to give reverse micelles. The possibilities for a tl. can be represented by the placement of the donor atom relative to the hydrophobic and hydrophilic regions of the molecule.
Most tl. have geometries dictated by the nature of the donor atom, which complicates the structure, e.g., $Ph_2P(CH_2)_3SO_3Na$; (menthyl)-$P[(CH_2)_8C_6H_4SO_3 \ Na]_2$; $P[(C_6H_4)(CH_2)_6C_6H_4SO_3Na]_3$, (binaphthyl)-$P_2[(C_6H_4)(CH_2)_6C_6H_4SO_3Na]_4$, or $P[C_5H_3N^+CH\{(CH_2)_2(SO_3^-)\}C_{12}H_{25}]_3$. Tls. have found application in *two-phase catalysis where they improve reaction rates for water insoluble substrates. Tls. may also be constructed to function with other solvent combinations, most notably fluorocarbon solvents (*fluorous-phase cat.). Fluorinated tensides can also act as solubilizers in *super-critical-fluid catalysis. **F** ligand surfactif;
G Tensidliganden.   B. HANSON

Ref.: *J.Mol.Catal.* **1995**, *98*, 117; *Oganometallics* **1994**, *13*, 3761; US 5.777.087 (1998); Cornils/Herrmann-2, p. 123.

terephthalic acid (TPA, *PTA) an important bulk chemical, manuf. from *p*-xylene by hom. *LPO *oxidation or co-oxidation (procs. of *Amoco, *Eastman, *Glitsch, *IFP, *Mobil, *Toray), ammonoxidation (*Lummus), *carbonylation of toluene (*Mitsubishi Gas), *hydrolysis of *DMT (*Hüls), or by *isomer-ization or *disproportionation reaction from phthalic acid anhydride or benzoic acid (*Henkel PTA processes I and II). *PTA* stands for purified terephthalic acid. **F** acide térephthalique; **G** Terephthalsäure. B. CORNILS

Ref.: *Hydrocarb.Proc.* **1972**, (9), 103.

termination reactions This term is a general description of the last reaction step in chain-growth *polymerizations. It leads to the end of the reaction cycle, when the growing polymer chain is broken off. In free-*radical polymerizations the growing polymer radical may be destroyed by a variety of procs., including termination by added substances, disproportionation (transfer of an H radical) and recoupling of two radicals depending on monomer type and polymerization conds. (*radical reactions [polymerization]). In *Ziegler-Natta catalysis the proposed polymerization mechanisms including the termination procs. are quite complex and not fully understood. The possible occurrence of a H transfer from the polymer chain to the metal center via a β-*agostic interaction leaving a metal hydride and a vinyl-terminated polymer chain has been discussed. The termination reaction can be increased by adding hydrogen leading to saturated chain ends. **F** réaction de termination; **G** Kettenabbruchreaktion.

W. KAMINSKY, O. PYRLIK

Ref.: Ullmann, 6th Ed.,1998 Electronic Release; J. Boor, *Ziegler-Natta Catalysts and Polymerizations*, Academic Press, New York 1979.

terpolymerization Terpolymers are *copolymers that are synth. with three different *monomers. The two major technically important prods. are *ABS and *EPDM. ABS *polymerization is cat. by free *radicals such as peroxides. Only one of the double bonds of each diene molecule takes part in polymerization, resulting in a saturated polymer backbone with a pendant double bond that facilitates the use of conventional sulfur curing (*vulcanization). Three types of ter*monomers* are used mainly: ethylidenenorbonene features the fastest curing rates, dicyclopentadiene the best processabilities, and *1,4-hexa-diene the best heat resistance. **F** terpolymeri-sation; **G** Terpolymerisation.

W. KAMINSKY, O. PYRLIK

Ref.: Ullmann, 6th Ed., Electrionic release 1998; *Macromol. Rev.* 1975, *10*, 1.

tertiary structure (of proteins) → sec-ondary structure

tetradentate phosphines *ligands con-taining four trivalent P atoms linked by a chain of carbon or hetero atoms. These lig-ands are able to *chelate or bridge one to four cat. active metals. Similarly to other polydentate *phosphine ligands, the coordina-tion properties are dependent on the elec-tron-donating or -withdrawing character of the substituents and on the geometry and chain length of the bridging units. Despite the great potential in stabilizing high and low *transition metal oxidation states and hydride and nitrogen *complexes, there are only a few reports on the use of tps. in cat. (e.g., in *hydroformylation). **F** phosphines quatre-den-taire; **G** vierzähnige Phosphane.   R. ECKL

Ref.: Pignolet; *Science* 1993, *260*, 1784; *Angew.Chem. Int.Ed.Engl.* 1994, *33*, 67.

Texaco *FCC process   for the processing of a wide range of both virgin and cracked gas oil fractions and deasphalted oils to light ole-finic hydrocarbons, *LPG, high-octane gaso-line, and petrochemical feedstocks. The con-cept includes *riser cracking of feed and re-cycle, segregation of efficient spent cat. strip-per, continuous *regeneration of cat., etc. (typical temp. >320 °C). **F** procédé FCC de Texaco; **G** Texaco FCC-Verfahren. B. CORNILS
Ref.: *Hydrocarb.Proc.* 1972, (9), 134 and 1978, (9), 109.

Texaco hydrogen finishing process   a variant of the *hydrotreating technology for paraffinic and naphthenic lubricating base oils or deoiled hard waxes. The proc. includes a fixed-bed, commercial hydrotreating cat. at 2.5–10 MPa . **F** procédé hydrogen finishing de

Texaco; **G** Texaco hydrogen finishing-Verfah-ren.   B. CORNILS
Ref.: *Hydrocarb.Proc.* 1984, (9), 89; McKetta-1, Vol. 2B, p. 347.

Texaco HyTex process   for the manuf. of high-purity hydrogen from *FCC and coker offgases, natural gas, etc., including CO *shift. **F** procédé HyTex de Texaco; **G** Texaco Hy-Tex-Verfahren.   B. CORNILS

Texaco propylene oxide process   deve-lopment for the manuf. of *PO by oxidation/co-oxidation of propylene and the *reaction carrier isobutane (cf. *oxirane proc.). The co-product(s) are *tert*-butanol (and/or isobu-tene). **F** procédé Texaco pour l'oxide de pro-pylène; **G** Texaco PO-Verfahren.   B. CORNILS

Texaco T-Star process   a variant of the *hydrotreating/*hydrocracking technique of heavy vacuum and heavy coker gas oils, lube oil extracts, etc., for fuel gases, naphta, low-sulfur middle distillates, etc., in ebullated-bed, two-phase fluid operation at 290–450 °C/1.5–20 MPa. The cat. is withdrawn periodically and carbon is burned off. **F** procédé T-Star de Texaco; **G** Texaco T-Star-Verfahren.

B. CORNILS

textural promoter   structural promoter, → promoter

texture (of catalysts)   the detailed topogra-phy of the void space in the particles of the ma-terial; cf. *pore structure. **E=F**; **G** Textur.
Ref.: Anderson/Boudart, Vol. 2, p. 171.   B. CORNILS

TG   → thermogravimetric methods

THD   → Gulf procs.

Theorell-Chance mechanism   a special case of an ordered, sequential bi-bi reaction, (see *enzyme mechanism) a two-substrate/two-product *enzymatic reaction in which the substrates enter in a specific sequence, react, then leave in a specific sequence:

If the concs. of the ternary *complexes EAB and EPQ are extremely low (the ternary reaction and expulsion of product are rapid compared to the other steps), then B and P will display competitive *product inhibition with respect to each other (in ordered sequential bi-bi, they would normally be noncompetitive): **F** mécanisme de Theorell-Chance; **G** Theorell-Chance-Mechanismus. P.S. SEARS

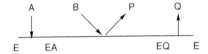

thermal analysis methods general term which comprise *calorimetry, *TDS, *thermogravimetry, differential thermal analysis, and similar methods such as evolved-gas analysis. Normally, the methods are destructive but well suited for powdered solids and allow fingerprint techniques. R. SCHLÖGL

thermal cracking → petroleum processing, *thermolysis

thermal decomposition a variant of the prepn. of (mainly *bulk) cats. by td. of metal-organic or metal-inorganic *precursors, e.g. the manuf. of *Adkins' cats. or preparing Cu/Zn carbonates (for the *WGSR) by td. of mixed hydrogencarbonates (cf. also *flame hydrolysis, *solid-state reactions). **F** décomposition thermale; **G** thermische Zersetzung. B. CORNILS

thermal desorption spectroscopy (TDS) This technique is applied for the determination of thermodynamic and kinetic parameters of sorption procs. by linear heating of samples with pre-adsorbed gases at various temps. and pursuing the partial press. changes.

The ex-situ UHV technique is destructive to the sample but requires multiple experiments for data analysis and yields no information about reaction and the gas-solid equilibria (cf. calorimetry). **E=F=G**. R. SCHLÖGL
Ref.: Ertl/Knözinger/Weitkamp, Vol. 2, p. 676.

thermitase a *subtilisin-like *serine protease produced by the thermophile *Thermoactinomyces vulgaris*. It has a surprisingly high ratio of esterase: amidase activity, and unlike most *proteases and *esterases it can accept the esters of extremely hindered alcohols such as *tert*-butyl alcohol. It has been used for the selective cleavage of *t*-butyl protecting groups in the synth. of *O*-glycopeptides. **E=F=G**. P.S. SEARS
Ref.: *Synlett* **1992**, 37.

Thermoanaerobacter, Thermoanaerobium alcohol dehydrogenases (EC 1.1.1.2)
The alcohol dehydrogenase from the ethanologenic thermophile *Thermoanaerobium brockii* (TADH) is an *NADP-dependent *enzyme, which cat. the *reduction of ketones and the *oxidation of secondary alcohols with high stereospecificity. It is specific for the *pro*-*R*-hydrogen of NADPH. *Hydride transfer generally occurs to the *re* face of the carbonyl to give the *S*-alcohol. An interesting reversal of stereochemistry is seen with smaller substrates, where the hydride is delivered to the *si* face to give the *R*-alcohol. A similar temp.-dependent phenomenom was observed with a secondary alcohol dehydrogenase from *Thermoanaerobacter ethanolicus* and was attributed to the relationship between enthalpy and entropy of activation. The enantiomeric excess of the reduced alcohols decreases when temp. increases. The general trend of reactivity for TADH is secondary alcohols >> linear ketones > cyclic ketones > primary alcohols. Several monocyclic ketones and the bicyclic ketones 4-norbornanone and bicyclo[3.2.0]-hept-2-en-6-one are substrates. The latter cpd. and derivatives are synthons for optically active natural products.

TADH uses *NADP(H) as a *coenzyme. The reduction of ketones can be coupled with the TADH, cat. oxidation of 2-propanol to regenerate the cofactor in a single enzyme system for synth. purposes. TADH is also a useful enzyme for the regeneration of NADP(H) in other systems. TADH exhibits extraordinary thermostability. The activity is unchanged at 65 °C and also shows good stability at 85 °C. Normal experimental ranges are from 10 to 50 °C. **E=F=G.**   C.-H. WONG

Ref.: *Enzyme Microb.Technol.* **1981**, *3*, 144; *J.Am. Chem.Soc.* **1985**, *107*, 4028 and **1986**, *108*, 162.

thermocatalysis a possible area of application of cat. in the future by introduction of solar energy into the energy of chemical cpds. as a source of tremendous heat which can be stored, in contrast to solar energy itself (solar-to-chemical energy). A reversible system under discussion is methane → *syngas → methane (similiar to the *Adam and Eve principle); cf. also *pyrolysis. **F** thermocatalyse; **G** Thermokatalyse.   B. CORNILS

thermodynamics in catalysis A cat. accelerates the reaction rate to attain the equil. For a given set of conds. the rates of both the forward and reverse reactions are accelerated by the cat., but cannot alter the position of that equil. The *rate of the reaction depends on the temp. through variation of the rate coefficient k, expressed as the *Arrhenius equ. This was derived for reversible reactions from thermodynamic considerations, viz. the van't Hoff relation (1):

$$\frac{d \ln K^{\ominus}}{dT} = \frac{\Delta_R H^{\ominus}}{RT^2}, \quad \text{where } K^{\ominus} = \frac{k_+}{k_-}$$

with $K^{\ominus}$ = standard equilibrium constant, k_+, k_- rate coefficients of the forward/reverse reaction, and $\Delta_R H^{\ominus}$ = standard reaction enthalpy (exothermic: $\Delta_R H^{\ominus} < 0$, endothermic $\Delta_R H^{\ominus} > 0$. The Arrhenius equ. is valid for single reactions. For complex reactions which are not quantitatively described in detail, only the overall rate coefficient results from the Arrhenius plot. The standard free-enthalpy change is expressed as equ. 2:

$$\Delta_R G^{\ominus} = -RT \ln K^{\ominus} = -RT \ln \frac{a_B^{\ominus}}{a_A^{\ominus}} \quad \text{for A} \rightleftarrows \text{B}$$

where $a_A^{\ominus}$, $a_B^{\ominus}$ are the standard activities. From this, the criterion arose from the thermodynamic point of view, that the reaction takes place if $\Delta_R G^{\ominus} < 0$ is fulfilled as a result of the reaction. A solid cat. cannot change $\Delta_R G^{\ominus}$ and hence does not change the ratio $a_B^{\ominus}/a_A^{\ominus}$. The ability of a cat. to increase a reaction rate can be described as the result of a decrease in the *activation energy of the reaction. Het. cat. reactions involve the following rate processes: *adsorption, formation and breakup of an activated complex, and *desorption of prods. In the case of hom. cat. reactions the activated *complex formed is energetically similar to that formed on solid cat. Each process is associated with an energy change and has its own *activation energy E, depicted as shown in the diagram of energy vs. reaction path, where ΔH_{ads} is the *adsorption enthalpy (exothermic), ΔH_{des} the *desorption enthalpy (endothermic), and $\Delta_R H$ is the reaction enthalpy, which is equal for both cat. and non-cat. reactions (Figure).

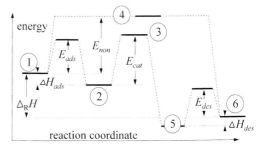

1: reactants; 2: adsorbed reactants; 3: activated complex, catalyzed reaction; 4: activated complex, *non-catalyzed* reaction; 5: adsorbed products; 6: products.

It is obvious that the activation energy E_{cat} of the cat.reaction is smaller than that of the non-cat. reaction E_{non}. According to the Arrhenius equ., the activation energy appears as an exponent, whereby even a slight change in it has a marked effect on the rate coefficient, and thus on the reaction rate. Because kinetic parameters (*activation energy, *pre-exponential factor) are related to the thermodynamics of a chemical reaction, they must fulfill the cond. of thermodynamic consistency: for a given cat. reaction mechanism, the sum of the activation energies for all the independent reaction paths from educts to prods. must be equal to the heat of the net equil. reaction. Additionally, the preexponential factors for all independent elementary steps are related to the standard entropy change for the net reaction. The thermodynamic consistency of kinetic parameters is an important criteria to check their physicochemical reliability. **F** thermodynamique en catalyse; **G** Thermodynamik in der Katalyse. P. CLAUS, D. HÖNICKE

Ref.: Smith, *Basic Chemical Thermodynamics*, Oxford University Press, Oxford 1982; van Santen/van Leeuwen.

thermofor process (TCC) a thermal cat. *cracking process, developed originally by Socony Vacuum Oil, for the manuf. of *LPG, gasoline, kerosene, etc. from gas oil fractions. The cats. consist today of acid-surfaced *alumosilicates mixed with suitable *zeolites; cf. *petroleum processing. A typical example is *Mobil's Airlift TCC proc. **F** procédé Thermofor; **G** Thermofor-Verfahren. B. CORNILS

Ref.: Ullmann, Vol. 10, p. 677 and Vol. A18, p. 62.

thermogravimetric methods (TG) are important for cat. research because of the possibility of reaction monitoring by weight changes of solid samples. The ex-situ method is destructive, has excellent in-situ capabilities, and is versatile for many *solid-state reaction problems. The data are integral and bulk- sensitive (cf. *calorimetry and also *temperature-programmed reactions).

Ref.: Ullmann, Vol.B6, p. 1. R. SCHLÖGL

thermolysis the spontaneous or directed cracking (thermal degradation) of molecules or materials under the influence of thermal energy and possibly cats. (*pyrolysis, *ther-

mocatalysis, e.g., procs. of *ATO heptaldehyde, *Chem. Systems PO; MDI procs. of *Atlantic Richfield or *Asahi). **F=G** thermolyse. B. CORNILS

thermoregulated ligands ligands which control the dependence of a cat.'s behavior on the temp. Typical trps. are *phosphines containing hydrophilic poly(ethylene glycol) (PEG) chains and an adjusted ratio between hydrophilic and hydrophobic groups such as $P[C_6H_5\text{-}O\text{-}(CH_2CH_2O)_nH]_3$. They exhibit an inverse-temp.-dependent solubility in water, i.e., their aqueous solution will undergo a phase separation on heating to a low critical solution temp. (cloud point, Cp) and become water soluble again on cooling to a temp. lower than the Cp (Figure; indicating the movement of the catalyst from the aqueous phase [Aq] into the organic phase [O] above the cloud point, the monophasic conversion S → P, and the return of the cat. into the aqueous phase below Cp). Thus, the *mobile* cat. transfers between the aqueous and the organic phase in response to temp. changes (general principle described earlier, see literature). Such a dramatic solubility change by changing temps. could be attributed to the H-bonds ex-

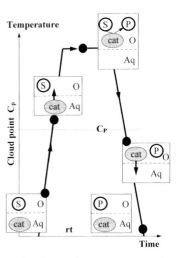

O = organic phase; Aq = aqueous phase; S = substrate; P = product

isting between PEG and water molecules and was found to be useful for cat. recovery. For example, in *aqueous biphasic *hydroformyl-ation of higher alkenes ($>C_6$) a higher reaction rate and a nearly quantitative cat. separation just by decantation can be provided by applying trl.-modified Rh cats. Thus, the problem of low convs. of reactants with low water solubility can be solved by the change from *mono*phasic reaction at a higher temp. to *bi*-phasic separation at a lower temp. **F** ligands thermorégulés; **G** thermoregulierte Liganden.

ZILIN JIN

Ref.: *J.Mol.Catal.* **1997**, *116*, 55; Cornils/Herrmann-2, p. 233.

thermostability of enzymes *Enzymes are somewhat delicate cats. Enzymes from mesophilic organisms (optimal growth temps. between 20 and 40 °C) typically unfold at relatively low temps. (~50 °C). Enzymes from extremophiles (optimal growth temps. 40–85 °C) and hyperthermophiles (optimal growth at >85 °C) are intrinsically more stable, many surviving boiling temps. for extended periods of time (β-glucosidase from *Pyrococcus furiosus* loses only 50 % of its activity after 13 h at 110 °C), though the known three 3D structures of proteins from hyperthermophilic organisms have not revealed any clear basis for such stability. In some cases, comparison of homologous proteins from mesophiles and thermophiles shows truncation of loops in the thermophiles, and incorporation of the N- and C-termini within secondary structural elements such as β-sheets.

At temps. much above 100 °C, covalent modifications such as deamination of asparagine residues, aspartimide formation, oxidation of a variety of residues (Cys, Met, His, Trp, Tyr), and backbone *hydrolysis can occur (cf. also *stability, enzyme). **F** thermostabilité des enzymes; **G** Thermostabilität von Enzymen.

P.S. SEARS

Ref.: *Bioorg.Med.Chem.* **1994**, *2*, 659; *Biochem.J.* **1996**, *317*, 1.

theta-value (Θ) is a steric parameter which (in *Tolman's theory) complements the χ- (electronic) value. Θ is a measure for the *steric* demand of *phosphine *ligands PR_3 and describes a cone angle (measured in degrees) which embraces all the atoms of the R substituents on the P atom, observed at a distance of 2.28 A from the P atom (Figure). Θ of PPh_3 is 145°; for $(2\text{-}MeC_4H_4)_3$ it is 194°.

B. CORNILS

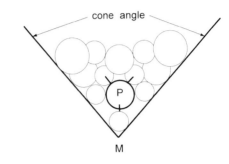

Ref.: Moulijn/van Leeuwen/van Santen, p. 206.

THF tetrahydrofuran

thiamine pyrophosphate is an *enzyme *cofactor derived from thiamine (vitamin B_1).

Thiamine pyrophosphate

The tp.-dependent enzymes are responsible for a variety of reactions that make or cleave C-C bonds, such as decarboxylases. The C2 of the coenzyme's thiazolium core is acidic and can be deprotonated to form the carbanion, which is nucleophilic and will attack carbonyl cpds., such as pyruvate (cf *pyruvate dehydrogenase). **E=F=G**.

P.S. SEARS

Thiele modulus a dimensionless number which represents the ratio of the intrinsic rate of chemical reaction, i.e., without any pore diffusion resistance, to the maximum diffusional transport rate inside the pore. For irreversible nth-order reactions the generalized Thiele modulus is defined by the equation

$$\Phi = L \sqrt{\frac{(n+1)}{2} \frac{k_s c_s^{n-1}}{D_{eff}}}$$

where k_s is the rate constant $[\mathrm{m}^{3(n-1)}/(\mathrm{kmol}^{(n-1)} \mathrm{s})]$ of reaction of order n, D_{eff} the effective diffusion coefficient, c_s the *surface conc. and L a characteristic length of the cat. particle (typically L is the cat. volume divided by the cat. outer *surface area; for spherical pellets $L = r_P/3$). For a first-order reaction it an expression follows which is analogous to the reaction modulus (*Hatta number) of fluid-fluid systems. If Φ is small, *intraparticle mass transport has no effect on the reaction rate. For large Φ, the educt conc. drops rapidly to zero on moving into the pore, so that diffusion strongly influences the reaction rate (under conds. of so-called strong pore resistance), and thus the cat. is not entirely used up. Hence, the *effectiveness factor η, which represents the ratio of the observed rate to the rate determined at outer surface conds. (i.e., without gradients), is strongly influenced. Whereas η approaches unity for $\Phi \ll 1$ at low cat. *activity or large *pore sizes, high *porosity, and small cat. particles, a good approximation for the effectiveness factor in the case of strong pore resistance ($\Phi > 5$) is $\eta = 1/\Phi$. This situation occurs typically with very active cats. or with small pore sizes, low porosity, and large cat. particles. **F** nombre de Thiele; **G** Thiele-Modul.   P. CLAUS, D. HÖNICKE

Thiele-Winter quinone aromatization

acid-cat. aromatization of p-quinone in the presence of *AAA which yields 1,2,4-triacetoxybenzene. **F** aromatisation Thiele-Winter de quinones; **G** Thiele-Winter Chinon-Aromatisierung.   B. CORNILS

Ref.: Krauch/Kunz.

thiolprotease   Ts. contain a *cat. triad (or sometimes a dyad, as in the *caspases) in the *active site composed of Asn, His and Cys, with Cys acting as a nucleophile. In the first step the *enzyme forms a covalent acyl *intermediate which is deacylated with water in a second distinct cat. step. In the *hydrolysis of peptide bonds, the *acylation is usually rate limiting; in the *hydrolysis of esters, deacylation of the acylated *enzyme is rate-limiting. In addition to water, other nucleophiles – for example alcohols, amines, or thiol groups – can also react with the acyl intermediate generated from an ester substrate and form new prods. of transacylation: esters, peptides, or thioesters. Acyl transfer reactions are often performed in media that contain high concs. of organic solvents to minimize competing hydrolysis. Examples of thioproteases are *papain, clostripain, and some *cathepsins. **E=F=G**.   P.S. SEARS

thiosubtilisin   a modified form of the *serine protease *subtilisin in which the *active-site nucleophile, a serine hydroxyl, has been modified (either chemically or via *site-directed mutagenesis) to cysteine thiol. The resulting *enzyme is useful for the *kinetically controlled synth. of peptides from amino acid, peptide, or glycopeptide esters. It has a greatly improved *aminolysis/*hydrolysis ratio as compared to the wild type enzyme. **E=F=G**.   P.S. SEARS

Ref.: *J.Am.Chem.Soc.* **1989**, *111*, 4513, **1987**, *109*, 808, **1995**, *117*, 819, and **1993**, *115*, 5893.

Thorpe nitrile addition   the *addition of nitriles to cpds. with active methylene groups, cat. by *alkoxides. In molecules with nitrile *and* methylene groups *cyclization occurs. **F** addition des nitriles selon Thorpe; **G** Thorpe Nitril-Addition.   B. CORNILS

Ref.: Krauch/Kunz.

three-phase catalysis (TPC)   special variant of het. cat., the cat., the reactant A, and the reactant B being in separate phases, or – depending on the definition – a *biphasic het.

catalysis or a variant of *phase transfer cat. **F** catalyse triphasique; **G** Dreiphasenkatalyse.

B. CORNILS

Ref.: *Angew.Chem.* **1979**, *91*, 464; K.D.P. Nigam, A. Schumpe, *Three-Phase Sparged Reactors*, Gordon and Breach 1996.

three-way catalyst (TWC) Cats. for otto engine exhaust are referred to as *three-way catalysts (TWC): They provide activity for three simultaneous reactions – CO and hc. *oxidation, and NOx (NO$_x$) *reduction. Early technology refrained from NOx reduction (oxidation catalysts) or performed reduction in an extra-converter, with air injection prior to the oxidation cat. (dual-bed converter, engine operated in fuel-rich regime). The TWC is linked to stoichiometric engine operation controlled by an oxygen *sensor (lambda sensor). It fails to reduce NOx under lean conds., and to oxidize CO and hcs. under fuel-rich ("rich") conds. The TWC is able to accumulate O or hcs., which permits its operation with the cyclic variation of the exhaust-gas O content imposed by the feedback control of the engine. After accumulating oxygen, it provides high oxidation convs. during excursion into the rich region; after accumulating hcs. it is able to reduce NOx during subsequent lean excursion. Overall, application of the TWC reduces the emissions of CO and hcs. by ca. 95 %, that of NOx by ca. 75 %. Since the cat. is *poisoned by lead, TWC technology has had a strong impact on the phaseout of leaded gasoline.

To minimize the press. drop, the TWC is usually applied in a monolith design. The cat. contains noble metal particles on a high-area *support modified by *additives to improve *stability and storage properties (cf. *automotive exhaust cats., *DeNOx reactions). **E=F**; **G** Dreiwegekatalysator.

M. MUHLER

Ref.: Ertl/Knözinger/Weitkamp, Ullmann, Vol. A16, p. 750; Thomas/Thomas, p. 577.

threonine aldolases The glycine-dependent tas. are *pyridoxal phosphate-dependent *enzymes which cat. the reversible *aldol re-

action of glycine with an aldehyde to form a β-hydroxy-α-amino acid. Two classes of such *aldolases have been discovered: the serine hydroxymethyltransferases (SHMT) and the *threonine aldolases. The natural reaction of the former is to create serine from glycine and formaldehyde, and the reaction uses the formaldeyde-activated tetrahydrofolate *cofactor, N^5,N^{10}-methylene tetrahydrofolate. SHMT accepts a variety of other aldehyde substrates, and the tetrahydrofolate *cofactor is not required in these reactions. When acetaldehyde is the acceptor, L-allo-threonine is formed, originally thought to be cat. by L-allothreonine aldolase (EC 4.1.2.6). SHMT, or L-allothreonine aldolase, has been used for the resolution of racemic *erythro* β-hydroxy α-amino acids to produce pure D-*erythro* isomers. The enzyme is selective for the *L*-configuration at the α-center, but in general displays poor *erythro-threo* discrimination.

D,L-*threo*
R = H, OH, Br, NO$_2$
D-*threo*

Threonine aldolases cat. the reversible *aldol reaction between glycine and acetaldehyde to give threonine and both D- and L-threonine aldolases have been reported. Hydroxy aldehydes gave complex product mixtures because of their interaction with free amino acids. This could be prevented by the protection of the OH groups and in this way several L-threonine and L-allothreonine derivatives with a protected OH group and C4 were prepared. Acceptor aldehydes with an oxygen functionality at the α- position of the aldehyde gave, in general, high *erythro/threo* ratios. **E=F=G**.

C.-H. WONG

Ref.: *Adv.Enzymol.* **1982**, *53*, 83; *Tetrahedron Lett.* **1995**, *36*, 4081; *J.Am.Chem.Soc.* **1997**, *119*, 11734.

THREOPHOS, threophos a potential tridentate *aminophosphinephosphinite *ligand (AMPP), derived from threonine (Figure).

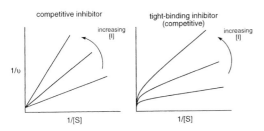

It is obtained in a three-stage synth. using the methyl ester hydrochloride of (2S,3R)-threonine as starting material. N-formylation and subsequent *reduction with LiAlH$_4$ yields the N-methylated 1,3-diol which is phosphinated with (C$_6$H$_5$)$_2$PCl/NEt$_3$.

Ni(O) *complexes obtained in situ from Ni (*cod)$_2$ and *chiral THREOPHOS are effective cats. for the *asymmetric hydrovinylation of cyclohexa-1,3-diene with ethylene in the presence of AlEt$_2$Cl (93 % *ee). Similar AMPPs are derived from other α-amino acids, e.g., (S)-phenylalanine, (S)-alanine, (R)-phenylglycine. O. STELZER

Ref.: *J.Org.Chem.* **1985**, *50*, 1781; Cornils/Herrmann-1, Vol. 2, p. 1029.

tight-binding inhibitors *enzyme inhibitors that bind with dissociation constants well below the useful conc. range of the enzyme (E) they inhibit ($K_I \ll [E]$). Addition of small amounts of inhibitor therefore "titrates" the enzyme, and in general the inhibitor conc. used is of the same order of magnitude as the enzyme conc.([I]~[E] and [I]≫K_I). This adds a level of difficulty to K_I determinations since methods for determining this parameter typically require that the inhibitor conc. be well known: it should not be significantly changed by binding to enzyme. This assumption fails for tbis., and so the *Lineweaver-Burk plot for a tight-binding inhibitor will be curved. **E=F=G**. P.S. SEARS

Ref.: *Methods Enzymol.* **1979**, *63*, 437 and **1995**, *249*, 144.

time of flight (TOF) is a detection method for mass analysis by pulsed ions with extreme transmission; even single-ion analysis is possible. The method can operate over a very wide range of masses. R. SCHLÖGL

Ref.: *Chem.Rev.* **1996**, *96*, 1307.

tin in catalysis Apart from Sn chloride, dialkyl-Sn oxide, SnMe$_4$, SnBu$_3$, etc. act as *Lewis acids (as, e.g.. in the *aldol reaction, *Mukaiyama aldol reaction, *PET synth.) Sn and Sn cpds. are used for *Stille cross-couplings, the *hydrocyanation (Sn/BPh$_3$), or the *Prins reaction. Hom. Pt-Sn cat. systems have a certain relevance in *carbonylation and *hydroformylation research. Sn as a dephlegmating *promoter (decreasing the *deactivation rate) is used for Pt-based cats. for *dehydrogenation (*UOP Oleflex proc.), *hydroesterification, or *cat. reforming (*UOP cat. reforming proc.; *Phillips STAR proc.). Pd-Sn cats. are proposed for *catalytic nitrate reduction and the hydrogenation of alkynes (*alloying cats.).

Sn *antimonate was the cat. for *BP's older version of the *ammoxidation proc. and is still an ingredient of the contact mass of the *Rochow synth. *Hüls' proc. for butene oxidation to acetic acid used an Sn-doped V cat. SnR$_4$ is proposed for *SOMC cats. The *Giese-Stork method has been developed in which the Sn halides produced are reduced in situ by Na cyanoborohydride to Sn$_3$H. Thus, only cat. amounts of toxic Sn cpds. are involved in the reaction cycle. **F** l'étain en catalyse; **G** Zinn in der Katalyse. B. CORNILS

Ref.: A.G. Davies, *Organotin Chemistry*, Wiley-VCH, Weinheim 1997; *Houben-Weyl/Methodicum Chimicum*, Vol. XIII/6 and Vol. E18, p. 746,827; *Acc. Chem.Res.* **1994**, *27*, 191.

TIP process total isomerization process, → Union Carbide procs.

Tishchenko reaction a special case of the *Cannizzaro reaction (aldehyde *disproportionation) with aldehydes with or without α-hydrogen atoms: from aldehydes, cat. with Na or Al *alkoxides, the *esters* of the corresponding acids and alcohols are formed (e.g., acetaldehyde → ethyl acetate, *Kvaerner proc.). **F** réaction de Tishtshenko; **G** Tischtschenko-Reaktion.   B. CORNILS

titania White rutile is the thermodynamic stable modification of titania. Its color changes to orange-yellow at high temps. It is very stable against acid and base treatment and possesses itself acid as well as basic properties. The *rutile structure can be described as a slightly distorted hexagonal close packing of oxygen ions. Half of the octahedral holes are filled by Ti^{4+} ions so that a space-centered tetragonal unit cell is formed. Each TiO_6 octahedron has two opposite, that are edges, common with neighboring octahedra.

Anatase is the second phase of t., metastable at rt. In the anatase structure each TiO_6 octahedron has four common edges with neighboring octahedra. It transforms rapidly into rutile at temps. above $870\,°C$. This *phase transition practically does not occur at temps. below $610\,°C$. It is reported that V impurities in anatase accelerate the rutilization above $550\,°C$. W impurities, on the other hand, lead to an increase in the temp. at which the phase transition to rutile occurs. A pronounced rutilization is also induced by small traces of Mo, if crystals of rutile are already present in the t. sample. Sulfate and phosphate impurities in anatase also lead to a stabilization of this metastable modification.

Both t. modifications, pure or as mixtures, are frequently used as supports for Mo-W-V catalysts for *selective cat. *reduction of NO_x (*SCR). Noble metals supported on t. exhibit *SMSI upon high-temp. reduction. The *SMSI modifies the cat. properties of noble metal cats. by a spread phase of t. suboxides (*Magnéli phases) on the *surface of the noble metal particles. This suboxide overlayer modifies the electronic properties of the cats. as well as the space available for the cat. reaction, which is thought to be the interface between metal and suboxide. Gold-colored TiO can be synth. from Ti and TiO_2 at temps. of about $1500\,°C$, while reduc-

tion of TiO_2 with H_2 at temps. of about $1000\,°C$ results in the formation of amethyst-colored Ti_2O_3. **E=F=G**.   G. MESTL
Ref.: Augustine, p. 165.

titania photocatalysis In this respect, cat. procs. in which the cat. is a photosensitive semiconductor are understood. *Photocatalytic reactions range from organic synth. to wastewater treatment and purification (cf. *catalytic wastewater treatment, *environmental cat.). The energy necessary for the cat. proc. is gained by the absorption of photons and the creation of electron-hole pairs, which directly or indirectly recombine. At surfaces, these electron-hole pairs may be trapped; thus, their lifetime is increased, so that the excited electron or hole may be transferred to an adsorbed electron acceptor or donor. This proc. activates the adsorbate molecule, which may then be transformed into the prod. before the back electron transfer occurs and the electron-hole pair recombines. Colloidal TiO_2 is known as a photocat. in which electrons are trapped within about 30 ps after excitation and holes within about 250 ns. The charge transfer to adsorbed molecules occurs between nano- and milliseconds. Usually adsorbed oxygen acts as electron acceptor, whereas adsorbed organic substrates play a role as electron donors and are thus oxidized during this proc. The adsorption of the organic molecule, if charged, depends on the surface point of zero charge (pzc). When the pH value of the reacting medium is higher than the pzc, the TiO_2 surface is negatively charged and anionic species cannot be adsorbed. When the pH is smaller than the TiO_2 pzc the surface is positively charged, anions are adsorbed, and may be converted photocat. **F** photocatalyse de titane; **G** Titan-Photokatalyse.   G. MESTL
Ref.: *Chem.Rev.* **1995**, 95, 735.

titanium as catalyst metal   1-in hom. cat.: Cat. amounts of Ti in the oxidation state IV (or sometimes III) and in a variety of different *ligand environments are effective in inducing a multitude of reactions in polymer and in synth. organic chemistry. *Ziegler cats. comprising trialkylaluminum reagents and TiCl$_4$ catalyze the *Ziegler-Natta *polymerization of α-olefins such as ethylene or propylene or *dimerizations (cf. *IFP Alphabutol proc.). Titanocenes (*metallocenes) cat. olefin polymerization and certain synth. organic reactions such as *reductions (cf. *Tebbe-[Grubbs] reagent). Ti(OR)$_4$ are efficient cats. in the *transesterification of carboxylic acid esters. Ti(OiPr)$_4$ is a key comp. in the *$tert$-butyl hydroperoxide-based *epoxidation of allylic alcohols, enantiomeric cat. versions being possible if tartaric acid esters are included as *chiral *ligands. Other chirally modified Ti complexes cat. *enantioselective *aldol additions, *Diels-Alder *cycloadditions, glyoxylate ene reactions, and *Grignard-type *additions of dialkyzinc reagents. *McMurry couplings and the *Mukaiyama aldol reaction are also cat. by Ti. **G** Titan als Katalysatormetall. M.T. REETZ

Ref.: G. Fink, R. Mülhaupt, H.H. Brintzinger (Eds.), *Ziegler Catalysts*, Springer, Berlin 1995; M. Schlosser (Ed.), *Organometallics in Synthesis*, Wiley, New York 1994 and 1999; M.T. Reetz, *Organotitanium Reagents in Organic Synthsis*, Springer, Berlin 1986; *Organic Reactions*, Vol. 43, p. 1.

2-in het. cat.: TiO$_2$ is a *support for het. cats. (*titania), e.g., for *Hüls' AA proc., the *Claus variant of *Rhône-Poulenc, the propylene oxide procs. of *Shell and *ARCO, or for *selective cat. reductions (e.g., Ti on *vanadia). It plays a role in *photocatalysis and is proposed for *hydrotreating reactions (Ni-Mo on *titania/*zirconia). *Ti silicalite (and Ti *zeolites) are applied in various procs. such as *Enichem's procvs. for *ammoxidations, hydroquinone, propylene oxide. *Perovskite (CaTiO$_3$) is used as electroceramic or as cat. in *fuel cells, for *catalytic combustion, and for *DeNOx reactions. Cf. also the other "Ti" keywords. B. CORNILS

titanium silicalite (TS-1) has the framework topology of *ZSM-5 and a typical framework composition of Ti$_{0.02}$Si$_{0.98}$O$_2$ but is less acidic. TS-1 cat. many highly *selective *oxidation reactions using environmentally benign hydrogen peroxide as oxidant (cf. *environmental cat.). Examples are the *Enichem hydroquinone and *Enichem PO procs. **E=F; G** Titansilikalit. P. BEHRENS

Ref.: Ertl/Knözinger/Weitkamp, Vol. 5, p. 2330.

TMA trimellitic (acid) anhydride

TMDI 2,2,4-trimethyl-1,6-hexane diisocyanate

tn 1,3-diaminopropyl (trimethylene diamine), H$_2$N-(CH$_2$)$_3$-NH-, a group in *complex cpds. or as a *ligand

TOF 1: *turnover frequency; 2: *time of flight

Tokuyama isopropanol process for the manuf. of isopropyl alcohol by *hydration of propylene over water-soluble silicotungsten acid cats. in aqueous phase at 270 °C/20 MPa. The aqueous cat. solution is recycled. **F** procédé Tokuyama pour l'isopropanol; **G** Tokuyama Isopropanol-Verfahren. B. CORNILS

Tolman concept Tolman developed a concept for estimating steric and electronic effects of *ligands, most prominently P(III) ligands, and thus the influence of ligands during cat. optimization and in order to interpret structure/reactivity relationships. Electronic effects are regarded as a result of transmission along chemical bonds; steric effects are regarded as a result of (usually nonbonding) forces between different parts of the molecule (cf. *chi- and *theta-values).

Increasing the size of the substituent (S) on the P atom of a ligand (L) will acc. to Tolman cause the following effects. 1: Opening of the SPS angle and the angles between L and other ligands on the metal center (M). 2: Reducing of the s character of the P lone pair, thus decreasing $^1J_{MP}$ and shifting the $\delta(^{31}P)$ to lower field. 3: Increasing the basicity of the lone pair of the P atom; 4: Increasing the rates of dissociative reactions and decreasing the

PR$_3$ Ligand; R=	χ	Θ
H		87
t-Bu	0	182
n-Bu	4	132
Ph	13	145
CH$_3$O	20	107
PhO	29	128
Cl	41	
F	55	
CF$_3$	59	
2-MeC$_6$H$_4$		194

rates of associative reactions. If the ligands become bulky enough, they can even interfere with the coordination of other ligands such as O$_2$ or CO which are usually strongly held. 5: Favoring ligands which are less crowded. 6: Favoring lower coordination numbers, possibly leading to M-M bond cleavage. 7: Favoring coordination of other ligands which are in competition for coordination sites.

Two parameters (χ,Θ) are used to describe the electronic and the steric effects. As χ is derived from differences in IR frequencies, the steric parameter Θ for symmetrical ligands is the apex angle of a cylindrical cone, centered 228 ppm from the center of the P atom, which touches the *van der Waals radii of the outermost atoms of the ligands attached to the P-atom. The Table lists some χ- and Θ-values. **F** règle de Tolman; **G** Tolmans Konzept.   F.E. KÜHN

Ref.: *J.Am.Chem.Soc.* **1970**, *92*, 2953,2956; **1974**, *96*, 53; *Chem. Rev.* **1977**, *77*, 313; *J.Organomet.Chem.* **1997**, *530*, 131.

Tolman cone angle

The cone angle Θ (cf. *theta-value) as a ligand property is defined as the opening angle of a cone originating at a metal bonded to a symmetrical phosphine PR$_3$ at a distance of 228 pm (the M-L bond length, Ni(0)-PR$_3$ distance) that just encompasses all *van der Waals spheres of the atoms in the ligand. In unsymmetrical *phosphines PR^{1}R^{2}R^3 average values are chosen; substituents of the ligands are arranged in rotameric conformations so as to minimize the Θ-value. The cone angle can be determined from X-

ray crystal structures or can be calcd. using theoretical models. The concept can be extended to all other kinds of ligands. **E=F=G**.

J. EPPINGER

Ref.: *Chem.Rev.* **1977**, *77*, 313; *Organometallics* **1989**, *8*, 1; *J.Am.Chem.Soc.* **1991**, *113*, 2520; *Inorg.Chem.* **1992**, *31*, 1286; *Adv.Organomet.Chem.* **1994**, *95*, 36; *Trans.Met.Chem.* **1995**, *26*, 1; Abel/Stone/Wilkinson, Vol.2, p.1012 (1987).

Tolman's 18–16 electron rule

predicts the *stability and *reactivity of *transition metal *complexes. These show a tendency to form complexes in which the metal has an effective atomic number corresponding to the next-higher inert gas. The basic premise is that 16- and 18-electron configurations are readily accessible to diamagnetic organometallic transition metal complexes. Tolman proposed 1: that diamagnetic organometallic complexes may exist in a significant conc. at moderate temps. only if the metal's valence shell contains 16 or 18 electrons; 2: that organometallic reactions (including cat. ones) proc. by elementary steps involving only *intermediates with 16 or 18 metal valence electrons. There are exceptions from the rule, such as MeAuPPh$_3$ (14 valence electrons) or [Co(CN)$_5$]$^{3-}$ (17e). **F** règle de Tolman de 18–16 électrons; **G** Tolmans 18–16-Elektronenregel.   T. STRASSNER

Ref.: *Chem.Soc.Rev.* **1972**, *1*, 337.

TON → turnover number

topa, topaquinone → copper amine oxidases

Topas

tradename of a *COC polymer from Celanese

topoisomerases

Two classes of isomerases change the extent of supercoiling of *DNA. The type I topoisomerases relax supercoiled DNA by nicking a single strand and passing the other through the break. The *enzyme forms a covalent adduct to the DNA via a phosphodiester linkage to an *active-site

tyrosine. No additional energy input is required to reform the DNA phosphodiester backbone.
In *prokaryotes, the type II ts. (DNA gyrases) supercoil DNA with concomitant *hydrolysis of *ATP. They appear to form a local loop in the DNA so that two DNA duplexes lie side by side, then break both strands of one DNA duplex, pass the adjacent DNA duplex through the gap, and religate. **E=F**; **G** Topoisomerasen. P.S. SEARS

Toprina process → BP protein proc.

Toray Hytoray process a gas-phase cat. proc. for the *hydrogenation (*dearomatization) of benzene to cyclohexane. **F** procédé Hytoray de Toray; **G** Toray Hytoray-Verfahren. B. CORNILS

Toray Isolen process *isomerization of xylenes at 400–500 °C over Al$_2$O$_3$-SiO$_2$ cats. To avoid side reactions another variant uses Pt-Al$_2$O$_3$ cats. under press. **F** procédé Isolen de Toray; **G** Toray Isolen-Verfahren. B. CORNILS

Toray lysine process manuf. of L-lysine through combined chemical and microbiological steps from D,L-α-amino-ε-caprolactam (from cyclohexane and nitrosyl chloride) and its *hydrolysis with an immobilized L-hydrolase. **F** procédé de Toray pour le L-lysine; **G** Toray Lysin-Verfahren. B. CORNILS
Ref.: Weissermel/Arpe.

Toray PNC (photonitrosation of cyclohexane) process for the *oximation of cyclohexane by photonitrosation to cyclohexanone oxime at <20 °C, initiated by UV light in the presence of HCl. **F** procédé PNC de Toray; **G** Toray PNC-Verfahren. B. CORNILS
Ref.: *Hydrocarb. Proc.* **1975**, (3), 83.

Toray Stex (styrene extraction) process extracts styrene from pyrolysis gasoline where it is present at 6–8 %. **F** procédé Stex de Toray; **G** Toray Stex-Verfahren. B. CORNILS

Toray terephthalic acid process for the manuf. of purified terephthalic acid (PTA) by co-oxidation of paraldehyde and *p*-xylene. **F** procédé Toray pour PTA; **G** Toray PTA-Verfahren. B. CORNILS

Toray/UOP Tatoray process for the gasphase manuf. of benzene and xylenes by *disproportionation/*transalkylation of toluene at 450–530 °C/2 MPa over CoO-MoO$_2$ cats. on alumosilicates/Al$_2$O$_3$. **F** procédé Tatoray de Toray/UOP; **G** Toray/UOP Tatoray-Verfahren. B. CORNILS
Ref.: Ertl/Knözinger/Weitkamp, Vol. 5, p. 2138.

Total/IFP process R.2.R for the manuf. of middle and light distillates from cracker feeds, using ultrastable *zeolites (USY) at 490–530 °C/0.15 MPa. The proc. is specially suited for revamping existing plants. **F** procédé R.2.R de Total/IFP; **G** R.2.R-Verfahren von Total/IFP. B. CORNILS
Ref.: *Appl.Catal.* **1994**, *115*, 178.

total isomerization process (TIP) → Union Carbide procs.

total surface area (TSA) a *BET variant, applied for the nonspecific determination of *surface areas (cf. *adsorption). R. SCHLÖGL

total turnover number (TTN) 1: → turnover number; 2-in *enzymatic reactions: The total turnover number is the total number of moles of product formed per mole of *enzyme or *cofactor, assuming a selectivity of 100 %. The relation between those conversions and realistic reactions in hom. cat. (with σ ≠ 100) is TTN = *TON x selectivity. **E=F=G**. B. CORNILS

toxicity in catalysis There are relatively few data on the t. of cats. Publications deal mainly with the Pt group metals in catalytic converters for cars (*automotive exhaust cats.) and the exposure of workers. **Ni**: In the *List of MAK and BAT Values 1998* (acc. to TLV and BEI [biogical exposure indices] values) Ni is classified in category 1, belonging to those cpds. which cause cancer in man and which can be assumed to make a significant contribution to cancer risk. There is a technical exposure limit (TRK = Technische Richtkonzentration) of 0.5 mg/m^3 for the inspirable fraction of Ni (as metal, NiS, etc.). For Ni cpds. in the form of inspirable droplets, the TRK value is 0.05 mg/m^3. Ni carbonyl is classified in category 2 (carcinogenic in animal studies); the TWA (time-weighted average) value is 0.05 ppm. Ni, its alloys and water-soluble salts are designated as "Sah" (danger of sensitization of the airways and the skin), and Ni carbonyl as "H" (danger of cutaneous absorption). Ni is an essential comp. of *metalloenzymes in urease and for methanogenesis. **Co**: In Germany the TRK value is 0.5 mg/m^3. The air threshold value varies between 0.05 (USA, Scandinavia) and 0.5 mg/m^3 (former USSR). There is a TLV-TWA (threshold limit value) of 0.1 mg/m^3 for Co carbonyl. There are no sound data on the toxicology of Co carbonyl (ACGIH 1991). Occupational exposure via the air has been proven by elevated levels in blood and urine. Co-containing metalloenzymes are essential for B12 metabolism. In cat. production, the EU classification of Co cpds. has to be taken into account. Both Co sulfate and Co chloride have been classified as highly potent carcinogens. **Pt**: Pt released from the cat. converters of cars is emitted predominantly in the elemental form in the range of ng/km. Pt is emitted in particulate form via mechanical rubbing of the cats. Asthma and allergies from Pt or Pt salts are known to occur. A short-term exposure value of 2 µg/m^3 is suggested. Pt cpds. (chloroplatinates) are designated as "Sah". Some Pt complexes are mutagenic. **Rh** and **Pd**: The analog Rh complexes are considerably less mutagenic; Pd shows no mutagenic potential. **Se**: The handling of Se-containg cpds. and the inspiration of SeO$_2$ are reasons for insomnia, nausea, etc. An MAK value of 0.1 mg/m^3 for Se cpds. (inspirable fraction) is cited; the TLV-TWA value is 0.2 mg/m^3. **Ti**: Inhaled Ti is biologically inert. **V**: At high concs. (5–150 mg/m^3) rhinitis

and chronic bronchitis are observed; an MAK value of 0.05 mg/m^3 is given for V$_2$O$_5$. The TLV-TWA value for ferrovanadium is 1 mg/m^3. **Be**: Be (via inhalation) is carcinogenic; the TLV-TWA value is 0.002 mg/M^3.

The MAK values (TLV-TWA values in [] (both in [mg/m^3]) of other elements in cat. application are: **Ba** 0.5 for dusts; **Zr** 5 [5] (as Zr); **Mo** 5 [10]; **Cu** 0.1–1 [1] (dust); **Ag** 0.01 [0.1]; **Cd** 0.05 [0.01]; **Hg** 0.1 [0.025] (organo-Hg cpds. are extremely toxic); **Sn** 2 [2]; **Te** 0.1; **As** 0.1 (for As$_2$O$_3$) [0.01]; **Sb** 0.5 [0.5]; **Pb** 0.1 [0.05]; **Os** OsO$_4$ is particularly poisonous because of its facile reduction to Os black (for comparison: the MAK value for HCN is 10 mg/m^3). **Al** [10]; **W** [5]; **Ta** [5]; **Y** [1]; **Cr** [0.5]; **Mn** [0.2]; **U** [0.2]. Many of these metals are incorporated in *metalloenzymes and are essential to the metabolism (Co, Zn, Fe, Mo, Ni, Cu, Mn, and Se).

The use of *phosphines as *ligands in hom. cat. can cause headache, coughing, or respiratory stress. Studies on mice showed weakly genotoxic effects. In the *List of MAK and BAT Values 1998* the value for phosphines is given as 0.1 ppm (corresponding to 0.14 mg/m^3); the TWA value is 0.3 ppm. Phosphine is classified by the EPA in group "D" (not classifiable as to human toxicity). **G** Toxizität in der Katalyse.   R. Schwabe, H. Greim

Ref.: *American Conference of Governmental Industrial Hygienists* (ACGIH) 1991 and 1999; *J.Trace Elem. Electrolytes Health Dis.* **1990**, *4* (ISS 1), 1; *Toxicol. Appl.Pharmacol.* **1992**, *115*, 137; *J.Soc.Occupat.Medicine* **1986**, *36*, 29; *Staub, Reinhalt. Luft* **1991**, *51*, 361; *MAK und BAT-Werte-Liste 1998*, DFG, Mitt. 34, Wiley-VCH, Weinheim 1998; *Toxicology Lett.* **1988**, *42*, 257; *Am.Ind.Hyg.Assoc.* **1975**, *36*, 452; *WHO Environmental Health Criteria* Nos. 24 (1982), 58 (1987), and 81 (1988); H. Greim (Ed.), *Occupational Toxicants*, Wiley-VCH, Weinheim (Germany) 1999; N.I. Sax, R.J. Lewis, *Sax's Dangerous Properties of Industrial Materials*, 3 Vols., 9th Ed., Wiley, New York 1998; *Cobalt News* **1998**, (4), 12.

Toyo Soda butanediol process   manuf. of *butanediol (BDO) by *hydrolysis of 1,4-dichloro-2-butene with Na formate and water at 110 °C. **F** procédé de Toyo Soda pour le BDO; **G** Toyo Soda BDO Verfahren.

Ref.: Winnacker/Küchler, Vol. 6, 40.   B. Cornils

TPA terephthalic acid (cf. also *PTA, purified terephthalic acid)

TPC → three-phase catalysis

TPD → temperature-programmed desorption

TPD process → Mobil procs.

TPN → temperature-programmed nitridation

TPO temperature-programmed oxidation, → hydrogen-oxygen titration

tpp, TPP 1: → triphenylphosphine, PPh_3; *thiamine pyrophosphate

TPPDS, TPPMS, TPPTS → triphenylphosphine di-, mono- and trisulfonate

TPR temperature-programmed reduction, → hydrogen-oxygen titration

TPRS → temperature-programmed reaction spectroscopy

TPS → temperature-programmed sulfidation

tpy tonnes per year

transaldolase (EC 2.2.1.2) Like *transketolase, ta. is an *enzyme in the oxidative pentose phosphate pathway. Ta., which operates through a *Schiff base intermediate, cat. the transfer of the C1-C3 aldol unit from *D-sedoheptulose 7-phosphate to glyceraldehyde-3-phosphate (*G3P) to produce D-fructose-6-phosphate and D-erythrose 4-phosphate.

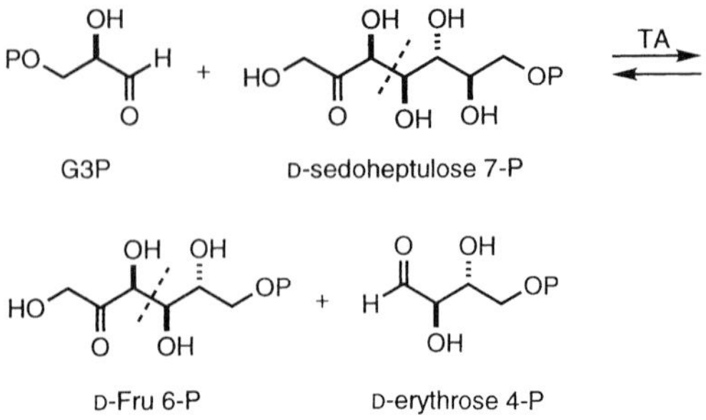

G3P   D-sedoheptulose 7-P

D-Fru 6-P   D-erythrose 4-P

Although it is commercially available, ta. has rarely been used in organic synth., and no detailed substrate specificity study has yet been performed. In one application, ta. was used in the synthesis of D-fructose from starch (cf. *pentose phosphate pathway). **E=F=G**.

C.-H. WONG

Ref.: *Carbohydr. Res.* **1985**, *143*, 288.

transalkylation transfer of alkyl groups between aromatic molecules, e.g., the conv. of toluene, blended with C_9 aromatics, to xylenes. The reaction is cat. by acids or by *alumosilicates. Procs. of interest which include t. are *Atlantic Richfield Xylene-Plus, *CD Cumene, *Ethyl Epal, *Lummus/UOP EB, *Mobil/Badger EB, 3GEB, LTD, LTI, TDP-3, or TransPlus, *Monsanto/Lummus EB, *Toray Tatoray, etc. **E=F**; **G** Transalkylierung.

B. CORNILS

Ref.: Ertl/Knözinger/Weitkamp, Vol. 5, p. 2136.

transamidation transfer of an amido group from one substrate to another. Similar to *transamination, it is an important reaction in biochemistry, mainly cat. by *enzymes. **F=E**; **G** Transamidierung.

B. DRIESSEN-HÖLSCHER

Ref.: *Chem. Listy* **1992**, *86*, 95.

transaminases are transferases, which cat. the *transamination reaction, i.e., the reaction between L-amino acids and 2-oxoglutaric acids acc. to L-amino acid + 2-oxoglutaric acid → L-glutamic acid + 2-oxo acid. **E=F=G**.

transamination transfer of an amino group from one substrate to another. Ts. are mainly cat. by *enzymes (*transaminases, *aminotransferases). Important t. reactions take place in the liver, where they guarantee in vivo the transformation of L-amino acids into oxocarboxylic acids as well as the reverse reaction. **F**=E; **G** Transaminierung.

B. DRIESSEN-HÖLSCHER

Transcat process   → Lummus procs.

transesterification (ester interchange) reaction producing esters by cat. alcoholysis of other esters acc. to R^1-$COOR^2$ + R^3-OH → R^1COOR^3 + R^2OH. Commercially t. is important for the manuf. of glycerol esters in fat chemistry or polyethylene terephthalate from *DMT. For the reaction kinetics cf. *Taft equation. **F** transestérification; **G** Umesterung.   B. CORNILS

trans effect   → electronic factors

transferases *enzymes which transfer groups of atoms (e.g., C_1 units, aldehyde or ketone units, phosphates, sugars) from a donor to an acceptor (cf. *enzymes). **F** transférases; **G** Transferasen.   P.S. SEARS

transfer hydrogenation *reduction of an organic substrate (acceptor A) by transfer of dihydrogen equivalents from a suitable donor molecule (DH_2) acc. to A + DH_2 → AH_2 + D. A typical example of th. is the *Meerwein-Ponndorf-Verley reduction of ketones to alcohols using isopropanol as the H source in the presence of a main group metal *alkoxide. The dehydrogenated donor, acetone, is continuously removed from the reaction mixture by distillation to shift the reversible reaction to the prod. side. The reverse reaction can also be of synth. value and is known as *Oppenauer oxidation. The *Eschweiler-Clarke methylation of primary amines utilizes formic acid as the H source. The thermodynamic stability and the volatility of the dehydrogenated prod. CO_2 make th. with this do-

nor quasi-irreversible. Other suitable donors include *hcs. (e.g., 1,3-cyclohexadiene, tetralin), dihydropyridines, *hydrazine, or hydrophosphoric acid. The reaction principle can be applied to a wide variety of functionalities such as NO_2, C=N, C=O, C=C, and to alkynes. Th. also provides an efficient methodology for hydrogenolytic *dehalogenation or debenzylation.

Many th. reactions are effectively cat. by typical het. or hom. *hydrogenation cats. *Chiral *complexes of Rh, Ir, and Ru allow highly enantioselective reductions with isopropanol or formic acid as H sources. Chiral Rh phosphine cats. reduce C=C bonds *chemoselectively whereas the selectivity of Ru amine systems follow the order C=N > C=O > C=C. Th. cat. by various dehydrogenases are of key importance for biological *redox couples involving *nicotinamides or flavins as hydrogen donors.

Th. is attractive for commercial applications on a small scale. It avoids the potential hazards of gaseous H_2 and allows the use of standard tank reactors (*catalytic reactors). **F** hydrogénation de transfert; **G** Transferhydrierung.   W. LEITNER

Ref.: *Chem.Rev.* **1985**, *85*, 129; *J.Am.Chem.Soc.* **1993**, *115*, 152; *Acc.Chem.Res.* **1997**, *30*, 97; A.R. Katrizky, O. Meth-Cohn (Eds.), *Best Synthetic Methods*, Academic Press, London 1990, Chapter 10; *J.Mol.Catal.A:* **1999**, *148*, 69.

transglycosidases   cat. the transfer of the glycosyl moiety from a glycoside to another acceptor with a minimal amount of *hydrolysis. Ts. have been found to be useful cats. for *glycosylation. For example, a β-fructofuranosidase from *Antherobacter* sp. K-1 has been used to transfer fructose from sucrose to the 6-position of the glucose residues of stevioside and rubusoside. A sucrase from *Bacillus subtilis* cat. the reversible transfer of fructose from sucrose to the 6-hydroxyl of a fructose unit at the nonreducing end of a levan chain. Several unnatural sucrose derivatives have been prepared by taking advantage of this process.

A transsialidase from *Trypanosoma cruzi* has been shown to transfer sialic acid reversibly to and from the 3-position of terminal β-Gal residues. Thus, sialic acid has been transferred to simple galactosides that are not substrates for sialyltransferases. Additionally, this *enzyme has been shown to resialylate the terminal galactose units of the cell-surface glycoproteins and glycolipids of sialidase-treated erythrocytes. **E=F=G**.   C.-H. WONG

transhydrocyanation → pseudomonas lipases; *oxylonitrilases

transition metal oxides are cpds. of a *transition metal and oxygen. Some of them are covalently bonded molecules; OsO_4, for example, appears as a gas at ambient conds., Mn_2O_7 is a liquid. However, most of them are solid. Their structures can be described as more or less dense packings of oxide anions, the interstices of which are occupied by metal cations. In spite of these structural models, the bonding is, however, never purely ionic, but mixed ionic-covalent, sometimes also featuring metallic bonding. As compared to oxides of non-metals and maingroup metals, tmos. have some special properties. Tmos. span a wide range of electrical properties from wide band-gap *semiconductors (TiO_2) through hopping semiconductors (Fe_3O_4) to metals (ReO_3) and superconductors ("High-T_c" or "ceramic" superconductors as $YBa_2Cu_3O_{7-x}$). Many compounds exhibit transitions from a non-metallic to a metallic state as a function of composition (e.g., *tungsten bronzes like $Na_x WO_3$, battery materials such as Li_xCoO_2), of temp. (e.g., VO_2), or of press. (V_2O_3).
Tmos. have many applications. Their adsorption properties are strongly influenced by their *surface chemistry. The variability in oxidation states, together with the possibilities of forming mixed-valency and non-stoichiometric cpds. give rise to important *redox cat. properties. In hom. cat., high oxidation state tmos. (e.g., *methyltrioxo-Re) can be used as cats. in olefin *metathesis and *oxidation.

F oxides des métaux de transition; **G** Übergangsmetalloxide.   P. BEHRENS
Ref.: P.A. Cox, *Transition Metal Oxides*, Clarendon, Oxford 1995.

transition metals 1-in het. catalysis: Tms. and their oxides play a dominant role in cat. The late tms. are used in their elemental form for the *activation of small molecules like O and H and the activation of C-H bonds (*C-H activation). The early tms. are used in their oxidic forms (mostly as suboxides) for *selective *oxidation and *hydrogenation reactions where the control of the oxidizing or reducing potential is more critical than with the late noble metals. Elemental metals are only rarely used in *bulk mass form in het. cat. (e.g., Pt *gauze for *oxidation of ammonia, *Raney Ni). Mostly, these metals are *supported on oxides or on *carbon. The prepn. of these finely divided forms of tms. today is mostly via *impregnation, *calcination-reduction sequences using simple inorganic metal cpds. (e.g., nitrates, halides, rarely *alkoxy *complexes or carbonyl complexes). In this way and by *tailoring the *metal-support interaction by choosing the right support material a great variety of *particle sizes, *shapes, and *dispersions can be achieved which all exhibit different cat. properties. Only in recent years has it been possible to begin to study the origin of this complex phenomenology by the advent of *single- crystalline support oxides and defined methods for metal *cluster prepn. and their *deposition.
Tm. oxides are usually used as mixtures in the form of multimetal multiphase oxide systems (MMO catalysts; cf. *mixed oxides). Typical applications are selective oxidation of propane to acrolein, formaldehyde synth. from methanol, or *ammoxidation to acylonitrile. The extreme chemical and structural complexity of technical systems which cannot be studied by simple oxide models up to now precludes a serious understanding of their mode of action. It is thought that the structural complexity is necessary to allow the reversible transfer of lattice O onto adsorbed

organic species and the replenishment of the vacancy by gas-phase O after the selective oxidation prod. has left the cat. *surface (*Mars-van Krevelen mechanism).

Coordinatively unsaturated metal oxygen polyhedra at surfaces are often energetically relaxed by dissociative adsorption of water, giving rise to surface OH groups. These groups can act as *solid acids. Terminal metal-O bonds exhibit basic properties. In this way all tm. oxide surfaces exposed to acidic/basic molecules behave as solid acids or bases. The strength and reactivity of these solid acids depends not only on the type of tm. (early tm.: acidic; late tm.: basic) but also on the details of the bonding geometry. In this way a vast selection of solid acids can be obtained either when the parent oxide is mixed with different tms. (*zirconia with Mn oxide as *superacid) or when a controlled *defect structure defines the disposition of terminating centers at the surface (modification of surface acidity by preparation conds. and post-synth. thermal treatment of tm. oxides).   R. SCHLÖGL

2-in hom. cat.: Tms. are the key to hom. cat. Molecular cats. may contain inner (f-elements) as well as outer (d-elements) tms. They form *complexes (or *organometallic cpds.) together with *ligands, which allow the fine-tuning of their cat. properties (*ligand tuning, *tailoring). The ability to change the oxidation state during a *cat. cycle and the formation of lower-energy *intermediates lead to a reduction of the activation barrier of cat. reactions. Additionally, tms. are moderate to strong *Lewis acids, depending on their oxidation state. Almost every tm. has found use in cat. procs.; therefore many of them are described individually. The applications of f-transition metals are described under *rare earth metals, while the d-elements are described under "*... as catalyst metal". Such entries cover the 3d-elements V, Cr, Co, Ni, Cu; the 4d-transition metals Mo, Ru, Rh, Pd, Ag, and 5d-elements such as W, Re, Os, Ir, Pt, Au, Hg. Some others are mentioned in the context of specific reactions, e.g., Zr under *metallocenes. **F** métaux de transition; **G** Übergangsmetalle.

T. STRASSNER

Ref.: J. Sinfelt, *Bimetallic catalysts: Discovery, Concepts and Applications*, Wiley, New York 1984; R.J. Madix (Ed.), *Surface Reactions*, Springer, Berlin 1994; Abel/Stone/Wilkinson; Beller/Bolm; Crabtree; Werner/Schreier; Wilkins.

transition state 1-general: The fundamental ts. concept states that for a reaction to proceed the reactant molecules must overcome a free energy barrier (Figure). Once the reactants (e.g. CO and O, $\Delta H = 67.6$ kcal/mol) have reached this state of highest free energy – the *transition state (or *activated complex, 25 kcal/mol; acc. to $[ES]^{\neq}$ in the Figure under 2: enzymatic), denoted by the symbol $\neq$ – they proceed on at a fixed rate (for the given reaction $\Delta G_P = \Delta G_{CO2}$ is ~5 kcal/mol). It is believed that there is a single potential energy surface on which the comps. will move; there is only one path on this *surface that requires the least energy for transformation. The position along this path is the *reaction coordinate. If a reaction proceeds through two or more steps, the one that has the highest free energy will often, but not always, be the rate-limiting step: in consecutive bi- and unimolecular reactions, for example, changes in conc. can shift the rate-limiting step from one to the other. For enzymatic reaction cf. *Eyring equ.

B. CORNILS

2-in enzymatic reactions: the reaction ts. is stabilized by the *enzyme to lower the activation energy. This is the main source of rate acceleration in enzymatic catalysis.

*Ts. theory can be applied to relate the first-order rate constants for the enzymatic (k_{cat}) and nonenzymatic (k) reactions to the corresponding equil. constants ($K^{\neq}_{cat}$ and $K^{\neq}$) for the formation of the ts. complex, that is $k_{cat}/k \sim K^{\neq}_{cat} K^{\neq}$.

According to the thermodynamic cycle, these equil. constants are related to the dissociation constants for the ts. (K_T) and for the substrate (K_S), so that $K_S K^{\neq} = K_T K_{cat}^{\neq}$. This simple

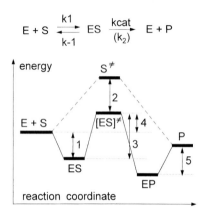

$$E + S \underset{k\text{-}1}{\overset{k1}{\rightleftharpoons}} ES \xrightarrow[(k_2)]{k_{cat}} E + P$$

1: ΔG_S; 2: ΔG_{ts}; 3: k_{cat}; 4: k_{cat}/K_m; 5: ΔG_P; $\Delta G_{ts} > \Delta G_S$ or ΔG_P; $S^{\neq}$ is the *uncat.* ts., $[ES]^{\neq}$ is the *cat.* ts.; ES = enzyme substrate complex, EP = enzyme product complex.

analysis concludes that the enzyme binds to the ts. $S^{\neq}$ more strongly than to the ground state S by a factor approximately equal to the rate acceleration; that is, $k_{cat}/k \sim K_S/K_T$ (cf. *transition state theory). **F** état de transition; **G** Übergangszustand. C.-H. WONG

Ref.: Gates; Mouljin/van Leeuwen/van Santen, p. 78; *Acc. Chem.Res.* **1999**, *32*, 127.

transition state analogs → organome-tallic model catalysts

transition state (shape) selectivity implies the restriction of the formation of certain molecules in the pores; cf. *microporous solids. F. JENTOFT

transition state theory (in enzyme catalysis) The fundamental concept that for a reaction to proceed the reactant molecules must overcome a free energy barrier has provided the basis for quantitative approaches to *enzyme kinetics. Once the reactants have reached this state of highest free energy – the *transition state (ts.) – they proceed to prods. at a fixed rate. Free energy contains both enthalpic and entropic terms. In general, the lower the activation energy, the faster the overall reaction will proceed. If a reaction proceeds through two or more steps, the one

that has the highest free energy will often, but not always, be the rate-limiting step: in consecutive bi- and unimolecular reactions, for example, changes in conc. can shift the rate-limiting step from one to the other.

The assumptions in ts. theory that the reactant ground state is in equilibrium with the ts., and that the ts. proceeds to products at a fixed rate, have led to the development of the *Eyring equ. (cf. *transition states). Ts. theory has proven to be an excellent, durable model with which to analyze basic principles of enzyme action. One role of enzymes can be considered to be the reduction in the free energy of activation by stabilizing the rate-limiting transition state. This reduction results in an acceleration in reaction rate. Enzymes accomplish this reduction by either reducing the enthalpy of activation, setting up more favorable interactions between substrates (an entropy effect), modifying interactions with solvent, or all of these (cf. also *intermediates). **F** théorie d'état de transition; **G** Theorie des Übergangszustandes. B. CORNILS

transketolase (EC 2.2.1.1) Tk. cat. the reversible transfer of the C1-C2 ketol unit from D-xylulose 5-phosphate to D-ribose 5-phosphate to generate *D-sedoheptulose 7-phosphate and glyceraldehyde-3-phosphate (*G3P). The *enzyme relies on two *cofactors for activity, *thiamine pyrophosphate (TPP) and Mg^{2+}. Tk. from baker's yeast is commercially available, and the enzyme has also been isolated from spinach.

β-Hydroxypyruvic acid (HPA), also, acts as a ketol donor which is transferred to an aldose acceptor. One valuable feature of HPA is that once the ketol unit is transferred by tk., CO_2 is lost, rendering the overall reaction irreversible. Other analogs of HPA are not substrates, but a wide range of aldehydes are ketol acceptors, including aliphatic, α,β-unsaturated, aromatic, and heterocyclic aldehydes (cf. also *pentose phosphate pathway; *Calvin cycle). **E=F=G**. C.-H. WONG

Ref.: *Tetrahedron Lett.* **1987**, *28*, 5525; *J.Chem.Soc., Perkin Trans.* **1993**, *1*, 165.

transmetallation exchange of the metal in organometallic cpds. or cats., usually applied in the synth. of otherwise inaccessible, or difficultly accessible derivatives. The general method works for M = Li to Cs, Be to Ba, Al, Ga, Sn, Pb, Bi, Se, Te, Zn, and Cd, acc. to $M + R\text{-}M^1 \rightarrow R\text{-}M + M^1$, e.g.,

$$Zn + (CH_3)_2Hg \longrightarrow (CH_3)_2Zn + Hg \downarrow$$
$$(Mg) \qquad\qquad\qquad\qquad (Mg)$$

$R\text{-}M^1$ should be of weakly exothermic or preferably endothermic in nature (e.g., $HgMe_2$, $\Delta H = +94$ kJ/mol). The equ. depends on the difference in the free enthalpies of formation $\Delta(\Delta G_f^o)$ of R-M and $R\text{-}M^1$, resp. Therefore, alkyl derivatives of noble metals are best suited for t.; cf. the t. with Zn, $\Delta H = -35$ kJ/mol). The equil. can be influenced by vapor press., solubilities, etc. of the prods. **F** transmétallation; **G** Transmetallierung.   W.A. HERRMANN

Ref.: C. Elschenbroich, A. Salzer, *Organometallics – A Concise Introduction*, VCH, Weinhein 1989.

transmethylation   → *S*-adenosylmethionine

transmission electron microscopy (TEM) uses transmitted and diffracted electrons and is similar to an *optical microscope, replacing optical by electromagnetic lenses. The data are local, integral and bulk-sensitive. The UHV ex-situ technique is destructive to the sample, but best suited for *single crystals. No information can be obtained about reactions and equilibria between gases and solids.   R. SCHLÖGL

Ref.: Niemantsverdriet, p. 168; Ullmann, Vol. B6, p. 229.

TransPlus process   → Mobil procs.

N-transribosylase cat. the transfer of the sugar moiety between two nucleosides. Two types of *enzymes have been identified: type I enzymes cat. the transfer of the sugar moiety between two purine bases, while type II cat. the transfer between any two bases. Like *nucleoside phosphorylases, *transribosylases are stereospecific for the β-anomer of the nucleo-

side prod. Thymidine and 2′-deoxycytidine are the best glycosyl donors, and a reasonable amount of variation in the acceptor bases is tolerated. **E=F=G**.   C.-H. WONG

Ref.: *TIBTECH* **1990**, *8*, 348.

transvinylation In the presence of Pd salts vinyl cpds. such as vinyl chloride and vinyl esters undergo oxyacylations in which the nucleophile at the vinyl group is replaced by a carboxylic residue. This reaction gained technical inmportance for the synth. of vinyl esters of higher carboxylic acids, such as vinyl esters of *Versatic or *Neo acids or divinyl adipate from adipic acid and vinyl acetates which are monomers for lacquers, etc. **E=F**; **G** Transvinylierung.   A. ECKERLE

TRAP, trap a *trans-*chelating chiral diphosphine *ligand bearing a $*C_2$ symmetric biferrocene framework with central and planar elements of chirality (Figure).

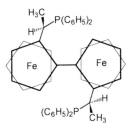

TRAP is generated by a multistep synth. from substituted ferrocenes. The Rh *complex of the *trans*-chelating TRAP with Et groups at P cat. the asymmetric *hydrogenation of dimethylitaconate (96 % *ee). Other applications of TRAP include *hydrosilylation and *Michael reactions.   O. STELZER

Ref.: *Tetrahedron Asymmetry* **1991**, *2*, 593 and **1995**, *6*, 2521; *Tetrahedron Lett.* **1995**, *36*, 5239; *J.Am.Chem. Soc.* **1992**, *114*, 8295.

trapping   → radical scavenger

trehalase, trehalose   → glycosyl phosphorylase

Treibs reaction allylic *acetoxylation of terminal alkenes by means of cat. active Hg acetate in acetic acid or other solvents. **F** réaction de Treibs; **G** Treibs-Reaktion. B. CORNILS

tren triaminotriethylamine ($[H_2NCH_2CH_2]_3N$) in *complex cpds. or as a *ligand

tribochemistry that branch of chemistry which investigates the effects exerted by mechanical energy upon chemical cpds., usually solids (e.g., cats.). When mechanical stress acts on a solid, deformations occur and finally the solid breaks apart. During such a proc. a whole series of chemical changes may occur in the material. Solid-state defects are created ranging from point defects to extended crystallographic shears. Cracks in material under mechanical activation may be generated with extremely high strain at the tip of the crack. The internal energy at the fissure may exceed the surface free energy of the material by a factor of 10^6. When the crack proceeds through the material, partial release of this high internal energy may lead to temps. at the tip of the crack as high as 10^5 K for extremely short periods of time. The mechanical generation of defects enhances the internal strain and stress in the solid and, thus, its internal free energy. This increased internal energy leads to an reinforcement of the material for solid-state reactions, for example, *sintering, cpd. formation, or *solid-solid wetting. The conc. of cracks, fissures, and of defects in the material rises and, thus, its internal strain changes during mechanical activation. Three different types of intrinsic strain are differentiated: first-order intrinsic strain is hom. distributed over a few X-ray-coherent areas: second-order intrinsic strain is limited to one primary crystallite, and third-order intrinsic strain is focused to atomic distances on the shortest.

To activate mechanically solids mills are usually used. Extended milling even may lead to recrystallization and sintering of particles. Of course, these procs. are strong functions of the material which is mechanically activated, the type of mill used, and the medium in the mill, e.g., an inert gas atmosphere, reactive gases, or liquids. **F** tribochimie; **G** Tribochemie. G. MESTL

Ref.: Heinicke et al., *Tribochemistry*, Akademie-Verlag, Berlin 1984.

tricarboxylic acid cycle → Krebs cycle

Tricat catalysts Tricat Industries Inc. was founded 1992 to provide *regeneration and presulfiding to provide services to *refineries that use *hydroprocessing cats. Tricat operates plants at McAlester (Oklahoma), Baltimore (Maryland) and Bitterfeld (Germany). J. KULPE

trickle-bed reactor continuous, vertical, cat. fixed-bed reactor where the liquid reactants trickles down over the bed while the gaseous reactants flow either co- or counter-currently (*catalytic reactors). **F** réacteur à ruissellement; **G** Riesel(bett)reaktor.
Ref.: McKetta-1, Vol. 46, p. 469. B. CORNILS

tridentate phosphines are *phosphines with three PR_2 groups (or other binding atoms, cf. *oxazoline) in the molecule thus being able to multiple bind *transition metals. (cf. *chelating ligands, *AMPP, *THREOPHOS) (Figure).

Left: linear type; Right: *tripod type

Similarly to other polydentate phosphine ligands, the coordination properties are dependent on the electron-donating or withdrawing character of the substituents and on the geometry and chain length of the bridging units. Most tps. synth. so far are powerful chelating ligands but there are also examples of bi- and trimetallic coordination modes. Their use in hom. cat. is almost essentially limited to mechanistic studies since the number of coordination sites and the dissociation rates are too low for sufficient cat. *activity.

In some cases reversible *arm-off dissociation of one of the phosphine moieties was observed during the cat. reaction. **F** phosphine tridentaire; **G** dreizähnige Phosphane. R. ECKL

Ref.: Hartley; Cornils/Herrmann-1, Chapter 3.2.13; *Organometallics* **1991**, *10*, 3798; *Tetrahedron Lett.* **1996**, *37*, 4933.

trien triethylenetetramine ($[H_2NCH_2CH_2-NHCH_2]_2$) in *complex cpds. or as a *ligand

triglycerides → interesterification

trimellitic acid (anhydride, TMA) intermediate, manuf. cat. by *oxidation of triethylbenzene (*Bofors proc.) or the *carbonylation of *m*-xylene (procs. of *Amoco, *Mitsubishi). B. CORNILS

trimerization *oligomerization of three molecules of, e.g., ethylene to 1-hexane, cat. by *Cr cpds. B. CORNILS

Triolefin process → Phillips procs.

triosephosphate isomerase (EC 5.3.1.1) a glycolytic *enzyme that cat. the interconversion of glyceraldehyde-3-phosphate (*G3P) and dihydroxyacetone phosphate (DHAP). The reaction proceeds through an enediol intermediate, and is cat. by an *active-site glutamate that acts as a cat. base, then acid. Ti. is considered to be a "perfect" enzyme, as the rate-determining step is the binding of G3P to the enzyme, and this rate is near the diffusional limit ($\sim 10^8 \, s^{-1}$). **E=F=G**. P.S. SEARS

Ref.: *Adv. Enzymol.* **1975**, *43*, 491; *Acc. Chem. Res.* **1977**, *10*, 105.

triphenylphosphine (TPP, tpp, PPh$_3$) the best-known (and the cheapest) organic *phosphine. X-ray studies reveal a distorted "propeller"-shaped molecule. TPPs form *complexes with a wide range of 3d, 4d, and 5d *transition metals M in different oxidation

states. The PPh$_3$ *ligand within the M-PPh$_3$ units displays a symmetric propeller-type arrangement of the C_6H_5 rings. TPP is manuf. commercially by the reaction of PCl_3 and C_6H_5Cl with Na.

TPP is widely employed for the prep. of *ligand-modified cats. and as a starting material for the prep. of mono- and multidentate phosphines containing Ph$_2$P moieties. The Rh(I) complex of TPP is an excellent cat. for *hydrogenations and *hydrosilylations (cf. *Vaska cpds.). The hydrido complex HRh(CO)(TPP)$_3$ is the well-known *hydroformylation cat. (cf. oxo procs. of *Union Carbide, *BASF, *Mitsubishi). HRuCl(TPP)$_3$ is extemely fast in the hydrogenation of terminal alkenes and of dienes/trienes to monoenes. Pd complexes of TPP serve in *carbonylations, *Heck reactions, and *Suzuki cross-couplings. Cat. *deactivation occurs through metal-assisted P-C cleavage, which requires *regeneration during the *cat. cycle after certain periods. TPP-stabilized Ru(II) alkylidenes initiate *living polymerizations of norbornene derivatives through *ROMP. **F=G** triphenylphosphan. O. STELZER

Ref.: *Acta Crystallog.* **1991**, *C47*, 345; *J. Chem. Soc., Dalton Trans.* **1991**, 653; Patai-2, Vol. 1, p. 9, 151; Cornils/Herrmann-1; *Angew. Chem. Int. Ed. Engl.* **1994**, *33*, 2379; *Chem. Rev.* **1995**, *95*, 2457; *J. Am. Chem. Soc.* **1992**, *114*, 3974.

triphenylphosphine di-*m*-sulfonate disodium (TPPDS) a water-soluble *ligand for hom. cat., prepd. by direct *sulfonation of *TPP with oleum at 58 °C in the presence of boric acid. The Pd(0)/TPPDS *complex has been employed for the activation of the Sn-C bond in organotrichlorostannanes RSnCl$_3$ in C-C coupling reactions. TPPDS is not used commercially. O. STELZER

Ref.: Cornils/Herrmann-2, p. 329; *J. Organomet. Chem.* **1995**, *502*, 177; *Angew. Chem. Int. Ed. Engl.* **1995**, *34*, 811; *Tetrahedron Lett.* **1995**, *18*, 3111.

triphenylphosphine mono-*m*-sulfonate sodium (TPPMS) a *ligand for hom. cat. It is moderately soluble in water and in organic phases and thus partly hydrophilic and partly lipophilic. TPPMS is prepared by direct *sulfonation of *TPP with oleum. TPPMS forms defined *complexes with Ag(I), Rh(I), Os(II), Pd(0), Pd(II), etc. Rh complexes of TPPMS or its Li salt are cats. for the *hydroformylation of higher alkenes using "solubilizing" agents, detergents, or *tenside ligands. Pd(0)/TPPMS complexes effect C-C couplings acc. to the *Suzuki coupling. A Pd cat. containing a *phosphonium salt based on TPPMS is used for *Kuraray's *hydrodimerization proc. Ru and Os complexes of TPPMS are emloyed as cats. for *regioselective *hydrogenations in *two-phase media.

O. STELZER

Ref.: *J.Chem.Soc.* **1958**, 276; Cornils/Herrmann-2, p. 42,71; *J.Cord.Chem.* **1991**, *24*, 43; *Organometallics* **1991**, *10*, 2126; *Nouv.J.Chim.* **1978**, *2*, 137; *J.Am.Chem. Soc.* **1990**, *112*, 4324; *J.Mol.Catal.A:* **1996**, *108*, 153, **1989**, *54*, 65 and **1997**, *116*, 167; *Chem.Eng.News* **1995**, 25; *Fett, Lipid.* **1996**, *98*, 393.

triphenylphosphine tri-*m*-sulfonate trisodium (TPPTS, tppts) a colorless solid, highly soluble in water (1.2 kg/L), and insoluble in most organic solvents. TPPTS is the prototype of water-soluble *phosphines, and is also used commercially as a *ligand (especially for Rh and Pd) in hom. cat.(*RCH/RP proc.). *Complexes with Mo, W, Mn, Fe, Ru, Co, Rh, Ir, Ni, Pd, Pt, Ag, and Au have been identified. It is manuf. by *sulfonation of *TPP with oleum (addition of boric acid minimizes the formation of TPPTS oxide), followed by an appropriate purification (Figure).

The Rh(I) complex of HRh(CO)(TPPTS)$_3$ is used as a cat. for industrial-scale *hydroformylation of lower alkenes (*Ruhrchemie/Rhône-Poulenc proc.). Higher alkenes are processed with the *SAPC method. C-C coupling reactions (*Suzuki, *Heck) are cat. by Pd(0)/TPPTS complexes. For *hydrocyanation of butadiene in biphasic systems Ni complexes are used. Biphasic *carbonylation are also cat. by Pd complexes. Rh(I)/TPPTS cpds. show remarkable cat. activity for the *hydrogenation of CO$_2$ to HCOOH in aqueous phase. Unsaturated aldehydes can be *hydrogenated by either Rh or Ru complexes. Co$_2$ (CO)$_6$(TPPTS)$_2$ is active in hexene hydroformylation in water. The selective *reduction of nitrobenzenes to anilines could be achieved using CO in the presence of Pd/TPPTS or Pd/ *BINAS complexes.

O. STELZER

Ref.: *J.Organomet.Chem.* **1990**, *389*, 85,103; **1991**, *403*, 221; **1994**, *482*, 45, and **1995**, *502*, 177; Cornils/Herrmann-1 and -2, p. 71; *Angew.Chem.Int.Ed.Engl.* **1995**, *34*, 811; *J.Mol.Catal.A:* **1996**, *110*, 189; **1997**, *116*, 131,217; *J.Catal.* **1990**, *121*, 327; *Tetrahedron Lett.* **1995**, *36*, 9305; *J.Chem.Soc.,Chem.Commun.* **1993**, 1465.

TRIPHOS, triphos serves as a *tridentate *ligand and forms stable *complexes with a variety of metals such as V, Cr, Co, Rh, or Pd, favoring *facial* coordination of one *transition metal.It is prep. by reaction of H$_3$CC(CH$_2$Cl)$_3$ with (C$_6$H$_5$)$_2$PM (M = Li,Na) (Figure).

*Complexes HM(TRIPHOS) and H$_3$M(TRIPHOS) (M = Rh, Ir) may be applied for cat. *hydrogenations, *hydrogenolysis, and *desulfurization. In Rh-cat. *hydroformylation the *tripod ligand TRIPHOS undergoes reversible *arm-off dissociation to generate *vacant coordination sites for incoming CO. The water- soluble derivative of TRIPHOS is SULPHOS (the methyl group of TRIPHOS substituted by a *p*-sulfonate sodium benzyl group) and is proposed for *hydroformylation, *hydrogenation, and *hydrogenolysis.

O. STELZER

Ref.: Patai-2, Vol. 1, p. 191; Cornils/Herrmann-2, p. 477; Cornils/Herrmann-1, Vol. 2, p. 969; *Organometallics* **1991**, *10*, 3798 and **1995**, *14*, 5458; *J.Am.Chem. Soc.* **1995**, *117*, 8567.

tripod ligands *ligands able to coordinate a metal center via three donor atoms in a tripod-type mode (cf. Figure under the keyword *tridentate phosphines). The best-investigated tls. are *tridentate phosphines but tripod amines and isocyanides have also been studied. The use of tls. in cat. is limited essentially to mechanistic studies and some special applications. One of the best known tl. is *TRIPHOS, which was applied in the development of the *hydrodesulfurization (*HDS) proc. and the hom. thiophene *hydrogenation cats. Some other tls. were also synth. and applied in *two-phase cat.; *chiral tls. have been used for *asymmetric hydrogenations. **F** ligands tripod; **G** Tripod-Liganden.

                                                    R. ECKL

Ref.: *J.Organomet.Chem.* **1997**, *539*, 67; *Organometallics* **1993**, *12*, 4949; Cornils/Herrmann-2.

tritox, TRITOX   the *alkoxide *ligand [OC(*tert*-butyl)$_3$]$^-$

Trojan horse inhibitors   → active site-directed inhibitors

Tropsch, Hans (1889–1935), professor of chemistry in Prague (Czech Republic), Mülheim (Germany), and Chicago (USA). Co-worker of Franz *Fischer and co-inventor of the *Fischer-Tropsch synthesis.   B. CORNILS

true activation energy   → apparent activation energy

trypsin   a digestive *serine protease produced in the pancreas of mammals as a virtually inactive *proenzyme (or *zymogen), trypsinogen. It is activated by the enteropeptidase (or trypsin)-cat. cleavage at lysine-15. Upon activation, the newly formed amino terminus moves from the surface to a buried position within the protein where it may form a *salt bridge with aspartate-194.
A polypeptide inhibitor of trypsin, frequently isolated from bovine pancreas (bovine pancreatic trypsin inhibitor, or BPTI), is a small (58 amino acid) protein with an extremely high affinity

for trypsin (K_D ~ 10^{-13} M). The inhibitor occupies the trypsin *active site, and has residues in what would seem to be an ideal position for *hydrolysis. **E=F=G**.   P.S. SEARS

tryptophan synthase (EC 4.2.1.20)   a heterotetrameric ($\alpha_2\beta_2$), bifunctional *enzyme from *Salmonella typhimurium*. The α- subunits cat. the *aldol cleavage of indole-3-glycerol phosphate to produce indole and *G3P while the *pyridoxal phosphate-containing β-subunits cat. the *condensation of indole with serine to produce tryptophan. The indole produced in the first reaction is believed to pass through a solvent-filled "channel" connecting the α- and β-subunit active sites. The β-subunits can also cat. the β-elimination and *deamination of serine to pyruvate, which has given a clue to the enzyme's mechanism of action: imine formation between the pyridoxal phosphate aldehyde and the amine of serine; β-elimination to produce the aminoacrylate (still bound to pyridoxal phosphate); addition of indole to the β-carbon; and hydrolysis of the imine. It also cat. the β-displacement reaction using β-chloroalanine. **E=F=G**.

                                    P.S. SEARS, C.-H. WONG

Ref.: *Appl.Biochem.Biotech.* **1986**, *13*, 147.

TS-1   → titanium silicalite

TSA   → total surface area

TSR   → two-state reactivity

T-Star process   → Texaco procs.

Tsuji-Trost reaction   → allylation

TTN   → total turnover number

tube wall reactor   → Synthane process

tungsta   tungsten oxide (e.g., WO$_3$) as a basis for cats.

tungsten as catalyst metal   W oxide- (and Mo-) based cats. have received considerable attention due to their high *metathesis activity in alkene conv., in both hom. and het. phases, but their reactivity is very similar and related to the ability of both these elements to form metal *carbene *complexes as *intermediates. W is mainly used in liquid-phase *metathesis of alkenes (e.g., procs. of *Phillips Triolefin or isoprene, *Shell FEAST, *Hüls Vestenamer, etc.), in comb. with Ni for *hydrotreating cats. (*HDS, e.g., WS_2 as cat.), and with V in the cats. for the reduction of NO in flue gas emissions (*stationary exhaust cat.), supported on *zirconia together with Pt for cats. for alkane *isomerization, as an oxoanion in liquid-phase applications (e.g., *Tokoyama proc.), and as a *support for Pt for the *hydrogenation of organics. In general it is found as a *promoter often used in several multicomp. *oxide cats. (cf. also *mixed oxide cats.).

W has chemical properties closely related to V and Mo and for this reason is often used to substitute partially for these elements in cat. formulations in order to tune the cat. behavior. For example, in industrial cats. based on *vanadia-*titania for the reduction of NO by NH_3/O_2 in flue gas, the partial substitution of V with W makes it possible to decrease the *activity, but increase the resistance to *deactivation by SO_2 (*SCR cats, *DeNOx).

W forms cpds. very analogous to those of Mo (for example, they both form dioxo-type species and upon reduction their oxides form closely related shear structures) and thus often the cat. behavior of materials based on these two elements is very similar, although often slightly less advantageous for tungsten-based cats. W and Mo also form analogous *polyoxometallates, e.g., the *Keggin-type $H_3PM_{12}O_{40}$ *heteropolyacids, where M can be W or Mo. Mixed W and Mo cpds. are also possible. Substitution of Mo with W improves the thermal stability, although the reactivity decreases. These or other types of related polyoxometallates, with substitution of the M metal with other elements such as V and Ti, have received increasing attention in recent years as oxidation cats. in both hom. and het. cat.

W, similarly to V and Mo, forms peroxo-type species (peroxotungstic acid and salts) in the liquid phase by reaction with *hydrogen peroxide and these species show excellent behavior in *epoxidation reactions. Examples are the manuf. of propylene oxide via *epoxidation of propylene (procs. cited under the keyword *propylene oxide) or *maleic anhydride (*Butakem tartaric acid), and the addition of H_2O_2 to e.g., allyl alcohol in presence of *Milas reagent (*Degussa glycerol, *Shell glycidol). W and Mo oxides are also known to form similar hydrogen bronzes (e.g., $WO_3H_{0.35}$, *tungsten bronze) which show interesting properties as fast ionic conductors (FICs). An interesting new direction of research is the prepn. of membranes based on these FIC cpds. for the preparation of *membrane reactors for *hydrogenation or *dehydrogenation reactions at relatively high oxygen activities. W containing enzymes are known, too. **F** le tungstène comme métal catalytique; **G** Wolfram als Katalysatormetall.

G. CENTI

Ref.: *Appl.Catal.A:* **1997**, *158*, 53; *Appl.Catal.A:* **1996**, *134*, 81; *J.Catal.* **1983**, *82*, 395; M.T. Pope, *Heteropoly Oxometalates*, Springer, Berlin 1983; *J.Mol.Catal.A:* **1995**, *95*, 147; *Houben-Weyl/Methodicum Chimicum*, Vol. XIII/7; *Chem.Rev.* **1996**, *96*, 2817; Adams/Cotton.

tungsten bronze Tbs. of general formula A_xWO_3 ($0 < x < 1$, usually $0.3 < x < 0.9$), are formed by the reaction of WO_3 or alkali or alkaline earth tungstenates with reducing agents like H or alkali metals, or electrochemically. They are very stable cpds. with different colors, always exhibiting a metallic "bronze" luster. The different structures (cubic, hexagonal or tetragonal) they exhibit rely on arrays of $[WO_{6/2}]$ octahedra, which share mainly corners and (sometimes) edges (*rhenium oxide based-structures). The voids of these host structures contain the guest cations. The metallic or *semi conducting properties of tungsten bronzes are caused by the introduction of the excess electrons from the reducing agent into the *conduction band of the WO_3 host structure.

The main use of tbs. is as pigment. The *redox reactivity of these cpds., which is coupled to insertion/de-insertion procs., gives rise to their use in *oxidation cat. **F** bronze au tungstène; **G** Wolframbronze.   P. BEHRENS

Ref.: *Acc.Chem.Res.* **1996**, *29*, 219.

tunicamycin a cpd. from *Streptomyces* which inhibits the *N*-acetylglucosamine-1-phosphotransferase that transfers *N*-acetylglucosamine phosphate (GlcNAc-P) from the nucleotide-activated sugar, UDP-GlcNAc, to dolichyl phosphate to form GlcNAc pyrophosphoryl dolichol (GlcNAc-PP-Dol). The formation of GlcNAc-PP-Dol is the first committed step in the synth. of the large oligosaccharide donor used by *oligosaccharyltransferase for the N-glycosylation of proteins in eukaryotic cells. T. mimics the two substrates, dolichyl phosphate and UDP-GlcNAc, as they are probably constrained within the active site. **E=F=G**.   P.S. SEARS

turnover frequency (TOF) The turnover frequency (the term was borrowed from *enzyme cat.) quantifies the specific *activity of a cat. center for a special reaction under defined reaction conds. by the number of molecular reactions or *cat. cycles occurring at the center per unit time. In practice, the experimentally determined volumetric rate of reaction [moles · volume^{-1} · time^{-1}] is divided by the number of centers [moles · volume^{-1}] yielding the TOF [time^{-1}]. The number of active centers per unit volume for well-defined hom. cat. reactions is calc. from the conc. of the cat. For het. cats. the number of active centers is derived from characterization methods for the *surface of the cat., predominantly sorption methods, based on the assumption of uniform activity for all *active sites. The reaction conds. have to be arranged in a way that ensures that the *rate of reaction is determined by the cat. and is not limited by transport phenomena. In this respect the TOF is inferior to an experimentally derived kinetic expression (cf., *reaction probability). The TOF is frequently used to compare different

cats. for a given reaction, enabling conclusions to be drawn about the type (e.g., *central atom) or structure (e.g., *ligands bound to the central atom) of the cat. and their relationship to the rate of reaction. For most relevant industrial applications the TOF is in the range between 10^{-2} and $10^2 s^{-1}$; for enzymes as cats. between 10^3 and 10^7 s^{-1}. The relationship between TOF and TON is TOF = TON · time^{-1}. **E=F**; **G** TOF, Wechselzahl.   C.D. FROHNING

Ref.: Thomas/Thomas; Ertl/Knözinger/Weitkamp, Vol. 3, p. 959.

turnover number (TON) specifies the maximum use that can be made of a cat. for a special reaction under defined reaction conds. by the number of molecular reactions or reaction cycles occurring at the reactive center up to the decay of *activity. In this respect, the TON represents the maximum yield of prods. attainable from a cat. center. The TON results from multiplication of the *turnover frequency TOF [time^{-1}] and the *lifetime of the cat. [time]. For hom. cat. reactions the TON may be calculated easily from the conc. of the cat. and the lifetime achieved and therefore it is unequivocally and frequently used to compare the efficiency of different hom. cats. For solid (heterogeneous) cats. the TON (like the TOF) depends on the accuracy of the method by which the number of *active sites per unit catalyst is determined. For fixed-bed operation the TON is inconvenient due to the inhomogeneous *deactivation in the cat. bed (for the different definitions, cf. *single turnover). For industrial applications the TON is in the range from 10^6 to 10^7. For the enzymatic definition of TON, cf. *catalytic constant and *total turnover number. **E=F=G**.   C.D. FROHNING

Ref.: Ertl/Knözinger/Weitkamp, Vol. 3, p. 958; *Chem. Rev.* **1995**, *95*, 661.

TWC → three-way catalyst, *automotive exhaust catalyst

Twitchell reaction the cleavage of *fats into the fatty acids and glycerol in presence of H$_2$SO$_4$/sulfonaphthenic acid as cat., and an emulsifier. **F** réaction de Twitchell; **G** Twitchell-Spaltung.   B. CORNILS

two-phase catalysis   the general case of a hom. cat. procedure in which the cat. is dissolved in one liquid *phase and the reactants and reaction prods. in another, immiscible phase (fluid-fluid two-phase cat.; for the case of two separate phases of which one is solid, cf. *immobilization). The liquids may be relatively nonpolar (organic), ionic/*nonionic, *molten salts, or aqueous. If the prods. formed during cat. prefer the non-cat. phase as a consequence of partition coefficients, the separation and thus the recycle of the cat. is very greatly facilitated (*mobile or liquid supports). Variants of handling of the two phases and thus isolating the prods. are known (*mass transfer of the reactants into the cat. phase and automatic separation of the phases during cat., *extraction of the reaction prods. with appropriate liquid extractants, etc.). Also known are examples of procs. with one-phase cat. coupled with biphase cat. separation (*aqueous-phase cat., *fluorous-phase cat., *thermoregulated ligands).
Two-phase cat. offers several advantages, such as easy cat. recycle without any chemical or thermal stress, the avoidance   of side reactions, and often a better *selectivity than a monophase reaction. Commercial examples are Shell's *SHOP proc.(organic/organic), the *Ruhrchemie/Rhône-Poulenc proc. (organic/ aqueous), *Kuraray telomerization proc. (organic/ aqueous), and *IFP Dimersol proc. (organic/*NAIL). For enzymatic tpc. cf. *bioreactors and *enzymatic cat. in organic solvents. **F** catalyse biphasique; **G** Zweiphasen-Katalyse.   B. CORNILS

Ref.: M.P. Allen, D.J. Tildesley, *Computer Simulation of Liquids*, Oxford Science, Oxford 1987; Cornils/ Herrmann-1, p. 577; Cornils/Herrmann-2; Horváth/ Joó; *Cattech* **1998**, *(6)*, 47; Reichardt.

two-phase liquid reaction   1: reactions in two-phase mixtures which secure an easy separation of cats. (*two-phase catalysis) and/ or parts of the reactants or reaction prods.

2: Reactions which need *phase transfer procedures or media to take place. **F** réactions en deux phases liquides; **G** Zweiphasen-Flüssig-reaktionen.   B. CORNILS

two-state reactivity   Neglection of spin multiplicity is a common dogma in transition metal cat. Recent studies of *bare metal ions indicate, however, that this view needs to be reconsidered. So, e.g., several oxidations mediated by *transition metals involve spin crossover as a mechanistc distributor. These results lead to formulate two-state reactivity (TSR) as a new, general mechanistic pattern in metal cat. The key feature of TSR is that it involves spin-inversion along the reaction coordinate as the rate-determining step. This behavior is unusual and does not obey classical *Arrhenius kinetics, thereby providing new concepts for cat. selectivity. TSR may be a fundamental factor for the broad versatility of *transition metal cats. Particular relevance concerns oxidation cat., because TSR allows for reactions which are otherwise spin-forbidden. **E=F=G**.   D. SCHRÖDER

Ref.: *J.Am.Chem.Soc.* **1995**, *117*, 11745; *Chem.Eur.J.* **1998**, *4*, 193.

TWR   tube wall reactor, → Synthane process

tyrosinase   a copper-containing *enzyme that cat. the *oxidation of *o*-diphenols to *ortho*-quinones using molecular oxygen. It also cat. the oxidation of monophenols to *ortho*-quinones. **E=F=G**.   C.-H. WONG

tyrosine hydroxylase   a pterin-containing monooxygenase which cat. the *oxidation of tyrosine to L-dopa using molecular oxygen and *NADPH. The *enzyme *phenylalanine monooxygenase (phenylalanine hydroxylase) proceeds through the same mechanism to form tyrosine from phenylalanine. **E=F=G**.

C.-H. WONG

tyrosine kinase → kinase, *protein kinase

tyrosine phenol lyase (EC 4.1.99.2) catalyzes the synth. of tyrosine from pyruvate, phenol, and ammonia. It also accepts β-chloroalanine as a substrate for β-displacement reactions. **E=F=G**.   C.-H. WONG

Ref.: *Appl.Biochem.Biotech.* **1986**, *13*, 147.

tyrosyl-tRNA synthetase   → nonribosomal peptide synthesis

U

Ube aminoundecanoic acid process
multi-step proc. for the manuf. of aminoundecanoic acid by 1: *aminative peroxidation of
cyclohexanone with H_2O_2 and ammonia to
1,1′-peroxydicyclohexylamine, 2: *pyrolysis of
the peroxide to 11-cyanoundecanoic acid, and
3: cat. *hydrogenation of the cyano- to the
aminoundecanoic acid. **F** procédé d'Ube
pour l'acide aminoundecanoique; **G** Ube-
Aminoundecansäure-Verfahren.   B. CORNILS

Ube dimethyl carbonate process two-
step proc. for the manuf. of dimethyl carbonate (DMC) by 1: generation of methyl nitrite
from methanol, NO and O_2, 2: the hom.
$PdCl_2$-cat. *carbonylation of methyl nitrite.
F procédé d'Ube pour le diméthyl carbonate;
G Ube Dimethylcarbonat-Verfahren.
                                          B. CORNILS
Ref.: *Catalysis Surveys from Japan* **1991**, *1*, 77.

Ube dimethyl oxalate process for the
manuf. of DMO by Pd-cat. *carbonylation of
butyl nitrite (oxidatively from methanol and
NO), similar to the Ube DMC proc. **F** procédé d'Ube pour le diméthyl oxalate; **G** Ube
Dimethyloxalat-Verfahren.   B. CORNILS
Ref.: *Catalysis Surveys from Japan* **1991**, *1*, 77.

Ube glycol process manuf. of ethylene
glycol by a two-step procedure from 1: *oxidative carbonylation of methanol to dimethyl
oxalate with a Pd cat. at 110 °C/9 MPa, and
2: *hydrogenation of the diester to glycol over
Ru cats. Methanol is recycled. **F** procédé
d'Ube pour le glycol; **G** Ube Glykol-Verfah-
ren.   B. CORNILS

Ube hydroquinone process for the manuf. of hydroquinone by cat. *oxidation of phenol with H_2O_2. **F** procédé d'Ube pour le hydroquinone; **G** Ube Hydrochinon-Verfahren.
                                          B. CORNILS

Ube oxamid process Proc. for the manuf.
of oxalic acid diamide (oxamide as sustained
fertilizer) by oxidative dimerisation of HCN
with aqueous $Cu(NO_3)_2$ solutions. **F** procédé
d'Ube pour l'oxamide; **G** Ube Oxamid-Ver-
fahren.   B. CORNILS

UCB Union Chimique Belge

**UCB (Alusuisse/UCB) maleic acid anhy-
dride process** for the manuf. of *maleic
anhydride by oxidation of benzene over
V_2O_5-based cats. at 360 °C/0.5 MPa. **F** procédé d'UCB pour l'anhydride maléique;
G UCB Maleinsäureanhydrid-Verfahren.
                                          B. CORNILS
Ref.: van Santen/van Leeuwen/Moulijn/Averill.

UCC → Union Carbide Corp.

UCT upper critical temperature, → lower
critical temperature

UDP → uridine diphosphate

UDP-*N*-acetylgalactosamine (UDPGal-
NAc) can be prepared from GalNAc-1-P
and *UTP using UDP-GalNAc pyrophosphorylase, or from UDP-GlcNAc using an
epimerase. An alternative procedure is based
on a UMP exchange reaction between UDP-
Glc and GalN-1-P, with cat. by UDP-glucose:
galactosylphosphate uridyltransferase (EC
2.7.7.12). Galactose-1-phosphate is the natural substrate for the *enzyme, but 2-deoxyga-
lactose-1-phosphate, and others are also

accepted. The UDP-GalN thus produced is acetylated with acetic anhydride in a subsequent step to give UDP-GalNAc. **E=F=G.**

C.-H. WONG

UDP-*N*-acetylglucosamine (UDP-GlcNAc) Two *enzymatic methods have been developed for the synth. of UDP-GlcNAc. The first involves a reaction between GlcNAc-1-phosphate and *UTP, cat. by UDP-GlcNAc pyrophosphorylase. The second procedure exploits UDP-Glc pyrophosphorylase to catalyze a *condensation between *UTP and glucosamine-1-phosphate (GlcN-1-P) to afford UDP-glucosamine. UDP-GlcN can then be selectively *N*-acetylated to provide UDP-GlcNAc. GlcN-1-P was synth. from GlcN by phosphorylation of the 6-position with *hexokinase to give GlcN-6-P, followed by a *phosphoglucomutase-cat. *isomerization to provide GlcN-1-P (Figure). **E=F=G.**

C.-H. WONG

E$_1$: hexokinase from yeast; E$_2$: pyruvate kinase; E$_3$: phosphoglucomutase; E$_4$: UDP-Glc pyrophosphorylase; E$_5$: inorganic pyrophosphatase

UDP-glucose (UDP-Glc) and UDP-galactose (UDP-Gal) UDP-glucose has been prepared from *UTP and glucose-1-phosphate with cat. by UDP-glucose pyrophosphorylase. UDP-Gal can be synth. in an analogous fashion using UDP-Gal pyrophosphorylase.

Additionally, UDP-Gal can be generated from UDP-Glc by epimerization of C 4 with UDP-glucose epimerase. Although the equil. for this reaction favors UDP-Glc, the reaction can be coupled to an in-situ glycosylation with galactosyltransferase to shift the equil. The latter proc. has been applied to a large-scale synth. of *N*-acetyllactosamine. UDP-Gal has been prepared on a gram scale from Gal-1-phosphate and UDP-Glc using UDP-Gal uridyltransferase, and generated in situ for glycosylation (Figure). **E=F=G.**

C.-H. WONG

UDP-glucose epimerase → UDP-glucose, UDP-galactose

UDP-glucose pyrophosphorylase → UDP-glucose, UDP-galactose

UDP-glucuronic acid (UDP-GlcUA) biosynth. by *oxidation of C 6 of UDP-Glc with UDP-Glc dehydrogenase, an NAD-dependent *enzyme. The *NAD *cofactor was regenerated with *lactate dehydrogenase in the presence of pyruvate for preparative synth. of *UDP-GlcA. **G** UDP-Glucuronsäure.

C.-H. WONG

UF → ultrafiltration

Ugi four-component condensation (4CC) a one-pot proton-cat. reaction of an isonitrile, a carbonyl cpd., an amine, and a carboxylic acid to yield α-acylaminocarboxylic acid amides. The characteristic feature is the α-addition of an iminium ion and the anion X of an acid HX to an isocyanide. A spontaneous rearrangement of the α-adduct gives the stable α-aminocarboxamide. Chiral amines undergo stereoselective 4CCs used for peptide synth. The 4CC can be extended to 7CC by combination with the Asinger condensation. **F** condensation d'Ugi par quatre composants; **G** Ugi Vierkomponentenkondensation. M. ECKERT, M. BELLER

Ref.: I. Ugi, *Isonitrile Chemistry*, Academic Press, New York 1971; Trost/Fleming, Vol. 2, p. 1083.

Ugine Kuhlmann process for the manuf. of *sulfuric acid by the *contact technique (V_2O_5 as cat.) acc. to $SO_2 + \frac{1}{2} O_2 \rightarrow SO_3$ at 0.5 MPa. **F** procédé d'Ugine Kuhlmann pour l'acide sulfurique; **G** Ugine-Kuhlmann Schwefelsäureverfahren. B. CORNILS

Ref.: Büchner/Schliebs/Winter/Büchel, p. 112.

UHMW-PE → ultrahigh molecular weight polyethylene

UHV ultra-high vacuum

UK Wesseling HP methanol process for the manuf. of methanol from *syngas at 350 °C/30 MPa over fixed-bed $ZnO\text{-}Cr_2O_3$ cats., one of the few remaining high-press. MeOH procs. **F** procédé UK Wesseling pour le méthanol à haute tension; **G** UK Wesseling Methanol-Hochdruckverfahren. B. CORNILS

Ullmann reaction the reaction of two aryl halides in the presence of stoichiometric amounts of Cu or Cu salts at elevated temps. to yield biaryl and copper halides. Aryl iodides are fairly reactive as starting materials, whereas aryl bromides and chlorides react in efficient yields only if they carry electron-withdrawing substituents. Functional groups are tolerated, too. In the presence of cat. amounts of Ni and Pd *complexes, the coupling reaction is performed under milder reaction conds. **F** reaction d'Ullmann; **G** Ullmann-Reaktion. C. BREINDL, M. BELLER

Ref.: J. Mulzer et al. (Eds.), *Organic Synthesis Highlights*, VCH, Weinheim 1991; *Angew.Chem.Int. Ed.Engl.* **1988**, *27*, 1113.

Ultee cyanhydrin synthesis *addition of anhydrous HCN to the carbonyl double bond of aldehydes or ketones under the influence of basic cats. at lower temps. **F** synthèse des cyanhydrines d'Ultee; **G** Ultee Cyanhydrin-Synthese. B. CORNILS

Ultracat process → Standard Oil (of Indiana) procs.

Ultracracking process → Standard Oil (of Indiana) procs.

ultrafiltration pressure-driven *membrane-based fractionation of molecules mainly by size, in which particles and dissolved macromolecules smaller than 0.1 μm and larger than about 2 nm (corresponding to a molecular weight range from 10^3 to 10^6 g · mol^{-1}) are retained while smaller molecules pass the *membrane. Typical values for the press. applied are 0.05 to 0.7 MPa. The membranes are made from organic polymers such as polysulfone or cellulose derivatives. For better stability against organic solutes and solvents polyaramide or polyamide (and recently inorganic materials like aluminum oxide) are also used. Typical applications of u. are concentration, clarification, diafiltration (*membranes), and purification of proteins or other biological macromolecules. **E=F=G**. U. KRAGL

Ref.: Ullmann, Vol. A16, p. 187.

Ultrafining process → Standard Oil (of Indiana) procs.

Ultraforming process → Standard Oil (of Indiana) procs.

ultrahigh molecular weight polyethylene (UHMW-PE) a high-density polyethylene (unbranched; cf. *density of PEs) with an extremely high molecular weight (3–6×10^6) and a density of only 940 kg/m^3. Crystallinity of UHMWPE is about 45 %, typically. UHMWPE shows a remarkable combination of properties: high abrasion resistance, chemical inertness, low friction, im-

pact toughness, good corrosion and radiation resistance, and acceptability in contact with foodstuffs.
*Polymerization of UHMWPE is carried out with an organometal cat.(*Ziegler-Natta, *Phillips) similar to that for conventional *high-density polyethylene (HDPE). Most UHMWPE is produced by a slurry process (batch or continuous), but either solution or bulk procs. are applicable. Main applications of UHMWPE are in chemical processing, the food and beverage industries, medical implants, and transportation. For example, UHMWPE can be used as a permanent solid lubricant to protect metal surfaces. **F** polyéthylène à densitée ultrahaute; **G** ultrahochmolekulares PE.   W. KAMINSKY, V. SCHOLZ

Ref.: *Brennst.Chem.* **1968**, *49*, 337; Ullmann, Vol. A21; Mark/Bikales/Overberger/Menges, Vol. 6, p. 490.

UltraKat process   → Amoco procs.

ultramicropores   → micropores

Ultra-Orthoflow process   → Kellogg procs.

ultraviolet spectroscopy   → UV spectroscopy

UMP   → uridine monophosphate

uncompetitive inhibition   An uncompetitive inhibitor gives a collection of parallel lines on a *Lineweaver-Burk plot or lines that intersect on the x-axis in an *Eadie-Hofstee plot, indicating an influence of the inhibitor on *both* K_m and V_{max} (Figure 1).

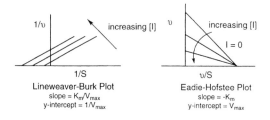

Lineweaver-Burk Plot
slope = K_m/V_{max}
y-intercept = $1/V_{max}$

Eadie-Hofstee Plot
slope = $-K_m$
y-intercept = V_{max}

One interpretation of ui. is that the inhibitor binds only to the *Michaelis complex and not to the *enzyme (Figure 2). **E=F=G**.   P.S. SEARS

$$E + S \underset{}{\overset{K_S}{\rightleftharpoons}} ES \xrightarrow{k_{cat}} E + P$$
$$+$$
$$I$$
$$K_I \updownarrow$$
$$ESI$$

Uncompetitive Inhibition

$$\upsilon = \frac{k_{cat}[E_o][S]}{[S](1+[I]/K_I)+[K_m]}$$

Unichema catalysts   → Synetix catalysts

Unicracking process   → 1: UOP Unocal procs.; 2: Union Oil procs.

Unidak process   → Union Oil procs.

Unifining process   → UOP procs.

uniform distribution of metals   → metal distribution

Union Carbide Coalcon process   former autothermal fluidized-bed, low-temp. carbonization and gasification of coal proc. under H_2 press. (formerly "agglomerating ash" development of Battelle). **F** procédé Coalcon de Union Carbide; **G** Union Carbide Coalcon Verfahren.   B. CORNILS

Ref.: Winnacker/Küchler, Vol. 5, p. 463.

Union Carbide Ethoxene process   for the oxidative *dehydrogenation of ethane at 300–400 °C/0.1–4 MPa in the gas phase over a proprietary cat. The economics of the process depend on the utilization of the byprod. acetic acid. **F** procédé Ethoxene de Union Carbide; **G** Union Carbide Ethoxene-Verfahren.

Ref.: McKetta-2, p. 810.   B. CORNILS

Union Carbide ethylene oxide process   early proc. for the manuf. of *EO by *oxidation of ethylene with air over Ag an cat. at 220–270 °C/1–3 MPa and *GHSVs of 2000–4500. The yield was 75 %. **F** procédé Union

Carbide pour l'EO; **G** Union Carbide EO-Verfahren.   B. CORNILS

Ref.: Ullmann, Vol. A10, p. 101.

Union Carbide IsoSiv process for the

non-cat. adsorptive isolation of paraffins by means of *molecular sieves. **F** procédé Isosiv de Union Carbide; **G** Union Carbide Isosiv-Verfahren.   B. CORNILS

Ref.: *Hydrocarb.Proc.* **1978**, (9), 213; Ullmann, Vol. A18, p. 83.

Union Carbide oxo (*LPO) process the

low-press. variant of *hydroformylation with Rh carbonyls, *ligand-modified with a large excess of *triphenylphosphine. The reaction between propylene and carefully cleaned *syngas takes place in stirred tank reactors at 95–100 °C/up to 1.8 MPa; the effluent product vapors pass demisting pads. A liquid effluent stream containing the cat., free phosphine ligand, aldehydes, and condensation prods., is taken from the reactor. The stream passes a let-down valve for press. release and enters a flash evaporator where the major part of the unconverted reactants is taken overhead and recycled. The liquid is freed from most of the aldehydes in the first distillation column. The cat. is concentrated in the second column under reduced press. and recycled to the reactor. The alkene conv. per pass is approx. 35 %. In a continuous bypass portions of the cat. cycle are simultaneously regenerated. No feed-

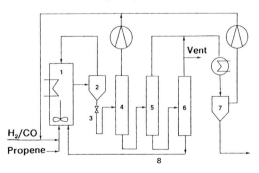

1 reactor; 2 separator; 3 let-down valve; 4 evaporator; 5,6 distillation columns; 7 gas separator; 8 catalyst recycle.

stocks other than propylene or butylenes are known. The UCC oxo proc. is spread worldwide (Figure). **F** procédé oxo LPO de UCC; **G** UCC LPO-Oxo-Verfahren.   C.D. FROHNING

Ref.: Winnacker/Küchler, Vol. 5, p. 555; Falbe-1; *J.Mol.Catal. A:* **1995**, *104*, 18.

Union Carbide sorbic acid process for-

mer two-step proc. for the manuf. of sorbic acid by 1: basic cat. *aldol condensation of *acetaldehyde to sorbic aldehyde, and 2: Ag-cat. *oxidation of 2,4-hexanedialdehyde to sorbic acid. **F** procédé Union Carbide pour l'acide sorbique; **G** Union Carbide Sorbinsäure-Verfahren.   B. CORNILS

Union Carbide syngas processes var-

iant of various procs. for the conv. of *syngas to valuable intermediates such as ethylene glycol, 1,2-propandiol, and glycerol at 125–350 °C/140–340 MPa (over promoted Rh cats.).   B. CORNILS

Union Carbide Total Isomerization (TIP)

process is an integration of Shell's *Hysomer with UCC's IsoSiv technology for upgrading of the *ON level of light gasoline fractions. The vapor-phase fixed-bed *hydroisomerization step uses a *dual-function cat. with a noble metal (Pt) on *zeolites/molecular sieve. **F** procédé TIP de Union Carbide; **G** Union Carbide TIP-Verfahren.   B. CORNILS

Ref.: Moulijn/van Leeuwen/van Santen, p. 33; Ertl/Knözinger/Weitkamp, Vol. 4, p. 2012; van Santen/van Leeuwen.

Union Carbide Unipol process the first

gas-phase *polymerization proc. of Union Carbide for *HDPE, today used also for prod. of *LLDPE and *PP. In the fluidized-bed reactor the synth. proceeds at press. of 1.5–2.5 MPa and temps. below 110 °C. The reactor is filled with a bed of dry polymer particles and fluidized by ethylene. Fluidization and heat removal require rapid recycling of ethylene through the reactor; as a result the conv. of ethylene in a single pass is only about 2 % for HDPE, but higher when using a *comonomer in the condensing mode (Figure 1). (*catalytic reactors).

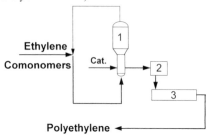

Polyethylene ◄───

1 fluidized-bed reactor; 2 product discharge tank; 3 granulator/extruder.

A solid *Ziegler or *Phillips cat. is fed continuously to the reactor above the distribution plate as dry powder or as concentrated slurry in high-boiling mineral oil. The polymer particles can grow to an average size of 400–700 μm and still retain their spherical shape. On average the *residence time of the cat. is 2½ – 4 h. As *co-catalyst AlEt₃ is fed with the *co-monomer into the ethylene stream below the distribution plate to activate the cat. Figure 2 shows the PP configuration. **F** procédé Unipol d'Union Carbide; **G** Unipol-Verfahren der Union Carbide.   W. KAMINSKY, C. STRÜBEL

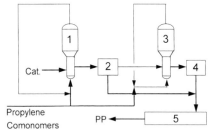

1 fluidized-bed reactor; 2,4 product discharge tank; 3 impact reactor; 5 granulator/extruder.

Ref.: US 4,011,382; *Chem.Eng.* **1972**, *79*(21), 104; McKetta-1, Vol. 41, p. 77.

Unionfining process   → Union Oil procs.

Union Oil Unicracking process   for the *hydrocracking/ *hydrotreatment of atmospheric and vacuum resids over fixed-bed cats. in one or two stages to remove high sulfur contents (*HDS), nitrogen (*HDN), and metals

(*HDM) at 260–425 °C/7–14 MPa. The proc. includes a *guard bed with Unicracking/*HDS cats. (non-noble metal on *molecular sieve) to remove particulate matter and residual salt. **F** procédé Unicracking/HDS de Union Oil; **G** Union Oil Unicracking-Verfahren.

Ref.: *Hydrocarb.Proc.* **1972**, (9), 146.   B. CORNILS

Union (Sun) Oil Unidak process   manuf. of naphthalene by cat. *hydrodealkylation of methylnaphthalenes at 600 °C/3–10 MPa over Co-Mo on *alumina. A steady addition of steam secures continuous cat. *regeneration. **F** procédé Unidak de Sun Oil; **G** Unidak-Verfahren der Sun Oil.   B. CORNILS

Union Oil Unionfining process   fixed-bed proc. for the accomplishment of *hydrodesulfurization, *hydrodenitrogenation, and *hydrogenation of a wide range of petroleum feedstocks over *Unifining cats. (today an UOP process). **F** procédé Unionfining de Union Oil; **G** Union Oil Unionfining-Verfahren.   B. CORNILS

Ref.: *Hydrocarb.Proc.* **1978**, (9), 154; *Hydrocarb.Proc.* **1996**, (11), 134.

Union Oil Unisar process   for the saturation of aromatics present in turbine fuel stocks, solvent naphtha, and other petroleum-derived feedstocks over single, fixed-bed reactors with multiple cat. beds located between suitable hydrogen quench trays. **F** procédé Unisar de Union Oil; **G** Union Oil Unisar-Verfahren.   B. CORNILS

Ref.: *Hydrocarb.Proc.* **1978**, (9), 155.

Unipol process   → Union Carbide procs.

Unisar process   → Union Oil procs.

unit operations of oil-processing industry   the configuration of the different unit operations and processes (which are mostly catalyzed); → petroleum processing, *refinery.

units of enzymatic activity is the amount of *enzyme required to catalyze the conversion of 1 μmol/min of a standard substrate.

P.S. SEARS

unnatural amino acids A *biosynthetic method for incorporating unnatural amino acids into protein has been developed which involves the replacement of the tRNA anticodon via chemical depurination at low pH followed by *ribonuclease A digestion and ligation with another oligonucleotide using *T4 RNA ligase and T4 polynucleotide *kinase, and synth. of a knot from single- or double-stranded *DNA. **F** acides aminés non naturels; **G** unnatürliche Aminosäuren. C.-H. WONG

Ref.: *Biochemistry* **1982**, *21*, 855; *J.Am.Chem.Soc.* **1991**, *113*, 2722.

Unocal process → UOP procs.

unsteady-state processes in catalysis In ussp. or periodic operation, the dynamic behavior of the cat. and the *cat. reactor is used as an additional parameter for proc. optimization. Frequency, amplitude, and form of the forced periodic oscillations become important operating variables for influencing conv., *selectivity, and yield. The special interest in periodic operation results from the fact that it can be possible to obtain improvements compared to the optimal *steady state in the admissible range for the control variables used, such as temp., press., conc., and flow rates. The cat. unsteady state can be controlled in two ways: by forced conc. oscillations or flow reversal in fixed beds by spatial regulation when input conds. of the reactor do not change within time but the cat. moves in a field of variable reactant concs. and/or temps., e.g., in a fluidized bed. The overall performance of chemical reactors can be considerably increased by unsteady operation for reactions with *substrate inhibition. The rates achievable under dynamic conds. are higher than the maximum steady-state values and depend on the sorption kinetics of the inhibiting species. Examples of increasing selectiv-

ities by ussp. are *partial *oxidations of *hcs., such as butane to maleic anhydride. A high prod. selectivity can be achieved over oxidized cats. in absence of gas-phase oxygen. The reduced cat. is then reoxidized periodically. **E=F=G.**

A. RENKEN

Ref.: Ertl/Knözinger/Weitkamp, Vol. 3, p. 1464; Y.S. Matros, *Catalytic Processes under Unsteady-state Conditions*, Elsevier, Amsterdam 1989; *Catal. Today* **1995**, *25*, 91 and **1988**, *3*, 175; *Int.Chem.Eng.* **1984**, *24*, 202; *AIChE J.* **1986**, *32*, 1612; *Chem. Eng.Sci.* **1991**, *46*, 1083; *Stud.Surf.Sci.Cat.* **1997**, *109*, 469; N.J. Themelios, *Transport and Chemical Rate Phenomena*, Gordon and Breach, London 1995.

unsupported catalysts For reasons given under the keyword *support, only a few industrial cats. are used in the form of ucs. (or *bulk cats.). This is specially true either when a cat. reaction is extremely fast and/or when no impurities of the reactants are present. Both reasons apply for the *oxidation of ammonia where unsupported Pt-Rh *gauze cats. are used. Other examples are the synth. of *ethylene oxide (e.g., over Ag granules) or the use of *skeletal (*Raney) cats. which can also be regared as ucs., or of the *Adams cat.

B. CORNILS

UOP one of the leading technology licensors and cat. suppliers for the *petroleum refining and processing, petrochemical, and gas processing industry. UOP is headquartered in Des Plaines (IL, USA).

J. KULPE

UOP Alkar process for the gas-phase *alkylation of benzene and ethylene over BF_3 on Al_2O_3 cats. at 290 °C/6 MPa to *EB. The proc. tolerates diluted ethylene streams, oxygen, and CO_2. **F** procédé Alkar de UOP; **G** UOP Alkar-Verfahren.

B. CORNILS

UOP Benfield process for the removal of CO_2, H_2S, COS, and RSH from gases. Although non-cat., an absorption stage with *DEA may be added as the HiPure version (cf. *Catacarb proc.). **F** procédé Benfield de UOP; **G** UOP Benfield-Verfahren. B. CORNILS

UOP BenSat process for the removal of benzene from light *reformate by cat. *hydrogenation to cyclohexane, either in a stand-alone option or integrated in the *isomerization features of the *UOP Penex process (Figure). **F** procédé BenSat de UOP; **G** UOP BenSat-Verfahren. B. CORNILS

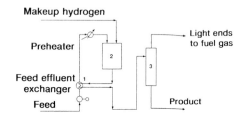

1 heat exchanger; 2 reactor (with noble metal cat.); 3 stabilizer.

Ref.: *Hydrocarb.Proc.* **1996**, (11), 94.

UOP BOC (BOC Isomax) process for cat. residual hydroconversion in fixed-bed operation (black oil conv.). The degree of S removal is said to be 70–80 %. **F** procédé BOC de UOP; **G** UOP BOC-Verfahren. B. CORNILS
Ref.: *Hydrocarb.Proc.* **1972**, (9), 139; *Sulfur* **1988**, *195*, 24; Ullmann, Vol. A18, p. 76.

UOP/BP Cyclar process for the manuf. of petrochemical-grade benzene, toluene, and xylenes (*BTX) by cat. *aromatization of propane and butanes over Ga-doped *zeolites in stacked radial-flow reactors combined with a *CCR unit. **F** procédé Cyclar de UOP/BP; **G** UOP/BP Cyclar-Verfahren. B. CORNILS
Ref.: *Hydrocarb.Proc.* **1997**, (3), 116.

UOP Butamer process for the cat. *isomerization of *n*- to *i*-butane in the presence of hydrogen over Pt on Al$_2$O$_3$. The H$_2$ suppresses the *polymerization of the minor amounts of alkenes and is doped with trace amounts of organic chloride (Figure; cf. *chlorinated alumina). **F** procédé Butamer de UOP; **G** UOP Butamer-Verfahren. B. CORNILS

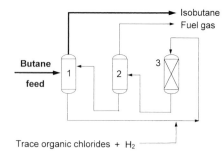

1 deisobutanizer; 2 stabilizer; 3 isomerization reactor.

Ref.: *Hydrocarb.Proc.* **1978**, (9), 168; *Oil Gas J.* **1958**, *56*(13), 73; Moulijn/van Leeuwen/van Santen, p. 59.

UOP catalytic cracking MSCC process for the conv. of gas oils and resid feedstocks to light olefins, LPG, petrochemical intermediates, naphthas, etc. using an ultra-short contact time reaction system with a proprietary cat.-feed scheme via novel injection nozzles. **F** procédé MSCC catalytic cracking de UOP; **G** UOP MSCC catalytic cracking-Verfahren. B. CORNILS
Ref.: *Hydrocarb.Proc.* **1996**, (11), 96.

UOP catalytic *reforming process for the upgrading of naphtha for use in gasoline blendstocks or for petrochemicals prod. by UOP's *CCR *platforming procs. The cats. are UOP's R-130 series (Sn-doped Pt). **F** procédé catalytic reforming de UOP; **G** UOP catalytic reforming-Verfahren. B. CORNILS
Ref.. *Hydrocarb.Proc.* **1996**, (11), 98.

UOP CCR technique → continuous catalyst generation

UOP cumene process for the manuf. of *cumene by gas-phase *alkylation of benzene and propylene/propane mixtures over H$_3$PO$_4$-SiO$_2$ (BF$_3$-promoted, cf. *SPA) cats. at 250 °C/2–4 MPa. The *lifetime of the cat. is higher if steam is present. Polyalkylates have to be *transalkylated in separate reactors. **F** procédé UOP pour le cumène; **G** UOP Cumol-Verfahren. B. CORNILS

UOP Cumox process   variant of the *Hock phenol process

UOP DeFine process-step   a one-step selective *hydrogenation to be incorporated into other procs., e.g., *UOP linear alkylbenzene proc.   B. CORNILS

UOP Detal process   for the manuf. of *linear alkylbenzenes (LABs) by *alkylation of benzene with higher alkenes such as 1-dodecene or internal alkenes over (probably) *zeolite cats. **F** procédé Detal de UOP; **G** UOP Detal-Verfahren.   B. CORNILS
Ref.: Ertl/Knözinger/Weitkamp, Vol. 5, p. 2130.

UOP Ethermax process   → Hüls/UOP procs.

UOP *FCC process   to convert selectively hydrocarbon feeds, from naphtha to bottoms, into high-octane gasoline, distillate and fuel oils, alkenes, etc. The reaction system represents an advanced, side-by-side reactor and *regenerator configuration with a short contact-time *riser with efficient contact between the cat. and *hcs. (preacceleration zone with elevated Optimix feed nozzles). Burning of the used cat. takes place with air. **F** procédé FCC de UOP; **G** UOP FCC-Verfahren.

                                    B. CORNILS
Ref.: *Hydrocarb.Proc.* **1978**, (9), 110 and **1996**, (11), 123.

UOP HDC Unibon process   single-stage cat. *hydrocracker proc. for vacuum gas oils to maximize the yield of distillate fuels. **F** procédé HDC Unibon de UOP; **G** UOP HDC Unibon-Verfahren.   B. CORNILS

UOP HF alkylation process   for the manuf. of highly branched high-octane gasolines (*motor fuel alkylates) by HF-cat. *alkylation of isoparaffins and alkenes at low temp. (Figure). **F** procédé HF alkylation de UOP; **G** UOP HF-Alkylierungs-Verfahren.

                                    B. CORNILS

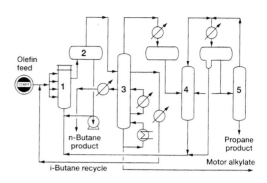

1 reactor; 2 acid settler; 3 isostripper; 4 depropanizer; 5 HF stripper.

Ref.: *Hydrocarb.Proc.* **1978**, (9), 176 and **1996**, (11), 93.

UOP Hydrobon process   for removing objectionable materials from petroleum distillates by cat. *hydrogenation or *dehydrosulfurization (e.g. S may be reduced below 0.5 ppm, improvement of smoke point). **F** procédé Hydrobon de UOP; **G** UOP Hydrobon-Verfahren.   B. CORNILS
Ref.: *Hydrocarb.Proc.* **1972**, (9), 160.

UOP Hydeal process   for the cat. *hydrodealkylation of alkyl aromatics, e.g., toluene → benzene. The yield is reported to be 98 %. **F** procédé Hydeal de UOP; **G** UOP Hydeal-Verfahren.   B. CORNILS

UOP Hydrar process   for the *hydrogenation of benzene and alkenes in order to increase the yield of raw materials for the chemistry by gas-phase *hydrogenation over Ni cats. in three succesive adiabatic fixed beds at 400–600 °C/3 MPa. **F** Procédé Hydrar de UOP; **G** UOP Hydrar-Verfahren.   B. CORNILS
Ref.: Winnacker-Küchler, Vol. 5, p. 224.

UOP hydrogen once-through (HOT) process   a modern version of UOP's *Penex proc. Thus the recycle compressor, the prod. separator, and associated heat exchangers can be omitted.   **F** procédé HOT de UOP; **G** UOP HOT-Verfahren.   B. CORNILS
Ref.: *Fuel Proc.Technol.* **1993**, 35, 183.

UOP Hypro process   an older proc. for the conv. of methane into hydrogen by cat. *cracking of methane without addition of steam. Coke deposits on the cat. must be burned off intermittently. **F** procédé Hypro de UOP; **G** UOP Hypro-Verfahren.

B. CORNILS

UOP Indirect Alkylation (InAlk) process   for the selective conv. of mainly C_4 alkanes to high-octane number, gasoline boiling range alkanes by a three-step procedure, 1: cat. C_4 alkane *dehydrogenation, 2: C_4 *zeolite cat. alkene *oligomerization to C_8 alkenes, and 3: *hydrogenation to C_8 alkanes (highly branched). **F** procédé InAlk de UOP; **G** UOP InAlk process.   B. CORNILS

UOP Isomar process   for the cat. adjustment of the equil. of isomeric xylenes in a combination of UOP *Parex and UOP *Isomar procs. **F** procédé Isomar de UOP; **G** UOP Isomar-Verfahren.   B. CORNILS

Ref.: *Hydrocarb.Proc.* **1977**, (11), 239 and **1997**, (3), 166.

UOP linear alkylbenzene process   to produce linear alkylbenzenes (*LAB) from C_{10}–C_{14} by *alkylating benzene with alkenes over a solid, het. cat. in fixed-bed operation. Since the feed comprise linear paraffins the proc. includes a *Pacol *dehydrogenation step and a *DeFine selective *hydrogenation for diolefins. **F** procédé UOP pour des alkylbenzènes lineaires; **G** UOP-Verfahren für lineare Alkylbenzole.   B. CORNILS

Ref.: *Hydrocarb.Proc.* **1997**, (3), 107.

UOP Lomax process   a forerunner of Chevron's *Isomax process

UOP Merox process   for the removal of mercaptans from gases, LPG, and lower-boiling fractions and for the *sweetening of gasolines by their in-situ conv. (aerobic oxidation) in the presence of metal cat. ("Merox cat.") to give insoluble, decantable disulfides (*de-

sulfurization). **F** procédé Merox de UOP; **G** UOP Merox-Verfahren.   B. CORNILS

Ref.: *Hydrocarb.Proc.* **1996**, (4), 128.

UOP/Norsk Hydro *MTO process   for the manuf. of mainly ethylene/propylene from methanol by fluidized-bed operations over *SAPO-34 (silicoaluminaphosphate). In a second fluidized bed the cat. is regenerated by burning off the coke (Figure). **F** procédé MTO de UOP/Norsk Hydro; **G** UOP/Norsk Hydro MTO process.   B. CORNILS

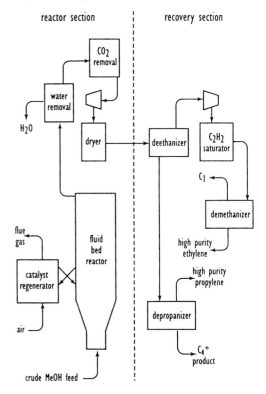

Ref.: *Ind.Catal.News* **1998**, *1*, 10.

UOP Oleflex process   for the cat. *dehydrogenation of propane (or butane) to produce polymer-grade propylene (and hydrogen) in a radial-flow reactor with *CCR based on Pt cats.(on *alumina, doped with Sn) at 550–620 °C/200–500 kPa. **F** procédé Oleflex de UOP; **G** UOP Oleflex-Verfahren.

B. CORNILS

Ref.: *Hydrocarb.Proc.* **1997**, (3), 158, McKetta-2, p. 809; Ertl/Knözinger/Weitkamp, Vol. 5, p. 2148; Moulijn/van Leeuwen/van Santen, p. 59.

UOP Pacol-Olex process for the gasphase *dehydrogenation of paraffins (C_6–C_{19}) at 500 °C/0.3 MPa over Sn-doped Pt-Al_2O_3 cats. Unconverted paraffins are separated by means of *molecular sieves. This proc. is advantageously incorporated into other procs. such as the *UOP linear alkylbenzene proc. **F** procédé Pacol-Olex de UOP; **G** UOP Pacol-Olex-Verfahren.   B. CORNILS

Ref.: Ertl/Knözinger/Weitkamp, Vol. 5, p. 2130; Farrauto/Bartholomew, p. 460.

UOP Parex process a non-cat. separation proc. for isomeric xylenes over *molecular sieves; advantageously to be combined with the *UOP Isomar proc. **F** procédé Parex de UOP; **G** UOP Parex-Verfahren.   B. CORNILS

Ref.: *Hydrocarb.Proc.* **1997**, (3), 166.

UOP Penex process for the hydrogenative upgrading of pentane and/or hexane fractions from refinery naphtha by *isomerization over a Pt cat. (on *alumina, HCl-activated; cf. *chlorinated alumina). **F** procédé Penex de UOP; **G** UOP Penex-Verfahren.   B. CORNILS

Ref.: *Hydrocarb.Proc.* **1978**, (9), 172 and **1996**, (11), 142; *Oil Gas J.* **1985**, May 27, 80.

UOP Platforming process for the *reforming of low-octane naphtha or for high yields of C_6–C_8 aromatics for petrochemical feedstocks. The installation includes a continuous *regeneration of withdrawn cat. under steady-state conds. The cat. is *bimetallic. **F** procédé Platforming de UOP; **G** UOP Platforming-Verfahren.   B. CORNILS

Ref.: *Hydrocarb.Proc.* **1978**, (9), 163; Winnacker-Küchler, Vol. 5, p. 97,103.

UOP polymer olefin process for the manuf. of tetrapropylene through *oligomerization of propylene over fixed-bed H_3PO_4-SiO_2 cat. (*SPA-1) at 200 °C/4–6 MPa. **F** procédé UOP pour des oléfines polymères; **G** UOP Polymerolefin-Verfahren.   B. CORNILS

UOP Q-Max process for the *alkylation of benzene with propylene to *cumene over liquid-phase *zeolite cats., which are non-corrosive and have (cat.) *lifetimes from five years. **F** procédé Q-Max de UOP; **G** UOP Q-Max-Verfahren.   B. CORNILS

Ref.. *Hydrocarb.Proc.* **1997**, (3), 121.

UOP RCD Unibon process for the deep *desulfurization of S-containing heavy fuel oils to produce low-sulfur residuals as heavy fuel oils or as feedstocks for *coking and *FCC units. The RCD proc. is fixed-bed two-stage (Figure). **F** procédé RCD Unibon de UOP; **G** UOP RCD Unibon-Verfahren.

B. CORNILS

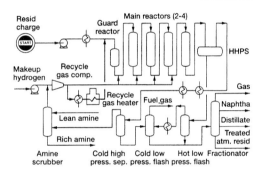

Ref.: *Hydrocarb.Proc.* **1978**, (9), 147 and **1996**, (11), 136.

UOP SafeCat technology the combination of cat. procs. with preceding adsorption/desorption towers for the removal of the high-boiling comps. coming from a *hydrotreater.

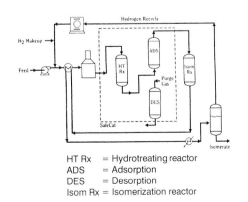

HT Rx   = Hydrotreating reactor
ADS   = Adsorption
DES   = Desorption
Isom Rx = Isomerization reactor

The intermittently operated towers act as *guard beds. An example is the extension of UOP's *TIP proc. and its integration with the SafeCat technology (Figure). **F** technique Safe-Cat de UOP; **G** UOP SafeCat-Technologie.
Ref.: *Fuel Proc.Technol.* **1993**, *35*, 183. B. CORNILS

UOP Selectox process variant of the *Claus proc. for the removal of H$_2$S from gas streams. **F** procédé Selectox de UOP; **G** UOP Selectox-Verfahren. B. CORNILS

UOP SMART styrene process The proc. (originally called Styro-Plus) converts ethylbenzene via *dehydrogenation over Fe cats. to *styrene. The hydrogen formed is oxidized over noble metal cats. to H$_2$O; this *oxidation partly ensures the heat balance of the proc. **F** procédé SMART de UOP; **G** UOP SMART-Verfahren. B. CORNILS

UOP Unifining process for the *hydrotreatment and removal of sulfur (*HDS) from pyrolysis of naphtha or heavy gas oils up to vacuum gas oils, and to prepare them as *platforming feed at 350 °C/3–6 MPa over Co-Mo or Ni-Mo cats. **F** procédé Unifining de UOP; **G** UOP Unifining-Verfahren. B. CORNILS

UOP Unocal (Unicracking) process for the hydrocracking of relatively heavy oil feedstocks into lighter transportation fuels under higher hydrogen press. (300–450 °C/5–20 MPa) in the presence of Ni-Mo or Ni-W cats. in fixed-bed, trickle-flow operation. **F** procédé Unocal de UOP; **G** UOP Unocal-Verfahren. B. CORNILS
Ref.: *NPRA Paper AM-91-12*, **1991**.

UPS → UV photoelectron spectroscopy

urease a urea-cleaving *metalloenzyme which contains Ni. The final prod. are carbamic acid and NH$_3$, thus removing ammonia from organisms. B. CORNILS

uridine diphosphate *N*-acetylgalactosamine → UDP-*N*-acetylgalactosamine

uridine triphosphate → nucleoside triphosphate, sugar nucleotide synthesis, *sugar nucleotide regeneration

Urushibara nickel special cats. for lab-scale *hydrogenation of double bonds and carbonyl groups based on Ni, Co, or Cu, and prepared by precipitation of the metals from aqueous solution. The cats. are activated by washing with acids or bases. **F** catalyseur d'Urushibara; **G** Urushibara-Katalysatoren.
Ref.: Augustine, p. 249. B. CORNILS

US Chemicals Co. (USI) VAM process for the manuf. of vinyl acetate from ethylene, acetic acid, and oxygen over a noble metal cat. in a fixed-bed tubular reactor. The technology is similar to the *Bayer-Hoechst proc. **F** procédé de USI pour l'acétate de vinyle; **G** USI-Vinyl-acetatverfahren. B. CORNILS
Ref.: *Hydrocarb.Proc.* **1972**, *51*(11), 141 and **1977**, *56*(11), 235.

USI → US Chemicals Co.

UTP → uridine triphosphate

UV near infrared spectroscopy (UV-NIR) is a non-destructive method with good in-situ capabilities for the determination of electronic structure and *redox states of solid samples (visible IR light). The data obtained are integral, *surface- and bulk-sensitive.
R. SCHLÖGL

UV-NIR → UV near infrared spectroscopy

UV photoelectron spectroscopy (UPS) is related to *XPS in that UV light is used instead of X-rays. The exciting energies are thus low and the photoemission is limited to valence electrons (electronic structure near the *Fermi edge).
The UHV, ex-situ method is non-destructive for clean solid *surfaces; the data are integral, bulk- and surface-sensitive with in-situ capabilities. R. SCHLÖGL
Ref.: Niemantsverdriet, p. 60.

UV-Vis spectroscopy (also electronic spectroscopy) the measurement of the absorption of light in the spectral range 190–800 nm ($50\,000$–$12\,500$ cm^{-1}). UV-Vis absorption spectra can provide information on cat. properties from electron transitions within the cat. and between the cat. and *adsorbed species. The electron transitions which can be observed include metal-centered transitions (trs.), charge-transfer trs., electron trs. involving defects, and *band-gap trs. Metal-centered trs. occur between orbitals within the same atom in *transition metals, in *rare earths, and in the main group elements. Charge-transfer trs. include ligand-to-metal (LMCT), metal-to-ligand (MLCT), and metal-to-metal (MMCT) trs., and trs. between molecular orbitals of inorganic or organic molecules. Band-gap trs. are observed as continuum absorption initiating at the band-gap energy. The high energy of UV-Vis trs. cause vibrational excitations which result in relatively broad bands.

UVs. can provide information on valence states and coordination of cat. *active species in both hom. and het. cats. Information on the electronic state of the active species and changes in the active species during cat. prepn. and reaction can assist in the determination of mechanisms and in optimization of treatment procedures as well as, for solid cats., choice of cat. *supports, support pretreatment, and cat. *precursor. The interaction of transition metal *complexes with support materials can be elucidated using information from UVs. *Active sites can be characterized and their interaction with adsorbed species can be observed, whether the adsorbed species is a reactant, an *intermediate, a *spectator species, or a cat. *poison. With the appropriate experimental setup, electronic properties of working cats. can be measured under reaction conds. (temps. from 100 to 800 K) which can give further insight into mechanism.

UVs. can be performed in transmission mode for hom. solutions and transparent solids or in diffuse reflectance (DR) mode for powder samples. Reference materials include MgO, $BaSO_4$, and more recently polytetrafluoroethylene; these materials can age and should be checked and replaced periodically. **E=F=G**.   R. JENTOFT

Ref.: Ertl/Knözinger/Weitkamp, Vol. 2, p. 641; Ullmann, Vol. B5, p. 383.

V

V8 protease (endoproteinase Glu-C; EC 3.4.21.19) a *serine protease from *Staphylococcus aureus* that cleaves specifically at the C-terminal side of glutamate residues, which has been useful in the site-selective cleavage of proteins. It has also been used for the *protease-cat. synth. of peptides from peptide esters. Due to its strong S1 site specificity, Vp. will not cleave most peptide bonds (thus limiting side reactions). **E=F=G.**   P.S. SEARS
Ref.: *J.Med.Chem.* **1992**, *35*, 3934.

VA, VAM, VINA vinyl acetate (VA and VINA) and vinyl acetate monomer (VAM)

vacant (coordination) sites In hom. cat., this term describes the status of coordinative unsaturation of metal centers (called free coordination sites), which is necessary to bring reactants together (also called *coordinatively unsaturated metal centers, cus). Normally the removal of ligands creates these vacant sites (cf., e.g., *ligand tuning, *photocatalysis). In crystallography vacancy is an imperfection in a crystal structure with missing atoms or ions. **F** site vacant (de coordination); **G** Leerstelle.   B. CORNILS
Ref.: Moulijn/van Leeuwen/van Santen, p. 106.

vacuum gas oil (VGO) at approx. 425–600 °C the highest-boiling fraction from vacuum distillation of petroleum (*petroleum processing), and the subject of various cat. procs. (e.g., *Chevron RDS and VGO, *Exxon GO Fining, *Gulf Gulfining, *Lummus LC-Fining, *Texaco T-Star, *Union Oil/UOP Unionfining, *UOP HDC and RCD, and the *hydrocracking variants of *Exxon, *IFP, *Kellogg, *VEBA.   B. CORNILS
Ref.: McKetta-1, Vol. 60, p. 304,317.

valence band → band theory

VAM synthesis cat. reaction of ethylene with *acetic acid to yield vinyl acetate monomer (VAM) (*acetoxylation, *Moiseev reaction). Originally a homogeneous proc., VAM (cat. by Hg salts) is now produced predominantly by vapor-phase procs. with het. Pd cats. The reactants are converted in tubular reactors (*catalytic reactors) on fixed-bed cats. at 140–180 °C/0.5–1.1 MPa to form VAM and water. Byprods. are CO_2, acetaldehyde, ethyl acetate, and condensation prods. (heavy ends). *STYs range from 200–800 g VAM per L cat. per h. The worldwide capacity in 1994 was 3.4×10^6 t/a; main producers are Celanese, Quantum, Union Carbide, DuPont, and BP (cf. procs. of *Bayer-Hoechst and *Halcon)., The Pd cat. is *supported on *activated carbon, *silica, *alumina, or *aluminosilicates. Additional *promoters such as Au, Cd, Pt, Rh, Ba, Cu, Mn, or Fe are applied together with alkaline salts such as CH_3COOK. In the *Wacker proc. VAM is obtained when the solvent water is substituted by acetic acid. **F** synthèse de l'acétate de vinyle monomère; **G** Vinylacetatsynthese.
          C. W. KOHLPAINTNER
Ref.: Parshall/Ittel, p. 200; Ertl/Knözinger/Weitkamp, Vol. 5, p. 2295; Farrauto/Bartholomew, p. 496.

vanadia vanadium oxide (V_2O_5), used as a base for cats.; cf. *vanadium oxides

vanadium as catalyst metal V is a widely used active element in several industrial cats., especially for *oxidation reactions or for environmental applications. *Bulk V_2O_5 is the cat. for SO_2 oxidation to SO_3 (*sulfuric acid, *contact proc., *Bayer double contact proc.); V supported on *titania is the active comp. of cats. for phthalic anhydride synth. or other acids (procs. of *Alusuisse, *BASF, *Lummus, *Hüls, *Mitsubishi, *Rhône-Progil, etc.) and for the *selective cat. reduction of NO to N_2 in the presence of NH_3/O_2. V-phosphorus oxides (*VPO) are the cat. for n-butane *selective oxidation of maleic anhydride; V in combination with Mo oxide is the cat. for acrolein oxidation to acrylic acid and benzene oxidation to maleic anhydride and in combination with Sb oxide for propane *ammoxidation to acrylonitrile. V also shows good cat. properties in the oxidative *dehydrogenation of light alkanes (in combination with Nb or Mg oxides or dispersed on *zeolite matrices), gas-phase oxidation of polycyclic aromatics (in combination with Fe and Mo oxides) and aldehyde formation from alkylaromatics (V supported on oxides or incorporated in *micro- or *mesoporous materials). V is also active in liquid-phase selective oxidation with *hydrogen peroxide or alkylperoxo *complexes. A recent development is the use in these reactions of solid cats. constituted by V incorporated in *MFI or *MeALPO zeotype materials, although their use is limited by leaching of the V during the cat. reaction.

The special cat. chemistry of V oxide and derived cats. results from various factors – 1: presence of partially filled d-orbitals, 2: possibility of different formal oxidation states (from two to five), and easy conv. between them (thus high *redox activity), 3: considerable extension of vanadium d-wave-functions (not involved in the metal-oxygen bonds) which allows *overlap with the p-orbital of π-symmetry, 4: redox potential relative to the anion valence-band edge favorable for bond formation, 5: easy rearrangement of coordination polyhedra due to the formation of shear planes which facilitate oxygen ion removal from the lattice, and 6: electronic state of the *surface lying near the *Fermi level, and either partially occupied by electrons or close enough to the filled *valence band to allow simultaneous donation and acceptance of

electrons to and from the interacting molecule. As a result of these chemical properties, the striking feature of V is its capability of formation of oxygenated, acidic-type prods. However, this oxidation ability may be considerably modified by the presence of a second comp. or by the possibility of isolating the V centers in an oxide matrix. It is thus possible to have fine tuning of its oxidation characteristics, which explains its wide use in several cat. formulations (*tailoring of cats.). Some *haloperoxidases contain V. **F** le vanadium comme métal catalytique; **G** Vanadium als Katalysatormetall.   G. CENTI

Ref.:*Appl.Catal.A:* **1997**, *157*, (issue Vanadia Catalysts); *Appl.Catal. A:* **1996**, *147*, 267; *Appl.Catal. A:* **1997**, *157*, 3; *Houben-Weyl/Methodicum Chimicum*, Vol. XIII/7; *Chem.Rev.* **1997**, *97*, 2707.

vanadium oxides play an important role in cat. (*V as cat. metal, *vanadia) for *partial *oxidations (e.g., butane → maleic anhydride, procs. of *Alusuisse LAR, *Lummus ALMA, *Mitsubishi, *SD, *UCB, *ammoxidation, or others) while the *SCR of NO_x is carried out over supported V_2O_5-WO_3/TiO_2 cats. The most stable V oxide is the orange V_2O_5, which easily form collodial solutions. The crystal structure is similar to MoO_3 (*Mo oxides). It is easily soluble in strong bases, forming vanadates, M_3VO_4. Upon acidification, *condensation to polyvanadic acids $H_{n+2}V_nO_{3n+1}$ and $H_{n-2}V_nO_{3n-1}$ occurs. The ability of V to build highly condensed polyoxometal acids is similar for Nb, Ta, Cr, Mo, and W. **F** oxides de vanadium; **G** Vanadiumoxide.   G. MESTL

van der Waals-London forces weakly attracting forces between molecules or between molecules and a *surface which are caused by momentarily existing dipoles or multipoles induced by fluctuations in the charge distribution. The forces are of the order of the condensation enthalpy and they decay rapidly with distance following a $1/r^6$ dependence. They are the forces which are also responsible for *physisorption of molecules/

atoms at surfaces. **F** forces de van der Waals-London; **G** van der Waals-London-Kräfte.

R. IMBIHL

VAPO   → V-modified *aluminophosphate

Vaska compounds   are called *complexes of general formula $MX(CO)L_2$ with M = Rh, Ir; X = CN, Cl, Br, N_3; L = *ligands such as *triphenylphosphine (PPh_3). Vaska's complex is the square planar d^8-Ir(I)Cl(PPh_3)$_2$. Vcs. cat. hom. a few different reactions, such as *hydrogenations, *oxidations (with O_2; phosphines → phosphine oxides, alkenes → *epoxides, aldehydes, alcohols, or ketones → carboxylic acids), or *isomerization reactions. The rate of isomerizations may be increased by UV radiation. **F** composés de Vaska; **G** Vaska-Verbindungen.

F. RAMPF

Ref.: Abel/Stone/Wilkinson, Vol. 5, p. 542.

VCC   → VEBA processes

VC, VCM   vinyl chloride, vinyl chloride monomer (VCM)

VCH   vinylcyclohexane

VEBA Combi Cracking (VCC) process   a combination of liquid-phase *catcracking (440–490 °C/25 MPa; over coal cats. [Fe] or metal- doped coal cats.) and gas-phase *hydrogenation (over *HDS cats.) of coal liquids, coal residues, or other resids (*coal hydrogenation). **F** procédé VCC de VEBA; **G** VEBA VCC-Verfahren.

B. CORNILS

Ref.: Oil Gas J. **1982**, *80*(12), 121; Ullmann, Vol. A18, p. 76.

VEBA ethanol process   for the manuf. of ethyl alcohol by *hydration of ethylene over fixed-bed H_3PO_4-SiO_2 cats. **F** procédé VEBA pour l'éthanol; **G** VEBA Ethanol-Verfahren.

B. CORNILS

VEBA *hydrocracking process   for the upgrading of heavy and extra-heavy crudes to high-quality *syncrude by hydrogen addition in liquid-phase operation and fixed-bed reactors. **F** procédé hydrocracking de VEBA; **G** VEBA Hydrocracking-Verfahren.

Ref.: Hydrocarb.Proc. **1996**, (11), 129.   B. CORNILS

VEB ISIS-Chemie L-dopa process   was developed for the synth. of *L-dopa, which is administered as a drug against morbus Parkinson. The pivotal step represented the hom. catalyzed asymmetric *hydrogenation of (Z)-α-benzoylamino-3-(3,4-dimethoxyphenyl)-2-propenoate over Rh(I) with *Ph-ß-glup as a chiral ligand at 0.01 MPa H_2/40 °C. The *chiral amino acid was obtained in 90 % *ee. The former VEB ISIS-Chemie produced L-dopa at a scale of 1 t/a. **E=F=G**. A. BÖRNER

Ref.: J.Mol.Catal. **1986**, *37*, 213; Chem.Tech. **1987**, *39*, 123.

VEB Leuna Aris process   This *aromatics isomerization* was developed for the *hydroisomerization of C_8 aromatics to xylenes at 400–450 °C/0.7–1.7 MPa. The cat. consisted of Pt on Al_2O_3/ mordenite. If desired, ethylbenzene can be *dealkylated by low Pt on pentasil to benzene. The activity of the cats. can be influenced by addition of basic nitrogen cpds. **F** procédé Aris de VEB Leuna; **G** VEB Leuna Aris-Verfahren.

A. BÖRNER

Ref.: Chem.Tech. **1978**, *30*, 407 and 626 and **1981**, *33*, 357.

vector (cloning)   A vector is a carrier. In molecular biology, vector refers to a carrier of a segemnt of nuclei acid, e.g., a cloning vector is a carrier for getting a piece of *DNA inside a cell. See also *cloning, expression. P.S. SEARS

Vegard's rule   This rule states that in hom. solid solutions, the lattice parameters of the members change continuously and linearly between those of the end-member cpds. Typical examples for solid solution series that obey Vegard's rule are found among metallic *alloys. Ionic cpds. also often follow Vr., whereas covalent cpds. often do not. **F** règle de Vegard; **G** Vegard's Regel.

P. BEHRENS

Vestenamer   → Hüls procs.

VFI *zeolite structural code for Virginia Five zeolites (e.g., *MCM materials)

VGO → vacuum gas oil

VIC → volatile inorganic compounds

Vilsmeier reaction between electron-rich aromatics and DMF under the influence of POCl₃ to form arylaldehydes, e.g., benzene → benzaldehyde. **F** réaction de Vilsmeier; **G** Vilsmeier-Reaktion. B. CORNILS

vinyl acetate (VA, VAM, VINA) important cat. manuf. monomer (cf. procs. of *Bayer, *Halcon, *USI), → VAM synthesis

vinylation 1-general: → olefination; 2: → Reppe reaction; 3: → Heck reaction

viologens 4,4′-bispyridium salts, → methyl viologen, *regeneration of NAD(P)H.
Ref.: P.M.S. Monk, *The Viologens*, Wiley, Chichester 1998.

visbreaking a milder versions of thermal *cracking for the bottoms of atmospheric distillation to lower their viscosity and pour point. Besides lower-boiling comps. tars are produced. **F=G=E**. B. CORNILS

vitamin A intermediates cf. procs. of *BASF and *Hoffmann-La Roche
Ref.: van Santen/van Leeuwen/Moulijn/Averill.

vitamin B₁₂ an *enzyme *cofactor. B₁₂ and other related Co *complexes have been used to cat. a number of C-C bond-forming reactions under reductive conds.
Co(I) in B₁₂ is a very good nucleophile and reacts rapidly with alkyl halides in an *oxidative addition reaction to form an organo-Co derivative. Homolytic cleavage of the C-C bond, by either *electrochemical or *photochemical methods, gives a carbon *radical which undergoes inter- or intramolecular *addition to an alkene. **E=F=G**. C.-H. WONG

Ref.: *Pure Appl.Chem.* **1983**, 55, 1791; B. Kräutler, B.T. Golding, D. Arigoni, *Vitamin B₁₂ and B₁₂-Proteins*, Wiley-VCH, Weinheim 1998.

vitrification thermal encapsulation (e.g., of spent cat.) in glassy materials; cf. proc. de l'*Atelier de Vitrification Marcoule (cf. *final deposition). **E=F**; **G** Schmelzflußeinkapselung. B. CORNILS

VOC → volatile organic compounds

void volume → bulk density of catalysts

volatile inorganic compounds (VICs) inorganic waste cpds. such as SO₂, NH₃, HF, Cl, CS₂, or H₂S, or metals such as Hg, Cd, Pb, Cr, As, Co, or Ni). They are treated similarly to *VOCs; the metals are separated in *guard beds. B. CORNILS

volatile organic compounds (VOCs) are emissions from a variety of different sources (e.g., solvents, reactants, reaction prods., etc.; from the paint industry, enameling operations, plywood manuf., printing industry, etc; cf. also *VICs). They are mostly oxidized catalytically (cf. *catalytic oxidation, *catalytic combustion) in gas cleanup procs. in the presence of Pt, Pd, Fe, Mn, Cu, or Cr cats. (cat. *afterburning). **E=F=G**.   B. CORNILS

Ref.: *Ind.Eng.Chem.Res.* **1987**, *26*, 2165; Ertl/Knözinger/Weitkamp, Vol. 4, p. 1664; Farrauto/Bartholomew, p. 640.

volcano plot   Plotting a reaction property (e.g., the rate of decomposition of formic acid, T_R, on various metal surfaces) versus the strength of adsorption (e.g., the heat of formation of metallic formates, ΔH^o_f) yields a volcano-shaped curve (Figure). The pronounced maximum of the "volcano" indicates that the adsorption may either be too *weak* (and the amount of adsorbed species too low to sustain a reaction) or too *strong* so that the species cannot leave the surface. The top of the volcano is reached from both sides: it has an intermediate value of the *steady-state surface *coverage (*Sabatier's principle).
Other visual presentations use the *Tanaka-Tamaru plot or the Sachtler-Fahrenfort plot. **F** diagramme de volcan; **G** Vulkandiagramm.
                                                                B. CORNILS

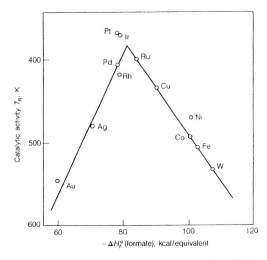

Ref.: *Z.Phys.Chem.* **1960**, *26*, 16; Ertl/Knözinger/Weitkamp, Vol. 1, p. 4f; Augustine, p. 12; Gates, p. 355; Thomas/Thomas, p. 29; van Santen/van Leeuwen.

volcano principle   This states that maximum cat. activity is shown by the material on which the reactants are *chemisorbed as weakly as possible, while still achieving high *surface *coverage. Any further increase in *adsorption strength causes the adsorbed reactants to be less reactive, while a decrease leads to lower rates because the surface is not being fully utilized. The plot of activity versus strength of reactant adsorption is therefore volcano-shaped (cf. *volcano plot, *catalytic activity). **F** principe de volcan; **G** Vulkanprinzip.
                                                                G.C. BOND

Vollmer diffusion   → mass transport in catalysis

von Heyden process   → Wacker procs.

VPO   1: vapor-phase oxidation; 2: → VPO catalysts

VPO catalysts   cats. containing vanadium, phosphorus, and oxygen, for, e.g., *amm(on)-oxidation, oxidation of butene to *maleic anhydride; i.e., vanadium phosphates (e.g., $V(IV)OHPO_4 \cdot \frac{1}{2} H_2O$) as a *precursor for cats.
                                                                B. CORNILS

VRDS   vacuum redium desulfurizer

VRDS process   → Chevron procs.

VRSD   vacuum residue short-path distillation

vulcanization   is – as defined in rubber technology – the *cross-linking of rubbers or polymers still containing double bonds by means of cross-linking agents such as *peroxides, catalysts, *radicals, etc. An example is the v. of polydiene rubbers with sulfur-cat. by Zn oxide (cf. *polybutadiene). **F** vulcanisation; **G** Vulkanisierung, Vulkanisation.
                                                                W. KAMINSKY

W

Wacker oxidation → Wacker processes

Wacker phthalic anhydride process
for the manuf. of *phthalic acid (anhydride) from o-xylene or naphthalene (or mixtures thereof) over a V_2O_5-based (ring-shaped) cat. through *oxidation with air in a multitubular reactor at 375–410 °C (former von Heyden proc.). **F** procédé Wacker pour le PA; **G** Wacker PA-Verfaren.   B. CORNILS

Ref.: *Hydrocarb.Proc.* **1997**, (3), 146.

Wacker processes to acetaldehyde
1: for the industrial *oxidation of ethylene to acetaldehyde cat. by a hom. two-comp. system consisting of $PdCl_2$ and $CuCl_2$ acc. to $H_2C=CH_2 + \frac{1}{2} O_2 \rightarrow CH_3CHO$. The principle of this proc. is based upon the observation by Phillips in 1894 that $PdCl_2$ reacts with ethylene in water yielding AcH and *Pd black. The commercial realization of this observation, the Wacker process, relies on the discovery that the Pd(0) formed can be reoxidized by molecular O_2 in the presence of $CuCl_2$. The $CuCl_2$ reoxidizes Pd(0) to Pd(II) and the Cu_2Cl_2 thus formed is readily converted to Cu(II) chloride by air or O_2. The mechanism of the reaction involves a nucleophilic attack by water on a (π-ethene)palladium *complex. Stereochemical studies by the use of (Z)-1,2-dideuterioethylene have shown that the attack by water occurs on the face of the ethylene opposite to that of Pd (*trans* attack). The 2-hydroxyethyl-Pd complex thus formed decomposes to AcH and Pd(0), a proc. which involves an intramolecular *hydrogen shift (β-hydride elimination-readdition).

Two modes of operation have been developed, the single-stage and the two-stage procs. In the single-stage proc. (Figure) a mixture of ethylene and molecular oxygen is fed into an aqueous solution of $PdCl_2$ and $CuCl_2$ at 120–130 °C/0.4 MPa. In the two-stage proc. ethylene is stoichiometrically oxidized in water to AcH by $PdCl_2$-$CuCl_2$ in the first reactor at 105–110 °C/1 MPa. AcH is removed and the aqueous solution consisting of $PdCl_2$/Cu_2Cl_2 is transferred to another reactor where the Cu(I) is reoxidized by air. Both procs. give yields of 95 %. Substitution of water by acetic acid yields vinyl acetate.

J.-E. BÄCKVALL

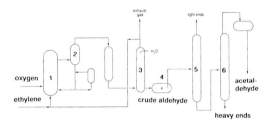

1 reactor; 2 separating vessel; 3 scrubber; 4 crude aldehyde tank; 5, 6 distillations.

Ref.: *J.Am.Chem.Soc.* **1894**, *16*, 255 and **1979**, *101*, 2411; *Angew.Chem.* **1959**, *71*, 176; *Chem.Ber.* **1962**, *95*, 1575; *Hydrocarb.Proc.* **1977**, 56(11), 117.

2: An obselete proc. using Hg salts as cats. for the *addition of water to acetylene (*mercury as catalyst metal). AcH synth. is effectively cat. by Hg_2^{2+} and Hg^{2+} salts. However, Hg^{2+} partially oxidizes the AcH to AcOH, and is itself reduced to metallic Hg. In a former Wacker proc. the metallic mercury was reoxidized to Hg^{2+} by Fe(III) sulfate. In a separate step, the Fe(II) was oxidized back to Fe(III) using nitric acid and was recycled. **F** procédé Wacker pour l'acetaldehyde; **G** Wacker-Acet-aldehydverfahren.   W.A. HERRMANN

Wagner-Meerwein rearrangement
*Lewis acid-cat. rearrangement involving

alkylcarbonium cations of terpenes. **F** réarangement de Wagner-Meerwein; **G** Wagner-Meerwein-Umlagerung. B. CORNILS
Ref.: Krauch/Kunz.

Wakamatsu reaction → amidocarbonylation

WAO wet air oxidation, → cat. waste water treatment

washcoat A w. is a thin layer of het. cat. applied on a structured *support such as *ceramic or metallic *monolithic *honeycombs, but also *foam structures and even tubes. The geometrical exchange area ranges from 0.5 to 5 m²/L. The w. is usually present as a single layer; for some applications up to three w. layers with different chemical compositions are used as well. The total thickness of the w. is between 10 and 200 μm. The w. is composed of porous particles of inorganic oxides, the diameter of which is between 1 to 30 μm. The voids between the w. particles are about 1 μm wide and constitute the w. *macroporosity. The porous particles have a dedicated *micro- and *mesoporosity giving an internal surface area between 50 and 300 m²/g w. Depending upon the support and on the requirements of the cat. function, a typical w. loading is between 50 and 400 g/L support volume. This yields a total internal surface area between 5000 and 40 000 m²/L.
The chemical composition of the w. is determined by its application; e.g., for *automotive exhaust cats. (cf. also *diesel engine emissions) a mixture of up to 20 different inorganic oxides is used. A common constituent of this type of w. is transitional *alumina in the γ-, δ- and θ-modification. Other ingredients comprise rare earth oxides such as doped Ce oxides, *zeolites, *transition metals, or alkaline earth group elements. The w. has different tasks: besides acting as a cat. directly it also serves as a provider of a high and stable internal surface area which is needed to guarantee a fine dispersion for the Pt, Pd, or Rh metal, each of which is used in these cats. The

w. also catches (similarly to *guard beds) some poisoning elements.
After prepn. of the w. it is then applied onto the support by submerging, pouring, pumping, sucking, or other procedures. Air blowing through the support removes the excess w. Finally, the washcoated support is subject to various thermal unit operations such as *drying, *calcining, or *reduction. Sometimes the precious metals are applied as a final step upon the washcoated supports by some kind of *incipient wetness impregnation. Currently, about 100 MM washcoated cats. are used annually worldwide, mainly in cat. exhaust gas emission control. **E=F=G**. E.S.J. LOX
Ref.: Ertl/Knözinger/Weitkamp, Vol. 3, p. 1402 and Vol. 4, p. 1675.

wastewater treatment (photocatalytic)
→ catalytic wastewater treatment, *titania photocatalysis

water-gas shift reaction (WGSR) In the presence of suitable cats. CO is converted by water (steam) to H_2 with CO_2 formed as a byprod.: $CO + H_2O \rightleftarrows H_2 + CO_2$, $\Delta H_{227\,°C} = -39.8$ kJ. The reaction is applied to adjust the CO/H_2 ratio to generate pure H or to remove CO from gas streams. Depending on the required CO conv. the reaction is carried out in one or more cat. beds. In the first stage (high-temp. conv. at 280–350 °C inlet temp., high-temp. shift, HTS) an Fe oxide/Cr oxide cat. is applied. As the equil. at high temp. is unfavorable for complete conv. a second stage operates at a lower temp. level (180–260 °C at the inlet, LTS) in the presence of Cu-containing cats. (or CuO-ZnO). Both stages together can reduce the content of CO to <0.5 wt% (cf.*Texaco HyTex proc.). Complete removal, for example of hydrogen for NH_3 synth., requires *methanation of the remainder to below 50 ppm. For the WGSR of sulfur-rich feedstocks cf. *sour gas shift and *Claus COS conversion.
Elemental Se is also proposed as a cat. in *ARCO's ethylurethane proc.to generate hydrogen in a *WGSR. The proc. was never ap-

plied commercially, probably due to the highly toxic intermediates H_2Se and $Se=C=O$. **F=G=E**. 　　　　　　　C.D. FROHNING

Ref.: Twigg, p. 283; Ertl/Knözinger/Weitkamp, Vol. 4, p. 1831; Cornils/Herrmann-1, p. 960; Farrauto/Bartholomew, p. 357.

water-soluble catalysts → aqueous-phase catalysis

water-soluble ligands

structurally defined as organic molecules bearing one or more polar functions and one or more heteroatoms with the ability to coordinate to metal centers. Wsls. are the supposition of water-soluble *complex cats. for an *aqueous-phase hom. catalysis. In general wsls. exhibit solubilities in water of >50 g/L.

Water-soluble aromatic amines and *phosphines are the predominant examples of wsls. Polar groups are either introduced during the synth. of wsls. or directly by derivatization of the water insoluble parent cpd. Both methods have been widely applied. Polar groups are represented by $-SO_3M$, $-COOM$, $-NR_3X$, PR_3X (M = Na, K, NH_4; X = Cl, Br, I; R = alkyl), or polyether chains $-CH_2(OCH_2CH_2)_n$ CH_3 (n = 12–120). These polar groups are introduced by various methods such as *aromatic substitution (*sulfonation, *carboxylation), quaternization, or *condensation reactions. Sulfonated derivatives of mono- or di*phosphines are the most widely applied, especially the *triphenylphosphine sulfonates *TPPMS and *TPPTS, the latter being the only example of commercial use. More than 50 defined metal complexes of TPPTS are known and characterized up to now.

TPPTS has been proposed and is applied in many cat. reactions such as *hydrogenation, *hydrocyanation, *carbonylation, C-C coupling reactions (*Heck and *Suzuki type), and *hydroformylation (*Ruhrchemie/Rhône-Poulenc proc.). In these reactions TPPTS offers the typical advantages of two-phase cat. where the wsl. retains the metal complex in the aqueous phase and the prod. is separated from the aqueous phase just by phase separation (*aqueous-phase catalysis). **F** ligands hydrosolubles; **G** wasserlösliche Liganden.

C. W. KOHLPAINTNER

Ref.: Angew.Chem.Int.Ed.Engl. **1993**, 32, 1524; Inorg. Synth. **1998**, 32, 1.

water-splitting → photoinduced oxidation/reduction of water

Watson-Crick base pairing rules

Complementarity between *DNA strands is determined by preferred *hydrogen bonding interactions between different bases.

Certain bases pair preferentially with each other. In *DNA, the four bases adenine (A), thymine (T), guanine (G), and cytosine (C) form A-T and G-C pairs. In *RNA, thymine is replaced with uracil (U). **E=F=G**.　　P.S. SEARS

WAX → wide-angle X-ray scattering

waxes

materials which are plastic solids (often crystalline and brittle) at ambient temps. and which become a low-viscosity liquid at elevated temps. The consistency is strongly temp.-dependent. Besides of insect, animal, mineral, or vegetable waxes there are synthetic waxes: polyalkylene (*polyethylene), *Fischer-Tropsch, and modified hydrocarbon waxes, which are mainly the only ones that are subject of cat. procs. The manuf. of petroleum-based waxes proceeds via *dewaxing. **F** cires; **G** Wachse.　　B. CORNILS

Ref: Kirk/Othmer-1, Vol. 25, p. 614; Kirk/Othmer-2, p. 1259; Mark/Bikales/Overberger/Menges, Vol. 17, p. 784.

Wax hydrofinishing process → BP procs.

weak metal-support interactions

(WMSI) → metal-support interactions

Weisz criterion evaluates the influence of the transport of intermediates from one *active site of a *bifunctional cat. to the second one on the overall reaction rate of a given complex reaction A → B → P. Suppose both reaction steps to proceed irreversibly on different *active sites separated by a distance r. If the intermediate cpd. B is a well-defined chemical species, the production rate of P is influenced not only by the reaction *rate of the second reaction step, but also by the velocity of the diffusional transport of B from the first active site to the second via the gaseous phase. The transport influence, however, will be of minor importance if the time t_D the intermediate needs to reach the second active site is shorter than the time t_R this site needs to transform B to P. Assuming the second reaction follows a first-order rate law with the rate constant k, there is thus no transport limitation if

$$\frac{t_D}{t_R} = \frac{r^2 k}{D_B} < 1$$

hold, wherein D_B denotes the gas diffusion coefficient of B in the reaction mixture (cf. also *ensembles). **E=F=G**. D. HESSE
Ref.: *Adv.Catal.* **1962**, *13*, 137.

Werner complexes → complexes

wet air oxidation (WAO) → catalytic wastewater treatment

WGSR → water-gas shift reaction

wheat germ acid phosphatase → phosphatases

Wheelabrator ARI LO-CAT II process for the removal of H_2S and the prod. of high purity S from gas streams by absorptive treatment with iron chelates. **F** procédé ARI LO-CAT II de Wheelabrator; **G** Wheelabrator ARI LO.CAT II-Verfahren. B. CORNILS
Ref.: *Hydrocarb.Proc.* **1996**, (4), 106.

whiskers single crystals from metals, oxides, carbides, nitrides, etc. They have been proposed (but not used commercially) as het. cats. **E=F=G**. B. CORNILS
Ref.: Ullmann (4th Ed.), Vol. 11, p. 383.

WHSV weight hourly space velocity, definition acc. to *GHSV or *LHSV but basing on weights.

wide angle X-ray scattering (WAX) identical to *XRD

Wilkinson Geoffrey (1921–1996), professor of inorganic chemistry at Harvard and Imperial College, London. He worked on *transition metal *complexes (low-temp./low-press. *hydrogenation with the *Wilkinson cat.), *ferrocenes, Rh-cat. reactions. Nobel laureate 1973 (together with E.O. *Fischer).
Ref.: *Angew.Chem.* **1974**, *86*, 664. W.A. HERRMANN

Wilkinson catalyst the square planar Rh(I) d^8-cpd. [Rh(Cl)(PPh$_3$)$_3$], first reported in 1965 and revealed to be the first effective homogeneous *hydrogenation cat. for alkenes and alkynes. Solutions of Wc. are sensitive to air and *hydroperoxides of solvents or substrates; *oxidative addition of diatomic molecules (e.g., H_2, O_2, Cl_2) yields penta- or hexacoordinated *complexes. The Wc. permits various transformations of unsaturated C-C bonds, e.g., *hydrogenations, *hydrosilylations, *transfer hydrogenations, and *dehydrogenations of alkenes or alkynes and furthermore *oxidations, *isomerizations, and *decarbonylation reactions. **F** catalyseur de Wilkinson; **G** Wilkinson-Katalysator.
F. RAMPF
Ref.: *J.Chem.Soc.A* **1966**, 1711; *Prog.Inorg.Chem.* **1981**, *28*, 63; Cornils/Herrmann-1, p. 201; Gates-1, p. 79.

Willgerodt reaction cat. conv. of aryl alkyl ketones to ω-arylalkylcarboxylic acid amides by means of ammonium polysulfide. **F=G=E**. B. CORNILS

Winkler, Clemens (1838–1904), professor in Freiberg (Germany). Developed, together with *Knietsch, the *contact process for H_2SO_4 manuf. by Pt cat. oxidation, $SO_2 \rightarrow SO_3$.   B. CORNILS

Ref.: Neufeldt.

Witten DMT peocess   This well-introduced proc. (developed by Chemische Werke Witten, Germany) consists of two-stage concurrent *oxidation/*esterification reactions. A mixture of *p*-xylene and recycle methyl *p*-toluate (MpT) is oxidized with air (no solvent) at 150–170 °C with Co-Mn cats. The oxidized mixture of *p*-toluic acid and terephthalic acid is esterified with methanol with the resultant *DMT and MpT being separated by distillation. MpT is recycled to the oxidizer and the DMT is further purified by distillation and crystallization. **F** procédé Witten pour le DMT; **G** Witten DMT-Verfahren.   B.L. SMITH

Wittig reaction   → olefination

WMSI   weak metal-support interactions, → metal-support interactions

W/O   water-in-oil (emulsions)

Wolffenstein-Böters hydroxynitration the simultaneous *oxidation and *nitration of aromatics with nitric acid in the presence of cat. amounts of Hg salts. **F** hydroxy-nitruration de Wolffenstein-Böters; **G** Wolffenstein-Böters Hydroxynitrierung.   B. CORNILS

Wolff-Kishner reduction   base-cat. conv. of ketones to the corresponding methylene cpd. via intermediate hydrazones at 150–200 °C. **E=F**; **G** Wolff-Kishner-Reduktion.
                                                           B. CORNILS

Ref.: *Organic Reactions*, Vol. 4, p. 375.

wood gas   manuf. by *pyrolysis of wood. The main constituents are CO_2, CO, and methane; wg. has little and only local importance (the same as peat gas) as raw material for *syngas. **F** gaz de bois; **G** Holzgas.
                                                           B. CORNILS

work function   the energy required to remove an electron from a solid and to bring it to an infinite distance from the *surface. The wf. is different for different materials and for a given material it depends on the crystallographic orientation of the surface plane. The wf. contains a bulk contribution which is hardly affected by *adsorption and a surface comp. which responds highly sensitively to *adsorption and to changes of the surface structure or structural imperfections. On the clean surface the surface comp. is given by the dipole layer that electrons form when they spill out into vacuum. Different orientations of a metal exhibit different wfs., obeying the Smulochowski principle which states that smoother, i.e., close-packed surfaces exhibit a higher wf. than rough surfaces. Upon adsorption the electrons at the surface are redistributed and a charge transfer from the surface to the adsorbate or vice versa occurs such that an adsorbate complex with a dipole moment develops which modifies the wf. according to the Helmholtz equ.   The integral wf. of a material can be measured with various techniques (e.g., with photoelectron spectroscopy, *atomic force microscopy [AFM], or *mirror electron microscopy [MEM]). **E=F=G**.
                                                           R. IMBIHL

Ref.: K. Christmann, *Introduction to Surface Physical Chemistry*, Steinkopff, Darmstadt 1991.

X/Y

XANES another term for NEXAFS (*near-edge X-ray absorption fine structure spectroscopy) or EXAFS (*extended X-ray absorption fine structure spectroscopy

xanthine oxidase (EC 1.2.3.2) a molybdenum iron-sulfur flavin hydroxylase that cat. the *hydroxylation of purine and pyrimidines and the *oxidation of benzaldehyde. Phenol, salicylate, melilotate and *p*-hydroxybenzoate hydroxylase are flavoproteins that cat. monohydroxylation of aromatic hydroxy cpds. utilizing dioxygen and reduced FAD.

Xanthine → O₂ H₂O₂ → Uric Acid (enol form)

Enantioselective oxidation of sulfide to sulfoxide can only be achieved with certain substrates and certain microorganisms. Although most *monooxygenases cat. this type of reaction, the enantioselectivity is often low. **E=F=G.** C.-H. WONG
Ref.: *Biochemistry* **1981**, *21*, 2490 and **1990**, *29*, 3101.

XAS → X-ray absorption spectroscopy

xerogels → sol-gel procs., *silica

XES → X-ray emission spectroscopy

XIS isomerization processes of xylenes to *p*-xylene

XIS process → Maruzen process

XPD 1: X-ray powder diffraction; cf. *XRD; 2: → X-ray photoelectron diffraction

XPEEM X-ray photoemission electron microscopy, a *PEEM using X-rays
Ref.: *J. Mol. Catal. A:* **1999**, *141*, 129.

XPS → X-ray photoelectron spectroscopy, also known as *ESCA

X-ray absorption fine structure (EXAFS) → extended X-ray absorption fine structure spectroscopy

X-ray absorption near edge structures → *XANES

X-ray absorption spectroscopy (XAS) is a term used for low energy *NEXAFS; basing on high photon flux from synchroton facilities as light sources. R. SCHLÖGL

X-ray diffraction (XRD, *XPD) used for the X-ray investigation of bulk phases, their analysis and phase transformation (basing on the comparison of the observed set of reflections of the cat. sample with those of pure reference phases). The ex-situ technique is non-destructive to the sample, very versatile, and universal (solids, liquids, powder or *single crystal) and has good in-situ capabilities. The data are integral and bulk-sensitive. R. SCHLÖGL
Ref.: Niemantsverdriet, p. 138; van Grieken/Markowicz.

X-ray emission spectroscopy (XES) for the determination of electronic structures of solids and of adsorbates by white X-ray photons. The ex-situ, UHV technique is non-destructive to the sample but requires a synchroton. The data are integral and *surface- as well as bulk-sensitive; they also have an excellent in-situ capability. R. SCHLÖGL

X-ray fluorescence spectroscopy (XRF) a multipurpose, non-destructive, ex-situ detection technique which is suitable for elemental chemical analysis. The data are integral and bulk-sensitive (cf. also *EDX and *EXAFS).

R. SCHLÖGL

Ref.: Niemantsverdriet, p. 174; Ullmann, Vol.B5, p. 675; van Grieken/Markowicz.

X-ray photoelectron diffraction (XPD) for the observation of local geometries of disordered *adsorbates by X-ray photons. The UHV ex-situ technique is non-destructive and requires *single crystals (model studies need a synchroton). The data are integral and *surface-sensitive. The method complements *LEED.

R. SCHLÖGL

X-ray photelectron spectroscopy (XPS) is one of the most frequently used ex-situ techniques in catalysis. The data obtained are integral, *surface- and bulk-sensitive (e.g., elemental composition, oxidation state of surfaces, etc.).

R. SCHLÖGL

Ref.: P.K. Gosh, *Introduction to Photoelectron Spectroscopy*, Wiley, New York 1983; Niemantsverdriet; Chastain; Ullmann, Vol.B6, p. 23; van Grieken/Markowicz.

XRD → X-ray diffraction

XRE → X-ray emission spectroscopy

XRF → X-ray fluorescence spectroscopy

Xylene Plus process → Atlantic Richfield procs.

xyliphos *ferrocene based *ligand for Ir *complexes, used for enantioselective *hydrogenations (cf. *Blaser-Heck reaction).

xylose isomerase → glucose isomerase

Yarsintez process → Snamprogetti procs.

yeast Many species of yeast have been used industrially for the prod. of *enzymes and other proteins, as well as smaller biochemicals. A number of yeasts have *GRAS status. *Saccharomyces (S.) cerevisiae* is an important species of yeast in the food industry. It can grow aerobically, and for this purpose is used as a leavening agent. It can also grow anaerobically via the fermentation of sugar to ethanol, and is used in ethanol production. A number of endogenous enzymes are isolated from *S.* spp., such as *lactase and alcohol dehydrogenase. *S.* spp. are used for expression of heterologous genes, as well. These yeasts are easily grown and a large amount of precedent is available regarding their genetic manipulation. As eukaryotic hosts, *S.* are better than bacterial hosts for correctly processing eukaryotic proteins. However, other yeasts are better producers. In addition, *S.* spp. are well known for heavily mannosylating proteins: they can add 150 or more mannose units per N-glycosylation site.

Another genus, *Pichia*, has been observed to secrete proteins to very high yields (>10g/L) and does not mannosylate proteins nearly as heavily as *S.* Cloning vectors are available for these and other yeast, including *Schizosaccharomyces, Kluyveromyces, Hansenula, and Yarrowia*. **F** levain; **G** Hefe. P.S. SEARS

Ref.: *Curr.Opin.Biotech.* **1996**, *7*, 517.

yeast alcohol dehydrogenase (EC 1.1.1.1.) The in-vivo role of (baker's and brewer's) yeast alcohol dehydrogenase (YADH) is to *reduce acetaldehyde and *oxidize ethanol. The purified *enzyme has limited applicability in organic synth., but readily reduces a variety of aldehydes and oxidizes acyclic primary alcohols with stoichiometric consumption of *NAD or *NADH. A substantial body of work has accumulated wherein the *enzyme is used in a whole-cell prepn., either baker's or brewer's yeast, primarily *Saccharomyces cerevisiae*. The characteristics of these whole-cell oxidations and reductions are very different from the purified enzyme, presumably due to the action of other *oxidoreductases, and may be superior for many applications.

Reactions are run at temps. at or below 30 °C, above which the enzyme is unstable and inactivates quickly. The pH for optimum reaction appears to be 8, but the enzyme may be used in the pH 6–8.5 range. Substrate or *product inhibition poses little or no problem with YADH. Organic solvents such as 20 % acetone, 30 % glycerol, and 30 % ethylene glycol are tolerated by the enzyme.

The *pro-R* hydride of the nicotinamide ring is transferred to the *re* face of the substrate carbonyl during reduction, and vice versa during oxidation. **E=F=G**.   C.-H. WONG

Z

Zeise William Christopher (1789–1847), apothecary and professor in Copenhagen (Denmark). He synthesized the first metal-olefin complex, the so-called *Zeise salt.

W.A. HERRMANN

Zeise salt known as the first organo-*transition metal *complex, $K[Pt^{II}Cl_3(\eta^2–C_2H_4)]$ · H_2O, and was described by *Zeise. He correctly determined the composition. The constitution as a π-complex of ethylene was first elucidated in 1953 by Chatt and Duncanson subsequently to the π-coordination model for $Ag(C_2H_4)^+$ by Dewar (*Chatt-Dewar-Duncanson model). **F** sel de Zeise; **G** Zeise-Salz. R. TAUBE

ZEKE spectroscopy ZEKE (*zero kinetic energy*) photoelectron spectroscopy (zs.) as a new method has found applications in cat. In zs. only the electrons of zero kinetic energy at threshold $E_{kin}(e^-) = 0$ are measured, acc. to $E^{neutral} + h\upsilon – IE = E^{ion} + E_{kin}(e^-)$. Zs. provides very high-resolution measurement of the energy states of cations, usually by pulsed field ionization of long-lived very high-n Rydberg states. ZEKE photodetachment of anions makes accessible the energy states of the corresponding neutral species, which in many cases is a reactive *intermediate.

Zs. has been applied to cat. interesting metal carbides and *metal oxide systems. Small V, Nb, and Y *clusters were studied since the geometric structures of Nb and Nb oxide *transition metal clusters may be models for reactive systems in het. cat. In order to take full advantage of the cluster-surface analogy it is necessary to establish the structure of metal clusters containing several transition metal atoms, in both the presence and absence of *ligands. From zs. the most stable structure for both Nb_3O and Nb_3O^+ was found to be planar with $*C_{2v}$ symmetry. The oxygen is bound with equal bond lengths to two Nb atoms. Two distinct Nb-Nb bond distances are present in the cluster. The ZEKE spectra observed are in perfect agreement with this structure; thus zs., combined with the sensible application of a suitable *ab-initio theory, reveals vibrational infomation on small transition metal clusters. In iron carbide, FeC_2^- has been studied. Acc. to zs. the most probable structure is linear, i.e., Fe-C-C, clearly indicating a cat. site. **E=F=G**.

K. MÜLLER-DETHLEFS

Ref.: I. Powis, T. Baer, C.-K. Ng (Ed.), *High Resolution Photoionisation and Photoelectron Studies*, Wiley, Chichester 1995, p. 21; *Chem.Rev.* **1994**, *94*, 1845; *Angew. Chem.Int.Ed.Eng.* **1998**, *110*, 1415; R.W. Field et al. (Eds.), *Nonlinear Spectroscopy for Molecular Structure Determination*, Blackwell, Oxford 1998, p. 167; E.W. Schlag, *ZEKE-Spectroscopy*, Cambridge University Press, Cambridge 1998.

zeolites *microporous solids with *pore sizes ranging from ca. 3 to ca. 7 A. Acc. to older definitions from mineralogy, zs. possess an aluminosilicate comp. As used today, the term z. is used in a wider sense and comprises many different oxidic compositions (e.g., aluminophosphates, metal silicates, etc.); the term *zeotypes* has been proposed to signify these zeolite-like microporous solids which are not aluminosilicates. The term z. is now applied to different materials, when they express zeolitic behavior such as the selective sorption (*molecular sieves) of small molecules (especially water), *ion exchange, large *surface areas and when they possess a crystalline structure, giving to rise to perfectly defined pore sizes.

From a structural point of view, most of the zs., especially those materials where this term is used in a narrower sense, possess a framework structure constructed from maingroup element atoms (so-called *T atoms as Al, Si, P) which are coordinated tetrahedrally by oxygen atoms. Within these frameworks, all the [TO$_4$] tetrahedra are corner-linked, so that the oxygen atoms are in two-fold coordination, bridging between the T atoms. Z. frameworks are open, i.e., they contain voids (*cavities). These voids can be channels (straight or sinusoidal) or cages of spherical or other shapes. Cages are usually interconnected by channels, so that zeolitic behavior becomes possible (*faujasite structure, zeolite Y, Figure).

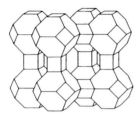

The size of the channels is depicted by the number n of T atoms surrounding the opening as n-membered rings. This gives rise to a classification into small-pore z. (access to the pore system via six- or eight-membered rings, diameter ca. 2.8 to 4 A, examples: sodalite, zeolite A; cf. Figure), middle-pore z. (access via ten-membered rings, diameter ca. 5 to 6 A, example: *ZSM-5, ZSM-11) and large-pore zeolites (access via 12-membered or larger rings, diameter larger than 7 A, examples: zeolite X and Y, UTD-1). Another classification of zeolites is based on the dimensionality of the pore system. It differentiates between structures with isolated cages (clathrate-like structures) and frameworks with one-, two- and three-dimensionally interconnected channels. To date, more than 150 different z. structural topologies, combining the structural characteristics of the pore systems just described, are known. These are assembled in the *International Zeolite Atlas*, where the topologies obtain specific three-letter codes (*zeolite codes).

The presence of pores which can be filled by molecular species renders zs. a class of the large family of *host-guest cpds. Zeolitic host frameworks can be neutral as in pure silica zeolites (zeosils) and in aluminophosphates, or they can be negatively charged when the average formal charge on the tetrahedral framework atoms is less than four, as is the case for aluminosilicate zs. In the former case, the voids may contain neutral guest molecules or ion pairs. In the latter case, the voids must contain cationic guest species (e.g., alkali or tetraalkylammonium ions) in order to balance the charge of the host framework. By special procedures, protons can be introduced as charge-compensating guest species, giving rise to strongly acidic sites.

Only a few zs. occur in Nature, e.g., *faujasite and analcime, where they were formed under hydrothermal conds. The artificial synth. of zs. imitates these conds. As all zeolites are metastable modifications with regard to denser, thermodynamically more stable *phases of similar composition, mild synth. conds. have to be applied to run the prepn. under kinetic control. Typical synth. parameters involve temps. from 100 to 200 °C, which are applied over several days to aqueous solutions containing sources of the T atom elements, which dissolve under hydrothermal conds. (e.g., pyrogenic or precipitated silica for Si, pseudoboehmite for Al), and a mineralizing agent (a base or fluoride ions). Under these circumstances, the great variety of structural topologies emerges through a proc. designated as structure-directed synth. Non-framework species present in the synth. solution are incorporated into the growing crystals and occupy their voids. These species, the so-called structure-directing agents (or *templates) such as hydrated inorganic cations (alkaline or alkaline earth) and organic cations (e.g., tetraalkylammonium ions) or molecules (typically amines), also control the nucleation step of the crystallization and thus exert a certain influence on the type of z. framework formed. Other factors influencing the outcome of a z. synth. are, e.g., ageing times of the prepns. before the hydrothermal treatment and agitation of the synth. batch.

Zs. are widely applied, with most of the mass produced going into their uses as *ion-exchangers and drying agents. The largest productivity however is achieved in their applications as cats. and cat. *supports. Zs. are uniform cats. in that the whole bulk volume is cat. active, not only the outer *surface. Proton-exchanged zeolites are strongly acidic, exhibiting *Brønsted as well as *Lewis acidity. Zs. are thus used as het. cats. in many acid-cat. procs., involving important large-scale commercial procs (cf. procs. *Mobil MTG, MTO, MOGD). In addition, zs. are also used in many small-scale procs. for the prod. of fine chemicals. Zs. ion-exchanged with Cs or loaded with CsO exhibit basic properties. More important is the use of *transition metal (Ti, V, Cr)-substituted z. which can act as *oxidation cats. The most successful z. in this realm is *titanium silicalite-1 (TS-1). Due to the molecular sieving properties of zs., the limited access to the cat. *active sites and the geometric constraints inside the z. voids, het. cat. by zs. is often highly selective with regard to the size and the shape of the reactants. This *shape-selective cat. is a special feature of zs. Depending on whether the influence of the z. framework is exerted on the entrance of the reactant molecules, the formation of a space-consuming *transition state, or the exit of prod. molecules, one differentiates between reactant selectivity, *transition-state selectivity and *product selectivity. Most zs. have to be modified after synth. before they can be applied in cats. Such modifications can include ion exchange of structure-directing inorganic cations, for example against lanthanide ions or protons, *calcination to remove organic structure-directing molecules from the voids, *activation (removal of water from the voids), or the *deposition of metal *clusters inside the voids. For cat. uses, zs. are usually compacted to pellets or another suitable form.

Examples of prepns. of zeolitic cats. are *alkylation/ *transalkylation cats. (*Albene EB, *CDTech/CD Cumene, *Dow cumene, *Lummus UOP styrene/EB, *Mobil/Badger EB and 3GEB, *Mobil TDI, *Mobil TPD-3, *UOP Detal, *UOP Q-Max), *isomerization cats. (*Mobil HAI/LPI and Isofin), *cyclization cats. (*Lonza nicotinamide, *UCC TIP), *SCR cats. (*Degussa Desonox), *hydrocracking/*FCC/*cat. cracking cats. (*Engelhardt FCC, *Houdry Houdresid, *IFP hydrocracking, *Mobil MDDW/LDW, *Mobil Selectoforming), and methanol conv. procs. such as *Mobil MTG and MTO or *UOP/Norsk Hydro MTO. Other reactions may also be based on zeolite cats. (*Deutsche Texaco MIBK, *Mobil dearomatization, *UOP/BP Cyclar, etc.). For zeolite-type cats. cf. *alum-(in)ophosphates, *SAPO, *MeALPO, *MCM cats., and also *molecular sieves, etc. **E=F**; **G** Zeolithe.   P. BEHRENS

Ref.: Ertl/Knözinger/Weitkamp, Vol. 1, p. 286 ff.; Thomas; C.R.A. Catlow (Ed.), *Modelling of Structure and Reactivity in Zeolites*, Academic Press, London 1992; R.M. Barrer, *Hydrothermal Chemistry of Zeolites*, Academic Press, London 1982; J. Fraissard, L. Petrakis (Eds.), *Acidity and Basicity of Solids*, Kluwer, Dordrecht 1999; W.M. Meier, D.H. Olson, Ch. Baerlocher, *Atlas of Zeolite Structure Types*, 4th Ed., Butterworth-Heinemann, 1996; *Zeolites* **1996**, *17*, 1.

zeolite catalytic action   Zeolites may act as het. cats. by the intrinsic reactivity of their frameworks, which is typically an acidic (*Brønsted or *Lewis-type) functionality. The substitution of special metal atoms into the framework may give rise to other reactivities; for example, Ti substitution in *titanium silicalites gives rise to *oxidation catalysis. In many cat. reactions, zs. act as cat. *supports; then, additional guest species, introduced by exchange of cations or by deposition from the gas or the solution phase, perform specific cat. functions, e.g., base or *redox cat.; *bifunctional cat. is also possible.

Independently of the type of reaction catalyzed, the cat. action of zs. involves two main characteristics: firstly, zs. are uniform het. catalysts, i.e., cat. can occur throughout the volume of a zeolite phase and is not – as for many other het. cats. – restricted to the outer *surface of the solid phase. Therefore, the number of *active sites is usually large, and their activity is usually similar. Secondly, zs.

Code	Material from which the code was derived	Isotypic materials
AFI	AlPO$_4$-5 (*AlPO$_4$-Fi*ve)	MgAPO-5, *SAPO-5, SSZ-24 (pure silica variant)
BEA	Zeolite *beta*	Tschernichite (naturally occurring mineral, found later than the synth. zeolite beta)
ERI	*Eri*onite	LZ-220, Linde-T, *ALPO-17
FAU	*fau*jasite (naturally occurring mineral, found before the synth. variants)	Zeolite X, zeolite Y, US-Y, SAPO-37
FER	*Fer*rierite	NU-23, ZSM-35, FU-9
LTA	Zeolite A (*Linde type A*)	ZK-4, gallophosphate, zeolite A, SAPO-42
MFI	ZSM-5 (*Mobil Fi*ve)	*Silicalite-1 (pure silica variant), *titanium silicalite-1 (TS-1)
MOR	*Mor*denite (naturally occurring mineral)	Zeolon, Ca-Q
VFI	VPI-5 (*Virginia Polytechnical Institute Fi*ve)	*MCM-9

exert geometric constraints on the reaction to be catalyzed. These constraints may act on the sorption of reactants, on the *transition state of a reaction to be cat., or on the *desorption of prods. Correspondingly, a distinction is to be made between reactant selectivity, restricted transition state selectivity, and product selectivity. These processes are summarized as *shape-selective catalysis.

P. BEHRENS

zeolite structural codes Zeolites and zeolite-type solids contain open frameworks built from corner-linked tetrahedra such as [AlO$_4$], [SiO$_4$] or [PO$_4$]. Their structures differ in the kind of connections between the tetrahedra, i.e., the topology of the framework. The known framework topologies of tetrahedral zeolite frameworks are compiled in the *International Zeolite Atlas*. Each topology has been assigned a specific three-letter code, which is (in some way that is often not very obvious) related to the name which was given to a certain material when it was first described (Table). As the topology of a certain structure is determined wholly by the linking of its tetrahedra (composition, bond angles and bond lengths, and other structural details are not considered), materials of very different composition, and quite different struc-

tures may actually appear under the same structural code.

P. BEHRENS

Ref.: W.M. Meier, D.H. Olson, Ch. Baerlocher, *Atlas of Zeolite Structure Types*, 4th Ed., Butterworth-Heinemann, 1996; *Zeolites* **1996**, *17*, 1; Ertl/Knözinger/Weitkamp, Vol. 1, p. 287.

zeolite-supported catalysts *Zeolites are suitable to anchor *transition metal particles on their *surface. In addition, zeolites exhibit a functional large internal surface which carries solid acid functions on Al-OH sites and which allows only the transport of certain sizes of molecules (molecular sieving effect, *molecular traffic control). The geometric constraints within the pore system create further constraints on possible reactions of activated molecules within the zeolite and affect the *selectivity (*shape selectivity, *reactant selectivity, *product selectivity) of reaction prods.

The generation of a transition metal–zeolite cat., in which the transition metal hydrogenates or dehydrogenates an organic reactant and the resulting prod. is further converted within the zeolite using the acidity of the zeolites is a characteristic example of a *multifunctional cat. In such a system several chemically different reaction paths are combined in one solid cat. and can be exerted in one proc. This strategy is the equivalent of a one-pot synth. in hom. chemistry. Areas of technological application are in the petrochemical reaction sequence.   R. SCHLÖGL

zeolon   a *mordenite based *zeolite.

Zeonex   → Nippon Zeon procs.

zeosils   → zeolites

zeotype   → zeolites

zeozymes   enzymes entrapped in zeolites; → ship-in-the-bottle cats.

zero emision automobile   → automotive exhaust catalysts, *DeNOx reactions

Zewail Ahmed H., Egyptian scientist and professor at the California Institute of Technology (USA); Nobel laureate in 1999 for the femtosecond probing of *transition states in chemical reactions.   B. CORNILS
Ref.: *J.Chem.Phys.* **1987**, 87, 2395.

Ziegler Karl (1898–1973), professor of chemistry at various German universities and head of the Max-Planck-Institut für Kohlenforschung at Mülheim (Germany). Worked on Li, K, and Al alkyls and discovered the low-press. (*high-density) polyethylene synth. (*Ziegler-Natta catalysis). Nobel laureate in 1963 (together with *Natta).   W.A. HERRMANN

Ziegler-Natta catalysis ZN cats. are formed by a cpd. of a *transition metal from the groups 4–8 (often a Ti cpd.) and a cpd. of an element of the groups 1, 2, or 3 (e.g., an AlR$_3$; *organoaluminum cpds.). The cat. species is assumed to be a binuclear *complex with a *vacant coordination site to which an alkene can bind and later insert into the Ti-alkyl bond. By repeating this cycle very often, long polymer chains are formed.

$(C_2H_5)_2TiCl_2 + Et_2AlCl →$

Commercially the *het.* ZN. cats. (nowadays prepared on MgCl$_2$ *supports) are widely used to produce several kinds of polyolefins. Since the discovery of *hom.* ZN. cats. based on *metallocenes such as $(C_2H_5)_2ZrCl_2$ and *methylalumoxan (MAO) not only has the mechanistical behavior been investigated, but also the activities are much higher than those of the het. systems. Thus, it has become possible to tailor the geometrical arrangement of the polymer by changing the *ligand structure of the metallocene (*tailoring of cats.). F catalyse selon Ziegler-Natta; G Ziegler-Natta-Katalyse.   W. KAMINSKY, M. VATHAUER
Ref.: Ullmann, Vol. A28, p. 506; *Angew.Chem.* **1952**, *64*, 323 and **1955**, *67*, 542; *J.Am.Chem.Soc.* **1955**, 77, 1708; *Adv.Organomet.Chem.* **1980**, *18*, 99; Mark/Bikales/Overberger/Menges, Vol. 17, p. 1027; Ertl/Knözinger/Weitkamp, Vol. 5, p. 2405.

Ziegler-Natta catalysts (generations)
During development of ZN. cats. various generations emerged. The first-generation was originally described by Ziegler et al. as a mixture of TiCl$_4$ and AlEt$_3$ or other alkylaluminums or alkyl-Al halides which acted as het. cats. In the second-generation cats. an ether washing extracted coprecipitated AlCl$_3$, thus preventing the formation of EtAlCl$_2$ as a cat. *poison. An additional washing with hydrocarbons removed adsorbed TiCl$_4$. The third generation consisted of cats. *supported on MgCl$_2$ cf. *high-mileage cats.).   W. KAMINSKY
Ref.: Ertl/Knözinger/Weitkamp, Vol. 5, p. 2405.

zinc as catalyst metal Zn is contained in many *enzymes, e.g., in some *acylases and *aldolases, *carbonic anhydrase, *HLAD, *carboxypeptidase, *aminopeptidase, or (together with Cu) in *superoxide dismutase (cf. *metalloenzymes). *Metalloproteases often utilize a Zn^{2+} ion as a *Lewis acid that is coordinated in the cat. step to the nucleophilic water (to increase its acidity) and to the carbonyl group of the scissile bond.

Zn can used be as a cat. active metal for various reactions (*Clemmensen, *Negishi, *Diels-Alder, *Friedel-Crafts, *Gattermnn, *hydrocyanation [Zn/BPh$_3$], *Mukayima aldol, *Reformatsky, *Reppe vinylation, *Simmons-Smith). Zn is constituent of various het. cats. for *aromatizations (as well as Ga-Zn on *zeolites, e.g., *UOP/BP Cyclar proc.) and in mixtures with other elements, e.g., Zn-Cr (zinc chromite) for *BASF's methanol proc., *dehydrogenations, and the *isobutyl oil synth., Zn-Cu (*Adkins type cat.) for *fat hydrogenation, Zn-Fe (zinc ferrite) for *oxidative *dehydrogenation, Zn-Cd for *selective hydrogenations (*Degussa allylic alcohol, *Henkel TPA), Zn-Pt for *catalytic reforming (*Phillips STAR proc.), or Zn-*alumina for HF *addition to double bonds. In the *Aldox proc. it acts as a *promoter. ZnBr$_2$ is used for the ene reaction in *Takasago's menthol proc. Zn may be an ingredient of the contact mass of the *Rochow synth. and is used for *vulcanization reactions. **F** le zinc comme métal catalytique; **G** Zink als Katalysatormetall.

B. CORNILS

Ref.: *Organic Reactions* Vol. 20; *Chem. Rev.* **1996**, *96*, 2375; *Acc. Chem. Res.* **1999**, *32*, 589; E. Erdik, *Organozinc Reagents in Organic Synthesis*, CRC Press, Boca Raton 1996.

zinc protease → metalloproteases

zirconia ZrO$_2$, a catalyst *support

zirconium as catalyst metal This has gained a certain importance during the last decade, e.g., in zirconocenes (*metallocenes) wherein the metal center is coordinated to two π-donor *ligands of the *cp type and two σ-donor ligands, such as H, Cl, CH$_3$, OR, or NR$_2$. Often, the cp ligands are linked by an -SiR$_2$- or -CH$_2$CH$_2$- bridge (*ansa-metallocenes) to reduce the fluxionality of the cat. active species. Such cpds. have found application in alkene *polymerization.

Besides these procs. zirconocenes and other Zr *complexes have been used as cats. in *hydrogenation, *hydrocyanation, and enantioselective carbomagnesation reactions, where 2,5-dihydrofurans and pyrroles are converted into *chiral homoallylic alcohols and amines. *Zirconia is used as a *support, especially in the sulfated form (*sulfated zirconia). Silicon-bonded Zr has been proposed for *surface organometal cats. **F** le zirconium comme métal catalytique; **G** Zirconium als Katalysatormetall.

W.R. THIEL

Ref.: *J. Catal.* **1997**, *172*, 24; *Cat. Today* **1989**, 5, 493 and **1994**, *20*, 219; *Houben-Weyl/Methodicum Chimicum*, Vol. E 18, p. 828 and Vol. XIII/7; van Santen/van Leeuwen/Moulijn/Averill.

zirconocenes metallocenes based on Zr

zymogen → proenzyme

ZSM-5 zeolite structural code for *MFI zeolites (spoken: Mobil Five).